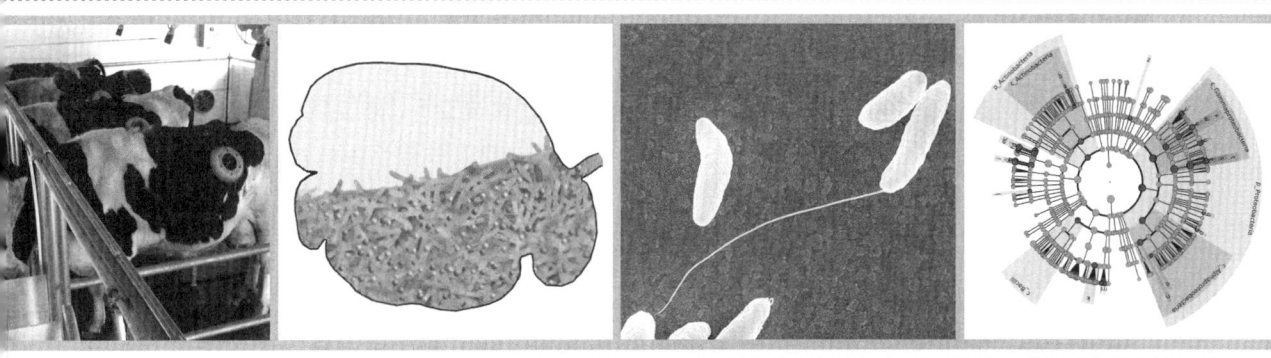

瘤胃微生物学

Improving Rumen Function

[澳] 克里斯·麦克斯威尼 (C. S. McSweeney),
[美] 罗德里克·麦基 (R. I. Mackie) 著

赵圣国 等 编译

王加启 主审校

中国农业科学技术出版社

图书在版编目（CIP）数据

瘤胃微生物学 /（澳）克里斯·麦克斯威尼（C. S. McSweeney），（美）罗德里克·麦基（R. I. Mackie）著；赵圣国等编译. --北京：中国农业科学技术出版社，2023.10（2025.4重印）

书名原文：Improving rumen function

ISBN 978-7-5116-6229-3

Ⅰ.①瘤… Ⅱ.①克…②罗…③赵… Ⅲ.①反刍动物-瘤胃微生物-研究 Ⅳ.①S823

中国国家版本馆 CIP 数据核字（2023）第 046399 号

Improving Rumen Function
ISBN：9781786763327

©Burleigh Dodds Science Publishing，2020，except the following: the contribution of Dr Stephanie A. Terry, Dr Carlos M. Romero, Dr Alex V. Chaves and Dr Tim A. McAllister in Chapter 15 is © Her Majesty the Queen in Right of Canada. All rights reserved.）

Improving Rumen Function 的中文翻译版由 Burleigh Dodds Science Publishing Limited 授权出版。

责任编辑	金　迪	
责任校对	贾若妍	李向荣
责任印制	姜义伟	王思文

出 版 者	中国农业科学技术出版社	
	北京市中关村南大街 12 号　邮编：100081	
电　　话	（010）82106625（编辑室）　（010）82109702（发行部）	
	（010）82109709（读者服务部）	
网　　址	https://castp.caas.cn	
经 销 者	各地新华书店	
印 刷 者	北京地大彩印有限公司	
开　　本	185 mm×260 mm　1/16	
印　　张	38.75　彩插　6 面	
字　　数	1 010 千字	
版　　次	2023 年 10 月第 1 版　2025 年 4 月第 2 次印刷	
定　　价	168.00 元	

◣━━ 版权所有·翻印必究 ━━◢

《瘤胃微生物学》
译校者名单

主 编 译：赵圣国

主 审 校：王加启

编译人员（按姓氏笔画排序）：

王佳堃　毛跃建　成艳芬　刘　晶　米见对

杨春蕾　杨　鼎　张美玲　罗玉衡　金　巍

庞小燕　赵圣国　黄小丹　董依然　焦金真

裴彩霞　谭　翠

顾　　问（按姓氏笔画排序）：

王之盛　龙瑞军　朱伟云　刘建新　谭支良

译者的话

"牛吃的是草,挤出的是奶",这句话形象体现了反刍动物转化饲草料成肉奶的能力,这种能力的主角就是瘤胃微生物。瘤胃微生物由细菌、古菌、真菌、原虫和病毒等组成,具有多样性高的特点,它们之间相互依赖或竞争,构成了瘤胃微生物生态系统。瘤胃微生物能将饲料中的木质纤维素、蛋白质、非蛋白氮、脂肪等消化、代谢和利用,为反刍动物提供可吸收利用的能量、氮、脂肪酸和维生素等营养素,对维持动物健康、促进生产性能具有决定性意义。

1966年美国微生物学家Robert Hungate教授出版了第一本瘤胃微生物书籍《瘤胃及其微生物》(*The Rumen and Its Microbes*),首次系统性阐述了瘤胃微生物分离培养、种类及生理特征,开启了瘤胃微生物研究的大门。进入21世纪后,随着测序和生物信息技术的快速发展,微生物组(Microbiome)成为国际前沿热点研究领域。将微生物组技术引入瘤胃微生物研究,让我们对瘤胃微生物种类与功能有了全新的认识。近年来,国际上瘤胃微生物组相关项目越来越多,比如欧洲RuminOmics、新西兰Hungate1000项目,我国的国家重点研发计划、国家自然科学基金中也设置了大量瘤胃微生物研究的项目,所以瘤胃微生物研究在国内外已进入快速发展轨道。

澳大利亚联邦科工组织(CSIRO)Chris McSweeney教授和美国伊利诺伊大学Roderick Mackie教授是享誉全球的瘤胃微生物学领域资深专家,更重要的是两位教授倾心培养中国研究生和青年学者,为中国瘤胃微生物学发展作出了贡献。欣喜地看到,两位教授2020年组织出版了一部权威著作 *Improving Rumen Function*,综述了瘤胃微生物领域的最新研究进展,填补了瘤胃微生物系统性书籍二十余年的空缺。为了让国内学者尽快地看到这本国际权威著作,我作为McSweeney教授的学生,决定组织其他曾在两位教授实验室学习过的中国学者,将这本英文著作翻译成中文。

本书介绍了瘤胃微生物研究技术方法,阐述了瘤胃各种微生物的特征与功能,论述了瘤胃微生物在饲料营养代谢、动物健康以及环境影响中的作用,提出了调控瘤胃微生物功能的营养策略，为我国反刍动物营养、消化道微生物领域的本科生、研究生、教师和科研工作者提供了有价值的参考。

本书书名"Improving Rumen Function"直译应为"促进瘤胃功能"，但为了更符合书的内容以及学科习惯，我们将书名翻译为"瘤胃微生物学"，希望能为我国瘤胃微生物学研究进入国际第一方阵贡献力量。

本书原著包含22个章节，内容丰富，翻译工作量大。感谢全体译者的无私奉献，没有大家一次次的认真校核，就没有高质量的书稿。感谢主审校和所有顾问专家的指导，正是你们的鼓励、建议和认可，才让本书成为一张瘤胃微生物领域的"名片"。同时，感谢张晓音、刘思佳、仲慧月、卜莹、李可馨、胡钰峰等研究生的大力协助，感谢出版社金迪博士的辛勤工作。特别感谢国家重点研发计划（2022YFD1301000）、中国农业科学院科技创新工程的资助。

本书的翻译专业性强、时间要求紧，对译者的学术水平和翻译水平提出了很高要求。中译本在内容上忠实于原著，力求准确、流畅，但由于知识水平的限制，在翻译过程中难免存在不妥之处，敬请读者批评指正。

谨以此书向已退休的 Chris McSweeney 和 Roderick Mackie 两位教授致敬！

赵圣国

中国农业科学院北京畜牧兽医研究所

序 言

从公元前 2500 年开始，大多数反刍动物物种逐渐被人类驯化。从此，被驯化的反刍动物随人类一起在全球迁移、扩散，它们为人类的生存和繁衍提供食物来源，为人类迁移和耕作提供畜力，以及提供用于制作衣服的纤维。反刍动物与人类的特殊关系源于它们能够将人类难以消化的碳水化合物和氮转化为肉和奶等优质蛋白质的能力，避免了人畜争粮问题。反刍动物能将人类无法消化的植物转化为动物可代谢的蛋白质和能量，这种能力主要归功于定植在其前胃（网胃、瘤胃和瓣胃）中的微生物。尽管很久以前人类就认识到了驯化这些动物的益处，但直到 20 世纪，"瘤胃"微生物在这一过程中的核心作用才被详细地记录。20 世纪中叶，随着一些瘤胃微生物得以分离，人类对厌氧发酵过程中生化反应的认识才被发表出来。1966 年，由 R. E. Hungate 编写的第一本关于瘤胃微生物学的专著《瘤胃及其微生物》出版，这部经典著作描述了当时已被鉴定的瘤胃微生物及其生理活动和营养需求，以及为培养和分离严格厌氧微生物而发展的技术。此外，书中也包含一些实用的内容，描述了提高反刍动物的生产力和防止瘤胃发酵异常等方面的基本知识。由此，人们也认识到瘤胃微生物学领域不能被独立研究，而需要密切联系农业科学。自 Hungate 编写的书出版后，瘤胃微生物学不断发展。1988 年，另一部名为《瘤胃微生物生态系统》（P. N. Hobson 编写）的专著出版，并于 1997 年出版了第 2 版（P. N. Hobson 和 C. S. Stewart 编写）。这两部书遵循了《瘤胃及其微生物》一书的基本形式，但邀请了更多相关专家撰写相关章节，以介绍新发现的微生物、生化反应及分析技术。

在过去的 25~30 年里，微生物学经历了一场革命，该领域已经从以培养为基础的技术发展到了分子方法，从而可在无需培养的情况下探究复杂的微生物生态系统。与之前研究瘤胃微生物学的专著一样，本专著旨在展示 21 世纪瘤胃分子生态学问世后发现的瘤胃微生物和生化途径的新知识。值得一提的是，目前瘤胃微生物学面临比以往任何时候都严峻的挑战，即如何面对反刍动物产业的重大问题。在本书中，我们特别关注了公众所关心的反刍动物生产对环境的负面影响，包括甲烷和氧化亚氮排放，以及尿液和粪便导致的土壤和水体富营养化。虽然环境问题在瘤胃研究中越来越受到重视，但提高反刍动物生产效率以及提供令人满意的、营养丰富的、健康的产品仍然非常重要，这些都可能直接受到瘤胃微生物生态系统的影响。

  一些被驯化的反刍动物已进化出了独特的解剖和生理特征，这使得多样的瘤胃微生物群能够有效利用蛋白质含量相对较低但木质素含量高的饲料，从而保证宿主的生存和繁衍。在20世纪，反刍动物生产效率的提高大多是通过提供高品质日粮和均衡配方的集约化养殖实现，但是这一模式与人类和单胃动物争夺优质日粮营养素，将受到环保人士、监管机构和消费者更为严格的关注。将来，人们需要更多地关注反刍动物将木质纤维素饲料转化为可供人类利用的优质蛋白的独特能力，以便在环境和经济角度体现反刍动物在可持续的养殖生产系统中的优势地位。因此，本书将为学生和研究人员提供支撑未来20~30年反刍动物高效生产中瘤胃微生物学基础知识。

C. S. McSweeney

R. I. Mackie

引 言

分析技术和基因组学的重大进展改变了人类对瘤胃微生物学的认识，这些认识对畜牧业生产至关重要，因为瘤胃的功能影响反刍动物的营养利用效率、废物排放（如甲烷和氧化亚氮）以及宿主健康。本书综述了我们对瘤胃微生物群落的了解，以及营养调控在优化瘤胃微生物群落功能以实现可持续畜牧业生产方面的作用。

第 1 章通过介绍瘤胃微生物群落定植和建立的最新进展开篇。本章综述了幼龄反刍动物胃肠道中微生物群落的建立，以及如何调控微生物群落，以促进动物健康和达到理想生产性能。同时，提供了关于改善动物健康和生产以及减少肠道甲烷排放的研究实例。本章的结尾介绍了未来研究的潜在领域，并列举更多信息来源。

第一部分 研究瘤胃微生物组的工具

第 2 章至第 4 章为本书的第一部分，概括了分析瘤胃微生物组的先进方法。

第 2 章综述了组学时代肠道微生物组是如何"复活"的。近年来，对哺乳动物肠道微生物组的研究主要是通过非培养的方法对基因进行分析。但是，越来越多的科学家对基于培养的研究重新产生了兴趣，因为培养能增加肠道微生物分离物的广度和深度，并能够获得微生物及其基因的功能信息。本章首先概述了难培养肠道微生物的常用培养方法，以及营养物质对微生物生长的影响。此外，我们也讨论了新的培养方法，包括利用现有丰富的宏基因组数据而建立起来的基于基因组、基因和抗体的分离策略。

第 3 章重点介绍瘤胃代谢组学，它是阐释和理解瘤胃功能和健康的工具。瘤胃是一个复杂的生态系统，对动物的健康和生产至关重要。瘤胃代谢组学研究为解析瘤胃中存在的代谢物以及影响瘤胃微生物组和代谢组的因素提供了重要数据。本章首先概述了瘤胃代谢组试验与分析技术及其影响因素。最后，作者指出了这个新兴研究领域的未来趋势，并对代谢组学分析技术进行了总结。

第 4 章是第一部分的最后一章，回顾了瘤胃微生物功能的数学模型。随着新技术和新数据类型不断被引入瘤胃研究，为了更好地利用现有大数据，需要构建数学模型阐明生物学功能并进行预测。第 4 章综述了在瘤胃整体水平上定量分析瘤胃微生物功能的方法，以及在数学建模中使用的单元和取样技术，这些可以用来预测瘤胃中的发酵和消化过程。

第二部分 瘤胃微生物群落

第 5 章至第 10 章为本书的第二部分，介绍瘤胃微生物群落中细菌、古菌、原虫、厌氧真菌、病毒以及瘤胃壁微生物的最新研究。

正如第 5 章中所指出的，微生物基因组测序对理解微生态产生了巨大的影响，这其中也包括对瘤胃中定植的细菌的种类、关系和功能的理解。自从 2003 年第一个瘤胃细菌的基因组公开，科学家们已得到超过 500 个分离菌株基因组和超过 5 000 个宏基因组组装的基因组，这些信息揭示了微生物通过群体互作发挥植物降解功能的机制。但是，这些机制仍然是不完整的，因为许多已知存在的微生物物种在基因组数据库中缺失，只有填补上这些缺失才可能全面了解瘤胃微生物组。第 5 章介绍了单菌基因组测序的作用和瘤胃微生物组参考基因组集，介绍了宏基因组数据在鉴定瘤胃中新基因组和关键功能的应用，以及应用基因组测序来揭示瘤胃微生物间的相互作用。

第 6 章介绍瘤胃产甲烷菌。本章回顾了瘤胃古菌的培养和分子生物学研究，讨论了目前产甲烷菌单菌基因组以及混合菌和宏基因组组装基因组信息，讲述了产甲烷菌功能的研究进展，最后总结了瘤胃产甲烷菌在甲烷产生和瘤胃功能方面的重要性。

第 7 章介绍了对瘤胃原虫的认识，包括它们的物种分类和种群组成，评估了瘤胃中原虫种群的生态扰动、相互作用，以及原虫对反刍动物营养、健康和排放的影响，讨论了瘤胃原虫培养的挑战，还介绍了一项调控瘤胃纤毛虫的实例。本章最后强调了当前研究的空白领域。

第 8 章讨论了厌氧真菌的生活周期、分类和形态特征。本章不仅综述了目前已知的单中心、多中心和球茎属的瘤胃厌氧真菌，以及它们的基因组和宏基因组，还介绍了瘤胃生态系统中真菌和其他成员之间的相互作用。本章最后强调了瘤胃厌氧真菌研究的重要性。

第 9 章指出，尽管瘤胃微生物生态学研究越来越多，但人类对包括病毒、质粒在内的瘤胃可移动基因组，及其对瘤胃功能的影响认识不够。已有证据显示，病毒种群与瘤胃微生物群落共存、并且捕食微生物。此外，质粒也常常与微生物存在内在联系。本章介绍了目前对瘤胃病毒种群和移动元件的认识，描述了可移动遗传元件的载体，例如细胞膜外囊泡，并探讨了可移动基因组对瘤胃功能的影响。

第 10 章是第二部分的最后一章，讨论了瘤胃壁微生物群落。瘤胃微生物通常根据其栖息地而被分为三大类：游离微生物、附着于饲料颗粒的微生物和瘤胃壁附着的微生物。在这三类微生物中，瘤胃壁附着微生物群落是最少被研究和了解的。瘤胃壁附着微生物在氧气清除、组织循环、尿素代谢和营养运输等过程发挥关键作用。近年来，核酸测序技术的发展使我们能够更好地探究该群落的组成和功能。本章总结了目前关于瘤胃壁微生物群落的知识，包括其多样性、生态学、功能以及对宿主生理的影响。

第三部分　营养代谢在瘤胃微生物和宿主互作中的作用

第 11 章至第 17 章为本书的第三部分，阐述了瘤胃代谢纤维和蛋白质等营养物质并产生能量、脂质和释放甲烷的作用。

第 11 章论述瘤胃中纤维的消化。由于反刍动物获得的大部分能量来自其共生菌群，因此饲料转化效率和肉奶的品质与瘤胃微生物组的变化和功能密切相关。本章概述了微生物群落在瘤胃木质纤维素降解中的作用及其机制，讨论了瘤胃球菌、纤维杆菌和普雷沃氏菌等已知菌以及拟杆菌等未知菌的研究进展。揭示了纤维小体、酶、细胞膜外囊泡、多糖利用位点和大型多模块酶的新研究进展，从而使人们更深入地了解参与瘤胃纤维消化的复杂微生物网络。

第 12 章介绍瘤胃中的蛋白质分解和氨同化。长期以来，人们就认识到反刍动物瘤胃的氮代谢与纤维的高效降解、采食量和生产性能密切相关。本章重点介绍氮代谢的最新研究进展，以及对瘤胃氨同化过程及其调控、蛋白分解和氮捕获的最新认识。瘤胃氮代谢的营养调控有助于提高反刍动物生产性能，同时最大限度地减少对环境的负面影响。

第 13 章介绍影响瘤胃能量代谢效率的因素。首先阐述了瘤胃发酵的主要途径，以及这些途径如何用于生产挥发性脂肪酸和三磷酸腺苷（ATP）等产物，以及代谢氢的清除。本章介绍瘤胃中甲烷的产生，以及如何调控甲烷减排来促进动物生产性能。本章还讨论了影响微生物生长效率的因素，以及瘤胃中能量代谢和氮代谢之间的相互作用，强调了提高发酵产能、调控挥发性脂肪酸组成和提高微生物生长效率的重要性。

第 14 章的主题是瘤胃脂质代谢及其在改善牛奶质量和监测动物健康中的作用。瘤胃脂质代谢在很大程度上决定了乳制品的脂肪酸组成，因此，乳脂肪酸可作为瘤胃健康甚至动物健康的指标。调控瘤胃代谢也有助于生产富含不饱和脂肪酸的乳制品，为了更好地理解乳制品中脂肪酸的来源，本章首先介绍了脂肪酸在瘤胃代谢、肠道消化、机体转运以及乳腺利用的过程，然后讨论了改善乳制品脂肪酸组成和增强人体健康的潜力，特别介绍了瘤胃不饱和脂肪酸的生物氢化抑制技术，最后探讨了利用乳脂肪酸组成变化作为瘤胃和动物健康的监测方法。

第 15 章分析了影响反刍动物温室气体产生的营养因素，介绍了调控肠道和粪便甲烷排放的营养调控启示。反刍动物是全球温室气体排放的重要贡献者。人们已经研究了如何减少胃肠道和粪便中甲烷的产生，但对其他温室气体研究较少。降低胃肠道甲烷的释放可能导致粪便中温室气体的意外增加。考虑到瘤胃的复杂性及其对粪便的影响，需要一个完整的技术体系来评估添加剂对甲烷排放的影响和可行性。本章总结了一系列可用于减少胃肠道和粪便中甲烷排放的营养策略，综述了包括替代电子受体、抑制剂、植物次生化合物和碳衍生材料等饲料添加剂在减少温室气体总排放的效果，评估了它们如何改变瘤胃和粪便微生物组。

第 16 章研究了宿主-瘤胃微生物组的互作及其对饲料转化效率、甲烷产生和其他生产性能的影响。虽然，我们对瘤胃微生物群落组成研究较多，但是随着微生物组这一理念的出现，如何从微生物组的角度阐明动物生产效率、健康、甲烷等废物排放的机制成为新热点。本章重点介绍了瘤胃微生物组在能量获取、甲烷排放中的作用，分享了宿主定向影响瘤胃微生物群落组成方面的最新发现。

第 17 章是第三部分的最后一章，讨论了瘤胃微生物调节机体免疫功能的作用。瘤胃及其微生物在为宿主提供能量、蛋白质、矿物质和维生素等关键营养物质，以及塑造奶牛免疫系统方面发挥着重要作用。瘤胃健康失调，如亚急性瘤胃酸中毒，会引起瘤胃生态失衡，导致上皮屏障功能紊乱和炎症反应，继而促进瘤胃和后肠来源的脂多糖（LPS）转运到血液中，从而影响机体代谢和免疫反应。本章总结了关于瘤胃健康、LPS 暴露及其在调节机体代谢组和肝脏健康中的作用，探讨长期瘤胃酸中毒对乳腺健康的影响，强调 LPS 在损害血乳屏障和乳腺组织方面的作用。

第四部分　优化瘤胃微生物功能的营养策略

第18章至第22章为本书的第四部分，探讨优化瘤胃功能的营养策略，涉及牧草、青贮饲料、谷物、植物次生化合物和益生菌等内容。

第18章重点介绍瘤胃微生物组在放牧饲养的反刍动物生产系统中的作用。牧草是反刍动物常用饲料，但饲喂牧草会产生更多温室气体，对环境产生负面影响，同时造成能量和氮的损失。本章讨论了放牧生产系统中瘤胃微生物组、宿主饲料转化效率和环境排放之间的关系，还综述了不同牧草和草场管理措施对瘤胃微生物的影响。

第19章分析了青贮饲料和精料在提高奶牛饲料转化效率和减少甲烷、氮排放中的作用。充分认识反刍动物转化人类不可食用生物质为高品质乳品的独特能力，构建减少温室气体和氮排放的可持续奶牛饲养策略。本章综述了在不影响奶牛生产性能（采食量、产奶量、饲料和氮转化效率、甲烷排放）的情况下，青贮植物种类（禾本科草、豆科草、玉米）、青贮植物生长阶段、粗饲料与精料的比例等降低碳氮排放的潜力。本章也分析了精料成分（脂类、碳水化合物和蛋白质）的作用，并且通过一个研究实例来介绍粉碎油菜籽在减少牧草青贮日粮导致的环境负面影响方面的潜力。

第20章介绍了精料营养调控提高饲料利用效率和减少甲烷排放的研究进展。首先重点介绍了日粮谷物种类，例如大麦、玉米、小麦、燕麦和高粱，然后讨论了影响产甲烷的日粮因素、以及淀粉和饲草在甲烷产生中的作用，分析了氢和甲烷产生的关系，综述了利用谷物提高饲料转化效率、减少甲烷产生的研究。本章还对参与谷物发酵的的细菌和古菌进行了总结。最后，本章重点介绍了影响饲料降解的重要因素即瘤胃滞留时间，并描述了瘤胃酸中毒等。

第21章讨论了植物次生产物及其对反刍动物营养和生产的有益作用。植物次生化合物是植物的次级代谢产物，在经过提取和浓缩后具有调节肠道微生物功能的作用。本章介绍了三大类植物次生化合物——精油、单宁和皂苷的组成、活性和对瘤胃功能、动物生产性能的影响。这些化合物的有益作用包括提高日粮蛋白和能量的利用效率、减少甲烷排放，从而增加产奶量或动物增重。本章探讨了目前在反刍动物日粮中使用植物提取物作为饲料添加剂的局限性，以及如何克服这些局限性。

第22章介绍了益生菌作为添加剂在反刍动物生产中的应用。当前，在维持动物健康和福利的同时，人们对提高日粮消化效率和生产性能的需求不断增加。益生菌是指适量使用且对宿主有益的活的微生物。本章介绍了反刍动物全生命周期中可使用益生菌的关键时期，以及益生菌的种类、使用方式和调节作用，总结了益生菌的益处和作用方式、以及在幼龄反刍动物中的应用潜力。最后，本章讨论了益生菌在成年反刍动物的饲料转化效率、甲烷产生、病原菌控制和增强免疫力等方面的作用。

（庞小燕译）

目 录

第 1 章 瘤胃微生物区系的定植和建立——调控生产性能和甲烷排放的窗口期 ……（1）
  1 前言 …………………………………………………………………………………（1）
  2 瘤胃微生物区系的建立 ……………………………………………………………（1）
  3 调控幼龄反刍动物胃肠道微生物区系以实现动物健康和高产 …………………（8）
  4 案例研究：促进健康和生产的早期调控策略 ……………………………………（9）
  5 案例研究：减少肠道甲烷排放的早期策略 ………………………………………（10）
  6 结论和展望 …………………………………………………………………………（11）
  7 更多信息 ……………………………………………………………………………（11）
  8 参考文献 ……………………………………………………………………………（13）

第一部分 研究瘤胃微生物组的工具

第 2 章 微生物培养的挑战：让消化道微生物在组学时代苏醒 ………………………（21）
  1 前言 …………………………………………………………………………………（21）
  2 培养方法及营养因素对微生物生长的影响 ………………………………………（22）
  3 基于基因组的消化道微生物定向分离技术 ………………………………………（26）
  4 基于分子的消化道微生物分离技术 ………………………………………………（29）
  5 基于抗体的消化道微生物分离技术 ………………………………………………（30）
  6 结论和展望 …………………………………………………………………………（32）
  7 参考文献 ……………………………………………………………………………（32）

第 3 章 瘤胃代谢组——探究和了解瘤胃功能和健康的有力工具 ……………………（40）
  1 前言 …………………………………………………………………………………（40）
  2 瘤胃代谢组：分析和提取技术 ……………………………………………………（41）
  3 影响瘤胃代谢组的因素 ……………………………………………………………（43）
  4 研究展望 ……………………………………………………………………………（45）
  5 结论 …………………………………………………………………………………（46）
  6 参考文献 ……………………………………………………………………………（46）

第 4 章 瘤胃微生物功能数学建模的方法 ………………………………………………（49）
  1 前言 …………………………………………………………………………………（49）
  2 瘤胃整体功能建模的方法 …………………………………………………………（50）
  3 量化瘤胃微生物功能 ………………………………………………………………（57）
  4 功能单元与取样技术 ………………………………………………………………（59）

  5 结论和展望 …………………………………………………………………… (62)
  6 更多信息 ……………………………………………………………………… (63)
  7 参考文献 ……………………………………………………………………… (64)

第二部分 瘤胃微生物群落

第 5 章 基因组测序与瘤胃微生物组 …………………………………… (69)
  1 前言 …………………………………………………………………………… (69)
  2 第一个瘤胃微生物基因组 …………………………………………………… (69)
  3 单菌基因组测序蕴含的能量 ………………………………………………… (70)
  4 构建瘤胃可培养微生物参考基因组集 ……………………………………… (72)
  5 宏基因组数据在新基因组构建中的应用 …………………………………… (73)
  6 瘤胃比较基因组与关键功能 ………………………………………………… (75)
  7 基因组是蛋白质组的蓝图 …………………………………………………… (76)
  8 微生物间的相互作用 ………………………………………………………… (77)
  9 结论 …………………………………………………………………………… (79)
  10 未来趋势 …………………………………………………………………… (83)
  11 更多信息 …………………………………………………………………… (84)
  12 参考文献 …………………………………………………………………… (84)

第 6 章 瘤胃古菌 …………………………………………………………… (93)
  1 前言 …………………………………………………………………………… (93)
  2 瘤胃中的古菌 ………………………………………………………………… (93)
  3 瘤胃中产甲烷菌的培养 ……………………………………………………… (97)
  4 利用分子生物学技术鉴定和定量瘤胃产甲烷菌 ………………………… (105)
  5 氢营养型产甲烷菌：*Methanobrevibacter ruminantium* M1 …………… (114)
  6 其他氢营养型产甲烷菌 …………………………………………………… (118)
  7 甲基营养型产甲烷菌 ……………………………………………………… (119)
  8 乙酸营养型产甲烷菌：*Methanosarcina* sp. CM1 ……………………… (123)
  9 结论 ………………………………………………………………………… (123)
  10 参考文献 …………………………………………………………………… (124)

第 7 章 瘤胃原虫 …………………………………………………………… (137)
  1 前言 ………………………………………………………………………… (137)
  2 瘤胃原虫的发现 …………………………………………………………… (137)
  3 瘤胃原虫分类和种群类型 ………………………………………………… (138)
  4 瘤胃原虫基因组学 ………………………………………………………… (141)
  5 原虫种群的变化 …………………………………………………………… (142)
  6 瘤胃内原虫的相互作用 …………………………………………………… (142)
  7 研究瘤胃原虫面临的挑战 ………………………………………………… (145)

8 原虫功能对反刍动物营养、健康和甲烷排放的影响 …………………… (146)
  9 研究例证：调控瘤胃纤毛虫 ……………………………………………… (148)
  10 未来趋势和结论 …………………………………………………………… (149)
  11 更多信息 …………………………………………………………………… (150)
  12 参考文献 …………………………………………………………………… (151)

第8章 瘤胃厌氧真菌 ……………………………………………………………… (158)
  1 前言 ………………………………………………………………………… (158)
  2 厌氧真菌的生命周期 ……………………………………………………… (158)
  3 厌氧真菌的分类学和形态学特征 ………………………………………… (159)
  4 厌氧真菌的属和种 ………………………………………………………… (161)
  5 单中心属 …………………………………………………………………… (164)
  6 多中心属 …………………………………………………………………… (168)
  7 球根属 ……………………………………………………………………… (171)
  8 厌氧真菌的基因组学 ……………………………………………………… (172)
  9 厌氧真菌的宏组学 ………………………………………………………… (187)
  10 瘤胃真菌与瘤胃生态系统中其他成员之间的交互作用 ………………… (189)
  11 结论 ………………………………………………………………………… (193)
  12 更多信息 …………………………………………………………………… (193)
  13 参考文献 …………………………………………………………………… (193)

第9章 瘤胃病毒和染色体外遗传元件 …………………………………………… (206)
  1 前言 ………………………………………………………………………… (206)
  2 染色体外元件 ……………………………………………………………… (206)
  3 瘤胃病毒 …………………………………………………………………… (210)
  4 移动体对瘤胃功能的作用和影响 ………………………………………… (218)
  5 结论和展望 ………………………………………………………………… (222)
  6 致谢 ………………………………………………………………………… (223)
  7 更多信息 …………………………………………………………………… (223)
  8 参考文献 …………………………………………………………………… (223)

第10章 瘤胃上皮附着微生物 …………………………………………………… (234)
  1 前言 ………………………………………………………………………… (234)
  2 瘤胃上皮附着微生物群落结构 …………………………………………… (234)
  3 瘤胃上皮附着微生物的定植 ……………………………………………… (237)
  4 影响瘤胃上皮附着微生物的因素 ………………………………………… (238)
  5 瘤胃上皮附着微生物对反刍动物生产性能的影响 ……………………… (239)
  6 挑战与展望 ………………………………………………………………… (243)
  7 结论 ………………………………………………………………………… (245)
  8 致谢 ………………………………………………………………………… (245)
  9 参考文献 …………………………………………………………………… (245)

第三部分　营养代谢在瘤胃微生物和宿主互作中的作用

第 11 章　瘤胃内纤维的消化 (255)
 1　前言 (255)
 2　木质纤维素类生物质 (255)
 3　碳水化合物活性酶 (257)
 4　瘤胃中原核生物降解纤维的策略 (259)
 5　目前的知识盲区 (266)
 6　改善纤维消化过程 (268)
 7　结论与未来发展趋势 (269)
 8　更多信息 (269)
 9　参考文献 (270)

第 12 章　瘤胃蛋白质分解和氨同化 (279)
 1　前言 (279)
 2　微生物氮代谢 (280)
 3　提高瘤胃氮代谢效率的方法 (291)
 4　结论 (292)
 5　未来展望 (293)
 6　更多信息 (293)
 7　参考文献 (293)

第 13 章　影响瘤胃能量代谢效率的因素 (302)
 1　前言 (302)
 2　瘤胃发酵的主要途径 (302)
 3　甲烷 (313)
 4　影响微生物生长效率的因素 (317)
 5　瘤胃能量与氮代谢的相互作用 (323)
 6　结论和未来趋势 (325)
 7　更多信息 (325)
 8　参考文献 (325)

第 14 章　阐明瘤胃脂质代谢以优化乳品质、促进人类健康并监测动物健康 (337)
 1　前言 (337)
 2　瘤胃脂质代谢及脂肪酸合成 (339)
 3　日粮和瘤胃脂肪酸向乳腺的转运以及乳腺中脂肪酸代谢 (342)
 4　乳腺内源性脂肪酸的代谢 (343)
 5　反刍动物脂肪酸对人类健康的影响 (344)
 6　增加多不饱和脂肪酸（PUFA）以改善牛奶脂肪酸结构 (346)
 7　过瘤胃脂质保护技术 (347)

8	瘤胃源的乳脂肪酸作为生物标志物监测动物健康	(350)
9	结论	(355)
10	参考文献	(355)

第15章　反刍动物温室气体产生的营养因素：对肠道和粪便甲烷排放的影响 (366)

1	前言	(366)
2	干酒糟及其可溶物（DDGS）	(372)
3	硝基化合物	(373)
4	植物次生化合物	(377)
5	碳衍生材料	(379)
6	微生物的氢利用	(381)
7	未来趋势及结论	(383)
8	更多信息	(384)
9	参考文献	(385)

第16章　瘤胃微生物与宿主间的相互作用及对饲料转化效率、甲烷产量和其他生产性能的影响 (396)

1	前言	(396)
2	瘤胃核心菌群、微生物组成的恢复和变化	(396)
3	依赖微生物组的特性	(398)
4	甲烷产生	(401)
5	氮化合物：利用和排放	(403)
6	微生物组和宿主遗传	(403)
7	参考文献	(404)

第17章　瘤胃对奶牛免疫功能的调节 (410)

1	前言	(410)
2	亚急性瘤胃酸中毒（SARA）在牛群中的发病率	(411)
3	瘤胃健康、代谢活动和紊乱	(412)
4	瘤胃健康和乳腺免疫系统	(415)
5	结论	(418)
6	致谢	(418)
7	补充信息	(418)
8	参考文献	(418)

第四部分　优化瘤胃微生物功能的营养策略

第18章　瘤胃微生物在反刍动物生产系统中的作用 (429)

1	前言	(429)
2	日粮和瘤胃微生物组	(430)
3	瘤胃纤维素降解	(431)

  4 瘤胃微生物组和饲料效率 ……………………………………………………… (436)
  5 瘤胃微生物群和甲烷产生 ……………………………………………………… (444)
  6 甲烷产生与剩余采食量 ………………………………………………………… (446)
  7 饲用植物对动物生产性能的影响 ……………………………………………… (448)
  8 结论 ……………………………………………………………………………… (457)
  9 展望 ……………………………………………………………………………… (457)
  10 参考文献 ……………………………………………………………………… (458)

第19章 优化瘤胃功能：青贮饲料和精料提高奶牛饲料转化率、减少甲烷和
     氮排放的作用 ……………………………………………………………… (472)
  1 前言 ……………………………………………………………………………… (472)
  2 青贮的作用：禾本牧草、豆科牧草和玉米 …………………………………… (473)
  3 精料的作用：脂质、碳水化合物和蛋白质 …………………………………… (481)
  4 案例研究：粉碎油菜籽对奶牛产奶量、乳脂组成和瘤胃甲烷排放的影响 … (493)
  5 总结和未来趋势 ………………………………………………………………… (496)
  6 补充材料 ………………………………………………………………………… (497)
  7 参考文献 ………………………………………………………………………… (498)

第20章 应用舍饲与谷物提高反刍动物饲料利用率并减少甲烷产生 ……………… (505)
  1 前言 ……………………………………………………………………………… (505)
  2 肉牛谷饲的类型 ………………………………………………………………… (506)
  3 谷物生产 ………………………………………………………………………… (508)
  4 影响反刍动物产甲烷的日粮因素 ……………………………………………… (509)
  5 淀粉和饲草在甲烷形成中的作用 ……………………………………………… (510)
  6 瘤胃 H_2 库和甲烷的产生 ……………………………………………………… (512)
  7 利用谷物提高饲料转化率并降低甲烷产量 …………………………………… (513)
  8 谷物发酵微生物学 ……………………………………………………………… (515)
  9 参与发酵的细菌与古细菌 ……………………………………………………… (516)
  10 过瘤胃时间 …………………………………………………………………… (518)
  11 饲料引起的酸中毒和其他负面影响 ………………………………………… (518)
  12 小结 …………………………………………………………………………… (520)
  13 更多信息 ……………………………………………………………………… (520)
  14 参考文献 ……………………………………………………………………… (521)

第21章 植物次生产物：在反刍动物可持续生产及营养中的有益作用 …………… (528)
  1 前言 ……………………………………………………………………………… (528)
  2 精油 ……………………………………………………………………………… (531)
  3 单宁 ……………………………………………………………………………… (538)
  4 皂苷 ……………………………………………………………………………… (547)
  5 未来趋势和结论 ………………………………………………………………… (553)
  6 参考文献 ………………………………………………………………………… (554)

第22章　益生菌作为反刍动物添加剂的应用 (568)

1 前言 (568)

2 益生菌对反刍动物生命周期中关键时期的靶向调控 (568)

3 定义、递送机制和管理 (571)

4 益生菌的益处和作用方式：幼龄反刍动物 (574)

5 益生菌的益处和作用方式：成年反刍动物的饲料效率 (577)

6 益生菌的益处和作用方式：甲烷生成 (581)

7 益生菌的益处和作用方式：病原体控制 (581)

8 益生菌的益处和作用方式：对免疫系统的影响 (583)

9 结论和未来趋势 (585)

10 致谢 (586)

11 更多信息 (586)

12 参考文献 (587)

第1章 瘤胃微生物区系的定植和建立——调控生产性能和甲烷排放的窗口期

Diego P. Morgavi、Milka Popova，法国国家农业食品与环境研究院，法国；
David Yañez-Ruiz，西班牙高等科研理事会，西班牙；
Evelyne Forano，法国国家农业食品与环境研究院，法国
（王佳堃译）

1 前言

反刍动物的胃肠道微生物与宿主密不可分。肠道共生微生物在宿主与周围环境的相互作用中发挥重要作用，它能够为宿主提供基本的营养、免疫和保护。与机体其他重要的"器官"相似，胃肠道微生物区系也经历从早期形成到成熟的一系列发展阶段。然而，与胚胎的形成过程不同，胃肠道微生物区系的发育从出生才开始，并且在生命早期动态演替（Savage，1977；Jami 等，2013；Rey 等，2014）。多种微生物种群在胃肠道生态系统中的定植和演替，对于微生物区系发挥正常的功能及微生物与宿主互作至关重要（Costello 等，2012）。

因为日粮、解剖结构和肠道理化状态对动物胃肠道微生物具有很强的选择性，所以环境是决定动物胃肠道微生物群落结构的强大驱动力（Ley 等，2008）。然而，一些偶然的和持续性的操作也会影响胃肠道微生物群落的组合，甚至可能对反刍动物产生持久性的影响（Yáñez-Ruiz 等，2010；Morgavi 等，2015；Moraïs 和 Mizrahi，2019）。本章综述了幼龄反刍动物瘤胃和后肠道微生物区系的建立及其调控技术，以期促进动物健康和获得理想表型。

2 瘤胃微生物区系的建立

由于瘤胃微生物区系组成直接影响宿主动物的消化和代谢，所以很多研究探究宿主动物从出生到成年瘤胃内微生物的定植规律。这些研究包括早期使用微生物纯培养方法开展的工作（Fonty 等，1983，1988），也包括近期使用高通量测序技术在犊牛、绵羊和山羊羔羊上开展的研究（Jami 等，2013；Rey 等，2014；Guzman 等，2015；Wang 等，2017b；Abecia 等，2018；Dias 等，2018）。新生反刍动物的瘤胃尚未发育成熟，这为调控瘤胃共生微生物，从而对成年反刍动物产生长期影响提供了特有的机会（Yáñez-Ruiz 等，2015）。

2.1 定植：从出生（不反刍）至功能齐全的瘤胃

最近的综述描述了当动物只采食固体饲料时，从出生至断奶及断奶后瘤胃中微生物群落的演替（Malmuthuge 等，2015；Yanez-Ruiz 等，2015；Meale，2017a）。一些功能种群以及存在于成年瘤胃中的种属，在反刍动物出生后不久，就以一种渐进的方式，以明确的顺序出现在瘤胃中。一些研究采用高通量测序和 qPCR 方法监测了犊牛从出生到断奶期间瘤胃细菌群落的建立过程（Jami 等，2013；Rey 等，2014；Guzman 等，2015）。研究发现，出生后的几天里瘤胃细菌组成变化迅速。1~3 日龄的犊牛，变形菌门（Proteobacteria）和链球菌属（*Streptococcus*）相关的序列丰富，随后被严格厌氧的细菌迅速取代（Jami 等，2013）。随着动物的生长，变形菌门逐渐被拟杆菌门（Bacteroidetes）取代，而厚壁菌门（Firmicutes）从幼年到成年一直存在（表1）。上述结果与早期使用纯培养技术的研究结果相一致，即好氧和兼性厌氧细菌先定植于瘤胃（Fonty 等，1987）。尽管如此，在成熟瘤胃中发挥重要功能的纤维素分解细菌和产甲烷菌等严格厌氧微生物在动物出生后的 1 d 或 2 d 就已存在于瘤胃中（Fonty 等，1987；Gagen 等，2012；Jami 等，2013；Guzman 等，2015）。

表1 宿主动物从出生到成年主要菌门的定植时间线（数值以平均百分比表示）

菌门	时间						
	3 d	7 d	14 d	28 d	42 d	6 个月	2 年
拟杆菌门	13.9~42.6	56.3~56.9	46.0~61.3	49.9~56.3	56.3~74	38.5~55.2	38.5~50.2
厚壁菌门	5.1~13.9	13.9~17.5	13.9~34.0	13.9~42.1	10~43.9	36.8~48.9	34.5~56.7
放线菌门	0.05~4.9	0.6~4.9	0.9~4.9	0.3~4.9	0.3~4.1	3	3
梭杆菌门	4.7~5.5	4.7~5.3	0.2~0.6	0.2~0.3	0.2~0.4	0.1	0.1
螺旋体门	0~0.4	0.1~0.4	0.4~2.6	0.4~0.9	0.4	0.7~1.2	0.9~2.5
纤维杆菌门	0~0.3	0~0.3	0.2~0.3	0.3~1.5	0.3~1.6	0.2~1.7	0.5~2.1
柔膜菌门	0	0.8	0.2	0.9	1.0	1.0~1.6	1.3~2.3

资料来源：Li 等，2012；Jami 等，2013；Rey 等，2014；Yáñez-Ruzi 等，2015；Abecia 等，2018。

在进食固体饲料前的 2~4 日龄，以及两周后，羔羊尚未成熟的瘤胃中就可以检测到产甲烷菌，而且产甲烷菌的丰度与成年羊的丰度相当（Fonty 等，1987；Morvan 等，1994）。羔羊出生 17 h 后置于无菌隔离器中饲养，虽然传统培养技术无法检测到产甲烷菌，但分子生物学方法却可以鉴定出丰度低但多样化的产甲烷菌群（以 *Methanobrevibacter* spp. 为主）（Gagen 等，2012）。近期以山羊为对象的一项研究也表明，在羔羊出生后的第 1 天，有活性的产甲烷菌已经定植于瘤胃中，其中甲烷杆菌目（Methanobacteriales）的甲烷短杆菌属（*Methanobrevibacter*）、甲烷球菌属（*Methanosphaera*）和甲烷马赛球菌目（Methanomassiliicoccales）的嗜甲烷念珠菌属（*Methanomethylophilus*）是丰度最高的 3 个属

(Wang 等，2017b)。甲烷杆菌目、甲烷微菌目（Methanomicrobiales）、甲烷八叠球菌目（Methanosarcinales）和甲烷马赛球菌目是瘤胃中常见的产甲烷菌目（Janssen 和 Kirs，2008）。这些产甲烷细菌在 1~2 周龄的犊牛中大量存在，但通过 qPCR 方法在成熟瘤胃中仅能检测到甲烷杆菌目和甲烷马赛球菌目（Friedman 等，2017）。基于生成甲烷的底物类型，Friedman 等认为早期的产甲烷菌群可能甲基营养型产甲烷活性强，由甲烷八叠球菌目执行甲基营养型产甲烷过程。随后，真核微生物也在瘤胃中定植。羔羊出生后的 8~10 d，瘤胃中可以检测到厌氧真菌（Fonty 等，1987）。早在动物定期采食固体饲料之前，瘤胃中就已存在分解纤维素的厌氧真菌。纤毛虫要在出生 2~3 周后才可以检测到，而且定植顺序是先内毛虫属（Entodinium）（15~20 d），之后是多腹甲纤毛虫属（Polyplastron）、真双毛虫属（Eudiplodinium）和前毛属（Epidinium）（20~25 d），最后是等毛虫属（Isotricha）（50 d）（Fonty 等，1988）。与细菌和古菌不同，如果新生动物在出生后立即母子分离，瘤胃中就无原虫定植（Fonty 等，1988；Chaucheyras-Durand 等，2019）。此外，复杂的微生物群落是纤毛虫定植的前提（Fonty 等，1983，1988）。图 1 展示了从哺乳期至断奶结束，羔羊瘤胃中微生物的主要定植事件。瘤胃壁上细菌的定植也与年龄有关，它们以一定的顺序逐渐丰富起来（Rieu 等，1990）。变形菌门，尤其是该门中的埃希氏杆菌属（Escherichia）是瘤胃壁上的优势菌（Jiao 等，2015；Wang 等，2017a）。而与瘤胃壁相关的这些变形菌门在瘤胃内的丰度随年龄增长而减少，厚壁菌门和拟杆菌门的丰度随年龄增长而增加（Jiao 等，2015）。

断奶前犊牛瘤胃食糜中的细菌区系与瘤胃壁黏附的细菌菌群存在较大差异。与瘤胃壁黏附的细菌相比，食糜中普雷沃氏菌属（Prevotella）丰度更高，拟杆菌门的丰度更低（Malmuthuge 等，2014）。

尽管本章的重心是瘤胃，但也会提及其他胃肠道部位的一些微生物信息。虽然瘤胃微生物随日龄的演替已经研究得相对深入，但关于网胃和瓣胃这些前胃的信息甚少。2018 年，Lei 等（2018）和 Yeoman 等（2018）分别分析了山羊羔羊出生后 3~56 d，犊牛出生至 21 d 的瘤胃液、网胃液和瓣胃液中微生物的组成。瘤胃变形菌门的相对丰度随年龄增长而逐渐减少，但拟杆菌门的相对丰度随年龄增长而增加。

对于后消化道，分娩后的最初几小时内也发现了极为多样的微生物（Alipour 等，2018）。厚壁菌门、变形菌门、放线菌门（Actinobacteria）和拟杆菌门是新生动物直肠中的优势菌群，但微生物群落结构在动物出生后的早期会迅速变化。有研究报道，出生后数小时内，大肠杆菌属（E. coli）和链球菌属率先在犊牛和羔羊的胃肠道所有区域定植（Smith，1965）。非培养的方法也证实，出生后 1 d，大肠杆菌属和链球菌属为直肠微生物中的优势菌属（Alipour 等，2018）。对于瘤胃（Jami 等，2013），推测这些兼性厌氧菌能够消耗氧气，为严格厌氧菌提供适宜的生存环境。乳酸杆菌（Lactobacilli）正是利用这种优势，在 1 日龄反刍动物的所有肠段内定植（Smith，1965）。变形菌门、厚壁菌门和拟杆菌门普遍存在于出生 3 周内的所有胃肠道部位（Malmuthuge 等，2014；Alipour 等，2018；Yeoman 等，2018）。厚壁菌门为幼龄反刍动物胃肠道远端（结肠和粪便）的优势菌群，而拟杆菌门在网胃、瘤胃、瓣胃和皱胃中丰度更高（Malmuthuge 等，2014；Yeoman 等，2018）。值得注意的是，厚壁菌门和拟杆菌门在内容物中的丰度更高，而变形菌门在胃壁

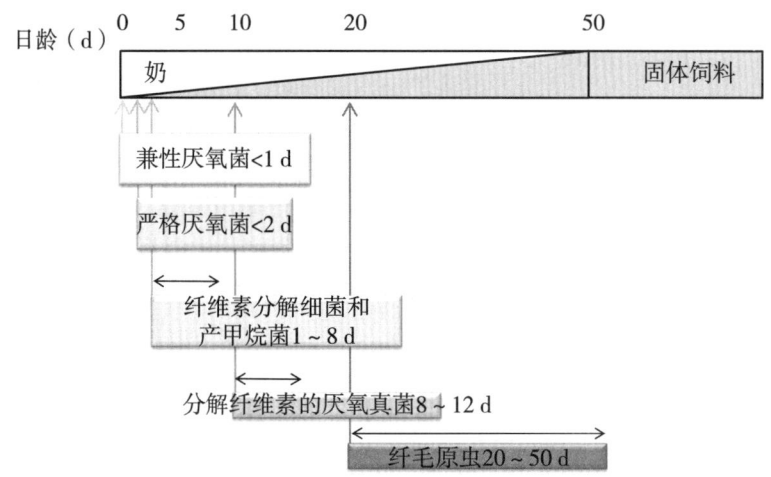

图1 羔羊瘤胃的微生物定植。通过微生物纯培养技术和分子生物学方法检测微生物群在瘤胃的定植。将羔羊与母畜共同饲养。箭头表示定植开始的时期。

黏膜中丰度更高（Yeoman等，2018），这应该是微氧等环境条件和黏蛋白等底物驱动了这种黏膜相关种群特有的生态位。几乎所有肠道部位的内容物样品中，微生物的丰富度和 α 多样性随着日龄的增长而增加，但黏膜微生物种群则没有这种变化趋势。尽管如此，小肠和大肠黏膜相关细菌种群的多样性要高于食糜（Malmuthuge 等，2014）。尽管到目前为止，胃后的肠道还没有受到太多的关注，但有证据表明，肠道中的微生物群对后期动物的健康和生长性能起着至关重要的作用。例如，初乳提供的被动免疫下降时，乳酸菌通过刺激宿主的适应性免疫系统，增强犊牛的主动免疫来增加免疫球蛋白 A（IgA）的生成（Corthésy 等，2007）。

2.2 宿主—微生物间互作

哺乳动物的胃肠道有多种非特异性和特异性保护机制使宿主与栖息的微生物共存（Hooper 等，2012）。营养素吸收、共生微生物耐受和致病微生物屏障间存在功能冲突，这需要一个复杂的物理、生化和细胞机制系统来保护胃肠上皮和宿主免受入侵（Kuhn 和 Stappenbeck，2013）。研究表明，免疫系统需要经历一个训练过程来应对微生物影响，这在生命早期阶段尤为重要（Collado 等，2012；Wu 和 Wu，2012）。可能因为瘤胃与肠道相比，上皮结构特殊，所以其对最初定植的微生物的"耐受"机制尚不清楚。瘤胃是多达 15 层细胞的角化上皮，没有系统的淋巴组织（Sharpe 等，1975）。这限制了免疫球蛋白等大分子的渗透。瘤胃的免疫平衡主要通过以下两种机制来实现：（i）Toll 样受体（toll-like receptors，TLRs）的信号传导（Malmuthuge 等，2012）；（ii）经唾液提供免疫球蛋白（主要是分泌型 IgA）（Williams 等，2009）。

（1）TLRs 是一个模式识别受体（PRRs）家族，通过充当传感器识别病原体和共生微生物菌表达的分子模式，在黏膜表面发挥关键作用。其他哺乳动物具有的 TLRs，奶牛也都有，如 TLR1-10 在奶牛瘤胃中都已经被鉴别到，并且被进行了表征（McGuire 等，

2006），而且有研究发现营养代谢紊乱（如酸中毒）与 TLR 的表达水平相关（Chen 等，2012）。但 TLR 是否参与生命早期的微生物定植却知之甚少。除 TLR-1 和 TLR-3 外，新生犊牛胃肠道中其他 TLR 的表达水平高于 6 月龄犊牛（Malmuthuge 等，2012）。最近，Abecia 等（2017）评估了山羊羔羊早期母乳喂养与人工喂养两种管理模式对瘤胃免疫反应的影响，发现 TLR1、TLR2、TLR5、TLR8 和 TLR10 表达存在日龄依赖性，表现出在 5~7 日龄时表达量增加，随后下降至稳定。TLRs 表达量增加的同时，挥发性脂肪酸（VFA）浓度和微生物定植量也增加。尽管随后微生物仍持续定植，但 TLRs 基因表达水平并未进一步提高。这与新生婴儿（Teran 等，2011）和新生犊牛（Malmuthuge 等，2012）血液中 TLR 表达水平随日龄增长而下调，但记忆性 T 细胞数量增加的报道相一致。丁酸盐也可以促进 TLR 活性的下调，丁酸盐通过稳定细胞间紧密连接蛋白信号传导增强屏障功能（Jiao 等，2017）。因此，为数不多的几项研究表明，随日龄增长 TLR 下调可能是宿主避免对微生物做出非必要炎症反应的机制之一。对于相同日龄的哺乳期犊牛结肠黏膜中 IL-8、IL-10 和紧密连接蛋白-4（claudin-4）的表达水平而言，喂生乳的犊牛要低于喂热处理乳的犊牛（Bach 等，2017）。喂生乳的犊牛结肠黏膜中乳酸杆菌的丰度更高，提示日龄增长诱导微生物发育得更加成熟，随之下调 TLR 表达。

（2）免疫球蛋白尤其是 IgA，主要通过唾液进入瘤胃。IgA 是唾液中主要的免疫球蛋白，也是瘤胃中的主要免疫球蛋白类型（Subharat 等，2015；Abecia 等，2017）。因为 IgA 的分泌组分更耐受蛋白酶，所以在瘤胃中 IgA 要比 IgG 稳定（Snoeck 等，2006）。犊牛胃肠道中，分泌型 IgA 包裹着共生细菌（Tsuruta 等，2012；Fouhse 等，2017），而且与未被 IgA 包裹的口腔微生物相比，IgA 包裹的口腔微生物与瘤胃微生物的相似性更高，这提示唾液 IgA 包裹可能是宿主影响共生微生物定植的机制（Fouhse 等，2017）。在某种程度上，这与前期针对特定微生物开发疫苗的工作矛盾。前期的研究提示唾液 IgA 对某些微生物有抑制作用。但近期 Donaldson 等（2018）的研究支持 Fouhse 等（2017）的观点。Donaldson 等（2018）呈现了 IgA 的特异性免疫识别促进了细菌在肠上皮细胞上的黏附，从而介导了微生物的稳定定植。这表明，除了清除病原体的作用外，IgA 还有助于宿主接纳共生微生物的定植。但这还需要在早期瘤胃定植中验证。

瘤胃定植还通过基因转录诱导肝脏反应，这可能与共生微生物耐受的形成有关（Li 等，2019）。将成年奶牛的瘤胃液移植给新生犊牛，肝脏主要上调的基因富集于免疫反应、抗炎反应和细胞信号传导（Li 等，2019）。

2.3 断奶与瘤胃微生物区系的稳定

瘤胃发育研究有两个核心的问题：一是哪些因素（日龄、断奶前日粮、断奶后期）驱动了反刍动物在断奶过程中有一个健康的、功能性的微生物群？二是瘤胃微生物从何日龄起稳定下来，可被认为是发育成熟了？在标准的养殖管理模式下，断奶过渡期可称为应激期，会减少采食量和降低生长速度。与断奶后的瘤胃一样，拟杆菌门、厚壁菌门和变形菌门也是断奶前犊牛瘤胃里的优势菌门，只是这些门的菌群相对丰度随日龄和发育阶段的不同而不同（Li 等，2012；Jami 等，2013；Meale 等，2016；Dias 等，2018）。一般而言，断奶后厚壁菌门的菌群相对丰度增加，而拟杆菌门的菌群相对丰度降低（Jami

等，2013；Meale 等，2016，2017b）。但是反刍动物的类型不同，甚至是同种动物不同的试验，这些菌门随时间的变化以及细菌科和属的演替顺序也会有所不同。管理模式、断奶日龄和饲料也都会影响微生物的演替（Meale 等，2016，2017a,b）。例如，6 周龄早期断奶的犊牛，瘤胃微生物的 β 多样性变化迅速，而 8 周龄断奶的犊牛，瘤胃微生物呈现更加渐进式的变化（Meale 等，2017b）。Meale 等认为与早期断奶相比，晚期断奶犊牛固体饲料采食量的逐渐增加可减少断奶过渡期的生理应激强度。此外，给断奶前犊牛喂奶的同时，补喂精料开食料可促进瘤胃巨型球菌（*Megasphaera*）、沙棘菌（*Sharpea*）和琥珀酸弧菌（*Succinivibrio*）等易发酵碳水化合物降解细菌的多样性，并有利于甲烷球形菌而非甲烷短杆菌的生长（Dias 等，2018）。在 Dias 等（2018）的研究中，真菌的相对丰度并没有随日粮或日龄显著变化，这可能是组内变异大以及日粮纤维含量低的原因。近期以山羊为对象的研究显示，断奶前后瘤胃微生物组成和代谢组明显不同，说明了日粮对瘤胃微生物功能的影响（Abecia 等，2018）。同样在山羊上观察到产甲烷菌的组成随日粮和年龄的变化而变化且不稳定，但在断奶时稳定下来（Wang 等，2017b）。断奶似乎是瘤胃微生物群落结构趋同于成年反刍动物的转折点。尽管如此，在较低的分类水平上，许多微生物类群是不同的（Dill-McFarland 等，2017）。从断奶转变为成年样的瘤胃微生物区系可能需要几个月甚至一年的时间（Fonty 等，1988；Dill-McFarland 等，2017）。

2.4 微生物活性与功能

目前为止的研究主要是以 16S（18S/ITS）rDNA 扩增子测序来监测瘤胃微生物的定植。非常有必要通过检测微生物活性或将功能性微生物一个个测定出来，以确定瘤胃建立了饲料降解和其他功能活性。前文已经提到，1~2 日龄反刍动物的瘤胃中已经存在纤维素分解细菌和产甲烷菌（Fonty 等，1987；Guzman 等，2015）。1 日龄起，瘤胃中就可以检测到利用饲料中植物纤维、蛋白质和淀粉等营养物质的酶活性（Rey 等，2012；Jia 等，2015）。这些酶的活性第 1 周不断增加，并在 1 个月左右，即微生物定植的初始阶段完成之后呈现出最大的活性（Jiao 等，2015）。这些变化可能部分反映了饮食从初乳到常乳或代乳粉，再逐渐转度为固体饲料的过程。如前文所述，早在出生后的 2~3 d 就可以检测到产甲烷活性（Morvan 等，1994；Friedman 等，2017）。纯培养和同位素标记试验也显示，羔羊瘤胃中氢营养活性的建立似乎也是有次序的。出生 20h 的羔羊瘤胃中已存在氢依赖的产乙酸过程，但是氢依赖的产甲烷过程要出生 30h 后才检测得到。氢利用型硫酸盐还原菌要出生第 3 天才定植（Morvan 等，1994）。通过监测未成熟羔羊瘤胃中乙酰辅酶 A（acetyl-CoA）合成酶和甲酰四氢叶酸（formyltetrahydrofolate）合成酶基因，探究了出生后还原型产乙酸菌的定植。鉴定出的潜在的产乙酸菌归属于布劳特氏菌属（*Blautia*）和毛螺菌科（Lachnospiraceae），产甲烷活性的建立并未从实质上影响这些羔羊的产乙酸菌多样性（Gagen 等，2012）。Li 等（2012）通过宏基因组学探究了反刍前的犊牛瘤胃微生物功能的多样性。在 14 日龄犊牛的瘤胃中检测到 8 000 多个推测的 Pfam 蛋白家族，鉴定出 60 多个糖苷水解酶家族，表明早在断奶前就已经有很高的碳水化合物代谢潜力（Li 等，2012）。尽管需要谨慎解读基于分类信息获得的功能代谢预测结果（Vieira-Silva 等，2016），但根据 16S rRNA 基因序列仍推断出 5~9 周龄犊牛的碳水化合物代谢增强（Meale

等，2017b）。

2.5 传递方式

传统上认为子宫是无菌环境，微生物在新生儿肠道中的定植始于分娩。通过母亲的阴道、粪便、初乳、皮肤和唾液以及环境传递给新生儿。然而，早在胎儿出生 20 min，就在瘤胃、肠道和胎粪中检测到了细菌和古细菌的 DNA（Guzman 等，2015），这提示瘤胃接种可能发生在出生前。婴儿和新生鼠的研究表明，这些微生物的肠道定植始于子宫（Ihekweazu 和 Versalovic，2018）。这使得反刍动物的研究结果间矛盾。一些研究表明出生前微生物就开始了反刍动物肠道的定植（Alipour 等，2018），而另一些研究报道的结果则相反（Malmuthug 和 Griebel，2018）。因为胎盘结构和母-胎"对话"在哺乳动物间差异很大，且反刍动物具有最完整的胎盘屏障，所以在解释来自人或其他物种的结果时应谨慎（Benirschke 等，2012）。另外在分析这类低微生物量样品时，一定要避免污染（Malmuthugg 和 Griebel，2018）。现有研究结果不足以证实或反驳反刍动物胃肠道微生物始于子宫。

无论胃肠道的细菌定植是否始于出生前，尚不清楚所有其他菌源的相对贡献。但肯定的是母亲是新生反刍动物微生物的重要供体。事实上，与母亲分开单独饲养的新生羔羊，瘤胃里没有纤维分解细菌、真菌和原虫（Fonty 等，1988）。新生双胞胎羔羊瘤胃产甲烷菌组成相似，也说明母亲是非常重要的微生物供体（Skillman 等，2004）。因为反刍，所以母羊口腔微生物可能额外重要。羔羊出生后母羊本能地舔舐就会将口腔微生物传递给羔羊。初乳和常乳也都是新生动物胃肠道定植细菌的供体。与单胃动物一样，吸乳的反刍动物，液体饲料经食管沟直接进入真胃（皱胃）。但这个沟闭合得并不完全，少量的乳会进入瘤胃，拟杆菌属（*Bacteroides*）和乳杆菌属（*Lactobacillus*）等能够利用乳营养成分，是仅饲喂乳的哺乳犊牛的优势微生物（Dias 等，2018）。产琥珀酸杆菌（*F. succinogenes*）能利用乳糖生长，这可以用来解释断奶前犊牛为何可以检测到大量纤维降解菌（Ghali 等，2017）。最近，研究人员探讨了来自母畜不同身体部位的微生物群落对新生儿胃肠道定植的重要性（Alipour 等，2018；Yeoman 等，2018）。Alipour 等（2018）的研究显示，母畜口腔的微生物群落与新生犊牛直肠的微生物群落最为相似。但是 Yeoman 等（2018）则发现，犊牛 1~21 日龄时，与初乳或阴道样本相比，母畜的乳房皮肤与犊牛的肠腔和黏膜共有的操作分类单元（OTUs）最多。奇怪的是，在奶牛的阴道样本中检测出纤维分解菌和产甲烷菌的序列，说明分娩时阴道微生物对犊牛肠道微生物定植是有作用的（Yeoman 等，2018）。除此之外，母畜阴道微生物对子代上呼吸道微生物的定植也起重要作用（Lima 等，2019）。

2.6 管理模式的影响

如果日粮补充固体饲料是瘤胃微生物定植的一个关键转折点，那么需要回答的一个重要问题就是新生反刍动物的饲养管理是否会改变微生物定植模式。很少有研究在促进或抑制某些影响微生物定植的因素时，比较未发育瘤胃中微生物定植，这些因素包括母本效应、隔离饲养的子代、微生物调节剂的使用等。

如前文所述，母畜对早期定植具有重要影响。最近的研究进一步阐释了管理体系如何影响母本效应（Abecia 等，2014b，2017，2018；Belanche 等，2015；Yeoman 等，2018）。反刍动物生产中主要有两套子代管理体系。商品化奶牛体系中，新生犊牛通常一出生就与母牛分离，饲喂代乳粉或全乳。相反，在肉牛和粗放型生产体系中，直至断奶才母子分离。因此，这两种体系意味着在乳的类型（全乳与代乳粉）以及是否与成年同种动物接触方面存在差异，许多情况下，这两套体系是混淆的。人工喂养的动物，其原虫定植模式与母畜喂养的动物不同（Abecia 等，2014a；Belanche 等，2015）。母羊自然哺乳与人工代乳粉喂养相比，留在母亲身边的山羊羔在其发育过程中瘤胃 pH 值始终较低（Abecia 等，2014a）。Abecia 等（2014a）推测群饲化学习（social feeding learning）可能使自然哺乳的羔羊在早期进食了更多的精料。De Paula Vieira 等（2012）也报道了群饲化学习，他们观察到，将断奶前犊牛与日龄大一点的牛一起饲养，可以刺激采食行为，促进断奶前后的生长。除学习行为外，与日龄大一些的同种动物一起饲养也会影响瘤胃微生物定植，如整个定植过程中，母羊哺乳的山羊羔要比人工喂养的羊羔细菌多样性丰富得多（Abecia 等，2017）。纤维杆菌门（Fibrobacteres）（瘤胃中主要的纤维素分解菌门之一）从出生后第 7 天起就存于自然喂养的幼畜体内；然而，直到第 28 天，它才在母子分离饲养的幼畜的瘤胃中定植，发酵琥珀酸生成丙酸的琥珀酸杆菌（*Succiniclasticum*）也有类似的情况。而且，在母畜哺乳喂养的幼畜瘤胃中挥发性脂肪酸（VFA）浓度更高，这表明微生物发酵在母畜哺乳喂养的幼畜瘤胃中建立得更快（Abecia 等，2014a）。但不清楚的是，这种自然哺喂和人工饲喂引起的微生物定植差异是否有长期效应，是否影响生长后期动物的消化能力。值得进一步研究母体对出生后幼畜微生物定植的影响，因为揭示这一影响有助于畜牧生产中开发出新的微生物定植策略。

3 调控幼龄反刍动物胃肠道微生物区系以实现动物健康和高产

前面章节中已经提到，很多因素影响幼龄反刍动物胃肠道微生物的定植。这其中尤以生产体系、牧场管理和日粮最为重要。好的管理模式通常在出生后第一周内降低疾病的发生，促进动物的快速生长，这显然会影响胃肠道微生物定植。例如，饲喂优质的初乳可以建立有益的黏膜相关细菌屏障，从而保护幼龄反刍动物免受肠道感染（Malmutjug 等，2015）。初乳有利于有益的双歧杆菌（*Bifidobacterium*）增殖，减少结肠中潜在病原菌和埃希氏杆菌属及志贺菌属（*Shigella*）中的微生物（Song 等，2019 年）。吮吸母乳和同种成年动物接触是确保早期自然获得功能性胃肠道微生物区系的最佳方式。然而，在目前的生产体系下，这种自然模式并不总是适用的，应该考虑采用早期的营养干预去促进（或防止）某些微生物群体的定植。一些特殊情况下，如果存在传播不良微生物种群的风险，则不推荐与母畜接触。鸟分枝杆菌亚种副结核（*Mycobacterium avium* subsp. *paratuberculosis*）是引起慢性细菌性肠炎（Johne's disease）的病原体，其感染风险，尤其是出生后 1 周内，随着新生动物与受感染成年动物共处天数的增加而上升（Burgess 等，2018）。反之，通过促进与某些动物接触以实现微生物传递，进而减少肠道甲烷排放或提高饲料效率等，这应该也是可行的（Shabat 等，2016；Difford 等，2018）。

日粮是公认的胃肠道微生物调节剂，对各年龄段的动物都有调节作用，但对幼龄反刍动物微生物区系的发育和微生物—宿主互作有着特定的调节作用（Bach 等，2017；Dill-McFarland 等，2019；Dong 等，2019）。牛奶加工过程中的差异，从宏观营养角度来看可以忽略不计，但却对某些肠道微生物种群和炎症指标产生很大影响。与巴氏杀菌和高温热处理的乳相比，生乳使乳酸杆菌数量增加，促炎白细胞介素的表达量降低，奶牛犊的生长性能提高（Bach 等，2017）。

除了上述提到的管理和日粮因素外，早期还应该考虑另外两种主要的营养干预措施来调节胃肠道微生物区系。一个是直接接种特定的微生物，另一个是补饲添加剂等化合物阻止或促进一些微生物群体的定植。这两种策略主要是为了促进实现目标表型。

4 案例研究：促进健康和生产的早期调控策略

给反刍动物饲喂活的微生物并不是一个新概念，大量关于直接饲喂微生物（direct-fed microbials, DFM）的研究已被报道（Martin 和 Nisbet，1992；Jeyanathan 等，2014）。Theodorou 等（1990）报道添加新美鞭菌属（*Neocallimastix* spp.）的厌氧瘤胃真菌可增加断奶时犊牛的采食量和活体重。Ziolecka 等（1984）和 Ziolecki 等（1984）报道补充瘤胃提取物在断奶期间可以提高犊牛的活体重，刺激瘤胃发育。早期新鲜瘤胃液接种已引起新的关注。最近的研究显示将成年动物的新鲜瘤胃液接种到新生幼畜的瘤胃中对生产性能有积极的影响（Zhong 等，2014；De Barbieri 等，2015b）。瘤胃液接种促进了绵羊和山羊羔羊断奶后的平均日增重、采食量、消化率和瘤胃发育。尽管接种后 5 月龄时瘤胃细菌的组成仍有所不同，但对日增重、采食量、消化率和瘤胃发育的影响可能消失（De Barbieri 等，2015a,b）。

作为特定的 DFM 制剂，给哺乳期犊牛饲喂乳酸菌可刺激反刍和瘤胃发育（Nakanishi 等，1993），乳酸菌或双歧杆菌可提高饲料转化率和增重（Abe 等，1995）。后一项研究中 DFMs 降低了腹泻率。Signorini 等（2012）的荟萃分析证实了乳酸菌对腹泻的作用。有趣的是，只有多菌株的 DFM，而且是在犊牛饲喂生乳的情况下才能观察到这种保护作用。乳酸菌的这些作用可部分归因于这些细菌对宿主免疫有正向的调节活性，这个内容在"宿主—微生物互作"小节中已进行阐述。瘤胃埃氏巨型球菌（*Megasphaera elsdenii*）利用乳酸生产丁酸盐，丁酸盐是胃肠道上皮黏膜可以利用的有益的能量化合物。埃氏巨型球菌也被用作 DFM，在幼龄反刍动物上使用，但效果并不一致。Muya 等（2015）发现该菌可提高采食量，改善瘤胃发育，表明上皮细胞代谢增强，消化终产物吸收增加。然而，Yohe 等（2018）使用相同菌株和相似的操作规程却未观察到该菌的作用。

腹泻是断奶前的反刍动物，尤其是奶用犊牛主要的健康问题（美国农业部，2018 年）。粪便微生物多样性低，特别是粪杆菌属的 *Faecalibacterium* spp.（已知它是一种具有抗炎活性的共生细菌）的相对丰度低，与腹泻发病率高度一致（Oikonomou 等，2013）。反之，*Faecalibacterium* spp. 与增重正相关。与未处理的对照组相比，使用普氏栖粪杆菌（*F. prausnitzii*）作为奶犊牛的 DFM 可将重度腹泻发病率降低一半，死亡率降低 2/3（Foditsch 等，2015）。

另一受关注的策略是早期使用活性酵母。尽管酵母在反刍动物营养中被广泛使用（Chaucheyras-Durand 等，2012），但在反刍前动物的日粮中应用酵母来改变微生物定植却是很新的想法。已有研究显示，酵母，尤其是作为 DFM 时可促进犊牛的生长，促进瘤胃和小肠发育，改善免疫和健康状态（Kim 等，2011；Alugongo 等，2017）。具体而言，酿酒酵母（Saccharomyces cerevisiae yeast，SCY）可以提高犊牛的干物质采食量、生长和饲料转化率（Lesmeister 和 Heinrichs，2004；Alugongo 等，2017），并降低腹泻发生率（Galvao 等，2005；Brewer 等，2014）。另外，瘤胃发酵（丁酸盐的生成量增加）和瘤胃乳头生长略有提高（Lesmeister 和 Heinrichs，2004）。处于应激状态的犊牛或犊牛养殖环境中存在大量致病因子时，更容易观察到酵母上述的效果。虽然关于反刍动物早期 SCY 的研究报道日渐增多，但关于 SCY 对微生物定植的影响还知之甚少。近期，Terré 等（2015）报道，用活性 SCY 饲喂幼龄雄性犊牛，会增加瘤胃中的纤维分解菌和白色瘤胃球菌（R. albus）的丰度，同时提高瘤胃 pH 值，但对其他细菌和原虫无影响。但也有研究显示，饲喂代乳粉的新生羔羊补饲活性酵母可增加瘤胃中产琥珀酸丝状杆菌（F. succinogenes）、真菌和原虫的定植（Chaucheyras-Durand 等，2019）。另有报道，SCY 增加粪便中乳酸杆菌丰度（Fomenky 等，2017）；增加断奶应激期的氧化应激和吞噬活性，表明 SCY 可增强先天免疫反应（Fomenky 等，2018）。一般来说，所有 DFM 和成年瘤胃微生物移植都还需要进一步开展整个微生物组的分析，需要在肠道免疫和后期效应等方面开展深入研究，对于新生犊牛管理模式尤为需要开展研究。如果想得出 DFM 和成年瘤胃微生物移植对幼龄反刍动物作用的一致结论，这些研究是必需要开展的。

5 案例研究：减少肠道甲烷排放的早期策略

调控发育中瘤胃的微生物定植以减少甲烷排放是另一关注点。为了这个目标，人们采用两种方法筛选一些作为饲料添加剂或饲料补充剂的化合物。一种方法是专门针对产甲烷菌，另一种方法是利用已报道的有多种作用机制的广谱抑制剂。针对产甲烷菌，补饲溴氯甲烷可改变古菌在山羊羔羊瘤胃中的定植，从而使甲烷排放量减少约 50%。分娩后母羊和羔羊同时补饲溴氯甲烷，停止补饲后，溴氯甲烷降低羔羊甲烷排放量的效果仍可持续 3 个月（Abecia 等，2013，2014a）。对于广谱抑制剂，人们评估了那些被证实有效改变成年动物微生物代谢和甲烷排放的脂类和植物提取物的作用效果。Debruyne 等（2018）评估了山羊出生前后椰子油中链脂肪酸的补饲效果。该处理通过抑制产甲烷菌的丰度和活性，减少了 4 周龄羔羊的体外甲烷排放量，但这会导致瘤胃发酵和细菌丰度下降，甚至抑制羔羊的日增重。出生后的处理是在 11 周龄时结束，但有些瘤胃乳头的变化在 28 周龄时仍可观察到，表明早期营养管理可以影响瘤胃乳头发育。Saro 等（2018）评估了绵羊羔羊早期联合添加大蒜油和亚麻籽油的效果，以及后期再次添加的作用。处理过程中产甲烷菌的组成发生变化，但停止处理后对产甲烷菌和甲烷排放量的作用都消失。相反，处理组和未处理组羔羊瘤胃细菌的组成在干预过程中和停止干预后，均有差异。停止干预后瘤胃和尿液代谢组图谱凸显了微生物和代谢物之间的相互作用，尤其是甲基化的化合物和甲烷马赛球菌科（Methanomassiliicoccaeae）产甲烷菌之间的互作。他们证明，长时间的早期干预会诱导瘤胃细菌组成改变，这种改变在干预停止后持续存在，但对古细菌和原虫的组成没有影响或影响极小。早期干预对甲烷生成的作用并不持久，表明该功能具有显著的恢复力。与此相似，Lyons等（2017）用亚麻籽油处理绵羊羔羊，发现细菌群落结构的差异持续存在，但细菌群落结构的差异与表型已不对应。

产甲烷菌在瘤胃生态系统中具有特定的生态位。抑制产甲烷菌时，氢的流向及其他电子供体会向丙酸盐等电子受体转移。这可以解释前文所有引用的研究中为何主要是在细菌群落中观察到营养代谢的重排和微生物种群的变化。虽然由于抑制剂类型不同可能对产甲烷菌组成的影响有所不同，产甲烷菌的丰度不一定受到影响，但产甲烷菌的多样性通常都是会改变的。亚麻籽和大蒜油降低绵羊羔羊甲烷马赛球菌目的相对丰度（Saro等，2018）。溴氯甲烷增加山羊羔羊甲烷短杆菌属的相对丰度，降低甲烷球菌属的相对丰度（Abecia等，2014b）。而亚麻籽增加绵羊羔羊甲烷球菌属的相对丰度（Lyons等，2017）。

6 结论和展望

反刍动物从出生起就不断接触各种来源的微生物。只有创建了适宜的环境，才能建立稳定的微生物区系。适宜的环境中日粮和胃肠道的成熟度间紧密互作。肠道微生物群落的演替和紧随其后的代谢功能的扩展是一个协调的过程，与胃肠道的解剖结构和生理发育相匹配。虽然瘤胃的变化主要发生在3~8周龄的不反刍向反刍的过渡阶段，及8周龄起的反刍阶段（Jiao等，2015），但0~3周的不反刍阶段，多样化的微生物群落的建立，尤其细菌在后消化道的定植，对动物健康和生长至关重要（Oikonomou等，2013；Bach等，2017）。为长期调节胃肠道微生物组成，在日粮、解剖结构和生理改变发生的时期进行干预更为有效（Dill-McFarland等，2019）。

多种策略能影响和调节早期胃肠道微生物群落（图2）。然而，如果在养殖场应用这些策略，还有一些突出的问题需要解决。进一步研究胃肠道微生物对动物表型的作用，将使我们了解对于幼龄反刍动物应该强化哪种微生物菌群结构。必须评估出生前胃肠道是否有定植，出生之前的阶段在成熟瘤胃微生物区系建立中的作用等。类似的，必需优化调节微生物的最佳窗口期。建立不同的菌群可能不是一次操作就可以解决的，想获得成年样的微生物菌群（Morgavi等，2015）或成功植入特定的微生物（Yohe等，2018），必须持续处理。更好地界定早期微生物群落在宿主微生物互作和免疫系统调节中的作用对于建立不同的菌群也很重要。未来几年，所有这些主题的预期成果可通过改善幼龄动物健康和提高效率来促进反刍动物产业的可持续发展。

7 更多信息

7.1 重要的会议

肠道微生物学研讨会（Gut Microbiology Symposium）：由法国国家农业科学研究院（INRA France）和阿伯丁大学Rowett研究所（Rowett Institute-Aberdeen University）联合

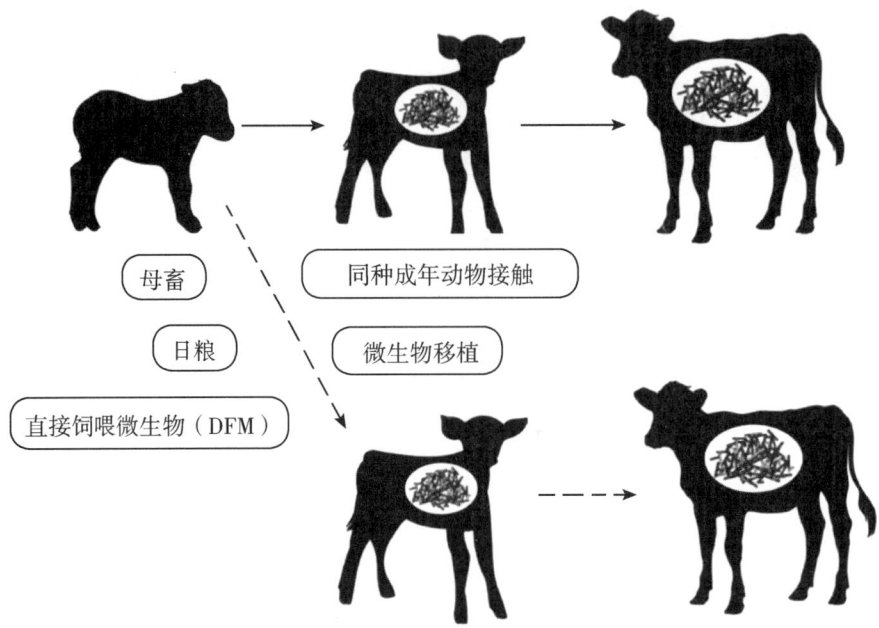

图2 通过饲养管理和有针对性的干预措施实现早期肠道微生物群的定植（虚线箭头），以促进健康和生产性能，例如提高饲料效率和减少甲烷排放。调节方法有：允许与含有理想微生物群的母畜或同种成年动物接触，从同种成年动物肠胃内容物移植微生物群，日粮调控和补充直接饲喂微生物（DFM）、益生菌和饲料添加剂。

举办，每两年一次。

胃肠功能大会（Congress on Gastrointestinal Function）：每两年在美国芝加哥举行一次。

动物科学学会的年会（Annual meetings from animal science societies）：例如欧洲动物科学联合会和美国奶业科学学会的年会。

犊牛智能饲养会议（https://smart-calf-rearing.com）。

7.2 主要的国际研究项目和网络平台

畜牧全球研究联盟—瘤胃微生物基因组学网（Rumen Microbial Genomics Network，Global Research Alliance-Livestock Group）是一个促进反刍动物微生物学家之间合作和交流的大型国际网络平台（http://global researchalliance.org/research/livestock/networks/rumen-microbial-genomics-network/）。

MASTER（通过工程技术和企业，将微生物组应用于可持续食品系统）是一个专注于来自不同环境包括反刍动物胃肠道的微生物组的表征和调节的H2020项目（http://www.master-h2020.eu/）。

8 参考文献

Abe, F., Ishibashi, N. and Shimamura, S. 1995. Effect of administration of bifidobacteria and lactic acid bacteria to newborn calves and piglets. J. Dairy Sci. 78 (12), 2838-46.

Abecia, L., Martín-García, A. I., Martínez, G., et al., 2013. Nutritional intervention in early life to manipulate rumen microbial colonization and methane output by kid goats postweaning. J. Anim. Sci. 91 (10), 4832-40.

Abecia, L., Ramos-Morales, E., Martínez-Fernandez, G., et al., 2014a. Feeding management in early life influences microbial colonisation and fermentation in the rumen of newborn goat kids. Anim. Prod. Sci. 54 (9), 1449-54.

Abecia, L., Waddams, K. E., Martinez-Fernandez, G., et al., 2014b. An antimethanogenic nutritional intervention in early life of ruminants modifies ruminal colonization by Archaea. Archaea 2014, 841463.

Abecia, L., Jiménez, E., Martínez-Fernandez, G., et al., 2017. Natural and artificial feeding management before weaning promote different rumen microbial colonization but not differences in gene expression levels at the rumen epithelium of newborn goats. PLoS ONE 12 (8), e0182235.

Abecia, L., Martínez-Fernandez, G., Waddams, K., et al., 2018. Analysis of the rumen microbiome and metabolome to study the effect of an antimethanogenic treatment applied in early life of kid goats. Front. Microbiol. 9, 2227.

Alipour, M. J., Jalanka, J., Pessa-Morikawa, T., et al., 2018. The composition of the perinatal intestinal microbiota in cattle. Sci. Rep. 8 (1), 10437.

Alugongo, G. M., Xiao, J. X., Wu, Z. H., et al., 2017. Review: Utilization of yeast of *Saccharomyces cerevisiae* origin in artificially raised calves. J. Anim. Sci. Biotechnol. 8, 34.

Bach, A., Aris, A., Vidal, M., et al., 2017. Influence of milk processing temperature on growth performance, nitrogen retention, and hindgut's inflammatory status and bacterial populations in a calf model. J. Dairy Res. 84 (3), 355-9.

Belanche, A., de la Fuente, G. and Newbold, C. J. 2015. Effect of progressive inoculation of fauna-free sheep with holotrich protozoa and total-fauna on rumen fermentation, microbial diversity and methane emissions. FEMS Microbiol. Ecol. 91 (3).

Benirschke, K., Burton, G. J. and Baergen, R. N. 2012. Pathology of the Human Placenta. Springer Science & Business Media, Heidelberg.

Brewer, M. T., Anderson, K. L., Yoon, I., et al., 2014. Amelioration of salmonellosis in pre-weaned dairy calves fed *Saccharomyces cerevisiae* fermentation products in feed and milk replacer. Vet. Microbiol. 172 (1-2), 248-55.

Burgess, T. L., Witte, C. L. and Rideout, B. A. 2018. Early-life exposures and Johne's disease risk in zoo ruminants. J. Vet. Diagn. Invest. 30 (1), 78-85.

Chaucheyras-Durand, F., Chevaux, E., Martin, C., et al., 2012. Use of yeast probiotics in ruminants: effects and mechanisms of action on rumen pH, fibre degradation, and microbiota according to the diet. In: Rigobelo, E. (Ed.), Probiotic in Animals. Intech Europe, Rijeka, Croatia, pp. 119-52.

Chaucheyras-Durand, F., Ameilbonne, A., Auffret, P., et al., 2019. Supplementation of live yeast based feed additive in early life promotes rumen microbial colonization and fibrolytic potential in lambs. Sci.

Rep. 9, 19216.

Chen, Y., Oba, M. and Guan, L. L. 2012. Variation of bacterial communities and expression of toll-like receptor genes in the rumen of steers differing in susceptibility to subacute ruminal acidosis. Vet. Microbiol. 159 (3-4), 451-9.

Collado, M. C., Cernada, M., Baüerl, C., et al., 2012. Microbial ecology and host-microbiota interactions during early life stages. Gut Microbes 3 (4), 352-65.

Corthésy, B., Gaskins, H. R. and Mercenier, A. 2007. Cross-talk between probiotic bacteria and the host immune system. J. Nutr. 137, 781S-90S.

Costello, E. K., Stagaman, K., Dethlefsen, L., et al., 2012. The application of ecological theory toward an understanding of the human microbiome. Science 336 (6086), 1255-62.

De Barbieri, I., Hegarty, R. S., Silveira, C., et al., 2015a. Programming rumen bacterial communities in newborn Merino lambs. Small Rumin. Res. 129, 48-59.

De Barbieri, I., Hegarty, R. S., Silveira, C., et al., 2015b. Positive consequences of maternal diet and post-natal rumen inoculation on rumen function and animal performance of Merino lambs. Small Rumin. Res. 129, 37-47.

Debruyne, S., Ruiz-Gonzalez, A., Artiles-Ortega, E., et al., 2018. Supplementing goat kids with coconut medium chain fatty acids in early life influences growth and rumen papillae development until 4 months after supplementation but effects on *in vitro* methane emissions and the rumen microbiota are transient. J. Anim. Sci. 96 (5), 1978-95.

De Paula Vieira, A., von Keyserlingk, M. A. G. and Weary, D. M. 2012. Presence of an older weaned companion influences feeding behavior and improves performance of dairy calves before and after weaning from milk. J. Dairy Sci. 95 (6), 3218-24.

Dias, J., Marcondes, M. I., Motta de Souza, S., et al., 2018. Bacterial community dynamics across the gastrointestinal tracts of dairy calves during preweaning development. Appl. Environ. Microbiol. 84 (9).

Difford, G. F., Plichta, D. R., Lovendahl, P., et al., 2018. Host genetics and the rumen microbiome jointly associate with methane emissions in dairy cows. PLoS Genet. 14 (10), e1007580.

Dill-McFarland, K. A., Breaker, J. D. and Suen, G. 2017. Microbial succession in the gastrointestinal tract of dairy cows from 2 weeks to first lactation. Sci. Rep. 7, 40864.

Dill-McFarland, K. A., Weimer, P. J., Breaker, J. D., et al., 2019. Diet influences early microbiota development in dairy calves without long-term impacts on milk production. Appl. Environ. Microbiol. 85 (2), e02141-02118.

Donaldson, G. P., Ladinsky, M. S., Yu, K. B., et al., 2018. Gut microbiota utilize immunoglobulin A for mucosal colonization. Science 360 (6390), 795-800.

Dong, L. -f., Ma, J. -N., Tu, Y., et al., 2019. Weaning methods affect ruminal methanogenic archaea composition and diversity in Holstein calves. J. Integr. Agric. 18 (5), 1080-92.

Foditsch, C., Pereira, R. V., Ganda, E. K., et al., 2015. Oral administration of Faecalibacterium prausnitzii decreased the incidence of severe diarrhea and related mortality rate and increased weight gain in preweaned dairy heifers. PLoS ONE 10 (12), e0145485.

Fomenky, B. E., Chiquette, J., Bissonnette, N., et al., 2017. Impact of *Saccharomyces cerevisiae* boulardii CNCMI-1079 and *Lactobacillus acidophilus* BT1386 on total lactobacilli population in the gastrointestinal tract and colon histomorphology of Holstein dairy calves. Anim. Feed Sci. Tech. 234, 151-61.

Fomenky, B. E., Chiquette, J., Lessard, M., et al., 2018. *Saccharomyces cerevisiae* var. boulardii

CNCM I-1079 and *Lactobacillus acidophilus* BT1386 influence innate immune response and serum levels of acute-phase proteins during weaning in Holstein calves. Can. J. Anim. Sci. 98 (3), 576-88.

Fonty, G., Gouet, P., Jouany, J. P., et al., 1983. Ecological factors determining establishment of cellulolytic bacteria and protozoa in the rumens of meroxenic lambs. J. Gen. Microbiol. 129 (1), 213-23.

Fonty, G., Gouet, P., Jouany, J. -P., et al., 1987. Establishment of the microflora and anaerobic fungi in the rumen of lambs. J. Gen. Microbiol. 133 (7), 1835-43.

Fonty, G., Senaud, J., Jouany, J. P., et al., 1988. Establishment of ciliate protozoa inthe rumen of conventional and conventionalized lambs: influence of diet and management conditions. Can. J. Microbiol. 34 (3), 235-41.

Fouhse, J. M., Smiegielski, L., Tuplin, M., et al., 2017. Host immune selection of rumen bacteria through salivary secretory IgA. Front. Microbiol. 8, 848.

Friedman, N., Jami, E. and Mizrahi, I. 2017. Compositional and functional dynamics of the bovine rumen methanogenic community across different developmental stages. Environ. Microbiol. 19 (8), 3365-73.

Gagen, E. J., Mosoni, P., Denman, S. E., et al., 2012. Methanogen colonisation does not significantly alter acetogen diversity in lambs isolated 17 h after birth and raised aseptically. Microb. Ecol. 64 (3), 628-40

Galvao, K. N., Santos, J. E. P., Coscioni, A., et al., 2005. Effect of feeding live yeast products to calves with failure of passive transfer on performance and patterns of antibiotic resistance in fecal *Escherichia coli*. Reprod. Nutr. Dev. 45 (4), 427-40.

Ghali, I., Sofyan, A., Ohmori, H., et al., 2017. Diauxic growth of *Fibrobacter succinogenes* S85 on cellobiose and lactose. FEMS Microbiol. Lett. 364 (15), 9.

Guzman, C. E., Bereza-Malcolm, L. T., De Groef, B., et al., 2015. Presence of selected methanogens, fibrolytic bacteria, and proteobacteria in the gastrointestinal tract of neonatal dairy calves from birth to 72 hours. PLoS ONE 10 (7), e0133048.

Hooper, L. V., Littman, D. R. and Macpherson, A. J. 2012. Interactions between the microbiota and the immune system. Science 336 (6086), 1268-73.

Ihekweazu, F. D. and Versalovic, J. 2018. Development of the pediatric gut microbiome: impact on health and disease. Am. J. Med. Sci. 356 (5), 413-23.

Jami, E., Israel, A., Kotser, A., et al., 2013. Exploring the bovine rumen bacterial community from birth to adulthood. ISME J. 7 (6), 1069-79.

Janssen, P. H. and Kirs, M. 2008. Structure of the archaeal community of the rumen. Appl. Environ. Microbiol. 74 (12), 3619-25.

Jeyanathan, J., Martin, C. and Morgavi, D. P. 2014. The use of direct-fed microbials for mitigation of ruminant methane emissions: a review. Animal 8 (2), 250-61.

Jiao, J., Li, X., Beauchemin, K. A., et al., 2015. Rumen development process in goats as affected by supplemental feeding v. grazing: age-related anatomic development, functional achievement and microbial colonisation. Br. J. Nutr. 113 (6), 888-900.

Jiao, J., Zhou, C., Guan, L. L., et al., 2017. Shifts in host mucosal innate immune function are associated with ruminal microbial succession in supplemental feeding and grazing goats at different ages. Front. Microbiol. 8, 1655.

Kim, M. H., Yang, J. Y., Upadhaya, S. D., et al., 2011. The stress of weaning influences serum levels of acute-phase proteins, iron-binding proteins, inflammatory cytokines, cortisol, and leukocyte subsets in Holstein calves. J. Vet. Sci. 12 (2), 151-7.

Kuhn, K. A. and Stappenbeck, T. S. 2013. Peripheral education of the immune system by the colonic microbiota. Semin. Immunol. 25 (5), 364-9.

Lei, Y., Zhang, K., Guo, M., et al., 2018. Exploring the spatial-temporal microbiota of compound stomachs in a pre-weaned goat model. Front. Microbiol. 9, 1846.

Lesmeister, K. E. and Heinrichs, A. J. 2004. Effects of corn processing on growth characteristics, rumen development, and rumen parameters in neonatal dairy calves. J. Dairy Sci. 87 (10), 3439-50.

Ley, R. E., Lozupone, C. A., Hamady, M., et al., 2008. Worlds within worlds: evolution of the vertebrate gut microbiota. Nat. Rev. Microbiol. 6 (10), 776-88.

Li, R. W., Connor, E. E., Li, C., et al., 2012. Characterization of the rumen microbiota of pre-ruminant calves using metagenomic tools. Environ. Microbiol. 14 (1), 129-39.

Li, W. L., Edwards, A., Riehle, C., et al., 2019. Transcriptomics analysis of host liver and meta-transcriptome analysis of rumen epimural microbial community in young calves treated with artificial dosing of rumen content from adult donor cow. Sci. Rep. 9 (1), 790.

Lima, S. F., Bicalho, M. L. S. and Bicalho, R. C. 2019. The Bos taurus maternal microbiome: role in determining the progeny early-life upper respiratory tract microbiome and health. PLoS ONE 14 (3), e0208014.

Lyons, T., Boland, T., Storey, S., et al., 2017. Linseed oil supplementation of lambs' dietin early life leads to persistent changes in rumen microbiome structure. Front. Microbiol. 8, 1656.

Malmuthuge, N. and Griebel, P. J. 2018. Fetal environment and fetal intestine are sterile during the third trimester of pregnancy. Vet. Immunol. Immunopathol. 204, 59-64.

Malmuthuge, N., Li, M., Fries, P., et al., 2012. Regional and age dependent changes in gene expression of toll-like receptors and key antimicrobial defence molecules throughout the gastrointestinal tract of dairy calves. Vet. Immunol. Immunopathol. 146 (1), 18-26.

Malmuthuge, N., Griebel, P. J. and Guan, L. 2014. Taxonomic identification of commensal bacteria associated with the mucosa and digesta throughout the gastrointestinal tracts of preweaned calves. Appl. Environ. Microbiol. 80 (6), 2021-8.

Malmuthuge, N., Griebel, P. J. and Guan, L. 2015. The gut microbiome and its potential role in the development and function of newborn calf gastrointestinal tract. Front. Vet. Sci. 2, 36.

Martin, S. A. and Nisbet, D. J. 1992. Effect of direct-fed microbials on rumen microbial fermentation. J. Dairy Sci. 75 (6), 1736-44.

McGuire, K., Jones, M., Werling, D., et al., 2006. Radiation hybrid mapping of all 10 characterized bovine toll-like receptors. Anim. Genet. 37 (1), 47-50.

Meale, S. J., Li, S., Azevedo, P., et al., 2016. Development of ruminal and fecal microbiomes are affected by weaning but not weaning strategy in dairy calves. Front. Microbiol. 7, 582.

Meale, S. J., Chaucheyras-Durand, F., Berends, H., et al., 2017a. From pre-to postweaning: transformation of the young calf's gastrointestinal tract. J. Dairy Sci. 100 (7), 5984-95.

Meale, S. J., Li, S. C., Azevedo, P., et al., 2017b. Weaning age influences the severity of gastrointestinal microbiome shifts in dairy calves. Sci. Rep. 7 (1), 198.

Moraïs, S. and Mizrahi, I. 2019. The Road Not Taken: the rumen microbiome, functional groups, and community states. Trends Microbiol. 27 (6), 538-49.

Morgavi, D. P., Rahahao-Paris, E., Popova, M., et al., 2015. Rumen microbial communities influence metabolic phenotypes in lambs. Front. Microbiol. 6, 1060.

Morvan, B., Dore, J., Rieu-Lesme, F., et al., 1994. Establishment of hydrogen-utilizing bacteria in the rumen of the newborn lamb. FEMS Microbiol. Lett. 117 (3), 249-56.

Muya, M. C., Nherera, F. V., Miller, K. A., et al., 2015. Effect of *Megasphaera elsdenii* NCIMB 41125 dosing on rumen development, volatile fatty acid production and blood beta-hydroxybutyrate in neonatal dairy calves. J. Anim. Physiol. Anim. Nutr. (Berl) 99 (5), 913-8.

Nakanishi, Y., Arave, C. W. and Stewart, P. H., 1993. Effects of feeding *Lactobacillus acidophilus* yogurt on performance and behavior of dairy calves. J. Dairy Sci. 76 (Suppl. 1), 244.

Oikonomou, G., Teixeira, A. G. V., Foditsch, C., et al., 2013. Fecal microbial diversity in pre-weaned dairy calves as described by pyrosequencing of metagenomic 16S rDNA. Associations of Faecalibacterium species with health and growth. PLoS ONE 8 (4), e63157.

Rey, M., Enjalbert, F. and Monteils, V. 2012. Establishment of ruminal enzyme activities and fermentation capacity in dairy calves from birth through weaning. J. Dairy Sci. 95 (3), 1500-12.

Rey, M., Enjalbert, F., Combes, S., et al., 2014. Establishment of ruminal bacterial community in dairy calves from birth to weaning is sequential. J. Appl. Microbiol. 116 (2), 245-57.

Rieu, F., Fonty, G., Gaillard, B., et al., 1990. Electron microscopy study of the bacteria adherent to the rumen wall in young conventional lambs. Can. J. Microbiol. 36 (2), 140-4.

Saro, C., Hohenester, U. M., Bernard, M., et al., 2018. Effectiveness of interventions to modulate the rumen microbiota composition and function in pre-ruminant and ruminant lambs. Front. Microbiol. 9, 1273.

Savage, D. C. 1977. Microbial ecology of the gastrointestinal tract. Annu. Rev. Microbiol. 31, 107-33.

Shabat, S. K., Sasson, G., Doron-Faigenboim, A., et al., 2016. Specific microbiome-dependent mechanisms underlie the energy harvest efficiency of ruminants. ISME J. 10 (12), 2958-72.

Sharpe, M., Latham, M. and Reiter, B. 1975. The immune response of the host animal to bacteriain the rumen and caecum. In: McDonald, I. W. and dan Warner, A. C. I. (Eds), Digestion and Metabolism in the Ruminant (1st edn.). The University of New England, Sydney, p. 149.

Signorini, M. L., Soto, L. P., Zbrun, M. V., et al., 2012. Impact of probiotic administration on the health and fecal microbiota of young calves: a meta-analysis of randomized controlled trials of lactic acid bacteria. Res. Vet. Sci. 93 (1), 250-8.

Skillman, L. C., Evans, P. N., Naylor, G. E., et al., 2004. 16S ribosomal DNA-directed PCR primers for ruminal methanogens and identification of methanogens colonising young lambs. Anaerobe 10 (5), 277-85.

Smith, H. W. 1965. The development of the flora of the alimentary tract in young animals. J. Pathol. Bacteriol. 90 (2), 495-513.

Snoeck, V., Peters, I. R. and Cox, E. 2006. The IgA system: a comparison of structure and function in different species. Vet. Res. 37 (3), 455-67.

Song, Y., Malmuthuge, N., Li, F. Y., et al., 2019. Colostrum feeding shapes the hindgut microbiota of dairy calves during the first 12 h of life. FEMS Microbiol. Ecol. 95 (1), 12.

Subharat, S., Shu, D., Zheng, T., et al., 2015. Vaccination of cattle with a methanogen protein produces specific antibodies in the saliva which are stable in the rumen. Vet. Immunol. Immunopathol. 164 (3-4), 201-7.

Teran, R., Mitre, E., Vaca, M., et al., 2011. Immune system development during early childhood in tropical Latin America: evidence for the age-dependent down regulation of the innate immune response. Clin. Immunol. 138 (3), 299-310.

Terré, M., Maynou, G., Bach, A., et al., 2015. Effect of *Saccharomyces cerevisiae* CNCM I-1077 supplementation on performance and rumen microbiota of dairy calves. Prof. Anim. Sci. 31 (2), 153-8.

Theodorou, M., Beever, D., Haines, M., et al., 1990. The effect of a fungal probiotic on intake and performance of early weaned calves. Anim. Prod. 50, 577.

Tsuruta, T., Inoue, R., Tsukahara, T., et al., 2012. Commensal bacteria coated by secretory immunoglobulin A and immunoglobulin G in the gastrointestinal tract of pigs and calves. Anim. Sci. J. 83 (12), 799-804.

USDA. 2018. Dairy 2014, "Health and Management Practices on U. S. Dairy Operations, 2014". Vol. # 696. 0218. USDA-APHIS-VS-CEAH-NAHMS, Fort Collins, CO.

Vieira-Silva, S., Falony, G., Darzi, Y., et al., 2016. Species-function relationships shape ecological properties of the human gut microbiome. Nat. Microbiol. 1 (8), 16088.

Wang, Z., Elekwachi, C., Jiao, J., et al., 2017a. Changes in metabolically active bacterial community during rumen development, and their alteration by rhubarb root powder revealed by 16s rRNA amplicon sequencing. Front. Microbiol. 8, 159.

Wang, Z., Elekwachi, C. O., Jiao, J., et al., 2017b. Investigation and manipulation of metabolically active methanogen community composition during rumen development in black goats. Sci. Rep. 7 (1), 422.

Williams, Y. J., Popovski, S., Rea, S. M., et al., 2009. A vaccine against rumen methanogens can alter the composition of archaeal populations. Appl. Environ. Microbiol. 75 (7), 1860-6.

Wu, H. J. and Wu, E. 2012. The role of gut microbiota in immune homeostasis and autoimmunity. Gut Microbes 3 (1), 4-14.

Yáñez-Ruiz, D. R., Macías, B., Pinloche, E., et al., 2010. The persistence of bacterial and methanogenic archaeal communities residing in the rumen of young lambs. FEMS Microbiol. Ecol. 72 (2), 272-8.

Yáñez-Ruiz, D. R., Abecia, L. and Newbold, C. J. 2015. Manipulating rumen microbiome and fermentation through interventions during early life: a review. Front. Microbiol. 6, 1133.

Yeoman, C. J., Ishaq, S. L., Bichi, E., et al., 2018. Biogeographical differences in the influence of maternal microbial sources on the early successional development of the bovine neonatal gastrointestinal tract. Sci. Rep. 8 (1), 3197.

Yohe, T. T., Enger, B. D., Wang, L., et al., 2018. Short communication: does early-life administration of a *Megasphaera elsdenii* probiotic affect long-term establishment of the organism in the rumen and alter rumen metabolism in the dairy calf? J. Dairy Sci. 101 (2), 1747-51.

Zhong, R. Z., Sun, H. X., Li, G. D., et al., 2014. Effects of inoculation with rumen fluid on nutrient digestibility, growth performance and rumen fermentation of early weaned lambs. Livest. Sci. 162, 154-8.

Ziolecka, A., Osinska, Z. and Ziolecki, A. 1984. The effect of Stabilized Rumen Extract on growth and development of calves. 1. Liveweight gain and efficiency of feed utilization. Z. Tierphysiol. Tierernahr. Futtermittelkd. 51 (1-2), 13-20.

Ziolecki, A., Kwiatkowska, E. and Laskowska, H. 1984. The effect of Stabilized Rumen Extract on growth and development of calves. 2. Digestive activity in the rumen and development of microflora in the rumen and faeces. Z. Tierphysiol. Tierernahr. Futtermittelkd. 51 (1-2), 20-31.

第一部分

研究瘤胃微生物组的工具

第 2 章 微生物培养的挑战：让消化道微生物在组学时代苏醒

Páraic Ó Cuív

昆士兰大学 Microba 生命科学与 Mater 研究所，澳大利亚

（王佳堃译）

1 前言

哺乳动物的消化道中定植着数量巨大且品种多样的微生物，这些微生物构成了宿主消化道微生态，并通过维持宿主营养、调节宿主免疫等多个方面来调控宿主健康。在 20 世纪，人们对于哺乳动物消化道微生物的认识主要来源于反刍动物瘤胃微生物的培养，瘤胃微生物与动物的健康和生产性能的相关性使得人们对于消化道微生物的功能有了更深层次的了解。随着人们对瘤胃微生物的不断探索，20 世纪 40 年代时，已有大量文献报道了用于研究消化道微生物的培养基及培养方法，以及各种已经被成功培养的消化道微生物，但微生物培养的局限性也极大限制了人们对消化道微生物的进一步探索。宏基因组学概念的出现打破了传统培养方法的限制，随着 DNA 测序技术和计算方法的不断创新，不依赖于培养的高通量测序技术促进了人们对于消化道微生物的深入探究，"消化道微生物组"已成为一个独立的研究领域。宏基因组学让我们对动物消化道微生物的多样性和功能有了前所未有的认识，并让我们以未培养的微生物和功能基因的形式进一步探索消化道微生物"暗箱"（Stewart 等，2018；Hess 等，2011；Nayfach 等，2019；Pasolli 等，2019）。

然而，一些研究也表明，仅采用不依赖培养的方式来研究微生物仍具有一定的局限性。例如，基于培养，我们能够更好地将微生物与各种生物学过程相关联，从而进一步阐明消化道微生物研究中的因果关系（Hoedt 等，2016；Davis 等，2012；Allison 等，1992）。另外，单一消化道微生物菌株的测序结果能够进一步丰富微生物组数据库，提高宏基因组数据的比对率和组装性能（Forster 等，2019；Zou 等，2019）。随着人类微生物组的兴起，人们对瘤胃生物学家们已经建立的培养方法，以及如何建立新的培养方法又重燃了兴趣。宏基因组学的局限性和微生物组的发展都为培养工作的重新兴起注入了新的活力，现在已经有许多种哺乳动物消化道微生物的培养组项目，包括人类［例如，人类肠道菌群参考序列数据库（Culturable Genome Reference）（Zou 等，2019）和人类胃肠细菌培养集（Human Gastrointestinal Bacteria Culture Collection）（Forster 等，2019）］、反刍动物［例如，Hungate1000 培养集（Hungate1000 collection）（Seshadri 等，2018）］和啮齿类动物［小鼠肠道细菌培养集（Mouse Intestinal Bacterial Collection）（Lagkouvardos 等，2016）］的消化道培养集。值得注意的是，近年来几项基于 16S rRNA 基因和宏基因组分

析的研究认为，人们可能夸大了哺乳动物消化道环境中许多微生物的不可培养性（Martiny 等，2019；Lagkouvardos 等，2017；Zehavi 等，2018；Lau 等，2016）。但这个说法仍存在争议，最近的一项研究表明，目前仍有超过70%的人类消化道微生物缺乏可培养的代表菌株（Almeida 等，2019）。

哺乳动物具有丰富的消化道微生物区系，主要包括病毒、细菌、古菌、真菌和原虫。大量研究表明，许多"挑剔"的消化道微生物在实验室培养条件下生长良好，但仍有一些微生物无法培养（He 等，2015），这表明我们对它们的某些基本的生物学特性仍然缺乏了解。培养技术的发展正在不断扩充消化道微生物培养集（Stewart 等，2012；Vartoukian 等，2010；Clavel 等，2017），宏组学技术也帮助我们认识和了解这些微生物的代谢潜力，从而为实验室条件下的分离培养提供理论指导。因此，微生物培养和不依赖培养方法的有机结合将为解析宿主—微生物的动态关系提供新思路。

本章概述了现有的和新兴的微生物分离策略，这些策略可广泛应用于配备了消化道微生物学常规设施和设备的实验室。此外，本章还重点讨论了营养因素对"挑剔"的消化道细菌和古菌可培养性的影响，以及如何基于新兴的宏组学、遗传学和抗体技术使消化道微生物组"活起来"。

2 培养方法及营养因素对微生物生长的影响

本节重点介绍了细菌和古菌培养方法的研究进展，对消化道真菌感兴趣的读者可参考Theodorou 等（2005）和 Haitjema 等（2014）的著作，这些著作全面概述了厌氧真菌的分离、培养和保存方法。目前，我们在培养"挑剔"的消化道细菌、古菌和真菌方面已经取得了重大突破，但在肠源性噬菌体和原虫的培养方面仍然存在巨大挑战。感兴趣的读者请参考 Gilbert 和 Klieve 的著作（Gilbert 等，2015）与 Newbold 等的综述（Newbold 等，2015），它们分别介绍了目前我们对瘤胃噬菌体和原虫已有的认知。

如何从瘤胃和其他消化道中分离和培养"挑剔"的细菌和古菌？McSweeney 等（2005）和 Joblin（2005）已经对微生物分离和培养的方法进行了全面系统的阐述。目前，在厌氧微生物学的研究中，使用最为广泛的技术仍然是基于 Hungate 和 Bryant 等（Eller 等，1971；Hungate 等，1969）的发明和报道，以及 Stewart 等（1997）的描述和总结。McSweeney 等（2005）报道的培养基配方及其衍生配方仍被广泛应用于科研工作中，能满足各种消化道细菌的营养需求。大多数微生物学家都对微生物培养基的制备和厌氧箱的使用了如指掌，在此就不做赘述。准备好培养基之后，在厌氧条件下接种微生物，随后将培养基转移至厌氧箱中，通过还原剂维持厌氧状态。初学者对厌氧培养基的配置方法和原理尚不熟悉，但在有经验研究人员的指导下可以很快掌握。厌氧培养箱的使用增加了操作的灵活性和实验条件的一致性，因此，严格遵守操作规程，防止氧气进入干扰培养箱内部的厌氧环境也尤其重要。例如，试剂和耗材必须确保无氧后方能放入厌氧培养箱。

厌氧微生物培养基主要包括氮源（如蛋白胨）、碳源（如葡萄糖）、微量营养素（如酵母膏、微量元素、维生素）、pH 缓冲盐溶液、氧气指示剂和还原剂。制备培养基时，按照培养基配方将所有组分置于可经受高压和高温的容器中混合（不包括不耐热成分和还原剂），利用氧气不易溶于水的性质，将培养基煮沸以达到快速除氧的目的，随后向培养基中通入二氧化碳或氮气直至冷却，在通入气体的过程中，用铝箔包裹容器口部，让容器内部形成正压以防止大气中的氧气进入。冷却后，将pH值调整至所需范围，添加还原剂进一步去除氧气。最后，在厌氧培养箱中将培养基按需分装后进行高压灭菌，培养基冷却后按需添加除菌后的不耐热成分。

配置培养基的另一种方法是分别制备培养基的某些组分，除菌后，在厌氧培养箱内将这些组分混合。这种方法可避免在高温高压时，糖和氨基酸发生美拉德反应或产生其他有毒物质（Einarsson等，1983）。常用的缓冲介质磷酸盐会生成有毒产物，从而抑制微生物生长，降低培养基中磷酸盐的浓度有利微生物的培养和新菌株的分离（Nyonyo，2013；Kenters等，2011；Tanaka等，2014）。此外，凝胶剂也会影响微生物的生长，常用的凝胶剂为细菌琼脂，由于其低成本而被广泛使用，但其他凝胶剂也能促进微生物的复苏和培养，例如结冷胶等（Tanaka等，2014；Kawasaki等，2017；Das等，2015）。

环境模拟培养基已被广泛用于分离"挑剔"的消化道微生物（Kenters等，2011；Rettedal等，2014）。虽然这种培养基能够支持多种微生物的生长，但一些数量较少或生长缓慢的微生物易被快速生长的微生物取代（Cray等，2013）。此外，还有许多复杂的因素都是微生物培养中所面临的挑战。例如，许多消化道微生物都为营养缺陷型，它们在培养基中生长所需的营养物质要通过外部环境来满足（Zengler等，2018；Mee等，2014；Sharma等，2019）。相反，一些细菌能够在培养基中生长，说明这些细菌是原养型细菌，或培养基能够完全满足这些细菌的营养需求。因此，在进行微生物培养时，需要根据目标微生物菌株的营养需要合理设计培养基，从而特异性促进目标微生物的生长，抑制非目标微生物的生长。下一节中将简要讨论常量和微量营养素对微生物生长的影响。

2.1 氮源

瘤胃中大部分氮以氨和蛋白质的形式存在，其中，氨是瘤胃微生物的主要氮源。许多瘤胃细菌都表现出特定的氨基酸营养缺陷或偏好，一些特定的氨基酸（如亮氨酸、甘氨酸）或小肽能够特异性促进某些细菌的生长（Richardson等，2013；Oliphant等，2019；Wallace等，1996）。因此，瘤胃微生物进化出了专门的氮排出和同化的策略。瘤胃微生物主要通过二肽基肽酶将多肽降解为二肽，然后进一步被二肽酶降解为氨基酸（Wallace等，1996）。众所周知，瘤胃微生物对氮的来源有特殊的偏好。例如，Pittman和Bryant（Pittman等，1964）报道，瘤胃普雷沃氏菌（*Prevotella ruminicola*）[后被归类为栖瘤胃拟杆菌（*Bacteroides ruminicola*）]更倾向于利用肽和氨态氮，而无法利用来源于游离氨基酸或其他小分子物质中的氮；同样，互养菌MFA1（*Synergistetes strain* MFA1）、布氏普雷沃氏菌B14（*Prevotella bryantii* B14）、月形单胞菌（*Selenomonas*）和链球菌属（*Streptococcus* spp.）也对肽的偏好性高于氨基酸（Atasoglu等，1998；Leong等，2016）；而其他细菌，例如嗜淀粉梭状芽孢杆菌（*Clostridium aminophilum*）、厌氧消化链球菌（*Peptostreptococcus anaerobius*）和纤维杆菌（*Fibrobacter* spp.）对氨基酸表现出更强的偏好性（Chen等，1989；Ling等，1995）。因此，氮源会影响某些微生物的生长，一方面，蛋白胨的来源可以影响特定细菌的生长速度和细胞特性（Gray等，2008；Hinton等，2016；

Schär-Zammaretti 等,2005),另一方面,游离氨基酸也能促进或抑制特定微生物的生长(Tramontano 等,2018)。

2.2 碳源

碳水化合物是大多数消化道微生物主要的能量来源。不同的碳水化合物具有不同的结构,因而消化道微生物利用碳水化合物的能力也各不相同,部分微生物对特定的碳水化合物表现出偏好性。McSweeney 等(2005)报道了一系列能够分离纤维分解菌、木聚糖分解菌、果胶分解菌和淀粉分解菌的选择性培养基,这些及其类似的培养基至今仍在被广泛使用。在消化道内,碳水化合物对微生物生长的影响取决于宿主日粮组成及其消化程度。微生物通常优先利用结构简单的碳水化合物,其次是结构复杂的碳水化合物。在消化道中,主导碳水化合物降解的微生物将结构复杂的碳水化合物分解代谢成为结构简单的寡糖或单糖,为其他微生物提供能量。微生物对碳水化合物的偏好性被广泛应用于微生物的分离。例如,含有复杂碳水化合物的培养基可用来分离降解复杂碳水化合物的微生物,通过快速转移代谢产物,不断富集降解复杂碳水化合物的微生物,减少降解寡糖和单糖微生物的数量。在以往的研究中,人们主要通过经验来判断不同碳水化合物对微生物富集的能力,如今,全基因组测序技术的发展、各种专业软件和服务器[例如 dbCAN2(Zhang 等,2018)和 SACCHARIS(Jones 等,2018)]的出现帮助人们更清楚地了解微生物对碳水化合物的选择性,从而更好地利用不同的碳水化合物来富集和分离目标微生物。

一些消化道微生物能够转变为非糖发酵或非蛋白质发酵类型,例如梭状芽孢杆菌(*Clostridia*)和链球菌(*Peptostreptococci*)等(Richardson 等,2013;Smith 等,1998),它们能够利用非碳水化合物中的碳。消化道中还存在着其他不依赖碳水化合物生长的微生物,包括已被报道的非酵解糖细菌[例如,考拉杆菌(*Phascolarctobacterium* spp.)(Watanabe 等,2012;Shigeno 等,2019)和 *Catenibacillus scindens*(Braune 等,2018)]。此外,产乙酸菌[例如,黏液真杆菌(*Eubacterium limosum*)(Kelly 等,2016)和从袋鼠(Tammar wallaby)体内分离的 TWA4 菌株(Gagen 等,2014)]和产甲烷菌也能利用非碳水化合物中的碳,且产甲烷菌能利用二氧化碳、甲酸盐、甲醇和甲胺等多种非碳水化合物作为碳源(Gaci 等,2014)。因此,微生物可以利用各种碳源,它们对碳源的偏好性会影响其在实验室中的可培养性。

2.3 微量营养素

许多消化道微生物的生长需要特定的微量营养素,包括维生素、脂质、元素和信号分子(Vartoukian 等,2010;Xavier 等,2017)。维生素和微量元素的重要性是众所周知的,因而,在配制培养基时,通常将其配制为标准溶液,需要时直接添加使用(McSweeney 等,2005)。一些消化道微生物对微量营养素表现出复杂的需求,这可以一定程度上解释它们难以在实验室培养的原因。此外,微量营养素的量极少,很难预测每种微生物对微量营养素的具体需求。因此,在培养基中添加一些细胞提取物(例如酵母提取物、牛肉提取物)能够为微生物提供微量营养素,对于一些"挑剔"的肠道微生物,还需要添加瘤胃液(McSweeney 等,2005)和粪便水(Ternan 等,2018),这些物质不仅能够提供微量营养素，也能够模拟微生物的生活环境，更加有利于微生物的培养。

直链和支链挥发性脂肪酸能促进消化道细菌的生长（Allison 等，1958；Bryant 等，1962），而非挥发性脂肪酸也具有类似功能。在培养基中添加苯丙酸或苯乙酸能促进白色瘤胃球菌 8（*Ruminococcus albus* 8）的生长及其消化纤维素的能力（Hungate 等，1982），白色瘤胃球菌 8 能利用苯乙酸合成苯丙氨酸，促进纤维素的降解，当培养基中加入苯乙酸时，白色瘤胃球菌 8 降解纤维素的速率与添加瘤胃液的培养基相当（Stack 等，1983）。二氧化碳也能促进一些重要消化道细菌的生长（Dehority 等，1971；Reilly 等，1980）。虽然，产甲烷菌通常利用氢气产生甲烷，但也有研究显示，在氢气含量低的情况下，甲烷球形菌 WGK6（*Methanosphaera* spp. WGK6）能够利用乙醇代替氢气合成甲烷（Hoedt 等 2016），这可能是由于微生物细胞内存在能够支撑这一反应的乙醇或乙醛脱氢酶，研究者在其他产甲烷微生物中也发现了这种乙醇和乙醛脱氢酶基因的同源基因，因此，这一现象并非偶然。有研究报导，在扩散室中分离到一些环境微生物，推测在扩散室中可能有促进微生物生长的物质产生。基于此，Lewis 和同事从密集生长的细菌平板中筛选到了一些生长缓慢的菌落，这些菌落的生长依赖于其他快速生长的菌落（Fenn 等，2017）。利用这种方法，他们筛选到了新的粪杆菌（*Faecalibacterium* sp.）和苏特氏菌（*Sutterella* sp.），这些细菌的生长依赖于其他细菌代谢产生的醌类物质。纯化的醌类物质可以在没有其他辅助细菌的情况下促进粪杆菌和苏特氏菌的生长，但不同的菌株表现出醌特异性。Lewis 等也通过这种方法筛选到能够利用其他辅助细菌产生的 γ-氨基丁酸（GABA）的菌株，当培养基中加入 GABA 时，该菌株能够在没有辅助菌株的情况下生长（Strandwitz 等，2019）。Vartoukian 等（2016）报导，在培养基中添加铁载体可以培养出新的口腔微生物菌株。

微生物对微量营养素的需要量是极少的。Tramontano 等（2018）研究了 90 种人体肠道细菌，发现特定的培养基成分和微生物代谢物的积累可能抑制细菌的生长。因此，在确保重要营养素供给的同时，也需要保证其浓度不会抑制微生物的生长。

2.4 还原剂

大部分反刍动物和其他哺乳动物的消化道微生物都是严格厌氧的，虽然，有些微生物可以耐受氧气［例如拟杆菌（*Bacteroides* spp.）、普雷沃氏菌（*Prevotella* spp.）］，但它们也只在厌氧条件下才能生长。半胱氨酸-盐酸和钛基还原剂都是常用的还原剂，半胱氨酸-盐酸使用方便，而钛基还原剂的还原力强（McSweeney 等，2005；Jones 等，1980；Zehnder 等，1976）。Raoult 等报道，当培养基中添加抗坏血酸和/或谷胱甘肽还原剂时，活泼瘤胃球菌（*Ruminococcus gnavus*）和坏死梭形杆菌（*Fusobacterium necrophorum*）可以在大气条件下生长（La Scola 等，2014）。随后，Dione 等（2016）进一步将这项研究扩展到了 623 种不同的细菌，包括 82 种严格厌氧和 9 种微好氧细菌，发现这些细菌均可以在添加抗坏血酸和谷胱甘肽的培养基中生长（Hwang 等，2012），尿酸、血红素和 α-酮戊二酸能够进一步提高这些微生物的可培养性，但分枝杆菌（*Mycobacteria*）无法在大气条件下生长。这些研究结果也被应用到对氧气高度敏感的产甲烷菌的培养中（Khelaifia 等，2016；Khelaifia 等，2013）。虽然，上述研究结果尚未被广泛采用，但它们已经被用于培养严格厌氧的消化道细菌，例如 *Dysosmobacter welbionis*（Le Roy 等，2019），这种方法使我们能够在有氧环境下对"挑剔"的厌氧菌进行常规的处理。

总而言之，影响微生物可培养性的因素有很多。采用低营养培养基和延长培养时间的方法，能够促进微生物之间的相互竞争或减少生长因子的浓度，防止微生物过度生长，从而有助于促进更多难以培养的细菌的生长（Vartoukian 等，2010；Pulschen 等，2017；Kato 等，2018；Bartelme 等，2019；Braun 等，2008）。培养组学建立在传统培养方法的基础上，通过广泛应用现有的各种培养基和培养条件，分离和培养许多新的消化道微生物菌株［图 1a（Ito 等，2019；Lagier 等，2016）］。然而，我们不能忽视常量和微量营养素，以及其他可变生长条件对微生物可培养性的影响，尽管已经有研究概述了应对这一挑战的策略［例如 Taguchi 阵列（Rao 等，2008）、因子设计］，系统地探究这些因素的具体作用将为现有菌株的代谢需求和生长潜力提供新的见解（Hwang 等，2012；Manzoor 等，2017），促进"挑剔"消化道微生物的体外培养，为将未培养的微生物带入大众视野提供了可能（Li 等，2018；Poeker 等，2019）。同时，小型菌落筛选机器人也将为微生物的培养提供技术支撑，该机器人首次实现了在厌氧培养箱内自动进行微生物菌落的采集。

3 基于基因组的消化道微生物定向分离技术

环境模拟培养基能够最大限度地促进微生物的生长并保留微生物的多样性，因而受到研究者们的青睐。然而，这些培养基只能分离群落中的主要微生物，对丰度低、生长慢的菌株分离效果较差。因此，针对目标微生物研发特异性的新型培养基，促进目标微生物生长的同时抑制非目标微生物生长，对于微生物的筛选意义重大。

DNA 测序的出现彻底改变了微生物学的发展进程，基于测序技术，研究者们首次对微生物的基因功能和代谢能力进行了合理的预测。随着 1995 年流感嗜血杆菌（*Haemophilus influenzae*）和生殖支原体（*Mycoplasma genitalium*）基因组信息的发布，微生物功能和代谢的预测信息得到了进一步扩充（Fraser 等，1995；Fleischmann 等，1995）。然而，直到 2005 年，Tyson 等（2005）才首次基于宏基因组测序数据将细菌从混合样本中分离出来。该研究采用宏基因组技术，从低多样性酸性矿山排水系统的生物膜样本中重建了部分微生物的基因组，并发现一种公认的固氮 nif 调节子，该调节子可能由硝化螺旋菌门（Nitrospirae）钩端螺旋体Ⅲ（Leptospirillum group Ⅲ）编码，该细菌谱系尚没有可培养代表菌属。因此，学者们利用上述研究结果，在无氮培养基中，结合梯度稀释富集培养法（dilution-to-extinction），从含有目的菌属丰度<10%的样本中分离出了单株 *Leptospirillum ferrodiazotrophum* 菌落。这在微生物研究中无疑是一个重大突破，该结果证实了利用基因组信息，分离和培养关键微生物单菌落的可行性。León 等（2014）结合宏基因组数据，在 2000 多个微生物菌落中分离出一种中度嗜盐细菌 *Spiribacter salinus*，该菌是中等盐度盐场中的核心微生物，其丰度高达原核生物群落的 15%。Wurch 等（2016）使用同样的方法，从微酸性的地热池样品中富集培养了共生纳米古菌（*Nanopusillus acidilobi*），该菌占微酸性地热池古菌群落的 7%，进一步通过细胞分选技术，成功地将该菌与其宿主［泉古菌门酸叶菌属（*Crenarchaeote Acidilobus*）］进行了共培养。同样，基于基因组信息的微生物筛选技术也被用于分离动物消化道中的细菌。例如，研究者在分析袋

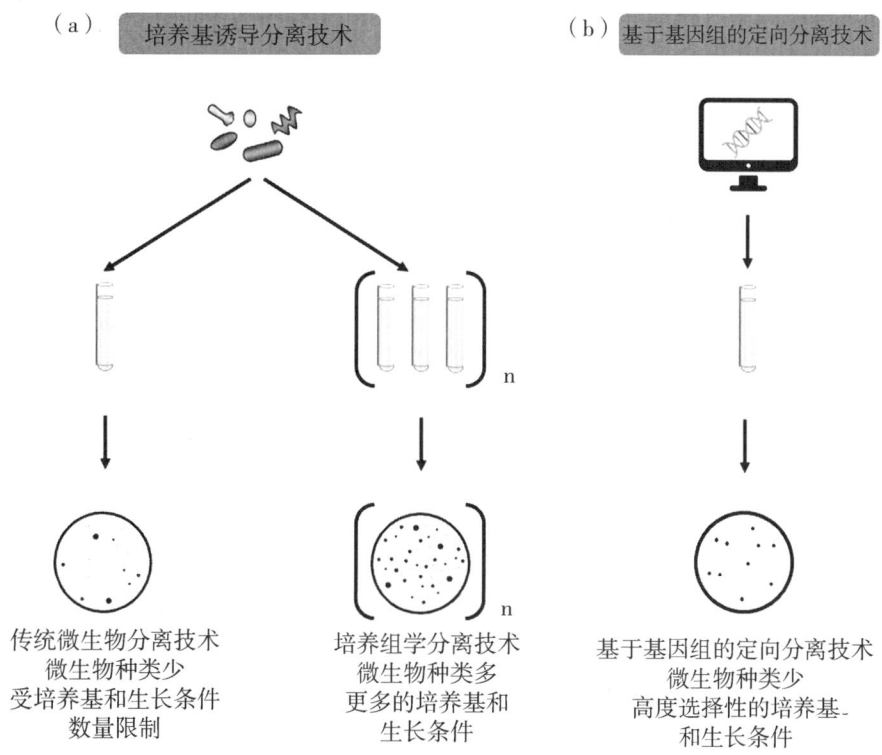

图 1 （a）传统的微生物分离方法基于前人的研究报道、资料和学者的相关经验，测试的培养基和培养条件相当有限，这使得分离到的微生物种类有限，难以筛选到目标微生物。培养组学建立在传统微生物分离方法的基础上，囊括了更多的培养基和生长条件（n），极大增加了可培养新菌株的数量。因此，培养组学能够分离和培养更多种类的微生物，但它不针对某一目标微生物。（b）基于基因组的定向分离技术是通过分析目标微生物的基因组数据来预测它们的营养和生长特性，然后根据目标微生物的特性研制特异的培养基。这些特异性培养基只允许目标微生物的生长，而非目标微生物无法在该培养基上生长。

鼠消化道内容物的宏基因组数据时，发现一个属于琥珀弧菌科（Succinivibrionaceae）的微生物，其丰度高达9%，进一步分析组装的基因组数据（大于2Mb），发现该细菌能够分别利用尿素和淀粉作为氮源和碳源，并耐受杆菌肽。基于这些研究结果，学者们采用半连续分批培养方法，在含有尿素、淀粉和杆菌肽的培养基上分离到了WG-1菌株。

虽然基于基因组定量分离微生物的方法已经取得很大的进展，但该方法仍然存在缺陷。基因组信息反映目标微生物的潜在功能和代谢能力，而并未反映目标微生物的真正代谢功能，这大大增加了微生物分离的难度。为了解决这一问题，Bomar等（2011）采用转录技术分析了药用水蛭（*Hirudo verbena*）消化道微生物区系基因的转录表达情况，发现该微生物区系主要由维氏气单胞菌（*Aeromonas veronii*）和一种类里氏杆菌（*Rikenella*-like bactcrium）构成。虽然，学者们已经报导了大量类里氏杆菌的测序数据，但仍未能成功分离培养该菌属。而Bomar等（2011）通过转录组分析，发现当类里氏杆菌快速增殖时，维氏气单胞菌已开始进入稳定期，进一步分析发现类里氏杆菌的一个操纵子表达量极高，这个操纵子上包含一个编码 GH18 家族成员的内切 β-N-乙酰氨基葡萄糖苷酶基因及其相邻的基因，这提示蛋白聚糖可能是该细菌的主要能量来源。为了验证这一点，学者们采用从牛颌下腺或猪胃腺分离的黏蛋白替代 Eggerth-Gagnon 培养基中的葡萄糖，经过 2 周培养，成功分离出维氏气单胞菌和类里氏杆菌，进一步用猪胃黏蛋白分离出了更多的菌株，大大增加了可培养的类里氏杆菌的种类。该方法的优势在于发现了该微生物在特定条件下的高表达基因，从而帮助我们更好地了解该菌在快速生长时的营养偏好性和代谢活性。

基因组信息不仅能够助力微生物的分离培养，还能用于优化临床和农业中关键细菌的培养条件。例如，基因组信息已被用于优化伯纳特氏立克次氏体（*Coxiella burnetii*）（Omsland 等，2009）和人消化道克里斯滕森菌（*Christensenella minuta*）（Xiao 等，2019）的培养策略。同样，基于基因组分析的结果，学者们也成功培养了一直以来无法培养的弗兰克氏菌株（*Frankia strain*），并探究了从结节细胞中分离的弗兰克氏菌属的各种性状（Gtari 等，2015）。

近期的研究主要集中在规模化和自动化的"基因型-表型"匹配分析，从而为微生物的分离提供更多的策略。McHardy 等研发了一种微生物表型分析工具，名为 Traitar（Weimann 等 2016），它可以根据微生物的基因组数据自动预测微生物的表型。Traitar 支持在线使用或下载使用，可预测包括碳源、细胞形态和酶活性等在内的 67 种表型。Traitar 采用已测序的微生物菌株的蛋白注释数据，利用全球传染病和流行病学在线网络（Global Infectious Disease and Epidemiology Online Network）的表型数据来构建表型分类模型，其对于微生物表型分类的结果与《伯杰氏古菌和细菌系统学手册》（*Bergey's Systematics of Archaea and Bacteria*）中的细菌表型一致，且该工具对于单菌株测序和宏基因组测序数据产生的不完整的基因组信息也具有良好的预测功能。Traitar 最强大的功能在于它能够比较多个基因组的表型，这项功能有助于分析多样性较低的微生物样本（例如富集培养物）。Traitar 还能够识别可能支撑特定表型的蛋白质，为更好地探索特定微生物的功能提供了机会。Feldbauer 等（2015）采用 Python 成功将比较"基因型-表型"算法（PICA）和"基因型-表型"关联及其学习算法（MacDonald 等，2010）应用到大规模基因组数据集和部分基因组序列，然而该算法预测仅限于 10 种表型，在分离肠道微生物时参考价值有限。

为了阐明消化道微生物的营养需求，学者们对基因组尺度的代谢模型也产生极大兴趣。读者可以参考 Santos 等（van der Ark 等，2017）以及 Sen 和 Orešič（Sen 等，2019）的综述，这些综述对支撑各种代谢模型的原理和实验策略进行了系统的讨论。简言之，学者们基于基因组信息，通过数学模型计算微生物的代谢网络，从而预测微生物营养偏好，为目的微生物定制适合其生长培养基提供理论基础。在迄今为止规模最大的此类研究中，Magnúsdóttir 等构建了 773 种人类消化道细菌的 AGORA 基因组尺度的代谢模型，为这些微生物的代谢能力和营养偏好提供了理论支撑（Magnúsdóttir 等，2017）。虽然基因组尺度的代谢模型功能强大，但注释错误和缺乏新代谢途径注释，使这种代谢模型的建立极易受到不准确因素的干扰，从而导致预测错误。例如，Tramontano 等（2018）研究表明，在 40 株 AGORA 菌株中只有 10 株菌能在其设计的培养基上生长，这个结果表明以基因组预测"挑剔"消化道细菌的营养需求仍存在着很大的局限性。Tramontano 等（2018）进一步改良了培养基，并成功培养了其余 30 株 AGORA 菌株，使 AGORA 模型得以改进。同样，学者们为普拉梭菌（*Faecalibacterium prausnitzii*）（Heinken 等，2014）和嗜黏蛋白阿克曼氏菌（*Akkermansia muciniphila*）（Ottman 等，2017）等重要的消化道微生物也设计了特异性的培养基，然而有些培养基并不是该微生物的最佳培养基，因此，这个模型还有待进一步完善。不同谱系的细菌在特定培养基上生长的能力表明，消化道细菌的营养需求可能不像想象的那么广泛。

那么，我们能从这些研究结果中得到什么结论呢？首先，与传统的方法和培养组学相比，多组学指导的分离策略可以用于富集或筛选目标微生物（图 1b，Gutleben 等，2018），然而，目前用该方法从消化道中分离的菌株仍较少。其次，目前分离微生物的效率低下，并且这些策略更适用于具有高丰度目标微生物的样本或能应用高度选择性培养策略的样本。综上所述，虽然学者们正尽力解决上述问题，但目前仍然缺乏基于基因组数据指导目标消化道微生物纯培养的理想工具和计算算法（Zhang 等，2019）。

4　基于分子的消化道微生物分离技术

宏基因组分析表明，反刍动物和人类的消化道微生物区系广泛受到基因水平转移的影响，一些研究也证实消化道环境中确实存在遗传元件在消化道细菌之间的转移。微生物在其生存环境中可能会经历选择性压力，这使它们更容易接受外源 DNA，从而为用分子方法分离微生物提供了可能。这个方法将为选择性地瞄准新的消化道微生物，并解决遗传可塑性消化道微生物稀少的问题提供了机会。

基于上述现象，ÓCuív 和他的同事研发了一种新的方法，称为宏亲本交配（Metaparental mating，MPM），该方法利用 RP4（RK2）介导的接合转移，将带有抗生素耐药性标记的重组 pEHR 模块载体引入复杂微生物区系的受体微生物中。RP4 接合转移系统的宿主广泛，可绕过宿主限制修饰系统，并易于实现高通量自动化。利用 MPM 方法和 pEHR 载体系统，他们证实采用分子方法可以选择性地分离出人类消化道中"挑剔"的梭状芽胞杆菌（*Clostridium*）IV、XIVa、XV 和 XVIII（ÓCuív 等，2015）。pEHR 质粒能够在宿主中保持稳定，这表明了它们的遗传可塑性，这一点是尤为重要的。从理论上讲，MpM 方法可用于任何合适的 RP4 载体，且 ÓCuív 与他的同事们随后也证明，宿主窄谱的肠道细菌载体 pJQ200sk（+）可以用于从人类粪便样本中回收大肠杆菌（*E. coli*）接合子（ÓCuív 等，2016）。除了肠细菌和拟杆菌属外，目前可用来分析消化道微生物功能的遗传可塑性消化道微生物很少。分离消化道微生物是一个耗时的过程，而 MPM 方法的一个关键优势在于，我们能够用成熟的遗传方法来鉴定分离菌株的特征。例如，ÓCuív 等利用 MPM 分离出一株遗传可塑性粪肠球菌（*Enterococcus faecalis*），该菌株具有强大的抗炎活性，这为探索抗炎活性的遗传基础提供了新的机会（ÓCuív 等，2018）。目前，学者们已经在常见的反刍动物和人类微生物中报道了多种可用的 RP4 载体，这些载体能够帮助加速遗传可塑性菌株的分离（Heap 等，2010；IIorn 等，2016；Mimce 等，2015）。

Ronda 等（2019）采用与 MPM 类似的方法，研发了一种模块化复制或整合载体，并提出了一种原位接合介导的消化道微生物宏基因组调控（Metagenomic Alteration of Gut microbiome by in situ Conjugation，MAGIC）方法，能在原位环境中实现对微生物的基因编辑。他们证实使用这种方法能够稳定编辑梭状芽胞杆菌XIVa、拟杆菌属和肠球菌属（*Enterococcus* spp.）等多种消化道细菌，从而验证了对微生物进行原位基因编辑的可行性。另外，Brophy等（2018）研发了一个宿主范围更窄的XPORT系统，它能够利用枯草芽胞杆菌（*Bacillus subtilis*）中的整合和接合原件（ICEBs1）将小型遗传元件转移到供体细菌中。XPORT系统由一个供体菌株构成，其供体菌株包含3个元件：一个要整合到受体细菌中的微小ICEBs1（mini-ICEBs1）、一个介导微小ICEBs1转移的IV型分泌系统，以及一个诱导接合的调控物质，这些元件都位于染色体上不同的位置。在部分受体菌株中，微小ICEBs1通常插入在细菌保守的亮氨酸转移RNA序列（tRNA）上，但在细菌缺少亮氨酸转移RNA时，微小ICEBs1会插入其他位点，或作为染色体以外的元件保留。与MPM方法类似，XPORT系统中的供体菌株是经过营养缺陷筛选的。同样，通过双亲本配对的方法，微小ICEBs1也被有效地转移到35种具有代表性的人类消化道和皮肤细菌中。此外，在土壤中应用MPM方法也能将微小ICEBs1转移到土壤细菌中。然而，Brophy等（2018）回收的大部分细菌都属于芽胞杆菌（*Bacilli*），并未检测到大肠杆菌（*E. coli*），说明了XPORT系统的宿主范围很窄。

与传统的分离培养方法相比，这些基于分子的微生物分离方法有如下优点：第一，由于引入了抗生素耐药性标记物，受体菌株能够在营养充足且添加合适抗生素的培养基上生长，进而可预先评估受体微生物对所选抗生素的耐受程度，进而筛选最优的基因转移条件；第二，通过分子方法筛选的菌株遗传可塑性更强，更有利于研究目标微生物的功能，然而，我们无法预先评估目标微生物的遗传可塑性，且克服野生菌株限制修饰系统仍然是一个挑战；第三，这些方法重燃了人们使用遗传学的方法来研究消化道微生物功能的兴趣，并促进了各种针对细菌和古菌遗传工具的研发（ÓCuív等，2015；García-Bayona等，2019；Enzmann等，2018），其中很多工具可以定制并应用于分离新的微生物，从而为加快宿主—微生物相互作用的研究提供了宝贵的遗传可塑的菌株资源。

5 基于抗体的消化道微生物分离技术

近几十年来，人们开始热衷于应用物理方法从混合群落中分离/富集目标微生物（Olsvik等，1994.；Palm等，2014）。基于免疫磁珠吸附的细胞分离技术最初是用于真核细胞的分离（例如，不同的血细胞），后来逐渐发展到微生物细胞。这项技术得以发展是因为微生物附着在磁珠上后仍然具有活性，且添加新鲜的培养基能够使其生长［图2a（Palm等，2014；Torensma等，1992；Armstrong等，2019）］。

该方法的首次实践成功证明了抗体介导的免疫磁珠吸附分离技术可用于分离临床相关的大肠杆菌K88+（*E. coli* K88+）菌株。Lund等（1988）将抗大肠杆菌K88+菌毛抗原的IgG2单克隆抗体固定在超顺磁珠上，并使用该磁珠在30 min内成功分离鉴定了样本中携带K88的细菌。此外，该方法能从不同O血清型大肠杆菌菌株的混合物中几乎特异性的分离大肠杆菌K88表达菌株。该领域的许多后续工作都是围绕临床病原菌的检测，包括

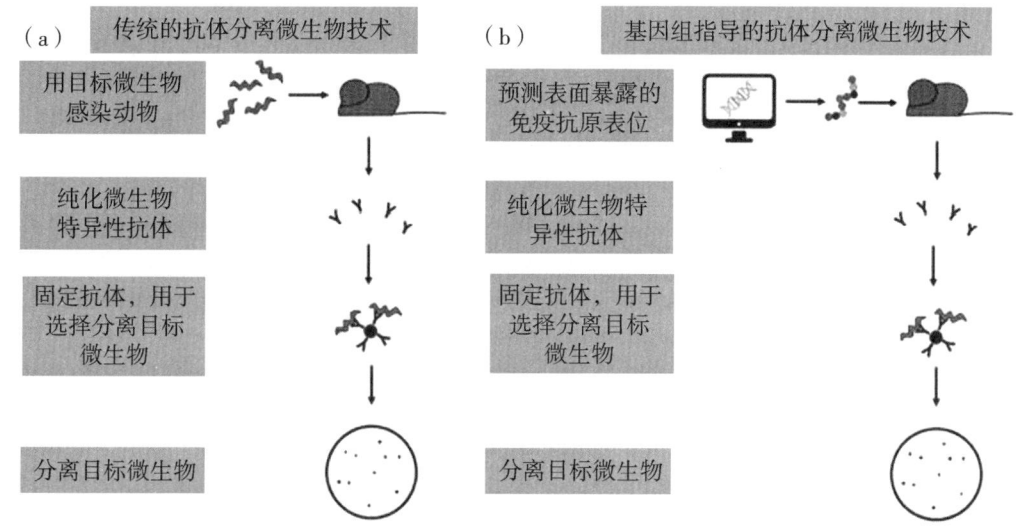

图 2 （a）传统的抗体分离目标微生物，用目标微生物免疫动物，获得并纯化目标微生物的特异性抗体，该抗体固定在磁珠上，用于在混合细胞群中物理分离目标微生物。这种方法可以分离出活的微生物，并能在合适的培养基中生长。（b）采用基因组指导的抗体分离目标微生物（反向基因组学），分析目标微生物的基因组数据，预测表面暴露的和其他可能的免疫抗原表位，表达抗原表位并用于免疫动物，获得并纯化相应的抗体用于目标微生物的分离。

李斯特菌（*Listeria*）、假单胞菌（*Pseudomonas*）和其他致病性大肠杆菌。Okrend 等（1992）发现可以将大肠杆菌 O157 的特异性 IgG 抗体包被到磁珠上，用磁珠选择性富集混合样品和肉类样品中的目标细菌，并且可以将分离的细菌接种到培养基上进行进一步鉴定。Okrend 等（1992）还报导鱼精蛋白可以阻止细菌和磁珠的非特异性结合，从而保证在分离大肠杆菌 O157：H7 时的一致性。

利用这些早期研究，Close 等（2013）采用 M13 丝状噬菌体展示的天然 scFV 抗体文库，鉴定了能与嗜酸乳杆菌 ATCC 4356（*Lactobacillus acidophilus* ATCC 4356）结合的抗体。这些抗体对嗜酸乳杆菌高表达的 S 层 A 蛋白具有高度特异性，并且不与包括其他 6 种乳酸菌（*Lactobacillus* spp.）在内的消化道细菌结合。他们进一步采用上述筛选的抗体，发现即使混合样品中目标细菌的丰度低于 0.2%，仍可以检测到目标细菌。基于这一研究结果，他们发现当嗜酸乳杆菌在样本中小于 0.2% 时，可使用荧光激活细胞分选法来分离嗜酸乳杆菌。该方法分离的单个菌细胞通过多重置换扩增测序，单个细胞的测序结果能达到 63% 的基因组覆盖率，而 50 个细胞的结果经过重头组装后，能获得 99.8% 的基因组覆盖率。最后，作者证实类似的方法可以应用在目标微生物在 10% 以下的复杂微生物模拟群落，从而揭示了无须培养单个细菌就可以获得菌种特异性抗体的可行性。

这些研究为有效地从复杂微生物群落中分离目标微生物的可行性提供了关键见解。和传统的分离培养方法相比，基于抗体的微生物分离技术具有更多优点，例如仅用一个抗体就能从多个样本中分离目标微生物，即使目标微生物的丰度较低也不受影响。此外，该技术还可应用于厌氧环境，并能够将细菌细胞与其他生长抑制因子或微生物进行物理分离。

尽管如此，这个方法仍然未被广泛应用，因为在过去，生产抗原特异性抗体所需的时间和成本投入远多于以上优点带来的效益。该技术的关键缺陷在于必须用纯的靶细胞群来生产特异性抗体，因为只有在没有其他抗原存在的情况下，针对所需抗原筛选出来的抗体才更具特异性（Lou 等，2001）。如今，这一技术难题已经被突破，我们能够通过大量的基因组数据筛选目标微生物所特有的抗原，随后采用候选抗原序列快速生产多肽，用于制备特异性的抗体，从而加快目标微生物的分离。Cross 等（2019）利用基因组信息指导筛选的抗体分离了单糖菌门（Saccharibacteria）和之前未培养的 SR1 细菌。随着目标抗体生产的相关技术限制的突破，基于抗体的分离方法可能会逐渐被用于新的微生物的分离（图 2b）。

6　结论和展望

本章概述了从哺乳动物消化道分离"挑剔"的细菌和古菌所面临的挑战。同时，还介绍了各种克服这些问题的新旧策略和方法。如要获得更详细的信息，可以查阅"参考文献"部分中列出的引文和最新的科学综述，以了解新兴的培养方法。

消化道微生物研究的变革使人们不断努力去阐明宿主—微生物区系的相互作用，及其对宿主表型的影响。不依赖于培养的研究更深入地揭示了消化道微生物区系的多样性和功能潜力。虽然仍有许多消化道微生物基因尚未被研究，随着核心基因的不断扩充和基因功能覆盖面的日益饱和，不依赖培养的方法可提供的信息将逐渐到达极限。而相比之下，尽管人们对培养消化道微生物的兴趣从未如此高涨，但纯培养的方法仍然相对落后。不言而喻，应利用瘤胃微生物学家的专业知识来更好地了解人类和试验动物的微生物区系。同样，反刍动物在为人类提供高质量营养方面起着重要作用，并且人们仍在不断探索如何减少反刍动物甲烷和氧化亚氮排放，从而减轻集约化养殖对环境的影响，因此，将微生物领域广泛应用的高通量技术应用于瘤胃微生物研究也十分重要。在后续研究中，不断了解和掌握微生物的营养偏好，基于此研发新的培养策略，进一步有效结合和利用宏组学技术和基因组数据，将为深入探究宿主—微生物互作机制提供新思路。

7　参考文献

Allison, M. J., Bryant, M. P. and Doetsch, R. N. 1958. Volatile fatty acid growth factor for cellulolytic cocci of bovine rumen. Science 128 (3322), 474-5.

Allison, M. J., Mayberry, W. R., McSweeney, C. S., et al., 1992. Synergistes jonesii, gen. nov., sp. nov.: a rumen bacterium that degrades toxic pyridinediols. Syst. Appl. Microbiol. 15 (4), 522-9.

Almeida, A., Nayfach, S., Boland, M., et al., 2019. A unified sequence catalogue of over 280, 000 genomes obtained from the human gut microbiome. bioRxiv, 762682.

Armstrong, H., Alipour, M., Valcheva, R., et al., 2019. Host immunoglobulin G selectively identifies pathobionts in pediatric inflammatory bowel diseases. Microbiome 7 (1), 1.

Atasoglu, C., Valdés, C., Walker, N. D., et al., 1998. De novo synthesis of amino acids by the ruminal bacteria *Prevotella bryantii* B14, *Selenomonas ruminantium* HD4, and *Streptococcus bovis* ES1. Appl. Environ. Microbiol. 64（8），2836-43.

Bartelme, R. P., Custer, J. M., Dupont, C. L., et al., 2019. The concentration of heterotrophic growth substrates in culture medium is a crucial determinant of the culturability of subsurface soil microbes. bioRxiv：726661.

Bomar, L., Maltz, M., Colston, S., et al., 2011. Directed culturing of microorganisms using metatranscriptomics. mBio 2（2），e00012-11.

Braun, P. R., Al-Younes, H., Gussmann, J., et al., 2008. Competitive inhibition of amino acid uptake suppresses chlamydial growth：involvement of the chlamydial amino acid transporter BrnQ. J. Bacteriol. 190（5），1822-30.

Braune, A. and Blaut, M. 2018. *Catenibacillus scindens* gen. nov., sp. nov., a C-deglycosylating human intestinal representative of the Lachnospiraceae. Int. J. Syst. Evol. Microbiol. 68（10），3356-61.

Brophy, J. A. N., Triassi, A. J., Adams, B. L., et al., 2018. Engineered integrative and conjugative elements for efficient and inducible DNA transfer to undomesticated bacteria. Nat. Microbiol. 3（9），1043-53.

Bryant, M. P. and Robinson, I. M. 1962. Some nutritional characteristics of predominant culturable ruminal bacteria. J. Bacteriol. 84, 605-14.

Chen, G. and Russell, J. B. 1989. More monensin-sensitive, ammonia-producing bacteria from the rumen. Appl. Environ. Microbiol. 55（5），1052-7.

Clavel, T., Lagkouvardos, I. and Stecher, B. 2017. From complex gut communities to minimal microbiomes via cultivation. Curr. Opin. Microbiol. 38, 148-55.

Close, D. W., Ferrara, F., Dichosa, A. E. K., et al., 2013. Using phage display selected antibodies to dissect microbiomes for complete de novo genome sequencing of low abundance microbes. BMC Microbiol. 13（1），270.

Cray, J. A., Bell, A. N., Bhaganna, P., et al., 2013. The biology of habitat dominance；can microbes behave as weeds? Microb. Biotechnol. 6（5），453-92.

Cross, K. L., Campbell, J. H., Balachandran, M., et al., 2019. Targeted isolation and cultivation of uncultivated bacteria by reverse genomics. Nat. Biotechnol. 37（11），1314-21.

Das, N., Tripathi, N., Basu, S., et al., 2015. Progress in the development of gelling agents for improved culturability of microorganisms. Front. Microbiol. 6, 698.

Davis, C. K., Webb, R. I., Sly, L. I., et al., 2012. Isolation and survey of novel fluoroacetate-degrading bacteria belonging to the phylum Synergistetes. FEMS Microbiol. Ecol. 80（3），671-84.

Dehority, B. A. 1971. Carbon dioxide requirement of various species of rumen bacteria. J. Bacteriol. 105（1），70-6.

Dione, N., Khelaifia, S., La Scola, B., et al., 2016. A quasiuniversal medium to break the aerobic/anaerobic bacterial culture dichotomy in clinical microbiology. Clin. Microbiol. Infect. 22（1），53-8.

Einarsson, H., Snygg, B. G. and Eriksson, C. 1983. Inhibition of bacterial growth by Maillard reaction products. J. Agric. Food Chem. 31（5），1043-7.

Eller, C., Crabill, M. R. and Bryant, M. P. 1971. Anaerobic roll tube media for nonselective enumeration and isolation of bacteria in human feces. Appl. Microbiol. 22（4），522-9.

Enzmann, F., Mayer, F., Rother, M., et al., 2018. Methanogens：biochemical background and biotechnological applications. AMB Express 8（1），1-.

Feldbauer, R., Schulz, F., Horn, M., et al., 2015. Prediction of microbial phenotypes based on comparative genomics. BMC Bioinformatics 16 (S1).

Fenn, K., Strandwitz, P., Stewart, E. J., et al., 2017. Quinones are growth factors for the human gut microbiota. Microbiome 5 (1), 161.

Fleischmann, R. D., Adams, M. D., White, O., et al., 1995. Whole-genome random sequencing and assembly of Haemophilus influenzae Rd. Science 269 (5223), 496-512.

Forster, S. C., Kumar, N., Anonye, B. O., et al., 2019. A human gut bacterial genome and culture collection for improved metagenomic analyses. Nat. Biotechnol. 37 (2), 186-92.

Fraser, C. M., Gocayne, J. D., White, O., et al., 1995. The minimal gene complement of Mycoplasma genitalium. Science 270 (5235), 397-403.

Gaci, N., Borrel, G., Tottey, W., et al., 2014. Archaea and the human gut: new beginning of an old story. World J. Gastroenterol. 20 (43), 16062-78.

Gagen, E. J., Wang, J., Padmanabha, J., et al., 2014. Investigation of a new acetogen isolated from an enrichment of the tammar wallaby forestomach. BMC Microbiol. 14, 314.

García-Bayona, L. and Comstock, L. E. 2019. Streamlined genetic manipulation of diverse bacteroides and parabacteroides isolates from the human gut microbiota. mBio 10 (4), e01762-19.

Gilbert, R. A. and Klieve, A. V. 2015. Ruminal viruses (Bacteriophages, Archaeaphages). In: Puniya, A. K., Singh, R. and Kamra, D. N. (Eds), Rumen Microbiology: From Evolution to Revolution. Springer, New Delhi India, pp. 121-41.

Gray, V. L., Muller, C. T., Watkins, I. D., et al., 2008. Peptones from diverse sources: pivotal determinants of bacterial growth dynamics. J. Appl. Microbiol. 104 (2), 554-65.

Gtari, M., Ghodhbane-Gtari, F., Nouioui, I., et al., 2015. Cultivating the uncultured: growing the recalcitrant cluster-2 Frankia strains. Sci. Rep. 5, 13112.

Gutleben, J., Chaib De Mares, M., van Elsas, J. D., et al., 2018. The multi-omics promise in context: from sequence to microbial isolate. Crit. Rev. Microbiol. 44 (2), 212-29.

Haitjema, C. H., Solomon, K. V., Henske, J. K., et al., 2014. Anaerobic gut fungi: advances in isolation, culture, and cellulolytic enzyme discovery for biofuel production. Biotechnol. Bioeng. 111 (8), 1471-82.

He, X., McLean, J. S., Edlund, A., et al., 2015. Cultivation of a human-associated TM7 phylotype reveals a reduced genome and epibiotic parasitic lifestyle. Proc. Natl. Acad. Sci. U. S. A. 112 (1), 244-9.

Heap, J. T., Kuehne, S. A., Ehsaan, M., et al., 2010. TheClosTron: mutagenesis in Clostridium refined and streamlined. J. Microbiol. Methods 80 (1), 49-55.

Heinken, A., Khan, M. T., Paglia, G., et al., 2014. Functional metabolic map of *Faecalibacterium prausnitzii*, a beneficial human gut microbe. J. Bacteriol. 196 (18), 3289-302.

Hess, M., Sczyrba, A., Egan, R., et al., 2011. Metagenomic discovery of biomass-degrading genes and genomes from cow rumen. Science 331 (6016), 463-7.

Hinton, A. 2016. Growth of Campylobacter incubated aerobically in fumaratepyruvate media or media supplemented with dairy, meat, or soy extracts and peptones. Food Microbiol. 58, 23-8.

Hoedt, E. C., Cuiv, P. Ó., Evans, P. N., et al., 2016. Differences down-under: alcohol-fueled methanogenesis by archaea present in Australian macropodids. ISME J. 10 (10), 2376-88.

Horn, N., Carvalho, A. L., Overweg, K., et al., 2016. A novel tightly regulated gene expression system for the human intestinal symbiont Bacteroides thetaiotaomicron. Front. Microbiol. 7 (1080), 1080.

Hungate, R. E. 1969. A roll tube method for cultivation of strict anaerobes. Methods in Microbiology 3 (Part

B), 117-32.

Hungate, R. E. and Stack, R. J. 1982. Phenylpropanoic acid: growth factor for Ruminococcus albus. Appl. Environ. Microbiol. 44 (1), 79-83.

Hwang, C. F., Chang, J. H., Houng, J. Y., et al., 2012. Optimization of medium composition for improving biomass production of *Lactobacillus plantarum* Pi06 using the Taguchi array design and the BoxBehnken method. Biotechnol. Bioprocess Eng. 17 (4), 827-34.

Ito, T., Sekizuka, T., Kishi, N., et al., 2019. Conventional culture methods with commercially available media unveil the presence of novel culturable bacteria. Gut Microbes 10 (1), 77-91.

Joblin, K. N. 2005. Methanogenic archaea. In: Makkar, H. P. S. and McSweeney, C. S. (Eds), Methods in Gut Microbial Ecology for Ruminants. Springer Netherlands, Dordrecht, pp. 47-53.

Jones, D. R., Thomas, D., Alger, N., et al., 2018. SACCHARIS: an automated pipeline to streamline discovery of carbohydrate active enzyme activities within polyspecific families and de novo sequence datasets. Biotechnol. Biofuels 11 (1), 27.

Jones, G. A. and Pickard, M. D. 1980. Effect of titanium (III) citrate as reducing agent on growth of rumen bacteria. Appl. Environ. Microbiol. 39 (6), 1144-7.

Kato, S., Yamagishi, A., Daimon, S., et al., 2018. Isolation of previously uncultured slow-growing bacteria by using a simple modification in the preparation of agar media. Appl. Environ. Microbiol. 84 (19), e00807-18.

Kawasaki, K. and Kamagata, Y. 2017. Phosphate-catalyzed hydrogen peroxide formation from agar, gellan, and kappa-carrageenan and recovery of microbial cultivability via catalase and pyruvate. Appl. Environ. Microbiol. 83 (21).

Kelly, W. J., Henderson, G., Pacheco, D. M., et al., 2016. The complete genome sequence of Eubacterium limosum SA11, a metabolically versatile rumen acetogen. Stand. Genom. Sci. 11, 26.

Kenters, N., Henderson, G., Jeyanathan, J., et al., 2011. Isolation of previously uncultured rumen bacteria by dilution to extinction using a new liquid culture medium. J. Microbiol. Methods 84 (1), 52-60.

Khelaifia, S., Lagier, J. C., Nkamga, V. D., et al., 2016. Aerobic culture of methanogenic archaea without an external source of hydrogen. Eur. J. Clin. Microbiol. Infect. Dis. 35 (6), 985-91.

Khelaifia, S., Raoult, D. and Drancourt, M. 2013. A versatile medium for cultivating methanogenic archaea. PLoS ONE 8 (4), e61563.

La Scola, B., Khelaifia, S., Lagier, J. C., et al., 2014. Aerobic culture of anaerobic bacteria using antioxidants: a preliminary report. Eur. J. Clin. Microbiol. Infect. Dis. 33 (10), 1781-3.

Lagier, J. C., Khelaifia, S., Alou, M. T., et al., 2016. Culture of previously uncultured members of the human gut microbiota by culturomics. Nat. Microbiol. 1, 16203.

Lagkouvardos, I., Overmann, J. and Clavel, T. 2017. Cultured microbes represent a substantial fraction of the human and mouse gut microbiota. Gut Microbes 8 (5), 493-503.

Lagkouvardos, I., Pukall, R., Abt, B., et al., 2016. The Mouse Intestinal Bacterial Collection (miBC) provides host-specific insight into cultured diversity and functional potential of the gut microbiota. Nat. Microbiol. 1 (10), 16131.

Lau, J. T., Whelan, F. J., Herath, I., et al., 2016. Capturing the diversity of the human gut microbiota through culture-enriched molecular profiling. Genome Med. 8 (1), 72-.

Le Roy, T., Van der Smissen, P., Paquot, A., et al., 2019. *Dysosmobacter welbionis* gen. nov., sp. nov., isolated from human faeces and emended description of the genus *Oscillibacter*. Int. J. Syst. Evol. Microbiol. .

Leong, L. E., Denman, S. E., Hugenholtz, P., et al., 2016. Amino acid and peptide utilization profiles of the fluoroacetate-degrading bacterium synergistetes strain MFA1 under varying conditions. Microb. Ecol. 71 (2), 494-504.

León, M. J., Fernández, A. B., Ghai, R., et al., 2014. From metagenomics to pure culture: isolation and characterization of the moderately halophilic bacterium *Spiribacter salinus* gen. nov., sp. nov. Appl. Environ. Microbiol. 80 (13), 3850-7.

Li, L., Zhang, X., Ning, Z., et al., 2018. Evaluating *in vitro* culture medium of gut microbiome with orthogonal experimental design and a metaproteomics approach. J. Proteome Res. 17 (1), 154-63.

Ling, J. R. and Armstead, I. P. 1995. The *in vitro* uptake and metabolism of peptides and amino acids by five species of rumen bacteria. J. Appl. Bacteriol. 78 (2), 116-24.

Lloyd, K. G., Steen, A. D., Ladau, J., et al., 2018. Phylogenetically novel uncultured microbial cells dominate earth microbiomes. mSystems 3 (5), e00055-18.

Lou, J., Marzari, R., Verzillo, V., et al., 2001. Antibodies in haystacks: how selection strategy influences the outcome of selection from molecular diversity libraries. J. Immunol. Methods 253 (1-2), 233-42.

Lund, A., Hellemann, A. L. and Vartdal, F. 1988. Rapid isolation of K88+ *Escherichia coli* by using immunomagnetic particles. J. Clin. Microbiol. 26 (12), 2572-5.

MacDonald, N. J. and Beiko, R. G. 2010. Efficient learning of microbial genotypephenotype association rules. Bioinformatics 26 (15), 1834-40.

Magnúsdóttir, S., Heinken, A., Kutt, L., et al., 2017. Generation of genome-scale metabolic reconstructions for 773 members of the human gut microbiota. Nat. Biotechnol. 35 (1), 81-9.

Manzoor, A., Qazi, J. I., Haq, I. U., et al., 2017. Significantly enhanced biomass production of a novel bio-therapeutic strain *Lactobacillus plantarum* (AS-14) by developing low cost media cultivation strategy. J. Biol. Eng. 11 (1), 17.

Martiny, A. C. 2019. High proportions of bacteria are culturable across major biomes. ISME J. 13 (8), 2125-8.

McSweeney, C. S., Denman, S. E. and Mackie, R. I. 2005. Rumen bacteria. In: Makkar, H. P. S. and McSweeney, C. S. (Eds), Methods in Gut Microbial Ecology for Ruminants. Springer, Dordrecht, pp. 23-37.

Mee, M. T., Collins, J. J., Church, G. M., et al., 2014. Syntrophic exchange in synthetic microbial communities. Proc. Natl. Acad. Sci. U. S. A. 111 (20), E2149-56.

Mimee, M., Tucker, A., Voigt, C., et al., 2015. Programming a human commensal bacterium, Bacteroides thetaiotaomicron, to sense and respond to stimuli in the murine gut microbiota. Cell Syst. 1 (1), 62-71.

Nayfach, S., Shi, Z. J., Seshadri, R., et al., 2019. New insights from uncultivated genomes of the global human gut microbiome. Nature 568 (7753), 505-10.

Newbold, C. J., de la Fuente, G., Belanche, A., et al., 2015. The role of ciliate protozoa in the rumen. Front. Microbiol. 6 (1313), 1313.

Nyonyo, T., Shinkai, T., Tajima, A., et al., 2013. Effect of media composition, including gelling agents, on isolation of previously uncultured rumen bacteria. Lett. Appl. Microbiol. 56 (1), 63-70.

Okrend, A. J. G., Rose, B. E. and Lattuada, C. P. 1992. Isolation of *Escherichia coli* 0157: H7 using

0157 specific antibody coated magnetic beads. J. Food Prot. 55 (3), 214-7.

Oliphant, K. and Allen-Vercoe, E. 2019. Macronutrient metabolism by the human gut microbiome: major fermentation by-products and their impact on host health. Microbiome 7 (1), 91-.

Olsvik, O., Popovic, T., Skjerve, E., et al., 1994. Magnetic separation techniques in diagnostic microbiology. Clin. Microbiol. Rev. 7 (1), 43-54.

Omsland, A., Cockrell, D. C., Howe, D., et al., 2009. Host cell-free growth of the Q fever bacterium Coxiella burnetii. Proc. Natl Acad. Sci. U. S. A. 106 (11), 4430-4.

Ottman, N., Davids, M., Suarez-Diez, M., et al., 2017. Genome-scale model and omics analysis of metabolic capacities of *Akkermansia muciniphila* reveal a preferential mucin-degrading lifestyle. Appl. Environ. Microbiol. 83 (18).

ÓCuív, P. Ó, Smith, W. J., Pottenger, S., et al., 2015. Isolation of genetically tractable most-wanted bacteria by metaparental mating. Sci. Rep. 5, 13282.

ÓCuív, P., Burman, S., Pottenger, S., et al., 2016. Exploring the bioactive landscape of the gut microbiota to identify metabolites underpinning human health. In: Beale, D. J., Kouremenos, K. A. and Palombo, E. A. (Eds), Microbial Metabolomics: Applications in Clinical, Environmental, and Industrial Microbiology. Springer International Publishing, Cham, pp. 49-82.

ÓCuív, P., Giri, R., Hoedt, E. C., et al., 2018. *Enterococcus faecalis* AHG0090 is a genetically tractable bacterium and produces a secreted peptidic bioactive that suppresses nuclear factor kappa B activation in human gut epithelial cells. Front. Immunol. 9 (790), 790.

Palm, N. W., de Zoete, M. R., Cullen, T. W., et al., 2014. Immunoglobulin A coating identifies colitogenic bacteria in inflammatory bowel disease. Cell 158 (5), 1000-10.

Pasolli, E., Asnicar, F., Manara, S., et al., 2019. Extensive unexplored human microbiome diversity revealed by over 150,000 genomes from metagenomes spanning age, geography, and lifestyle. Cell 176 (3), 649-662. e20.

Pittman, K. A. and Bryant, M. P. 1964. Peptides and other nitrogen sources for growth of Bacteroides ruminicola. J. Bacteriol. 88, 401-10.

Poeker, S. A., Lacroix, C., de Wouters, T., et al., 2019. Stepwise development of an *in vitro* continuous fermentation model for the murine caecal microbiota. Front. Microbiol. 10 (1166), 1166.

Pope, P. B., Smith, W., Denman, S. E., et al., 2011. Isolation of Succinivibrionaceae implicated in low methane emissions from Tammar wallabies. Science 333 (6042), 646-8.

Pulschen, A. A., Bendia, A. G., Fricker, A. D., et al., 2017. Isolation of uncultured bacteria from Antarctica using long incubation periods and low nutritional media. Front. Microbiol. 8, 1346.

Rao, R. S., Kumar, C. G., Prakasham, R. S., et al., 2008. The Taguchi methodology as a statistical tool for biotechnological applications: a critical appraisal. Biotechnol. J. 3 (4), 510-23.

Reilly, S. 1980. The carbon dioxide requirements of anaerobic bacteria. J. Med. Microbiol. 13 (4), 573-9.

Rettedal, E. A., Gumpert, H. and Sommer, M. O. A. 2014. Cultivation-based multiplex phenotyping of human gut microbiota allows targeted recovery of previously uncultured bacteria. Nat. Commun. 5, 4714.

Richardson, A. J., McKain, N. and Wallace, R. J. 2013. Ammonia production by human faecal bacteria, and the enumeration, isolation and characterization of bacteria capable of growth on peptides and amino acids. BMC Microbiol. 13, 6.

Ronda, C., Chen, S. P., Cabral, V., et al., 2019. Metagenomic engineering of the mammalian gut microbiome in situ. Nat. Methods 16 (2), 167-70.

Schär-Zammaretti, P., Dillmann, M. L., D'Amico, N., et al., 2005. Influence of fermentation medium composition on physicochemical surface properties of *Lactobacillus acidophilus*. Appl. Environ. Microbiol. 71 (12), 8165-73.

Sen, P. and Orešič, M. 2019. Metabolic modeling of human gut microbiota on a genome scale:an overview. Metabolites 9 (2), 22.

Seshadri, R., Leahy, S. C., Attwood, G. T., et al., 2018. Cultivation and sequencing of rumen microbiome members from the Hungate1000 Collection. Nat. Biotechnol. 36 (4), 359-67.

Sharma, V., Rodionov, D. A., Leyn, S. A., et al., 2019. B-vitamin sharing promotes stability of gut microbial communities. Front. Microbiol. 10 (1485), 1485.

Shigeno, Y., Kitahara, M., Shime, M., et al., 2019. *Phascolarctobacterium wakonense* sp. nov., isolated from common marmoset (Callithrix jacchus) faeces. Int. J. Syst. Evol. Microbiol. 69 (7), 1941-6.

Smith, E. A. and Macfarlane, G. T. 1998. Enumeration of amino acid fermenting bacteria in the human large intestine:effects of pH and starch on peptide metabolism and dissimilation of amino acids. FEMS Microbiol. Ecol. 25 (4), 355-68.

Stack, R. J., Hungate, R. E. and Opsahl, W. P. 1983. Phenylacetic acid stimulation of cellulose digestion by *Ruminococcus albus* 8. Appl. Environ. Microbiol. 46 (3), 539-44.

Stewart, C. S., Flint, H. J. and Bryant, M. P. 1997. The rumen bacteria. In:Hobson, P. N. and Stewart, C. S. (Eds), The Rumen Microbial Ecosystem. Springer Netherlands, Dordrecht, pp. 10-72.

Stewart, E. J. 2012. Growing unculturable bacteria. J. Bacteriol. 194 (16), 4151-60.

Stewart, R. D., Auffret, M. D., Warr, A., et al., 2018. Assembly of 913 microbial genomes from metagenomic sequencing of the cow rumen. Nat. Commun. 9 (1), 870.

Strandwitz, P., Kim, K. H., Terekhova, D., et al., 2019. GABAmodulating bacteria of the human gut microbiota. Nat. Microbiol. 4 (3), 396-403.

Tanaka, T., Kawasaki, K., Daimon, S., et al., 2014. A hidden pitfall in the preparation of agar media undermines microorganism cultivability. Appl. Environ. Microbiol. 80 (24), 7659-66.

Ternan, N. G., Moore, N. D., Smyth, D., et al., 2018. Increased sporulation underpins adaptation of Clostridium difficile strain 630 to a biologically-relevant faecal environment, with implications for pathogenicity. Sci. Rep. 8 (1), 16691.

Theodorou, M. K., Trinci, A. P. J. and Brookman, J. 2005. Anaerobic fungi. In:Makkar, H. P. S. and McSweeney, C. S. (Eds), Methods in Gut Microbial Ecology for Ruminants. Springer Netherlands, Dordrecht, pp. 55-66.

Torensma, R., Visser, M. J., Aarsman, C. J., et al., 1992. Monoclonal antibodies that detect live salmonellae. Appl. Environ. Microbiol. 58 (12), 3868-72.

Tramontano, M., Andrejev, S., Pruteanu, M., et al., 2018. Nutritional preferences of human gut bacteria reveal their metabolic idiosyncrasies. Nat. Microbiol. 3 (4), 514-22.

Tyson, G. W., Lo, I., Baker, B. J., et al., 2005. Genome-directed isolation of the key nitrogen fixer *Leptospirillum ferrodiazotrophum* sp. nov. from an acidophilic microbial community. Appl. Environ. Microbiol. 71 (10), 6319-24.

van der Ark, K. C. H., van Heck, R. G. A., Martins Dos Santos, V. A. P., et al., 2017. More than just a gut feeling:constraint-based genomescale metabolic models for predicting functions of human intestinal microbes. Microbiome 5 (1), 78.

Vartoukian, S. R., Adamowska, A., Lawlor, M., et al., 2016. *In vitro* cultivation of 'Unculturable' oral bacteria, facilitated by community culture and media supplementation with siderophores. PLoS ONE 11 (1), e0146926.

Vartoukian, S. R., Palmer, R. M. and Wade, W. G. 2010. Strategies for culture of 'unculturable' bacteria. FEMS Microbiol. Lett. 309 (1), 1-7.

Wallace, R. J. 1996. Ruminal microbial metabolism of peptides and amino acids. J. Nutr. 126 (4 Suppl.), 1326S-34S.

Watanabe, Y., Nagai, F. and Morotomi, M. 2012. Characterization of *Phascolarctobacterium succinatutens* sp. nov., an asaccharolytic, succinate-utilizing bacterium isolated from human feces. Appl. Environ. Microbiol. 78 (2), 511-8.

Weimann, A., Mooren, K., Frank, J., et al., 2016. From genomes to phenotypes: traitar, the microbial trait analyzer. mSystems 1 (6).

Wurch, L., Giannone, R. J., Belisle, B. S., et al., 2016. Genomics-informed isolation and characterization of a symbiotic Nanoarchaeota system from a terrestrial geothermal environment. Nat. Commun. 7, 12115-.

Xavier, J. C., Patil, K. R. and Rocha, I. 2017. Integration of biomass formulations of genome-scale metabolic models with experimental data reveals universally essential cofactors in prokaryotes. Metab. Eng. 39, 200-8.

Xiao, H., Liu, B., Yong, J., et al., 2019. Quantitative analysis and medium components optimizing for culturing a fastidious bacterium Christensenella minuta. bioRxiv: 632836.

Zehavi, T., Probst, M. and Mizrahi, I. 2018. Insights into culturomics of the rumen microbiome. Front. Microbiol. 9, 1999.

Zehnder, A. J. and Wuhrmann, K. 1976. Titanium (III) citrate as a nontoxic oxidationreduction buffering system for the culture of obligate anaerobes. Science 194 (4270), 1165-6.

Zengler, K. and Zaramela, L. S. 2018. The social network of microorganisms-how auxotrophies shape complex communities. Nat. Rev. Microbiol. 16 (6), 383-90.

Zhang, A. N., Mao, Y., Wang, Y., et al., 2019. Mining traits for the enrichment and isolation of not-yet-cultured populations. Microbiome 7 (1), 96.

Zhang, H., Yohe, T., Huang, L., et al., 2018. dbCAN2: a meta server for automated carbohydrate-active enzyme annotation. Nucleic Acids Res. 46 (W1), W95-W101.

Zou, Y., Xue, W., Luo, G., et al., 2019. 1,520 reference genomes from cultivated human gut bacteria enable functional microbiome analyses. Nat. Biotechnol, 37 (2), 179-85.

第3章 瘤胃代谢组——探究和了解瘤胃功能和健康的有力工具

Tom F. O'Callaghan,爱尔兰 Teagasc Moorepark 食品研究所;

Eva Lewi,英国德夫尼什

(张美玲译)

1 前言

瘤胃中栖息的微生物包括细菌、古菌、真菌和原生动物,组成了复杂的生态系统(de Almeida 等,2018)。近年来,人们对于肠道微生物在宿主生长和健康中重要作用的关注度不断提升。迄今为止,大多数肠道微生物组的研究以人为研究对象,但很多结论也适用于动物。实际上,肠道菌群已经被认为是人和畜禽动物体内的一个"内分泌器官"(O'Hara 和 Shanahan,2006,O'Callaghan 等,2016),其原因在于肠道菌群不仅能够影响肠道本身的功能,还可以影响到全身系统的功能,包括宿主的新陈代谢、大脑和行为、肝功能、心血管系统、肠道神经系统和免疫系统(O'Callaghan 等,2016)。

目前,全世界约有75亿人口,但是人口数量还在不断增加,全球人口对粮食的需求量在不断增加。瘤胃使反刍动物能够消化很多人体无法消化的物质,例如富含纤维素的食物。这些代谢活动的消化产物不但可以为瘤胃微生物提供丰富的代谢底物,还可以为人类提供营养丰富的食品,比如牛奶和肉类(Saleem 等,2013)。因此,深入了解瘤胃微生物的组成及其功能,对于提高畜禽动物的生产效率、健康状况、可持续性和实现生产力最大化至关重要。

代谢组(Metabolome)是指活细胞或者活体中所有内源性或者外源性的小分子代谢物的集合(Wishart,2005)。代谢组学(Metabolomics)是一个新兴的研究领域,主要利用核磁共振(NMR)、液/气色谱(LC/GC)和质谱(MS)等技术对小分子物质和代谢物进行定性和定量分析(Wishart,2008)。代谢组学通常被用于生物医学、营养学和作物研究。近年来,代谢组学在畜禽研究和畜禽动物健康监测领域的作用也日益突出——主要用于动物健康评估、疾病诊断和生物产品定性(Goldansaz 等,2017),也逐渐用于体增重、牛奶质量和健康等生产性状潜在生物标志物的鉴定(do Prado 等,2018)。生物标志物对于打击食品造假和提高食品安全性有重要作用。与传统化学检测分析法相比,代谢组学的优点是可以一次性对大量的代谢物进行高通量分析。此外,代谢组学还具有检测无创、省时、样品制备成本更加低廉的优点。然而,代谢组学分析所需的仪器尚未得到广泛应用,且购买和维护费用较为昂贵,同时需要大量数据资源来解释和利用结果。

本章主要介绍目前已知的瘤胃代谢组学研究,并重点介绍可能影响瘤胃代谢物组成和功能的因素。

1.1 靶向与非靶向代谢组学的比较

代谢组分析分为靶向和非靶向两种方法，每种方法都有各自的优点和缺点。靶向代谢组（Targeted metabolomics）是指对样品中预设的代谢物进行定量分析。因此，在生物标记物的识别和验证以及典型代谢通路分析等的应用中，对于要关注的样品和分子物质有预先的了解是必要的。靶向代谢组能够减少数据分析和代谢谱解析的时间，但它的缺点是只能检测有限、已知的代谢物。非靶向代谢组（Untargeted metabolomics）对于代谢物的检测没有预设性，只要在检测仪器灵敏度、检测范围、提取方法允许的条件下，任何代谢物都有可能被检测到，因此非靶向代谢组更有可能在样品中检测到新的化合物和生物标志物（Cajka 和 Fiehn，2016）。但是，非靶向代谢组的代谢谱确认和数据分析需要花费更多的时间，此外，它无法像靶向代谢组一样对代谢物进行定量分析。

2 瘤胃代谢组：分析和提取技术

科学家们一直致力于瘤胃代谢组的组成分析。早在 2013 年，Saleem 等人就利用一系列技术描述瘤胃代谢组的特征。这项研究成果成功构建了瘤胃代谢组数据库（www.rumendb.ca），该数据库是一个全面的可访问的网络资源，包含了超过 200 种能够识别和量化的瘤胃代谢物以及它们各自的结构和浓度。

瘤胃代谢组学分析流程如图 1 所示。代谢组学主要的分析方法包括核磁共振（NMR）和质谱分析（MS）。其中还包括多种不同的分析技术，如 ^{1}H-NMR、^{13}C-NMR、^{15}N-NMR 和 ^{31}P-NMR。虽然核磁共振是应用最为广泛的技术之一，但是其他技术，如气相色谱质谱联用（GC-MS）、液相色谱质谱联用（LC-MS）、毛细管电泳质谱联用（CE-MS）也被用于代谢组分析（Markley 等，2017）。各种技术都有其优缺点，质谱技术的灵敏度更高，可检测到的化合物数量多；核磁共振更适合分析含量丰富的化合物，可重复性高且能够定量，样品制备步骤也较少（Markley 等，2017）。2008 年，Wishart 等人讨论了 NMR、GC-MS 和 LC-MS 光谱在代谢组分析中的优缺点，也正是由于每种方法都有各自特异的代谢物检测水平，我们往往需要综合使用多种方法对样品的代谢组特征进行全面分析。

2013 年，Saleem 等联合使用核磁共振波谱、电感耦合等离子体质谱（ICP-MS）、GC-MS、直接流动注射质谱（DFI-MS）和脂质组学来分析瘤胃代谢组，共鉴定了 246 种代谢物。利用核磁共振技术，他们在瘤胃液中鉴定并定量了 50 种化合物，其中 98% 的可见峰都可以与一种化合物对应。使用 GC-MS 方法，他们鉴定并定量了 28 种极性代谢物，其中 8 种化合物无法被核磁共振检测到；但是，核磁共振检测到了 GC-MS 无法检测到的 30 种化合物。60% 的 GC-MS 可见峰能够被识别。而直接流动注射质谱（DFI-MS）可以对 116 种代谢物进行定量，其中，98 种化合物或化合物种类是难以用 GC-MS 或 NMR 检测到的。气相色谱-飞行时间质谱法也已成功用于瘤胃代谢组的检测，并能鉴定出 165 种代谢物（Sunet 等，2015）。这些研究都表明为了更全面地了解一个样品的代谢物组成，多种分析方法的联用是非常有必要的。

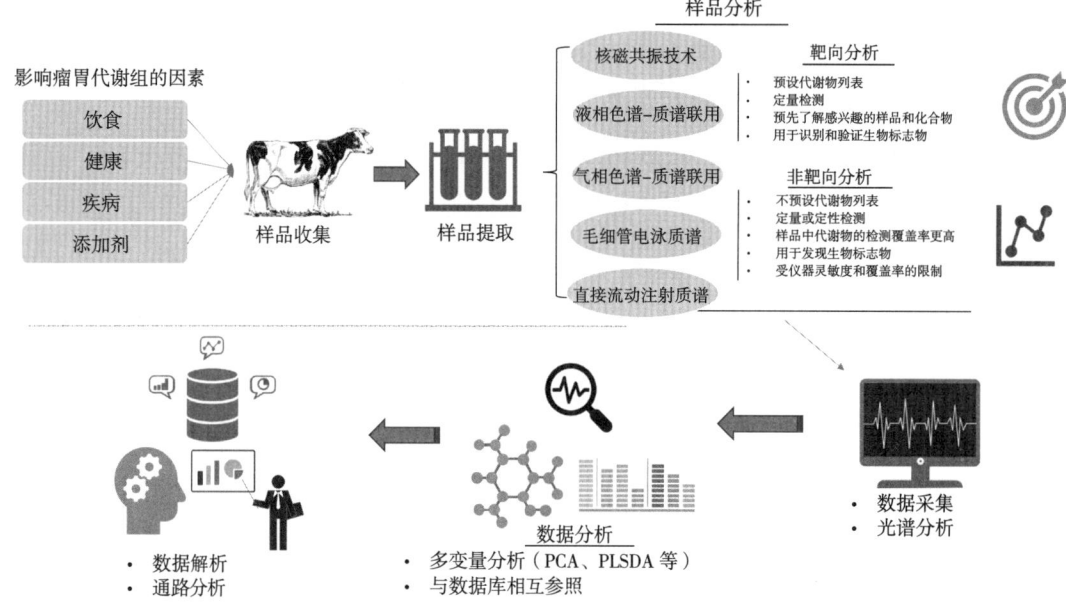

图 1　瘤胃代谢组学分析方法示例

2018 年，Almeida 等人使用液相色谱-高分辨质谱法检测了包括液-液萃取、固相萃取、原始 QuEChERS、缓冲 QuEChERS 和酸碱 QuEChERS 技术在内的不同样品提取技术对化合物鉴定的影响。该研究总共确定了 1 882 个分子的特征，其中只有 3.56% 的分子与全球天然产物分子网络数据库中的分子匹配（Global Natural Product Social Molecular Networking Database）。其中，液-液萃取能够在样品中提取到最多的具有中等极性、非极性和疏水性特征的化合物。作者还发现 pH 值对代谢物提取结果会有一定影响。根据溶剂选择的差异，固相萃取技术主要提取到的是中低极性的化合物，而 QuEChERS 方法对于提取样品中低极性的化合物也较为有效。

2.1　分析代谢组学数据的工具

代谢组收集到的数据需要进行进一步的分析和解析。MetaboAnalyst（http://www.metaboanalyst.ca）是分析和解析代谢组学数据的强有力工具，它可以处理由 MS 和 NMR 产生的代谢组数据，这个综合性的网络应用程序还支持进行数据分析和数据可视化操作，包括单变量和多变量方法，如主成分分析（PCA）、偏最小二乘判别式分析（PLS-DA）、热图聚类和机器学习方法（Xia 和 Wishart，2016；Chong 等，2018）。此外，用于多变量数据分析的 SIMCA 和用于核磁共振分析的 Chenomx 也可以与代谢物参考数据库（Metabolite Reference Libraries）联合使用，以识别和测量核磁共振光谱中的可见谱。Bioconductor（https://www.bioconductor.org/）是一个用于生物信息学的开源软件，R 语言中包含多种软件包（package），如"Metab"就是可以分析 GC-MS 高通量代谢组数据的软件包（Aggio 等，2011）。此外，还有多种工具用于分析不同代谢物的富集途径，包括京都基因和基因组百科全书（KEGG，https://www.genome.jp/kegg/）；小分子通路数据库

（SMPDB，http://smpdb.ca/）等。还有一些免费的数据库可以帮助识别和分析代谢组学，包括但不限于：牛瘤胃代谢组数据库（The Bovine Rumen Metabolome Database，http://www.rumendb.ca）、牛代谢组数据库（The Bovine Metabolome Database，http://www.cowmetdb.ca）、牛奶成分数据库（The Milk Composition Database）（Foroutan 等，2019）、人类代谢组数据库（The Human Metabolome Database）（Wishart 等，2013），The Exposome-Explorer（Neveu 等，2017），Phenol-Explorer（Rothwell 等，2013），Food DB（http://foodb.ca/），小分子通路数据库（Small Molecule Pathway Database）（Jewison 等，2014），有毒暴露物数据库（The Toxic Exposome Database）（Wishart 等，2015）、酵母代谢组数据库（The Yeast Metabolome Database）（Jewison 等，2012）、人类尿液代谢组（The Human Urine Metabolome）（Bouatra 等，2013）、人类血清代谢组（The Human Serum Metabolome）（Psychogios 等，2011）等。

3 影响瘤胃代谢组的因素

目前，大多数基于瘤胃代谢组的研究，都会探究动物日粮对瘤胃代谢组组成的影响。据报道，瘤胃液的化学成分反映了瘤胃微生物组和日粮之间的相互作用（Saleem 等，2013）。世界各地都有与当地气候、环境、土壤条件以及奶牛的营养和能量需求相匹配的多种饲养模式。

世界上大多数养殖系统采用室内全混合日粮（TMR）饲喂系统，而小部分气候温和、降水充足且土地肥沃的地区采用牧场或"草饲"养殖系统，如爱尔兰和新西兰。2013 年，Saleem 等研究了奶牛日粮中大麦比例（0、15%、30% 和 45%）的增加对瘤胃代谢组的影响，结果发现大麦对瘤胃代谢组的组成有显著影响，其中葡萄糖、丙酸盐、苯乙酸盐、丁酸盐、次黄嘌呤、尸胺、甲胺、腐胺、精氨酸、缬氨酸、丝氨酸和 L-丙氨酸等化合物受影响最大，并且含量随着大麦比例的增加而增加。这些结果表明，富含淀粉的日粮增加了游离葡萄糖的可利用性，它们可以作为细菌生长的底物增加挥发性脂肪酸的产量，从理论上看，对动物产量的提升是有益的。Zhang 等（2017）发现高精饲料饲喂的牛瘤胃中丁酸和丙酸的含量更高，而这对于小母牛的能量代谢和瘤胃的发育具有积极作用。然而，Ametaj 等（2010）还研究了谷物比例的增加对奶牛以及对瘤胃中代谢物的影响，结果发现大麦含量高于 30% 会增加奶牛瘤胃中有毒代谢物的含量。Saleem 等的研究（2012）也证实了这些结果，日粮中大麦比例超过 30% 会增加瘤胃中几种有毒、炎症和非天然化合物的浓度。山羊瘤胃代谢组的研究也报道了高谷物日粮对于动物健康的负面影响（Hua 等，2017）。

同样是研究日粮类型，Zhang 等（2017）的研究证明，改变日粮粗料与精料的比例会显著地改变瘤胃中的氨基酸、脂类、有机酸和碳水化合物的浓度。粗饲料类型也被证明会影响瘤胃代谢组，从而影响有机酸、胺和氨基酸的水平（Zhao 等，2014）。Sun 等（2015）发现分别饲喂苜蓿干草与玉米秸秆的奶牛瘤胃代谢组有显著差异，其中高品质的苜蓿干草会增加瘤胃中氨基酸、肽和碳水化合物的含量。

在一项综合日粮系统的比较研究中，O'Callaghan 等（2018）使用核磁共振检测了牧场和室内全混合日粮饲喂模式（TMR）对瘤胃和牛奶代谢物的影响。以牧草为基础的饮食能够增加异位酸、氨基酸和 P-甲酚的浓度，而 TMR 饲养则会使糖、胆碱和 3-苯丙酸盐浓度增加。这项研究还阐明了瘤胃代谢物与生牛乳之间的关联，即瘤胃液和牛奶中含有大量相同的代谢物。P-甲酚是瘤胃中产生的一种代谢物（Martin，1982；Carlson 和 Breeze，1984），近年来在人们评价牧场牛乳和乳制品的感观品质方面，这个指标也受到广泛关注。P-甲酚是一种有芳香气味的物质，含量较低时，它会使乳制品含有谷仓或类似动物的气味（Lopez 和 Lindsay，1993），而对于不习惯于牧场来源的牛奶和乳制品口味的人群，这种味道就属于不良气味。Sun 等（2015，2017）研究发现很多代谢物可以在不同的生物体液中，包括瘤胃液、血清、牛奶和尿液中被检测到。他们将瘤胃代谢组学与血清、牛奶、尿液和乳腺代谢谱相结合，以研究泌乳与非泌乳动物之间的代谢差异，结果发现在泌乳和非泌乳牛中一共有 33 种代谢物是共同变化的，其中 5 条代谢途径包括糖异生途径、丙酮酸代谢途径、三羧酸循环（TCA）途径、甘油酯代谢途径和天冬氨酸代谢途径被富集。O'Callaghan 等（2018）发现了多种可以作为"草饲"乳制品生物标志物的候选代谢物，如牛奶中的二甲基砜和马尿酸，并证明 NMR 代谢组学与多变量分析能够区分牧场和 TMR 饲养模式下的瘤胃代谢组和牛奶代谢组，并说明人们在食品真实性意识不断提高的背景下，NMR 代谢组学未来可以作为一种验证食品来源的工具。值得一提的是，在该研究中，作者还使用 16S rRNA 基因测序法检测了瘤胃微生物组，可能是由于试验饲喂周期较短，作者并未发现不同饮食之间瘤胃微生物组存在显著差异，但瘤胃代谢组的差异表明短期内的饮食变化虽然不会显著改变微生物组的组成，但会改变瘤胃微生物的功能，进而导致瘤胃代谢组的变化。

瘤胃代谢组学还可以用于寻找影响动物平均日增重的关键因素。Artegoitia 等（2017）确认了 33 种与公牛的平均日增重差异有关的代谢物。作者通过对牛的瘤胃代谢组进行分析，确定了亚油酸和 α-亚麻酸代谢以及芳香族氨基酸的生物合成相关功能途径的改变与平均日增重的变化有关。

之前的研究发现给反刍动物添加活酵母有益于提高动物饲料效率和生长性能，这种效应与瘤胃微生物组的改变如促纤维素分解细菌丰度的增加有关，此外，还与氧清除功能的增加（Calsamiglia 等，2005）和瘤胃乳酸浓度的降低有关（Dias 等，2018；Robinson，2010）。使用 LC-MS 进行瘤胃代谢组学研究，可以更好地探究活酵母有益效应的潜在作用机制。Ogunade 等（2019）研究发现，添加活酵母增加了 4-环己烷二酮和吡喃葡萄糖苷的浓度，减少了苏糖酸、黄苷、脱氧胆酸、月桂酰肉碱、甲氧基苯甲酸和十五烷烯苯甲酸的浓度。此外，作者还发现瘤胃微生物组的改变与氨基酸和能量代谢途径中的几种代谢产物呈现正相关关系，而这可能是添加酵母能够提高饲料效率和性能的原因。

亚急性瘤胃酸中毒（SARA）是饲喂高精饲料时常见的代谢性疾病。SARA 对牛的影响包括日粮摄入量减少、产奶量下降、牛乳脂肪含量降低、瘤胃 pH 值下降、生物胺和挥发性脂肪酸的积累以及蹄叶炎。之前有研究报道瘤胃中有多种与 SARA 有关的化合物，包括甲胺、亚硝胺、乙胺和乙醇（Ametaj 等，2010）。Xue 等（2018）将硫胺素作为缓解 SARA 的添加剂并探究其对瘤胃代谢组的影响。结果表明，饲喂 SARA 诱导日粮会增加瘤胃中的 VFAs、丙酮酸、乳酸和生物胺的浓度。而在 SARA 诱导日粮中添加硫胺素能够减少生物胺的含量,增加丙酮酸甲酸裂解酶的活性,抑制乳酸的产生,进而来缓解SARA。

4 研究展望

尽管瘤胃代谢组学是一个新兴的研究领域,但它在反刍动物代谢、疾病和日粮干预作用机制相关的化合物/代谢物检测方面的作用是深远的。了解影响瘤胃代谢组的因素能够为我们提供新型的干预策略,有效防止瘤胃功能受损和动物生产力下降。

当然,仪器和提取方法的选择也会影响代谢物的检出率,尤其是在非靶向代谢组学中,很多检测到的峰谱依然难以确认其组成。有研究发现,在某些情况下观察到的1 800多个峰谱中,只有约3.5%的峰的组成可以被确定(de Almeida 等,2018)。未来通过公开可用的交互式网络工具来进行数据集成,增强和提高非靶向代谢组学中峰谱的识别和鉴定比例,对于进一步了解瘤胃代谢组以及确定重要生理状况的新型生物标志物十分有益,这将有助于人们制定积极的干预策略,以提高反刍动物及其奶、肉制品的生产力和可持续性,增加动物的健康和福利。迄今为止,大多数研究都分析了饮食、饮食干预或健康状况是如何影响瘤胃代谢组。鉴于瘤胃可能对全身其他器官产生影响,其他因素如品种、遗传异质性和哺乳期对瘤胃代谢物及其功能的研究也是十分有意义的。

当今全球农业面临的主要挑战之一是其对全球变暖的影响。据报道,反刍动物甲烷排放的影响是农业对环境影响的重要部分。甲烷占温室气体总量的16%(Wallace 等,2015),在某些地区,反刍动物的甲烷排放率高达37%(Cottle 等,2011)。此外,反刍动物产生的甲烷代表着能量的损失,而这些能量原本可以用于生长和生产牛奶(Tapio 等,2017)。任何减少农业温室气体排放的努力都将有益于环境并改变消费者对该行业的看法。迄今为止,人们围绕瘤胃微生物组,在理解和减少甲烷排放方面做了大量工作。目前已经确定了与甲烷产生相关的微生物,包括古菌(Archaea)、纤毛类原生动物(Ciliate protozoa)、厌氧真菌(Anaerobic fungi)、琥珀酸弧菌(Succinovibrionaceae)和普雷沃氏菌(*Prevotella*)(Wallace 等,2015;Tapio 等,2017)。此外,代谢途径如甲酸盐代谢途径、氢和甲烷生成途径的改变也与甲烷排放有内在联系(Tapio 等,2017)。Martinez-Fernandez 等(2016)证明了氯仿能够减少牛体内甲烷的产生。作者发现,通过将[H]重新转化为还原性微生物终产物,可以减少30%~35%甲烷的生成,而这种变化不会对瘤胃功能产生不良影响。但是由于氯仿的毒性及其对环境影响,它并不能在养殖条件下使用。

Martinez-Fernandez 等(2018)在热带牧草中添加3-Nitrooxypropanol(3-NOP)和氯仿饲喂牛,并探究其对甲烷和氢的产量、瘤胃代谢物和微生物群落结构的影响。作者证明3-NOP 具有与氯仿相似的抑制产甲烷的作用。补充这些物质会使瘤胃中的各种代谢物和产甲烷菌发生变化,作者认为甲烷生成抑制是由于[H]不再转化为CH_4,而是转化形成其他对宿主有益的还原终产物。虽然减少甲烷排放是有益的,但同时也要注意甲烷的生成能够避免瘤胃中H_2的过度积累,H_2过度积累将通过改变瘤胃的pH值影响瘤胃的纤维消化能力(Matthews 等,2019)。因此,我们需要采用综合的方法来解决这一问题。虽然目前用代谢组学来探究瘤胃中甲烷生成动力学还较为困难,但是它也为我们理解饮食、微生物组、瘤胃代谢和代谢物的产生与牛体内甲烷产生之间的联系提供了可能。深入了解这些因素之间的相互作用，将有利于人类通过开发新的策略来减少养殖业整体的碳排放。

5 结论

正如我们之前所介绍的，多种技术可以用于代谢组学分析，而每种技术都有各自的优势和局限性；提取方法的不同也会影响代谢物的检出率。为了更全面地了解代谢组，多种技术方法联合使用依然是一个有效的策略。

为了更好地了解瘤胃的代谢物组成和功能，人们已经开展了大量的研究；然而，我们对于瘤胃这个器官的了解还远远不够。瘤胃代谢物的检测可以确定饮食对反刍动物代谢、饲料效率、牛奶成分和质量以及对动物健康的影响。通过技术创新提高代谢物的检测灵敏度对于代谢组的鉴定和解析非常重要，同时也使得鉴定更多新的代谢物成为可能，进一步加深了我们对反刍动物代谢和生产的了解。

6 参考文献

Aggio, R., Villas-Bôas, S. G. and Ruggiero, K. 2011. Metab: an R package for highthroughput analysis of metabolomics data generated by GC-MS. Bioinformatics, 27 (16), 2316-8.

Ametaj, B. N., Zebeli, Q., Saleem, F., et al., 2010. Metabolomics reveals unhealthy alterations in rumen metabolism with increased proportion of cereal grain in the diet of dairy cows. Metabolomics, 6 (4), 583-94.

Artegoitia, V. M., Foote, A. P., Lewis, R. M., et al., 2017. Rumen fluid metabolomics analysis associated with feed efficiency on crossbred steers. Scientific Reports 7 (1), 2864.

Bouatra, S., Aziat, F., Mandal, R., et al., 2013. The human urine metabolome. PLoS ONE 8 (9), e73076.

Cajka, T. and Fiehn, O. 2016. Toward merging untargeted and targeted methods in mass spectrometry-based metabolomics and lipidomics. Analytical Chemistry 88 (1), 524-45.

Calsamiglia, S., Castillejos, L. and Busquet, M. 2005. Alternatives to antimicrobial growth promoters in cattle. In: Smith, P. and Wiseman, J. (Eds), Recent Advances in Animal Nutrition. Nottingham University Press, Nottingham, UK, 129-67.

Carlson, J. R. and Breeze, R. G. 1984. Ruminal metabolism of plant toxins with emphasis on indolic compounds. Journal of Animal Science 58 (4), 1040-9.

Chong, J., Soufan, O., Li, C., et al., 2018. MetaboAnalyst 4.0: towards more transparent and integrative metabolomics analysis. Nucleic Acids Research 46 (W1), W486-94.

Cottle, D. J., Nolan, J. V. and Wiedemann, S. G. 2011. Ruminant enteric methane mitigation: a review. Animal Production Science 51 (6).

De Almeida, R. T. R., Do Prado, R. M., Porto, C., et al., 2018. Exploring the rumen fluid metabolome using liquid chromatography-highresolution mass spectrometry and Molecular Networking. Scientific Reports 8 (1), 17971.

Dias, A. L. G., Freitas, J. A., Micai, B., et al., 2018. Effect of supplemental yeast culture and dietary starch content on rumen fermentation and digestion in dairy cows. Journal of Dairy Science 101 (1), 201-21.

Do Prado, R. M., Porto, C., Nunes, E., et al., 2018. Metabolomics and agriculture: what can be done? mSystems 3 (2), e00156-17.

Foroutan, A., Guo, A. C., Vazquez-Fresno, R., et al., 2019. Chemical composition of commercial cow's milk. Journal of Agricultural and Food Chemistry 67 (17), 4897-914.

Goldansaz, S. A., Guo, A. C., Sajed, T., et al., 2017. Livestock metabolomics and the livestock metabolome: a systematic review. PLoS ONE 12 (5), e0177675.

Hua, C. F., Tian, J., Tian, P., et al., 2017. Feeding a high concentration diet induces unhealthy alterations in the composition and metabolism of ruminal microbiota and host response in a goat model. Frontiers in Microbiology 8, 138.

Jewison, T., Knox, C., Neveu, V., et al., 2012. YMDB: the yeast metabolome database. Nucleic Acids Research 40 (Database issue), D815-20.

Jewison, T., Su, Y., Disfany, F. M., et al., 2014. Smpdb 2. 0: big improvements to the small molecule pathway database. Nucleic Acids Research 42 (Database issue), D478-84.

Lopez, V. and Lindsay, R. C. 1993. Metabolic conjugates as precursors for characterizing flavor compounds in ruminant milks. Journal of Agricultural and Food Chemistry 41 (3), 446-54.

Markley, J. L., Brüschweiler, R., Edison, A. S., et al., 2017. The future of NMR-based metabolomics. Current Opinion in Biotechnology 43, 34-40.

Martin, A. K. 1982. The origin of urinary aromatic compounds excreted by ruminants: 3. The metabolism of phenolic compounds to simple phenols. British Journal of Nutrition 48 (3), 497-507.

Martinez-Fernandez, G., Denman, S. E., Yang, C., et al., 2016. Methane inhibition alters the microbial community, hydrogen flow, and fermentation response in the rumen of cattle. Frontiers in Microbiology 7, 1122.

Martinez-Fernandez, G., Duval, S., Kindermann, M., et al., 2018. 3-NOP vs. Halogenated compound: methane production, ruminal fermentation and microbial community response in forage fed cattle. Frontiers in Microbiology 9, 1582.

Matthews, C., Crispie, F., Lewis, E., et al., 2019. The rumen microbiome: a crucial consideration when optimising milk and meat production and nitrogen utilisation efficiency. Gut Microbes 10 (2), 115-32.

Neveu, V., Moussy, A., Rouaix, H., et al., 2017. Exposome-Explorer: a manually-curated database on biomarkers of exposure to dietary and environmental factors. Nucleic Acids Research 45 (D1), D979-84.

O'Callaghan, T. F., Ross, R. P., Stanton, C., et al., 2016. The gut microbiome as a virtual endocrine organ with implications for farm and domestic animal endocrinology. Domestic Animal Endocrinology 56 (Suppl.), S44-55.

O'Callaghan, T. F., Vázquez-Fresno, R., Serra-Cayuela, A., et al., 2018. Pasture feeding changes the bovine rumen and milk metabolome. Metabolites 8 (2), 27.

Ogunade, I., Schweickart, H., McCoun, M., et al., 2019. Integrating 16S rRNA sequencing and LC-MS-based metabolomics to evaluate the effects of live yeast on rumen function in beef cattle. Animals 9 (1), 28.

O'Hara, A. M. and Shanahan, F. 2006. The gut flora as a forgotten organ. EMBO Reports 7 (7), 688-93.

Psychogios, N., Hau, D. D., Peng, J., et al., 2011. The human serum metabolome. PLoS ONE 6

(2), e16957.

Robinson, P. 2010. Yeast products for growing and lactating ruminants: a literature summary of impacts on rumen fermentation and performance. Cooperative Extension Specialist Department of Animal Science, University of California, Davis, CA, p. 95616.

Rothwell, J. A., Perez-Jimenez, J., Neveu, V., et al., 2013. Phenol-Explorer 3. 0: a major update of the Phenol-Explorer database to incorporate data on the effects of food processing on polyphenol content. Database: the Journal of Biological Databases and Curation 2013, bat070.

Saleem, F., Ametaj, B. N., Bouatra, S., et al., 2012. A metabolomics approach to uncover the effects of grain diets on rumen health in dairy cows. Journal of Dairy Science 95 (11), 6606-23.

Saleem, F., Bouatra, S., Guo, A. C., et al., 2013. The bovine ruminal fluid metabolome. Metabolomics 9 (2), 360-78.

Sun, H. Z., Wang, D. M., Wang, B., et al., 2015. Metabolomics of four biofluids from dairy cows: potential biomarkers for milk production and quality. Journal of Proteome Research 14 (2), 1287-98.

Sun, H. Z., Shi, K., Wu, X. H., et al., 2017. Lactationrelated metabolic mechanism investigated based on mammary gland metabolomics and 4 biofluids' metabolomics relationships in dairy cows. BMC Genomics 18 (1), 936.

Tapio, I., Snelling, T. J., Strozzi, F., et al., 2017. The ruminal microbiome associated with methane emissions from ruminant livestock. Journal of Animal Science and Biotechnology 8, 7.

Wallace, R. J., Rooke, J. A., Mckain, N., et al., 2015. The rumen microbial metagenome associated with high methane production in cattle. BMC Genomics 16, 839.

Wishart, D. S. 2005. Metabolomics: the principles and potential applications to transplantation. American Journal of Transplantation: Official Journal of the American Society of Transplantation and the American Society of Transplant Surgeons 5 (12), 2814-20.

Wishart, D. S. 2008. Metabolomics: applications to food science and nutrition research. Trends in Food Science and Technology 19 (9), 482-93.

Wishart, D. S., Jewison, T., Guo, A. C., et al., 2013. HMDB 3. 0—the human metabolome database in 2013. Nucleic Acids Research 41 (Database issue), D801-7.

Wishart, D., Arndt, D., Pon, A., et al., 2015. T3DB: the toxic exposome database. Nucleic Acids Research 43 (Database issue), D928-34.

Xia, J. and Wishart, D. S. 2016. Using MetaboAnalyst 3. 0 for comprehensive metabolomics data analysis. Current Protocols in Bioinformatics 55 (14), 14. 10. 1-14. 10. 91.

Xue, F., Pan, X., Jiang, L., et al., 2018. GC-MS analysis of the ruminal metabolome response to thiamine supplementation during high grain feeding in dairy cows. Metabolomics: Official Journal of the Metabolomic Society 14 (5), 67.

Zhang, J., Shi, H. T., Wang, Y. J., et al., 2017. Effect of dietary forage to concentrate ratios on dynamic profile changes and interactions of ruminal microbiota and metabolites in holstein heifers. Frontiers in Microbiology 8, 2206.

Zhao, S., Zhao, J., Bu, D., et al., 2014. Metabolomics analysis reveals large effect of roughage types on rumen microbial metabolic profile in dairy cows. Letters in Applied Microbiology 59 (1), 79-85.

第4章 瘤胃微生物功能数学建模的方法

André Bannink、Soumya Kar、Dirkjan Schokker、Jan Dijkstra

瓦格宁根大学,荷兰

(王佳堃译)

1 前言

瘤胃作为消化器官,在反刍动物饲料消化过程中扮演非常重要的角色。因为瘤胃中栖息着大量的微生物,才使得反刍动物能够发酵纤维和非纤维饲料获取发酵产物。也正因为瘤胃及其微生物的存在,反刍动物可以利用营养价值低,甚至是其他动物无法食用的饲料满足自身的维持和生产需要。瘤胃的消化作用主要归功于瘤胃中微生物的发酵作用。瘤胃中的微生物包括细菌、原虫和真菌。此外,由于瘤胃内部极端的缺氧环境,古细菌(产甲烷菌)也是其中一员。尽管不同反刍动物品种间,瘤胃微生物种群的组成极其复杂、功能极为多样,但不同物种、不同区域和不同管理模式下,都存在一个核心微生物组(Henderson 等,2015)。这个核心微生物组也并非一成不变。核心微生物的波动主要由日粮引起。一方面,日粮决定了瘤胃中微生物可利用底物的数量、类型和质地;而另一方面,日粮与宿主互作(反刍、瘤胃收缩与瘤胃内容物移动和外排)影响瘤胃内食糜流动,影响食糜的物理和化学状态,而食糜的流动和它的理化状态又决定了微生物的活性(Zebeli 等,2012;Bannink 等,2016)。瘤胃内的理化因素主要有瘤胃内容物的粒度分布、纤维层厚度、液体体积、水和颗粒流速、流体渗透性、酸度和氧化还原电位,以及二氧化碳、甲烷和氢气的压力。半个多世纪以来,研究人员建模时都兼顾到了多数这些理化因素,但对氧化还原电位(Dijkstra 等,2020)、气压(Van Lingen 等,2017)和渗透性(López 等,2003)的研究甚少。近年来分子生物学与应用生物学相关技术的发展,使深入研究瘤胃内环境成为可能。2018 年 Ungerfeld 和 Newbold 主编的电子书中收录了多篇关于该领域最新研究进展的高质量综述。而且他们在书评中展望了功能基因表达、理化原理应用、瘤胃热动力学和新的动力测定方法等微生物多组学技术(multi-omics)联用带来的变革。

随着新技术和新数据类型的不断引入,为了充分利用收集到的数据,必须建立数学模型来捕获其中的生物信息,并在瘤胃整体水平上去预测功能。目前已有许多瘤胃整体水平上的模型,但是模型基于的组织层面和所采用的建模理念大相径庭。系统动力学机制模型方法仍是目前最为复杂的建模方法。图1列出了系统动力学机制模型所涉及的各级组织层面信息。1987 年和 1992 年 Baldwin 等和 Dijkstra 等分别构建了预测瘤胃功能的动力学模型

(Baldwin 等, 1987; Dijkstra 等, 1992)。很多研究团队基于这两个模型又衍生出了新的模型。而这些动力学模型是迄今为止最精细的建模方法。Bannink 等 2016 年的综述 (Bannink 等, 2016) 中引用了 Tedeschi 等 (2014) 的文章, Tedeschi 等指出其他模型采用的方法相对简单,它们采用静态的或者较少的机制算法,而更多的应用经验值(取决于建模的实际目的,简单可能更适宜)。最近,研究人员把热动力学驱动力对瘤胃微生物的影响(Van Lingen 等, 2016)和瘤胃氢动力对瘤胃微生物活力的影响(Wang 等, 2016; Van Lingen 等, 2019)引入模型中。用这两项指标反映瘤胃氢气压力对发酵终产物形成的热动力学调控作用。但是所有这些瘤胃模型都没有整合多组学技术的成果。虽然有些模型量化了瘤胃微生物功能,以期实现瘤胃整体功能的预测,但这些量化并未与多组学方法的量化相结合,仍各自独立。

本章讨论了量化瘤胃微生物群或对瘤胃微生物进行数学建模的理念和方法;讨论了从瘤胃整体水平上数学建模预测微生物功能的重要性。瘤胃作为器官,只有在整体水平建模才能更好地预测瘤胃的发酵和消化状况。

2 瘤胃整体功能建模的方法

2.1 反刍动物宿主与瘤胃内容物之间的相互作用

瘤胃内容物与反刍动物宿主之间存在大量的相互作用(图1)。基因与环境因素使得反刍动物个体间在采食和反刍、瘤胃壁形态和生理、瘤胃收缩和食糜经网胃行至瓣胃和皱胃的节律方面都存在差异。反刍动物宿主本身可以调控液体和食糜颗粒在瘤胃内的滞留时间,所以对不同品质日粮有不同的适应能力。慢速降解的饲料颗粒,在瘤胃内的滞留时间更长,使微生物有更充足的时间去降解,但慢速降解的饲料多了,占瘤胃的体积也随之增大。唾液的产生量和挥发性脂肪酸(VFA)的吸收情况,决定了瘤胃的缓冲能力,而瘤胃的缓冲能力是影响瘤胃微生物活力的另一个决定性因素。瘤胃壁高度适应瘤胃的酸性环境,这证实了瘤胃壁在确保瘤胃正常发酵中的调节作用。如果没有上述缓冲机制,瘤胃会酸化,微生物活力会减弱,进而降低饲料的消化率和营养价值。氮和磷经瘤胃壁和唾液从血液回流至瘤胃也是宿主和瘤胃环境间另一个非常重要的相互作用。

目前相对复杂一点的瘤胃模型都会考虑前文提到的宿主与瘤胃微生物之间的互作(Bannink 等, 2016)。这些模型旨在量化整个瘤胃的功能,预测饲料降解情况及微生物蛋白质、VFA、氨和甲烷等发酵产物的产生量。Tedeschi 等综述指出,尽管有些复杂的瘤胃模型采用了高度机制化的表示方法,但如实际应用中的蛋白质评估系统,这些应用性强的模型远没有那么复杂。这些应用型模型采用一般假设处理瘤胃排空速度和瘤胃内环境等理化因素(Tedeschi 等, 2014)。为了防止瘤胃氮缺乏,应用型模型也会考虑计算出的瘤胃氮平衡不能低于阈值。而机制模型试图描述氮循环,并以此来预测瘤胃发酵状态。应用型模型基本忽略了宿主和瘤胃内环境的相互作用(图1)。虽然复杂的机制模型考虑了这些互作关系,但机制模型的许多参数是没有表型数据的,用的是宿主的平均值,这限制了个体的参数化。Bannink 等(2016)的综述有进一步关于互作关系以及如何将这些互作关系纳入瘤胃模型的讨论。Tedeschi 等（2014）的综述中也介绍了许多介于应用型和机制模型间的瘤胃模型。

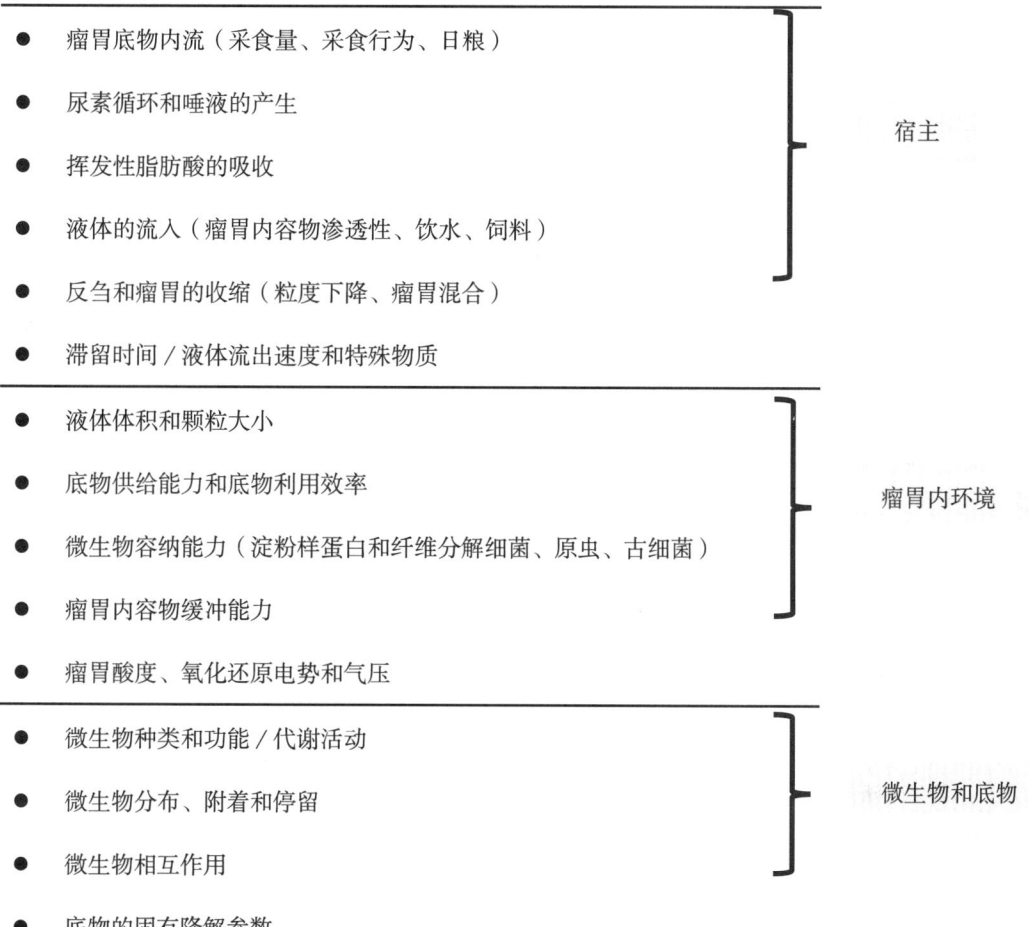

图1 与瘤胃功能相关的各级层次示意图，从瘤胃整体水平到多组学数据收集。

2.2 瘤胃内环境

在瘤胃层面，影响微生物活性的各种理化因素都是数学建模的关键参数。这些因素包括瘤胃中氢作为微生物代谢的热力学驱动力对产甲烷菌活性的影响（Wang 等，2016；Van Lingen 等，2016，2019），宿主通过调节瘤胃上皮组织活性（Bannink 等，2012），调节瘤胃内饲料颗粒分布、粒径和瘤胃液外排动力（Gregorini 等，2015），实现对瘤胃酸度的调控作用。这些参数会对微生物代谢产生直接影响，量化这些参数有利于准确预测营养策略对瘤胃功能的影响以及瘤胃功能的宿主特异性。尽管这些瘤胃内环境因素非常重要，但本章不做深入讨论，本章将把重点放在瘤胃微生物群落组成和功能的多样性及其在发酵过程中对瘤胃功能的影响上。

2.3 微生物功能类别

在现有的瘤胃整体模型中，关于微生物活性有两个重要的参数，分别是淀粉分解活性和纤维分解活性。淀粉分解活性与糖和淀粉等易降解的碳水化合物的利用有关，一般认为这部分碳水化合物粒径小或存在于液相中，在瘤胃中滞留时间短。纤维分解活性则与粒径大、滞留时间长的结构性碳水化合物的利用有关。尽管饲料并不能绝对归入结构性和非结构性的碳水合化物，而且很多微生物可以利用多种底物，但不同类群的微生物都可以利用结构性和非结构性碳水化合物生长繁殖。这就使得在种水平上讨论微生物是淀粉分解菌还是纤维分解菌存在某种随意性。除细菌外，瘤胃中还存在大量的原虫。原虫除了直接利用饲料，还捕食细菌，从而促进瘤胃内循环（Newbold 等，2015）。原虫不容易随瘤胃内容物外排，因此在瘤胃内滞留的时间比细菌长得多。尽管原虫具有如此重要的作用，但是迄今为止，近乎所有的瘤胃模型都没有把原虫的活性与淀粉分解细菌和纤维分解细菌的活性并提（Tedeschi 等，2014）。唯有 Dijkstra（1994）与 Hook 等（2017）的模型量化了瘤胃中原虫的动态变化和滞留情况。Nagorcka 等（2000）采用了更详尽的方法表示不同的微生物功能类别。他们使用一系列公认的化学方法测量了不同类别微生物的 VFA 产量，再根据特定底物下，每一类别微生物的 VFA 贡献度，赋予微生物权重。这种方法并没有进一步细化模型中功能性微生物的分类，而是改变了 VFA 的化学计量方式。Dougherty 等（2017）将该方法引入瘤胃模型来指导澳大利亚的牛肉生产。在该模型中也同时引入了乳酸利用细菌。但遗憾的是，目前该模型的方程和假设尚未发表，这限制了与其他瘤胃模型的比较，但是在构建新的系统动力学机制模型时，研究人员总是会将这个方法与以往的方法并提（Tedeschi 等，2014；Bannink 等，2016）。

乳酸作为微生物代谢的终产物，Mills 等（2014）采用了一种更详尽的、更动态的方法来研究营养供应时瘤胃中乳酸的代谢情况。他们在 Dijkstra 模型（1994）的基础上引入了除纤维分解菌和原虫外的乳酸生成菌和乳酸利用菌的动态变化，并用乳酸利用菌来取代淀粉分解菌。该模型引入了特定的参数研究瘤胃中细菌和原虫对乳酸的竞争，以及两类微生物的动态互作。随着测量反刍动物肠道甲烷热度的增加，越来越多的人试图定量甲烷产量。几乎所有的模型都是把瘤胃氢平衡产生的氢净值完全转化为甲烷来计算甲烷产量。仅 2019 年，Van Lingen 等才把产甲烷菌作为一种功能微生物类别引入构建的模型中（Van Lingen 等，2019）。

综上所述，在瘤胃数学模型中通常的微生物功能类别是淀粉分解菌和纤维分解菌。更专业的模型会细分乳酸产生菌、乳酸利用菌、原虫和产甲烷菌。但是没有一个瘤胃模型进一步细化到每个属，或者多组学技术获得的具体功能。因为底物的瘤胃降解参数信息很容易获取（Li 等，2018；Dijkstra 等，1992），所以现有的瘤胃模型通常细化到饲养试验报道的底物类型和相关的微生物活性，而不区分微生物群落中的特定微生物属。

2.4 微生物互作关系

不同种类微生物间的互作是瘤胃微生物种群功能的重要特点。因为微生物的组成和互作与底物降解和发酵、微生物数量、VFA、氢和甲烷等发酵终产物的形成密切相关，所以原则上，模型设计者都想在模型中细化不同的微生物类群以及它们之间的互作关系。对微生物互作关系进行建模意味着一类微生物的活动与另一类微生物存在相互作用。瘤胃微生物组的研究表明瘤胃中存在大量不同种类的微生物，微生物会联合起来完成饲料降解等复杂的工作（Belanche 等，2019）。与此同时，日粮影响着微生物群落的组成与多样性（Henderson 等，2015）。然而，建模必须注意到瘤胃微生物种群存在极强的功能冗余，也就是说多物种间存在功能重叠。有研究表明，微生物群落组成和多样性显著变化时，并不改变 pH 值和 VFA 摩尔比等瘤胃发酵关键指标，这显然表明瘤胃微生物功能冗余（Weimer，2015）。Belanche 等（2019）研究也发现绵羊由干草-精料饲养转换为黑麦草放牧后，瘤胃微生物组发生了极大的变化。泌乳期奶牛瘤胃细菌丰度比干奶期低，这可能是因为与干奶期相比，泌乳期奶牛进食量大，且饲料中精料比例高（Dieho 等，2017）（图2）。产犊后精料供给量的增加，奶牛瘤胃细菌丰度会短期内显著下降。精料增加得越快，细菌丰度下降得越严重。这种下降仍是暂时性的。但 Dieho 等（2017）发现微生物组成的显著改变与饲料的瘤胃降解动力并不一致。尽管瘤胃中详细的微生物群落组成研究为揭示微生物如何适应不同的瘤胃环境和日粮提供了精辟的见解，但仍未深入到微生物之间的互作，也没能给出瘤胃建模时如何量化微生物的功能。为了解决这一问题，除了研究瘤胃中不同属微生物的丰度外，还需要对特定的功能进行定量。鸟枪法宏基因组测序技术便于获取瘤胃微生物的功能信息。但相比于 16S/18S/ITS 扩增子测序，瘤胃微生物宏基因组的数据偏少。

按照现有瘤胃模型的方法，最常划分的微生物功能包括淀粉分解活性（利用可溶性或易降解碳水化合物）和纤维分解活性（利用结构性碳水化合物）。如此简化瘤胃功能，是因为能量是决定微生物生长的最主要的因素，微生物可获得的能量主要来自这些底物。但其他模型的功能划分也是很好理解的。如构建瘤胃酸中毒模型时考虑乳酸生成菌和乳酸利用菌，构建瘤胃氮利用效率模型时考虑蛋白分解菌和原虫，或者引入瘤胃上皮黏附微生物来代表这类微生物在瘤胃中的功能。

微生物在瘤胃中是游离的，还是黏附在饲料颗粒上的，与功能也密切相关。通常机制型模型会将这两种类型的微生物都考虑在内（Tedeschi 等，2014；Bannink 等，2016）。瘤胃模型包含细菌与原虫的功能（Dijkstra，1994）、乳酸生成菌与乳酸利用菌（Mills 等，2014；Dougherty 等，2017）、氢敏感性（底物降解速度快、微生物生长迅速）与氢不敏感性代谢（底物降解速度慢、微生物生长缓慢）（Van Lingen 等，2017），储能多糖的合成能力（Dijkstra 等，1992）等微生物活性信息越多，越可以深入区分微生物种类。目前的瘤胃模型大多默认不同瘤胃功能是互相独立的，比如认为淀粉分解与纤维分解是两个独立的过程。只有微生物类别之间确实存在物质交换，或者不同微生物之间共用一种底物时，模型才会引入微生物之间的依赖关系。例如当不同类别微生物之间共用一种氮源时，可以用氮的可用性来表示微生物类别之间的依赖关系（Li 等，2019）。Dijkstra 等（1992）和 Dijkstra（1994）模拟了另一种与底物相关的相互作用。原虫和淀粉分解菌会吞食易发酵的糖和淀粉，并将其储存在细胞内，防止其他微生物快速发酵这些碳水化合物，从而减轻瘤胃酸化，防止瘤胃氢分压的增长。当然也可以利用微生物间的直接关系与相互作用建模，如原虫捕食细菌（Dijkstra，1994）；甲烷合成依赖细菌与原虫产生的氢，但合成的甲烷又会影响细菌甚至原虫的发酵模式，反过来改变瘤胃的氢分压（Van Lingen 等，2019）。假如某些相互作用很难被表示（如没有共用的底物或直接关系），那么建模者总是会直接或间接地假定不同类别瘤胃微生物间功能独立（Russell 等，1992）。

除 Reichl 和 Baldwin（1976）的线性规划模型、Mills 等（2014）和 Dougherty 等（2017）的模型外，一般瘤胃整体模型中并未纳入微生物在属甚至种水平的功能。Reichl 和 Baldwin 根据底物特异性、发酵速率以及生长的营养需求，将 10 多个优势瘤胃微生物种分成 8 类纳入模型。Mills 等（2014）和 Dougherty 等（2017）为了研究瘤胃的乳酸动态和酸度，分别以埃氏巨球形菌（*Megasphaera elsdenii*）和牛链球菌（*Streptococcus bovis*）作为瘤胃中主要的乳酸生成菌和乳酸利用菌纳入模型。除了 1994 年 Dijkstra 的模型考虑了原虫和细菌共用碳源外，目前微生物在模型中均是独立的，瘤胃模型均未考虑在相同的糖、淀粉、半纤维素和纤维素碳源下，微生物可能有着多种功能。1994 年 Dijkstra 的模型是其 1992 年模型的细化版。很多研究者试图利用组学技术将瘤胃的功能与具体的微生物一一对应，获得了大量新的、不同类型的数据（图3）。这些数据让我们对不同瘤胃环境下瘤胃微生物组成的变化和推测的功能（尽管这方面的数据也很有限）有了新的了解（Denman 等，2018；Stewart 等，2018，2019），但目前的数据还不能指征微生物的功能，不能应用于瘤胃整体功能模型的数学预测。对这些新数据的合理分析可能会得出一些全新的概念，将有助于构建瘤胃整体功能模型。如果将这些概念通过指征瘤胃功能的方式纳入当前的瘤胃功能模型中，可能会大大扩展模型的适用性，并应用于多组学联用。下文将讨论文献中报道的一些概念。

2.5 微生物群落动力学

功能冗余和回弹等瘤胃微生物功能相关的术语，已被用于复杂生态系统的研究中。Weimer（2015）认为设计一种包含多种生态位和多种微生物种群的瘤胃生态系统是可行的。他的一个主要观点是瘤胃微生物群落存在强大的功能冗余，这意味着即使微生物的丰富度或多样性降低，由于微生物种间的功能重叠，瘤胃的发酵特性和消化能力可能并不受影响（Weimer，2015）。回弹是瘤胃微生物种群的一个关键特征，瘤胃微生物可以在饲料、饮水和日粮组成变化造成的强扰动后恢复原有状态（Weimer，2015；Scheffer 等，2015）。Dieho 等（2017）观察到产后奶牛干物质和精料摄入量的增加，在一定程度上降低了细菌群落的丰度（图2a；基于"系统发育树多样性"的丰富度度量，Faith，1992）。产后奶牛精料摄入从每日 1kg 干物质快速增长至 11kg 的过程中，瘤胃细菌丰度迅速下降（图2a；实心符号）。但这种现象在一段时间后基本消失。尽管初始的精料摄入量不同，但一段时间后，各组奶牛精料摄入量不再有显著差异，瘤胃细菌丰度也不再有差异，这表明微生物种群具有恢复力。与此同时，微生物丰度的变化并未影响饲料的原位降解率，这表明微生物种群功能冗余。随后 Dieho 等（未发表）探讨了干奶期补饲精料的影响，发现在干奶期的最后 28 d 每日给奶牛补饲 3kg 的精料（干物质），干奶期结束时，精料补饲组奶牛的瘤胃微生物丰度未发生改变，这可能是因为每日总的饲料干物质摄入量较低，仅为 12kg 所致。产后精料进食量从每日 1kg 增至 9kg 的过程中，干奶期补饲精料并未影响精料的进食，但在泌乳早期，供应同样的泌乳日粮的情况下，干奶期补饲精料的奶牛瘤胃细菌

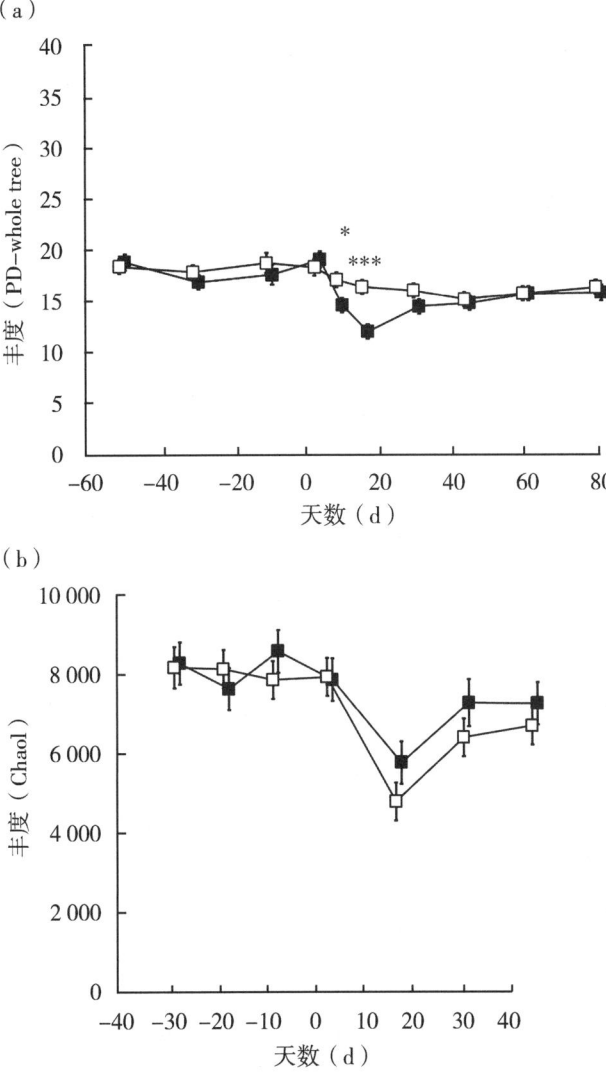

图 2 (a) 奶牛的瘤胃微生物丰度（■□；基于"系统发育多样性全树"的丰富度；Chao，1984）。产犊后精料供给量快速（1.0 kg DM/d；■，n=6）和平稳（0.25 kg DM/d；□，n=6）增加，值代表最小二乘均值±标准误差（ *** 代表 $P<0.001$，* 代表 $P<0.05$）（转载自 Dieho 等，2017）。(b) 干奶期对照日粮（0.0 kg DM/d 浓缩物，□；n=5）和补充精料的细菌群落丰富度（■□；基于"Chao1"的丰富度度量；Chao，1984）（3.0 kg 精料 DM/d，■；n=4）。值代表最小二乘平均值±标准误差。资料来源于 Dieho 等未发表的结果。

丰度高于未补饲精料的奶牛（图 2b；基于"Chao1"的丰富度；Chao，1984）。这可能表明，干奶期末期瘤胃微生物可以为产犊后精料摄入量的快速增加做好准备；当然，这还需更多结论性结果的证实。这些结果表明，伴随饲喂制度的变化，瘤胃微生物丰度也会出现瞬时的变化，其中饲喂水平可影响瘤胃微生物，但可发酵有机物快速和大量摄入对瘤胃微生物的影响更强。尽管采食量和日粮组成的改变（从干奶期低采食量、少精料的状态转变成泌乳期高采食量、高精料的状态）会导致细菌多样性轻微地改变，但最终微生物丰度会恢复到干扰前的状态（图2a和图2b）。在这两项研究中，瘤胃微生物种群表现出相当强的回弹性。这体现在尽管细菌多样性发生变化，饲料的瘤胃降解率仍保持不变。Belanche等（2019）基于系统生物学方法，发现当日粮由干草/精料变为黑麦草放牧时，微生物组成及关键微生物属的丰度会发生永久的变化。他们用"可塑性"这个术语来表示瘤胃微生物区系适应日粮和瘤胃发酵条件的强烈变化而不丧失其功能的能力。一般来说，应对日粮的彻底改变，虽然短期内瘤胃微生物的功能和代谢活动会发生改变，但随着时间的推移，瘤胃微生物区系似乎是复杂而稳定的（O'Callaghan等，2018）。

现有的瘤胃整体功能模型均未涵盖上述所观察到的瘤胃微生物动态变化和行为的特征；甚至连概念都未引入（Dougherty等，2017；Li等，2018；Van Lingen等，2019）。然而，在瘤胃模型中表示这些概念是很有意义的。尽管人们设计出了基因组尺度代谢模型（genome-scale metabolic models，GEMs），而且也应用该模型描述特定的物种，以及两两物种间的互作关系，但对于像瘤胃这种微生物群落，模型还是很少涉及群落的动态行为。瘤胃微生物群的功能互补和功能冗余，使瘤胃微生物的动态行为更为复杂。理论上讲，Scheffer等（2015）提出功能互补促进整个生态系统演变，而功能冗余则促进生态系统从演变中复原，微生物通过这种生物多样性的保险效应抵抗外界环境的扰动。功能冗余是体型小的生物群体进化的机制，不易在体型大的生物群体中发生。尽管许多物种具有相同的功能（也利用相同的底物），但面对胁迫时，它们做出的反应可能不同。对于体型大的生物群体，它们在生态系统中的功能越独特，它们的丰富度就越低，功能冗余也就越罕见。它们一但从生态系统中消失，生态系统的功能会发生剧烈改变。虽然Scheffer等（2015）用遍布世界的4 000多种潜水甲虫阐释了功能冗余这一概念，但这一理论性概念可能仅适用于瘤胃这种体型小的和体型大的微生物共存的复杂生态系统。瘤胃生态系统中细菌是体型小的微生物，原虫则是体型大的微生物。基于此，研究原虫在瘤胃生态系统中的作用和变化应该是很有意思的（Newbold等，2015）。原虫的体型较细菌大得多，是细菌的20~100倍，原虫可以占到瘤胃微生物总生物量的一半（Bainbridge等，2018）。瘤胃生态系统的功能冗余和回弹在细菌群落中得以很好的体现（Weimer，2015）。但对于体型较大的原虫而言，它的底物利用率低，它在瘤胃里的功能很容易受环境扰动的影响（Dijkstra，1994）。尽管原虫具有抵抗外排和隔离的能力，并能通过反刍动物之间接触水平转移，但扰动可以（暂时）削弱它们在瘤胃中的作用，从而使瘤胃建立新的平衡，改变微生物生态系统整体的功能。古细菌可能与原虫一样，因为它们在进食后，不仅活性低于细菌，而且活性会延迟（Van Lingen等，2017）。因为原虫和古细菌在瘤胃生态系统中起着关键作用，所以有必要通过调节底物发酵速度和氢动力，改变瘤胃的整体发酵模式，制造扰动，解析扰动对原虫和古细菌的影响和作用机制（Newbold等，2015）。因为瘤胃发酵模式决定了宿主从瘤胃吸收的生糖和生酮物质的比率以及甲烷的排放量，所以瘤胃发酵模式是瘤胃整体功能模型需要预测的一个重要反应变量（Dijkstra，1994；Huws等，2018；Van Lingen等，2019）。对于功能冗余和回弹是否也适用于古细菌和原虫，以及日粮变化和特定补充剂饲喂等突然的扰动是否会引起瘤胃生态系统功能的变化，将是非常有意思的研究内容。通过育种评估不同瘤胃发酵模式下的扰动易感性可能也非常有意义。例如，育种筛选出的低甲烷产量的母羊，它的瘤胃更小，固体和液体食糜在瘤胃中的滞留时间更短（Goopy 等，2014）。如果可以证实上述瘤胃功能的变化，将其纳入瘤胃整体功能模型，那么首先要了解这些变化与功能的关系。目前的研究经常以试验描述性的方式报道瘤胃状态之间的差异，但并不了解不同瘤胃状态下的生态系统功能，以及如何在瘤胃整体功能模型中量化消化率、酸度、微生物合成效率、瘤胃内循环、发酵终产物 VFA 的吸收等功能。研究瘤胃和分析瘤胃微生物的行为可能需要全新的试验设计。

3 量化瘤胃微生物功能

2018 年 Denman 等在综述中讨论了利用多种组学技术表征瘤胃微生物功能，他们的结论是，通过（计算）建模整合组学获得的数据，可以构建瘤胃特定微生物（代谢）模型。例如 GEMs 专注于用数学表达生物体的代谢能力。该方法已应用于细菌系统（Oberhardt 等，2009），可用于表征表型、演示代谢工程和研究物种（物种间）相互作用（Freilich 等，2011）。然而，与其行为特征有关，每一种微生物都有其自身的活动，这就使事情变得更为复杂（Wintermute 和 Silver，2010）。此外，在生物学中，微生物适应"混合"培养的能力至关重要。整合生物学的应用有望使人们不再局限于实验室人工培养的微生物，而去专注于阐释微生物行为（生物学）的自然过程。如本章第一节所讨论的瘤胃整体功能（图 1），整合生物学是指从生命体各级水平出发，整合各学科（异质）数据的跨学科研究。因此，这些各级水平可能包括反刍动物宿主本身的生产性能、体重、采食量、饲料消化率、行为（反刍、采食行为、分泌唾液）以及特定组织中的宿主 DNA 和基因表达。原则上，它还应该包括如微生物组成（DNA）、微生物基因表达（RNA 和蛋白质）、酶活（蛋白质）、酶活检测（转化率和流量）和代谢的生物指纹图谱（如代谢物）等瘤胃微生物生态系统功能数据。现有的瘤胃模型是基于消化生理学原理，以传统的方式"自上而下"开发的（即从整个瘤胃出发，向下囊括一些额外的细节和复杂性），而不是"自下而上"的（即通过多组学技术，从最细节的分子水平出发，通过分析、整合和关联，试图掌握整个瘤胃功能）。但是，同时从这两个角度研究"复杂"的生物学行为可能会产生创新的方法，去更准确地揭示生物学机制，找到量化微生物功能的方法。得益于此，现有的瘤胃整体数学模型的预测能力会得到提升，同时也可能会开发出新的应用策略，解决前一节中讨论的微生物组成的动态变化问题。而目前还不清楚如何在宽泛的营养条件和状态下，将这些在体的瘤胃生态系统信息，在瘤胃整体水平上纳入数学模型。这些模型试图在瘤胃整体层面去预测功能（Bannink 等，2016），当然无法表征多组学技术获得的细节数据（Chong 和 Xia，2017）。只有解决了这个矛盾，才能将两种方法结合到一起。图 3 给出了检测瘤胃功能所涉及的宿主和微生物信息收集相关的各级生物学层面和技术。这些收集到的数据可以在瘤胃层面，也可以在宿主层面与瘤胃整体的表型数据进行关联（Huws 等，2018）。瘤胃层面的表型数据主要有发酵速率、瘤胃外排速率和微生物合成速率，宿主层面的表型数据主要有全肠道消化率、生产性能和饲料效率。无论是直接监测到的，还是多组学分析的瘤胃微生物种群特征似乎都还不能在现有的瘤胃整体功能模型中预测微生物功能变化会导致瘤胃功能如何改变。通过改进的和更精细的表型分析将微生物变化转化为瘤胃功能变化（Huws 等，2018）是将这些多组学数据应用于现有预测模型的先决条件，我们应该在未来的研究中更多地关注这些工作。可以研究瘤胃微生物和瘤胃代谢产物之间的变化和联系，从而了解微生物变化对瘤胃功能产生的影响。这些微生物变化可能是断奶和母畜对初始细菌组成的影响（Abecia 等，2018），也可能是调节瘤胃原虫数量的影响。因为原虫体内共生细菌，而且这部分细菌的组成与外界环境中的组成不同，调节瘤胃

图3 瘤胃宿主和微生物组以及相关生物学功能的示意图。宿主（浅灰色底和白色底）和微生物组（深灰色底和白色底）都可以影响瘤胃功能。对于宿主和微生物组，描述了几个相关的生物学层面，即 DNA、RNA、肽/蛋白质和代谢物。在所有这些不同的生物学层面之下，描述了分子技术和由此产生的活动/结果。代谢物处于中心，因为这是宿主和微生物组之间相互作用的整合。**SNP**，单核苷酸多态性；**MS**，质谱；**GC**，气相色谱；**NMR**，核磁共振；**LC**，液相色谱；**CE**，毛细管电泳；**WMGS**，全基因组测序；**ITS**，内部转录间隔。

原虫数量有助于了解细菌与原虫之间的互作关系（Levy 和 Jami，2018）。虽然有这些结果，但模拟瘤胃微生物组的功能还是面临不少挑战的。归纳这些挑战，Huws 等（2018）认为反刍动物宿主的表型通常没有得到很好的描述，这阻碍了量化与预测的进程（尤其是作为功能器官的瘤胃）；可以预见组学有助于数学模型和瘤胃指标的开发，并可能彻底改变我们对瘤胃微生物组的认知。

此外，我们可以监测到瘤胃中存在微生物功能极短暂的影响和瞬时的变化，但现有的瘤胃整体数学模型还不能呈现这些信息。这些模型采用了确定性的方法，并没有给瘤胃微生物留出"代谢记忆"的空间。模型不会呈现应激或偶然扰动的后果，也不会保留长期的瘤胃状态信息。相反，这些模型采用了静态的概念，基本上忽略了微生物功能的时间变化。通常，试验至少设置两周的预饲期，让瘤胃微生物适应新的日粮，进入试验状态（Hristov 等，2019）。这也就表明短期动态变化原则上与瘤胃功能和日粮改变及宿主有关。瘤胃微生物短期的功能性和代谢性活动确实发生了变化（O'Callaghan 等，2018），当试图理解和预测瘤胃动力和功能，或是限制瘤胃微生物功能冗余、互补和回弹时，可以证实日粮和宿主相关因素导致微生物出现了这种短期的变化。

应该探究日粮、宿主和微生物之间的联系（Chong 和 Xia，2017；Huws 等，2018；Denman 等，2018），并在不久的将来应用于实际生产。但值得注意的是，Tedeschi 等（2014）和 Bannink 等（2016）的综述中都指出，只有将这些结果或这些联系转换为模型中指征微生物功能的术语，才能应用这些结果改进当前的瘤胃整体数学预测模型。多组学研究的描述性和技术驱动性本质提供了前所未有的信息量和细节层级。因为现有的瘤胃整体预测模型也主要是基于描述性知识发展而来，所以组学研究本身的描述性本质不一定有问题。但是，要让结果的分析服务于现有的瘤胃预测模型是很具挑战性的，而对这种分析的需求比可靠的、详细的表型更迫切。面对挑战，确定多组学观测的功能单元是非常重要的，这将在下一节中讨论。因为瘤胃整体模型需要依据瘤胃中底物和微生物的存在情况（Dijkstra 等，1992；Li 等，2018）或瘤胃微生物蛋白的总合成效率或净合成效率（Dijkstra 等，1998）去量化底物的降解速率，所以通过数据分析聚焦哪一层级或解决哪一层级的功能是另一挑战。多组学数据分析本身是无法实现这种量化的。

4 功能单元与取样技术

多组学技术已经用于分析瘤胃壁、瘤胃食糜或瘤胃液、瘤胃微生物等目标器官的样本。在这方面，多组学技术可以"自下而上"的方式，深入揭示和定量瘤胃微生物的功能（Huws 等，2018；Denman 等，2018）。尽管表征相同的生物体，但总体而言，瘤胃整体功能数学预测模型所采用的功能单元与多组学分析所采用的单元是不同的。图4呈现了两者在瘤胃组织（图4a）、瘤胃微生物（图4b）和瘤胃液体（图4c）取样时的差异。克服这个差异，才能直接将获得的组学数据及其分析的结果应用于瘤胃整体功能建模，这就需要将这些样本的结果（本质上只提供了一个浓度）分别扩展到表征瘤胃壁、整个瘤胃微生物群体和瘤胃液的层面。多组学技术缺少另一必要的信息就是"流量"，即进出瘤胃壁（黏膜和浆膜侧）和通过食道和网胃进出瘤胃的信息。预测性瘤胃模型首先就是要量化流量,而且要量化瘤胃微生物功能变化和日粮进食对流量的影响,而不是仅给出浓度值。为了实现瘤胃流量动态预测,瘤胃整体功能的动态机制模型将酶降解与微生物合成能力描述为瘤胃中可利用底物和微生物的功能。这意味着需要量化蛋白质或酶活力,这些取决于浓度的指标(Bannink 等,2016)。瘤胃样本的 DNA、RNA 和代谢物等参数也可以间接地反映瘤胃的功能。为了实现"自下而上"和"自上而下"方法与瘤胃功能量化相匹配,就需要在含有微生物的内容物中,在具有酶活性和瘤胃壁(上皮和肌肉组织)反应的宿主中,以及在化合物和液体(包括瘤胃内容物的缓冲和血液中氮和磷的回流)流入和流出的瘤胃器官中找到蛋白质和酶活力层级的共同点(图 1 和图 3)。除了需要弥合所需功能单元和实际产出单元的差异外,瘤胃内容物本身的取样过程就是一个需要考虑的关键的方法学问题(图 4b,图 4c)。基因型(Li 等,2019)、日龄(Jami 等,2013)、日粮、瘤胃内容物的取样阶段(Mohammed 等,2012)、宿主的营养水平、一天内的采样时间、实时瘤胃 pH 值以及动物的个体差异(Wemer 等,2010)等因素可以影响微生物群落分析的结果(Weimer,2015)。动物行为、瘤胃的形态和生理导致了动物个体差异。所有这些因素都可能对试验设计和动物选择产生重要影响。与试验设计相关的另一组因素是取样程序,消化阶段的取样位点、样品处理、使用的分离方法和分析方法。任一差异都会影响荟萃分析等数据间的比较分析。取瘤胃内容物(如饲喂前取样)不一定反映瘤胃功能的日平均状态。Van Lingen 等(2017)在日饲喂两次的过程中,对瘤胃进行多次采样,发现瘤胃代谢产物浓度和瘤胃微生物组成随进食后时间的变化要强于亚麻籽油补饲的作用。因此

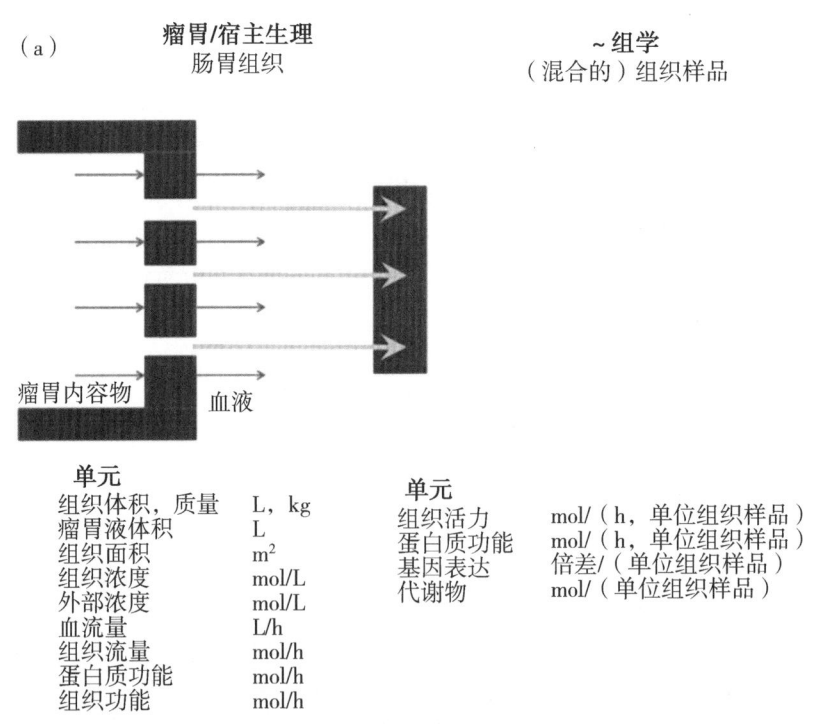

(b) 瘤胃微生物　　　　　　　　　　~组学
　　　　　　　　　　　　（筛选的/混合的）瘤胃内容物样品

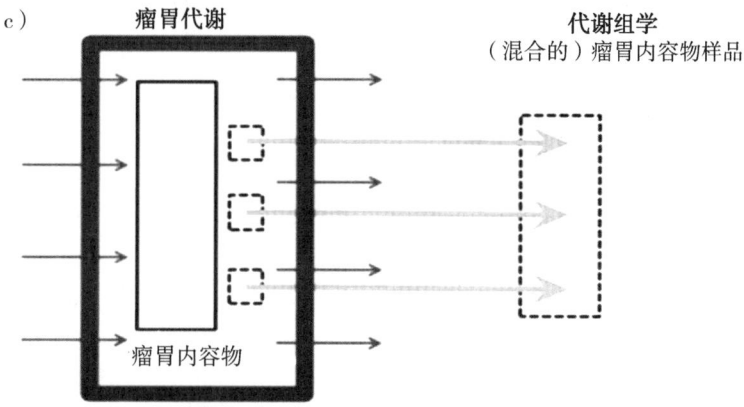

用于瘤胃功能的单元
瘤胃内容物　　　　　　L，kg
瘤胃液体积　　　　　　L
食糜质量　　　　　　　kg
食糜表面积　　　　　　m²
游离的/吸附的瘤胃液　　kg
食糜流出　　　　　　　/h
瘤胃液流出　　　　　　/h
微生物活力　　　　　　mol/（h·kg微生物
　　　　　　　　　　　质量·mol底物）
微生物质量　　　　　　kg/kg内容物
底物量　　　　　　　　mol/kg内容物

用于瘤胃样品的单元
DNA　g/（kg微生物质量，单位样品）
RNA　g/（kg微生物质量，单位样品）
蛋白质　mol/（kg微生物质量，单位样品）

(c) 瘤胃代谢　　　　　　　　　　代谢组学
　　　　　　　　　　　　（混合的）瘤胃内容物样品

用于瘤胃内容物的单元
瘤胃内容物　　　　　　L，kg
瘤胃液体积　　　　　　L
食糜质量　　　　　　　kg
食糜表面积　　　　　　m²
游离的/吸附的瘤胃液　　kg
食糜/瘤胃液流出量　　　/h
微生物活力　　　　　　mol/（h·kg微生物质
　　　　　　　　　　　量·mol底物）
微生物量　　　　　　　kg/kg内容物
底物量　　　　　　　　mol/kg内容物
代谢物量　　　　　　　mol/kg内容物

用于瘤胃样品的单元
代谢物　mol/（kg微生物质量，单位样品）
或者
mol/L瘤胃液

图4 用于瘤胃（左侧）和瘤胃样品（右侧）的单元差异示意图。(a) 瘤胃壁组织，(b) 瘤胃微生物群，(c) 瘤胃液。

证明，单个时间点取样可能无法反映瘤胃功能的日平均状态。而且，很久以前（Martin 和 Michalet-Doreau，1995；Huhtanen 等，1992）人们就知道一日内瘤胃微生物的酶活变化剧烈，食糜黏附性细菌需要在进食后很长一段时间才能达到最大降解能力。当结果来自单个取样时间点的样本时，在做结论和概括结果时这种日变化是不可以忽略的，必须加以考虑。

5 结论和展望

无论是量化瘤胃整体功能的经典方法（"自上而下"），还是使用多组学技术的"下一代"方法（"自下而上"），在研究瘤胃生物学方面都有各自的优点和局限性。经典建模方法，至少复杂的模型是通过描述瘤胃发酵过程的基本化学和物理原理来预测瘤胃功能，采用动态的方法来量化底物和微生物种群之间的互作（图 1）。通常，模型也表征宿主生理过程和瘤胃内环境等更高层级的信息。此外，基于营养研究，底物化学分类描述了区分微生物功能和量化微生物底物可用性的方法（Bannink 等，2016）。相反，"下一代"方法通过收集瘤胃样本的多组学数据，关注分子特征。这种方法是技术驱动性和描述性的，但得益于细节层级和数据量。通过瘤胃微生物组的宏基因组测序或瘤胃代谢物的表

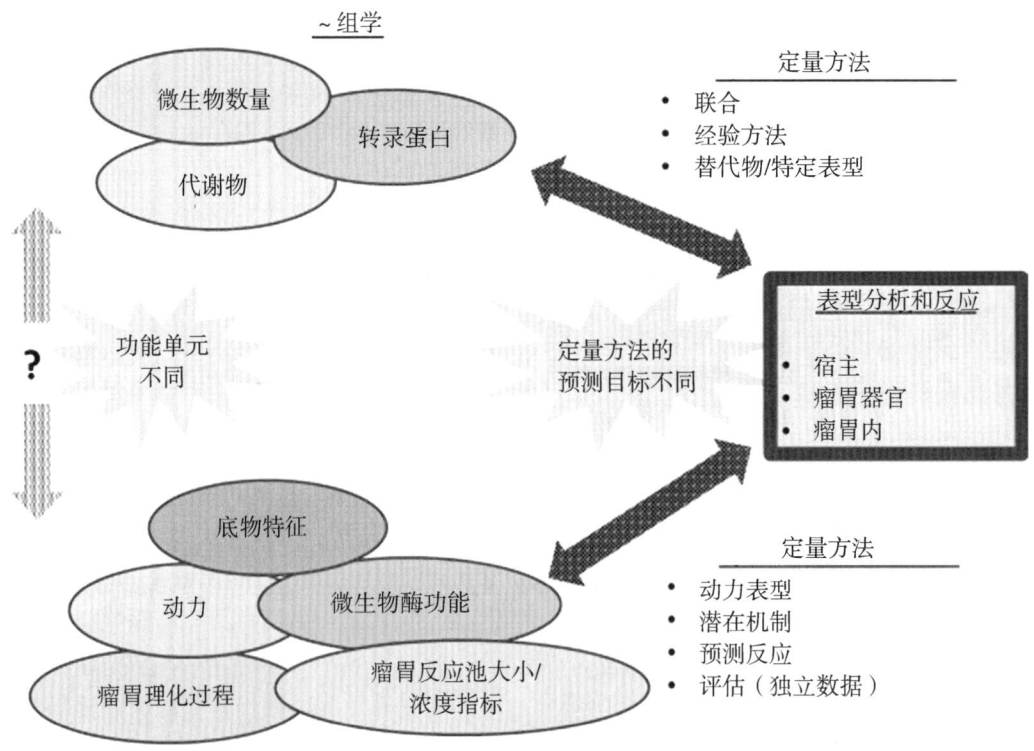

图 5　多组学方法（自下而上）和全瘤胃建模方法（自上而下）示意图。
示意图展示了两种方法使用的单位、涉及层面以及定量目标的差异。

征，可以实现微生物与表型、微生物与目标反应的关联。虽然这两种方法都关注瘤胃，但它们关注的内容不同，所以我们在前面的章节中可以看到这两种方法不匹配的地方。两种方法在使用的功能单元和量化的功能方面存在差异，并且量化的目的也不同，一个是寻找关联，一个是基于机制去预测。图5试图去说明两者之间的不同和分歧。可以预见，如果多组学方法可以提供与现有瘤胃整体功能数学模型中使用的元素直接相关或可能相关的功能特征，这两种方法在未来的研究中可能会相互受益。因为与目前基于底物类型区分微生物功能相比，多组学方法能更详细地揭示微生物种群功能，所以瘤胃整体功能模型可能会受益于多组学方法。改善现状的首要一步是技术标准化，以使研究具有可比性，这样就可以创建包含瘤胃两个"部分"数据的规模型集中式数据库。这个数据库不仅需要包含试验参数和宿主特征的信息，还应该包括瘤胃功能的细节信息。

总之，如果多组学的结果及其衍生出的知识与目前的瘤胃整体模型的过程和元素更相近的话，两种方法的知识交流将更频繁和顺畅。已经证实这些模型对于解释观察到的降解和发酵、瘤胃外排和吸收以及微生物合成等瘤胃功能和微生物活性的变化具有重要意义。量化这种关系将使建模者吸纳新元素和新概念，使现有的瘤胃整体功能预测模型受益于瘤胃微生物功能的深入了解和细化。为了进一步整合组学和瘤胃整体建模方法，必须解决两种方法所用功能单元不同以及定量方法的预测目标不同的问题。

6 更多信息

有几个研究项目正试图通过组学技术量化瘤胃微生物组的功能。另外，通过全球研究联盟（GRA）（https://globalresearchalliance.org）及其相关的瘤胃微生物和基因组学网（https://globalresearchalliance.org/research/livestock/networks/rumen-microbial-genomics-network）、饲料和营养网（https://globalresearchalliance.org/research/livestock/networks/feed-nutrition-network）组建了全球研究联盟。

读者可参阅以下信息：
- Huws 等 2018 年撰写了一篇与瘤胃微生物和基因组学网密切相关的如何使用组学数据量化瘤胃功能的立场性文章（https://www.frontiersin.org/articles/10.3389/fmicb.2018.02161/full）；
- Ungerfeld 和 Newbold 作为客座编辑在 *Frontiers of Microbiology* 上刊登的一期如何设计瘤胃代谢途径的专刊（https://www.frontiersin.org/articles/10.3389/fmicb.2017.02627/full）；
- Henderson 等 2015 年公布的全球瘤胃普查结果（https://www.nature.com/articles/srep14567）；
- CEDERS 项目关于通过清单和牧场核算方法获取的日粮对牧场温室气体排放影响的结果，以及其他 ERAGAS 项目的结果（https://www.eragas.eu/en/eragas/Research-projects）；
- 通过 GRA 畜牧研究小组进行的，包括瘤胃微生物和基因组学网以及饲料和营养网的活动和成果报告在内的各种交流（https://globalresearchalliance.org/research/live-

stock)。

7 参考文献

Abecia, L., Martínez-Fernandez, G., Waddams, K., et al., 2018. Analysis of the rumen microbiome and metabolome to study the effect of an antimethanogenic treatment applied in early life of kid goats. Frontiers in Microbiology 9, 2227.

Bainbridge, M. L., Saldinger, L. K., Barlow, J. W., et al., 2018. Alteration of rumen bacteria and protozoa through grazing regime as a tool to enhance the bioactive fatty acid content of bovine milk. Frontiers in Microbiology 9, 904.

Baldwin, R. L., Thornley, J. H. M. and Beever, D. E. 1987. Metabolism of the lactating cow. II. Digestive elements of a mechanistic model. Journal of Dairy Research 54 (1), 107-31.

Bannink, A., Gerrits, W. J. J., France, J., et al., 2012. Variation in rumen fermentation and the rumen wall during the transition period in dairy cows. Animal Feed Science and Technology 172 (1-2), 80-94.

Bannink, A., van Lingen, H. J., Ellis, J. L., et al., 2016. The contribution of mathematical modeling to understanding dynamic aspects of rumen metabolism. Frontiers in Microbiology 7, 1820.

Belanche, A., Kingston-Smith, A. H., Griffith, G. W., et al., 2019. A multi-kingdom study reveals the plasticity of the rumen microbiota in response to a shift from non-grazing to grazing diets in sheep. Frontiers in Microbiology 10, 122.

Chao, A. 1984. Non-parametric estimation of the number of classes in a population. Scandinavian Journal of Statistics 11, 265-70.

Chong, J. and Xia, J. 2017. Computational approaches for integrative analysis of the metabolome and microbiome. Metabolites 7 (4), 62.

Denman, S. E., Morgavi, D. P. and McSweeney, C. S. 2018. Review: The application of omics to rumen microbiota function. Animal: an International Journal of Animal Bioscience 12 (s2), s233-45.

Dieho, K., van den Bogert, B., Henderson, G., et al., 2017. Changes in rumen microbiota composition and in situ degradation kinetics during the dry period and early lactation as affected by rate of increase of concentrate allowance. Journal of Dairy Science 100 (4), 2695-710.

Dijkstra, J. 1994. Simulation of the dynamics of protozoa in the rumen. British Journal of Nutrition 72 (5), 679-99.

Dijkstra, J., Neal, H. D., Beever, D. E., et al., 1992. Simulation of nutrient digestion, absorption and outflow in the rumen: model description. Journal of Nutrition 122, 2239-56.

Dijkstra, J., France, J. and Davies, D. R. 1998. Different mathematical approaches to estimating microbial protein supply in ruminants. Journal of Dairy Science 81 (12), 3370-84.

Dijkstra, J., van Gastelen, S., Dieho, K., et al., 2020. Rumen pH and redox potential measured by rumen sensors: data and interpretation for key rumen metabolic processes. Animal: an International Journal of Animal Bioscience 14 (s1), s176-86.

Dougherty, H. C., Kebreab, E., Evered, M., et al., 2017. The AusBeef model for beef production: I. Description and evaluation. Journal of Agricultural Science 155 (9), 1442-58.

Faith, D. P. 1992. Conservation evaluation and phylogenetic diversity. Biological Conservation 61 (1), 1-10.

Freilich, S., Zarecki, R., Eilam, O., et al., 2011. Competitive and cooperative metabolic interactions in bacterial communities. Nature Communications 2, 589.

Goopy, J. P., Donaldson, A., Hegarty, R., et al., 2014. Low-methane yield sheep have smaller rumens and shorter rumen retention time. British Journal of Nutrition 111 (4), 578-85.

Gregorini, P., Beukes, P., Waghorn, G., et al., 2015. Development of an improved representation of rumen digesta outflow in a mechanistic and dynamic model of a dairy cow, Molly. Ecological Modelling 313, 293-306.

Henderson, G., Cox, F., Ganesh, S., et al., 2015 Rumen microbial community composition varies with diet and host, but a core microbiome is found across a wide geographical range. Scientific Reports 5, 14567.

Hook, S. E., France, J. and Dijkstra, J. 2017. Further assessment of the protozoal contribution to the nutrition of the ruminant animal. Journal of Theoretical Biology 416, 8-15.

Hristov, A. N., Bannink, A., Crompton, L. A., et al., 2019. Invited review: Nitrogen in ruminant nutrition: a review of measurement techniques. Journal of Dairy Science 102 (7), 5811-52.

Huhtanen, P., Dakowski, P. and Vanhatalo, A. 1992. Composition, digestibility and particle-associated enzyme activities in rumen digesta as influenced by particle size and time after feeding. Journal of Animal and Feed Sciences 1 (3-4), 223-35.

Huws, S. A., Creevey, C. J., Oyama, L. B., et al., 2018. Addressing global ruminant agricultural challenges through understanding the rumen microbiome: past, present, and future. Frontiers in Microbiology 9, 2161.

Jami, E., Israel, A., Kotser, A., et al., 2013. Exploring the bovine rumen bacterial community from birth to adulthood. ISME Journal 7 (6), 1069-79.

Levy, B. and Jami, E. 2018. Exploring the prokaryotic community associated with the rumen ciliate protozoa population. Frontiers in Microbiology 9, 2526.

Li, M. M., White, R. R. and Hanigan, M. D. 2018. An evaluation of Molly cow model predictions of ruminal metabolism and nutrient digestion for dairy and beef diets. Journal of Dairy Science 101 (11), 9747-67.

Li, F., Li, C., Chen, Y., et al., 2019. Host genetics influence the rumen microbiota and heritable rumen microbial features associate with feed efficiency in cattle. Microbiome 7 (1), 92.

López, S., Hovell, F. D. D., Dijkstra, J., et al., 2003. Effects of volatile fatty acid supply on their absorption and on water kinetics in the rumen of sheep sustained by intragastric infusions. Journal of Animal Science 81 (10), 2609-16.

Martin, C. and Michalet-Doreau, B. 1995. Variations in mass and enzyme activity of microorganisms: effect of barley and buffer supplements. Journal of the Science of Food and Agriculture 67 (3), 407-13.

Mills, J. A. N., Crompton, L. A., Ellis, J. L., et al., 2014. A dynamic mechanistic model of lactic acid metabolism in the rumen. Journal of Dairy Science 97 (4), 2398-414.

Mohammed, R., Stevenson, D. M., Weimer, P. J., et al., 2012. Individual animal variability in ruminal bacterial communities and ruminal acidosis in primiparous Holstein cows during the periparturient period. Journal of Dairy Science 95 (11), 6716-30.

Nagorcka, B. N., Gordon, G. L. R. and Dynes, R. A. 2000. Towards a more accurate representation of fermentation in mathematical models of the rumen. In: McNamara, J. P., France, J. and Beever, D. E. (Eds), Modelling Nutrient Utilization in Farm Animals. CAB International, Wallingford, UK, pp.

37-48.

Newbold, C. J., de la Fuente, G., Belanche, A., et al., 2015. The role of ciliate protozoa in the rumen. Frontiers in Microbiology 6, 1313.

Oberhardt, M. A., Palsson, B. Ø and Papin, J. A. 2009. Applications of genome-scale metabolic reconstructions. Molecular Systems Biology 5, 320.

O'Callaghan, T. F., Vazquez-Fresno, R., Serra-Cayuela, A., et al., 2018. Pasture feeding changes the bovine rumen and milk metabolome. Metabolites 8 (2), 27.

Reichl, J. R. and Baldwin, R. L. 1976. A rumen linear programming model for evaluation of concepts of rumen microbial function. Journal of Dairy Science 59 (3), 439-54.

Russell, J. B., O'Connor, J. D., Fox, D. G., et al., 1992. A net carbohydrate and protein system for evaluating cattle diets: I. Ruminal fermentation. Journal of Animal Science 70 (11), 3551-61.

Scheffer, M., Vergnon, R., van Nes, E. H., et al., 2015. The evolution of functionally redundant species: evidence from beetles. PLoS ONE 10 (10), e0137974.

Stewart, R. D., Auffret, M. D., Warr, A., et al., 2018. Assembly of 913 microbial genomes from metagenomic sequencing of the cow rumen. Nature Communications 9 (1), 870.

Stewart, R. D., Auffret, M. D., Warr, A., et al., 2019. Compendium of 4,941 rumen metagenome-assembled genomes for rumen microbiome biology and enzyme discovery. Nature Biotechnology 37 (8), 953-61.

Tedeschi, L. O., Cavalcanti, L. F. L., Fonseca, M. A., et al., 2014. The evolution and evaluation of dairy cattle models for predicting milk production: an agricultural model intercomparison and improvement project (AgMIP) for livestock. Animal Production Science 54 (12), 2052-67.

Ungerfeld, E. M. and Newbold, C. J. 2018. Editorial: Engineering rumen metabolic pathways: where we are, and where are we heading. Frontiers in Microbiology 8, 2627.

Van Lingen, H. J., Plugge, C. M., Fadel, J. G., et al., 2016. Thermodynamic driving force of hydrogen on rumen microbial metabolism: a theoretical investigation. PLoS ONE 11 (10), e0161362.

Van Lingen, H. J., Edwards, J. E., Vaidya, J. D., et al., 2017. Diurnal dynamics of gaseous and dissolved metabolites and microbiota composition in the bovine rumen. Frontiers in Microbiology 8, 425.

Van Lingen, H. J., Fadel, J. G., Moraes, L. E., et al., 2019. Bayesian mechanistic modeling of thermodynamically controlled volatile fatty acid, hydrogen and methane production in the bovine rumen. Journal of Theoretical Biology 480, 150-65.

Wang, Y., Janssen, P. H., Lynch, T. A., et al., 2016. A mechanistic model of hydrogen-methanogen dynamics in the rumen. Journal of Theoretical Biology 393, 75-81.

Weimer, P. J. 2015. Redundancy, resilience and host specificity of the ruminal microbiota: implications for engineering improved ruminal fermentations. Frontiers in Microbiology 6, 296.

Weimer, P. J., Stevenson, D. M., Mantovani, H. C., et al., 2010. Host specificity of the ruminal bacterial community in the dairy cow following near-total exchange of ruminal contents. Journal of Dairy Science 93 (12), 5902-12.

Wintermute, E. H. and Silver, P. A. 2010. Dynamics in the mixed microbial concourse. Genes and Development 24 (23), 2603-14.

Zebeli, Q., Aschenbach, J. R., Tafaj, M., et al., 2012. Invited review: Role of physically effective fiber and estimation of dietary fiber adequacy in high-producing dairy cattle. Journal of Dairy Science 95 (3), 1041-56.

第二部分

瘤胃微生物群落

第5章 基因组测序与瘤胃微生物组

Jessica C. A. Friedersdorff、Benjamin J. Thomas，英国女王大学；

Sara E. Pidcock，英国亚伯大学；

Francesco Rubino、Christopher J. Creevey，英国女王大学

（赵圣国译）

1 前言

基因组测序的出现对生物学研究产生了重要影响，通过预测蛋白质的结构和功能，促进了我们对物种分类、系统发育和进化关系的理解。微生物基因组测序推动了瘤胃微生物组学的研究进展，不仅有助于比较动物体内微生物的种类、功能和代谢差异，还有助于比较世界各地不同条件下（如日粮变化和反刍动物物种）微生物的变化。

在全基因组测序出现以前，微生物功能特征主要通过形态学、生物化学等来确定（Bryant，1959；Leahy 等，2013）。测序可将细菌和古菌群落与重要瘤胃功能相关联，使用 16S rRNA 基因确定微生物分类后，通过与基因组关联，便可将微生物种类与功能联系起来（Wilkinson 等，2018）。

尽管这种方法能提供微生物功能信息，但很大程度上依赖于数据库的完整性，但事实上数据库中瘤胃微生物信息有限，且没有考虑到那些具有重要功能但丰度低的微生物（Delgado 等，2019）。"组学"技术的发展，丰富了对瘤胃微生物群落多样性、结构和功能的认识（Kingston-Smith 等，2013；Morgavi 等，2013；Creevey 等，2014），这对于提高反刍动物生产性能和环境保护都至关重要。

本章综述了基于基因组测序的研究报道，这将有助于我们理解微生物及其在瘤胃生态系统中的功能和作用，有助于提出相应营养调控策略。主要要点总结如图 1 所示。

2 第一个瘤胃微生物基因组

第一个被基因组测序和分析的瘤胃微生物是 2003 年发表的产琥珀酸沃廉（弧）菌 DSMZ1740 [*Wolinella* (*Vibrio*) *succingenes* DSMZ1740]（Baar 等，2003），晚于细菌流感嗜血杆菌 Rd（*Haemophilus influenzae* Rd）（Fleischmann 等，1995）和生殖支原体（*Mycoplasma genitalium*）（Fraser 等，1995）基因组测序 8 年。虽然产琥珀酸沃廉菌（*W. succinogenes*）最初是从牛的瘤胃中分离出来的，但其 16S rRNA 基因和蛋白质序列与人类病原体幽门螺杆菌（*Helicobacter pylori*）和空肠弯曲杆菌（*Campylobacter jejuni*）相似，而且产琥珀酸沃廉菌（*W. succinogenes*）在反刍动物宿主或人中均未表现出致病性（Baar

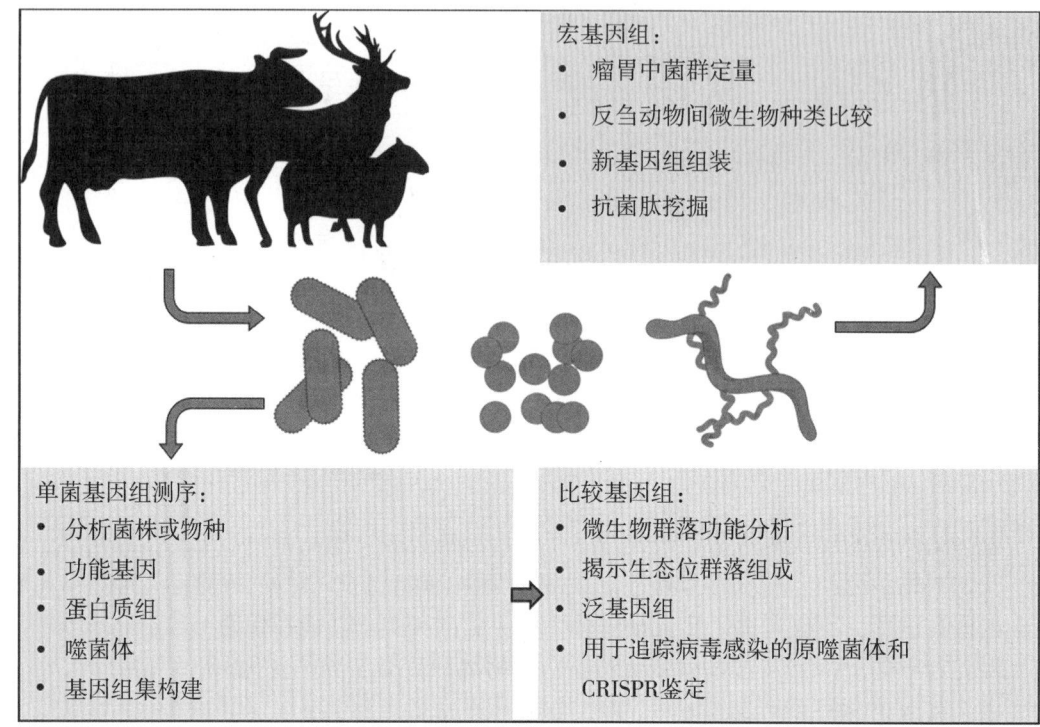

图 1 本章要点

等，2003）。

对产琥珀酸沃廉菌进行基因组测序的主要原因是该细菌同时在动物和人体内定植。对该基因组编码的基因和酶分析表明，产琥珀酸沃廉菌在生长过程中通过厌氧呼吸将延胡索酸加氢还原为琥珀酸，但由于缺乏葡萄糖激酶编码基因和葡萄糖转运系统，该菌不能发酵利用葡萄糖，而其他碳水化合物代谢酶的存在，提示该细菌可能使用糖酵解酶来代替糖异生（Baar 等，2003）。

通过对产琥珀酸沃廉菌与远亲微生物基因组进行比较，发现该菌多个基因来自水平转移。产琥珀酸沃廉菌所拥有的代谢通路和基因，在同一纲分类水平上的其他微生物基因组中并不常见，表明产琥珀酸沃廉菌可能有更多更独特的功能特性（Baar 等，2003）。

从 2003 年开始，可培养瘤胃微生物基因组开始增加（尽管很少超过 10 个）。直到 2018 年，随着 Hungate 项目收集的近 500 个基因组的发表，瘤胃微生物基因组信息有了最大幅度的增加（Seshadri 等，2018）（图 2）。未被培养瘤胃微生物基因组从 2011 年开始出现（Hess 等，2011），数量也迅速增加，现在它们的数量已经超过了可培养微生物基因组的 10 倍以上（图 2）。

3 单菌基因组测序蕴含的能量

虽然研究微生物的表型非常重要，但是微生物基因组测序可以揭示更多基因和基础表

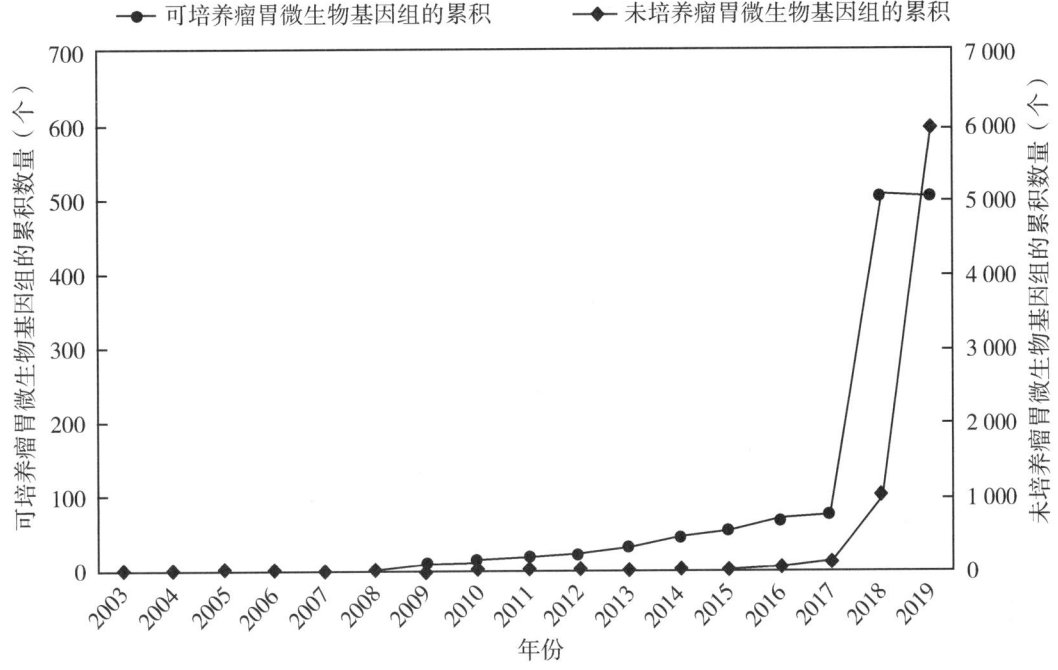

图 2 瘤胃微生物基因组累计数量的变化

型信息。在已知表型和功能的微生物生态位中,对分离菌株进行基因组测序,可以对提出的假设进行验证,并将观察到的表型与基因组信息关联起来(Creevey 等,2014)。

优势纤维分解细菌产琥珀酸丝状杆菌(*Fibrobacter succinogenes*)是利用基因组来阐明瘤胃微生物功能机制的典型例子。将产琥珀酸丝状杆菌 S85(*F. succinogenes* S85)在各种条件下进行培养,测定纤维素酶活性(Weimer,1993)。研究发现该菌对纤维素具有依赖性并具有快速降解结晶纤维素的能力,降解速度明显快于其他瘤胃细菌(Weimer,1993)。研究人员测定该菌多糖水解和利用情况,结果表明,该菌仅利用纤维素降解产物,而不利用其他多糖降解产物。多年后基因组测序显示,该菌缺乏进一步分解或利用纤维素以外其他多糖降解产物所需的酶(Suen 等,2011)。

第一个完整的产甲烷菌基因组来自于瘤胃甲烷短杆菌 M1(*Methanobrevibacter ruminantium* M1),该基因组为认识产甲烷菌代谢途径提供了重要指导。该基因组具有大量独有基因以及一个完整原噬菌体和大量表面黏附相关基因,该菌比来自其他环境同一科古菌拥有更大的基因组(Leahy 等,2010)。此外,该基因组提供了甲烷生成相关的基因信息,为了解产甲烷菌产甲烷机制提供了重要信息。通过分析古菌基因组,可鉴定古菌细胞膜蛋白编码基因,为发现抗原和设计新型抗体疫苗用于甲烷减排提供了重要信息(Leahy 等,2010,2013)。

单菌基因组信息对于认识微生物功能有重要指导意义,而大量单菌基因组信息将有助于通过比较基因组学方法更深入理解微生物群落的生态功能。

4　构建瘤胃可培养微生物参考基因组集

随着基因组测序技术的快速发展，人们需要构建完整的瘤胃微生物基因（组）数据库，以便让新瘤胃微生物测序数据获得更好功能注释。通过分离培养的方法，从瘤胃中分离微生物，提取 DNA 并进行基因组测序，能够为瘤胃微生物组提供真实可靠的基因数据。Hungate 1000 项目正是建立在上述思路基础上，联合全球多个国家，从瘤胃中分离和测序了独特的微生物，以创建一个高质量的瘤胃微生物参考基因组集（Seshadri 等，2018）。

该参考基因组集结合了 91 个已经公开的可培养微生物基因组，以及另外 410 个使用 Illumina 和 PacBio 测序的基因组，包含了多个优势瘤胃微生物基因组。该项目于 2018 年结束，共 21 个国家参与，最终获得 501 个基因组，并以"Hungate Collection"形式公开（Seshadri 等，2018）。这些微生物大多数分离自牛、羊、鹿等反刍动物，但是一部分微生物分离自非反刍动物，如马、鹅和猪，但这些微生物也在瘤胃中存在，因此被加入到该基因组集中（Seshadri 等，2018）。

这个参考基因组集的优势在于提供了根据基因组溯源微生物的数据库，这些基因组来自分离的单菌，基因组质量高，DNA 污染少。根据基因组信息提出的科学假设可以通过体外培养的方法进行证实（Creevey 等，2014）。全球范围内的大规模合作增加了不同反刍动物样本的多样性和覆盖面，使人们对瘤胃微生物组有了全面的了解。根据参考基因组集，还有助于通过排除已知微生物序列，从而实现对宏基因组数据中新微生物基因组的组装（Wilkinson 等，2018）。

这个参考基因组集的主要缺点是依赖于微生物分离培养，但是很多微生物都无法分离培养，并且参考基因组集中包含的微生物多样性仍不足（Attwood 等，2008b）。因此，在利用这些参考基因组进行分析时要考虑这些局限性，并采取一些方法来克服局限性，例如，在对参考基因组集进行基于序列相似性搜索时，将阈值适当调低。

尽管如此，Hungate 1000 项目研究成果对瘤胃微生物组学研究产生了巨大的影响（Seshadri 等，2018），促进了瘤胃微生物组分析新工具的开发（Wilkinson 等，2018），增加了对瘤胃氢替代代谢途径的深入了解（Greening 等，2019），鉴定了产甲基化合物的微生物（Kelly 等，2019），增加了对瘤胃微生物抗生素耐药性的认识（Sabino 等，2019），揭示了丁酸弧菌等菌群的多糖降解能力（Palevich 等，2019），并解析了瘤胃微生物间的信息交流（Won 等，2020）。

通过基因组测序鉴定的基因，并不能确定其真实功能，只能认为是具有潜在功能的基因（Creevey 等，2014）。然而，目前瘤胃中仅有约 23% 微生物是可培养的，大部分微生物难以分离培养，因此可培养微生物基因组有一定局限性（Zehavi 等，2018）。然而，培养组学的出现有助于获得更多可培养微生物，从而有助于最大程度上获得更多可培养微生物基因组。

5 宏基因组数据在新基因组构建中的应用

随着测序通量的增加和成本的降低，使得从宏基因组数据中组装基因组成为可能。早期从宏基因组测序项目中，组装出了一些完整的微生物基因组（Tyson 等，2004）。虽然最初仅能从一些非常简单的微生物群落中取得成功，但这为复杂计算算法的发展奠定了基础，新算法使得从宏基因组数据（Alneberg 等，2014；Kang 等，2015；Gregor 等，2016）甚至复杂瘤胃微生物群中重组基因组获得可能（Hess 等，2011；Pope 等，2012；Svartström 等，2017；Stewart 等，2018）。

首先对宏基因组数据进行组装，然后将原始序列与组装的序列进行比对，估算覆盖度（即每个组装的 DNA 序列或 contig 在原始序列中出现的次数），然后计算一定长度的核苷酸序列的频率（通常称为长度 k，这些核苷酸序列的总称为 k-mers）。k-mer 的重要性源于先前的一项研究，该研究表明可以识别出特定物种的 k-mer 谱（Karlin 等，1998）用于物种分类组装。因此，根据宏基因组数据，就可能计算组装的所有 contigs 的 k-mer 频率，并将它们进行分类。

分类的过程被称为分箱（binning），是从宏基因组数据集重构基因组方法中最重要的步骤之一。因此，所有被分箱到一起的 contigs，被认为来自同一生物体，代表一个宏基因组组装的基因组（Metagenomically Assembled Genome，MAG），在早期文献中有时称为分类分箱（taxonomic binning）。另外，还可以计算多个样本间 contigs 的丰度，如果它们都来自同一生物体，那么这些 contig 的丰度分布具有相关性，据此将这些 contig 作为同一个基因组（Alneberg 等，2014）。

利用宏基因组数据组装基因组的优势是可以获得难以分离培养微生物的基因信息（Creevey 等，2014），据此揭示微生物潜在的新功能，并更好地了解微生物对整个瘤胃功能的贡献。

近年来，将瘤胃宏基因组数据进行分箱用于基因组功能研究的报道越来越多（Hess 等，2011；Pope 等，2012；Svartström 等，2017；Stewart 等，2018；Stewart 等，2019b），目前产生了 5 000 多个宏基因组组装的基因组，这比单菌基因组的数量多的多（图 2）。对这些 MAGs 进行评估后发现，未分离细菌基因组的构建仍是很大的挑战（Creevey 等，2014）。目前，通常是利用 16S rRNA 基因测序分析细菌种类的组成［如全球瘤胃普查（Global Rumen Census）］。

由于 16S rRNA 基因的可变性和保守性双特征，从宏基因组数据中组装这些基因非常困难，因此，当前构建的大多数瘤胃 MAGs 中都缺失了 16S rRNA 基因。除了 16S rRNA 基因外，我们应该利用更多的其它基因，来解决 MAGs 的物种分类注释问题。

我们可以通过比较单菌基因组、MAGs 和 16S rRNA 基因 3 种分析方法揭示的微生物相对丰度变化，从而找到新的有待研究的重要微生物（表 1）。虽然 Hungate 项目（Seshadri 等，2018）、MAGs 组装项目（Stewart 等，2019b）和全球瘤胃普查项目（16S 测序）（Henderson 等，2015）中揭示的排名前三的细菌科是普雷沃氏菌科、瘤胃球菌科和毛螺菌科，但它们的排名顺序并不一致（表 1）。这些差异可能与宿主的日粮、品种或地理位置有关。

表1 全球瘤胃普查（Global Rumen Cencus）项目、Hungate 项目（Seshadri 等，2018）和宏基因组组装基因组（MAGs）项目（Stewart 等，2019b）中排名前20的细菌科的基因组相对丰度（括号中为基因组数量）（Henderson 等，2015）。

科水平分类	Hungate 项目（%）	MAGs 项目（%）	Global Rumen Census 项目（%）
普雷沃氏菌科（Prevotellaceae）	7（36）	16（521）	35.29
瘤胃球菌科（Ruminococcaceae）	8（38）	35（1 111）	17.53
毛螺菌科（Lachnospiraceae）	32（162）	20（640）	16.51
副尊苔科（Paraprevotellaceae）	0（0）	0（0）	5.13
韦荣氏菌科（Veillonellaceae）	6（30）	1（25）	3.34
纤维杆菌科（Fibrobacteraceae）	0（2）	1（42）	3.10
琥珀酸弧菌科（Succinivibrionaceae）	1（7）	1（16）	2.73
艰难杆菌科（Mogibacteriaceae）	0（0）	0（0）	2.04
克里斯滕森菌科（Christensenellaceae）	0（0）	0（0）	1.37
梭菌科（Clostridiaceae）	3（13）	1（35）	1.29
丹毒丝菌科（Erysipelotrichaceae）	3（13）	9（291）	1.28
氨基酸球菌科（Acidaminococcaceae）	1（4）	1（43）	1.14
拟杆菌科（Bacteroidaceae）	4（18）	0（0）	< 1
螺旋体科（Spirochaetaceae）	1（6）	2（52）	< 1
厌氧绳菌科（Anaerolinaceae）	0（0）	0（0）	< 1
棒状杆菌科（Corynebacteriaceae）	0（2）	0（2）	< 1
脱硫糖卵黄科（Dethiosulfovibrionaceae）	0（0）	0（0）	< 1
乳酸杆菌科（Lactobacillaceae）	2（10）	1（22）	< 1
链球菌科（Streptococcaceae）	8（39）	0（6）	< 1
厌氧支原体科（Anaeroplasmataceae）	0（0）	0（0）	< 1

瘤胃微生物 MAGs 数据带来了新的微生物认识。例如，变形菌门（Proteobacteria）等新类群被认为是全球牛瘤胃的"核心"微生物（Stewart 等，2019b），最令人吃惊的是，研究人员发现了大量新的纤维素相关多糖利用位点、碳水化合物酶（CAZymes）基因，包括几十万个与碳水化合物代谢相关的新基因（Stewart 等，2019b）。

毫无疑问，MAGs 将有助于我们进一步了解瘤胃微生物群落的功能，并为未来生物酶工程的发展提供巨大的生物资源（Svartström 等，2017）。随着参考基因组数据的不断增加，宏基因组数据的分类率已经达到 50%~70%，这表明未来瘤胃宏基因组研究可能不需要从头组装和注释，而是直接对测序数据与参考基因组或基因集比对、获取注释信息

(Stewart 等，2019b）。

长读长测序的出现，使得未来更有可能组装出更高质量的 MAGs。该技术已经被应用于宏基因组组装研究（Stewart 等，2019b），并获得了完整的 16S rRNA 基因，从而有助于准确鉴定微生物物种种类，有助于揭示生产效率、动物健康和环境相关的瘤胃微生物群落及其功能（Wilkinson 等，2018）。

6 瘤胃比较基因组与关键功能

Hungate 1000 项目的比较基因组分析显示，在 410 个基因组中，仅有 2% 以上的基因与多糖降解有关，其中大多数属于多样性高的糖基水解酶家族（Seshadri 等，2018）。通过对参与植物结构性碳水化合物降解和代谢基因多样性分析，可以预测微生物群落在不同发酵途径中的作用。由于这些基因组来自分离的单菌，所以这些基因功能都可以进行试验验证，从而为优化瘤胃功能提供可能的指导策略。有趣的是，这项分析还发现了一些与瘤胃内日粮营养素分解不相关的功能，比如拟杆菌（Bacteroidetes）基因组中发现了参与宿主多聚糖降解的糖基水解酶。Seshadri 等（2018）推测，这可能与宿主唾液分泌物中 N 连接糖蛋白的分解代谢有关。

与碳水化合物代谢相关基因不同，人们对瘤胃微生物氮代谢相关基因的认识仍然很少。然而，瘤胃微生物的氮代谢直接关系到宿主的营养供给，瘤胃微生物将不同氮源转化成自身蛋白质，在宿主肠道中被消化吸收，为反刍动物提供优质蛋白质（Wallace 等，1997）。因此，了解瘤胃中蛋白质分解和代谢途径对于提高饲料利用率是非常重要的。

通过基因组、转录组和蛋白质组分析，研究人员阐明了栖瘤胃普雷沃氏菌 23（*Prevotella ruminicola* 23）的氮代谢途径（Kim 等，2017）。普雷沃氏菌（*Prevotella*）是瘤胃微生物群落中的优势菌，也是人类肠道和口腔微生物群落的成员，因此研究普雷沃氏菌（*Prevotella*）的氮代谢途径具有重要意义。当细菌可获得充足的氨时，会上调氨转运蛋白，但当氨不足时，细菌则会上调多肽和多胺转运体蛋白，这种适应性变化显示出微生物对不同氮源的适应性（Kim 等，2017）。

微生物对氨基酸和肽的降解也是氮代谢的重要过程。虽然参与这一过程的微生物在瘤胃微生物群中只占一小部分，但降解产生的氨对环境有害，这是因为过量的氨以尿素的形式从宿主体内排出，同时也意味着宿主氮的损失。产氨菌在这一过程尤为重要，大量的研究旨在利用产氨菌的基因组和转录组更好地了解它们的功能（Friedersdorff 等，2019），以便提出高效氮利用调控策略。

细菌基因组具有可变性，特别是在竞争激烈的复杂微生物群落中。细菌基因总是被转移出、转移进入或突变，细菌基因的遗传多样性和转移性是可以被预测的（Andreani 等，2017）。随着细菌基因组数据越来越丰富，开发新的分析技术越来越重要（Kislyuk 等，2011）。其中一项研究就是"泛基因组"研究（Tettelin 等，2005），因为任何单一基因组都不能代表该物种的遗传组成。同一个物种不同菌株之间共有的一组基因被称为核心基因组，而其他基因被称为附属基因组，核心和附属基因组共同构成了该物种的全基因组。

利用微生物物种的泛基因组有助于对该物种生物学功能的清晰认识，从整体上对该物种进行理解。例如，对菌株间基因组成的分析，可以发现抗性基因的转移规律（Subedi 等，2019）。瘤胃细菌含有大量同工酶，表明瘤胃细菌基因组的附属基因很多，这会有助于微生物降解和代谢各种各样的底物（Rubino 等，2017）。

瘤胃微生物经常协同参与代谢，比如生物氢化作用，即微生物将多不饱和脂肪酸转化为饱和脂肪酸的过程（Mosley 等，2002）。溶纤维丁酸弧菌（*Butyrivibrio fibrisolvens*）能够将亚油酸（18:2）生物氢化为顺-9，反-11共轭亚油酸，然后解蛋白丁酸弧菌（*B. proteoclasticus*）将顺-9，反-11共轭亚油酸还原为硬脂酸（18:0）（Ramos-Morales 等，2013）。

微生物群落中新菌株的加入，往往会增加泛基因组中新基因的数量。如果新菌株的加入能给泛基因组增加大量基因，那么群落基因组就可以称为开放的（open）。相反，如果增加的新基因数量有限，那么整个基因组就是封闭的（closed）（Tettelin 等，2008）。丁酸弧菌（*Butyrivibrio*）的基因组是非常开放的，该属一些物种的核心基因不到3%，大部分核心基因属于管家基因，大量的附属基因行使着广泛的功能。溶纤维丁酸弧菌（*B. fibrisolvens*）的核心基因主要由翻译、核糖体结构和生物发生相关的基因组成（70%），附属基因主要行使碳水化合物代谢、复制、重组和修复等一系列功能（Pidcock 等，2019）。

同一物种不同菌株的基因组变异是常见现象（Bentley，2009），所以即便是来自同一物种的菌株也可能会呈现不同的代谢功能（Hussain 等，2016）。所以，从菌株水平认识微生物基因组及其功能对农业生产（如反刍动物饲料效率提升）有重要指导意义。

7 基因组是蛋白质组的蓝图

蛋白质组这个术语是在20世纪90年代出现的，用来描述由基因组编码的各种蛋白质（Wilkins 等，1996）。蛋白质组学是其他组学平台的补充，提供了对生物体功能的更全面认识。基因组学为蛋白质组学提供了翻译蛋白的蓝图（Tyers 和 Mann，2003）。人类基因组中大约有20 235个基因编码了10万种蛋白质（Zhang 等，2013）。

DNA测序为蛋白质分析提供了基础，利用质谱数据搜索蛋白质和核苷酸序列数据库可以识别蛋白质（Yates，2000）。鸟枪蛋白质组学是最常用的蛋白质组学方法，它将蛋白质水解为肽段通过质谱进行鉴定。随后，通过肽谱匹配算法搜索序列数据库，确定蛋白质种类（Carvalho 等，2016）。鸟枪法一词，与鸟枪基因组测序相似，是将大分子（如核酸、蛋白质）变成碎片，然后进行检测与分析（Yates，1998；Zhang 等，2013）。

近年来，蛋白质组学被应用于瘤胃微生物组研究，比如分析特定时间点微生物群蛋白质组的变化（Wilmes 和 Bond，2004）。瘤胃微生物宏蛋白质组学虽然还不完善，但近年来在瘤胃微生物研究越来越多（Deusch 和 Seifert，2015；Snelling 和 Wallace，2017；Hart 等，2018）。研究发现，与宏基因组和16S rRNA基因测序方法相比，瘤胃宏蛋白质组揭示的微生物种类多样性具有一定的一致性（Hart 等，2018）。蛋白质组学未来将是瘤胃微生物功能研究的重要手段，有助于对瘤胃功能的认识。

8 微生物间的相互作用

细菌基因组不仅能体现细菌在瘤胃中的功能和作用，而且还可以表征细菌与其他微生物的相互作用。在竞争条件下，微生物形成了特异性的生态位，并且与特异的碳水化合物活性酶有关，Rubino 等（2017）研究了微生物群体中蛋白质（由（Schlüter 等，2009）定义的"亚型"）的遗传变异。该研究调查了梭菌属（*Clostridium*）和普雷沃氏菌属（*Prevotella*）细菌，发现它们虽然具有相同的生理代谢能力，但在植物细胞壁降解和利用相关蛋白质多样性方面呈现出明显的差异。这种差异或变化可能与植物饲料在瘤胃中分解利用的过程相关（Huws 等，2016），这表明瘤胃微生态的变化受瘤胃环境的影响而发生适应性变化（Barrett 和 Schluter，2008）。

微生物产生和分泌的抗菌化合物由于具有调控剂的应用潜力，深受人们的关注（Williams，2009）。瘤胃中微生物之间存在竞争关系，许多次级代谢物在其中发挥作用。瘤胃内还分布着多样性很高的病毒，其中一些病毒能消灭特定微生物。尽管病毒占据了瘤胃微生物群落很大的一部分，但对病毒的研究仍比较少（Gilbert 等，2020）。噬菌体是一种以细菌为食的病毒，在瘤胃液中发现的噬菌体浓度为 $10^7 \sim 10^8$ 个/mL（Klieve 和 Bauchop，1988）。

噬菌体需要特定的细菌种类甚至是菌株进行自我复制。一旦与细胞表面结合，噬菌体将其基因组（通常储存在衣壳中）注入宿主细胞。噬菌体利用细菌产生更多的噬菌体，最终产生破坏细胞的溶解酶，导致细胞死亡，以排出噬菌体感染更多细胞。这种循环对细菌细胞有害，因此会导致细菌各种防御机制的进化，最典型的是 CRISPR-Cas 系统。噬菌体基因组也能被整合到细菌基因组中，依靠细菌繁殖而进行复制。

8.1 细菌基因组中的病毒和原噬菌体

在未获得细菌基因组之前，研究人员主要通过诱导来确定噬菌体的存在。比如用诱变剂丝裂霉素 C 处理瘤胃细菌，以破坏细菌细胞并诱导原噬菌体裂解循环并形成病毒，并使用显微镜观察了这一过程（Klieve 等，1989）。当获得了细菌基因组序列后，则可避开培养，通过与已知病毒基因进行比对从而确定是否存在原噬菌体或原病毒基因，或者是否感染过噬菌体或病毒。如果从细菌基因组中发现一个完整的原噬菌体基因组，那么就表明噬菌体已经整合到细菌基因组中。对产甲烷菌和醋酸菌的基因组序列进行分析，发现了原病毒基因片段。例如，在黏液真杆菌 SA11（*Eubacterium limosum* SA11）基因组中发现了一个 55kb 的原病毒基因（Kelly 等，2016），在甲烷短杆菌 SM9（*Methanobrevibacter millerae* SM9）基因组中发现了一个长度为 49kb 的原病毒基因，在甲烷杆菌 BRM9（*Methanobacterium formicicum* BRM9）基因组中发现了一个长度为 37kb 的原病毒基因。

虽然瘤胃中存在大量的病毒和噬菌体，但它们的功能仍不清楚。目前仅对病毒 Φmru 有了初步的认识（Attwood 等，2008a）。在对瘤胃甲烷短杆菌 M1（*Methanobrevibacter ruminantium* M1）基因组分析时，发现了一个长约 40 kb 的高 GC 含量的片段，经鉴定是一个完整的原病毒基因组，具有重要功能（如整合、表达结构蛋白和 DNA 复制），该完整的原病毒命名为φmru（Attwood 等，2008a）。φmru 还编码一种独特的内异肽酶 PeiR，当这种酶被加入瘤胃甲烷短杆菌 M1（*M. ruminantium* M1）的培养物中时，会裂解细菌细胞，并减少甲烷产生，这表明重组病毒酶在产甲烷菌调控中具有应用潜力（Leahy 等，2010；Altermann 等，2018）。通过在瘤胃微生物基因组序列中寻找噬菌体功能基因，或许能找到缓解甲烷排放的调控剂。

8.2 利用 CRISPR 序列揭示噬菌体与细菌的相互作用

规律间隔成簇短回文重复序列（CRISPR）与细菌和古菌 Cas 酶一起形成对抗病毒感染的防御机制。通过在 CRISPR 阵列中将外源病毒核酸的短区域整合到细菌基因组中，这个小 RNA 序列可以与酶复合物一起与外来核酸靶向结合（Karginov 和 Hannon，2010）。现有基因组序列中 CRISPR 阵列和噬菌体间隔区表明，细菌或古菌与噬菌体或古菌病毒之间存在相互作用。这种 CRISPR 序列在瘤胃甲烷短杆菌 M1（*Methanobrevibacter ruminantium* M1）（Attwood 等，2008a）、甲烷短杆菌 SM9（*M. millerae* SM9）（Kelly 等，2016）、甲酸甲烷杆菌 BRM9（*Methanobacterium formicium* BRM9）（Kelly 等，2014）和产乙酸菌黏液真杆菌 SA11（*Eubacterium limosum* SA11）（Kelly 等，2016）中存在，但缺乏功能研究。

在 Hungate 项目的基因组中，241 个基因组被鉴定出含有 CRISPR 区域。然而，在病毒数据库中，只有 2.7% 的间隔子与病毒数据库匹配（Seshadri 等，2018），揭示了 31 种微生物与 83 种病毒之间的新关系。26% 的间隔序列与瘤胃微生物序列相似，而约 61% 间隔序列与人体肠道微生物序列相似，这凸显了目前公共数据库中病毒序列来源的偏差性。迄今为止，从瘤胃分离出来的噬菌体只有 5 个被测序（Gilbert 等，2017），因此未来噬菌体和病毒种群的基因组测序，对增加瘤胃内病毒和微生物间相互作用认识具有重要意义。

8.3 微生物之间的竞争

为了生存或维持生态位，几乎所有生命体都进化出了对抗竞争物种或感染生物（例如细菌、真菌、寄生虫或病毒）的防御系统。抗菌化合物是微生物群中弱势微生物的竞争性生存利器，也是细胞先天免疫的一部分（Aminov，2010）。抗菌肽（AMPs）是一类重要的抗菌次级代谢物，也常被称为先天免疫肽或宿主防御肽。AMP 可以说是最广泛的天然抗菌防御系统，存在于几乎所有生命中，从单细胞到多细胞生物，包括细菌、古菌、真菌、植物、昆虫、两栖动物、鱼类、爬行动物和哺乳动物（Thomas，2019）。

细菌素由细菌基因组中 AMPs 核糖体编码，对相关菌株具有杀灭作用。在瘤胃细菌中发现了编码 I 类细菌素的基因簇，如广谱抗生素丁型弧菌素（Butyrivibriocin OR79A）（Kalmokoff 等，1999）和牛链球菌 HC5（*Streptococcus bovis* HC5）中的羊毛硫细菌素（Bovicin HC5）（Mantovani 等，2002）。羊毛硫抗生素（Lantibiotics）常常翻译后被修饰，而 II 类细菌素在结构和抗菌活性方面具有更高的多样性（Ditu 等，2014），不需要翻译后修饰（Yang 等，2014）。在瘤胃细菌中也检测到 II 类细菌素，如链球菌（*Streptococcus* spp.）（Bovicin 255）（Whitford 等，2001）和白色瘤胃球菌 7（*Ruminococcus albus* 7）中的细菌素（Albusin B），它们能抑制黄色瘤胃球菌 FD-1（*R. flavefaciens* FD-1）的生长（Chen 等，2004）。

来自可培养瘤胃细菌基因组越来越多，Hungate 1000 项目和宏基因组数据中，已经发现了大量编码细菌素和防御代谢物的基因簇（Oyama 等，2017a）。通过对 224 种瘤胃细菌和 5 种古菌的基因组分析，发现了来自 33 个菌株的 46 个细菌素基因簇（Azevedo 等，2015）。虽然 Hungate 项目增加了抗菌相关基因的数据量，但瘤胃微生物中仍有大量的抗菌基因资源未被挖掘（Oyama 等，2017b）。

瘤胃细菌和古菌的竞争关系，应将真核生物（如厌氧真菌和纤毛虫）和日粮纤维同时考虑进去（Oyama 等，2017b），因为纤维分解就是各种微生物物种之间争夺共享资源和相互对抗的过程。瘤胃中也存在宿主防御的其他次生代谢产物（Agarwal 等，2015），它们有助于形成一个复杂的生物防御生态系统。将瘤胃中发现的微生物或微生物代谢物作为益生菌或调控剂，比如将抗菌肽作为添加剂调控瘤胃微生物群，将是很重要的一个方向（Wang 等，2016；Liu 等，2017）。

8.4 瘤胃微生物耐药基因

抗性基因存在于有抗生素环境的微生物群落中，抗性基因使微生物自身发生适应性进化（Huttner 等，2013；Blair 等，2015；Brown 和 Wright，2016），迫使它们进化出对抗抗生素的耐药性，以确保生存（Manges 等，2001；Aminov 和 Mackie，2007；Denning 等，2017）。人为过度使用抗生素可能会进一步加剧耐药性的发展（Shea，2003；Fischbach 和 Walsh，2009；Moran，2017）。分析不同环境下的耐药性有助于未来尤其是临床重要微生物的耐药性解决策略的开发（Huttner 等，2013；O'Neill，2016；Martens 和 Demain，2017）。

由于养殖人员与反刍动物长期接触，瘤胃微生物耐药基因（ARGs）可能会转移到人类病原菌中。反刍动物瘤胃和粪便中存在耐药基因（Flint 和 Stewart，1987），并且在使用抗生素治疗期间，牛消化道内耐药基因及丰度会发生变化（Kanwar 等，2014）。由于瘤胃微生物生态系统的多样性（Choudhury 等，2015；Henderson 等，2015），产生新型抗菌化合物的可能性很高，事实上，绵羊瘤胃是耐药基因最大的储存库（Hitch 等，2018）。对 435 个瘤胃微生物基因组中耐药性基因分析发现，编码四环素耐药性的基因丰度很高，表明四环素耐药 *tet*（W）基因正处于正向选择压力下（Sabino 等，2019）。

9 结论

大量分析工具使得瘤胃微生物基因组和宏基因组的研究越来越深入。表 2 列出了一些用于分析瘤胃微生物数据尤其是基因组的工具。一般来说，分析步骤包括：数据质量评估、基因组/宏基因组组装、组装结果验证和统计、基因组分箱、基因或开放阅读框（ORF）预测、功能和基因组注释、基因组比较、分类和系统发育分析、差异分析和病毒分析（CRISPR 和噬菌体）。

虽然表 2 汇总了常用的分析工具，但随着生物信息学的发展，更好的工具将会陆续出现。此外，对于相同的分析或过程，通常有多个工具，而每个工具都有自己的优缺点或特定的使用条件。我们建议读者在开始大规模的生物信息学分析之前仔细分析每种工具的优缺点，并利用许多在线（和离线）课程来学习分析流程（例如欧洲生物信息研究所宏基因组培训课程：https://www.ebi.ac.uk/training/online/topic/metagenomics）。

分析工具通常被放在软件包或流程包中，以便快速和自动执行。初学人员通常喜欢图形界面或自动化分析，而不是命令行操作。Galaxy 是一个很好的例子，它既提供图形界面，又允许构建自动化的工作流程。许多机构都在本地安装了 Galaxy，供研究人员自己或分享给其他人使用（https://usegalaxy.org/）。

使用自动化分析的缺点是：对于用户来说，分析流程是一个暗箱，只有输入数据和输出结果，很难控制或理解分析过程。此外，一些工具的更新，或新工具的出现，都可能导致自动化分析过时。表 2 列出了一些用于分析瘤胃微生物数据的自动化分析范例。

表 2　瘤胃微生物组分析工具

方法	工具	参考文献和用途	应用案例参考文献
数据质控	Trimmomatic	Bolger 等（2014）；一个很灵活的质控工具，修剪和过滤序列，去除测序接头	Stewart 等（2018）
	Sickle	Joshi 和 Fass（2011）；根据 read 质量和长度使用滑动窗口的方法对 read 和接头剪裁	Svartström 等（2017）
	FLASH	Magoč 和 Salzberg（2011）；在组装前将短 reads 进行末端配对的方法	Shi 等（2014）
	BBDuK（BBTools 组件）	Bushnell（sourceforge.net/projects/bbmap/）；修剪和过滤，利用 K-mers 去除接头	Stewart 等（2018）
	poRe	Watson 等（2015）；处理 MinION Nanopore 测序数据的 R 包	Stewart 等（2019b）
组装	Velvet	Zerbino 和 Birney（2008）；使用 K-mers 和 de Bruijn 图对长 contig 和基因组进行组装	Seshadri 等（2018），Hess 等（2011）
	ALLPATHS	Butler 等（2008）；对鸟枪法测序读长（25-50bp）进行基因组组装	Seshadri 等（2018）
	HGAP	Chin 等（2013）；利用长测序读长和单分子实时数据（SMRT）组装基因组	Seshadri 等（2018）
	MEGAHIT	Li 等（2015）；适合复杂的、大数据量的宏基因组数据组装	Stewart 等（2018，2019b）
	MetaSPAdes	Nurk 等（2017）；宏基因组数据组装	Stewart 等（2018）
	Ray	Boisvert 等（2010）；混合测序平台得到的短读长 reads 组装	Svartström 等（2017）
	Newbler	Margulies 等（2005）；鸟枪法测序数据组装	Svartström 等（2017），Pope 等（2012）
	Canu	Koren 等（2017）；PacBio 或 nanopore 测序得到的长 reads 组装	Stewart 等（2019b）

(续表)

方法	工具	参考文献和用途	应用案例参考文献
组装后基因组验证及统计	Bowtie	Langmead 等（2009）；把短读长 reads（25~50 bp）与组装的基因组比对。对长一点的 reads 使用 Bowtie2（Langmead 和 Salzberg，2012）	Hess 等（2011）
	Burrows-Wheeler Aligner（BWA）	Li（2013）；将 reads 与 contigs、scaffolds 或基因组比对	Stewart 等（2018，2019b），Seshadri 等（2018）
	SamTools	Li 等（2009）；一套用于处理序列比对/MAP（SAM）文件格式的工具，将 SAM 文件格式转换为 BAM 文件格式	Stewart 等（2018，2019b）
	CheckM	Parks 等（2015）；一种通过计算基因组完整度和污染度来评估基因组质量的工具	Stewart 等（2018，2019b），Svartström 等（2017），Seshadri 等（2018）
	nanopolish	Simpson, J（https://github.com/jts/nanopolish）；提高 nanopore 长读长 reads 数据的质量	Stewart 等（2019b）
	Racon	Vaser 等（2017）；快速识别模式，从 nanopore 长读长 reads 数据组装高质量基因组	Stewart 等（2019b）
分箱	MetaBAT2	Kang 等（2019）；一种用于对宏基因组数据进行基因组分箱的工具	Stewart 等（2018，2019b）
	dRep	Olm 等（2017）；一种对分箱基因组做比较，去除低质量基因组的工具	Stewart 等（2018，2019b）
	CONCOCT	Alneberg 等（2014）；根据序列覆盖度和组成，对 contigs 进行分箱的工具	Svartström 等（2017）
	PhyloPythiaS	Patil 等（2011）；一种对宏基因组数据进行物种分类的分类器	Pope 等（2012）
基因或 ORF 的预测	Prodigal	Hyatt 等（2010）；基因预测，与其他基因预测工具相比，具有更好的基因结构预测、翻译起始位点识别和低假阳性	Seshadri 等（2018），Stewart 等（2019b），Pope 等（2012）
	MetaGeneMark	Zhu 等（2010）；一种对微生物鸟枪法测序组装的基因组中基因预测的工具	Hess 等（2011）
	GeneMark.hmm	Borodovsky 等（2003）；一种使用隐马尔可夫模型（Hidden Markov Models）预测原核生物基因的工具	Pope 等（2012）
	MetaGene	Noguchi 等（2006）；一种使用密码子频率和 GC 含量预测鸟枪法测序得到的基因组中基因的工具	Pope 等（2012）
	GenePRIMP	Pati 等（2010）；一个提高原核生物基因组中基因预测的程序	Seshadri 等（2018）
	微生物基因组注释流程（MGAP）	Huntemann 等（2015）；一种综合性工具，包含质量控制、结构注释和功能注释的功能，包括用于基因预测的 Prodigal 和 GenePrimp，用于查找 tRNAs 的 tRNAscan 等，这是美国能源部（DOE）联合基因组研究所（JGI）使用的标准方法	Seshadri 等（2018）

(续表)

方法	工具	参考文献和用途	应用案例参考文献
功能和基因组注释	Prokka	Seemann（2014）；一个可注释全部原核生物基因组的工具，集合了用于基因预测的 Prodigal 和用于蛋白质家族搜索的 HMMER 等工具	Stewart 等（2019b），Svartström 等（2017）
	IMG/M	Chen 等（2019）；一个在线分析和注释基因组及宏基因组数据的软件包，结合了来自 JGI 和用户上传的数据，并使用了大量的工具，如 HMMER 和 BLAST 来搜索基因和蛋白质数据库	Seshadri 等（2018），Pope 等（2012）
	USEARCH	Edgar（2010）；一个全局和局部搜索的工具	Shi 等（2014）
	Diamond	Buchfink 等（2015）；一种允许大量蛋白质序列与参考数据库快速比对的工具	Stewart 等（2018，2019b），Svartström 等（2017）
	dbCAN/dbCAN2	Yin 等（2012）；一种在线与碳水化合物活性酶（CAZymes）蛋白质序列进行注释的工具	Stewart 等（2018，2019b）
	Profile Hidden Markov Model	Eddy（1998），Mistry 等（2013）；使用工具例如 HMMER 或 Pfam-Scan，搜索符合隐马尔可夫模型（Hidden Markov Model）数据库，例如 CAZy 或 pfam	Seshadri 等（2018），Hess 等（2011），Stewart 等（2018，2019b），Svartström 等（2017），Pope 等（2012）
	MAGpy	Stewart 等（2019a）；一个注释和描述 MAGs 的软件包，包括 CheckM，Prodigal，BLASTP，Diamond，Pfam-Scan，PhyloPhlAn 和 sourmash	Stewart 等（2018，2019b）
	tRNAscan-SE	Lowe 和 Eddy（1997）；预测 tRNA 基因	Stewart 等（2019b）
	barrnap	Seeman, T（https://github.com/tseemann/barrnap）；预测 rRNA 基因	Stewart 等（2019b）
	RNAmmer	Lagesen 等（2007）；使用隐马尔可夫模型（Hidden Markov Model）预测 rRNA 基因	Svartström 等（2017）
	SINA	Pruesse 等（2012）；根据 SILVA 数据库对 rRNA 基因进行多序列比对，并利用最后共同祖先信息进行物种分类	Svartström 等（2017）
基因组比较、分类和系统发育分析	PULpy	Stewart 等（2018）；发现多糖利用位点	Stewart 等（2019b）
	sourmash	Titus Brown 和 Irber（2016）；利用 MinHash 绘制 DNA 草图，并与其他基因组进行比较	Stewart 等（2018）
	MEGAN4	Huson 等（2011）；用于宏基因组、宏转录组、宏蛋白质组和宏分类（rRNA）数据的物种分类和功能分析的工具	Svartström 等（2017）
	MUMer	Kurtz 等（2004）；快速比对完整的基因组或基因组草图	Stewart 等（2019b）
	Minimap2	Li（2018）；用于将长 DNA 片段或 mRNA 序列比对到参考基因组或参考序列上	Stewart 等（2019b）

(续表)

方法	工具	参考文献和用途	应用案例参考文献
基因组比较、分类和系统发育分析	FastANI	Jain 等（2018）；计算基因组间平均核苷酸一致性（ANI）	Stewart 等（2019b）
	PhyloPhlAn	Segata 等（2013）；一种为基因组进行进化树分析和物种分类分析的工具	Stewart 等（2018），Svartström 等（2017）
	RAxML	Stamatakis（2014）；一种系统发育分析和构建系统发育树的软件	Seshadri 等（2018）
	Interactive Tree of Life（iTOL）	Letunic 和 Bork（2007）；一种对系统发育树进行可视化展示的工具	Seshadri 等（2018，2019b）
	FigTree	Rambaut, A（http://tree.bio.ed.ac.uk/software/figtree/）；用于图形化查看由 BEAST 系统发育分析软件输出的文件	Svartström 等（2017），Stewart 等（2018，2019b）
	GraPhlAn	Asnicar 等（2015）；微生物基因组和宏基因组大数据集的简单可视化	Stewart 等（2018，2019b）
	Kraken	Wood 和 Salzberg（2014）；利用基于 K-mer 搜索快速地对宏基因组序列进行物种分类鉴定	Stewart 等（2018）
	CD-HIT	Fu 等（2012）；一种将序列聚类并去冗余的工具	Stewart 等（2019b）
差异分析	DESeq2	Love 等（2014）；用于定量和差异分析的 R 包，主要用于 RNA 测序和基因表达的数据	Stewart 等（2019b），Shi 等（2014）
	Metastats	White 等（2009）；一种用于分析两组样本的微生物丰度差异的工具	Seshadri 等（2018）
病毒的相互作用；CRISPR 和噬菌体	IMG/M 和 IMG/VR 数据库	Chen 等（2019）；检测 CRISPR，然后在特定病毒数据库中进行搜索（Paez-Espino 等，2017）	Seshadri 等（2018）

10 未来趋势

随着测序成本的降低，微生物基因组数量和质量正在大幅增加。虽然大量基因组研究来自人类遗传学和医学领域（Cheifet，2019），但是，微生物基因组测序分析也很重要。研究重点可能会从分离菌株测序转向宏基因组测序转变，以尽可能获得更多基因组。然而，如前所述，基于宏基因组的物种分类可能无法与已有微生物分类系统相对应，并且缺乏分离菌种的确证性证据。根据 MAGs 的基因信息预测培养基的方法可能有助于获得菌株（Song 等，2008）。目前，培养组或单细胞分选测序（Huws 等，2018）逐渐成熟，很有可能成为下一代瘤胃微生物基因组分析的重要方法。

随着全球人口的不断增长，反刍动物将在营养性食物供应方面发挥重要作用。然而，反刍动物数量的增加对环境和全球变暖的影响不容忽视。我们需要更好地认识养殖方式对反刍动物健康和饲料转化效率的影响，以及胃肠道甲烷和其他废物排放，而要做到这一点，需要对瘤胃微生物群有充分全面的认识（Huws 等，2018）。现有瘤胃微生物基因数据集对于认识上述问题起到了重要作用，但是仍然不够。应当建立全面的瘤胃微生物基因组数据库，包含相关代谢途径，从而用于回答哪些物种促进了这种功能，这些途径如何促进了瘤胃整体功能等问题，或者回答什么是增强瘤胃功能的理想微生物群落（Friedersdorff 等，2019）。因此，有理由认为，基因组资源开发与共享将是未来全球瘤胃微生物基因组研究的重点。

11 更多信息

美国能源部联合基因组研究所（JGI）综合微生物基因组和微生物组（IMG/M）[The Joint Genome Institute（JGI），Integrated Microbial Genomes and Microbiomes] 资源平台（https://img.JGI.doe.gov/）提供了在JGI测序的微生物基因组和微生物组注释、分析，尤其包含 Hungate 项目瘤胃微生物组基因组数据。

瘤胃微生物相关机构包括：

瘤胃微生物基因组（Rumen Microbial Genomics，RMG）（http://www.rmgnetwork.org/）。

全球研究联盟（Global Research Alliance，GRA）家畜研究小组的温室气体研究（https://globalresearchalliance.org/research/livestock/）。

12 参考文献

Afgan, E., Baker, D., Batut, B., et al., 2018. The Galaxy platform for accessible, reproducible and collaborative biomedical analyses:2018 update. Nucleic Acids Research 46（W1），W537-44.

Agarwal, N., Kamra, D. N. and Chaudhary, L. C. 2015. Rumen microbial ecosystem of domesticated ruminants. In:Puniya, A. K., Singh, R. and Kamra, D. N. (Eds), Rumen Microbiology:from Evolution to Revolution. Springer, New Delhi, India, pp. 17-30.

Alneberg, J., Bjarnason, B. S., de Bruijn, I., et al., 2014. Binning metagenomic contigs by coverage and composition. Nature Methods 11 (11), 1144-6.

Altermann, E., Schofield, L. R., Ronimus, R. S., et al., 2018. Inhibition of rumen methanogens by a novel archaeal lytic enzyme displayed on tailored bionanoparticles. Frontiers in Microbiology 9, 2378.

Aminov, R. I. 2010. A brief history of the antibiotic era:lessons learned and challenges for the future. Frontiers in Microbiology 1, 134.

Aminov, R. I. and Mackie, R. I. 2007. Evolution and ecology of antibiotic resistance genes. FEMS Microbiology Letters 271 (2), 147-61.

Andreani, N. A., Hesse, E. and Vos, M. 2017. Prokaryote genome fluidity is dependent on effective population size. The ISME Journal 11 (7), 1719-21.

Asnicar, F., Weingart, G., Tickle, T. L., et al., 2015. Compact graphical representation of phylogenetic data and metadata with GraPhlAn. PeerJ 3, e1029.

Attwood, G. T., Kelly, W. J., Altermann, E. H., et al., 2008a. Analysis of the Methanobrevibacter ruminantium draft genome:understanding methanogen biology to inhibit their action in the rumen. Australian

Journal of Experimental Agriculture 48 (2), 83.

Attwood, G. T., Kelly, W. J., Altermann, E. H., et al., 2008b. Application of rumen microbial genome information to livestock systems in the postgenomic era. Australian Journal of Experimental Agriculture 48 (7), 695.

Azevedo, A. C., Bento, C. B. P., Ruiz, J. C., et al., 2015. Distribution and genetic diversity of bacteriocin gene clusters in rumen microbial genomes. Applied and Environmental Microbiology 81 (20), 7290-304.

Baar, C., Eppinger, M., Raddatz, G., et al., 2003. Complete genome sequence and analysis of Wolinella succinogenes. Proceedings of the National Academy of Sciences of the United States of America 100 (20), 11690-5.

Barrett, R. D. H. and Schluter, D. 2008. Adaptation from standing genetic variation. Trends in Ecology and Evolution 23 (1), 38-44.

Bentley, S. 2009. Sequencing the species pan-genome. Nature Reviews. Microbiology 7 (4), 258-9.

Blair, J. M. A., Webber, M. A., Baylay, A. J., et al., 2015. Molecular mechanisms of antibiotic resistance. Nature Reviews. Microbiology 13 (1), 42-51.

Boisvert, S., Laviolette, F. and Corbeil, J. 2010. Ray: simultaneous assembly of reads from a mix of high-throughput sequencing technologies. Journal of Computational Biology 17 (11), 1519-33.

Bolger, A. M., Lohse, M. and Usadel, B. 2014. Trimmomatic: a flexible trimmer for Illumina sequence data. Bioinformatics 30 (15), 2114-20.

Borodovsky, M., Mills, R., Besemer, J., et al., 2003. Prokaryotic gene prediction using GeneMark and GeneMark. hmm. Current Protocols in Bioinformatics 1 (1), 4. 5. 1-16.

Brown, E. D. and Wright, G. D. 2016. Antibacterial drug discovery in the resistance era. Nature 529 (7586), 336-43.

Bryant, M. P. 1959. Bacterial species of the rumen. Bacteriological Reviews 23 (3), 125-53.

Buchfink, B., Xie, C. and Huson, D. H. 2015. Fast and sensitive protein alignment using DIAMOND. Nature Methods 12 (1), 59-60.

Butler, J., MacCallum, I., Kleber, M., et al., 2008. ALLPATHS: de novo assembly of whole-genome shotgun microreads. Genome Research 18 (5), 810-20.

Carvalho, P. C., Lima, D. B., Leprevost, F. V., et al., 2016. Integrated analysis of shotgun proteomic data with PatternLab for proteomics 4. 0. Nature Protocols 11 (1), 102-17.

Cheifet, B. 2019. Where is genomics going next? Genome Biology 20 (1), 17.

Chen, J., Stevenson, D. M. and Weimer, P. J. 2004. Albusin B, a bacteriocin from the ruminal bacterium *Ruminococcus albus* 7 that inhibits growth of *Ruminococcus flavefaciens*. Applied and Environmental Microbiology 70 (5), 3167-70.

Chen, I. -M. A., Chu, K., Palaniappan, K., et al., 2019. IMG/M v. 5. 0: an integrated data management and comparative analysis system for microbial genomes and microbiomes. Nucleic Acids Research 47 (D1), D666-77.

Chin, C. S., Alexander, D. H., Marks, P., et al., 2013. Nonhybrid, finished microbial genome assemblies from long-read SMRT sequencing data. Nature Methods 10 (6), 563-9.

Choudhury, P. K., Salem, A. Z. M., Jena, R., et al., 2015. Rumen microbiology: an overview. In: Puniya, A. K., Singh, R. and Kamra, D. N. (Eds), Rumen Microbiology: from Evolution to Revolution. Springer, New Delhi, India, pp. 3-16.

Ciccarelli, F. D., Doerks, T., von Mering, C., et al., 2006. Toward automatic reconstruction of a highly resolved tree of life. Science 311 (5765), 1283-7.

Creevey, C. J., Doerks, T., Fitzpatrick, D. A., et al., 2011. Universally distributed single-copy genes indicate a constant rate of horizontal transfer. PLoS ONE 6 (8), e22099.

Creevey, C. J., Kelly, W. J., Henderson, G., et al., 2014. Determining theculturability of the rumen bacterial microbiome. Microbial Biotechnology 7 (5), 467-79.

Delgado, B., Bach, A., Guasch, I., et al., 2019. Whole rumen metagenome sequencing allows classifying and predicting feed efficiency and intake levels in cattle. Scientific Reports 9 (1), 11.

Denning, D. W., Perlin, D. S., Muldoon, E. G., et al., 2017. Delivering on antimicrobial resistance agenda not possible without improving fungal diagnostic capabilities. Emerging Infectious Diseases 23 (2), 177-83.

Deusch, S. and Seifert, J. 2015. Catching the tip of the iceberg - evaluation of sample preparation protocols for metaproteomic studies of the rumen microbiota. Proteomics 15 (20), 3590-5.

Ditu, L. M., Chifiriuc, M., Pelinescu, D., et al., 2014. Class I and II bacteriocins: structure, biosynthesis and drug delivery systems for the improvement of their antimicrobial activity. International Journal of Clinical Pharmacology and Therapeutics 11 (2), 121-7.

Eddy, S. R. 1998. Profile hidden Markov models. Bioinformatics 14 (9), 755-63.

Edgar, R. C. 2010. Search and clustering orders of magnitude faster than BLAST. Bioinformatics 26 (19), 2460-1.

Fischbach, M. A. and Walsh, C. T. 2009. Antibiotics for emerging pathogens. Science 325 (5944), 1089-93.

Fleischmann, R. D., Adams, M. D., White, O., et al., 1995. Wholegenome random sequencing and assembly of Haemophilus influenzae Rd. Science 269 (5223), 496-512.

Flint, H. J. and Stewart, C. S. 1987. Antibiotic resistance patterns and plasmids of ruminal strains of Bacteroidesruminicola and Bacteroides multiacidus. Applied Microbiology and Biotechnology 26 (5), 450-5.

Fraser, C. M., Gocayne, J. D., White, O., et al., 1995. The minimal gene complement of Mycoplasma genitalium. Science 270 (5235), 397-403.

Friedersdorff, J., Creevey, C. and Kingston-Smith, A. 2019. Characterising the genomes and transcriptomes of hyper ammonia producing bacteria from the rumen. Access Microbiology 1 (1A).

Fu, L., Niu, B., Zhu, Z., et al., 2012. CD-HIT: accelerated for clustering the next-generation sequencing data. Bioinformatics 28 (23), 3150-2.

Gilbert, R. A., Kelly, W. J., Altermann, E., et al., 2017. Toward understanding phage: host interactions in the rumen; complete genome sequences of lytic phages infecting rumen bacteria. Frontiers in Microbiology 8, 2340.

Gilbert, R. A., Townsend, E. M., Crew, K. S., et al., 2020. Rumen virus populations: technological advances enhancing current understanding. Frontiers in Microbiology 11, 450.

Greening, C., Geier, R., Wang, C., et al., 2019. Diverse hydrogen production and consumption pathways influence methane production in ruminants. The ISME Journal 13 (10), 2617-32.

Gregor, I., Dröge, J., Schirmer, M., et al., 2016. PhyloPythiaS+: a self-training method for the rapid reconstruction of low-ranking taxonomic bins from metagenomes. PeerJ 4, e1603.

Hart, E. H., Creevey, C. J., Hitch, T., et al., 2018. Meta-proteomics of rumen microbiota indicates niche compartmentalisation and functional dominance in a limited number of metabolic pathways between abundant bacteria. Scientific Reports 8 (1), 10504.

Henderson, G., Cox, F., Ganesh, S., et al., 2015 Rumen microbial community composition varies with diet and host, but a core microbiome is found across a wide geographical range. Scientific Reports 5, 14567.

Hess, M., Sczyrba, A., Egan, R., et al., 2011. Metagenomic discovery of biomass-degrading genes and genomes from cow rumen. Science 331 (6016), 463-7.

Hitch, T. C. A., Thomas, B. J., Friedersdorff, J. C. A., et al., 2018. Deep sequence analysis reveals the ovine rumen as a reservoir of antibiotic resistance genes. Environmental Pollution 235, 571-5.

Huntemann, M., Ivanova, N. N., Mavromatis, K., et al., 2015. The standard operating procedure of the DOE-JGI Microbial Genome Annotation Pipeline (MGAP v. 4). Standards in Genomic Sciences 10, 86.

Huson, D. H., Mitra, S., Ruscheweyh, H. J., et al., 2011. Integrative analysis of environmental sequences using MEGAN4. Genome Research 21 (9), 1552-60.

Hussain, M. H., Zainol, S., Ming Chong, N. F., et al., 2016 Characterisation of Klebsiella pneumoniae Xylanase and increment of its activity in heterologous expression system. Borneo Journal of Resource Science and Technology 6 (1), 1-9.

Huttner, A., Harbarth, S., Carlet, J., et al., 2013. Antimicrobial resistance: a global view from the 2013 World Healthcare-Associated Infections Forum. Antimicrobial Resistance and Infection Control 2, 31.

Huws, S. A., Edwards, J. E., Creevey, C. J., et al., 2016. Temporal dynamics of the metabolically active rumen bacteria colonizing fresh perennial ryegrass. FEMS Microbiology Ecology 92 (1).

Huws, S. A., Creevey, C. J., Oyama, L. B., et al., 2018. Addressing global ruminant agricultural challenges through understanding the rumen microbiome: past, present, and future. Frontiers in Microbiology 9, 2161.

Hyatt, D., Chen, G. L., Locascio, P. F., et al., 2010. Prodigal: prokaryotic gene recognition and translation initiation site identification. BMC Bioinformatics 11, 119.

Jain, C., Rodriguez-R, L. M., Phillippy, A. M., et al., 2018. High throughput ANI analysis of 90K prokaryotic genomes reveals clear species boundaries. Nature Communications 9 (1), 5114.

Joshi, N. A. and Fass, J. N. 2011. Sickle: a sliding-window, adaptive, quality-based trimming tool for FastQ files (Version 1.33) [Software]. Available at https://github.com/najoshi/sickle.

Kalmokoff, M. L., Lu, D., Whitford, M. F., et al., 1999. Evidence for production of a new lantibiotic (butyrivibriocin OR79A) by the ruminal anaerobe *Butyrivibrio fibrisolvens* OR79: characterization of the structural gene encoding butyrivibriocin OR79A. Applied and Environmental Microbiology 65 (5), 2128-35.

Kang, D. D., Froula, J., Egan, R., et al., 2015. MetaBAT, an efficient tool for accurately reconstructing single genomes from complex microbial communities. PeerJ 3, e1165.

Kang, D. D., Li, F., Kirton, E., et al., 2019. MetaBAT 2: an adaptive binning algorithm for robust and efficient genome reconstruction from metagenome assemblies. PeerJ 7, e7359.

Kanwar, N., Scott, H. M., Norby, B., et al., 2014. Impact of treatment strategies on cephalosporin and tetracycline resistance gene quantities in the bovine fecal metagenome. Scientific Reports 4, 5100.

Karginov, F. V. and Hannon, G. J. 2010. The CRISPR system: small RNA-guided defense in bacteria and archaea. Molecular Cell 37 (1), 7-19.

Karlin, S., Campbell, A. M. and Mrázek, J. 1998. Comparative DNA analysis across diverse genomes.

Annual Review of Genetics 32, 185-225.

Kelly, W. J., Leahy, S. C., Li, D., et al., 2014. The complete genome sequence of the rumen methanogen *Methanobacterium formicicum* BRM9. Standards in Genomic Sciences 9, 15.

Kelly, W. J., Henderson, G., Pacheco, D. M., et al., 2016. The complete genome sequence of *Eubacterium limosum* SA11, a metabolically versatile rumen acetogen. Standards in Genomic Sciences 11, 26.

Kelly, W. J., Leahy, S. C., Kamke, J., et al., 2019. Occurrence and expression of genes encoding methyl-compound production in rumen bacteria. Animal Microbiome 1 (1), 15.

Kim, J. N., Méndez-García, C., Geier, R. R., et al., 2017. Metabolic networks for nitrogen utilization in *Prevotella ruminicola* 23. Scientific Reports 7 (1), 7851.

Kingston-Smith, A. H., Davies, T. E., Rees Stevens, P., et al., 2013. Comparative metabolite fingerprinting of the rumen system duringcolonisation of three forage grass (*Lolium perenne* L.) varieties. PLoS ONE 8 (11), e82801.

Kislyuk, A. O., Haegeman, B., Bergman, N. H., et al., 2011. Genomic fluidity: an integrative view of gene diversity within microbial populations. BMC Genomics 12, 32.

Klieve, A. V. and Bauchop, T. 1988. Morphological diversity of ruminal bacteriophages from sheep and cattle. Applied and Environmental Microbiology 54 (6), 1637-41.

Klieve, A. V., Hudman, J. F. and Bauchop, T. 1989. Inducible bacteriophages from ruminal bacteria. Applied and Environmental Microbiology 55 (6), 1630-4.

Koren, S., Walenz, B. P., Berlin, K., et al., 2017. Canu: scalable and accurate long-read assembly via adaptive k-mer weighting and repeat separation. Genome Research 27 (5), 722-36.

Kurtz, S., Phillippy, A., Delcher, A. L., et al., 2004. Versatile and open software for comparing large genomes. Genome Biology 5 (2), R12.

Lagesen, K., Hallin, P., Rødland, E. A., et al., 2007. RNAmmer: consistent and rapid annotation of ribosomal RNA genes. Nucleic Acids Research 35 (9), 3100-8.

Langmead, B. and Salzberg, S. L. 2012. Fast gapped-read alignment with Bowtie 2. Nature Methods 9 (4), 357-9.

Langmead, B., Trapnell, C., Pop, M., et al., 2009. Ultrafast and memory-efficient alignment of short DNA sequences to the human genome. Genome Biology 10 (3), R25.

Leahy, S. C., Kelly, W. J., Altermann, E., et al., 2010. The genome sequence of the rumen methanogen Methanobrevibacter ruminantium reveals new possibilities for controlling ruminant methane emissions. PLoS ONE 5 (1), e8926.

Leahy, S. C., Kelly, W. J., Ronimus, R. S., et al., 2013. Genome sequencing of rumen bacteria and archaea and its application to methane mitigation strategies. Animal 7 (Suppl. 2), 235-43.

Letunic, I. and Bork, P. 2007. Interactive Tree Of Life (iTOL): an online tool for phylogenetic tree display and annotation. Bioinformatics 23 (1), 127-8.

Li, H. 2013. Aligning sequence reads, clone sequences and assembly contigs with BWAMEM. Available at: http://arxiv.org/abs/1303.3997 (accessed on 12 September 2019).

Li, H. 2018. Minimap2: pairwise alignment for nucleotide sequences. Bioinformatics 34 (18), 3094-100.

Li, H., Handsaker, B., Wysoker, A., et al., 2009. The Sequence Alignment/Map format and SAMtools. Bioinformatics 25 (16), 2078-9.

Li, D., Liu, C. M., Luo, R., et al., 2015. MEGAHIT: an ultra-fast singlenode solution for large and complex metagenomics assembly via succinct de Bruijn graph. Bioinformatics 31 (10), 1674-6.

Liu, Q., Yao, S., Chen, Y., et al., 2017. Use of antimicrobial peptides as a feed additive for juvenile goats. Scientific Reports 7 (1), 12254.

Love, M. I., Huber, W. and Anders, S. 2014. Moderated estimation of fold change and dispersion for RNA-seq data with DESeq2. Genome Biology 15 (12), 550.

Lowe, T. M. and Eddy, S. R. 1997. TRNAscan-SE: a program for improved detection of transfer RNA genes in genomic sequence. Nucleic Acids Research 25 (5), 955-64.

Magoč, T. and Salzberg, S. L. 2011. FLASH: fast length adjustment of short reads to improve genome assemblies. Bioinformatics 27 (21), 2957-63.

Manges, A. R., Johnson, J. R., Foxman, B., et al., 2001. Widespread distribution of urinary tract infections caused by amultidrugresistant *Escherichia coli* clonal group. The New England Journal of Medicine 345 (14), 1007-13.

Mantovani, H. C., Hu, H., Worobo, R. W., et al., 2002. Bovicin HC5, a bacteriocin from *Streptococcus bovis* HC5. Microbiology 148 (11), 3347-52.

Margulies, M., Egholm, M., Altman, W. E., et al., 2005. Genome sequencing in microfabricated high-density picolitre reactors. Nature 437 (7057), 376-80.

Martens, E. and Demain, A. L. 2017. The antibiotic resistance crisis, with a focus on the United States. The Journal of Antibiotics 70 (5), 520-6.

Mistry, J., Finn, R. D., Eddy, S. R., et al., 2013. Challenges in homology search: HMMER3 and convergent evolution of coiled-coil regions. Nucleic Acids Research 41 (12), e121.

Moran, D. 2017. Antimicrobial resistance in animal agriculture: understanding user attitudes andbehaviours. The Veterinary Record 181 (19), 508-9.

Morgavi, D. P., Kelly, W. J., Janssen, P. H., et al., 2013. Rumen microbial (meta) genomics and its application to ruminant production. Animal 7 (Suppl. 1), 184-201.

Mosley, E. E., Powell, G. L., Riley, M. B., et al., 2002. Microbial biohydrogenation of oleic acid to trans isomers *in vitro*. Journal of Lipid Research 43 (2), 290-6.

Noguchi, H., Park, J. and Takagi, T. 2006. MetaGene: prokaryotic gene finding from environmental genome shotgun sequences. Nucleic Acids Research 34 (19), 5623-30.

Nurk, S., Meleshko, D., Korobeynikov, A., et al., 2017. metaSPAdes: a new versatile metagenomic assembler. Genome Research 27 (5), 824-34.

Olm, M. R., Brown, C. T., Brooks, B., et al., 2017. dRep: a tool for fast and accurate genomic comparisons that enables improved genome recovery from metagenomes through de-replication. The ISME Journal 11 (12), 2864-8.

Oyama, L. B., Crochet, J. A., Edwards, J. E., et al., 2017a. Buwchitin: a ruminal peptide with antimicrobial potential against Enterococcus faecalis. Frontiers in Chemistry 5, 51.

Oyama, L. B., Girdwood, S. E., Cookson, A. R., et al., 2017b. The rumen microbiome: an underexplored resource for novel antimicrobial discovery. NPJ Biofilms and Microbiomes 3, 33.

O'Neill, J. 2016. Tackling Drug-Resistant Infections Globally: Final Report and Recommendations. Review of Antimicrobial Resistance. HM Government andWellcome trust, London, UK.

Paez-Espino, D., Chen, I. A., Palaniappan, K., et al., 2017. IMG/VR: a database of cultured and uncultured DNA Viruses and retroviruses. Nucleic Acids Research 45 (D1), D457-65.

Palevich, N., Kelly, W. J., Leahy, S. C., et al., 2019. Comparative genomics of rumen Butyrivibrio uncovers a continuum of polysaccharide-degrading capabilities. Applied and Environmental Microbiology 86

(1).

Parks, D. H., Imelfort, M., Skennerton, C. T., et al., 2015. CheckM: assessing the quality of microbial genomes recovered from isolates, single cells, and metagenomes. Genome Research 25 (7), 1043-55.

Pati, A., Ivanova, N. N., Mikhailova, N., et al., 2010. GenePRIMP: a gene prediction improvement pipeline for prokaryotic genomes. Nature Methods 7 (6), 455-7.

Patil, K. R., Haider, P., Pope, P. B., et al., 2011. Taxonomic metagenome sequence assignment with structured output models. Nature Methods 8 (3), 191-2.

Pidcock, S., Skvortsov, T., Santos, F., et al., 2019. Understanding the evolution and metabolic capabilities of the Butyrivibrio group. Access Microbiology 1 (1A), (1A).

Pope, P. B., Mackenzie, A. K., Gregor, I., et al., 2012. Metagenomics of the Svalbard reindeer rumen microbiome reveals abundance of polysaccharide utilization loci. PLoS ONE 7 (6), e38571.

Pruesse, E., Peplies, J. and Glöckner, F. O. 2012. SINA: accurate high-throughput multiple sequence alignment of ribosomal RNA genes. Bioinformatics 28 (14), 1823-9.

Ramos-Morales, E., Martínez-Fernández, G., Abecia, L., et al., 2013. Garlic derived compounds modify ruminal fatty acid biohydrogenation and induce shifts in the *Butyrivibrio* community in continuous-culture fermenters. Animal Feed Science and Technology 184 (1-4), 38-48.

Rubino, F., Carberry, C., Waters, S. M., et al., 2017. Divergent functional isoforms drive niche specialisation for nutrient acquisition and use in rumen microbiome. The ISME Journal 11 (6), 1510.

Sabino, Y. N. V., Santana, M. F., Oyama, L. B., et al., 2019. Characterization of antibiotic resistance genes in the species of the rumen microbiota. Nature Communications 10 (1), 5252.

Schlüter, H., Apweiler, R., Holzhütter, H. G., et al., 2009. Finding one's way in proteomics: a protein species nomenclature. Chemistry Central Journal 3, 11.

Seemann, T. 2014. Prokka: rapid prokaryotic genome annotation. Bioinformatics 30 (14), 2068-9.

Segata, N., Börnigen, D., Morgan, X. C., et al., 2013. PhyloPhlAn is a new method for improved phylogenetic and taxonomic placement of microbes. Nature Communications 4, 2304.

Seshadri, R., Leahy, S. C., Attwood, G. T., et al., 2018. Cultivation and sequencing of rumen microbiome members from the Hungate1000 Collection. Nature Biotechnology 36 (4), 359-67.

Shea, K. M. 2003. Antibiotic resistance: what is the impact of agricultural uses of antibiotics on children's health? Pediatrics 112 (1 Pt 2), 253-8.

Shi, W., Moon, C. D., Leahy, S. C., et al., 2014. Methane yield phenotypes linked to differential gene expression in the sheep rumen microbiome. Genome Research 24 (9), 1517-25.

Snelling, T. J. and Wallace, R. J. 2017. The rumen microbial metaproteome as revealed by SDS-PAGE. BMC Microbiology 17 (1), 9.

Song, H., Kim, T. Y., Choi, B. K., et al., 2008. Development of chemically defined medium for *Mannheimia succiniciproducens* based on its genome sequence. Applied Microbiology and Biotechnology 79 (2), 263-72.

Stamatakis, A. 2014. RAxML version 8: a tool for phylogenetic analysis and post-analysis of large phylogenies. Bioinformatics 30 (9), 1312-3.

Stewart, R. D., Auffret, M. D., Warr, A., et al., 2018. Assembly of 913 microbial genomes from metagenomic sequencing of the cow rumen. Nature Communications 9 (1), 870.

Stewart, R. D., Auffret, M. D., Snelling, T. J., et al., 2019a. MAGpy: a reproducible pipeline for the downstream analysis of metagenome-assembled genomes (MAGs). Bioinformatics 35 (12), 2150-2.

Stewart, R. D., Auffret, M. D., Warr, A., et al., 2019b. Compendium of 4,941 rumen metagenome-assembled genomes for rumen microbiome biology and enzyme discovery. Nature Biotechnology 37 (8), 953-61.

Subedi, D., Kohli, G. S., Vijay, A. K., et al., 2019. Accessory genome of the multi-drug resistant ocular isolate of Pseudomonas aeruginosa PA34. PLoS ONE 14 (4), e0215038.

Suen, G., Weimer, P. J., Stevenson, D. M., et al., 2011. The complete genome sequence of *Fibrobacter succinogenes* S85 reveals a cellulolytic and metabolic specialist. PLoS ONE 6 (4), e18814.

Svartström, O., Alneberg, J., Terrapon, N., et al., 2017. Ninety-nine de novo assembled genomes from the moose (Alces alces) rumen microbiome provide new insights into microbial plant biomass degradation. The ISME Journal 11 (11), 2538-51.

ettelin, H., Masignani, V., Cieslewicz, M. J., et al., 2005. Genome analysis of multiple pathogenic isolates of *Streptococcus agalactiae*: implications for the microbial 'pan-genome'. Proceedings of the National Academy of Sciences of the United States of America 102 (39), 13950-5.

Tettelin, H., Riley, D., Cattuto, C., et al., 2008. Comparative genomics: the bacterial pan-genome. Current Opinion in Microbiology 11 (5), 472-7.

Papke, R. T. 2009. A critique of prokaryotic species concepts. Methods in Molecular Biology 532, 379-95.

Thomas, B. J. 2019. Eukaryotic and bacterial antimicrobial peptides. In: John Wiley & Sons Ltd (Ed.), eLS. John Wiley & Sons, Ltd, Chichester, UK.

Titus Brown, C. and Irber, L. 2016. sourmash: a library for MinHash sketching of DNA. The Journal of Open Source Software 1 (5).

Tyers, M. and Mann, M. 2003. From genomics to proteomics. Nature 422 (6928), 193-7.

Tyson, G. W., Chapman, J., Hugenholtz, P., et al., 2004. Community structure and metabolism through reconstruction of microbial genomes from the environment. Nature 428 (6978), 37-43.

Vaser, R., Sović, I., Nagarajan, N., et al., 2017. Fast and accurate de novo genome assembly from long uncorrected reads. Genome Research 27 (5), 737-46.

Wallace, R. J., Onodera, R. and Cotta, M. A. 1997. Metabolism of nitrogen-containing compounds. In: Hobson, P. N. and Stewart, C. S. (Eds), The Rumen Microbial Ecosystem (2ndedn.). Blackie Academic & Professional, London, UK, pp. 283-328.

Wang, S., Zeng, X., Yang, Q., et al., 2016. Antimicrobial peptides as potential alternatives to antibiotics in food animal industry. International Journal of Molecular Sciences 17 (5).

Watson, M., Thomson, M., Risse, J., et al., 2015. poRe: an R package for the visualization and analysis of nanopore sequencing data. Bioinformatics 31 (1), 114-5.

Weimer, P. J. 1993. Effects of dilution rate and pH on the ruminal cellulolytic bacterium *Fibrobacter succinogenes* S85 in cellulose-fed continuous culture. Archives of Microbiology 160 (4), 288-94.

White, J. R., Nagarajan, N. and Pop, M. 2009. Statistical methods for detecting differentially abundant features in clinical metagenomic samples. PLoS Computational Biology 5 (4), e1000352.

Whitford, M. F., McPherson, M. A., Forster, R. J., et al., 2001. Identification of bacteriocin-like inhibitors from rumen *Streptococcus* spp. and isolation and characterization of bovicin 255. Applied and Environmental Microbiology 67 (2), 569-74.

Wilkins, M. R., Pasquali, C., Appel, R. D., et al., 1996. From proteins to proteomes: large scale protein identification by two-dimensional electrophoresis and amino acid analysis. Nature Biotechnology 14 (1), 61-5.

Wilkinson, T. J., Huws, S. A., Edwards, J. E., et al., 2018. Cowpi: a rumen microbiome focussed version of the picrust functional inference software. Frontiers in Microbiology 9, 1095.

Williams, P. G. 2009. Panning for chemical gold: marine bacteria as a source of new therapeutics. Trends in Biotechnology 27 (1), 45-52.

Wilmes, P. and Bond, P. L. 2004. The application of two-dimensional polyacrylamide gel electrophoresis and downstream analyses to a mixed community of prokaryotic microorganisms. Environmental Microbiology 6 (9), 911-20.

Won, M. Y., Oyama, L. B., Courtney, S. J., et al., 2020. Can rumen bacteria communicate to each other? Microbiome 8 (1), 23.

Wood, D. E. and Salzberg, S. L. 2014. Kraken: ultrafast metagenomic sequence classification using exact alignments. Genome Biology 15 (3), R46.

Yang, S. C., Lin, C. H., Sung, C. T., et al., 2014. Antibacterial activities of bacteriocins: application in foods and pharmaceuticals. Frontiers in Microbiology 5, 241.

Yates, J. R. 1998. Database searching using mass spectrometry data. Electrophoresis 19 (6), 893-900.

Yates, J. R. 2000. Mass spectrometry. Trends in Genetics 16 (1), 5-8.

Yin, Y., Mao, X., Yang, J., et al., 2012. dbCAN: a web resource for automated carbohydrate-active enzyme annotation. Nucleic Acids Research 40 (Web Server issue), W445-51.

Zehavi, T., Probst, M. and Mizrahi, I. 2018. Insights into culturomics of the rumen microbiome. Frontiers in Microbiology 9, 1999.

Zerbino, D. R. and Birney, E. 2008. Velvet: algorithms for de novo short read assembly using de Bruijn graphs. Genome Research 18 (5), 821-9.

Zhang, Y., Fonslow, B. R., Shan, B., et al., 2013. Protein analysis by shotgun/bottom-up proteomics. Chemical Reviews 113 (4), 2343-94.

Zhu, W., Lomsadze, A. and Borodovsky, M. 2010. Ab initio gene identification in metagenomic sequences. Nucleic Acids Research 38 (12), e132.

第6章 瘤胃古菌

Graeme T. Attwood、Sinead C. Leahy，新西兰皇家农科院和新西兰温室气体研究中心；
William J. Kelly，新西兰 Donvis 有限公司

(金巍译)

1 前言

饲养反刍动物生产奶、肉和毛是一项重要的农业生产活动，对世界上许多国家的经济做出了重大贡献。植物性饲料原料决定反刍动物的生产性能，这些植物饲料在反刍动物的瘤胃和网胃中降解并发酵生成挥发性脂肪酸（VFAs）、二氧化碳（CO_2）和甲烷（CH_4）。甲烷是由产甲烷菌通过相对简单的代谢机制，利用有限的几种底物产生的。甲烷主要在瘤胃中产生并通过嗳气排出体外。仅有约10%的甲烷是由后肠发酵产生的（Murray等，1978）。家畜的甲烷排放是温室气体（GHG）排放的一个重要来源，占世界温室气体总排放量的14%左右。反刍动物的甲烷排放与全球变暖和气候变化有关。人们越来越关注如何通过减少甲烷排放来减少动物摄入饲料能量的损失，同时降低反刍动物饲养对环境的影响。本章将总结瘤胃古菌的培养研究和古菌群落组成的分子研究。本章还将通过从产甲烷菌纯培养菌株的基因组以及混合富集培养物和瘤胃样品的宏基因组数据组装基因组中的基因估测产甲烷菌的功能。

2 瘤胃中的古菌

产甲烷菌属于广古菌门（Euryarchaeota）古菌，是一个具有细菌外观的古老微生物谱系，但它们在系统进化发育上是不同的。目前，广古菌门分为4个纲（甲烷杆菌纲 Methanobacteria、甲烷球菌纲 Methanococci、甲烷微菌纲 Methanomicrobia 和甲烷火菌纲 Methanopyri），7个目（甲烷杆菌目 Methanobacteriales、甲烷球菌目 Methanococcales、甲烷微菌目 Methanomicrobiales、甲烷胞菌目 Methanocellales、甲烷八叠球菌目 Methanosarcinales、甲烷火菌目 Methanopyrales 和甲烷马赛球菌目 Methanomassiliicoccales）。近期，在深古菌门（Bathyarchaeot）（Evans 等，2015）及新描述的 Candidatus "Methanofastidiosa"（Nobu 等，2016）和佛斯特拉古菌门（Verstraetearchaeota）（Vanwonterghem 等，2016）发现了潜在的甲烷代谢通路，暗示甲烷产生能力的分布不仅局限于广古菌门。然而，对全球反刍动物的调查（在后面将详细讨论）表明瘤胃甲烷的形成仅限于广古菌门的产甲烷菌（Henderson 等，2015）。

通常在富含碳，但硝酸盐、硫酸盐和氧化铁的浓度很低的厌氧环境中发现产甲烷菌。在这些环境中，产甲烷菌是碳的最终还原者，仅使用几种简单化合物作为生长的能量来源。在硝酸盐、硫酸盐、氧气或氧化铁存在的环境中，产甲烷菌竞争不过能利用上述几种化合物的微生物。与生成甲烷相比，利用上述几种化合物作为末端电子受体可以产生更高的自由能，因此可以产生更多的ATP和菌体量。

反刍动物可以采食多种多样的植物性饲料，包括禾本科和豆科牧草（通过放牧鲜饲或制作青贮和干草保存）、谷物（玉米、大麦、燕麦和小麦）、饲料作物（十字花科芥属、紫花苜蓿、柳条、银合欢）、农业副产品（棕榈油渣、葡萄渣、甘蔗渣、稻草）或上述的组合。这些植物饲料原料富含碳营养素，而硫酸盐、铁和硝酸盐的浓度通常很低（Kennedy和Milligan，1978；Hungate，1966；Jamieson，1959）。因此，一碳化合物转化成甲烷的代谢过程在瘤胃代谢链末端的还原过程中占主导地位。一旦饲料进入瘤胃就会被微生物定植。定植的微生物主要是细菌，它们分泌的酶会攻击和分解各种植物成分。结构性多糖（纤维素、半纤维素、果胶）被分解成低聚糖和单糖，然后通过几种不同的发酵方式转化成挥发性脂肪酸（VFAs，主要是乙酸、丙酸和丁酸，还有少量的甲酸、戊酸、琥珀酸、乳酸和支链挥发性脂肪酸）、二氧化碳和氢气。非结构性多糖（淀粉、果糖）和

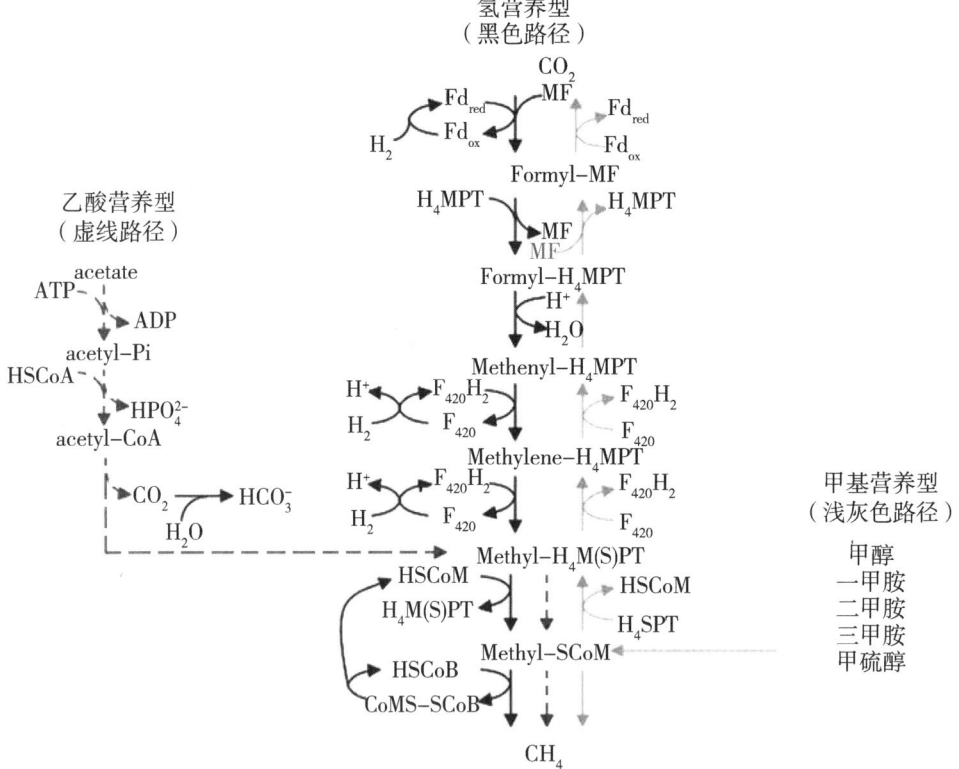

图1 产甲烷菌中的产甲烷途径。MF，甲烷呋喃；Formyl-MF，甲酰甲烷呋喃；H_4MPT，四氢甲烷蝶呤；F420，8-羟基-f-脱氮黄素；CoM，辅酶M（2-磺酰亚砜磺酸盐）；CoB，辅酶B（7-巯基庚酰基苏氨酸磷酸）（Thauer，1998）。

可溶性糖类也会被快速发酵成 VFAs。甲基化合物，如甲胺、甲硫化物和甲醇，在瘤胃中也可能分别从植物脂类、含硫氨基酸或果胶的降解中释放出来（Patterson 和 Hespell，1979；Pol 和 Demeyer，1988）。与其他富含碳的厌氧环境不同，产生的 VFAs 在瘤胃中不会被其他共存的微生物利用，因为将 VFAs 氧化分解成更小的化合物和氢气的微生物生长缓慢，而且瘤胃周转率太高，这样缓慢的代谢过程无法建立（Janssen，2010）。瘤胃中产生的 VFAs 通过瘤胃上皮被吸收，而后被反刍动物代谢利用获得能量并用于生长。

这些发酵过程的几种终产物可以作为产甲烷菌生长的能量来源，如氢气、甲酸、乙酸、短链的醇、甲胺和甲硫化物（Zinder，1993）。在形成甲烷的过程中，作为末端电子受体的碳可能来自二氧化碳、甲酸、甲醇、甲基化合物或乙酸的羧基。由于这些底物是其他降解更复杂有机化合物微生物的发酵产物，因此产甲烷菌通常与这些微生物形成稳定的关系，以便于底物在两种微生物间转移。产甲烷菌有四种公认的甲烷生成模式（Thauer 等，2008）：①氢营养（消耗氢气）途径，通过 7 个酶促反应过程利用氢气将二氧化碳还原成甲烷；②依赖氢气的甲基营养（消耗甲基化合物）途径，利用氢气将甲基化合物还原成甲烷；③不依赖氢气的甲基营养途径，将甲基化合物分解为甲烷和二氧化碳；④乙酸营养（乙酸裂解）途径，将乙酸裂解生成甲烷和二氧化碳（图 1）。

表 1 瘤胃内甲烷生成反应的自由能和竞争氢气利用的途径*

甲烷生成途径	反应	$\Delta G^{0'}$（kJ/mol）*
氢营养型	$4H_2+CO_2 \rightarrow CH_4+2H_2O$	-135
不依赖氢的甲醇营养型	$4CH_3OH \rightarrow 3CH_4+CO_2+2H_2O$	-105
依赖氢气的甲醇营养型	$CH_3OH+H_2 \rightarrow CH_4+H_2O$	-113
甲胺	$4CH_3-NH_2+2H_2O \rightarrow 3CH_4+CO_2+4NH_3$	-75
二甲胺	$2(CH_3)_2-NH+2H_2O \rightarrow 3CH_4+CO_2+2NH_3$	-73
三甲胺	$4(CH_3)_3-N+6H_2O \rightarrow 9CH_4+3CO_2+4NH_3$	-74
二甲基硫	$2(CH_3)_2-S+2H_2O \rightarrow 3CH_4+CO_2+2H_2S$	-49
乙酸营养型	$CH_3COO^-+H^+ \rightarrow CH_4+CO_2$	-33

资料来源：改编自 Liu 和 Whitman（2008）。

这四种甲烷代谢途径产生的可利用自由能不同（表 1）。定量分析表明，氢气是生成甲烷的主要能量来源（Hungate 等，1970；Hungate，1975），尽管利用甲酸生成的甲烷的量可能占瘤胃中甲烷总生成量的 18%（Hungate 等，1970）。甲胺和甲醇也可以被甲烷八叠球菌目的甲基营养型产甲烷菌、甲烷杆菌目的甲烷球形菌（Liu 和 Whitman，2008）以及新的第 7 个产甲烷菌目（Methanomassiliicoccales）的产甲烷菌利用。甲基化合物对瘤胃甲烷产生的贡献还没有被直接测定，但是研究已经发现灌注到绵羊瘤胃中的甲醇被迅速完全地转化成甲烷（Pol 和 Demeyer，1988），暗示瘤胃中可利用的甲基化合物限制了甲烷生成甲基营养途径。在全球范围内对能够还原甲醇和甲胺的瘤胃产甲烷菌丰度的估测表明，约 22% 的瘤胃产甲烷菌可能是甲基营养型（Henderson 等，2015）。虽然乙酸在瘤胃中的浓度很高（约 60 mmol/L），但利用乙酸生成的甲烷可以忽略不计，因为瘤胃液在瘤胃中的保留时间通常不超过一天，而乙酸营养型产甲烷菌数量翻倍时间为 1~2 d，因此无法通过快速增殖来克服被排出瘤胃的问题。在上文提到的全球瘤胃普查中，能够利用乙酸生成甲烷的产甲烷菌（甲烷八叠球菌属 Methanosarcina 和甲烷鬃菌属 Methanosaeta）极为罕见，占瘤胃总古菌的比例不到 0.015%（Henderson 等，2015）。虽然已经从瘤胃中分离出了利用乙酸的产甲烷菌（如甲烷八叠球菌），但这些菌通常是在甲醇或甲胺等底物存在的情况下分离的（Patterson 和 Hespell，1979）。甲烷八叠球菌利用甲醇或甲胺作为底物生长比利用氢气生长得更快（Hutten 等，1980；Sowers 等，1984），推测当这些底物在瘤胃中存在时，它们可以快速地生长。当瘤胃的排空速率急剧降低时，甲烷八叠球菌的数量也会剧增（Rowe 等，1979），很可能是排空速率降低，滞留时间延长允许它们利用乙酸进行生长。

尽管氢气是瘤胃中甲烷生成的主要能量来源，但在某些条件下，其他的氢营养微生物可以竞争氢气（表2）。同型乙酸生成菌通过乙酸生成途径利用氢气还原二氧化碳生成乙酸，已经从瘤胃中分离获得了一些菌株（Genthner 等，1981；Rieu–Lesme 等，1996；Greening 和 Leedle，1989）。分子检测结果显示，同型乙酸生成菌在瘤胃中含量很低（Morvan 等，1994，1996；Henderson 等，2010）。同型乙酸菌的氢气阈值为 350~700 nM，而氢营养型产甲烷菌则低得多，氢气阈值在 20~75 nM（Breznak 和 Kane，1990；Mackie 和 Bryant，1994；Cord-Ruwisch 等，1988；Kim，2012）。瘤胃中溶解的氢气浓度根据日粮、饲喂时间和瘤胃周转速率不同而变化较大，一般在 400~3 400 nM（Janssen，2010）。

表2 瘤胃中与产甲烷竞争氢反应的吉布斯自由能和阈值

反应	反应	ΔG（kJ/mol H_2）*	氢阈值（ppm）**
乙酸生成	$4H_2+2HCO_3^-+H^+\rightarrow CH_3COO^-+4H_2O$	-2.2	520~700
甲烷生成	$4H_2+CO_2\rightarrow CH_4+2H_2O$	-16.9	28~100
硫酸盐还原	$4H_2+SO_4^{2-}+H^+\rightarrow HS^-+4H_2O$	-21.1	9
富马酸还原	H_2+富马酸$\rightarrow$琥珀酸	-84	0.02
硝酸盐还原	$4H_2+NO_3^-+2H^+\rightarrow NH_4^++3H_2O$	-125.5	0.02

资料来源：*Ungerfeld 和 Kohn（2006），**Cord Ruwisch 等（1988）。

通常瘤胃中的氢气浓度高于产甲烷菌生长需要的阈值，但低于乙酸产生菌需要的阈值。此外，通过氢营养途径生成甲烷产生的自由能（-135 kJ/mol）大于生成乙酸产生的自由能（-71.6 kJ/mol）。这些因素共同确定了产甲烷菌在瘤胃中占据主导地位（Cord-Ruwisch 等，1988；Ungerfeld 和 Kohn，2006；Janssen，2010）。硫酸盐还原菌（如脱硫脱硫弧菌；Huisingh 等，1974；Howard 和 Hungate，1976）和硝酸盐还原菌（如反刍月形单胞菌 Selenomonas ruminantium 和产琥珀酸沃廉菌 Wolinella succinogenes，Henderson，1980；Martin 和 Park，1996；Wolin 等，1961；Tanner 等，1981；Unden 等，1980；Jones，1972；Iwamoto 等，2002；Bokranz 等，1983）也是氢气的潜在竞争者。然而，如上所述，硫酸盐和硝酸盐的浓度通常太低,所以这些过程在瘤胃中不占主导地位。在瘤胃中,产琥珀酸沃廉菌等微生物可能会将富马酸还原成琥珀酸(Wolin 等,1961),但富马酸主要是细胞内的中间代谢物,一般情况下无法达到可以利用氢气的浓度。类似地,脱毒脱氮杆菌等微生物还原牧草中的硝基化合物(如 3-硝基丙醇、3-硝基丙酸)也仅可能起到很小的作用(Anderson 等,1993,1996,2000)。

3 瘤胃中产甲烷菌的培养

产甲烷菌是严格的厌氧菌,有非常特殊的生长要求(McAllister 等,1996)。早期科学家们尝试从各种环境中分离培养产甲烷菌,但均受到了阻碍,因为无法达到并维持一个足够低氧化还原电势的厌氧环境供产甲烷菌在体外生长。直到 1940 年,当一种厌氧琼脂培养方法被开发出来后,才得以在体外培养中获得产甲烷菌(Barker,1940)。Robert Hungate 完善了预先还原培养基的制备(Bryant,1972;Hungate,1950,1969),并开发了在厌氧培养基的制备过程中排除氧气的方法、厌氧培养基的灭菌方法以及在维持氧化还原电势低于 -330 mV 的培养基中无菌接种和传代厌氧微生物的方法。这些方法中涉及使用二氧化碳-碳酸氢盐-碳酸盐缓冲液来维持接近中性的 pH 值,使用半胱氨酸和硫化钠作为还原剂,采用刃天青作为氧化/还原指示剂。对于产甲烷菌培养管上的橡胶塞,应采用无菌操作方法打开,以防止污染或氧气进入培养管中。完美地使用这些技术要求太高,因此在 1974 年之前只有少数的产甲烷菌纯菌被分离获得。开发出可以在加压空气中培养产甲烷菌的方法是分离和培养这些微生物的一大进步,从根本上消除了微生物污染或被氧化的机会(Balch 和 Wolfe,1976;Balch 等,1979)。培养物的纯度也是一个问题。从沉积物样品中获得的首批产甲烷菌培养物之一——奥氏甲烷芽孢杆菌(*Methanobacillus omelianskii*)(Barker,1936a,b),尽管经过了几位科研人员的详细研究(Barker,1940,1941;Barker 等,1940;Johns 和 Barker,1960;Wolin 等,1963,1964a,b;Wood 等,1965;Knight 等,1966),但后来被发现仍是一个混合培养物。这个混合培养物最终被纯化,分离出一株产甲烷菌(被命名为 *Methanobacterium* M. o. H,后被重命名为 *Methanobacterium bryantii*(Bryant 等,1967)和一株脂肪酸氧化共养菌——沃氏共养单胞菌(*Syntrophomonas wolfei*)(McInerney 等,1979,1981a)。在产甲烷菌分离培养历程中,产甲烷菌培养物中含有其他微生物是一个常见的问题,这反映了它们需要和与他们紧密联系的微生物共生,这些共生的微生物为它们提供底物或生长因子。

关于瘤胃古菌的知识源于对各种反刍动物瘤胃产甲烷菌的培养(表 3)。Beijer(1952)首次描述了从山羊瘤胃液中分离出一株甲烷八叠球菌属(*Methanosarcina*)和一株甲烷杆菌属(*Methanobacterium*)产甲烷菌。Oppermann 等(1957 年)描述了用甲酸盐富集牛的瘤胃液产生了一个混合培养物,其中含有一株甲酸甲烷杆菌(*Methanobacterium formicium*),而用乙酸盐富集了一种短小、丰满的杆菌,被认为是一株形态变异的索氏甲烷杆菌(*Methanobacterium sohngenii*)。随后,又分离出一株与索氏甲烷杆菌相似的产甲烷菌,被描述为索氏甲烷丝菌(*Methanothrix soehngenii*)(Huser 等,1982)。Nelson 等(1958)也描述了一些源于牛瘤胃液的利用丁酸和戊酸产生甲烷的稳定富集培养物。然而，第一个被分离、纯化和详细描述的瘤胃产甲烷菌是反刍兽甲烷杆菌（*Methanobacterium ruminantium*）（Smith 和 Hungate，1958）。该属的名称后来被改为甲烷短杆菌（Balch 等，1979）。由于最初获得的菌株未能在保存中复活，目前该菌种模式菌株 *Methanobrevibacter ruminantium* M1T（DSM 1093）是由 Bryant（1965）从牛的瘤胃内容物中重新分离出来的。在世界各地的瘤胃微生物实验室中，如何让纯化的产甲烷菌培养物存活并让它们在菌种保藏库中长期保存是一个仍然有待解决的问题。

自早期培养实验以来，已从世界许多地方的牛羊瘤胃内容物中分离出许多甲烷菌，并在一定程度上进行了描述（表3）。在美国，奶牛和肉牛在反刍动物养殖业中占主导地位，人们对描述牛体内甲烷特别感兴趣。

Paynter 和 Hungate（1968）从牛瘤胃中分离出一个产甲烷菌培养物，被命名为运动甲烷杆菌（*Methanobacterium mobilis*），因为它有一根单极性鞭毛，可以让细胞运动。这一产甲烷菌培养物后来被重新分类为运动甲烷微菌（*Methanomicrobium mobile*）（Balch 等，1979）。Patterson 和 Hespell（1979）的研究结果显示在瘤胃内容物中甲烷八叠球菌（*Methanosarcina*）的浓度介于 $10^5 \sim 10^6$ 个/mL，同时证明了最初从沉积物中分离得到的巴氏甲烷八叠球菌 MS（*Methanosarcina barkeri* MS）可以利用三甲胺和甲胺（胆碱的分解产物）生长，但瘤胃源的反刍兽甲烷短杆菌 M1（*Mbb. ruminantium* M1）不能利用。他们认为甲烷八叠球菌可能是瘤胃中利用甲胺的主要微生物。McInerney 等（1981b）在从牛瘤胃液中富集的脱硫弧菌菌株 G11 和一株甲烷八叠球菌的混合培养物中检测到一种厌氧丁酸降解菌，其形态类似于沃氏共养单胞菌（*Syntrophomonas wolfei*）。这一发现解释了这一混合培养物利用丁酸生成甲烷的原因（Nelson 等，1958）。但是，在富集培养物中甲烷八叠球菌作为主要的氢利用菌是不寻常的，因为沃氏共养单胞菌典型的共生菌是亨氏甲烷螺菌（*Methanospirillum hungatei*），同时一般甲烷短杆菌在瘤胃液中的含量更高。Lovely 等（1984）分离出类似反刍兽甲烷短杆菌的菌株 10-16B 和 RMB-1，这两个菌株能够合成辅酶 M（CoM），而菌株 RMB-2、RMB-3 和 RMB-4 不能合成 CoM，需要在培养基中添加 CoM 才能生长。与能够自身合成 CoM 的菌株相比，不能自身合成 CoM 的菌株生长速率更慢，营养需求更复杂，因此被认为在瘤胃中占据不同的生态位。从高倍稀释的牛瘤胃内容物中检测到另外 6 种产甲烷菌菌株（Z3、Z4、Z6、Z8、ZA-4、ZA-10），这些菌株具有甲烷短杆菌属菌株的形态和生理特征（Miller 等，1986）。其中 4 个菌株需要 CoM，2 个不需要。这 6 个菌株的生长要么依赖于异丁酸、异戊酸、2-甲基丁酸和戊酸的混合物，要么受这些挥发性脂肪酸的诱导，这表明它们依赖瘤胃中其他微生物为其提供必需生长因子。然而，所有菌株均未与反刍兽甲烷短杆菌 M1 抗血清发生反应，表明它们代表了不同于甲烷短杆菌菌株的更广泛的免疫多样性。

大约在同一时间，在新西兰开始研究与瘤胃真菌相关的甲烷菌的特征，瘤胃真菌是由 Colin Orpin 发现的（Orpin，1975）。Bauchop 和 Mountfort（1981）分离出一种类似甲烷短杆菌的产甲烷菌（命名为 RA-1）菌落，这一产甲烷菌与厌氧真菌 *Neocalimastix frontalis* PN-1 培养在一起。当共培养时，由于 RA-1 可以利用真菌产生的氢气和甲酸，真菌表现出更高的纤维素降解程度和速率。

表 3 来源于反刍动物的产甲烷菌

培养物	菌株名称（保藏中心登录号）	使用的底物*	来源	参考文献
氢营养型产甲烷菌				
Methanobrevibacter olleyae	KM1H5-1P^T（DSM）	H_2+CO_2, HCOOH	羊瘤胃，澳大利亚	Rea 等（2007）
Methanobrevibacter olleyae	OCP	H_2+CO_2	牛瘤胃，澳大利亚	Rea 等（2007）
Methanobrevibacter olleyae	AK-87	H_2+CO_2	牛瘤胃，澳大利亚	Rea 等（2007）
Methanobrevibacter boviskoreani	JH1	H_2+CO_2, HCOOH, EtOH*	牛瘤胃（Hanwoo 牛），韩国	Lee 等（2013a）
Methanobrevibacter millerae	ZA-10^T（DSM 16643）	H_2+CO_2, 少量 HCOOH	牛瘤胃，美国	Rea 等（2007）
Methanobrevibacter olleyae	YLM1	H_2+CO_2*, HCOOH*	羊瘤胃，新西兰	Skillman 等（2004）
Methanobrevibacter olleyae	BU1	足量 H_2+CO_2	水牛瘤胃，印度	Joshi（2018）
Methanobrevibacter olleyae	GO1	足量 H_2+CO_2	山羊瘤胃，印度	Joshi（2018）
Methanobrevibacter ruminantium	M1^T（DSM 1093）	H_2+CO_2, HCOOH	牛瘤胃，新西兰	Bryant（1965）
Methanobrevibacter ruminantium	0-16B, RMB-1, -2, -3, -4	H_2+CO_2, HCOOH	牛瘤胃，美国	Lovely 等（1984）
Methanobrevibacter ruminantium	MF2	H_2+CO_2	羊瘤胃，法国	Chaucheyras 等（1995）
Methanobrevibacter ruminantium	YE286	H_2+CO_2*, HCOOH*	牛瘤胃，澳大利亚	澳大利亚 GOLD 计划 ID Gp0035230
Methanobrevibacter smithii	GMS-01	ND	山羊瘤胃，印度	Gupta 和 Chaudhary（2010）
Methanobrevibacter smithii（*millerae*）	SM9	H_2+CO_2*, HCOOH*	羊瘤胃，新西兰	未发表

(续表)

培养物	菌株名称（保藏中心登录号）	使用的底物*	来源	参考文献
Methanobrevibacter sp.	AbM4	$H_2+CO_2^*$, $HCOOH^*$, $EtOH+CO_2^*$	羊皱胃，新西兰	未发表
Methanobrevibacter sp.	YE 301	H_2+CO_2	牛瘤胃，澳大利亚	Gilbert 等（2010）
Methanobrevibacter sp.	YE 304	H_2+CO_2	牛瘤胃，澳大利亚	Gilbert 等（2010）
Methanobrevibacter sp.	YE315	$H_2+CO_2^*$, $HCOOH^*$	牛瘤胃，澳大利亚	GOLD project ID Gp0118019
Methanobrevibacter sp.	RA-1	H_2+CO_2, 少量 HCOOH	羊瘤胃，新西兰	Bauchop 和 Mountfort（1981）
Methanobrevibacter sp.	MB-9	H_2+CO_2, HCOOH, 少量 2PrOH	羊瘤胃，日本	Tokura 等（1999）
Methanobrevibacter spp.	Z3, Z4, Z6, Z8, ZA-4	H_2+CO_2, 少量 HCOOH	牛瘤胃，美国	Miller 等（1986）
Methanobrevibacter thaueri	ISO4-G16	ND	羊瘤胃，新西兰	Jeyanathan（2010）
Methanoculleus bourgensis	KOR-2	H_2+CO_2, HCOOH	牛瘤胃，韩国	Battumur 等（2019）
Methanomicrobium mobile	Strain 1, BP^T（DSM 1539）	H_2+CO_2, HCOOH	牛瘤胃，美国	Paynter 和 Hungate（1968）
Methanomicrobium mobile	BRM16	H_2+CO_2, HCOOH	牛瘤胃，新西兰	Jarvis 等（2000）
Methanobacterium sp.	ND	H_2+CO_2, HCOOH	山羊，美国	Beijer（1952）
Methanobacterium bryantii	YE 299	H_2+CO_2	牛瘤胃，澳大利亚	Gilbert 等（2010）
Methanobacterium formicicum	BRM9	H_2+CO_2, HCOOH	牛瘤胃，新西兰	Jarvis 等（2000）

(续表)

培养物	菌株名称 (保藏中心登录号)	使用的底物*	来源	参考文献
Methanobacterium formicicum	SM1	足量 H_2+CO_2	绵羊瘤胃，印度	Joshi 等（2018）
Methanobacterium formicicum	ND	足量 HCOOH	牛瘤胃，美国	Oppermann 等（1957）
Methanocorpusculum aggregans	BU5	足量 H_2+CO_2	水牛瘤胃，印度	Joshi 等（2018）
氢依赖甲基营养型产甲烷菌				
Methanosphaera sp.	ISO3-F5	$H_2 + CO_2 + MeOH + CH_3COOH$	羊瘤胃，新西兰	Jeyanathan（2010）
Methanosphaera sp.	BMS	H_2+MeOH	牛瘤胃（婆罗门牛），澳大利亚	Hoedt（2017）
Methanomassiliicoccales sp.	ISO4-H5	$H_2 + MeOH^*$, MMA^*, DMA^*, TMA^*	羊瘤胃，新西兰	Jeyanathan（2010）
Methanomassiliicoccales sp.	ISO4-G1	$H_2 + MeOH^*$, MMA^*, DMA^*, TMA^*	羊瘤胃，新西兰	Jeyanathan（2010）
Methanomassiliicoccales sp.	ISAO4-G11	ND	羊瘤胃，新西兰	Jeyanathan（2010）
Thermoplasmatales-associated lineage C (TALC)	BRNA1	$H_2 + MeOH^*$, MMA^*, DMA^*, TMA^*	牛瘤胃，澳大利亚	Genbank CP002916.1, 2011
Rumen Cluster C (RCC)	LGM-AF04	ND	山羊瘤胃，中国	Jin 等（2014）
乙酸营养型产甲烷菌				
Methanosarcina	ND	乙酸盐	山羊，美国	Beijer（1952）

(续表)

培养物	菌株名称 (保藏中心登录号)	使用的底物*	来源	参考文献
Methanosarcina barkeri	CM1	H_2+CO_2, MeOH, CH_3COOH, MMA, TMA	牛瘤胃,新西兰	Jarvis 等 (2000)
Methanosarcina sp.	ND	MeOH, MMA, TMA, CH_3COOH, H_2+CO_2	牛瘤胃,美国	McInerney 等 (1981b)
Methanothrix (*Methanobacterium*) *sohngenii*	未报告菌株名称	富含 CH_3COOH	牛瘤胃,美国	Oppermann 等 (1957)
瘤胃源共培养或富集培养物				
Methanobrevibacter ruminantium co-culture with Neocallimastix	Yaktz 1-7	ND	西藏牦牛瘤胃	Wei 等 (2016)
Methanobrevibacter ruminantium co-culture with Piromyces	Yak G18	ND	西藏牦牛瘤胃	Wei 等 (2017)
Methanobrevibacter sp. Z8 in co-culture with Piromyces	F2	ND	山羊瘤胃,中国	Jin 等 (2011)
Methanobrevibacter thaueri CW in co-culture with Piromyces	F1	ND	山羊瘤胃,中国	Jin 等 (2011)
Methanomassiliicoccales enrichment	瘤胃 M1 环境簇	富含 CH_3COOH, HCOOH, TMA; MeOH* MMA*	牛瘤胃,澳大利亚	Söllinger 等 (2016)
Methanomassiliicoccales enrichment	瘤胃 M2 GIT 簇	富含 CH_3COOH, HCOOH, TMA	牛瘤胃,澳大利亚	Söllinger 等 (2016)

(续表)

培养物	菌株名称 (保藏中心登录号)	使用的底物*	来源	参考文献
Methanomassiliicoccales enrichment	1R26 GIT 簇	富含 CH_3COOH, $HCOOH$, TMA; $MeOH^*$	牛瘤胃，丹麦	Noel 等 (2016)
Rumen Cluster C phylotype in co-culture with a *Piromyces* sp.	LGM-AFM04	ND	山羊瘤胃，中国	Jin 等 (2011)
非瘤胃源产甲烷菌，但也存在于瘤胃中				
Methanosphaera sp.	WGK6	H_2+MeOH, $EtOH+MeOH$	西方灰色袋鼠，澳大利亚	Hoedt 等 (2016)
Methanobrevibacter sp. 1Y co-culture with *Piromyces* sp.	1Y	ND	水牛粪便	Jin 等 (2011)
Methanobrevibacter gottschalkii	HO^T	H_2+CO_2	马粪便	Miller 和 Lin (2002)
Methanobrevibacter thaueri	CW^T	H_2+CO_2	牛粪便	Miller 和 Lin (2002)
Methanobrevibacter wolinii	SH^T (DSM 11976)	H_2+CO_2	绵羊粪便，美国	Miller 和 Lin (2002)
Methanobrevibacter woeseis	GS^T	H_2+CO_2, $HCOOH$	鹅粪便	Miller 和 Lin (2002)

*从基因组推断；ND，未检测。

在后来的培养研究中，发现饲喂精料日粮的反刍动物瘤胃产甲烷菌数量为每克瘤胃内容物 $10^7 \sim 10^9$ 个，而放牧反刍动物的瘤胃产甲烷菌数量为每克瘤胃内容物 $10^9 \sim 10^{10}$ 个（Joblin，2005）。史密斯甲烷短杆菌 SM9（*Mbb. smithii* SM9）分离自饲喂新鲜牧草的绵羊（Joblin，1999）。从绵羊皱胃中分离获得一株命名为 AbM4 的甲烷短杆菌，这是研究奥斯特线虫（*Ostertagia circumcincta*）对皱胃环境影响的一部分（Simcock 等，1999）。作为对新西兰反刍动物产甲烷调查项目的一项内容，Jarvis 等（2000）从新西兰黑麦草-白三叶草牧场放牧的牛中分离出 3 株产甲烷菌，并将其分别分类为甲酸甲烷杆菌 *Mb. formicicum*（BRM9）、运动甲烷微菌 *Mm. mobile*（BRM16）和巴氏甲烷八叠球菌 *Ms. barkeri*（CM1）。Skillman 等（2004）从羔羊的瘤胃中分离出一株奥利亚甲烷短杆菌 *Mbb. olleyae* YLM1。后来从绵羊瘤胃中成功分离出一株瘤胃甲烷球形菌（*Methanosphaera* sp. ISO3-F5）和 4 个混合培养物，每个培养物均含有一种产甲烷菌（Jeyanathan，2010）。其中一个混合培养物被命名为 ISO4-G16，含有一种与 *Mbb. thaueri* 进化关系很近的产甲烷菌，而其余 3 个培养物（ISO4-G1、ISO4-G11、ISO4-H5）中的产甲烷菌均与甲烷马赛球菌目（Methanomassiliicocacales）密切相关。随后，一株甲烷马赛球菌目菌株被从 ISO4-H5 培养物中分离出来，这株菌的唯一共培养菌株为溶糊精琥珀酸弧菌（*Succinivibrio dextrinisolvens*），被命名为 H5（Li，2016）。

在反刍动物源产甲烷菌的分离工作中，澳大利亚研究人员获得 3 株利用甲酸的产甲烷菌，这些产甲烷菌被鉴定为奥利亚甲烷短杆菌（*Mbb. olleyae*，绵羊源菌株 KM1H5-1P^T、牛源菌株 OCP 和 AK-87）。研究人员同时将先前 Miller 等（1986）分离的牛源菌株 ZA-10^T 分类为一个新种 *Mbb. millerae*（Rea 等，2007）。牛源甲烷短杆菌 YE301（GQ906575，与 *Mbb. smithii* ATCC 35061 相似性 98%）和 YE304（GQ906576，与 *Mbb. smithii* 相似性 98%），以及一株布氏甲烷杆菌 *Mb. bryantii* YE299（GQ906568，与 *Mb. bryantii* DSM 863 相似性 99%）在布里斯班的实验室被分离获得，并被用于筛选对产甲烷菌具有抑制活性的微生物（Gilbert 等，2010）。从 4 头饲喂低质量干草 28 d 的杂交牛的混合瘤胃液中一株甲烷短杆菌菌株 YE315 被分离出来（Dianne Ouwerkerk，结果未发表；GOLD project ID Gp0118019）。从澳大利亚牛的瘤胃内容物中，富集分离获得一株热原体目相关谱系 C（TALC）菌株 BRNA1（Genbank CP002916.112011；Denman 等，结果未发表）。从鸡的粪便中分离获得一株相似的菌株（Padmanabha 等，2013）。袋鼠虽然不是反刍动物，但具有发酵型前胃，也是产甲烷菌培养物的来源。从澳大利亚西部灰袋鼠的前胃内容物中获得一株甲烷球形菌菌株（*Methanosphaera* sp. WGK6），这株菌具有将乙醇作为还原物还原二氧化碳生成甲烷的独特能力（Hoedt 等，2016，2018）。另一株甲烷球形菌菌株（*Methanosphaera* sp. BMS）被从昆士兰州加顿的婆罗门牛中分离获得（Hoedt，2017；基因组序列未发表；JGI IMG 基因组 ID 2651869595）。

其他国家也分离获得了很多产甲烷菌菌株。反刍兽甲烷短杆菌 MF2（*Mbb. ruminantium* MF2）是法国克莱蒙特·费兰蒂斯 INRA 实验室从绵羊瘤胃中分离出的（Chaucheyras 等，1995）。甲烷短杆菌菌株 MB-9（与 *Mbb. ruminatium* M1 相似性 97%）分离自日本绵羊瘤胃纤毛虫的富集培养物（Tokura 等，1999）。史密斯甲烷短杆菌 GMS-01（*Mbb. smithii* GMS-01）分离自印度山羊瘤胃（Gupta 和 Chaudhary，2010）。最近，从水牛、山羊和绵羊中分离到一些新的产甲烷菌菌株，这些菌株在进化关系上分别与

Mbb. olleyae、*Mb. subterraneum*、*Methanocorpusculum aggregans* 和 *Mb. Formicicum* 相近（Joshi 等，2018）。与先前对新西兰绵羊的研究类似，来自中国的研究发现与反刍兽甲烷短杆菌（*Mbb. ruminantium*-like）类似的菌株 Yak-G18 和 Yaktz 1-7 分别存在于厌氧真菌 *Piromyces* 和 *Neocallimastix frontalis* 培养液中，这些共培养物分离自西藏牦牛瘤胃液（Wei 等，2016，2017）。中国科研工作者还从中国的多种食草动物瘤胃液或粪便中分离到厌氧真菌与产甲烷菌自然共生的共培养物（Jin 等，2011）。他们对源于海门山羊瘤胃液的厌氧真菌富集培养物的深入研究发现，一个属于瘤胃古菌 C 簇（RCC）的新产甲烷菌种在厌氧真菌（*Piromyces. sp.*）的继代培养液中长久存在（Jin 等，2014）。从韩国本地牛（HanWoo；*Bos taurus Corenae*）的瘤胃中分离到一株与甲烷短杆菌 AbM4（*Methanobrevibacter* sp. AbM4）进化关系高度相近的产甲烷菌，这株菌被分类成为一个新的甲烷短杆菌种 *Mbb. boviskoreani* JH1（Lee 等，2013a）。从韩国的荷斯坦公牛瘤胃中也分离到一种罕见的瘤胃产甲烷菌 *Methanoculleus bourgensis* KOR-2（Battumur 等，2019）。

前文描述的 TALC BRNA1 菌株和甲烷马赛球菌目的富集培养物 ISO4-G1、ISO4-G11、ISO4-H5 等菌株和培养物是利用含有三甲胺的培养基从奥地利棕色瑞士牛瘤胃液中富集分离来的（Söllinger 等，2016）。对 16S rRNA 基因序列的分析表明，富集培养物中含有一种与 *Mms. Luminyensis* 和 *Ca. M. intestinalis* 相似性 96% 的微生物，另外一种微生物与 *Ca. M. intestinalis* 有 95% 的相似性，与 *Ca. Mms. alvus* 有 91% 的相似性。另一个从牛瘤胃内容物富集的培养物含有一株甲烷马赛球菌目菌株 *Candidatus Methanomethylophilus* sp. 1R26，这株菌可利用三甲胺和甲醇生成甲烷（Noel 等，2016）。

虽然 *Mbb. gottschalkii* HO^T（分离自马粪便）、*Mbb. thaueri* CW^T（分离自牛粪便）、*Mbb. woesei* GS^T（分离自鹅粪）和 *Mbb. wolinii* SH^T（分离自绵羊粪便）是已知的栖息在瘤胃内的产甲烷菌，但是已经获得的培养物均是非瘤胃来源的（Rea 等，2007；Miller 和 Lin，2002）。

尽管有大量关于产甲烷菌分离和富集的报道，但只有少数瘤胃产甲烷菌菌种得到了有效的描述（Smith 和 Hungate，1958；Bryant，1965；Jarvis 等，2000；Rea 等，2007；Lee 等，2013a），并且可以从商业菌库中获得（表 3）。这是由于许多产甲烷菌培养物难以纯化，并且它们在长期储存过程中会失去活性。几种常用的瘤胃产甲烷菌储存方法包括：加入 10% 甘油或 5% 二甲基亚砜，低温冷冻 -80℃ 以下，或冷冻干燥（Tindall，2007），然而，它们在储存期间存活率低的原因还不完全清楚。

4 利用分子生物学技术鉴定和定量瘤胃产甲烷菌

上述基于培养的早期研究在很大程度上依赖于通过检测表型特征来鉴定产甲烷菌。一直到分子生物学方法和 DNA 分析技术的出现，才实现了通过杂交和测序技术鉴定产甲烷菌。David Stahl 实验室开展了早期的分子检测工作，研究人员使用从瘤胃样本中提取的 RNA，并将其吸附在过滤器上，然后使用标记的寡核苷酸杂交探针对其进行检测，该探针特异性地结合产甲烷菌的 16S rRNA（Lin 等，1997）。这一方法首次提供了瘤胃样本中古菌总数的定量数据，并用于描述在多种动物体内发现的产甲烷菌类型。根据反刍动物物种的不同，古菌小亚基核糖体RNA（SSU rRNAs）占瘤胃样本总SSU rRNAs的0.3%~3.3%（Lin等，1997），占原虫总SSU rRNAs的1.46%（Sharp等，1998）。甲烷杆菌科（Methanobacteriaceae）是牛和山羊瘤胃中最丰富的产甲烷菌群，而甲烷微菌目在绵羊瘤胃中最丰富（Lin等，1997）。原虫共生菌中99.2%的古菌属于甲烷杆菌科，暗示这一产甲烷菌群作为瘤胃原虫共生菌的重要性（Sharp等，1998）。甲烷八叠球菌目（Methanosarcinales）以前被认为是仅次于甲烷杆菌科的第二大菌群，但实际上仅占总古菌SSU rRNA的很小一部分。

标记探针的另一种用途是荧光原位杂交（FISH），这一技术将探针与核糖体RNA杂交，可以让微生物群落中的目标菌群可视化。Soliva等（2003）利用FISH研究月桂酸和肉豆蔻酸混合物对瘤胃产甲烷菌的影响，并成功地鉴定出了甲烷球菌目（Methanococcales）产甲烷菌，这一产甲烷菌目在以前的基于PCR分析中没有被发现过。Valle等（2015）使用甲烷菌辅因子F420（F420）自发荧光和激光共聚焦扫描显微镜鉴定瘤胃产甲烷菌，并确定其在自由生活、生物膜和与原虫共生的微环境中的空间分布。产甲烷菌16S rRNA特异性探针（如Arch915）可以与一些缺乏F_{420}的细胞结合，有助于鉴定出新甲烷马赛球菌目产甲烷菌（*Methanomassiliicoccales*）。以甲烷生成途径中特异性甲基辅酶M还原酶（mcr）基因的RNA为靶点的探针可检测甲烷八叠球菌（*Methanosarcin*）细胞，其信号强度与16S rRNA杂交相关。然而，该探针未能与大多数激发了F_{420}自然荧光（蓝绿光）的瘤胃产甲烷菌杂交，可能原因是不同产甲烷菌种属之间的细胞壁通透性具有差异。研究发现产甲烷菌被整合到微生物生物膜中，并且产甲烷菌与瘤胃原虫存在外共生和内共生现象。

随着聚合酶链式反应（PCR）技术和商用热循环仪的发展，使16S rRNA基因扩增、克隆和测序成为可能（Böttger，1989）。这极大地促进了产甲烷菌的鉴定和定量分析，并使得以前未培养的产甲烷菌得以被发现。例如，Whitford等（2001）从牛瘤胃中鉴定到属于甲烷杆菌科（Methanobacteriaceae）和甲烷八叠球菌科（Methanosarcinaceae）的未培养物种，并对代谢标记基因 *mcrA* 基因也进行了靶向PCR扩增和测序（Tatsuoka等，2004；Ozutsumi等，2012；Poulsen等，2013；Shi等，2014；Denman等，2007；Popova等，2011）。

瘤胃中一些产甲烷菌与原虫密切相关，通过细胞内或细胞外的形式与原虫共生（Vogels等，1980；Finlay等，1994）。通过从瘤胃内容物中分离单个原虫细胞并对产甲烷菌16S rRNA基因进行PCR扩增（Irbis和Ushida，2004；Regensbogenova等，2004），或通过分离整个原虫群落，然后进行16S rRNA或mcrA基因测序的方法（Chagan等，1999；Tokura等，1999；Tymensen等，2012；Xia等，2014；Valle等，2015），已鉴定出产甲烷菌-原虫共生体。

氢气是原虫细胞中氢酶体的代谢产物，产甲烷菌可以利用氢气生成甲烷，因此产甲烷菌通过与原虫紧密连接获得氢气而获益，而原虫通过产甲烷菌利用氢气降低氢分压受益，低氢分压有利于原虫产生更多的代谢产物（Müller，1993；Finlay等，1994；Akhmanova等，1998）。研究发现，虽然与原虫相关的产甲烷菌主要属于甲烷短杆菌属（*Methanobrevibacter*），但是仍然有一小部分甲烷马赛球菌目（Methanomassiliicoccales）、甲烷球形菌属（*Methanosphaera*）和甲烷微菌属（*Methanomicrobium*）产甲烷菌与原虫相关。与特定瘤胃原虫相关的产甲烷菌种类变化较大（Finlay 等，1994；Lloyd 等，1996），与不同原虫相关的产甲烷菌多样性可能相似（Belanche 等，2014）。

PCR 技术的出现也催生了几种针对产甲烷菌 16S rRNA 基因的新型微生物"指纹"技术。变性梯度凝胶电泳（DGGE）和温度梯度凝胶电泳（TGGE）已用于分离在变性剂浓度梯度（DGGE）或温度梯度（TGGE）中变性顺序不同的产甲烷菌 16S rRNA 基因扩增子。Nicholson 等（2007）使用 PCR-TGGE 技术检测了牛和羊的瘤胃甲烷菌，然而 Yu 等（2008）发现，脂肪补充剂的添加改变了产甲烷菌的 PCR-DGGE 图谱，增加了斯氏甲烷球形菌（*Ms. stadtmanae*）的丰度，并减少了 *Methanobrevibacter sp.* AbM4. 的丰度。Zhou 等（2010）利用 PCR-DGGE 分析了饲喂生长期日粮和肥育期日粮牛的瘤胃产甲烷菌菌群，发现产甲烷菌菌群从以反刍兽甲烷短杆菌（*Mbb. ruminantium*）为主的菌群转变成了由不同产甲烷菌组成的混合菌群。末端片段长度多态性（TRFLP）技术是将 16S rRNA 基因 PCR 扩增子利用限制性内切酶消化成不同长度的片段，然后对标记的末端片段长度进行分析，这一技术也已应用于描述奶牛不同胃肠道部位的产甲烷菌（Frey 等，2010）以及与瘤胃原虫相关的产甲烷菌（Belanche 等，2014）。虽然这些技术有助于发现产甲烷菌群落的变化，但它们有一些局限性。其中包括难以解释多个条带是否来自相同或不同的基因组，缺乏对每个种系型的定量信息，以及需要对条带进行切割测序或对克隆片段测序以获得更明确的物种鉴定信息，现在这些技术已被高通量测序技术取代。

16S rRNA 基因克隆文库被广泛用于描绘各种条件下瘤胃古菌群落（Ohene Adjei 等，2007；Regensbogenova 等，2004；Shin 等，2004；Skillman 等，2006；Tajima 等，2001；Tokura 等，1999；Wright 等，2004、2006、2007 和 2008）。Janssen 和 Kirs（2008）分析总结了这些研究，结果显示，大多数瘤胃产甲烷菌归属于两个已知的属：甲烷短杆菌属（*Methanobrevibacter* spp.）和甲烷微菌属（*Methanomicrobium* spp.）。重要的是，系统发育进化树上出现了一个大的、未描述的分支，这一分支先后被称为瘤胃古菌 C 簇（RCC）或 *Thermoplasmatales*-associated lineage C（TALC）。这个未培养产甲烷菌分支 16S rRNA 基因激起了研究人员分离培养这一分支菌株的热情（Jeyanathan，2010；Denman 等，2011，结果未发表，Genbank CP002916.1；Li 等，2016a）。最近，在人类肠道和环境样本中也检测到与这一分支相似的产甲烷菌，这一分支被命名为产甲烷菌的一个新目"甲烷马赛球菌目"（Methanomassiliicoccales）。

使用基因克隆文库的缺点是 PCR 反应使 DNA 呈指数级放大，因此特定微生物的丰度只能相对于利用相同引物扩增的其他微生物的丰度来表达。产甲烷菌的绝对数量与在瘤胃中观察到的微生物密度无关。实时定量 PCR 技术（qRT-PCR）被用于对目标基因进行定量。通过与参考基因相比，可以实现对目标基因的绝对定量，该方法已被用于量化古菌菌群的绝对丰度（Hook 等，2009；Ohene Adjei 等，2008；Zhou 等，2009 和 2010），还被用于比较不同饲料效率动物体内的微生物丰度变化（Zhou，2009）。该技术的一个特殊优势是能够在食糜样品中检测到少量古菌（Frey 等，2010），但其高成本和低通量限制了其在微生物生态学研究中的应用。

16S rRNA 基因克隆和测序技术的低通量也限制了可分析的克隆数量。然而，DNA 测序技术的快速发展使扩增子文库的高通量测序成为可能，这大大增加了每个样本中可分析的序列数量。对 PCR 引物进行标记化也使得在一次测序运行中可以对多个样本同时进行测序。目前古菌特异性引物已通过各种测序平台应用于瘤胃微生物区系的扩增和测序（Kittelmann 等，2013、2014；Shi 等，2014；Henderson 等，2015；Seedorf 等，2015；Myer 等，2016；Lee 等，2012；Li 等，2016a；de la Fuente 等，2014 年；Snelling 等，2014；McCabe 等，2015；Fouts 等，2012；Indugu 等，2016）。研究者对这些高通量技术已经进行了比较，还与其他微生物定量方法进行了比较（Myer 等，2016；Snelling 等，2014；de la Fuente 等，2014；Li 等，2016a；Indugu 等，2016）。这些技术已经被用于描述与牛和羊瘤胃相关的古菌区系（Seedorf 等，2015；Fouts 等，2012），比较高产甲烷和低产甲烷绵羊之间的古菌区系（Kittelmann 等，2013、2014；Shi 等，2014），分析微生物区系与瘤胃代谢物的相关性（Lee 等，2012），还用于检测限饲牛体内的细菌和古菌区系（McCabe 等，2015）。Henderson 等（2015）的里程碑式的全球瘤胃普查研究采用了这种方法，发现在世界各地饲养的反刍动物体内产甲烷菌区系相似，主要受宿主动物日粮的影响。瘤胃中最丰富的产甲烷菌是氢营养型的甲烷短杆菌，主要是反刍兽甲烷短杆菌（*Mbb. ruminantium*）以及它的近似菌种（称为"RO 簇"）和 *Mbb. gottschalkii* 及近似菌种（称为"SGMT 簇"），这两个产甲烷菌簇平均占瘤胃总古菌的 74%。一种甲基营养型产甲烷菌甲烷球形菌属（Methanosphaera）和另外两种亲缘较远的甲基营养型产甲烷菌科甲烷马赛球菌科（*Methanomassiliicoccaceae*）大约占瘤胃总古菌的 16%。利用标记物基因（marker gene）分析鉴定的一些产甲烷菌尚没有培养物，因此它们的功能仍然未知。通过利用标记物基因的丰度信息，并根据已培养的亲缘关系最近的产甲烷菌的代谢信息，对每一个产甲烷菌菌群进行生理学功能分析，可以推断出瘤胃中的大多数古菌（约 78%）是氢营养型产甲烷菌，而其他古菌（约 22%）可能通过甲基营养途径利用氢气还原甲基化合物（如甲醇或甲胺）进行生长（Henderson 等，2015），该研究还证实，世界范围内的瘤胃古菌群落几乎完全由产甲烷菌组成。在 16S rRNA 基因文库中检测到两个未培养的古菌菌群：Qld26 菌群（Wright 等，2006，2007）和 *Crenarchaeota* 的一个分支簇（Shin 等，2004）。Qld26 菌群的系统进化发育关系表明，它具有氢营养型甲烷生成模式，但归属 *Crenarchaeota* 菌群的生理学功能尚不清楚。全球瘤胃普查研究还发现，一些被标注为"其他"或没有 BLAST 匹配的序列，在每一个单一样本中的平均丰度小于 0.04%，最大相对丰度小于 1%（Henderson 等，2015）；因此，这些菌群不可能对整个瘤胃产甲烷菌菌群具有重要贡献。

16S rRNA 基因扩增子数据的分析需要全面且精心整理的参考数据库。目前已经开发了一个数据库，用于在菌种水平上对来自瘤胃、动物和人类肠道产甲烷菌的 16S rRNA 基因扩增子数据进行分类学鉴定（Seedorf 等，2014），该数据库称为 RIM-DB（瘤胃和肠道产甲烷菌数据库），包含 2 379 个几乎全长的非冗余 16S rRNA 基因序列。该数据库与 QIIME 兼容，也可与 SILVA 数据库结合使用。

表 4 瘤胃产甲烷菌基因组序列和宏基因组组装基因组基因组

瘤胃源培养物基因组

基因组/样品名（保藏中心）	测序项目	IMG 基因组 ID	基因组大小*	基因数量*	来源	状态	参考文献
Methanobacterium formicicum BRM9	AgR[1] (PGGRC[2])	2630968343	2 449 987	2 452	牛	草图	Kelly 等 (2014)
Methanobrevibacter boviskoreani JH1	KRIBB[3]	2551306657	2 045 801	1 799	牛	永久草图	Lee 等 (2013b)
Methanobrevibacter millerae ZA-10 (DSM 16643)	JGI[4] (Hungate 1000[5])	2593339167	2 725 667	2 467	牛	永久草图	Seshadri 等 (2018)
Methanobrevibacter millerae SM9	AgR (PGGRC)	2654587756	2 543 538	2 365	绵羊	草图	Kelly 等 (2016b)
Methanobrevibacter olleyae 1H5-1P (DSM 16632)	JGI (Hungate1000)	2593339150	2 122 444	1 854	牛	永久草图	Seshadri 等 (2018)
Methanobrevibacter olleyae YLM1	AgR (PGGRC)	2681812966	2 201 192	1 882	绵羊	草图	Kelly 等 (2016b)
Methanobrevibacter ruminantium M1 (DSM 1093)	AgR (PGGRC)	646311943	2 937 203	2 283	牛	完成图	Leahy 等 (2010)
Methanobrevibacter ruminantium YE286	Macrogen[6]	2524023089	1 816 712	1 728	牛	草图	Leahy 等 (2010)
Methanobrevibacter sp. AbM4	AgR (NZAGRC)	2540341183	1 998 189	1 716	绵羊	完成图	Leahy 等 (2010)
Methanobrevibacter sp. YE315	Queensland govt[7]	2667527400	2 273 296	1 982	牛	草图	Seshadri 等 (2018)
Methanomicrobium mobile BP (DSM1539)	JGI (Hungate1000)	2571042922	1 711 791	1 686	牛	永久草图	Seshadri 等 (2018)
Methanosphaera sp. BMS	U. Queensland[8]	2651869595	2 868 093	2 258	牛	草图	Li 等 (2016b)
Methanogenic archaeon ISO4-H5	AgR (NZAGRC[9])	2660238307	1 937 883	1 890	绵羊	草图	Kelly 等 (2016c)
Unclassified archaeon ISO4-G1	AgR (NZAGRC)	2667527398	1 593 503	1 549	绵羊	草图	Kelly 等 (2016c)
Thermoplasmatales archaeon BRNA1	CSIRO[10]	2565956561	1 461 105	1 577	牛	完成图	Lambie 等 (2015)
Methanosarcina barkeri CM1	AgR (PGGRC)	2634166439	4 501 171	3 878	牛	草图	Lambie 等 (2015)

(续表)

基因组/样品名（保藏中心）	测序项目	IMG 基因组 ID	基因组大小*	基因数量*	来源	状态	参考文献
Methanosarcina sp. DSM 11855	JGI (Hungate1000)	2595698250	3 100 602	2 732	绵羊	永久草图	Seshadri 等 (2018)
非瘤胃源菌株基因组							
Methanobrevibacter gotschalki HO^T (DSM 11977)	JGI (KMG-IV[11])	2788546169	1 878 029	1 898	马粪	草图	
Methanobrevibacter gotschalki HO^T (DSM 11977)	CSIRO	2622736425	1 879 371	1 889	马粪	草图	
Methanobrevibacter gotschalki PG (DSM 11978)	JGI (Hungate1000)	2654588138	1 864 477	1 942	猪粪	草图	Seshadri 等 (2018)
Methanobrevibacter thaueri CW^T (DSM 11995)	CSIRO	2622736427	2 244 956	2 184	牛粪	草图	
Methanobrevibacter woesei GS^T (DSM 11979)	CSIRO	2622736428	1 544 134	1 628	鹅粪	草图	
Methanosphaera stadtmanae MCB-3 (DSM 3091)	Max Planck Institute[12]	637000163	1 767 403	1 592	人粪	已完成	Fricke 等 (2006)
Methanosphaera stadtmanae WGK6	454 Life Sciences[13]	2595698213	1 729 155	1 643	鹅肠道	草图	Hoedt 等 (2016)
Methanobrevibacter wolinii SH^T (DSM 11976)	JGI (Hungate1000)	2558860120	2 041 814	1 747	绵羊粪	永久草图	Seshadri 等 (2018)
瘤胃样品的宏基因组装基因组							
Methanomassiliicoccales archaeon RumEn M1	U. Vienna[14]	2667527221	2 121 026	1 926	牛	草图	Söllinger 等 (2016)
Methanomassiliicoccales archaeon RumEn M2	U. Vienna	2667527222	1 280 797	1 225	牛	草图	Söllinger 等 (2016)
Candidate *Methanomethylophilus* sp. hRUG898	Roslin Institute[15]	2799112866	1 388 681	1 459	牛	草图	Stewart 等 (2018)

(续表)

基因组/样品名（保藏中心）	测序项目	IMG 基因组 ID	基因组大小*	基因数量*	来源	状态	参考文献
Candidate Methanomethylophilus sp. RUG779	Roslin Institute	2799112865	1 262 884	1 400	牛	草图	Stewart 等 (2018)
Methanobrevibacter sp. RUG011	Roslin Institute	2799112914	2 484 127	2 252	牛	草图	Stewart 等 (2018)
Methanobrevibacter sp. RUG012	Roslin Institute	2799112907	2 366 395	2 113	牛	草图	Stewart 等 (2018)
Methanobrevibacter sp. RUG018	Roslin Institute	2799112916	2 431 573	2 363	牛	草图	Stewart 等 (2018)
Methanobrevibacter sp. RUG020	Roslin Institute	2799112915	2 285 716	2 036	牛	草图	Stewart 等 (2018)
Methanobrevibacter sp. RUG031	Roslin Institute	2799112917	2 324 332	2 214	牛	草图	Stewart 等 (2018)
Methanobrevibacter sp. RUG076	Roslin Institute	2799112919	1 863 219	1 669	牛	草图	Stewart 等 (2018)
Methanobrevibacter sp. RUG092	Roslin Institute	2799112922	2 324 420	2 154	牛	草图	Stewart 等 (2018)
Methanobrevibacter sp. RUG120	Roslin Institute	2799112903	2 575 445	2 281	牛	草图	Stewart 等 (2018)
Methanobrevibacter sp. RUG121	Roslin Institute	2799112904	2 386 750	2 220	牛	草图	Stewart 等 (2018)
Methanobrevibacter sp. RUG186	Roslin Institute	2799112911	2 487 973	2 342	牛	草图	Stewart 等 (2018)
Methanobrevibacter sp. RUG201	Roslin Institute	2799112913	1 920 279	1 753	牛	草图	Stewart 等 (2018)
Methanobrevibacter sp. RUG236	Roslin Institute	2799112912	1 908 294	2 016	牛	草图	Stewart 等 (2018)

(续表)

基因组样品名（保藏中心）	测序项目	IMG 基因组 ID	基因组大小*	基因数量*	来源	状态	参考文献
Methanobrevibacter sp. RUG256	Roslin Institute	2799112920	1 806 492	1 825	牛	草图	Stewart 等 (2018)
Methanobrevibacter sp. RUG338	Roslin Institute	2799112921	2 281 773	2 175	牛	草图	Stewart 等 (2018)
Methanobrevibacter sp. RUG492	Roslin Institute	2799112909	2 038 501	2 127	牛	草图	Stewart 等 (2018)
Methanobrevibacter sp. RUG526	Roslin Institute	2799112910	1 740 466	2 013	牛	草图	Stewart 等 (2018)
Methanobrevibacter sp. RUG545	Roslin Institute	2799112908	2 220 727	2 146	牛	草图	Stewart 等 (2018)
Methanobrevibacter sp. RUG648	Roslin Institute	2799112924	2 291 898	2 360	牛	草图	Stewart 等 (2018)
Methanobrevibacter sp. RUG780	Roslin Institute	2799112905	1 767 289	1 913	牛	草图	Stewart 等 (2018)
Methanobrevibacter sp. RUG787	Roslin Institute	2799112906	1 597 431	1 832	牛	草图	Stewart 等 (2018)
Methanobrevibacter sp. RUG823	Roslin Institute	2799112918	1 424 219	1 638	牛	草图	Stewart 等 (2018)
Methanobrevibacter sp. RUG833	Roslin Institute	2799112923	2 477 776	2 872	牛	草图	Stewart 等 (2018)
Methanobrevibacter sp. RUG761	Roslin Institute	2799112880	1 638 250	1 752	牛	草图	Stewart 等 (2018)

注：* 组装。[1] 新西兰皇家农业科学院；[2] 新西兰畜牧业温室气体研究联盟；[3] 韩国生物科学与生物技术研究所；[4] 美国能源部联合基因组研究所；[5] Hungate 1 000项目，瘤胃微生物参考基因组集；[6] 千年基因公司；[7] 昆士兰政府；[8] 昆士兰大学；[9] 新西兰农业温室气体研究中心；[10] 澳大利亚联邦科工组织；[11] 千种微生物基因组第四阶段项目；[12] 德国马克斯·普朗克研究所；[13] 454 生命科学公司；[14] 奥地利维也纳大学；[15] 英国苏格兰罗斯林研究所农学院兰肉牛和绵羊研究中心。

随着 16S rRNA 基因扩增子高通量分析（NGS）技术迅速发展，这些大规模并行测序技术使瘤胃宏基因组 DNA 的深度鸟枪法测序成为可能。Hess 等（2011）首次将高通量测序应用于瘤胃，这一研究描绘了牛瘤胃中生物质降解酶基因和附着在植物纤维上的微生物基因组。该研究对 268 Gb 的宏基因组 DNA 进行了测序和分析，鉴定出 27 755 个可能的碳水化合物活性酶基因，重要的是，证明了利用瘤胃宏基因组数据重新组装未培养的微生物基因组的可行性。NGS 技术的容量和通量的增长已经使检测低丰度的序列成为可能，包括瘤胃古菌的序列。这个高通量、深度测序方法还获得了足够数量的产甲烷菌基因，基于这些基因可以重建甲烷菌的代谢通路，还可以根据新西兰高产和低产甲烷绵羊瘤胃样品中的转录本丰度估算基因表达活性（Shi 等，2014）。这项研究证实在高产甲烷和低产甲烷的动物中，产甲烷菌和甲烷生成途径中的基因丰度相似，但在高产甲烷的绵羊中，氢营养型甲烷生成途径基因的转录显著增加。这些结果表明了瘤胃原驻产甲烷菌的甲烷生成对氢气生成的响应。据推测，产甲烷菌基因表达的变化可能受绵羊瘤胃大小、饲料颗粒在瘤胃滞留时间和/或食糜的排空速率间接控制（Shi 等，2014），还可能与乳酸的生成和利用代谢有关（可以导致氢气和甲烷生成的减少）（Kamke 等，2016）。Wirth 等（2018）利用半导体测序技术（Ion Torrent）对 10 头荷斯坦奶牛的瘤胃液样品进行了宏基因组和转录组测序。尽管获得的序列数据量很低，但他们报告了一个由 47 个细菌属和一个单一的古菌属——甲烷短杆菌属（*Methanobrevibacter*）组成的核心微生物组。

最近，从 43 头苏格兰牛的 800 Gb 瘤胃宏基因组序列数据中重新组装了 913 个细菌和古菌基因组草图（Stewart 等，2018）。这项研究结合了宏基因组分箱技术和 DNA 交联方法，通过物理方式将 DNA 分子紧密地连接在一起，从而捕获长基因组片段，保证基因序列的连续性。这些宏基因组重组基因组（MAG）（被称为瘤胃未培养基因组（RUGs））中的许多代表了以前未测序的菌株和物种，总体上包含的 CAZYme 基因数量是 Hess 等（2011）研究中相应基因数量的两倍以上。重新组装的基因组还包括 28 个瘤胃未培养的产甲烷菌基因序列，其中 17 个与 *Mbb. gottschalkii* SGMT 分支相关，8 个与 *Mbb. ruminantium* 相关 RO 分支相关，1 个与从袋鼠中分离到的产甲烷菌菌株 WGK6 相关，2 个与 *Candidatus Methanomethylophilus* 密切相关。MAGs 的出现提高了公共数据库中瘤胃微生物基因组的覆盖率，提升了瘤胃生态学研究中宏基因组序列识别和分类的潜力。NGS 技术大大降低了 DNA 测序的成本，并可实现更多的产甲烷菌基因组测序。目前在 JGI GOLD 数据库中，有 17 个瘤胃产甲烷菌培养物来源的基因组序列，8 个非瘤胃产甲烷菌培养物来源的基因组，但有瘤胃源的 16S rRNA 基因序列，25 个瘤胃宏基因组数据组装的产甲烷菌 MAGs（表 4）。瘤胃产甲烷菌培养物来源的基因组主要由甲烷短杆菌属的代表菌株基因组组成，这也反映了这个属在瘤胃中的高丰度。然而，也有来源于数量较少的产甲烷菌菌群的基因组，包括甲烷杆菌科 Methanobacteriaceae（*Methanosphaera*，*Methanomicrobium*，*Methanobacterium*）、甲烷八叠球菌目 Methanosarcinales（*Methanosarcina*）和新命名的甲烷马赛球菌目 Methanomassiliicoccales 的三个基因组。以下章节描述了从基因组序列分析中获得的瘤胃产甲烷菌的主要特征。首先利用 *Mbb. ruminantium* M1 作为模型描述了氢营养型产甲烷菌，以 *Methanosphaera* 和 *Methanomassiliicoccales* sp. ISO4-H5 为例，对甲基营养型产甲烷菌进行了基因组分析。

5 氢营养型产甲烷菌：*Methanobrevibacter ruminantium* M1

氢营养型产甲烷菌是瘤胃中最主要的产甲烷菌。*Methanobrevibacter ruminantium* $M1^T$（DSM 1093）是一种氢营养型产甲烷菌，是在不同日粮条件下、不同反刍动物瘤胃中发现的甲烷短杆菌 OR 分支的代表菌株（Henderson 等，2015）。归属于 *Mbb. ruminantium* 的 16S rRNA 基因序列在全球瘤胃普查分析的样本中，平均占瘤胃古菌的 27%。*Mbb. ruminantium* $M1^T$是研究的最全面深入的瘤胃产甲烷菌菌株之一，由于 M1 培养物易于在实验室中进行常规培养（Smith 和 Hungate，1958；Bryant，1965），所以该培养物被选为第一个用于基因组测序的瘤胃产甲烷菌（Leahy 等，2010）。*Mbb. ruminantium* 的基因组信息证实该产甲烷菌具有氢营养型甲烷生成模式，具有编码利用氢气和二氧化碳以及甲酸生成甲烷所需的所有酶和大多数辅助因子（图1）。M1 没有编码甲基辅酶还原酶Ⅱ（*mcr*Ⅱ或*mrt*）的基因，这种酶是一种甲基辅酶还原酶Ⅰ（*mcr*）的同工酶（Reeve 等，1997），可以在高氢分压下介导甲烷生成，暗示 M1 适应在瘤胃低氢分压环境下生长。

M1 基因组中，编码氢营养甲烷生成途径的基因高度保守（图2），反映了该途径在瘤胃产甲烷菌能量生成中的重要性。利用 M1 基因组基因，通过与载体结合的中间产物和独特的辅助因子推导出甲烷生成途径，M1 可以通过氢化酶从氢气或者通过甲酸脱氢酶从甲酸获得还原当量。M1 包含 5 个氢化酶基因、两个膜结合能量转换 ［Ni-Fe］氢化酶（ehaABCDEFGHIJKLMNOPQR、ehbABCDEFGHIJKLMNOPQ）、一个 F_{420} 非还原性氢化酶（mvhABDG）、一个 F_{420} 还原性氢化酶（frhABB2DDG）和一个氢生成亚甲基四氢甲烷蝶呤脱氢酶（hmd），这些编码的酶均位于细胞质中。甲烷生成途径起始于甲烷呋喃（methanofuran，MF）结合二氧化碳，然后被含钨的甲酰甲烷呋喃脱氢酶（formylmethanofuran dehydrogenase，fwdABCDEFGH）催化还原生成甲酰甲烷呋喃。然后甲酰转移酶（ftr）将甲酰基转移到四氢甲烷蝶呤上，并催化生成甲酰基四氢甲烷蝶呤（Formyl-H_4MPT）。接着，N^5-甲酰基四氢甲烷蝶呤被 N^5，N^{10}-次甲基四氢甲烷蝶呤环水解酶（mch）催化水解为 N^5，N^{10}-次甲基四氢甲烷蝶呤。随后，在亚甲基四氢甲烷蝶呤脱氢酶（mtd）的催化下，次甲基四氢甲烷蝶呤被还原为亚甲基四氢甲烷蝶呤，并进一步还原为甲基四氢甲烷蝶呤，这两种反应均使用还原的 F_{420} 作为还原剂。第二种亚甲基四氢甲烷蝶呤脱氢酶（hmd）直接将亚甲基四氢甲烷蝶呤还原与氢气氧化连接起来，而不使用还原的 F_{420}。亚甲基四氢甲烷蝶呤还原酶（mer）利用还原的 F_{420} 将亚甲基四氢甲烷蝶呤还原为甲基四氢甲烷蝶呤。

然后，在四氢甲烷蝶呤 S-甲基转移酶（mtrABCDEFGH；mtrA2，mtrH2）的催化下，甲基从 N^5-甲基四氢甲烷蝶呤转移到辅酶 M（HS-CoM）上。甲基辅酶 M 还原酶（mcrABCDG；mrtAGDB）催化甲基辅酶 M 最终的还原，这个过程涉及两种辅酶：B9（HS-HTP）和 F_{430}。B9 在甲基辅酶 M 还原为甲烷以及 HS-CoM 和 HS-HTP（CoM-S-S-HTP）的混合二硫化物过程中充当电子供体。在甲基还原酶反应中形成的异二硫化物（CoM-S-S-HTP），在氢依赖的异二硫化物还原酶系统（hdrABCD/E）作用下，还原裂解再生成 HS-CoM 和 HS-HTP。

M1 基因组的一个非常不寻常的特征是它编码两个 NADPH 依赖的 F_{420} 脱氢酶（*nplG*1，

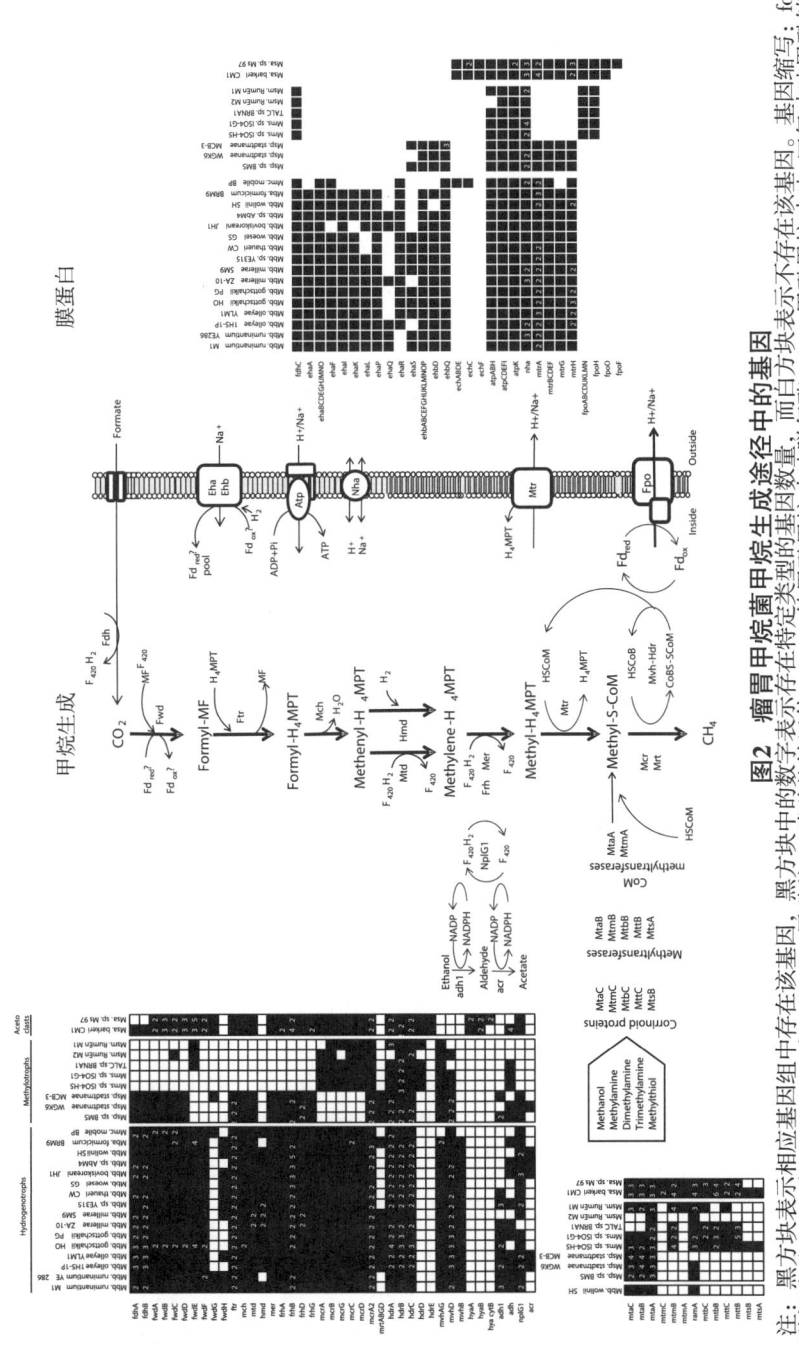

图2 瘤胃甲烷菌甲烷生成途径中的基因

注：黑方块表示相应基因组中存在该基因，白方块中的数字表示存在特定类型的基因数量，而白方块表示不存在该基因。基因缩写：fdh，甲酸脱氢酶（EC1.2.99.-）；fdhC，甲酸盐亚硝酸盐转运体；fwd，钨甲酰甲烷呋喃脱氢酶（EC1.2.99.-）；ftr，甲酰甲烷呋喃—四氢甲烷蝶呤脱甲酰基转移酶（EC2.3.1.101）；mch，次甲基四氢甲烷蝶呤环化水解酶（EC3.5.4.27）；mtd，亚甲基四氢甲烷蝶呤还原酶（EC1.5.99.9）；hmd，生成 H_2 的亚甲基四氢甲烷蝶呤脱氢酶（EC1.12.98.2）；mer，亚甲基四氢甲烷蝶呤还原酶（EC1.5.99.11），辅酶F420还原氢化酶；frh，辅酶F420还原氢化酶；fno，还原态辅酶F420:NADP氧化还原酶；mcr，甲基辅酶M还原酶；hdr，CoB-CoM异二硫化物还原酶（EC1.6.99.-）；mvh，甲基紫精还原酶；acr，NADP依赖乙醇脱氢酶；adh，乙醇脱氢酶；mtaA，甲醇：辅酶M甲基转移酶；mtaB，甲醇—钴胺素甲基转移酶；mtaC，甲醇类咕啉蛋白；mtbB，二甲胺甲基转移酶；mtbC，类咕啉蛋白；nplG1，还原态铁氧还蛋白；mtbA，甲胺：辅酶M甲基转移酶；mtmC，一甲胺类咕啉蛋白；mtmB，一甲胺甲基转移酶；mtmC，一甲胺类咕啉蛋白；mttB，三甲胺甲基转移酶；mttC，三甲胺类咕啉蛋白；VA型H+-甲胺酶；类咕啉甲基转移酶；mtsB，甲硫醇类活化蛋白；mtsA，甲基硫：辅酶M甲基转移酶；ech，膜结合氢化酶；fpo，F420H2脱氢酶/NADH醌氧化还原酶；atp，能量转换酶；hya，甲基苯丙哚还原蛋白；nha，Na+/K+/阳离子/H+逆向转运蛋白；H4MPT，四氢甲烷蝶呤；F420，辅因子F420；MF，甲烷呋喃；CoBS-SCoM，CoB-CoM异二硫化物；Fd，铁氧还蛋白。

2）基因和三个 NADP 依赖的乙醇脱氢酶（adh1，2，3）基因。这些基因编码的酶允许一些非瘤胃源产甲烷菌利用乙醇或者异丙醇生长。这一生物过程将 NADP$^+$ 依赖的醇类氧化过程与利用 F_{420} 因子将次甲基四氢甲烷蝶呤还原为甲基四氢甲烷蝶呤过程偶联（Berk 和 Thauer，1997）。尽管之前没有关于 M1 利用乙醇或甲醇生长的报道（Smith 和 Hungate，1958），但研究表明，在 H_2+CO_2 含量有限的情况下，乙醇和甲醇会刺激 M1 的生长，但这些醇在缺少氢气时不能支持 M1 生长（Leahy 等，2010）。由于 M1 缺乏甲醇利用基因（mta），似乎 M1 的 Adh 酶将乙醇的氧化与 F_{420} 的还原关联起来，从而增加细胞内还原的 F_{420} 量，并通过保留用于生成还原的 F_{420} 的氢气来刺激 M1 的生长。

与其他产甲烷菌一样，M1 具有特殊的生长要求，包括培养基中需要添加乙酸盐、2-甲基丁酸盐和辅酶 M（CoM）（Bryant 等，1971），M1 的基因组序列能够解释这些生长要求。在 M1 的细胞碳生物合成过程中需要乙酸参与，乙酸先被 acs 和 acsA 活化，生成乙酰辅酶 A，然后还原羧化为丙酮酸（porABCDEF）。由于缺少一个编码高丝氨酸激酶的基因，M1 不能从苏氨酸合成异亮氨酸，所以支链挥发性脂肪酸 2-甲基丁酸是异亮氨酸生物合成所必需的。辅酶 M 生物合成途径（comADE）中所需的 3 个基因在 M1 中缺失，这意味着它需要外源供应的辅酶 M 来生长。这些生长需求可能反映了 M1 对瘤胃环境的适应。其他瘤胃产甲烷菌能够合成辅酶 M，因此 M1 似乎放弃了自己的辅酶 M 生物合成途径，依赖于从周围环境中摄取这种辅助因子。相似地，瘤胃中能够产生大量的乙酸和支链挥发性脂肪酸，因此 M1 可以从环境中获得这些化合物而不需要自身合成，这在生物学上也是有意义的。M1 有一个基因编码未确定的转运蛋白 SSS 家族蛋白（mru1786），该蛋白可能具有乙酸钠转运蛋白的功能，协助摄取乙酸。

M1 的胞外膜作为抵抗外部环境的屏障，介导瘤胃代谢物和共在微生物的相互作用。瘤胃产甲烷菌的细胞表面蛋白质和多糖也引起了研究人员的兴趣，因为它们是能够开发减少反刍动物甲烷生成的抗产甲烷菌疫苗的潜在靶标（Leahy 等，2010）。M1 细胞的超微结构研究表明，细胞壁由三层组成；粗糙不规则的外层（推测由细胞壁糖聚合物、细胞壁相关蛋白和可能的其他成分组成）、由致密的新合成的假肽聚糖组成的薄的电子致密内层以及由假肽聚糖组成的较厚的电子密度较低的中间层（Zeikus 和 Bowen，1975；Miller，2001 年；Graham 和 Beveridge，1994）。在 M1 三层细胞壁之外，存在一系列大的黏附素样蛋白质，它们通过不同的细胞锚定结构域连接，包括一个含有蛋白质的 LPxTG 结构域和与假肽聚糖结合的多黏附素、C 端和 C 端跨膜锚定蛋白（Leahy 等，2010）。其中一些黏附素非常大，长度超过 5 000 个氨基酸（aa）的有两种，长度超过 3 000 个氨基酸的有 11 种。这些非常大的蛋白质数量以及合成它们所需的细胞资源表明，它们在 M1 中发挥着重要的功能，可能介导 M1 与环境和其他瘤胃微生物的相互作用。事实上，M1 与降解木聚糖、产氢气的细菌蛋白质分裂丁酸弧菌（*Butyrivibrio proteoclasticus*）共培养，M1 中 6 种黏附素样蛋白表达被上调。对共培养物的显微镜观察结果显示，M1 和 *B. proteolaticus* 细胞共聚集在一起（Leahy 等，2010）。其中一种黏附素（Mru_1499）含有 3 个细菌免疫球蛋白样 1（Big_1）结构域，通常在细菌黏附素中发现（Bodelón 等，2013）。Mru_1499 随后被证明在 M1 中起到黏附素的作用，并能够与多种瘤胃原虫（包括 *Epidinium* 和 *Entodinium*）以及细菌 *B. proteoclasticus* 结合（Ng 等，2016）。M1 基因组中编码着 60 多种黏附素样蛋白，其中一些包含 Big_1 结构域，尽管这些黏附素样蛋白的作用尚不清楚，但是似乎它们对瘤胃中的产甲烷菌群生态很重要。与其他氢营养型产甲烷菌一样，M1 需要氢气才能生长，而 Mru_1499 介导的与产氢气瘤胃微生物相互作用的研究表明，至少其中一些蛋白质参与了促进 M1 与其他产氢气细菌和原虫的相互作用。似乎它还可能与其他瘤胃微生物或宿主发生相互作用。

在 M1 的基因组中，约 70 Kb 的基因组区域在测序数据中所占比例过高。随后发现该区含有原噬菌体。该区域占基因组序列比例过高的原因可能是为获得 M1 基因组 DNA 而进行的大规模培养过程中原噬菌体被切除和复制造成的。原噬菌体区域在一个 62kb 的富集 GC（39%GC 含量）区域上含有 70 个 ORFs（mru0256-0325），并被命名为 Ø-mru。检测到编码整合、DNA 复制、DNA 包装、噬菌体衣壳、裂解性和溶原性功能的不同模块，并在裂解性模块内鉴定出一个编码假定的裂解酶——内异肽酶（endoisopeptidase PeiR，mru0320）的基因。PeiR 代表了一种新的酶，因为它与公共数据库中的任何序列都没有显著的同源性。随后，重组 PeiR 裂解 M1 细胞的能力在纯培养试验中被证实。目前已经将这个酶与聚羟基脂肪酸酯纳米颗粒结合，并发现其在多个产甲烷菌体外培养试验中能够抑制甲烷生成（Altermann 等，2018）。

M1 的一个未预见的新特征是存在两个大蛋白（mru0068 和 mru0351），显示了非核糖体肽合成酶（nonribosomal peptide synthetases，NRPSs）的独特结构域。NRPSs 可以产生种类繁多的小分子天然产物，具有如抗菌肽、铁载体、免疫抑制剂或抗肿瘤药物等生物应用价值（Amoutzias 等，2008），M1 的 NRPSs 是在古菌中首次报道。由 mru0068 编码的 NRPSs 预计包含两个模块，每个模块均包含缩合、腺苷酸化和硫酰化结构域。第一模块中缩合结构域的存在通常与制造 N-酰化肽的 NRPs 有关（Fischbach 和 Walsh，2006）。第二个模块后是一个末端硫酯酶结构域，该结构域被认为是从最后的硫酰化结构域释放肽。*Mru0068* 被编码两种丝氨酸磷酸酶（*mru0066*、*mru0071*）、一种抗 σ 因子拮抗剂（*mru0067*）和一种 MatE-efflux 家族蛋白（*mru0069*）的基因包围，这四种基因可能参与环境感应、NRPS 表达调节和 NRP 输出的过程。*Mru0068* 编码的全长蛋白质序列与来自沃氏共养单胞菌沃氏亚种 Göttingen 菌株（*Syntrophomonas wolfei* subsp. *Wolfei* strain *Göttingen*）的假定 NRPS 蛋白序列一致，而这株革兰氏阳性细菌与产甲烷菌共生（McInerney 等，1979）。

第二个 NRPS 基因（*mru0351*）包含 4 个模块和一个硫酯酶结构域。*mru0351* 的下游是另一个 MatE-efflux 蛋白家族（*mru0352*），可能参与 NRP 的输出。另外，位于基因组其他地方的较小基因簇（*mru0513-0516*）似乎编码 NRPS 相关功能。该簇包括磷酸泛酰巯基乙胺基转移酶（*mru0514*），其通过将磷酸泛酰巯基乙胺基添加到硫酰化结构域内保守的丝氨酸上来启动 NRPS。酰基转移酶（*mru0512*）可能参与到 NRP 酰化的过程中。丝氨酸磷酸酶（*mru0515*）、抗 σ 因子拮抗剂（*mru0513*）和抗 σ 调节因子丝氨酸/苏氨酸蛋白激酶（*mru0516*）可能在感知环境和 NRPS 调节中发挥作用。尽管每个 NRP 的产物未知，但对腺苷酰化结构域氨基酸序列的分析预测出了 10 个对底物结合和催化作用重要的残基。水平基因转移分析显示，这些基因可能起源于细菌（Darkhorse；Podell 和 Gaasterland，2007）。

6 其他氢营养型产甲烷菌

6.1 *Methanobrevibacter olleyae* YLM1

Mbb. olleyae YLM1 是从羔羊瘤胃中分离出来的（Skillman 等，2004）。YLM1 基因组在大小和总基因含量上与 *Mbb. ruminantium* M1 相当，表明这两种氢营养型产甲烷菌的基本代谢是相似的。YLM1 包含一个原噬菌体，仅 40 Kb，比 Ø-mru 稍小，编码 2 个规律间隔成簇短回文重复序列（CRISPR）。它有 64 个大的黏附素样蛋白，可介导与其他瘤胃微生物的相互作用。YLM1 有两个基因组区域，预测在细胞功能中起重要作用：一个是 10 个基因的插入子，编码 CoB-CoM 异二硫键还原酶（*hdrABC*）、参与辅酶 M 产生的酶（*comADE*），甲烷合成标志蛋白 16、黏附素样蛋白和两个假设蛋白；第二个是第 9 基因的插入子，编码一组甲酸脱氢酶、氢化酶成熟蛋白、一个 ATP 酶和甲基辅酶 M 还原酶 Ⅱ（*mrtBDGA*）操纵子，这个操纵子被认为能够让产甲烷菌在高氢气浓度下生长。能够合成自身 CoM 和在高氢气浓度下生长的能力可能让 YLM1 在瘤胃中占据与 M1 稍有不同的生态位。

6.2 *Methanobrevibacter* sp. AbM4

AbM4 最初是从绵羊的皱胃内容物中分离出来的，它与 *Mbb. wolinii* SH 最相似（16S rRNA 基因相似性 95%）。它被选为 *Mbb. woliniif* 分支的代表菌株进行基因组测序，结果显示其基因组小于 M1（2.0 Mb vs 2.93 Mb）。它编码的开放阅读框（ORFs）较少（1671 vs 2217），G+C 含量也较低（29% vs 33%），但与 *Mbb. wolinii* SH^T（JGI 项目编号：Gp0047017）的基因组大小相似（2.0 Mb）。全基因组共线性分析发现 AbM4 和 SH^T 基因组共线性非常高，含有许多相同的编码甲烷生成通路的基因（图 2），表明这两个菌株的代谢高度相似。AbM4 和 SH^T 仅使用 McrI 系统进行最终的甲基 CoM 还原，似乎适应了在低氢气水平下生长。AbM4 能够在甲醇上生长，但基因组没有甲醇-钴胺素甲基转移酶基因（*MtaABC*），而 SH^T 基因组编码了甲醇-类咕啉甲基转移酶（T523DRAFT_00271）和甲基钴（Ⅲ）甲醇特异性类咕啉蛋白：辅酶 M 甲基转移酶（T523DRAFT_00274）。与 M1 相同，AbM4 编码两个 NADP 依赖的 F_{420} 脱氢酶基因（*AbM4_0649* 和 *AbM4_1626*）和 3 个醇脱氢酶基因（*AbM4_1002*、*AbM4_1297* 和 *AbM4_1629*）（Leahy 等，2013），表明 AbM4 可以使用醇作为降低甲烷生成潜能的替代来源。当培养基同时添加 20 mM 浓度的乙醇和甲醇时，AbM4 能够在没有氢气的情况下生长，这一特性可以用来开发高通量、基于微量滴定板的生物测定法，筛选甲烷生成抑制化合物（Weimar 等，2017）。AbM4 具有完整的 CoM 生物合成途径，但不包含原噬菌体或任何 NRPS 基因。然而，它含有一个大的 CRISPR 区域和几个 Ⅰ 型和 Ⅱ 型限制修饰系统元件。不寻常的是，AbM4 具有两个 DNA 指导的 RNA 聚合酶 β′和 β″亚单位，它们连接在一起，这一特征以前仅在一些嗜热古菌中被发现。AbM4 编码黏附素样蛋白的基因较少，暗示它与 M1 占据的瘤胃生态位不同。

6.3 *Methanobrevibacter millerae* SM9

SM9 是从一只采食鲜牧草日粮的绵羊中分离得到的,其 16S rRNA 基因与 *Mbb. millerae* ZA-10^T（DSM 16643）相似性 99%。ZA-10^T 也有基因组序列（JGI 基因组 ID 2593339167）。这两株菌是 *Mbb. gottschalkii* 分支（*Mbb. gottschalkii*, *Mbb. millerae* 和 *Mbb. thaueri*；Janssen 和 Kirs, 2008）的代表性菌株。全球瘤胃普查中, *Mbb. gottschalkii* 分支在瘤胃中占优势地位, 平均丰度占总古菌 rRNA 基因序列的 42.4%（Henderson 等, 2015）。SM9 和 ZA-10^T 基因组高度共线性。这两株菌的甲烷生成基因与 *Mbb. ruminantium* M1 高度相似。然而, 它们的不同之处在于, SM9 和 ZA-10^T 具有甲基辅酶 M 还原酶基因（*mrtAGDB*）、两个拷贝的 F_{420} 依赖的亚甲基四氢甲烷蝶呤脱氢酶（*mtd*）和另一系列甲酸脱氢酶基因（*fdhAB*）。SM9 和 ZA-10 具有合成 CoM（*comABCDE*）和钴胺素的基因。虽然它们没有合成生物素的基因, 但拥有摄取生物素的 BioY 转运体。这两个基因组编码的蛋白与来源于 *Lactobacillus plantarum* 的胞外单宁酶（TanALp）具有高度同源性, 因此在含植物单宁较高的日粮中可能具有优势。SM9 和 ZA-10^T 都编码 NRPSs 的基因, 但 SM9 有 2 个, ZA-10^T 有 3 个, 尽管它们的 NRPS 与 M1 中不同的 NRPSs 表现出弱同源性, 但它们没有同源性。这些特性让 *Mbb. ruminantium* 和 *Mbb. millerae* 分支不同, 这可能让它们占据独立的瘤胃生态位, 并解释为什么在瘤胃的 16S rRNA 基因研究中总是发现这两个分支。

6.4 *Methanobacterium formicicum* BRM9

甲酸甲烷杆菌 BRM9 是从一头新西兰黑麦草/三叶草牧场放牧饲养的黑白花牛的瘤胃中分离出来的, 其 16S rRNA 基因序列与 *Mbb. formicicum* DSM 1535 相似性 99.8%（Jarvis 等, 2000）。BRM9 是氢营养型产甲烷菌, 仅利用甲酸、氢气和二氧化碳生成甲烷, 是一种非运动性的短杆菌, 但在生长后期往往变得更长且不规则。BRM9 编码甲烷生成途径的基因与甲烷杆菌科的其他产甲烷菌相似, 只是没有编码 [Fe]-氢化酶/脱氢酶（*hmd*）的基因, 它利用甲基辅酶 M 还原酶系统（*mrtAGDB*）。BRM9 特别显著的特征是, 它具有许多酶, 这些酶被预测参与氧化应激反应, 包括超氧化物歧化酶、过氧化氢酶/过氧化物酶和过氧化还原酶（烷基过氧化氢还原酶）。它还具有 3 个四氢嘧啶合成基因（*ectABC*）, 生成的四氢嘧啶（1,4,5,6-tetrahydro-2-methyl-4-pyrimidin ecarboxylic acid）具有调节细胞渗透压的功能, 帮助生物体在渗透胁迫下生存。BRM9 似乎可以非常精密地监测氧化还原电位、氧气和能量水平变化, 因为它有大量编码组氨酸激酶/反应调节蛋白信号转导系统元件的基因, 可感知这些环境条件变化。BRM9 似乎还编码几种氮同化机制：它有两种氨转运蛋白, 一个是谷氨酰胺合成酶（GS）/谷氨酸合成酶（谷氨酰胺/2-氧谷氨酸转氨酶, GOGAT）途径；令人惊讶的是, 它还有一个 *nif* 操纵子, 编码了一个固氮酶和一个固氮酶辅因子生物合成基因。

7 甲基营养型产甲烷菌

甲基营养型甲烷合成途径, 包含从甲醇、甲胺和甲硫醇生成甲烷。甲烷杆菌目、甲烷八叠球菌目和甲烷马赛球菌目中的产甲烷菌都有这种代谢途径。对前文提到的两个新的甲烷产生菌门 Bathyarchaeota 和 Verstraetearchaeota 的基因组序列分析显示，其均含有甲基营养型甲烷代谢机制（Evans 等，2015；Vanwonterghem 等，2016）。瘤胃中主要的甲基化合物是甲醇和甲胺。在喂饲干草和谷物的牛瘤胃中甲醇的浓度约为 0.8 mM（Vantcheva 等，1970），这些甲醇被认为是从日粮果胶的脱甲氧基化生成的，这一化学反应是由来源于细菌的果胶甲基酯酶（PME；EC3.1.1.11）催化的。*Lachnospira multipara*（Silley，1985，1986）和 *Butteriviribrio* 属和 *Prevotella* 属的细菌均含有这种酯酶。一甲胺、二甲胺和三甲胺主要是来源于植物磷脂酰胆碱的降解（Dawson 和 Hemington，1974），磷脂酰胆碱被瘤胃微生物降解成胆碱，胆碱再被进一步降解成甲胺。饲喂谷类日粮的奶牛瘤胃液中，甲胺浓度约为 85 μM（Ametaj 等，2010）。饲喂不同含量大麦日粮牛的瘤胃中，甲胺浓度范围是 28.5~703 μM（牛瘤胃代谢组数据库，2018）。

瘤胃中甲胺是如何产生的，目前相关信息很少。研究表明，标记的胆碱注入瘤胃后，被瘤胃微生物迅速代谢成三甲胺（TMA），标记的甲基最终沉积在甲烷中（Broad 和 Dawson，1976；Neill 等，1978）。最近的一项研究发现，瘤胃甲烷马赛球菌目菌群与尿中氧化三甲胺（TMAO）浓度存在负相关（Morgavi 等，2015），这可能是由于甲烷马赛球菌将 TMA 转化成了甲烷，改变了 TMA 进入肝脏被氧化成为 TMAO 的代谢路径。

7.1 属于甲烷马赛球菌目（Methanomassiliicoccales）的产甲烷菌

Tajima 等（2001）在研究牛瘤胃中古菌多样性的过程中，首次发现了一组与热源体古菌（*Thermoplasma*）相近，但系统进化距离又较远的未培养古菌的序列。Wright 等（2004）还从西澳大利亚州、澳大利亚昆士兰州的绵羊瘤胃（Wright 等，2006）以及安大略省和加拿大爱德华王子岛的饲喂马铃薯的牛（Wright 等，2007）中观察到了这些新的热源体古菌相关序列。Janssen 和 Kirs（2008）对所有可以得到的产甲烷菌 16S rRNA 基因序列进行荟萃分析最终发现，与热源体古菌相关的大量未培养瘤胃古菌占全球数据集的近 16%，并提出了瘤胃 C 簇（RCC）的名称。在后来的研究中，发现与热源体古菌相关的序列在来自中国青藏高原牦牛和黄牛的产甲烷菌克隆文库（Huang 等，2012）中占优势地位，而属于甲烷马赛球菌科的 *Mmc*. Group 10 和 *Mmc*. Group 4 两个菌群的成员在牦牛和藏羊的 16S rRNA 基因序列中占优势，*Mmc*. Group 12 在引进牛和杂交羊中占优势（Huang 等，2016）。在中国晋南牛中，热源体古菌相关 16S rRNA 基因序列广泛分布于瘤胃上皮、瘤胃固相和液相部分（Pei 等，2010）。在各种反刍动物的瘤胃样品中也发现了热源体古菌相关序列，在某些情况下是瘤胃中最优势的产甲烷菌菌群（King 等，2011；Gu 等，2011；Franzolin 等，2012；Jeyanathan 等，2011；Chen 等，2009）。

到目前为止，RCC-热源体古菌相关序列的所有信息都是从 16S rRNA 基因文库中推断出来的，但正如前面提到的，从牛瘤胃中富集得到一个属于热源体古菌相关谱系 C（TALC）组的菌株 BRNA1（Denman 等，2011；数据未发表），从鸡的粪便中获得一个纯培养物（*Methanoplasma gallocaecorum* DOK-1，Padmanabha 等，2013），这两株菌均利用氢气将甲胺和甲醇还原为甲烷。尽管 BRNA1 菌株是一个多菌株富集培养物的一部分，但已经从混合培养物的宏基因组数据中重组装了基因组草图，基因回补证实其能够产生甲烷。Poulsen 等（2013）发现，牛瘤胃中的热源体古菌是甲基营养型产甲烷菌，并且在泌乳牛中添加菜籽油可减少其数量。使用宏转录组学方法研究发现，热源体古菌 16S rRNA 和 mcr 基因转录下调，伴随着利用甲胺生成甲烷的途径中相关酶的 mRNAs 降低，这一结果暗示这一菌群利用甲胺。体外培养试验证实了甲基营养型甲烷生成途径。在体外培养试验中，向培养基中添加甲胺，增强了热源体古菌的生长和甲烷的生成。随后，利用瑞士褐牛瘤胃液三甲胺富集培养物宏基因组序列重组了两个甲烷马赛球菌目产甲烷菌基因组（Söllinger 等，2016）。RumEn M1 MAG 由 182 个 contigs 组成，估计基因组大小为 2.21Mbp，并且它的 16S rRNA 基因序列证实它是甲烷马赛球菌科的一个新成员。第二个 MAG，RumEn M2，包含 18 个 contigs，估计的基因组大小为 1.28 Mbp，是甲烷马赛球菌目胃肠道（GIT）簇的一个新成员。利用三甲胺和甲醇生成甲烷的牛瘤胃内容物的富集培养物中含有另一个甲烷马赛球菌目的成员 "*Candidatus Methanomethylophilus*" sp. 1R26（Noel 等，2016）。最近在体试验发现，给牛饲喂产甲烷菌抑制剂 3-硝酸酯丙醇（3-nitrooxy propanol，3-NOP）降低了瘤胃中甲烷马赛球菌的丰度（Martinez Fernandez 等，2018）。同时发现，三甲胺在饲喂 3-NOP 的动物瘤胃中积累（1.2 mmol/L），对照动物的瘤胃中三甲胺浓度为 0.33 mmol/L，这证实了甲烷马赛球菌在三甲胺代谢中的作用。

大约在同一时间，Dridi 等（2012）从人类粪便中分离出一株与热源体古菌相关的菌株，并将其命名为 *Methanomassiliicoccus luminyensis*，而 Borrel 等（2012）对另一个 RCC 相关的菌株——*Candidatus Methanomethylophilus alvus Mx*1201 的基因组进行了测序。Paul 等（2012 年）将所有热源体古菌相关的 16S rRNA 基因和 mcrA 基因序列，来自高等白蚁和千足虫的富集培养物以及 *Mms. luminyensis* 聚集到一起，并提出了产甲烷菌的第 7 个目，即甲烷马赛球菌目 Methanomassiliicoccales。

Jeyanathan 等（2011）在之前提到的产甲烷菌培养工作中，产生了 3 个仅含有 RCC 产甲烷菌的混合富集培养物。ISO4-H5 混合培养物包含一个单一甲烷马赛球菌菌株和一个共生细菌 *Succinivibrio dextrinisolvens*，对该混合培养物进行了宏基因组测序。将其中的甲烷马赛球菌相关基因序列重组进入了一个含有 47 个 contigs 的 scaffold 中，获得了基因组完成图（Li 等，2016b）。*Methanomassiliicoccales* sp. ISO4-H5 的基因组大小为 1.9 Mb，G+C 含量 54%，与 Mx1201、B10 和 BRNA1 的基因组相似。ISO4-H5 基因组编码依赖氢还原甲基底物的甲基营养型甲烷生成途径，但是没有将甲基底物氧化成二氧化碳的能力。通过基因组预测发现，ISO4-H5 可以利用氢气、甲醇、一甲胺、二甲胺、三甲胺和 3-甲硫基丙酸甲酯，ISO4-H5 比瘤胃中其他甲基营养型产甲烷菌具有更高的代谢多样性。

与其他甲烷马赛球菌描述的相同，ISO4-H5 利用异二硫化物还原酶（*hdrABC*）和甲基紫精氢化酶（*mvhADG*）以及氢化酶生成的还原当量完成 CoM 的循环利用，这一过程与 F_{420} 脱氢酶 Fpo 样复合物偶联以产生膜电位和 ATP。ISO4-H5 没有编码 CoM 生物合成的基因，这表明其已适应瘤胃环境，在瘤胃环境中，其他产甲烷菌可产生 CoM。它还缺少编码辅因子 F_{420} 合成的基因，这与显微观察结果一致，即它在 420 nm 的激发光下不会发出蓝绿荧光。ISO4-H5 可产生吡咯赖氨酸，并具有特异的氨基酰-tRNA 合成酶，可使终止密码子 UAG 通读。在 46 个 ISO4-H5 基因中发现了表明吡咯赖氨酸插入的框内密码子，包括许多编码甲胺和甲醇利用的基因。

Jeyanathan 等（2011）对获得的 ISO4-G1 富集培养物也进行了基因组测序，基因组是一个单个的 1.6 Mb 环形染色体，GC 含量为 55.5%，预测的编码蛋白质的基因 1 501 个，但不包含任何可识别的质粒、原噬菌体或 CRISPR 序列（Kelly 等，2016c）。与 ISO4-H5 一样，ISO4-G1 以甲醇和甲胺为底物，通过依赖氢气的甲基营养型甲烷生成途径产生能量。它也缺乏合成 CoM 和色氨酸的基因，但编码许多转运蛋白，包括 15 个 ABC 转运蛋白，这些转运蛋白被预测参与 Fe^{3+} 或铁载体的摄取。ISO4-G1 拥有一个完整的吡咯赖氨酸生物合成操纵子（AUP07_0651-AUP07_0654）和特异的氨酰-tRNA 合成酶，以及 25 个编码吡咯赖氨酸蛋白质的基因，其中 9 个是一/二/三甲胺：类咕啉甲基转移酶。ISO4-G1 还编码一个 NRPS，被预测包含一个吡咯赖氨酸残基，但其功能尚不清楚。

ISO4-H5 和 ISO4-G1 的甲基化合物甲烷生成途径可能每生成两个甲烷分子仅泵送一个离子穿过细胞膜，以形成跨膜离子梯度，而利用相同代谢途径的 *Msp. stadtmanae* 泵送两个离子。因此，据预测，ISO4-H5 和 ISO4-G1 的 ATP 产量要低得多，因此生长速度要慢于 *Methanosphaera* spp.，这与 ISO4-H5 和 ISO4-G1 在培养液中极低的细胞密度一致。然而，甲烷形成的不同化学计量比（表1）意味着 ISO4-H5 和 ISO4-G1 可能具有较低的氢气利用阈值。与瘤胃中主要的甲基营养竞争者 *Methanosphaera* spp. 相比，它们能够在更低的氢气浓度下生长。这在生态学上区分了 *Methanomassiliicoccales* 和 *Methanosphaera*。当反刍动物采食或反刍时，瘤胃中大范围的发酵条件可能为这两组甲基营养产甲烷菌在瘤胃中的生长和共存提供了机会。

7.2 *Methanosphaera* sp. BMS

Methanosphaera sp. BMS 是从澳大利亚婆罗门牛的瘤胃内容物的甲醇富集培养物中分离出的（Hoedt，2017）。生长试验结果显示，BMS 是一种依赖氢气的甲基营养型产甲烷菌，仅利用甲醇，不利用甲胺、乙酸、甲酸、丙醇或氢气/二氧化碳。BMS 没有利用乙醇替代氢气来支持甲烷生成的能力，与在从袋鼠上获得的菌株上观察到的结果一致（*Methanosphaera* sp. *WGK6*；Hoedt 等，2016）。BMS 的基因组（2.9 Mb，2204 个蛋白质编码序列）比人类和袋鼠来源的甲烷球形菌菌株的基因组（均为约 1.7 Mb）大。KEGG 通路分析甲烷的代谢结果显示，BMS 编码氢驱动的甲醇还原所需的所有基因，并编码 II 型 *mrt* 系统以及辅酶 M 甲基转移酶亚基 *mtrABC*。从多种宏基因组数据集中重组了 *Methanosphaera* sp. MAGs，对这些基因组进行分析发现，依赖氢气将甲醇还原为甲烷似乎是所有 *Methanosphaera* 基因组的共同特征。然而，如前所述，对袋鼠源 *Methanosphaera* 菌株 WGK6 的培养试验结果显示，它可以将乙醇作为还原甲醇生成甲烷的唯一还原力来源，这个过程需要醇和醛脱氢酶的参与（Hoedt 等，2016）。这些醇和醛脱氢酶同源基因也在从低产甲烷绵羊宏基因组组装而成的 *Methanosphaera* sp. MAG（SHI1033）中检测到（Shi 等，2014），但在任何其他 MAG 或 *Methanosphaera* 属 BMS 菌株以及 *Methanosphaera stadtmanae* DSMZ 3091T 基因组中均未检测到。这就提出了一个有趣的可能性，即甲烷球形菌的乙醇驱动甲烷生成作用是低产甲烷胃肠道环境的一个普遍特征。

8 乙酸营养型产甲烷菌：*Methanosarcina* sp. CM1

在瘤胃中利用乙酸生成甲烷的量很少（Hungate 等，1970），因为能够利用乙酸生成甲烷的产甲烷菌（通过裂解乙酸的甲烷生成途径）在这种环境中极为罕见（<0.015%；Henderson 等，2015）。虽然有这些发现，但已经有甲烷八叠球菌曾多次从瘤胃中分离出来的例证（Patterson 和 Hespell，1979；McInerney 等，1981；Jarvis 等，2000），因此，虽然它们在瘤胃菌群中的比例很少，但并不妨碍对它们的培养。甲烷八叠球菌 CM1（16S rRNA 基因与 *Ms. barkeri* MST DSM 800 相似性 99%）是从采食黑麦草/白三叶草的新西兰黑白花奶牛的瘤胃中分离出来的。CM1 在肉汤培养基中生长成大的细胞聚集体，其形态特征与 *Ms. Barkeri* 相似。它可以利用氢气及氧气、乙酸、甲醇和甲胺生长并产生甲烷。尽管它编码甲酸脱氢酶基因，但它不利用甲酸。CM1 基因组比其他瘤胃产甲烷菌大得多（4.5 Mb；3656 个预测基因），编码乙酸营养、氢营养和甲基营养型甲烷生成途径。CM1 不编码［Fe］-氢化酶/脱氢酶（hmd）或 mrt。虽然 CM1 编码了 CoM 生物合成途径，但该途径与其他已测序的瘤胃甲烷杆菌不同。CM1 的几个特点使它成为瘤胃中特殊的存在。它编码吡咯赖氨酸生物合成酶（pylSBCD）和一种特异的吡咯赖氨酸氨基酰 tRNA 合成酶。基因预测显示它的甲基转移酶氨基酸序列中含有吡咯赖氨酸。

预测其细胞表面由 S 层蛋白质组成，而不是在其他甲烷杆菌目中发现的假肽聚糖。它有一个糖苷水解酶家族 18（GH18；几丁质酶）蛋白质，预测从细胞内分泌，并可能介导与含有几丁质的瘤胃厌氧真菌的相互作用。尽管 CM1 有一个编码鞭毛（*flaB-flaJ*）的操纵子和相关的趋化基因，但 CM1 可能不是一个运动的细胞，这表明这些鞭毛蛋白在 CM1 中可能有不同的用途。与 *Mb. formicicum* BRM9 相似，CM1 包含两个不同的 *nif* 操纵子，它们编码固氮酶和固氮酶辅因子生物合成基因，这些酶和辅因子在固氮中起作用。尽管 CM1 的菌群数量不高，并且在大多数情况下在瘤胃中几乎检测不到，然而，CM1 具有利用多种底物的能力，这标志着 CM1 是一个多面手，能够占据很多不同的生境。然而，当利用果胶诱导瘤胃中的甲醇产生时，甲烷八叠球菌可以被富集和分离（Pol 和 Demeyer，1988）。

9 结论

对反刍动物甲烷排放引起全球变暖的认识，重新引起了人们对瘤胃产甲烷菌研究的兴趣，并试图找到降低其活性的方法。本章对瘤胃微生态学研究的总结表明，目前已经有了很多瘤胃产甲烷菌培养物，但只有少数被成功纯化，并且其中仅有少部分被充分描述并保存在国际公认的、可获得的菌种保藏中心。这迫切需要新的举措，如 Hungate 1000 项目（Seshadri 等，2018），获得更多的瘤胃产甲烷菌培养物，并进行详细的表型和基因组描述。需要更全面的产甲烷菌菌株目录，以便更好地描述其生理特性，并进一步了解不同甲烷形成模式的细节。尽管产甲烷菌基因组序列在某种程度上解释了为什么在瘤胃中存在多个产甲烷菌类型，但还需要进行大量工作。在菌株水平上，阐明它们的生长特征以及它们如何响应能源物质氢气浓度变化，以及在较小程度上，对不同水平短链醇作出响应。此外，了解瘤胃中氢产生菌和甲基化合物产生菌与产甲烷菌的互作关系对于确定种间氢转移的作用以及该过程控制甲烷生成的程度至关重要。

甲烷生成途径中酶的大多数生化过程已经在非瘤胃产甲烷菌中完成，主要是 *Methanobacterium thermoautotrophicum* 和 *Methanosarcina barkeri*（Thauer 等，2008）。重要的是，至少要确认瘤胃产甲烷菌甲烷生成的生化过程与其他地方描述的机制和途径相同，最好是充分阐明主要瘤胃产甲烷菌群的酶学。通过对特定产甲烷菌基因敲除，从而研究基因功能。在非瘤胃古菌中，已经开发出了基因改造系统，这些系统随后成为古菌微生物学研究的主要工具（Metcalf 等，1997；Rother 和 Metcalf，2005）。目前尚不存在瘤胃产甲烷菌的基因操纵系统，但如果要在确定大量未知功能的产甲烷菌基因的功能方面取得真正进展，就需要这些系统。瘤胃产甲烷菌中存在一些基因操作系统的原材料，如质粒、噬菌体和CRISPR-Cas 系统，因此需要共同努力使这些基因元件发挥作用，以揭示瘤胃产甲烷菌生物学特征的细节。

未来下一代高通量测序技术无疑将在描述瘤胃产甲烷菌菌群过程中发挥重要作用。下一代测序技术已经使培养的产甲烷菌菌种的基因组测序成为常规手段，并增强了我们对产甲烷菌标记基因进行登记和分析的能力。他们还提供了强大的新工具，以便越来越深入地检索甲烷宏基因组序列，从而允许通过宏基因组重组直接从瘤胃样本获取基因组信息。预计这些新方法将产生大量的标记基因、基因组和宏基因组序列数据，这些数据需要在产甲烷菌生物学背景下进行解释，获得的认识和理解才能应用于瘤胃甲烷减排。在专用数据库中，按分类学和功能类别构建的注释良好的参考基因和全基因组库，对于准确解释这些新数据集至关重要。不断增加的数据量还要求极大地扩展计算能力以及更好的分析软件，以便识别、提取和分析这些新数据集。虽然有非常大的挑战，但这些新数据集提供的机会也是巨大的，预示着未来对瘤胃产甲烷菌生物学的认识将有重大突破。

10 参考文献

Akhmanova, A., Voncken, F., et al., 1998. Ahydrogenosome with a genome. Nature 396 (6711), 527-8.

Altermann, E., Schofield, L. R., Ronimus, R. S., et al., 2018. Inhibition of rumen methanogens by a novel archaeal lytic enzyme displayed on tailored bionanoparticles. Frontiers in Microbiology 9, 2378.

Ametaj, B. N., Zebeli, Q., Saleem, F., et al., 2010. Metabolomics reveals unhealthy alterations in rumen metabolism with increased proportion of cereal grain in the diet of dairy cows. Metabolomics 6 (4), 583-94.

Amoutzias, G. D., Van de Peer, Y. and Mossialos, D. 2008. Evolution and taxonomic distribution of non-ribosomal peptide and polyketide synthases. Future Microbiology 3 (3), 361-70.

Anderson, R. C., Rasmussen, M. A. and Allison, M. J. 1993. Metabolism of the plant toxins nitropropionic acid and nitropropanol by ruminal microorganisms. Applied and Environmental Microbiology 59 (9), 3056-61.

Anderson, R. C., Rasmussen, M. A. and Allison, M. J. 1996. Enrichment and isolation of a nitropropanol-metabolizing bacterium from the rumen. Applied and Environmental Microbiology 62 (10), 3885-6.

Anderson, R. C., Rasmussen, M. A., Jensen, N. S., et al., 2000. *Denitrobacterium detoxificans* gen. nov., sp. nov., a ruminal bacterium that respires on nitrocompounds. International Journal of Systematic and Evolutionary Microbiology 50 (2), 633-8.

Balch, W. E. and Wolfe, R. S. 1976. New approach to the cultivation of methanogenic bacteria: 2-mercaptoethanesulfonic acid (HS-CoM) -dependent growth of *Methanobacterium ruminantium* in a pressurized atmosphere. Applied and Environmental Microbiology 32 (6), 781-91.

Balch, W. E., Fo, G. E., Magrum, L. J., et al., 1979. Methanogens: reevaluation of a unique biological group. Microbiological Reviews 43 (2), 260-96.

Barker, H. A. 1936a. On the biochemistry of the methane fermentation. Archives of Microbiology 7 (1-5), 404-19.

Barker, H. A. 1936b. Studies upon the methane-producing bacteria. Archives of Microbiology 7 (1-5), 420-38.

Barker, H. A. 1940. Studies upon the methane fermentation. IV. The isolation and culture of *Methanobacterium omelianski*. Antonie van Leeuwenhoek 6, 201-20.

Barker, H. A. 1941. Studies of the methane fermentation. V. Biochemical activities of *Methanobacterium omelianskii*. Journal of Biological Chemistry 137, 153-67.

Barker, H. A., Ruben, S. and Kamen, M. D. 1940. The reduction of radioactive carbon dioxide by methane-producing bacteria. Proceedings of the National Academy of Sciences of the United States of America 26 (6), 426-30.

Battumur, U., Lee, M., Bae, G. S., et al., 2019. Isolation and characterization of a new Methanoculleus bourgensis strain KOR-2 from the rumen of Holstein steers. Asian Australasian Journal of Animal Sciences 32 (2), 241-8.

Bauchop, T. and Mountfort, D. O. 1981. Cellulose fermentation by a rumen anaerobic fungus in both the absence and the presence of rumen methanogens. Applied and Environmental Microbiology 42 (6), 1103-10.

Beijer, W. H. 1952. Methane fermentation in the rumen of cattle. Nature 170 (4327), 576-7.

Belanche, A., de la Fuente, G. and Newbold, C. J. 2014. Study of methanogen communities associated with different rumen protozoal populations. FEMS Microbiology Ecology 90 (3), 663-77.

Berk, H. and Thauer, R. K. 1997. Function of coenzyme F_{420}-dependent NADP reductase in methanogenic archaea containing an NADP-dependent alcohol dehydrogenase. Archives of Microbiology 168 (5), 396-402.

Bodelón, G., Palomino, C. and Fernández, L. Á 2013. Immunoglobulin domains in *Escherichia coli* and other enterobacteria: from pathogenesis to applications in antibody technologies. FEMS Microbiology Reviews 37 (2), 204-50.

Bokranz, M., Katz, J., Schröder, I., et al., 1983. Energy metabolism and biosynthesis of *Vibrio succinogenes* growing with nitrate or nitrite as terminal electron acceptor. Archives of Microbiology 135 (1), 36-41.

Borrel, G., Harris, H. M. B., Tottey, W., et al., 2012. Genome sequence of "*Candidatus Methanomethylophilus alvus*" Mx1201, a methanogenic archaeon from the human gut belonging to a seventh order of methanogens. Journal of Bacteriology 194 (24), 6944-5.

Böttger, E. C. 1989. Rapid determination of bacterial ribosomal RNA sequences by direct sequencing of enzymatically amplified DNA. FEMS Microbiology Letters 53 (1-2), 171-6.

Bovine Rumen Metabolome Database. 2018. Available at: http://www.rumendb.ca (accessed on 29 June 2018).

Breznak, J. A. and Kane, M. D. 1990. Microbial H_2/CO_2 acetogenesis in animal guts: nature and nutritional significance. FEMS Microbiology Reviews 7 (3-4), 309-13.

Broad, T. E. and Dawson, R. M. 1976. Role of choline in the nutrition of the rumen protozoon Entodinium caudatum. Journal of General Microbiology 92 (2), 391-97.

Bryant, M. P. 1965. Rumen methanogenic bacteria. In: Dougherty, R. W., Allen, R. S., Burroughs, R. W., Jacobson, N. L. and McGilliard, A. D. (Eds), Physiology of Digestion of the Ruminant. Butterworths Publishing, Inc., Washington DC, pp. 411-8.

Bryant, M. P. 1972. Commentary on the Hungate technique for culture of anaerobic bacteria. American Journal of Clinical Nutrition 25 (12), 1324-8.

Bryant, M. P., Wolin, E. A., Wolin, M. J., et al., 1967. Methanobacillus omelianskii, a symbiotic association of two species of bacteria. Archives of Microbiology 59 (1), 20-31.

Bryant, M. P., Tzeng, S. F., Robinson, I. M., et al., 1971. Nutritional requirements of methanogenic bacteria. In: Pohland, F. G. (Ed.), Anaerobic Biological Treatment Processes. Advances in Chemistry Series 105. American Chemical Society, Washington DC, pp. 23-40.

Chagan, I., Tokura, M., Jouany, J. P., et al., 1999. Detection of methanogenic archaea associated with rumen ciliate protozoa. Journal of General and Applied Microbiology 45 (6), 305-8.

Chaucheyras, F., Fonty, G., Bertin, G., et al., 1995. In vitro H_2 utilization by a ruminal acetogenic bacterium cultivated alone or in association with an Archaea methanogen is stimulated by a probiotic strain of *Saccharomyces cerevisiae*. Applied and Environmental Microbiology 61 (9), 3466-7.

Cheng, Y. F., Mao, S. Y., Liu, J. X., et al., 2009. Molecular diversity analysis of rumen methanogenic archaea from goat in eastern China by DGGE methods using different primer pairs. Letters in Applied Microbiology 48 (5), 585-92.

Cord-Ruwisch, R., Seitz, H. J. and Conrad, R. 1988. The capacity of hydrogenotrophic anaerobic bacteria to compete for traces of hydrogen depends on the redox potential of the terminal electron acceptor. Archives of Microbiology 149 (4), 350-7.

Dawson, R. M. C. and Hemington, N. L. 1974. Digestion of grass lipids and pigments in the sheep rumen. British Journal of Nutrition 32 (2), 327-40.

De la Fuente, G., Belanche, A., Girwood, S. E., et al., 2014. Pros and cons of Ion-Torrent next generation sequencing versus Terminal Restriction Fragment Length Polymorphism T-RFLP for studying the rumen bacterial community. PLoS ONE 9 (7), e101435.

Denman, S. E., Tomkins, N. W. and McSweeney, C. S. 2007. Quantitation and diversity analysis of ruminal methanogenic populations in response to the antimethanogenic compound bromochloromethane. FEMS Microbiology Ecology 62 (3), 313-22.

Denman, S. E., Evans, P., Bragg, L., et al., 2011. Thermoplasmatales-like gut symbionts are pyrrolysine-dependent-methanogens. Genbank CP002916.1, unpublished.

Dridi, B., Fardeau, M. L., Ollivier, B., et al., 2012. *Methanomassiliicoccus luminyensis* gen. nov., sp. nov., a methanogenic archaeon isolated from human faeces. International Journal of Systematic and Evolutionary Microbiology 62 (8), 1902-7.

Evans, P. N., Parks, D. H., Chadwick, G. L., et al., 2015. Methane metabolism in the archaeal phylum Bathyarchaeota revealed by genome-centric metagenomics. Science 350 (6259), 434-8.

Evans, P. N., Boyd, J. A., Leu, A. O., et al., 2019. An evolving view of methane metabolism in the Archaea. Nature Reviews. Microbiology 17 (4), 219-32

Finlay, B. J., Esteban, G., Clarke, K. J., et al., 1994. Some rumen ciliates have endosymbiotic methanogens. FEMS Microbiology Letters 117 (2), 157-61.

Fischbach, M. A. and Walsh, C. T. 2006. Assembly-line enzymology for polyketide and non-ribosomal peptide antibiotics: logic, machinery and mechanisms. Chemical Reviews 106 (8), 3468-96.

Fouts, D. E., Szpakowski, S., Purushe, J., et al., 2012. Next generation sequencing to define prokaryotic and fungal diversity in the bovine rumen. PLoS ONE 7 (11), e48289.

Franzolin, R., St-Pierre, B., Northwood, K., et al., 2012. Analysis of rumen methanogens diversity in water buffaloes (Bubalus bubalis) under three different diets. Microbial Ecology 64 (1), 131-9.

Frey, J. C., Pell, A. N., Berthiaume, R., et al., 2010. Comparative studies of microbial populations in the rumen, duodenum, ileum and faeces of lactating dairy cows. Journal of Applied Microbiology 108 (6), 1982-93.

Fricke, W. F., Seedorf, H., Henne, A., et al., 2006. The genome sequence of *Methanosphaera stadtmanae* reveals why this human intestinal archaeon is restricted to methanol and H_2 for methane formation and ATP synthesis. Journal of Bacteriology 188 (2), 642-58.

Genthner, B. R., Davis, C. L. and Bryant, M. P. 1981. Features of rumen and sewage sludge strains of *Eubacterium limosum*, a methanol - and $H_2 - CO_2$ - utilizing species. Applied and Environmental Microbiology 42 (1), 12-9.

Gilbert, R. A., Ouwerkerk, D., Zhang, L. H., et al., 2010. *In vitro* detection and primary cultivation of bacteria producing materials inhibitory to ruminal methanogens. Journal of Microbiological Methods 80 (2), 217-8.

Graham, L. L. and Beveridge, T. J. 1994. Structural differentiation of the *Bacillus subtilis* 168 cell wall. Journal of Bacteriology 176 (5), 1413-21.

Greening, R. C. and Leedle, J. A. Z. 1989. Enrichment and isolation of *Acetitomaculum ruminis*, gen. nov., sp. nov.: acetogenic bacteria from the bovine rumen. Archives of Microbiology 151 (5), 399-406.

Gu, M. J., Alam, M. J., Kim, S. H., et al., 2011. Analysis of methanogenic archaeal communities of rumen fluid and rumen particles from Korean black goats. Animal Science Journal = Nihon Chikusan Gakkaiho 82 (5), 663-72.

Gupta, A. and Chaudhary, U. B. 2010. Isolation and characterization of *Methanobrevibacter smithii* GMS-01 from rumen of goats. Indian Veterinary Journal 87, 1009-12.

Henderson, C. 1980. The influence of extracellular hydrogen on the metabolism of Bacteroides ruminicola, *Anaerovibrio lipolytica* and *Selenomonas ruminantium*. Microbiology 119 (2), 485-91.

Henderson, G., Naylor, G. E., Leahy, S. C., et al., 2010. Presence of novel, potentially homoacetogenic bacteria in the rumen as determined by analysis of formyltetrahydrofolate synthetase sequences from ruminants. Applied and Environmental Microbiology 76 (7), 2058-66.

Henderson, G., Cox, F., Ganesh, S., et al., 2015. Rumen microbial community composition varies with diet and host, but a core microbiome is found across a wide geographical range. Scientific Reports 5, 14567.

Hess, M., Sczyrba, A., Egan, R., et al., 2011. Metagenomic discovery of biomass-degrading genes and genomes from cow rumen. Science 331 (6016), 463-7.

Hoedt, E. C. 2017. Functional and comparative studies of members of the genus *Methanosphaera*, and their adaptations to the gut environment. PhD Thesis. The University of Queensland, Brisbane, Australia.

Hoedt, E. C., ÓCuív, P. Ó, Evans, P. N., et al., 2016. Differences down-under: alcohol-fuelled methanogenesis by archaea present in Australian macropodids. ISME Journal 10 (10), 2376-88.

Hoedt, E. C., Parks, D. H., Volmer, J. G., et al., 2018. Culture-and metagenomics-enabled analyses of the *Methanosphaera* genus reveals their monophyletic origin and differentiation according to genome size. ISME Journal 12 (12), 2942-53.

Hook, S. E., Northwood, K. S., Wright, A. D. G., et al., 2009. Long-term monensin supplementation does not significantly affect the quantity or diversity of methanogens in the rumen of the lactating dairy cow. Applied and Environmental Microbiology 75 (2), 374-80.

Howard, B. H. and Hungate, R. E. 1976. Desulfovibrio of the sheep rumen. Applied and Environmental Microbiology 32 (4), 598-602.

Huang, X. D., Tan, H. Y., Long, R. J., et al., 2012. Comparison of methanogen diversity of yak (Bos grunniens) and cattle (Bos taurus) from the Qinghai-Tibetan Plateau, China. BMC Microbiology 12, 237.

Huang, X. D., Martinez-Fernandez, G., Padmanabha, J., et al., 2016. Methanogen diversity in indigenous and introduced ruminant species on the Tibetan Plateau. Archaea 2016, 5916067.

Huisingh, J., McNeill, J. J. and Matrone, G. 1974. Sulfate reduction by a *Desulfovibrio* species isolated from sheep rumen. Applied Microbiology 28 (3), 489-97.

Hungate, R. E. 1950. The anaerobic mesophilic cellulolytic bacteria. Bacteriological Reviews 14 (1), 1-49.

Hungate, R. E. 1966. The Rumen and Its Microbes. Academic Press, New York, NY and London, UK.

Hungate, R. E. 1969. A roll-tube method for cultivation of strict anaerobes. Methods in Microbiology 3B, 117-32.

Hungate, R. E. 1975. The rumen microbial ecosystem. Annual Review of Ecology and Systematics 6 (1), 39-66.

Hungate, R. E., Smith, W., Bauchop, T., et al., 1970. Formate as an intermediate in the rumen fermentation. Journal of Bacteriology 102, 384-97.

Huser, B. A., Wurmann, K. and Zehnder, A. J. B. 1982. *Methanothrix soehngenii* gen. nov. sp. nov., a new acetotrophic non-hydrogen-oxidizing methane bacterium. Archives of Microbiology 132 (1), 1-9.

Hutten, T. J., Bongaerts, H. C. M., Van der Drift, C., et al., 1980. Acetate, methanol and carbon dioxide as substrates for growth of *Methanosarcina barkeri*. Antonie van Leeuwenhoek 46 (6), 601-10.

Iino, T., Tamaki, H., Tamazawa, S., et al., 2013. *Candidatus Methanogranum caenicola*: a novel methanogen from the anaerobic digested sludge, and proposal of *Methanomassiliicoccaceae* fam. nov. and *Methanomassiliicoccales* ord. nov., for a methanogenic lineage of the class Thermoplasmata. Microbes and Environments 28 (2), 244-50.

Indugu, N., Bittinger, K., Kumar, S., et al., 2016. A comparison of rumen microbial profiles in dairy cows as retrieved by 454 Roche and Ion Torrent (PGM) sequencing platforms. PeerJ 4, e1599.

Irbis, C. and Ushida, K. 2004. Detection of methanogens and proteobacteria from a single cell of rumen ciliate protozoa. Journal of General and Applied Microbiology 50 (4), 203-12.

Iwamoto, M., Asanuma, N. and Hino, T. 2002. Ability of *Selenomonas ruminantium*, *Veillonella parvula*, and *Wolinella succinogenes* to reduce nitrate and nitrite with special reference to the suppression of ruminal methanogenesis. Anaerobe 8 (4), 209-15.

Jamieson, N. D. 1959. Rumen nitrate metabolism and the changes occurring in the composition of the rumen volatile fatty acids of grazing sheep. New Zealand Journal of Agricultural Research 2 (2), 314-28.

Janssen, P. H. 2010. Influence of hydrogen on rumen methane formation and fermentation balances through microbial growth kinetics and fermentation thermodynamics. Animal Feed Science and Technology 160 (1-2), 1-22.

Janssen, P. H. and Kirs, M. 2008. Structure of the archaeal community of the rumen. Applied and Environmental Microbiology 74 (12), 3619-25.

Jarvis, G. N., Strompl, C., Burgess, D. M., et al., 2000. Isolation and identification of ruminal methanogens from grazing cattle. Current Microbiology 40 (5), 327-32.

Jeyanathan, J. 2010. Investigation of rumen methanogens in New Zealand livestock. PhD Thesis. Massey University, Palmerston North, New Zealand.

Jeyanathan, J., Kirs, M., Ronimus, R. S., et al., 2011. Methanogen community structure in the rumens of farmed sheep, cattle and red deer fed different diets. FEMS Microbiology Ecology 76 (2), 311-26.

Jin, W., Cheng, Y. F., Mao, S. Y., et al., 2011. Isolation of natural cultures of anaerobic fungi and indigenously associated methanogens from herbivores and their bioconversion of lignocellulosic materials to methane. Bioresource Technology 102 (17), 7925-31.

Jin, W., Cheng, Y. F., Mao, S. Y., et al., 2014. Discovery of a novel rumen methanogen in the anaerobic fungal culture and its distribution in the rumen as revealed by realtime PCR. BMC Microbiology 14, 104.

Joblin, K. N. 1999. Ruminal acetogens and their potential to lower ruminant methane emissions. Australian Journal of Agricultural Research 50 (8), 1307-13.

Joblin, K. N. 2005. Methanogenic archaea. In: Makkar, H. P. S. and McSweeney, C. (Eds), Methods in Gut Microbial Ecology for Ruminants. Kluwer Academic Publishers, the Netherlands, pp. 47-53.

Johns, A. T. and Barker, H. A. 1960. Methane formation: fermentation of ethanol in the absence of carbon dioxide by *Methanobacillus omelianskii*. Journal of Bacteriology 80, 837-41.

Jones, G. A. 1972. Dissimilatory metabolism of nitrate by the rumen microbiota. Canadian Journal of Microbiology 18 (12), 1783-7.

Joshi, A., Lanjekar, V., Dhakephalkar, P. K., et al., 2018. Cultivation of multiple genera of hydrogenotrophic methanogens from different environmental niches. Anaerobe 50, 64-8.

Kamke, J., Kittelmann, S., Soni, P., et al., 2016. Rumen metagenome and metatranscriptome analyses of low methane yield sheep reveals a Sharpea-enriched microbiome characterised by lactic acid formation and utilisation. Microbiome 4 (1), 56.

Kelly, W. J., Leahy, S. C., Li, D., et al., 2014. The complete genome sequence of the rumen methanogen *Methanobacterium formicicum* BRM9. Standards in Genomic Sciences 9 (1), 15.

Kelly, W. J., Li, D., Lambie, S. C., et al., 2016a. Draft genome sequence of therumen methanogen *Methanobrevibacter olleyae* YLM1. Genome Announcements 4 (2).

Kelly, W. J., Pacheco, D. M., Li, D., et al., 2016b. The complete genome sequence of the rumen methanogen *Methanobrevibacter millerae* SM9. Standards in Genomic Sciences 11, 49.

Kelly, W. J., Li, D., Lambie, S. C., et al., 2016c. Complete genome sequence of *Methanogenic archaeon* ISO4-G1, a member of the *Methanomassiliicoccales*, isolated from a sheep rumen. Genome Announcements 4 (2).

Kennedy, P. M. and Milligan, L. P. 1978. Quantitative aspects of the transformations of sulphur in sheep. British Journal of Nutrition 39 (1), 65-84.

Kim, C. C. -H. 2012. Identification of rumen methanogens, characterization of substrate requirements and measurement of hydrogen thresholds. MSc Thesis. Massey University, Palmerston North, New Zealand.

King, E. E., Smith, R. P., St-Pierre, B., et al., 2011. Differences in the rumen methanogens populations of lactating Jersey and Holstein dairy cows under the same diet regimen. Applied and Environmental Microbiology 77 (16), 5682-7.

Kittelmann, S., Seedorf, H., Walters, W. A., et al., 2013. Simultaneous amplicon sequencing to explore co-occurrence patterns of bacterial, archaeal and eukaryotic microorganisms in rumen microbial communities. PLoS ONE 8 (2), e47879.

Kittelmann, S., Pinares-Patinö, C. S., Seedorf, H., et al., 2014. Two different bacterial community types are linked with the low-methane emission trait in sheep. PLoS ONE 9 (7), e103171.

Knight, M., Wolfe, R. S. and Elsden, S. R. 1966. The synthesis of amino acids by *Methanobacterium omelianskii*. Biochemical Journal 99 (1), 76-86.

Lambie, S. C., Kelly, W. J., Leahy, S. C., et al., 2015. The complete genome sequence of the rumen methanogen *Methanosarcina barkeri* CM1. Standards in Genomic Sciences 10, 57.

Leahy, S. C., Kelly, W. J., Altermann, E. H., et al., 2010. The genome sequence of the rumen methanogen *Methanobrevibacter ruminantium* reveals new possibilities for controlling ruminant methane emissions. PLoS ONE 5 (1), e8926.

Leahy, S. C., Kelly, W. J., Li, D., et al., 2013. The complete genome sequence of *Methanobrevibacter sp.* AbM4. Standards in Genomic Sciences 8 (2), 215-27.

Lee, H. J., Jung, J. Y., Oh, Y. K., et al., 2012. Comparative survey of rumen microbial communities and metabolites across one caprine and three bovine groups, using bar-coded pyrosequencing and ^{1}H nuclear magnetic resonance spectroscopy. Applied and Environmental Microbiology 78 (17), 5983-93;

Lee, J. H., Kumar, S., Lee, G. H., et al., 2013a. *Methanobrevibacter boviskoreani* sp. nov., isolated from the rumen of Korean native cattle. International Journal of Systematic and Evolutionary Microbiology 63 (11), 4196-201.

Lee, J. H., Rhee, M. S., Kumar, S., et al., 2013b. Genome sequence of *Methanobrevibacter* sp. strain JH1, isolated from rumen of Korean native cattle. Genome Announcements 1 (1). pii:e00002-13.

Li, Y. 2016. Comparative genomics of rumen methanogens. PhD Thesis. Massey University, Palmerston North, New Zealand. Available at:http://hdl. handle. net/10179/10953.

Li, F., Henderson, G., Sun, X., et al., 2016a. Taxonomic assessment of rumen microbiota using totalRNA and targeted amplicon sequencing approaches. Frontiers in Microbiology 7, 987.

Li, Y., Leahy, S. C., Jeyanathan, J., et al., 2016b. The complete genome sequence of the methanogenic archaeon ISO4-H5 provides insights into the methylotrophic lifestyle of a ruminal representative of the *Methanomassiliicoccales*. Standards in Genomic Sciences 11 (1), 59.

Lin, C., Raskin, L. and Stahl, D. A. 1997. Microbial community structure in gastrointestinal tracts of domestic animals:comparative analyses using rRNA targeted oligonucleotide probes. FEMS Microbiology Ecology 22 (4), 281-94.

Liu, Y. and Whitman, W. B. 2008. Metabolic, phylogenetic, and ecological diversity of the methanogenic Archaea. Annals of the New York Academy of Sciences 1125, 171-89.

Lloyd, D., Williams, A. G., Amann, R., et al., 1996. Intracellular prokaryotes in rumen ciliate protozoa: detection by confocal laser scanning after in situ hybridization with fluorescent 16S rRNA probes. European Journal of Protistology 32 (4), 523-31.

Lovely, D. R., Greening, R. C. and Ferry, J. G. 1984. Rapidly growing rumen methanogenic organism that synthesizes coenzyme M and has a high affinity for formate. Applied and Environmental Microbiology 48 (1), 81-7.

Mackie, R. I. and Bryant, M. P. 1994. Acetogenesis and the rumen: syntrophic relationships. In: Drake, H. L. (Ed.), Acetogenesis. Chapman and Hall, New York, NY, pp. 331-64.

Martin, S. A. and Park, C. M. 1996. Effect of extracellular hydrogen on organic acid utilization by the ruminal bacterium *Selenomonas ruminantium*. Current Microbiology 32 (6), 327-31.

Martinez-Fernandez, G., Duval, S., Kindermann, M., et al., 2018. 3-NOP vs. halogenated compound: methane production, ruminal fermentation and microbial community response in forage fed cattle. Frontiers in Microbiology 9, 1582.

McAllister, T. A., Cheng, K. J., Okine, E. K., et al., 1996. Dietary, environmental and microbiological aspects of methane production in ruminants. Canadian Journal of Animal Science 76 (2), 231-43.

McCabe, M. S., Cormican, P., Keogh, K., et al., 2015. Illumina MiSeq phylogenetic amplicon sequencing shows a large reduction of an uncharacterised *Succinivibrionaceae* and an increase of the *Methanobrevibacter gottschalkii* clade in feed restricted cattle. PLoS ONE 10 (7), e0133234.

McInerney, M. J., Bryant, M. P. and Pfennig, N. 1979. Anaerobic bacterium that degrades fatty acids in syntrophic association with methanogens. Archives of Microbiology 122 (2), 129-35.

McInerney, M. J., Bryan, M. P., Hespell, R. B., et al., 1981a. *Syntrophomonas wolfei* gen nov., sp. nov., an anaerobic, syntrophic, fatty acid-oxidizing bacterium. Applied and Environmental Microbiology 41 (4), 1029-39.

McInerney, M. J., Mackie, R. I. and Bryant, M. P. 1981b. Synthrophic association of a butyrate-degrading bacterium and Methansarcina enriched from bovine rumen fluid. Applied and Environmental Microbiology 41 (3), 826-8.

Metcalf, W. W., Zhang, J. K., Apolinario, E., et al., 1997. A genetic system for Archaea of the genus *Methanosarcina*: liposome-mediated transformation and construction of shuttle vectors. Proceedings of the National Academy of Sciences of the United States of America 94 (6), 2626-31.

Miller, T. L. 2001. Genus II. Methanobrevibacter. In: Boone, D. R. and Castenholz, R. W. (Eds), Bergey's Manual of Systematic Bacteriology Vol. 1.: The Archaea and the Deeply Branching and Phototrophic Bacteria. Springer, New York, NY, pp. 218-26.

Miller, T. L. and Lin, C. 2002. Description of *Methanobrevibacter gottschalkii* sp. nov., *Methanobrevibacter thaueri* sp. nov., *Methanobrevibacter woesei* sp. nov. and *Methanobrevibacter wolinii* sp. nov. International Journal of Systematic and Evolutionary Microbiology 52 (3), 819-22.

Miller, T. L., Wolin, M. J., Zhao, H. X., et al., 1986. Characteristics of methanogens isolated from bovine rumen. Applied and Environmental Microbiology 51 (1), 201-2.

Morgavi, D. P., Rathahao-Paris, E., Popova, M., et al., 2015. Rumen microbial communities influence metabolic phenotypes in lambs. Frontiers in Microbiology 6, 1060.

Morvan, B., Dore, J., Rieu-Lesme, F., et al., 1994. Establishment of hydrogen-utilizing bacteria in the rumen of the newborn lamb. FEMS Microbiology Letters 117 (3), 249-56.

Morvan, B., Bonnemoy, F., Fonty, G., et al., 1996. Quantitative determination of H_2-utilizing acetogenic and sulfate-reducing bacteria and methanogenic archaea from digestive tract of different mammals. Current Microbiology 32 (3), 129-33.

Müller, M. 1993. The hydrogenosome. Journal of General Microbiology 139 (12), 2879-89.

Murray, R. M., Bryant, A. M. and Leng, R. A. 1978. Methane production in the rumen and lower gut of sheep given Lucerne chaff: effect of level of intake. British Journal of Nutrition 39 (2), 337-45.

Myer, P. R., Kim, M. S., Freetly, H. C., et al., 2016. Evaluation of 16S rRNA amplicon sequencing using two next-generation sequencing technologies for phylogenetic analysis of the rumen bacterial community in steers. Journal of Microbiological Methods 127, 132-40.

Neill, A. R., Grime, D. W. and Dawson, R. M. 1978. Conversion of choline methyl groups through trimethylamine into methane in the rumen. Biochemical Journal 170 (3), 529-35.

Nelson, W. O., Oppermann, R. A. and Brown, R. E. 1958. *In vitro* studies on methanogenic rumen bacteria. II. Fermentation of butyric and valeric acid. Journal of Dairy Science 41 (4), 545-51.

Ng, F., Kittelmann, S., Patchett, M. L., et al., 2016. An adhesin from hydrogen-utilizing rumen methanogen *Methanobrevibacter ruminantium* M1 binds a broad range of hydrogen-producing microorganisms. Environmental Microbiology 18 (9), 3010-21.

Nicholson, M. J., Evans, P. N. and Joblin, K. N. 2007. Analysis of methanogen diversity in the rumen using temporal temperature gradient gel electrophoresis: identification of uncultured methanogens. Microbial Ecology 54 (1), 141-50.

Nobu, M. K., Narihiro, T., Kuroda, K., et al., 2016. Chasing the elusive Euryarchaeota class WSA2: genomes reveal a uniquely fastidious methyl-reducing methanogen. ISME Journal 10 (10), 2478-87.

Noel, S. J., Højberg, O., Urich, T., et al., 2016. Draft genome sequence of '*Candidatus Methanomethylophilus*' sp. 1R26, enriched from bovine rumen, a methanogenic archaeon belonging to the Methanomassiliicoccales order. Genome Announcements 4 (1), e01734-15.

Ohene-Adjei, S., Teather, R. M., Ivan, M., et al., 2007. Post-inoculation protozoan establishment and association patterns of methanogenic archaea in the ovine rumen. Applied and Environmental Microbiology 73 (14), 4609-18.

Ohene-Adjei, S., Chaves, A. V., McAllister, T. A., et al., 2008. Evidence of increased diversity of methanogenic archaea with plant extract supplementation. Microbial Ecology 56 (2), 234-42.

Oppermann, R. A., Nelson, W. O. and Brown, R. E. 1957. *In vitro* studies on methanogenic rumen bacteria. Journal of Dairy Science 40 (7), 779-88.

Orpin, C. G. 1975. Studies on the rumen flagellate, *Neocallimastix frontalis*. Journal of General Microbiology 91 (2), 249-62.

Ozutsumi, Y., Tajima, K., Takenaka, A., et al., 2012. The mcrA gene and 16S rRNA gene in the phylogenetic analysis of methanogens in the rumen of faunated and unfaunated cattle. Animal Science Journal = Nihon Chikusan Gakkaiho 83 (11), 727-34.

Padmanabha, J., Liu, J., Kurekci, C., et al., 2013. A methylotrophic methanogen isolate from the *Thermoplasmatales affiliated* RCC clade may provide insight into the role of this group in the rumen. Proceedings of the 5th Greenhouse Gases and Animal Agriculture Conference, Dublin, Ireland, 23-26 June 2013. Cambridge University Press, Cambridge, p. 259.

Patterson, J. A. and Hespell, R. B. 1979. Trimethylamine and methylamine as growth substrates for rumen bacteria and *Methanosarcina barkeri*. Current Microbiology 3 (2), 79-83.

Paul, K., Nonoh, J. O., Mikulski, L., et al., 2012. *Methanoplasmatales*, Thermoplasmatales-related archaea in termite guts and other environments, are the seventh order of methanogens. Applied and Environmental Microbiology 78 (23), 8245-53.

Paynter, M. J. B. and Hungate, R. E. 1968. Characterization of *Methanobacterium mobilis*, sp. n., isolated from the bovine rumen. Journal of Bacteriology 95 (5), 1943-51.

Pei, C. X., Mao, S. Y., Cheng, Y. F., et al., 2010. Diversity, abundance and novel 16S rRNA gene sequences of methanogens in rumen liquid, solid and epithelium fractions of Jinnan cattle. Animal 4 (1), 20-9.

Podell, S. and Gaasterland, T. 2007. Darkhorse: a method for genome-wide prediction of horizontal gene transfer. Genome Biology 8 (2), R16.

Pol, A. and Demeyer, D. I. 1988. Fermentation of methanol in the sheep rumen. Applied and Environmental Microbiology 54 (3), 832-4.

Popova, M., Marti, C., Eugène, M., et al., 2011. Effect of fibre- and starch-rich finishing diets on methanogenic Archaea diversity and activity in the rumen of feedlot bulls. Animal Feed Science and Technology 166-167, 113-21.

Poulsen, M., Schwab, C., Borg Jensen, B. B., et al., 2013. Methylotrophic methanogenic *Thermoplasmata* implicated in reduced methane emissions from bovine rumen. Nature Communications 4, 1428.

Rea, S., Bowman, J. P., Popovski, S., et al., 2007. *Methanobrevibacter millerae* sp. nov. and *Methanobrevibacter olleyae* sp. nov., methanogens from the ovine and bovine rumen that can utilize formate for growth. International Journal of Systematic and Evolutionary Microbiology 57 (3), 450-6.

Reeve, J. N., Nolling, J., Morgan, R. M., et al., 1997. Methanogenesis: genes, genomes, and who's on first? Journal of Bacteriology 179 (19), 5975-86.

Regensbogenova, M., McEwan, N. R., Javorsky, P., et al., 2004. A re-appraisal of the diversity of the methanogens associated with the rumen ciliates. FEMS Microbiology Letters 238 (2), 307-13.

Rieu-Lesme, F., Morvan, B., Collins, M. D., et al., 1996. A new H_2/CO_2-using acetogenic bacterium from the rumen: description of *Ruminococcus schinkii* sp. nov. FEMS Microbiology Letters 140 (2-3), 281-6.

Rother, M. and Metcalf, W. W. 2005. Genetic technologies for Archaea. Current Opinion in Microbiology 8 (6), 745-51.

Rowe, J. B., Loughnan, M. L., Nolan, J. V., et al., 1979. Secondary fermentation in the rumen of a sheep given a diet based on molasses. British Journal of Nutrition 41 (2), 393-7.

Seedorf, H., Kittelmann, S., Henderson, G., et al., 2014. RIM-DB: a taxonomic framework for community structure analysis of methanogenic archaea from the rumen and other intestinal environments. PeerJ 2, e494.

Seedorf, H., Kittelmann, S. and Janssen, P. H. 2015. Few highly abundant operational taxonomic units dominate within rumen methanogenic archaeal species in New Zealand sheep and cattle. Applied and Environmental Microbiology 81 (3), 986-95.

Seshadri, R., Leahy, S. C., Attwood, G. T., et al., 2018. Cultivation and sequencing of rumen microbiome members from the Hungate1000 Collection. Nature Biotechnology 36 (4), 359-67.

Sharp, R., Ziemer, C. J., Stern, M. D., et al., 1998 Taxon-specific associations between protozoal and methanogen populations in the rumen and a model rumen system. FEMS Microbiology Ecology 26 (1), 71-8.

Shi, W., Moon, C. D., Leahy, S. C., et al., 2014. Methane yield phenotypes linked to differential gene expression in the sheep rumen microbiome. Genome Research 24 (9), 1517-25.

Shin, E. C., Choi, B. R., Lim, W. J., et al., 2004. Phylogenetic analysis of archaea in three fractions of cow rumen based on the 16S rDNA sequence. Anaerobe 10 (6), 313-9.

Silley, P. 1985. A note on the pectinolytic enzymes of *Lachnospira multiparus*. Journal of Applied Bacteriology 58 (2), 145-9.

Silley, P. 1986. The production and properties of a crude pectin lyase from *Lachnospira multiparus*. Letters in Applied Microbiology 2 (2), 29-31.

Simcock, D. C., Joblin, K. N., Scott, I., et al., 1999. Hypergastrinaemia, abomasal bacterial population densities and pH in sheep infected with *Ostertagia circumcincta*. International Journal for Parasitology 29 (7), 1053-63.

Skillman, L. C., Evans, P. N., Naylor, G. E., et al., 2004. 16S ribosomal DNA-directed PCR primers for ruminal methanogens and identification of methanogens colonising young lambs. Anaerobe 10 (5), 277-85.

Skillman, L. C., Evans, P. N., Strömpl, C., et al., 2006. 16S rDNA directed PCR primers and detection of methanogens in the bovine rumen. Letters in Applied Microbiology 42 (3), 222-8.

Smith, P. H. and Hungate, R. E. 1958. Isolation and characterization of *Methanobacterium ruminantium* n. sp. Journal of Bacteriology 75 (6), 713-8.

Snelling, T. J., Genç, B., McKain, N., et al., 2014. Diversity and community composition of methanogenic Archaea in the rumen of Scottish Upland sheep assessed by different methods. PLoS ONE 9 (9), e106491.

Soliva, C. R., Hindrichsen, I. K., Meile, L., et al., 2003. Effects of mixtures of lauric and myristic acid on rumen methanogens and methanogenesis *in vitro*. Letters in Applied Microbiology 37, 35-9.

Söllinger, A., Schwab, C., Weinmaier, T., et al., 2016. Phylogenetic and genomic analysis of Methanomassiliicoccales in wetlands and animal intestinal tracts reveals clade-specific habitat preferences. FEMS Microbiology Ecology 92 (1), fiv149.

Sowers, K. R., Baron, S. F. and Ferry, J. G. 1984. *Methanosarcina acetivorans* sp. nov., an acetotrophic methane producing bacterium isolated from marine sediments. Applied and Environmental Microbiology 47 (5), 971-8.

Stewart, R. D., Auffret, M. D., Warr, A., et al., 2018. Assembly of 913 microbial genomes from metagenomic sequencing of the cow rumen. Nature Communications 9 (1), 870.

Tajima, K., Nagamine, T., Matsui, H., et al., 2001. Phylogenetic analysis of archaeal 16S rRNA libraries from the rumen suggests the existence of a novel group of archaea not associated with known methanogens. FEMS Microbiology Letters 200 (1), 67-72.

Tanner, A. C. R., Badger, S., Lai, C. H., et al., 1981. *Wolinella* gen. nov., *Wolinella succinogenes* (*Vibrio succinogenes* Wolin et al.) comb. nov., and description of *Bacteroides gracilis* sp. nov., *Wolinella recta* sp. nov., *Campylobacter concisus* sp. nov. International Journal of Systematic and Evolutionary Microbiology 31, 432-45.

Tatsuoka, N., Mohammed, N., Mitsumori, M., et al., 2004. Phylogenetic analysis of methyl coenzyme-M reductase detected from the bovine rumen. Letters in Applied Microbiology 39 (3), 257-60.

Thauer, R. K. 1998. Biochemistry of methanogenesis: a tribute to Marjory Stephenson. 1998 Marjory Stephenson Prize Lecture. Microbiology 144 (9), 2377-406.

Thauer, R. K., Kaster, A. K., Seedorf, H., et al., 2008. Methanogenic archaea: ecologically relevant differences in energy conservation. Nature Reviews Microbiology 6 (8), 579-91.

Tindall, B. J. 2007. Vacuum-drying and cryopreservation of prokaryotes. In: Day, J. G. and Stacey, G. N. (Eds), Cryopreservation and Freeze – Drying Protocols. Methods in Molecular Biology (vol. 368). Humana Press, Inc., Totowa, NJ.

Tokura, M., Tajima, K. and Ushida, K. 1999. Isolation of *Methanobrevibacter* sp. as a ciliateassociated ruminal methanogen. Journal of General and Applied Microbiology 45 (1), 43-7.

Tymensen, L. D., Beauchemin, K. A. and McAllister, T. A. 2012. Structures of free-living and protozoa-associated methanogen communities in the bovine rumen differ according to comparative analysis of 16S rRNA and mcrA genes. Microbiology 158 (7), 1808-17.

Unden, G., Hackenberg, H. and Kröger, A. 1980. Isolation and functional aspects of the fumarate reductase involved in the phosphorylative electron transport of *Vibrio succinogenes*. Biochimica et Biophysica Acta 591 (2), 275-88.

Ungerfeld, E. M. and Kohn, R. A. 2006. The role of thermodynamics in the control of ruminal fermentation. In: Sejrsen, K., Hvelplund, T. and Nielsen, M. O. (Eds), Ruminant Physiology. Wageningen Academic Publishers, Wageningen, pp. 55-85.

Valle, E. R., Henderson, G., Janssen, P. H., et al., 2015. Considerations in the use of fluorescence in situ hybridization (FISH) and confocal laser scanning microscopy to characterize rumen methanogens and define their spatial distributions. Canadian Journal of Microbiology 61 (6), 417-28.

Vantcheva, Z. M., Prodhan, K. and Hemken, R. W. 1970. Rumen methanol *in vivo* and *in vitro*. Journal of Dairy Science 53 (10), 1511-4.

Vanwonterghem, I., Evans, P. N., Parks, D. H., et al., 2016. Methylotrophic methanogenesis discovered in the archaeal phylum Verstraetearchaeota. Nature Microbiology 1, 16170.

Vogels, G. D., Hoppe, W. F. and Stumm, C. K. 1980. Association of methanogenic bacteria with rumen ciliates. Applied and Environmental Microbiology 40 (3), 608-12.

Wei, Y. Q., Long, R. J., Yang, H., et al., 2016. Fiber degradation potential of natural co-cultures of *Neocallimastix frontalis* and *Methanobrevibacter ruminantium* isolated from yaks (Bos grunniens) grazing on the Qinghai Tibetan Plateau. Anaerobe 39, 158-64.

Wei, Y. Q., Yang, H. J., Long, R. J., et al., 2017. Characterization of natural co – cultures of Piromyces with *Methanobrevibacter ruminantium* from yaks grazing on the Qinghai-Tibetan Plateau: a microbial consortium with high potential in plant biomass degradation. AMB Express 7 (1), 160.

Weimar, M. R., Cheung, J., Dey, D., et al., 2017. Development of multiwell-plate methods using pure cultures of methanogens to identify new inhibitors for suppressing ruminant methane emissions. Applied and Environmental Microbiology 83 (15), e00396-17.

Whitford, M. F., Teather, R. M. and Forster, R. J. 2001. Phylogenetic analysis of methanogens from the bovine rumen. BMC Microbiology 1, 5.

Wirth, R., Kádár, G., Kakuk, B., et al., 20018. The planktonic core microbiome and core functions in the cattle rumen by next generation sequencing. Frontiers in Microbiology 9, 2285.

Wolin, M. J., Wolin, E. A. and Jacobs, N. J. 1961. Cytochrome-producing anaerobic vibrio, *Vibrio succinogenes*, sp. n. Journal of Bacteriology 81, 911-7.

Wolin, M. J., Wolin, E. A. and Wolfe, R. S. 1963. ATP-dependent formation of methane from methylcobalamin by extracts of *Methanobacillus omelianskii*. Biochemical and Biophysical Research Communications

12 (6), 464-8.

Wolin, E. A., Wolfe, R. S. and Wolin, M. J. 1964a. Viologen dye inhibition of methane formation by *Methanobacillus omelianskii*. Journal of Bacteriology 87 (5), 993-8.

Wolin, M. J., Wolin, E. A. and Wolfe, R. S. 1964b. The cobalamin product of the conversion of methyl-cobalamin to CH4 by extracts of *Methanobacillus omelianskii*. Biochemical and Biophysical Research Communications 15 (5), 420-3.

Wood, J. M., Allam, A. M., Brill, W. J., et al., 1965. Formation of methane from serine by cell-free extracts of *Methanobacillus omelianskii*. Journal of Biological Chemistry 240 (12), 4564-9.

Wright, A. D. G., Williams, A. J., Winder, B., et al., 2004. Molecular diversity of rumen methanogens from sheep in Western Australia. Applied and Environmental Microbiology 70 (3), 1263-70.

Wright, A. D., Toovey, A. F. and Pimm, C. L. 2006. Molecular identification of methanogenic archaea from sheep in Queensland, Australia reveal more uncultured novel archaea. Anaerobe 12 (3), 134-9.

Wright, A. D. G., Auckland, C. H. and Lynn, D. H. 2007. Molecular diversity of methanogens in feedlot cattle from Ontario and Prince Edward Island, Canada. Applied and Environmental Microbiology 73 (13), 4206-10.

Wright, A. D. G., Ma, X. and Obispo, N. E. 2008. Methanobrevibacter phylotypes are the dominant methanogens in sheep from Venezuela. Microbial Ecology 56 (2), 390-4.

Xia, Y., Kon, Y. H., Seviour, R., et al., 2014. Fluorescence in situ hybridization probing of protozoal *Entodinium* spp. and their methanogenic colonizers in the rumen of cattle fed alfalfa hay or triticale straw. Journal of Applied Microbiology 116 (1), 14-22.

Yu, Z., García-González, R., Schanbacher, F. L., et al., 2008. Evaluations of different hypervariable regions of archaeal 16S rRNA genes in profiling of methanogens by Archaea-specific PCR and denaturing gradient gel electrophoresis. Applied and Environmental Microbiology 74 (3), 889-93.

Zeikus, J. G. and Bowen, V. G. 1975. Comparative ultrastructure of methanogenic bacteria. Canadian Journal of Microbiology 21 (2), 121-9.

Zhou, M., Hernandez-Sanabria, E. and Guan, L. L. 2009. Assessment of the microbial ecology of ruminal methanogens in cattle with different feed efficiencies. Applied and Environmental Microbiology 75 (20), 6524-33.

Zhou, M., Hernandez-Sanabria, E. and Luo Guan, L. L. 2010. Characterization of variation in rumen methanogenic communities under different dietary and host feed efficiency conditions, as determined by PCR-denaturing gradient gel electrophoresis analysis. Applied and Environmental Microbiology 76 (12), 3776-86.

Zinder, S. H. 1993. Physiological ecology of methanogens. In: Ferry, J. G. (Ed.), Methanogenesis: Ecology, Physiology, Biochemistry and Genetics. Chapman and Hall, New York, NY, pp. 128-206.

第7章 瘤胃原虫

Sharon A. Huws，英国女王大学；
Cate L. Williams，阿伯里斯特威斯大学；
Neil R. McEwan，英国罗伯特高登大学
（刘晶译）

1 前言

Gruby 等（1843）首次报道了瘤胃原虫的存在。此后，人们进行了大量关于瘤胃原虫功能的研究，由于瘤胃原虫研究的复杂性，人们对瘤胃原虫的认识远不及对瘤胃细菌的理解。瘤胃原虫主要属于纤毛门（Ciliophora）中的全毛目（Holotrich）和内毛目（Entodiniomorphid）。虽然瘤胃原虫的数量比瘤胃细菌少，但是它们体积大，占瘤胃微生物生物量的 5%~50%。反刍动物有特定的瘤胃原虫群（A 型、B 型、O 型、K 型），宿主基因组和原虫捕食行为（例如原生动物摄食其他原生动物）可能与原虫群落的形成有关，但是具体原因尚不清楚。瘤胃原虫的细胞核有两种，即小核和大核，小核为生殖细胞，不参与 RNA 转录，大核通过 RNA 转录进行生长。

瘤胃原虫的存在对瘤胃细菌、真菌和古菌有重要的影响。例如，许多瘤胃原虫的选择性捕食导致瘤胃微生物群落发生改变。瘤胃产甲烷菌与原虫，尤其是全毛虫关系密切，纤毛虫具有氢小体，氢小体可以影响能量代谢的末端反应导致氢的释放，从而使产甲烷菌更有效地利用氢生成甲烷。

瘤胃原虫与日粮中的植物纤维密切相关，前毛虫（*Epidinium*）会积极寻找、吞噬并储存植物叶绿体。这一现象的原因尚不清楚，但可能是由于植物叶绿体含有丰富的脂质和蛋白质，能满足原虫的能量需求。内毛目中的许多原虫还能通过糖基水解酶降解纤维，这些糖基水解酶很可能是通过瘤胃细菌水平基因转移获得的。体内驱除原虫试验和体内试验荟萃分析证实，瘤胃原虫可以增加纤维分解和甲烷排放，降低瘤胃微生物蛋白质生成和动物平均日增重，对宿主生长和环境有一定负面影响。需要指出的是，驱除所有的瘤胃原虫并不是最好的甲烷减排策略，但是去除全毛虫可能是合适的，因为全毛虫在甲烷排放中的作用较大，而在日粮纤维消化中的作用较小。鉴于驱除所有原虫的挑战性，开发针对特定原虫的驱除新技术将更有意义。

2 瘤胃原虫的发现

1676 年，安东尼·范·列文虎克（Antonie Van Leeuwenhoek）的开创性工作，使得游

离原虫被发现。由于瘤胃样本的难以获得性以及认识有限性,科学家们没有在第一时间研究瘤胃中的原虫,直到1843年瘤胃原虫才被报道(Gruby和Delafond,1843)。瘤胃具有温暖(39℃)、厌氧、含大量食物颗粒并且规律性收缩外排的特征。瘤胃类似于微生物恒定反应器,原虫传代时间慢(5~15.7h/代)(Warner,1962;Karnati等,2007)且流速为$2×10^{11}$个/d,因此原虫在瘤胃内生存并不容易。它们通常会附着在瘤胃内容物上(Orpin,1985)或者瘤网状胃壁上(Abe和Iriki,1989),以避免从瘤胃内流出。原虫也能够根据外排流速或底物变化调整传代时间,以保证自身生存(Dehority,1998,2004;Karnati等,2007)。

早期研究表明,小部分被描述为原虫的细胞实际上是另一类稀有瘤胃真核生物—厌氧真菌产生的游动孢子(Orpin,1975),它们的密度非常低。现在人们已清楚地认识到,瘤胃原虫密度远低于瘤胃细菌(原虫,10^4~10^6个/mL;细菌,10^9~10^{11}个/mL)。由于原虫体型较大,能根据宿主饮食而变化,所以它们占瘤胃微生物生物量的5%~50%(Williams和Coleman,1992;Denton等,2015)。

3 瘤胃原虫分类和种群类型

瘤胃原虫主要是纤毛虫,属于纤毛虫门;而鞭毛虫,如毛滴虫属(*Trichomonas* spp.)、单尾滴虫属(*Monocecromonas* spp.)和唇鞭虫属(*Chilomastix* spp.)等数量极少(Williams和Coleman,1992;图1)。瘤胃纤毛虫与骆驼前肠(Kubesy和Dehority,2002)及其他草食动物(如马和犀牛)后肠(Moon-van der Staay等,2014)的纤毛虫一起构成独特的系统发育簇,根据形态特征,瘤胃纤毛虫被分为全毛目和内毛目(Kamra,2005;图1)。

全毛目有4个科:等毛科(Isotrichhidae)、盔毛科(Blepharocorythidae)、布契利科(Buetschliidae)和亚等毛滴虫科(Paraisotrichidae),等毛科分布更广泛,占纤毛虫种群的40%(Williams和Coleman,1992)。等毛科优先利用可溶性碳水化合物,有一定的耐氧性,拥有覆盖整个细胞表面且长度均匀的纤毛,纤毛围绕前庭和柔性膜融合(Williams和Coleman,1992;Dehority,2003;图2)。肠等毛虫(*Isotricha intestinalis*)、前缘等毛虫(*Isotricha prostoma*)和反刍厚毛虫(*Dasytricha ruminantium*)是等毛虫科主要的原虫,其他种类原虫丰度较低(Hobson和Stewart,1997;图1,图2)。

内毛目主要由头毛科(Ophryscolecidae)组成(图1),还包括 *Cycloposthiidae Parentodium* spp. 和 *Rhinozetidae Rhinozeta* spp.(Imai,1998)。

门　Ciliophora
　亚门　Intramacronucleata
　　纲　直口纲(Litostomatea)
　　　亚纲　Trichostomatia
　　　　目　全毛目(Vestibuliferida)
　　　　　科　等毛科(Isotrichidae)

　　　　属　（*Aviisotricha*）
　　　　属　厚毛虫属（*Dasytricha*）
　　　　属　等毛虫属（*Isotricha*）
　目　内毛目（Entodiniomorphida）
　　亚目　（Entodiniomorphina）
　　　科　头毛科（Ophryoscolecida）
　　　　亚科　内毛亚科（Entodiniinae）
　　　　　属　内毛属（*Entodinium*）
　　　　亚科　双毛亚科（Diplodiniinae）
　　　　　属　双毛属（*Diplodinium*）
　　　　　　原始纤毛属（*Eodinium*）
　　　　　　真双毛属（*Eudiplodinium*）
　　　　　　硬甲属（*Ostracodinium*）
　　　　　　甲属（*Enoploplastron*）
　　　　　　后毛属（*Metadinium*）
　　　　　　鞘甲属（*Elytroplastron*）
　　　　　　多甲属（*Polyplastron*）
　　　　亚科　头毛亚科（Ophryoscolecinae）
　　　　　属　前毛属（*Epidinium*）
　　　　　　前甲属（*Epiplastron*）
　　　　　　后毛属（*Opisthotrichum*）
　　　　　　头毛属（*Ophryscolex*）
　　　　　　美头属（*Caloscolex*）

图1　瘤胃纤毛虫分类（改编自 Imai，1998）

内毛虫与全毛虫在形态方面存在一定差异，内毛虫的纤毛在前端和/或后端聚集，并根据纤毛带、骨板数量、尾部突起和纤毛在细胞表面的排列等特征被分为不同的种类（Dehority，2003；图2）。内毛虫有一层厚厚的外膜，不如全毛虫灵活，这是对瘤胃内恶劣环境的一种适应性进化（Williams，Coleman，1992）。瘤胃内毛虫包括内毛亚科（Entodiniiae）、双毛亚科（Diplodiniiae）、头毛亚科（Ophryoscolecinae）、前毛亚科（Epidiniinae）、后毛亚科（Opisthotrichinae）、美头亚科（Caloscolecinae）（Williams，Coleman，1992；Henderson 等，2015；图1，图2）。由于内毛虫种内变异较大，在种水平上鉴定内毛虫很困难。尾刺内毛虫（*Entodinium caudatum*）只有在法氏囊存在的情况下，才产生尾棘（Hobson，Stewart，1997）。

瘤胃普查（Rumen Census）项目（Henderson 等，2015）研究了不同地理位置、宿主和饮食条件下反刍动物瘤胃微生物组成变化，是迄今为止最全面的瘤胃微生物研究之一。该研究利用测序技术，针对所有细菌、古菌、纤毛虫、真菌和病毒，分析768个瘤胃样本的微生物群落（Henderson 等，2015）。结果发现，99%以上的原虫来自于以下12个属：

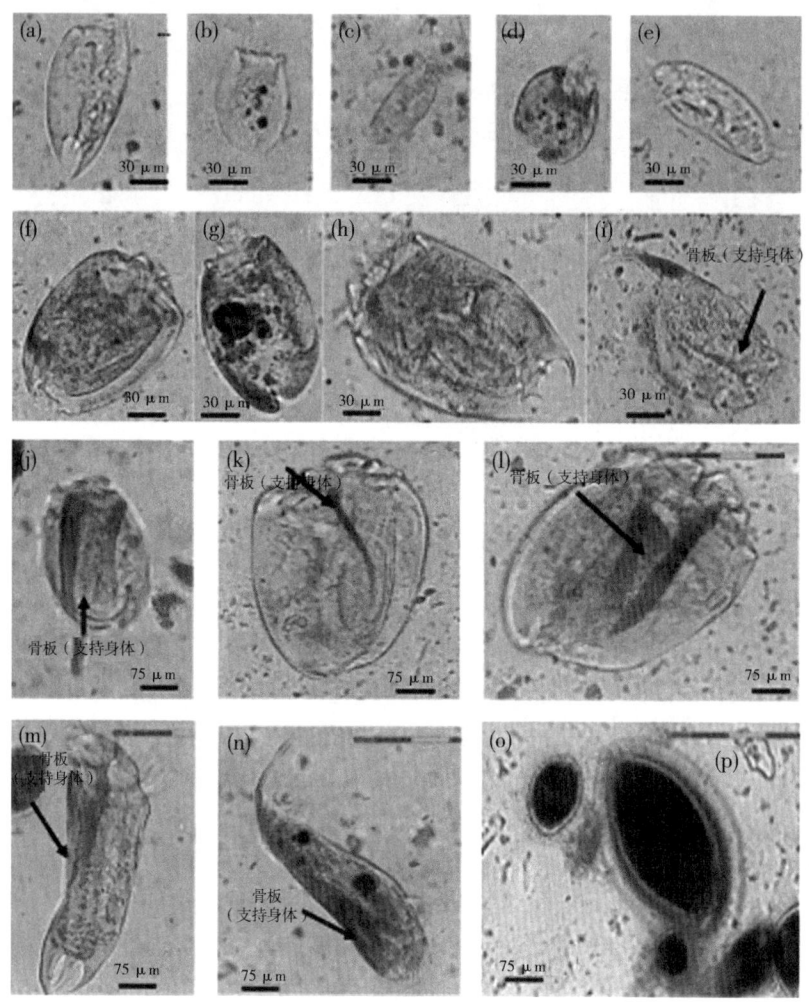

图 2 鲁哥氏碘液染色瘤胃纤毛虫的光学显微镜照片。(a) ~ (e) 是内毛属 (*Entodinium* spp.), (f) ~ (h) 是双毛属 (*Diplodium* spp.), (i) 单甲属 (*Eremoplastron* spp.), (j) 硬甲属 (*Ostracodinium* spp.), (k) 真双毛属 (*Endiplodinium* spp.), (l) 后毛属 (*Metadinium* spp.), (m) 和 (n) 前毛属 (*Epidimiu* spp.), (o) 厚毛虫属 (*Dasytricha* spp.), (p) 等毛虫属 (*Isotricha* spp.)。图像均来自 Williams (2018),图中原虫均属于 B 型种群 (彩图 1)。

内毛属、前毛属、甲属 (*Enoploplastron*)、头毛属 (*Ophryscolex*)、无甲亚属 (*Anoplodinium*)/双毛属、原始纤毛属 (*Eremoplastron*)/双甲属 (*Diploplastron*)、真双毛属、后毛属、硬甲属、多甲属 (*Polyplastron*)、厚毛虫属和等毛虫属,其中内毛属和前毛属丰度分别为 32.2% 和 16.5% (Henderson 等,2015)。此外,原虫在地理位置上的变异比细菌要明显得多 (Henderson 等,2015)。

在基因组分析技术出现之前,人们就发现,反刍动物瘤胃拥有一类独特的原虫种群

（Eadie，1962；Williams 和 Coleman，1992；图 1）。瘤胃原虫种群被分为 A 型、B 型、O 型和 K 型，其中 A 型和 B 型的原虫种类组成最多样化，是反刍动物瘤胃主要种群类型。某些原虫，如双毛属和头毛属可能会消失，体积较大的内毛虫在不同动物之间的变化最大，而全毛虫则相对稳定（Eadie，1962；Williams 和 Coleman，1992）。

反刍动物拥有特定瘤胃原虫种群类型，瘤胃原虫随地理位置变化的原因尚不清楚。据推测，动物的遗传、生理、饮食、传播（原虫的定植和排出）和纤毛虫之间的拮抗都是影响瘤胃纤毛虫群落建立与丰度变化的重要因素。实际上，体积较大的原虫可以内吞体积较小的原虫，这就是大原虫和小原虫很少同时存在的原因（Williams 和 Coleman，1992）。Eadie（1967）发现多泡多甲虫（*Polyplastron multivesiculatum*）早于前毛属、*Endiplodinium maggii*、牛单甲虫（*Eremoplastron bovis*）和硬甲属出现。因此，多泡多甲虫可被用于分析反刍动物是否具有 A 型或 B 型原虫种群。

表 1 牛、绵羊和山羊的原虫种群类型

A 型	B 型	O 型	K 型（仅在牛中存在）
内毛属	内毛属	内毛属	内毛属
多泡多甲虫	前毛属 （不含 *E. tricaudaum*）		水牛鞘甲虫 （*Elytroplastron bubali*）
邻近双甲虫 （*Diploplastron affine*）	*Eudiplodinium maggi*		喙状单甲虫 （*Eremoplastron rostratum*）
头毛属（牛、羊）	单甲属		双毛属
硬甲属（牛、羊）	硬甲属（羊）		
单甲属（牛中很少）	头毛属（山羊）		
双毛属	双毛属		
甲属	甲属（牛中很少）		
原始纤毛属	双毛属		
	原始纤毛属		

资料来源：改编自 Williams 和 Coleman（1992）。

4 瘤胃原虫基因组学

瘤胃原虫有两种细胞核：小核和大核。在非瘤胃源的纤毛虫中，小核是生殖核，不表达基因，大核具有编码酶的基因，支持营养生长（Prescott，1994）。一些非反刍动物纤毛虫仅含有单基因的染色体，即染色体上只有一个基因，两侧是端粒。瘤胃纤毛虫的染色体被称为中染色体，长度相对较短，但是比单基因染色体长得多（Thomas 等，2004）。小核和大核均富含 AT 碱基，编码区对 A 或 T 密码子表现出强烈的偏好性，并且只使用有限的密码子（McEwan 等，2000a，b）。由于碱基堆积过程中的相互作用，低 GC 含量的 DNA 比高 GC 含量的 DNA 更不稳定。所以，富含 AT 的基因组不利于测序，这成为原虫基因组测序的限制因素（Chen 等，2013）。

虽然大核已经被成功分离了，但是由于小核尺寸与细菌相似，所以小核很难被纯化（Young 等，2015）。最近，尾刺内毛虫的大核基因组草图被公布，这是第一个瘤胃原虫大核基因组序列草图（Park 等，2018）。利用嗜热四膜虫（*Tetrahymena thermophila*）进行基因预测训练，结果显示尾刺内毛虫大核基因组与单细胞生物 *Oxytricha trifallax* 相似。但是鉴于反刍动物肠道共生纤毛虫与其他环境的游离原虫有显著差异，所以这种基因注释结果并不理想。瘤胃原虫参考序列的缺乏易导致瘤胃元基因组数据集信息不全面，以及基因注释的覆盖率低。瘤胃厌氧真菌 *Orpinomyces* 的 AT 含量在 80%~85%，在进行基因组分析时注释率较低。尽管如此，基于基因测序的方法仍然是认识瘤胃原虫功能的重要方法，为未来其他原虫的大核测序提供基础。

5　原虫种群的变化

动物采食 6 h 后，瘤胃内毛虫数量减少 50% 以上，到 24 h，原虫恢复至采食前的数量（Williams 和 Coleman，1992）。这是由于进入瘤胃内的唾液和水增加，对微生物进行了稀释，然而丰富的营养物质使原虫快速生长，并很快恢复数量（Williams 和 Coleman，1992）。动物无论每天采食多少次，这种变化模式几乎都是一致的。但当动物一天只采食一次，原虫数量恢复的时间则更长一些（Williams 和 Coleman，1992）。

瘤胃全毛目和内毛目原虫有着各自独立的节律变化。Purser 和 Moir（1959）指出，动物采食后内毛虫数量先下降后上升，而 Purser（1961）在后期的试验指出全毛虫数量在采食时达到峰值，然后逐渐减少。动物采食前和采食过程中，网胃收缩和可溶性糖刺激全毛虫的迁移，这一现象主要发生在以细菌为食并能分解植物成分（纤维素、半纤维素、果聚糖、果胶、淀粉、不溶性糖、蛋白质和脂类）的原虫身上。

虽然原虫是瘤胃微生态系统中一个天然的、正常的组成部分，但是动物在没有原虫的情况下也可以生存。通过隔离饲养动物（Belanche 等，2015）或者补充 2%（W/V）硫酸铜和二辛基硫代丁二酸钠均可以实现瘤胃内去除原虫（Aban 和 Bestil，2016）。虽然细菌和真菌能分解植物纤维，但是去除原虫后，动物对纤维的总消化率显著下降，并且大肠和盲肠内出现纤维代偿消化现象（Demeyer，1981）。

6　瘤胃内原虫的相互作用

6.1　瘤胃细菌与原虫相互作用

Bryant 和 Small（1960）研究发现，去原虫犊牛接种瘤胃总内容物（含原虫）后，瘤胃细菌数从 6.0×10^9 个/mL 增加到 13.6×10^9 个/mL。这一结果说明，原虫对瘤胃细菌有相当大的影响。部分原虫可以吞噬细菌，释放氢气，增加甲烷的产生。细菌可能是原虫含氮化合物最重要的来源，内毛目和全毛目原虫在吞噬的细菌种类上表现明显的偏好性，内毛目吞噬反刍单胞菌（*Selenomonas ruminantium*）和丁酸弧菌（*Butyrivibrio fibrisolvens*）的速度比摄取其他细菌的速度快得多（Coleman，1964）。在吞噬瘤胃细菌和真菌的过程中，原虫中特定活性酶发挥作用。E. maggii 产生胞外和胞内几丁质酶用于分解真菌细胞壁（Miltko 等，2012）。瘤胃原虫转录组结果也表明蛋白酶（B 和 F）呈现溶菌酶活性，参与细菌细胞壁的分解（Williams，2018）。细菌种类影响原虫的吞噬速度，大肠杆菌（Escherichia coli）和产气克雷伯氏菌（Klebsiella aerogenes）被尾刺内毛虫快速吞噬，而奇异变形杆菌（Proteus mirabilis）似乎对原虫吞噬有一定抵抗力（White，1969）。原虫除了影响细菌数量和多样性，还拥有特定的胞内共生菌（Levy 和 Jami，2018）。

6.2 瘤胃真菌与原虫相互作用

Newbold 等（2015）的荟萃分析表明，动物去原虫后，瘤胃纤维分解微生物（包括真菌）明显减少，这可能是由于瘤胃内纤维分解是一个复杂的过程，需要包括瘤胃原虫在内的多种微生物协同作用。因此，瘤胃去原虫后可能影响纤维分解微生物群体功能，进而影响纤维分解效率。然而，Hsu 等（1991）发现，瘤胃去原虫后，真菌游动孢子数量增加，这可能是消除了原虫的捕食和营养竞争而导致的。Morgavi 等（1994）发现瘤胃原虫具有几丁质酶活性，可以降解瘤胃真菌细胞壁中的几丁质。

6.3 瘤胃产甲烷菌与原虫相互作用

瘤胃原虫含有氢化酶，能释放氢气，为产甲烷菌提供底物合成甲烷，这意味着原虫与产甲烷菌之间有着强烈的相互合作关系（Fenchel 和 Finlay，2006；Belanche 等，2014）。最近的一项研究表明，与瘤胃原虫相关的主要产甲烷菌是氢营养型的甲烷短杆菌（Methanobrevibacter）（Tymensen 等，2012；Belanche 等，2014；Wang 等，2017）。与内毛虫相比，全毛虫有不同的胞内共生产甲烷菌（Belanche 等，2014），可能是因为全毛虫的氢化酶活性更强（Paul 等，1990）或者全毛虫在碳水化合物过剩时快速合成糖原，导致氢气产量增加（Hall，2011；Denton 等，2015）。Belance 等（2014）研究发现，与内毛虫相比，去原虫山羊接种全毛虫后甲烷排放量更大。

6.4 瘤胃植物与原虫相互作用

瘤胃微生物种群对日粮有明显、快速的反应，原虫也不例外（Williams 和 Coleman，1992；Newbold 等，2015；Nikkhah，2016）。体型较大的内毛虫，如前毛属和多甲属通常偏爱颗粒物，与植物纤维紧密联系（Huws 等，2009，2012，2018）。前毛属常在黑麦草中游动，并在游动过程中吞噬植物叶绿体（Akin 和 Amos，1979；Huws 等，2018；图3）。Bauchop（1980）通过体外试验发现，前毛属能快速定植到植物表面，分解叶肉组织，这说明纤毛虫是叶肉组织的主要分解微生物。

据推测，瘤胃原虫在淀粉、纤维素、纤维二糖、果胶和半纤维素的分解中均发挥作用（Wright，1960；Bailey 等，1962；Coleman，1978，1985；Forsberg 等，1984；Orpin，1984；Veira，1986；Varel 和 Dehority，1989；Jouany 和 Ushida，1999；Wang 和 McAllister，2002；Newbold 等，2015）。事实上，有人曾提出在植物细胞壁降解过程中，瘤胃细菌和真菌贡献80%的降解活性，原虫贡献20%的降解活性（Wang 和 McAllister，2002；Newbold 等，

2015)。然而，受分离与培养的限制性，我们可能低估了原虫的作用（Wang 和 McAllister，2002）。无论如何，许多瘤胃原虫比如 *E. maggii*、*Ostracodinium dilobum*、*Metadinium affine*、牛真双毛虫（*Endiplodinium bovis*）、有尾头毛虫（*Ophryoscolex caudatus*）、多泡多甲虫、肠等毛虫和前缘等毛虫已经被证实具有潜在的纤维分解、半纤维分解和果胶分解活性（Wang 和 McAllister，2002），但是尾刺内毛虫、*Diplodinium pentacanthum*、*Endoploplastron triloricatum*、*Ophryoscolex tricoronatus*、薄硬甲虫（*Ostracodinium gracile*）则没有这些活性（Dehority，1993）。

关于瘤胃原虫碳水化合物酶活性的研究大多集中在 1996—2000 年（Guttmacher 和 Collins，2003；Konstantinidis 等，2006），大多数研究都是利用纯培养或比较去原虫前后变化而获得信息。这些研究证明了原虫参与瘤胃内纤维分解（Jouany 和 Senaud，1979）。酶谱研究显示，前毛属内存在多种植物细胞壁多糖分解酶。Bailey 和 Gaillard（1965）研究发现，尾刺内毛虫能产生阿拉伯呋喃木糖苷酶、内切-1,4-β-木聚糖酶和内切-1,4-β-木糊精酶，水解植物中的三种半纤维素。Bailey（1962）研究证明，尾刺内毛虫可以水解小麦木聚糖，释放阿拉伯糖、木二糖和木糖，牛单甲虫产生木聚糖酶活性、木糊精酶活性、纤维糊精酶活性和胞外-1,4-β-葡聚糖酶活性，*E. maggii* 分解纤维素和果胶（Hungate，1942，1943；Bailey 和 Clarke，1963；Coleman，1978，1986）。

随着测序技术发展，研究人员在瘤胃原虫尤其是较大的内毛虫中发现了纤维分解基因（Devillard 等，1999；Ricard 等，2006；Wereszka 等，2004，2006）。Devillard 等（1999）建立了多泡多甲虫的宏基因组库，成功鉴定 cDNA 文库中糖基水解酶家族中的 11 个木聚糖酶，通过构建 11 个木聚糖酶催化结构域的系统发育树发现，木聚糖酶与黄色瘤胃球菌（*Ruminococcus flavefaciens*）和短小芽孢杆菌（*Bacillus pimilus*）的亲缘关系较近。这增加了原虫通过水平基因转移，从摄取的细菌中获得糖基水解酶的可能性（Devillard 等，1999；Friedman 和 Ely，2012）。最近，从尾刺内毛虫中分离到了一个谷氨酸脱氢酶 cDNA 基因序列，该基因也可能是纤毛虫通过水平基因转移而得到的（Newbold 等，2005）。

Ricard 等（2006）利用瘤胃纤毛虫约 4 000 个表达序列标签（Expressed Sequence Tags，ESTs）研究了从细菌向纤毛虫水平基因转移情况。该研究选择一些常见的、数量多的纤毛虫：简单内毛虫（*Entodinium simplex*）、尾刺内毛虫、*E. maggii*、中间后毛虫（*Metadinium medium*）、*Diploplastron affine*、多泡多甲虫、*E. ecaudatum*、前缘等毛虫、肠等毛虫和反刍厚毛虫，构建了广覆盖度的 cDNA 文库和各类群的系统发育树，揭示了系统发育树上的聚类。在候选的水平转移基因中，与碳水化合物代谢相关的酶约占 3/4，其中 35%的酶是糖基水解酶。内毛虫和全毛虫的酶谱存在显著差异，全毛虫表达烯醇化酶、果糖激酶和葡萄糖激酶，而内毛虫不表达这些酶。另一方面，内毛虫表达纤维二糖磷酸化酶、纤维素酶、木聚糖酶、果胶裂解酶、天冬氨解氨酶和硝基还原酶，而全毛虫缺乏这些酶。事实上，全毛虫中唯一的糖基水解酶（GHs）并不能溶解纤维，这支持了全毛虫不分解纤维的观点。Ricard 等（2006）提出，水平基因转移是原虫适应高碳水化合物营养环境的进化学方法，原虫中有 4.1%的序列表达标签（Expressed Sequence Tags，ESTs）是由水平转移获得的。

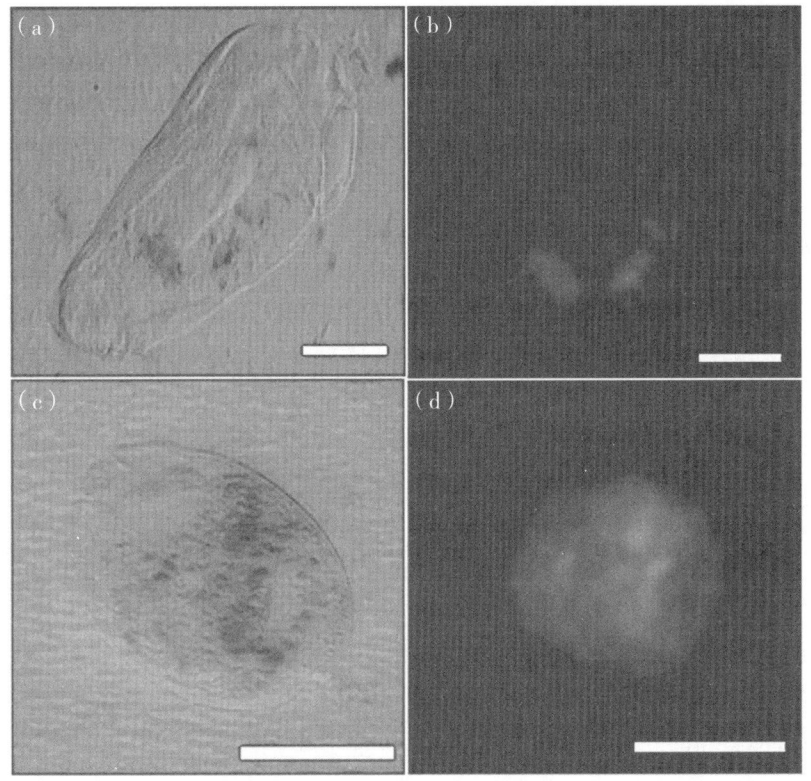

图 3 前毛虫和内毛虫的显微成像。（a）前毛虫光学显微镜图像（比例尺 10 μm），内含胞内叶绿体，分离自饲喂草 2 h 后的肉牛瘤胃；（b）前毛虫荧光图像（比例尺 10 μm），内含胞内叶绿体，分离自饲喂草 2 h 后的肉牛瘤胃；（c）内毛虫光学显微镜图像（比例尺 10 μm），内含胞内叶绿体，分离自饲喂草 2 h 后的肉牛瘤胃；（d）内毛虫光学显微镜图像（比例尺 10 μm），内含胞内叶绿体，分离自饲喂草 2 h 后的肉牛瘤胃（彩图 2）。

在快速生长的微生物中，密码子倾向于发生中性突变或者自然选择，从而实现最佳翻译效率（Rocha，2004；Behura 和 Severson，2013；Brandis 和 Hughes，2016）。McEwan 等（2000b）发现瘤胃纤毛虫遵循使用"通用"密码子编码，但不使用 TAA、TAG 或者 TGA 编码氨基酸，表明原虫具有不寻常的密码子模式。AT 含量较高的生物体首选终止密码子是 TAA。尽管纤毛虫使用了 61 个"通用"密码子，但是对某些密码子表现出明显的偏爱，比如纤毛虫更偏爱利用 AGA 编码精氨酸，而精氨酸在其他许多生物体中的使用率通常是最低的。此外，瘤胃纤毛虫在 mRNA 的 3′端非翻译区内不使用"通用的"多聚腺苷酸信号，而是使用不常见的核苷酸模式（Destables 等，2005）。

7 研究瘤胃原虫面临的挑战

瘤胃原虫在遗传学和纯培养试验中面临许多挑战。很多与遗传学研究相关的问题已经在上一部分被讨论过了：双核结构、不寻常的密码子使用模式、非标准的多聚腺苷酸化等。在体外培养试验中，不能将瘤胃原虫不与或仅与一种细菌进行培养。许多研究单一种类瘤胃原虫的试验都是将单一原虫引入去除原虫的体内，从而达到单原虫状态，以便后续研究。使用显微操纵仪挑选出有足够形态相同性和单一物种的原虫细胞，然后将这些原虫接种反刍动物。动物被接种前必须保证瘤胃内无原虫，如果动物接种前有过原虫，那么瓣胃（反刍动物前胃的第三个腔）也必须保证无原虫（Michalowski 等，1991）。然而，当不同的原虫物种具有相似的形态结构时，则无法保证动物接种的是单一原虫物种。即使接种了一个原虫物种，接种动物也必须与其他反刍动物隔离饲养，避免被其他原虫物种污染。

Coleman 等（1976）详细介绍了 6 种原虫（*E. triloricatum*、*E. maggii*、*D. affine*、内毛虫、*Diplodinium monacanthum*、*D. pentacanthum*）的培养方法，建立了利用 ^{14}C 标记黑麦草的 α-[^{14}C]纤维素分析纤维素酶活性的方法。除 *D. monacanthum* 外，其余 5 种原虫均表现纤维素酶活性。真正区分原虫和细菌的纤维素酶活性是极具挑战性的，离心去除细菌后，无细胞提取液仍然表现纤维素分解活性，少数细菌也有在原虫中存在的可能。

显微操作方法挑选单细胞也可以用于原虫体外培养。一些盐溶液，如 Hungate 盐溶液（Hungate，1942）、"caudatum"盐溶液（Coleman 等，1972）、"simplex"型溶液（Coleman 等，1972）和"人工瘤胃液"（Michalowski 等，1999）也被用于原虫体外培养。但是，显微操作的方法通常依赖于接种形态相似的细胞，而这些细胞实际上可能不是单一物种。即使成功建立了原虫的体外培养体系，原虫也可能会在几轮细胞分裂后死亡，这可能是因为缺乏细胞间的结合，无法产生小核，培养物只能依靠无性繁殖，导致原虫有限的稳定性和寿命，但是这只是一种推断，目前还没有被证实。

8 原虫功能对反刍动物营养、健康和甲烷排放的影响

Gruby 和 Delafond（1843）提出假设：瘤胃原虫一定在宿主营养代谢过程中发挥作用。后期研究表明，瘤胃原虫确实在反刍动物中发挥重要作用，但是也不是必要的（Newbold 等，2015；Li 等，2018）。本章节前部分基于实验室研究对原虫功能进行了讨论，本部分将对活体动物研究的结果进行讨论。驱除原虫的研究提供了评估瘤胃原虫对动物生产影响的数据。近期发表的基于驱除原虫试验的荟萃分析，增强了对于瘤胃原虫在宿主表型中作用的理解。

8.1 瘤胃原虫与氮利用率

瘤胃原虫的蛋白代谢研究数据比较有限并且结论经常是矛盾的，例如，Forsberg（1984）和 Coleman（1983）发现，在中性条件下，原虫酪蛋白酶几乎没有活性，而其他人则发现具有酪蛋白酶活性（Brock 等，1982）。有证据表明，内毛虫会产生"硫醇"和"羧基"类型的弱活性蛋白水解酶（Coleman，1983；Forsberg，1984）。必须指出的是，早期的研究无法区分原虫摄入细菌源的酶活性和原虫本身的酶活性，研究结果可能没有最近的研究结果那么可信（Hess 等，2011）。

许多研究表明，去除原虫可以增加瘤胃氮保留量（Eugène 等，2004；Newbold 等，

2015；Li 等，2018），但是也有研究发现没有影响（Nguyen 等，2016）。造成结果差异的原因有很多，可能是宿主品种不同、饮食不同、驱虫时间不同等，这使得荟萃分析在真实全面分析原虫功能方面更有优势。目前已有三篇荟萃分析文献，这些分析基于不同数量的研究，排除或纳入了某些因素，所有分析均表明，去除原虫会导致瘤胃氨氮浓度增加（Eugène 等，2004；Newbold 等，2015；Li 等，2018）。这可能是由于原虫对细菌蛋白质脱氨，而原虫不利用氨，导致氨释放至瘤胃液，引起氨氮浓度增加（Ivan 等，1991；Demeyer 和 Fievez，2004）。瘤胃原虫能够吞噬蛋白质和碳水化合物等大分子，促进微生物氮从瘤胃流入十二指肠（Bach 等，2005）。然而，由于原虫被选择性地保留在瘤胃中，它们对蛋白总供应量的贡献约占 11%（Shabi 等，2000）。

8.2 瘤胃原虫与纤维消化

瘤胃去除原虫后，有机物降解率显著降低。纤维消化是一个复杂的过程，需要几种不同的纤维分解微生物共同合作，原虫负责纤维定植和纤维消化的初始阶段（Newbold 等，2015）。虽然全毛虫纤维分解活性低于内毛虫（可能是相对丰度不同导致的），但是全毛虫确实积极参与了营养代谢过程。动物日粮含有较高水平可溶性碳水化合物时，全毛虫密度大大增加。一般来说，全毛虫很大程度上受宿主日粮的影响。然而，需要指出的是，关于全毛虫代谢活性的研究只包括了本章节之前提到的 3 个主要的全毛虫物种，这些物种参与可溶性糖和非结构性多糖的利用，并利用趋化性向营养物质迁移。这类原虫也与日粮植物密切相关，但是原虫降解植物细胞壁多糖的能力可能有限（Hobson 和 Stewart，1997）。去除原虫对整个微生物群落也有一定影响，真菌浓度大大降低，而纤维分解菌数量显著增加（Newbold 等，2015）。去除原虫对饲料转化率的影响目前存在争议。Williams 和 Coleman（1992）报道了去除原虫能提高饲料转化率和平均日增重。去除原虫的另一个显著效果是流向十二指肠的蛋白质增加，这是由于细菌没有被原虫消化，增加了瘤胃细菌种群多样性和数量（Williams 和 Coleman，1992；Newbold 等，2015）。

8.3 瘤胃原虫与反刍动物产品品质

目前原虫在脂质代谢中的作用存在争议。原虫富含对人类健康有益的多不饱和脂肪酸（PUFA）（Huws 等，2009），这不是因为原虫能够直接参与脂质代谢过程，而是因为它们倾向利用富含 PUFA 的叶绿体，导致自身 C18：3 n-3 含量很高（Huws 等，2009；图3）。原虫吞噬叶绿体后，部分叶绿体被摄入的细菌代谢，导致共轭亚油酸（CLA）和顺-9，反-11 十八碳烯酸在原虫体内的积累，这两种酸均对人类健康有益（Lourenco 等，2010）。Huws 等（2009，2012）研究发现，瘤胃原虫可以有效保护瘤胃内的多不饱和脂肪酸不被脂解和生物氢化，补充日粮叶绿体能促进 PUFA 向十二指肠和肉奶制品中转移。近期，Francisco 等（2019）的一项研究表明，瘤胃原虫数量与肉类有益脂肪酸沉积呈正相关。然而，瘤胃原虫多被保留在瘤胃内，流向十二指肠的原虫及其多不饱和脂肪酸很少（Huws 等，2012）。

8.4 瘤胃原虫与甲烷排放

农业生产中的大部分甲烷是由反刍动物产生的（每年大约释放 14% 的大气甲烷）（Huws 等，2018）。甲烷主要是在前肠内产生，主要是由瘤胃古菌（产甲烷菌）产生。许多产甲烷菌都与瘤胃原虫有关，有的产甲烷菌附着在原虫细胞表面，而有的在原虫胞内共生。这种相互作用高度依赖瘤胃内的气体成分，当瘤胃中氮气浓度较高而氢气浓度较低时，产甲烷菌与纤毛虫有很强的相关性；当瘤胃中氢气浓度较高时，这种关联性降低（Stumm 等，1982）。研究表明瘤胃产甲烷菌和原虫间的联系是进化适应的结果，以确保它们在瘤胃氢气浓度低时可以获得氢气。所以，以瘤胃原虫为靶标可作为降低甲烷产量的一种调控方式。去除原虫虽然不影响产甲烷菌数量，但是可以降低甲烷产量（Finlay 等，1994）。原虫数量和甲烷排放量之间存在线性关系，这为通过调控瘤胃原虫来缓解甲烷排放提供了基础（Newbold 等，2015）。

9 研究例证：调控瘤胃纤毛虫

目前，瘤胃原虫功能的数据表明，减少瘤胃原虫数量或者完全去除瘤胃原虫可能是促进反刍动物生产性能和降低环境影响的有效措施。瘤胃原虫通过唾液实现在动物之间的传播（Becker 和 Hsiung，1929；Yáñez-Ruiz 等，2015）。幼年反刍动物出生后两周内，便可在瘤胃内观察到纤毛虫，并且小纤毛虫在大型原虫出现之前定植（Eadie，1962）。因此，可以让小牛或者羔羊在早期（通常在出生后 24~48 h，接触母体初乳后）与其他反刍动物（有正常的原虫群）分开，单独饲养，从而建立无原虫动物群体。另外，也可以通过化学处理，去除反刍动物后期定植的原虫。虽然这两种方法在研究中均比较常见，但是在动物生产过程中都不是特别适用，因此，需要其他方法来调控原虫群（减少或者清除）。

离子载体（包括莫能菌素）是减少原虫数量、提高反刍动物生产效率、降低环境影响的有效添加剂（Nagaraja 等，1997）。离子载体能够促进反刍动物生长。已有研究证明莫能菌素可以抑制草履虫储存小泡内的蛋白水解（Fok 和 Ueno，1987）和食物液泡溶酶体融合（Gautier 等，1994）。莫能菌素可能对不同原虫有相似的作用方式，但是这一论断还未得到充分证实。此外，瘤胃原虫能适应离子载体，从而降低离子载体作用能力（Nagaraja 等，1997）。鉴于原虫的适应性，以及 2006 年欧盟禁止离子载体用于畜牧生产，因此必须找到一种原虫调控替代方法。

饲料中禁止使用离子载体后，人们把大量的注意力放在了植物源化合物上面。皂苷被人们大量研究，它由一个苷元或皂苷元通过糖苷键与一个或多个糖基连接而成（Francis 等，2002）。不同来源的皂苷具有不同程度的效应（Wallace 等，2002），从苜蓿中提取的皂苷可以改变体外瘤胃发酵（Lu 等，1987），减少原虫数量和甲烷产量（Patra 和 Saxena，2009）。然而，反刍动物日粮中添加皂苷的研究结果显示，皂苷效果差异较大，有的研究可以减少 6.7% 的甲烷排放（Santoso 等，2004），有的研究甚至没有任何效果（Ślisiński 等，2002；Holtshausen 等，2009）。人们认为甲烷产量的减少是丙酸产量增加和原虫数量减少的结果（Hristov 等，1999）。皂苷的不同来源和不同浓度有可能是造成结果间差异的原因，Hess等人在研究中使用了无患子果实的提取物，其皂苷含量为75%，远远高于丝兰提取物的含量（Śliśinski等，2002；Goel和Makkar，2012）。在这些研究中，日粮种类更多样，包括牧草、饲草以及精料（Goel和Makkar，2012）。

单宁是天然存在的植物次生代谢物，其作用方式与皂苷类似，具有抗菌作用，能够影响瘤胃发酵途径。单宁有两种类型——缩合单宁和水解单宁，这两种单宁都可能对动物产生不利或有益的影响。富含缩合单宁的植物或其提取物主要被用于降低毒性风险（Beauchemin等，2008）。人们认为，单宁的减甲烷作用可能是一种或两种方式造成的：直接影响瘤胃产甲烷菌或者通过减少饲料降解从而降低氢产量（Tavendale等，2005）。单宁对瘤胃纤维降解的负面影响可能是由纤维分解菌减少和生成单宁-纤维复合体而导致的（McSweeney等，2001；Makkar等，1995）。与皂苷的研究类似，在不同研究之间，单宁对原虫影响的研究结果也是不一致的、矛盾的（Benchaar等，2008；Goel和Makkar，2012）。

其他饲料添加剂，如精油，也可以在一定程度上减少瘤胃原虫数量（Benchaar等，2008）。目前精油对动物影响的作用机制仍缺乏清晰的认识。虽然有一些研究证明精油具有显著的效果，但是这些研究无法被复制（Patra，2011）。此外，精油使用还存在监管和安全问题，较高浓度的精油才能产生显著效果，这可能导致毒性作用以及成本增加。

有一家销售植物性饲料添加剂的公司（Delacon），产品主要用于提高反刍动物生产和健康性能，但是因为监管、管理和盈利问题，这些添加剂仍没有被广泛应用。全毛虫与产甲烷菌关系密切，因此应以定向去除全毛虫为目标。而内毛虫因具有纤维分解能力，应该尽量被保留。

10 未来趋势和结论

预计到2025年，全球人口将达到98亿人左右，提供充足且富含营养的食物将成为社会面临的主要问题之一（UN，2017）。因此，畜牧业提供的动物性食物的需求量会大幅增加，同时可持续性发展和环境保护压力越来越大。解决这一问题的方法是最大限度地提高瘤胃功能和效率，但要实现这一目标，首先必须对瘤胃微生物群有一个完整的认识。目前，我们对瘤胃原虫的认识非常有限，因此，使用最新的测序技术和分子技术进行研究是至关重要的。这些方法为进一步阐明原虫在瘤胃内的作用提供重要帮助，指导更好地调控微生物种群。与纯培养方法相比，宏基因组学、代谢组学、宏蛋白质组学和宏转录组学等元组学技术提供了研究整个微生物种群的机会。这些方法有助于开展荟萃分析，深入阐明微生物群内以及微生物群与宿主或营养素间的相互作用，尤其是避免了原虫培养的困难（Newbold等，2015）。基因数据库中瘤胃原虫序列仍较少（Park等，2018），在生物信息学分析过程中，原虫分类与注释可能存在错误或不足，导致结果不可靠以及一些代谢过程被忽视（Comtet-Marre等，2017）。因此，充分利用高通量测序和组学技术对瘤胃微生物尤其是原虫进行功能注释和基因组测序是很重要的。更多、更有力的数据是开发调控动物生产效率和环境效应调控技术的关键，这需要开发微生物组生物标记物，揭示原虫与产甲烷菌的关系和宿主与微生物组的交关作用。

在原虫与其他微生物群之间相互作用方面，特别是产甲烷菌与原虫间的关系方面，还存在文献空白。定植在瘤胃原虫上的产甲烷菌在丰度和组成方面随原虫属不同而有所差异，例如，较小的内毛虫与较大的多泡多甲虫相比，甲烷产量更高（Ranilla等，2007）。

应深入研究宿主基因型与微生物群尤其是与产甲烷相关的微生物物种之间的关系。Difford等（2018）研究了宿主基因和微生物群对甲烷产量的影响，甲烷产量13%的变异归因于瘤胃微生物群，21%的变异归因于宿主遗传。总之，瘤胃原虫研究是一项具有挑战性的工作，这也限制了我们对原虫作用的认识。但是，瘤胃原虫研究仍非常有必要，这对理解原虫的功能、开发针对特定瘤胃原虫来改善宿主表型和环境影响的方法非常重要。

11 更多信息

关键研究

- Hobson, P. N. and Stewart, C. S. 1997. The Rumen Microbial Ecosystem (2nd edn.). Blackie Academic and Professional, UK.
- Huws, S.A., Creevey, C.J., Oyama, L.B., Mizrahi, I., Denman, S.E., Popova, M., Muñoz-Tamayo, R., Forano, E., Waters., S.M., Hess, M., Tapio, I., Smidt, H., Krizsan, S.J., Yáñez-Ruiz, D.R., Belanche, A., Guan, L., Gruninger, R.J., McAllister, T.A., Newbold, C.J., Roehe, R., Dewhurst, R.J., Snelling, T.J., Watson, M., Suen, G., Hart, E.H., Kingston-Smith, A.H., Scollan, N.D., do Prado, R.M., Pilau, E.J., Mantovani, H.C., Attwood, G.T., Edwards, J.E., McEwan, N.R., Morrisson, S., Mayorga, O.L., Elliott, C. and Morgavi, D.P. 2018. Addressing global ruminant agricultural challenges through understanding the rumen microbiome: past, present and future. Front. Microbiol. 9, 1-33.
- Kamra, D.N. and Agarwal, N. 2003. Techniques in Rumen Microbiology. Indian Veterinary Research Institute, India.
- Newbold, C.J., de la Fuente, G., Belanche, A., Ramos-Morales, E. and McEwan, N.R. 2015. The role of ciliate protozoa in the rumen. Front. Microbiol 6.
- Russell, J.B. 2002. Rumen Microbiology. James B. Russell, Ithaca, NY.
- Williams, A.G. and Coleman, G.S. 1992. The Rumen Protozoa (1st edn.). Springer-Verlag, New York, NY.

关键会议/期刊

- Conference: INRA-Rowett conference on gastrointestinal tract function held biannually between the Rowett Institute, Aberdeen University and INRA-Theix, France.
- Conference: Congress on Gastrointestinal Function held biannually in the University of Chicago, Illinois.
- Journals: Microbiome; Animal Microbiome.

12 参考文献

Aban, M. L. and Bestil, L. C. 2016. Rumen defaunation: determining the level and frequency of *Leucaena leucocephala* Linn. forage. Int. J. Food. Eng. 2 (1), 55-60.

Abe, M. and Iriki, T. 1989. Mechanism whereby holotrich ciliates are retained in the reticulo-rumen of cattle. Br. J. Nutr. 62 (3), 579-87.

Akin, D. E. and Amos, H. E. 1979. Mode of attack on Orchard grass leaf blades by rumen protozoa. Appl. Environ. Microbiol. 37 (2), 332-8.

Bach, A., Calsamiglia, S. and Stern, M. D. 2005. Nitrogen metabolism in the rumen. J. Dairy Sci. 88 (Suppl. 1), E9-E21.

Bailey, R. W. and Clarke, R. T. J. 1963. Carbohydrase activity of tureen *Entodinium* spp. from sheep on a starch-free diet. Nature 198 (4882), 787.

Bailey, R. W. and Gaillard, B. D. 1965. Carbohydrases of the rumen ciliate *Epidinium ecaudatum* (Crawley). Hydrolysis of plant hemicellulose fractions and beta-linked glucose polymers. Biochem. J. 95 (3), 758-66.

Bailey, R. W., Clarke, R. T. J. and Wright, D. E. 1962. Carbohydrases of the rumen ciliate *Epidinium ecaudatum* (Crawley). Biochem. J. 83 (3), 517-23.

Bauchop, T. 1980. Scanning Electron Microscopy in the Study of Microbial Digestion of Plant Fragments in the Gut. New York Academic Press, London, UK.

Beauchemin, K. A., Kreuzer, M., O'Mara, F., et al., 2008. Nutritional management for enteric methane abatement: a review. Aust. J. Exp. Agric. 48 (2), 21-7.

Becker, E. R. and Hsiung, T. S. 1929. The method by which ruminants acquire their fauna of infusoria, and remarks concerning experiments on the host-specificity of these protozoa. Proc. Natl. Acad. Sci. U. S. A. 15 (8), 684-90.

Behura, S. K. and Severson, D. W. 2013. Codon usage bias: causative factors, quantification methods and genome-wide patterns: with emphasis on insect genomes. Biol. Rev. Camb. Philos. Soc. 88 (1), 49-61.

Belanche, A., De La Fuente, G. and Newbold, C. J. 2014. Study of methanogen communities associated with different rumen protozoal populations. FEMS. Microbiol. Ecol. 90 (3), 663-77.

Belanche, A., De La Fuente, G. and Newbold, C. J. 2015. Effect of progressive inoculation of fauna-free sheep with Vestibuliferida protozoa and total-fauna on rumen fermentation, microbial diversity and methane emissions. FEMS. Microbiol. Ecol. 91 (3).

Benchaar, C., McAllister, T. A. and Chouinard, P. Y. 2008. Digestion, ruminal fermentation, ciliate protozoal populations, and milk production from dairy cows fed cinnamaldehyde, quebracho condensed tannin, or *Yucca schidigera* Saponin extracts. J. Dairy Sci. 91 (12), 4765-77.

Brandis, G. and Hughes, D. 2016. The selective advantage of synonymous codon usage bias in *Salmonella*. PLoS Genet. 12 (3), e1005926.

Brock, F. M., Forsberg, C. W. and Buchanan-Smith, J. G. 1982. Proteolytic activity of the rumen microorganisms and effects of proteinase inhibitors. Appl. Environ. Microbiol. 44, 561-9.

Bryant, M. P. and Small, N. 1960. Observations on the ruminal microorganisms of isolated and inoculated calves. J. Dairy Sci. 43 (5), 654-67.

Chen, Y. C., Liu, T., Yu, C. H., et al., 2013. Effects of GC bias in next-generations-sequencing data on de novo genome assembly. PLoS ONE 8 (4).

Coleman, G. S. 1964. The metabolism of *Escherichia coli* and other bacteria by *Entodinium caudatum*. J. Gen. Microbiol. 37 (2), 209-23.

Coleman, G. S. 1978. The metabolism of cellulose, glucose and starch by the tureen ciliate protozoon *Endiplodinium maggii*. J. Gen. Microbiol. 107 (2), 359-66.

Coleman, G. S. 1983. Hydrolysis of fraction 1 leaf protein and casein by rumen entodiniomorphid protozoa. J. Appl. Bacteriol. 55 (1), 111-8.

Coleman, G. S. 1985. The cellulose content of 15 species of entodiniomorphid protozoa, mixed bacteria and plant debris isolated from the ovine rumen. J. Agric. Sci. 164, 349-60.

Coleman, G. S. 1986. The distribution of carboxymethyl cellulase between fractions taken from the rumens of sheep containing no protozoa or one of five different protozoal populations. J. Agric. Sci. 106 (1), 121-7.

Coleman, G. S., Davies, J. I. and Cash, M. A. 1972. The cultivation of rumen ciliates *Epidinium ecaudatum caudatum* and *Polyplastron multivesiculatum in vitro*. J. Gen. Microbiol. 73 (3), 509-21.

Coleman, G. S., Laurie, J. I., Bailey, J. E., et al., 1976. The cultivation of cellulolytic protozoa isolated from the rumen. J. Gen. Microbiol. 95 (1), 144-50.

Comtet-Marre, S., Parisot, N., Lepercq, P., et al., 2017. Metatranscriptomics reveals the active bacterial and eukaryotic fibrolytic communities in the rumen of dairy cow fed a mixed diet. Front. Microbiol. 8 (67), 67.

Dehority, B. A. 1993. Microbial ecology of cell wall fermentation. In: Jung, H. G., Buxton, D. R., Hatfield, R. D. and Ralph, J. (Eds), Forage Cell Wall Structure and Digestibility. Wiley, pp. 425-53.

Dehority, B. A. 1998. Generation times of *Epidinium caudatum* and *Entodinium caudatum*, determined in vitro by transferring at various time intervals. J. Anim. Sci. 76 (4), 1189-96.

Dehority, B. A. 2003. Rumen Microbiology. Nottingham University Press, Nottingham, UK.

Dehority, B. A. 2004. In vitro determination of generation times for *Entodinium exiguum*, *Ophryoscolex purkynjei* and *Endiplodinium maggii*. J. Eukaryot. Microbiol. 51 (3), 333-8.

Demeyer, D. I. 1981. Rumen microbes and digestion of plant cell walls. Agric. Environ. 6 (2-3), 295-337.

Demeyer, D. and Fievez, V. 2004. Is the synthesis of rumen bacterial protein limited by the availability of pre-formed amino acids and/or peptides? Br. J. Nutr. 91 (2), 175-6.

Denton, B. L., Diese, L. E., Firkins, J. L., et al., 2015. Accumulation of reserve carbohydrate by rumen protozoa and bacteria in competition for glucose. Am. Soc. Microbiol. 81 (5), 1832-8.

Destables, E., Thomas, N. A., Boxma, B., et al., 2005. The 3' untranslated region of mRNAs from the ciliate Nyctotherus ovalis. Acta Protozolog. 44, 231-6.

Devillard, E., Newbold, C. J., Scott, K. P., et al., 1999. A xylanase produced by the rumen anaerobic protozoan *Polyplastron multivesiculatum* shows close sequence similarity to family 11 xylanases from grampositive bacteria. FEMS Microbiol. Lett. 181 (1), 145-52.

Difford, G. F., Plichta, D. R., Lovendahl, P., et al., 2018. Host genetics and the rumen microbiome jointly associate with methane emissions in dairy cows. PLoS Genet. 14 (10), e1007580.

Eadie, J. M. 1962. Inter-relationships between certain rumen ciliate protozoa. Microbiology 29 (4), 579-88.

Eadie, J. M. 1967. Studies on the ecology of certain rumen ciliate protozoa. J. Gen. Microbiol. 49 (2),

175-94.

Eugène, M., Archimede, H. and Sauvant, D. 2004. Quantitative meta-analysis on the effects of defaunation of the rumen on growth, intake and digestion in ruminants. Livest. Prod. Sci. 85 (1), 81-97.

Fenchel, T. and Finlay, B. J. 2006. The diversity of microbes: resurgence of the phenotype. Philos. Trans. R. Soc. Lond. B Biol. Sci. 361 (1475), 1965-73.

Finlay, B. J., Esteban, G., Clarke, K. J., et al., 1994. Some rumen ciliates have endosymbiotic methanogens. FEMS Microbiol. Lett. 117 (2), 157-61.

Fok, A. K. and Ueno, M. S. 1987. Ionophores and weak bases inhibit phagolysosomal proteolysis in Paramecium. Eur. J. Cell Biol. 45 (1), 145-50.

Forsberg, C. W., Lovelock, L. K., Krumholz, L., et al., 1984. Protease activities of rumen protozoa. Appl. Environ. Microbiol. 47 (1), 101-10.

Francis, G., Kerem, Z., Makkar, H. P., et al., 2002. The biological action of saponins in animal systems: a review. Br. J. Nutr. 88 (6), 587-605.

Francisco, A. E., Santos-Silva, J. M., V Portugal, A. P., et al., 2019. Relationship between rumen ciliate protozoa and biohydrogenation fatty acid profile in rumen and meat of lambs. PLoS ONE 14 (9), e0221996.

Friedman, R. and Ely, B. 2012. Codon usage methods for horizontal gene transfer detection generate an abundance of false positive and false negative results. Curr. Microbiol. 65 (5), 639-42.

Gautier, M. C., Garreau de Loubresse, N., Madeddu, L., et al., 1994. Evidence for defects in membrane traffic in Paramecium secretory mutants unable to produce functional storage granules. J. Cell Biol. 124 (6), 893-902.

Goel, G. and Makkar, H. P. 2012. Methane mitigation from ruminants using tannins and saponins. Trop. Anim. Health. Pro. 44 (4), 729-39.

Gruby, D. and Delafond, H. M. O. 1843. Recherches ser des animalcules se développant en grand nombre dans l'estomacet dans les intestins, pedant la digestion des animaux herbivores et carnivores. Compt. Rend. Acad. Sci. 17, 1304-8.

Guttmacher, A. E. and Collins, F. S. 2003. Welcome to the genomic era. New Engl. J. Med. 349 (10), 996-8.

Hall, M. B. 2011. Isotrichid protozoa influence conversion of glucose to glycogen and other microbial products. J. Dairy Sci. 94 (9), 4589-602.

Henderson, G., Cox, F., Ganesh, S., et al., 2015. Rumen microbial community composition varies with diet and host, but a core microbiome is found across a wide geographical range. Sci. Rep. UK 5, 14567.

Hess, H. D., Beuret, R. A., Lötscher, M., et al., 2004. Ruminal fermentation, methanogenesis and nitrogen utilization of sheep receiving tropical grass hay-concentrate diets offered with *Sapindus saponaria* fruits and *Cratylia argentea* foliage. Anim. Sci. 79 (1), 177-89.

Hess, M., Sczyrba, A., Egan, R., et al., 2011. Metagenomic discovery of biomass-degrading genes and genomes from cow rumen. Science 331 (6016), 463-7.

Hobson, P. N. and Stewart, C. S. 1997. The Rumen Microbial Ecosystem (2nd edn.). Blackie Academic and Professional, UK.

Holtshausen, L., Chaves, A. V., Beauchemin, K. A., et al., 2009. Feeding saponin-containing *Yucca schidigera* and *Quillaja saponaria* to decrease enteric methane production in dairy cows. J. Dairy Sci. 92

(6), 2809-21.

Hristov, A. N., McAllister, T. A., Van Herk, F. H., et al., 1999. Effect of *Yucca schidigera* on ruminal fermentation and nutrient digestion in heifers. J. Anim. Sci. 77 (9), 2554-63.

Hungate, R. E. 1942. The culture of *Endiplodinium neglectum* with experiments on the digestion of cellulose. Biol. Bull. 83 (3), 303-19.

Hungate, R. E. 1943. Further experiments on cellulose digestion by the protozoa in the rumen of cattle. Biol. Bull. 84 (2), 157-63.

Hsu, J. T., Fahey, G. C., Merchen, N. R., et al., 1991. Effects of defaunation and various nitrogen supplementation regimens on microbial numbers and activity inthe rumen of sheep. J. Anim. Sci. 69 (3), 1279-89.

Huws, S. A., Kim, E. J., Kingston-Smith, A. H., et al., 2009. Rumen protozoa are rich in polyunsaturated fatty acids due to the ingestion of chloroplasts. FEMS Microbiol. Ecol. 69 (3), 461-71.

Huws, S. A., Lee, M. R. F., Kingston-Smith, A. H., et al., 2012. Ruminal protozoal contribution to the duodenal flow of fatty acids following feeding of steers on forages differing in chloroplast content. Br. J. Nutr. 108 (12), 2207-14.

Huws, S. A., Creevey, C. J., Oyama, L. B., et al., 2018. Addressing global ruminant agricultural challenges through understanding the rumen microbiome: past, present and future. Front. Microbiol. 9, 1-33.

Imai, S. 1998. Phylogenetic taxonomy of rumen ciliate protozoa based on their morphology and distribution. J. Appl. Anim. Res. 13 (1-2), 17-36.

Ivan, M., Charmley, L. L., Neill, L., et al., 1991. Metabolic changes in the rumen following protozoal inoculation of fauna-free sheep fed a corn silage diet supplemented with casein or soybean meal. Ann. Rech. Vet. 22 (2), 227-38.

Jouany, J. P. and Senaud, J. 1979. Role of rumen protozoa in the digestion of food cellulosic materials. Ann. Rech. Vet. 10 (2-3), 261-3.

Jouany, J. P. and Ushida, K. 1999. The role of protozoa in feed digestion-review. Asian Australas. J. Anim. Sci 12 (1), 113-28.

Kamra, D. N. 2005. Rumen microbial ecosystem. Curr. Sci. India 89 (1), 124-35.

Karnati, S. K. R., Sylvester, J. T., Noftsger, S. M., et al., 2007. Assessment of ruminal bacterial populations and protozoal generation time in cows fed different methionine sources. J. Dairy Sci. 90 (2), 798-809.

Konstantinidis, K. T., Ramette, A. and Tiedje, J. M. 2006. The bacterial species definition in the genomic era. Philos. Trans. R. Soc. Lond. B Biol. Sci. 361 (1475), 1929-40.

Kubesy, A. A. and Dehority, B. A. 2002. Forestomach ciliate Protozoa in Egyptian dromedary camels (Camelus dromedarius). Zootaxa 51 (1), 1-12.

Levy, B. and Jami, E. 2018. Exploring the prokaryotic community associated with the rumen ciliate protozoa population. Front. Microbiol. 9 (9), 2526.

Li, Z., Deng, Q., Liu, Y., et al., 2018. Dynamics of methanogenesis, ruminal fermentation and fiber digestibility in ruminants following elimination of protozoa: a meta-analysis. J. Anim. Sci. Biotechnol. 9 (9), 89.

Lourenco, M., Ramos-Morales, E. and Wallace, R. J. 2010. The role of microbes in rumen lipolysis and biohydrogenation and their manipulation. Animal 4 (7), 1008-23.

Lu, C. D., Tsai, L. S., Schaefer, D. M., et al., 1987. Alteration of fermentation in continuous culture of mixed rumen bacteria by isolated alfalfa saponins. J. Dairy Sci. 70 (4), 799-805.

Makkar, H. P. S., Blümmel, M. and Becker, K. 1995. *In vitro* effects of and interactions between tannins and saponins and fate of tannins in the rumen. J. Sci. Food Agric. 69 (4), 481-93.

McEwan, N. R., Eschenlauer, S. C. P., Calza, R. E., et al., 2000a. The 3' untranslated region of messages in the rumen protozoan *Entodinium caudatum*. Protist 151 (2), 139-46.

McEwan, N. R., Gatherer, D., Eschenlauer, S. C. P., et al., 2000b. An unusual codon usage pattern in the ciliatefamily Ophryoscolecidae and its implications for determining the source of cloned DNA. Anaerobe 6 (1), 21-8.

McSweeney, C. S., Palmer, B., McNeill, D. M., et al., 2001. Microbial interactions with tannins: nutritional consequences for ruminants. Anim. Feed Sci. Tech. 91 (1-2), 83-93.

Michałowski, T., Muszyński, P. and Landa, I. 1991. Factors influencing the growth of rumen ciliates Endiplodinium maggii in vitro. Acta. Protozool. 25, 419-26.

Michałowski, T., Harmeyer, J. and Bełżecki, G. 1999. The importance of washing the omasum for successful defaunation of sheep. J. Anim. Feed. Sci. 8 (4), 611-9.

Miltko, R., Belezecki, G. and Michalowski, T. 2012. Chitinolytic enzymes of the rumen ciliate *Endiplodinium maggii*. Folia Microbiol. (Praha) 57 (4), 317-9.

Moon-van der Staay, S. Y., van der Staay, G. W. M., Michałowski, T., et al., 2014. The symbiotic intestinal ciliates and the evolution of their hosts. Eur. J. Protistol. 50 (2), 166-73.

Morgavi, D. P., Sakurada, M., Tomita, Y., et al., 1994. Presence in rumen bacterial and protozoal populations of enzymes capable of degrading fungal cell walls. Microbiology (Reading, Engl.) 140 (3), 631-6.

Nagaraja, T. G., Newbold, C. J., Van Nevel, C. J., et al., 1997. Manipulation of ruminal fermentation. In: Hobson, P. N. and Stewart, C. S. (Eds), The Rumen Microbial Ecosystem. Blackie Academic and Professional, London, UK, pp. 523-632.

Newbold, C. J., McEwan, N. R., Calza, R. E., et al., 2005. An NAD+-dependent glutamate dehydrogenase cloned from the ruminal ciliate protozoan, *Entodinium caudatum*. FEMS Microbiol. Lett. 247 (2), 113-21.

Newbold, C. J., de la Fuente, G., Belanche, A., et al., 2015. The role of ciliate protozoa in the rumen. Front. Microbiol. 6, 1313.

Nguyen, S. H., Li, L. and Hegarty, R. S. 2016. Effects of rumen protozoa of Braham heifers and nitrate on fermentation and *in vitro* methane production. Asian-Australas. J. Anim. Sci. 29 (6), 807-13.

Nicholson, M. J., Theodorou, M. K. and Brookman, J. L. 2005. Molecular analysis of the anaerobic rumen fungus Orpinomyces – insights in an AT – rich genome. Microbiology (Reading, Engl.) 151 (1), 121-33.

Nikkhah, A. 2016. Rumen microbial protein synthesis in multigrain and barley fed lactating dairy cows. J. Anim. Sci. 94 (suppl_2), 160.

Orpin, C. G. 1975. Studies on the rumen flagellate Neocallimastix frontalis. J. Gen. Microbiol. 91 (2), 249-62.

Orpin, C. G. 1984. The role of ciliate protozoa and fungi in the rumen digestion of plant cell walls. Anim. Feed. Sci. Tech. 10 (2-3), 121-43.

Orpin, C. G. 1985. Association of rumen ciliate populations with plant particles *in vitro*. Microb. Ecol. 11 (1), 59-69.

Park, T., Wijeratne, S., Meulia, T., et al., 2018. Draft macronuclear genome sequence of the ruminal ciliate *Entodinium caudatum*. Microbiol. Res. Ann. 7 (1), e00826-18.

Patra, A. K. 2011. Effects of essential oils on rumen fermentation, microbial ecology and ruminant production. Asian J. Anim. Vet. Adv. 6 (5), 416-28.

Patra, A. K. and Saxena, J. 2009. The effect and mode of action of saponins on the microbial populations and fermentation in the rumen and ruminant production. Nutr. Res. Rev. 22 (2), 204-19.

Paul, R. G., Williams, A. G. and Butler, R. D. 1990. Hydrogenosomes in the rumen Entodiniomorphid ciliate *Polyplastron multivesiculatum*. J. Gen. Microbiol. 136 (10), 1981-9.

Prescott, D. M. 1994. The DNA of ciliated protozoa. *Microbiol*. Rev. 58 (2), 233-67.

Purser, D. B. 1961. A diurnal cycle for Vestibuliferida protozoa of the rumen. Nature 190, 831-2.

Purser, D. and Moir, R. 1959. Ruminal flora studies in the sheep. IX. The effect of pH on the ciliate population of the rumen *in vivo*. Aust. J. Agric. Res. 10 (4).

Quail, M. A., Smith, M., Coupland, P., et al., 2012. A tale of three next generation sequencing platforms: comparison of Ion Torrent, Pacific Biosciences and Illumina MiSeq sequencers. BMC Genomics 13, 341.

Ranilla, M. J., Jouany, J. P. and Morgavi, D. P. 2007. Methane production and substrate degradation by rumen microbial communities containing single protozoal species in vitro. Lett. Appl. Microbiol. 45 (6), 675-80.

Ricard, G., McEwan, N. R., Dutilh, B. E., et al., 2006. Horizontal gene transfer from bacteria to rumen ciliates indicates adaptation to their anaerobic, carbohydrates-rich environment. BMC Genomics 7, 22.

Rocha, E. P. C. 2004. Codon usage bias from tRNA's point of view: redundancy, specialization and efficient decoding for translational optimization. Genome Res. 14 (11), 2279-86.

Santoso, B., Mwenya, B., Sar, C., et al., 2004. Effects of supplementing galacto-oligosaccharides, Yucca schidigera or nisin on rumen methanogenesis, nitrogen and energy metabolism in sheep. Livest. Prod. Sci. 91 (3), 209-17.

Shabi, Z., Tagari, H., Murphy, M. R., et al., 2000. Partitioning of amino acids flowing to the abomasum into feed, bacterial, protozoal, and endogenous fractions. J. Dairy Sci. 83 (10), 2326-34.

Śliwiński, B. J., Soliva, C. R., Machmüller, A., et al., 2002. Efficacy of plant extracts rich in secondary constituents to modify rumen fermentation. Anim. Feed Sci. Tech. 101 (1-4), 101-14.

Stumm, C. K., Gijzen, H. J. and Vogels, G. D. 1982. Association of methanogenic bacteria with ovine rumen ciliates. Br. J. Nutr. 47 (1), 95-9.

Tavendale, M. H., Meagher, L. P., Pacheco, D., et al., 2005. Methane production from *in vitro* rumen incubations with Lotus pedunculatus and Medicago sativa, and effects of extractable condensed tannin fractions on methanogenesis. Anim. Feed Sci. Tech. 123-124, 403-19.

Thomas, N. A., Regensbogenova, M., de Graaf, R. M., et al., 2004. Diversity in the length of macronuclear chromosomes in the phylum Ciliophora: rumen ciliates and Nyctotherus-a case study. Reprod. Nutr. Dev. 44 (Suppl. 31).

Tymensen, L. D., Beauchemin, K. A. and McAllister, T. A. 2012. Structures of free-living and protozoa-associated methanogen communities in the bovine rumen differ according to comparative analysis of 16S rRNA and mcrA genes. Microbiology (Reading, Engl.) 158 (7), 1808-17.

UN. 2017. 2017 Revision of World Population Prospects: Key Findings and Advance Tables.

van Zijderveld, S. M., Fonken, B., Dijkstra, J., et al., 2011. Effects of a combination of feed additives on methane production, diet digestibility and animal performance in lactating dairy cows. J. Dairy Sci. 94 (3), 1445-54.

Varel, V. H. and Dehority, B. A. 1989. Ruminal cellulolytic bacteria and protozoa from bison, cattle-bison hybrids and cattle fed three alfalfa-corn diets. Appl. Environ. Microbiol. 55 (1), 148-53.

Veira, D. M. 1986. The role of ciliate protozoa in nutrition of the ruminant. J. Anim. Sci. 63 (5), 1547-60.

Wallace, R. J., McEwan, N. R., McIntosh, F. M., et al., 2002. Natural products as manipulators of rumen fermentation. Asian Australas. J. Anim. Sci 15 (10), 1458-68.

Wang, Y. and McAllister, T. A. 2002. Rumen microbes, enzymes and feed digestion-a review. Asian Australas. J. Anim. Sci 15 (11), 1659-76.

Wang, Z., Elekwachi, C. O., Jiao, J., et al., 2017. Investigation and manipulation of metabolically active methanogen community composition during rumen development in black goats. Sci. Rep. 7 (1), 422.

Warner, A. C. I. 1962. Some factors influencing the rumen microbial population. J. Gen. Microbiol. 28 (1), 129-46.

Wereszka, K., McIntosh, F. M., Michałowski, T., et al., 2004. A cellulase produced by the rumen anaerobic protozoan *Epidinium ecaudatum* has an unusual pH optimum. Endocyt. Cell Res. 15, 561-9.

Wereszka, K., Michałowski, T., Newbold, C. J., et al., 2006. Xylanolytic activity of the rumen protozoan Diploplastron affine. J. Anim. Feed Sci. 15, S43-6.

White, R. W. 1969. Viable bacteria inside the rumen ciliate *Entodinium caudatum*. J. Gen. Microbiol. 56 (3), 403-8.

Williams, A. G. 1986. Rumen vestibuliferida ciliate protozoa. Am. Soc. Microbiol. 50 (1), 25-49.

Williams, C. L. 2018. Prospecting the rumen protozoa for hydrolytic enzymes and their evolutionary origins. Ph. D. Thesis. Aberystwyth University.

Williams, A. G. and Coleman, G. S. 1992. The Rumen Protozoa (1st edn.). Springer-Verlag, New York, NY.

Wright, D. E. 1960. Pectic enzymes in rumen protozoa. Arch. Biochem. Biophys. 86, 251-4.

Yáñez-Ruiz, D. R., Abecia, L. and Newbold, C. J. 2015. Manipulating rumen microbiome and fermentation through interventions during early life: a review. Front. Microbiol. 6 (6), 1133.

Young, K. W., Thomas, N. A., Duval, S. M., et al., 2015. Isolation of macro-nuclear DNA from the rumen ciliate *Entodinium caudatum*. Endocyt. Cell Res. 26, 50-5.

第8章 瘤胃厌氧真菌

Matthias Hess，美国加利福尼亚大学戴维斯分校；
KaterinaFliegerová，捷克科学院动物生理学与遗传学研究所；
Shyam Paul，印度农业研究理事会家禽研究所；
Anil KumarPuniya，印度农业研究理事会国家乳业研究所
（成艳芬译）

1 前言

新美鞭毛菌门（Neocallimastigomycota）的成员已被证实广泛分布于草食性哺乳动物的瘤胃和后肠内，在纤维饲料消化过程中起关键作用（Gruninger等，2014）。这些严格厌氧的真菌最先定植在摄入的植物上，它们的菌丝尖端具有高浓度的纤维素分解酶，能够作为"生物撬棍"分解纤维物质，从而增加其他纤维素分解微生物获取植物碳水化合物的途径。尽管厌氧真菌在宿主肠道微生物区系中仅占很少部分（7%~9%），它们能释放植物组分中超过50%的可发酵糖，因此厌氧真菌在纤维饲料的消化过程中发挥着关键作用（Theodorou等，1996）。研究表明，驱除绵羊消化道内的厌氧真菌后，绵羊的饲料采食量大大降低（Gordon和Phillips，1993）。

2 厌氧真菌的生命周期

厌氧真菌通过无性繁殖，它们的生命周期包括可运动鞭毛（游动孢子）阶段，以及无运动性的营养繁殖（菌体）阶段（Mountfort，1987）。处于无运动性阶段时，瘤胃厌氧真菌定殖和降解纤维饲料，因此它们能在瘤胃纤维消化过程中起作用。瘤胃厌氧真菌的生命周期始于生殖体（孢子囊）分化成游动孢子，这也是对瘤胃液中可溶性碳水化合物、血红素浓度增加，以及其他相关卟啉物质释放的响应（Orpin和Greenwood，1986）。单鞭毛或者多鞭毛真菌游动孢子的丰度在采食后30~60 min达到峰值（Orpin和Joblin，1997）。虽然游动孢子能持续运动数小时（Lowe等，1987），但通常在孢子囊释放后的30 min内它们就附着并包裹在植物碎片上（Heath等，1986）。可运动的游动孢子受到可溶性糖和酚类物质梯度变化而产生趋化现象，从而在摄入纤维的潜在附着位点聚焦（Wubah和Kim，1996）。一旦附着，它们的体积增大，鞭毛脱落，通过加厚细胞壁形成包囊，但仍保持像变形虫一样的运动能力。包囊发芽参与鞭毛产生部位的相反极性端的芽管产生，单孢子囊（单中心）和多孢子囊（多中心）的瘤胃真菌的包囊发芽不同。单中心真菌进行内源性孢子萌发，在此期间细胞核在包囊内分裂，包

囊产生的游动孢子囊和无核假根系统，可以充当锚点和营养界面（Ho 和 Barr，1995）。在游动孢子发生后，剩余的菌体发生自溶（Lowe 等，1987）。多中心真菌展示出外源性包囊萌发的能力，在此过程中细胞核迁移至假根系统，从而形成多个孢子囊（Barr 等，1989）。单中心厌氧真菌的生命周期为 14~32 h，具体阶段包括：游动孢子附着、包囊萌发、菌体发育、繁殖阶段、游动孢子释放（Lowe 等，1987；Ho 等，1996）。多中心厌氧真菌既能通过产生游动孢子进行繁殖，又可通过营养生长进行繁殖，因为菌丝具有细胞核，因此能够无限生长。对于两个球根菌属（*Caecomyces* 和 *Cyllamyces*），其细胞核存在于吸附器官和孢囊梗，孢囊梗与外源性发育一致，这两个菌属的菌体发育不像单中心属那样严格，但是比多中心属更严格。

除了生物学的相关性，游动孢子数量还被作为琼脂滚管法测定瘤胃液中厌氧真菌丰度的指标（Bauchop，1979；Joblin，1981）。与其他方法一样，这种方法仅提供一个估算，因为游动孢子的丰度在 24 h 内有相当大的波动，而每个孢子囊的游动孢子之间的关系仍未被完全理解。尽管该方法存在局限性，但其优势在于能够观察到形态学表型，便于初步鉴定和分类，并且已使用该方法成功分离了不同厌氧真菌菌属（Phillips 和 Gordon，1988，1995；Borneman 等，1989）。滚管技术的另一种替代方法是在厌氧培养箱中接种琼脂平板（Borneman 等，1989）。考虑到已从草食动物不同胃肠道区域、胃肠道或粪便、家养动物或野生草食动物中分离出的厌氧真菌不同，这些差异可能是采用不同分离技术造成。因此应该考虑将这些技术结合起来，以最大限度地提高分离工作的成功率，并最准确地描述从这些生态位中分离出的真菌。

许多途径能促进真菌在动物之间的转移，如唾液、粪便、气溶胶（Lowe 等，1987；Orpin，1989）。举例而言，Milne 等报道在 39℃保存 8 h 的绵羊唾液、在 39℃保存 128 d 的干绵羊粪便仍可能分离出厌氧真菌（Milne 等，1989）。据报道，粪便中含有大量的真菌孢子囊，当粪便干燥时，真菌孢子囊的数量下降非常缓慢，并可维持活力长达 10 个月（Theodorou 等，1990）。据观察，干燥粪便真菌存活率高于潮湿条件下储存的粪便中真菌的存活率，Trinci 等（1988）猜测干燥处理刺激抗性结构的形成（Trinci 等，1988）。同时，潮湿条件下大量的细菌生长会抑制真菌的活性与生长（McGranaghan 等，1999），以上两点可以解释低湿度条件下真菌存活率增加的现象。厌氧真菌具有形成耐氧结构的能力，如 *Anaeromyces* 的多室孢子（Brookman 等，2000；Ozkose 等，2001），这种结构可以解释厌氧真菌存在于垃圾填埋场（McDonald 等，2012）、深海沉积物（Nagahama 和 Nagano，2012）和沼气反应器的现象（Haitjema 等，2014）。这也解释了反刍动物胃肠道内维持无厌氧真菌的难度，尽管孢子形成和后续萌发的过程目前仍不清楚（Gruninger 等，2014）。

3 厌氧真菌的分类学和形态学特征

瘤胃真菌及其分类学体系的研究历史持续了近一个世纪。20 世纪初，瘤胃液中的这些鞭毛微生物被首次描述，尽管观察到的细胞比瘤胃液中培养的真正鞭毛原虫小（Jensen 和 Hammond，1964），它们仍被归类为原虫（Liebetanz，1910；Braune，1913）。直到 1975 年，Colin Orpin 描述了它们生命周期的阶段（Orpin，1975），发现几丁质是它们细胞壁主要的结构多糖（Orpin，1977a），并将其正确地重新分类为第一个种的厌氧真菌（Orpin，1976，1977b）。在那个时代，真菌学界认为真菌在自然界中具有高度的氧化性，在缺氧的情况下无法代谢碳水化合物（Foster，1949；Vavra 和 Joyon，1966），这与 Orpin 对真菌的重新分类相矛盾。

在 1980 年，瘤胃真菌被归类为壶菌纲（Chytridiomycetes）、小壶菌目（Spizellomycetales）（Barr，1980），后续也根据其 18S rRNA 基因序列进一步证实了这一理论（Dore 和 Stahl，1991；Bowman 等，1992；Li 和 Heath，1992）。尽管有分子证据，但是瘤胃真菌独特的表型特征，如严格厌氧、线粒体缺失、氢体和多鞭毛游动孢子的存在（Li 等，1993）在其他小壶菌目（Spizellomycetales）中没有发现，这引出了对 Barr 分类体系（Barr，1980，1988）的质疑。这种分歧导致壶菌纲（Chytridiomycetes）出现一个新的目，即新美鞭菌目（Neocallimastigales），其中仅有新美鞭菌科（Neocallimastigaceae）一个科，该科仅包含厌氧真菌（Li 等，1993）。

"组装真菌生命树"项目采用多基因方法破译了低级真菌界成员间的进化发育关系，在真菌系统学方面取得了重大进展（James 等，2006）。"六基因系统发育"的组合包括 4 个来自 rRNA 操纵子（即 18S rRNA，28S rRNA，ITS）的基因和两个蛋白编码基因（即 EF1α，RNA 聚合酶 Ⅱ 最大亚基 RPB1，以及其第二大亚基 RPB2）。"六基因系统发育"的组合与瘤胃真菌独特的形态特征最终造成厌氧真菌从壶菌门（Chytridiomycota）分离，并形成一个新的菌门，即新美鞭菌门（Neocallimastigomycota）。该菌门在纲、目、科水平上分别由新美鞭菌纲（Neocallimastigomycetes）、新美鞭菌目（Neocallimastigales）、新美鞭菌科（Neocallimastigaceae）组成（Hibbett 等，2007）。与"六基因系统发育"相比，基于三个细胞核核糖体区域（即 ITS，LSU 和 SSU）和一个蛋白编码基因区域（即 RPB1）的分类（Schoch 等，2012），以及基于 46 个缓慢进化和 107 个中速进化的直系同源蛋白编码基因的系统发育学分类（Ebersberger 等，2012）均表明具有游动孢子的含几丁质真菌是单系起源的。Tedersoo 等（2018）充分证明了厌氧真菌是一个独立的菌门，并且在更高的分类水平上提出了命名的变化，从而将它们引入真菌亚界（表1）。基于分子系统发育、分化时间和单系准则，Tedersoo 等提出了新亚界 Chytridiomyceta，其中包含三个门，分别名为壶菌门（Chytridiomycota）、单毛壶菌门（Monoblepharomycota）和新美鞭菌门（Neocallimastigomycota）（Tedersoo 等，2018）。

为了确定厌氧真菌最接近的共同祖先，Wang 等（2018）利用 27 个新美鞭菌门（Neocallimastigomycota）厌氧真菌的基因组和转录组数据计算它们的分化时间。他们的分析结果表明，厌氧真菌大约在（73.5±5）百万年前开始分化，这与食草性哺乳动物的进化时间相一致。

表 1　厌氧真菌的常用分类

分类水平	Tedersoo 等（2018）	Hibbett 等（2007）	NCBI 分类法
界	Fungi	Fungi	Fungi
亚界	Chytridiomyceta[a]	—	—

(续表)

分类水平	Tedersoo 等（2018）	Hibbett 等（2007）	NCBI 分类法
门	Neocallimastigomycota[b]	Neocallimastigomycota[b]	Chytridiomycota[c]
亚门	Neocallimastigomycotina[d]		
纲	Neocallimastigomycetes[e]	Neocallimastigomycetes[e]	Neocallimastigomycetes[e]
目	Neocallimastigales[f]	Neocallimastigales[f]	Neocallimastigales[f]
科	Neocallimastigaceae[g]	Neocallimastigaceae[g]	Neocallimastigaceae[g]

注：[a] 亚界：Chytridiomycota Tedersoo et al. subkgd. nov.（Tedersoo 等，2018）。
[b] 门：Neocallimastigomycota M. J. Powell, phylum nov.（Hibbett 等，2007）。
[c] 门：Chytridiomycota（Barr，2001）。
[d] 亚门：Neocallimastigomycotina Tedersoo et al. subphyl. nov.（Tedersoo 等，2018）。
[e] 纲：Neocallimastigomycetes M. J. Powell, class. nov.（Hibbett 等，2007）。
[f] 目：Neocallimastigales（Li 等，1993）。
[g] 科：Neocallimastigaceae（Li 等，1993）。

4 厌氧真菌的属和种

迄今为止，已有 18 个厌氧真菌属被描述，但是非培养研究表明至少存在 25 个厌氧真菌属。由于厌氧真菌分类存在难题，被描述的菌种数量仅在 31~41 个。考虑到厌氧真菌的准确分类具备挑战性，以及一些已描述的物种存在冗余的可能性，因此明智的做法是认可较低的数据为正确数据。随着更加先进的分离培养方法的发展，以及特异性针对厌氧真菌的新分子技术，未来厌氧真菌将会有更加精确的分类。

形态特征是厌氧真菌分类的关键，其菌属的定义基于单中心或者多中心菌体的形成、丝状或者球状的假根、孢子囊的形状，以及是否形成单鞭毛或者多鞭毛的游动孢子。但是，形态学方法研究厌氧真菌的系统分类存在诸多困难，主要表现在广泛的形态学变异、孢子囊和假根结构的多型现象、单中心属和单鞭毛属形态特征的相似性、无法产生孢子囊、一些多中心真菌不存在游动孢子发生等。表 2 总结了目前公认厌氧真菌属的关键形态学特征。

迄今为止，大多数培养试验和非培养试验使用基于 ITS 的序列。近期研究已证实，在多个厌氧真菌属的种级分化中，内部转录间隔（ITS）序列分析不如基于核糖体大亚基（LSU）的序列分析，但是目前用于比较的 LSU 序列筛选的数量仍然有限。因此，与形态学无关的基于 DNA 分析的分子方法代表了一个非常强大的工具，能够阐明其他烦琐的分类系统，以及厌氧真菌个体间的分化和关系。近期，Hanafy 等（2019）进行了 LSU 序列的系统发育分析，并将目前已知的 18 个瘤胃真菌属的进化关系进行可视化展示（图 1）。

表 2　目前公认的厌氧真菌属的主要形态学特征

属（参考文献）	游动孢子/菌体形态	其他特点
Neocallimastix（Heath 等，1983）	多鞭毛/单中心，丝状	管状或者膨大的假根在孢子囊颈部以下，孢子囊位于不分支的或者分支的孢子囊柄上
Piromyces（Barr 等，1989）	单鞭毛/单中心，丝状	双鞭毛或者四鞭毛游动孢子，假根有/无孢子囊下膨大，成熟孢子囊通常有隔膜
Oontomyces（Dagar 等，2015）	单鞭毛/单中心，丝状	中间假根膨大，孢子囊在末端不形成短尖，长孢子囊柄可以通过明显的缢缩与假根菌丝体分离
Buwchfawromyces（Callaghan 等，2015）	单鞭毛/单中心，丝状	具有扭曲假根的大量假根系统，孢子囊顶端无突出物，隔膜可见，细胞核位于孢子囊，在孢子囊柄或假根中未观察到细胞核
Pecoramyces（Hanafy 等，2017）	单鞭毛/单中心，丝状	双鞭毛游动孢子，内源性和外源性游动孢子囊发育，单个末端孢子囊，孢子囊不分支，孢子囊下通常形成隆起或卵杯状膨大。大量无核假根系统缺乏假根膨大或缢缩
Liebetanzomyces（Joshi 等，2018）	单鞭毛/单中心，丝状	内源性和外源性孢子囊发育，大量无核的大量无核假根系统，单个顶生孢子囊，孢子囊在长短不等的孢子囊柄上有隔膜，有时在孢子囊下方形成卵杯状结构或者表现为囊肿状结构。孢子囊和假根结构在不同底物中具有典型的多形性
Feramyces（Hanafy 等，2018a，b）	多鞭毛/单中心，丝状	具有宽和狭窄菌丝，大量高度分支的假根形成，宽菌丝在不规则的间隔处缢缩，每个菌体由单个的顶生孢子囊，偶有假间隔孢子囊，孢子囊通常卷曲或宽而扁平，内源性和外源性游动孢子囊发育过程中，游动孢子通常在孢子囊下方形成一个突出的，或者卵杯状的膨大，游动孢子通过孢子囊顶孔释放，释放的同时保持完整孢子囊壁保持完整或整个孢子囊脱离
Agriosomyces（Hanafy 等，2019）	单鞭毛/单中心，丝状	内源性和外源性游动孢子囊发育，膨大的假根在紧缩的孢子囊颈部以下，孢子囊膨大
Aklioshbomyces（Hanafy 等，2019）	单鞭毛/单中心，丝状	双鞭毛三鞭毛状游动孢子，内源性和外源性游动孢子囊发育，乳头状孢子囊，偶有假间隔的内源性孢子囊，不分支的孢子囊柄
Capellomyces（Hanafy 等，2019）	单鞭毛/单中心，丝状	内源性和外源性游动孢子囊发育，不分支的孢子囊表现出囊下膨大，游动孢子通过顶孔释放

（续表）

属（参考文献）	游动孢子/菌体形态	其他特点
Ghazallomyces (Hanafy 等, 2019)	多鞭毛/单中心，丝状	内源和外源性游动孢子囊发育，高度分支的假根，不分支的孢子囊，孢子囊颈部狭窄，孢子囊通过顶孔释放
Joblinomyces (Hanafy 等, 2019)	单鞭毛/单中心，丝状	双鞭毛游动孢子，内源和外源性游动孢子囊发育，孢子囊的长度不同，具隔膜的多形性孢子囊的顶孔释放，形成空杯状孢子囊
Khoyollomyces (Hanafy 等, 2019)	单鞭毛/单中心，丝状	内源性和外源性游动孢子囊发育，高度分支的假根，分支的孢子囊柄具有 2~4 个孢子囊，游动孢子通过宽大的顶孔释放
Tahromyces (Hanafy 等, 2019)	单鞭毛/单中心，丝状	双鞭毛或四鞭毛状游动孢子，内源性和外源性游动孢子囊发育，分支的假根，短而膨大的孢子囊柄，孢子囊具有隔膜，孢子囊颈部缢缩
Anaeromyces (Breton 等, 1990)	单鞭毛/多中心，丝状	孢子囊具长尖形（短尖）的顶端，可以位于直立的、单独的、不分支的孢子囊柄上，菌丝高度分支，通常有许多缢缩（香肠状外观），有时有根状外观
Orpinomyces (Barr 等, 1989)	多鞭毛/多中心，丝状	多核根状菌丝，菌丝有大量分支，较宽的菌丝在近距离有紧密缩的点（串珠状或香肠状外观）
Caecomyces (Gold 等, 1988)	单鞭毛/单中心，球状	双鞭毛或三鞭毛状游动孢子，营养阶段无发达的分支假根系统，有球形或卵球形主体（附着器官或吸附器官），管状孢子囊柄和球茎状假根，细胞核通常存在于孢子囊和营养细胞中
Cyllamyces (Ozkose 等, 2001)	单鞭毛/多中心，球状	双鞭毛或三鞭毛状游动孢子，具有多个孢子囊的无假根的分支孢子囊柄，细胞核存在于球茎附着器官，它可以生在一个伸长或分支的孢子囊柄上

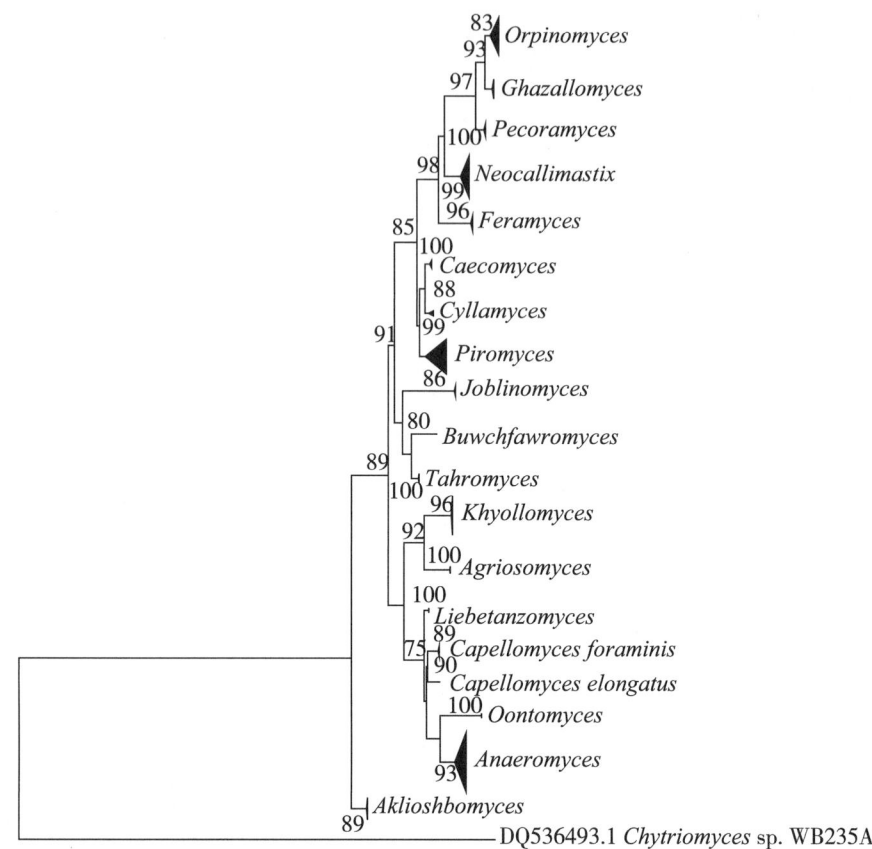

图1 基于 LSU 序列所有已知瘤胃厌氧真菌的进化关系。资料来源：经 Hanafy 等（2019）许可改编。使用最大可能法在 MAFFT 中对序列进行对比（Nakamura 等，2018），在 MEGA7 中构建系统发育树（Kumar 等，2016）。对于自引支持度超过 70% 的节点，将显示来自 100 个重复的支持率。

5 单中心属

5.1 Neocallimastix

Neocallimastix 是最早被描述的厌氧真菌属，其最初被 Braune（1913）分类为鞭毛原虫，并命名为 Callimastix frontalis。1966 年，该菌种更名为 Neocallimastix frontalis，但仍被分类为原虫（Vavra 和 Joyon，1966），直到 Orpin（1975）将其重新分类为真菌。后来，Heath 等（1983）对 N. frontalis 进行了更为详细的描述。但是，该菌株与 Orpin 最初研究的菌株不同，因此 Orpin 将他分离培养的真菌更名为 N. patriciarum（Orpin 和 Munn，1986）。Neocallimastix 广泛分布于反刍动物和非反刍草食动物中，是至今研究最充分的厌氧真菌。

菌种：多年来，已有 7 种属于 *Neocallimastix* 属（MB#25486）的不同菌种被报道，包括 *Neocallimastix frontalis*（Heath 等，1983），*Neocallimastix patriciarum*（Orpin 和 Munn，1986），*Neocallimastix joyonii*（Breton 等，1989），*Neocallimastix hurleyensis*（Webb 和 Theodorou，1991），*Neocallimastix variabilis*（Ho 等，1993b），*Neocallimastix cameroonii*（Ariyawansa 等，2015），*Neocallimastix californiae*（Li 等，2016）。研究发现，*Neocallimastix joyonii* 与 *Orpinomyces* 相同，因此该菌株更名为 *Orpinomyces joyonii*（Li 等，1991）。1991 年，Wubah 等对 *N. patriciarum* 和 *N. frontalis* 的模式培养菌进行了直接比较，并发现它们在形态和培养上是无法区分的（Wubah 等，1991）。Ho 和 Barr 后来提出将 *Neocallimastix patriciarum*，*N. variabilis* 与 *N. frontalis* 等同，并且他们质疑 *N. hurleyensis* 是否为一个独立的菌种（Ho 和 Barr，1995）。2017 年，3 个位点的系统发育分析为 *N. hurleyensis* 事实上应该等同于 *N. frontalis* 提供了分子证据（Wang 等，2017），这意味着 *N. frontalis*，*N. cameroonii* 和 *N. californiae* 是当前可从 *Neocallimastix* 菌属成功分离培养的仅有 3 个菌种。

5.2 *Piromyces*

瘤胃单鞭毛原虫在 100 多年前首次被报道（Liebetanz，1910），但是直到 20 世纪 70 年代末这些微生物才被认为是真菌（Orpin，1977a）。Orpin 将他分离的菌种命名为 *Piromonas communis*，并将其分类为 *Piromonas* 属。为了强调真菌的亲缘关系，*Piromonas* 后来被更名为 *Piromyces*（Gold 等，1988），并在第二年发表了对 *P. communis* 的详细描述（Barr 等，1989）。在厌氧真菌中，*Piromyces* 似乎是异质性最大的一个菌属，但是，由于系统发育分析和对比表明 *Piromyces* 菌属的分离物是多源的（Wang 等，2017），该菌属可能需要进行详细修订。

菌种：在 *Piromyces* 属（MB#25332），目前已有 9 个菌种被描述，即 *P. communis*（Barr 等，1989），*P. mae*（Li 等，1990），*P. dumbonicus*（Li 等，1990），*P. rhizinflatus*（Breton 等，1991），*P. spiralis*（Ho 等，1993c），*P. minutus*（Ho 等，1993d），*P. polycephalus*（Chen 等，2002），*P. irregularis*（Ariyawansa 等，2015）和 *P. finnis*（Li 等，2016）。另有一个菌种（即 *P. cryprodigmaticus*）也已通过 16S rRNA 扩增子测序被检测到。

5.3 *Oontomyces*

Dagar 等（2015）从印度骆驼前胃首次分离出该菌属的可培养代表菌，其形态学上与 *Piromyces* 属的成员相似。但是遗传学分析表明，该菌属与多中心 *Anaeromyces* 分支的成员更接近。目前，与该菌属相同或者相似的序列仅在骆驼样品中被报道，表明该真菌可能是骆驼特有的。

菌种：在 *Oontomyces* 属（MB#550795），目前仅有一个菌种 *Oontomyces anksri* 被描述（Dagar 等，2015）。

5.4 *Buwchfawromyces*

早先，*Buwchfawromyces* 属是用 SK2 表示的一个分支，其中仅包含不可培养的厌氧真菌（Kittelmann 等，2012；Koetschan 等，2014）。在 2015 年获得可培养菌种后，该菌属更名为 *Buwchfawromyces*（Callaghan 等，2015）。这个菌属中的代表菌在形态学上与 *Piromyces* 相似，并已从奶牛、绵羊、马的粪便中分离获得。但是，从遗传学角度它们与 *Piromyces* 不同，反而构成 *Anaeromyces* 的一个姐妹分支。

菌种：*Buwchfawromyces eastonii*（Callaghan 等，2015）是目前 *Buwchfawromyces*（MB# 550797）中唯一具有特征的代表菌。

5.5 *Pecoramyces*

Pecoramyces 早先被标记为 *Orpinomyces* 5（Kittelmann 等，2012），随后更名为 *Orpinomyces* sp. C1A（Youssef 等，2013），于 2017 年重新被分类为 *Pecoramyces ruminantium*（Hanafy 等，2017）。在形态学上，该菌属的唯一代表菌与其他厌氧真菌成员相似，但是遗传学上它是独立的，从而构成一个具有很强自引支持度的单系群体，并且是 *Orpinomyces* 属的姐妹分支。

菌种：*Pecoramyces ruminantium*（Hanafy 等，2017）是目前具有 *Pecoramyces*（MB# 552530）特点的唯一代表菌。

5.6 *Liebetanzomyces*

早先，该菌属是用 SP4 表示的分支（Paul 等，2018），但是在同年该分支第一个可培养代表菌被发现后，该菌属更名为 *Liebetanzomyces*（Joshi 等，2018）。该菌属从山羊瘤胃食糜中分离的可培养菌株在形态学上与其他单中心菌属相似，但是在遗传学上截然不同，从而构成 *Anaeromyces* 的一个姐妹分支。

菌种：在 *Liebetanzomyces*（MB#554794）中，目前仅 *Liebetanzomyces polymorphus* 被描述（Joshi 等，2018）。

5.7 *Feramyces*

Feramyces 是从环境序列中检测到的不可培养厌氧真菌的分支（Liggenstoffer 等，2010），后来被分类为 AL6（Kittelmann 等，2012）。2018 年，该菌属中首个也是目前唯一可培养分离菌被描述，该菌属更名为 *Feramyces*（Hanafy 等，2018a）。该菌属的菌种分离自野生的、未驯化的巴巴里绵羊和黇鹿的粪便样品和瘤胃食糜。*Feramyces* 的形态学代表菌与其他厌氧真菌属相似，但是在遗传学上是唯一的，从而构成以 *Neocallimastix-Pecoramyces-Orpinomyces* 分支为基础的的独立分支。

菌种：在 *Feramyces*（MB#823650）中，目前仅 *Feramyces austinii* 被描述（Hanafy 等，2018a）。

5.8 *Agriosomyces*

该菌属分离自欧洲盘羊和野生的波尔山羊粪便（Hanafy 等，2019）。基于 LSU 序列，*Agriosomyces* 构成 *Khoyollomyces* 的一个姐妹分支，与 *Oontomyces-Anaeromyces-Liebetanzomyces* 分支聚类。基于 ITS1 区域的系统发育分析显示，该菌属与 *Ghazallomyces-*

Orpinomyces-Pecoramyces 聚类为姐妹分支。

菌种：在 *Agriosomyces*（MB#830737）中，目前仅 *Agriosomyces longus* 被描述（Hanafy 等，2019）。

5.9 *Aklioshbomyces*

该菌属在形态学上与 *Oontomyces* 和 *Feramyces*（即假夹层内生孢子囊），以及 *Piromycesmae*（即乳头状孢子囊）有某些相似点，但是在系统发育学上该菌属与其他菌属相当疏远，从而构成目前所有可培养厌氧真菌的一个姐妹分支（Hanafy 等，2019）。

菌种：目前为止，在 *Aklioshbomyces*（MB#830735）中，仅一种真菌（*Aklioshbomyces papillarum*）被描述。

5.10 *Capellomyces*

该菌属包含从野生波尔山羊和家养山羊粪便分离的两个代表菌。形态学特征与 *Piromyces rhizinflatus*（Ho 和 Barr，1995）和 *Neocallimastix frontalis*（Barr 等，1995）有某些共同点，但是在遗传学上它们是唯一的。LSU 序列将该真菌属归入 *Oontomyces-Anaeromyces-Liebetanzomyces* 属分支。

菌种：在 *Capellomyces*（MB#830739）中，目前已有两个菌种被描述，分别名为 *Capellomyces foraminis* 和 *Capellomyces elongatus*（Hanafy 等，2019）。

5.11 *Ghazallomyces*

该菌属仅在近期被描述，Hanafy 等（2019）从野生的印度梅花鹿的粪便中分离出一个代表菌。该菌属构成一个具有很强自引支持度的单系聚类，作为 *Orpinomyces* 和 *Pecoramyces* 的一个姐妹分支。

菌种：在 *Ghazallomyces*（MB#830733）中，目前仅 *Ghazallomyces constrictus* 被详细描述（Hanafy 等，2019）。

5.12 *Joblinomyces*

该菌属目前包含 Hanafy 等（2019）从驯养的山羊和绵羊粪便中分离出的一个菌种（*Joblinomyces apicalis*）代表。展现出的游动孢子释放模式和孢子囊形态与 *Piromyces minutus* 相似（Ho 和 Barr，1995），但是在遗传学上是唯一的。基于 ITS 1 序列，该菌属构成与 *Neocallimastix*、*Feramyces*、*Agriosomyces*、*Ghazallomyces*、*Orpinomyces* 和 *Pecoramyces* 有远亲关系的一个独立分支。基于 LSU 序列，*Joblinomyces* 构成 *Buwchfawromyces* 和 *Tahromyces* 的姐妹分支。

菌种：在 *Joblinomyces*（MB#830867）中，目前仅 *Joblinomyces apicalis* 被描述（Hanafy 等，2019）。

5.13 *Khoyollomyces*

该菌属是 Liggenstoffer 等（2010）首次从环境序列中描述的一个分支，后续被标记为

AL1（Kittelmann 等，2012），根据从驯养的马和动物园饲养的斑马粪便中分离出的首个可培养代表菌，该菌属被更名为 *Khoyollomyces*（Hanafy 等，2019）。基于 ITS 1 序列分析，该菌属构成一个独立分支。基于 LSU 序列，*Khoyollomyces* 作为 *Agriosomyces* 属的姐妹分支，构成一个具有很强自引支持度的单系聚类。

菌种：在 *Khoyollomyces*（MB＃830741）中，目前仅 *Khoyollomyces ramosus* 被描述（Hanafy 等，2019）。

5.14 *Tahromyces*

近期，Hanafy 等（2019）从野生的尼尔吉里塔尔羊粪便中分离出代表菌，进而对该菌属进行了描述。形态学上，该菌属与 *Piromyces*、*Buwchfawromyces* 和 *Neocallimastix* 相似，但是遗传学上该菌属是唯一的，构成一个具有自引支持度的单系聚类，与 *Buwchfawromyces* 形成姐妹分支。

菌种：在 *Tahromyces*（MB＃830865）中，目前仅 *Tahromyces munnarensis* 被描述（Hanafy 等，2019）。

5.15 单中心菌属的形态学

具有单中心的真菌，包括单鞭毛的（*Piromyces*、*Oontomyces*、*Buwchfawromyces*、*Pecoramyces*、*Liebetanzomyces*、*Aklioshbomyces*、*Agriosomyces*、*Capellomyces*、*Joblinomyces*、*Khoyollomyces*、*Tahromyces*），或多鞭毛的（*Neocallimastix*、*Feramyces*、*Ghazallomyces*）球型到宽椭球形（直径 4~13 μm）游动孢子，它们可以内源性或者外源性萌发。形成双鞭毛游动孢子是偶发性的（6%~9%）（Hanafy 等，2017，2019），对于 *Piromyces* 甚至可以观察到 4 个鞭毛（Barr 等，1989）。多鞭毛游动孢子能携带 7~17 根鞭毛。鞭毛长度为 15~37 μm，与原包囊细胞分离。单中心属在营养阶段无核，高度分支，孢子囊为球形、卵球形、椭圆形、棒状、三角形、梨形、心形、卵形、近圆柱形或者不规则形状（长 40~185 μm，宽 20~100 μm）。主要假根可以呈管状，或者在孢子囊以下膨大。在 *Oontomyces*（Dagar 等，2015）和 *Khoyollomyces*（Hanafy 等，2019）可以观察到卵球形到亚卵球形间假根膨胀。宽大的菌丝可表现出不规则的多重缢缩。孢子囊可位于不分支或者分支的孢子囊柄上。孢子囊柄长度不同（15~600 μm），通常是卷曲或者宽而扁平，经常在孢子囊下方形成突起或卵杯状膨胀（Hanafy 等，2017，2018a，b；Joshi 等，2018）。在成熟的游动孢子囊中，通常可以看到分隔游动孢子囊和孢子囊柄的隔膜（Heath 等，1983；Callaghan 等，2015；Joshi 等，2018）。孢子囊和假根结构在不同底物上具有典型的多形性（Joshi 等，2018）。

6 多中心属

6.1 *Orpinomyces*

Barr 等（1989）将 *Orpinomyces bovis* 描述为该菌属的模式菌种，同年 Breton 等（1989）描述了 *Neocallimastix joyonii*。由于形态描述的高度相似性，该菌株被认为等同于

Barr 的菌株，在两年后被归入菌属 *Orpinomyces*，且更名为 *Orpinomyces joyonii*（Li 等，1991）。

菌种：在 *Orpinomyces*（MB#25326）中，目前已描述 3 个菌种，分别是 *O. bovis*（Barr 等，1989），*O. joyonii*（Li 等，1991）和 *O. intercalaris*（Ho 等，1994）。但是，*Orpinomyces bovis* 后续被证实等同于 *O. joyonii*。

6.2 *Anaeromyces*

1990 年，有两篇文章描述了一个新的多中心厌氧真菌属，分别是从奶牛瘤胃中分离的 *Anaeromyces mucronatus*（Breton 等，1990）和从阉牛瘤胃分离的 *Ruminomyces elegans*（Ho 等，1990）。基于某些形态学差异，*R. elegans* 最初被描述为 *Anaeromyces* 的一个新菌种，并更名为 *Anaeromyces elegans*（Ho 等，1993a）。随后，Ho 和 Barr（1995）对 *A. elegans* 的分类表示质疑。因为没有 *A. elegans* 的培养物或 DNA 可用于进一步阐明其信息，目前仍不确定 *A. elegans* 是否等同于 *A. mucronatus*。有观点认为这两个菌属是等同的，且 *Anaeromyces mucronatus* 为有效的菌种名。

菌种：在 *Anaeromyces*（MB#27188）中，目前已描述了 4 个菌种，包括 *A. mucronatus*（Breton 等，1990）、*A. polycephalus*，*A. robustus*（Li 等，2016）和 *A. contortus*（Hanafy 等，2018b）。2012 年，根据其 ITS 序列，有观点认为将之前描述的 *Piromyces polycephalus*（Chen 等，2002）更名为 *Anaeromyces polycephalus*（Kirk，2012）。但是，由于该菌种在形态学和遗传学上都与 *Piromyces* 和 *Anaeromyces* 的成员不同，因此它可能不属于这两个菌属，应该被重新分类。

另一个在 *Anaeromyces* 属重新分类的菌种是 *A. robustus*，它与 *Oontomyces* 共同构成 *Anaeromyces* 的姐妹分支。但是，对其部分 28S rRNA 序列分析表明，*A. robustus* 是一个与 *Anaeromyces* 属无关的独立群体（Li 等，2016）。

6.3 多中心菌属的形态学

多中心菌体的真菌具有单鞭毛（*Anaeromyces*）或多鞭毛（*Orpinomyces*）的球形或者椭圆形（直径 8～16 μm）的游动孢子。多中心游动孢子可携带 10～25 根鞭毛（长 30～50 μm）。营养阶段由密集的多核假根状菌丝组成，菌丝普遍分支，比单中心菌种菌丝更大。这些菌丝高度分支、很大、通常有大量的缢缩，因此呈串珠状或香肠状的外观（Barr 等，1989；Breton 等，1990；图 2）。据观察，*Anaeromyces* 的菌丝具有易损的、根状的外观（Breton 等，1990）。孢子囊具有球形、近球形、椭球形或者不规则形状（长 30～120 μm，宽 8～80 μm），*Anaeromyces* 具有典型的尖形（短尖形）顶端（Breton 等，1990）。*Orpinomyces* 的孢子囊是远端的，形成于单个或者分支的孢子囊柄复合体的顶端（Li 等，1991）或者作为菌丝间隔中的较小膨胀，或者作为菌丝的侧枝生长（Ho 等，1994）。*Anaeromyces* 的孢子囊可位于直立的、孤立的、不分支的孢子囊柄（长 5～100 μm）上，孢子囊柄从菌丝的侧面或者远端产生（Breton 等，1990）。但是，某些培养物无法产生成熟的孢子囊，游动孢子很少，因此很难通过形态学手段进行分类（Ho 和 Bauchop，1991）。

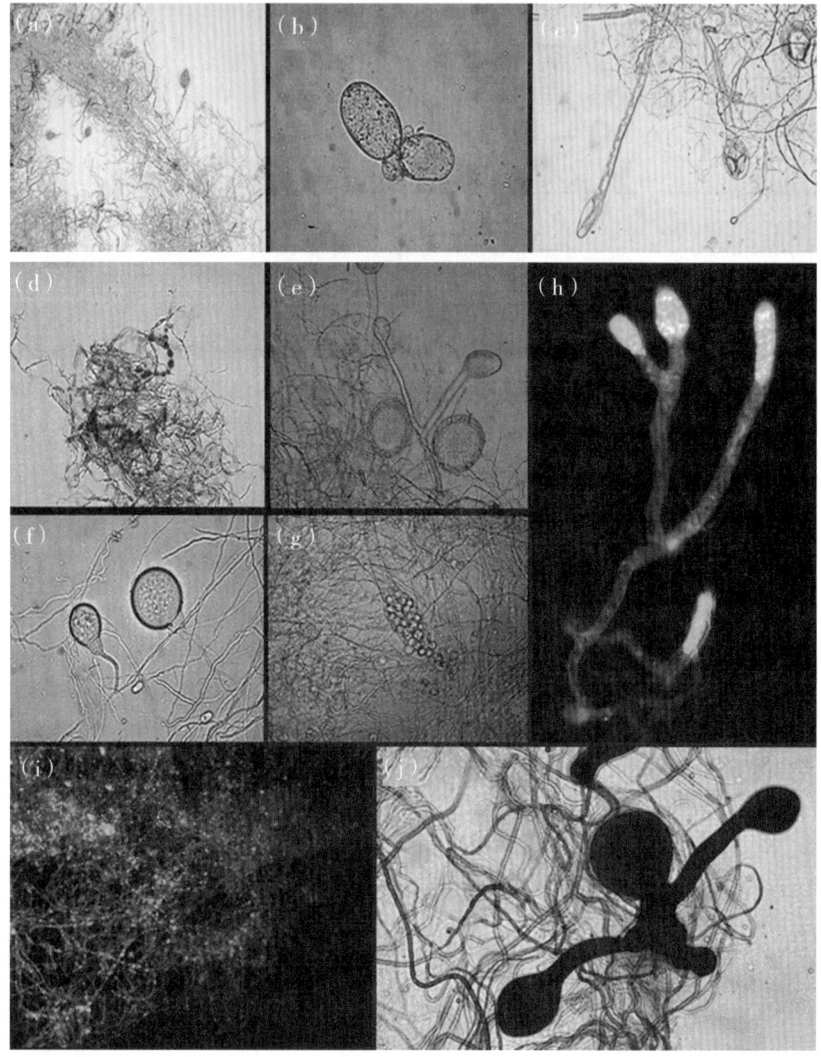

图 2　光镜下厌氧真菌的特征。（a）*Anaeromyces*：位于不分支孢子囊柄的孢子囊具尖形（短尖形）的顶端；（b）*Caecomyces*：球根状的假根；（c）*Piromyces*：长的、无分支的孢子囊柄；（d）*Anaeromyces*：具有大量缢缩的菌丝，呈香肠状或串珠状外观；（e）*Piromyces*：分叉的孢子囊柄；（f）*Piromyces*：在孢子囊下方的孢子囊柄形成一个卵杯状膨胀；（g）*Piromyces*：游动孢子从孢子囊释放；（h）*Piromyces*：分支的孢子囊柄，细胞核集中在顶端（DAPI 染色）；（i）*Orpinomyces*：普遍分支的菌丝上存在密集的多核根状菌丝体（DAPI 染色）；（j）*Piromyces*：孢子囊形状不规则（卢戈氏染色）（彩图 3）。

7 球根属

7.1 *Caecomyces*

表现出该菌属形态学的菌株最初被描述为 *Sphaeromonas*。Liebetanz（1910）最先将 *Sphaeromonas* 这一属名用于鞭毛原虫，后来 Orpin（1976）使用该属名描述与 Liebetanz 所描述的培养物相似的具有游动孢子的真菌分离物。为了强调这些分离菌属于厌氧真菌，Gold 等（1988）建立了一个新的菌属 *Caecomyces*，*Sphaeromonas communis* 更名为 *Caecomyces communis*。

菌种：在 *Caecomyces*（MB#25287）中，目前已经描述了 4 个菌种，包括 *C. communis*（Gold 等，1988），*C. equi*（Gold 等，1988），*C. sympodialis*（Chen 等，2007）和 *C. churrovis*（Henske 等，2017）。Ho 和 Barr（1995）质疑 *C. equi* 作为一个独立菌种的分类，但是由于缺少对 *C. equi* 培养物的进一步试验验证，很难确定该菌种是否与 *C. communis* 等同。

7.2 *Cyllamyces*

能够形成球状菌丝体的 *Cyllamyces*，该菌属的首个代表菌是从威尔士的奶牛新鲜粪便中分离出来的（Ozkose 等，2001）。在家牛和水牛的瘤胃液和新鲜粪便（Sridhar 等，2007），野牛和羚羊的粪便（Liggenstoffer 等，2010）中也有关于 *Cyllamyces* sp. 的报道。据报道，高能量日粮可提高奶牛 *Cyllamyces* 相对丰度（Kumar 等，2015）。但是，根据现有的知识，*Cyllamyces* 并不是普遍存在于草食动物的一个菌属。

菌种：在 *Cyllamyces*（MB#28540）中，目前仅 2 个菌种被描述，分别是 *C. aberensis*（Ozkose 等，2001）和 *C. icaris*（Sridhar 等，2014）。

7.3 球根属的形态学

球根形态是单中心 *Caecomyces* 和多中心 *Cyllamyces* 的典型特征，两个菌属都具有圆形或椭圆形（*Caecomyces*）（直径 7~9 μm；图 2）的单鞭毛游动孢子。偶尔有 2~3 条鞭毛（长 20~30 μm），*Caecomyces sympodialis* 的游动孢子甚至可以是四鞭毛的（Chen 等，2007）。*Caecomyces* 的营养阶段不存在发达的分支假根系统，由球形或者卵球体（附着或吸附）、管状孢子囊柄、球根状假根组成。细胞核通常存在于孢子囊（直径 22~33 μm）和营养体中（Gold 等，1988）。菌体发育可能在单孢子囊或者多孢子囊阶段终止（Gold 等，1988）。*Cyllamyces* 的营养阶段包含球根状吸附器官（直径 30~50 μm），无假根，有多个球形或者卵球形孢子囊（直径 12~15 μm；Ozkose 等，2001）。孢子囊可生于单个伸长的或分支的孢子囊柄（长 85 μm）。即使作者将 *Cyllamyces* 菌体描述为多中心的，其菌体发育仍被认为是单中心多孢子囊的，因为细胞核存在于菌体的营养部分（球根状吸附器官和孢子囊柄），并且不断产生大量的孢子囊（Ozkose 等，2001）。

8 厌氧真菌的基因组学

厌氧真菌被认为是木质纤维素分解过程的关键角色，它们通过物理穿透和各种细胞壁降解酶共同完成。分子生物学和计算生物学技术的进步促进了实验和技术标准的建立，使我们现在能够破译复杂的真菌基因组和单个真菌的生物学功能，而不需要广泛的生物信息学网络和训练有素的专业人员。真菌基因组中异常高的 A 碱基和 T 碱基含量、大量的非编码基因间区，真菌基因组内的基因重复，以及厌氧真菌未知的倍数性和复杂的生命周期，使得对获得的数据进行核酸测序和分析变得相当困难（Brownlee，1989；Chen 等，2006；Youssef 等，2013；Gruninger 等，2014）。

在过去，这些困难都是通过无差别测序解决，只有少数研究机构能负担得起测序成本。遇到的障碍以及如何解决的一个很好的例子是 *Piromyces* sp. E2 的基因组草图，*Piromyces* sp. E2 是一种高产的生物质降解菌。2011 年，美国能源部联合基因组研究所利用桑格数据进行了初步组装（https://genome.jgi.doe.gov/PirE2_1/PirE2_1.info.html）。6 年后，Illumina 的短长度测序技术生成的数据得以完成初步组装，为解释 *Piromyces* sp. E2 的生物学和分子机制提供了有用的见解（Haitjema 等，2017）。

最近，PacBio 开发的"单分子实时"（SMRT）测序技术为基因组测序开辟了新的可能性。SMRT 技术产生长读长时伴随的高出错率可以通过 Illumina 修正 SMRT 数据解决。通过不同类型的"下一代测序"（NGS）技术的结合，成功获得了多种瘤胃厌氧真菌的基因组，如 *Pecoramyces ruminantium*（原先分类为 *Orpinomyces* sp. 菌株 C1A；Youssef 等，2013；Hanafy 等，2017），*Anaeromyces robustus*，*Piromyces finnis* 和 *Neocallimastix californiae*（Haitjema 等，2017）。

表 3 列出了目前瘤胃厌氧真菌基因组测序情况。这些基因组研究除了 *Pecoramyces* 的代表菌 *Pecoramyces ruminantium* 外，目前仅限于 *Orpinomyces* 的成员，证明了新美鞭菌科（Neocallimastigaceae）的成员使用纤维小体的猜想。据报道，纤维小体是大型多蛋白复合物，能够加强一系列植物生物质降解酶，从而提高生物质水解（Artzi 等，2017）。

8.1 单细胞基因组学

自 20 世纪 90 年代以来，基于 PCR 的方法以及先进的等温 DNA 扩增技术使得单细胞测序成为解读人类细胞基因组变异的有力工具（Zhang 等，1992；Dean 等，2001）。不久后，单细胞基因组学（SCG）就成功用于古菌和细菌的基因组测序（Raghunathan 等，2005；Zhang 等，2006；Podar 等，2013）。甚至组装了一种不可培养的瘤胃微生物（Hess 等，2011）和海洋单细胞真核生物（Yoon 等，2011；Strassert 等，2018）的基因组草图。真菌 SCG 有关的难题与传统测序方法相同（例如，高倍数性、线粒体基因组、多染色体、转座子和广泛的 GC 变化）。但是，对于 SCG，高度坚固的真菌细胞壁是一个更严重的问题，因为它阻碍了对 DNA 模板的获取，导致在下游扩增过程中模板数量有限（Ahrendt 等，2018）。为了克服这一障碍，Ahrendt 等开发了一种稳健捕获和从头组装的实验方法，使得需氧真菌单细胞基因组组装的完成度达到 88%，在测序前结合目标生物的多个细胞时，可使得完整度提高至约97%（Ahrendt 等，2018）。这种强大的技术应用到厌氧真菌基因组研究仅仅是时间问题。

表3　目前可获取的厌氧真菌基因组统计

微生物名称	基因组大小[a]（碱基对）	基因数量[a]	参考文献
Anaeromyces robustus	71 685 009	12 832	Haitjema 等（2017）
Caecomyces churrovis	165 495 782	15 009	Henske 等（2017）
Neocallimastix californiae	193 495 782	20 219	Haitjema 等（2017）
Pecoramyces ruminantium（原名 *Orpinomyces* sp. C1A）	100 954 185	18 936	Youssef 等（2013）
Piromyces finnis	56 455 805	10 992	Haitjema 等（2017）
Piromyces sp. E2	71 019 055	14 648	Haitjema 等（2017）

注：[a] https://genome.jgi.doe.gov/neocallimastigomycota/.

8.2　转录组学

转录组和蛋白组研究中，生物个体的基因表达和蛋白质组成在一套确定的代谢条件下测定，并相互比较，相比传统基因组学研究测定生物体的整体代谢潜力，转录组和蛋白组能够提供一个更具功能性的方式。应用于真核生物，包括应用于厌氧真菌时，这两种方法额外的好处是它们可提供内含子组装和更长的重复读长，这两者是真菌基因组组装的主要难题，已经过时（Gruninger 等，2014）。

以杂交为基础测定细菌基因表达谱的方法在20世纪90年代末已被广泛使用，但是在真菌学方面使用较少（Nilsson 等，2019）。基于功能的杂交检测，如 CAZyChip, FibroChip, 和 GeoChip 被开发用于快速定量功能的关键酶表达，即使是在复杂的微生物群落也可进行（Tu 等，2014；Abot 等，2016；Comtet-Marre 等，2018）。这些检测方法包含探针，能允许量化那些在序列上足够相似，且可以与连接在阵列矩阵上探针结合的基因。由于关键基因可能没有足够的亲和力，目的基因需要有相对较高的相似性是一个显著的缺点。随着核酸测序和序列分析成本的不断降低，这些杂交阵列方法将变得不那么常见。

Anaeromyces robustus、*Neocallimastix californiae* 和 *Piromyces finnis* 的转录组分析表明，*N. californiae* 和 *P. finnis* 的木聚糖水解酶活力比商业的 *Aspergillus* 酶混合物提高了3倍，约2%的转录本编码木质纤维素糖苷水解酶（GH）和其他碳水化合物活性酶（CAZymes；Solomon 等，2016）。与 *Trichoderma* 和 *Aspergillus* 相比，上述3种厌氧真菌都表达了更多的半纤维素酶（即 GH10）和果胶降解酶，使它们能去除半纤维素和果胶，从而在获取植物细胞内富含能量的碳水化合物方面更有优势。有趣的是，*A. robustus* 更偏好水解葡萄糖，而不是更复杂的碳水化合物，这种对简单糖类的倾向表明一旦植物纤维中复杂的成分被其他微生物消化，*A. robustus* 对植物的降解更为重要（Solomon 等，2016）。

除了被分类为纤维素酶和半纤维素酶的转录编码 GH 家族外，转录组学还表明了这些真菌表达的辅助酶（AAs）和羧酸酯酶（CE）基因数量的增加。这也解释了为什么 Neocallimastigomycota 门的成员，包括 *Pecoramyces ruminantium* 能够降解一系列富含木质素，且未经前处理的饲草（Couger 等，2015；Solomon 等，2016）。Wang 等结合 Roche 和 Illumina 的 NGS 技术鉴定了瘤胃厌氧真菌 *Neocallimastix patriciarum* W5 在以稻秸为底物时积极表达的 288 个独特的类 GH 重叠群（Wang 等，2011）。这些类 GH 重叠群可编码多种纤维素酶、半纤维素酶，包括 GH10 家族成员，几丁质酶和包含非催化的锚定结构域（NCDDs）的开放性阅读框，后者则是纤维小体的特征（Gilmore 等，2015）。

Gruninger 等（2018）鉴定了大量的 CAZymes，分别占 *Anaeromyces mucronatus*、*Neocallimastix frontalis*、*Orpinomyces joyonii* 和 *Piromyces rhizinflata* 总转录本的 8.1%，9%，11.2% 和 8.9%。此外，Gruninger 等能辨认 12 个 CAZy 家族和 10 个碳水化合物结合模块（CBMs），与需氧真菌、非瘤胃细菌和瘤胃细菌转录组相比，上述 3 种瘤胃厌氧真菌的转录组丰度增加了 2 倍以上（表 4）。在这些厌氧真菌转录组中，富集的 CAZyme 家族包括 GH6、GH11、GH48、CE1，以及 CBM1、CBM26、和 CBM29 的丰度甚至提高了 5 倍。更引人注意的是，这些表达谱中，纤维素特异性的 CBM10 和几丁质结合的 CBM18 的转录本丰度提高了 20 倍以上（表 4）。

由于真菌纤维小体 NCDDs 与细菌纤维小体 NCDDs 相似性较低，因此前人认为真菌纤维小体是独立进化的，但是通过水平基因转移，真菌在瘤胃生态系统从共存的细菌获取了一些有益的特性（如催化域）（Haitjema 等，2017）。*Caecomyces churrovis* 是一种缺乏大量假根系统的厌氧真菌，这与其他已知肠道真菌的特点不同，通过 Illumina 的短读长数据，我们首次深入了解了 *Caecomyces churrovis* 的基因表达（Henske 等，2017）。Henske 等证明，在 *C. churrovis* 的 CAZyme 转录本中，含有 NCDDs 的转录本比例较低（15%），而已报道具有假根的厌氧真菌有约 30% 的 CAZyme 转录本与至少一种 NCDD 相关。基于这一现象，Henske 等（2017）猜想 *C. churrovis* 更依赖 CAZymes 降解植物碳水化合物，而不是基于纤维小体的生物降解策略。

虽然最近的基因组和转录组研究为许多厌氧真菌的分子机制提供了重要的解释，但是数据采集、分析和报告的差异使得很难直接比较这些数据。尽管存在这些难题，这些研究的结果已经证实了真菌的 CBMs 和真菌含有催化结构域的 CAZymes 在瘤胃厌氧生态系统降解复杂植物物质中发挥关键作用。所有厌氧真菌基因组的 GH 家族中，GH5、GH6、GH9、GH45 和 GH48 属于纤维素酶。有趣的是，这些 GH 的代表菌通常也与纤维小体特异性 CBMs 相关，部分研究已提供了充分的细节（图 3 和表 4）。GH10 和 GH11 也是如此，预计它们的目标是木糖，且是一组被归类为低聚糖降解的 GH（表 4）。虽然厌氧真菌使用的所有半纤维素酶库总是包含 GH10 和 GH11 的成员，一组额外 GH 家族在真菌个体半纤维素成分降解过程中略有变化（图 3）。这些额外的半纤维素酶属于 GH8、GH26 和 GH53，它们分别属于包含处理木聚糖内切酶、木聚糖酶以及内切-1,4-β-半乳聚糖酶活力的家族。

表 4 厌氧真菌的碳水化合物活性酶组成

主要已知活力[c]		Gruninger 等 (2018)			Solomon 等 (2016)[h] /Haitjema 等 (2017)[i]					Wang 等 (2011)[j]	Henske 等 (2017)[k]	Couger 等 (2015)[h]
		Anaeromyces[d] mucranatus	Neocallimastix[d] frontalis	Orpinomyces[d] joyonii	Piromyces[d] rhizinflata	Anaeromyces robustus	Neocallimastix californiae	Pecoramyces ruminantium	Piromyces finnis	Piromyces sp. E2	Neocallimastix patriciarium W5	Caecomyces churrovis	Pecoramyces ruminantium
糖苷水解酶 (GH) 家族[a]													
纤维素酶													
GH5	纤维素酶	24[e]	40[e]	31[e]	49[e]	22/10	44/33	45/24	25/14	NR/9	20	19 (1)	36
GH6	内切葡聚糖酶	12[f]	35[f]	31[f]	27[f]	3/12	18/9	46/32	1/17	NR/22	35	27 (8)	18
GH7	内切葡聚糖酶	0	0	0	0	NR/NR	NR/NR	NR/NR	NR/NR	NR/NR	0	0	NR
GH9	内切葡聚糖酶	14[e]	13[e]	11[e]	14[e]	14/8	24/13	14	8	NR/11	12	29 (11)	21
GH44	内切葡聚糖酶	0	0	0	0	NR/NR	NR/NR	NR/NR	NR/NR	NR/NR	NR	NR	NR
GH45	内切葡聚糖酶	14[e]	22[e]	16[e]	20[e]	13/7	23/15	13/11	7/9	NR/5	14	26 (12)	16
GH48	内切纤维素酶	6[f]	22[f]	19[f]	16[f]	7/6	15/16	14/6	313	NR/11	12	25 (5)	17
合计		70	132	108	126	59/43	124/86	132/81	44/64	–/58	93	126 (37)	108
半纤维素酶													
GH8	内切木聚糖酶	1	2	4	2	1/2	4/1	1/0	2/1	NR/1	NR	2 (1)	2
GH10	内切-1,4-β-木聚糖酶	10[e]	35[e]	23[e]	25[e]	14/6	60/24	30/11	24/12	NR/4	21	15 (2)	28
GH11	木聚糖酶	24[f]	41[f]	23[f]	49[f]	24/11	47/9	43/15	11/10	NR/27	11	123 (12)	24

（续表）

	主要已知活力[c]	Gruninger 等 (2018)				Solomon 等 (2016)[h]/Haijema 等 (2017)[i]					Wang 等 (2011)[j]	Henske 等 (2017)[k]	Couger 等 (2015)[h]
		Anaeromyces mucranatus[d]	Neocallimastix frontalis[d]	Orpinomyces joyonii[d]	Piromyces rhizinflata[d]	Anaeromyces robustus	Neocallimastix californiae	Pecoramyces ruminantium	Piromyces finnis	Piromyces sp. E2	Neocallimastix patriciarum W5	Caecomyces churrovis	Pecoramyces ruminantium
GH12	内切葡聚糖酶和木聚糖水解酶	0	0	0	0	NR/NR	NR/NR	NR/NR	NR/NR	NR/NR	0	NR	NR
GH26	β-甘露聚糖酶和木聚糖酶	3	10	1	7	NR/2	NR/12	NR/0	NR/2	NR/4	4	NR	7
GH28	半乳糖醛酸酶	0	0	1	0	0/NR	0/NR	0/NR	0/NR	NR/NR	NR	NR	5
GH53	内切-1,4-β-半乳糖苷酶	1	2	1	2	NR/1	NR/2	NR/0	NR/1	NR/0	1	NR	2
合计		39	90	53	85	39/22	111/48	74/26	37/26	–/36	37	126 (15)	68
解支链酶													
GH51	α-L-阿拉伯呋喃糖苷酶	0	0	0	0	NR/NR	NR/NR	NR/NR	NR/NR	NR/NR	0	NR	NR
GH54	α-L-阿拉伯呋喃糖苷酶	0	0	0	0	NR/NR	NR/NR	NR/NR	NR/NR	NR/NR	0	NR	NR
GH62	α-L-阿拉伯呋喃糖苷酶	0	0	0	0	NR/NR	NR/NR	NR/NR	NR/NR	NR/NR	0	NR	NR
GH67	α-葡萄糖醛酸酶	1	0	0	1	NR/NR	NR/NR	NR/NR	NR/NR	NR/NR	0	NR	1
GH78	α-L-鼠李糖苷酶	1	1	1	1	NR/NR	NR/NR	NR/NR	NR/NR	NR/NR	NR	NR	1

（续表）

主要已知活力[c]		Gruninger 等（2018）				Solomon 等（2016）[h]/Haitjema 等（2017）[i]					Wang 等（2011）[j]	Henske 等（2017）[k]	Couger 等（2015）[h]
		Anaeromyces mucranatus[d]	Neocallimastix frontalis[d]	Orpinomyces joyonii[d]	Piromyces rhizinflata[d]	Anaeromyces robustus	Neocallimastix californiae	Pecoramyces ruminantium	Piromyces finnis	Piromyces sp. E2	Neocallimastix patriciarium W5	Caecomyces churrovis	Pecoramyces ruminantium
合计		2	1	1	2	—	—	—	—	—	0	0	—
低聚糖分解酶													
GH1	β-葡糖苷酶和其他 β-连接二聚体	5	11	11	6	10/NR	14/NR	16/NR	4/NR	NR/NR	7	20（0）	9
GH2	β-葡糖苷酶和其他 β-连接二聚体	2	2	1	1	NR/0	NR/6	NR/0	NR/0	NR/1	NR	NR	1
GH3	主要地 β-葡糖苷酶	13	18	15	10	16/3	30/3	17/2	13/3	NR/2	10	16（3）	17
GH29	α-L-岩藻糖苷酶	0	0	0	0	NR/NR	NR/NR	NR/NR	NR/NR	NR/NR	0	0	NR
GH35	β-半乳糖苷酶	0	0	0	0	NR/NR	NR/NR	NR/NR	NR/NR	NR/NR	NR	NR	NR
GH38	α-甘露糖苷酶	1	1	1	1	NR/NR	NR/NR	NR/NR	NR/NR	NR/NR	NR	NR	1
GH39	β-木糖苷酶	2	4	6	3	4/5	7/6	0/1	1/2	NR/3	NR	3（2）	9
GH42	β-半乳糖苷酶	0	0	0	0	NR/NR	NR/NR	NR/NR	NR/NR	NR/NR	NR	NR	NR
GH43	阿拉伯糖酶和木糖苷酶	24[e]	37[e]	30[e]	18[e]	15/8	32/14	24/6	7/10	NR/7	20	59（10）	32

（续表）

主要已知活力[c]		Gruninger 等 (2018)				Solomon 等 (2016)[h]/Haitijema 等 (2017)[i]					Wang 等 (2011)[j]	Henske 等 (2017)[k]	Couger 等 (2015)[h]
		Anaeromyces mucranatus[d]	*Neocallimastix frontalis*[d]	*Orpinomyces joyonii*[d]	*Piromyces rhizinflata*[d]	*Anaeromyces robustus*	*Neocallimastix californiae*	*Pecoramyces ruminantium*	*Piromyces finnis*	*Piromyces sp.* E2	*Neocallimastix patriciarium* W5	*Caecomyces churrovis*	*Pecoramyces ruminantium*
GH52	β-木糖苷酶	0	0	0	0	NR/NR	NR/NR	NR/NR	NR/NR	NR/NR	NR	NR	NR
	合计	47	73	64	39	45/16	83/29	57/9	25/15	-/13	37	98 (15)	-
辅助酶（AA）家族[b]													
AA1	漆酶	0	0	0	0	NR/NR	NR/NR	NR/NR	NR/NR	NR/NR	NR	NR	NR
AA2	锰过氧化物酶和木质素过氧化物酶	0	0	0	0	NR/NR	NR/NR	NR/NR	NR/NR	NR/NR	NR	NR	NR
AA3	纤维二糖脱氢酶和芳基醇氧化酶	0	0	0	0	NR/NR	NR/NR	NR/NR	NR/NR	NR/NR	NR	NR	NR
AA4	香草醇氧化酶	1	2	1	1	NR/NR	NR/NR	NR/NR	NR/NR	NR/NR	NR	NR	NR
AA5	半乳糖氧化酶和乙醇氧化酶	0	0	0	0	NR/NR	NR/NR	NR/NR	NR/NR	NR/NR	NR	NR	NR
AA6	1,4-苯醌还原酶	1	1	1	2	NR/NR	NR/NR	NR/NR	NR/NR	NR/NR	NR	NR	NR
AA7	葡寡糖氧化酶和壳寡糖氧化酶	0	0	0	0	NR/NR	NR/NR	NR/NR	NR/NR	NR/NR	NR	NR	NR
AA8	铁还原酶	0	0	0	0	NR/NR	NR/NR	NR/NR	NR/NR	NR/NR	NR	NR	NR

第8章 瘤胃厌氧真菌

(续表)

主要已知活力[c]		Gruninger 等 (2018)				Solomon 等 (2016)[h]/Haitjema 等 (2017)[i]					Wang 等 (2011)[j]	Henske 等 (2017)[k]	Couger 等 (2015)[h]
		Anaeromyces mucranatus[d]	Neocallimastix frontalis[d]	Orpinomyces joyonii[d]	Piromyces rhizinflata[d]	Anaeromyces robustus	Neocallimastix californiae	Pecoramyces ruminantium	Piromyces finnis	Piromyces sp. E2	Neocallimastix patriciarium W5	Caecomyces churrovis	Pecoramyces ruminantium
AA9 (原先 GH61)	铜依赖裂解性多糖单加氧酶 (LPMOs)	0	0	0	0	NR/NR	NR/NR	NR/NR	NR/NR	NR/NR	NR	NR	NR
AA10 (原先 CBM33)	铜依赖裂解性多糖单加氧酶 (LPMOs)	0	0	0	0	NR/NR	NR/NR	NR/NR	NR/NR	NR/NR	NR	NR	NR
AA11	铜依赖裂解性多糖单加氧酶 (LPMOs)	0	0	0	0	NR/NR	NR/NR	NR/NR	NR/NR	NR/NR	NR	NR	NR
AA12	吡咯并喹啉醌依赖氧化还原酶	0	0	0	0	NR/NR	NR/NR	NR/NR	NR/NR	NR/NR	NR	NR	NR
AA13	铜依赖裂解性多糖单加氧酶 (LPMOs)	0	0	0	0	NR/NR	NR/NR	NR/NR	NR/NR	NR/NR	NR	NR	NR
合计		2	3	2	3	–	–	–	–	–	–	–	–
羧基酯酶 (CE) 家族[b]													
CE1	乙酰木聚糖酯酶和肉桂酰酯酶	22[f]	29[f]	21[f]	17[f]	NR/13	NR/25	NR/20	NR/10	NR/16	NR	NR	67
CE2	乙酰木聚糖酯酶	2[e]	8[e]	3[e]	5[e]	NR/NR	NR/NR	NR/NR	NR/NR	NR/NR	NR	NR	4
CE3	乙酰木聚糖酯酶	6	8	2	5	NR/0	NR/1	NR/1	NR/0	NR/1	NR	NR	9

· 179 ·

(续表)

主要已知活力[c]	Gruninger 等 (2018)				Solomon 等 (2016)[h]/Haitjema 等 (2017)[i]					Wang 等 (2011)[j]	Henske 等 (2017)[k]	Couger 等 (2015)[h]
	Anaeromyces mucranatus[d]	Neocallimastix frontalis[d]	Orpinomyces joyonii[d]	Piromyces rhizinflata[d]	Anaeromyces robustus	Neocallimastix californiae	Pecoramyces ruminantium	Piromyces finnis	Piromyces sp. E2	Neocallimastix patriciarum W5	Caecomyces churrovis	Pecoramyces ruminantium
CE4 乙酰木聚糖酯酶和儿丁质脱乙酰酶	34[e]	53[e]	32[e]	49[e]	53/1	85/0	48/1	41/3	NR/0	NR	NR	66
CE5 乙酰木聚糖酯酶和角质酶	0	0	0	0	NR/NR	NR/NR	NR/NR	NR/NR	NR/NR	NR	NR	NR
CE6 乙酰木聚糖酯酶	8[e]	12[e]	12[e]	19[e]	NR/8	NR/8	NR/4	NR/4	NR/7	NR	NR	10
CE7 乙酰木聚糖酯酶和头孢菌素-C脱乙酰酶	0	0	0	0	NR/NR	NR/NR	NR/NR	NR/NR	NR/NR	NR	NR	4
CE8 果胶甲酯酶	4	3	2	9	NR/NR	NR/NR	NR/NR	NR/NR	NR/NR	NR	NR	5
CE9 N-乙酰氨基葡萄糖6-磷酸脱乙酰酶	0	0	0	0	NR/NR	NR/NR	NR/NR	NR/NR	NR/NR	NR	NR	3
CE11 UDP-3-O-酰基 N-乙酰氨基葡萄糖脱乙酰酶	0	0	0	0	NR/NR	NR/NR	NR/NR	NR/NR	NR/NR	NR	NR	NR
CE12 果胶乙酰酯酶和乙酰木聚糖酯酶	8	4	6	8	NR/NR	NR/NR	NR/NR	NR/NR	NR/NR	NR	NR	4
CE13 果胶乙酰酯酶	0	0	0	0	NR/NR	NR/NR	NR/NR	NR/NR	NR/NR	NR	NR	NR

（续表）

	主要已知活力[c]	Gruninger 等 (2018)				Solomon 等 (2016)[h] /Haitjema 等 (2017)[i]					Wang 等 (2011)[j]	Henske 等 (2017)[k]	Couger 等 (2015)[h]
		Anaeromyces mucranatus[d]	Neocallimastix frontalis[d]	Orpinomyces joyonii[d]	Piromyces rhizinflata[d]	Anaeromyces robustus	Neocallimastix californiae	Pecoramyces ruminantium	Piromyces finnis	Piromyces sp. E2	Neocallimastix patriciarium W5	Caecomyces churrovis	Pecoramyces ruminantium
CE14	N-乙酰-1-D-肌糖-2-氨基-2-脱氧-α-D-吡喃葡萄糖苷脱乙酰酶	1	1	1	1	NR/NR	NR/NR	NR/NR	NR/NR	NR/NR	NR	NR	1
CE15	4-O-甲基-葡糖醛酸甲基酯酶	3	4	3	3	NR/2	NR/4	NR/0	NR/1	NR/0	NR	NR	2
CE16	乙酰酯酶	6	9	3	14	NR/2	NR/3	NR/1	NR/1	NR/0	NR	NR	12
	合计	94	131	85	130	53/26	85/41	48/27	41/19	0/24	—	—	187
多糖裂解酶 (PL) 家族[b]													
PL1	果胶酸裂解酶和果胶酸裂解酶	6	9	8	13	NR/1	NR/0	NR/0	NR/0	NR/0	NR	NR	21
PL3	果胶酸裂解酶	1	5	5	10	NR/NR	NR/NR	NR/NR	NR/NR	NR/NR	NR	NR	9
PL4	鼠李聚糖半乳糖醛酸内切裂解酶	3	4	3	7	NR/1	NR/1	NR/1	NR/1	NR/1	NR	NR	4
PL9	果胶酸裂解酶	0	0	0	0	NR/NR	NR/NR	NR/NR	NR/NR	NR/NR	NR	NR	1
PL10	果胶酸裂解酶	0	0	0	0	NR/NR	NR/NR	NR/NR	NR/NR	NR/NR	NR	NR	1
PL11	鼠李聚糖半乳糖醛酸裂解酶	0	1	0	0	NR/NR	NR/NR	NR/NR	NR/NR	NR/NR	NR	NR	1

(续表)

		Gruninger 等 (2018)					Solomon 等 (2016)[h]/Haitjema 等 (2017)[i]					Wang 等 (2011)[j]	Henske 等 (2017)[k]	Couger 等 (2015)[h]
	主要已知活力[c]	Anaeromyces mucranatus[d]	Neocallimastix frontalis[d]	Orpinomyces joyonii[d]	Piromyces rhizinflata[d]	Anaeromyces robustus	Neocallimastix californiae	Pecoramyces ruminantium	Piromyces finnis	Piromyces sp. E2	Neocallimastix patriciarium W5	Caecomyces churrovis	Pecoramyces ruminantium
PL22	低聚半乳糖醛酸裂解酶	0	0	0	0	NR/NR	NR/NR	NR/NR	NR/NR	NR/NR	NR	NR	5
	合计	10	19	16	30	−/2	−/1	−/1	−/2	−/1	−	−	42
碳水化合物结合模块 (CBM) 家族[b]													
CBM1	纤维素和几丁质	81[f]	106[f]	85[f]	119[f]	NR/47	NR/26	NR/24	NR/46	NR/16	1	NR	NR
CBM2	纤维素, 几丁质和木聚糖	2	2	1	3	NR/NR	NR/NR	NR/NR	NR/NR	NR/NR	NR	NR	NR
CBM6	纤维素和 β-1,4-木聚糖	10[e]	11[e]	15[e]	5[e]	NR/4	NR/5	NR/4	NR/6	NR/5	1	NR	NR
CBM10	纤维素	275[g]	424[g]	283[g]	390[g]	NR/3	NR/8	NR/10	NR/10	NR/0	16	NR	NR
CBM12	几丁质	3	1	1	2	NR/NR	NR/NR	NR/NR	NR/NR	NR/NR	NR	NR	NR
CBM13	半乳糖, 甘露糖, 木聚糖	27[e]	41[e]	35[e]	24[e]	NR/17	NR/38	NR/15	NR/15	NR/6	NR	NR	NR
CBM18	几丁质	95[g]	193[g]	105[g]	159[g]	NR/1	NR/3	NR/1	NR/0	NR/4	NR	NR	NR
CBM20	淀粉	3	4	0	0	NR/NR	NR/NR	NR/NR	NR/NR	NR/NR	NR	NR	NR
CBM21	淀粉	6[e]	11[e]	5[e]	6[e]	NR/NR	NR/NR	NR/NR	NR/NR	NR/NR	NR	NR	NR

（续表）

	主要已知活力[c]	Gruninger 等 (2018)				Solomon 等 (2016)[h]/Haitjema 等 (2017)[i]					Wang 等 (2011)[j]	Henske 等 (2017)[k]	Couger 等 (2015)[h]
		Anaeromyces mucranatus[d]	Neocallimastix frontalis[d]	Orpinomyces joyonii[d]	Piromyces rhizinflata[d]	Anaeromyces robustus	Neocallimastix californiae	Pecoramyces ruminantium	Piromyces finnis	Piromyces sp. E2	Neocallimastix patriciarium W5	Caecomyces churrovis	Pecoramyces ruminantium
CBM22	木聚糖和混合的 β-1,3/β-1,4-葡聚糖	2	1	1	0	NR/0	NR/1	NR/0	NR/1	NR/0	NR	NR	NR
CBM25	淀粉	2[e]	3[e]	8[e]	1[e]	NR/NR	NR/NR	NR/NR	NR/NR	NR/NR	NR	NR	NR
CBM26	淀粉	6[f]	11[f]	22[f]	2[f]	NR/NR	NR/NR	NR/NR	NR/NR	NR/NR	NR	NR	NR
CBM29	甘露聚糖/葡甘露聚糖	7[f]	9[f]	9[f]	16[f]	NR/1	NR/8	NR/9	NR/5	NR/2	NR	NR	NR
CBM32	半乳糖，乳糖和多聚半乳糖醛酸	0	3	0	2	NR/NR	NR/NR	NR/NR	NR/NR	NR/NR	NR	NR	NR
CBM35	木聚糖和甘露聚糖，β-半乳聚糖	3	3	2	8	NR/2	NR/6	NR/0	NR/2	NR/2	NR	NR	NR
CBM36	木聚糖和低聚木糖	0	1	0	0	NR/NR	NR/NR	NR/NR	NR/NR	NR/NR	NR	NR	NR
CBM48	糖原	6	8	5	6	NR/NR	NR/NR	NR/NR	NR/NR	NR/NR	NR	NR	NR
CBM50	壳五糖	2	4	4	5	NR/NR	NR/NR	NR/NR	NR/NR	NR/NR	NR	NR	NR
CBM52	β-1,3-葡聚糖	3	6	4	5	NR/1	NR/1	NR/0	NR/1	NR/0	NR	NR	NR
CBM61	β-1,4-半乳聚糖	3	1	1	2	NR/NR	NR/NR	NR/NR	NR/NR	NR/NR	NR	NR	NR

(续表)

	主要已知活力[c]	Gruninger 等 (2018)[b]				Solomon 等 (2016)[h]/Haitjema 等 (2017)[i]					Wang 等 (2011)[j]	Henske 等 (2017)[k]	Couger 等 (2015)[h]
		Anaeromyces mucranatus[d]	Neocallimastix frontalis[d]	Orpinomyces joyonii[d]	Piromyces rhizinflata[d]	Anaeromyces robustus	Neocallimastix californiae	Pecoramyces ruminantium	Piromyces finnis	Piromyces sp. E2	Neocallimastix patriciarium W5	Caecomyces churrovis	Pecoramyces ruminantium
CBM63	纤维素	8[e]	5[e]	5[e]	8[e]	NR/1	NR/3	NR/2	NR/2	NR/0	NR	NR	NR
CBM66	果聚糖	1	0	1	0	NR/NR	NR/NR	NR/NR	NR/NR	NR/NR	NR	NR	NR
CBM67	L-鼠李糖	1	1	1	1	NR/NR	NR/NR	NR/NR	NR/NR	NR/NR	—	—	—
合计		546	849	593	764	–/77	99	–/65	–/88	–/35	—	—	—

注：[a] Allgaier 等 (2010)。
[b] Gruninger 等 (2018)。
[c] 碳水化合物结合模块的主要底物。
[d] 检测到的转录本数量。
[e] 碳水化合物活性酶家族和碳水化合物结合模块家族，来自厌氧真菌转录组的该家族转录本丰度是需氧真菌的 2 倍以上。
[f] 碳水化合物活性酶家族和碳水化合物结合模块家族，来自厌氧真菌转录组的该家族转录本丰度是需氧真菌的 5 倍以上。
[g] 碳水化合物活性酶家族和碳水化合物结合模块家族，来自厌氧真菌转录组的该家族转录本丰度是需氧真菌的 20 倍以上。
[h] 检测到的转录本数量。
[i] 检测到的具有非催化锚定结构域的基因数量。
[j] 编码碳水化合物活性酶家族的表达基因数量。
[k] 检测到碳水化合物结合结构域的转录本数量（包含至少一个锚定结构域的转录本数量）。

NR＝未报道。

尽管这些 CAZymes 无处不在，更为详细的转录组研究揭示了以前通常被忽略的 AAs、CEs 和多糖裂解酶（PLs）的重要性。这些酶通过分解和调整晶体纤维素核心周围的难降解组分，从而机械性地削弱纤维素结构，增强目前研究报道较多的 GH 的催化活性（Ochiai 等，2007；Gilbert，2010；Makela 等，2018）。在 *A. mucranatus*、*N. frontalis*、*O. joyonii* 和 *P. rhizinflata* 中鉴定出分类为 AA4 和 AA6 的转录本（Gruninger 等，2018），而 CEs（即 CE1、CE4 和 CE6）来自 *Pecoramyces ruminantium*（Couger 等，2015）、*A. robustus*、*N. californiae*、*P. finnis* 的转录组，以及 *Piromyces* sp. E2 的基因组（Solomon 等，2016；Haitjema 等，2017）。据报道，PLs 中的 PL1、PL3 和 PL4 家族存在于 *A. mucranatus*、*N. frontalis*、*O. joyonii*、*P. rhizinflata*（Gruninger 等，2018），以及 *Pecoramyces ruminantium*，此外，PL5 和 PL9 也在 *Pecoramyces ruminantium* 的转录本中检测到（Couger 等，2015）。在 *A. robustus*、*N. californiae*、*Pecoramyces ruminantium*、*P. finnis* 和 *Piromyces* sp. E2 的基因组中也检测到了含有 NCDD 的 PL4 成员（Haitjema 等，2017）。图 3 显示了厌氧真菌在不同转录组和基因组中检测到的最丰富的 CAZyme 家族。表 4 列出了从每种瘤胃厌氧真菌个体鉴定出的不同 CAZyme 家族的详细分类。

基因表达是一个仍知之甚少的复杂且严格调控的过程。非编码 RNA 链的转录及与编码链的相互作用，是在生命的所有领域中调控基因表达控制的主要机制（Wagner 和 Simons，1994；Katayama 等，2005；Donaldson 和 Saville，2012；Britto-Kido Sde 等，2013）。基于反义的表达调控能迅速适应进化压力（Yan 和 Wang，2012），但是更重要的是它能实现类开关响应（Pelechano 和 Steinmetz，2013），这也解释了为什么天然的反义转录本（NATs）在真菌基因组中被发现，并且在它们的代谢过程中扮演重要角色。最近的一项研究显示，在早期瘤胃厌氧真菌（即 *A. robustus*，*N. californiae* 和 *P. finnis*）中 NATs 介导的基因调控是保守的（Solomon 等，2018）。尽管在这 3 个研究样本中普遍存在，但是 NATs 的丰度低于其他真菌，因此作者推断这可能反映了厌氧真菌的早期分化，以及与细菌水平基因转移的高效率。在比较分析中鉴别的 NATs 调控过程约 10% 与木质纤维素降解有关，表明为了充分了解厌氧真菌的生物量降解表型，以及其他可能的表现，可能需要对相应的 NATs 图谱进行全面分析。

据我们所知，目前还没有旨在生成全球蛋白质图谱或者鉴定厌氧真菌未知酶的独立蛋白质组学研究。目前生成的蛋白质图谱仅用于确定已经鉴定的 CAZyme 家族或者溶质转运载体，这些 CAZyme 或载体参与 *N. patriciarium* W5、*A. robustus*、*N. californiae*、*P. ruminantium* 和 *P. finnis* 碳水化合物代谢（Wang 等，2011；Seppala 等，2016；Solomon 等，2016）。Wang 等通过质谱法验证了 *N. patriciarum* W5 中属于纤维素酶（即 GH6，GH9，GH45 和 GH48）、半纤维素酶（即 GH10 和 GH11），以及其他低聚糖降解酶家族（即 GH1，GH3 和 GH43）的蛋白分泌（Wang 等，2011）。*A. robustus*、*N. californiae* 和 *P. finnis* 分泌相同的纤维素酶和半纤维素酶（Solomon 等，2016），为这些微生物使用相似的蛋白质来解决复杂植物聚合物的抗性这一猜想提供了支持。

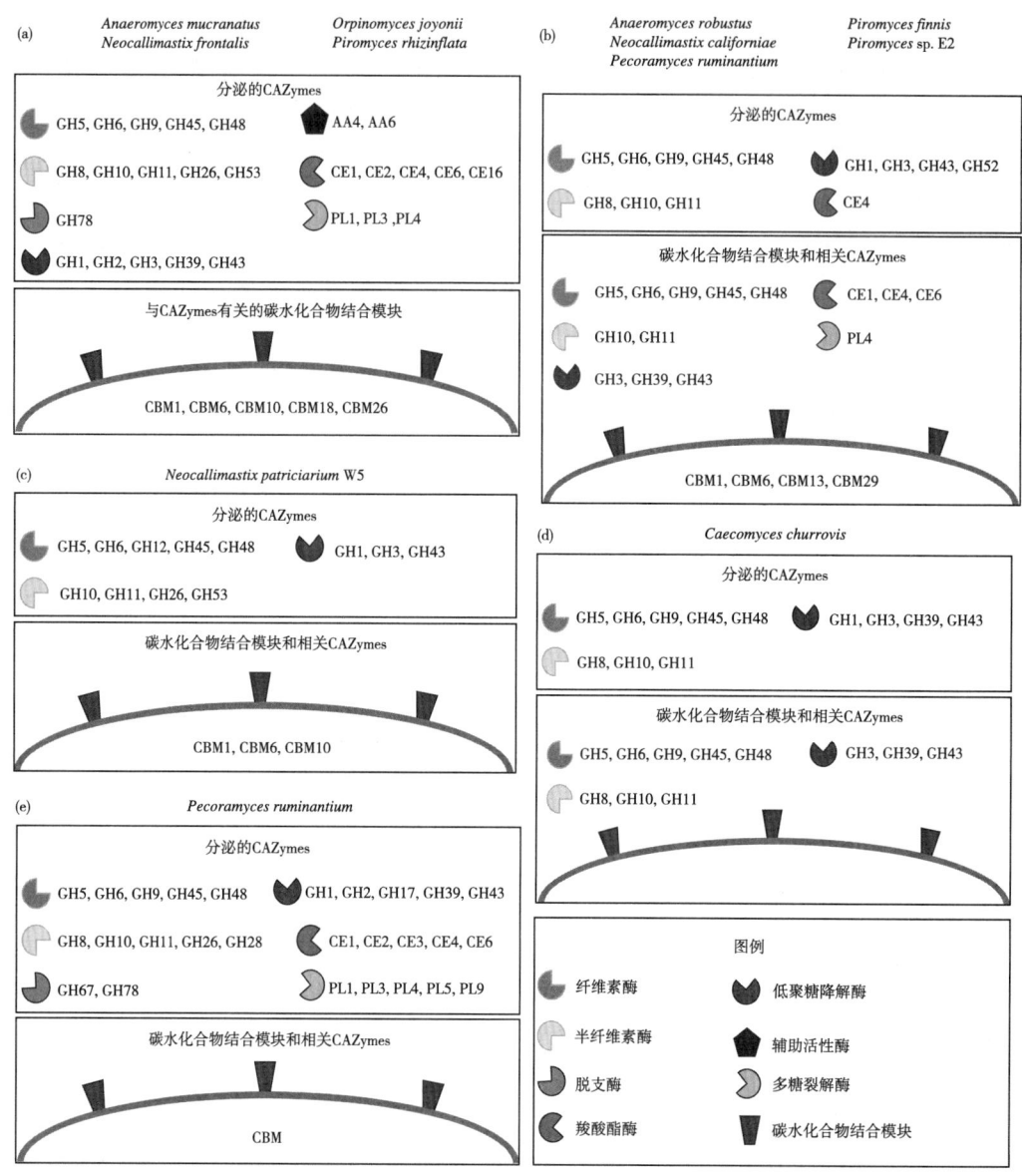

图3 瘤胃厌氧真菌碳水化合物活性酶（CAZyme）库。(a) *Anaeromyces mucranatus*、*Neocallimastix frontalis*、*Orpinomyces joyonii* 和 *Piromyces rhizinflata* 表达谱中检测到的CAZymes转录本（Gruninger等，2018）。仅报道这4种微生物转录组中检测到的5个丰度最高的CAZyme家族。(b) *A. robustus*、*N. californiae*、*Pecoramyces ruminantium* 和 *P. finnis* 的表达谱（Solomon等，2016），*A. robustus*、*N. californiae*、*Pecoramyces ruminantium*、*P. finnis* 和 *Piromyces* sp. E2 的基因组（Haitjema等，2017）中检测到的CAZymes。在 *A. robustus*、*N. californiae*、*Pecoramyces ruminantium* 和 *P. finnis* 的转录组，以及所有5种微生物（即 *A. robustus*、*N. californiae*、*Pecoramyces ruminantium*、*P. finnis* 和 *Piromyces* sp. E2）的基因组中，仅报道检测到的5个最丰富的CAZyme家族。(c) *Neocallimastix patriciarium* W5 表达谱中检测到的CAZymes（Wang等，2011）。仅报道检测到的5个最丰富的CAZyme家族。(d) *Caecomyces churrovis* 的表达谱中检测到的CAZymes（Henske等，2017）。仅报道检测到的5个最丰富的CAZyme家族。(e) *Pecoramyces ruminantium* 的表达谱中检测到的CAZymes（Couger等，2015）。仅报道检测到的5个最丰富的CAZyme家族。

9 厌氧真菌的宏组学

9.1 宏基因组学

利用测序技术更好地理解特定环境下微生物群体的装配和潜在功能，已经成为微生物生态学的标准方法。宏基因组学分析流程已经完善地建立了许多年，从 16S rRNA 基因的高变区生成的扩增子，有助于通过相对低的成本产生原核生物种群的丰度谱。尽管鸟枪法宏基因组学可以从生成的短读长数据中组装古菌、细菌，甚至是病毒的全基因组，扩增子测序通常仍是研究微生物菌群及其对外部因素潜在响应的第一步。

尽管扩增子测序具有局限性，如 PCR 偏好和有限的分辨率，当不得不计划后续试验和分析方法用于更复杂的鸟枪法宏基因组或转录组的研究时，扩增子测序获得的数据仍然很有价值，因为鸟枪法宏基因组或转录组是更为资源密集型的（Turaev 和 Rattei，2016；Staley 和 Sadowsky，2018）。对于真菌菌群组成，细胞核 rRNA 操纵子的 ITS 区域通常被用于标记基因的选择（Schoch 等，2012）。厌氧真菌的分子生态学研究主要集中在 ITS1 区域，并为各种草食动物消化道真菌的多样性和结构提供了信息（Dill-McFarland 等，2019；Mura 等，2019）。

当只对有限数量的已知真菌感兴趣时，探针特异性方法，如 Denman 等（2008）开发的自动核糖体基因间隔区分析已经足够，并且能够长时间追踪目标真菌的丰度。近期，Dollhofer 等（2016）发表了基于 PCR 的方法，能够允许快速且相对廉价地进行真菌菌群组装，以及对基于 18S rRNA、28S rRNA 基因和编码 GH5 内切葡聚糖酶基因的纤维素分解活力进行初步评估。尽管真菌扩增子领域中有这些进展，但是主要的难点仍是所有真菌分类标记基因可实现的分辨率相当低。随着培养品种数量的增加，以及这些标记基因参考数据库的增加，该难题可能最终被克服。

鸟枪法宏基因组学是将环境 DNA 直接测序，然后组装成较长的传染性 DNA 片段（重叠群），最终形成基因组草图的方法，该方法为更全面地了解任何给定环境的微生物组功能潜力提供了机会。形态学和基因组的特征阻碍了厌氧真菌个体以及自然丰度低的厌氧真菌基因组信息的获取和分析，样品制备技术如尼龙袋技术的孔径大小阻碍了体积较大、生长较慢的厌氧真菌在尼龙袋内样品上的附着，这可能是宏基因组数据中真菌基因组组装信息较少的重要因素。从宏基因组数据中成功组装较长真核细胞片段的研究偶有报道，但这种情况很少（Sharon 等，2013；Kantor 等，2015，2017；Quandt 等，2015；Mosier 等，2016；Raveh-Sadka 等，2016）。为了从不同微生物群落的宏基因组数据中组装真核生物片段，加州大学伯克利分校的班菲尔德实验室开发了一种基于 $k\text{-}mer$ 的方法，称为 *EukRep*，以识别来自不同环境样本的数据集的真核生物序列组装，从而提高基因预测的质量和进一步的合并策略（West 等，2018）。他们在先前发表的 268 Gbp 瘤胃宏基因组数据（Hess 等，2011）测试这种方法时，*EukRep* 没有发现任何真核生物组装，表明取样方法的重要性（未发表的数据）。在这个特例中，将磨碎的植物材料放置在一个孔径为 50 μm 的尼龙袋中，在瘘管奶牛瘤胃内培养 72 h，使瘤胃原核生物能够进入并定殖于磨碎的生物质中，但这可能导致瘤胃厌氧真菌对纤维的定殖不充分，从而无法在生成的宏基因组中检测其遗传物质。

9.2 宏转录组学

与单个真菌分离物的表达谱分析类似，许多与真菌宏基因组相关的生物信息学难题可以通过宏转录组学方法来解决。但是到目前为止，对瘤胃相关真菌菌群表达谱的深入研究仅仅是完整菌群表达谱的副产品，而原核生物转录本构成了生成读长的主要部分。在一篇代表性论文中，Qi 等（2011）报道了由 2.8 Gbp 宏转录组测序数据组装、平均长度为 310 个碱基对（bp）的 59 129 个组装序列。这些数据是对两头麝牛固态瘤胃内容物总 mRNA 进行聚腺化 mRNA 富集后生成的，聚腺化 mRNA 是真核生物 RNA 加工的特征（Stewart，2019）。在预测编码蛋白质的读长中，约 14.4%属于真菌，其中约一半来自新美鞭菌门（Neocallimastigomycota）的成员。进一步分析发现了各种 CAZymes 的存在，包括属于纤维素酶的 GH（即 GH5、GH6、GH7、GH9、GH45 和 GH48），属于内切纤维素酶的 GH（即 GH8、GH10、GH11、GH26 和 GH28），属于脱支酶的 GH（即 GH67 和 GH78），属于低聚糖降解酶的 GH（即 GH1、GH2、GH3 和 GH38），以及多种 CEs、PLs 和 CBMs，例如 CBM1、CBM6、CBM10、CBM13、CBM18 和 CBM29。Qi 等（2011）还鉴定了几个含有多种 CAZyme 结构域的高表达基因，在大多情况下一个 CBM10 与一个 GH6 或 GH48 结合，表明厌氧真菌利用含有 GH6 或 GH48 催化结构域的纤维小体。总之，这项工作提供了第一手的组学证据，证明真菌菌群在复杂碳水化合物中发挥重要作用，真菌菌群利用酶的多模块机制，促进瘤胃生态系统中难降解植物组织的分解。

在瘤胃发生活性降解的情况下，饲喂后 1 h 两头奶牛瘤胃混合菌群的转录组数据中仅有总读长的约 0.12%来自瘤胃真菌（Dai 等，2015）。这些真菌的读长大多来自 *Piromyces* 或 *Neocallimastix*，分别占真菌读长的 54%和 41%。尽管真菌的转录本丰度较低，但它们约占已鉴定的 GH48 总量的 14%，表明它们是这种蛋白的主要生产者。此外，也检测到了真菌转录本编码的纤维素酶（即 GH6、GH9、GH48），半纤维素酶（即 GH10 和 GH11），β-葡糖苷酶（GH1 和 GH3），以及 CBM6 和纤维小体特异性的 CBM10。类似的研究使用 Illumina 双端测序技术生成了混合菌群 mRNA 的表达谱，结果表明泌乳奶牛的厌氧真菌在活跃的自由漂浮菌群中约占 7.5%，其中 *Neocallimastigacea* 的成员对纤维素酶转录本的基因表达谱数量具有显著贡献（Sollinger 等，2018）。这些发现表明，厌氧真菌在瘤胃生态系统的植物生物量降解作用被低估。早先认为真菌通过机械和酶学方法使难分解生物量更容易获取，随后被 *Prevotellacea* 等微生物定殖，这些微生物的基因组具有丰富的与半纤维素、淀粉和蛋白质降解相关的基因，而不是与纤维素降解相关的基因（Flint 等，2012；Accetto 和 Avgustin，2015），混合瘤胃 mRNA 的这两个宏转录组中真菌纤维素酶转录本丰度增加，表明真菌在纤维素降解中存在重要作用。然而，更详细地了解植物纤维降解过程中上调表达的真菌基因以及以真菌为目标的宏组学方法将是至关重要的，如 Foster 等的研究（Qi 等，2011）。否则，我们只能继续研究真菌的冰山一角。

10 瘤胃真菌与瘤胃生态系统中其他成员之间的交互作用

10.1 厌氧真菌和产甲烷菌之间的交互作用

在瘤胃中厌氧真菌与多种微生物共存，其中包括细菌、原虫、产甲烷菌和噬菌体。除了产甲烷菌和噬菌体，这些微生物产生的酶能够用于饲料物质的分解和发酵，从而产生多种终产物，如挥发性脂肪酸（主要是乙酸、丙酸、丁酸）、氢气和二氧化碳。与产甲烷菌共同生长时，厌氧真菌的发酵产物从电子沉积产物（即乙醇和乳酸）转变为更多的还原性产物（即乙酸和甲酸，Theodorou 等，1996）。在厌氧真菌和产甲烷菌共培养中，乙酸是主要产物，二氧化碳产量增加，而乳酸和乙醇产量降低（Bauchop 和 Mountfort，1981）。除了发酵的变化外，由于去除了发酵抑制中间体（即乙醇、甲酸、乳酸），产甲烷菌共培养显著增加真菌生物量。

厌氧真菌的主要代谢产物包括乙酸、甲酸、氢气和二氧化碳。据报道厌氧真菌能够介导种间氢转移，并通过甲烷生成作用在能量上有利地处理电子，因此产甲烷菌和厌氧真菌表面存在物理联系也就不足为奇了（Edwards 等，2017；Li 等，2019）。许多研究发现，产甲烷菌与厌氧真菌之间存在共生关系（Cheng 等，2009；Jin 等，2011，2014；Leis 等，2014）。一类属于不可培养的古菌菌群"Rumen Cluster C"的新瘤胃产甲烷菌菌种随后被归类为 *Methanomassiliicoccus*（Seedorf 等，2014；Paul 等，2015），且这类菌种与厌氧真菌培养物有关（Jin 等，2014）。Jin 等（2011）报道，*Methanobrevibacter* spp. 是 *Piromyces*、*Anaeromyces*、*Neocallimastix* 相关的主要产甲烷菌群。Leis 等（2014）建立了 *Caecomyces communis* 和 Methanobacteriales 产甲烷菌的共生培养物。Sun 等（2014）还获得了 3 种 *Methanobrevibacter olleyae* 与 *Piromyces*、*Neocallimastix* 和 *Caecomyces* 的共培养物，以及一种 *Neocallimastix/Methanobrevibacter thaueri* 的共培养物。这种共生作用能够促进真菌生长（Cheng 等，2009；Li 等，2019），提高底物利用（Bootten 等，2011）、酶产生（Teunissen 等，1992）、发酵产物产量（Cheng 等，2013），以及对有毒羧基离子载体的抗性（Stewart 和 Richardson，1989）。但是，这些真菌与原虫和某些瘤胃细菌存在竞争关系，特别是竞争同一底物的细菌。相反，某些缺乏木质纤维素分解机制的瘤胃细菌可以从厌氧真菌中受益，因为厌氧真菌的假根物理性分解后，增加了细菌对木质纤维素内部的黏附和利用。

10.2 饲料成分对瘤胃真菌菌群的影响

在反刍动物采食的牧草中添加谷物是提高日粮能量密度和可利用碳水化合物采食的常用手段，但是日粮中添加易发酵的含淀粉精料对厌氧真菌的影响不同。添加谷物后瘤胃厌氧真菌数量略低（Orpin，1977a；Gordon，1985；Grenet 等，1989），而在体外试验中，高粱青贮中添加玉米提高了厌氧真菌的降解能力（Akin 和 Windham，1989）。此外，在干草日粮中添加主要为谷物的精料可显著提高绵羊瘤胃内真菌游动孢子的数量（大于 20 倍），但是真菌生物量仅提高了 1~2 倍（Faichney 等，1997）。这些表观差异的可能解释是仅 *Neocallimastix*、*Piromyces*、*Orpinomyces* 3 个菌属的部分厌氧真菌能够产生淀粉酶，因而具有发酵淀粉的能力（Phillips 和 Gordon，1988，1995；Yanke 等，1993；Mountfort，1994）。McAllister 等（1993）记录了具有淀粉分解能力的厌氧真菌对谷物的降解过程，其结果发现瘤胃微生物组中似乎发生了复杂的交互作用。目前，没有关于对瘤胃真菌菌群影响的普遍结论。饲喂游离脂质对反刍动物瘤胃发酵不利、延缓纤维降解（Jenkins，1993），鉴于目前含油脂的油籽粕使用越来越多，这是一个需要考虑的重要因素。厌氧真菌作为瘤胃微生物群体中的一部分，日粮中添加脂质对厌氧真菌同样有不利影响。据报道，添加菜籽油显著减少真菌数量，但是具体机制尚未阐明（Fonty 和 Grenet，1994）。Elliott 等（1987）和 Calderon-Cortes（1989）发现，大麦秸秆日粮中补充葵花籽粕导致绵羊瘤胃真菌数量减少到可检测水平以下。而当日粮中添加棉籽粕时，绵羊瘤胃真菌的游动孢子数量减少到可检测水平以下，并且无法检测到真菌的 DNA。饲喂中链脂肪酸钙（C6-12）减少绵羊瘤胃真菌游动孢子的数量，而长链脂肪酸盐（C12-14）对厌氧真菌没有影响（Ushida 等，1991），表明至少可以通过化学预处理缓解油籽粕长链脂肪酸的抑制作用。

10.3　厌氧真菌对宿主利用中间代谢产物的影响

研究表明，采食难降解木质纤维素日粮后，动物消化道厌氧真菌种群增加，这归因于这些真菌具有穿透、定殖木质化组织和溶解植物细胞壁的独特能力（Akin 等，1983；Grenet 等，1989），从而促进细菌随后的植物发酵。动物采食细长、坚硬的植物茎时，肠道真菌很丰富，而采食软叶日粮时真菌数量较少（Grenet 等，1989）。绵羊瘤胃内驱除真菌会降低秸秆消化，而将真菌重新引入瘤胃系统后，这种现象得以恢复（Calderon-Cortes 等，1989）。当厌氧真菌存在时，体内消化率从 3% 增加到 8%（Elliott 等，1987；Gordon 和 Phillips，1993；Kumar 等，2004；Dayananda 等，2007；Tripathi 等，2007；Sehgal 等，2008；Saxena 等，2010）。*Neocallimastix* sp. 和 *Piromonas* sp. 在降解植物组织碎片方面均优于 *Caecomyces* sp.，这可能是因为丝状假根穿透坚硬的植物组织比球状假根更有效（Orpin，1989），而在混合瘤胃细菌培养过程中添加 *Neocallimastix* 可使小麦秸秆的降解率提高 15%（Hillaire 和 Jouany，1989）。上述结果表明，厌氧真菌极有可能在纤维性饲料的降解过程中发挥重要作用，并且使其他不消化的碳水化合物能够被宿主动物进一步代谢利用。

厌氧真菌也有助于宿主动物的蛋白质供应，不仅通过产生蛋白水解酶，也可以作为瘤胃内合成的微生物蛋白，从而流入真胃和肠道被宿主消化和吸收。与细菌不同，厌氧真菌是蛋白酶阳性的，可以通过假根渗透饲料的蛋白部分，因此它们在降解纤维相关蛋白或者单宁蛋白复合物方面发挥重要作用（Wallace 和 Munro，1986；Gordon 等，1995）。但是，真菌对瘤胃蛋白水解的贡献尚未确定（Bonnemoy 等，1993）。瘤胃真菌也被证实具有氨肽酶活性（Michel 等，1993）。单中心厌氧真菌 *Neocallimastix*、*Piromyces* 和 *Caecomyces* 的大部分蛋白质成分在绵羊肠道被消化和吸收（Gulati 等，1988，1989）。尽管厌氧真菌对宿主微生物蛋白的总体贡献很小，但是真菌来源的蛋白质量高，且易于宿主获取（Faichney 等，1997）。由于厌氧真菌可以利用氨作为唯一氮源（Lowe 等，1985；Guliye 和 Wallace，2007），可以通过直接接种菌株或通过日粮补充剂刺激它们的活力，从而调控瘤胃真菌菌群，在这两种情况下均可以增加其对宿主高质量微生物蛋白的供应。

厌氧真菌的发酵受到底物和其他微生物的影响（Theodorou 等，1996；Sirohi 等，2012）。当使用葡萄糖作为底物时，厌氧真菌的主要发酵产物是乙酸、乙醇、甲酸、乳酸、琥珀酸、二氧化碳和氢气（Lowe 等，1987）。多中心厌氧真菌产生的乳酸少于单中心真菌（Borneman 等，1989；Phillips 和 Gordon，1995），*Piromyces* 不产生乳酸（Ho 等，1996）。当使用葡萄糖和木糖作为底物时，*Neocallimastix* sp. 产生甲酸、乙酸、乳酸和乙醇（Lowe 等，1987）。Borneman 等（1989）得出的结论是，发酵产物积累伴随着底物利用。厌氧真菌的主要发酵产物是甲酸、乙酸、D（-）乳酸、乙醇、二氧化碳和氢气。除了从驴盲肠分离出的 *Piromyces* 菌株无法产生乳酸，其余厌氧真菌无论其来源如何，均能产生乳酸（Julliand 等，1998）。

在饲喂 30 min 内，厌氧真菌的游动孢子释放，并附着在植物细胞和受损的表面上（Edwards 等，2008）。附着后，游动孢子首先穿透植物组织，然后孢子分叉形成网状的假根结构（Dollhofer 等，2015；Cheng 等，2018）。真菌游动孢子对木质化组织中的酚酸（对香豆酸、阿魏酸、丁香酸）具有趋化反应（Wubah 和 Kim，1996），并破坏木质素使酚酸从细胞壁释放。厌氧真菌游动孢子的运动对植物物质释放的可溶性糖和酚酸有化学倾向（Orpin 和 Bountiff，1978）。游动孢子附着在植物组织后，鞭毛脱落，形成包囊。芽管在原始鞭毛的位置产生，随后发展出一个分支的假根系统，通过酶和假根的共同作用可以穿透植物组织（Ho 等，1988）。这种特性使厌氧真菌能在其他微生物之前快速侵蚀进入消化道的植物组织（Edwards 等，2008；Cheng 等，2018）。高浓度的酚类单体对瘤胃真菌有毒性，尽管真菌对这些酸的抵抗力远高于其他瘤胃微生物（Akin 和 Rigsby，1987；Paul 等，2003）。瘤胃真菌能产生胞外阿魏酸酯酶和对香豆酰酯酶，这些酶可以使羟基肉桂酸从阿拉伯木聚糖中释放出来（Borneman 等，1990）。McSweeney 等（1994）研究了 *Neocallimastix patriciarium* 消化典型木质素化合物和木质化植物的能力，发现该真菌不会降解典型的木质素化合物，也不会溶解从茅草中分离出来的酸性洗涤剂木质素。该真菌能够裂解耦合木质素和多糖的酯键，但是未能裂解木质素和多糖间的醚键。Akin 和 Benner（1988）、Chesson（1993）也得出结论，瘤胃真菌能够将其溶解，但不能将其降解为最终代谢产物。

10.4 通过调控瘤胃真菌菌群提高木质纤维素降解

在反刍动物和非反刍动物生产中，已有应用厌氧真菌作为直接饲喂微生物，从而提高低品质牧草利用的研究，无真菌绵羊补饲真菌培养物后秸秆日粮采食量提高 40%（Gordon 和 Phillips，1998）。体外纤维消化和瘤胃发酵分析结果也显示，添加真菌培养物对提高乙酸、体外干物质消化率、分配系数值、微生物生物量合成水平具有促进作用（Sirohi 等，2013）。据报道，在反刍动物日粮中添加厌氧真菌培养物能够提高采食量、动物生长速率、饲料转化效率、产奶量（Lee 等，2000；Dey 等，2004；Paul 等，2004；Tripathi 等，2007；Sehgal 等，2008；Saxena 等，2010；Paul 等，2011；Gao 等，2013；Kumar 等，2015）。此外，饲喂厌氧真菌对幼龄反刍动物（Sehgal 等，2008）和缺乏厌氧真菌的绵羊（Elliott 等，1987；Gordon 和 Phillips，1993）的效果更好。研究表明，应用厌氧真菌作为直接饲喂微生物可以通过提高瘤胃发酵特性（pH 值、挥发性脂肪酸、氨态氮）、微生物菌群，以及纤维素酶的活性，从而提高饲料消化率。相反，单独添加厌氧真菌分泌的酶不会改变瘤胃发酵，这突出了利用活菌作为反刍动物饲料添加剂的重要性（Lee 等，2000）。但是，考虑到这些严格厌氧真菌在培养和使用时的困难，通过饲喂厌氧真菌提高瘤胃内纤维消化的可行性仍存在问题。

干草日粮或牧草的硫含量是调控瘤胃真菌菌群的重要因素（Akin 等，1983；Gordon，1985）。当日粮中每千克有机物包含 1.0 g 硫或者更低时，采食热带牧草干草（*Digitaria pentzii*）的绵羊瘤胃内不存在厌氧真菌（Akin 等，1983）。在用于制作干草的草场上施用硫肥（Akin 等，1983）或者在低硫干草中添加含硫添加剂（Gordon 等，1983）后，瘤胃内厌氧真菌数量显著提高。草场施用硫肥同样使反刍动物的平均自由采食量提高 38%（Akin 等，1983；Gordon，1985）。另一种低硫的热带牧草干草也导致瘤胃内无法检测到厌氧真菌（Morrison 等，1990）。此外，低硫秸秆日粮中可检测到瘤胃厌氧真菌菌群，但数量较低（Gordon 等，1983；Gulati 等，1985；Weston 等，1988）。草场硫含量降低与瘤胃真菌数量减少之间的关系在黑麦草草场并不适用（Millard 等，1987），绵羊采食未施肥干草（包含 0.9 g 硫/kg DM）后瘤胃厌氧真菌数量高于采食施肥草场干草（2~4 g 硫/kg DM）。施肥草场中附加的硫主要存在于硫酸盐和非蛋白质组分，这与 Akin 等（1983）使用的施肥 *D. pentzii* 中硫的分布相似，其附加硫主要存在于可溶性非蛋白成分。低硫牧草中硫的类型和分布可能与总硫含量对瘤胃真菌菌群大小的影响同样重要。低硫马唐属（*Digitaria*）牧草日粮中添加蛋氨酸或者单质硫（Gordon，1985），谷物秸秆中添加蛋氨酸（Gordon 等，1983；Gulati 等，1985）或者硫酸盐（Weston 等，1988）同样有益于瘤胃中的厌氧真菌。与未添加硫酸盐的干草相比，针茅日粮补充硫酸盐可提高真菌数量（Morrison 等，1990）。重要的是，体外培养的厌氧真菌需要还原硫（Orpin 和 Greenwood，1986；Gordon 和 Phillips，1995），表明在厌氧真菌可利用前，需要将瘤胃内补充的硫酸盐还原。迄今为止，可用于刺激瘤胃厌氧真菌的饲料添加剂（单质硫、硫酸盐或蛋氨酸）的硫含量可能对所有的瘤胃微生物区系都有影响。有研究比较日粮中两种有机硫营养素巯基-1-丙酸（MPA）、3-巯基-1-丙磺酸（MPS）与无机硫添加剂在牛上的效果，其结果表明这些有机硫提高牛的氮利用和微生物蛋白产量，但令人惊讶的是，这是由于微生物发酵木质纤维素效率的普遍提高，而不是由于对真菌的特定刺激（McSweeney 和 Denman，2007）。针对厌氧真菌的含硫添加剂研究成果尚待鉴定。

厌氧真菌在不同反刍动物之间转移的可行性已得到证实（Orpin，1989）。Lee 等（2000）报道，山羊源的 *Orpinomyces* 菌株接种到绵羊瘤胃，可提高绵羊的营养物质消化率。同样地，Paul 等（2004）报道，一种优秀的厌氧真菌（*Piromyces* sp. FNG5，从野生蓝牛粪便中分离）具有比水牛分离的厌氧真菌更强的木质纤维素分解活性，饲喂这种厌氧真菌的培养物能够提高水牛的消化率。瘤胃内添加分离的真菌显著增加挥发性脂肪酸的浓度。同时，饲喂真菌提高纤维素酶和半纤维素酶的活性以及细菌和真菌总数。瘤胃内容物中的羧甲基纤维素酶、木聚糖酶、微晶纤维素酶、乙酰酯酶、阿魏酸酯酶和蛋白酶的活性也有所提高。Gordon 等（2000）注意到，给绵羊添加非内源性厌氧真菌（从牛中分离的 *Piromyces* sp. CS15）比从绵羊中分离的厌氧真菌具有更高的纤维素分解活性，导致自由采食量增加 12%。

11　结论

尽管瘤胃厌氧真菌的生态重要性已被普遍认可,但是针对这类微生物的研究很少。最近发展起来的先进分子技术提高了对真菌的检测和分类能力,并首次深入了解了厌氧瘤胃真菌的真正系统发育和功能多样性。目前分离至今未知分类的真菌的步伐正在加快,以及对真菌的鉴定、酶学,以及过去认为厌氧真菌不存在的酶库的研究进展也在加快,这为厌氧真菌的研究开辟了新的可能。上述工作有望激励科学界更多地关注目前仍知之甚少的微生物群体,以及它们在瘤胃生态系统的作用。

12　更多信息

我们对厌氧真菌的了解仍然相对有限,但是有几篇文章已经很好地介绍这些有趣的微生物的生物学特点。此外,组学技术的最新进展为这些难降解生物质降解菌的分子过程提供了一些重要的见解。厌氧真菌网络(Anaerobic Fungi Network,https://anaerobicfungi.org/)是了解厌氧真菌领域最新进展的一个非常好的网站。该网站包含大多数厌氧真菌研究人员的名单。由 Joan Edwards 博士整理的厌氧真菌的基本介绍(https://youtu.be/VEis-RWzNyfI)、生命周期的总结(https://youtu.be/x8jJbkT7t3o)和分类学信息(https://youtu.be/vMhb4QL5zRQ)可以在 Youtube 上找到,从而使我们能够快速了解该领域。另一个关于厌氧真核生物的有用资源则是一个研究主题网站,其中包含关于栖息于草食动物肠道内厌氧真菌和其他真核生物的开放获取的同行评议文章。该研究主题网站的 URL 是 https://www.frontiersin.org/research-topics/9250。在 de Vries 等(2018)创办的 *Fungal Genomics* 中可以找到优秀的真菌基因组操作指南,包括厌氧真菌的基因组学。近期,Wang 等(2019)利用比较基因组学方法阐明了厌氧真菌出现的时间点,以及水平基因转移在厌氧真菌进化和特殊代谢功能上的潜在作用。Elshahed 教授团队的工作大大扩展了厌氧真菌系统发育学的研究(Hanafy 等,2019)。最后,强烈推荐 Orpin(1975)、Gordon 和 Philipps(1998)这几位研究者的标志性论文。

13　参考文献

Abot, A., Arnal, G., Auer, L., et al., 2016. CAZyChip:dynamic assessment of exploration of glycoside hydrolases in microbial ecosystems. BMC Genomics 17,671.

Accetto, T. and Avgustin, G. 2015. Polysaccharide utilization locus and CAZyme genome repertoires reveal diverse ecological adaptation of *Prevotella* species. Syst. Appl. Microbiol. 38(7),453-61.

Ahrendt, S. R., Quandt, C. A., Ciobanu, D., et al., 2018. Leveraging single-cell genomics to expand the fungal tree of life. Nat. Microbiol. 3(12),1417-28.

Akin, D. E. and Benner, R. 1988. Degradation of polysaccharides and lignin by ruminal bacteria and fungi. Appl. Environ. Microbiol. 54(5),1117-25.

Akin, D. E. and Rigsby, L. L. 1987. Mixed fungal populations and lignocellulosic tissue degradation in the bovine rumen. Appl. Environ. Microbiol. 53（9）, 1987-95.

Akin, D. E. and Windham, W. R. 1989. Influence of diet on rumen fungi. In: Nolan, J. V., Leng, R. A. and Demeyer, D. I. （Eds）, The Role of Protozoa and Fungi in Ruminant Digestion. Penumbul Books, Armidale, New South Wales, Australia, pp. 75-82.

Akin, D. E., Gordon, G. L. and Hogan, J. P. 1983. Rumen bacterial and fungal degradation of Digitaria pentzii grown with or without sulfur. Appl. Environ. Microbiol. 46（3）, 738-48.

Ariyawansa, H. A., Hyde, K. D., Jayasiri, S. C., et al., 2015. Fungal diversity notes 111-252—taxonomic and phylogenetic contributions to fungal taxa. Fungal Divers. 75（1）, 27-274.

Artzi, L., Bayer, E. A. and Morais, S. 2017. Cellulosomes: bacterial nanomachines for dismantling plant polysaccharides. Nat. Rev. Microbiol. 15（2）, 83-95.

Barr, D. J. S. 1980. An outline for the reclassification of the Chytridiales, and for a new order, the Spizellomycetales. Can. J. Bot. 58（22）, 2380-94.

Barr, D. J. S. 1988. How modern systematics relates to the rumen fungi. Biosystems 21（3-4）, 351-6.

Barr, A. J. 2001. Chytridiomycota. In: McLaughlin, D. and Spatafora, J. W. （Eds）, The Mycota Ⅶ-Systematics and Evolution-Part A. Springer-Verlag, Berlin, pp. 93-112.

Barr, D. J. S., Kudo, H., Jakober, K. D., et al., 1989. Morphology and development of rumen fungi: *Neocallimastix* sp., *Piromyces communis*, and *Orpinomyces bovis* gen. nov., sp. nov. Can. J. Bot. 67（9）, 2815-24.

Barr, D. J. S., Yanke, L. J., Bae, H. D., et al., 1995. Contributions on the morphology and taxonomy of some rumen fungi from Canada. Mycotaxon 54, 203-14.

Bauchop, T. 1979. Rumen anaerobic fungi of cattle and sheep. Appl. Environ. Microbiol. 38（1）, 148-58.

Bauchop, T. and Mountfort, D. O. 1981. Cellulose fermentation by a rumen anaerobic fungus in both the absence and the presence of rumen methanogens. Appl. Environ. Microbiol. 42（6）, 1103-10.

Bonnemoy, F., Fonty, G., Michel, V., et al., 1993. Effect of anaerobic fungi on the ruminal proteolysis in gnotobiotic lambs. Reprod. Nutr. Dev. 33（6）, 551-5.

Bootten, T. J., Joblin, K. N., McArdle, B. H., et al., 2011. Degradation of lignified secondary cell walls of Lucerne （*Medicago sativa* L.） by rumen fungi growing in methanogenic co-culture. J. Appl. Microbiol. 111（5）, 1086-96.

Borneman, W. S., Akin, D. E. and Ljungdahl, L. G. 1989. Fermentation products and plant cell wall-degrading enzymes produced by monocentric and polycentric anaerobic ruminal fungi. Appl. Environ. Microbiol. 55（5）, 1066-73.

Borneman, W. S., Hartley, R. D., Morrison, W. H., et al., 1990. Feruloyl and p-coumaroyl esterase from anaerobic fungi in relation to plant cell wall degradation. Appl. Microbiol. Biotechnol. 33（3）, 345-51.

Bowman, B. H., Taylor, J. W., Brownlee, A. G., et al., 1992. Molecular evolution of the fungi: relationship of the Basidiomycetes, Ascomycetes, and Chytridiomycetes. Mol. Biol. Evol. 9（2）, 285-96.

Braune, R. A. 1913. Untersuchungen über die im Wiederkäuermagen vorkommenden Protozoen. Arch. Protistenkunde 32, 111-70.

Breton, A., Bernalier, A., Bonnemoy, F., et al., 1989. Morphological and metabolic characterization of a new species of strictly anaerobic rumen fungus: *Neocallimastix joyonii*. FEMS Microbiol. Lett. 58（2-3）, 309-14.

Breton, A., Bernalier, A., Dusser, M., et al., 1990. *Anaeromyces mucronatus* nov. gen., nov. sp. A new strictly anaerobic rumen fungus with polycentric thallus. FEMS Microbiol. Lett. 58 (2), 177-82.

Breton, A., Dusser, M., Gaillard-Martinie, B., et al., 1991. *Piromyces rhizinflata* nov. sp., a strictly anaerobic fungus from faeces of the Saharian ass: a morphological, metabolic and ultrastructural study. FEMS Microbiol. Lett. 66 (1), 1-8.

Britto-Kido Sde, A., Neto, J. R. F., Pandolfi, V., et al., 2013. Natural antisense transcripts in plants: a review and identification in soybean infected with Phakopsora pachyrhizi SuperSAGE library. Sci. World J. 2013:219798.

Brookman, J. L., Ozkose, E., Rogers, S., et al., 2000. Identification of spores in the polycentric anaerobic gut fungi which enhance their ability to survive. FEMS Microbiol. Ecol. 31 (3), 261-7.

Brownlee, A. G. 1989. Remarkably AT-rich genomic DNA from the anaerobic Fungus Neocallimastix. Nucleic Acids Res. 17 (4), 1327-35.

Calderon-Cortes, J. F., Elliott, R. and Ford, C. W. 1989. Influence of rumen fungi on the nutrition of sheep fed forage diets. In: Nolan, J. V., Leng, R. A. and Demeyer, D. I. (Eds), The Roles of Protozoa and Fungi in Ruminant Digestion. Penumbul Books, Armidale, Australia, pp. 181-7.

Callaghan, T. M., Podmirseg, S. M., Hohlweck, D., et al., 2015. *Buwchfawromyces eastonii* gen. nov., sp. nov.: a new anaerobic fungus (Neocallimastigomycota) isolated from buffalo faeces. MycoKeys 9, 11-28.

Chen, Y. C., Hseu, R. S. and Chien, C. Y. 2002. *Piromyces polycephalus* (Neocallimastigaceae), a new rumen fungus. Nova Hedwigia 75 (3), 409-14.

Chen, H., Hopper, S. L., Li, X. L., et al., 2006. Isolation of extremely AT-rich genomic DNA and analysis of genes encoding carbohydrate-degrading enzymes from Orpinomyces sp. strain PC-2. Curr. Microbiol. 53 (5), 396-400.

Chen, Y. C., Tsai, S. D., Cheng, H. L., et al., 2007. *Caecomyces sympodialis* sp. nov., a new rumen fungus isolated from Bos indicus. Mycologia 99 (1), 125-30.

Cheng, Y. F., Edwards, J. E., Allison, G. G., et al., 2009. Diversity and activity of enriched ruminal cultures of anaerobic fungi and methanogens grown together on lignocellulose in consecutive batch culture. Bioresour. Technol. 100 (20), 4821-8.

Cheng, Y. F., Jin, W., Mao, S. Y., et al., 2013. Production of citrate by anaerobic fungi in the presence of co-culture methanogens as revealed by (1) H NMR spectrometry. Asian-Australas. J. Anim. Sci. 26 (10), 1416-23.

Cheng, Y., Shi, Q., Sun, R., et al., 2018. The biotechnological potential of anaerobic fungi on fiber degradation and methane production. World J. Microbiol. Biotechnol. 34 (10), 155.

Chesson, A. 1993. Mechanistic models of forage cell wall degradation. In: Jung, H. G., Buxton, D. R., Hatfield, R. D. and Ralph, J. (Eds), Forage Cell Wall Structure and Digestibility. American Society of Agronomy, Madison, WI, pp. 347-76.

Comtet-Marre, S., Chaucheyras-Durand, F., Bouzid, O., et al., 2018. FibroChip, a functional DNA microarray to monitor cellulolytic and hemicellulolytic activities of rumen microbiota. Front. Microbiol. 9, 215.

Couger, M. B., Youssef, N. H., Struchtemeyer, C. G., et al., 2015. Transcriptomic analysis of lignocellulosic biomass degradation by the anaerobic fungal isolate Orpinomyces sp. strain C1A. Biotechnol. Biofuels 8, 208.

Dagar, S. S., Kumar, S., Griffith, G. W., et al., 2015. A new anaerobic fungus (*Oontomyces anksri* gen. nov., sp. nov.) from the digestive tract of the Indian camel (*Camelus dromedarius*). Fungal Biol. 119 (8), 731-7.

Dai, X., Tian, Y., Li, J., et al., 2015. Metatranscriptomic analyses of plant cell wall polysaccharide degradation by microorganisms in the cow rumen. Appl. Environ. Microbiol. 81 (4), 1375-86.

Dayananda, T. L., Nagpal, R., Puniya, A. K., et al., 2007. Biodegradation of urea-NH_3 treated wheat straw using anaerobic rumen fungi. J. Anim. Feed Sci. 16 (3), 484-9.

Dean, F. B., Nelson, J. R., Giesler, T. L., et al., 2001. Rapid amplification of plasmid and phage DNA using Phi 29 DNA polymerase and multiply-primed rolling circle amplification. Genome Res. 11 (6), 1095-9.

Denman, S. E., Nicholson, M. J., Brookman, J. L., et al., 2008. Detection and monitoring of anaerobic rumen fungi using an ARISA method. Lett. Appl. Microbiol. 47 (6), 492-9.

de Vries, R. P., Tsang, A. and Grigoriev, I. V. (Eds). 2018. Fungal Genomics: Methods and Protocols. Humana Press. Available at: https://www.springer.com/us/book/9781493978038.

Dey, A., Sehgal, J. P., Puniya, A. K., et al., 2004. Influence of an anaerobic fungal culture (Orpinomyces sp.) administration on growth rate, ruminal fermentation and nutrient digestion in calves. Asian Australas. J. Anim. Sci 17 (6), 820-4.

Dill-McFarland, K. A., Weimer, P. J., Breaker, J. D., et al., 2019. Diet influences early microbiota development in dairy calves without long-term impacts on milk production. Appl. Environ. Microbiol. 85 (2):

Dollhofer, V., Podmirseg, S. M., Callaghan, T. M., et al., 2015. Anaerobic fungi and their potential for biogas production. Adv. Biochem. Eng. Biotechnol. 151, 41-61.

Dollhofer, V., Callaghan, T. M., Dorn-In, S., et al., 2016. Development of three specific PCR-based tools to determine quantity, cellulolytic transcriptional activity and phylogeny of anaerobic fungi. J. Microbiol. Methods 127, 28-40.

Donaldson, M. E. and Saville, B. J. 2012. Natural antisense transcripts in fungi. Mol. Microbiol. 85 (3), 405-17.

Dore, J. and Stahl, D. A. 1991. Phylogeny of anaerobic rumen Chytridiomycetes inferred from small subunit ribosomal RNA sequence comparisons. Can. J. Bot. 69 (9), 1964-71.

Ebersberger, I., de Matos Simoes, R., Kupczok, A., et al., 2012. A consistent phylogenetic backbone for the fungi. Mol. Biol. Evol. 29 (5), 1319-34.

Edwards, J. E., Kingston-Smith, A. H., Jimenez, H. R., et al., 2008. Dynamics of initial colonization of nonconserved perennial ryegrass by anaerobic fungi in the bovine rumen. FEMS Microbiol. Ecol. 66 (3), 537-45.

Edwards, J. E., Forster, R. J., Callaghan, T. M., et al., 2017. PCR and omics based techniques to study the diversity, ecology and biology of anaerobic fungi: insights, challenges and opportunities. Front. Microbiol. 8, 1657.

Elliott, R., Ash, A. J., Calderon-Cortes, F., et al., 1987. The influence of anaerobic fungi on rumen volatile fatty acid concentrations *in vivo*. J. Agric. Sci. 109 (1), 13-7.

Faichney, G. J., Poncet, C., Lassalas, B., et al., 1997. Effect of concentrates in a hay diet on the contribution of anaerobic fungi, protozoa and bacteria to nitrogen in rumen and duodenal digesta in sheep. Anim. Feed Sci. Technol. 64 (2-4), 193-213.

Flint, H. J., Scott, K. P., Duncan, S. H., et al., 2012. Microbial degradation of complex carbohydrates in the gut. Gut Microbes 3 (4), 289-306.

Fonty, G. and Grenet, E. 1994. Effects of diet on the fungal population of the digestive tract of ruminants. In: Mountfort, D. O. and Orpin, C. G. (Eds), Anaerobic Fungi: Biology, Ecology and Function. Marcel Dekker, New York, pp. 229-39.

Foster, J. 1949. Chemical Activities of the Fungi. Academic Press, New York.

Gao, A. W., Wang, H. R., Yang, J. L., et al., 2013. The effects of elimination of fungi on microbial population and fiber degradation in sheep rumen. Appl. Mech. Mater. 295-298, 224-31.

Gilbert, H. J. 2010. The biochemistry and structural biology of plant cell wall deconstruction. Plant Physiol. 153 (2), 444-55.

Gilmore, S. P., Henske, J. K. and O'Malley, M. A. 2015. Driving biomass breakdown through engineered cellulosomes. Bioengineered 6 (4), 204-8.

Gold, J. J., Heath, I. B. and Bauchop, T. 1988. Ultrastructural description of a new chytrid genus of caecum anaerobe, *Caecomyces equi* gen. nov., sp. nov., assigned to the Neocallimasticaceae. Biosystems 21 (3-4), 403-15.

Gordon, G. L. R. 1985. The potential for manipulation of rumen fungi. In: Leng, R. A., Barker, J. S. F., Adams, D. B. and Hutchinson, K. J. (Eds), Biotechnology and Recombinant DNA Technology in the Animal Production Industries. Reviews in Rural Science No. 6. University of New England, Armidale, Australia, pp. 124-8.

Gordon, G. L. R. and Phillips, M. W. 1993. Removal of anaerobic fungi from the rumen of sheep by chemical treatment and the effect on feed consumption and *in vivo* fibre digestion. Lett. Appl. Microbiol. 17 (5), 220-3.

Gordon, G. L. R. and Phillips, M. W. 1995. New approach for the manipulation of anaerobic fungi in the rumen. In: Corbett, J. L., Choct, M., Nolan, J. V. and Rowe, J. B. (Eds), Recent Advances in Animal Nutrition in Australia. University of New England, Armidale, Australia, pp. 108-15.

Gordon, G. L. and Phillips, M. W. 1998. The role of anaerobic gut fungi in ruminants. Nutr. Res. Rev. 11 (1), 133-68.

Gordon, G. L. R., Gulati, S. K. and Ashes, J. R. 1983. Influence of low-sulphur straw on anaerobic fungal numbers in a sheep rumen. In: Proceedings of the Nutritional Society of Australia.

Gordon, G. L. R., Wong, H. and Phillips, M. V. 1995. In vitro degradation of 14C lignocellulose by polycentric and monocentric ruminal anaerobic fungi is inhibited differently by phenolic monomers. Ann. Zootech. 44 (Suppl. 1), 152-.

Gordon, G. L. R., Phillips, M. W., Rintoul, A. J., et al., 2000. Increased intake of fibrous feed by sheep orally dosed with a culture of an elite non-indigenous anaerobic gut fungus. Asian Aust. J. Anim. Sci. 13 (143).

Grenet, E., Breton, A., Barry, P., et al., 1989. Rumen anaerobic fungi and plant substrate colonization as affected by diet composition. Anim. Feed Sci. Technol. 26 (1-2), 55-70.

Gruninger, R. J., Puniya, A. K., Callaghan, T. M., et al., 2014. Anaerobic fungi (phylum Neocallimastigomycota): advances in understanding their taxonomy, life cycle, ecology, role and biotechnological potential. FEMS Microbiol. Ecol. 90 (1), 1-17.

Gruninger, R. J., Nguyen, T. T. M., Reid, I. D., et al., 2018. Application of transcriptomics to compare the carbohydrate active enzymes that are expressed by diverse genera of anaerobic fungi to degrade plant cell wall carbohydrates. Front. Microbiol. 9, 1581.

Gulati, S. K., Ashes, J. R., Gordon, G. L. R., et al., 1985. Possible contribution of rumen fungi to fibre digestion in sheep. Proc. Nutr. Soc. Aust. 10, 96.

Gulati, S. K., Ashes, J. R. and Gordon, G. L. R. 1988. Digestibility of sulphur amino acids in rumen fungal species. Proc. Nutr. Soc. Aust. 13 (133).

Gulati, S. K., Ashes, J. R., Gordon, G. L. R., et al., 1989. Nutritional availability of amino acids from the rumen anaerobic fungus *Neocallimastix* sp. LM1 in sheep. J. Agric. Sci. 113 (3), 383-7.

Guliye, A. Y. and Wallace, R. J. 2007. Effects of aromatic amino acids, phenylacetate and phenylpropionate on fermentation of xylan by the rumen anaerobic fungi, *Neocallimastix frontalis* and *Piromyces* communis. J. Appl. Microbiol. 103 (4), 924-9.

Haitjema, C. H., Solomon, K. V., Henske, J. K., et al., 2014. Anaerobic gut fungi: advances in isolation, culture, and cellulolytic enzyme discovery for biofuel production. Biotechnol. Bioeng. 111 (8), 1471-82.

Haitjema, C. H., Gilmore, S. P., Henske, J. K., et al., 2017. A parts list for fungal cellulosomes revealed by comparative genomics. Nat. Microbiol. 2, 17087.

Hanafy, R. A., Elshahed, M. S., Liggenstoffer, A. S., et al., 2017. *Pecoramyces ruminantium*, gen. nov., sp. nov., an anaerobic gut fungus from the feces of cattle and sheep. Mycologia 109 (2), 231-43.

Hanafy, R. A., Elshahed, M. S. and Youssef, N. H. 2018a. *Feramyces austinii*, gen. nov., sp. nov., an anaerobic gut fungus from rumen and fecal samples of wild Barbary sheep and fallow deer. Mycologia 110 (3), 513-25.

Hanafy, R. A., Johnson, B., Elshahed, M. S., et al., 2018b. *Anaeromyces contortus*, sp. nov., a new anaerobic gut fungal species (Neocallimastigomycota) isolated from the feces of cow and goat. Mycologia 110 (3), 502-12.

Hanafy, R., Lanjekar, V., Dhakephalkar, P., et al., 2019. Seven new Neocallimastigomycota genera from fecal samples of wild, zoo-housed, and domesticated herbivores: description of *Ghazallomyces constrictus* gen. nov., sp. nov., *Aklioshbomyces papillarum* gen. nov., sp. nov., *Agriosomyces longus* gen. nov., sp. nov., *Capellomyces foraminis* gen. nov., sp. nov. and *Capellomyces elongatus* sp. nov., *Joblinomyces apicalis* gen. nov., sp. nov., *Khoyollomyces ramosus* gen. nov., sp. nov., and *Tahromyces munnarensis* gen. nov., sp. nov. bioRxiv.

Heath, I. B., Bauchop, T. and Skipp, R. A. 1983. Assignment of the rumen anaerobe Neocallimastix frontalis to the Spizellomycetales (Chytridiomycetes) on the basis of its polyflagellate zoospore ultrastructure. Can. J. Bot. 61 (1), 295-307.

Heath, I. B., Kaminskyj, S. G. and Bauchop, T. 1986. Basal body loss during fungal zoospore encystment: evidence against centriole autonomy. J. Cell Sci. 83 (1), 135-40.

Henske, J. K., Gilmore, S. P., Knop, D., et al., 2017. Transcriptomic characterization of *Caecomyces churrovis*: a novel, non-rhizoid-forming lignocellulolytic anaerobic fungus. Biotechnol. Biofuels 10, 305.

Hess, M., Sczyrba, A., Egan, R., et al., 2011. Metagenomic discovery of biomass-degrading genes and genomes from cow rumen. Science 331 (6016), 463-7.

Hibbett, D. S., Binder, M., Bischoff, J. F., et al., 2007. A higher-level phylogenetic classification of the Fungi. Mycol. Res. 111 (5), 509-47.

Hillaire, M. C. and Jouany, J. P. 1989. Effects of rumen anaerobic fungi on the digestion of wheat straw and the end products of microbial metabolism, studies in a semicontinuous *in vitro* system. In: Nolan, J. V., Leng, R. A. and Demeyer, D. I. (Eds), The Role of Protozoa and Fungi in Ruminant Digestion.

Penumbul Books, Armidale, Australia, pp. 269-71.

Ho, Y. W. and Barr, D. J. S. 1995. Classification of anaerobic gut fungi from herbivores with emphasis on rumen fungi from Malaysia. Mycologia 87 (5), 655-77.

Ho, Y. W. and Bauchop, T. 1991. Morphology of three polycentric rumen fungi and description of a procedure for the induction of zoosporogenesis and release of zoospores in cultures. J. Gen. Microbiol. 137 (1), 213-7.

Ho, Y. W., Abdullah, N. and Jalaludin, S. 1988. Colonization of guinea grass by anaerobic rumen fungi in swamp buffalo and cattle. Anim. Feed Sci. Technol. 22 (1-2), 161-71.

Ho, Y. W., Bauchop, T., Abdullah, N., et al., 1990. *Ruminomyces elegans* gen. et sp. nov, a polycentric anaerobic rumen fungus from cattle. Mycotaxon 38, 397-405.

Ho, Y. W., Barr, D. J. S., Abdullah, N., et al., 1993a. Anaeromyces, an earlier name for Ruminomyces. Mycotaxon 47, 283-4.

Ho, Y. W., Barr, D. J. S., Abdullah, N., et al., 1993b. *Neocallimastix variabilis*, a new species of anaerobic fungus from the rumen of cattle. Mycotaxon 46, 241-58.

Ho, Y. W., Barr, D. J. S., Abdullah, N., et al., 1993c. *Piromyces spiralis*, a new species of anaerobic fungus from the rumen of goat. Mycotaxon 47, 285-93.

Ho, Y. W., Barr, D. J. S., Abdullah, N., et al., 1993d. A new species of Piromyces from the rumen of deer in Malaysia. Mycotaxon 47, 285-93.

Ho, Y. W., Barr, D. J. S., Abdullah, N., et al., 1994. *Orpinomyces intercalaris*, a new species of polycentric anaerobic rumen fungus from cattle. Mycotaxon 50, 139-50.

Ho, Y. W., Abdullah, N. and Jalaludin, S. 1996. Microbial colonisation and degradation of some fibrous crop residues in the rumen of goats. Asian Australas. J. Anim. Sci 9 (5), 519-24.

James, T. Y., Kauff, F., Schoch, C. L., et al., 2006. Reconstructing the early evolution of Fungi using a six-gene phylogeny. Nature 443 (7113), 818-22.

Jenkins, T. C. 1993. Lipid metabolism in the rumen. J. Dairy Sci. 76 (12), 3851-63.

Jensen, E. A. and Hammond, D. M. 1964. A morphological study of trichomonads and related flagellates from the bovine digestive tract. J. Protozool. 11 (3), 386-94.

Jin, W., Cheng, Y. F., Mao, S. Y., et al., 2011. Isolation of natural cultures of anaerobic fungi and indigenously associated methanogens from herbivores and their bioconversion of lignocellulosic materials to methane. Bioresour. Technol. 102 (17), 7925-31.

Jin, W., Cheng, Y. F., Mao, S. Y., et al., 2014. Discovery of a novel rumen methanogen in the anaerobic fungal culture and its distribution in the rumen as revealed by real-time PCR. BMC Microbiol. 14 (1), 104.

Joblin, K. N. 1981. Isolation, enumeration, and maintenance of rumen anaerobic fungi in roll tubes. Appl. Environ. Microbiol. 42 (6), 1119-22.

Joshi, A., Lanjekar, V. B., Dhakephalkar, P. K., et al., 2018. *Liebetanzomyces polymorphus* gen. et sp. nov., a new anaerobic fungus (Neocallimastigomycota) isolated from the rumen of a goat. MycoKeys 40 (40), 89-110.

Julliand, V., Riondet, C., de Vaux, A., et al., 1998. Comparison of metabolic activities between *Piromyces citronii*, an equine fungal species, and *Piromyces communis*, a ruminal species. Anim. Feed Sci. Technol. 70 (1-2), 161-8.

Kantor, R. S., van Zyl, A. W., van Hille, R. P., et al., 2015. Bioreactor microbial ecosystems for thiocyanate and cyanide degradation unravelled with genome-resolved metagenomics. Environ. Microbiol. 17 (12), 4929-41.

Kantor, R. S., Huddy, R. J., Iyer, R., et al., 2017. Genome-resolved metaomics ties microbial dynamics to process performance in biotechnology for thiocyanate degradation. Environ. Sci. Technol. 51 (5), 2944-53.

Katayama, S., Tomaru, Y., Kasukawa, T., et al., 2005. Antisense transcription in the mammalian transcriptome. Science 309 (5740), 1564-6.

Kirk, P. M. 2012. Anaeromyces polycephalus (Chen YC, Chien CY, Hseu RS) Flieg., K. Voigt & P. M. Kirk. IF550012. Index Fungorum 1, 1.

Kittelmann, S., Naylor, G. E., Koolaard, J. P., et al., 2012. A proposed taxonomy of anaerobic fungi (class Neocallimastigomycetes) suitable for large-scale sequence-based community structure analysis. PLoS ONE 7 (5), e36866.

Koetschan, C., Kittelmann, S., Lu, J., et al., 2014. Internal transcribed spacer 1 secondary structure analysis reveals a common core throughout the anaerobic fungi (Neocallimastigomycota). PLoS ONE 9 (3), e91928.

Kumar, B. M., Puniya, A. K., Singh, K., et al., 2004. *In vitro* degradation of cell-wall and digestibility of cereal straws treated with anaerobic ruminal fungi. Indian J. Exp. Biol. 42 (6), 636-8.

Kumar, S., Sehgal, J. P., Puniya, A., et al., 2015. Growth performance and fibre utilization of Murrah male buffalo calves fed wheat straw based complete feed blocks incorporated with superior anaerobic fungal zoospores (*Neocallimastix* sp. GR-1). Indian J. Anim. Sci. 85 (3), 275-81.

Kumar, S., Stecher, G. and Tamura, K. 2016. MEGA7: molecular evolutionary genetics analysis version 7. 0 for bigger datasets. Mol. Biol. Evol. 33 (7), 1870-4.

Lee, S. S., Ha, J. K. and Cheng, K. J. 2000. Influence of an anaerobic fungal culture administration on in vivo ruminal fermentation and nutrient digestion. Anim. Feed Sci. Technol. 88 (3-4), 201-17.

Leis, S., Dresch, P., Peintner, U., et al., 2014. Finding a robust strain for biomethanation: Anaerobic fungi (Neocallimastigomycota) from the Alpine ibex (Capra ibex) and their associated methanogens. Anaerobe 29, 34-43.

Li, J. and Heath, I. B. 1992. The phylogenetic relationships of the anaerobic chytridiomycetous gut fungi (Neocallimasticaceae) and the Chytridiomycota. I. Cladistic analysis of rRNA sequences. Can. J. Bot. 70 (9), 1738-46.

Li, J., Heath, I. B. and Bauchop, T. 1990. *Piromyces mae* and *Piromyces dumbonica*, two new species of uniflagellate anaerobic chytridiomycete fungi from the hindgut of the horse and elephant. Can. J. Bot. 68 (5), 1021-33.

Li, J., Heath, I. B. and Cheng, K. J. 1991. The development and zoospore ultrastructure of a polycentric chytridiomycete gut fungus, Orpinomyces joyonii comb. nov. Can. J. Bot. 69 (3), 580-9.

Li, J., Heath, I. B. and Packer, L. 1993. The phylogenetic relationships of the anaerobic chytridiomycetous gut fungi (Neocallimasticaceae) and the Chytridiomycota. II. Cladistic analysis of structural data and description of Neocallimasticales ord. nov. Can. J. Bot. 71 (3), 393-407.

Li, G. J., Hyde, K. D., Zhao, R. L., et al., 2016. Fungal diversity notes 253-366: taxonomic and phylogenetic contributions to fungal taxa. Fungal Divers. 78 (1), 1-237.

Li, Y., Li, Y., Jin, W., et al., 2019. Combined genomic, transcriptomic, proteomic, and physiological characterization of the growth of *Pecoramyces* sp. F1 in monoculture and co-culture with a syntrophic methanogen. Front. Microbiol. 10, 435.

Liebetanz, E. 1910. Die parasitischen Protozoen des Wiederkauermagens. Arch. Prot. 19, 19-90.

Liggenstoffer, A. S., Youssef, N. H., Couger, M. B., et al., 2010. Phylogenetic diversity and community structure of anaerobic gut fungi (phylum Neocallimastigomycota) in ruminant and non-ruminant herbivores. ISME J. 4 (10), 1225-35.

Lowe, S. E., Theodorou, M. K., Trinci, A. P. J., et al., 1985. Growth of anaerobic rumen fungi on defined and semi-defined media lacking rumen fluid. Microbiology 131 (9), 2225-9.

Lowe, S. E., Theodorou, M. K. and Trinci, A. P. J. 1987. Growth and fermentation of an anaerobic rumen fungus on various carbon sources and effect of temperature on development. Appl. Environ. Microbiol. 53 (6), 1210-5.

Makela, M. R., Dilokpimol, A., Koskela, S. M., et al., 2018. Characterization of a feruloyl esterase from *Aspergillus terreus* facilitates the division of fungal enzymes from carbohydrate esterase family 1 of the carbohydrate-active enzymes (CAZy) database. Microb. Biotechnol. 11 (5), 869-80.

McAllister, T. A., Dong, Y., Yanke, L. J., et al., 1993. Cereal grain digestion by selected strains of ruminal fungi. Can. J. Microbiol. 39 (4), 367-76.

McDonald, J. E., Houghton, J. N. I., Rooks, D. J., et al., 2012. The microbial ecology of anaerobic cellulose degradation in municipal waste landfill sites: evidence of a role for fibrobacters. Environ. Microbiol. 14 (4), 1077-87.

McGranaghan, P., Davies, J. C., Griffith, G. W., et al., 1999. The survival of anaerobic fungi in cattle faeces. FEMS Microbiol. Ecol. 29 (3), 293-300.

McSweeney, C. S. and Denman, S. E. 2007. Effect of sulfur supplements on cellulolytic rumen micro-organisms and microbial protein synthesis in cattle fed a high fibre diet. J. Appl. Microbiol. 103 (5), 1757-65.

McSweeney, C. S., Dulieu, A., Katayama, Y., et al., 1994. Solubilization of lignin by the ruminal anaerobic fungus Neocallimastix patriciarum. Appl. Environ. Microbiol. 60 (8), 2985-9.

Michel, V., Fonty, G., Millet, L., et al., 1993. *In vitro* study of the proteolytic activity of rumen anaerobic fungi. FEMS Microbiol. Lett. 110 (1), 5-9.

Millard, P., Gordon, A. H., Richardson, A. J., et al., 1987. Reduced ruminal degradation of ryegrass caused by sulphur limitation. J. Sci. Food Agric. 40 (4), 305-14.

Milne, A., Theodorou, M. K., Jordan, M. G. C., et al., 1989. Survival of anaerobic fungi in feces, in saliva, and in pure culture. Exp. Mycol. 13 (1), 27-37.

Morrison, M., Murray, R. M. and Boniface, A. N. 1990. Nutrient metabolism and rumen micro-organisms in sheep fed a poor-quality tropical grass hay supplemented with sulphate. J. Agric. Sci. 115 (2), 269-75.

Mosier, A. C., Miller, C. S., Frischkorn, K. R., et al., 2016. Fungi contribute critical but spatially varying roles in nitrogen and carbon cycling in acid mine drainage. Front. Microbiol. 7, 238.

Mountfort, D. O. 1987. The rumen anaerobic fungi. FEMS Microbiol. Lett. 46 (4), 401-8.

Mountfort, D. O. 1994. Regulatory constraints in the degradation and fermentation of carbohydrates by anaerobic fungi. In: Mountfort, D. O. and Orpin, C. G. (Eds), Anaerobic Fungi: Biology, Ecology and Function. Marcel Dekker, New York, pp. 147-68.

Mura, E., Edwards, J., Kittelmann, S., et al., 2019. Anaerobic fungal communities differ along the horse digestive tract. Fungal Biol. 123 (3), 240-6.

Nagahama, T. and Nagano, Y. 2012. Cultured and uncultured fungal diversity in deepsea environments. Prog. Mol. Subcell. Biol. 53, 173-87.

Nakamura, T., Yamada, K. D., Tomii, K., et al., 2018. Parallelization of MAFFT for large-scale multiple sequence alignments. Bioinformatics 34 (14), 2490-2.

Nilsson, R. H., Anslan, S., Bahram, M., et al., 2019. Mycobiome diversity: high - throughput sequencing and identification of fungi. Nat. Rev. Microbiol. 17 (2), 95-109.

Ochiai, A., Itoh, T., Kawamata, A., et al., 2007. Plant cell wall degradation by saprophytic *Bacillus subtilis* strains: gene clusters responsible for rhamnogalacturonan depolymerization. Appl. Environ. Microbiol. 73 (12), 3803-13.

Orpin, C. G. 1975. Studies on the rumen flagellate *Neocallimastix frontalis*. J. Gen. Microbiol. 91 (2), 249-62.

Orpin, C. G. 1976. Studies on the rumen flagellate *Sphaeromonas communis*. J. Gen. Microbiol. 94 (2), 270-80.

Orpin, C. G. 1977a. The occurrence of chitin in the cell walls of the rumen organisms *Neocallimastix frontalis*, *Piromonas communis* and *Sphaeromonas communis*. J. Gen. Microbiol. 99 (1), 215-8.

Orpin, C. G. 1977b. The rumen flagellate *Piromonas communis*: its life-history and invasion of plant material in the rumen. J. Gen. Microbiol. 99 (1), 107-17.

Orpin, C. G. 1989. Ecology of rumen anaerobic fungi in relation to the nutrition of the host animal. In: Nolan, J. V., Leng, R. A. and Demeyer, D. I. (Eds), The Role of Protozoa and Fungi in Ruminant Digestion. Penumbul Books, Armidale, Australia, pp. 29-37.

Orpin, C. G. and Bountiff, L. 1978. Zoospore chemotaxis in the rumen phycomycete *Neocallimastix frontalis*. J. Gen. Microbiol. 104 (1), 113-22.

Orpin, C. G. and Greenwood, Y. 1986. The role of haems and related compounds in the nutrition and zoosporogenesis of the rumen chytridiomycete *Neocallimastix frontalis* H8. Microbiology. 132 (8), 2179-85.

Orpin, C. G. and Joblin, K. N. 1997. The rumen anaerobic fungi. In: Hobson, P. N. and Stewart, C. S. (Eds), The Rumen Microbial Ecosystem. Blackie Academic and Professional, pp. 140-95.

Orpin, C. G. and Munn, E. A. 1986. *Neocallimastix patriciarum* sp. nov., a new member of the Neocallimasticaceae inhabiting the rumen of sheep. Trans. Br. Mycol. Soc. 86 (1), 178-81.

Ozkose, E., Thomas, B. J., Davies, D. R., et al., 2001. *Cyllamyces aberensis* gen. nov sp. nov., a new anaerobic gut fungus with branched sporangiophores isolated from cattle. Can. J. Bot. 79 (6), 666-73.

Paul, S. S., Kamra, D. N., Sastry, V. R. B., et al., 2003. Effect of phenolic monomers on biomass and hydrolytic enzyme activities of an anaerobic fungus isolated from wild nil gai (*Baselophus tragocamelus*). Lett. Appl. Microbiol. 36 (6), 377-81.

Paul, S. S., Kamra, D. N., Sastry, V. R. B., et al., 2004. Effect of administration of an anaerobic gut fungus isolated from wild blue bull (*Boselaphus tragocamelus*) to buffaloes (Bubalus bubalis) on *in vivo* ruminal fermentation and digestion of nutrients. Anim. Feed Sci. Technol. 115 (1-2), 143-57.

Paul, S. S., Deb, S. M., Punia, B. S., et al., 2011. Effect of feeding isolates of anaerobic fungus *Neocallimastix* sp. CF 17 on growth rate and fibre digestion in buffalo calves. Arch. Anim. Nutr. 65 (3), 215-28.

Paul, S. S., Deb, S. M., Dey, A., et al., 2015. 16S rDNA analysis of archaea indicates dominance of Methanobacterium and high abundance of Methanomassiliicoccaceae in rumen of Nili-Ravi buffalo. Anaerobe 35 (B), 3-10.

Paul, S. S., Bu, D., Xu, J., et al., 2018. A phylogenetic census of global diversity of gut anaerobic fungi and a new taxonomic framework. Fungal Divers. 89 (1), 253-66.

Pelechano, V. and Steinmetz, L. M. 2013. Gene regulation by antisense transcription. Nat. Rev. Genet. 14 (12), 880-93.

Phillips, M. W. and Gordon, G. L. R. 1988. Sugar and polysaccharide fermentation by rumen anaerobic fungi from Australia, Britain and New Zealand. Biosystems 21 (3-4), 377-83.

Phillips, M. W. and Gordon, G. L. R. 1995. Carbohydrate fermentation by three species of polycentric ruminal fungi from cattle and water buffalo in tropical Australia. Anaerobe 1 (1), 41-7.

Podar, M., Makarova, K. S., Graham, D. E., et al., 2013. Insights into archaeal evolution and symbiosis from the genomes of a nanoarchaeon and its inferred crenarchaeal host from Obsidian Pool, Yellowstone National Park. Biol. Direct 8, 9.

Qi, M., Wang, P., O'Toole, N., et al., 2011. Snapshot of the eukaryotic gene expression in muskoxen rumen-a metatranscriptomic approach. PLoS ONE 6 (5), e20521.

Quandt, C. A., Kohler, A., Hesse, C. N., et al., 2015. Metagenome sequence of *Elaphomyces granulatus* from sporocarp tissue reveals Ascomycota ectomycorrhizal fingerprints of genome expansion and a Proteobacteria-rich microbiome. Environ. Microbiol. 17 (8), 2952-68.

Raghunathan, A., Ferguson Jr., H. R., et al., 2005. Genomic DNA amplification from a single bacterium. Appl. Environ. Microbiol. 71 (6), 3342-7.

Raveh-Sadka, T., Firek, B., Sharon, I., et al., 2016. Evidence for persistent and shared bacterial strains against a background of largely unique gut colonization in hospitalized premature infants. ISME J. 10 (12), 2817-30.

Saxena, S., Sehgal, J. P., Puniya, A. K., et al., 2010. Effect of administration of rumen fungi on production performance of lactating buffaloes. Benefic. Microbes 1 (2), 183-8.

Schoch, C. L., Seifert, K. A., Huhndorf, S., et al., 2012. Nuclear ribosomal internal transcribed spacer (ITS) region as a universal DNA barcode marker for Fungi. Proc. Natl Acad. Sci. U. S. A. 109 (16), 6241-6.

Seedorf, H., Kittelmann, S., Henderson, G., et al., 2014. RIM-DB: a taxonomic framework for community structure analysis of methanogenic archaea from the rumen and other intestinal environments. PeerJ 2, e494.

Sehgal, J. P., Jit, D., Puniya, A. K., et al., 2008. Influence of anaerobic fungal administration on growth, rumen fermentation and nutrient digestion in female buffalo calves. J. Anim. Feed Sci. 17 (4), 510-8.

Seppala, S., Solomon, K. V., Gilmore, S. P., et al., 2016. Mapping the membrane proteome of anaerobic gut fungi identifies a wealth of carbohydrate binding proteins and transporters. Microb. Cell Fact. 15 (1), 212.

Sharon, I., Morowitz, M. J., Thomas, B. C., et al., 2013. Time series community genomics analysis reveals rapid shifts in bacterial species, strains, and phage during infant gut colonization. Genome Res. 23 (1), 111-20.

Sirohi, S. K., Singh, N., Dagar, S. S., et al., 2012. Molecular tools for deciphering the microbial community structure and diversity in rumen ecosystem. Appl. Microbiol. Biotechnol. 95 (5), 1135-54.

Sirohi, S. K., Choudhury, P. K., Dagar, S. S., et al., 2013. Isolation, characterization and fibre degradation potential of anaerobic rumen fungi from cattle. Ann. Microbiol. 63 (3), 1187-94.

Sollinger, A., Tveit, A. T., Poulsen, M., et al., 2018. Holistic assessment of rumen microbiome dynamics through quantitative metatranscriptomics reveals multifunctional redundancy during key steps of anaerobic feed degradation. mSystems 3 (4).

Solomon, K. V., Haitjema, C. H., Henske, J. K., et al., 2016. Early-branching gut fungi possess a large, comprehensive array of biomass-degrading enzymes. Science 351 (6278), 1192-5.

Solomon, K. V., Henske, J. K., Gilmore, S. P., et al., 2018. Catabolic repression in early-diverging anaerobic fungi is partially mediated by natural antisense transcripts. Fungal Genet. Biol. 121, 1-9.

Sridhar, M., Kumar, D., Anandan, S., et al., 2007. Occurrence and prevalence of Cyllamyces genus-a putative anaerobic gut fungus in Indian cattle and buffaloes. Curr. Sci. 92 (10), 1356-8.

Sridhar, M., Kumar, D. and Anandan, S. 2014. *Cyllamyces icaris* sp. nov., a new anaerobic gut fungus with nodular sporangiophores isolated from Indian water buffalo (Bubalus bubalis). Int. J. Curr. Res. Aca. Rev. 2, 7-24.

Staley, C. and Sadowsky, M. J. 2018. Practical considerations for sampling and data analysis in contemporary metagenomics-based environmental studies. J. Microbiol. Methods 154, 14-8.

Stewart, M. 2019. Polyadenylation and nuclear export of mRNAs. J. Biol. Chem. 294 (9), 2977-87.

Stewart, C. S. and Richardson, A. J. 1989. Enhanced resistance of anaerobic rumen fungi to the ionophores monensin and lasalocid in the presence of methanogenic bacteria. J. Appl. Bacteriol. 66 (1), 85-93.

Strassert, J. F. H., Karnkowska, A., Hehenberger, E., et al., 2018. Single cell genomics of uncultured marine alveolates shows paraphyly of basal dinoflagellates. ISME J. 12 (1), 304-8.

Sun, M., Jin, W., Li, Y., et al., 2014. Isolation and identification of cellulolytic anaerobic fungi and their associated methanogens from holstein cow. Wei Sheng Wu Xue Bao 54 (5), 563-71.

Tedersoo, L., Sánchez-Ramírez, S., Kõljalg, U., et al., 2018. High-level classification of the Fungi and a tool for evolutionary ecological analyses. Fungal Divers. 90 (1), 135-59.

Teunissen, M. J., Kets, E. P. W., Op den Camp, H. J. M., et al., 1992. Effect of coculture of anaerobic fungi isolated from ruminants and non-ruminants with methanogenic bacteria on cellulolytic and xylanolytic enzyme activities. Arch. Microbiol. 157 (2), 176-82.

Theodorou, M. K., Gill, M., King-Spooner, C., et al., 1990. Enumeration of anaerobic chytridiomycetes as thallus-forming units: novel method for quantification of fibrolytic fungal populations from the digestive tract ecosystem. Appl. Environ. Microbiol. 56 (4), 1073-8.

Theodorou, M. K., Zhu, W. Y., Rickers, A., et al., 1996. Biochemistry and ecology of anaerobic fungi. In: Howard, D. H. and Miller, J. D. (Eds), Human and Animal Relationships. Springer Berlin Heidelberg, Berlin, Heidelberg, pp. 265-95.

Trinci, A. P., Lowe, S. E., Milne, A., et al., 1988. Growth and survival of rumen fungi. Biosystems 21 (3-4), 357-63.

Tripathi, V. K., Sehgal, J. P., Puniya, A. K., et al., 2007. Effect of administration of anaerobic fungi isolated from cattle and wild blue bull (*Boselaphus tragocamelus*) on growth rate and fibre utilization in buffalo calves. Arch. Anim. Nutr. 61 (5), 416-23.

Tu, Q., Yu, H., He, Z., et al., 2014. GeoChip 4: a functional gene-array-based high-throughput environmental technology for microbial community analysis. Mol. Ecol. Resour. 14 (5), 914-28.

Turaev, D. and Rattei, T. 2016. High definition for systems biology of microbial communities: metagenomics gets genome-centric and strain-resolved. Curr. Opin. Biotechnol. 39, 174-81.

Ushida, K., Jouany, J. P. and Demeyer, D. I. 1991. Physiological aspects of digestion and metabolism in ruminants. In: Proceedings of the Seventh International Symposium on Ruminant Physiology, New York, pp. 625-54.

Vavra, J. and Joyon, L. 1966. Etude sur la changes in the chemical composition of a cambial morphologie, le cycle evolutif et la position syscell during its differentiation into xylem and phloem tematique de Callimastix cyclops, Weissenberg, tissue in trees. Protistologica 2, 5-16.

Wagner, E. G. and Simons, R. W. 1994. Antisense RNA control in bacteria, phages, and plasmids. Annu. Rev. Microbiol. 48, 713-42.

Wallace, R. J. and Munro, C. A. 1986. Influence of the rumen anaerobic fungus *Neocallimastix frontalis* on the proteolytic activity of a defined mixture of rumen bacteria growing on a solid substrate. Lett. Appl. Microbiol. 3 (2), 23-6.

Wang, T. Y., Chen, H. L., Lu, M. J., et al., 2011. Functional characterization of cellulases identified from the cow rumen fungus *Neocallimastix patriciarum* W5 by transcriptomic and secretomic analyses. Biotechnol. Biofuels 4, 24.

Wang, X., Liu, X. and Groenewald, J. Z. 2017. Phylogeny of anaerobic fungi (phylum Neocallimastigomycota), with contributions from yak in China. Antonie Leeuwenhoek 110 (1), 87-103.

Wang, Y., Youssef, N., Couger, M. B., et al., 2018. Comparative genomics and divergence time estimation of the anaerobic fungi in herbivorous mammals. bioRxiv, 401869.

Wang, Y., Youssef, N. H., Couger, M. B., et al., 2019. Molecular dating of the emergence of anaerobic rumen fungi and the impact of laterally acquired genes. mSystems 4, e00247-19.

Webb, J. and Theodorou, M. K. 1991. *Neocallimastix hurleyensis* sp. nov., an anaerobic fungus from the ovine rumen. Can. J. Bot. 69 (6), 1220-4.

West, P. T., Probst, A. J., Grigoriev, I. V., et al., 2018. Genome-reconstruction for eukaryotes from complex natural microbial communities. Genome Res. 28 (4), 569-80.

Weston, R. H., Lindsay, J. R., Purser, D. B., et al., 1988. Feed intake and digestion responses in sheep to the addition of inorganic sulfur to a herbage diet of low sulfur content. Aust. J. Agric. Res. 39 (6), 1107-19.

Wubah, D. A. and Kim, D. S. H. 1996. Chemoattraction of anaerobic ruminai fungi zoospores to selected phenolic acids. Microbiol. Res. 151 (3), 257-62.

Wubah, D. A., Fuller, M. S. and Akin, D. E. 1991. Studies on Caecomyces communis: morphology and development. Mycologia 83 (3), 303-10.

Yan, B. and Wang, Z. 2012. Long noncoding RNA: its physiological and pathological roles. DNA Cell Biol. 31 (Suppl. 1), S34-41.

Yanke, L. J., Dong, Y., McAllister, T. A., et al., 1993. Comparison of amylolytic and proteolytic activities of ruminal fungi grown on cereal grains. Can. J. Microbiol. 39 (8), 817-20.

Yoon, H. S., Price, D. C., Stepanauskas, R., et al., 2011 Single-cell genomics reveals organismal interactions in uncultivated marine protists. Science 332 (6030), 714-7.

Youssef, N. H., Couger, M. B., Struchtemeyer, C. G., et al., 2013. The genome of the anaerobic fungus Orpinomyces sp. strain C1A reveals the unique evolutionary history of a remarkable plant biomass degrader. Appl. Environ. Microbiol. 79 (15), 4620-34.

Zhang, L., Cui, X., Schmitt, K., et al., 1992. Whole genome amplification from a single cell: implications for genetic analysis. Proc. Natl Acad. Sci. U. S. A. 89 (13), 5847-51.

Zhang, K., Martiny, A. C., Reppas, N. B., et al., 2006. Sequencing genomes from single cells by polymerase cloning. Nat. Biotechnol. 24 (6), 680-6.

第9章 瘤胃病毒和染色体外遗传元件

Rosalind Ann Gilbert、Diane Ouwerkerk，
澳大利亚昆士兰大学，昆士兰政府和昆士兰农业与
食品创新联盟，农业和渔业部
（焦金真译）

1 前言

在20世纪50年代，马文·布莱恩特（Marvin Bryant）指出瘤胃是一种自然的微生物栖息地，"似乎比许多其他微生物栖息地更容易受到微生物生态学家的关注和分析"（Bryant 1959）。直到今天，无论是从经济还是环境角度，学者对于它的研究兴趣仍然高涨。目前，已知草食动物的瘤胃中定植有原核生物（细菌、古细菌）和真核生物（原生动物、真菌）等多种微生物群落，它们负责高效发酵植物饲料，这些微生物的遗传和功能（酶促）特征被广泛关注，用以研发改善瘤胃功能和提高饲料效率，并降低排泄物对环境影响的饲养策略。

尽管瘤胃微生物生态学是持续的研究重点，然而对瘤胃内可移动基因组的了解相对匮乏，包括影响可移动基因组的主要因素（病毒和质粒）以及可移动基因组对瘤胃功能的影响。病毒种群已经被证实与瘤胃内的微生物种群存在共生；但是，对它们的共生关系仍缺乏基本的生物学和遗传学信息。更令人惊讶的是，关于非病毒外染色体元件（质粒）的信息更少，这些信息通常与瘤胃微生物种群内在地联系在一起。本章旨在综述当前对瘤胃病毒群体和外染色体元件的理解，并描述移动遗传元件（MGEs）的载体，例如细胞外膜囊泡（MVS）。此外，也关注了基于瘤胃和其他微生物生态系统的进展，探讨了可移动基因组对瘤胃功能的潜在影响。

2 染色体外元件

染色体外元件，是指在细胞中未整合到染色体中的遗传元件。染色体外元件出现在真核生物和原核生物细胞中，可能存在自我复制。染色体外元件和稳定的染色体之间持续交换，构成了基因组的重要特征，这有助于细胞的遗传能力和生物体之间的基因转移——水平基因转移（HGT）。在这方面，染色体外元件也可以被称为移动遗传元件，并且统称为可移动基因组（Koonin和Wolf 2008），其中包括的所有遗传元件有助于基因的水平转移和微生物的进化（Brussow等，2004；Koonin和Dolja，2014；Croucher等，2016；Krupovic等，2019）。

水平基因转移有四种主要策略：①转化：细胞从环境中获取外源性 DNA 的自然能力；②转导：将 DNA 从一个细胞转移到另一个细胞，例如病毒；③共轭：通过共轭（配对）系统从供体细胞到受体细胞的接触依赖性单向转移；④融合：将两个细胞或细胞与囊泡（例如外膜囊泡）连接（Bellanger 等，2014；Johnson 和 Grossman，2015；Guedon 等，2017；Rowan-Nash 等，2019）。所有这些策略可能会发生在菌群密集且多样的瘤胃微生物生态系统中（Morrison，1996），其中染色体外元件扮演着不可或缺的重要角色。基于前期大量研究集中在瘤胃的大背景下，"染色体外元件"包括了所有有助于可移动基因组的元素：①病毒基因组（完整和残留的病毒基因组）；②质粒；③其他的遗传物质，如细胞外膜囊泡中的 DNA（图 1）。

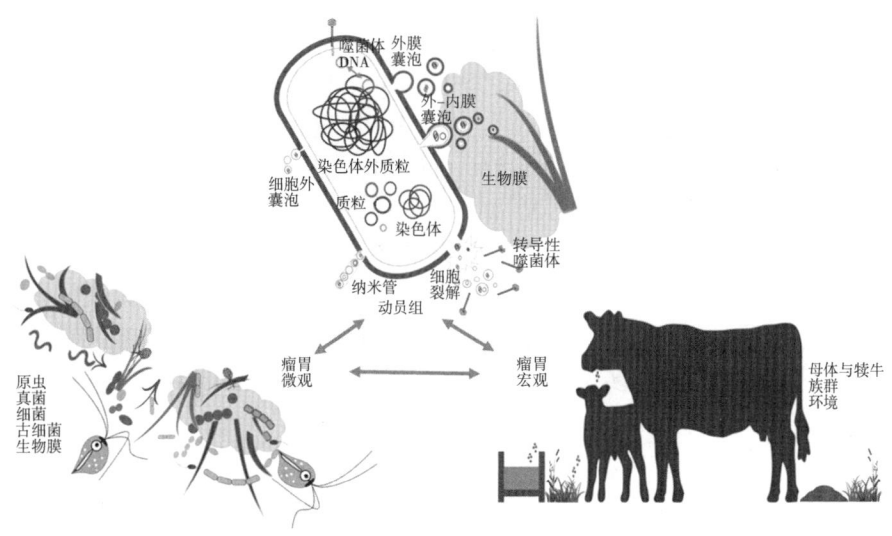

图 1　瘤胃可移动基因组：宏观和微观相互作用的解释

在亚微观遗传水平上，可移动基因组可以通过重组、染色体整合、转导等机制促进基因转移。可移动基因组包括了微生物病毒（如溶解性的、溶原性的和转导性的噬菌体）、质粒（染色体外的和游离的）、大型染色质以及囊泡（外膜囊泡和内外膜囊泡）。可移动基因组的成分可能在微生物细胞溶菌作用后释放并被生物膜所利用。在微观水平，瘤胃的原核生物（细菌、古生菌）和真核生物（真菌和原虫）可以产生可移动基因组元素，且存在转移的可能性。在宏观水平，可移动基因组的成分转移的方式如下：通过奶牛和犊牛之间的交互（如唾液、气溶胶）；一群个体成员之间的相互作用（如唾液、气溶胶、粪便等）；与环境的相互作用（如食物、水）。

2.1　病毒基因组

在瘤胃中找到的可以更好地理解染色体外元件的是原核生物的病毒，它们通常被描述为独立对象（病毒和噬菌体），而不是被称为染色体外元件。细胞内病毒基因组或前噬菌体可以作为游离体在细胞中被发现（两者都整合到群体中的染色体和染色体外元件）或作为质粒（仅在染色体外）（Deutsch 等，2018）。从这种意义上来说，病毒基因组也可以被描述为整合和结合元件（ICE）（Delavat 等，2017）。噬菌体能够整合新的 DNA 进入它们的基因组，也被描述为转导噬菌体（Toussaint 和 Rice，2017）。然而，ICE 的更广泛定义还包括模块化遗传结合转座子等元素，通常包含的基因用于结合（IV 型分泌系统松弛酶或用于复制的蛋白质起始和易位）和整合（酪氨酸整合酶，丝氨酸或 DDE 家族）（Liu 等，2019），并且通常来自有缺陷的噬菌体（Bobay 等，2014）。尽管具有选择性优势的前噬菌体可以避免突变并保持活力，选择压力倾向作用于噬菌体，如染色体外和整合的噬菌体序列可以代表宿主的遗传和复制。因此，前噬菌体可能对于使它们有缺陷的突变的累积更敏感（Winstanley 等，2009；Koonin 和 Dolja，2014）。

在瘤胃细菌基因组中可以很容易地检测到有缺陷的前噬菌体（Gilbert 和 Klieve，2015）。它们以类似前噬菌体的形式存在；然而，无论是单个基因还是整个基因模块，通常是那些编码结构蛋白（外壳蛋白和尾蛋白），可能会丢失并且仅保留整合和复制蛋白。可以使用检测前噬菌体的工具如 PHASTer 和 PhiSpy 来识别原核生物基因序列中有缺陷的前噬菌体（Akhter 等，2012；Arndt 等，2016）。然而，使用这些工具需要尽量更全面地注释前噬菌体相关基因并确定预期的噬菌体基因模块，以确定所鉴定的前噬菌体是否具有完整的噬菌体相关基因（Gilbert 等，2017）。然而，有缺陷的前噬菌体可能是形成新 ICE 的基础，因此有助于瘤胃细菌之间的水平基因转移。确定病毒（完整的或有缺陷的）在瘤胃中对水平基因转移的贡献程度是相当难的；然而，考虑到它们出现的高频率，预计它们的影响将是巨大的（Seshadri 等，2018）。鉴于瘤胃病毒是研究最广泛的对瘤胃运动有贡献的遗传因素，第 3 节进一步综述了瘤胃病毒研究领域的最新成果。

2.2 质粒

Teather（1982）首次报道了与瘤胃细菌分离物相关的质粒，他发现了 *Butyrivibrio fibrisolvens* ATCC19171 的典型菌株和 6 种新的 *B. fibrisolvens* 牛分离菌都含有质粒。该报告还表明，这些质粒可能对赋予这种常见瘤胃细菌属的竞争适应性很重要，并可能影响其发酵单糖和多糖以及降解蛋白质的能力。随后进一步的研究表明，在瘤胃球菌属 *Ruminococcus* 的其他分离株中检测到了几种质粒（Asmundson 和 Kelly，1987）。随着 DNA 测序的出现，报道了瘤胃球菌属、新月形单胞菌属、丁酸弧菌属和普雷沃氏菌（后来重新分类作为拟杆菌属）等分离株的质粒序列（Champion 等，1988；Attwood 和 Brooker，1992；Hefford 等，1993；May 等，1996；Ogata 等，1996；Ohara 等，1998）。这些序列显示存在复制和功能蛋白模块，以及通常无法识别的辅助蛋白（Ohara 等，1998），这正如对自我复制的染色体外元件所预期的那样（Thomas 等，2017）。还有一些关于在瘤胃拟杆菌菌株之间转移编码抗生素（四环素）抗性的质粒研究（Flint 等，1988）。这些研究的大部分是由当时的生物技术发展推动的，旨在挖掘参与碳水化合物分解的新酶，并开发用于基因转化瘤胃细菌的工具，以提高纤维分解的效率（Morrison，1996；Flint，1997）。

虽然人们对遗传转化的兴趣已经减弱，但对瘤胃细菌质粒的兴趣仍在继续。随着基因组测序技术的发展，许多瘤胃细菌分离株的质粒序列已经被报道，包括白色瘤胃球菌（Suen 等，2011；Dassa 等，2014）、纤维降解菌丁酸弧菌（Kelly 等，2010；Palevich 等，2017），以及抗性淀粉降解菌小猪双歧杆菌（Jung 等，2018）的原始菌株。自从 20 世纪

90年代最初尝试分离和测序质粒以来，基因组注释也有了显著的改进，这些新鉴定的质粒携带的基因正在被赋予功能特性。例如，与 *B. choerinum* FMB-1 相关的质粒携带一个与抗生素耐药性相关的利胆醇外排泵基因（Jung 等，2018）。几种白色瘤胃球菌菌株的基因组测序表明，在该菌种中发现的质粒，例如白色瘤胃球菌菌株 7，含有 dockerins 的基因（Dassa 等，2014），这是形成纤维素降解多酶纤维素体复合物所必需的（Nash 等，2016）。相比之下，两个与 *Butyrivibrio proteolaticus* B316 相关的质粒似乎编码对瘤胃存活重要的基因，但对植物多糖降解没有明显贡献（Kelly 等，2010）。

有趣的是，*Butterivibrio hungatei* MB2003 基因组中相对较大的比例（7.7%）不是直接与染色体相关的，因为它包含四个复制子，一个单一染色体（3 143 784 bp）、一个较小的类铬或次级染色体（91 776 bp）、一个大质粒（144 470 bp）和一个小质粒（6 284 bp）（Palevich 等，2017）。在这些染色质外成分中发现了类色素编码细胞内分解碳水化合物的基因（β-葡萄糖苷酶、β-半乳糖苷酶和多糖脱乙酰酶）。以前在原核生物中已经描述过类铬（Harrison 等，2010），它们不同于较小的质粒，具有与主染色体相似的 GC 含量，但含有利于维持和复制的质粒基因。随着对与瘤胃微生物基因组相关质粒的注释和特征的关注越来越多，可以预期，关于这些染色体外元件对瘤胃功能的影响将有更深入的了解。

宏基因组测序可以作为表征和理解质粒在瘤胃中作用的一种重要方法，有研究已对从奶牛瘤胃液中纯化的质粒部分进行了测序（Kav 等，2012）。此研究开发了从瘤胃液中纯化质粒的新方法，检测了携带质粒的瘤胃微生物的类型，并测定了质粒基因的功能。结果表明，瘤胃细菌携带质粒的可能性高于古细菌和真核生物，与质粒相关的优势菌门分布在厚壁菌门（47%）、拟杆菌门（22%）、变形菌门（20%）和放线菌门（9%）。利用 SEED 对质粒相关基因进行功能分析（Overbeek 等，2005），并与质粒数据库进行比较，结果表明氨基酸、DNA 代谢、辅助因子、维生素、细胞壁和荚膜、碳水化合物、呼吸和蛋白质代谢等系统具有较高的表达量。这一功能分析表明，瘤胃细菌的质粒可能含有包括复制和侧向基因转移在内的质粒相关基因以及辅助功能基因等，通常与瘤胃中的微生物存活和底物利用有关（包括细胞壁糖基转移酶、荚膜多糖、蛋白质代谢酶和碳水化合物降解酶）。

最近，随着宏基因组学和生物信息学的发展，已经从牛瘤胃样本中获得了宏基因组组装基因组以及质粒序列（Stewart 等，2018）。尽管获得和分析质粒不是本研究的重点，但已鉴定出两个质粒重叠群，它们与白色瘤胃球菌 7 质粒（pRUMAL02）相似（Dassa 等，2014），其中也包括 dockerin 蛋白的基因。此外，这些质粒重叠群似乎在其他四个组装基因组中共享，包括 3 个梭菌属和 1 个黄色瘤胃球菌（Stewart 等，2018）。虽然瘤胃球菌属也被归为梭菌纲，但这些结果表明质粒为负责瘤胃碳水化合物分解的关键蛋白质的跨菌种转移提供了机制。

2.3 其他染色质外成分和膜泡

迄今为止，除病毒和质粒以外，瘤胃微生物通过细胞外元素摄取遗传物质的能力仍是未知的，仍处于冷门的研究领域。这可能部分归因于瘤胃液本身的性质：大量微生物系统发育和表型的多样性，以及这种多样性所带来的物理复杂性；固相和液相组分之间存在分层，生物膜的形成和胞外多糖黏液的产生，以及与瘤胃上皮的相互作用（Bryant，1959；

Russell 和 Rychlik，2001；Huws 等，2018）。已经证实，瘤胃内的微生物细胞存在溶解现象，其内容物可以释放到瘤胃液中（Wells 和 Russell，1996），其中的蛋白质成分被其他微生物快速吸收（Leng 和 Nolan，1984），溶解产生的核酸成分在瘤胃中被快速降解（Russell 和 Wilson，1988）。因此，外膜囊泡等物理结构可能在保护核酸免受瘤胃降解方面发挥重要作用（Klieve 等，2005）。通过这种方式，外膜囊泡可能在染色体外元件的运输中发挥主要作用，并促进瘤胃中的基因转移。

外膜囊泡可能不仅包含 DNA，还包含蛋白质和脂质（Willms 等，2016；Clarke，2018；Volgers 等，2018）。外膜囊泡是球形双层结构，大小从 20~400 nm 不等（Toyofuku 等，2019）。外膜囊泡通常由革兰氏阴性细菌形成，它们会从细胞表面脱落或起泡，但从细胞表面释放外膜囊泡的能力在所有原核生物（细菌、古细菌）和真核生物（真菌和原虫）中都是保守的（Deatherage 和 Cookson，2012）。微生物可以利用外膜囊泡获取营养，抵御其他微生物，实现细胞间信息交流，抵抗宿主免疫系统的压力，并有助于生物膜的形成（Ellis 和 Kuehn，2010；Pope 等，2011）。它们也可以用来防御病毒或噬菌体感染，外膜囊泡上的受体为质粒附着提供位点；此外，外膜囊泡还可以输出和携带病毒颗粒以及细胞外 DNA（Toyofuku 等，2019）。

有研究表明，瘤胃真菌（如 *Sphaeromonas communis*）也产生外膜囊泡（Gaillard 等，1989），但对瘤胃外膜囊泡的大多数研究都集中在细菌外膜囊泡运输和对饲料降解有关的酶的能力上，例如对白色瘤胃球菌（Kim 等，2001）和产琥珀酸丝状杆菌的研究（Groleau 和 Forsberg，1981；Gong 和 Forsberg，1993；Arntzen 等，2017）。栖瘤胃拟杆菌也能产生外膜囊泡，但它们在植物降解或 DNA 转移中的作用尚不清楚（Huws 等，2018）。可能存在更多种类的瘤胃细菌能够产生外膜囊泡，这些外膜囊泡有助于饲料分解，也可积极促进瘤胃内染色体外元件的运输和转移。

3　瘤胃病毒

病毒可能通过饲料、水、气溶胶和动物间传播（例如唾液、接触粪便物质；图1），瘤胃中发现的大多数病毒是那些主动感染和在微生物中复制的病毒。鉴于瘤胃中最主要的微生物是细菌（Hungate，1966），瘤胃液中存在的绝大多数病毒是感染细菌噬菌体病毒（Adams 等，1966；Hoogenraad 等，1967）。瘤胃中发现的古菌群也被古菌病毒感染（Baresi 和 Bertani 1984；Attwood 等，2008），有时也被称为噬菌体或古噬菌体（Abedon 和 Murray，2013）。考虑到其他微生物生态系统的发现（Ghabrial 等，2015；Grybchuk 等，2018），瘤胃真核生物种群（原生动物和真菌）也可能被病毒感染。然而，真核病毒还没有从瘤胃中分离出来，它们的存在只能通过对瘤胃病毒群的宏基因组分析，检测到原生动物病毒中的相关基因（Berg Miller 等，2012；Anderson 等，2017）。

瘤胃噬菌体研究的进展紧随研究技术的进步与发展。随着透射电子显微镜（TEM）的出现，研究人员开始检测驯养的食草动物（牛、羊）和驯鹿的瘤胃样本中病毒是否存在，并对其病毒样颗粒进行形态解析（Adams 等，1966；Hoogenraad 和 Hird，1970；Tarakanov，1972）。病毒分类的流程随着研究的深入不断完善，这些研究第一次确定了瘤胃是以尾状病毒目尾状噬菌体 Caudovirales 为主，包括长尾噬菌体 Siphoviridae、肌尾噬菌体 Myoviridae 和短尾噬菌体 Podoviridae。

从以山羊瘤胃液为接种液的厌氧发酵反应器中采集微生物样品，通过透射电镜观察到了长尾噬菌体 Siphoviridae、肌尾噬菌体 Myoviridae 和短尾噬菌体 Podoviridae 的存在（图2）。虽然利用透射电子显微镜有时很难区分断尾噬菌体的小尾巴结构，这项研究仍然发现了属于复层噬菌体科 Tectiviridae、覆盖噬菌体科 Corticoviridae 和微病毒科 Microviridae 的无尾病毒颗粒，这些病毒也会感染细菌（Ackermann 和 Prangishvili，2012）。作者团队之前通过透射电镜观察到牛瘤胃液中存在这些细小粒子；然而，这些噬菌体目前还没有从瘤胃中分离出来。同时，丝状噬菌体科 Inoviridae 在瘤胃液也可被观察到；然而，由于它们类似于断裂的噬菌体尾巴和细菌胞外丝状结构（细毛和鞭毛），很难区分这些缺少头部结构、处于粗提阶段的又长又细的噬菌体。

在以透射电镜为主要研究方法的阶段，研究人员也开始分离和纯化侵染瘤胃液中的噬菌体，并将这些噬菌体进行分离培养。由于分离单个噬菌体需要使用细菌宿主易受噬菌体感染（De Jong 等，2019），且宿主的范围较为有限，选择微生物宿主用于噬菌体分离培养是一种可行的方式。相较于采用多种瘤胃寄主细菌菌株的广谱方式，大多数分离研究集中在使用特定的细菌作为宿主，这一策略在 Gilbert 和 Klieve（2015）的综述中得到较为系统的阐明。此外，大多数分离的噬菌体被用作细菌宿主，用于研发噬菌体疗法来控制有害的瘤胃微生物（如牛链球菌），或通过生物技术方法用于植物的解毒，如 Bacteroides ruminantium 和 Bacteroides ruminicola 等（Gregg 等，1994；Tarakanov，1994；Wong 等，2003）。

迄今为止，从瘤胃分离出的大多数噬菌体是 Siphoviridae，感染来自瘤胃的链球菌菌株，即牛链球菌（又称马链球菌）（Schlegel et al.，2003）。这些研究是在几个不同的国家进行的，分离的噬菌体来源于饲喂多种日粮的多种反刍动物（牛，包括奶牛，绵羊）（Adams 等，1966；Brailsford 和 Hartman，1968；Iverson 和 Millis，1976a；Tarakanov，1976；Styriak 等，1989；Klieve 和 Bauchop，1991）。噬菌体的宿主的范围通常很窄，每个噬菌体只能感染有限数量的链球菌宿主菌株（Klieve 等，1999），虽然也有例外，只有一份报告显示噬菌体可感染 10 株马链球菌中的 5 株（Styriak 等，1994）。此外，从细菌的角度，马链球菌 2B 比其他菌株的链球菌（Iverson 和 Millis，1976a）似乎更容易受到噬菌体感染（Klieve 等，1999；Gilbert 等，2017）。

随着检测瘤胃噬菌体的分子和遗传学方法的出现，基于分离培养探索瘤胃噬菌体特性的研究越来越少。尽管如此，运用分离培养的噬菌体可以很好地解析噬菌体在瘤胃中的生物学特性。早期的培养研究表明，瘤胃噬菌体感染链球菌不仅取决于宿主范围，还影响其特性，如生长速率（从噬菌体感染到宿主细胞裂解的复制时间）和暴发大小（感染和复制后释放的颗粒数量）等。其他的基于噬菌体分离培养的研究明确了瘤胃液中病毒粒子衰减率（Orpin 和 Munn，1974；Tarakanov，1976；Swain，1999）、噬菌体耐药性的发展规律（Klieve 和 Bauchop，1991）以及噬菌体与瘤胃上皮细胞的相互作用（Styriak 等，1991）。

随着分子生物学的出现，基于 TEM 和培养的研究逐渐被基于 DNA 的方法所替代，用

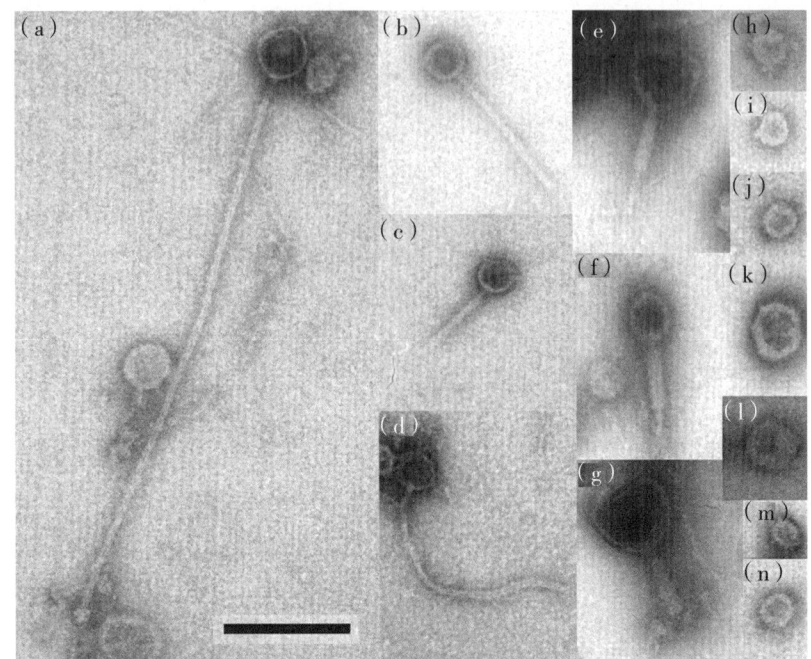

图2 常见的瘤胃病毒形态。包括 Siphoviridae（a~d）、Myoviridae（e~g）、Podoviridae（h，i）和潜在的无尾噬菌体（j~n），如皮质病毒科和微病毒科（比例尺 200 nm）。所有的颗粒都是从一个体外发酵系统中收集的样品（Infors）分离纯化，这个体外发酵系统来源于山羊的瘤胃液。噬菌体颗粒用 2.5%的戊二醛固定，用 1%钼酸铵 pH 7.0 染色。所有图像用 Joel 1400 透射电子显微镜以相同的倍数进行拍摄。图片由 Kathy Crew 博士提供。

以获得和解析任何一个样本或特定时间点的瘤胃噬菌体种群的整体图像或快照（Klieve 和 Gilbert，2005）。基于透射电镜研究推测每毫升牛瘤胃液中至少有 5×10^7 个噬菌体（Paynter 等，1969），每毫升牛和羊瘤胃液中有 $2 \times 10^7 \sim 1 \times 10^8$ 个噬菌体（Klieve 和 Bauchop，1988），牛的最高估计为每毫升瘤胃液中含有 $>10^9$ 个噬菌体颗粒（Ritchie 等，1970）。根据噬菌体 DNA 浓度推算的噬菌体种群数略高于以往的透射电镜研究，文献报道的噬菌体数分别为 3×10^9 个/mL 和 1.6×10^{10} 个/mL 瘤胃液（Klieve 和 Swain，1993）。这些方法还可以根据基因组长度获得噬菌体群体的图谱或指纹图谱，并显示在瘤胃液中持续存在的两个主要组成部分：一是广泛的基因组长度范围（30~200 kb），代表许多不同噬菌体的 DNA 基因组，包括尾噬菌体的预期基因组大小；二是单个或多个噬菌体产生的 DNA 离散带，可能代表噬菌体产生的大量繁殖（Swain 等，1996）。

这些研究还首次表明，日粮及其成分可能会影响瘤胃内噬菌体数量，即饲喂天然牧草的绵羊的噬菌体数量显著高于饲喂切碎干草（燕麦：苜蓿为 70：30）基础日粮的绵羊。他们还研究了个体之间和不同反刍动物物种（绵羊、牛和山羊）之间噬菌体总数的差异（Swain 等，1996）。一天内绵羊瘤胃内噬菌体数量并非总是恒定的，而是随着采食而变化的，采食 4 h 后总噬菌体数量增加，采食 8~10 h 后噬菌体数量达到最大值。这些模式与之前在牛的瘤胃细菌中观察到的模式相似（Leedle 等，1982）。由此推测，噬菌体数量随着细菌总数的变化而波动。噬菌体数量在细菌数量达到峰值后约 2 h 达到峰值，这种时间延迟被认为是噬菌体复制和宿主细胞裂解需要一定的时间的结果。

与此同时，有研究利用体外培养，通过改变日粮成分或添加饲料添加剂来刺激噬菌体裂解或从瘤胃中清除噬菌体颗粒，以此来调控瘤胃噬菌体数量（Swain，1999）。体外检测的化合物包括次生植物化合物（皂苷、芦丁、儿茶素、槲皮素和单宁酸）、植物激素苯乙酸、离子载体抗生素莫能菌素和蒙脱土、膨润土等。以上化合物均不能够增加噬菌体的数量。膨润土可能会沉淀噬菌体颗粒，将它们从瘤胃液中结合并去除。类似地，单宁酸也减少了噬菌体的数量，可能是通过结合噬菌体颗粒的蛋白质成分。随后对绵羊进行的动物试验（Swain 1999）表明，在 24 h 内，瘤胃内添加单宁酸可以降低噬菌体的浓度（图 3）。这些研究为解析日粮成分如何在化学上或物理上与瘤胃内噬菌体颗粒相互作用提供了深入的见解。尽管有大量研究关注植物化合物对瘤胃细菌的影响（Hart 等，2008；Huws 等，2009；Belanche 等，2012；Wang 等，2019），其对瘤胃噬菌体种群的影响尚不明晰。

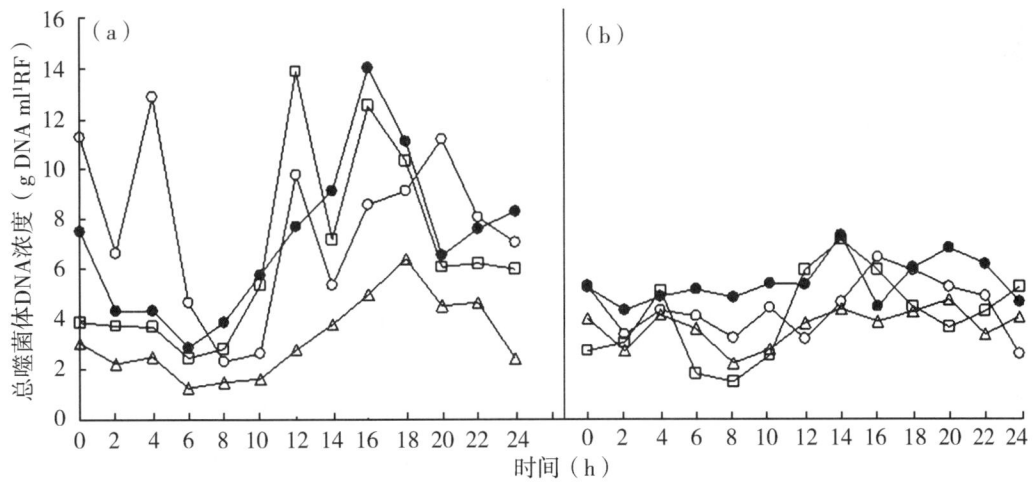

图 3 典型的次级植物化合物单宁酸对瘤胃噬菌体种群的影响。在 24h 内采集山羊瘤胃液测定总噬菌体 DNA 浓度（μg mL-1 RF）。(a) 4 只绵羊没有在瘤胃内添加单宁酸；(b) 同样的四只绵羊在试验第 4 小时（09:00）于瘤胃内添加单宁酸（50mL 的 8% wt/vol. 溶液）。所有的羊均在试验的第 4 小时（09:00）饲喂 1 次燕麦：苜蓿糖比例为 70:30 的日粮。(○) 羊 A，(△) 羊 B，(□) 羊 C，(●) 羊 D。改编自 Swain（1999）。

细菌基因组测序技术的进步也提升了我们对瘤胃噬菌体的认识。这是由于常见的噬菌体复制策略可以导致噬菌体基因组 DNA 整合到宿主细胞的基因组中。全基因组测序技术可以有效地将这种噬菌体遗传物质纳入细菌的总基因组序列数据中，基于此的分析可以明晰噬菌体相关遗传物质（Koonin 和 Wolf，2008）。虽然用于描述噬菌体复制策略的术语有时不一致，并正在审查（Hobbs 和 Abedon，2016），但噬菌体感染宿主细胞并将其基因组纳入宿主细胞，从而产生稳定的、可遗传状态的能力，通常被称为溶原性，这种复制周期被称为溶原性周期（图 3）。经历溶原过程的噬菌体为温和噬菌体，寄生在宿主细胞中的噬菌体基因组被称为前噬菌体（Zabriskie，1964；Howard-Varona 等，2017）。前噬菌体要么将其 DNA 整合到宿主染色体中，要么以质粒的形式在细胞质中循环，这些状态是可互换的（Deutsch 等，2018）。20 世纪 70 年代，瘤胃中首次发现了溶原性噬菌体，在使用化学诱导剂丝裂霉素 C 刺激马链球菌后，检测到完整的噬菌体样颗粒（Tarakanov，1974）。另外一项研究检测了来自 5 个不同属的 38 株瘤胃细菌，结果显示，使用丝裂霉素 C 诱导后，9 株（23.7%）可以产生完整的噬菌体样颗粒（Klieve 等，1989）。在通过载体状态和慢性感染等抑制溶原性后，噬菌体可以引起持续感染，维持并与微生物宿主共存（Weinbauer，2004；Clokie 等，2011），这在瘤胃细菌（如瘤胃拟杆菌）中也得到了证实（Klieve 等，1991）。虽然可以通过细菌基因组学来提示这些状态的存在（即是否存在完整的前噬菌体序列），但仍需要进一步的基于培养的实验方法来验证这些复制方式的存在（图 4）。

迄今为止，瘤胃噬菌体基因组序列在公开的遗传数据库中仅有少量报道。这些序列包括感染瘤胃微生物属（链球菌、拟杆菌和瘤胃球菌）的溶菌噬菌体，以及长尾噬菌体 *Siphoviridae* 和短尾噬菌体 *Podoviridae* 家族的噬菌体（Gilbert 等，2017）。在这些噬菌体中，链球菌噬菌体 φSb01 最初来源于牛瘤胃液，其生物学特征最先被明确（Klieve 和 Bauchop，1991）。该噬菌体具有典型的长尾噬菌体形态，缺乏溶原性的遗传模块（图 5）。在厌氧培养中，φSb01 能在感染宿主菌株 *S. equinus* 2B 后 3 h 内裂解，并在感染培养物持续孵育后形成抗噬菌体细胞。抗噬菌体的细胞改变其生长为聚集生长，形成厚的多糖胶囊（Klieve 和 Bauchop，1991）。φSb01 噬菌体颗粒在瘤胃液中被降解（图 5），颗粒衰减率为每小时 36.6%~53.0%（Swain，1999）。

瘤胃微生物（细菌和古菌）全基因组测序技术的发展迅速扩大了噬菌体的检测速度。在公开的培养瘤胃细菌和古菌基因组序列中，基于前噬体序列的流行率调查研究发现，大多数瘤胃细菌属的菌株基因组中均存在与噬菌体相关的序列（Berg Millere 等，2012；Gilbert 和 Cliff，2015；Seshadri 等，2018）。然而，这些噬菌体序列是否可以编码完整的、活的噬菌体颗粒，仍然需要大量的实验证明。虽然已知存在感染产甲烷菌的裂解噬菌体（Meile 等，1989；Nölling，1993；Weidenbach 等，2017；Wolf 等，2019），迄今为止只有一篇文章描述了一种溶解古细菌病毒感染了瘤胃产甲烷菌（Baresi 和 Bertani，1984）。据报道，在产甲烷的瘤胃古菌中，包括栖瘤胃甲烷短杆菌 *M. ruminantium* M1、甲烷短杆菌 *Methanobrevibacter* sp. JH1 和 *Methanobrevibacter formicicum* BRM9 等的基因组序列中检测到几种噬菌体（Attwood 等，2008；Leahy 等，2013；Kelly 等，2014）。所有这些噬菌体都是有尾噬菌体；然而，它们很难在体外被诱导或培养。尽管很难获得完整的噬菌体颗粒，但古菌前噬菌体编码的蛋白质已经被试验表达，并表明可以特异性水解产甲烷古生菌的假肽聚糖细胞壁成分（Altermann 等，2018）。

细菌和古菌的全基因组测序也揭示了噬菌体感染的防御系统的存在，包括限制修改（RM）系统和成簇的、规律间隔的短回文重复序列（CRISPR）和 CRISPR 相关蛋白（Cas）。RM 系统可使外来 DNA 发生特异性和非特异性的甲基化和/或裂解（Koonin 等，2017），从而防止噬菌体被感染。在多种瘤胃菌株中发现了 RM 系统，包括链球菌、瘤胃

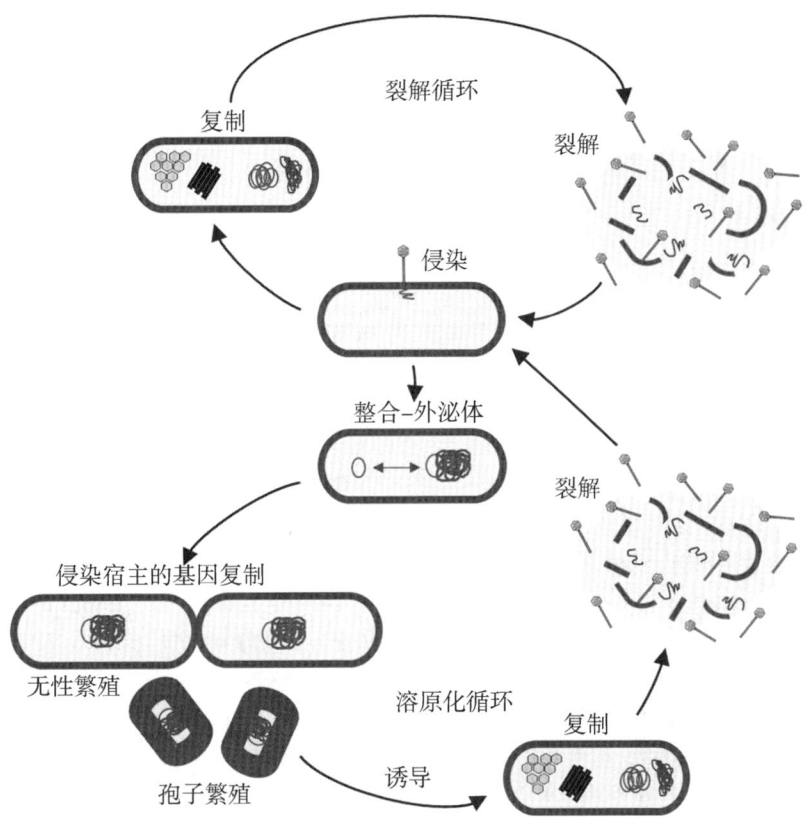

图 4 瘤胃中发生的噬菌体复制周期的示意图。噬菌体吸附和侵染后，可能会立即开始裂解循环，噬菌体基因组被转录、复制和产生噬菌体蛋白，使成熟的噬菌体颗粒得以组装，并在宿主细胞裂解后释放。溶原化循环包括建立稳定的、可遗传的状态，噬菌体基因组形成一个外泌体，或整合到宿主染色体上，或环化，这一状态被称为前噬菌体，并在无性繁殖或孢子繁殖期间保留。大多数噬菌体基因在前噬菌体中被转录抑制，但在裂解循环中经诱导后，这种抑制被移除，使得噬菌体能够复制并裂解宿主细胞以释放成熟的噬菌体颗粒。

球菌、拟杆菌属、巨型瘤胃菌、光冈菌属、月形单胞菌属和密螺旋体等（Styriak 等，1998；Piknova 等，2004；Pristas 和 Piknova，2005；Gilbert 等，2017）。进一步的研究表明 RM 系统的类型和丰度存在菌种特异性差异，这表明某些种类的瘤胃细菌可能比其他种类更容易受到噬菌体感染。

在几项关于瘤胃细菌基因组的研究中也发现了 CRISPR-Cas 系统的存在（Attwood 等，2008；Berg Miller 等，2012；Kelly 等，2014；Gilbert，2017；Seshadri 等，2018）。在古菌和细菌中发现了 CRISPR-Cas 系统（Sorek 等，2008），并为微生物提供了对传入 DNA 的免疫，随着时间的推移，通过获取进入 CRISPR 位点的侵入性核酸的短段（称为"间隔段"），从而建立了免疫力（Barrangou，2013）。以上研究提供了一种机制来识别（通过间隔区域的同源性）和切割（通过 Cas 蛋白）传入的 DNA，包括病毒 DNA。对瘤胃细菌

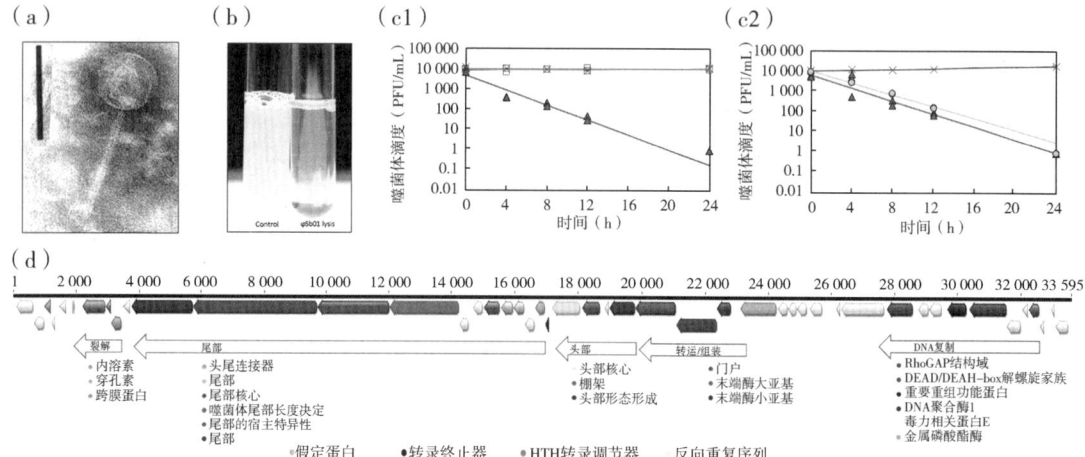

图5 基于培养和遗传学技术解析的瘤胃噬菌体 φSb01 感染瘤胃细菌马粪肠球菌 2B 的特征。(a) 透射电子显微镜观测 φSb01 的典型长尾噬菌体形态（比例尺 100nm）；(b) 马粪肠球菌 2B 厌氧条件下纯培养生长 6h 后（对照）和感染 φSb01 后 3h 内（φSb01 裂解）；(c) φSb01 颗粒的存活率测定，用蚀斑试验测定剩余的活体噬菌体颗粒的滴度［每毫升空斑形成单位（PFU）及其回归曲线］。(c1) 分别在绵羊瘤胃液（▲）、热处理（高压灭菌107℃，103KPa，45min）瘤胃液（□）和无瘤胃液对照培养基（×）的复制管中培养孵育；(c2) 分别在绵羊瘤胃液（▲）、澄清，0.22μm 过滤（低蛋白残留过滤器 HV，Millipore）的瘤胃液（●）和无瘤胃液的对照培养基（×）；(d) φSb01 的全基因组序列显示了模块化的基因排列和相关功能基因的聚类（彩图4）。改编自 Gilbert 等（2017）。

基因组中 CRISPR 序列的系统研究表明，在所检测的 410 个基因组中，其中的 241 个包含 6 344 个 CRISPR 间隔序列（Seshadri 等，2018）。通过检索 IMG/VR 数据库（Paez-Espino 等，2017），发现 83 个病毒操作分类单元（OTUs）和 31 个宿主之间存在关联，只有一个相同的间隔序列出现在肠杆菌目和假单胞菌目的分离株之间。这些结果进一步表明，瘤胃噬菌体种群具有高度多样性和遗传异质性，只有极少数培养菌株对高度相关的噬菌体产生免疫力或受到其感染。

病毒宏基因组学的发展进一步加深了我们对瘤胃噬菌体种群的理解，其数据集被称为病毒组。宏基因组学包括获取环境样本（包括病毒）中所有遗传物质的序列数据。然而，由于其相当小的基因组长度，病毒往往在宏基因组数据集中相对不足。病毒宏基因组学通常涉及对环境样本进行预处理，以浓缩或"富集"病毒颗粒，并降低污染、非病毒 DNA 的浓度（Edwards 和 Rohwer，2005；Thurber 等，2009）。除非引入特定步骤来克隆和测序 ssDNA，并转录和扩增这些富集病毒样本中包含的任何病毒 RNA（Edwards 和 Rohwer，2005；Bolduc 等，2012），这些 ssDNA 和 RNA 病毒通常不会被 DNA 测序技术检测到。这可能会妨碍检测到 RNA 病毒，包括可能会感染瘤胃相关的真核生物（真菌、原虫）的分体病毒科 Partitivirridae 和 Totivirridae（Nibert 等，2009；Goodman 等，2011），以及感染细菌和古细菌的 Cystoviridae 和 Leviviridae（Krupovic 等，2011）。病毒宏基因组学的优势在于克服了传统培养为基础的方法局限性，发现了新的病毒，并揭示了自然界中瘤胃和其他微生物生态系统中病毒的多样性和丰度（Paez-Espino 等，2016）。

迄今为止，病毒宏基因组已用于奶牛（Berg Miller 等，2012；Ross 等，2013）、水牛（Parmar 等，2016）、肉牛（Anderson 等，2017）和山羊（Namonyo 等，2018）中相关病毒的研究。从奶牛（Dinsdale 等，2008）、绵羊（Yutin 等，2015）和驼鹿（Solden 等，2018）的瘤胃宏基因组中也发现了多种病毒。除了病毒宏基因组外，还可以利用独立注释的 kmers（短核苷酸序列）等病毒特征进行病毒组研究。有研究基于此方式对奶牛的瘤胃病毒组进行了研究（Willner 等，2009），发现瘤胃病毒种群与其他环境中发生的病毒种群截然不同。尽管日粮、品种、病毒样品制备和测序技术（454-pyrosequencing、Illumina HiSeq 和 Ion torrent）存在差异，但所有这些研究都表明，瘤胃病毒种群以有尾噬菌体目 Caudovirales 的尾部噬菌体种群为主。其中最多的是长尾噬菌体科 Siphoviridae，其次是肌尾噬菌体科 Myoviridae 和短尾噬菌体科 Podoviridae。这些发现与早期基于 TEM 的瘤胃病毒的研究结果一致，即尾部噬菌体为牛、绵羊和山羊瘤胃样本中的优势病毒（图 2）。迄今为止，大多数病毒宏基因组学研究都使用有限数量的瘤胃样本来探索病毒多样性（Berg Miller 等，2012；Parmar 等，2016；Namonyo 等，2018），并保持日粮、品种和泌乳期等试验条件不变（Ross 等，2013）。只有最近的一项研究采集了大量样本，并纳入了更多的试验变量，如日粮的影响（Anderson 等，2017）。然而，尽管存在这些局限性，高通量测序和生物信息学技术所获得的信息深度无疑超过了以往所有用于描述瘤胃病毒种群组成的技术。病毒宏基因组学和不富集病毒的宏基因组学研究使病毒基因的鉴定、功能分类和分类分配以及部分和/或接近完整的病毒基因组的组装成为可能。

瘤胃中含有大量且高度多样化的病毒种群。尽管使用基于基因的方法探究不同病毒类型的物种丰富度，如 Phage Communities from Contig Spectrum（PHACCS）和 CatchAll（Angly 等，2005；Bunge 等，2012），所得的结果存在巨大差异，然而所有的研究都认为存在数千种不同的病毒类型（Berg Miller 等，2012；Ross 等，2013；Anderson 等，2017）。我们可以通过增加测序深度和组装更完整的病毒 contigs 等测序技术的改进来进行更为系统的病毒组的研究。

瘤胃病毒通常含有高丰度的基因，这些基因被分类在由噬菌体、前噬菌体、转座元件和质粒组成的 SEED 子系统中（Aziz 等，2008；Berg Miller 等，2012）。这些病毒功能基因包括与病毒复制相关的功能基因（DNA 复制和控制蛋白）和有助于病毒颗粒形成的结构基因（头部和尾部蛋白），以及细胞裂解蛋白（包括溶血素和内肽酶）（Ross 等，2013）。有趣的是，最近一项对驼鹿瘤胃微生物和病毒宏基因组的研究也进行了代谢重建，并使用宏蛋白质组学检测病毒基因的表达（即病毒蛋白的产生）（Solden 等，2018）。虽然检测到的大多数（80%）病毒蛋白功能未知，仍有一些检测到的蛋白可以归类为结构蛋白（如衣壳蛋白）（Solden 等，2018）。

除了 3 个主要的尾状病毒科，瘤胃病毒宏基因组还检测到非尾部病毒科，如复层噬菌体科 Tectivirridae（Berg Miller 等，2012；Ross 等，2013；Solden 等，2018）。尽管这一病毒以前曾在瘤胃液中观察到，但从未被成功培养过。瘤胃病毒宏基因组学研究也发现了与巨型病毒（如 Mimivirus）高度相关的基因（Berg Miller 等，2012；Anderson 等，2017），它们通常在瘤胃原虫种群中感染和复制。常用的从瘤胃液样品中纯化病毒颗粒的方法，包括通过 0.22 μm 过滤器过滤，可以排除这些非常大的病毒（*Acanthamoeba polyphaga* mimivirusis 病毒直径为 400 nm）。目前尚不清楚这些巨型病毒基因是否实际上是来自较小的、新的瘤胃病毒的同源基因，这些病毒是否在基因数据库中不存在。在绵羊瘤胃中也发现了新型病毒噬菌体的存在，它们是寄生于 Mimivirridae 科的巨型病毒（Yutin 等，2015）。宏基因组学正在迅速扩大我们对于与瘤胃微生物种群相关的病毒类型的理解。

在无法进行微生物-病毒共培养的情况下，前述的 CRISPR 间隔物可以用于研究瘤胃病毒-宿主相互作用。这可以通过比较病毒宏基因组序列或组装的病毒 contigs 与 CRISPR 间隔序列数据库或从细菌基因组序列派生的 CRISPR 间隔序列的同源性来实现（Berg Miller 等，2012；Anderson 等，2017）。这种方法有时在瘤胃病毒种群和已知的 CRISPR 间隔区域之间显示出很少的相似性。然而，最近的一项病毒宏基因组研究从同一个瘤胃液样本中组装了微生物基因组，使用了基于 CRISPR 的方法将 113 个病毒 contigs（来自研究中获得的 1 907 个>10 kb 病毒）与 4 个门的微生物宿主关联起来（Seshadri 等，2018）。本研究通过宏基因组学方法表明，瘤胃中的病毒捕食可影响饲料分解的主要微生物种群，包括参与复杂碳水化合物（木糖聚糖、半纤维素聚合物和单糖，如木糖）降解的微生物种群。

4 移动体对瘤胃功能的作用和影响

通过与其他微生物生态系统（如土壤、水和其他肠道生态系统）类比，可以推测到瘤胃中移动体的潜在作用（Wommac 等，1996；Srinivasiah 等，2008；Letarov，2012；Roux 和 Brum，2019；Shkoporov 和 Hill，2019）。然而，瘤胃环境中的一些物理属性会使得微生物群落（细菌、古细菌、真菌和原虫）以及与这些微生物相关的病毒种群与其他环境中的显著不同。

瘤胃环境与其中微生物种群的生长所需的条件十分契合（Bryant，1959），整体是一个大型的肠腔，具有恒定的温度（大约 39℃）和厌氧环境。瘤胃中的营养物质（饲料，包括容易发酵的糖和复杂的碳水化合物）和水的稳定供应，加上大量缓冲唾液的作用，维持了其相对恒定的接近中性（微酸性）的 pH 值。瘤胃发酵的产物和微生物本身可以流入后消化道或者被瘤胃上皮吸收入血（Mackie 等，2002），这样可以保持瘤胃微生物种群的活跃生长，正如在水生环境的上层中一样，既不会像在土壤环境中可能出现脱水现象，也不会受到阳光的影响（Weinbauer 等，1999，Weinbauer，2004）。在除了类似于其他宿主相关的肠道环境外，瘤胃的较大体积、反刍行为、相对较长的停留时间和草食性造就了瘤胃微生物菌群的独特性。

在影响瘤胃功能众多的移动体中，病毒和质粒是迄今为止研究最多的。病毒和质粒在瘤胃微生物生态系统中的作用包括微生物细胞裂解、水平基因转移、瘤胃微生物种群的调节、表型性状的改变以及对动物宿主如免疫的潜在影响。

4.1 细胞的裂解作用

研究表明，病毒裂解细胞是许多不同环境中细菌死亡的主要原因，原核生物病毒对营养循环有重大影响（Weinbauer 和 Höfle，1998；Weinbauer，2004；Clokie 等，2011）。就瘤胃而言，病毒（噬菌体）裂解长期以来被认为是导致瘤胃内细菌裂解的原因（Jarvis，

1968；Nolan 和 Leng，1972；Kliev 等，1989）。从反刍动物营养和蛋白质效率的角度来看，噬菌体介导的细菌裂解是一种负面影响，蛋白质在瘤胃中循环利用，而不是进入下部肠道以供动物吸收利用（Leng 和 Nolan，1984）。然而从微生物生态学的角度来看，事实可能恰好相反，噬菌体介导的细菌裂解和蛋白质的瘤胃内循环实际上对瘤胃微生物群产生了积极有利的影响。这种对整个生态系统碳和养分循环的积极贡献已在水生（如海洋）生态系统中得到了证实（Fuhrman，1999；Roux 和 Brum，2019）。噬菌体介导的细胞裂解可为瘤胃内的其他生物体提供蛋白质（包括酶）、DNA 和细菌代谢产物（单糖、降解的植物化合物）。

在绵羊瘤胃中曾观察到裂解噬菌体活性的暴发和总噬菌体数量的变化，这可能正好对应着细菌的裂解现象（Klieve 和 Swain，1993；Swain 等，1996）。最近的一项研究表明，病毒的侵染可能有助于瘤胃中的微生物种间互利（营养循环）（Solden 等，2018），病毒会感染并溶解参与复杂植物化合物降解的关键微生物种群。此外，作为同一研究的部分内容所进行的蛋白质组学研究表明，最常表达的病毒蛋白质是与结构噬菌体成分（如头部蛋白质）相关的蛋白质（Solden 等，2018），而不是参与细菌细胞壁裂解的噬菌体编码酶（如内异肽酶、赖氨酸和胆碱）。在此基础上仍需要利用多种技术和分析手段（如蛋白质组学、病毒和微生物宏基因组学以及瘤胃氨水平）进行进一步研究，从而更全面地了解病毒介导的裂解对瘤胃微生物种群的影响程度。

4.2 水平基因转移

在所有微生物生态系统中，移动体能够促进生物间遗传物质的转移，确保基因组和表型的可塑性，维持微生物多样性，并相互适应（Koonin 和 Wolf，2008；Jørgensen 等，2014；Hülter 等，2017）。鉴于这些基因转移所带来的严重的临床影响和人类病原体的出现，微生物类群的毒素基因和抗生素耐药基因的转移一直是水平基因转移研究的重点（Aminov，2011；De Sordi 等，2019；Dunivin 等，2019）。有研究证明，毒素基因可以转移到噬菌体基因组中，一个典型的例子是噬菌体 CEβ 和 DEβ 将神经毒素转移到肉毒梭菌中（Eklund 等，1971；Eklund 等，1972）。此外，毒素表达所需的大多数调节因子和辅助因子也由噬菌体编码；因此，噬菌体感染和噬菌体基因组 DNA 整合到细菌染色体中形成前噬菌体（溶原转化）对于易感梭菌菌株的"毒素转化至关重要"（Brussow 等，2004）。

在不同的微生物类群中，移动体单元通过转移并形成稳定的遗传状态而顺利完成水平基因转移的能力存在显著差异。某些物种如化脓性链球菌，含有大量 MGE，测序菌株的总基因组中多达 10% 与前噬菌体相关，因此通常被认定为是多溶源性的（Brussow 等，2004）。在某些情况下，微生物获得致病性的过程可能需要多重 MGE 的水平基因转移。例如，虽然大多数大肠杆菌菌株都是良性的共生肠道细菌，但不排除一些菌株可能通过水平基因转移转化为强致病菌。根据这些菌株毒力因子及其引起的疾病可以对其进行分类，如肠致病性大肠杆菌（EPEC）、肠出血性大肠杆菌（EHEC）、产志贺毒素大肠杆菌（STEC）、肠侵袭性大肠杆菌（EIEC）、弥漫黏附性大肠杆菌（DAEC）和尿致病性大肠杆菌（UPEC）（Kaper 等，2004；Blount，2015）。与这些菌株相关的水平基因转移毒力因子包括致病基因、转座子、质粒和噬菌体（Brussow 等，2004）。虽然噬菌体编码的 AB 型毒素（志贺毒素基因 *stx*1 和 *stx*2），可抑制蛋白质合成并导致严重疾病，但它需要多种致病因子的组合，以使得强毒力大肠杆菌菌株能够在肠道上皮定居并产生毒素，肠细胞脱落位点（LEE）基因组促进了菌株附着和病变易化的形成（Hazen 等，2015）。而质粒编码基因，包括 α-溶血素（hlyA），对 O157：H7 等菌株的毒力有增强作用（Johnson 和 Nolan，2009）。在一些大肠杆菌菌株中，质粒相关基因也编码 IV 型菌毛，这赋予了其在肠上皮细胞表面的局部黏附能力和细胞毒性（Lang 等，2018）。

虽然瘤胃移动体的主要作用之一是水平基因转移（Morrison 1996），可以推测多种不同属的瘤胃微生物会发生遗传交换进而导致菌株变异。由于瘤胃微生物生态系统的构成过于复杂，目前鲜有实验证据证明这一点。尽管噬菌体、质粒和染色质外等移动体已被证明归属于几个不同的瘤胃微生物属中，例如链球菌、瘤胃球菌、拟杆菌和甲烷杆菌（Klieve 等，1989；Attwood 等，2008；Gilbert 等，2017），尚未有研究表明这些 MGE 携带的基因是否存在于其他瘤胃微生物中。不同的瘤胃微生物也携带有相同功能的基因，如编码分解复杂碳水化合物酶的基因（Seshadri 等，2018）。这使得微生物间特定基因的转移很难通过实验的手段进行验证。

此外，还可以使用基于共享序列的预测网络研究水平基因转移（Shapiro 和 Putonti，2018）。从整合病毒基因组和微生物基因组的瘤胃宏基因组学数据建立网络（Solden 等，2018），可以提供遗传联系和潜在基因转移的信息，却无法具体确定特定的单个基因的转移信息。最近的一项基于病毒组的研究表明，瘤胃病毒编码的基因（糖苷水解酶）增强了瘤胃中复杂碳水化合物的分解能力（Anderson 等，2017）。然而，Solden 等，（2018）对病毒和微生物宏基因组的研究发现，参与复杂碳水化合物分解的基因通常与瘤胃病毒无关。值得注意的是，对微生物和病毒进化的研究表明，许多较小的病毒（基因组大小为 5～50 kb）只有在很少的情况下会从宿主中获取基因，并且往往不会在其基因组中携带非必需的辅助基因（Forterre 和 Prangishvili，2013）。目前迫切需要从瘤胃中获取更多原核生物病毒和染色质外的成分（质粒和前噬菌体），并将其序列整合进遗传数据库中，以便更好地识别这些遗传成分所携带的可能会转移的基因单元。

4.3 微生物菌群的调节

瘤胃移动体的特定组成部分，如质粒，可能起到传递基因的作用，这可能给受体微生物带来有益的选择优势，从而促进载体或宿主微生物的增殖（Teather，1982）。同样，病毒衍生的染色质外成分（前噬菌体）可能会提高宿主微生物的竞争力或对环境的适应能力（Winstanley 等，2009）。这不仅发生在原核生物中，也发生在真核生物（真菌、原虫）中；例如，受真菌病毒感染的酵母菌和真菌（包括酵母菌、*Hanseniaspora*、*Ustilago* 和 *Zygosaccharomyces*）可能会增加其产生"致死毒素"的能力（Ghabrial 等，2015），即使大多数真菌病毒只会导致无明显症状的感染（Sato 等，2018）。然而，病毒对瘤胃内厌氧真菌或原虫的影响尚未见报道。

另一种途径是通过大规模的宿主细胞裂解来调节微生态系统中的细菌种群，其中常见的方式就是噬菌体侵染。这种噬菌体调控系统中最丰富或最主要的微生物种群的现象，曾被描述为"推翻垄断者"(Weinbauer,2004)或"由上及下的控制"(Sieradzki 等,2019),用以维持种群多样性,即保证其他的竞争微生物类群的共存。这种现象最常发生在营养相对有限的水生生态系统中,在这些生态系统中,微生物会定期暴发,其中可能涉及裂解和溶原噬菌体的活动(Wommack 和 Colwell,2000;Suttle,2007;Howard Varona 等,2017)。

在瘤胃中,营养限制通常不像在淡水和海洋环境中那么关键;但噬菌体数量也呈现出随采食量变化而波动的规律(Swain 等,1996)。研究表明,相当一部分优势瘤胃细菌属可能被溶原噬菌体感染,而溶原噬菌体可在原噬菌体诱导下进行增殖(Klieve 等,1989);因此,溶原性噬菌体也可能在调节瘤胃细菌种群的相对丰度方面发挥作用。然而,就大多数瘤胃微生物属而言,有关宿主与病毒相互作用的基本信息仍不清楚,如宿主范围(病毒可感染的物种数量)、暴发大小(受感染细胞释放的病毒颗粒数量)和瘤胃条件下噬菌体颗粒降解的速率等。因此,很难准确计算出新病毒感染的发生率,也很难预测病毒对瘤胃微生物的种群调节能到何种程度。

4.4 对表型性状和生长性能的影响

迄今为止,关于移动体对瘤胃微生物种群表型特征和生长性能影响的研究很少。然而,在其他微生物生态系统中,胞外元素(如质粒)被证明具有抗药性等特征(Heuer 和 Smalla,2012;Hülter 等,2017),并可直接影响和改善微生物的生存能力和持久性。噬菌体感染还可导致细菌产生细菌毒素(Kuhl 等,2012;Penadés 等,2015),并可促进蓝藻的光合作用和产氧(Sieradzki 等,2019),以及提高细菌宿主对其他噬菌体感染的抵抗力(噬菌体重复感染)(Refardt,2011)。

在瘤胃细菌马链球菌 S. equinus 中,噬菌体感染可通过形成厚的多糖荚膜(Klieve 和 Bauchop,1991)促进具有改变生长习性的抗噬菌体菌株的增殖,这可能因为多糖荚膜掩盖了细胞表面的噬菌体受体从而保护细胞免受其他噬菌体的侵染。值得注意的是,细菌和古细菌形成胞外多糖荚膜的能力通常有助于生物膜的形成(Orell 等,2013;Turnbull 等,2016)。饲料中的植物材料在瘤胃中的聚集导致了细菌生物膜的形成,而噬菌体颗粒与瘤胃中这些生物膜相互作用可能有助于形成这些生物膜的现象,这或许会是一个新的研究方向。如今,宏基因组学似乎难以通过检测与噬菌体受体相关的位点突变等方法来检测抗噬菌体细菌表型的变化(De Sordi 等,2019)。这就需要使用简化的群落或单个微生物菌株进行进一步的体外研究,以充分了解瘤胃中移动体影响微生物生长发育的程度。

4.5 对反刍动物宿主的影响

移动体与微生物之间固有的、不可分割的联系,会随着反刍动物中微生物种群的定植和更替而改变(Fonty 等,1987;DillMcFarland 等,2017;Dias 等,2018),移动体(如病毒种群)在瘤胃发育的早期阶段便会出现。迄今为止,尚不清楚移动体是否能对反刍动物宿主的生理状态产生直接的影响。

人类肠道相关微生物群落的最新研究表明,包括噬菌体在内的病毒可能渗透肠道上皮，并对宿主免疫产生影响（Barr，2019；De Sordi 等，2019；Shkoporov 和 Hill，2019）。有研究证明噬菌体颗粒有多种免疫调节形式，并可与人类肠道上皮相关的免疫细胞（如树突状细胞）发生相互作用（Keen 和 Dantas，2018）。此外，噬菌体介导的细菌裂解后，脂多糖（LPS）和其他细菌免疫原的释放可能会激发微生物的天然免疫机制（Duerkop 和 Hooper，2013）。也有研究认为，噬菌体颗粒附着于黏蛋白糖蛋白的能力，即 BAM 模型（噬菌体黏附黏液），能够使噬菌体在肠道黏液层中积累，并形成富含噬菌体的保护层，从而减少细菌对肠道上皮的渗透和定植（Barr 等，2013）。

目前，移动体与宿主相互作用被认为是一个重要的新兴研究领域。但相对较大的瘤胃体积、瘤胃液的蛋白水解性质、瘤胃上皮与人类肠道的生理差异等因素是否会对反刍动物宿主免疫能力产生潜在的影响，仍需进一步的研究。

5 结论和展望

总而言之，当前对瘤胃移动体的研究并不是高度热门领域。尽管移动体的单个成分，如瘤胃病毒，被认为是瘤胃的互利共生成分，但一般认为移动体不是需要重点研究的对象。然而，随着更多研究的开展，研究人员在进一步探究瘤胃移动体对瘤胃微生物群平衡和功能的潜在影响时，发现其所占据的地位和产生的影响是重要且深远的（Kav 等，2012；Anderson 等，2017；Solden 等，2018）。

因此，瘤胃病毒和细胞外成分的未来研究趋势或将围绕两个主要焦点：①探究上述问题的答案，②开发实际的应用。成功实现这两个目标的关键取决于技术的发展。随着测序技术、生物信息学和蛋白质组学的发展，现在的研究比以往任何时候都更适合描述和解析瘤胃中病毒和染色质外成分的影响。当与基于培养的技术结合使用时，这些技术所提供的信息预计将会远远超过在测序前和生物信息学时代获得的有关瘤胃相关病毒和染色体外成分的生命周期和生物学的信息量。

随着抗生素在农业生产中的禁用，需要一些新的方法对有害的细菌种群进行生物学调控。目前，已经研发出一种使用完整病毒（噬菌体疗法）或病毒编码酶（噬菌体酶制剂）的噬菌体疗法，常被用于控制园艺生产、家禽和水产养殖业中的病原体，从而逐步解决生产和管理中的难题（Monk 等，2010；Chan 等，2013；Fernández 等，2018；Seal 等，2018）。在反刍动物生产体系的背景下，尤其是集约化养殖的奶业生产一些疾病如乳腺炎、致病性大肠杆菌菌株的侵染和甲烷排放等问题均可通过基于噬菌体的疗法来解决（Klieve 和 Hegarty 1999；Raya 等，2011；Dias 等，2013；Gilbert 等，2015；Gutiérrez 等，2019）。靶向微生物细胞壁（如卵磷脂和赖氨酸、内肽酶和尾穗蛋白）的完整性，开发应用噬菌体编码的酶，是一种可以有效减少和控制有害瘤胃微生物种群的强有力的新途径。

随着技术的更新，未来有关瘤胃移动体的研究工作将更加热门。基因数据库中瘤胃特定微生物高质量注释序列的增加以及其中病毒和质粒相关信息的扩展显得尤为重要。新的参考序列数据将大大增强目前有限的赋予新基因元件相关序列同源性的能力。只有同步探究这些问题，才能揭示移动体对瘤胃功能的系统影响。

6 致谢

感谢昆士兰农业和渔业部 Kathy Crew 博士在透射电子显微镜方面的帮助。此外,本章特别致敬 Athol Klieve 博士,感谢他在 30 多年里对瘤胃微生物学和瘤胃噬菌体领域做出的重大贡献。

7 更多信息

可通过以下出版物获得原核生物病毒和染色体外成分的信息:Brussow et al.(2004)、Weinbauer(2004)、Ackermann and Prangishvili(2012)、Bellanger et al.(2014)、Delavat et al.(2017)、Keen and Dantas(2018)、Roux and Brum(2019)以及 Shkoporov and Hill(2019)。有关生物信息学资源的其他信息和链接,请访问:https://www.ncbi.nlm.nih.gov/genome/viruses/ 和 http://www.isvm.org/resources.html。

关于瘤胃及其微生物以及瘤胃病毒和质粒的更多具体信息,请参见以下出版物:Kav et al.(2012)、Gilbert and Klieve(2015)、Gilbert et al.(2017)、Huws et al.(2018)和 Seshadri et al.(2018)。相关瘤胃微生物学方面的国际研究工作、联系和合作项目,请访问:https://globalresearchalliance.org/research/livestock/networks/rumen-microbial-genomics-network/。

8 参考文献

Abedon, S. T. and Murray, K. L. 2013. Archaeal viruses, not archaeal phages: an archaeological dig. Archaea 2013, 1-10.

Ackermann, H. W. and Prangishvili, D. 2012. Prokaryote viruses studied by electron microscopy. Arch. Virol. 157 (10), 1843-9.

Adams, J. C., Gazaway, J. A., Brailsford, M. D., et al., 1966. Isolation of bacteriophages from the bovine rumen. Experientia 22 (11), 717-8.

Akhter, S., Aziz, R. K. and Edwards, R. A. 2012. PhiSpy: a novel algorithm for finding prophages in bacterial genomes that combines similarity- and composition-based strategies. Nucleic Acids Res. 40 (16), e126.

Altermann, E., Schofield, L. R., Ronimus, R. S., et al., 2018. Inhibition of rumen methanogens by a novel archaeal lytic enzyme displayed on tailored bionanoparticles. Front. Mirobiol. 9, 2378.

Aminov, R. I. 2011. Horizontal gene exchange in environmental microbiota. Front. Mirobiol. 2, 158.

Anderson, C. L., Sullivan, M. B. and Fernando, S. C. 2017. Dietary energy drives the dynamic response of bovine rumen viral communities. Microbiome 5 (1), 155.

Angly, F., Rodriguez-Brito, B., Bangor, D., et al., 2005. PHACCS, an online tool for estimating the structure and diversity of uncultured viral communities using metagenomic information. BMC Bioinformatics 6, 41.

Arndt, D., Grant, J. R., Marcu, A., et al., 2016. PHASTER: a better, faster version of the PHAST phage search tool. Nucleic Acids Res. 44 (W1), W16-21.

Arntzen, M. Ø., Várnai, A., Mackie, R. I., et al., 2017. Outer membrane vesicles from *Fibrobacter succinogenes* S85 contain an array of carbohydrate-active enzymes with versatile polysaccharide-degrading capacity. Environ. Microbiol. 19 (7), 2701-14.

Asmundson, R. V. and Kelly, W. J. 1987. Isolation and characterization of plasmid DNA from *Ruminococcus*. Curr. Microbiol. 16 (2), 97-100.

Attwood, G. T. and Brooker, J. D. 1992. Complete nucleotide sequence of a *Selenomonas ruminantium* plasmid and definition of a region necessary for its replication in *Escherichia coli*. Plasmid 28 (2), 123-9.

Attwood, G. T., Kelly, W. J., Altermann, E. H., et al., 2008. Analysis of the *Methanobrevibacter ruminantium* draft genome: understanding methanogen biology to inhibit their action in the rumen. Aust. J. Exp. Agric. 48 (2), 83-8.

Aziz, R. K., Bartels, D., Best, A. A., et al., 2008. The RAST server: rapid annotations using subsystems technology. BMC Genomics 9, 75.

Baresi, L. and Bertani, G. 1984. Isolation of a bacteriophage for a methanogenic bacterium. Admin. Ann. Meet. Americ Soc. Microbiol. 28, 133.

Barr, J. J. 2019. Missing a phage: unravelling tripartite symbioses within the human gut. mSystems 4 (3).

Barr, J. J., Auro, R., Furlan, M., et al., 2013. Bacteriophage adhering to mucus provide a non-host-derived immunity. PNAS 110 (26), 10771-6.

Barrangou, R. 2013. CRISPR-Cas systems and RNA-guided interference. Wiley Interdisc. Rev. RNA 4 (3), 267-78.

Belanche, A., Doreau, M., Edwards, J. E., et al., 2012. Shifts in the rumen microbiota due to the type of carbohydrate and level of protein ingested by dairy cattle are associated with changes in rumen fermentation. J. Nutr. 142 (9), 1684-92.

Bellanger, X., Payot, S., Leblond-Bourget, N., et al., 2014. Conjugative and mobilizable genomic islands in bacteria: evolution and diversity. FEMS Microbiol. Rev. 38 (4), 720-60.

Berg Miller, M. E., Yeoman, C. J., Chia, N., et al., 2012. Phage-bacteria relationships and CRISPR elements revealed by a metagenomic survey of the rumen microbiome. Environ. Microbiol. 14 (1), 207-27.

Blount, Z. D. 2015. The unexhausted potential of *E. coli*. eLife 4, e05826.

Bobay, L. M., Touchon, M. and Rocha, E. P. C. 2014. Pervasive domestication of defective prophages by bacteria. PNAS 111 (33), 12127-32.

Bolduc, B., Shaughnessy, D. P., Wolf, Y. I., et al., 2012. Identification of novel positive-strand RNA viruses by metagenomic analysis of archaea-dominated Yellowstone Hot Springs. J. Virol. 86 (10), 5562-73.

Brailsford, M. D. and Hartman, P. A. 1968. Characterisation of *Streptococcus durans* bacteriophages. Can. J. Microbiol. 14 (4), 397-402.

Brussow, H., Canchaya, C. and Hardt, W. D. 2004. Phages and the evolution of bacterial pathogens: from genomic rearrangements to lysogenic conversion. Microbiol. Mol. Biol. Rev. 68 (3), 560-602.

Bryant, M. P. 1959. Bacterial species of the rumen. Bacteriol. Rev. 23 (3), 125-53.

Bunge, J., Woodard, L., Bohning, D., et al., 2012. Estimating population diversity with CatchAll. Bioinformatics 28 (7), 1045-7.

Champion, K. M., Helaszek, C. T. and White, B. A. 1988. Analysis of antibiotic susceptibility and extra-chromosomal DNA content of *Ruminococcus albus* and *Ruminococcus flavefaciens*. Can. J. Microbiol. 34 (10), 1109-15.

Chan, B. K., Abedon, S. T. and Loc-Carrillo, C. 2013. Phage cocktails and the future of phage therapy. Future Microbiol. 8 (6), 769-83.

Clarke, A. J. 2018. The 'hole' story of predatory outer-membrane vesicles. Can. J. Microbiol. 64 (9), 589-99.

Clokie, M. R. J., Millard, A. D., Letarov, A. V., et al., 2011. Phages in nature. Bacteriophage 1 (1), 31-45.

Croucher, N. J., Mostowy, R., Wymant, C., et al., 2016. Horizontal DNA transfer mechanisms of bacteria as weapons of intragenomic conflict. PLoS Biol. 14 (3), e1002394.

Dassa, B., Borovok, I., Ruimy-Israeli, V., et al., 2014. Rumen cellulosomics: divergent fiber-degrading strategies revealed by comparative genome-wide analysis of six ruminococcal strains. PLoS ONE 9 (7), e99221.

Deatherage, B. L. and Cookson, B. T. 2012. Membrane vesicle release in bacteria, eukaryotes, and archaea: a conserved yet underappreciated aspect of microbial life. Infect. Immun. 80 (6), 1948-57.

De Jonge, P. A., Nobrega, F. L., Brouns, S. J. J., et al., 2019. Molecular and evolutionary determinants of bacteriophage host range. Trends Microbiol. 27 (1), 51-63.

Delavat, F., Miyazaki, R., Carraro, N., et al., 2017. The hidden life of integrative and conjugative elements. FEMS Microbiol. Rev. 41 (4), 512-37.

De Sordi, L., Lourenço, M. and Debarbieux, L. 2019. The battle within: interactions of bacteriophages and bacteria in the gastrointestinal tract. Cell Host Microbe 25 (2), 210-8.

Deutsch, D. R., Utter, B., Verratti, K. J., et al., 2018. Extrachromosomal DNA sequencing revealsepisomal prophages capable of impacting virulence factor expression in *Staphylococcus aureus*. Front. Microbiol. 9, 1406.

Dias, R. S., Eller, M. R., Duarte, V. S., et al., 2013. Use of phages against antibiotic-resistant *Staphylococcus aureus* isolated from bovine mastitis. J. Anim. Sci. 91 (8), 3930-9.

Dias, J., Marcondes, M. I., Motta de Souza, S., et al., 2018. Bacterial community dynamics across the gastrointestinal tracts of dairy calves during preweaning development. Appl. Environ. Microbiol. 84 (9), e02675-17.

Dill-McFarland, K. A., Breaker, J. D. and Suen, G. 2017. Microbial succession in the gastrointestinal tract of dairy cows from 2 weeks to first lactation. Sci. Rep. 7, 40864.

Dinsdale, E. A., Edwards, R. A., Hall, D., et al., 2008. Functional metagenomic profiling of nine biomes. Nature 452 (7187), 629-32.

Duerkop, B. A. and Hooper, L. V. 2013. Resident viruses and their interaction with the immune system. Nat. Immunol. 14 (7), 654-9.

Dunivin, T. K., Choi, J., Howe, A. and Shade, A. 2019. RefSoil+: a reference database for genes and traits of soil plasmids. mSystems 4 (1), e00349-18.

Edwards, R. A. and Rohwer, F. 2005. Viral metagenomics. Nat. Rev. Microbiol. 3 (6), 504-10.

Eklund, M. W., Poysky, F. T., Reed, S. M., et al., 1971. Bacteriophage and the toxigenicity of *Clostridium botulinum* type C. Science 172 (3982), 480-2.

Eklund, M. W., Poysky, F. T. and Reed, S. M. 1972. Bacteriophage and the toxigenicity of *Clostridium botulinum* type D. Nat. New Biol. 235 (53), 16-7.

Ellis, T. N. and Kuehn, M. J. 2010. Virulence and immunomodulatory roles of bacterial outer membrane vesicles. Microbiol. Mol. Biol. Rev. 74 (1), 81-94.

Fernández, L., Gutiérrez, D., Rodríguez, A., et al., 2018. Application of bacteriophages in the agro-food sector: a long way toward approval. Front. Cell. Infect. Microbiol. 8, 296.

Flint, H. J. 1997. The rumen microbial ecosystem—some recent developments. Trends Microbiol. 5 (12), 483-8.

Flint, H. J., Thomson, A. M. and Bisset, J. 1988. Plasmid-associated transfer of tetracycline resistance in *Bacteroides ruminicola*. Appl. Environ. Microbiol. 54 (4), 855-60.

Fonty, G., Gouet, P., Jouany, J. P., et al., 1987. Establishment of the microflora and anaerobic fungi in the rumen of lambs. J. Gen. Microbiol. 133 (7), 1835-43.

Forterre, P. and Prangishvili, D. 2013. The major role of viruses in cellular evolution: facts and hypotheses. Curr. Opin. Virol. 3 (5), 558-65.

Fuhrman, J. A. 1999. Marine viruses and their biogeochemical and ecological effects. Nature 399 (6736), 541-8.

Gaillard, B., Breton, A. and Bernalier, A. 1989. Study of the nuclear cycle of four species of strictly anaerobic rumen fungi by fluorescence microscopy. Curr. Microbiol. 19 (2), 103-7.

Ghabrial, S. A., Caston, J. R., Jiang, D., et al., 2015. 50-plus years of fungal viruses. Virology 479-480, 356-68.

Gilbert, R. A. and Klieve, A. V. 2015. Ruminal viruses (bacteriophages, Archaeaphages). In: Puniya, A. K., Singh, R. and Kamra, D. N. (Eds), Rumen Microbiology: From Evolution to Revolution. Springer, New Delhi, pp. 121-41. https://doi-org.ezproxy.library.uq.edu.au/10.1007/978-81-322-2401-3_9.

Gilbert, R. A., Ouwerkerk, D. and Klieve, A. V. 2015. Phage therapy in livestock methane amelioration. In: Malik, P. K., Bhatta, R., Takahashi, J., Kohn, R. A. and Prasad, C. S. (Eds), Livestock Production and Climate Change. CABI Publishing, Wallingford, Oxfordshire, UK, pp. 318-35.

Gilbert, R. A., Kelly, W. J., Altermann, E., et al., 2017. Toward understanding phage: host interactions in the rumen; complete genome sequences of lytic phages infecting rumen bacteria. Front. Mirobiol. 8, 2340.

Gong, J. and Forsberg, C. W. 1993. Separation of outer and cytoplasmic membranes of *Fibrobacter succinogenes* and membrane and glycogen granule locations of glycanases and cellobiase. J. Bacteriol. 175 (21), 6810-21.

Goodman, R. P., Ghabrial, S. A., Fichorova, R. N., et al., 2011. Trichomonasvirus: a new genus of protozoan viruses in the family Totiviridae. Arch. Virol. 156 (1), 171-9.

Gregg, K., Kennedy, B. G. and Klieve, A. V. 1994. Cloning and DNA sequence analysis of the region containing AttP of the temperate phage φAR29 of *Prevotella ruminicola* AR29. Microbiology (Reading, Engl.) 140 (8), 2109-14.

Groleau, D. and Forsberg, C. W. 1981. Cellulolyticactivity of the rumen bacterium *Bacteroides succinogenes*. Can. J. Microbiol. 27 (5), 517-30.

Grybchuk, D., Akopyants, N. S., Kostygov, A. Y., et al., 2018. Viral discovery and diversity in trypanosomatid protozoa with a focus on relatives of the human parasite Leishmania. PNAS 115 (3), E506-15.

Guedon, G., Libante, V., Coluzzi, C., et al., 2017. The obscure world of integrative and mobilizable elements, highly widespread elementsthat pirate bacterial conjugative systems. Genes (Basel) 8 (11), 337.

Gutiérrez, D., Fernández, L., Rodríguez, A., et al., 2019. Role of bacteriophages in the implementation of a sustainable dairy chain. Front. Mirobiol. 10, 12-.

Harrison, P. W., Lower, R. P. J., Kim, N. K. D., et al., 2010. Introducing the bacterial 'chromid': not a chromosome, not a plasmid. Trends Microbiol. 18 (4), 141-8.

Hart, K. J., Yanez-Ruiz, D. R., Duval, S. M., et al., 2008. Plant extracts to manipulate rumen fermentation. Anim. Feed Sci. Tech. 147 (1-3), 8-35.

Hazen, T. H., Kaper, J. B., Nataro, J. P., et al., 2015. Comparative genomics provides insight into the diversity of the attaching and effacing *Escherichia coli* virulence plasmids. Infect. Immun. 83 (10), 4103-17.

Hefford, M. A., Teather, R. M. and Forster, R. J. 1993. The complete nucleotide sequence of a small cryptic plasmid from a rumen bacterium of the genus *Butyrivibrio*. Plasmid 29 (1), 63-9.

Heuer, H. and Smalla, K. 2012. Plasmids foster diversification and adaptation of bacterial populations in soil. FEMS Microbiol. Rev. 36 (6), 1083-104.

Hobbs, Z. and Abedon, S. T. 2016. Diversity of phage infection types and associated terminology: the problem with 'lytic or lysogenic'. FEMS Microbiol. Lett. 363 (7), fnw047.

Hoogenraad, N. J. and Hird, F. J. R. 1970. Electron-microscopic investigation of the flora of sheep alimentary tract. Aust. J. Biol. Sci. 23 (4), 793-808.

Hoogenraad, N. J., Hird, F. J. R., Holmes, I., et al., 1967. Bacteriophages in rumen contents of sheep. J. Gen. Virol. 1 (4), 575-6.

Howard-Varona, C., Hargreaves, K. R., Abedon, S. T., et al., 2017. Lysogeny in nature: mechanisms, impact and ecology of temperate phages. ISME J. 11 (7), 1511-20.

Hülter, N., Ilhan, J., Wein, T., et al., 2017. An evolutionary perspective on plasmid lifestyle modes. Curr. Opin. Microbiol. 38, 74-80.

Hungate, R. E. 1966. The Rumen and Its Microbes. Academic Press, New York.

Huws, S. A., Kim, E. J., Kingston-Smith, A. H., et al., 2009. Rumen protozoa are rich in polyunsaturated fatty acids due to the ingestion of chloroplasts. FEMS Microbiol. Ecol. 69 (3), 461-71.

Huws, S. A., Creevey, C. J., Oyama, L. B., et al., 2018. Addressing global ruminant agricultural challenges through understanding the rumen microbiome: past, present, and future. Front. Mirobiol. 9, 2161.

Iverson, W. G. and Millis, N. F. 1976a. Characterisation ofStreptococcus bovis bacteriophages. Can. J. Microbiol. 22 (6), 847-52.

Iverson, W. G. and Millis, N. F. 1976b. Lysogeny inStreptococcus bovis. Can. J. Microbiol. 22 (6), 853-7.

Jarvis, B. D. W. 1968. Lysis of viable rumen bacteria in bovine rumen fluid. Appl. Microbiol. 16 (5), 714-23.

Johnson, C. M. and Grossman, A. D. 2015. Integrative and conjugative elements (ICEs): what they do and how they work. Annu. Rev. Genet. 49, 577-601.

Johnson, T. J. and Nolan, L. K. 2009. Pathogenomics of the virulence plasmids of *Escherichia coli*. Microbiol. Mol. Biol. Rev. 73 (4), 750-74.

Jørgensen, T. S., Kiil, A. S., Hansen, M. A., et al., 2014. Current strategies for mobilome research.

Front. Mirobiol. 5, 750.

Jung, D. H., Chung, W. H., Seo, D. H., et al., 2018. Complete genome sequence of *Bifidobacterium choerinum* FMB-1, a resistant starch-degrading bacterium. J. Biotechnol. 274, 28-32.

Kaper, J. B., Nataro, J. P. and Mobley, H. L. T. 2004. Pathogenic *Escherichia coli*. Nat. Rev. Microbiol. 2 (2), 123-40.

Kav, A. B., Sasson, G., Jami, E., et al., 2012. Insights into the bovine rumen plasmidome. PNAS 109, 5452-7.

Keen, E. C. and Dantas, G. 2018. Close encounters of three kinds: bacteriophages, commensal bacteria, and host immunity. Trends Microbiol. 26 (11), 943-54.

Kelly, W. J., Leahy, S. C., Altermann, E., et al., 2010. The glycobiome of the rumen bacterium *Butyrivibrio proteoclasticus* B316 (T) highlights adaptation to a polysaccharide-rich environment. PLoS ONE 5 (8), e11942.

Kelly, W. J., Leahy, S. C., Li, D., et al., 2014. The complete genome sequence of the rumen methanogen *Methanobacterium formicicum* BRM9. Stand. Genom. Sci. 9, 15.

Kim, Y. S., Singh, A. P., Wi, S. G., et al., 2001. Cellulosomelike structures in ruminal cellulolytic bacterium *Ruminococcus albus* F-40 as revealed by electron microscopy. Asian Australas. J. Anim. Sci 14 (10), 1429-33.

Klieve, A. V. and Bauchop, T. 1988. Morphological diversity or ruminal bacteriophages from sheep and cattle. Appl. Environ. Microbiol. 54 (6), 1637-41.

Klieve, A. V. and Bauchop, T. 1991. Phage resistance and altered growth habit in a strain of *Streptococcus bovis*. FEMS Microbiol. Lett. 64 (2-3), 155-9.

Klieve, A. V. and Gilbert, R. A. 2005. Bacteriophage populations. In: Makkar, H. P. S. and McSweeney, C. S. (Eds), Methods in Gut Microbial Ecology for Ruminants. Springer Netherlands, Dordrecht, pp. 129-37.

Klieve, A. V. and Hegarty, R. 1999. Opportunities for biological control of ruminal methanogenesis. Aust. J. Agric. Res. 50 (8), 1315-9.

Klieve, A. V. and Swain, R. A. 1993. Estimating ruminal bacteriophage numbers using pulsed field gel electrophoresis and laser densitometry. Appl. Environ. Microbiol. 59 (7), 2299-303.

Klieve, A. V., Hudman, J. F. and Bauchop, T. 1989. Inducible bacteriophages from ruminal bacteria. Appl. Environ. Microbiol. 55 (6), 1630-4.

Klieve, A. V., Gregg, K. and Bauchop, T. 1991. Isolation andcharacteristics of lytic phages from *Bacteroides ruminicola ss brevis*. Curr. Microbiol. 23 (4), 183-7.

Klieve, A. V., Heck, G. L., Prance, M. A., et al., 1999. Genetic homogeneity and phage susceptibility of ruminal strains of *Streptococcus bovis* isolated in Australia. Lett. Appl. Microbiol. 29 (2), 108-12.

Klieve, A. V., Yokoyama, M. T., Forster, R. J., et al., 2005. Naturally occurring DNA transfer system associated with membrane vesicles in cellulolytic *Ruminococcus* spp. of ruminal origin. Appl. Environ. Microbiol. 71 (8), 4248-53.

Koonin, E. V. and Dolja, V. V. 2014. Virus world as an evolutionary network of viruses and capsidless selfish elements. Microbiol. Mol. Biol. Rev. 78 (2), 278-303.

Koonin, E. V. and Wolf, Y. I. 2008. Genomics of bacteria and archaea: the emerging dynamic view of the prokaryotic world. Nucleic Acids Res. 36 (21), 6688-719.

Koonin, E. V., Makarova, K. S. and Wolf, Y. I. 2017. Evolutionary genomics of defense systems in archaea and bacteria. Annu. Rev. Microbiol. 71, 233-61.

Krupovic, M., Prangishvili, D., Hendrix, R. W., et al., 2011. Genomics of bacterial and archaeal viruses: dynamics within the prokaryotic virosphere. Microbiol. Mol. Biol. Rev. 75 (4), 610-35.

Krupovic, M., Makarova, K. S., Wolf, Y. I., et al., 2019. Integrated mobile genetic elements in Thaumarchaeota. Environ. Microbiol. 21 (6), 2056-78.

Kuhl, S., Abedon, S. T. and Hyman, P. 2012. Diseasescaused by phages. In: Hyman, P. and Abedon, S. T. (Eds) Bacteriophages in Health and Disease. CAB International, Wallingford, Oxfordshire, UK, pp. 21-32.

Lang, C., Fruth, A., Holland, G., et al., 2018. Novel type of pilus associated with a Shiga-toxigenic *E. coli* hybrid pathovar conveys aggregative adherence and bacterial virulence. Emerg. Microbes Infect. 7 (1), 203.

Leahy, S. C., Kelly, W. J., Li, D., et al., 2013. The complete genome sequence of *Methanobrevibacter* sp. AbM4. STD Genomics Sci. 8 (2), 215-27.

Leedle, J. A. Z., Bryant, M. P. and Hespell, R. B. 1982. Diurnalvariations in bacterial numbers and fluid parameters in ruminal contents of animals fed low-or high-forage diets. Appl. Environ. Microbiol. 44 (2), 402-12.

Leng, R. A. and Nolan, J. V. 1984. Nitrogen metabolism in the rumen. J. Dairy Sci. 67 (5), 1072-89.

Letarov, A. V. 2012. Bacteriophages as a part of thehuman microbiome. In: Hyman, P. and Abedon, S. T. (Eds), Bacteriophages in Health and Disease. CAB International, Wallingford, Oxfordshire, UK, pp. 6-20.

Liu, M., Li, X., Xie, Y., et al., 2019. ICEberg 2.0: an updated database of bacterial integrative and conjugative elements. Nucleic Acids Res. 47 (D1), D660-5.

Mackie, R. I., McSweeney, C. S. and Klieve, A. V. 2002. Microbial ecology of the ovine rumen. In: Freer, M. and Dove, H. (Eds), Sheep Nutrition. CAB International, pp. 71-94.

May, T., Kocherginskaya, S. A., Mackie, R. I., et al., 1996. Complete nucleotide sequence of a cryptic plasmid, pbaw301, from the ruminai anaerobe *Ruminococcus flavefaciens* R13e2. FEMS Microbiol. Lett. 144 (2-3), 221-7.

Meile, L., Jenal, U., Studer, D., et al., 1989. Characterization of ψM1, a virulent phage of *Methanobacterium thermoautotrophicum* Marburg. Arch. Microbiol. 152 (2), 105-10.

Monk, A. B., Rees, C. D., Barrow, P., et al., 2010. Bacteriophage applications: where are we now? Lett. Appl. Microbiol. 51 (4), 363-9.

Morrison, M. 1996. Do rumen bacteria exchange geneticmaterial. J. Dairy Sci. 79 (8), 1476-86.

Namonyo, S., Wagacha, M., Maina, S., et al., 2018. A metagenomic study of the rumen virome in domestic caprids. Arch. Virol. 163 (12), 3415-9.

Nash, M. A., Smith, S. P., Fontes, C. M., et al., 2016. Single versus dual-binding conformations in cellulosomal cohesin-dockerin complexes. Curr. Opin. Struct. Biol. 40, 89-96.

Nibert, M. L., Woods, K. M., Upton, S. J., et al., 2009. Cryspovirus: a new genus of protozoan viruses in the family Partitiviridae. Arch. Virol. 154 (12), 1959-65.

Nolan, J. V. and Leng, R. A. 1972. Dynamic aspects of ammonia and urea metabolism in sheep. Br. J. Nutr. 27 (1), 177-94.

Nölling, J., Groffen, A. and de Vos, W. M. 1993. φF1 and φF3, two novel virulent, archaeal phages infecting different thermophilic strains of the genus *Methanobacterium*. J. Gen. Microbiol. 139 (10),

2511-6.

Ogata, K., Aminov, R. I., Nagamine, T., et al., 1996. Structural organization of pRAM4, a cryptic plasmid from *Prevotella ruminicola*. Plasmid 35 (2), 91-7.

Ohara, H., Miyagi, T., Kaneichi, K., et al., 1998. Structural analysis of a new cryptic plasmid pAR67 isolated from *Ruminococcus albus* AR67. Plasmid 39 (1), 84-8.

Orell, A., Frols, S. and Albers, S. V. 2013. Archaeal biofilms: the great unexplored. Annu. Rev. Microbiol. 67, 337-54.

Orpin, C. G. and Munn, E. A. 1974. The occurrence of bacteriophages in the rumen and their influence on rumen bacterial populations. Experientia 30 (9), 1018-20.

Overbeek, R., Begley, T., Butler, R. M., et al., 2005. The subsystems approach to genome annotation and its use in the project to annotate 1000 genomes. Nucleic Acids Res. 33 (17), 5691-702.

Paez-Espino, D., Eloe-Fadrosh, E. A., Pavlopoulos, G. A., et al., 2016. Uncovering Earth's virome. Nature 536 (7617), 425-30.

Paez-Espino, D., Chen, I. M. A., Palaniappan, K., et al., 2017. IMG/VR:a database of cultured and uncultured DNA Viruses and retroviruses. Nucl. Acids Res. 45 (D1), D457-65.

Palevich, N., Kelly, W. J., Leahy, S. C., Altermann, E., Rakonjac, J. and Attwood, G. T. 2017. The complete genome sequence of the rumen bacterium *Butyrivibrio hungatei* MB2003. Stand. Genom. Sci. 12, 72.

Parmar, N. R., Jakhesara, S. J., Mohapatra, A., et al., 2016. Rumen virome: an assessment of viral communities and their functions in the rumen of an Indian Buffalo. Curr. Sci. 111 (5), 919-25.

Paynter, M. J. B., Ewert, D. L. and Chalupa, W. 1969. Some morphological types of bacteriophages in bovine rumen contents. Appl. Microbiol. 18 (5), 942-3.

Penadés, J. R., Chen, J., Quiles-Puchalt, N., et al., 2015. Bacteriophage-mediated spread of bacterial virulence genes. Curr. Opin. Microbiol. 23, 171-8.

Piknova, M., Filova, M., Javorsky, P., et al., 2004. Different restriction and modification phenotypes in ruminal lactate-utilizing bacteria. FEMS Microbiol. Lett. 236 (1), 91-5.

Pope, P. B., Totsika, M., Aguirre de Carcer, D., et al., 2011. Muramidases found in the foregut microbiome of the tammar wallaby can direct cell aggregation and biofilm formation. ISME J. 5 (2), 341-50.

Pristas, P. and Piknova, M. 2005. Underrepresentation of short palindromes in *Selenomonas ruminantium* DNA:evidence for horizontal gene transfer of restriction and modification systems? Can. J. Microbiol. 51 (4), 315-8.

Raya, R. R., Oot, R. A., Moore-Maley, B., et al., 2011. Naturally resident and exogenously applied T4-like and T5-like bacteriophages can reduce *Escherichia coli* O157:H7 levels in sheep guts. Bacteriophage. 1 (1), 15-24.

Refardt, D. 2011. Within-host competition determines reproductive success of temperate bacteriophages. ISME J. 5 (9), 1451-60.

Ritchie, A. E., Robinson, I. M. and Allison, M. J. 1970. Rumen bacteriophage: survey of morphological types. In:Favard, P. (Ed.), Microscopie Electronique (vol. 3). SocieteFrancaise de Microscopie Electronique, Paris, pp. 333-4.

Ross, E. M., Petrovski, S., Moate, P. J., et al., 2013. Metagenomics of rumen bacteriophage from thirteen lactating dairy cattle. BMC Microbiol. 13, 242.

Roux, S. and Brum, J. R. 2019. A viral reckoning:viruses emerge as essential manipulators of global ecosystems. Environ. Microbiol. Rep. 11 (1), 3-8.

Rowan-Nash, A. D., Korry, B. J., Mylonakis, E., et al., 2019. Cross-domain and viral interactions in the microbiome. Microbiol. Mol. Biol. Rev. 83 (1), e00044-18.

Russell, J. B. and Rychlik, J. L. 2001. Factors that alter rumen microbial ecology. Science 292 (5519), 1119-22.

Russell, J. B. and Wilson, D. B. 1988. Potential opportunities and problems for genetically altered rumen microorganisms. J. Nutr. 118 (2), 271-9.

Sato, Y., Caston, J. R. and Suzuki, N. 2018. The biologicalattributes, genome architecture and packaging of diverse multi-component fungal viruses. Curr. Opin. Virol. 33, 55-65.

Schlegel, L., Grimont, F., Ageron, E., et al., 2003. Reappraisal of the taxonomy of the *Streptococcus bovis/Streptococcus equinus* complex and related species: description of *Streptococcus gallolyticus* subsp. gallolyticus subsp. nov., *S. gallolyticus* subsp. macedonicus subsp. nov. and *S. gallolyticus* subsp. Pasteurianus subsp. nov. Int. J. Syst. Evol. Microbiol. 53 (3), 631-45.

Seal, B. S., Drider, D., Oakley, B. B., et al., 2018. Microbialderived products as potential new antimicrobials. Vet. Res. 49 (1), 66.

Seshadri, R., Leahy, S. C., Attwood, G. T., et al., 2018. Cultivation and sequencing of rumen microbiome members from the Hungate1000 Collection. Nature Biotech. 36, 359-67.

Shapiro, J. W. and Putonti, C. 2018. Gene co-occurrence networks reflect bacteriophage ecology and evolution. mBio 9 (2), e01870-17.

Shkoporov, A. N. and Hill, C. 2019. Bacteriophages of the human gut: the 'Known Unknown' of the microbiome. Cell Host Microbe 25 (2), 195-209.

Sieradzki, E. T., Ignacio-Espinoza, J. C., Needham, D. M., et al., 2019. Dynamic marine viral infections and major contribution to photosynthetic processes shown by spatiotemporal picoplankton metatranscriptomes. Nat. Comm. 10 (1), 1169.

Solden, L. M., Naas, A. E., Roux, S., et al., 2018. Interspecies crossfeeding orchestrates carbon degradation in the rumen ecosystem. Nat. Microbiol. 3 (11), 1274-84.

Sorek, R., Kunin, V. and Hugenholtz, P. 2008. CRISPR — awidespread system that provides acquired resistance against phages in bacteria and archaea. Nat. Rev. Microbiol. 6 (3), 181-6.

Srinivasiah, S., Bhavsar, J., Thapar, K., et al., 2008. Phages across the biosphere: contrasts of viruses in soil and aquatic environments. Res. Microbiol. 159 (5), 349-57.

Stewart, R. D., Auffret, M. D., Warr, A., et al., 2018. Assembly of 913 microbial genomes from metagenomic sequencing of the cow rumen. Nat. Commun. 9 (1), 870.

Styriak, I., Kmet, V. and Spanova, A. 1989. Isolation and characterisation of two rumen *Streptococcus bovis* bacteriophages. Microbiol. 12, 317-22. Styriak, I., Galfi, P. and Kmet, V. 1991. Preliminary observations of interaction between bacteriophages and *Streptococcus bovis* bacteria on ruminal epithelium primoculture. Vet. Microbiol. 29 (3-4), 281-7.

Styriak, I., Spanova, A., Montagova, H., et al., 1994. Isolation and characterization of a new ruminal bacteriophage lytic to *Streptococcus bovis*. Curr. Microbiol. 28 (6), 355-8.

Styriak, I., Pristas, P. and Javorsky, P. 1998. Lack of surface receptors not restrictionmodification system determines F4 phage resistance in *Streptococcus bovis* II/1. Folia Microbiol. 43, 35-8.

Suen, G., Stevenson, D. M., Bruce, D. C., et al., 2011. Complete genome of the cellulolytic ruminal bacterium *Ruminococcus albus* 7. J. Bacteriol. 193 (19), 5574-5.

Suttle, C. A. 2007. Marine viruses--major players in the global ecosystem. Nat. Rev. Microbiol. 5 (10), 801-12.

Swain, R. A. 1999. Factors Affecting Bacteriophages of the Rumen Ecosystem. Ph. D. Thesis. University of New England.

Swain, R. A., Nolan, J. V. and Klieve, A. V. 1996. Natural variability and diurnal fluctuations within the bacteriophage population of the rumen. Appl. Environ. Microbiol. 62 (3), 994-7.

Tarakanov, B. V. 1972. The electron-microscopy examination of the microflora of reindeer rumen. Microbiology 41, 862-70 (translated from Mikrobiologiya 41, 862-70).

Tarakanov, B. V. 1974. Lysogenic cultures of *Streptococcus bovis* isolated from rumen of cattle and sheep. Microbiology 43, 375-7 (translated from Mikrobiologiya 43, 375-7).

Tarakanov, B. V. 1976. Biological properties of *Streptococcus bovis* bacteriophages isolated from lysogenic cultures and sheep rumen. Microbiology 45 (4), 695-700 (translated from Mikrobiologiya 45, 695-700).

Tarakanov, B. V. 1994. Regulation of microbial processes in the rumen by bacteriophages of *Streptococcus bovis*. Microbiology 63, 373-8 (translated from Mikrobiologiya 63, 657-67).

Teather, R. M. 1982. Isolation of plasmid DNA from *Butyrivibrio fibrisolvens*. Appl. Environ. Microbiol. 43 (2), 298-302.

Thomas, C. M., Thomson, N. R., Cerdeño-Tárraga, A. M., et al., 2017. Annotation of plasmid genes. Plasmid 91, 61-7.

Thurber, R. V., Haynes, M., Breitbart, M., et al., 2009. Laboratory procedures to generate viral metagenomes. Nat. Protoc. 4 (4), 470-83.

Toussaint, A. and Rice, P. A. 2017. Transposable phages, DNA reorganization and transfer. Curr. Opin. Microbiol. 38, 88-94.

Toyofuku, M., Nomura, N. and Eberl, L. 2019. Types and origins of bacterial membrane vesicles. Nat. Rev. Microbiol. 17 (1), 13-24.

Turnbull, L., Toyofuku, M., Hynen, A. L., et al., 2016. Explosive cell lysis as a mechanism for the biogenesis of bacterial membrane vesicles and biofilms. Nat. Comm. 7, 11220.

Volgers, C., Savelkoul, P. H. M. and Stassen, F. R. M. 2018. Gram-negative bacterial membrane vesicle release in response to the host-environment: different threats, same trick? Crit. Rev. Microbiol. 44 (3), 258-73.

Wang, B., Ma, M. P., Diao, Q. Y., et al., 2019. Saponin-induced shiftsin the rumen microbiome and metabolome of young cattle. Front. Microbiol. 10, 356,

Weidenbach, K., Nickel, L., Neve, H., et al., 2017. *Methanosarcina spherical* virus, a novel archaeal lytic virus targeting Methanosarcina strains. J. Virol. 91 (22), e00955-17.

Weinbauer, M. G. 2004. Ecology of prokaryotic viruses. FEMS Microbiol. Rev. 28 (2), 127-81.

Weinbauer, M. G. and Höfle, M. G. 1998. Significance of viral lysis and flagellate grazing as factors controlling bacterioplankton production in a eutrophic lake. Appl. Environ. Microbiol. 64 (2), 431-8.

Weinbauer, M. G., Wilhelm, S. W., Suttle, C. A., et al., 1999. Sunlight-induced DNA damage and resistance in natural viral communities. Aquat. Microb. Ecol. 17, 111-20.

Wells, J. E. and Russell, J. B. 1996. Why do many ruminal bacteria dies and lyse so quickly? J. Dairy Sci. 79 (8), 1487-95.

Willms, E., Johansson, H. J., Mäger, I., et al., 2016. Cells release subpopulations of exosomes with distinct molecular and biological properties. Sci. Rep. 6, 22519.

Willner, D., Thurber, R. V. and Rohwer, F. 2009. Metagenomic signatures of 86 microbial and viral metagenomes. Environ. Microbiol. 11 (7), 1752-66.

Winstanley, C., Langille, M. G. I., Fothergill, J. L., et al., 2009. Newly introduced genomic prophage islands are critical determinants of *in vivo* competitiveness in the Liverpool Epidemic Strain of *Pseudomonas aeruginosa*. Genome Res. 19 (1), 12-23.

Wolf, S., Fischer, M. A., Kupczok, A., et al., 2019. Characterization of the lytic archaeal virus Drs3 infecting *Methanobacterium formicicum*. Arch. Virol. 164 (3), 667-74.

Wommack, K. E. and Colwell, R. R. 2000. Virioplankton: viruses in aquatic ecosystems. Microbiol. Mol. Biol. Rev. 64 (1), 69-114.

Wommack, K. E., Hill, R. T., Muller, T. A., et al., 1996. Effects of sunlight on bacteriophage viability and structure. Appl. Environ. Microbiol. 62 (4), 1336-41.

Wong, C. M., Klieve, A. V., Hamdorf, B. J., et al., 2003. Family of shuttle vectors for ruminal Bacteroides. J. Mol. Microbiol. Biotechnol. 5 (2), 123-32.

Yutin, N., Kapitonov, V. V. and Koonin, E. V. 2015. A new family of hybrid virophages from an animal gut metagenome. Biol. Direct 10, 19.

Zabriskie, J. B. 1964. The role of temperate bacteriophage in the production of erythrogenic toxin by group A Streptococci. J. Exp. Med. 119, 761-80.

第 10 章 瘤胃上皮附着微生物

Mi Zhou, 加拿大阿尔伯塔大学；
刘军花, 南京农业大学, 中国；
Le Luo Guan, 加拿大阿尔伯塔大学

(焦金真译)

1 前言

在过去的几十年里，随着核酸测序方法的进步，人们对瘤胃内微生物组成和功能进行了大量的研究。这些研究旨在了解瘤胃微生物在调控宿主生理的作用，并基于此研发调控微生物活性的措施。瘤胃微生物通常分为三大类（图1）：游离于瘤胃液中的浮游微生物（约占30%）；附着于饲料颗粒的微生物（约占70%）；附着于瘤胃壁的微生物（通常也叫上皮附着微生物，占1%～5%）。

尽管上皮附着微生物只占整个瘤胃微生物群的很小部分，但它们在维持宿主生产性能和健康状况方面发挥着重要作用。瘤胃壁氧气含量丰富，血液中的尿素经瘤胃壁可以扩散至胃内，因此我们认为，上皮附着微生物的主要功能是水解尿素和清除氧气（Cheng 等，1979；Cheng 和 Wallace，1979）。上皮附着微生物可以通过消化从瘤胃上皮脱落的细胞，参与瘤胃上皮组织循环过程。它们还可以作为瘤胃内环境与宿主互作的媒介。因此，充分了解上皮附着微生物在宿主生理过程中的作用是很必要的。然而，由于技术短板，相较于瘤胃液与饲料颗粒中的瘤胃微生物，我们对瘤胃上皮附着微生物群落的了解有限。在接下来的章节中，我们将集中讨论瘤胃上皮附着微生物的最新进展、面临的挑战及未来的趋势。

2 瘤胃上皮附着微生物群落结构

瘤胃上皮附着微生物主要包括细菌和古菌，而最近的一篇论文揭示了瘤胃上皮附着微生物群落中也存在着原虫和真菌（Ishaq 等，2017）。其组成和多样性受多种因素的影响，包括品种、日粮组成、健康状况和地理位置等（例如 Petri 等，2013；Wetzels 等，2017；Bi 等，2018；Li 等，2019a）。瘤胃上皮附着微生物的变化可能会对反刍动物的产奶量、乳品质、饲料效率和甲烷产量等产生潜在影响（例如 Kong，2016；Mann 等，2018；Li 等，2019b；Neubauer 等，2019）。鉴于此，利用营养手段调控瘤胃上皮附着微生物可能有助于提高动物生产效率。本章将围绕以下几个方面对于瘤胃上皮附着微生物进行讨论：

(1) 瘤胃上皮附着微生物组的组成和功能；
(2) 瘤胃上皮附着微生物的定植；

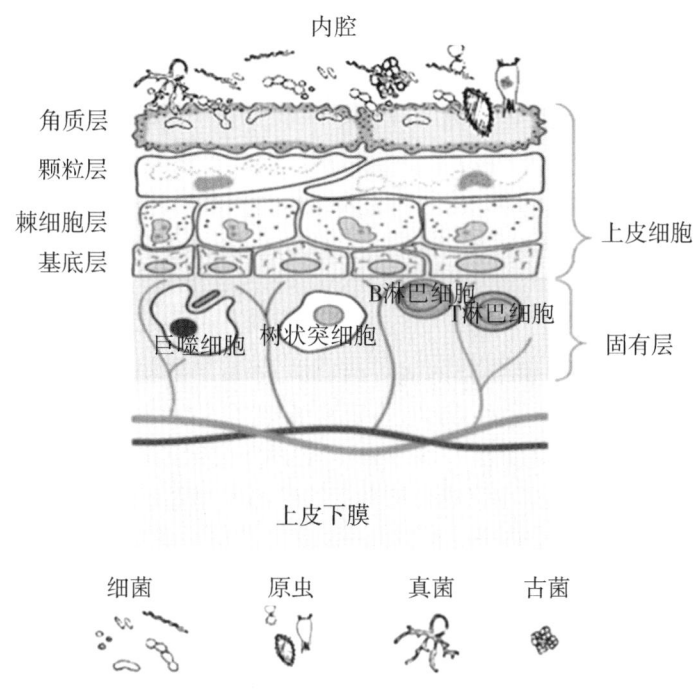

图 1　瘤胃上皮附着微生物区系图解（Garcia 等，2017）

（3）影响瘤胃上皮附着微生物的因素；
（4）瘤胃上皮附着微生物对反刍动物生产性能的影响。

2.1　瘤胃上皮附着细菌组成和功能

细菌是瘤胃上皮附着微生物群的主要菌群，约占瘤胃上皮附着微生物总 rRNA 的 90%（Nagaraja，2016）。基于培养的研究表明，绵羊瘤胃上皮每平方厘米平均附着 $1.2×10^7$ CFU 细菌，以丁酸弧菌和拟杆菌属为优势菌属（Dehority 和 Grubb，1981）。Wallace 等（1979）发现，采食干草的绵羊每克瘤胃上皮附着 $4.4×10^7$ 到 $2.2×10^8$ CFU 细菌。Mueller 等（1984b）研究表明，附着于瘤胃上皮的细菌群落与瘤胃内容物的细菌群落没有分类学上的差异。然而，Sadet 等（2007）基于 PCR 变性梯度凝胶电泳（PCR-DGGE）图谱，研究发现 5 月龄羔羊瘤胃上皮细菌群落与瘤胃内容物的细菌群落存在显著差异。通过 16S rDNA 基因克隆技术发现，食草的绵羊瘤胃内以厚壁菌门（Firmicutes）、拟杆菌门（Bacteroidetes）和变形菌门（Proteobacteria）为优势门，瘤胃上皮中的变形菌门比例（14%）高于瘤胃内容物中的比例（Sadet-Bourgeteau 等，2010）。基于 PCR-DEGG 和实时荧光定量 PCR 技术的研究进一步证实，厚壁菌门、变形菌门和拟杆菌门在肉牛瘤胃上皮附着微生物中占主导地位（Chen 等，2011）。

随着测序技术在山羊、肉牛和奶牛上的逐步应用，越来越多的证据显示，瘤胃上皮附着细菌的含量最高的是厚壁菌门，其次是拟杆菌门和变形菌门（Petri 等，2013；Mao 等，2015；Liu 等，2015，2016）。这些研究也进一步证实瘤胃上皮附着细菌中的变形菌门的相对丰度远高于瘤胃内容物的相对丰度，这可能与它们在反刍动物体内发挥的水解尿素、清除氧气和循环上皮组织的功能紧密相关。Wetzels 等（2017）利用 Illumina MiSeq 测序技术发现，当奶牛处于亚急性瘤胃酸中毒（SARA）的时候，变形菌门（相对丰度 45.2%）为优势门，其次是厚壁菌门（33.7%）和拟杆菌门（15.9%）。Petri 等（2013）基于 454 焦磷酸测序技术发现，在高谷物日粮条件下，肉牛瘤胃上皮微生物中 *Atopobium*、*Desulfocurvus*、*Fervidicola*、*Lactobacillus* 和 *Olsenella* 是最具优势的属。

Liu 等（2015）利用 454 测序技术研究发现，山羊瘤胃上皮的细菌群落在属水平上主要是丁酸弧菌属（占所有序列的 11.01%）、脱硫叶菌属（5.51%）、分支杆菌属（4.33%）和普氏菌属（4.01%）。Mao 等（2015）采用 Illumina 测序技术发现丁酸弧菌属是泌乳期奶牛瘤胃上皮中最主要的优势菌。在 56 日龄的绵羊中，瘤胃上皮附着微生物的优势菌包括普雷沃氏菌属（22.65%）、丁酸弧菌属（13.86%）、弯曲菌属（6.91%）、密螺旋体属（6.16%）、RC9 属（5.51%）和脱硫叶菌属（2.99%）（Liu 等，2017）。Wetzels 等（2017）报道，在饲喂高精料日粮的奶牛瘤胃上皮中，弯曲菌属（15.5%）、金氏菌属（7.8%）、脱硫叶菌属（4.7%）和单胞菌属（4.2%）是主要的优势菌。这些结果表明，瘤胃上皮附着微生物的群落组成可能受动物品种、日粮组成和地理位置的影响。同样值得注意的是，不同的测序技术也可能导致结果的差异。确切地了解微生物群落的组成以及它们是如何受日粮、宿主和环境的影响，需要进一步规范相关的研究方法。

尽管早期的研究表明附着在瘤胃上皮的细菌参与水解尿素、清除氧气和瘤胃上皮组织循环（Cheng 等，1979；Cheng 和 Wallace，1979），但是其具体功能在很大程度上是未知的。基于瘤胃上皮附着细菌的组成和功能预测的研究表明，其可能参与 VFA 吸收（Chen 等，2011）、上皮细胞增殖、疾病发生、蛋白质代谢和能量代谢等过程（Mao 等，2015；Liu 等，2017；Lin 等，2019）。但是，目前仍然缺少直接的证据表明反刍动物瘤胃上皮附着微生物的具体功能。在未来，需要通过培养技术、宏基因组和宏转录组等技术来更深层次地探究瘤胃上皮附着细菌具体的功能。

2.2 瘤胃上皮附着古菌的组成与功能

古菌是一种定植在瘤胃内的特异性微生物。甲烷短杆菌（61.6%）、甲烷微菌属（14.9%）和 Methanomassiliicoccales（以前称为瘤胃 C 簇，15.8%）被认为是瘤胃内产甲烷的三大优势菌（Nagaraja，2016）。大多数产甲烷菌利用 H_2 和 CO_2 合成 CH_4，而甲基营养型产甲烷菌（如 Methanomassiliicoccales）主要利用甲基化合物和 H_2 合成 CH_4。与细菌生态系统相比，瘤胃上皮古菌相关的信息是非常有限的。Janssen 和 Kirs（2008）发现在瘤胃上皮中古菌占所有微生物 rRNA 的 3%~5%。Shin 等（2004）应用 16S rDNA 测序，发现在韩国奶牛瘤胃中，甲烷微菌科是瘤胃上皮古菌中的优势菌，占比达 95%；而甲烷短杆菌科是瘤胃内容物中的优势菌。Pei 等（2010）利用 16S rRNA 测序技术比较了中国济南地区肉牛瘤胃液相、固相和附着于瘤胃上皮的产甲烷菌的多样性和丰富度。研究发现甲烷短杆菌属是附着于饲料颗粒上（77.2%）和瘤胃上皮（77.8%）的优势菌。通过实时荧光定量 PCR，他们还报道了瘤胃上皮附着的产甲烷菌群丰度显著高于附着在饲料颗粒上的，并认为这可能是由于瘤胃上皮周转速度较慢引起的（Pei 等，2010）。然而，其他的研究表明与瘤胃内容物和瘤胃液中相比，奶牛瘤胃上皮相关的古菌群落多样性较低（De Mulder 等，2017），且数量较少（Liu 等，2016；Wang 等，2017）。瘤胃古菌多样性也因宿主种类而异。在济南地区的肉牛瘤胃上皮中古菌主要为甲烷短杆菌（Pei 等，2010）。Shin 等（2004）发现韩国的奶牛瘤胃中古菌主要为甲烷微菌。Wang 等（2017）发现在黑山羊的瘤胃上皮中古菌主要为甲烷球形菌属（大约 20%）。有研究认为古菌群落中 *Mbb. gottschalkii* 与 *Mbb. ruminantium* 的比例可能与宿主甲烷的排放量紧密相关（Danielsson 等，2017），但还未在其他研究中被证实为瘤胃上皮古菌群落的关键特征。附在瘤胃上皮的古菌的组成和功能仍不甚清楚。

3 瘤胃上皮附着微生物的定植

如前所述，成年反刍动物的瘤胃上皮拥有一个复杂的微生物系统，在出生时瘤胃是一个无菌的环境（Nagaraja，2016）。幼龄反刍动物瘤胃的发育过程中很重要的一个过程就是瘤胃内容物和上皮附着微生物的定植与建立（Yanez-Ruiz 等，2015）。新生反刍动物的瘤胃微生物定植遵循一个典型的生长顺序，即细菌在出生后立即在瘤胃液中定植，然后在接下来的 36~48 h 内定植于瘤胃上皮（Yanez-Ruiz 等，2015）。定植的液相细菌促进了随后的真菌定植，其次是瘤胃液中的原虫（Yanez-Ruiz 等，2015）。瘤胃中液相饲料颗粒附着和上皮附着的微生物定植最终形成了一个复杂的系统。一旦形成，与内容物相关的微生物通常是稳定的，只有在营养物质发生变化时才会发生改变（Stewart 等，1988），而瘤胃上皮附着微生物群落的多样性似乎随着年龄的增长而变化（Mueller 等，1984a；Rieu 等，1990）。生命早期是调控瘤胃微生物群，包括附着在瘤胃上皮的微生物群的关键阶段。

许多研究表明，瘤胃上皮的细菌群落不同于瘤胃内容物的细菌群落。它们可能发挥一些特定的功能，并影响幼龄反刍动物瘤胃的发育和免疫（主要是屏障功能）（Malmuthuge 等，2012，2014）。鉴于瘤胃上皮微生物在宿主代谢、天然免疫和瘤胃发育中发挥的特殊功能，在幼龄阶段瘤胃上皮附着微生物的定植应该值得更多的关注。

Mueller 等（1984a）利用扫描电子显微镜技术，在 1~10 周龄的羔羊瘤胃上皮中鉴定到了 24 种与上皮附着细菌相关的形态类型，但是在较大一些的羔羊和成年绵羊中仅有 7 种类型。Malmuthuge 等（2014）发现在断奶前 3 周的雄性奶牛牛犊的瘤胃上皮细菌群落与瘤胃内容物中的细菌群落存在显著差异。研究发现，与瘤胃内容物细菌群落相比，瘤胃上皮中的普雷沃氏菌属的丰度更低，拟杆菌属的丰度更高。瘤胃上皮细菌群落的发育与断奶前犊牛先天免疫相关的一些关键基因的 mRNA 表达密切相关（Malmuthuge 等，2014）。这些表明应该更多地关注幼龄反刍动物瘤胃上皮细菌群落的定植过程。

Jiao 等（2015）基于 16S rRNA 的 Miseq 测序技术发现，瘤胃附着上皮细菌的组成和多样性在山羊的发育期（0~70 d）发生变化。他们发现，在山羊瘤胃发育的早期，厚壁菌门、拟杆菌门和变形菌门是瘤胃上皮附着细菌群落的主要优势菌，厚壁菌门和拟杆菌门的比例随着年龄的增长而增加，而变形菌门的比例随着年龄的增长而减少。在属水平上，来自母体阴道、皮肤、乳房或者环境中的大肠杆菌（80.79%）是新生羔羊瘤胃上皮的优势菌属。在 42 日龄和 70 日龄，丁酸弧菌属和弯曲菌属变成了瘤胃上皮的优势菌，显著地高于瘤胃内容物中的丰度，这表明这两个属在引入固体饲料后可能在宿主功能中发挥关键作用。除此之外，他们还报道了丁酸弧菌属、弯曲菌属和脱硫杆菌属可能参与羔羊瘤胃形态和功能的发育过程。该研究表明提供固体的饲料是影响羔羊瘤胃上皮附着微生物群落定植和建立的主要因素。了解掌握这3个属细菌的具体功能和与宿主之间的相互作用需要更进一步地探究。

Liu等（2015）探究了添加固体开食料对湖羊瘤胃上皮细菌群落建立的影响，发现与只喂养羊奶相比，添加固体开食料显著改变了瘤胃上皮细菌的结构和组成。脱硫杆菌属、霍华德菌属、拟杆菌属、共营养球菌属、椎体杆菌属、双歧杆菌属和巨球菌的相对丰度较高，而弯曲杆菌属和斯诺格拉菌属的相对丰度较低。所有的这些研究均表明，瘤胃上皮附着细菌的定植与建立都与年龄相关（在2月龄建立）。他们还指出，许多因素（母体影响、单独饲养、液体或固体饲料）都可能影响瘤胃上皮附着微生物群落的定植和建立。值得注意的是，年龄的影响受饮食的影响，瘤胃上皮的微生物建立可能是饮食和动物的生长相互作用的结果。这些是未来早期干预需要考虑的重要因素。

4 影响瘤胃上皮附着微生物的因素

长期以来，饮食一直被认为是改变瘤胃内容物微生物群落生态的重要驱动因素之一。高饲草和高谷物日粮可以显著改变瘤胃微生物群落，这一变化通常认为是对不同日粮类型的适应。

近期一些研究探讨了日粮组成对上皮附着微生物群落的影响，但结果不尽相同。AlZahal等（2017）研究表明，当奶牛从高饲草日粮转变为高谷物日粮时，瘤胃上皮附着微生物的多样性显著降低，纤维杆菌属（由5.3%降至1.3%）和瘤胃球菌科（由7.5%降至2.3%）的丰度降低，普雷沃氏菌属的丰度增加（由16.1%降至19.2%）。Wetzels等（2017）发现，当日粮中逐渐增加精料的比例时，弯曲杆菌属（由20.05%降至15.15%）和金氏杆菌属（由12.46%降至7.31%）两个优势菌的丰度显著降低。Petri等（2018）发现，当用优质的干草替代精料饲喂荷斯坦奶牛，优质干草使得瘤胃上皮的优势菌从厚壁菌门转变成了变形门菌，同时降低了微生物的多样性。他们认为，这种变化是由于弯曲杆菌的增殖（大约5倍），这可能与优质干草中的蛋白质和非蛋白氮含量增加有关。这两项研究都强调了弯曲杆菌属在瘤胃上皮附着细菌群落中的潜在重要作用，值得进一步研究。

无论宿主物种、地理范围和饮食的不同，瘤胃固相和液相微生物的种群都相对稳定（Henderson等，2015）。然而，上皮附着微生物菌群因宿主物种而不同。据报道，与内容物微生物相比，肉牛上皮微生物的多样性较低（Li等，2012）。这与奶牛瘤胃上皮微生物的多样性高于内容物微生物形成对比（Malmutjug等，2012；AlZahal等，2017）。AlZahal等（2017）认为，奶牛瘤胃上皮附着微生物的高多样性可能表明它们参与了更为复杂的生物过程。

不同反刍动物种间的上皮附着微生物在主要优势菌方面存在差异。在门水平上，变形杆菌是山羊（Jiao等，2015）和肉牛（Kong等，2016）的主要优势菌。拟杆菌门是成年山羊的主要优势菌（Zhang 等，2017a，b）。在奶牛从出生到成年的过程中，观察到拟杆菌门和厚壁菌门替换了变形菌门（Jami 等，2013）。在探究瘤胃上皮微生物种群时，也发现了个体交互差异。McCann 等（2016）观察到瘤胃酸中毒引起的上皮微生物的跨个体变化。研究发现患有瘤胃酸中毒的牛的上皮微生物中，金氏菌和固氮弧菌数量减少，反刍杆菌数量增加（Wetzels 等，2016）。有人认为，这种上皮微生物群的改变可能对瘤胃功能产生负面影响，导致反刍动物瘤胃 pH 值发生更大幅度的下降，酸中毒的严重程度增加。De Mulder 等（2017）还报道，在泌乳中期的荷斯坦-黑白花奶牛中，上皮微生物群的 α 多样性和 β 多样性指数的个体差异高于液相和固相微生物。这些发现表明，与瘤胃内容物微生物种群相比，宿主遗传因素可能对与宿主密切相关的上皮微生物群有更强的影响。

对非反刍动物的研究表明，不同的生长促进剂或有毒的生长抑制剂是影响上皮微生物定植的关键因素（Schluter 和 Foster，2012）。但是，宿主效应对微生物的影响通常被其他因素（如日粮和环境）所掩盖。为了更好地考虑宿主效应，一个潜在的研究方向是将上皮微生物种群与宿主遗传标记（如从基因封闭的群体中识别的 SNP）联系起来。成对或者三个一组对反刍动物进行的研究也可能是阐明宿主效应的另一种选择。

根据瘤胃普查项目，瘤胃内容物微生物群落在地理上有所不同（Henderson 等，2015）。同一种牛的瘤胃上皮微生物群落也因地理区域而异。以荷斯坦奶牛为例，Liu 等（2016）的研究发现，厚壁菌门在中国的荷斯坦奶牛上皮附着微生物中占 80% 以上。然而，Petri 等（2018）的一项研究发现，厚壁菌门在奥地利的牛瘤胃上皮群落中所占比例不超过 50%。这些研究中的总体上皮附着微生物特征也有所不同。虽然宿主物种相同（荷斯坦奶牛），处于相似的哺乳期，并且提供了具有相似能量值的日粮，但微生物图谱的这种差异与之前报告的微生物群的差异是符合的（Henderson 等，2015）。地理位置的差异可能归因于复杂的综合效应，如温度、周围环境、不同的日粮组成和不同的饲养管理。此外，不同实验室的取样技术和样品处理方式不同，也可能导致多项研究中报告不同的结果。表 1 总结了从这些研究中观察到的瘤胃上皮附着微生物物种分类学组成。

5 瘤胃上皮附着微生物对反刍动物生产性能的影响

5.1 乳成分

乳成分与瘤胃发酵和瘤胃上皮吸收的有效养分有关（Matthews 等，2019）。瘤胃中的微生物氮代谢对于向宿主提供用于产奶的微生物蛋白质是至关重要的（Tadele 和 Amha，2015）。挥发性脂肪酸（VFAs）也会影响乳脂成分（Hurtaud 等，1993）。大多数研究探讨了瘤胃微生物与牛奶成分之间的关系，主要集中在固相和液相微生物（Jami 等，2014；Bainbridge 等，2016；Zhang 等，2017a，b）。结果发现，通过调节瘤胃微生物，使酪蛋白或乳清蛋白相对于瘤胃中的非蛋白氮（N）占有更大比例，最终可以提高乳蛋白产量（Tadele 和 Amha，2015）。

表 1 不同研究、不同方法揭示瘤胃上皮附着微生物群落的综述

研究	物种	日粮	生长阶段	方法	主要种群[a]
Chen 等 (2011)	肉用培育母牛	玉米对照组：97% 干草，3% 精料；处理组：由 60∶40、40∶60、25∶75、15∶85、8∶92 粗精比过渡	8 月龄	16S rRNA 变性梯度凝胶电泳	细菌：变形菌门、厚壁菌门、拟杆菌门
Petri 等 (2013)	安格斯母牛	牧草：95% 草干草，5% 添加物；混合牧食：60% 大麦青贮，30% 大麦粒，10% 添加物；高谷物：9% 大麦青贮，81% 大麦粒，10% 添加物	体重：(308±35) kg	16S rRNA 变性梯度凝胶电泳	细菌：厚壁菌门、变形菌门、拟杆菌门、放线菌门
Liu 等 (2015)	波尔×长江三角洲白山羊	干草：81% 羊草，15% 苜蓿；高谷物：30% 羊草，45% 玉米粉，20% 小麦粉，1.1% 大豆	2~3 年	16S rRNA 基因 MiSeq 测序	细菌：厚壁菌门、变形菌门、拟杆菌门、放线菌门、螺旋菌门
Jiao 等 (2015)	山羊	1~40 d：0.5L 羊奶，0.12 kg 开食料；40~70 d：每顿 0.06 kg 鲜草，0.17 kg 开食料，每顿 0.04 kg 鲜草	1~70 日龄	16S rRNA 基因 MiSeq 测序	细菌：变形菌门、厚壁菌门、拟杆菌门、放线菌门、梭杆菌门
Wetzels 等 (2017)	荷斯坦奶牛	玉米对照组：50% 青贮料，50% 二次割草甸干草；处理组：20% 青贮饲料，20% 二次割草甸干草，60% 精料	体重：(710±118) kg	16S rRNA 基因 MiSeq 测序	细菌：变形菌门、厚壁菌门、拟杆菌门、互养菌门、迷踪菌门
Abecia 等 (2014b)	山羊	苜蓿自由采食；NAT：随母生活；ART：代乳品	1, 3, 7, 14, 21, 28 日龄	16S rRNA 焦磷酸测序	细菌：变形菌门、厚壁菌门、蓝细菌门
De Mulder 等 (2017)	荷斯坦-弗里斯奶牛	粗精比 70∶30	泌乳中期	16S rRNA 基因 MiSeq 测序	细菌：厚壁菌门、变形菌门、拟杆菌门、纤维杆菌、放线菌门、螺旋菌门、TM7；古菌：产甲烷菌

（续表）

研究	物种	日粮	生长阶段	方法	主要种群[a]
AlZahal 等（2017）	荷斯坦奶牛	高粗料：粗精比为77：23；高谷物：粗精比为49：51	泌乳期	16S rRNA 基因 MiSeq 测序	细菌：拟杆菌门，厚壁菌门，变形菌门，软壁菌门，SR1，螺旋菌门，纤维杆菌门，蓝细菌门
Mann 等（2018）	荷斯坦奶牛	玉米对照组：50%青贮料，50%二次割草甸干草；处理组：60%精料，40%粗料	非泌乳（3～4胎）	宏转录组	细菌：变形菌门，厚壁菌门，拟杆菌门，螺旋菌门，放线菌门；古菌：甲烷球菌，甲烷短杆菌
Frutos 等（2018）	美利奴羔羊	43.3%大麦，15.0%玉米，23.7%豆粕，15.0%大麦秸秆，3.0%维生素-矿物质预混料	育肥期	实时定量 PCR	细菌：报道了总细菌。普雷沃氏菌，反刍月形单胞菌，产甲烷菌
Li 等（2019b）	荷斯坦公牛	玉米组织组：35.3%淀粉，25.3%中性洗涤纤维；玉米颗粒组：42.7%淀粉，15.1%中性洗涤纤维	1～17周	宏转录组	细菌：变形菌门，厚壁菌门，放线菌门，拟杆菌门，梭杆菌门
Lin 等（2019）	湖羊	玉米对照组：仅母乳；处理组：每日两次喂食开食料	10～56日龄	rRNA 基因 MiSeq TruSeq 宏基因组测序	细菌：拟杆菌门，厚壁菌门，放线菌门，变形菌门，螺旋菌门，软皮菌门；原生动物：内纤毛虫，瘤胃纤毛虫，未分类的毛口虫，头毛虫，双毛虫，等毛虫
Neubauer 等（2019）	荷斯坦奶牛	玉米对照组：50：50 干草：青贮牧草；处理组：精料65%，粗料35%	非泌乳期	MiSeq 测序	细菌：变形菌门，厚壁菌门，放线菌门，互养菌门

[a] 根据每个研究中的相对丰度列出菌种类型，仅呈现丰度高于1%的微生物种群。

一些瘤胃上皮附着细菌（厚壁菌门、变形菌门、放线菌门和拟杆菌门）具有脲酶基因（Jin 等，2017；Mann 等，2018），并可能在瘤胃氮代谢中发挥作用。从瘤胃上皮微生物群落中观察到的独特的尿素分解菌和脲酶基因可能暗示着其与内容物微生物群落的功能差异（Jin 等，2017）。目前，上皮尿素分解菌的特异性功能、脲酶活性以及上皮附着微生物在多大程度上影响瘤胃氮代谢尚不清楚。宏基因组学和转录组学等先进的组学方法可能有助于揭示上皮附着微生物在整体氮代谢中的功能，但这依赖于准确分离瘤胃上皮附着微生物。

除了乳蛋白，乳脂也是牛奶生产中另一个经济上重要的品质性状。牛羊等的乳房组织脂肪酸从头合依赖于瘤胃 VFAs 的转运（Palmquist，2006）。VFA 的吸收和转运与瘤胃上皮生理，如上皮细胞表面和转运蛋白的活性密切相关（O'Shea 等，2016）。有证据表明瘤胃上皮细菌丰度与瘤胃总 VFAs 浓度有关（Chen 等，2012），提示该群落可能在调节 VFA 吸收方面发挥作用。探索活性转运蛋白的作用以及运用宏转录组技术鉴定相应 VFAs 的活性转运途径及其与瘤胃上皮附着活性微生物群的关系，应该引起更进一步的重视（Mann 等，2018）。

5.2 饲料效率

据报道，瘤胃上皮附着微生物也是影响宿主饲料效率的一个因素。Kong（2016）比较了不同剩余采食量（RFI）肉牛的上皮微生物，RFI 是衡量肉牛饲料转化率的常用指标（低 RFI 表明效率较高）。研究发现，弯曲杆菌科和奈瑟氏菌科在低剩余采食量组（L-RFI）的上皮中的细胞活性明显高于高剩余采食量组（H-RFI）的上皮（分别为 5.8% vs 2.4%，10.8% vs 1.0%）。研究表明，这两种微生物所具有的较高的氧化酶活性可能使 L-RFI 动物具有较高的氧清除能力，从而维持较好的厌氧环境进行微生物厌氧发酵，进而提高能量利用效率。Liang 等（2017）研究了湖羊羔羊育肥过程中的上皮附着微生物菌群，发现 H-RFI 动物中，溶纤维丁酸弧菌和大肠杆菌的比例较高。据报道，较高丰度的溶纤维丁酸弧菌，与较低的丙酸浓度和较高的乙丙比相关联，这与较低的饲料转化率相吻合。结果表明，大肠杆菌数量越高，H-RFI 动物瘤胃内 pH 值和 VFA 浓度波动越大（Liang 等，2017）。有研究表明，上皮附着微生物对宿主饲料转化率的影响可能依赖于除饲料降解的其他功能。应用宏基因组学或宏转录组的方法来探明微生物活性途径将有助于更好地解析可能影响宿主饲料转化率的具体机制以及上皮微生物在多大程度上影响牛的饲料效率。

5.3 甲烷排放

研究发现瘤胃上皮的氧浓度远高于瘤胃管腔内的氧浓度。在瘤胃上皮微生物群中通常有大量厌氧古菌。浮游生物群落的 *Mbb. gottschalkii*：*Mbb. ruminantium* 比例被认为与宿主甲烷排放呈正相关（Danielsson 等，2017），但这并没有得到其他研究的验证。

虽然没有直接证据表明瘤胃上皮微生物群影响甲烷排放，但可能存在值得研究的联系，例如瘤胃上皮微生物群在调节管腔［H］利用率中的作用。如果发现标志性的上皮微生物与管腔［H］浓度间接地与甲烷生成有关，我们可能能够开发出新的甲烷减排方式，即除内容物微生物群外，还可以专注于调控这些上皮微生物。

有间接证据表明宿主对古菌群落有影响。在瘤胃内容物移植的研究中，Zhou（2018）等发现古菌群落不受捐赠者表型的影响，大多数动物的古菌群落都恢复了原始状态。他们认为瘤胃上皮尤其是古菌可能具有重建微生物群的内在潜力。Roehe 等（2016）报道了宿主对甲烷排放的影响，根据宿主后代群体对瘤胃古菌种群和宿主甲烷排放量进行了排名。他们还提出，瘤胃古菌的丰度受到宿主基因的控制。Difford 等（2018）研究了奶牛甲烷排放的相关因素，发现宿主基因是导致这一过程的首要因素，其次是细菌和古菌。作者提出了通过基因选择动物来调节瘤胃微生物群的潜力。这些研究强调有必要单独研究瘤胃上皮古菌，以制定更好的甲烷减排策略。

6 挑战与展望

6.1 微生物功能的深入解析

尽管近年来的研究已经应用基于多组学的方法揭示了瘤胃微生物群的组成和功能（例如 Stewart 等，2018；Li 等，2019），但对瘤胃上皮附着微生物的认识仍然十分有限。Mann 等（2018）采用宏转录组方法研究了瘤胃上皮细菌的基因表达和功能潜力。他们发现，半乳糖、淀粉、蔗糖和能量代谢相关基因在上皮微生物中高度表达。这一结果表明瘤胃上皮微生物通过跨瘤胃上皮运输提供宿主相关代谢产物的积极作用。他们首次报道了牛瘤胃上皮微生物的固氮作用，还观察到与氧化应激有关的基因的高水平表达。他们还确定了上皮微生物中活跃的古菌和真菌群落，尽管它们的功能仍不甚清楚。

毫无疑问，利用多组学方法可以获得更多关于瘤胃上皮附着微生物的描述性数据，但必须进行准确的分析，以鉴定微生物群落的"真正和关键"成员。样本收集方法（Paz等，2016）、DNA/RNA 提取步骤（Villegas-Rivera 等，2013；Henderson 等，2015）、测序方法和用于数据处理的生物信息模块（Neves 等，2017）等因素都会影响识别瘤胃上皮附着微生物菌群的结果。此外，用于数据解释的数据库也至关重要。由 Seshadri 等（2018）组装的宏基因组是从 Hungate1000 数据库中检索到的，该数据库对瘤胃微生物高度专门化，并宣称涵盖了约 3/4 的属级微生物分类群（JGI Hungate 库）。然而，瘤胃内容物微生物和上皮微生物之间存在成分和功能差异。只有将序列片段正确地匹配到真正的表型才能定义准确的微生物功能，但目前可用的工具还无法实现这一结果。

最近，Wilkins 等（2019）开发了一种更有效的检索宏基因组组装基因组（MAGs）的方法。这些方法为瘤胃微生物学家提供了一个新的方向，以确定瘤胃上皮微生物 MAGs，从而为以后的研究提供更好的分辨率。除了鉴定微生物的组成之外，研究微生物的代谢产物可能会直接证明微生物活动如何影响宿主的性能。尽管目前还没有基于代谢物的瘤胃微生物研究，但基于蛋白质组的瘤胃液微生物研究（Hart 等，2018）为今后从代谢物/酶/蛋白质等方面研究瘤胃微生物提供了方向，使我们能够完整描述其功能及其对宿主的影响。不过，值得注意的是，微生物代谢产物随着时间的推移是动态变化的。从这些样品中获得的代谢组是否具有代表性，是否足以反映瘤胃内环境的真实情况仍是一个问题。

6.2 对宿主—微生物互作的深入理解

由于与宿主的距离较近，瘤胃上皮微生物被认为与宿主有密切的相互作用。Chen 等（2012）报道，在耐酸中毒肉牛的瘤胃上皮细菌数量与 *TLR*4 基因表达量呈正相关，但在酸中毒敏感动物中未见这种关系。Liu 等，（2015）发现 *TLR*2 基因的表达与 10 种上皮细菌类群有关，而 *TLR*4 基因表达与厌氧绳菌科的 1 个类群有关。这些结果提示，上皮细菌的丰度可能刺激宿主基因的表达。今后的研究应探讨宿主基因表达与上皮微生物之间的关系，以更好地阐明微生物定植如何影响宿主。

脂多糖（LPS）存在于革兰氏阴性菌（GNB）的外层（Wang 和 Quinn，2010）。瘤胃液内 GNB 的裂解与瘤胃 LPS 水平的升高有关（Nagaraja 等，1978；Gozho 等，2005，2007）。瘤胃上皮 GNB 释放的 LPS 被认为是具有强烈促炎潜能的内毒素，从而引起瘤胃组织损伤（Steele 等，2011），改变 TLRs 的表达和瘤胃上皮紧密连接结构（Liu 等，2015；McCann 等，2016）。

在这些瘤胃上皮相关的 GNB 中，坏死梭形杆菌（*Fusobacterium necrosisphorum*）是目前已知的最被了解的菌种之一，是一种常见于瘤胃上皮的耐氧厌氧细菌，参与饲料的降解、乳酸和上皮蛋白的代谢（Li 等，2019a）。在正常瘤胃条件下，坏死梭形杆菌主要作为饲料降解菌发挥作用。然而，在患有 SARA 的动物中，它成为一种条件致病菌（Berg 和 Scanlan，1982），在受到角质化不全和宿主长期 SARA 条件影响的瘤胃上皮黏膜上增殖（Okada 等，1999；Takayama 等，2000）。它还可以穿透瘤胃上皮，转移到血液中，侵入肝脏，引起脓肿（Nagaraja 和 Titgemeyer，2007；Steele 等，2009；Tadepalli 等，2009）。了解共生上皮微生物与宿主之间的相互作用至关重要，这将为今后在高谷物饲料和反刍动物代谢紊乱条件下预防坏死梭形杆菌的发病机制提供依据。

在固相和液相微生物方面，宿主—微生物相互作用已被广泛研究。这些相互作用有助于解释肉牛（Zhou 等，2009；Hernandez-Sanabria 等，2012；Myer 等，2015）和奶牛（Jami 等，2014；Jewell 等，2015）宿主饲料利用率、甲烷排放量（Carberry 等，2014；Shi 等，2014）和产奶量（Jami 等，2014；Jewell 等，2015）的变化。然而，上皮微生物群落在解释宿主表型中的作用还不是很清楚。直到最近，Mann 等，（2018）研究并报道了在上皮微生物中一系列微生物基因的表达。这些活性基因的产物可能作为未来识别其在影响上皮细胞形态、分子运输、信号传导等方面作用的靶点，为微生物-宿主互作提供直接证据。

6.3 高效的调控策略

最近有许多研究通过调控瘤胃微生物来提高发酵效率，营造更健康的瘤胃环境。这种在成年期间改变瘤胃微生物的研究并没有表现出一致的、长期的效果。Abecia 等（2014a，b）建议将早期瘤胃内容物微生物群作为减少甲烷排放的长期效果的目标。由于上皮微生物暴露在更强的宿主相关选择压力下，与其他生态位瘤胃微生物在功能上存在差异，进一步研究宿主对上皮微生物的调控作用可能有助于提高早期生命干预成功的机会。

需要对已有的不同研究产生的多组学数据集进行 Meta 分析，为新靶点如关键谱系类型、关键通路、关键酶等提供线索。目前，我们正在努力将所有不同类型的数据合理地分配到一个单一的分析平台，以便更好地解释来自多个研究的代谢组，并提供更准确的研究间比较。益生菌的应用也可能是未来调控上皮微生物的一个方向，这需要更多的试验证据来支持这种方法。

6.4　样本收集和数据处理的改进

虽然多组学技术可以对采集到的样品的核酸（DNA 和 RNA）和代谢产物提供高分辨率的分析，但研究上皮群落的主要障碍是鉴定真正的上皮微生物菌群。与通过管道或瘤胃插管方法容易获得瘤胃液和消化液样本不同，上皮样本相对较难获得。Van Niekerk 等（2018）开发了一种通过内镜采集瘤胃组织的方法，该方法允许在不插入瘤胃管的情况下重复采样。这种方法使我们能够更高效、更人性化地采集幼年和成年反刍动物的瘤胃组织样本，同时也可以研究同一动物接受不同处理后，其上皮微生物菌群的变化。

从瘤胃上皮组织中去除非黏附微生物也是样品处理的技术问题。采取适当的冲洗步骤，去除上皮样本中的任何非粘附物质是必要的（Chen 等，2011）。然而，由于微生物 mRNA 的半衰期可短至几秒（Laalami 等，2014），因此应探索快速而有效的去除非粘附微生物的方法，以确保样品的完整性。

在分析数据时，也应考虑诸如商业试剂盒中的空白对照的数据（Becker 等，2016；Thoendel 等，2017）和从宏组学数据中识别的宿主数据（Brown 等，2019）的干扰，以避免产生假阳性结果。适当选择反应试剂盒以及数据处理途径，可能有助于减少干扰。

7　结　论

目前分子生物学方法已经能够详细鉴定不同条件下的瘤胃上皮微生物组成及其变化，为明确它们在协同浮游微生物和饲料附着微生物以消化饲料，以及与宿主组织相互作用以转运营养物质和促进免疫功能方面的作用提供了机会。目前已被应用于瘤胃内容物菌群分析的先进的多组学方法也应被运用于瘤胃上皮微生物的研究，以了解其功能和对宿主生物学的贡献。综合分析现有数据集，建立合适的数据解释模型，以及大样本集的参与，有助于填补当前的知识空白，并为今后瘤胃上皮健康和功能的研究提出更好的建议。

8　致　谢

感谢阿尔伯塔省农林部（2018F095R）和 NSERC（加拿大自然科学与工程技术研究理事会）研发基金的资助。

9　参考文献

Abdel Rahman, S. 1966. Comparative study of the urease in the rumen wall and rumen content. Nature 209

(5023), 618-9.

Abdoun, K., Stumpff, F. and Martens, H. 2006. Ammonia and urea transport across the rumen epithelium: a review. Anim. Health Res. Rev. 7 (1-2), 43-59.

Abecia, L., Waddams, K. E., Martínez-Fernandez, G., et al., 2014a. An antimethanogenic nutritional intervention in early life of ruminants modifies ruminal colonization by Archaea. Archaea 2014, 841463.

Abecia, L., Ramos-Morales, E., Martínez-Fernandez, G., et al., 2014b. Feeding management in early life influences microbial colonisation and fermentation in the rumen of newborn goat kids. Anim. Prod. Sci. 54 (9), 1449-54.

Adjei-Fremah, S., Ekwemalor, K., Worku, M., et al., 2018. Probiotics and ruminant health. In: Enany, S. (Ed.), Probiotics: Current Knowledge and Future Prospects. IntechOpen, pp. 131-49. Chapter 8.

AlZahal, O., Li, F., Guan, L. L., et al., 2017. Factors influencing ruminal bacterial community diversity and composition and microbial fibrolytic enzyme abundance in lactating dairy cows with a focus on the role of active dry yeast. J. Dairy Sci. 100 (6), 4377-93.

Anders, S. and Huber, W. 2010. Differential expression analysis for sequence count data. Genome Biol. 11 (10), R106.

Bainbridge, M. L., Cersosimo, L. M., Wright, A. D., et al., 2016. Rumen bacterial communities shift across a lactation in Holstein, Jersey and Holstein × Jersey dairy cows and correlate to rumen function, bacterial fatty acid composition and production parameters. FEMS Microbiol. Ecol. 92 (5), fiw059.

Becker, L., Steglich, M., Fuchs, S., et al., 2016. Comparison of six commercial kits to extract bacterial chromosome and plasmid DNA for MiSeq sequencing. Sci. Rep. 6, 28063.

Berg, J. N. and Scanlan, C. M. 1982. Studies of *Fusobacterium necrophorum* from bovine hepatic-abscesses-biotypes, quantitation, virulence, and antibiotic susceptibility. Am. J. Vet. Res. 43 (9), 1580-6.

Bharanidharan, R., Arokiyaraj, S., Kim, E. B., et al., 2018. Ruminal methane emissions, metabolic, and microbial profile of Holstein steers fed forage and concentrate, separately or as a total mixed ration. PLoS ONE 13 (8), e0202446.

Bi, Y., Zeng, S., Zhang, R., et al., 2018. Effects of dietary energy levels on rumen bacterial community composition in Holstein heifers under the same forage to concentrate ratio condition. BMC Microbiol. 18 (1), 69.

Brown, S. M., Chen, H., Hao, Y., et al., 2019. MGS-Fast: metagenomic shotgun data fast annotation suing microbial gene catalogs. GIGA Sci. 8 (4), giz020.

Carberry, C. A., Kenny, D. A., Kelly, A. K., et al., 2014. Quantitative analysis of ruminal methanogenic microbial populations in beef cattle divergent in phenotypic residual feed intake (RFI) offered contrasting diets. J. Anim. Sci. Biotechnol. 5 (1), 41.

Chen, Y. H., Penner, G. B., Li, M. J., et al., 2011. Changes in bacterial diversity associated with epithelial tissue in the beef cow rumen during the transition to a high-grain diet. Appl. Environ. Microbiol. 77 (16), 5770-81.

Chen, Y., Oba, M. and Guan, L. L. 2012. Variation of bacterial communities and expression of Toll-like receptor genes in the rumen of steers differing in susceptibility to subacute ruminal acidosis. Vet. Microbiol. 159 (3-4), 451-9.

Cheng, K. J. and Wallace, R. J. 1979. The mechanism of passage of endogenous urea through the rumen wall and the role ofureolytic epithelial bacteria in the urea flux. Br. J. Nutr. 42 (3), 553-7.

Cheng, K. J., Mccowan, R. P. and Costerton, J. W. 1979. Adherent epithelial bacteria in ruminants and their roles in digestive-tract function. Am. J. Clin. Nutr. 32 (1), 139-48.

Cho, S. J., Cho, K. M., Shin, E. C., et al., 2006. 16S rDNA analysis of bacterial diversity in three fractions of cow rumen. J. Microbiol. Biotechn. 16, 92-101.

Danielsson, R., Dicksved, J., Sun, L., et al., 2017. Methane production in dairy cows correlates with rumen methanogenic and bacterial community structure. Front. Microbiol. 8, 226.

Dehority, B. A. and Grubb, J. A. 1981. Bacterial population adherent to the epithelium on the roof of the dorsal rumen of sheep. Appl. Environ. Microbiol. 41 (6), 1424-7.

De Mulder, T., Goossens, K., Peiren, N., et al., 2017. Exploring the methanogen and bacterial communities of rumen environments: solid adherent, fluid and epimural. FEMS Microbiol. Ecol. 93 (3), fiw251.

Difford, G. F., Plichta, D. R., Lovendahl, P., et al., 2018. Host genetics and the rumen microbiome jointly associate with methane emissions in dairy cows. PLoS Genet. 14 (10), e1007580.

Frutos, J., Andres, S., Yanez-Ruiz, D., et al., 2018. Early feed restriction of lambs modifies ileal epimural microbiota and affects immunity parameters during the fattening period. Animal 12 (10), 2115-22.

Garcia, M., Bradford, B. J. and Nagaraja, T. G. 2017. Invited review: ruminal microbes, microbial products, and systemic inflammation. Prof. Anim. Sci. 33 (6), 635-50.

Gloor, G. B., Macklaim, J. M., Pawlowsky-Glahn, V., et al., 2017. Microbiome dataset are compositional: and this is not option. Front. Microbiol. 8.

Gozho, G. N., Plaizier, J. C., Krause, D. O., et al., 2005. Subacute ruminal acidosis induces ruminal lipopolysaccharide endotoxin release and triggers an inflammatory response. J. Dairy Sci. 88 (4), 1399-403.

Gozho, G. N., Krause, D. O. and Plaizier, J. C. 2007. Ruminal lipopolysaccharide concentration and inflammatory response during grain-induced subacute ruminal acidosis in dairy cows. J. Dairy Sci. 90 (2), 856-66.

Granja-Salcedo, Y. T., Ramirez-Ui, R. A., Machado, E. G., et al., 2017. Studies on bacterial community composition are affected by the time and storage method of the rumen content. PLoS ONE 12 (4), e0176701.

Hart, E. H., Creevey, C. J., Hitch, T., et al., 2018. Meta-proteomics of rumen microbiota indicates niche compartmentalization and functional dominance in a limited number of metabolic pathways between abundant bacteria. Sci. Rep. 8 (1), 10504.

Henderson, G., Cox, F., Ganesh, S., et al., 2015 Rumen microbial community composition 40 The rumen wall microbiota communit varies with diet and host, but a core microbiome is found across a wide geographical range. Sci. Rep. 5, 14567.

Hernandez-Sanabria, E., Goonewardene, L. A., Wang, Z., et al., 2012. Impact of feed efficiency and diet on adaptive variations in the bacterial community in the rumen fluid of cattle. Appl. Environ. Microbiol. 78 (4), 1203-14.

Hristov, A. N. and Ropp, J. K. 2003. Effect of dietary carbohydrate composition and availability on utilization of ruminal ammonia nitrogen for milk protein synthesis in dairy cows. J. Dairy Sci. 86 (7), 2416-27.

Hu, F., Xue, Y., Guo, C., et al., 2018. The response of ruminal fermentation, epithelium-associated microbiota, and epithelial barrier function to severe feed restriction in pregnant ewes. J. Anim. Sci. 96 (10), 4293-305.

Hurtaud, C., Rulquin, H. and Verite, R. 1993. Effect of infused volatile fatty acids and caseinate on milk composition and coagulation in dairy cows. J. Dairy Sci. 76 (10), 3011-20.

Ishaq, S. L., AlZahal, O., Walker, N., et al., 2017. An investigation into rumen fungal and protozoal diversity in three rumen fractions, during high-fiber or grain-induced sub-acute ruminal acidosis conditions, with or without active dry yeast supplementation. Front. Microbiol. 8, 1943.

Jami, E., Israel, A., Kotser, A., 2013. Exploring the bovine rumen bacterial community from birth to adulthood. ISME J. 7 (6), 1069-79.

Jami, E., White, B. A. and Mizrahi, I. 2014. Potential role of the bovine rumen microbiome in modulating milk composition and feed efficiency. PLoS ONE 9 (1), e85423.

Janssen, P. H. andKirs, M. 2008. Structure of the archaeal community of the rumen. Appl. Environ. Microbiol. 74 (12), 3619-25.

Jewell, K. A., McCormick, C. A., Odt, C. L., et al., 2015. Ruminal bacterial community composition in dairy cows is dynamic over the course of two lactations and correlates with feed efficiency. Appl. Environ. Microbiol. 81 (14), 4697-710.

Jiao, J., Huang, J., Zhou, C., et al., 2015. Taxonomic identification of ruminal epithelial bacterial diversity during rumen development in goats. Appl. Environ. Microbiol. 81 (10), 3502-9.

Jin, D., Zhao, S. G., Zheng, N., et al., 2017. Differences in ureolytic bacterial composition between the rumen digesta and rumen wall based on ureC gene classification. Front. Microbiol. 8, 385.

Kong, R. 2016. Contribution of the rumen epithelial transcriptome and microbial community to variation in beef cattle feed efficiency. MSc Thesis.

Laalami, S., Zig, L. and Putzer, H. 2014. Initiation of mRNA decay in bacteria. Cell. Mol. Life Sci. 71 (10), 1799-828.

Li, M., Zhou, M., Adamowicz, E., et al., 2012. Characterization of bovine ruminal epithelial bacterial communities using 16S rRNA sequencing, PCR-DGGE, and qRT-PCR analysis. Vet. Microbiol. 155 (1), 72-80.

Li, F., Hitch, T. C. A., Chen, Y., et al., 2019a. Comparative metagenomic and metatranscriptomic analyses reveal the breed effect on the rumen microbiome and its associations with feed efficiency in beef cattle. Microbiome 7 (1), 6.

Li, W., Gelsinger, S., Edwards, A., et al., 2019b. Transcriptome analysis of rumen epithelium and meta-transcriptome analysis of rumen epimural microbial community in young calves with feed induced acidosis. Sci. Rep. 9 (1), 4744.

Liang, Y. S., Li, G. Z., Li, X. Y., et al., 2017. Growth performance, rumen fermentation, bacteria composition, and gene expressions involved in intracellular pH regulation of rumen epithelium in finishing Hu lambs differing in residual feed intake phenotype. J. Anim. Sci. 95 (4), 1727-38.

Lin, L., Xie, F., Sun, D., et al., 2019. Ruminal microbiome-host crosstalk stimulates the development of the ruminal epithelium in a lamb model. Microbiome 7 (1), 83.

Liu, J. H., Bian, G. R., Zhu, W. Y., et al., 2015. High-grain feeding causes strong shifts in ruminal epithelial bacterial community and expression of toll-like receptor genes in goats. Front. Microbiol. 6, 167.

Liu, J. H., Zhang, M. L., Zhang, R. Y., et al., 2016. Comparative studies of the composition of bacterial microbiota associated with the ruminal content, ruminal epithelium and in thefaeces of lactating dairy cows. Microb. Biotechnol. 9 (2), 257-68.

Liu, J., Bian, G., Sun, D., et al., 2017. Starter feeding altered ruminal epithelial bacterial communities and some key immune-related genes' expression before weaning in lambs. J. Anim. Sci. 95 (2), 910-21.

Lovell, D., Pawlowsky-Glahn, V., Egozcue, J. J., et al., 2015. Proportionality: a valid alternative to correlation for relative data. PLoS Comput. Biol. 11 (3), e1004075.

Malmuthuge, N., Li, M. J., Chen, Y. H., et al., 2012. Distinct commensal bacteria associated with ingesta and mucosal epithelium in the gastrointestinal tracts of calves and chickens. FEMS Microbiol. Ecol. 79 (2), 337-47.

Malmuthuge, N., Griebel, P. J. and le Guan, L. 2014. Taxonomic identification of commensal bacteria associated with the mucosa and digesta throughout the gastrointestinal tracts of preweaned calves. Appl. Environ. Microbiol. 80 (6), 2021-8.

Mann, E., Wetzels, S. U., Wagner, M., et al., 2018. Metatranscriptome sequencing reveals insights into the gene expression and functional potential of rumen wall bacteria. Front. Microbiol. 9, 43.

Mao, S., Zhang, M., Liu, J., et al., 2015. Characterising the bacterial microbiota across the gastrointestinal tracts of dairy cattle: membership and potential function. Sci. Rep. 5, 16116.

Matthews, C., Rispie, F., Lewis, E., et al., 2019. The rumen microbiome: a crucial consideration when optimizing milk and meat production and nitrogen utilization efficiency. Gut Microbes 19, 115-32.

McCann, J. C., Luan, S., Cardoso, F. C., et al., 2016. Induction of subacute ruminal acidosis affects the ruminal microbiome and epithelium. Front. Microbiol. 7, 701.

Mueller, R. E., Asplund, J. M. and Iannotti, E. L. 1984a. Successive changes in the epimural bacterial community of young lambs as revealed by scanning electron-microscopy. Appl. Environ. Microbiol. 47 (4), 715-23.

Mueller, R. E., Iannotti, E. L. and Asplund, J. M. 1984b. Isolation and identification of adherent epimural bacteria during succession in young lambs. Appl. Environ. Microbiol. 47 (4), 724-30.

Myer, P. R., Smith, T. P. L., Wells, J. E., et al., 2015. Rumen microbiome from steers differing in feed efficiency. PLoS ONE 10 (6), e0129174.

Nagaraja, T. G. 2016. Microbiology of the rumen. Rumenology, 39-61.

Nagaraja, T. G. and Titgemeyer, E. C. 2007. Ruminal acidosis in beef cattle: the current microbiological and nutritional outlook. J. Dairy Sci. 90 (Suppl. 1), E17-38.

Nagaraja, T. G., Bartley, E. E., Fina, L. R., et al., 1978. Evidence of endotoxins in the rumen bacteria of cattle fed hay or grain. J. Anim. Sci. 47 (1), 226-34.

Neubauer, V., Humer, E., Mann, E., et al., 2019. Effects clay mineral supplementation on particle-associated and epimural microbiota, and gene expression in the rumen of cows fed high-concentrate diet. Anaerobe 59, 38-48.

Neves, A. L. A., Li, F., Ghoshal, B., et al., 2017. Enhancing the resolution of rumen microbial classification from metatranscriptomic data using Kraken and Mothur. Front. Microbiol. 8, 2445.

O'Hara, E., Neves, A. L. A., Song, Y., et al., 2019. The role of gut microbiome in cattle production and health: driver or passenger?

Okada, Y., Kanoe, M., Yaguchi, Y., et al., 1999. Adherence of Fusobacterium necrophorum subspecies necrophorum to different animal cells. Microbios 99 (393), 95-104.

O'Shea, E., Waters, S. M., Keogh, K., et al., 2016. Examination of the molecular control of ruminal epithelial function in response to dietary restriction and subsequent compensatory growth in cattle. J. Anim.

Sci. Biotechnol. 7, 53.

Palmquist, D. L. 2006. Milk fat: origin of fatty acids and influence of nutritional factors thereon. In: Fox P. F. and McSweeney, P. L. H. (Eds), Advanced Dairy Chemistry Volume 2 Lipids. Springer, Boston, MA, pp. 43-92.

Paz, H. A., Anderson, C. L., Muller, M. J., et al., 2016. Rumen bacterial community composition in holstein and jersey cows is different under same dietary condition and is not affected by sampling method. Front. Microbiol. 7, 1206.

Pei, C. X., Mao, S. Y., Cheng, Y. F., et al., 2010. Diversity, abundance and novel 16S rRNA gene sequences of methanogens in rumen liquid, solid and epithelium fractions of Jinnan cattle. Animal 4 (1), 20-9.

Petri, R. M., Schwaiger, T., Penner, G. B., et al., 2013. Changes in the rumen epimural bacterial diversity of beef cattle as affected by diet and induced ruminal acidosis. Appl. Environ. Microbiol. 79 (12), 3744-55.

Petri, R. M., Kleefisch, M. T., Metzler-Zebeli, B. U., et al., 2018. Changes in the rumen epithelial microbiota of cattle and host gene expression in response to alterations in dietary carbohydrate composition. Appl. Environ. Microbiol. 84 (12).

Rainard, P. and Foucras, G. 2018. A critical appraisal of probiotics for mastitis control. Front. Vet. Sci. 5, 251.

Rey, M., Enjalbert, F., Combes, S., et al., 2014. Establishment of ruminal bacterial community in dairy calves from birth to weaning is sequential. J. Appl. Microbiol. 116 (2), 245-57.

Rieu, F., Fonty, G., Gaillard, B., et al., 1990. Electron-microscopy study of the bacteria adherent to the rumen wall in young conventional lambs. Can. J. Microbiol. 36 (2), 140-4.

Roehe, R., Dewhurst, R. J., Duthie, C. A., et al., 2016. Bovine host genetic variation influences rumen microbial methane production with best selection criterion for low methane emitting and efficiently feed converting hosts based on metagenomic gene abundance. PLoS Genet. 12 (2), e1005846.

Sadet, S., Martin, C., Meunier, B., et al., 2007. PCR-DGGE analysis reveals a distinct diversity in the bacterial population attached to the rumen epithelium. Animal 1 (7), 939-44.

Sadet-Bourgeteau, S., Martin, C. and Morgavi, D. P. 2010. Bacterial diversity dynamics in rumen epithelium of wethers fed forage and mixed concentrate forage diets. Vet. Microbiol. 146 (1-2), 98-104.

Schluter, J. and Foster, K. R. 2012. The evolution of mutualism in gut microbiota via host epithelial selection. PLoS Biol. 10 (11), e1001424.

Seshadri, R., Leahy, S. C., Attwood, G. T., et al., 2018. Cultivation and sequencing of rumen microbiome members from the Hungate1000 Collection. Nat. Biotechnol. 36 (4), 359-67.

Shi, W., Moon, C. D., Leahy, S. C., et al., 2014. Methane yield phenotypes linked to differential gene expression in the sheep rumen microbiome. Genome Res. 24 (9), 1517-25.

Shin, E. C., Choi, B. R., Lim, W. J., et al., 2004. Phylogenetic analysis of archaea in three fractions of cow rumen based on the 16S rDNA sequence. Anaerobe 10 (6), 313-9.

Steele, M. A., AlZahal, O., Hook, S. E., et al., 2009. Ruminal acidosis and the rapid onset of ruminal parakeratosis in a mature dairy cow: a case report. Acta Vet. Scand. 51, 39.

Steele, M. A., Croom, J., Kahler, M., et al., 2011. Bovine rumen epithelium undergoes rapid structural adaptations during grain-induced subacute ruminal acidosis. Am. J. Physiol. Regul. Integr. Comp. Physiol. 300 (6), R1515-23.

Stewart, C. S., Fonty, G. and Gouet, P. 1988. The establishment of rumen microbial comm-unities. Anim.

Feed Sci. Tech. 21 (2-4), 69-97.

Stewart, R. D., Auffret, M. D., Warr, A., et al., 2018. Assembly of 913 microbial genomes from metagenomic sequencing of the cow rumen. Nat. Commun. 9 (1), 870.

Tadele, Y. and Amha, N. 2015. Use of different non protein nitrogen sources in ruminant nutrition: a review. Adv. Life Sci. Technol. 29, 100-6.

Tadepalli, S., Narayanan, S. K., Stewart, G. C., et al., 2009. *Fusobacterium necrophorum*: a ruminal bacterium that invades liver to cause abscesses in cattle. Anaerobe 15 (1-2), 36-43.

Takayama, Y., Kanoe, M., Maeda, K., et al., 2000. Adherence of *Fusobacterium necrophorum* subsp. necrophorum to ruminal cells derived from bovine rumenitis. Lett. Appl. Microbiol. 30 (4), 308-11.

Thoendel, M., Jeraldo, P., Greenwood-Quaintance, K. E., et al., 2017. Impact of contaminating DNA in whole-genome amplification kits used for metagenomic shotgun sequencing for infection diagnosis. J. Clin. Microbiol. 55 (6), 1789-801.

van Niekerk, J., Middeldorp, M. and Steele, M. 2018. Technical Note: the development of a methodology for ruminal and colon tissue biopsying of young Holstein dairy calves. J. Dairy Sci. 101 (8), 7212-18.

Villegas-Rivera, G., Vargas-Cabrera, Y., Gonzalez-Silva, N., et al., 2013. Evaluation of DNA extraction methods of rumen microbial populations. World J. Microbiol. Biotechnol. 29 (2), 301-7.

Wallace, R. J., Cheng, K. J., Dinsdale, D., et al., 1979. An independent microbial flora of the epithelium and its role in the ecomicrobiology of the rumen. Nature 279 (5712), 424-6.

Wang, X. and Quinn, P. J. 2010. Endotoxins: lipopolysaccharides of gram-negative bacteria. Subcell. Biochem. 53, 3-25.

Wang, Z., Elekwachi, C. O., Jiao, J., et al., 2017. Investigation and manipulation of metabolically active methanogen community composition during rumen development in black goats. Sci. Rept 7 (1), 422.

Wetzels, S. U., Mann, E., Matzler-Zebeli, B. U., et al., 2016. Epimural indicator phylotypes of transiently-induced subacute ruminal acidosis in dairy cattle. Front. Microbiol. 7, 274.

Wetzels, S. U., Mann, E., Pourazad, P., et al., 2017. Epimural bacterial community structure in the rumen of Holstein cows with different responses to a long-term subacute ruminal acidosis diet challenge. J. Dairy Sci. 100 (3), 1829-44.

Wilkins, L., Ettinger, C., Josping, G., et al., 2019. Metagenome-assembled genomes provide new insights into microbial diversity. Sci. Rep. 9 (1), 3059-65.

Yanez-Ruiz, D. R., Abecia, L. and Newbold, C. J. 2015. Manipulating rumen microbiome and fermentation through interventions during early life: a review. Front. Microbiol. 6, 1133.

Zhang, J., Xu, C., Huo, D., et al., 2017a. Comparative study of the gut microbiome potentially related to milk protein in Murrah buffaloes (Bubalus bubalis) and Chinese Holstein cattle. Sci. Rep. 7, 42189.

Zhang, R., Ye, H., Liu, J., et al., 2017b. High-grain diets altered rumen fermentation and epithelial bacterial community and resulted in rumen epithelial injuries of goats. Appl. Microbiol. Biotechnol. 101 (18), 6981-92.

Zhou, M., Hernandez-Sanabria, E. and Guan, L. L. 2009. Assessment of the microbial ecology of ruminal methanogens in cattle with different feed efficiencies. Appl. Environ. Microbiol. 75 (20), 6524-33.

Zhou, M., Chen, Y., Griebel, P. J., et al., 2014. Methanogen prevalence throughout the gastrointestinal tract of pre-weaned dairy calves. Gut Microbes 5 (5), 628-38.

Zhou, M., Peng, Y. J., Chen, Y., et al., 2018. Assessment of microbiome changes after rumentransfaunation: implications on improving feed efficiency in beef cattle. Microbiome 6 (1), 62.

第三部分

营养代谢在瘤胃微生物和宿主互作中的作用

第11章 瘤胃内纤维的消化

Adrian E. Naas、Phillip B. Pope

挪威生命科学大学，挪威

（毛跃建、董依然译）

1 前言

草食反刍动物依赖共生的瘤胃微生物，将植物纤维物质转化为微生物蛋白和挥发性脂肪酸（McCann 等，2014）。瘤胃为微生物提供了适宜且相对稳定的生存环境，瘤胃微生物通过分解植物饲料中的细胞壁成分，在维持其自身生长繁殖的同时，还能够满足宿主营养需求。这些瘤胃微生物利用碳水化合物活性酶（CAZymes）来降解植物细胞壁中复杂且难以降解的碳水化合物，以此获取其生长所必需的能量。纤维素由于其自身的结构特点使其难以被降解，由于植物细胞壁是由纤维素和半纤维素通过复杂的交联嵌合形成，因此酶解植物细胞壁更加困难。然而瘤胃中的细菌、原虫和真菌利用不同的降解机制、通过分泌CAZymes 高效降解木质纤维素。

目前，对牛瘤胃中纤维素降解的研究主要集中在厚壁菌门（Firmicutes）和纤维杆菌门（Fibrobacters）中少数被分离的菌株，即白色瘤胃球菌（*Ruminococcus albus*）、黄色瘤胃球菌（*R. flavefaciens*）和产琥珀酸丝状杆菌（*Fibrobacter succinogenes*）（Hungate，1950，1960；Russell 等，2009）。未培养分子生物学术研究显示，分离菌株在瘤胃中的丰度往往很低；瘤胃微生物其实主要以未被培养的厚壁菌门和拟杆菌门为主（Stevenson 和 Weimer，2007；Konietzny 等，2014）。这些未知的微生物不仅是新的 CAZymes 的潜在来源，也可能蕴藏着糖酵解机制之外的新知识。目前的糖酵解机制是建立在瘤胃生物质和纤维素降解的经典观点之上的。从已知的培养微生物中所了解到的 CAZyme 结构和机制包括纤维小体，分泌的游离纤维素酶和多糖利用位点（PULs）。此外，某些微生物种中纤维素降解的机制仍然未知。

考虑到共生菌群决定反刍动物主要的能量生成，因此饲料的转化效率、牛肉及牛奶等产品的品质都与瘤胃菌群的动态变化和功能密切相关。在本章中，我们将对微生物群落在瘤胃木质纤维素降解中的作用进行概述，让大家了解不同微生物在做什么，以及它们降解木质纤维素的机制。

2 木质纤维素类生物质

反刍动物的解剖结构和饮食行为表明，其生长需要消耗大量的植物多糖。为了探讨植物纤维在瘤胃中的具体降解机制，我们首先简要地介绍植物细胞壁的组成（图1）。

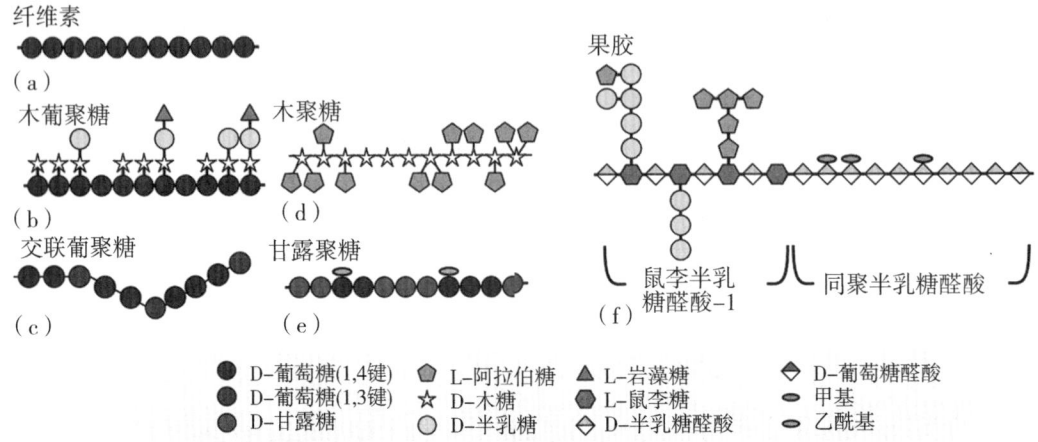

图1 植物多糖结构和组成的类型。包括（a）纤维素、（b-e）半纤维素和（f）果胶。具体底物包括（a）纤维素、（b）木葡聚糖、（c）交联葡聚糖、（d）木聚糖、（e）甘露聚糖、（f）果胶（包括鼠李糖醛酸）（彩图5）。

植物细胞壁是由多糖、木质素、脂类、矿物质和糖蛋白组成的复杂细胞外基质，它们为细胞提供机械强化保护。细胞壁由不同的层组成：中间片层、初生细胞壁和次生细胞壁。它们均由细胞原生质体依次分泌（Gibson，2012）。初生细胞壁由纤维素纤维组成，纤维素嵌入在果胶和半纤维素的基质中，如木葡聚糖、木聚糖和葡甘聚糖（Scheller 和 Ulvskov，2010）。这层是有柔性的，因此细胞可以生长。当其生长停止的时候，次生细胞壁就会通过修改初生细胞壁或者沉积另外第二层的方式产生。次生细胞壁和初生细胞壁最大的区别在于它掺入了木质素。木质素是交联酚类化合物的基质，为细胞壁提供额外的机械强度（Zhong 和 Ye，2015）。维管植物利用细胞壁木质化的细胞作为机械组织，让它们能够长得很高，有利于竞争阳光。淀粉虽然从生物学来说并不是构成细胞壁的结构多糖的一部分，但它是一种由α-D 葡萄糖单元组成的重要储存多糖，经常被反刍动物摄入。

2.1 纤维素

细胞壁的主要成分是纤维素，通常占木质纤维素生物量的 25%~51.4%（Isikgor 和 Becer，2015）。纤维素是 D-吡喃葡萄糖苷（葡萄糖）分子以 β-1,4 糖苷键组成的大分子（图1a）。单个葡萄糖分子相互旋转180°，形成重复单位——纤维二糖。纤维二糖的天然分子结构使其在较长的低聚物中具有很大的潜力形成氢键。当聚合度（DP）大于7时，纤维二糖对其他低聚物的亲和力很高，以至于分子聚集并在水溶剂中无法溶解（Brown，2004）。植物细胞壁中的天然纤维素呈一种叫做纤维素 I_β 的结晶形态，其中 18~24 条纤维素链是由细胞质膜上的纤维素合成酶复合物产生，然后通过平行排列的链之间的大量氢键来形成稳定的微纤（Schneider 等，2016）。初生纤维聚集到较大的、具有结晶和较少结晶（无定形）区域的微纤结构中，然后在植物细胞壁中再被半纤维素和木质素相互连接。

2.2 半纤维素

'半纤维素'是一个广义的术语，它描述了嵌入在植物细胞壁中不同的杂多聚体，其详细的结构和丰度可因植物种类而异。传统上来说，从植物生物质中提取果胶和木质素后，采用碱法从纤维素中分离出来的其余多糖，被命名为'半纤维素'（Scheller 和 Ulvskov，2010）。最近，Scheller 和 Ulvskov（2010）建议将"半纤维素"一词重新定义为在其骨架中轴中都有 β-1,4-糖苷键的细胞壁多糖。它们包括木葡聚糖、交联的 β-(1,3,1,4)-葡聚糖、木聚糖、甘露聚糖、葡甘露聚糖和半乳甘露聚糖。

木葡聚糖存在于除轮藻类以外所有的植物细胞壁中（Scheller 和 Ulvskov，2010），由含有或不含木糖取代（分别称为单体 X 和 G）的 β-1,4 连接的葡萄糖骨架组成。木糖残基反过来又可以被半乳糖（简称 L）取代，或者在草本植物中被岩藻酰化的半乳糖取代（简称 F）（图 1b）。X 和 G 在骨架和支链中的比例和分布模式因植物种类而异，双子叶植物中 XXXG 这样的重复单位比较常见，而茄科植物则 XXGG 较为常见，其中 X 可以是未取代的木糖（X），也可以是取代的半乳糖或岩藻酰化的半乳糖（L 或 F）（Attia 和 Brumer，2016）。另一种具有葡萄糖骨架的半纤维素是交联 β-葡聚糖。它存在于草本植物的细胞壁中，由 β-1,4-连接的葡萄糖三聚体和四聚体组成，通过 β-1,3-糖苷键连接在一起（图 1c）。

木聚糖是主链骨架中含有 β-1,4 糖苷键连接的木糖单元的半纤维素。它们通常是双子叶植物次生细胞壁中占主导地位的非纤维素多糖，通常有 α-1,2 连接的葡萄糖醛酸糖和 4-O-甲基葡萄糖醛酸糖取代，被称为葡萄糖醛酸木聚糖。在单子叶植物中，阿拉伯木聚糖在初生细胞壁中占据主导地位，更多地被阿拉伯糖取代（Scheller 和 Ulvskov，2010；图 1d）。

第三类半纤维素含有由 β-1,4 糖苷键连接的甘露糖骨架，在所有细胞壁中都有不同数量的存在。在甘露聚糖和半乳甘露聚糖中，骨架只含有甘露糖，以及各种半乳糖单元的取代。葡萄糖甘露聚糖在其骨架中既有甘露糖，也有葡萄糖，以不重复的模式由 β-1,4 键连接而成（图 1e）。

3 碳水化合物活性酶

木质纤维素类生物质是地球上最丰富的生物质，其复杂的结构使其具有很强的抗微生物侵袭破坏的能力。在自然界中，碳水化合物的结构有着无穷尽的变化，一个还原性六聚糖可以形成 1 012 个可能的线性和支链异构体（Laine，1994）。然而，如果不能释放并利用植物细胞壁碳水化合物中固定的碳，那么异养生物将无法获得能量。因此，自然界进化出了一系列的工具来克服这种碳水化合物分解的难题。

作用于碳水化合物多糖组装和分解的酶和辅助蛋白统称为碳水化合物活性酶，简称 CAZymes（Lombard 等，2014）。1999 年，CAZymes 数据库（www.cazy.org）正式上线，成为 CAZyme 信息的中心存储库，包括序列、三维结构和生化数据。该数据库目前根据作用方式将蛋白质家族分为 6 类：用于分解的糖苷水解酶（GHs）、碳水化合物酯酶

(CEs) 和多糖裂解酶 (PLs); 用于合成的糖基转移酶 (GTs); 帮助酶靶向到其底物的碳水化合物结合模块 (CBMs); 以及其他与 CAZymes 共同作用的的辅助活性酶 (AA), 包括氧化还原酶。CAZymes 根据其氨基酸序列分为不同的家族, 进而反映其三维结构和折叠 (Henrissat, 1991; Cantarel 等, 2009)。由于碳水化合物底物的数量大大超过折叠的数量, 酶在普通折叠的基础上进一步进化, 因此在同一家族中可以存在几种酶的特异性。同样的, 在不同的家族中也可以发现相同的酶特异性, 典型的趋同进化。CAZymes 通常是多模块的, 可以包含来自不同家族的多个结构域, 使得一个蛋白质序列可以被分到多个家族。本章的其余部分将概述各类 CAZymes 的功能。

3.1 糖苷水解酶 (GHs)

CAZymes 数据库中序列数和家族数最多的一类是 GHs, 共有 166 个家族, 超过 70 万条序列, 反映了可利用碳水化合物底物的巨大差异 (Lombard 等, 2014)。GHs 催化糖苷键的水解, 其反应可以通过两种不同的机制进行, 即保留糖苷键的异构体构型或者反转 (Koshland, 1953)。该反应由酶中两个保守的氨基酸残基催化, 通常是谷氨酸或天冬氨酸 (McCarter 和 Withers, 1994)。它们作为一般的酸 (质子供体) 和基底起作用, 其侧链的空间位置与机制的类型有关。

GHs 具有多种特异性: 作用于线性多糖的骨架、以结晶底物为靶、或作为脱支酶只攻击特定半纤维素的特定基团。作用在聚合物上的 GHs 既可以是内切作用, 也可以是外切作用, 分别指酶是攻击多糖内部的糖苷键, 还是攻击链末端的糖苷键。通常, 外切酶在不与底物分离的情况下, 可以逐步地进行若干次水解反应。它们对多糖的还原端或非还原端都具有特异性 (Davies 和 Henrissat, 1995; Barr 等, 1996)。在纤维素降解方面, 内切纤维素酶 (目前在 14 个 GH 家族中有发现, 但通常属于 GH5、GH9 或 GH45) 和外切纤维素酶 (通常是 GH6、GH7、GH48; 但有些 GH9 被报道为持续性内切纤维素酶, 释放纤维二糖) 协同合作来降解结晶底物 (Wood 和 McCrae, 1979; Kostylev 和 Wilson, 2012)。在这个过程中, 内切纤维素酶在无定形区域内裂解纤维素链, 为释放纤维二糖的持续性内切纤维素酶 (纤维二糖水解酶) 形成链末端。

3.2 碳水化合物结合模块 (CBMs)

碳水化合物结合模块 (CBMs) 是 CAZymes 的非催化模块, 但能帮助酶靶向到一种特定的碳水化合物上 (Boraston 等, 2004)。第一个被表征的 CBMs 是纤维素结合模块, 它被发现是用来促进里氏木霉 (*Trichoderma reesei*) 的纤维二糖水解酶 I 和 II 以及来源于粪肥纤维单胞菌 (*Cellulomonas fimi*) 的两种纤维素酶结合到纤维素上的 (Van Tilbeurgh 等, 1986; Tomme 等, 1988; Gilkes 等, 1988)。随后, CBMs 的碳水化合物靶点被证明几乎涵盖了所有已知的碳水化合物, 包括植物细胞壁中的纤维素和半纤维素 (McCartney 等, 2004; Lombard 等, 2014)。CBMs 根据其氨基酸序列分为家族, 迄今已有 84 个不同的家族, 可分为 3 个功能大类 (Gilbert 等, 2013)。类型 A 以结晶多糖的表面为靶点; 类型 B 以碳水化合物链的内部为靶点; 类型 C 则以聚糖链的末端为靶点。

CBMs 被认为是通过提高底物附近酶的浓度来提高其所结合的催化结构域的效率

（McCartney 等，2004）。保持催化域与其底物更接近，催化的发生概率就更大。这一点在低底物浓度条件时更为重要，比如有研究观察到纤维素酶中存在纤维素结合的 CBMs 时，在高底物浓度时对酶效率的影响较小，甚至是负面影响（Várnai 等，2013）。尽管 CBMs 通常针对提高催化域附近的底物浓度，但也有 CBMs 与不同底物结合的例子，使得位于其结合的聚糖附近的目标多糖能够被催化（Hervé 等，2010）。

3.3 其他 CAZyme 家族

辅助活性酶（AA）包括漆酶、纤维二糖脱氢酶、铜自由基加氧酶和利用碳水化合物氧化机制的各种酶（Lombard 等，2014）。以往认为多糖裂解单加氧酶（LPMOs）是有助于破坏几丁质结晶结构的 CBMs，但研究表明黏质沙雷氏菌的 Cbp21 酶可以引入断链，在结晶几丁质表面生成氧化的链末端（Vaaje-Kolstad 等，2010）。随后从细菌和真菌中发现有降解纤维素活性的 LPMOs，促使需要在 CAZymes 数据库中建立单独的类，即辅助活性酶（Forsberg 等，2011；Quinlan 等，2011；Levasseur 等，2013）。目前，在 13 个 AA 家族中，LPMOs 被分为 9、10、11 和 13 等四个 AA 家族。这些酶受到广泛关注，因为它们能提高 GHs 的活性，从而有助于提高鸡尾酒酶的整体效率。重要的是，LPMOs 可能在解决前面所讨论的 GHs 与底物结合方面起至关重要的作用，因为它们能够打破结晶环境中的糖苷键，从而提高 GHs 降解底物的效率。尽管有多种瘤胃分离菌的基因组编码了 AA10 中代表基因（Seshadri 等，2018），但 LPMOs 活性尚未在瘤胃中被检测到，迄今仅在好氧微生物中有过报道。

CEs 作为脱支酶发挥作用，对复杂多糖进行脱 O-或脱 N-酰基酯修饰。在去除这些酯基修饰后，使得 GHs 在复杂多糖中更容易进入其作用位点（Cantarel 等，2009）。PLs 利用 β-消除作用来切割糖醛酸多糖的糖苷键。这使得新的非还原端上的糖在 C4 和 C5 间是不饱和双键，而新的还原端则处于饱和状态（Garron 和 Cygler，2010）。作为 CAZymes 数据库中唯一的合成代谢的酶类，糖基转移酶利用糖磷酸酯在"活化"糖与其他糖类、脂类或蛋白质之间形成糖苷键（Lairson 等，2008）。糖基以相同或反转的方式转移到底物的亲核基团上。

4 瘤胃中原核生物降解纤维的策略

几十年来对各种生态系统（如土壤、海洋、动物相关）的研究表明，所有糖酵解微生物都依赖于 CAZymes 的作用。然而，这些 CAZymes 是如何被微生物利用的，可能存在较大个体差异。在本节中，我们将描述瘤胃细菌和厌氧真菌所使用的各种机制。

4.1 纤维小体

纤维小体是最普遍的纤维素酶解机制，与好氧真菌分泌的游离酶一起，是微生物纤维素降解的两种主要方式之一（Wilson，2011）。纤维素酶的纤维小体常见于几种厌氧细菌和厌氧真菌中，其首次在一种厌氧嗜热土壤细菌——热纤梭菌（*Clostridium thermocellum*）中被发现（Lamed 等，1983；Bayer 等，2008）。纤维小体是一种多模块的酶复合物，使细胞能够黏附到结晶纤维素上，在超微结构上降解纤维素（Bayer 和 Lamed，1986）。纤维小体的内切、外切纤维素酶和半纤维素酶除了含有催化结构域外，还含有锚定结构域，这有利于对接大的非催化支架蛋白亚基上的粘连素结构域（Bayer 等，1994；Yaron 等，1995）。热纤梭菌（cipA）的初级支架蛋白亚单位包含 9 个为了结合锚定连接酶亚基的粘连素结构域，1 个与结晶纤维素结合的 CBM3 模块，1 个与细胞壁锚定支架蛋白中 II 型粘连素结合的 C-端锚定结构域。锚定的支架蛋白含有 S 层同源结构域（SLH），将纤维小体固定在细胞表面，使细胞接近纤维素酶水解释放出来的可溶性纤维糊精（Bayer 等，2008）。3 种不同的锚定支架蛋白允许多达 63 种不同的纤维小体附着在热纤梭菌的单个复合体中。纤维小体的模块化结构使内切、外切纤维素酶紧密地结合在一起，并将它们与底物结合在一起，从而在纤维素降解中发挥协同作用（Krauss 等，2012）。

目前热纤梭菌在瘤胃中未被发现，但是已知的纤维素降解菌——黄色瘤胃球菌和白色瘤胃球菌，会部分利用纤维小体来促进植物生物质在瘤胃中的降解（Ohara 等，2000；Ding 等，2001）。与热纤梭菌（*C. thermocellum*）不同的是其纤维小体的模块组成，其中黄色瘤胃球菌（*R. flavefaciens*）有一个特别精细的系统，编码大量的锚定蛋白，包括新的 CBMs（Dassa 等，2014；Venditto 等，2016）。黄色瘤胃球菌的纤维小体蛋白在菌株间存在差异，基因组分析表明，锚定蛋白的数量在 53~223（Seshadri 等，2018）。此外，粘连素和锚定蛋白间的相互作用大部分具有菌株特异性（Israeli-Ruimy 等，2017）。已经证实，黄色瘤胃球菌 FD-1 的纤维小体多达 14 个酶亚基，被组装在 4 个不同的支架蛋白上（图 2a）。与黄色瘤胃球菌相比，白色瘤胃球菌（*R. albus*）含有较低的锚定蛋白编码基因，在 3 株测序的菌株中，2 株仅含有 1 个粘连素编码基因，第 3 株则不含相应的粘连素编码基因。这表明存在一种迄今为止尚未被发现的、具有新的类似粘连蛋白结构域的支架蛋白类型，或者说纤维素降解细菌缺乏利用'完整'的纤维小体机制（Dassa 等，2014）。

通过基因组测序发现，纤维小体在厌氧真菌基因组和厌氧细菌基因组存在很大差异（Haitjema 等，2017）。在所有 5 个测序的厌氧真菌基因组中都发现的大支架蛋白（ScaA）同源基因，与细菌支架蛋白没有序列相似性。有趣的是，来自 3 个属的肠道真菌的锚定结构域均能与含有粘连素模体的 ScaA 片段结合。因此，作者推测，在其原生环境中，真菌纤维小体实际上可能是由来自不同真菌物种的酶组成的复合物，不像细菌锚定蛋白—黏蛋白之间的相互作用那样具有高度物种特异性。

4.2 分泌型酶

微生物通过分泌游离的纤维素酶来降解纤维素是第二种降解纤维素的模式，主要存在于好氧真菌和细菌中（图 2b）。这种降解机制在嗜温丝状真菌里氏木霉（*Trichoderma reesei*）中已有较为深入的研究。里氏木霉最初是在第二次世界大战期间，从所罗门群岛腐烂的美国陆军装备中分离得到的，是目前最主要的工业纤维素酶生产菌（Bischof 等，2016；Reese，1956）。

里氏木霉在纤维素上生长时会分泌大量纤维素酶到培养液中，使纤维素迅速降解为葡萄糖（Sheir-Neiss 和 Montenecourt，1984）。分泌的酶包括内切葡聚糖酶（EG、Cel5A、Cel5B、Cel12A、Cel45A）、非还原和还原端纤维二糖水解酶（CBHI/GH6 和 CBHII/

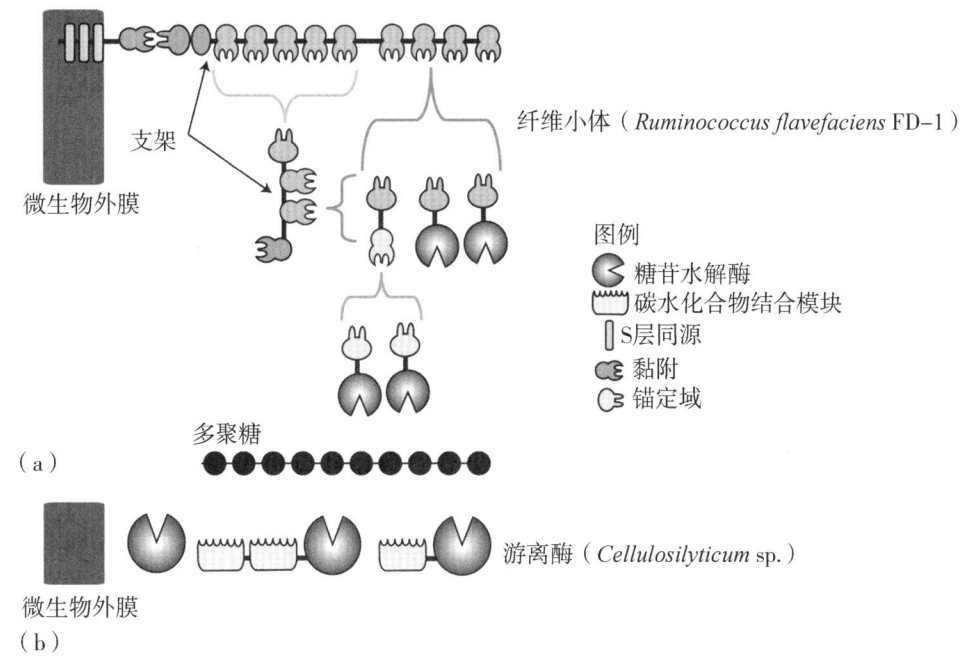

图 2 来源于瘤胃微生物的纤维小体（a）和游离酶（b）

Cel7A）、β-葡萄糖苷酶Ⅰ（GH3）和 AA9 LPMOs（之前称为 GH61；Saloheimo 等，1997；Wilson，2009；Westereng 等，2011；Li 等，2016）。所有这些酶协同工作，内切葡聚糖酶（含或不含 CBM1 结构域）攻击纤维素无定形区域的 β-1,4 糖苷键，为从链两端攻击的 CBHs 创造链末端，破坏结晶结构。从 CBHs 中释放的纤维二糖被降解为葡萄糖单体，然后被细胞吸收。AA9 型的 LPMOs 在纤维素结晶区引入氧化断裂，为 CBH 创造更多的链末端，并可能从木质素等非酶供体获得还原力（Westereng 等，2015）。其他真菌纤维素酶系统也利用分泌酶，但未对这些酶系进行深入研究（Wilson，2008）。有几种细菌也利用分泌型游离酶降解纤维素，降解方式与真菌相似（Wilson，2011）。目前对于褐色嗜热裂孢菌（*Thermobifida fusca*）降解纤维素机制的研究已经较为深入，它利用 GH5、GH6 和 GH9 内切纤维素酶，含有 CBM2 纤维素结合结构域的 GH6 和 GH48 外切纤维素酶，以及两个 AA10 型 LPMOs 协同来降解纤维素（Gomez Del Pulgar 和 Saadeddin，2013；Forsberg 等，2014）。

分泌型游离酶在草食动物瘤胃中的作用尚不清楚。例如，基于基因序列预测，认为黄色瘤胃球菌和瘤胃真菌除了利用纤维小体以外，还利用分泌酶（Dassa 等，2014）。基因预测表明，有几个纤维素降解厚壁菌微生物（Firmicutes）利用分泌型的外切（GH48）和内切（GH5、GH9）纤维素酶，例如腔隙杆菌属（*Lachnoclostridium*）、溶纤维素菌属（*Cellulosilyticum*），瘤胃梭菌属（*Ruminoclostridium*）及瘤胃球菌属（*Ruminococcus*）中的某些种（Cai 等，2010；Seshadri 等，2018）。另外基因测序结果表明，放线菌门的纤维素单胞菌（*Cellulomonas* sp.）也是通过分泌游离的内切和外切纤维素酶共同降解纤维素。此外，测序结果表明，许多种群还可利用游离酶降解半纤维素和淀粉，如瘤胃中的优势微生物溶纤维丁酸弧菌（*Butyrivibrio fibrisolvens*）（Seshadri 等，2018）。

4.3 多糖利用位点

大部分定植在肠道微生物生态系统的细菌都属于拟杆菌门（Tajima 等，1999；Hold 等，2002；Flint 等，2008）。这些革兰氏阴性细菌具有广泛的碳水化合物降解能力，与最早在人类肠道细菌 *Bacteroidetes thetaiotaomicron* 中被发现和鉴定的许多被称为 PULs 的基因簇有关（Bjursell 等，2006；Martens 等，2008）。*B. thetaiotaomicron* 编码 101 个 susC/D 基因对，使它能利用种类繁多的多糖和与宿主相关的黏液素生长。此外，转录组学分析表明当存在特定底物时，PULs 转录表达将会上调（Martens 等，2011）。PULs 不仅存在于人类肠道细菌，而且普遍存在于环境和食草动物（包括反刍动物）的肠道微生物组中的拟杆菌中（McBride 等，2009；Pope 等，2010，2012；Naas 等，2014；Dodd 等，2010；Terrapon 等，2015；Accetto 和 Avguštin，2015；Mackenzie 等，2015；Güllert 等，2016；Rosewarne 等，2014）。PULs 主要作用于淀粉、半纤维素和果胶，在牛瘤胃中由大量的普雷沃氏菌科（Prevotellaceae）编码，可能有助于从半纤维素基质中释放纤维素（Stewart 等，2018b；Solden 等，2018）。尤其是 *Prevotella bryantii* 所含的一种降解木聚糖的 PUL，是最早采用培养组学方法在人体肠道外进行详细研究的 PULs 之一（Dodd 等，2010）。

PULs 被定义为共存的类 SusC/SusD 基因、糖苷水解酶（GHs）和糖类转运蛋白（图3a）。类 SusC/类 SusD 蛋白的命名来源于在淀粉利用系统（Sus）中对它们的首次描述（Reeves 等，1996，1997）。Sus 基因簇包含 8 个基因，被命名为 *susRABCDEFG*，包含细胞降解和输入淀粉所需的所有酶（Foley 等，2016）。SusR 作为一种内膜跨膜蛋白，是基因簇的转录调控因子，具有识别淀粉降解产物麦芽糖的功能（D'elia 和 Salyers，1996b）。SusR 能上调表达基因簇中其余部分，使细胞能够对淀粉的存在做出响应。外膜脂蛋白 SusD、E 和 F 能促进淀粉与细胞表面的结合，使外膜脂质上锚定的 α-淀粉酶 SusG 能够靠近其底物（Shipman 等，1994，2000；Koropatkin 和 Smith，2010）。SusD 对于利用大于 DP4 的麦芽低聚糖作为底物的生长至关重要，与其单个淀粉结合域被破坏的突变体相比，SusD 能显著提高对麦芽糖的感知能力（Cameron 等，2014）。SusE 和 SusF 的结合位点和表达对转录的激活并非必需，但是它们通过底物依赖性的方式能提高其生长速度。SusG 释放的麦芽寡糖通过依赖 TonB 的外膜孔蛋白 SusC 进入周质空间，在那里它们被 SusB α-葡萄糖苷酶和 SusA 支链淀粉酶进一步水解（D'Elia 和 Salyers，1996a；Reeves 等，1996）。由此产生的葡萄糖被进一步输送到细胞中进行发酵。

每个 PULs 共存的碳水化合物活性酶（CAZymes）数量有所差异，这与预测的底物复杂性有关。预测的 CAZymes 数量范围从 *B. thetaiotaommicron* 果胶半乳聚糖 PULs 中的 2 种酶到果胶鼠李半乳聚糖 II PULs 中的 32 种酶（Cameron 等，2012）。目前已有几种 PULs 被广泛研究，通过对生长实验中单个基因的生物化学分析，揭示了酶在与果聚糖、卟啉、木葡聚糖、木聚糖和 α-甘露聚糖等底物的连续多糖解构活动过程中的复杂相互作用（Sonnenburg 和 Zheng，2010；Hehemann 等，2012；Larsbrink 等，2014；Cuskin 等，2015；Rogowski 等，2015）。PULs 通常含有针对一种特定多糖的酶，但是驯鹿瘤胃中尚未被培养的拟杆菌属编码的 PULs 所表征的 CAZymes 被证明可降解甘露聚糖、木聚糖、木葡聚糖和

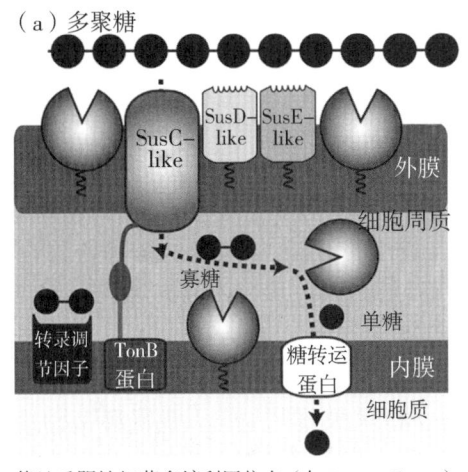 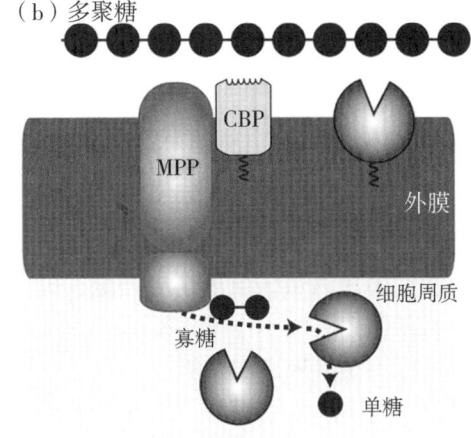

革兰氏阴性细菌多糖利用位点（如 *Prevotella* sp.）　　　革兰氏阳性细菌多糖利用位点（如 *Blautia* sp.）

图 3　革兰氏阴性（a）和革兰氏阳性（b）细菌多糖利用位点（PULs）。革兰氏阴性 PULs 的典型特征是类 SusC 脂蛋白、类 SusD 脂蛋白和糖苷水解酶的基因组共定位；而革兰氏阳性菌群的编码 ABC 转运蛋白的碳水化合物结合蛋白（CBP）和渗透蛋白（MPP）被编码。这两种方式都包括胞外（膜附着）和胞内 CAZymes 将复杂的多糖转化为单糖。

β-葡聚糖（Mackenzie 等，2015）。

目前，普遍认为拟杆菌 PULs 仅针对可溶性聚糖起作用（Koropatkin 等，2012）。然而，有研究表明土壤拟杆菌门中的 *Flavobacterium johnsoniae* 可以通过 PUL 与分泌的多模块几丁质酶共同作用有效地降解微晶纤维素同源物几丁质（McBride 等，2009；Larsbrink 等，2016）。此外，在几种草食动物肠道宏基因组中，PULs 还与假定的纤维素酶有联系（Pope 等，2010，2012；Dai 等，2012；Naas 等，2014）。但是，还没有通过分离获得的拟杆菌门代表性微生物可以通过 PULs 降解微晶纤维素。

4.4　源于革兰氏阳性菌的 PULs

尽管 PULs 长期以来一直被认为与革兰氏阴性菌（尤其是拟杆菌门）相关，近期针对厚壁菌门的研究表明，在革兰氏阳性菌中也存在类似降解细胞壁包裹的植物纤维的方式（La Rosa 等，2019；图 3b）。对人体肠道微生物 *Roseburia* 的 PULs 的基因组、转录组和蛋白质组详细分析表明，多基因位点编码和表达锚定在细胞壁上的必要的 CAZymes、结合蛋白、ABC 转运蛋白、转录调节因子和存在于细胞质的寡糖降解 CAZymes，它们将复杂的半纤维素转化为单糖。此外，对于上述 PULs 的每个组分的详尽生化特性表征也证实了关于它们代谢功能的预测（La Rosa 等，2019）。目前，反刍动物微生物中还没有关于革兰氏阳性菌 PULs 的详细报道。但是针对 Hungate 1000 基因组的分析表明，不同门包括厚壁菌门（如 *Blautia schinkii* DSM 10518）和放线菌门（如 *Bifidobacterium longum* AGR2137）都存在革兰氏阳性菌源 PULs（Seshadri 等，2018）。

4.5 产琥珀酸丝状杆菌与外膜囊泡（OMVs）

产琥珀酸丝状杆菌是奶牛瘤胃中主要的纤维素降解菌种之一，其纤维素降解机制与经典纤维素降解方式存在差异（Suen 等，2011）。产琥珀酸丝状杆菌表现出比其他种类的瘤胃细菌具有更强的降解植物纤维素的能力，因此被认为是瘤胃中关键的纤维降解菌群之一（Dehority 和 Scott，1967）。但是，它的基因组缺乏纤维素体组分锚定蛋白和粘连蛋白，同时它不编码任何外切纤维素酶，而已知的外切纤维素酶必须包含游离酶和纤维小体。产琥珀酸丝状杆菌只利用纤维素作为碳源，但它具有高度多样化的 CAZymes 和酶活性。产琥珀酸丝状杆菌编码的 31 种内切纤维素酶也不包含与纤维素结合相关的 CBMs，进一步证明酶分泌可能不是其采用的机制。

目前，科学家已经提出了几种关于产琥珀酸丝状杆菌纤维素降解机制的模型，其中部分研究所支持的观点是细胞利用纤维黏液蛋白和菌毛附着在纤维素上，使底物靠近与外膜结合的纤维素内切酶（Burnet 等，2015）。产琥珀酸丝状杆菌还产生含有 CAZymes 的外膜囊泡（OMVs），当它们从细胞中被释放后作用于植物纤维（Arntzen 等，2017；Fig. 4a）。OMVs 存在于各种生态系统中的革兰氏阴性菌中，其主要发挥水平基因转移、生物膜形成、生物通讯和生物分子传递在内的作用（Kulp 和 Kuehn，2010；Elhenawy 等，2014；Roier 等，2016）。来自产琥珀酸丝状杆菌的 OMVs 含有纤维–黏液蛋白、纤维素酶和半纤维素酶，并能降解纤维素、果胶和半纤维素。这些 OMVs 对于植物纤维的降解目前还存在争议，部分研究声称它们的产生是由于细胞的老化所致（Gaudet 和 Gaillard，1987）。然而也有证据表明，它们可能在纤维素降解中具有生物学作用（Forsberg 等，1981），使用产琥珀酸丝状杆菌 OMVs 对柳枝草进行预处理后，商业纤维素酶"鸡尾酒"的后续糖化作用增加了 2.4 倍（Arntzen 等，2017）。已有研究表明，来自人类肠道的其他能够降解植物纤维的拟杆菌门也会产生分解纤维的 OMVs（Elhenawy 等，2014），因此也有假设提出产琥珀酸丝状杆菌 OMVs 可以通过破坏木质纤维素基质的复杂结构来增加细菌对纤维素底物的利用与降解（Arntzen 等，2017）。

4.6 拟杆菌门、IX 型分泌系统和多模块 CAZymes

科学家针对尚未分离培养的瘤胃拟杆菌目微生物提出了不同的非经典纤维降解机制（Naas 等，2018）。被称为"*Candidatus* MH11"的新科包含一系列不编码纤维小体或类 PULs 系统的种群，它们利用 IX 型分泌系统（T9SS）分泌的多模块纤维素酶和半纤维素酶。T9SS 对土壤中拟杆菌降解结晶底物至关重要，例如 *Cytophaga hutchinsonii* 和 *F. johnsoniae* 分别利用 T9SS 分泌的大纤维素酶和多模几丁质酶来降解纤维素和几丁质。T9SS 对纤维降解的重要性首次被描述为 *C. hutchinsonii* 有氧滑行机制的一部分，它们具有纤维素降解能力，但缺乏外切纤维素酶以及锚定蛋白和粘连蛋白模块（Xie 等，2007）。这些细胞利用其滑动的能力附着并穿梭于纤维素上，据猜测这种滑动的能力可以通过定位更容易接触并利用的底物来帮助降解纤维素。拟杆菌门的滑行运动也在和它们具有较远亲缘关系的 *F. johnsoniae* 中得到了研究，通过 T9SS 在功能上与结晶底物的降解和分泌联系起来（Sato 等，2010；Ji 等，2012；Zhu 和 McBride，2014；McBride 和 Nakane，2015）。

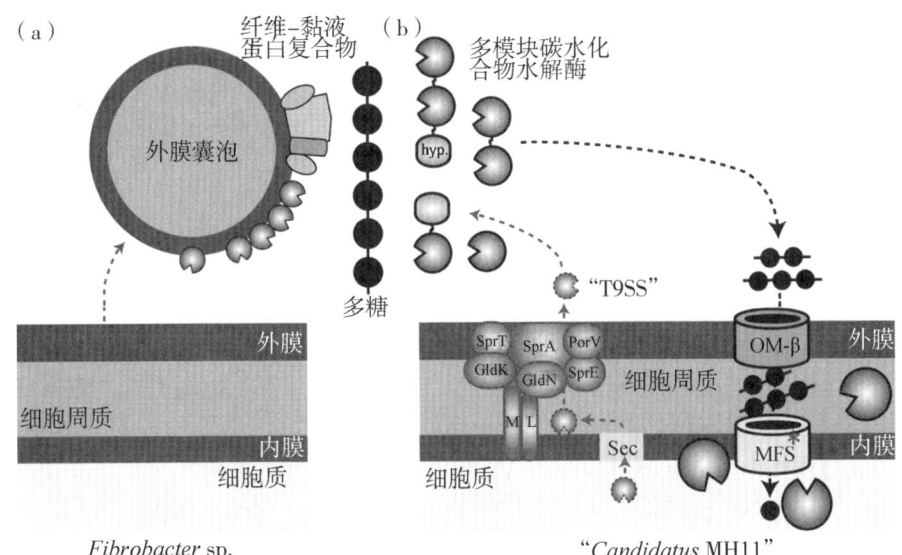

图4 产琥珀酸丝状杆菌（a）和尚未分离培养的拟杆菌科"Candidatus MH11"（b）的纤维素降解预测机制。产琥珀酸丝状杆菌产生外膜囊，包含多个 CAZymes 和纤维-黏液蛋白复合物用于利用各种纤维素、半纤维素和果胶。隶属于 Ca. MH11 的"Candidatus Paraporphyromonas polyenzyme"编码 IX 型分泌系统（T9SS）组分，该系统预计可促进多模块 CAZymes 的分泌。在 Ca. MH11 种群中 T9SS CAZymes 是游离的还是附着于细胞壁尚不清楚。

除了 T9SS 的重要性，多模块 CAZymes 最近也被证明广泛存在于细菌基因组中，其假定的作用对象包含纤维素和纤维素同系物几丁质（Talamantes 等, 2016）。例如嗜热微生物 Caldicellulosiruptor 分泌的一种多模块纤维素酶（CbCelA：GH9/CBM3c/CBM3b/CBM3b/GH48）由于其假设的内/外协同活性，在纤维素降解中发挥重要作用，其缺失能够显著降低细菌在纤维素中的生长速度（Yi 等, 2013; Young 等, 2014）。Brunecky 等的一项研究表明，从培养物上清液中纯化的 CbCelA 可能由于其域间协同作用，其活性优于商品化的外切-内切-纤维素酶混合物（Brunecky 等, 2013）。在拟杆菌门中，F. johnsoniae 的几丁质作用位点（ChiUL）的主要几丁质酶 FjChiA 与 CbCelA 相似，二者都含有 GH18 几丁质酶的一个内切-/外切作用侧翼对，其几丁质/纤维素结合域位于中间区域（Larsbrink 等, 2016）。这两种类似的酶作用于难降解底物都表现出高活性，表明了 GHs 的多模块性是一种有效的底物降解方式，也可能存在于在其他具有生物降解功能的生物中。FjChiA 也被证明是通过 T9SS 分泌的，基因敲除突变证明这种酶对几丁质代谢和细胞生长至关重要（Kharade 和 McBride, 2014）。

牛瘤胃内隶属于 Ca. MH11 科暂时命名为"Candidatus Paraporphyromonas polyenzymogenes"菌株的基因预测显示，该菌具有纤维素降解的潜力，基因组中包含超过 100 个 CAZyme 结构域其中有 17 个假定的纤维素酶、GH3 β-葡萄糖苷酶和 GH94 纤维二糖磷酸化酶（Naas 等, 2018）。值得注意的是，多个基因 accessibility 编码的 CAZymes 为多模块，包含一个特定的 C 端结构域（CTD）作用于目标蛋白用于通过 T9SS 进行分泌（图 4b）。

在绵羊瘤胃固有的6个 *Ca.* MH11 分支中也发现了类似的纤维素酶基因信息，表明 T9SS 分泌的多模块酶在瘤胃纤维素和半纤维素降解中发挥了作用。此外，对该菌株部分 CAZymes 生物化学和结构分析进一步证明这种未被培养的细菌的纤维素水解表型的预测，并证明其对线性聚合物（如无定型和结晶纤维素以及混合连接 β-葡聚糖）的降解活性（Naas 等，2018）。

5 目前的知识盲区

5.1 真核生物

虽然原核生物在数量上在瘤胃微生物组占主导地位，但真核生物也对纤维消化有重要贡献。本书第 2.8 部分（瘤胃纤毛原虫）和第 2.9 部分（瘤胃厌氧真菌）详细探讨了真核生物种群及其对瘤胃功能的贡献。由于瘤胃原虫的细胞体积相对较大，它们约占瘤胃微生物生物量的 20%（Huws 等，2018）。瘤胃厌氧真菌最早报道于 20 世纪 70 年代（Bauchop，1979；Orpin，1975），对于它们利用 CAZymes 来降解植物纤维组织已有较为深入的了解（Borneman 等，1989）。然而，对于瘤胃原虫和厌氧真菌仍然缺乏基因组层面的信息，很大程度上是因为它们难以分离、纯化培养，且由于测序技术的限制对未培养代表物种的基因组测序和注释尚有不足。2018 年研究者公布了瘤胃原虫（*Entodinium caudatum* MZG-1）的巨核基因组序列草图（Park 等，2018），有代表性的真菌基因组有 *Anaeromyces*、*Neocallimastix*、*Orpinomyces* 和 *Piromyces* 属报道（Haitjema 等，2017）。基于未培养的宏基因组数据的获取和生物信息学处理技术的快速发展，使得从未培养的厌氧真菌和原虫群体中构建代表性基因组更加便捷。例如：长片段测序技术（Oxford Nanopore、PacBio）正在提高真核基因组的组装完整性（DíazViraqué 等，2019），而基因组组装软件（EukRep）最近被设计成能够从复杂的微生物群落构建真核基因组（West 等，2018）。目前对于瘤胃原核微生物降解植物纤维组织的机制已经很全面，但是关于真核微生物（瘤胃原虫与厌氧真菌）如何在瘤胃内降解植物组织的具体机制，目前尚未被全面解析。一般来说，原虫利用多模块纤维素酶和半纤维素酶的协同作用降解植物纤维组织，同时尽管粘连蛋白结构域不具有序列上的同源性，瘤胃厌氧真菌中也存在纤维小体，其与细菌纤维小体在结构上具有相似性（Steenbakkers 等，2001；Fanutti 等，1995；Ljungdahl，2009；Haitjema 等，2017）。

5.2 瘤胃中存在多少 CAZymes 和多糖多样性？

存在于动物饲料中的植物多糖通常包含纤维素、半纤维素、淀粉或果胶。然而，在这些植物多糖中（例如半纤维素，果胶）存在丰富多样的多糖组成结构，许多尚未被解析。在较高的组成水平上，草、谷物和豆类等家养反刍动物的常见饲料来源以纤维素为主，其纤维素含量根据植物细胞定位（初级、次级等）占 15%～50%。交联多糖（例如半纤维素和果胶）在结构上更为复杂，可能含有支链或甲基、乙酰基等基因。草和谷物最常见的是木聚糖（20%～40%）和混合连接葡聚糖（10%～30%），而许多双子叶植物（如豆类）含有更高水平的木葡聚糖（20%～25%）、甘露聚糖（5%～10%）和果胶（20%～35%）（Mertens，2003；Vogel，2008；Pattathil 等，2015）。在放牧动物中，我们对其植物纤维组成的了解要少得多；然而，用于估计和/或绘制特定植物样品中"碳组成"的植物微阵列方法最近已被用于描述野生反刍动物如驼鹿和驯鹿的饮食特征（Mackenzie 等，2015；Solden 等，2018）。这些方法依赖于单克隆抗体或 CBMs（Moller 等，2007），再次证明复合纤维在放牧动物广泛的饮食来源中占主导地位，包括各种结构的甘露聚糖、木聚糖、木葡聚糖、半乳聚糖、阿拉伯聚糖和果胶（Mackenzie 等，2015；Solden 等，2018）。虽然已经鉴定了许多不同纤维物质的结构（Voiniciuc 等，2018；Wood 等，2017），但是反刍动物饮食中仍存在许多未表征的物质结构，未来仍需要进一步解析多糖固有的多样性。

瘤胃中的土著纤维降解微生物已经进化出与多聚糖多样性所匹配的丰富的 CAZymes 库，包括分解多糖骨架的内切和外切酶以及辅助酶（如 CEs、PLs 和其他脱支酶）。仅仅是主要来源于人类肠道的 964 个拟杆菌基因组的 CAZyme 图谱，就有超过 13 500 个 PULs 被分组并且估计存在大约几千种酶的组合，共同参与降解自然界中发现的各种多聚糖结构（Lapébie 等，2019）。对于瘤胃单一种群（如 *R. flavefaciens*）的分析已经显示了它们对多糖的非凡识别能力（Venditto 等，2016）；然而，对于含有数百种瘤胃纤维分解微生物的群落中许多是未培养的微生物，在这一群落层面的 CAZyme 图谱和组合的计算还尚未开展。因此，很明显还存在大量多聚糖和 CAZyme 多样性有待瘤胃微生物学家对其进行进一步研究。

5.3 功能组学研究能告诉我们什么？

得益于基因测序和生物信息学技术的迅猛发展，瘤胃微生物学家能够从瘤胃中构建成千上万个微生物基因组，同时基于功能性 RNA 和基于肽表达的研究也经历了转变。在过去几年，几种微阵列芯片已经被开发在转录水平上用于量化细菌、原虫和真菌种群菌群结构及功能分析（Abot 等，2016；Comtet-Marre 等，2018）。总的来说，尽管基于已知的 CAZymes 数据设计的这种芯片无法检测尚未进行基因组样品采样的单个群体的特定活性，这些方法已经显示出源于原核和真核的各种纤维素酶和半纤维素酶的高表达值。

定量转录组方法已被用于分析整个瘤胃微生物组的基因表达模式，并表明相似的分类群（如普氏菌科、琥珀弧菌科、纤维杆菌科）和 CAZymes 在多项研究中的普遍性（Söllinger 等，2018；Comtet-Marre 等，2017）。出乎意料的是这些研究还表明厌氧真菌和原虫贡献了大部分的纤维素和半纤维素降解酶转录本（Söllinger 等，2018；Comtet-Marre 等，2017；Qi 等，2011）。宏基因组、宏转录组或宏蛋白质组等新的多组学方法结合正在进一步提高分辨率，以具体识别单个群体及其纤维降解的机理（Solden 等，2018）。例如我们的多组学分析显示 *Candidatus Paraporphyromonas polyenzymogenes* 可在蛋白质组学数据中检测到，并在瘤胃中进行培养的植物生物质样品中得以富集，这表明复杂碳水化合物的消化是由新型 *Ca.* MH11 家族成员采用非传统的基于 T9SS 的糖分解机制完成（图 4b；Naas 等，2018）。基于阿拉斯加驼鹿基因组的宏蛋白质组数据显示超过 90% 检测到的 CAZymes 都表达自 PULs，因此多组学重申了隶属于拟杆菌的 PULs 对瘤胃纤维消化至关重要（Solden 等，2018）。初步研究表明这些方法可加深对微生物降解植物纤维的理解，但迄今为止只观察到了起作用的更大范围的群落动态变化中的一小部分，这些动态变化会随时间（如进食前后），以及动物宿主的个体因素不断变化。

6 改善纤维消化过程

瘤胃微生物群长期以来一直被视为潜在的增加纤维消化，提高动物生产力和健康，并减少甲烷排放的研究对象。在很大程度上，试图通过补充活性培养物（Chiquette 等，2007；Praesteng 等，2013；Krause 等，2001）、化学物质（Chalupa，1977）或外源性CAZymes（Beauchemin 等，2003）来直接促进特定的纤维分解物种种群，达到提高瘤胃纤维消化的目的，但是效果并不理想（Moraïs 和 Mizrahi，2019）。长期以来，人们一直认为饮食是形成两侧对称动物（尤其是反刍动物）的肠道微生物群落的主要驱动因子（Spor 等，2011；Henderson 等，2015），新的证据证实宿主的遗传学特征也同样重要。无论是肉牛（Li 和 Guan，2017）还是奶牛（Jami 和 Mizrahi，2012），即使在同样的环境下，采食相同的日粮，它们的瘤胃微生物群的个体差异也是存在的。最近，全基因组关联研究（GWAS）已经确定了可遗传的瘤胃细菌（Sasson 等，2017；Li 等，2019），并且证明奶牛的遗传变异可以导致微生物基因、种群丰度、宿主饲料转化效率和甲烷产量的差异（Roehe 等，2016；Difford 等，2018）。而且 Li 等（2019）最近的研究表明，可遗传的微生物群落还与位于牛基因组中单核苷酸多态性相关，这些单核苷酸多态性位于牛饲料转化效率的已知数量性状基因座，进一步强调了育种策略可用于操纵或选择有益的微生物群落。

于此，研究者又提出了一个新的问题：我们能否通过将宿主遗传学与微生物组功能以及饮食中特定的多糖图谱联系起来，来改善反刍动物的纤维消化？最终的想法是，我们可以理论上为特定的牛品种订制具有特定多糖结构的饮食，使它们与宿主微生物群落的酶活性或机理相匹配。然而截至目前，还未有反刍动物 GWAS 从深层功能水平阐明（多个）微生物种群中表达的代谢酶或途径如何与宿主基因型以及特定的聚糖结构相关联。此外，大多数已鉴定的可遗传种群被归为没有可培养的菌株、基因组或代谢信息的分类群。因此，我们对牛和微生物基因组之间的相互作用及其表达的酶或代谢途径如何影响纤维消化这一"共生功能体"还缺乏更深入的理解。从技术和经济角度来看，应对这样的挑战在历史上都是"越界"的；然而，随着分子生物学技术的迅速发展，研究数千个瘤胃微生物基因组，绘制动物饲料中消耗的聚糖结构，厘清饲料—肠道微生物组—宿主遗传学之间极其复杂的相互作用成为可能（Stewart 等，2018a；Seshadri 等，2018）。

因此，我们认为研究动物及其共生微生物群落（共生功能体）的高维多物种分子表型成为一种可能。这包括其基因组信息、基因组转录情况，以及这些基因在酶和相互作用的生化反应方面"产生"了什么。最终，通过研究所获得的目标基因、高效酶能够商业化，并促进遗传育种的发展，以此提高粗饲料纤维物质的降解率。

7 结论与未来发展趋势

牛瘤胃是"世界上最大的生物反应器",其为共生微生物群落提供了生存的场所(Weimer 等,2009),使宿主能够高效地利用植物纤维物质来满足自身的能量需求。培养和未培养技术的迅猛发展,提高了我们对瘤胃微生物组中存在的遗传和代谢多样性的理解。自 2010 年以来,我们对瘤胃微生物降解植物纤维的机制的了解已经从游离酶和纤维小体扩展到革兰氏阴性和阳性菌的 PULs、OMVs 和 T9SS 分泌的多模块酶。此外,最近更新的数据库资源将有助于该学科的进一步发展。例如,作为 Hungate1000 的一部分,410 个瘤胃分离菌的基因组的测序和注释提供了宝贵的资源,由此可以快速分析每个物种的 CAZyme 图谱并对它们的纤维分解潜力和策略进行评估(Seshadri 等,2018)。在短短的两年内,用于从瘤胃微生物群落构建种群基因组的宏基因组方法已经从数百个基因组提高到数千个(Stewart 等,2018b;Solden 等,2018),以此提供了大量有待挖掘的基因组和候选 CAZymes 数据。近一步深入挖掘这些基因资源很可能会发现与本章概述内容不同的 CAZymes 和纤维降解机制。

尽管这些基因组资源具有潜在的价值,但它们目前并没有解决瘤胃微生物学仍然需要克服的问题,即我们的认知是基于有详细全面研究的细菌和古菌建立起来的,而对真核生物和病毒群体几乎一无所知。瘤胃微生物学的这些有待研究的内容在动物的消化和气体排放方面有重要的作用。但是由于它们的不可培养性和基因组的复杂性,目前对这些方面还知之甚少。希望新技术,如长片段测序技术(Oxford nanopore)(Stewart 等,2018a),分别针对病毒和真核生物的特殊的宏基因组分箱软件 VirSorter(Roux 等,2015)(Emerson 等,2018)和 EukRep(West 等,2018)有助于解决目前面临的病毒和真核生物的挑战。此外,需要协同努力,将纯培养以及生化和酶学方法纳入未来揭开新的纤维分解机制的努力中,因为没有确凿的生化结果证据,我们将无法真正阐明其代谢功能。

8 更多信息

瘤胃微生物基因组学(RMG)网络成立于 2011 年,为全球使用微生物基因组学方法了解瘤胃微生物组(包括纤维降解)的研究人员合作网络提供了一个论坛(http://www.rmgnetwork.org)。关注胃肠道微生物和糖降解机制相互作用的会议包括:

- 阿伯丁大学 Rowett 研究所和法国国家农业研究所(INRA)举办的肠道微生物 Rowett-INRA 联合研讨会(Rowett-INRA Joint Symposium on Gut Microbiology)。
- 胃肠功能大会(The Congress on Gastrointestinal Function)(每两年在美国芝加哥举行一次)。
- 聚糖转化糖类活性酶戈登研究会议(The Gordon Research Conference on Carbohydrate-Active Enzymes for Glycan Conversions)(每两年在美国举行一次)。
- 糖类生物工程会议(The Carbohydrate Bioengineering Meeting)(每两年在欧洲举行一次)。

9 参考文献

Abot, A., Arnal, G., Auer, L., et al., 2016. CAZyChip: dynamic assessment of exploration of glycoside hydrolases in microbial ecosystems. BMC Genomics 17, 671.

Accetto, T. and Avguštin, G. 2015. Polysaccharide utilization locus and CAZyme genome repertoires reveal diverse ecological adaptation of *Prevotella* species. Syst. Appl. Microbiol. 38 (7), 453–61.

Arntzen, M. Ø., Várnai, A., Mackie, R. I., et al., 2017. Outer membrane vesicles from *Fibrobacter succinogenes* S85 contain an array of Carbohydrate-Active Enzymes with versatile polysaccharide-degrading capacity. Environ. Microbiol. 19 (7), 2701–14.

Attia, M. A. and Brumer, H. 2016. Recent structural insights into the enzymology of the ubiquitous plant cell wall glycan xyloglucan. Curr. Opin. Struct. Biol. 40, 43–53.

Barr, B. K., Hsieh, Y. L., Ganem, B., et al., 1996. Identification of two functionally different classes of exocellulases. Biochemistry 35 (2), 586–92.

Bauchop, T. 1979. Rumen anaerobic fungi of cattle and sheep. Appl. Environ. Microbiol. 38 (1), 148–58.

Bayer, E. A. and Lamed, R. 1986. Ultrastructure of the cell surface cellulosome of *Clostridium thermocellum* and its interaction with cellulose. J. Bacteriol. 167 (3), 828–36.

Bayer, E. A., Morag, E. and Lamed, R. 1994. The cellulosome – a treasure – trove for biotechnology. Trends Biotechnol. 12 (9), 379–86.

Bayer, E. A., Lamed, R., White, B. A., et al., 2008. From cellulosomes to cellulosomics. Chem. Rec. 8 (6), 364–77.

Beauchemin, K. A., Colombatto, D., Morgavi, D. P., et al., 2003. Use of exogenous fibrolytic enzymes to improve feed utilization by ruminants. J. Anim. Sci. 81, E37–47.

Bischof, R. H., Ramoni, J. and Seiboth, B. 2016. Cellulases and beyond: the first 70 years of the enzyme producer *Trichoderma reesei*. Microb. Cell Fact. 15 (1), 106.

Bjursell, M. K., Martens, E. C. and Gordon, J. I. 2006. Functional genomic and metabolic studies of the adaptations of a prominent adult human gut symbiont, *Bacteroides thetaiotaomicron*, to the suckling period. J. Biol. Chem. 281 (47), 36269–79.

Boraston, A. B., Bolam, D. N., Gilbert, H. J., et al., 2004. Carbohydrate-binding modules: fine-tuning polysaccharide recognition. Biochem. J. 382 (3), 769–81.

Borneman, W. S., Akin, D. E. and Ljungdahl, L. G. 1989. Fermentation products and plant cell wall-degrading enzymes produced by monocentric and polycentric anaerobic ruminal fungi. Appl. Environ. Microbiol. 55 (5), 1066–73.

Brown, R. M. 2004. Cellulose Structure and Biosynthesis: what is in Store for the 21st Century? J. Polym. Sci. A Polym. Chem. 42 (3), 487–95.

Brunecky, R., Alahuhta, M., Xu, Q., et al., 2013. Revealing nature's cellulase diversity: the digestion mechanism of *Caldicellulosiruptor bescii* CelA. Science 342 (6165), 1513–6.

Burnet, M. C., Dohnalkova, A. C., Neumann, A. P., et al., 2015. Evaluating models of cellulose degradation by *Fibrobacter succinogenes* S85. PLoS ONE 10 (12), e0143809.

Cai, S., Li, J., Hu, F. Z., et al., 2010. *Cellulosilyticum ruminicola*, a newly described rumen bacterium that possesses redundant fibrolytic-protein-encoding genes and degrades lignocellulose with multiple carbohydrate-borne fibrolytic enzymes. Appl. Environ. Microbiol. 76 (12), 3818-24.

Cameron, E. A., Maynard, M. A., Smith, C. J., et al., 2012. Multidomain carbohydrate-binding proteins involved in Bacteroides thetaiotaomicron starch metabolism. J. Biol. Chem. 287 (41), 34614-25.

Cameron, E. A., Kwiatkowski, K. J., Lee, B. H., et al., 2014. Multifunctional nutrient-binding proteins adapt human symbiotic bacteria for glycan competition in the gut by separately promoting enhanced sensing and catalysis. mBio 5 (5), e01441-14.

Cantarel, B. L., Coutinho, P. M., Rancurel, C., et al., 2009. The Carbohydrate-Active EnZymes database (CAZy): an expert resource for glycogenomics. Nucleic. Acids. Res. 37, D233-8.

Chalupa, W. 1977. Manipulating rumen fermentation. J. Anim. Sci. 45, 585-99.

Chiquette, J., Talbot, G., Markwell, F., et al., 2007. Repeated ruminal dosing of *Ruminococcus flavefaciens* NJ along with a probiotic mixture in forage or concentrate-fed dairy cows: effect on ruminal fermentation, cellulolytic populations and in sacco digestibility. Can. J. Anim. Sci. 87 (2), 237-49.

Comtet-Marre, S., Parisot, N., Lepercq, P., et al., 2017. Metatranscriptomics reveals the active bacterial and eukaryotic fibrolytic communities in the rumen of dairy cow fed a mixed diet. Front. Microbiol. 8, 67.

Comtet-Marre, S., Chaucheyras-Durand, F., Bouzid, O., et al., 2018. FibroChip, a functional DNA microarray to monitor cellulolytic and hemicellulolytic activities of rumen microbiota. Front. Microbiol. 9, 215.

Cuskin, F., Lowe, E. C., Temple, M. J., et al., 2015. Human gut Bacteroidetes can utilize yeast mannan through a selfish mechanism. Nature 517 (7533), 165-9.

D'Elia, J. N. and Salyers, A. A. 1996a. Contribution of a neopullulanase, a pullulanase, and an alpha-glucosidase to growth of *Bacteroides thetaiotaomicron* on starch. J. Bacteriol. 178 (24), 7173-9.

D'Elia, J. N. and Salyers, A. A. 1996b. Effect of regulatory protein levels on utilization of starch by *Bacteroides thetaiotaomicron*. J. Bacteriol. 178 (24), 7180-6.

Dai, X., Zhu, Y., Luo, Y., et al., 2012. Metagenomic insights into the fibrolytic microbiome in yak rumen. PLoS ONE 7 (7), e40430.

Dassa, B., Borovok, I., Ruimy-Israeli, V., et al., 2014. *Rumen cellulosomics*: divergent fiber-degrading strategies revealed by comparative genome-wide analysis of six ruminococcal strains. PLoS ONE 9 (7), e99221.

Davies, G. and Henrissat, B. 1995. Structures and mechanisms of glycosyl hydrolases. Structure 3 (9), 853-9.

Dehority, B. A. and Scott, H. W. 1967. Extent of cellulose and hemicellulose digestion in various forages by pure cultures of rumen bacteria. J. Dairy Sci. 50 (7), 1136-41.

Díaz-Viraqué, F., Pita, S., Greif, G., et al., 2019. Nanopore sequencing significantly improves genome assembly of the eukaryotic protozoan parasite *Trypanosoma cruzi*. Genome Biol. Evol. 11 (7), 1952-7.

Difford, G. F., Plichta, D. R., Løvendahl, P., et al., 2018. Host genetics and the rumen microbiome jointly associate with methane emissions in dairy cows. PLoS Genet. 14 (10), e1007580.

Ding, S. Y., Rincon, M. T., Lamed, R., et al., 2001. Cellulosomal scaffoldin-like proteins from *Ruminococcus flavefaciens*. J. Bacteriol. 183 (6), 1945-53.

Dodd, D., Moon, Y. H., Swaminathan, K., et al., 2010. Transcriptomic analyses of xylan degradation by *Prevotella bryantii* and insights into energy acquisition by xylanolytic Bacteroidetes. J. Biol. Chem. 285 (39), 30261-73.

Elhenawy, W., Debelyy, M. O. and Feldman, M. F. 2014. Preferential packing of acidic glycosidases and proteases into Bacteroides outer membrane vesicles. mBio 5 (2), e00909-14.

Emerson, J. B., Roux, S., Brum, J. R., et al., 2018. Host-linked soil viral ecology along a permafrost thaw gradient. Nat. Microbiol. 3 (8), 870-80.

Fanutti, C. C., Ponyi, T. T., Black, G. W. G., et al., 1995. The conserved noncatalytic 40-residue sequence in cellulases and hemicellulases from anaerobic fungi functions as a protein docking domain. J. Biol. Chem. 270 (49), 29314-22.

Flint, H. J., Bayer, E. A., Rincon, M. T., et al., 2008. Polysaccharide utilization by gut bacteria: potential for new insights from genomic analysis. Nat. Rev. Microbiol. 6 (2), 121-31.

Foley, M. H., Cockburn, D. W. and Koropatkin, N. M. 2016. The Sus operon: a model system for starch uptake by the human gut Bacteroidetes. Cell. Mol. Life Sci. 73 (14), 2603-17.

Forsberg, C. W., Beveridge, T. J. and Hellstrom, A. 1981. Cellulase and xylanase release from *Bacteroides succinogenes* and its importance in the rumen environment. Appl. Environ. Microbiol. 42 (5), 886-96.

Forsberg, Z., Vaaje-Kolstad, G., Westereng, B., et al., 2011. Cleavage of cellulose by a CBM33 protein. FEBS Lett. 20 (9), 1479-83.

Forsberg, Z., Mackenzie, A. K., Sørlie, M., et al., 2014. Structural and functional characterization of a conserved pair of bacterial cellulose-oxidizing lytic polysaccharide monooxygenases. Proc. Natl Acad. Sci. U. S. A. 111 (23), 8446-51.

Garron, M. L. and Cygler, M. 2010. Structural and mechanistic classification of uronic acid-containing polysaccharide lyases. Glycobiology 20 (12), 1547-73.

Gaudet, G. and Gaillard, B. 1987. Vesicle formation and cellulose degradation in *Bacteroides succinogenes* cultures: ultrastructural aspects. Arch. Microbiol. 148 (2), 150-4.

Gibson, L. J. 2012. The hierarchical structure and mechanics of plant materials. J. R. Soc. Interface 9 (76), 2749-66.

Gilbert, H. J., Knox, J. P. and Boraston, A. B. 2013. Advances in understanding the molecular basis of plant cell wall polysaccharide recognition by carbohydrate-binding modules. Curr. Opin. Struct. Biol. 23 (5), 669-77.

Gilkes, N. R., Warren, R. A., Miller, R. C., et al., 1988. Precise excision of the cellulose binding domains from two *Cellulomonas fimi* cellulases by a homologous protease and the effect on catalysis. J. Biol. Chem. 263 (21), 10401-7.

Gomez Del Pulgar, E. M. and Saadeddin, A. 2013. The cellulolytic system of *Thermobifida fusca*. Crit. Rev. Microbiol. 7828, 1-12.

Güllert, S., Fischer, M. A., Turaev, D., et al., 2016. Deep metagenome and metatranscriptome analyses of microbial communities affiliated with an industrial biogas fermenter, a cow rumen, and elephant feces reveal major differences in carbohydrate hydrolysis strategies. Biotechnol. Biofuels 9, 121.

Haitjema, C. H., Gilmore, S. P., Henske, J. K., et al., 2017. A parts list for fungal cellulosomes revealed by comparative genomics. Nat. Microbiol. 2, 17087.

Hehemann, J. H., Kelly, A. G., Pudlo, N. A., et al., 2012. Bacteria of the human gut microbiome catabolize red seaweed glycans with carbohydrateactive enzyme updates from extrinsic microbes. Proc. Natl Acad. Sci. U. S. A. 109 (48), 19786-91.

Henderson, G., Cox, F., Ganesh, S., et al., 2015. Rumen microbial community composition varies with diet and host, but a core microbiome is found across a wide geographical range. Sci. Rep. 5, e14567.

Henrissat, B. 1991. A classification of glycosyl hydrolases based on amino acid sequence similarities. Biochem. J. 280 (2), 309-16.

Hervé, C., Rogowski, A., Blake, A. W., et al., 2010. Carbohydrate-binding modules promote the enzymatic deconstruction of intact plant cell walls by targeting and proximity effects. Proc. Natl Acad. Sci. U. S. A. 107 (34), 15293-8.

Hold, G. L., Pryde, S. E., Russell, V. J., et al., 2002. Assessment of microbial diversity in human colonic samples by 16S rDNA sequence analysis. FEMS Microbiol. Ecol. 39 (1), 33-9.

Hungate, R. E. 1950. The anaerobic mesophilic cellulolytic bacteria. Bacteriol. Rev. 14 (1), 1-49.

Hungate, R. E. 1960. Symposium: selected topics in microbial ecology. I. Microbial ecology of the rumen. Bacteriol. Rev. 24 (4), 353-64.

Huws, S. A., Creevey, C. J., Oyama, L. B., et al., 2018. Addressing global ruminant agricultural challenges through understanding the rumen microbiome: past, present, and future. Front. Microbiol. 9, 2161.

Isikgor, F. H. and Becer, C. R. 2015. *Lignocellulosic biomass*: a sustainable platform for production of bio-based chemicals and polymers. Polym. Chem. 6 (25), 4497-559.

Israeli-Ruimy, V., Bule, P., Jindou, S., et al., 2017. Complexity of the *Ruminococcus flavefaciens* FD-1 cellulosome reflects an expansion of family-related protein-protein interactions. Sci. Rep. 7, 42355.

Jami, E. and Mizrahi, I. 2012. Composition and similarity of bovine rumen microbiota across individual animals. PLoS One 7 (3), e33306.

Ji, X., Xu, Y., Zhang, C., et al., 2012. A new locus affects cell motility, cellulose binding, and degradation by *Cytophaga hutchinsonii*. Appl. Microbiol. Biotechnol. 96 (1), 161-70.

Kharade, S. S. and McBride, M. J. 2014. *Flavobacterium johnsoniae chitinase* ChiA is required for chitin utilization and is secreted by the type IX secretion system. J. Bacteriol. 196 (5), 961-70.

Konietzny, S. G., Pope, P. B., Weimann, A., et al., 2014. Inference of phenotype-defining functional modules of protein families for microbial plant biomass degraders. Biotechnol. Biofuels 7 (1), 124.

Koropatkin, N. M. and Smith, T. J. 2010. SusG: a unique cell-membrane-associated alphaamylase from a prominent human gut symbiont targets complex starch molecules. Structure 18 (2), 200-15.

Koropatkin, N. M., Cameron, E. A. and Martens, E. C. 2012. How glycan metabolism shapes the human gut microbiota. Nat. Rev. Microbiol. 10 (5), 323-35.

Koshland, D. E. 1953. Stereochemistry and the mechanism of enzymatic reactions. Biol. Rev. 28 (4), 416-36.

Kostylev, M. and Wilson, D. B. 2012. Synergistic interactions in cellulose hydrolysis. Biofuels 3 (1), 61-70.

Krause, D. O., Bunch, R. J., Conlan, L. L., et al., 2001. Repeated ruminal dosing of *Ruminococcus* spp. does not result in persistence, but changes in other microbial populations occur that can be measured with quantitative 16S-rRNA-based probes. Microbiology 147 (7), 1719-29.

Krauss, J., Zverlov, V. V. and Schwarz, W. H. 2012. In vitro reconstitution of the complete *Clostridium thermocellum* cellulosome and synergistic activity on crystalline cellulose. Appl. Environ. Microbiol. 78 (12), 4301-7.

Kulp, A. and Kuehn, M. J. 2010. Biological functions and biogenesis of secreted bacterial outer membrane vesicles. Annu. Rev. Microbiol. 64, 163-84.

Laine, R. A. 1994. A calculation of all possible oligosaccharide isomers both branched and linear yields 1. 05 x 10 (12) structures for a reducing hexasaccharide: the Isomer Barrier to development of single-method saccharide sequencing or synthesis systems. Glycobiology 4 (6), 759-67.

Lairson, L. L., Henrissat, B., Davies, G. J., et al., 2008. Glycosyltransferases: structures, functions, and mechanisms. Annu. Rev. Biochem. 77, 521-55.

Lamed, R., Setter, E., Kenig, R., et al., 1983. The cellulosome: a discrete cell surface organelle of *Clostridium thermocellum* which exhibits separate antigenic, cellulose-binding and various cellulolytic activities. Biotechnol. Prog. 13, 163-81.

Lapébie, P., Lombard, V., Drula, E., et al., 2019. Bacteroidetes use thousands of enzyme combinations to break down glycans. Nat. Commun. 10 (1), 2043.

La Rosa, S. L., Leth, M. L., Michalak, L., et al., 2019. The human gut Firmicute *Roseburia intestinalis* is a primary degrader of dietary β-mannans. Nat. Commun. 10 (1), 905,

Larsbrink, J., Rogers, T. E., Hemsworth, G. R., et al., 2014. A discrete genetic locus confers xyloglucan metabolism in select human gut Bacteroidetes. Nature 506 (7489), 498-502.

Larsbrink, J., Zhu, Y., Kharade, S. S., et al., 2016. A polysaccharide utilization locus from *Flavobacterium johnsoniae* enables conversion of recalcitrant chitin. Biotechnol. Biofuels 9, 260.

Levasseur, A., Drula, E., Lombard, V., et al., 2013. Expansion of the enzymatic repertoire of the CAZy database to integrate auxiliary redox enzymes. Biotechnol. Bioeng. 6 (1), 41.

Li, F. and Guan, L. L. 2017. Metatranscriptomic profiling reveals linkages between the active rumen microbiome and feed efficiency in beef cattle. Appl. Environ. Microbiol. 83 (9), e00061-17.

Li, C., Lin, F., Li, Y., et al., 2016. A β-glucosidase hyper-production Trichoderma reesei mutant reveals a potential role of cel3D in cellulase production. Microb. Cell Fact. 15 (1), 151.

Li, F., Li, C., Chen, Y., et al., 2019. Host genetics influence the rumen microbiota and heritable rumen microbial features associate with feed efficiency in cattle. Microbiome 7 (1), 92.

Ljungdahl, L. G. 2009. A life with acetogens, thermophiles, and cellulolytic anaerobes. Annu. Rev. Microbiol. 63, 1-25.

Lombard, V., Golaconda Ramulu, H., Drula, E., et al., 2014. The carbohydrate-active enzymes database (CAZy) in 2013. Nucleic Acids Res. 42, D490-5.

Mackenzie, A. K., Naas, A. E., Kracun, S. K., et al., 2015. A polysaccharide utilization locus from an uncultured Bacteroidetes phylotype suggests ecological adaptation and substrate versatility. Appl. Environ. Microbiol. 81 (1), 187-95.

Martens, E. C., Chiang, H. C. and Gordon, J. I. 2008. Mucosal glycan foraging enhances fitness and transmission of a saccharolytic human gut bacterial symbiont. Cell Host Microbe 4 (5), 447-57.

Martens, E. C., Lowe, E. C., Chiang, H., et al., 2011. Recognition and degradation of plant cell wall polysaccharides by two human gut symbionts. PLoS Biol. 9 (12), e1001221.

McBride, M. J. and Nakane, D. 2015. Flavobacterium gliding motility and the type IX secretion system. Curr. Opin. Microbiol. 28, 72-7.

McBride, M. J., Xie, G., Martens, E. C., et al., 2009. Novel features of the polysaccharide-digesting gliding bacterium *Flavobacterium johnsoniae* as revealed by genome sequence analysis. Appl. Environ. Microbiol. 75 (21), 6864-75.

McCann, J. C., Wickersham, T. A. and Loor, J. J. 2014. High-throughput methods redefine the rumen microbiome and its relationship with nutrition and metabolism. Bioinform. Biol. Insights 8, 109-25.

McCarter, J. D. and Withers, S. G. 1994. Mechanisms of enzymatic glycoside hydrolysis. Curr. Opin. Struct. Biol. 4 (6), 885-92.

McCartney, L., Gilbert, H. J., Bolam, D. N., et al., 2004. Glycoside hydrolase carbohydrate-binding modules as molecular probes for the analysis of plant cell wall polymers. Anal. Biochem. 326 (1), 49-54.

Mertens, D. 2003. Nutritional implications of fiber and carbohydrate characteristics of corn silage and alfalfa hay. Proceedings of the California Animal Nutrition Conference, Fresno, CA, pp. 94-107.

Moller, I., Sørensen, I., Bernal, A. J., et al., 2007. High-throughput mapping of cell-wall polymers within and between plants using novel microarrays. Plant J. 50 (6), 1118-28.

Moraïs, S. and Mizrahi, I. 2019. Islands in the stream: from individual to communal fiber degradation in the rumen ecosystem. FEMS Microbiol. Rev. 43 (4), 362-79.

Naas, A. E., Mackenzie, A. K., Mravec, J., et al., 2014. Do rumen Bacteroidetes utilize an alternative mechanism for cellulose degradation? mBio 5 (4), e01401-14.

Naas, A. E., Solden, L. M., Norbeck, A. D., et al., 2018. 'Candidatus Paraporphyromonas polyenzymogenes' encodes multi-modular cellulases linked to the type IX secretion system. Microbiome 6 (1), 44.

Ohara, H., Karita, S., Kimura, T., et al., 2000. Characterization of the cellulolytic complex (cellulosome) from *Ruminococcus albus*. Biosci. Biotechnol. Biochem. 64 (2), 254-60.

Orpin, C. G. 1975. Studies on the rumen flagellate Neocallimastix frontalis. J. Gen. Microbiol. 91 (2), 249-62.

Park, T., Wijeratne, S., Meulia, T., et al., 2018. Draft macronuclear genome sequence of the ruminal ciliate *Entodinium caudatum*. Microbiol. Resour Announc. 7 (1), e00826-18.

Pattathil, S., Hahn, M. G., Dale, B. E., et al., 2015. Insights into plant cell wall structure, architecture, and integrity using glycome profiling of native and AFEXTM-pre-treated biomass. J. Exp. Bot. 66 (14), 4279-94.

Pope, P. B., Denman, S. E., Jones, M., et al., 2010. Adaptation to herbivory by the Tammar wallaby includes bacterial and glycoside hydrolase profiles different to other herbivores. Proc. Natl Acad. Sci. U. S. A. 107 (33), 14793-8.

Pope, P. B., Mackenzie, A. K., Gregor, I., et al., 2012. Metagenomics of the *Svalbard reindeer* rumen microbiome reveals abundance of polysaccharide Utilization Loci. PLoS One 7 (6), e38571.

Præsteng, K. E., Pope, P. B., Cann, I. K., et al., 2013. Probiotic dosing of *Ruminococcus flavefaciens* affects rumen microbiome structure and function in reindeer. Microb. Ecol. 66 (4), 840-9.

Qi, M., Wang, P., O'Toole, N., et al., 2011. Snapshot of the eukaryotic gene expression in muskoxen rumen—a metatranscriptomic approach. PLoS One 6 (5), e20521.

Quinlan, R. J., Sweeney, M. D., Lo Leggio, L., et al., 2011. Insights into the oxidative degradation of cellulose by a copper metalloenzyme that exploits biomass components. Proc. Natl Acad. Sci. U. S. A. 108 (37), 15079-84.

Reese, E. T. 1956. A microbiological process report: enzymatic hydrolysis of cellulose. Appl. Microbiol. 4 (1), 39-45.

Reeves, A. R., D'elia, J. N., Frias, J., et al., 1996. A *Bacteroides thetaiotaomicron* outer membrane protein that is essential for utilization of maltooligosaccharides and starch. J. Bacteriol. 178 (3), 823-30.

Reeves, A. R., Wang, G. R. and Salyers, A. A. 1997. Characterization of four outer membrane proteins that play a role in utilization of starch by *Bacteroides thetaiotaomicron*. J. Bacteriol. 179 (3), 643-9.

Roehe, R., Dewhurst, R. J., Duthie, C. A., et al., 2016. Bovine host genetic variation influences rumen microbial methane production with best selection criterion for low methane emitting and efficiently feed converting hosts based on metagenomic gene abundance. PLoS Genet. 12 (2), e1005846.

Rogowski, A., Briggs, J. A., Mortimer, J. C., et al., 2015. Glycan complexity dictates microbial resource allocation in the large intestine. Nat. Commun. 6, 7481.

Roier, S., Zingl, F. G., Cakar, F., et al., 2016. A novel mechanism for the biogenesis of outer membrane vesicles in Gramnegative bacteria. Nat. Commun. 7, 10515.

Rosewarne, C. P., Pope, P. B., Cheung, J. L., et al., 2014. Analysis of the bovine rumen microbiome reveals a diversity of Sus-like polysaccharide utilization loci from the bacterial phylum Bacteroidetes. J. Ind. Microbiol. Biotechnol. 41 (3), 601-6.

Roux, S., Enault, F., Hurwitz, B. L., et al., 2015. VirSorter: mining viral signal from microbial genomic data. PeerJ 3, e985.

Russell, J. B., Muck, R. E. and Weimer, P. J. 2009. Quantitative analysis of cellulose degradation and growth of cellulolytic bacteria in the rumen. FEMS Microbiol. Ecol. 67 (2), 183-97.

Saloheimo, M., Nakari-SetaLa, T., Tenkanen, M., et al., 1997. cDNA Cloning of a *Trichoderma reesei* cellulase and demonstration of endoglucanase activity by expression in yeast. Eur. J. Biochem. 249 (2), 584-91.

Sasson, G., Kruger Ben-Shabat, S., Seroussi, E., et al., 2017. Heritable bovine rumen bacteria are phylogenetically related and correlated with the cow's capacity to harvest energy from its feed. mBio 8 (4), e00703-17.

Sato, K., Naito, M., Yukitake, H., et al., 2010. A protein secretion system linked to bacteroidete gliding motility and pathogenesis. Proc. Natl Acad. Sci. U. S. A. 107 (1), 276-81.

Scheller, H. V. and Ulvskov, P. 2010. Hemicelluloses. Annu. Rev. Plant Biol. 61, 263-89.

Schneider, R., Hanak, T., Persson, S., et al., 2016. Cellulose and callose synthesis and organization in focus, what's new? Curr. Opin. Plant Biol. 34, 9-16.

Seshadri, R., Leahy, S. C., Attwood, G. T., et al., 2018. Cultivation and sequencing of rumen microbiome members from the Hungate1000 Collection. Nat. Biotechnol. 36 (4), 359-67.

Sheir-Neiss, G. and Montenecourt, B. S. 1984. Characterization of the secreted cellulases of Trichoderma reesei wild type and mutants during controlled fermentations. Appl. Microbiol. Biotechnol. 20 (1), 46-53.

Shipman, J. A., Cho, K. H., Siegel, H. A., et al., 1999. Physiological characterization of SusG, an outer membrane protein essential for starch utilization by *Bacteroides thetaiotaomicron*. J. Bacteriol. 181 (23), 7206-11.

Shipman, J. A., Berleman, J. E. and Salyers, A. A. 2000. Characterization of four outer membrane proteins involved in binding starch to the cell surface of *Bacteroides thetaiotaomicron*. J. Bacteriol. 182 (19), 5365-72.

Solden, L. M., Naas, A. E., Roux, S., et al., 2018. Interspecies crossfeeding orchestrates carbon degradation in the rumen ecosystem. Nat. Microbiol. 3 (11), 1274-84.

Söllinger, A., Tveit, A. T., Poulsen, M., et al., 2018. Holistic assessment of rumen microbiome dynamics through quantitative metatranscriptomics reveals multifunctional redundancy during key steps of anaerobic feed degradation. mSystems 3 (4), e00038-18.

Sonnenburg, E. D., Zheng, H., Joglekar, P., et al., 2010. Specificity of polysaccharide use in intestinal Bacteroides species determines diet-inducedmicrobiota alterations. Cell 141 (7), 1241-52.

Spor, A., Koren, O. and Ley, R. 2011. Unravelling the effects of the environment and host genotype on the gut microbiome. Nat. Rev. Microbiol. 9 (4), 279-90.

Steenbakkers, P. J., Li, X. L., Ximenes, E. A., et al., 2001. Noncatalytic docking domains of cellulosomes of anaerobic fungi. J. Bacteriol. 183 (18), 5325-33.

Stevenson, D. M. and Weimer, P. J. 2007. Dominance of *Prevotella* and low abundance of classical ruminal bacterial species in the bovine rumen revealed by relative quantification real-time PCR. Appl. Microbiol. Biotechnol. 75 (1), 165-74.

Stewart, R. D., Auffret, M. D., Warr, A., et al., 2018a. The genomic and proteomic landscape of the rumen microbiome revealed by comprehensive genome-resolved metagenomics. BioRxiv.

Stewart, R. D., Auffret, M. D., Warr, A., et al., 2018b. Assembly of 913 microbial genomes from metagenomic sequencing of the cow rumen. Nat. Commun. 9 (1), 870.

Suen, G., Weimer, P. J., Stevenson, D. M., et al., 2011. The complete genome sequence of *Fibrobacter succinogenes* S85 reveals a cellulolytic and metabolic specialist. PLoS ONE 6 (4), e18814.

Tajima, K., Aminov, R. I., Nagamine, T., et al., 1999. Rumen bacterial diversity as determined by sequence analysis of 16S rDNA libraries. FEMS Microbiol. Ecol. 29 (2), 159-69.

Talamantes, D., Biabini, N., Dang, H., et al., 2016. Natural diversity of cellulases, xylanases, and chitinases in bacteria. Biotechnol. Biofuels 9, 133.

Terrapon, N., Lombard, V., Gilbert, H. J., et al., 2015. Automatic prediction of polysaccharide utilization loci in Bacteroidetes species. Bioinformatics 31 (5), 647-55.

Tomme, P., Van Tilbeurgh, H., Pettersson, G., et al., 1988. Studies of the cellulolytic system of *Trichoderma reesei* QM 9414. Analysis of domain function in two cellobiohydrolases by limited proteolysis. Eur. J. Biochem. 170 (3), 575-81.

Vaaje-Kolstad, G., Westereng, B., Horn, S. J., et al., 2010. An oxidative enzyme boosting the enzymatic conversion of recalcitrant polysaccharides. Science 330 (6001), 219-22.

Van Tilbeurgh, H., Tomme, P., Claeyssens, M., et al., 1986. Limited proteolysisof the cellobiohydrolase I from *Trichoderma reesei*. FEBS Lett. 204 (2), 223-7.

Várnai, A., Siika-Aho, M. and Viikari, L. 2013. Carbohydrate-binding modules (CBMs) revisited: reduced amount of water counterbalances the need for CBMs. Biotechnol. Biofuels 6 (1), 30.

Venditto, I., Luis, A. S., Rydahl, M., et al., 2016. Complexity of the *Ruminococcus flavefaciens* cellulosome reflects an expansion in glycan recognition. Proc. Natl Acad. Sci. U. S. A. 113 (26), 7136-41.

Vogel, J. 2008. Unique aspects of the grass cell wall. Curr. Opin. Plant. Biol. 11 (3), 301-7.

Voiniciuc, C., Pauly, M. and Usadelb, B. 2018. Monitoring polysaccharide dynamics in the plant cell wall. Plant Physiol. 176 (4), 2590-600.

Weimer, P. J., Russell, J. B. and Muck, R. E. 2009. Lessons from the cow: what the ruminant animal can teach us about consolidated bioprocessing of cellulosic biomass. Bioresour. Technol. 100 (21), 5323-31.

West, P. T., Probst, A. J., Grigoriev, I. V., et al., 2018. Genomereconstruction for eukaryotes from complex natural microbial communities. Genome Res. 28 (4), 569-80.

Westereng, B., Ishida, T., Vaaje-Kolstad, G., et al., 2011. The putative endoglucanase PcGH61D from *Phanerochaete chrysosporium* is a metal-dependent oxidative enzyme that cleaves cellulose. PLoS ONE 6 (11), e27807.

Westereng, B., Cannella, D., Wittrup Agger, J., et al., 2015. Enzymatic cellulose oxidation is linked to lignin bylongrange electron transfer. Sci. Rep. 5, 18561.

Wilson, D. B. 2008. Three microbial strategies for plant cell wall degradation. Ann. N. Y. Acad. Sci. 1125, 289-97.

Wilson, D. B. 2009. Aerobic microbial cellulase systems. In: Himmel, M. E. (Ed.), Biomass Recalcitrance: Deconstructing the Plant Cell Wall for Bioenergy. Blackwell Publishing Limited Company.

Wilson, D. B. 2011. Microbial diversity of cellulose hydrolysis. Curr. Opin. Microbiol. 14 (3), 259-63.

Wood, T. M. and McCrae, S. I. 1979. Synergism between enzymes involved in the solubilization of native cellulose. Adv. Chem. 181, 181-209.

Wood, I. P., Pearson, B. M., Garcia-Gutierrez, E., et al., 2017. Carbohydrate microarrays and their use for the identification of molecular markers for plant cell wall composition. Proc. Natl Acad. Sci. U. S. A. 114 (26), 6860-5.

Xie, G., Bruce, D. C., Challacombe, J. F., et al., 2007. Genome sequence of the cellulolytic gliding bacterium *Cytophaga hutchinsonii*. Appl. Environ. Microbiol. 73 (11), 3536-46.

Yaron, S., Morag, E., Bayer, E. A., et al., 1995. Expression, purification and subunit-binding properties of cohesins 2 and 3 of the *Clostridium thermocellum* cellulosome. FEBS Lett. 360 (2), 121-4.

Yi, Z., Su, X., Revindran, V., et al., 2013. Molecular and biochemical analyses of CbCel9A/Cel48A, a highly secreted multi-modular Cellulase by *Caldicellulosiruptor bescii* during growth on crystalline Cellulose. PLoS One 8 (12), e84172.

Young, J., Chung, D., Bomble, Y. J., Het al., 2014. Deletion of Caldicellulosiruptor bescii CelA reveals its crucial role in the deconstruction of lignocellulosic biomass. Biotechnol. Biofuels 7 (1), 142.

Zhong, R. and Ye, Z. H. 2015. Secondary cell walls: biosynthesis, patterned deposition and transcriptional regulation. Plant Cell Physiol. 56 (2), 195-214.

Zhu, Y. and McBride, M. J. 2014. Deletion of the *Cytophaga hutchinsonii* type IX secretion system gene sprP results in defects in gliding motility and cellulose utilization. Appl. Microbiol. Biotechnol. 98 (2), 763-75.

第 12 章 瘤胃蛋白质分解和氨同化

Jeffrey L. Firkins，美国俄亥俄州立大学；
Roderick I. Mackie，美国伊利诺伊大学
（赵圣国译）

1 前言

数十年来，人们一直在研究瘤胃蛋白水解，保证含氮前体物的充足性，以支持微生物蛋白质的产生，同时最大限度地减少过度的蛋白水解，避免蛋白氮在营养、经济和环境方面的浪费（Schwab 和 Broderick，2017）。研究者提出需要充足的瘤胃降解蛋白（RDP）使得瘤胃甚至后肠中纤维降解达到最优化，以确保高采食量、生产力和饲养效率。动物产品中未被利用的过量氮也会降低动物的能量利用效率（Rced 等，2017）。但是，由于对于微生物功能（Huws 等，2018；Reed 等，2015）和调控（Foskolos 和 Moorby，2018）仍有许多不清楚的地方，难以做到反刍动物氮利用率的大幅提高。如果氮排泄量仅略低于氮摄入量，则很难通过降低氮摄入量实现氮排泄量大幅下降。

微生物蛋白水解（Walker 等，2005）、氨同化（Pengpeng 和 Tan，2013）以及氨基酸和核酸的生物合成（Morrison 和 Mackie，1997；walker 等，2005）已被综述，其中大部分来自模式细菌或同位素标记的混合微生物。关于氨基酸生物合成的认识大多来自大肠杆菌（llackmann 和 Firkins，2015b），但对瘤胃细菌的类似研究较滞后。本综述将按照图 1 中描

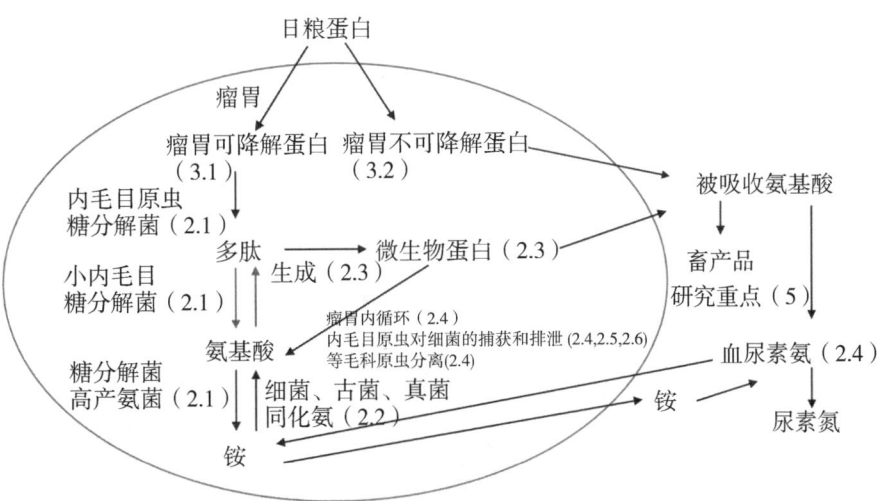

图 1 瘤胃氮代谢在日粮蛋白质向动物产品转化中的作用
（括号中的数字与正文中的章节编号相对应）

绘的整体框架，重点关注宏基因组学阐述的氮代谢新知识。对微生物蛋白质分解和合成机理的深入理解，有望帮助人们提升营养学研究水平，进而提升反刍动物生产性能，同时最大限度地减少环境负面影响。

2 微生物氮代谢

2.1 瘤胃蛋白水解和脱氨

Walker 等（2005）综述了关于瘤胃微生物蛋白酶、肽酶和脱氨酶的研究进展。尽管纤维分解菌和古菌的蛋白酶活性较低，但其他糖分解菌和原虫蛋白酶种类较多，分为胞内、胞壁和胞外酶。肽酶催化多肽分解为二肽和氨基酸。氨氮浓度偏低时，会导致奶牛瘤胃消化率、采食量和生产性能下降（Schwab 和 Broderick，2017）。在采食低品质牧草时，肉牛也需要补充充足的瘤胃可降解蛋白（RDP），以提高微生物蛋白质的合成效率。瘤胃内氨浓度与微生物蛋白质合成效率的关系（除了 RDP 本身）反映了碳水化合物的发酵能力和支链挥发性脂肪酸生长因子的利用率（Roman-Garcia 等，2016）。与肠道蛋白水解相比，瘤胃蛋白水解并不高。事实上，瘤胃未降解蛋白（RUP）的肠道消化率通常超过 80% 甚至 90%。外源蛋白酶似乎可以提高微生物对纤维（Colombatto 和 Beauchemin，2009）和淀粉（Ferraretto 等，2015）的分解能力。Griswold 等（2003）研究了在人工瘤胃连续培养时，通过添加尿素增强蛋白质水解的能力。

日粮粗蛋白降低至 15%（甚至 12%）以下时，会增加内源尿素氮转化为瘤胃细菌氮的比例（Reynolds 和 Kristensen，2008；Batista 等，2017）。通过调节瘤胃上皮尿素的转运不仅能调控瘤胃氨含量，也能调控尿素氮循环（Patra 和 Aschenbach，2018）。瘤胃细菌的脲酶活性与尿素和氨氮密切相关，脲酶基因种类很多，但大部分来自未知微生物，对于瘤胃细菌中脲酶的调控方式需要更加深入的研究（Mann 等，2018）。

研究者对微生物蛋白酶的特征进行了总结（Walker 等，2005），并与某些可培养细菌、原虫和真菌进行了关联分析（Hartinger 等，2018）。然而，宏基因组学方法以及 Hungate 1000 等测序项目（Seshadri 等，2018），为研究不同条件下蛋白酶表达机制提供了非常好的条件。例如，随着多年生黑麦草在瘤胃孵育时间的增加（1 h 与 4 h），纤维黏附细菌蛋白酶、肽酶和 AA 透性酶基因丰度增加（Mayorga 等，2016）。通过荧光原位杂交，人们发现了附着在大麦和玉米粒上的典型蛋白水解菌（Xia 等，2016）。但是，附着在颗粒物上的蛋白水解微生物的功能仍值得后续研究。因此，瘤胃取样（尤其是从口腔插管取样）时，必须有足够多的颗粒物（Paz 等，2016）。

瘤胃中真核生物比如植物、纤毛虫、真菌的蛋白酶种类和功能存在多样性（Walker 等，2005）（Hartinger 等，2018），近年来对真菌蛋白酶的研究越来越引起重视（Edwards 等，2017）。

与真菌相比，原虫需要与细菌共生（Fondevila 和 Dehority，2001），共生原因也越来越受关注（Park 和 Yu，2018a）。Hristov 和 Touanv（2005）综述了原虫水解蛋白方面的研究，提出超过一半的捕获性菌体蛋白由原虫分泌，半胱氨酸和天冬氨酸蛋白酶是原虫的主要蛋白酶，且氨肽酶活性高于脱氨酶活性（Forsberg 等，1984）。Wallace 和 McPherson（1987）提出内毛虫具有较高的蛋白水解和脱氨基能力。Diaz 等（2014）发现在特定条件下，会加剧等毛科原虫的自解。参与蛋白分解的纤毛虫数量下降，会导致瘤胃内氨浓度下降（Hlackmann 和 Firkins，2015b）。

活纤毛虫通常被用于蛋白水解的研究，因为它们通过维持细胞器的运行和囊泡运输而不是自溶，去参与日粮的消化以及内源性蛋白的转化（Diaz 等，2014）。与其他环境原虫相比，瘤胃纤毛虫泛素辅助的蛋白水解方式是相同的（Liu 等，2013）。与草履虫和四膜虫相似，瘤胃原虫通过吞噬体与溶酶体的配合，实现消化作用（Diaz 等，2014）。相较于原虫裂解液，原虫活细胞中的蛋白水解较慢，并且，低 pH 值会抑制原虫生长进而更缓慢水解蛋白。

瘤胃中植物蛋白和部分微生物蛋白脱氨过程，能为微生物特别是纤维分解菌提供氨和生长因子（如支链挥发性脂肪酸）（Walker 等，2005；Fikins 和 Yu，2015）。当蛋白质作为唯一底物时，蛋白水解菌生长缓慢（Wiallace 等，1997），相较于碳水化合物，氨基酸发酵过程产生的 ATP 较少，因此，加入碳水化合物更有利于蛋白的合成（Hackmann 和 Firkins，2015b）。

高产氨细菌（HAB）于 1980 年被发现，其在瘤胃的脱氨过程中起着重要作用（Walker 等，2005）。HAB 在瘤胃内数量较多且种类丰富。解糖型 HAB 主要依靠小肽而不是氨基酸进行生长（Wallace 等，2004；Leong 等，2016）。体外试验表明，抑制甲烷生成不会提高氢向微生物氨基酸的转化（Ungerfeld 等，2019）。因此，脱氨与还原氢代谢之间的关系仍不清楚（Hino 和 Russell，1985）。尽管大多数 HAB 分离自富含蛋白和肽的培养基，但是它们仍能利用碳水化合物（Bento 等，2015）。部分 HAB 虽然脱氨酶活性较高，但蛋白酶活性较低，对莫能菌素敏感（Shen 等，2018）。由于瘤胃内糖分解菌和产甲烷菌的影响作用，饲喂相同日粮的动物之间蛋白分解可能有所不同（Firkins 等，2007），并且蛋白分解的贡献难以量化（Rychlik 和 Russell，2000）。

2.2 氨同化

氨是日粮蛋白质和非蛋白氮（尿素和氨基酸）分解的主要终产物，也是瘤胃细菌合成蛋白质的主要氮源（Cotta 和 Russell1，1997；Morrison 和 Mackie，1997；Hallace 等，1997）。事实上，氨是许多细菌和古菌生长的优先氮源。60%~80% 的细菌氮来自氨氮（Mackie 和 White，1990），其他氮来自二肽、三肽以及氨基酸。瘤胃微生物氮在有机物中含量约 10%（Czerkawski，1976；Fessenden 等，2017），而微生物生长情况会影响氮含量（Hackmann 和 Firkins，2015b）。细菌蛋白质的合成受氨同化效率的影响很大，但我们对瘤胃细菌氨同化机制和调节的认识仍然有限。

2.2.1 铵转运

铵转运（Amt）蛋白属于膜蛋白家族，功能是实现铵跨膜转运。在细菌和古菌中，Amt 转运铵进入胞内，用于铵同化。在高氨浓度条件下，铵以被动扩散形式满足微生物的氮需求。质子化和带正电的铵以去质子化和气态形式存在，即氨。气态氨能够通过膜扩散并变成质子化铵。在 35℃ 下，铵的 pka 为 8.95，在生理 pH 值（6.5~7.5）下，氨气仅占

1%（Martinelle 和 Hlaggstrom，1997）。因此，微生物细胞需要转运铵，事实上铵转运系统在瘤胃和人体细菌中非常常见。

气态氨扩散是氨跨膜运输的重要形式，当铵浓度很低时，转运蛋白 AmtB 开始发挥主要作用（Soupene 等，1998，2002；van Heeswijk 等，1996；Winkler，2006）。即便在缺乏铵转运促进剂条件下，大肠杆菌、枯草芽孢杆菌、谷氨酸杆菌、肠沙门氏菌和酿酒酵母仍然能够在高浓度的铵/氨条件下很好的生长（Detsch 和 Stulke，2003；Meier-Wagner 等，2001；Soupene 等，1998）。混合瘤胃微生物胞内铵浓度比胞外高时，表明铵发生了主动转运（Russell 和 Strobel，1987）。当铵的浓度过低时，则需要转运铵通过胞质膜进入胞内。铵类似物如甲基铵和乙基铵被用于研究铵跨膜转运研究，但应谨慎使用。主要原因在于，转运蛋白对其配体的选择性非常高，特别是对于像铵这样的小分子，导致不同类似物间的扩散程度不同，可能会使研究结果产生偏差（Kleiner，1982；Kleiner 和 Castorph，1982；Stevenson 和 Silver，1977）。铵进入细胞质后通过铵同化转化成谷氨酸和谷氨酰胺，这两个氨基酸是细胞内氮循环的关键代谢中间体，谷氨酸是细胞中最丰富的代谢物，在大肠杆菌胞内浓度为 96 mmol/L。谷氨酸通过 α-酮戊二酸直接将氮代谢与碳代谢联系起来（Bennett 等，2009）。不管氮底物条件如何，细菌主要将铵转化成谷氨酸和谷氨酰胺，然后用于含氮化合物包括氨基酸、嘌呤、嘧啶等的合成。

2.2.2 铵同化的酶促反应

目前关于肠道细菌铵同化和调控认识主要来自变形杆菌（大肠杆菌、克雷伯氏菌、沙门氏菌）和芽孢杆菌（革兰氏阳性菌典型菌），但这些细菌并不一定能反映肠道优势细菌如拟杆菌门和厚壁菌门（Reitzer，2003；Van Heeswi 等，2013）。枯草芽孢杆菌中铵同化的调控比较经典（Gunka 和 Commichau，2012）。细胞铵浓度和氮水平不同，产生不同的铵利用途径（图2），包括高亲和力途径和低亲和力途径。高亲和力途径是在低铵浓度下发生，包括三种酶即 ATP 依赖型谷氨酰胺合成酶（GS）、谷氨酸合成酶（GOGAT）和铵转运蛋白（AmtB）。相比之下，低亲和力途径在高铵浓度下发生，由 NAD（P）H 氧化型谷氨酸脱氢酶（GDH）组成。在枯草芽孢杆菌中，GS/GOGAT 系统仅负责铵的同化，而 GDH 负责谷氨酸分解代谢的逆反应。这两个途径在能量消耗和催化机制方面具有差异。在大肠杆菌中，转录调控和翻译后修饰的巧妙调控协调了两种途径的贡献。

由于高亲和力途径比低亲和力途径更需要能量，所以微生物细胞需要在最小化能量消耗的情况下维持生长。变形菌等肠道微生物同样通过这两种途径利用氮，成为经典模式（Van Heeswijk 等，2013）。

2.2.3 调控机制

利用革兰氏阴性菌和革兰氏阳性菌模型，研究人员提出了一种调控模型（Van Heeswijk 等，2013）。氨同化基因表达通过 NRI/NRII 双组分系统和 Nac 进行调节，Nac 本身也受 NRI/NRII 的转录调控。通过修饰 GS 和 GDH 实现酶促的快速抑制或激活，酶活性受调节蛋白 GlnB（也称为 P-II）、ATase（腺苷酸转移酶/腺苷酸去除酶）和 UTase（尿苷酰转移酶/尿苷酰去除酶）的调节。在枯草芽孢杆菌中，转录调节因子包括 TnrA 和 GlnR，它们通过蛋白质与功能酶的相互作用来调节转录（Fisher，1999）。通过调节酶促反应及其调节途径，细胞将胞外铵转移到细胞内谷氨酸和谷氨酰胺中。

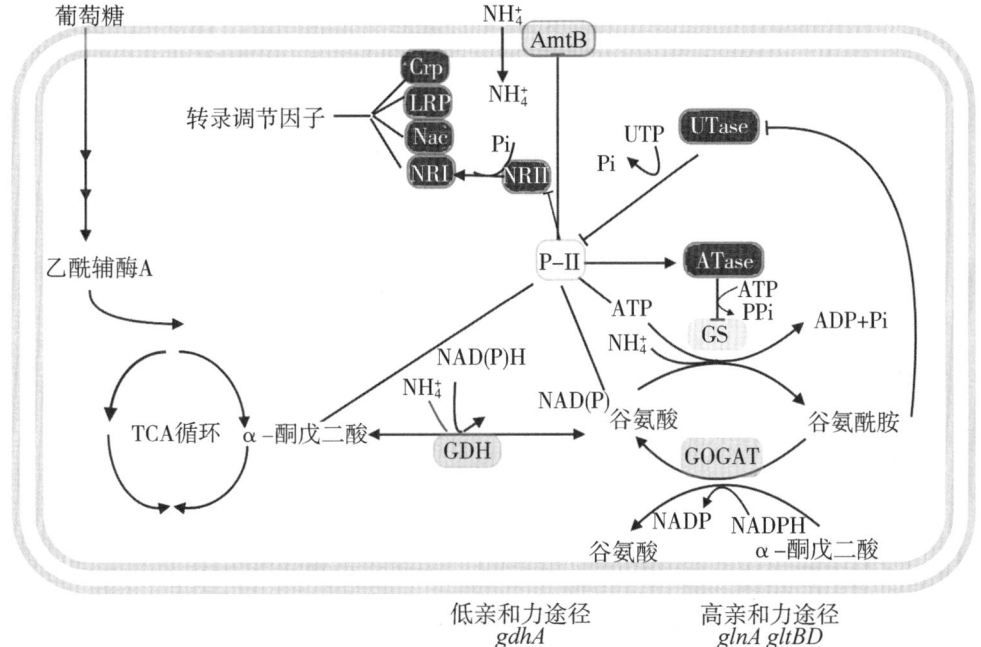

图 2 基于大肠杆菌的氨同化途径

低亲和力和高亲和力途径包括谷氨酸脱氢酶（GDH，基因 *gdhA*）、谷氨酰胺合成酶（GS，基因 *glnA*）、谷氨酸合成酶（GOGAT，基因 *gltB*）和铵转运蛋白（AmtB，基因 *amtB*）。调节因子包括蛋白活性相关的 UTase、ATase 和 GlnB（P-II），以及转录调节因子 NRI、NRII 和 Nac，标深灰色底纹的蛋白在拟杆菌中没有发现序列同源性（Iakiviak，2018）。

在大肠杆菌中，蛋白受 NRI/NRII 系统以及 Nac、CRP-CAMP、IHF、LrD 和 ArgR 的转录调控（Van Heeswijk 等，2013），传感蛋白 NRII 与铵结合后发生磷酸化，随后将磷酸基团转移给反应调节蛋白 NRI，磷酸化的 NRI 增加 *glnA*、*glnk*、*amtB*、*nac* 和其他基因的转录（Magasanik，1989）。Nac 在没有辅助因子或共价修饰的情况下仍能抑制 *gdhA* 转录。通过增加 *gltBD* 的转录来应对 Lrp 氨基酸的缺乏，并通过 Crp-CAMP 抑制 *gltBD* 表达和调节 *glnA* 的基础表达来应对能量（ATP）的缺乏（Van Heeswiik 等，2013）。

在厚壁菌、枯草芽孢杆菌模型中，已经确定了三个转录调节因子，即 GInR、TnrA 和 GItC（Fisher，1999；Fisher 和 wray，2002；Schumacher 等，2015；Wray 等，2001）。GInR 的转录活性受 GS 结合介导，与 DNA 稳定结合，并受 pH 值的影响。相反，当与谷氨酰胺合成酶（GS）结合时，TnrA 是失活的。此外，TnrA 与膜上 Glnk 和 AmtB 结合后，从 DNA 中分离。最后，细胞快速成长时，GltC 负责激活谷氨酸合成酶的转录以满足对谷氨酸的需求（Gunka 和 Commichau，2012）。P-II 蛋白是重要调节蛋白，能感应代谢物水平并调节酶转录和活性。P-II（GlnB）和 P-II 样蛋白（Glnk）由 *glnB* 和 *glnk* 编码，*glnk* 通常与 *amtB* 相邻（Arcondeguy 等，2001；Blauwkamp 和 Ninfa，2002；Detsch 和 Stulke，2003；Forchhanmer，2008；Van Heeswijk 等，1996）。P-II 蛋白是同

源三聚体，拥有 a-酮戊二酸和 ATP 的结合位点，以及脲酶中谷氨酰胺尿苷酰化结合位点。此外，分枝杆菌（mycobacteria）P-Ⅱ 蛋白可以进行腺苷酸化修饰，而蓝藻中 P-Ⅱ 蛋白进行磷酸化修饰（Forchharmer，2008；Gunka 和 Commichau，2012；Williams 等，2013）。变形菌 AmtB、ATase、NRⅡ 和脲酶以及枯草芽孢杆菌 TnrA 直接与 P-Ⅱ 结合。P-Ⅱ 通过直接插入转运通道三聚体 AmtB 中，抑制铵转运，P-Ⅱ 的脲酶尿苷酰化能抑制其调节作用（Reitzer，2003）。

变形杆菌Ⅰ型谷氨酰胺合成酶（GS-Ⅰ）活性受 ATase 共价修饰调节。ATase 诱导十二聚体 GS-Ⅰ 的一个亚基腺苷酸化后失活，由于腺苷酸化状态不同（0~12 之间），GS-Ⅰ 活性多样变化。ATase 还能通过多肽去腺苷酸化激活 GS-Ⅰ。P-Ⅱ 对 ATase 的调节活性取决于其通过 UTase 对 P-Ⅱ 进行尿苷酰化或脱尿苷酰化的能力，UTase 尿苷酰化和去尿苷酰化活性受谷氨酰胺和其他小分子的影响（图3）。最后，氨同化基因的转录也受到 P-Ⅱ 与 NRⅡ 的相互作用的影响。当能量和氮充足时，代谢物 ATP 和 alpha-酮戊二酸调控 P-Ⅱ 和 NRⅡ 之间的相互作用，导致自磷酸化水平下降。将肠道细菌调控机制扩展到瘤胃中优势菌普雷沃氏菌（*Prevotella*）和瘤胃球菌（*Ruminococcus*）是不适用的，因为这两类菌都缺失铵同化调节因子同源物，即双组分系统（NRⅠ/NRⅡ）、共价修饰（ATase/UTase）和转录因子。此外，在不同氮利用率下拟杆菌属的优势酶活性研究结果存在矛盾。

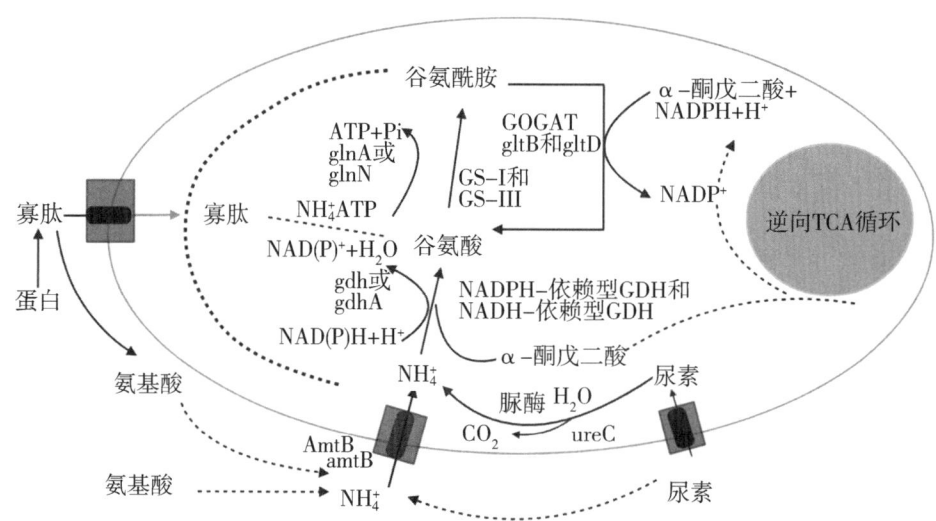

图3 白色瘤胃球菌（*Rumminococcus albus*）8 的氮代谢途径，glnA、glnN、gdh、gdhA、amtB、gltBD、ureC 代表基因名称（Kim 等，2014）

2.2.4 瘤胃细菌氨同化

人结肠拟杆菌具有与上述模型不同的氨同化和调节途径，体现在氮限制下 GDH 活性增加、铵过量下 GS 转录增加，以及新谷氨酰胺合成酶（GSⅢ）（Kim 等，2014，2017）。肠道拟杆菌内谷氨酰胺合成酶和谷氨酰胺脱氢酶的基因功能仍不清楚。尽管对拟杆菌内铵同化的研究不多，但在与经典途径不同的新机制方面已经有了进展。

研究者对两种瘤胃细菌：*Ruminococcus albus*（厚壁菌门内的革兰氏阳性菌，具有植物细胞壁降解活性；kim 等，2014）和 *Prevotella ruminicola*（拟杆菌门内的革兰氏阴性细菌；Kim 等，2017）的研究，总结了氨同化和氮代谢相关进展。

对白色瘤胃球菌 8 基因组序列（3.8Mb）分析，揭示了许多编码参与氮代谢和氨同化的重要酶基因（图 3）。研究人员鉴定了 NADH-依赖型 GDH（*gdh*）、NADPH-依赖型 GDH（*gdhA*）、高亲和性铵转运蛋白（*amtB*）、调节蛋白（*glnk*）、脲酶、两种 GS（Ⅰ型 GS）的基因 [*glnA*] 和Ⅲ型 GS [*glnN*]），以及 GOGAT 两个亚基（大亚基 [*gltB*] 和小亚基 [*gltD*]）。白色瘤胃球菌 8 以氨、尿素、肽或氨基酸作为氮源生长，通过分析氮代谢基因转录丰度和酶活性，揭示了氮利用情况。白色瘤胃球菌 8 利用氨和尿素，具有相似的生长规律。白色瘤胃球菌 8 也能够利用肽，尽管生长量低于首选氮源（尿素和氨），但最大生长率没有变化，这是白色瘤胃球菌 8 使用肽作为氮源的第一份试验报告。白色瘤胃球菌 8 还产生几种肽转运蛋白和肽酶，用于摄取和利用肽，包括三肽 ABC 转运蛋白、二肽 ABC 转运蛋白和 25 个参与肽代谢的基因。这些结果与瘤胃中氮利用方式一致，由于游离氨基酸迅速脱氨基产生氨，所以氨基酸常以肽形式转运至细菌细胞内，以便节约能量。

Prevotella ruminicola 23 不分解纤维素，但可以降解半纤维素和果胶以及宿主的蛋白多糖。*P. ruminicola* 23 可以利用氨和肽（优选较大的肽，高达 2kDa）作为生长的氮源。*P. ruminicola* 23 具有广泛和高活性的二肽基肽酶。图 4 展示了 *P. ruminicola* 23 响应氮源和铵浓度的转录组变化、蛋白质组变化和铵同化酶活性变化（图 4）。

P. ruminicola 23 可以利用肽和铵作为生长氮源，但不能利用氨基酸（Kim 等，2017）。当该细菌以肽为氮源时，铵同化途径未被激活。蛋氨酸作为甲基或甲硫醇（CH_3S）供体而不是氮源，对 *P. ruminicola* 23 的生长至关重要。

P. ruminicola 23 对氮变化的响应与变形杆菌如大肠杆菌或沙门氏菌不同（Kim 等，2017）。表明它们在氮代谢基因转录调控和酶活性大小方面有所不同（图 3）。先前的研究表明，*P. ruminicola* 具有 NADP-（合成代谢）和 NAD-依赖型（分解代谢）GDH 活性。相应地，在过量的铵浓度下，*gdhA* 转录丰度增高，NADP-GDH 活性增强。在铵限制条件下，细菌的主要铵同化途径为 GS-GOGAT 途径。然而，在非限制性铵条件下，*P. ruminicola* 23 被高度诱导。*P. ruminicola* 23 可以在铵底物条件下利用高底物亲和力的 GS-GOGAT 进行生长，这种模式在白色瘤胃球菌 8 中也有。据我们所知，这是第一次报道在非限制性铵条件下，细菌使用 GDH 和 GS-GOGAT 两种途径进行铵同化。研究报道 GS-GOGAT 系统需要 ATP，当氨同化依赖于 GDH 时，GS-GOGAT 的下调会减少能量浪费，GDH 和 GS-GOGAT 途径共同维持用于氨基酸合成的谷氨酸含量。当氨同化通过 GDH 途径进行时，GS-GOGAT 途径将从"耗能"下调为"节能"，GDH 和 GS-GOGAT 途径也通过维持谷氨酸的含量来进行氨基酸的合成。当瘤胃内氮浓度达到生长抑制浓度时（正常铵浓度范围为 4～70 mmol/L），*P. ruminicola* 23 仍可进行生长。这种铵同化模式将使 *P. ruminicola* 23 保持快速生长，优先竞争性利用铵作为氮源，这可能是普雷沃氏菌作为瘤胃优势细菌原因之一。

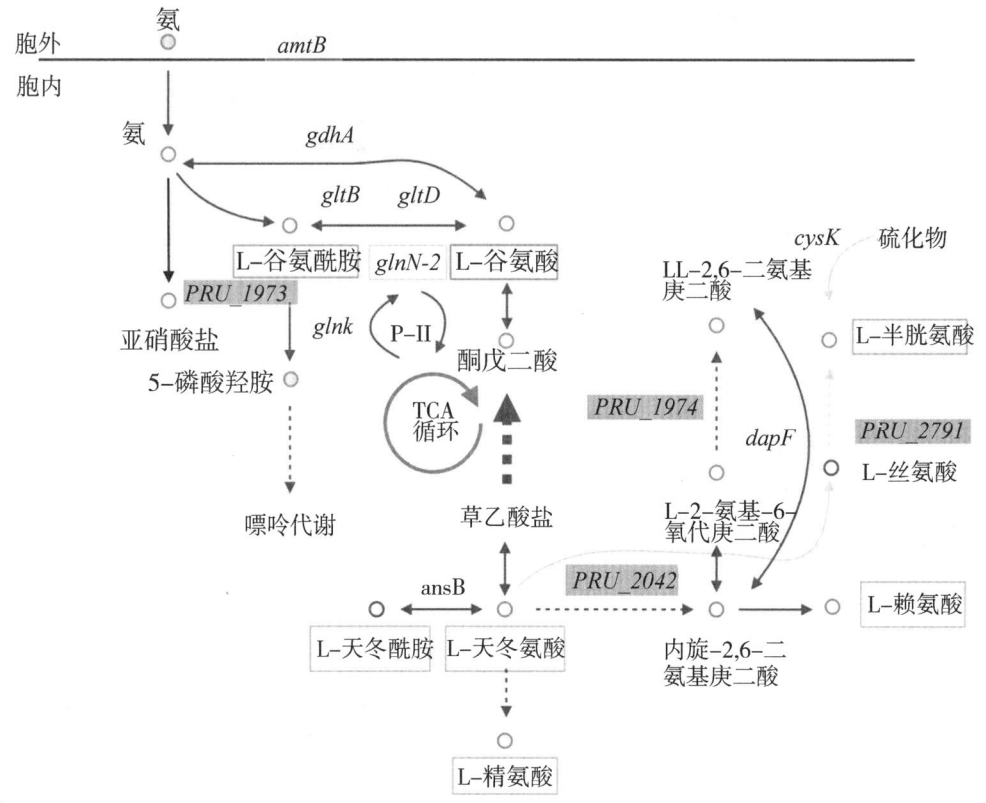

关键基因/位点	编码蛋白	途径
amtB	铵转运蛋白	
glnK	氮调控蛋白 P-Ⅱ	
gdhA	谷氨酸脱氢酶（NADP-依赖型）	谷氨酸、谷氨酰胺代谢
glnN-2	谷氨酰胺合成酶Ⅲ-2型	谷氨酸、谷氨酰胺代谢
gltB	谷氨酸合成酶，大亚基	谷氨酸、谷氨酰胺代谢
gltD	谷氨酸合成酶，小亚基	谷氨酸、谷氨酰胺代谢
PRU_1974	O-乙酰高丝氨酸氨丙基转移酶	赖氨酸合成
cysK	半胱氨酸合成酶 A	半胱氨酸合成
dapF	二氨基庚二酸异构酶	赖氨酸合成
PRU_1974	同源转氨酶	赖氨酸合成
PRU_2042	二氨基庚二酸脱氢酶	赖氨酸合成
PRU_1973	谷氨酰胺转氨酶	谷氨酸代谢
asnB	天冬酰胺合成酶	天冬酰胺合成

图 4　普雷沃氏菌 23（*P. ruminicola* 23）的氮代谢网（Kim 等，2017）。当 *P. ruminicola* 23 在高铵浓度下生长时，GDH 和 CS/GOGAT 的氨同化效率达到最大。化合物用灰色圆圈表示，含氮物用绿色表示，含硫物用蓝色表示。红色箭头代表氨诱导途径，黄色箭头代表氨基酸或肽诱导途径（彩图 6）。

基因组和蛋白组结果表明当氮浓度不被限制时，GSⅢ-2 是参与铵同化主要的酶。铵能诱导 GOGAT 基因转录丰度提高 22 倍，所以，酶 GSⅢ-2-GOGAT 互作在 *P. ruminicola* 23 的铵同化和循环过程起重要作用。铵转运体基因（*amtB*）在铵浓度足够高时转录丰度增加，因此，*P. ruminicola* 23 对环境内铵浓度敏感且受其浓度调控。当铵浓度较低时，细菌诱导肽和多胺 ABC 转运蛋白的生成。研究表明，以二肽或三肽形式转运多肽（每个氨基酸消耗 1/2 或 1/3 ATP）相较于转运单个游离氨基酸（每个氨基酸消耗 1 个 ATP）更节能。当铵浓度足够高时，氮调控蛋白（P-Ⅱ）诱导 *glnK* 的转录丰度增加。最新研究表明，P-Ⅱ 蛋白存在信号转运机制：该机制发生在翻译完成后，胞内 α-酮戊二酸通过调控铵转运蛋白通道影响铵浓度，导致氮浓度发生波动，从而诱导 ATP 酶活发生变化。*amtB* 基因的上游和下游都有 *glnk* 基因，该基因连锁高度保守且与蛋白产物功能高度相关。

当铵浓度足够高时，*P. ruminicola* 23 通过 NADP-GDH 和 GSⅢ-2-GOGAT 途径进行铵同化。*P. ruminicola* 23 通过产生谷氨酸、维持胞内谷氨酸/谷氨酰胺比例、嘌呤/嘧啶比例以及合成细胞壁来适应瘤胃内高铵浓度。相反，当铵浓度较低时，*P. ruminicola* 23 仍可利用 GDH/GS-GOGAT 途径去进行铵同化和生成氨基酸。

2.2.5 古菌氮代谢

对古菌的氮代谢研究相较于细菌较少，产甲烷菌作为一种古菌，它利用氮合成氨基酸、嘌呤和嘧啶，并且铵同化途径与细菌相同。

全基因组测序为古菌种类的研究提供了新方法，有助于研究产甲烷菌氮代谢途径。史密斯产甲烷短杆菌（Methanobrevibacter smithii）作为人体肠道内的优势菌，可优先利用铵作为氮源。*M. smithii* 含有一个铵转运基因（*AmtB*；MSM0234）及两条铵同化途径：①ATP 依赖型谷氨酰胺合成酶-谷氨酸合成酶途径，该途径对铵具有高亲和力；②非 ATP 依赖型谷氨酸脱氢酶途径，该途径对铵亲和力较弱（Samuel 等，2007）。Hansen 等（2011）利用含氨培养基对 5 株 *M. smithii* 进行 RNA 测序，结果表明 GS-GOGAT 和 GDH 途径在测序菌种中均被表达，其中 0.4%~1.21% 的测序序列来源于氨同化酶基因。

对海沼甲烷球菌（*Methanococcus maripaludis*）和马氏甲烷球菌（*Methanosarcina mazei*）的研究进一步丰富和扩展了对产甲烷菌氮代谢调控方面的认知（Leigh 和 Dodsworth，2007）。这两种菌种通过 GS/GOGAT 途径进行氨同化以及固氮，两者均含有固氮操纵子。氮浓度调控 *GlnA* 和固氮基因表达，也调控 GS 和固氮酶活性。对海沼甲烷球菌和马氏甲烷球菌的研究，发现了新型转录抑制因子 NrpR，揭示 *GlnK* 可以直接调控 GS，另外 *NifI* 可以直接调控固氮酶基因表达（Leigh 和 Dodsworth，2007）。

2.2.6 瘤胃厌氧真菌

Lowe 等（1985）研究发现瘤胃真菌可利用氨生成氨基酸，Orpin 和 Greenwood（1986）也证实了氨基酸（谷氨酸、丝氨酸和蛋氨酸）可以促进新美鞭菌（*Neocallimastix patriciarum*）生长。目前对瘤胃内真菌氨基酸代谢以及可合成氨基酸的真菌种类研究较少（Theodorou 等，1994）。通过同位素富集法，测定了不同培养基中 $^{15}NH_4Cl$ 的代谢变化，探究单鞭毛菌（*Piromyces communis*）和新美鞭菌（*Neocallimastix frontalis*）的氨基酸合成过程（Atasoglu 和 Wallace，2002）。当加入 1g/L 胰蛋白酶到培养基中后，单鞭毛菌和新美鞭菌可利用氨基酸氮进行氨基酸合成。肽/氨基酸浓度较高时，氨基酸合成减慢。当在培养基中加入胰蛋白酶酶解酪蛋白或氨基酸时，赖氨酸合成减慢，并且赖氨酸可能抑制瘤胃内真菌生长。将来还需要采用更先进的方法去进一步研究氮源对瘤胃发酵和代谢的影响。

2.2.7 纤毛虫

纤毛虫作为瘤胃内体积最大且含量最多的原虫，主要分为两大类，即等毛科原虫（前庭目）和内毛（内毛目）虫，这两类原虫在形态结构和代谢方面均不相同。等毛科原虫主要捕获微粒，等毛科原虫细胞质囊泡中含有产甲烷菌（Williams 和 Coleman，1997）。细菌为原虫生长提供主要氮源，细菌被吞噬后，在消化泡中被原虫消化。将大肠杆菌中的氨基酸进行 ^{14}C 标记，随后与内毛虫进行培养，结果发现部分被标记的氨基酸与原虫蛋白进行结合，其余形成游离氨基酸或被排放出去。细菌的核酸碱基可以结合到原虫核酸内，以核苷酸形式进行转移（Williams 和 Coleman，1997）。

瘤胃内的等毛科原虫主要为等毛虫属和多毛虫属，与瘤胃细菌相比，它们体积小且流动性强，为椭圆型，纤毛分布于细胞表面。等毛虫属和多毛虫属都可以吞噬瘤胃细菌和非瘤胃细菌，细菌被吞噬后，迅速失活随后被降解。等毛科原虫不仅通过吞噬和分解细菌来获取氮源，也通过外界环境来获取氨基酸，这些氨基酸被直接同化，也可被分解为其他含氮物（Williams 和 Coleman，1997）。

2.3 微生物蛋白合成

游离的氨基酸可以提高微生物生长效率（Hackmann 和 Firkins，2015b），反刍动物大部分微生物蛋白质来自氨同化合成的氨基酸，但常在检测时被低估（Ahvenjärvi 等，2018）。为了提高对微生物蛋白向十二指肠流动研究的准确性和精密性，我们对微生物蛋白合成研究进展进行了综述（Firkins 等，2007；haringer 等，2018）。然而，现阶段对体内微生物蛋白合成研究结果存在较大差异（White 等，2016），所以结果准确性和精密性将有助于揭示奶牛和肉牛中瘤胃降解蛋白质（RDP）与碳水化合物互作机制（Galyean 和 Tedeschi，2014；Van Amburgh 等，2015；White 等，2017b）。氨浓度必须足够高，才能有利于微生物蛋白合成（Schwab 和 Broderick，2017），但最佳氨浓度很难被预测（Firkins 等，2007），因为氨浓度与微生物需要量密切相关（Ahvenjärvi 和 Huhtanen，2018）。氨基酸氮浓度会成为微生物蛋白合成效率的限制因素，尤其是碳水化合物量多时，因此促进氨同化和增加游离氨基酸可以提高微生物生长效率（Firkins 等，2007）。因此，需将氮代谢与碳水化合物代谢进行结合研究。随着氨的减少，ATP 依赖型氨同化效率增加（第 2.2 节），ATP 产生途径也受到影响（Hackmann 等，2017），用于细胞生长的能量比例增加（Hackmann 和 Firkins，2015b）。

通常用合成量和分解量的差值来评估微生物蛋白产量。过瘤胃流通速率的提高会增加原虫外流和自解（Firkins 等，2007），但对其研究有限，尤其是等毛属纤毛虫（也称全毛虫科）（Diaz 等，2014）（Firkins 和 Yu，2015）。早期研究认为原虫蛋白不外流，这其实导致肠道内微生物氨基酸（特别是赖氨酸）供应量预测值存在误差（Sok 等，2017；Fessenden 等，2017）。有报道指出，瘤胃液相和固相细菌氨基酸组成不同。细菌蛋白流量会随着过瘤胃未降解纤维的增加而增加（Sauvant 和 Nozière，2016），过瘤胃氮和循环氮都与蛋白质分解、氨同化有联系，这方面内容将在随后进行讨论。

2.4 瘤胃内微生物蛋白氮循环

微生物蛋白质合成是通过自身细胞生长补充瘤胃蛋白外流损失的重要过程。细胞裂解比氨同化更浪费能量（Firkins等，2007），原虫被认为会导致日粮和微生物蛋白质的过度水解和浪费（Hristov和Jouany，2005）。但是，在体内很难评估氮循环情况，主要是由于$^{15}NH_4$的添加和混匀误差较大。研究人员利用^{15}N标记饲料和严谨取样方法（Ahvenjärvi等，2018），发现奶牛瘤胃中大多数原虫蛋白来源于细菌蛋白。奶牛瘤胃内氮循环率约为22%，低于绵羊中的氮循环率。瓣胃中15%的微生物氮来自原虫氮（Fessenden等，2019；Sok等，2017），而不是所谓原虫氮全部在瘤胃中循环。事实上原虫氮中40%是在瘤胃中循环的，主要来自等毛虫。

瘤胃内氮循环量可能被过高估计，原因包括：①细菌细胞被原虫捕获消化量估计过高；②原虫数量估计过高。在研究原虫捕获细菌时，常常使原虫处于饥饿状态，所以过高估计了原虫对细菌的捕获量。Williams和Coleman（1992）给混合内毛目原虫分别饲喂10、52、260、1 670个米粒/原虫细胞时，大肠杆菌的消失率分别下降0%、0%、18%和68%。在比较瘤胃内氮循环量时，也要考虑动物是否去原虫，因为原虫与细菌真菌的相互作用会导致结果发生偏差（Newbold等，2015）。

有人认为50%原虫量在瘤胃中循环，其实这个数据被估计的偏高（Wenner等，2018），原因包括未考虑细菌污染（Sylvester等，2005；Firkins等，2007）、标记不合理（Sok等，2017）以及错误认为所有原虫在瘤胃循环（事实上主要是等毛虫）（Diaz等，2014）。Karnati等（2007）根据消化残渣在瘤胃内滞留时间，通过细胞计数方法确定了奶牛原虫（主要是内毛虫）的消长时间。有研究认为，原虫在瘤胃中滞留量要比向瓣胃中排放量更多（Hook等，2007），但事实上该研究忽略了采样时瓣胃内水过量进入而导致的稀释作用，以及皱胃酸液倒流导致的原虫细胞自溶问题（Firkins和Yu，2006）。对于低采食量的绵羊来说，原虫可能占到微生物生物量的50%（Firkins等，2007），但对于奶牛而言可能只占25%（Ahvenjärvi等，2018）。

原虫捕获细菌、真菌游动孢子和部分较小原虫（Hartinger等，2018），但是无法完全消化它们（Hristov和Jouany，2005），导致微生物蛋白合成效率降低。真菌作为重要的纤维降解微生物（Edwards等，2017），原虫对真菌的拮抗作用可能对宿主不利。原虫对功能性细菌如纤维分解菌的选择性捕食（Park和Yu，2018b），可能会影响纤维消化率（Newbold等，2015）。瘤胃宏转录组学研究证实了原虫和真菌在影响纤维消化中的重要性（Comtet-Marre等，2017），这项研究建议重新考虑对原虫的传统功能认识，比如消耗O_2或快速摄取淀粉。通过选择性抑制等毛虫定植可能有利于瘤胃氮循环（Newbold等，2015），但是疫苗等抑制方法仍不完善（Hartinger等，2018）。

对原虫的控制要进行综合考虑。例如，原虫数量与单位干物质采食量产生甲烷量呈正相关，同时也与纤维消化率和干物质采食量呈负相关（Guyader等，2014），所以抑制甲烷时要考虑干物质采食量变化（Ungerfeld，2018）。同样，奶牛日粮蛋白质含量（可能RDP）与干物质采食量呈正相关（Zanton，2016），在低蛋白日粮条件下，原虫可能通过摄取更多细菌氮和尿素氮进行适应性生长（Oelker等，2009），而随着日粮蛋白含量的增加，原虫利用尿素氮会越来越少，如2.1节所述。

对奶牛而言，通过增加血液尿素氮水平来弥补日粮粗蛋白质的不足，这可能不是最佳的方法，因为采食量可能会被降低，而且十二指肠食糜比日粮具有更有利于动物生产的氨基酸组成（Schwab和Broderick，2017）。如果十二指肠微生物蛋白量过低，会降低动物生产效率，必须通过补充瘤胃不可降解蛋白才能弥补。但是，与瘤胃降解蛋白相比，瘤胃不可降解蛋白不仅价格贵，而且消化率变异大，还能导致血液尿素氮向肠道循环与排放增多（Batista等，2017）。肉牛在血液尿素氮转化为瘤胃细菌蛋白方面能力突出。

2.5 通过案例进一步认识原虫介导的蛋白质分解和瘤胃内微生物蛋白循环

Williams和Coleman（1992）报道，单一原虫清除细菌的速度比体外混合原虫快1.5~17.6倍。因此，Belanche等（2012）建立了一套复杂的体外取样方法，用来评估瘤胃微生物蛋白的循环利用。研究发现，孔径20μm筛上保留的原虫具有更强的细菌裂解能力（图5），主要包括内毛虫尤其是有尾内毛虫（Ishaq等，2017；Kittelmann等，2015）。分子生物学技术的发展，为原虫计数提供了更准确的方法，但是传统计数结合显微镜观察（尤其是实时观察结合自动计算）的方法，在原虫体积估算方面仍有前途，事实上原虫体积与代谢能力有关系（Wenner等，2018）。另外，有必要研究不同碳源条件下，原虫裂解细菌的能力变化（Ye等，2018；Diaz等，2014）。

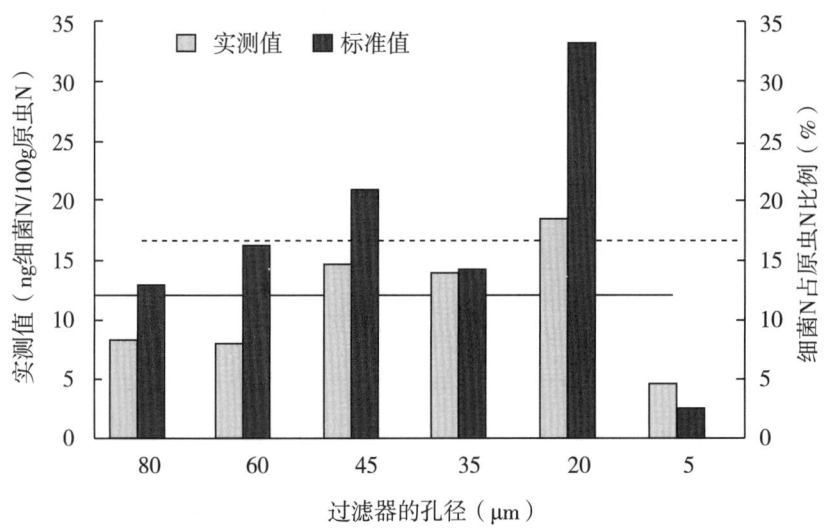

图5 不同大小原虫对细菌的裂解（Belanche等，2012）。
实线是12.0ng细菌N/100g原虫N，虚线为16.7%。

体外方法揭示了牛瘤胃原虫种群丰度、群落结构和功能活性对细菌裂解能力的影响。纤毛虫一般分为四种群落类型（Kittelmann等，2016），与原核生物相比具有更稳定的群落组成，受日粮影响较小（Mizrahi和Jami，2018）。但是，日粮的改变也会影响某些原虫。例如，椰子油能抑制除内毛虫外的大多数原虫（Reveneau等，2012）。内毛虫作为优势原虫，在细胞大小、生态位和捕获细菌能力方面的个体差异很大（Williams和Coleman，

2.6 利用测序方法进行原虫体内或体外培养研究的案例

原虫个体的培养研究有助于作用机制方面的挖掘。在尖尾内毛虫和有尾前毛虫的基因组中,存在瘤胃典型细菌和古菌基因序列,尤其是变形菌门的细菌,这类菌能够抵抗原虫蛋白分解,可能是原虫的内共生体(Park 和 Yu,2018a,b)。

许多原虫方面的科学问题仍有待于研究。例如,纤毛虫是为超级产氨菌提供肽和氨基酸,还是捕获超级产氨菌(Firkins 等,2007)?原虫数量或 18S rRNA 基因拷贝与饲料利用效率的关系仍不明确(Delgado 等,2019)。但是,纤毛虫与细菌古菌的相互作用是复杂的,只有结合转录组学和系统发育分析才能更好地阐明互作机制。另外,原虫基因数据库也不健全,后续需要增加原虫基因组的测序数据量(Park 等,2018)。

3 提高瘤胃氮代谢效率的方法

3.1 瘤胃降解蛋白的测定

饲料蛋白质水解是蛋白质代谢体系的重要内容。尽管各种体外和原位方法被用于蛋白质水解研究(Stern 等,1997;Schwab 和 Broderick,2017),但这些方法都有一定局限性。例如,标准化的蛋白酶可以被用于评估饲料相对 RDP 值,但是瘤胃中蛋白酶活性也受日粮影响。微生物富含蛋白质并且持续不断地定植到饲料中,而饲料蛋白质降解较快;最终结果是低估了其降解速度。对于高蛋白日粮而言,这种影响较小。然而对于那些有利于细菌定植且蛋白质含量相对较低的高纤维性饲料,通常利用细菌标记物如 ^{15}N(Kamoun 等,2014)和定量 PCR(Paz 等,2014)进行测定。在体外使用与尼龙袋孔径类似的滤纸,可实现对可溶性蛋白质的降解动力学研究。蛋白过瘤胃率的预测通常不会太准确(Firkins 等,1998),目前还没有一种最优方法实现对瘤胃可降解蛋白质的准确测定。更多细节参见 Hristov 等(2019)的研究报道。

3.2 日粮或添加剂对瘤胃蛋白质水解的影响

日粮组成多样性较高,并可能与饲养条件相互作用发挥效应。调控青贮制作方法,包括提高干物质含量(Hartinger 等,2018)和使用菌剂(Muck 等,2018),有助于减少青贮中蛋白质分解。过热处理饲料会增加蛋白质的过瘤胃量,但是降低其肠道利用率(Hartinger 等,2018;Schwab 和 Broderick,2017)。估算 RDP 和 RUP 肠道消化率的方法与饲料类型有关(Liebe 等,2018)。此外,每个氨基酸的瘤胃利用率并不相同(White 等,2017a)。因此有必要继续深入研究以提高估算瘤胃蛋白质分解对肠道蛋白质供应影响的准确性和精密度。

研究人员利用了各种饲料添加剂来限制瘤胃蛋白水解,包括精油、单宁和皂苷(Hartinger 等,2018)。有些靶向蛋白酶或脱氨酶,有些靶向微生物种群(Calsamiglia 等,2007)。不同批次添加剂的活性成分含量不同,且受贮藏时间或瘤胃适应性影响,因此应通过长期试验确定添加剂的最佳使用剂量。研究人员评估了精油的抑制甲烷作用（Benchaar 和 Greathead，2011），同时精油对产氨菌有抑制作用，所以有必要评估其对脱氨活性的影响（McIntosh 等，2003；Guyader 等，2017；Martinez-Fernandez 等，2016）。

皂苷和单宁也可以减少瘤胃中蛋白质水解，这可能与微生物种群变化有关。皂苷广泛应用于抑制原虫（Newbold 等，2015），但原虫对皂苷的反应并不总是一致（Patra 和 Saxena，2009）。特定皂苷或其衍生物可能对原虫抑制更持久有效（Ramos-Morales 等，2017a，b），然而有必要对高产动物进行适应性和有效性验证，但是抗原虫药物还未被批准用于减少肠道甲烷的产生（Hristov 等，2013）。山羊可以耐受单宁，但单宁会导致绵羊干物质采食量减少（Min 和 Solaiman，2018）。食草动物可能对单宁具有耐受性。单宁也会影响蛋白质或 NDF 的消化率（Patra 和 Saxena，2011）。虽然研究者们已经测试了多种浓缩单宁，但还需要更全面的评估（Hartinger 等，2018）。

离子载体最初是为了抑制原虫，后来被用于抑制超级产氨菌（Hartinger 等，2018），从而减少蛋白质水解。离子载体被长期使用时，易产生瘤胃适应性，从而不再抑制原虫数量（Schären 等，2017）。相反，超级产氨菌通常对离子载体敏感（Bento 等，2015；Shen 等，2018），从而不易发生长期使用导致的适应性（Ye 等，2018）。莫能菌素通常对革兰氏阳性细菌的抑制作用更强，而厚壁菌门细菌进化了适应性细胞壁结构以维持自身生长（Schären 等，2017）。这些作者发现莫能菌素未引起瘤胃氨浓度的变化，实际上瘤胃氨浓度除了受蛋白质水解和脱氨作用的影响外，还受微生物蛋白质合成和血液中尿素转入的影响。与交叉试验设计相比，连续饲养试验更能体现离子载体效果（Sauer 等，1998；Hartinger 等，2018）。Firkins 和 Yu（2015）发现莫能菌素的效果在肉牛中的一致性比奶牛高，这与添加的有效剂量有关，并且莫能菌素可改变采食行为和瘤胃微生态。

研究人员对中链脂肪酸和不饱和脂肪酸的抗原虫效果进行了研究（Firkins 和 Yu，2015），发现不饱和脂肪能改变微生物群落组成（Enjalbert 等，2017）。因此，研究人员应该充分考虑脂肪酸对微生物群落各种细菌而不仅仅是丁酸弧菌的影响（Hackmann 和 Firkins，2015a）。例如，定量 PCR 检测表明，氢化棕榈油增加了 *Prevotella bryantii* 的丰度（Vargas-Bello-Pérez 等，2016），该菌具有较高的肽酶活性（Hartinger 等，2018）。普雷沃氏菌和丁酸弧菌是奶牛瘤胃上皮优势细菌类群（Mann 等，2018），具有很强的蛋白水解（Walker 等，2005）以及尿素水解能力（Bickhart 和 Weimer，2018）。研究者对瘤胃壁细菌群落进行宏转录组分析，发现了参与氨基酸代谢的高表达基因，可能在蛋白水解中发挥重要作用（Mann 等，2018）。

4 结论

瘤胃具有广泛的蛋白质水解活性。在限制瘤胃蛋白质分解以减少氮排放时，应考虑淀粉和纤维的瘤胃降解特性、微生物蛋白质外流和干物质采食量。增加瘤胃微生物蛋白和过瘤胃蛋白是减少日粮蛋白质浪费的重要手段。原虫利用的大部分氨基酸来自捕食的细菌蛋白，但对宿主氮利用率的影响不大，除非大部分原虫蛋白被排放到瘤胃而不是外流。氨同化通常是谷氨酸脱氢酶和谷氨酰胺合成酶-谷氨酸合成酶完成的，基因组测序揭示了铵态

氮对 *R. albus* 8 和 *P. ruminicola* 23 氨同化酶表达的影响。宏基因组学分析表明营养素影响微生物蛋白酶、肽酶和转运蛋白表达。瘤胃微生物氮代谢的营养调控将有助于实现动物氮利用率的进一步提升。

5 未来展望

长期以来，人们对瘤胃蛋白质水解认识越来越多，但大部分研究来自单一菌株，缺少更多对其他细菌的研究，并且缺少对各类反刍动物在不同条件下的研究。RDP 的估计忽略了定植微生物蛋白的影响，因此我们需要将微生物学和营养学结合起来，以提高 RDP 的估计准确性。

蛋白质分解中，哪些来自饲料蛋白质水解，哪些来自微生物蛋白质循环，这些应更好的区分和明确，原虫在这些活动中发挥着重要作用，所以本章进行了阐述。除了驱除原虫方法外，我们还需要更多的方法调控原虫数量和活性，同时考虑日粮、瘤胃和全消化道消化率等影响。不同原虫呈现不同的蛋白水解特征，纤毛虫如何直接（捕食微生物）和间接（碳水化合物代谢、氧清除等）影响瘤胃微生物群功能仍有待进一步研究。新的研究方法如体外培养、宏基因组学和宏转录组学应被采用，并且应将蛋白分解、氨同化和氮循环进行系统性研究，提高反刍动物日粮蛋白质转化为动物产品氮的效率。

6 更多信息

在 P. N. Hobson 和 C. S. Stewart 合著的里程碑式著作《瘤胃微生物生态系统》中，有关于瘤胃蛋白质分解介绍的章节，后续研究者对相关内容进行了更新（Walker 等，2005），Ungerfeld 和 Newbold 的综述文章也有相关描述（doi：10.3389/fmicb.2017.02627）。通过美国国家科学院、工程学院和医学院的网站（https://www.nap.edu/topic/276/agriculture）可以获得肉牛、羊和奶牛的营养需求信息。澳大利亚（CSIRO；ISBN 9780643092624）、荷兰（DVE/OEB；doi：10.1017/S0021859610000912）、法国（INRA；doi.org/10.3920/978-90-8686-292-4）和北欧（doi.org/10.3920/978-90-8686-718-9）国家也有类似信息。《乳品科学百科全书》（ISBN 978-0-12-374407-4）和北美会议摘要（https://spac.adsa.org/index.asp）中也有相关信息。

7 参考文献

Ahvenjärvi, S. and Huhtanen, P. 2018. Effects of intraruminal urea-nitrogen infusions on feed intake, nitrogen utilization, and milk yield in dairy cows. Journal of Dairy Science 101 (10), 9004-15.

Ahvenjärvi, S., Vaga, M., Vanhatalo, A., et al., 2018. Ruminal metabolism of grass silage soluble nitrogen fractions. Journal of Dairy Science 101 (1), 279-94.

Arcondeguy, T., Jack, R. and Merrick, M. 2001. P (II) signal transduction proteins, pivotal players in microbial nitrogen control. Microbiology and Molecular Biology Reviews 65 (1), 80-105.

Atasoglu, C. and Wallace, R. J. 2002. De novo synthesis of amino acids by the ruminal anaerobic fungi, *Piromyces communis* and *Neocallimastix frontalis*. FEMS Microbiology Letters 212 (2), 243-7.

Batista, E. D., Detmann, E., Valadares Filho, S. C., et al., 2017. The effect of CP concentration in the diet on urea kinetics and microbial usage of recycled urea in cattle: a meta-analysis. Animal 11 (8), 1303-11.

Belanche, A., de la Fuente, G., Moorby, J. M., et al., 2012. Bacterial protein degradation by different rumen protozoal groups. Journal of Animal Science 90 (12), 4495-504.

Benchaar, C. and Greathead, H. 2011. Essential oils and opportunities to mitigate enteric methane emissions from ruminants. Animal Feed Science and Technology 166-167, 338-55.

Bennett, B. D., Kimball, E. H., Gao, M., et al., 2009. Absolute metabolite concentrations and implied enzyme active site occupancy in *Escherichia coli*. Nature Chemical Biology 5 (8), 593-9.

Bento, C. B. P., de Azevedo, A. C., Detmann, E., et al., 2015. Biochemical and genetic diversity of carbohydrate-fermenting and obligate amino acid fermenting hyper-ammonia-producing bacteria from Nellore steers fed tropical forages and supplemented with casein. BMC Microbiology 15, 28.

Bickhart, D. M. and Weimer, P. J. 2018. Symposium review: host-rumen microbe interactions may be leveraged to improve the productivity of dairy cows. Journal of Dairy Science 101 (8), 7680-9.

Blauwkamp, T. A. and Ninfa, A. J. 2002. Physiological role of the GlnK signal transduction protein of Escherichia coli: survival of nitrogen starvation. Molecular Microbiology 46 (1), 203-14.

Bowen, M. K., Poppi, D. P. and McLennan, S. R. 2017. Efficiency of rumen microbial protein synthesis in cattle grazing tropical pastures as estimated by a novel technique. Animal Production Science 57 (8), 1702-12.

Calsamiglia, S., Busquet, M., Cordozo, P. W., et al., 2007. Invited review: essential oils as modifiers of rumen microbial fermentation. Journal of Dairy Science 90 (6), 2580-95.

Colombatto, D. and Beauchemin, K. A. 2009. A protease additive increases fermentation of alfalfa diets by mixed ruminal microorganisms *in vitro*. Journal of Animal Science 87 (3), 1097-105.

Comtet-Marre, S., Parisot, N., Lepercq, P., et al., 2017. Metatranscriptomics reveals the active bacterial and eukaryotic fibrolytic communities in the rumen of dairy cow fed a mixed diet. Frontiers in Microbiology 8, 67.

Cotta, M. A. and Russell, J. B. 1997. Digestion of nitrogen in the rumen: a model for metabolism of nitrogen compounds in gastrointestinal environments. In: Mackie, R. I. and White, B. A. (Eds), Gastrointestinal Microbiology (vol. 1). Springer, New York, pp. 380-412.

Czerkawski, J. W. 1976. Chemical composition of microbial matter in the rumen. Journal of the Science of Food and Agriculture 27 (7), 621-32.

Delgado, B., Bach, A., Guasch, I., et al., 2019. Whole rumen metagenome sequencing allows classifying and predicting feed efficiency and intake levels in cattle. Scientific Reports 9 (1), 11.

DeMoll, E. 1993. Nitrogen and phosphorus metabolism in methanogens. In: Ferry, J. G. (Ed.), Methanogenesis: Ecology, Physiology, Biochemistry and Genetics. Chapman and Hall, New York, pp. 471-89.

Detsch, C. and Stulke, J. 2003. Ammonium utilization in Bacillus subtilis: transport and regulatory functions of NrgA and NrgB. Microbiology 149 (11), 3289-97.

Diaz, H. L., Barr, K. N., Godden, K. R., et al., 2014. Eukaryotic inhibitors or activators elicit responses to chemosensory compounds by ruminal isotrichid and entodiniomorphid protozoa. Journal of Dairy Science 97 (4), 2254-69.

Edwards, J. E., Forster, R. J., Callaghan, T. M., et al., 2017. PCR and omics based techniques to study the diversity, ecology and biology of anaerobic fungi: insights, challenges and opportunities. Frontiers in Microbiology 8, 1657.

Enjalbert, F., Combes, S., Zened, A., et al., 2017. Rumen microbiota and dietary fat: a mutual shaping. Journal of Applied Microbiology 123 (4), 782-97.

Ferraretto, L. F., Crump, P. M. and Shaver, R. D. 2015. Effect of ensiling time and exogenous protease addition to whole-plant corn silage of various hybrids, maturities, and chop lengths on nitrogen fractions and ruminal *in vitro* starch digestibility. Journal of Dairy Science 98 (12), 8869-81.

Fessenden, S. W., Hackmann, T. J., Ross, D. A., et al., 2017. Ruminal bacteria and protozoa composition, digestibility, and amino acid profile determined by multiple hydrolysis times. Journal of Dairy Science 100 (9), 7211-26.

Fessenden, S. W., Hackmann, T. J., Ross, D. A., et al., 2019. Rumen digestion kinetics, microbial yield, and omasal flows of nonmicrobial, bacterial, and protozoal amino acids in lactating dairy cattle fed fermentation by-products or urea as a soluble nitrogen source. Journal of Dairy Science 102 (4), 3036-52.

Firkins, J. L. and Yu, Z. 2006. Characterisation and quantification of the microbial populations in the rumen. In: Sejrsen, K., Hvelplund, T. and Nielsen, M. O. (Eds), Ruminant Physiology, Digestion, Metabolism and Impact of Nutrition on Gene Expression, Immunology and Stress. Wageningen Academic Publishers, Wageningen, The Netherlands, pp. 19-54.

Firkins, J. L. and Yu, Z. 2015. How to use data on the rumen microbiome to improve our understanding of ruminant nutrition. Journal of Animal Science 93 (4), 1450-70.

Firkins, J. L., Allen, M. S., Oldick, B. S., et al., 1998. Modeling ruminal digestibility of carbohydrates and microbial protein flow to the duodenum. Journal of Dairy Science 81 (12), 3350-69.

Firkins, J. L., Yu, Z. and Morrison, M. 2007. Ruminal nitrogen metabolism: perspectives for integration of microbiology and nutrition for dairy. Journal of Dairy Science 90 (E. Suppl.), E1-E16.

Fisher, S. H. 1999. Regulation of nitrogen metabolism in *Bacillus subtilis*: vive la difference! Molecular Microbiology 32 (2), 223-32.

Fisher, S. H. and Wray, L. V. 2002. Mutations in the *Bacillus subtilis* glnRA operon that cause nitrogen source-dependent defects in regulation of TnrA activity. Journal of Bacteriology 184 (16), 4636-9.

Fondevila, M. and Dehority, B. A. 2001. *In vitro* growth and starch digestion by Entodinium exiguum as influenced by the presence or absence of live bacteria. Journal of Animal Science 79 (9), 2465-71.

Forchhammer, K. 2008. PII signal transducers: novel functional and structural insights. Trends in Microbiology 16 (2), 65-72.

Forsberg, C. W., Lovelock, L. K. A., Krumholz, L., et al., 1984. Protease activities of rumen protozoa. Applied and Environmental Microbiology 47 (1), 101-10.

Foskolos, A. and Moorby, J. M. 2018. Evaluating lifetime nitrogen use efficiency of dairy cattle: a modelling approach. PLoS One 13 (8), e0201638.

Galyean, M. L. and Tedeschi, L. O. 2014. Predicting microbial protein synthesis in beef cattle: relationship to intakes of total digestible nutrients and crude protein. Journal of Animal Science 92 (11), 5099-111.

Griswold, K. E., Apgar, G. A., Bouton, J., et al., 2003. Effects of urea infusion and ruminal degradable protein concentration on microbial growth, digestibility, and fermentation in continuous culture. Journal of Animal Science 81 (1), 329-36.

Gunka, K. and Commichau, F. M. 2012. Control of glutamate homeostasis in *Bacillus subtilis*: a complex interplay between ammonium assimilation, glutamate biosynthesis and degradation. Molecular Microbiology 85 (2), 213-24.

Guyader, J., Eugene, M., Noziere, P., et al., 2014. Influence of rumen protozoa on methane emission in ruminants: a meta-analysis approach. Animal 8 (11), 1816-25.

Guyader, J., Ungerfeld, E. M. and Beauchemin, K. A. 2017. Redirection of metabolic hydrogen by inhibiting methanogenesis in the rumen simulation technique (RUSITEC). Frontiers in Microbiology 8, 393.

Hackmann, T. J. and Firkins, J. L. 2015a. Electron transport phosphorylation in rumen butyrivibrios: unprecedented ATP yield for glucose fermentation to butyrate. Frontiers in Microbiology 6, 622.

Hackmann, T. J. and Firkins, J. L. 2015b. Maximizing efficiency of rumen microbial protein production. Frontiers in Microbiology 6, 465.

Hackmann, T. J., Ngugi, D. K., Firkins, J. L., et al., 2017. Genomes of rumen bacteria encode atypical pathways for fermenting hexoses to short-chain fatty acids. Environmental Microbiology 19 (11), 4670-83.

Hansen, E. E., Lozupone, C. A., Rey, F. E., et al., 2011. Pan-genome of the dominant human gut-associated archaeon, *Methanobrevibacter smithii*, studied in twins. Proceedings of the National Academy of Sciences of the United States of America 108, 4599-606.

Hartinger, T., Gresner, N. and Südekum, K. H. 2018. Does intra-ruminal nitrogen recycling waste valuable resources? A review of major players and their manipulation. Journal of Animal Science and Biotechnology 9, 33.

Hino, T. and Russell, J. B. 1985. Effect of reducing-equivalent disposal and NADH/NAD on deamination of amino acids by intact rumen microorganisms and their cell extracts. Applied and Environmental Microbiology 50 (6), 1368-74.

Hook, S. E., France, J. and Dijkstra, J. 2017. Further assessment of the protozoal contribution to the nutrition of the ruminant animal. Journal of Theoretical Biology 416, 8-15.

Hristov, A. N. and Jouany, J.P. 2005. Factors affecting the efficiency of nitrogen utilization in the rumen. In: Hristov, A. N. and Pfeffer, E. (Eds), Nitrogen and Phosphorus Nutrition of Cattle and Environment. CAB International, Wallingford, UK, pp. 117-66.

Hristov, A. N., Oh, J., Firkins, J. L., et al., 2013. Special topics—mitigation of methane and nitrous oxide emissions from animal operations: I. A review of enteric methane mitigation options. Journal of Animal Science 91 (11), 5045-69.

Hristov, A. N., Bannink, A., Crompton, L. A., et al., 2019. Invited review: nitrogen in ruminant nutrition: a review of measurement techniques. Journal of Dairy Science 102 (7), 5811-52.

Huws, S. A., Creevey, C. J., Oyama, L. B., et al., 2018. Addressing global ruminant agricultural challenges through understanding the rumen microbiome: past, present, and future. Frontiers in Microbiology 9, 2161.

Iakiviak, M. 2018. Analysis of the ammonium assimilation pathways of the human colonic bacterium, Bacteroides thetaiotaomicron. PhD Thesis. University of Illinois, Urbana, IL.

Ishaq, S. L., AlZahal, O., Walker, N., et al., 2017. An investigation into rumen fungal and protozoal diversity in three rumen fractions, during high-fiber or grain-induced sub-acute ruminal acidosis conditions, with or without active dry yeast supplementation. Frontiers in Microbiology 8, 1943.

Jin, D., Zhao, S., Zheng, N., et al., 2018. Urea metabolism and regulation by rumen bacterial urease in ruminants-a review. Annals of Animal Science 18 (2), 303-18.

Kamoun, M., Ammar, H., Théwis, A., et al., 2014. Comparison of three 15N methods to correct for microbial contamination when assessing in situ protein degradability of fresh forages. Journal of Animal Science 92 (11), 5053-62.

Karnati, S. K. R., Sylvester, J. T., Noftsger, S. M., et al., 2007. Assessment of ruminal bacterial populations and protozoal generation time in cows fed different methionine sources. Journal of Dairy Science 90 (2), 798-809.

Kim, J. N., Henriksen, E. D., Cann, I. K., et al., 2014. Nitrogen utilization and metabolism in *Ruminococcus albus* 8. Applied and Environmental Microbiology 80 (10), 3095-102.

Kim, J. N., Mendez-Garcia, C., Geier, R. R., et al., 2017. Metabolic networks for nitrogen utilization in *Prevotella ruminicola* 23. Scientific Reports 7 (1), 7851.

Kittelmann, S., Devente, S. R., Kirk, M. R., et al., 2015. Phylogeny of the intestinal ciliates, including *Charonina ventriculi*, and comparison of microscopy and 18S rRNA gene pyrosequencing for rumen ciliate community structure analysis. Applied and Environmental Microbiology 81 (7), 2433-44.

Kittelmann, S., Pinares-Patino, C. S., Seedorf, H., et al., 2016. Natural variation in methane emission of sheep fed on a lucerne pellet diet is unrelated to rumen ciliate community type. Microbiology 162 (3), 459-65.

Kleiner, D. 1982. Ammonium (methylammonium) transport by *Klebsiella pneumoniae*. Biochimica et Biophysica Acta 688 (3), 702-8.

Kleiner, D. and Castorph, H. 1982. Inhibition of ammonium (methylammonium) transport in *Klebsiella pneumoniae* by glutamine and glutamine analogues. FEBS Letters 146 (1), 201-3.

Leigh, J. A. and Dodsworth, J. A. 2007. Nitrogen regulation in Bacteria and Archaea. Annual Review of Microbiology 61, 349-77.

Leong, L. E. X., Denman, S. E., Hugenholtz, P., et al., 2016. Amino acid and peptide utilization profiles of the fluoroacetate-degrading bacterium Synergistetes Strain MFA1 under varying conditions. Microbial Ecology 71 (2), 494-504.

Liebe, D. M., Firkins, J. L., Tran, H., et al., 2018. Technical note: methodological and feed factors affecting measurement of protein A, B, and C fractions, degradation rate, and intestinal digestibility of rumen-undegraded protein. Journal of Dairy Science 101 (9), 8046-53.

Liu, X., Shi, F. and Gong, J. 2013. Variations and evolution of polyubiquitin genes from ciliates. European Journal of Protistology 49 (1), 40-9.

Lowe, S. E., Theodorou, M. K., Trinci, A. P. J., et al., 1985. Growth of anaerobic rumen fungi on defined and semi-defined media lacking rumen fluid. Journal of General Microbiology 131, 2225-9.

Mackie, R. I. and White, B. A. 1990. Recent advances in rumen microbial ecology and metabolism: potential impact on nutrient output. Journal of Dairy Science 73 (10), 2971-95.

Magasanik, B. 1989. Regulation of transcription of the glnALG operon of Escherichia coli by protein phosphorylation. Biochimie 71 (9-10), 1005-12.

Mann, E., Wetzels, S. U., Wagner, M., et al., 2018. Metatranscriptome sequencing reveals insights into the gene expression and functional potential of rumen wall bacteria. Frontiers in Microbiology 9, 43.

Martinelle, K. and Haggstrom, L. 1997. On the dissociation constant of ammonium: effects of using an incorrect pKa in calculations of the ammonia concentration in animal cell cultures. Biotechnology Techniques 11

(8), 549-51.

Martinez-Fernandez, G., Denman, S. E., Yang, C., et al., 2016. Methane inhibition alters the microbial community, hydrogen flow, and fermentation response in the rumen of cattle. Frontiers in Microbiology 7, 1122.

Mayorga, O. L., Kingston-Smith, A. H., Kim, E. J., et al., 2016. Temporal metagenomic and metabolomic characterization of fresh perennial ryegrass degradation by rumen bacteria. Frontiers in Microbiology 7, 1854.

McIntosh, F. M., Williams, P., Losa, R., et al., 2003. Effects of essential oils on ruminal microorganisms and their protein metabolism. Applied and Environmental Microbiology 69 (8), 5011-4.

Meier-Wagner, J., Nolden, L., Jakoby, M., et al., 2001. Multiplicity of ammonium uptake systems in Corynebacterium glutamicum: role of Amt and AmtB. Microbiology 147 (1), 135-43.

Min, B. R. and Solaiman, S. 2018. Comparative aspects of plant tannins on digestive physiology, nutrition and microbial community changes in sheep and goats: a review. Journal of Animal Physiology and Animal Nutrition 102 (5), 1181-93.

Mizrahi, I. and Jami, E. 2018. Review: the compositional variation of the rumen microbiome and its effect on host performance and methane emission. Animal 12 (s2), s220-32.

Morrison, M. and Mackie, R. I. 1997. Biosynthesis of nitrogen-containing compounds. In: Mackie, R. I. and White, B. A. (Eds), Gastrointestinal Ecosystems and Fermentations (vol. 1). Chapman and Hall, New York, NY, pp. 424-69.

Muck, R. E., Nadeau, E. M. G., McAllister, T. A., et al., 2018. Silage review: recent advances and future uses of silage additives. Journal of Dairy Science 101 (5), 3980-4000.

Newbold, C. J., de la Fuente, G., Belanche, A., et al., 2015. The role of ciliate protozoa in the rumen. Frontiers in Microbiology 6, 1313.

Oelker, E. R., Reveneau, C. and Firkins, J. L. 2009. Interaction of molasses and monensin in alfalfa hay- or corn silage-based diets on rumen fermentation, total tract digestibility, and milk production by Holstein cows. Journal of Dairy Science 92 (1), 270-85.

Orpin, C. G. and Greenwood, Y. 1986. Nutritional and germination requirements of the rumen chytridiomycete Neocallimastix frontalis. Transactions-British Mycological Society 86, 103-9.

Park, T. and Yu, Z. 2018a. Aerobic cultivation of anaerobic rumen protozoa, Entodinium caudatum and Epidinium caudatum. Journal of Microbiological Methods 152, 186-93.

Park, T. and Yu, Z. 2018b. Do ruminal ciliates select their preys and prokaryotic symbionts? Frontiers in Microbiology 9, 1710.

Park, T., Wijeratne, S., Meulia, T., et al., 2018. Draft macronuclear genome sequence of the ruminal ciliate Entodinium caudatum. Microbiology Resource Announcements 7 (1), e00826-18.

Patra, A. K. and Aschenbach, J. R. 2018. Ureases in the gastrointestinal tracts of ruminant and monogastric animals and their implication in urea-N/ammonia metabolism: a review. Journal of Advanced Research 13, 39-50.

Patra, A. K. and Saxena, J. 2009. The effect and mode of action of saponins on the microbial populations and fermentation in the rumen and ruminant production. Nutrition Research Reviews 22 (2), 204-19.

Patra, A. K. and Saxena, J. 2011. Exploitation of dietary tannins to improve rumen metabolism and ruminant production. Journal of the Science of Food and Agriculture 91 (1), 24-37.

Paz, H. A., Klopfenstein, T. J., Hostetler, D., et al., 2014. Ruminal degradation and intestinal digestibility of protein and amino acids in high-protein feedstuffs commonly used in dairy diets. Journal of Dairy Science 97 (10), 6485-98.

Paz, H. A., Anderson, C. L., Muller, M. J., et al., 2016. Rumen bacterial community composition in Holstein and Jersey cows is different under same dietary condition and is not affected by sampling method. Frontiers in Microbiology 7, 1206.

Pengpeng, W. and Tan, Z. 2013. Ammonia assimilation in rumen bacteria: a review. Animal Biotechnology 24 (2), 107-28.

Ramos-Morales, E., de la Fuente, G., Duval, S., et al., 2017a. Antiprotozoal effect of saponins in the rumen can be enhanced by chemical modifications in their structure. Frontiers in Microbiology 8, 399.

Ramos-Morales, E., de la Fuente, G., Nash, R. J., et al., 2017b. Improving the antiprotozoal effect of saponins in the rumen by combination with glycosidase inhibiting iminosugars or by modification of their chemical structure. PLoS One 12 (9), e0184517.

Reed, K. F., Casper, D. P., France, J., et al., 2015. Prediction of nitrogen efficiency in dairy cattle: a review. CAB Reviews 10 (1), 1.

Reed, K. F., Bonfá, H. C., Dijkstra, J., et al., 2017. Estimating the energetic cost of feeding excess dietary nitrogen to dairy cows. Journal of Dairy Science 100 (9), 7116-26.

Reitzer, L. 2003. Nitrogen assimilation and global regulation in *Escherichia coli*. Annual Review of Microbiology 57, 155-76.

Reveneau, C., Karnati, S. K. R., Oelker, E. R., et al., 2012. Interaction of unsaturated fat or coconut oil with monensin in lactating dairy cows fed twelve times daily. I. Protozoal abundance, nutrient digestibility, and microbial protein flow to the omasum. Journal of Dairy Science 95 (4), 2046-60.

Reynolds, C. K. and Kristensen, N. B. 2008. Nitrogen recycling through the gut and the nitrogen economy of ruminants: an asynchronous symbiosis. Journal of Animal Science 86 (E. Suppl.), E293-305.

Roman-Garcia, Y., White, R. R. and Firkins, J. L. 2016. Meta-analysis of postruminal microbial nitrogen flows in dairy cattle. I. Derivation of equations. Journal of Dairy Science 99 (10), 7918-31.

Russell, J. B. and Strobel, H. J. 1987. Concentration of ammonia across cell membranes of mixed rumen bacteria. Journal of Dairy Science 70 (5), 970-6.

Rychlik, J. L. and Russell, J. B. 2000. Mathematical estimations of hyper-ammonia producing ruminal bacteria and evidencefor bacterial antagonism that decreases ruminal ammonia production. FEMS Microbiology Ecology 32 (2), 121-8.

Samuel, B. S., Hansen, E. E., Manchester, J. K., et al., 2007. Genomic and metabolic adaptations of *Methanobrevibacter smithii* to the human gut. Proceedings of the National Academy of Sciences of the United States of America 104 (25), 10643-8.

Sauer, F. D., Fellner, V., Kinsman, R., et al., 1998. Methane output and lactation response in Holstein cattle with monensin or unsaturated fat added to the diet. Journal of Animal Science 76 (3), 906-14.

Sauvant, D. and Nozière, P. 2016. Quantification of the main digestive processes in ruminants: the equations involved in the renewed energy and protein feed evaluation systems. Animal 10 (5), 755-70.

Schären, M., Drong, C., Kiri, K., et al., 2017. Differential effects of monensin and a blend of essential oils on rumen microbiota composition of transition dairy cows. Journal of Dairy Science 100 (4), 2765-83.

Schumacher, M. A., Chinnam, N. B., Cuthbert, B., et al., 2015. Structures of regulatory machinery reveal novel molecular mechanisms controlling B. subtilis nitrogen homeostasis. Genes and Development 29 (4), 451-64.

Schwab, C. G. and Broderick, G. A. 2017. A 100-year Review: protein and amino acid nutrition in dairy cows. Journal of Dairy Science 100 (12), 10094-112.

Seshadri, R., Leahy, S. C., Attwood, G. T., et al., 2018. Cultivation and sequencing of rumen microbiome members from the Hungate1000 Collection. Nature Biotechnology 36, 359.

Shen, J., Yu, Z. and Zhu, W. 2018. Insights into the populations of proteolytic and amino acid-fermenting bacteria from microbiota analysis using *in vitro* enrichment cultures. Current Microbiology 75 (11), 1543-50.

Sok, M., Ouellet, D. R., Firkins, J. L., et al., 2017. Amino acid composition of rumen bacteria and protozoa in cattle. Journal of Dairy Science 100 (7), 5241-9.

Soupene, E., He, L., Yan, D., et al., 1998. Ammonia acquisition in enteric bacteria: physiological role of the ammonium/methylammonium transport B (AmtB) protein. Proceedings of the National Academy of Sciences of the United States of America 95 (12), 7030-4.

Soupene, E., Lee, H. and Kustu, S. 2002. Ammonium/methylammonium transport (Amt) proteins facilitate diffusion of NH_3 bidirectionally. Proceedings of the National Academy of Sciences of the United States of America 99 (6), 3926-31.

Stern, M. D., Bach, A. and Calsamiglia, S. 1997. Alternative techniques for measuring nutrient digestion in ruminants. Journal of Animal Science 75 (8), 2256-76.

Stevenson, R. and Silver, S. 1977. Methylammonium uptake by *Escherichia coli*: evidence for a bacterial $NH4^+$ transport system. Biochemical and Biophysical Research Communications 75 (4), 1133-9.

Sylvester, J. T., Karnati, S. K., Yu, Z., et al., 2005 Evaluation of a real-time PCR assay quantifying the ruminal pool size and duodenal flow of protozoal nitrogen. Journal of Dairy Science 88 (6), 2083-95.

Theodorou, M. K., Davies, D. R. and Orpin, C. G. 1994. Nutrition and survival of anaerobic fungi. In: Mountfort, D. O. and Orpin, C. G. (Eds), Anaerobic Fungi: Biology, Ecology and Function. Marcel Dekker, New York, pp. 107-28.

Ungerfeld, E. M. 2018. Inhibition of rumen methanogenesis and ruminant productivity: a meta-analysis. Frontiers in Veterinary Science 5, 113.

Ungerfeld, E. M., Aedo, M. F., Martínez, E. D., et al., 2019. Inhibiting methanogenesis in rumen batch cultures did not increase the recovery of metabolic hydrogen in microbial amino acids. Microorganisms 7 (5), 115.

Van Amburgh, M. E., Collao-Saenz, E. A., Higgs, R. J., et al., 2015. The Cornell Net carbohydrate and Protein System: updates to the model and evaluation of version 6.5. Journal of Dairy Science 98 (9), 6361-80.

Van Heeswijk, W. C., Hoving, S., Molenaar, D., et al., 1996. An alternative PII protein in the regulation of glutamine synthetase in *Escherichia coli*. Molecular Microbiology 21 (1), 133-46.

Van Heeswijk, W. C., Westerhoff, H. V. and Boogerd, F. C. 2013. Nitrogen assimilation in *Escherichia coli*: putting molecular data into a systems perspective. Microbiology and Molecular Biology Reviews 77 (4), 628-95.

Vargas-Bello-Pérez, E., Cancino-Padilla, N., Romero, J. and Garnsworthy, P. C. 2016. Quantitative analysis of ruminal bacterial populations involved in lipid metabolism in dairy cows fed different vegetable oils. Animal 10 (11), 1821-8.

Walker, N. D., Newbold, C. J. and Wallace, R. J. 2005. Nitrogen metabolism in the rumen. In: Pfeffer, E. and Hristov, A. (Eds), Nitrogen and Phosphorus Nutrition of Cattle. CABI Publishing, Cambridge, MA, pp. 71–115.

Wallace, R. J. and McPherson, C. A. 1987. Factors affecting the rate of breakdown of bacterial protein in rumen fluid. The British Journal of Nutrition 58 (2), 313–23.

Wallace, R. J., Onodera, R. and Cotta, M. A. 1997. Metabolism of nitrogen-containing compounds. In: Hobson, P. N. and Stewart, C. S. (Eds), The Rumen Microbial Ecosystem (2nd edn.). Chapman and Hall, New York, NY, pp. 283–328.

Wallace, R. J., Chaudhary, L. C., Miyagawa, E., et al., 2004. Metabolic properties of *Eubacterium pyruvativorans*, a ruminal 'hyper-ammonia producing' anaerobe with metabolic properties analogous to those of Clostridium kluyveri. Microbiology 150 (9), 2921–30.

Wenner, B. A., Wagner, B. K. and Firkins, J. L. 2018. Using video microscopy to improve quantitative estimates of protozoal motility and cell volume. Journal of Dairy Science 101 (2), 1060–73.

White, R. R., Roman-Garcia, Y. and Firkins, J. L. 2016. Meta-analysis of postruminal microbial nitrogen flows in dairy cattle. II. Approaches to and implications of more mechanistic prediction. Journal of Dairy Science 99 (10), 7932–44.

White, R. R., Kononoff, P. J. and Firkins, J. L. 2017a. Technical note: methodological and feed factors affecting prediction of ruminal degradability and intestinal digestibility of essential amino acids. Journal of Dairy Science 100 (3), 1946–50.

White, R. R., Roman-Garcia, Y., Firkins, J. L., et al., 2017b. Evaluation of the 2001 dairy NRC and derivation of new prediction equations. 2. Rumen degradable and undegradable protein. Journal of Dairy Science 100 (5), 3611–27.

Williams, A. G. and Coleman, G. S. 1992. The Rumen Protozoa. Springer, New York. Williams, A. G. and Coleman, G. S. 1997. The rumen protozoa. In: Hobson, P. N. and Stewart, C. S. (Eds), The Rumen Microbial Ecosystem (2nd edn.). Blackie Academic and Professional, London, pp. 73–139.

Williams, K. J., Bennett, M. H., Barton, G. R., et al., 2013. Adenylylation of mycobacterial Glnk (PII) protein is induced by nitrogen limitation. Tuberculosis 93 (2), 198–206.

Winkler, F. K. 2006. Amt/MEP/Rh proteins conduct ammonia. Pflugers Archiv 451 (6), 701–7.

Wray, L. V., Zalieckas, J. M. and Fisher, S. H. 2001. *Bacillus subtilis* glutamine synthetase controls gene expression through a protein-protein interaction with transcription factor TnrA. Cell 107 (4), 427–35.

Xia, Y., Kong, Y., Huang, H., et al., 2016. *In situ* identification and quantification of protein-hydrolyzing ruminal bacteria associated with the digestion of barley and corn grain. Canadian Journal of Microbiology 62 (12), 1063–7.

Ye, D., Karnati, S. K. R., Wagner, B., et al., 2018. *Essential oil* and monensin affect ruminal fermentation and the protozoal population in continuous culture. Journal of Dairy Science 101 (6), 5069–81.

Zanton, G. I. 2016. Analysis of production responses to changing crude protein levels in lactating dairy cow diets when evaluated in continuous or change-over experimental designs. Journal of Dairy Science 99 (6), 4398–410.

第13章 影响瘤胃能量代谢效率的因素

Emilio M. Ungerfeld，智利农业研究所；
Timothy J. HAckmAnn，加州大学戴维斯分校，美国
（黄小丹译）

1 前言

自新石器时代以来，反刍动物就与人类紧密联系在一起，它们为人类提供肉、奶、毛并成为重要的役力和驮运工具。反刍动物可以利用人类无法利用的纤维和非蛋白氮（N），这也是它们被驯化的一个关键原因。反刍动物通过瘤胃中的细菌、原虫、真菌、产甲烷菌和噬菌体组成的复杂微生物群落对摄入的饲料进行消化和发酵，而由饲料发酵产生的挥发性脂肪酸（VFA）被宿主动物吸收并用作能量、葡萄糖和脂肪的来源。从瘤胃流出的微生物细胞则在胃肠道后部被消化，为反刍宿主动物提供主要的氨基酸来源。

虽然反刍动物的营养适应性使它们能够在为人类提供有用产品方面发挥核心作用，但这是以能氮利用效率低下为代价的。而瘤胃微生物群落调控的目的与能量、氮或脂肪酸代谢或解毒有关（Nagaraja等，1997；Lorenco等，2010）。在本章中，我们讨论了影响瘤胃能量利用效率的几个因素，并提出了提高瘤胃能量利用效率的研究方向。

2 瘤胃发酵的主要途径

2.1 挥发性脂肪酸的产生

在瘤胃中，紧随碳水化合物的消化之后是微生物的能量交换。微生物分解代谢（发酵）产生ATP提供能量，以驱动微生物合成代谢（生长）。发酵可以被定义为一种不完全的氧化过程，在这一过程中，最终的电子受体是碳水化合物。在瘤胃中，纤维素和淀粉等复杂碳水化合物水解释放的单糖发酵生成以乙酸、丙酸和丁酸为主的VFA，以及二氧化碳（CO_2）和甲烷（CH_4）等气体（图1）。在这个过程中会产生氢气（H_2）、甲酸、琥珀酸、乳酸，但不会积累，因为它们会迅速代谢成为发酵终产物。从能量的角度看，发酵与负的吉布斯自由能变化有关（ΔG），其中部分吉布斯自由能可被微生物细胞保存产生ATP，用于维持细胞生长过程中的合成代谢反应（Czerkawski，1986；Russel和Wallace，1997）。

在过去，糖酵解（EMP）被认为是瘤胃发酵中葡萄糖代谢的主要途径（Russell和Wallace，1997）。Wallnöfer等（1966）在体外培养标记葡萄糖和原位标记棉花纤维素后进

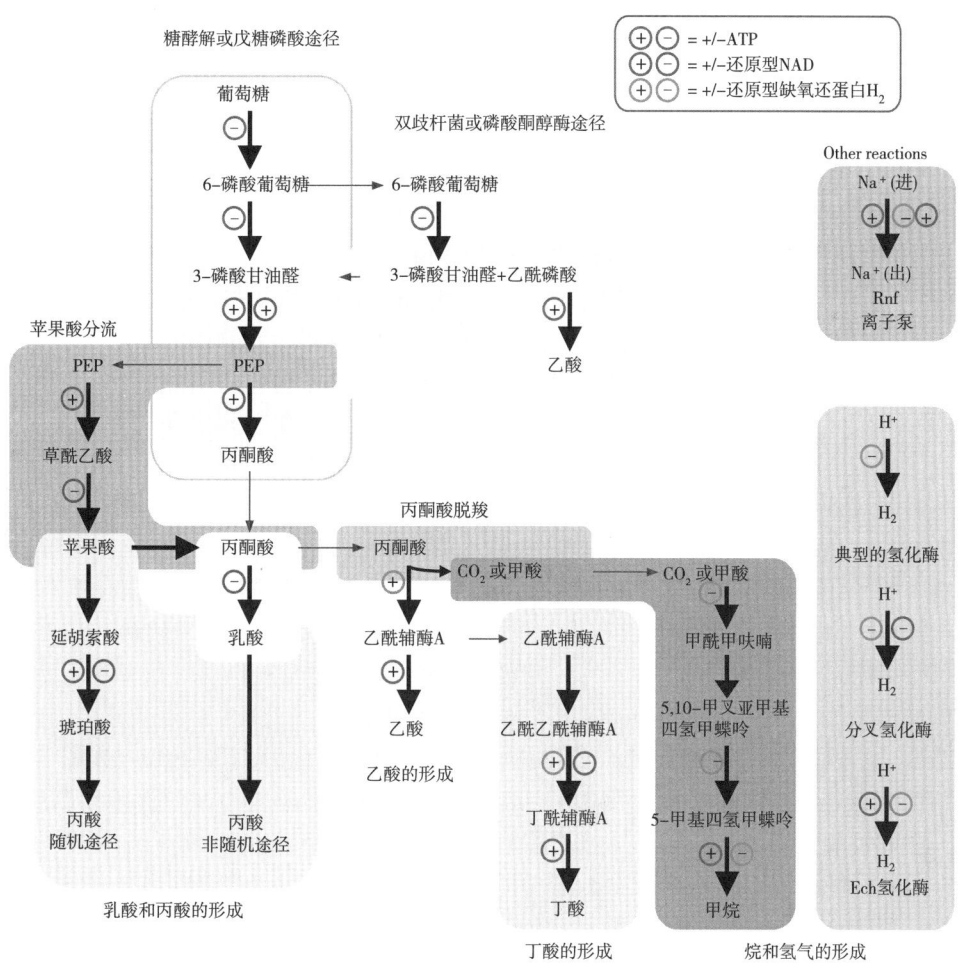

图1 瘤胃细菌和产甲烷菌的发酵途径概述。途径已被简化,每个箭头可能代表多种反应。详见 Hackmann 等（2017）。不同颜色代表参与的不同反应,红色：ATP 的生成或水解；蓝色：NAD 的还原或氧化；绿色：铁氧还蛋白或氢气（H_2）的形成。甲烷形成途径由 6 种瘤胃产甲烷菌（*Methanobrevibacter boviskoreani* JH‑1、*Methanobrevibacter millerae* ZA‑10、*Methanobrevibacter olleyae* 1H5‑1P、*Methanobrevibacter ruminantium* M1、*Methanomicrobium mobile* 1、*Methanosarcina* sp. Ms 97）编码。一种产甲烷菌（*Methanosarcina* sp. Ms 97）不编码从甲酸盐形成甲烷的途径。缩写：–P = 磷酸盐，–3P = 3‑磷酸盐，–6P = 6‑磷酸盐，CoA = 辅酶 A，CoM = 辅酶 M，H_4MPT = 四氢甲蝶呤或四氢糖蝶呤，MFR = 甲呋喃，PEP = 磷酸烯醇式丙酮酸（彩图 7）。

行的初步实验表明,糖酵解是葡萄糖的主要分解代谢途径。后来对瘤胃细胞游离提取物的研究结果也支持糖酵解作为葡萄糖和糖酵解中间体的主要代谢途径（Hamar 和 Borchers,1967）。

最近,对 48 种完整度为 99.5% 的瘤胃细菌的基因组检测表明,27% 的细菌没有或不完全编码糖酵解途径（Hackmann 等,2017）。此外,11% 的细菌编码除糖酵解外的多种葡萄糖分解途径。显然，每种葡萄糖分解代谢途径量化的重要性不仅取决于微生物基因组中是否存在编码必需酶的基因，还取决于微生物的群落组成、基因表达以及通过具有活性酶的每种分解代谢途径的碳流量。而进一步的生化实验也表明，一些瘤胃细菌缺少某些糖酵解酶（Scardovi，1965；Kelly 等，2010）。

鉴于针对碳水化合物代谢相关研究的缺乏，我们建议加强对动物饲养中实际使用的更复杂的以及具有代表性的底物进行研究，而不是纯碳水化合物。例如 Wallnöfer 等（1966）通过使用纯底物的体外发酵实验发现，只有 10% 左右的纤维素被降解，产生相对较低的乙酸与丙酸比，而粗料发酵通常会产生较高的乙酸与丙酸比（Janssen，2010）。正如 Hackmann 等（2017）对瘤胃细菌编码的各种途径所提示，与纯底物相比，动物饲料中碳水化合物降解所释放的单糖和寡糖的复杂混合物在不同速率下的分解代谢可能涉及更多样化的分解代谢途径。碳在不同途径中的分配可能受到各种单糖和寡糖的释放率、pH 值和其他变量的影响。

糖酵解过程中形成的丙酮酸是一个中心分支点，在这里，乙酸、丙酸和丁酸的产生途径发生分歧（Russell 和 Wallace，1997）。在乙酸和丁酸盐形成的第一步，糖酵解过程中形成的丙酮酸在丙酮酸氧化还原酶的催化下被氧化脱羧为乙酰辅酶 A、二氧化碳和铁氧还蛋白（图 1）。此外，在丙酮酸脱氢酶的催化下，它还可以脱羧为乙酰辅酶 A，二氧化碳和 NADH（Hackmann 等，2017）。另一种可能则是通过丙酮酸-甲酸裂解酶催化反应，脱羧成甲酸酯和乙酰辅酶 A（Asanuma 等，1999b）（图 1）。

然后，乙酰辅酶 A 通过磷酸乙酰转移酶和乙酸激酶转化为乙酸（Russell 和 Wallace，1997）（图 2）。某些细菌（硒单胞菌和 *Mitsuokella*）不编码磷酸乙酰转移酶和乙酸激酶，相反，它们可能使用琥珀酰辅酶 A：乙酸辅酶 A 转移酶和琥珀酸辅酶 A 连接酶（图 2）。瘤胃细菌还可能通过双歧途径产生乙酸（Hackmann 等，2017）（图 1）。

乙酰辅酶 A 也可以转化为丁酸盐（Russell 和 Wallace，1997）（图 1）。首先，两分子乙酰辅酶 A 发生缩合反应，生成乙酰乙酰辅酶 A，然后还原为 β-羟基丁酰辅酶 A。然后 β-羟基丁酰辅酶 A 被脱水成巴豆酰辅酶 A，巴豆酰辅酶 A 被还原成丁酰辅酶 A。在该途径的最后一步，丁酰辅酶 A 转化为丁酰磷酸盐，最后转化为丁酸盐（图 2）。（Asanuma 等，2003；Hackmann 等，2017）（图 2）。

丙酮酸和磷酸烯醇丙酮酸可以通过琥珀酸或乳酸作为中间体的两种不同途径代谢为丙酸（图 1）。它们的区别在于丙酮酸中碳的随机化（Russell 和 Wallace，1997）。在随机化（琥珀酸）途径中，丙酮酸或磷酸烯醇丙酮酸被羧基化成草酰乙酸。草酰乙酸还原为苹果酸，苹果酸脱水为富马酸。富马酸随后被还原为琥珀酸，NADH 或还原的铁氧还蛋白作为细胞内电子供体（Hackmann 等，2017）。外部氢气可被吸收以还原 NAD^+ 或氧化铁氧还蛋白来提供代谢氢（[H]）进而还原富马酸为琥珀酸（Henderson，1980；Asanuma 和 Hino，2000）。然后琥珀酸被激活为琥珀酰辅酶 A，经过链重排和脱羧生成丙酸（Russell 和 Wallace，1997）。琥珀酸到琥珀酰辅酶 A 的激活在能量上是无效的，因为它可以在途径的最后一步偶联到丙酰辅酶 A 转化到丙酸。一些微生物可释放琥珀酸作为最终产物，然后再被琥珀酸利用者吸收转化为丙酸。

第 13 章 影响瘤胃能量代谢效率的因素

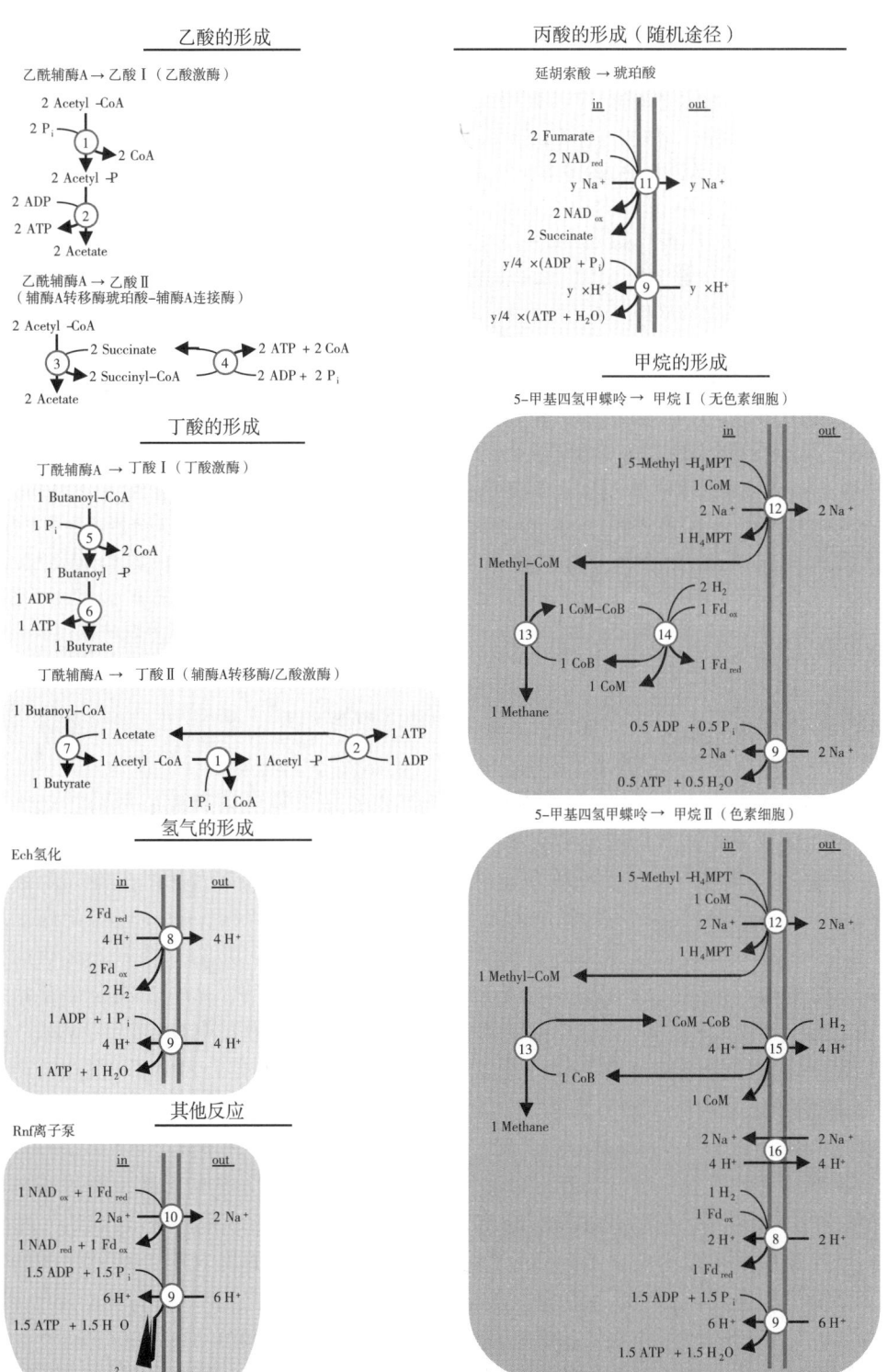

图 2 发酵过程中产生 ATP 的几个反应。未显示 EMP 途径和苹果酸分流中的反应。丙酸盐形

成中的系数 y 对于普雷沃氏菌属（编码 Nqr NADH 脱氢酶）为 4，而对于硒单胞菌属（编码 Ndh NADH 脱氢酶）为 0。丙酸盐形成的反应假设：①丙酮酸盐分解代谢仅形成还原的铁氧还蛋白（不是甲酸盐）和②每 2mol 丙酸盐形成 1mol 乙酸盐。丁酸盐形成的反应假定：①丙酮酸分解代谢仅形成还原的铁氧还蛋白（不是甲酸盐）和②不形成乙酸盐。如果未显示，则假定另一种反应（例如由反转运蛋白催化的反应）平衡了 Na^+ 和 H^+。参见文本和 Hackmann 等（2017）了解详情。甲烷形成的反应是由 6 种瘤胃产甲烷菌（*Methanobrevibacter boviskoreani* JH-1、*Methanobrevibacter millerae* ZA-10、*Methanobrevibacter olleyae* 1H5-1P、*Methanobrevibacter ruminantium* M1、*Methanomicrobium mobile* 1、*Methanosarcina* sp. Ms 97）编码的反应。搜索的数据库 ID 是 Hackmann 等（2017）的数据库 ID 及以下：K00577~K00584 和 pfam04210 用于 Mtr 甲基转移酶；Mcr 还原酶的 K00399~K00402 和 K03421~K03422；K03388~K03390 用于 HdrABC 还原酶；K14126~K14128 用于 Mvh 氢化酶；K08264 和 K08265 用于 HdrDE 氢化酶；用于 Vho 氢化酶的 COG0374、COG2864 和 COG1740；和 pfam00999 用于 Nha 反转运蛋白。这些反应由 KEGG（Kanehisa 等，2017）和 Thauer 等（2008）定义。*Methanomicrobium mobile* 1 不编码反应 14，也不清楚是什么反应替代。反应：1. 磷酸乙酰转移酶；2. 乙酸激酶；3. 琥珀酰辅酶 A：乙酸辅酶 A 转移酶；4. 琥珀酸-CoA 连接酶（ADP 形成）；5. 磷酸丁酰转移酶；6. 丁酸激酶；7. 丁酰辅酶 A：乙酸辅酶 A 转移酶；8. Ech 氢化酶；9. ATP 合酶；10. Rnf 铁氧还蛋白-NAD+氧化还原酶；11. NADH 脱氢酶和延胡索酸还原酶/琥珀酸脱氢酶；12. Mtr 甲基转移酶；13. Mcr 还原酶；14. HdrABC 还原酶/Mvh 氢化酶；15. HdrDE 还原酶/Vho 氢化酶；16. Nha 反转运蛋白。缩写：–P=磷酸盐，CoA=辅酶 A，CoB=辅酶 B，CoM=辅酶 M，Fdox=氧化铁氧还蛋白，Fdred=还原铁氧还蛋白，H4MPT=四氢甲蝶呤或四氢糖蝶呤，NADox=氧化 NAD，NADred=还原 NAD，和 Pi=无机磷酸盐。

在丙酸非随机、直接还原途径中，丙酮酸被还原为乳酸（图 1）。乳酸被激活为乳酸辅酶 A，后者被脱水为丙烯酰辅酶 A。然后丙烯酰辅酶 A 被还原为丙酰辅酶 A，最后丙酰辅酶 A 水解生成丙酸（Russell 和 Wallace，1997）。乳酸可以作为发酵的最终产物被一些微生物释放，并被乳酸利用者吸收，转化为丙酸、乙酸和少量的丁酸（Satter 和 Esdale，1968；Gill 等，1986）。对"非适应性"动物饲喂快速发酵饲料后，乳酸可作为发酵的最终产物积累，引起瘤胃 pH 值降低，导致乳酸酸中毒（Nagaraja 和 Titgemeyer，2007）。

半纤维素和果胶属于杂多糖，也是植物中结构性碳水化合物的一部分，含有戊糖单体，如木糖和阿拉伯糖（Scheller 和 Ulvskov，2010）。结果表明，木糖在混合瘤胃培养物中可通过转酮醇酶和转醛糖酶途径代谢为己糖中间体，然后通过糖酵解产生 VFA，并在较小程度上通过磷酸酮醇酶裂解（Pazur 等，1958；Wallnöfer 等，1966）（图 1）产生甘油醛-3 磷酸和乙酰磷酸。瘤胃普雷沃氏菌（*Prevotella ruminicola*）B14 和 S23、反刍月形单胞菌（*Selenomonas ruminantium*）D 和产琥珀酸丝状杆菌（*Fibrobater succinogenes*）S85 具有转酮醇酶活性，但只有反刍月形单胞菌 D 和瘤胃普雷沃氏菌 S23 具有磷酸酮醇酶活性（Matte 等，1992）。在半纤维素溶解菌丁酸弧菌（*Butyrivibrio fibrisolvens*）的一些菌株中也发现了磷酸酮醇酶活性（Marounek 和 Petr，1995）。瘤胃球菌具有结构性转酮醇酶活性，无磷酸酮醇酶活性（Thurston 等，1994）。戊糖的代谢与己糖一样可产生中间体，它们在微生物合成代谢中可作为氨基酸和核酸的前体（Marounek 和 Petr，1995）。

在瘤胃中，蛋白质水解产生的许多氨基酸都被脱氨，而不是直接并入微生物蛋白中，这一过程由大量具有低脱氨活性的细菌、少量但具有高脱氨活性的细菌以及原生动物进行（Wallace，1996；Wallace 等，1997；Hartinger 等，2018）。虽然可以推测由脱氨作用和氨基酸对之间的耦合氧化和还原反应释放的碳骨架代谢产生能量，但通过纯培养或混合培养的瘤胃细菌发酵氨基酸似乎很少提供能量供微生物生长（Russell 和 Wallace，1997；Wallace 等，1997）。

酰基脂肪酸在瘤胃中可被微生物脂肪酶水解产生脂肪酸和甘油，而糖脂和磷脂分别产生糖和磷酸盐（Harfoot 和 Hazlewood，1997）。由此释放的甘油可以通过糖酵解产生能量，最终代谢为丙酸（Bergner 等，1995；Krueger 等，2010；Avila 等，2011）。饱和脂肪酸在瘤胃中基本上是稳定的。尽管溶纤维丁酸弧菌（*B. fibrisolvens*）的还原酶似乎与膜相关，但不饱和脂肪酸的微生物生物氢化迄今尚未被证明能产生能量（Jenkins 等，2008）。

由氢气（H_2）还原二氧化碳形成甲烷（CH_4）被认为是瘤胃甲烷生成的主要途径（Hungate，1967），其次是甲酸（Hungate 等，1970）（图1）。氢气和甲酸都是碳水化合物的发酵产物（见2.3，代谢氢的处理）。在氢营养型的甲烷生成过程中，二氧化碳转移到载体上，然后碳逐渐还原为甲烷。第一个载体是甲烷呋喃（MFR），然后是四氢甲烷蝶呤（在无细胞色素的产甲烷菌中）或四氢肉瘤蝶呤（在有细胞色素的产甲烷菌中）和辅酶M。甲烷不是由甲酸直接形成的，相反，甲酸在还原为甲烷之前会被氧化成二氧化碳（Thauer 等，2008）。甲酸盐的氧化可以由产甲烷菌本身或细菌完成。而甲基营养型的产甲烷菌则可以利用果胶和甲胺释放的甲醇以及甲基化硫化合物作为甲烷的前体（Enzmann 等，2018）。近期研究表明，甲基营养型产甲烷菌在瘤胃甲烷生成中具有重要的定量意义（Söllinger 等，2018）；而甲基营养型产甲烷菌的富集可能与试验中添加的菜籽饼有关，因为菜籽饼中含有果胶（Jeong 等，2014），果胶在瘤胃中发生甲酯水解，产生甲醇（Pol 和 Demeyer，1988）。

2.2 三磷酸腺苷（ATP）的生成

各种发酵反应的负 ΔG 与 ATP 的产生是耦合的（表1）。糖酵解中每摩尔葡萄糖产生 2 mol ATP（Czerkawski，1986），与有氧代谢相同。当乙酸形成时，在乙酰辅酶 A 转化为乙酸的过程中，通过底物水平的磷酸化产生额外的 2 个 ATP（图2），导致每摩尔发酵的葡萄糖总共有 4 mol ATP（表1）。

当通过随机途径形成丙酸盐（或琥珀酸盐）时，富马酸盐还原成琥珀酸盐时会产生额外的 ATP（图2）。这一 ATP 是通过电子传递磷酸化（ETP）形成的。形成的近似数字是每摩尔葡萄糖（即葡萄糖）含有 1.5 mol ATP。对于普雷沃氏菌（*Prevotella*）来说，每摩尔丙酸盐需要 0.75 mol ATP，对于月形单胞菌（*Selenomonas*）来说，则需要 0.5 mol ATP（Hackmann 等，2017）。如果 ETP 产生 1.5 个 ATP，那么总产率是每葡萄糖 3.5 个 ATP（表1）。这些 ATP 产生的化学计量尚未通过实验确定。相反，它们是给定离子泵类型（NADH 脱氢酶和 Rnf；图3；参见 2.3 代谢氢的处理）由这些瘤胃细菌的基因组编码。这也与针对琥珀酸沃里氏菌（*Woline succinogenes*）的试验结果一致。在这种瘤胃细菌中，每 2 mol 富马酸还原为琥珀酸，产量约为 1 mol ATP（由 ~$1H^+/1e^-$、$2e^-$/富马酸和假设为 $4H^+$/ATP 的化学计量比得出）（Kröger 等，2002）。

表1　在各种反应（行）中每个发酵产物（列）产生的近似 ATP 数（mol/mol 葡萄糖）

	乙酸	丙酸	丁酸	乳酸	甲烷
葡萄糖→6-磷酸葡萄糖	−1	−1	−1	−1	0
6-磷酸葡萄糖→3-磷酸甘油醛	−1	−1	−1	−1	0
3-磷酸甘油醛→PEP	+2	+2	+2	+2	0
PEP→丙酮酸	+2	0	+2	+2	0
PEP→草酰乙酸	0	+2	0	0	0
乙酰辅酶A→乙酸	+2	0	0	0	0
延胡索酸→琥珀酸	0	+1	0	0	0
丁酰辅酶A→丁酸	0	0	+1	0	0
Rnf 离子泵	0	+0.5	+0.5	0	0
Ech 氢化酶	0	0	+1	0	0
5-甲基四氢甲蝶呤	0	0	0	0	+0.5
合计	+4	+3.5	+4.5	+2	+0.5

如果通过直接还原途径产生乳酸或丙酸，ETP 似乎不会产生额外的 ATP（Thauer 等，1977；Seeliger 等，2002）。因此，总产量是每发酵 1 mol 葡萄糖产生 2 mol ATP。

当丁酸形成时，额外的 ATP 通过两种方式产生。首先，每发酵 1 mol 葡萄糖（或产生丁酸）就会在丁酰辅酶 A 转化为丁酸的过程中通过底物水平的磷酸化产生 1 molATP（图 2）。其次，在大多数丁酸弧菌属、假丁酸弧菌中，ETP 最多可产生 1.5 个 ATP（图 2）。在其他细菌中，由于一个离子泵（Ech）未编码，只生成 0.5 个 ATP。产生 ATP 的数量尚未通过试验确定，这也是这些细菌所拥有的离子泵类型所预算的（Hackmann 和 Firkins，2015a；Hackmann 等，2017；Schoelmerich 等，2020）。如果 ETP 产生 1.5 个 ATP，那么每摩尔葡萄糖的总产量是 4.5 mol ATP（表 1）。

甲烷生成仅通过 ETP 产生 ATP（图 2）。没有细胞色素的产甲烷菌（包括甲烷短杆菌属和甲烷微菌属），每摩尔甲烷大约产生 0.5 mol ATP。含有细胞色素的产甲烷菌（甲烷八叠球菌属），每摩尔甲烷生成约 1.5 mol ATP（Thauer 等，2008）。由于大多数瘤胃产甲烷菌属于甲烷短杆菌（Henderson 等，2015），瘤胃内每摩尔甲烷生成约 0.5 mol ATP（表 1）。

很少在瘤胃细菌试验中测定每摩尔己糖发酵的 ATP 产量，但可以通过估算完成（表 1）。表 1 总结了一个估值，在此我们还应该考虑到每摩尔己糖发酵的 ATP 产量受多种因素影响，例如通过随机和非随机途径产生丙酸的比例（Russell 和 Wallace，1997），离子泵对每对电子的 H^+ 或 Na^+ 挤压的化学计量学（Schuchmann 和 Müller，2014），ATP 合酶的化学计量学，生化途径的变化，如丙酸随机途径中的丙酮酸羧化（Hackmann 等，2017），丙酮酸脱羧成乙酰辅酶 A（Hackmann 和 Firkins，2015a），焦磷酸磷酸化（Roberton 和

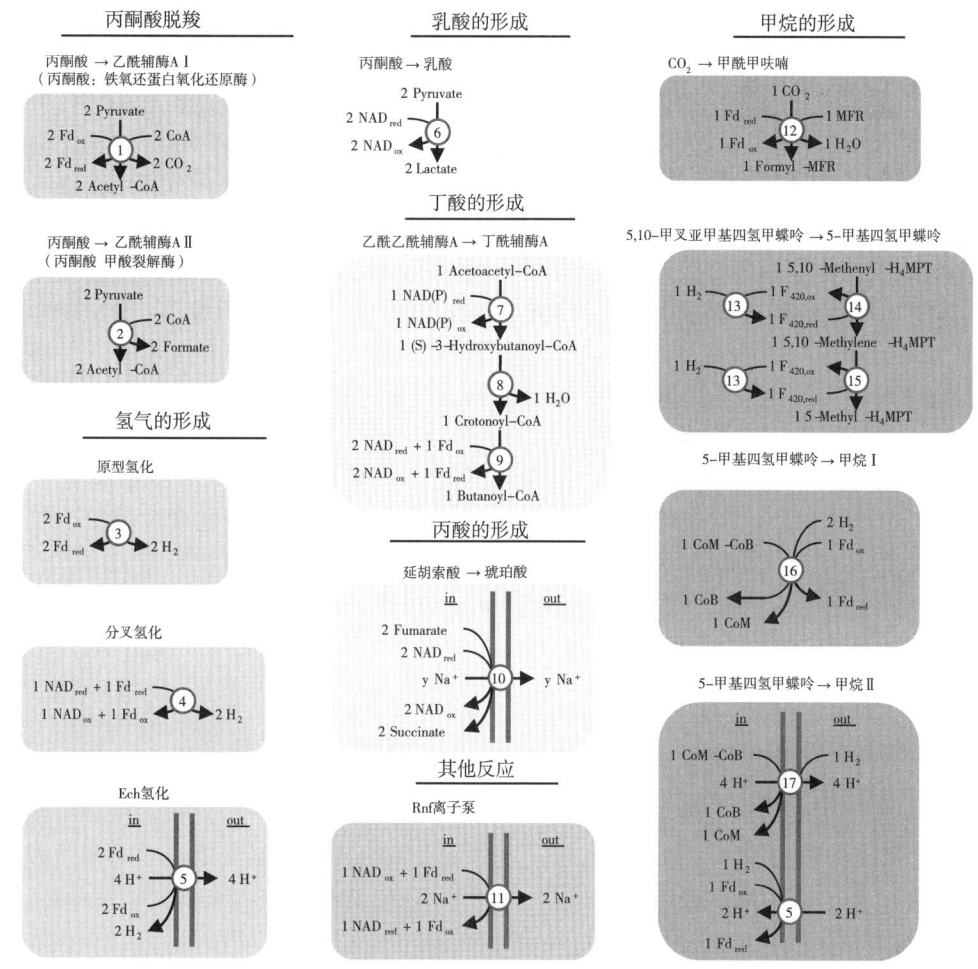

图3 发酵过程中的一些反应会产生或消耗代谢氢（[H]）。EMP 途径和苹果酸分流中的反应没有显示。详见正文、图 2 和 Hackmann 等（2017）。氢化酶的反应来自 Sondergaard 等（2016）和 Greening 等（2019）。甲烷生成反应是图 2 中描述的 6 种瘤胃产甲烷菌编码的反应，检索的数据库源自 Hackmann 等（2017），其中，K00200~K00203、K00205、K11260 和 K11261，用于甲酰基-MFR 脱氢酶；K00440~K00443 用于辅酶 F420 氢化酶；K00319 用于 5, 10-亚甲基-H4MPT 脱氢酶；K00320 用于 5, 10-亚甲基-H4MPT 还原酶。以下反应源自 KEGG（Kanehisa 等，2017）和 Thauer 等（2008）。反应：1. 丙酮酸：铁氧还蛋白氧化还原酶；2. 丙酮酸甲酸裂解酶；3. 原型氢化酶；4. 分叉氢化酶；5. Ech 氢化酶；6. 乳酸脱氢酶；7. 3-羟基丁酰辅酶 A 脱氢酶；8. 烯酰辅酶 A 水合酶；9. 丁酰辅酶 A 脱氢酶；10. NADH 脱氢酶和延胡索酸还原酶/琥珀酸脱氢酶；11. Rnf 铁氧还蛋白-NAD+氧化还原酶；12. 甲酰基-MFR 脱氢酶；13. 辅酶 F420 氢化酶；14. 5,10-亚甲基-H4MPT 脱氢酶；15. 5,10-亚甲基-H4MPT 还原酶；16. HdrABC 还原酶/Mvh 氢化酶；17. HdrDE 还原酶/Vho 氢化酶。缩写：CoA=辅酶 A、CoB=辅酶 B、CoM=辅酶 M、$F_{420,ox}$=氧化型辅酶 F_{420}、$F_{420,red}$=还原型辅酶 F_{420}、Fd_{ox}=氧化型铁氧还蛋白、Fd_{red}=还原型铁氧还蛋白、H4MPT=四氢甲蝶呤（产甲烷菌中不含细胞色素）或四氢糖蝶呤（在具有细胞色素的产甲烷菌中），MFR=甲呋喃，NAD(P)$_{ox}$=氧化的 NAD(P)，NAD(P)$_{red}$=还原的 NAD(P)。

Glucina，1982；Petzel 等，1989），碳水化合物进入细胞的运输类型（Russell 和 Wallace，1997），以及微生物合成代谢中发酵中间体的利用。

ATP 的生成对反刍动物的微生物蛋白产生至关重要，而微生物蛋白是反刍动物氨基酸最重要的来源（Wallace 等，1997）。也就是说，每摩尔底物（如葡萄糖）最大限度地产生 ATP 是否总能充分提高瘤胃微生物生物量，这一点值得怀疑。由于发酵反应的负 ΔG 的较大比例耦合到 ATP 的产生，净 ΔG 的绝对值下降，反应可以接近平衡。如果发生这种情况，逆反应的速率接近正向反应的速率，净反应的速率趋向于零，减缓发酵并最终停止发酵。这一概念的一个例子就是有细胞色素和没有细胞色素的产甲烷菌。有细胞色素的产甲烷菌每摩尔甲烷产生更多的 ATP，但它们的产甲烷必须以更低的 ΔG 来产生更多的 ATP。这意味着它们的氢气阈值更高，在极低的氢气压力环境中，它们的生长在热力学上是不可行的（Thauer 等，2008）。

如果某些发酵中间体（如乳酸）作为最终产物积累起来，每摩尔发酵单糖产生的 ATP 的摩尔数就会减少（Kohn 和 Boston，2000）。尽管与 VFA 产量相比，每单位发酵产乳酸的底物获得的 ATP 较低，但应该注意的是，产乳酸并不一定对产乳酸者不利。在快速发酵饲料中，乳酸产生比 VFA 产生快，因此单位时间内可产生更多的 ATP。相反，当能源匮乏时，最佳策略是通过产生乙酸和丙酸来最大化每摩尔发酵底物的 ATP 生成量（Russell 和 Wallace，1997）。

2.3 代谢氢的处理

在糖酵解和丙酮酸氧化脱羧成乙酰辅酶 A 过程中，[H] 被转移到氧化辅助因子，如 NAD^+ 和氧化的铁氧还蛋白（图 1，图 3）。碳水化合物的分解代谢需要还原的辅助因子被再次氧化和 [H] 的重新处理（Wolin 等，1997）。在瘤胃和其他厌氧微生物生态系统中，电子可以通过氢化酶从被还原的辅助因子转移到质子（H^+）中形成氢气（Czerkawski，1986；图 3）。三分之二的细菌和古菌基因组编码氢化酶，这也证明了氢气作为细胞间中间体在瘤胃发酵 [H] 流动中的核心作用（Greening 等，2019）。

产甲烷菌可利用氢气将二氧化碳还原为甲烷（Stewart 等，1997；图 3）甲烷生成使得瘤胃中可保持较低浓度的氢气，这允许被还原的辅助因子再次氧化，在热力学上有利于氢气的生成（Janssen，2010；van Lingen 等，2016）。产氢菌和氢营养菌之间的紧密距离对种间氢气转移非常重要（Leng，2014；Wolf 等，2016）。

一些试验研究了产氢菌的发酵产物和生长，以及同一产氢菌和氢营养菌（包括产甲烷菌和琥珀酸盐或丙酸盐生成者）的共同培养，指出了种间氢气转移在瘤胃发酵中发挥的重要作用。氢营养菌的存在有利于产氢菌产生更多的氧化发酵产物，并刺激发酵以及微生物的生长（Chung，1976；Chen 和 Wolin，1977；Bauchop 和 Mountfort，1981；Marvin-Sikkema 等，1990；Pavlostathis 等，1990）。同单一培养的白色瘤胃球菌相比，与可利用氢气的琥珀酸丝状杆菌共培养的白色瘤胃球菌依赖铁氧还蛋白的氢化酶转录减少了约两个数量级（Geier 等，2016；格林等，2019）。

[H] 的流动可以通过产生跨膜电化学梯度与能量守恒相耦合。NAD^+ 在 Rnf 离子泵的催化下氧化还原的铁氧还蛋白，可以产生一个 H^+ 或钠离子（Na^+）跨膜电化学梯度，可

以通过 ATP 酶驱动 ATP 生成（图 3；也见第 2.2 节）。同样，在铁还蛋白的催化下，H^+ 氧化还原铁还蛋白：H^+ 还原酶（Ech 氢化酶）形成氢气，也产生 H^+ 或 Na^+ 的跨膜电化学梯度（图 3）。通常在存在氢营养菌的厌氧环境中，后者需要较低的氢气压力才能在热力学上可行（Buckel 和 Thauer，2013）。

铁氧还原蛋白具有很低的标准还原电位（Eh°′），通常还原率超过 90%，在 Eh°′低至 −500 mV 的反应中可作为电子给体。如果不是通过适当的电子供体（如 NADH、NADPH、H_2、甲酸、丙酮酸或产甲烷菌中还原的 F420）和电子受体（如 NAD^+、$NADP^+$、巴豆酰辅酶 A、乙酰辅酶 A 和产甲烷菌中的异二硫化物 CoM-CoB）的组合与放能反应耦合的话，铁氧还原蛋白与 NADH 或 H_2 的再生是吸能的。这些铁氧还原蛋白还原的机制被称为基于黄素的电子分叉，因为它们是由含有黄素腺嘌呤二核苷酸（FAD）或单核核苷酸（FMN）的细胞质酶复合物催化的（Buckel 和 ThAuer，2013）。吸能和放能氧化还原反应的耦合在热力学上允许一个电子对分裂成一个 Eh°′较大的电子和另一个 Eh°′较小的电子，从而以损失前者为代价，来增加后者的还原电位（Buckel 和 Thauer，2018b）。在被称为聚合的逆反应中，高和低的 Eh′还原的辅因子提供电子来减少分叉的辅因子（Buckel 和 Thauer，2018a）。

瘤胃细菌埃氏巨球型菌（*Megasphaera elsdenii*）从 NADH 分叉两个电子，铁氧还原蛋白（吸能反应）和巴豆酰辅酶 A（放能反应），生成还原的铁还蛋白和丁酰基辅酶 A（Chowdhury 等，2015）。白色瘤胃球菌具有电子分叉的铁还蛋白/NAD-氢化酶和铁还蛋白依赖的氢化酶，其相对活性取决于氢气压力。hydS 基因与依赖铁氧还原蛋白的氢化酶共转录，含有氢气可以结合的 H-簇和 PAS 结构域，该结构域被认为可以通过触发信号传递来感知氢气压力（Zheng 等，2014）。

据 Greening 等（2019）报道，在培养的 501 个瘤胃细菌基因组中，最为富集的氢化酶是聚合 A3 组［FeFe］-氢化酶，它氧化 NADH，将铁蛋白还原为氢气。与此一致的另一个重要发现是，绵羊瘤胃中 54% 的氢化酶转录本对应于 A3 组［FeFe］-氢化酶，其中大多数属于产氢梭菌基因组（Greening 等，2019）。关于氢营养电子分叉氢化酶丰度的类似发现此前在人类结肠中也有报道（Wolf 等，2016）。Greening 等（2019）的研究重点指出了电子分叉在瘤胃［H］动态中发挥着核心作用。蛋白质组学试验可能有助于确认瘤胃中不同类型的氢化酶转录本的翻译。

被还原的胞内辅助因子中的代谢氢以及氢气和甲酸也可以转移到二氧化碳以外的电子受体上，如丙酸形成的草酰乙酸和富马酸或丁酸形成的乙酰乙酰辅酶 A 和巴豆酰辅酶 A（Henderson，1980；Asanuma 等，1999a；Asanuma 和 Hino，2000；Greening 等，2019）。琥珀酸和乳酸除了是丙酸形成的还原中间体外，还是细胞间电子载体（Stewart 等，1997；Asanuma 等，1999a）。代谢氢也可以提供给无机电子受体，如硝酸盐和硫酸盐。硝酸盐和硫酸盐的还原在热力学上比甲烷生成更为有利，但［H］进入这些途径通常会受到这些底物可用性的限制（Ungerfeld 和 Kohn，2006）。

乙酸和丁酸的生成可导致［H］的净生成（表 2）。而丙酸、乳酸和甲烷的生成可导致［H］的净消耗。在具有产甲烷功能的体外分批和连续培养中，甲烷似乎是主要的［H］库，丙酸排在第二位（Ungerfeld，2015b）。严格地说，还没有确定这在体内是否正确，因为目前还没有已发表的体内试验。然而，基于对已发表奶牛试验（Cabezas-Garcia等，2017）的meta分析，快速估算得出，甲烷作为［H］库肯定超过丙酸，至少在该分析中用到的动物和日粮中是这样（未显示计算结果）。

表2 发酵过程中产生的代谢氢（［H］）（mol/mol 葡萄糖）

	乙酸	丙酸	丁酸	乳酸	甲烷
A. 减少NAD					
3-磷酸甘油醛→PEP	+2	+2	+2	+2	0
草酰乙酸→苹果酸	0	-2	0	0	0
丙酮酸→乙酰辅酶A	0	0	0	0	0
丙酮酸→乳酸	0	0	0	-2	0
延胡索酸琥珀酸	0	-2	0	0	0
乙酰乙酰辅酶A→丁酰辅酶A	0	0	-3	0	0
Rnf离子泵	0	+1	+1	0	0
CO_2→甲酰甲呋喃	0	0	0	0	0
5,10-甲叉亚甲基四氢甲蝶呤→5-甲基四氢甲蝶呤	0	0	0	0	0
5-甲基四氢甲蝶呤→甲烷	0	0	0	0	0
总计	+2	-1	0	0	0
B. 减少铁氧化还原蛋白或H_2					
3-磷酸甘油醛→PEP	0	0	0	0	0
草酰乙酸→苹果酸	0	0	0	0	0
丙酮酸→乙酰辅酶A	+2	0	+2	0	0
丙酮酸→乳酸	0	0	0	0	0
延胡索酸琥珀酸	0	0	0	0	0
乙酰乙酰辅酶A→丁酰辅酶A	0	0	+1	0	0
Rnf离子泵	0	-1	-1	0	0
CO_2→5-甲基四氢甲蝶呤	0	0	0	0	-1
5,10-甲叉亚甲基四氢甲蝶呤→5-甲基四氢甲蝶呤	0	0	0	0	-2
5-甲基四氢甲蝶呤→甲烷	0	0	0	0	-1
总计	+2	-1	+2	0	-4

3 甲烷

3.1 重要性

在瘤胃中形成的甲烷通过打嗝和排气被释放到大气中。因此，甲烷中存在的燃烧热并没有被动物吸收和利用，从而导致摄入总能量的2%~12%的损失（Johnson和Johnson，1995）。反刍动物营养学家很早就发现甲烷的能量损失是导致低效瘤胃发酵的主要原因之一。有人提出，如果甲烷中的[H]可以转向有用的能量汇，动物的生成力很可能会提高（Czerkawski和Breckenridge，1975；Davies等，1982；Martin和Macy，1985）。有关抑制瘤胃甲烷生成的体外（Bauchop，1967）和体内（Clapperton，1974；Cole和McCroskey，1975；Czerkawski和Breckenridge，1975）研究已经进行了几十年。

最近，由于对气候变化的日益关注，人们对控制瘤胃中甲烷的形成和肠道甲烷向大气排放更加重视（Moss等，2000）。大气中甲烷的增加约占自工业革命以来以二氧化碳当量表示的温室气体总排放量（每种温室气体总量按其变暖潜力加权的总和）增量的20%。据估计，约30%人为排放的甲烷来源于牲畜肠道发酵和粪便分解（Saunois等，2016a）。在全球范围内，以二氧化碳当量计算，肠道甲烷只占温室气体总人为排放的6%或更少。然而，减少甲烷排放对改善温室气体排放具有战略意义，原因如下：①在过去十年中，大气中甲烷浓度的增加比二氧化碳的增加快得多，可能主要来自农业（尽管主要来源和汇的变化存在不确定性）（Saunois等，2016b）；②甲烷全球变暖潜力是二氧化碳的28倍，这意味着减少甲烷的排放将比同等数量的二氧化碳产生更大的影响；③相对于二氧化碳，甲烷在大气中的寿命要短得多（9年 vs 30年），这使得甲烷成为实现缓解气候变化短期影响更具吸引力的目标（Montzka等，2011；Saunois等，2016b）。

综上，我们有必要了解和控制瘤胃中甲烷的产生。读者可以参考过去和当前关于控制反刍动物甲烷排放研究的优秀综述（Beauchemin等，2008；Eckard等，2010；Martin等，2010；Morgavi等，2010）。在此，我们将重点讨论瘤胃发酵以及甲烷生成的生物化学过程，以及在瘤胃中抑制甲烷生成对微生物代谢的影响。

3.2 甲烷产量和VFA谱

主要挥发性脂肪酸的形成途径决定了乙酸和丁酸的产生与[H]净释放。如果乙酸通过双歧途径与乳酸一起形成，则不会释放[H]，尽管通过该途径形成的瘤胃乙酸的比例尚不清楚。另一方面，丙酸的产生需要[H]的净加入，这导致丙酸与产甲烷竞争[H]，并与甲烷的产生呈负相关（Janssen，2010）。这表明，在瘤胃发酵中形成的VFA与每单位有机物（OM）发酵产生的甲烷量密切相关。Wolin等（1997）总结了各种乙酸、丙酸和丁酸摩尔百分比的例子中，每摩尔发酵己糖理论上预期的甲烷摩尔数。

从理论上讲，3种主要挥发性脂肪酸（乙酸、丙酸和丁酸）的生成的化学计量学允许较大的波动范围，其中一些可能与不生成甲烷有关。例如，在理论上，丙酸和丁酸的摩尔生成比为2∶1，没有形成乙酸（因此乙酸和丙酸的摩尔比等于零），将导致理论上的"低甲烷"发酵曲线。在另一个极端，理论发酵曲线只有乙酸作为VFA和所有［H］并入甲烷，乙酸与丙酸的摩尔生成比实际上是+∞。然而事实上，在73个不同的试验中，乙酸和丙酸的体内摩尔浓度比在（4~1.5）∶1的范围内变化较小（Ungerfeld，2013）。

因此，尽管VFA产生的化学计量学可以导出十分宽泛的VFA谱，但这需要严格的发酵控制。在用标记VFA测定VFA产生流动的体内试验中，乙酸和丙酸、乙酸和丁酸之间VFA相互转换的反向流动的绝大多数测量值相对接近，表明VFA和热力学控制之间接近于热力学平衡，这可以解释为什么在瘤胃VFA谱中观察到的范围往往相对较窄（Ungerfeld和Kohn，2006）。

VFA谱的热力学控制理论可为其观察到的相对化学计量数较窄的范围提供一个可能的解释。然而，热力学平衡本身并不能解释VFA比率的变化，这些变化与饮食的变化以及与甲烷生成的程度有关。Janssen（2010）基于微生物生长数学模型Monod函数，提出了基于瘤胃氢气浓度和产甲烷菌生长速率的模型，以解释高精料日粮如何导致发酵从乙酸向丙酸转变。建立了产甲烷菌生长受瘤胃流出速率、pH值和产甲烷抑制剂影响的模型。根据Monod模型，产甲烷菌生长速率的变化会反过来影响氢气浓度。该模型还考虑了氢气产量的日变化（Janssen，2010）。下面将讨论瘤胃流出速率、瘤胃pH值和产甲烷抑制剂的变化如何通过产甲烷菌的生长速率和氢气浓度影响甲烷的产生和VFA谱。

高精料日粮提高了瘤胃流出率，这使得产甲烷菌有较大的生长速率。由Monod函数可知，以H_2作为能源生长的产甲烷菌生长速度越快，氢气浓度越高。反过来，根据不同VFA产生的化学计量学，从热力学上讲，增加H_2浓度有利于从乙酸转向丙酸的产生（Janssen，2010）。与这一理论预测一致的是，在恒温器中增加流出速率导致发酵向更多的丙酸和更少的甲烷的转变，尽管在该试验中随着流出速率的增加丁酸比乙酸减少得更多（Isaacson等，1975）。在连续培养中，在低稀释率和高稀释率下，［H］在甲烷、氢气和VFA中的回收率也有类似的结果（Stanier和Davies，1981）。

同样，产甲烷菌对pH值下降也很敏感，瘤胃pH值降低会降低其最大生长速率。根据Monod函数，这将导致氢气浓度增加，以保持相同的实际生长速率。增加氢气浓度将再次促进发酵向氢气协作途径的转变，如产生丙酸。甲烷生成抑制剂同样被假定通过类似的机制提高氢气浓度，并将促进发酵转向丙酸产生（Janssen，2010）。与甲烷生成抑制剂的预测效果一致，对甲烷生成抑制分批和连续培养（Ungerfeld，2015b）以及体内（Ungerfeld，2018）实验均指向较高的氢气积累。尽管有报道指出甲烷生成抑制批次的发酵转向丙酸，但没有连续培养的研究报道（Ungerfeld，2015b）。

体外分批培养试验发现粗料与精料的比例对乙酸与丙酸的比例的影响与pH值无关（Russell，1998）。Janssen（2010）也讨论了底物发酵性本身是如何影响氢气和甲烷的产生以及VFA分布。瘤胃环境的时空异质性导致部分生态位氢气产生和甲烷的产生紧密耦合，氢气浓度保持在较低水平，而快速发酵可能导致局部和瞬间的氢气浓度过高。时间变化的一个特别重要的例子是饲喂后的高发酵率和氢气产量（van Lingen等，2017；Söllinger等，2018），会导致氢气浓度增加，暂时抑制氢气释放过程，刺激丙酸生成。氢气产生和利用之间的不耦合被解释为氢气的快速演化超过了产甲烷菌吸收所有氢气的能力（Rooke等，2014）。Walker和Monk（1971）的一个有趣的试验似乎支持了这一观点：这些研究人员发现，在体外瘤胃分批培养中增加标记葡萄糖的剂量约30 000倍，导致乙酸与丙酸的比例从24.3下降到2.42。在该研究中没有报道pH值，因此不能确定乙酸与丙酸的比例的变化仅仅是发酵速率提高的结果，还是受pH值下降的影响。

发酵曲线的短期变化更有可能是由微生物群落内不同微生物的活动和微生物细胞内不同发酵途径的相对变化而引起，而不是微生物群落组成的变化。另外，由饮食变化引起的发酵曲线的长期变化则更有可能主要是由微生物群落组成的变化引起（Janssen，2010）。

[H]在瘤胃内流动的动力学和热力学模型（Kohn and Boston，2000；Offner and Sauvant，2006；van Lingen等，2016，2019）或描述了特定[H]在甲烷生成抑制瘤胃发酵中的热力学可行性（Ungerfeld，2015a）。进一步了解[H]流动的控制可能需要在体内测量辅助因子的氧化状态，特别是$NADH/NAD^+$对和铁氧还原蛋白，不同类型的氢化酶的活性，以及溶解的氢气梯度的大小。在不同的饮食和饲喂后的不同时间对动物进行测定是很重要的。关于瘤胃微生物细胞内pH值和还原以及氧化辅助因子浓度的报道目前很少，而且仅限于体外培养的$NADH/NAD^+$对（Hino和Russell，1985）。这些测定显然在试验上具有一定的挑战性。此外，有研究还提出若能更好地反映动物和日粮因素，如瘤胃流出率和颗粒大小，将有助于改进现有模型（van Lingen等，2019）。

3.3 抑制瘤胃发酵中甲烷的生成

通过抑制瘤胃中甲烷的产生，可以改善反刍动物生成对环境的影响，同时提高反刍动物的能量利用效率。然而，对于后一个方面，在没有改变饲料或动物的情况下，以特定的化学抑制剂为靶点的瘤胃甲烷生成试验的meta分析表明，抑制瘤胃甲烷生成并不能持续提高产奶效率或生长效率（Ungerfeld，2018）。

动物生产力的提高是选取控制瘤胃产甲烷方法的重要指标，因此了解为什么减少甲烷的能量损失并不总是转化为生产力的提高，这点十分重要。

（1）假设日粮干物质消化率为70%，根据总能量损失为2%~12% GE（Johnson和Johnson，1995），以甲烷为单位的能量损失约为摄入可消化能（DE）的3%~17%。在大多数实验中，抑制甲烷产生一直是适度的，例如30%，这可能代表摄入可消化能增加了1%~5%。忽略尿液能量损失和考虑维持和产奶的代谢到净能量转换因子为0.6，生长和增肥的代谢到净能量转换因子为0.2，抑制甲烷产生30%可获得的最大净能量增益约为3%，从而提高生产力。人们认为，很少有试验能够发现这种差异。

（2）体外试验在[2H]平衡中一致表明，VFA、甲烷和氢气中的[2H]回收率严重下降。除氢气外，当甲烷生成受到抑制时，可能会积累其他非典型[H]池，它们是具有功能性产甲烷作用的瘤胃发酵中间产物，不具有营养价值，如甲酸盐（Ungerfeld，2015b）。氢气的积累在甲烷未形成过程中所节省的能量中占相当大的比例。在体内试验中，即使使用相同的抑制剂和相似的日粮和动物，氢气积累对抑制甲烷生成的反应也存在很大差异（图4）。

（3）还需要进一步研究抑制甲烷生成对消化和发酵有机物量的影响。二氢积累表明[H]的处理中断和辅助因子的再氧化，这可以抑制发酵（Wolin等，1997）。当甲烷生成在体外受到抑制时，有抑制发酵的迹象（Ungerfeld，2015b），这与批次培养实验中观察到

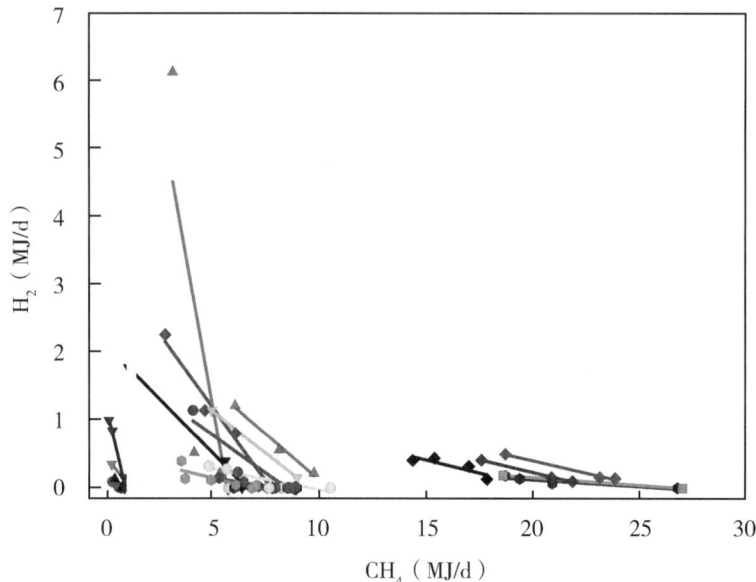

图4 在13项研究（包括20项实验或饮食）中，随着甲烷（CH_4）产量减少，氢气（H_2）能量损失的响应。H_2（MJ）= 1.62（±0.42；$P<0.001$）+exp（$P=0.081$）−0.30（±0.17；$P=0.085$）CH_4（MJ）+exp × CH_4（$P=0.004$）；N=67，$R^2=0.81$（$P=0.002$）。资料来源：Hristov 等（2015），Johnson 等（1972，1974），Lopes 等（2016），Martinez-Fernandez 等（2016，2017，2018），Mitsumori 等（2014），Olijhoek 等（2016），Veneman 等（2015），Vyas 等（2016，2018）。

的 NADH 与 NAD⁺ 比值的增加（Hino 和 Russell，1985）一致。

在某些动物体内抑制甲烷生成并不影响原位 DM 表观消化率（诺兰等，2010；Martinez-fernandez 等，2014）。但并非所有（Martinez-Fernandez 等，2018）试验都与总 VFA 浓度呈负相关（Ungerfeld，2018）。原位表观消化率和 VFA 浓度的测量都不能准确地反映消化和发酵过程，因为表观消化率没有考虑微生物生物量，VFA 浓度不仅受 VFA 产生的影响，还受 VFA 吸收、传代、微生物生物量和瘤胃体积的变化等的影响（Dijkstra 等，1993；Storm 等，2012；Hall 等，2015）。显然，[H] 流动的中断必然会停止发酵，但问题是，在不妨碍发酵的情况下，辅因子的氧化状态能在多大程度上被改变，以及哪些替代途径可以与甲烷生成相同的速率合并 [H]。

（4）抑制瘤胃产甲烷可能导致瘤胃发酵的变化，而这种变化可能并不总是符合动物的需要。例如，抑制甲烷生成可能促使瘤胃发酵从乙酸到丙酸的改变（Janssen，2010），这已在分批培养试验中得到证实（Ungerfeld，2015b）。如果抑制甲烷生成也会增加体内丙酸的产量，那么不缺乏产糖前体的动物可能对额外的丙酸没有反应，除非使饮食产生更多酮，以调整其适应抑制产甲烷的干预。由于受到其他营养物质供应的限制，更多的能量和葡萄糖的供应可能不会转化为受其他营养物质供应限制的更高的动物生成力。此外，丙酸作为一种饱腹感信号（Allen 等，2009），这似乎与减少动物摄入量从而抑制甲烷生成相一致（Ungerfeld，2018）。

很明显，为了最高生成力，仅抑制甲烷的产生是不够的，必须将 [H] 转到限制每种动物生成力的发酵产物上。反过来，这就需要对具有功能性甲烷生成的瘤胃和甲烷生成被抑制的瘤胃发酵的生成和利用 [H] 流的控制有透彻的了解。

将 [H] 引入有用的汇中这一方式取得的部分成效，这反映在氢气积累减少和一些感兴趣的产品的浓度增加，一些体外分批培养试验已经说明了这一点。苹果酸或富马酸是丙酸生成随机途径的中间产物，添加苹果酸或富马酸可以在甲烷生成被抑制时部分缓解氢气积累，有时还会增加最终丙酸的浓度（Mohammed 等，2004；Tatsuoka 等，2008；Ebrahimi 等，2011）。其他研究表明在添加了活性琥珀酸或丙酸生成生物体的分批培养中，可以提高丙酸、琥珀酸或丁酸的产量，减少甲烷的生成（Mamuad 等，2014；Kim 等，2016）。在抑制甲烷生成的批培养物中添加还原性乙酸降低了氢气的积累（Nollet 等，1997；Le Van 等，1998；Lopez 等，1999）。相比之下，丙酸菌体内剂量并不影响总甲烷排放（Vyas 等，2014a，b，2015）。在一项体内试验中，Martinez-Fernandez 等（2017）通过添加间苯三酚降低了氯仿抑制甲烷生成的阉牛的氢气排放。间苯三酚在还原为乙酸时明显加入了氢气。

4 影响微生物生长效率的因素

微生物利用发酵过程中产生的 ATP 为其生长提供能量，但其生长效率远非完美。提高这种效率是非常重要的，因为瘤胃中生长的微生物产生的蛋白质占动物消化所需蛋白质的 70%（Broderick 等，2010）。从理论上讲，微生物应该以高且几乎恒定的效率生长。从生化途径，我们可以计算出在微生物细胞中合成大分子需要多少 ATP。20 世纪 70 年代首次进行的这些计算表明，微生物的生长效率（YATP）应为每摩尔 ATP 约 30 g 干物质（Stouthamer，1973）。改变计算中的一些假设（如蛋白质是由葡萄糖和氨合成的还是由预先形成的氨基酸合成的），对计算效率影响不大。

在现实生活中，瘤胃微生物的生长效率低且多变。事实上，它们似乎从理论效率的 1/3 增长到 2/3（图 5）。大多数研究表明效率低是在纯培养中进行的（Russell 和 Wallace，1997）。然而，现有证据表明，混合培养（Isaacson 等，1975）和活体培养（Kennedy 和 Milligan，1978）的效率同样低。重要的是，生长效率不仅低于理论水平，而且变化很大，甚至在试验中也是如此（图 5）。在这些和其他报告 YATP 的试验中，没有测量 ATP 的生成（Bauchop 和 Elsden，1960；de Vries 等，1973），而是假设每个 VFA 产生 ATP 的化学计量常数进行估算。因此，报告的 YATP 值实际上对应于微生物生物量与 VFA 和甲烷产量不同线性组合的比率。如第 2.2 节所述，每摩尔 VFA 产生的 ATP 受到许多因素的影响，这些因素在不同的试验和处理中可能有所不同，因此，实际 YATP 的变化可能比报道的还要高。YATP 对乙酸、乳酸和乙醇生成的评估可能更准确，其中 ATP 是由底物水平的磷酸化产生的，但对丙酸、丁酸和甲烷的偏差可能是重要的，其中也会发生电子传递磷酸化。

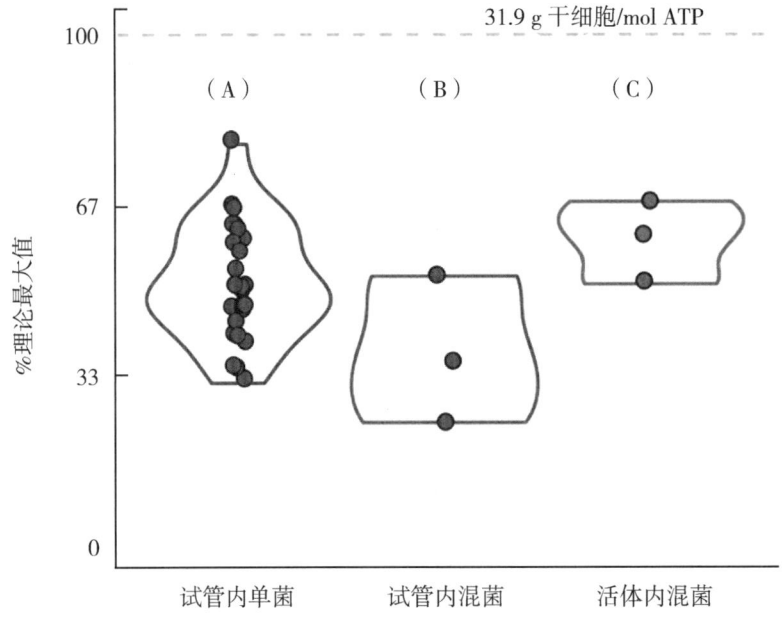

图 5 瘤胃微生物的生长效率。数值是实际生长效率（g 干细胞/mol ATP），表示为理论最大效率的百分比 [31.9 g 干细胞/mol ATP；（Stouthamer，1973）]。Russell 和 Wallace（1997）总结了体外单个细菌的数据（A）。体外混合细菌的数据（B）来自 Isaacson 等（1975）。体内混合微生物的数据（C）来自 Kennedy 和 Milligan（1978）。

4.1 能量汇

尽管人们早就认识到微生物的生长效率低,但其根本原因一直难以确定。我们可以推断,低效率的原因是微生物引导 ATP 远离生长而流向其他能量汇(图6)。虽然可能存在几种能量汇,但最近的研究表明,其中 3 种对混合瘤胃微生物是重要的(Hackmann 等,2013a;Teixeira 等,2017)。这些能量吸收是维持、糖原积累和能量溢出(图6)。

4.1.1 维持

维持指的是细胞的"基本代谢功能调节",主要包括维持细胞膜上的离子平衡,以及在较小程度上维持大分子的更替。尽管机动性占比较小,但它也是是维持的另一个组成部分(Russell 和 Cook,1995)。即使维持是微生物生存所必需的,但从牛和其他反刍动物的生成角度来看,维持是一种浪费,因为它的净产物是热量(而不是微生物细胞)。

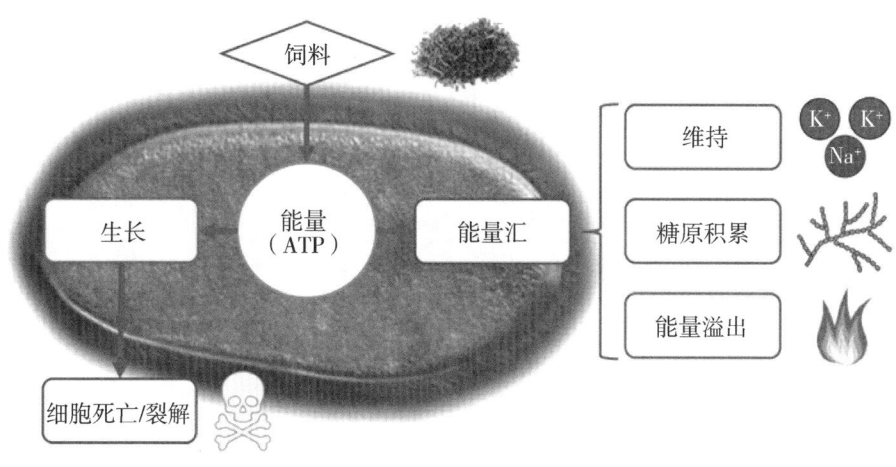

图 6 导致微生物低效生长的因素

饲料发酵释放的一些 ATP 直接用于生长,但大量用于能量汇,例如维持、糖原积累和能量溢出。细胞的死亡和裂解也会降低效率。

维持是影响生长效率的重要因素,特别是当微生物生长速率较低时。在 5%/h 的低生长速率下,维持消耗占混合瘤胃细菌总能量消耗的 30% 以上。相比之下,在 20%/h 的高速增长下,它只占 10%(Russell,2007A)。这是因为维持是一种固定成本,当生成率较低时,固定成本的重要性会相应增加。原生动物比细菌有更长的生成时间,这意味着在总能量消耗中比例更高的维持能量消耗(Newbold 等,2015)。虽然维持能量是一种非增长的固定成本,但有证据表明维持需求不是恒定的(Russell 和 Cook,1995)。例如,快速发酵的底物(葡萄糖)的维持能力可能高于缓慢发酵的底物(纤维二糖)(Thurston 等,1993)。为了更好地理解这些机制,如果可能的话,设计降低维持成本的方法,将对微生物低速率生长的广泛反刍动物生成系统产生积极的影响。

4.1.2 糖原累积

当细胞有多余的碳水化合物时,微生物就会积累糖原(Preiss 和 Romeo,1990)。在混合瘤胃微生物培养中，也积累了少量的葡萄糖和麦芽糖（Hackmann 等，2013b）。积累糖原似乎是储存额外碳水化合物的一种经济方法。然而，在这个过程中细胞会失去 ATP，除非细胞在没有水解其糖原的情况下排出瘤胃，糖原在十二指肠消化，以葡萄糖的形式被吸收。在糖原合成过程中，细胞每吸收 1 mol 葡萄糖就要消耗 2 mol ATP。如果细胞随后降解糖原，它只能恢复相当于每摩尔葡萄糖的 1 mol ATP（以葡萄糖-1-磷酸盐的形式）（Nelson 和 Cox，2017）。因为糖酵解过程中每摩尔葡萄糖只产生 2 mol ATP（表 1），用于糖原积累的 ATP 量可能很重要。

4.1.3　能量溢出

能量溢出表明分解代谢和合成代谢之间不耦合。在能量溢出的过程中，微生物细胞以热量的形式消耗多余的能量。当分解代谢产生的 ATP 多于细胞生长、维持和糖原积累时，细胞就会释放能量。当微生物细胞对过量的可发酵碳水化合物做出反应时，或者当微生物生长所需的营养物，如氮，变得有限时，微生物细胞会溢出能量（Russell，2002；Hackmann 和 Firkins，2015b）。

能量溢出是通过离子、糖原或海藻糖的无用循环来实现的（Russell 和 Cook，1995；Hackmann 和 Firkins，2015b）。例如，在牛链球菌中，H^+ 在膜上循环（Russell，2002）。牛链球菌能量溢出可由氨（NH_4^+）作为唯一氮源存在的过量葡萄糖触发。在这种细菌中，激活能量溢出的细胞内信使似乎是 ATP。细胞膜的通透性降低，这使得 H^+ 循环无效（Russell 和 Cook，1995）。混合瘤胃微生物能量溢出的机制尚未确定。通过混合培养了解不同底物能量溢出的可能大小和机制，对于理解瘤胃微生物在体内能量溢出的程度以及在何种条件下溢出具有重要意义。

4.1.4　汇的比较

维持、糖原积累和能量溢出在分批培养实验中被证明是重要的（Hackmann 等，2013a；Teixeira 等，2017）。最近的试验比较了构成混合微生物的两个最大群体细菌和原生动物的碳汇（Teixeira 等，2017）中的汇。当给予葡萄糖时，原生动物的反应是积累大量的糖原（图 7）。这在①测量糖原和②计算其合成能产生多少热量时很明显。相比之下，细菌合成的糖原相对较少，但其能量溢出很重要。在两组中，维持也很重要（图 7）。

从热力学的观点来看，维持、糖原积累、能量溢出和微生物生长都涉及散热的物理化学过程。人们普遍认为，热量的产生可能会随着生长速度的增加而增加，因为在生长速度更快的细胞中，在单位时间和单位体积内可能会发生更多的化学反应。然而，在不同生长速率和限制葡萄糖条件下连续培养的反刍链球菌和瘤胃拟杆菌（*Bacteroides*）则不是这样，除非生长速率足够高导致葡萄糖积累。通过限制葡萄糖，发现热量的产生似乎与维持有关，而不是与生长有关（Russell，1986）。

4.2　影响增长效率的其他因素

4.2.1　细胞组成

Stouthamer（1973）计算出在大肠杆菌中，对细胞生长的最大需求对应于氨基酸与蛋白质的聚合。

微生物细胞合成蛋白质的能量消耗高，其原因有两方面：一是组装氨基酸构建蛋白质

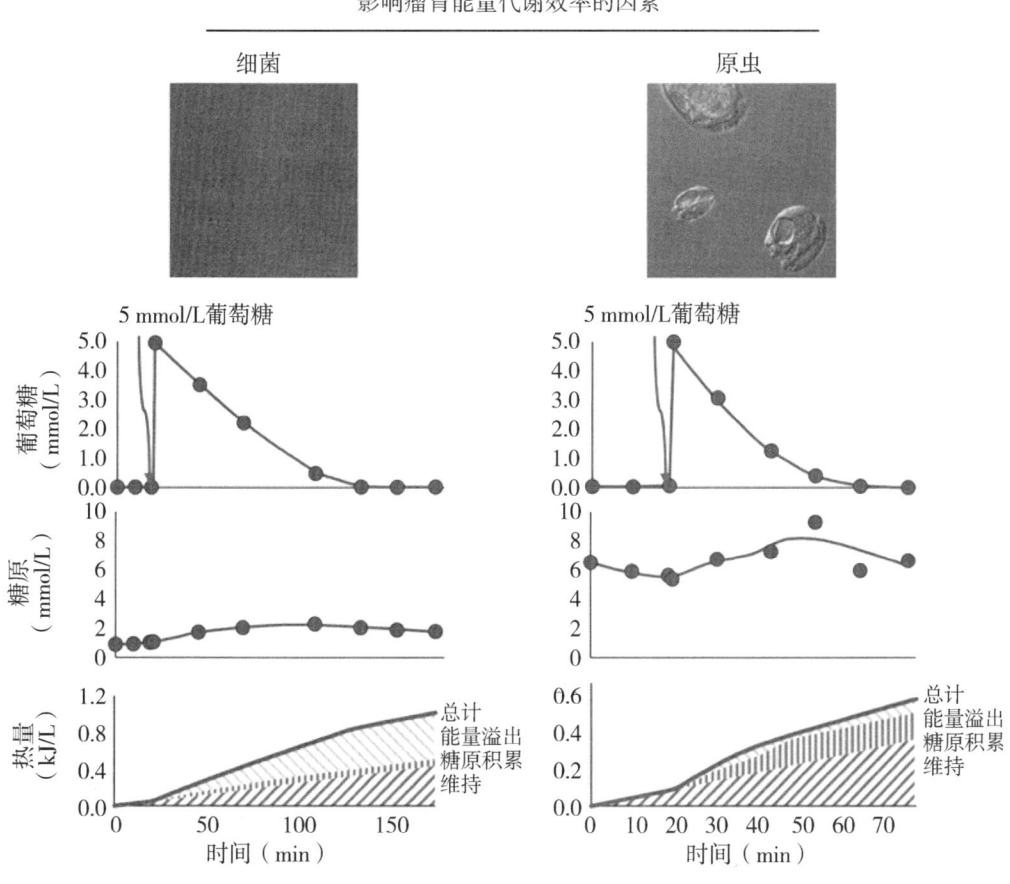

图 7　瘤胃细菌与原生动物不同能量汇的重要性。细菌和原生动物被分离并与葡萄糖（5 mmol/L）一起生长。能量溢出和维持对细菌很重要，而糖原和维持对原生动物很重要。改编自 Teixeira 等（2017）。

所需的 ATP 量高，二是蛋白质含量高。多糖和 RNA 在数量上也是重要的组成部分，其比例随生长阶段的不同而有很大的变化；然而，据估计在大肠杆菌中蛋白质合成的 ATP 成本更高（Russell 和 Cook，1995）。但是，在混合瘤胃细菌和原生动物中，糖原的积累及其合成的能量成本也可能是相当大的（Hackmann 和 Firkins，2015b）。

4.2.2　细胞裂解

虽然非生长能量吸收很重要，但导致效率低下的另一个因素是细胞死亡和裂解，导致微生物生物量的周转（图 5）。原生动物的捕食、自溶和噬菌体都是细胞裂解的原因。在瘤胃中，微生物蛋白的降解量很高，高达 50% 的微生物蛋白可降解为非蛋白氮（Wells 和 Russell，1996；Oldick 等，2000）。降解后的微生物蛋白中只有部分氨基酸和非蛋白氮被重新加入新合成的微生物蛋白中，说明氮的利用效率低下。此外，瘤胃内微生物蛋白降解也意味着能量利用效率低下，由于 ATP 必须用于蛋白质的再合成，即使是蛋白质分解后未降解的氨基酸，也可以用来合成新的微生物蛋白（Czerkawski，1986）。

由于低氮日粮的初步试验似乎指出了提高微生物蛋白产量和氮利用效率的方法,已提出消除原生动物是减少细胞裂解的一种手段。然而,对高产动物的研究显示消除原生动物对微生物氮生成的益处方面还不够明确(Firkins 等,2007)。此外,原生动物群落的组成可能会影响细菌的捕食和细胞裂解,因为全毛虫的细菌捕食活性低于内毛虫。目前,人们正在研究几种控制原生动物的方法,但目前还没有一种实用且安全的用于生成环境的灭虫方法(Newbold 等,2015)。同时,增加瘤胃外流率已被建议作为一种机制来增加细胞在裂解前从瘤胃排出的比例(Wells 和 Russel,1996),尽管应该考虑到更大的通过利率可能会导致更低的消化率(Løvendahl 等,2018)。

导致效率低下的另一个因素可能是细胞内代谢物(Bond 等,1998)、纤维多糖(Wells 等,1995)和麦芽多糖的释放(Matulova 等,2001;Nouaille 等,2005)。虽然在纯培养中观察到这种释放,但其对瘤胃混合微生物的重要性尚不清楚。

4.2.3 利用能源如碳源来促进细胞生长

碳在能量底物中的比例被纳入细胞生物量很大程度上取决于发酵过程中 ATP 的产生。如果每摩尔发酵的能量底物产生的 ATP 相对较低,大部分能量源将用于能量生成,但如果 ATP 产生高,能量源中的大部分碳可以被吸收到细胞成分中(Russell 和 Cook,1995)。从数学上讲,由 ATP 生成的变化引起的发酵产物和微生物合成代谢之间的碳分配的变化会影响每克有机物降解产生的微生物生物量,而不是每摩尔 ATP(YATP)产生的微生物生物量,因为微生物生物量的形成和 ATP 产生的方向是相同的。

4.3 提高微生物生长效率策略的展望

我们需要在比分批培养更复杂的试验模型中描述每个能量吸收的相对重要性。在葡萄糖上连续培养的试验已经证实,混合细菌的维持是重要的(Isaacson 等,1975)。未来还需要其他碳水化合物的试验,如淀粉或纤维,以扩大对活体动物的认识。当下只有分批培养被用来测量混合微生物中的能量溢出(Hackmann 和 Firkins,2015b)。需要更多的试验来评估和量化瘤胃能量溢出的重要性。

体内试验已经测量了混合细菌中的糖原(Jouany 和 Thivend,1972b;McAllan 和 Smith,1974;Leedle 等,1982)和原生动物(Jouany 和 Thivend,1972a)。这些试验证实了糖原是由混合微生物积累的。然而,这些试验还没有确定积累的数量有多重要。为了回答这个问题,未来的试验可以在不同的条件下提供不同同位素标记的碳水化合物,测量糖原中所占的比例。

提高微生物效率的一种方法是改进评估奶牛日粮的模型。尽管这些模型被营养学家广泛使用,但它们的某些方面仍然很粗糙。例如,美国 NRC(2001)假设微生物效率是一个恒定值(根据瘤胃降解氮进行调整)。这使得应用营养学家在提高效率方面缺乏有用的指导意见。

根据最近对能量汇的测量,可以构建更好的模型。这些改进的模型将帮助营养学家制定提高生产效率的日粮模式。

5 瘤胃能量与氮代谢的相互作用

瘤胃中的能量和氮代谢密切相关（Nocek 和 Russell，1988）。蛋白质、核酸等含氮化合物的合成需要利用分解代谢过程中产生的负 ΔG，而分解反应又需要酶的参与，酶的合成是 N 代谢的一部分。在本节中，我们将从 3 个不同方面讨论瘤胃能量和氮代谢之间的相互作用：①甲烷生成抑制对微生物氨基酸和蛋白质合成的影响；②氮源对能量溢出的影响；③能源与氮源供应的同步。

5.1 抑制瘤胃甲烷生成对微生物合成氨基酸的影响

混合瘤胃微生物群将培养基中的 NH_4^+ 结合到碳链中，合成用于蛋白质合成的氨基酸。胺化中每摩尔 NH_4^+ 的加入，需要加入一摩尔［2H］，通常由 NADH 或 NADPH 提供。相反的反应，脱氨，释放［2H］为 NADH 或 NADPH（Wallace 等，1997）。抑制瘤胃甲烷生成会破坏电子迁移，氢气积累和还原电位（Eh）降低就是证据（Czerkawski，1986；Sauer 和 Teather，1987）。据推测，抑制甲烷生成可以刺激［H］的胺化处理，并增加微生物蛋白的合成（Ungerfeld 等，2007）。此外，在瘤胃分批培养物中抑制甲烷生成提高了 $NADH/NAD^+$ 比值，抑制了氨基酸发酵（Russell 和 Jeraci，1984；Hino 和 Russel，1985）。在瘤胃细菌和原生动物的细胞提取物中，NADH 抑制了更多还原态氨基酸的脱氨作用，而氧化剂亚甲基蓝则刺激了脱氨作用（Hino 和 Russell，1985）。

分批和连续培养抑制瘤胃甲烷生成持续降低了主要 VFA 和气体中［2H］的回收率，有人提出，未解释的［2H］掺入可能部分是由于胺化作用的增加（Ungerfeld，2015b）。然而，最近的研究结果无法证明瘤胃分批培养物中微生物氨基酸合成的［2H］掺入量的净增加（Ungerfeld 等，2019）。在混合瘤胃分批培养物中抑制甲烷生成对微生物氮和其他细胞组分产量的影响并不一致，有些研究表明会导致其增加（Ungerfeld 等，2007；Guo 等，2009），有些研究表明会导致其减少（Russell 和 Martin，1984）。

在体内，抑制瘤胃甲烷生成对瘤胃微生物蛋白生成的影响还很少。Nolan 等（2010）观察到，当使用硝酸盐适度抑制甲烷生成时，通过尿液中嘌呤衍生物估计的微生物蛋白产量会增加。此外，当硝酸盐适度抑制甲烷的产生时，Wang 等（2018）观察到瘤胃中微生物氮含量增加。

Martinez-Fernandez 等（2016）研究了在体内抑制瘤胃甲烷生成对氨基酸代谢的影响，他们发现用氯仿抑制甲烷的产生，提高了饲喂混合日粮的牛瘤胃液中几种氨基酸的浓度，以及干草饲料中几乎所有氨基酸的浓度。他们的结果表明蛋白质水解增加（Martinez-Fernandez 等，2016），由于抑制游离氨基酸脱氨的另一种可能性似乎与抑制甲烷生成时两种日粮中支链脂肪酸的增加不一致。Martinez-Fernandez 等（2016）还发现，在干草日粮中抑制甲烷生成时，瘤胃液中嘌呤代谢物肌苷和次黄嘌呤浓度增加，这可能是由于饲料核酸的更大降解和/或微生物细胞的更大溶解。游离氨基酸浓度的增加也可能是由于氨基酸加入微生物蛋白和/或更多的微生物细胞裂解所致。

Martinez-Fernandez 等（2018）研究表明，氯仿或 3-硝基羟丙醇（3NOP）抑制瘤胃

甲烷生成时，甲基胺和其他甲基化合物的浓度增加。虽然甲基胺可以由氨基酸的脱羧作用形成，但其浓度的增加似乎更有可能是由于甲基营养产甲烷菌抑制了对甲基胺的利用。这与他们在第七产甲烷菌目中观察到的 3NOP 的减少相一致，尽管第七产甲烷菌目不受氯仿的影响（Martinez-Fernandez 等，2018）。到目前为止，体外和体内的结果都不支持胺化是甲烷生成的一种替代品，但还需要更多的研究证明，特别是体内试验。

5.2 氮对能量溢出的影响

纤维素分解型（Atasoglu 等，2001）和非纤维素分解瘤胃细菌（Atasoglu 等，1998）对添加氨基酸和多肽的反应是通过降低由 NH_4^+ 衍生的细菌蛋白的比例实现的。当添加氨基酸时，牛链球菌发酵过量葡萄糖的生长速度更快，每摩尔葡萄糖产生的蛋白质也比仅在 NH_4^+ 上生长时更多。补充氨基酸降低了每毫克细胞蛋白质产生的热量，这被解释为由于分解代谢和合成代谢之间的更紧密耦合而导致的能量溢出的减少（Russell，1993）。类似地，Bond 和 Russell（1996）估计，与生长在葡萄糖和氨氮源上的牛链球菌相比，生长在葡萄糖和氨基酸源上的牛链球菌以热量的形式释放的能量大约多出 2 倍。在生长于快速可用碳水化合物来源的瘤胃培养物中，氨基酸 N 的缺乏引发了从细胞成分合成中分离出来的 ATP 水解的能量溢出（Russell 和 Cook，1995；Van Kessel 和 Russell，1996；Russe，2007 b）。这一现象在不同动物饮食的体内有多重要还有待阐明。

5.3 同步性

瘤胃氮的缺乏会影响碳水化合物的代谢，反之亦然。N 利用率低会影响碳水化合物的消化，碳水化合物利用率不足会降低 NH_4^+ 进入微生物蛋白的量。此外，如果碳水化合物提供的能量有限，氨基酸可以作为一种能量来源发酵，而不是纳入蛋白质（Nocek 和 Russell，1988）。Newbold 和 Rust（1992）通过混合瘤胃分批培养物研究了碳水化合物和氮供应的异步性，得出结论，在氮和能量失衡停止后，培养物迅速恢复。Valkeners 等（2006）给牛饲喂相同配方的日粮，但通过改变上午和下午日粮的成分比例，在保持总体配方不变的情况下，造成了 3 种不同程度的瘤胃可降解有机物和氮的不平衡。平衡日粮时，早上饲喂后瘤胃 NH_4^+ 浓度峰值较大，反之，NH_4^+ 浓度维持在极低浓度，且不平衡程度最大。但对十二指肠微生物氮流量、瘤胃 NDF 降解和尿氮排泄均无影响。

通过日粮配方来同步营养的可利用性通常没有改善微生物蛋白生成或动物性能（Cole 和 Todd，2008；Hall 和 Huntington，2008）。动物具有改善日粮引起的瘤胃营养不平衡的机制。宿主动物通过尿素循环和内源性蛋白质向瘤胃贡献氮。因此，除了饮食之外，还必须考虑动物因素对营养平衡的影响（Hall 和 Huntington，2008）。

在畜牧系统中，平衡能量和氮的额外挑战是饲料摄入量的测量，以及对食用饲料化学成分及其时空变化的了解（Hersom，2008）。在粗放的放牧系统中，动物每天补充一次或更少的频率，可能会对营养同步作出更频繁的补充反应。然而，与每周补充一次棉籽粕相比，每天补充棉籽粕没有任何优势（Huston 等，1999）。每隔 24 h 或 48 h 用豆粕或玉米蛋白粉补充以玉米秸秆为基础的低质量饲料，不能表现出较频繁的蛋白质补充的一致优势（Collins 和 Pritchard，1992）。Farmer 等（2004）报道，在放牧低质量牧草的牛中，补饲频率与补给量的蛋白质含量和组成相互作用。当尿素氮含量低于 30%时，可以将补给量减少到每周 3 d，而不影响生产性能。总的来说，似乎很少有情况下同步提供能量和氮有助于提高瘤胃功能或动物性能。

6 结论和未来趋势

目前，我们亟须对瘤胃能量代谢进行研究，以填补对反刍动物营养重要方面的知识空白，如对瘤胃 VFA 和 CH_4 产量，以及微生物蛋白产生效率的调控。这些领域的进展可以改进营养模型，以优化动物饲料营养。而更深入地了解瘤胃发酵，也有助于优化特定干预措施的效果（如抑制甲烷生成），或设计新的干预措施来改善瘤胃功能。我们应将瘤胃及其微生物群落与宿主视为一个相互作用的系统。为了明晰研究目的，微生物代谢和生态学以及宿主生理的研究可以进行，但不应该被视为截然不同和孤立的。因此，综合的、系统的研究方法是必要的，因为大多数针对瘤胃代谢某一特定方面的干预将同时影响瘤胃微生物或动物系统的其他部分。此外，还需要加强对通过主要代谢途径的碳和 [H] 流进一步量化并研究其物理化学控制基础。而越来越多的强大的组学技术与同位素研究相结合将为未来瘤胃代谢研究提供有力工具。

7 更多信息

胃肠功能大会：https://www.congressgastro function.org/.
《瘤胃研究导论》（An Introduction to Rumen Studies），帕加马出版社，英国埃克塞特。
温室气体和畜牧业会议：http://www.ggaa2019.org/.
反刍动物生理学国际研讨会：https://www.isrp2019.com/.
瘤胃微生物基因组网络：http://www.rmgnetwork.org/.
RuminOmics 项目：http://www.ruminomics.eu/.
Russell, J. B. 2002. Rumen Microbiology and Its Role in Ruminant Nutrition. James B. Russell, Ithaca, NY. https://www.ars.usda.gov/research/software/.download/?softwareid=409.
Russell, J. B. and Wallace, R. J. 1997. Energy-yielding and energy-consuming reactions. In：Hobson, P. N. and Stewart, C. S. (Eds), The Rumen Microbial Ecosystem (2nd edn.). Blackie Academic & Professional, London, UK.
Rowett-INRA Conference：https://fems-microbiology.org/opportunities/11th-rowett-inra-conference-gut-microbiology-no-longer-forgotten-organ-11-14-june-2018-uk/.

8 参考文献

Allen, M. S., Bradford, B. J. and Oba, M. 2009. Board-invited review：the hepatic oxidation theory of the control of feed intake and its application to ruminants. Journal of Animal Science 87 (10), 3317-34.

Asanuma, N. and Hino, T. 2000. Activity and properties of fumarate reductase in ruminal bacteria. Journal of General and Applied Microbiology 46 (3), 119-25.

Asanuma, N., Iwamoto, M. and Hino, T. 1999a. Effect of the addition of fumarate on methane production by ruminal microorganisms *in vitro*. Journal of Dairy Science 82 (4), 780-7.

Asanuma, N., Iwamoto, M. and Hino, T. 1999b. The production of formate, a substrate for methanogenesis, from compounds related with the glyoxylate cycle by mixed ruminal microbes. Animal Science Journal 70 (2), 67-73.

Asanuma, N., Kawato, M., Ohkawara, S., et al., 2003. Characterization and transcription of the genes encoding enzymes involved in butyrate production in *Butyrivibrio fibrisolvens*. Current Microbiology 47 (3), 203-7.

Atasoglu, C., Valdes, C., Walker, N. D., et al., 1998. De novo synthesis of amino acids by the ruminal bacteria *Prevotella bryantii* B14, *Selenomonas ruminantium* HD4, and *Streptococcus bovis* ES1. Applied and Environmental Microbiology 64 (8), 2836-43.

Atasoglu, C., Newbold, C. J. and Wallace, R. J. 2001. Incorporation of [^{15}N] ammonia by the cellulolytic ruminal bacteria *Fibrobacter succinogenes* BL2, *Ruminococcus albus* SY3, and *Ruminococcus flavefaciens* 17. Applied and Environmental Microbiology 67 (6), 2819-22.

Avila, J. S., Chaves, A. V., Hernandez-Calva, M., et al., 2011. Effects of replacing barley grain in feedlot diets with increasing levels of glycerol on *in vitro* fermentation and methane production. Animal Feed Science and Technology 166-167, 265-8.

Bauchop, T. 1967. Inhibition of rumen methanogenesis by methane analogues. Journal of Bacteriology 94 (1), 171-5.

Bauchop, T. and Elsden, S. R. 1960. The growth of micro-organisms in relation to their energy supply. Journal of General Microbiology 23, 457-69.

Bauchop, T. and Mountfort, D. O. 1981. Cellulose fermentation by a rumen anaerobic fungus in both the absence and the presence of rumen methanogens. Applied and Environmental Microbiology 42 (6), 1103-10.

Beauchemin, K. A., Kreuzer, M., O'Mara, F., et al., 2008. Nutritional management for enteric methane abatement: a review. Australian Journal of Experimental Agriculture 48 (2), 21-7.

Bergner, H., Kijora, C., Ceresnakova, Z., et al., 1995. *In vitro* studies on glycerol transformation by rumen microorganisms. Archiv fur Tierernahrung 48 (3), 245-56.

Bond, D. R. and Russell, J. B. 1996. A role for fructose 1, 6-diphosphate in the ATPase-mediated energy-spilling reaction of *Streptococcus bovis*. Applied and Environmental Microbiology 62 (6), 2095-9.

Bond, D. R., Tsai, B. M. and Russell, J. B. 1998. The diversion of lactose carbon through the tagatose pathway reduces the intracellular fructose 1, 6-bisphosphate and growth rate of *Streptococcus bovis*. Applied Microbiology and Biotechnology 49 (5), 600-5.

Broderick, G. A., Huhtanen, P., Ahvenjärvi, S., et al., 2010. Quantifying ruminal nitrogen metabolism using the omasal sampling technique in cattle-a meta-analysis. Journal of Dairy Science 93 (7), 3216-30.

Buckel, W. and Thauer, R. K. 2013. Energy conservation via electron bifurcating ferredoxin reduction and proton/Na (+) translocating ferredoxin oxidation. Biochimica et Biophysica Acta 1827 (2), 94-113.

Buckel, W. and Thauer, R. K. 2018a. Flavin-based electron bifurcation, a new mechanism of biological energy coupling. Chemical Reviews 118 (7), 3862-86.

Buckel, W. and Thauer, R. K. 2018b. Flavin-based electron bifurcation, ferredoxin, flavodoxin, and anaerobic respiration with protons (Ech) or NAD$^+$ (Rnf) as electron acceptors: a historical review. Frontiers in Microbiology 9, 401.

Cabezas-Garcia, E. H., Krizsan, S. J., Shingfield, K. J., et al., 2017. Between-cow variation in digestion and rumen fermentation variables associated with methane production. Journal of Dairy Science 100 (6), 4409-24.

Clapperton, J. L. 1974. The effect of trichloroacetamide, chloroform and linseed oil given into the rumen of sheep on some of the end-products of rumen digestion. British Journal of Nutrition 32 (1), 155-61.

Cole, N. A. and McCroskey, J. E. 1975. Effects of hemiacetal of chloral and starch on the performance of beef steers. Journal of Animal Science 41 (6), 1735-41.

Cole, N. A. and Todd, R. W. 2008. Opportunities to enhance performance and efficiency through nutrient synchrony in concentrate-fed ruminants. Journal of Animal Science 86 (14 Suppl.), E318-33.

Collins, R. M. and Pritchard, R. H. 1992. Alternate day supplementation of corn stalk diets with soybean meal or corn gluten meal fed to ruminants. Journal of Animal Science 70 (12), 3899-908.

Czerkawski, J. W. 1986. An Introduction to Rumen Studies. Pergamon Press, Exeter, UK.

Czerkawski, J. W. and Breckenridge, G. 1975. New inhibitors of methane production by rumen micro-organisms. Experiments with animals and other practical possibilities. British Journal of Nutrition 34 (3), 447-57.

Chen, M. and Wolin, M. J. 1977. Influence of CH_4 production by Methanobacterium ruminantium on the fermentation of glucose and lactate by *Selenomonas ruminantium*. Applied and Environmental Microbiology 34 (6), 756-9.

Chowdhury, N. P., Kahnt, J. and Buckel, W. 2015. Reduction of ferredoxin or oxygen by flavin-based electron bifurcation in *Megasphaera elsdenii*. The FEBS Journal 282 (16), 3149-60.

Chung, K. T. 1976. Inhibitory effects of H_2 on growth of *Clostridium cellobioparum*. Appliedand Environmental Microbiology 31 (3), 342-8.

Davies, A., Nwaonu, H. N., Stanier, G., et al., 1982. Properties of a novel series of inhibitors of rumen methanogenesis; *in vitro* and *in vivo* experiments including growth trials on 2, 4-bis (trichloromethyl) -benzo [1, 3] dioxin-6-carboxylic acid. The British Journal of Nutrition 47 (3), 565-76.

De Vries, W., Van Wyck-Kapteyn, W. M. and Stouthamer, A. H. 1973. Generation of ATP during cytochrome-linked anaerobic electron transport in propionic acid bacteria. Journal of General Microbiology 76 (1), 31-41.

Dijkstra, J., Boer, H., Van Bruchem, J., et al., 1993. Absorption of volatile fatty acids from the rumen of lactating dairy cows as influenced by volatile fatty acid concentration, pH and rumen liquid volume. The British Journal of Nutrition 69 (2), 385-96.

Ebrahimi, S. H., Mohinia, M., Singhala, K. K., et al., 2011. Evaluation of complementary effects of 9, 10-anthraquinone and fumaric acid on methanogenesis and ruminal fermentation *in vitro*. Archives of Animal Nutrition 65 (4), 267-77.

Eckard, R. J., Grainger, C. and De Klein, C. A. M. 2010. Options for the abatement of methane and nitrous oxide from ruminant production: a review. Livestock Science 130 (1-3), 47-56.

Enzmann, F., Mayer, F., Rother, M., et al., 2018. Methanogens: biochemical background and biotechnological applications. AMB Express 8 (1), 1-.

Farmer, C. G., Woods, B. C., Cochran, R. C., et al., 2004. Effect of supplementation frequency and supplemental urea level on dormant tallgrass-prairie hay intake and digestion by beef steers and prepartum performance of beef cows grazing dormant tallgrass-prairie. Journal of Animal Science 82 (3), 884-94.

Firkins, J. L., Yu, Z. and Morrison, M. 2007. Ruminal nitrogen metabolism: perspectives for integration of microbiology and nutrition for dairy. Journal of Dairy Science 90 (Suppl. 1), E1-E16.

Geier, R. R., Kwon, I. H., Cann, I. K., et al., 2016. Interspecies hydrogen transfer and its effects on global transcript abundance in *Ruminococcus albus*, a predominant fiber-degrading species in the rumen. The FASEB Journal 30, 1102.

Gill, M., Siddons, R. C., Beever, D. E., et al., 1986. Metabolism of lactic acid isomers in the rumen of silage-fed sheep. The British Journal of Nutrition 55 (2), 399-407.

Greening, C., Geier, R., Wang, C., et al., 2019. Diverse hydrogen production and consumption pathways influence methane production in ruminants. The ISME Journal 13 (10), 2617-32.

Guo, W. S., Schaefer, D. M., Guo, X. X., et al., 2009. Use of nitrate-nitrogen as a sole dietary nitrogen source to inhibit ruminal methanogenesis and to improve microbial nitrogen synthesis *in vitro*. Asian-Australasian Journal of Animal Sciences 22 (4), 542-9.

Hackmann, T. J. and Firkins, J. L. 2015a. Electron transport phosphorylation in rumen butyrivibrios: unprecedented ATP yield for glucose fermentation to butyrate. Frontiers in Microbiology 6, 622.

Hackmann, T. J. and Firkins, J. L. 2015b. Maximizing efficiency of rumen microbial protein production. Frontiers in Microbiology 6, 465.

Hackmann, T. J., Diese, L. E. and Firkins, J. L. 2013a. Quantifying the responses of mixed rumen microbes to excess carbohydrate. Applied and Environmental Microbiology 79 (12), 3786-95.

Hackmann, T. J., Keyser, B. L. and Firkins, J. L. 2013b. Evaluation of methods to detect changes in reserve carbohydrate for mixed rumen microbes. Journal of Microbiological Methods 93 (3), 284-91.

Hackmann, T. J., Ngugi, D. K., Firkins, J. L., et al., 2017. Genomes of rumen bacteria encode atypical pathways for fermenting hexoses to short-chain fatty acids. Environmental Microbiology 19 (11), 4670-83.

Hall, M. B. and Huntington, G. B. 2008. Nutrient synchrony: sound in theory, elusive in practice. Journal of Animal Science 86 (14) (Suppl.), E287-92.

Hall, M. B., Nennich, T. D., Doane, P. H., et al., 2015. Total volatile fatty acid concentrations are unreliable estimators of treatment effects on ruminal fermentation *in vivo*. Journal of Dairy Science 98 (6), 3988-99.

Hamar, D. and Borchers, R. 1967. Glycolytic pathway in rumen microorganisms. Journal of Animal Science 26 (3), 654-7.

Harfoot, C. G. and Hazlewood, G. P. 1997. Lipid metabolism in the rumen. In: Hobson, P. N. and Stewart, C. S. (Eds), The Rumen Microbial Ecosystem. Blackie Academic & Professional, London, UK, pp. 382-426.

Hartinger, T., Gresner, N. and Sudekum, K. H. 2018. Does intra-ruminal nitrogen recycling waste valuable resources? A review of major players and their manipulation. Journal of Animal Science and Biotechnology 9, 33.

Henderson, C. 1980. The influence of extracellular hydrogen on the metabolism of *Bacteroides rurninicola*, *Anaerovibrio lipolytica* and *Selenomonas ruminantiurn*. Journal of General Microbiology 119 (2), 485-91.

Henderson, G., Cox, F., Ganesh, S., et al., 2015. Rumen microbial community composition varies with diet and host, but a core microbiome is found across a wide geographical range. Scientific Reports

5, 14567.

Hersom, M. J. 2008. Opportunities to enhance performance and efficiency through nutrient synchrony in forage-fed ruminants. Journal of Animal Science 86 (14 Suppl.), E306-17.

Hino, T. and Russell, J. B. 1985. Effect of reducing-equivalent disposal and NADH/NAD on deamination of amino acids by intact rumen microorganisms and their cell extracts. Applied and Environmental Microbiology 50 (6), 1368-74.

Hristov, A. N., Oh, J., Giallongo, F., et al., 2015. An inhibitor persistently decreased enteric methane emission from dairy cows with no negative effect on milkproduction. Proceedings of the National Academy of Sciences of the United States of America. Proceedings of the National Acadademy of Sciences of the United States of America 112 (34), 10663-8.

Hungate, R. E. 1967. Hydrogen as an intermediate in the rumen fermentation. Archiv für Mikrobiologie 59 (1), 158-64.

Hungate, R. E., Smith, W., Bauchop, T., et al., 1970. Formate as an intermediate in the bovine rumen fermentation. Journal of Bacteriology 102 (2), 389-97.

Huston, J. E., Lippke, H., Forbes, T. D. A., et al., 1999. Effects of supplemental feeding interval on adult cows in western Texas. Journal of Animal Science 77 (11), 3057-67.

Isaacson, H. R., Hinds, F. C., Bryant, M. P., et al., 1975. Efficiency of energy utilization by mixed rumen bacteria in continuous culture. Journal of Dairy Science 58 (11), 1645-59.

Janssen, P. H. 2010. Influence of hydrogen on rumen methane formation and fermentation balances through microbial growth kinetics and fermentation thermodynamics. Animal Feed Science and Technology 160 (1-2), 1-22.

Jenkins, T. C., Wallace, R. J., Moate, P. J., et al., 2008. Board-invited review: recent advances in biohydrogenation of unsaturated fatty acids within the rumen microbial ecosystem. Journal of Animal Science 86 (2), 397-412.

Jeong, H. S., Kim, H. Y., Ahn, S. H., et al., 2014. Optimization of enzymatic hydrolysis conditions for extraction of pectin from rapeseed cake (*Brassica napus* L.) using commercial enzymes. Food Chemistry 157, 332-8.

Johnson, D. E. 1972. Effects of a hemiacetal of chloral and starch on methane production and energy balance of sheep fed a pelleted diet. Journal of Animal Science 35 (5), 1064-8.

Johnson, D. E. 1974. Adaptational responses in nitrogen and energy balance of lambs fed a methane inhibitor. Journal of Animal Science 38 (1), 154-7.

Johnson, K. A. and Johnson, D. E. 1995. Methane emissions from cattle. Journal of Animal Science 73 (8), 2483-92.

Jouany, J. P. and Thivend, P. 1972a. Évolution postprandiale de la composition glucidique des corps microbiens du rumen en fonction de la nature des glucides du régime. I. — les protozoaires. Annales de Biologie Animale Biochimie Biophysique 12 (4), 673-7.

Jouany, J. P. and Thivend, P. 1972b. Évolution postprandiale de la composition glucidique des corps microbiens du rumen en fonction de la nature des glucides du régime. II. — les bactéries. Annales de Biologie Animale Biochimie Biophysique 12 (4), 679-83.

Kanehisa, M., Furumichi, M., Tanabe, M., et al., 2017. KEGG: new perspectives on genomes, pathways, diseases and drugs. Nucleic Acids Research 45 (D1), D353-61.

Kelly, W. J., Leahy, S. C., Altermann, E., et al., 2010. The glycobiome of the rumen bacterium Butyrivibrio proteoclasticus B316 (T) highlights adaptation to a polysaccharide-rich environment. PLoS ONE 5 (8), e11942.

Kennedy, P. M. and Milligan, L. P. 1978. Effects of cold exposure on digestion, microbial synthesis and nitrogen transformations in sheep. The British Journal of Nutrition 39 (1), 105-17.

Kim, S. H., Mamuad, L. L., Kim, D. W., et al., 2016. Fumarate reductase-producing enterococci reduce methane production in *in vitro* rumen fermentation. Journal of Microbiology and Biotechnology 26 (3), 558-66.

Kohn, R. and Boston, R. 2000. The role of thermodynamics in controlling rumen metabolism. In: McNamara, J. P., France, J. and Beever, D. E. (Eds), Modelling Nutrient Utilization in Farm Animals. CABI, Wallingford, UK, pp. 11-24.

Kröger, A., Biel, S., Simon, J., et al., 2002. Fumarate respiration of *Wolinella succinogenes*: enzymology, energetics and coupling mechanism. Biochimica et Biophysica Acta 1553 (1-2), 23-38.

Krueger, N. A., Anderson, R. C., Tedeschi, L. O., et al., 2010. Evaluation of feeding glycerol on free-fatty acid production and fermentation kinetics of mixed ruminal microbes in vitro. Bioresource Technology 101 (21), 8469-72.

Leedle, J. A., Bryant, M. P. and Hespell, R. B. 1982. Diurnal variations in bacterial numbers and fluid parameters in ruminal contents of animals fed low-or high-forage diets. Applied and Environmental Microbiology 44 (2), 402-12.

Leng, R. A. 2014. Interactions between microbial consortia in biofilms: a paradigm shift in rumen microbial ecology and enteric methane mitigation. Animal Production Science 54 (5), 519-43.

Le Van, T. D., Robinson, J. A., Ralph, J., et al., 1998. Assessment of reductive acetogenesis with indigenous ruminal bacterium populations and Acetitomaculum ruminis. Applied and Environmental Microbiology 64 (9), 3429-36.

Lopes, J. C., de Matos, L. F., Harper, M. T., et al., 2016. Effect of 3-nitrooxypropanol on methane and hydrogen emissions, methane isotopic signature, and ruminal fermentation in dairy cows. Journal of Dairy Science 99 (7), 5335-44.

Lopez, S., McIntosh, F. M., Wallace, R. J., et al., 1999. Effect of adding acetogenic bacteria on methane production by mixed rumen microorganisms. Animal Feed Science and Technology 78 (1-2), 1-9.

Lourenco, M., Ramos-Morales, E. and Wallace, R. J. 2010. The role of microbes in rumen lipolysis and biohydrogenation and their manipulation. Animal 4 (7), 1008-23.

Løvendahl, P., Difford, G. F., Li, B., et al., 2018. Review: Selecting for improved feed efficiency and reduced methane emissions in dairy cattle. Animal 12 (s2), s336-49.

Mamuad, L., Kim, S. H., Jeong, C. D., et al., 2014. Effect of fumarate reducing bacteria on *in vitro* rumen fermentation, methane mitigation and microbial diversity. Journal of Microbiology 52 (2), 120-8.

Marounek, M. and Petr, O. 1995. Fermentation of glucose and xylose in ruminal strains of *Butyrivibrio fibrisolvens*. Letters in Applied Microbiology 21 (4), 272-6.

Martin, S. A. and Macy, J. M. 1985. Effects of monensin, pyromellitic diimide and 2-bromoethanesulfonic acid on rumen fermentation *in vitro*. Journal of Animal Science 60 (2), 544-50.

Martin, C., Morgavi, D. P. and Doreau, M. 2010. Methane mitigation in ruminants: from microbe to the farm scale. Animal 4 (3), 351-65.

Martínez-Fernández, G., Abecia, L., Arco, A., et al., 2014. Effects of ethyl-3-nitrooxy propionate and 3-nitrooxypropanol on ruminal fermentation, microbial abundance, and methane emissions in sheep.

Journal of Dairy Science 97 (6), 3790-9.

Martinez-Fernandez, G., Denman, S. E., Yang, C., et al., 2016. Methane inhibition alters the microbial community, hydrogen flow, and fermentation response in the rumen of cattle. Frontiers in Microbiology 7, 1122.

Martinez-Fernandez, G., Denman, S. E., Cheung, J., et al., 2017. Phloroglucinol degradation in the rumen promotes the capture of excess hydrogen generated from methanogenesis inhibition. Frontiers in Microbiology 8, 1871.

Martinez-Fernandez, G., Duval, S., Kindermann, M., et al., 2018. 3-NOP vs. Halogenated compound: methane production, ruminal fermentation and microbial community response in forage fed cattle. Frontiers in Microbiology 9, 1582.

Marvin-Sikkema, F. D., Richardson, A. J., Stewart, C. S., et al., 1990. Influence of hydrogen-consuming bacteria on cellulose degradation by anaerobic fungi. Applied and Environmental Microbiology 56 (12), 3793-7.

Matte, A., Forsberg, C. W. and Gibbins, A. M. V. 1992. Enzymes associated with metabolism of xylose and other pentoses by *Prevotella* (Bacteroides) *ruminicola* strains, *Selenomonas ruminantium D*, and *Fibrobacter succinogenes* S85. Canadian Journal of Microbiology 38 (5), 370-6.

Matulova, M., Delort, A. M., Nouaille, R., et al., 2001. Concurrent maltodextrin and cellodextrin synthesis by *Fibrobacter succinogenes* S85 as identified by 2D NMR spectroscopy. European Journal of Biochemistry 268 (14), 3907-15.

McAllan, A. B. and Smith, R. H. 1974. Carbohydrate metabolism in the ruminant. Bacterial carbohydrates formed in the rumen and their contribution to digesta entering the duodenum. The British Journal of Nutrition 31 (1), 77-88.

Miller, T. L. and Jenesel, S. E. 1979. Enzymology of butyrate formation by *Butyrivibrio fibrisolvens*. Journal of Bacteriology 138 (1), 99-104.

Mitsumori, M., Matsui, H., Tajima, K., et al., 2014. Effect of bromochloromethane and fumarate on phylogenetic diversity of the formyltetrahydrofolate synthetase gene in bovine rumen. Animal Science Journal = Nihon Chikusan Gakkaiho 85 (1), 25-31.

Mohammed, N., Lila, Z. A., Ajisaka, N., et al., 2004. Inhibition of ruminal microbial methane production by β-cyclodextrin iodopropane, malate and their combination *in vitro*. Journal of Animal Physiology and Animal Nutrition 88 (5-6), 188-95.

Montzka, S. A., Dlugokencky, E. J. and Butler, J. H. 2011. Non-CO_2 greenhouse gases and climate change. Nature 476 (7358), 43-50.

Morgavi, D. P., Forano, E., Martin, C., et al., 2010. Microbial ecosystem and methanogenesis in ruminants. Animal 4 (7), 1024-36.

Moss, A. R., Jouany, J. P. and Newbold, J. 2000. Methane production by ruminants: its contribution to global warming. Annales de Zootechnie 49 (3), 231-53.

Nagaraja, T. G. and Titgemeyer, E. C. 2007. Ruminal acidosis in beef cattle: the current microbiological and nutritional outlook. Journal of Dairy Science 90 (Suppl. 1), E17-38.

Nagaraja, T. G., Newbold, C. J., Van Nevel, C. J., et al., 1997. Manipulation of rumen fermentation. In: Hobson, P. N. and Stewart, C. S. (Eds), The Rumen Microbial Ecosystem (2nd edn.). Blackie Academic & Professional, London, UK, pp. 524-632.

Nelson, D. L. and Cox, M. M. 2017. Lehninger Principles of Biochemistry. W. H. Freeman, Macmillan

Learning, New York.

Newbold, J. R. and Rust, S. R. 1992. Effect of asynchronous nitrogen and energy supply on growth of ruminal bacteria in batch culture. Journal of Animal Science 70 (2), 538-46.

Newbold, C. J., De La Fuente, G., Belanche, A., et al., 2015. The role of ciliate protozoa in the rumen. Frontiers in Microbiology 6, 1313.

Nocek, J. E. and Russell, J. B. 1988. Protein and energy as an integrated system. Relationship of ruminal protein and carbohydrate availability to microbial synthesis and milk production. Journal of Dairy Science 71 (8), 2070-107.

Nolan, J. V., Hegarty, R. S., Hegarty, J., et al., 2010. Effects of dietary nitrate on fermentation, methane production and digesta kinetics in sheep. Animal Production Science 50 (8), 801-6.

Nollet, L., Demeyer, D. I. and Verstraete, W. 1997. Effect of 2-bromoethanesulfonic acid and *Peptostreptococcus* productus ATCC 35244 addition on stimulation of reductive acetogenesis in the ruminal ecosystem by selective inhibition of methanogenesis. Applied and Environmental Microbiology 63 (1), 194-200.

Nouaille, R., Matulova, M., Delort, A. M., et al., 2005. Oligosaccharide synthesis in *Fibrobacter succinogenes* S85 and its modulation by the substrate. The FEBS Journal 272 (10), 2416-27.

NRC. 2001. Nutrient Requirements of Dairy Cattle. National Research Council, Washington DC.

Offner, A. and Sauvant, D. 2006. Thermodynamic modeling of ruminal fermentations. Animal Research 55 (5), 343-65.

Oldick, B. S., Firkins, J. L. and Kohn, R. A. 2000. Compartmental modeling with nitrogen-15 to determine effects of degree of fat saturation on intraruminal N recycling. Journal of Animal Science 78 (9), 2421-30.

Olijhoek, D. W., Hellwing, A. L. F., Brask, M., et al., 2016. Effect of dietary nitrate level on enteric methane production, hydrogen emission, rumen fermentation, and nutrient digestibility in dairy cows. Journal of Dairy Science 99 (8), 6191-205.

Pavlostathis, S. G., Miller, T. L. and Wolin, M. J. 1990. Cellulose fermentation by continuous cultures of *Ruminococcus albus* and *Methanobrevibacter smithii*. Applied Microbiology and Biotechnology 33 (1), 109-16.

Pazur, J. H., Shuey, E. W. and Georgi, C. E. 1958. The conversion of d-xylose into volatile organic acids by rumen bacteria. Archives of Biochemistry and Biophysics 77 (2), 387-94.

Petzel, J. P., Mcelwain, M. C., Desantis, D., et al., 1989. Enzymic activities of carbohydrate, purine, and pyrimidine metabolism in the Anaeroplasmataceae (class Mollicutes). Archives of Microbiology 152 (4), 309-16.

Pol, A. and Demeyer, D. I. 1988. Fermentation of methanol in the sheep rumen. Applied and Environmental Microbiology 54 (3), 832-4.

Preiss, J. and Romeo, T. 1990. Physiology, biochemistry and genetics of bacterial glycogen synthesis. In: Rose, A. H. and Tempest, D. W. (Eds), Advances in Microbial Physiology (vol. 30). Academic Press, pp. 183-238.

Roberton, A. M. and Glucina, P. G. 1982. Fructose 6-phosphate phosphorylation in Bacteroides species. Journal of Bacteriology 150 (3), 1056-60.

Rooke, J. A., Wallace, R. J., Duthie, C. A., et al., 2014. Hydrogen and methane emissions from beef cattle and their rumen microbial community vary with diet, time after feeding and genotype. The British Journal of Nutrition 112 (3), 398-407.

Russell, J. B. 1986. Heat production by ruminal bacteria in continuous culture and its relationship to maintenance energy. Journal of Bacteriology 168 (2), 694-701.

Russell, J. B. 1993. Effect of amino acids on the heat production and growth efficiency of *Streptococcus bovis*: balance of anabolic and catabolic rates. Applied and Environmental Microbiology 59 (6), 1747-51.

Russell, J. B. 1998. The importance of pH in the regulation of ruminal acetate to propionate ratio and methane production in vitro. Journal of Dairy Science 81 (12), 3222-30.

Russell, J. B. 2002. Rumen Microbiology and Its Role in Ruminant Nutrition. James B. Russell, Ithaca, NY.

Russell, J. B. 2007a. Can the heat of ruminal fermentation be manipulated to decrease heat stress? Proceedings of the Southwest Nutrition Conference, pp. 109-15.

Russell, J. B. 2007b. The energy spilling reactions of bacteria and other organisms. Journal of Molecular Microbiology and Biotechnology 13 (1-3), 1-11.

Russell, J. B. and Cook, G. M. 1995. Energetics of bacterial growth: balance of anabolic and catabolic reactions. Microbiological Reviews 59 (1), 48-62.

Russell, J. B. and Jeraci, J. L. 1984. Effect of carbon monoxide on fermentation of fiber, starch, and amino acids by mixed rumen microorganisms *in vitro*. Applied and Environmental Microbiology 48 (1), 211-7.

Russell, J. B. and Martin, S. A. 1984. Effects of various methane inhibitors on the fementation of amino acids by mixed rumen microorganisms *in vitro*. Journal ofAnimal Science 59 (5), 1329-38.

Russell, J. B. and Wallace, R. J. 1997. Energy-yielding and energy-consuming reactions. In: Hobson, P. N. and Stewart, C. S. (Eds), The Rumen Microbial Ecosystem (2nd edn.). Blackie Academic & Professional, London, UK, pp. 246-82.

Satter, L. D. and Esdale, W. J. 1968. *In vitro* lactate metabolism by ruminal ingesta. Applied Microbiology 16 (5), 680-8.

Sauer, F. D. and Teather, R. M. 1987. Changes in oxidation reduction potentials and volatile fatty acids production by rumen bacteria when methane synthesis is inhibited. Journal of Dairy Science 70 (9), 1835-40.

Saunois, M., Bousquet, P., Poulter, B., et al., 2016a. The global methane budget 2000-2012. Earth System Science Data 8 (2), 697-751.

Saunois, M., Jackson, R. B., Bousquet, P., et al., 2016b. The growing role of methane in anthropogenic climate change. Environmental Research Letters 11 (12).

Scardovi, V. 1965. The fructose-6-phosphate shunt as a peculiar pattern of hexose degradation in the genus Bifidobacterium. Annals of Microbiology and Enzymology 15, 19-29.

Scheller, H. V. and Ulvskov, P. 2010. Hemicelluloses. Annual Review of Plant Biology 61, 263-89.

Schoelmerich, M. C., Katsyv, A., Dönig, J., et al., 2019. Energy conservation involving 2 respiratory circuits. Proceedings of the National Academy of Sciences of the United States of America 117, 1167-73.

Schuchmann, K. and Müller, V. 2014. Autotrophy at the thermodynamic limit of life: a model for energy conservation in acetogenic bacteria. Nature Reviews. Microbiology 12 (12), 809-21.

Seeliger, S., Janssen, P. H. and Schink, B. 2002. Energetics and kinetics of lactate fermentation to acetate and propionate via methylmalonyl-CoA or acrylyl-CoA. FEMS Microbiology Letters 211 (1), 65-70.

Söllinger, A., Tveit, A. T., Poulsen, M., et al., 2018. Holistic assessment of rumen microbiome dynamics through quantitative metatranscriptomics reveals multifunctional redundancy during key steps of anaerobic feed degradation. mSystems 3 (4).

Sondergaard, D., Pedersen, C. N. and Greening, C. 2016. HydDB: A web tool for hydrogenase classification and analysis. Scientific Reports 6, 34212.

Stanier, G. and Davies, A. 1981. Effects of the antibiotic monensin and an inhibitor of methanogenesis on *in vitro* continuous rumen fermentations. The British Journal of Nutrition 45 (3), 567-78.

Stewart, C. S., Flint, H. J. and Bryant, M. P. 1997. The rumen bacteria. In: Hobson, P. N. and Stewart, C. S. (Eds), The Rumen Microbial Ecosystem (2nd edn.). Blackie Academic & Professional, London, UK, pp. 10-72.

Storm, A. C., Kristensen, N. B. and Hanigan, M. D. 2012. A model of ruminal volatile fatty acid absorption kinetics and rumen epithelial blood flow in lactating Holstein cows. Journal of Dairy Science 95 (6), 2919-34.

Stouthamer, A. H. 1973. A theoretical study on the amount of ATP required for synthesis of microbial cell material. Antonie van Leeuwenhoek 39 (3), 545-65.

Tatsuoka, N., Hara, K., Mikuni, K., et al., 2008. Effects of the essential oil cyclodextrin complexes on ruminal methane production in vitro. Animal Science Journal 79 (1), 68-75.

Teixeira, C. R. V., Lana, R. P., Tao, J., et al., 2017. Comparing the responses of rumen ciliate protozoa and bacteria to excess carbohydrate. FEMS Microbiology Ecology 93 (6).

Thauer, R. K., Jungermann, K. and Decker, K. 1977. Energy conservation in chemotrophic anaerobic bacteria. Bacteriological Reviews 41 (1), 100-80.

Thauer, R. K., Kaster, A. K., Seedorf, H., et al., 2008. Methanogenic archaea: ecologically relevant differences in energy conservation. Nature Reviews. Microbiology 6 (8), 579-91.

Thiele, J. H. and Zeikus, J. G. 1988. Control of interspecies electron flow during anaerobic digestion: significance of formate transfer versus hydrogen transfer during syntrophic methanogenesis in flocs. Applied and Environmental Microbiology 54 (1), 20-9.

Thurston, B., Dawson, K. A. and Strobel, H. J. 1993. Cellobiose versus glucose utilization by the ruminal bacterium *Ruminococcus albus*. Applied and Environmental Microbiology 59 (8), 2631-7.

Thurston, B., Dawson, K. A. and Strobel, H. J. 1994. Pentose utilization by the ruminal bacterium *Ruminococcus albus*. Applied and Environmental Microbiology 60 (4), 1087-92.

Ungerfeld, E. M. 2013. A theoretical comparison between two ruminal electron sinks. Frontiers in Microbiology 4, 319.

Ungerfeld, E. M. 2015a. Limits to dihydrogen incorporation into electron sinks alternative to methanogenesis in ruminal fermentation. Frontiers in Microbiology 6, 1272.

Ungerfeld, E. M. 2015b. Shifts in metabolic hydrogen sinks in the methanogenesis-inhibited ruminal fermentation: a meta-analysis. Frontiers in Microbiology 6, 37.

Ungerfeld, E. M. 2018. Inhibition of rumen methanogenesis and ruminant productivity: a meta-analysis. Frontiers in Veterinary Science 5, 113.

Ungerfeld, E. M. and Kohn, R. A. 2006. The role of thermodynamics in the control of ruminal fermentation. In: Sejrsen, K., Hvelplund, T. and Nielsen, M. O. (Eds), Ruminant Physiology. Wageningen Academic Publishers, Wageningen, The Netherlands, pp. 55-85.

Ungerfeld, E. M., Rust, S. R. and Burnett, R. 2007. Increases in microbial nitrogen production and efficiency *in vitro* with three inhibitors of ruminal methanogenesis. Canadian Journal of Microbiology 53 (4),

496-503.

Ungerfeld, E. M., Aedo, M. F., Martínez, E. D., et al., 2019. Inhibiting methanogenesis in rumen batch cultures did not increase the recovery of metabolic hydrogen in microbial amino acids. Microorganisms 7 (5), 115.

Valkeners, D., Théwis, A., Amant, S., et al., 2006. Effect of various levels of imbalance between energy and nitrogen release in the rumen on microbial protein synthesis and nitrogen metabolism in growing double-muscled Belgian Blue bulls fed a corn silage-based diet. Journal of Animal Science 84 (4), 877-85.

Van Kessel, J. S. and Russell, J. B. 1996. The effect of amino nitrogen on the energetics of ruminal bacteria and its impact on energy spilling. Journal of Dairy Science 79 (7), 1237-43.

Van Lingen, H. J., Plugge, C. M., Fadel, J. G., et al., 2016. Thermodynamic driving force of hydrogen on rumen microbial metabolism: a theoretical investigation. PLoS ONE 11 (10), e0161362.

Van Lingen, H. J., Edwards, J. E., Vaidya, J. D., et al., 2017. Diurnal dynamics of gaseous and dissolved metabolites and microbiota composition in the bovine rumen. Frontiers in Microbiology 8, 425.

Van Lingen, H. J., Fadel, J. G., Moraes, L. E., et al., 2019. Bayesian mechanistic modeling of thermodynamically controlled volatile fatty acid, hydrogen and methane production in the bovine rumen. Journal of Theoretical Biology 480, 150-65.

Veneman, J. B., Muetzel, S., Hart, K. J., et al., 2015. Does dietary mitigation of enteric methane production affect rumen function and animal productivity in dairy cows? PLoS ONE 10 (10), e0140282.

Vyas, D., McGeough, E. J., Mcginn, S. M., et al., 2014a. Effect of *Propionibacterium* spp. on ruminal fermentation, nutrient digestibility, and methane emissions in beef heifers fed a high-forage diet. Journal of Animal Science 92 (5), 2192-201.

Vyas, D., McGeough, E. J., Mohammed, R., et al., 2014b. Effects of Propionibacterium strains on ruminal fermentation, nutrient digestibility and methane emissions in beef cattle fed a corn grain finishing diet. Animal 8 (11), 1807-15.

Vyas, D., Alazezeh, A., McGnn, S. M., et al., 2015. Enteric methane emissions in response to ruminal inoculation of Propionibacterium strains in beef cattle fed a mixed diet. Animal Production Science 56 (7).

Vyas, D., McGinn, S. M., Duval, S. M., et al., 2016. Effects of sustained reduction of enteric methane emissions with dietary supplementation of 3-nitrooxypropanol on growth performance of growing and finishing beef cattle. Journal of Animal Science 94 (5), 2024-34.

Vyas, D., Alemu, A. W., McGinn, S. M., et al., 2018a. The combined effects of supplementing monensin and 3-nitrooxypropanol on methane emissions, growth rate, and feed conversion efficiency in beef cattle fed high-forage and high-grain diets. Journal of Animal Science 96 (7), 2923-38.

Vyas, D., McGinn, S. M., Duval, S. M., et al., 2018b. Optimal dose of 3-nitrooxypropanol for decreasing enteric methane emissions from beef cattle fed high-forage and high-grain diets. Animal Production Science 58 (6).

Walker, D. J. and Monk, P. R. 1971. Fate of carbon passing through the glucose pool of rumen digesta. Applied Microbiology 22 (5), 741-7.

Wallace, R. J. 1996. Ruminal microbial metabolism of peptides and amino acids. The Journal of Nutrition 126 (4 Suppl.), 1326S-34S.

Wallace, R. J., Onodera, R. and Cotta, M. A. 1997. Metabolism of nitrogen-containing compounds. In: Hobson, P. N. and Stewart, C. S. (Eds), The Rumen Microbial Ecosystem (2nd edn.). Blackie Academic & Professional, London, UK, pp. 283-328.

Wallnöfer, P., Baldwin, R. L. and Stagno, E. 1966. Conversion of C-labeled substrates to volatile fatty acids by the rumen microbiota. Applied Microbiology 14 (6), 1004-10.

Wang, R., Wang, M., Ungerfeld, E. M., et al., 2018. Nitrate improves ammonia incorporation into rumen microbial protein in lactating dairy cows fed a low-protein diet. Journal of Dairy Science 101 (11), 9789-99.

Wells, J. E. and Russell, J. B. 1996. Why do many ruminal bacteria die and lyse so quickly? Journal of Dairy Science 79 (8), 1487-95.

Wells, J. E., Russell, J. B., Shi, Y., et al., 1995. Cellodextrin efflux by the cellulolytic ruminal bacterium *Fibrobacter succinogenes* and its potential role in the growth of nonadherent bacteria. Applied and Environmental Microbiology 61 (5), 1757-62.

Wolf, P. G., Biswas, A., Morales, S. E., et al., 2016. H_2 metabolism is widespread and diverse among human colonic microbes. Gut Microbes 7 (3), 235-45.

Wolin, M. J., Miller, T. L. and Stewart, C. S. 1997. Microbe-microbe interactions. In: Hobson, P. N. and Stewart, C. S. (Eds), The Rumen Microbial Ecosystem (2nd edn.). Blackie Academic & Professional, London, UK, pp. 467-91.

Zheng, Y., Kahnt, J., Kwon, I. H., et al., 2014. Hydrogen formation and its regulation in *Ruminococcus albus*: involvement of an electron-bifurcating [FeFe] -hydrogenase, of a non-electron-bifurcating [FeFe] -hydrogenase, and of a putative hydrogen-sensing [FeFe] -hydrogenase. Journal of Bacteriology 196 (22), 3840-52.

第14章 阐明瘤胃脂质代谢以优化乳品质、促进人类健康并监测动物健康

Veerle Fievez、Nympha De Neve、Lore Dewanckele

比利时根特大学

(黄小丹译)

1 前言

牛乳中含有3.5%~5.0%的乳脂,以乳脂球的形式分散在水相中。它由乳脂球膜包裹一个相对较大的三酰基甘油酯/甘油三酯(TAG)核心组成,而乳脂球膜是由磷脂(PL)组成的的薄三层膜结构(Bernard 等,2018)。因此,乳脂主要由甘油三酯(97.5%~99%)、少量双甘油酯(<1%)、磷脂(PL)、非酯化脂肪酸(NEFA)、胆固醇和单甘油酯,以及无或仅微量的(<0.1%)胆固醇酯组成(Castro-Gómez 等,2014;Glasser 等,2007)。

受瘤胃微生物活性以及乳腺代谢的影响,牛乳乳脂肪酸含量较为丰富。典型的牛乳脂肪酸组成如表1所示。其中,奇链和支链脂肪酸(OBCFA)主要源自瘤胃微生物从头合成脂肪酸,而一系列反式和顺式不饱和脂肪酸(UFA)则是由(多)不饱和脂肪酸(PUFA)的不完全生物氢化形成,这也导致了乳脂中饱和脂肪酸(SFA)含量较高(约70%)。本章详细讨论了瘤胃脂质代谢的脂解和生物氢化作用这两个方面及其重要意义。

为了更好地了解乳制品中脂肪酸的来源,本章第一部分讨论了瘤胃代谢、肠道消化、脂肪酸的乳腺转运和代谢。随后讨论了如何改善乳制品中脂肪酸的组成以促进人类健康。在此,我们特别关注的是早期广泛报道的关于如何保护不饱和脂肪酸不受瘤胃生物氢化的研究(Kliem 和 Shingfield,2016;Gebreyowhans 等,2019)。最后,本章讨论了如何利用由瘤胃脂质代谢变化驱动乳脂肪酸组成的波动来监测动物健康。

表1 德国牛乳中夏季和冬季乳脂的脂肪酸组成 (wt %)[1]

脂肪酸	夏季	冬季
C4:0	3.79	3.85
C5:0	0.02	0.02
C6:0	2.10	2.37
C7:0	0.02	0.03
C8:0	1.19	1.39

(续表)

脂肪酸	夏季	冬季
C9:0	0.02	0.04
C10:0	2.44	3.03
C10:1	0.27	0.27
C11:0	0.04	0.06
C12:0	2.98	3.57
C12:1	0.08	0.09
C13:0 *iso*	0.14	0.13
C13:0 *anteiso*	0.02	—
C13:0	0.08	0.10
C14:0 *iso*	0.16	0.10
C14:0	9.75	11.1
C14:1	1.08	1.07
C15:0 *iso*	0.43	0.29
C15:0 *anteiso*	0.74	0.50
C15:0	1.35	1.17
C16:0 *iso*	0.30	0.22
C16:0	23.5	30.3
C16:1	2.00	2.03
C17:0 *iso*	0.65	0.55
C17:0 *anteiso*	0.55	0.52
C17:0	0.72	0.64
C17:1	0.39	0.36
C18:0 *iso*	0.05	0.08
C18:0	10.6	9.42
trans-4 C18:1	0.02	0.01
trans-5 C18:1	0.02	0.01
trans-6/7/8 C18:1	0.26	0.21
trans-9 C18:1	0.27	0.17
trans-10 C18:1	0.29	0.25
trans-11 C18:1	3.82	1.11
trans-12 C18:1	0.33	0.27

(续表)

脂肪酸	夏季	冬季
trans-13/14 C18:1	0.34	0.24
trans-15 C18:1	0.51	0.38
trans-16 C18:1	0.26	0.19
trans（总）C18:1	6.12	2.84
cis-9 C18:1	19.4	17.3
cis-11 C18:1	0.55	0.53
cis-12 C18:1	0.10	0.14
cis-13 C18:1	0.06	0.08
cis-15 C18:1	0.02	0.03
C18:2（总）	2.17	2.18
C18:3 n-3	0.61	0.42
C19:0	0.09	0.05
C20:0	0.16	0.15
C20:1	0.31	0.22
C20:4	0.09	—
C22:0	0.07	0.06
C22:1	0.03	—
C24:0	0.04	0.04

注：trans-，反式；cis-，顺式。

2 瘤胃脂质代谢及脂肪酸合成

2.1 日粮脂质的脂解和生物氢化作用

泌乳奶牛的日粮通常含有4%~5%的粗脂肪（干物质基础）。反刍动物日粮中脂质的主要来源为牧草和精料，主要含18碳不饱和脂肪酸（即α-亚麻酸，C18:3 n-3；亚油酸，C18:2 n-6；油酸，顺式-9 C18:1）（Ferleay 等，2017）。另外，脂质添加剂也可用来增加脂质含量。牧草中主要脂类是糖脂，而精料中的大多数脂类则以三酰甘油的形式存在。采食后，日粮中的脂质被水解，游离脂肪酸被释放到瘤胃中。α-亚麻酸、亚油酸和油酸首先通过顺式异构化转化为反式脂肪酸中间体，然后双键氢化，进而被瘤胃微生物群落转化为饱和脂肪酸（Harfoot 和 Hazlewood，1997；Shingfield 和 Wallace，2014）。这个过程被称为生物氢化。

瘤胃微生物群落主要由厌氧菌（10^{10}个/mL）、原虫（10^7个/mL）和厌氧真菌（10^6个/mL）组成（Jenkins 等，2008；Buccioni 等，2012）。瘤胃生物氢化主要是由与原生动物共生的细菌引起的，而原生动物和真菌的贡献可以忽略不计（Lourenco 等，2010；Buccioni 等，2012）。体外试验研究指出丁酸弧菌属（*Butyrivibrio* spp.）是主要的生物氢化细菌（Kepler 等，1966；Maia 等，2007；Wallace 等，2007）。然而，在一些体内试验研究中，生物氢化过程的中间产物或最终产物与瘤胃中或鼻腔入口处丁基弧菌的丰度之间几乎没有相关性（Zened 等，2016；Zhu 等，2016；Kairenius 等，2017）。这表明，其他未培养的微生物也可能参与瘤胃生物氢化过程（Huws 等，2011；Toral 等，2016）。大量关联性分析表明，普氏菌属（*Prevotella*）、毛螺菌科（Lachnospiraceae incertae sedis）、未分类的拟杆菌科（unclassified Bacteroidales）、梭状芽孢杆菌科（Clostridiales）和瘤胃球菌科（Ruminococcaceae）的未培养细菌均可能在瘤胃生物氢化中发挥作用。细菌对多不饱和脂肪酸（PUFA）进行生物氢化可降低其毒性（Maia 等，2007，2010；Fukuda 等，2009）。目前多不饱和脂肪酸抑菌作用机制尚未明晰，但关键靶点可能在细菌细胞膜以及发生在膜内和膜上的各种基本过程。细菌更倾向饱和脂肪酸来合成膜，因为不饱和脂肪酸中的双键会改变分子的形状，破坏脂质双分子层结构（Keweloh 和 Heipieper，1996）。

大量的体内和体外研究阐明了 C18:2 n-6、C18:3 n-3 和顺式-9 C18:1 的瘤胃生物氢化途径。在正常瘤胃条件下，C18:2 n-6 主要异构化为顺式-9，反式-11 共轭亚油酸，共轭亚油酸进一步氢化为反式-11-十八烯酸，最终为硬脂酸（图1，实线箭头）。C18:3 n-3 的主要生物氢化途径为顺式-9、反式-11、顺式-15 共轭亚麻酸（CLnA），反式-11、顺式-15-十八碳二烯酸和反式-11-十八烯酸为中间产物（图2，实线箭头），而油酸主要在瘤胃中直接氢化为硬脂酸（图3，实线箭头）。然而，C18:2 n-6、C18:3 n-3 和顺式-9 C18:1 的瘤胃生物氢化也可能导致其他几个次要的脂肪酸中间体的形成，例如反式-9、反式-11 共轭亚油酸、反式-10 C18:1 和顺式-12 C18:1（图1 至图3，虚线箭头）。在特定的瘤胃条件下，一些替代途径可能变得更重要，进而成为监测瘤胃状况的指标。例如，当饲喂高淀粉日粮时，C18:3 n-3 和 C18:2 n-6 可能通过替代途径转化，导致瘤胃中反式-10 中间体的积累增加，尤其是反式-10 C18:1（Shingfield 和 Griinari，2007）。然而，这些变化在不同反刍动物之间存在差异。研究指出添加葵花籽油的高淀粉日粮会导致奶牛瘤胃反式-11 到 反式-10 的转变，但山羊的这种转变较为少见（Toral 等，2016）。另一方面，富含极长链 n-3 PUFA 的海洋脂质可有效抑制牛和山羊 C18 脂肪酸生物氢化的最后一步，增加了反式-11 C18:1、顺式-9，反式-11 共轭亚油酸和反式-10 C18:1 的含量（Toral 等，2016）。此外，除了图1 至图3 中所示的路径外，其他替代途径也可能存在，近期在使用稳定同位素的体外试验中发现了许多其他的生物氢化中间体（氧化氘和^{13}C 标记的脂肪酸；Honkanen 等，2016；Toral 等，2018a，2019）。

2.2 瘤胃细菌合成脂肪酸

细菌合成饱和脂肪酸是由两种类型的脂肪酸合成酶介导的，即直链和支链脂肪酸合成酶（Kaneda，1991）。偶数碳的直链脂肪酸的从头合成是以丙二酰辅酶 A（CoA）与乙酰

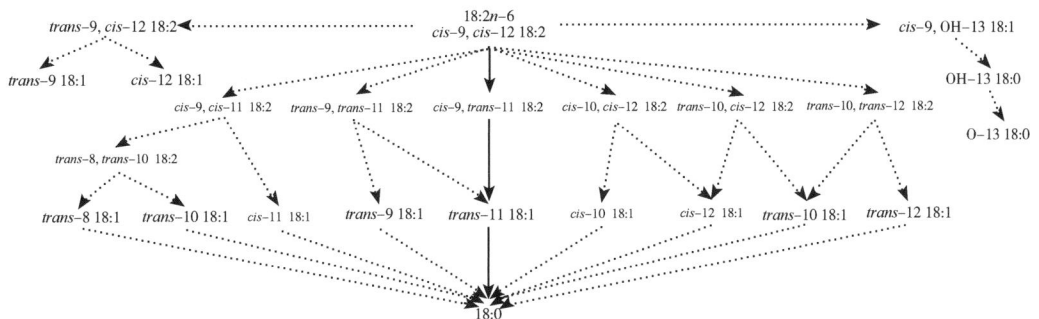

图 1 瘤胃 C18:2 n-6 脂肪酸代谢途径（Shingfield 和 Wallace，2014）。实线箭头表示瘤胃正常条件下主要的生物氢化途径，虚线箭头表示次要中间产物的形成。*trans*-：反式；*cis*-：顺式。

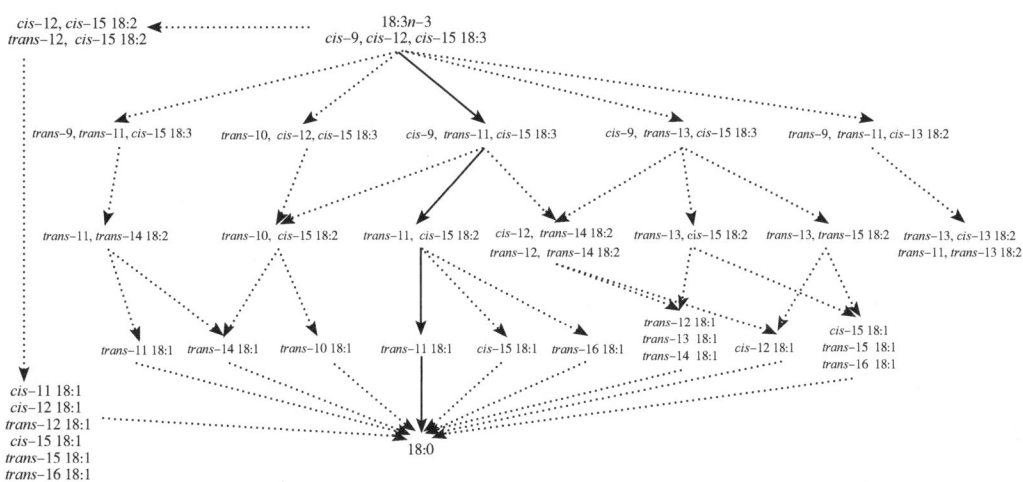

图 2 瘤胃 C18:3 n-3 脂肪酸代谢途径（Ferlay 等，2017）。实线箭头表示瘤胃正常条件下主要的生物氢化途径，虚线箭头表示次要中间产物的形成。*trans*-：反式；*cis*-：顺式。

辅酶 A 作为引物的重复缩合完成的，最终产物是棕榈酸（Fulco，1983）。当使用丙酰辅酶 A 而不是乙酰辅酶 A 作为引物时，形成线性奇链脂肪酸（Fulco，1983；Kaneda，1991）。支链脂肪酸可以分为 3 个系列：偶数异构脂肪酸（如异构 C14:0，异构 C16:0），奇数异构脂肪酸（如异构 C15:0，异构 C17:0）和奇数反式异构脂肪酸（如反式异构 C15:0，反式异构 C17:0）（Kaneda，1977），引物分别为异丁基辅酶 A、异戊酰辅酶 A 和 2-甲基丁基辅酶 A（Kaneda，1977；Annous 等，1997）。大多数植物饲料中只含有微量支链脂肪酸（Diedrich 和 Henschel，1990），由于细菌能从头合成支链脂肪酸，因此消化道含有大量共栖微生物的动物组织中含有支链脂肪酸（Keeney 等，1962）。更重要的是，瘤胃细菌对应其特有的奇数和支链脂肪酸谱，这似乎在很大程度上是由乙酰辅酶 A 酰基载体蛋白转酰基酶底物特异性决定的（Kaneda，1991）。因此，瘤胃奇数和支链脂肪酸谱的变化可以用

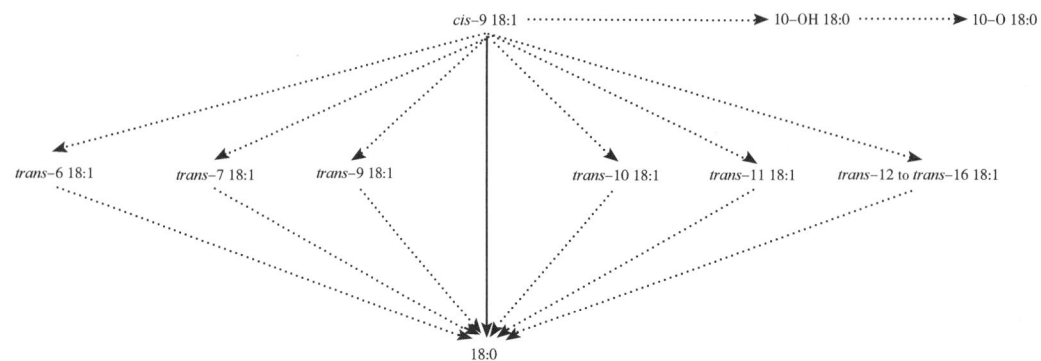

图3 瘤胃顺式-9 C18:1脂肪酸代谢途径（Shingfield 和 Wallace，2014）。实线箭头表示瘤胃正常条件下主要的生物氢化途径，虚线箭头表示次要中间产物的形成。trans-：反式；cis-：顺式。

来反映瘤胃中特定细菌种群相对丰度的变化。

3 日粮和瘤胃脂肪酸向乳腺的转运以及乳腺中脂肪酸代谢

日粮中脂肪酸可过在瘤胃进行代谢，同时瘤胃脂肪酸流入十二指肠。乳脂中这些十二指肠脂肪酸的表观回收/利用率取决于多种因素，包括消化率、脂肪酸代谢（合成和氧化）、牛的生理状态（能量平衡）、脂肪酸运输的血脂类别和十二指肠脂肪酸流量，在肠道流量较低时往往表现出较高的转移效率（Chilliard 等，2000）。另外，一些脂肪酸还会在乳腺中部分转化。

3.1 脂肪酸的肠道消化

约由70%的短链脂肪酸在瘤胃中被吸收，而中链和长链脂肪酸在到达小肠前的吸收可以忽略不计（Noble，1978）。到达小肠的脂肪酸在胶体相中溶解后在空肠被吸收。虽然在反刍动物中，由于瘤胃脂肪分解，主要是游离脂肪酸流向小肠，但这些脂肪酸仍需要在一定程度上"释放"。事实上，游离脂肪酸因其疏水性，往往附着在食糜的小颗粒上。这些疏水的脂肪酸要在水环境中"溶解"，必须形成胶束。在反刍动物和单胃动物中，胶束形成效率是小肠吸收脂肪酸的限制因素。长链脂肪酸、饱和脂肪酸（如硬脂酸）的胶束形成效率一般低于短链脂肪酸和不饱和脂肪酸。因此，与非反刍动物相比，反刍动物在脂肪酸吸收方面形成了许多重要的差异特性，包括两种胆盐组成的差异（牛磺酸结合的胆盐比甘氨酸结合的胆盐多）和胶束形成中主要的两亲性物质。这使得饱和脂肪酸能够被有效吸收（Bauman 和 Lock，2006）。溶血卵磷脂是反刍动物小肠中最重要的两亲性物质，而单甘油酯和胆盐主要在单胃动物中发挥这一作用。与这两种组分相比，溶血卵磷脂是较好的硬脂酸乳化剂。溶血卵磷脂（溶血磷脂酰胆碱）是卵磷脂（一种磷脂）通过磷脂酶（在胰腺产生）的作用在十二指肠形成的。此外，受采食量及过瘤胃的影响，脂肪酸供给小肠的特征是缓慢而持续地释放相对少量的脂肪酸。因此，反刍动物吸收饱和脂肪酸的能力远高于非反刍动物（Bauman 和 Lock，2006）。脂肪酸在肠道的平均表观吸收率为 0.83，而微生物脂肪酸吸收率更高（Schmidely 等，2008）。

近期的研究表明，外源性乳化剂注入真胃已被证明可以改善脂肪酸的吸收，这意味着脂肪酸的溶解是其从肠腔转移到血液的主要限制点（Prom 和 Lock，2019）。此外，过量补充日粮脂肪反而会抑制脂肪酸的消化率，且当 C18:0 的摄入量（或从瘤胃流出）高于 C16:0 时，这种影响更大（Boerman 等，2015，2017；Rico 等，2017）。

3.2 脂肪酸在血液中运输并转运到乳腺

游离脂肪酸被吸收后，在肠上皮细胞中酯化为三脂酰甘油和磷脂并运输，首先在淋巴中，然后在血液中。它们主要以乳糜微粒的形式存在（Demeyer 和 Doreau，1999），在极低密度脂蛋白（VLDL）中也有少量出现。这两种脂蛋白主要由甘油三酯（TAG）组成，此外还有磷脂（PL）、胆固醇酯（CE）和游离脂肪酸（NEFA）（按重要性顺序递减）（Bruss，1997）。乳中的脂类在组成上有较大不同，主要源于选择性的特定脂肪酸，以及不同脂肪酸输送到乳腺的有效性。事实上，由于乳腺脂蛋白脂肪酶对这些组分的亲和力较低，在胆固醇酯和磷脂中富集的脂肪酸很难转移到乳脂中（Annison 等，1967；Shennan 和 Peaker，2000）。脂类的这种分布差异（至少是部分差异）可以解释为多不饱和脂肪酸（PUFA）富集乳脂的困难性，例如 C18:3 n-3，多不饱和脂肪酸主要存在于血浆胆固醇酯中（Loor 等，2002b；Tyburczy 等，2009），然而，C20:5 n-3（二十碳五烯酸，EPA）和 C22:6 n-3（二十二碳六烯酸，DHA）主要存在于磷脂中。此外，与十二指肠脂肪酸组成相比，脂肪组织的调动可能有助于调节牛奶脂肪酸组成（Jorjong 等，2014），特别是脂肪组织中富含较长链的脂肪酸。对于奇数和支链脂肪酸而言，链长为 17 个碳原子的奇数和支链脂肪酸在合成代谢过程中比链长为 13~15 个碳原子的奇数和支链脂肪酸优先，而 C17:0 与 C15:0 在脂肪组织（2∶1）与牛乳（1∶2）中比例正好相反也支持这一结论（Craninx 等，2008）。因此，在泌乳的第 1 周，奶牛正在调动脂肪时，含有 17 个碳原子的奇数和支链脂肪酸增加。相反，其他短链和中链脂肪酸，较短的奇数和支链脂肪酸（链长为 14~15 个碳原子）在乳脂中被稀释（Craninx 等，2008）。

4 乳腺内源性脂肪酸的代谢

乳腺内源性脂肪酸代谢包括短链和中链脂肪酸的合成、脂肪酸的去饱和以及一定程度上的链延伸。本节仅介绍与人类健康（特别是不饱和脂肪酸）和动物健康（奇链和支链脂肪酸以及反式脂肪酸）相关的内源性牛奶脂肪酸合成的一般原则和一些最新研究，尤其是关于牛的研究。而主要的乳腺脂肪生成途径在不同物种之间具有差异，例如，山羊和牛乳腺中脂质代谢相关基因 mRNA 编码的丰度，以及与脂肪酸和脂质合成相关的酶活性，都显示出了较强的物种特异性（Bernard 等，2017）。

4.1 乳腺内源性奇链和支链脂肪酸的代谢

乳腺中脂肪酸的合成主要由脂肪酸合成酶（FASN）和乙酰辅酶 A 羧化酶

（ACACA）催化。在此过程中，由乙酰辅酶 A 生成的丙二酰辅酶 A 作为延伸底物，以乙酰辅酶 A 为引物生成短中偶数链脂肪酸，当丙酰辅酶 A 代替乙酰辅酶 A 作为前体，牛奶中会出现（微量）$C_{5:0}$、$C_{7:0}$、$C_{9:0}$、$C_{11:0}$ 和 $C_{13:0}$ 脂肪酸，而后者的微量脂肪酸也会在十二指肠内容物中检测到（Dodds 等，1981；Massart-Leën 等，1983）。此外，瘤胃添加丙酸不仅提高了乳脂中原奇链脂肪酸的含量，而且还提高了 $C_{15:0}$ 和 $C_{17:0}$ 的含量（Rigout 等，2003；Maxin 等，2011；French 等，2012）。可以说，这一过程增加 $C_{15:0}$ 和 $C_{17:0}$ 从十二指肠的转移，类似于乳脂中棕榈酸的双重来源。研究指出，乳腺中约一半的 $C_{16:0}$ 以及至少 1/3 的 $C_{15:0}$ 和 $C_{17:0}$ 来自脂肪酸的合成（Vlaeminck 等，2015）。

另外，通过（标记）异戊酸或 2-甲基丁酸酯的研究发现乳腺脂肪酸合成酶似乎并没有将异戊酰辅酶 A、2-甲基丁基辅酶 A 和异丁基辅酶 A 延伸至异奇链、前奇链或异偶数链脂肪酸（Verbeke 等，1959；French 等，2012。然而，异式十七碳脂肪酸（Dewhurst 等，2007）和反异式十七碳脂肪酸（Vlaeminck 等，2006）回收率高于 1.0。虽然通过两个碳原子的脂肪酸延伸产生超过 16 个碳的脂肪酸似乎与线性偶数链和奇数链脂肪酸的牛奶脂肪酸合成无关，但 Vlaeminck 等（2015）研究发现了"瘤胃后"两个碳原子的脂肪酸延伸促进异式十七碳脂肪酸和反异式十七碳脂肪酸的证据。然而，从血浆甘油三酯中异构 $C_{17:0}$ 和反式异构 $C_{17:0}$ 相对于十二指肠内容物的富集程度可得出异构 $C_{15:0}$ 和反式异构 $C_{15:0}$ 向 $C_{17:0}$ 对应脂肪酸的部分转化似乎也先于乳腺发生（在十二指肠上皮细胞）（Vlaeminck 等，2015）。此外，奇链脂肪酸可以通过 $\Delta 9$-去饱和酶活性进一步代谢，但只有 $C_{17:0}$ 到 $C_{17:1}$ 的转换有重要意义（Fievez 等，2003）。

4.2　乳腺内源性饱和脂肪酸和单不饱和脂肪酸的代谢

硬脂酰辅酶 A 去饱和酶（SCD）是牛乳腺中的一种重要酶。它通过在饱和碳链的碳原子 9 和碳原子 10 之间引入双键，将饱和脂肪酸转化为单不饱和脂肪酸（MUFA），例如将 $C_{18:0}$ 的饱和度降至顺式-9 $C_{18:1}$（Annison 等，1967）。然而，它也可以催化 MUFA 的去饱和，特别是反式-11 $C_{18:1}$，生成顺式-9，反式-11 共轭亚油酸（Griinari 等，2000）。

4.2.1　乳腺内源性多不饱和脂肪酸的代谢

在反刍动物的各种组织中，除乳腺外，两个碳原子的脂肪酸伸长产生链长超过 16 个碳的脂肪酸是一个非常常见的过程（Moore 和 Christie，1981）。

5　反刍动物脂肪酸对人类健康的影响

据欧洲 13 个国家的一项队列研究统计，乳制品约占膳食总脂肪摄入量的 14%（西班牙）~40%（德国）（Hulshof 等，1999），可见反刍动物脂肪酸在西方饮食中的重要性。由于生物氢化作用，反刍动物产品的脂肪酸结构以饱和脂肪酸为主，其中主要是 $C_{18:0}$ 脂肪酸（Shingfield 和 Wallace，2014）。$C_{18:0}$ 脂肪酸是关系到人类健康的中性脂肪酸，它对人类健康既不是消极的，也不是积极的。然而，瘤胃中不饱和 18 碳脂肪酸还原到 $C_{18:0}$ 涉及一系列 $C_{18:1}$、$C_{18:2}$ 和 $C_{18:3}$ 的中间产物以不同位置和构型的双键积累。在这些不饱和脂肪酸中,有一些可能会对人类健康产生积极或消极的影响,这取决于不饱和脂肪酸的结构。

5.1 反刍动物反式脂肪酸对人类健康的影响

通常牛奶或乳制品中主要的反式脂肪酸是反式-11 C18:1(占总 C18:1 异构体的 25%~75%)(Lock 和 Bauman,2004)。Kuhnt 等(2006)研究指出,膳食中 19%~24% 的反式-11 C18:1 通过 Δ9-去饱和酶内源性转化为顺式-9,反式-11 共轭亚油酸,且该共轭亚油酸异构体被证明具有抗癌和抗动脉硬化作用(Lock 和 Bauman,2004)。然而,根据人类流行病学的研究,其余未转化的反式-11 C18:1 可能会增加癌症风险(优势比从 0.30~3.69;Field 等,2009)。然而,已有的动物试验和体外研究并不支持这一观点(Field 等,2009)。此外,在人类 T 细胞中的反式-11 C18:1 已被证明具有减少细胞因子的作用(白细胞介素-2 和肿瘤坏死因子-α),因为没有发生生物转化,所以并不是顺式-9、反式-11 共轭亚油酸的作用(Jaudszus 等,2012)。这表明,除顺式-9、反式-11 共轭亚油酸外,反式-11 C18:1 也可能具有健康促进作用(Field 等,2009)。Kuhnt 等(2016)在其最新综述中总结了几项与反刍动物反式脂肪酸相关的人类干预研究,与人群的平均消耗量相比,摄入 3 倍或 10 倍反刍动物反式脂肪酸并不会对人群健康产生负面影响。研究中得出的对血脂产生负面影响的反刍动物反式脂肪酸摄入量非常高且不合理(11~12 g/d,相当于每天摄入大约 490 g 奶酪、1 200 mL 牛奶、430 g 酸奶和 80 g 黄油,Kuhnt 等,2016)。

当给奶牛饲喂高淀粉日粮时,乳脂可能从以反式-11 C18:1 为主转变为更高比例的反式-10 C18:1(如 Conte 等,2018)。而 C18:1 的单个反式异构体可能具有其特定的性质(Ferlay 等,2017)。流行病学研究表明,工业反式脂肪酸可能影响血清胆固醇和脂蛋白代谢,从而增加冠心病的患病风险(Kuhnt 等,2016)。而工业脂肪酸中反式-10 C18:1 的含量较高,因此富含反式-10 C18:1 的牛奶可能会对人类健康产生负面影响。然而,对于目前还没有相关的人体干预研究,只有两项动物试验研究将含有反式-10 C18:1 与含有反式-11 C18:1 的牛奶或黄油进行了比较。Roy 等(2007)发现,与富含反式-11 C18:1 的黄油相比,日粮中添加反式-10 C18:1 的家兔血浆中总胆固醇和低密度脂蛋白胆固醇浓度增加,主动脉中脂质沉积增加。此外,在大鼠试验中,用富含反式-10 C18:1 的乳脂处理倾向于增加甘油三酯的浓度,而用反式-11 C18:1 和顺式-9,反式-11 共轭亚油酸倾向于降低 TAG 的浓度(Anadon 等,2010)。尽管这些动物研究表明,含有反式-10 C18:1 的牛奶(产品)对人类健康有潜在的负面风险,但将动物研究的结果推论至人类时仍需谨慎。

5.2 共轭亚油酸和 n-3 脂肪酸对人体健康的影响

乳制品中含有多种不同几何构型的共轭亚油酸异构体,其双键位置在 7~9 和 12~14 之间,其中顺式-9,反式-11 共轭亚油酸是主要异构体(~90%)(Ferlay 等,2017)。近年来,大量研究指出共轭亚油酸对健康具有积极作用,如抗癌、抗肥胖和抗炎等作用(Dhiman 等,2005;Ferlay 等,2017)。然而,几乎所有的研究使用的都是共轭亚油酸异构体的混合物,包括等量的顺式-9、反式-11、反式-10、顺式-12 异构体(80%~

95%）和其他次要的共轭亚油酸异构体。近期研究指出反式共轭亚油酸异构体，无论是反式-9，反式-11 或二者混合，均对健康有益，如抗癌、抗炎、抗血小板聚集和预防低胆固醇血症及脂肪肝等（Kim 等，2016）。总之，摄入共轭亚油酸已被证明可以改善人体健康（Dhiman 等，2005）。然而，个别共轭亚油酸异构体的代谢作用及其相互作用仍需要更多的研究来明晰。

当提到多不饱和脂肪酸，我们对长链脂肪酸（LCFA）的 n-3 和 n-6 系列进行了区分，LCFA 是由必需的亚麻酸（C18:3 n-3）和亚油酸（C18:2 n-6）的延伸和去饱和形成的，并可进一步转化为二十烷类化合物。研究指出从花生四烯酸（C20:4 n-6，AA）中提取的 n-6 二十烷醇类化合物具有促炎特性，而从 EPA 中提取的 n-3 二十烷醇类化合物具有抗炎作用（Wall 等，2010）。由于由前体形成的 n-3 和 n-6 系列 LCFA 是由同一组酶催化的，所以它们之间存在竞争。因此，建议采用较低的 n-6 与 n-3 比率（1:1~1:4）（Wall 等，2010）。相比之下，在西方饮食中（高 C18:3 n-3），这个比例估计介于 15:1~20:1（Simopoulos，2001）。荷兰的一项研究指出，乳脂对必需 n-3 多不饱和脂肪酸 C18:3 n-3 的摄入贡献显著（5.3%~15.7%），但对必需 n-6 多不饱和脂肪酸 C18:2 n-6 的摄入贡献较小（1.8%~3.6%）（van Valenberg 等，2013）。几种极长链 n-3 和 n-6 多不饱和脂肪酸主要通过乳脂的摄入（van Valenberg 等，2013），其 n-6 与 n-3 的比值为 2.28，在理想范围内。唯一推荐摄入的长链多不饱和脂肪酸是 EPA 和 DHA，据估计 10%~15% 的每日 EPA 摄入量来自乳脂（van Valenberg 等，2013），然而乳脂对 DHA 摄入量的贡献微乎其微（Meyer 等，2003；Sioen 等，2006；Astorg 等，2004）。

6 增加多不饱和脂肪酸（PUFA）以改善牛奶脂肪酸结构

尽管摄入牛奶脂肪酸可以带来一些健康益处（Gómez-Cortés 等，2018），但人们对乳制品脂肪的普遍看法仍是负面的，因为它的饱和脂肪酸含量高，会增加心血管疾病的风险。牧草可以说是农民容易获得的廉价 n-3 脂肪酸来源，以牧草为基础的生产系统是提高牛奶中共轭亚油酸和 n-3 脂肪酸含量以促进健康的有效策略（Gebreyowhans 等，2019；Elgersma，2015）。事实上，当采用以牧草为基础的生产方式（最多 50% 以谷物为基础的饲料）时，通过牛奶和乳制品摄入的共轭亚油酸（包括人类从反式-11 C18:1 转化而来的共轭亚油酸）从每天 350~550 mg 增加到每天 707~1 107 mg（van Wijlen 和 Colombani，2010）。然而，由于有限的草地和气候的不适宜性，这种以草为基础的生产系统往往是不可行的。

以油或油籽的形式添加 n-3 或 n-6 多不饱和脂肪酸，可以被视为一种替代品，或者与补饲结合来调节瘤胃脂质代谢。由于瘤胃中广泛的瘤胃生物氢化作用，日粮中添加油或油籽只能使牛奶中少量的多不饱和脂肪酸富集（Gebreyowhans 等，2019；Shingfield 等，2013）。因此，人们尝试通过植物次生代谢物和微生物的添加提高瘤胃食糜、牛奶和肌内脂肪中多不饱和脂肪酸的浓度。

Toral 等（2018b）在综述中指出这种添加剂改变牛奶多不饱和脂肪酸成分的潜力有限。活性成分和剂量的多样性（不同植物次生化合物，包括单宁、含氧脂肪酸和皂苷等）和反刍动物物种差异，基础日粮组成以及处理时间都可能是导致结果不一致的因素。因此，使多不饱和脂肪酸在牛奶中富集的最有效方法是过瘤胃保护技术。

Toral 等（2018b）最近对未包被的脂质添加剂对瘤胃代谢的影响进行了较好的总结，本章将进一步关注过瘤胃保护技术。这类保护技术可以分为两种类型（Jenkins 和 Bridges，2007）：一是修饰脂肪酸的结构以抑制细菌异构酶；二是包埋在保护基质中。

表 2 给出了较为常见的技术汇总（Gadeyne 等，2017），以及多不饱和脂肪酸从饮食到牛奶的相应转化效率。相比之下，在瘤胃后灌注脂肪酸的研究中，C18:2 n-6 和 C18:3 n-3 的转化效率可达到 49% 和 50%（Shingfield 等，2013），而反式-10、顺式-12 共轭亚油酸和 C22:6 n-3 的转化效率分别在 22%（de Veth 等，2004）和 25%（Shingfield 等，2013）。

7 过瘤胃脂质保护技术

本节以 Gadeyne 等（2017）的报道为基础，其中 7.1 至 7.5 节逐字引自 Gadeyne 等（2017），而 7.6 节在 Gadeyne 等（2017）研究结果的基础上编写。

7.1 脂肪酸结构的改变：脂肪酸钙盐

LCFA 的钙盐是由脂肪酸的游离羧基和钙离子之间通过离子键形成的钙皂。Palmquist 和 Jenkins（1987）首次提出了通过保护性脂肪酸钙以防止瘤胃生物氢化，这可能是由于钙盐的不溶性［钙盐的解离常数（pKa）介于 4.5~6］使得脂肪酸得以在避免瘤胃微生物干扰的情况下有效过瘤胃（Sukhija 和 Palmquist，1990），脂肪酸钙在真胃的酸性环境中可再次解离，使得其在小肠中被吸收。然而，若瘤胃 pH 值降低到 6.3 以下，脂肪酸钙仍可能发生解离，使得脂肪酸被细菌异构酶所利用（Chalupa 等，1986；Van Nevel 和 Demeyer，1996）。事实上，pKa 也依赖于钙皂中脂肪酸的不饱和度（Sukhija 和 Palmquist，1990），这意味着在给定的瘤胃 pH 值下，随着不饱和脂肪酸浓度的增加，解离也会增加。目前，关于不饱和脂肪酸钙盐过瘤胃保护的研究结果并不一致，但大多数都报道了不完全的保护策略（表 2）。

7.2 脂肪酸结构的改变：脂肪酰基酰胺

脂肪酰基酰胺由通过酰胺与胺结合的脂肪酸组成。包括氨基酸（Fotouhi 和 Jenkins，1992）、非酸性伯胺（Jenkins，1996）或氨（Cummings 和 Forrest，1997）这可能有助于保护不饱和脂肪酸以避免瘤胃生物氢化作用，如含简单的酰胺保护补充剂的生产过程需要游离 FA 作为前体。但在将日粮多不饱和脂肪酸转移到牛奶这方面，脂肪酰基酰胺的效果同纯油相比不会更好（表 2）。

7.3 耐菌外壳包被：醛处理

1975 年，Scott 和 Hills（1975）提出了一种通过包封蛋白-醛反应产物来保护不饱和脂肪酸的方法。在添加醛之前，脂类首先需要使用蛋白质（如酪蛋白、明胶或其他植物、

表2 目前描述最多或最有前景的过瘤胃脂肪技术中多不饱和脂肪酸（PUFA）的保护机制与不足之处（文献摘录汇总）

保护技术	保护机制（+）缺点（−）	脂质来源	过瘤胃多不饱和脂肪酸评估	转运[b]（%）包被的添加剂	转运[b]（%）未包被的添加剂	参考
钙盐	+封闭游离脂肪酸羧基端	亚麻籽油	18:3n−3	0.67	—	Chouinard 等, 1998
		亚麻籽油	18:3n−3	1.2	—	Sultana 等, 2008
	−由于分离保护受损	亚麻籽油	18:3n−3	1.9	1.5[c]	Cortes 等, 2010
		鱼油	22:6n−3	6.0	3.3[d]	Castaneda-Gutierrez 等, 2007b
	−制多不饱和脂肪酸保护	高多不饱和脂肪酸棕榈油	18:2n−6	13.2	—	Theurer 等, 2009
		共轭亚油酸	t10 c12 18:2	1.9~7.2	—	de Veth 等, 2005e
	−需要游离脂肪酸	共轭亚油酸	t10 c12 18:2	3.2	0.0[f]	de Veth 等, 2005
		大豆油	18:2n−6	6.5	6.9	Lundy 等, 2004
脂肪酰基酰胺	+封闭羧基端	菜籽油	18:2n−6	17	18	Loor 等, 2002a
		大豆油	18:2n−6	5.5	6.9	Lundy 等, 2004
		共轭亚油酸	t10 c12 18:2	7.1	0.0[f]	Perfield 等, 2004
甲醛	+封装在甲醛−蛋白质基质中	大豆/菜籽油	18:2n−6	25~44	—	Gulati 等, 2005e
		棉籽油	18:2n−6	43	—	Gulati 等, 2005e
		大豆/亚麻籽油	18:3n−3	19~24	—	Gulati 等, 2005e
	−需要有毒物质保护	大豆/鱼油	22:n−3	10~14	—	Gulati 等, 2005e
	−目标反应	亚麻籽油	18:3n−3	13	3.0[g]	Sterk 等, 2012
	−价格高	共轭亚油酸	t10 c12 18:2	7.0	0.0[f]	de Veth 等, 2005
		共轭亚油酸	t10 c12 18:2	6.9~8.9	—	Gulati 等, 2006

(续表)

保护技术	保护机制（+）缺点（-）	脂质来源	过瘤胃多不饱和脂肪酸评估	转运[b]（%）包被的添加剂	转运[b]（%）未包被的添加剂	参考
脂质复合凝胶	+包封再在凝胶蛋白质中 -含有大量液体	大豆油	18:2n-6	46~9（16~30）[i]	22-37	Carroll 等，2006
		大豆亚麻籽油	18:3n-3	81~25[i]（9~43）[i]	21	Heguy 等，2006
		大豆亚麻籽油	18:3n-3	13~9	—	van Vuuren 等，2010
		油菜籽油	18:2n-6	11~5	13	Kliem 等，2016
脂质包裹	+封装在更高熔点的脂质基质中	共轭亚油酸	t10 c12 18:2	7.9	0.0[f]	Perfield 等，2004
		共轭亚油酸	t10 c12 18:2	5.1	0.0[f]	Castaneda-Gutierrez 等，2007a
		共轭亚油酸	t10 c12 18:2	4.8	—	Moallem 等，2010
	-低有效荷载	共轭亚油酸	t10 c12 18:2	6.3	0.0[f]	Odens 等，2007
	-过瘤胃释放低	共轭亚油酸	t10 c12 18:2	2.4~5.8	—	Pappritz 等，2011
		共轭亚油酸	t10 c12 18:2	4.9	—	Schwarz 等，2009
		海藻油	22:6n-3	1.0	—	Stamey 2012
		海藻油	22:6n-3	2.0~3.4	—	Stamey 等，2012
		Echium 油	18:4n-3	3.2~3.4	—	Bainbridge 等，2015
酪氨酸酶交联乳剂	+包封在酚质中 -蛋白质提取 -需要酚类介入	亚麻籽油	t10 c12 18:2	4.0	—	Gadeyne 等，2016

注：[a] 使用油中最显著的 PUFA 进行评估；[b] 转移计算为（牛奶中的 g PUFA）/（饮食中的 g PUFA）×100，其中假设脂肪含有 90%（w/w）的 FA；[c] 全亚麻籽；[d] 瘤胃输注鱼油，与治疗无统计学差异；[e] 早期研究的电子摘要；[f] No t10 c12 18:2 在对照处理的牛奶中测得；[g] 挤出整粒亚麻籽；[h] 没有描述体内牛奶数据的科学文献；[i] 参考文献中报告的 Net 传输效率；[j] 使用来自参考的数据进行计算会导致不切实际的高转移。资料来源：改编自 Gadeyne 等（2017）。

鱼、肉或油籽蛋白质）进行乳化，以确保脂类在蛋白质中的均匀分布，并可以进一步使用喷雾干燥处理，以获得涂层固体颗粒。然而，甲醛是一种有毒产品，其在欧盟的使用受到严格的限定（2011/391/EU）。尽管甲醛处理被认为是迄今为止最有效的技术，但受成本高、饲料化学处理接受度差以及最终动物产品中可能存在残留的影响，目前应用仍然有限（Doreau 等，2015；Palmquist 和 Jenkins，2017）。如表 2 所示，Scott 和 Hills（1975）提出的方法在体外和体内均能有效地防止瘤胃生物氢化。

7.4 耐菌外壳包被：脂质体凝胶包被

近期的一种方法提及了含有氨基酸和脂类的复合凝胶的过瘤胃保护作用。Rosenberg 和 De Peters（2010）指出，蛋白质在水相中的脂滴分散可以通过热诱导的凝胶化作用防止瘤胃降解。与醛处理相反，蛋白质的交联不是由甲醛等二价连接剂诱导的，而是通过蛋白质的凝胶化。还原糖可以通过美拉德反应在基质中额外交联蛋白质。而将脂质嵌入乳清或血液蛋白的基质中，既具有提升此类侧流的优势，又同时产生瘤胃旁路脂质。但凝胶状乳剂通常含水量较高，保质期很短，这可能会导致凝胶和封闭的脂质在储存期间变质（van Vuuren 等，2010）。然而，由于凝胶是在高温下制备的，人们往往认为复合凝胶在货架期是相对稳定的（Weinstein 等，2015）。Carroll 等（2006）首次报道了乳清蛋白凝胶复合物有提高牛乳脂中多不饱和脂肪酸含量的功效（表 2）。Heguy 等（2006）发现，饲喂大豆/亚麻籽油的乳清蛋白分离凝胶复合物可增加血浆和乳脂中的多不饱和脂肪酸含量，降低反式脂肪酸含量。最近的研究还证实了该凝胶的长期效果（10 周）（van Vuuren 等，2010）。

7.5 耐菌外壳包被：脂质体包被

前一节描述的技术依赖于不同类型的蛋白质交联来实现过瘤胃保护。在其他配方中，活性化合物被包裹在脂质微胶囊中，要么被嵌入脂质基质中；要么被配制成小球体，然后被脂质包裹（Desai 和 Jin Park，2005；Wu 和 Papas，1997）。通常，涂层由至少高于其包膜基质的高熔点脂肪组成（Lorenzon，2015；Meade 等，1999），但外涂层的组成不同，会导致保护效率的差异（表 2）。

7.6 耐菌外壳包被：酪氨酸酶交联乳剂

另外一项研究探索了脂质保护技术涉及酪氨酸酶交联蛋白乳剂（Gadeyne 等，2015）。通过体内试验证实了酪氨酸酶交联乳剂包裹的反式-10，顺式-12 共轭亚油酸从日粮转移到了牛奶中（Gadeyne 等，2016）。然而，在市售的过瘤胃保护产品中，其转移效率并不高于反式-10，顺式-12 共轭亚油酸（表 2）。因此，需要进一步研究来优化当前酶基方法，这种方法在环境温度和压力下可以进行无溶剂加工，并且有可能使工业和农业产品增值。

8 瘤胃源的乳脂肪酸作为生物标志物监测动物健康

瘤胃日粮脂质的代谢和脂肪酸从头合成的变化可能表明瘤胃菌群的变化甚至紊乱，最终还可能导致亚急性瘤胃酸中毒（SARA）。血液中急性期蛋白的增加预示着紊乱的瘤胃内环境（pH）往往与炎症息息相关（Gozho 等，2005，2007），进一步可能导致采食量下降和生产损失（Plaizier 等，2008）。瘤胃 pH 值的降低也可能影响纤维消化（Plaizier 等，2008）或抑制乳脂（Dewanckele 等，2019）。亚急性瘤胃酸中毒的其他后果还包括腹泻、蹄叶炎和跛行，以及肝脓肿。一些特定的反式脂肪酸，尤其是反式-10 中间体，奇链和异构脂肪酸已经被建议作为（早期）SARA 生物标志物。

8.1 瘤胃细菌与反式-10 中间体的积累有关

反式-10 的积累通常与瘤胃中乳酸利用型埃氏巨型球菌的丰度增加有关（Weimer 等，2010；Dewanckele 等，2018）。此外，体外（Mickdam 等，2016）和体内（Khafipour 等，2009；Fernando 等，2010；Plaizier 等，2017）试验均发现，谷物诱导的瘤胃酸中毒期间，埃氏巨型球菌丰度增加，这可能与该情况下较高的乳酸有关。

基于瘤胃微生物组成与牛乳脂肪酸（Pitta 等，2018；Dewanckele 等，2019）或瘤胃脂肪酸组成（Zened 等，2016；Dewanckele 等，2018）的相关性分析，揭示了反式-10 中间体与酸氨基球菌属（*Acidaminococcus* spp.）、丹毒丝菌属（*Bulleidia* spp.）、双歧杆菌属（*Bifidobacterium* spp.）、肉食杆菌属（*Carnobacterium* spp.）、脱硫弧菌属（*Desulfovibrio* spp.）、小杆菌属（*Dialister* spp.）、真杆菌属（*Eubacterium* spp.）、乳杆菌属（*Lactobacillus* spp.）、欧陆森氏菌属（*Olsenella* spp.）、夏普氏菌属（*Sharpea* spp.）、共营养球菌属（*Syntrophococcus* spp.）和未分类的科杆菌科（unclassified Coriobacteriaceae）、毛螺菌科（Lachnospiraceae）以及瘤胃菌科（Ruminococcaceae）之间的正相关性。其中一些属在亚急性瘤胃酸中毒期间也有增加，如双歧杆菌属（Mao 等，2013）、乳杆菌属（Petri 等，2013；Mickdam 等，2016；Plaizier 等，2017）、欧陆森氏菌属（Petri 等，2013；zening 等，2013）、夏普氏菌属（Petri 等，2013；Plaizier 等，2017）和共营养球菌属（Petri 等，2013）。此外，乳酸菌属和夏普氏菌属是瘤胃中重要的乳酸生产者（Sharpe 等，1973；Kamke 等，2016），这可以进一步解释它们与亚急性瘤胃酸中毒的关联。

Devillard 等（2007）和 McIntosh 等（2009）观察到从人肠道分离的丙酸杆菌（*Proonibacterium freudenreichii*）参与从 C18:2n-6 生成反式-10，顺式-12 共轭亚油酸。与此一致的是，Wallace 等（2007）也发现从绵羊瘤胃中分离的痤疮丙酸杆菌推动 C18:2n-6 转化为反式-10，顺式-12 共轭亚油酸（McKain 等，2010；Dewanckele 等，2017）。前期未发表数据也发现一株可促进 C18:3n-3 异构化为反式-10，顺式-12，顺式-15 共轭亚麻酸的人菌株痤疮丙酸杆菌（L. Dewanckele，J. Jeyanathan，B. Vlaeminck，V. Fievez，未发表数据）。丙酸杆菌产生丙酸，而有些菌株还能发酵乳酸（Bryant，1959）。高谷物日粮可能导致从主要生产乙酸向丙酸和乳酸的转变（Balch 和 Rowland，1957），这可能支持丙酸杆菌和亚急性瘤胃酸中毒之间的联系。然而，没有研究报道该菌属在亚急性瘤胃酸中毒期间丰度增加，这可能是由于该菌属在瘤胃内容物中的丰度相对较低所致（Kim 等，2002；Shingfield 等，2012）。

8.2 瘤胃细菌及其奇链和支链脂肪酸

Vlaeminck（2006）等已经详细概述了奇数和支链脂肪酸（OBCFA）及其对应的瘤胃微生物。奇链脂肪酸（C15:0 和 C17:0）是通过丙酸或戊酸延伸形成的，而支链脂肪酸的前体（异构 C13:0，异构 C14:0，异构 C15:0，异构 C16:0，异构 C17:0，异构 C18:0，反式异构 C13:0，反式异构 C15:0，反式异构 C17:0）是支链氨基酸（缬氨酸、亮氨酸和异亮氨酸）及其相应的分支短链羧酸（异丁酸、异戊酸和 2-甲基丁酸）。瘤胃细菌对应的奇数和支链脂肪酸组成似乎很大程度上取决于微生物的脂肪酸合成酶活性，而不是前体的可用性（Vlaeminck 等，2006）。因此，离开瘤胃的奇数和支链脂肪酸组成的变化在一定程度上可以反映瘤胃中特定细菌种群相对丰度的变化。因此，固相菌群中异构脂肪酸的比例较高，反映了纤维素降解菌中的富集，而在液相菌群中，较高比例的反异式 C15:0 可能指示果胶和糖发酵细菌的富集（表3）（Vlaeminck 等，2006；Bessa 等，2009）。

不同瘤胃细菌对应的奇数和支链脂肪酸组成及其主要发酵底物和最终产物的详细信息见表 3。白色瘤胃球菌（*Ruminococcus albus*）、溶纤维丁酸弧菌（*Butyrivibrio fibrisolvens*）和黄色瘤胃球菌（*Ruminococcus flavefaciens*）被认为是典型的瘤胃纤维降解菌，其对应的奇数和支链脂肪酸组成含有较高比例的偶链和/或奇链异式脂肪酸（表3）。这些纤维菌主要将纤维素、半纤维素和果胶发酵成乙酸、丁酸、氢（H_2）和二氧化碳（CO_2）。另外，以嗜淀粉反刍杆菌（*Ruminobacter amylophilus*）、反刍月形单胞菌（*Selenomonas ruminantium*）、牛链球菌（*Streptococcus bovis*）和解淀粉琥珀单胞菌（*Succinomonas amylolytica*）为代表的降解淀粉细菌对应的的奇数和支链脂肪酸谱显示支链脂肪酸比例较低，特别是异构支链脂肪酸；但富含线性奇链脂肪酸和/或反异构支链脂肪酸（表3）。后者似乎在糖或果胶发酵细菌中特别重要，如普雷沃氏菌属（*Prevotella* spp.）、多对毛螺菌（*Lachnospira multiparus*）和溶糊精琥珀酸弧菌（*Succinovibrio dextrinosolvens*）（表3）。淀粉和糖消化细菌主要发酵糖、淀粉和肽类以生成丙酸、丁酸、乙酸、乳酸、H_2 和 CO_2。反刍真杆菌和牛链球菌主要生成乳酸，不合成乙酸。这些细菌在酸中毒的发生中起重要作用。这其中最重要的奇数和支链脂肪酸是异构 C15:0 和反式异构 C15:0。埃氏巨型球菌也发挥着重要作用（Nagaraja 和 Lechtenberg，2007），其于细胞壁中主要脂肪酸是 C15:0（表3）。

8.3 瘤胃细菌与反式-10 中间体的积累有关，这其中是否与炎症相关的奇链脂肪酸丰富而异构脂肪酸缺乏相关

大多数瘤胃细菌与反式-10 中间体的积累有关，这其中富含奇链脂肪酸而异构脂肪酸少见。例如，酸氨球菌属、脱硫弧菌属、小杆菌属和共营养球菌属以及琥珀弧菌科（包括琥珀弧菌属）等革兰氏阴性菌。众所周知，革兰氏阴性菌的细胞壁外膜中含有内毒素，如脂多糖（LPS），它们可以以其游离形式作为免疫原性化合物（Hurley，1995）。这些内毒素在细菌生长的对数期和稳定阶段大量释放，也可在细胞解体和裂解后释放（Nagaraja 等，1978a，b；赫尔利，1995；Plaizier 等，2012）。然而，脂多糖的效力因菌种而异，因此，它的促炎反应可能也因菌种而异（Ghaffari 等，2017）。此外，埃氏巨型球菌产生的

第14章 阐明瘤胃脂质代谢以优化乳品质、促进人类健康并监测动物健康

表3 参与瘤胃碳水化合物发酵的重要细菌的主要底物（Harfoot 和 Hazlewood, 1997）、发酵最终产物（Russell 和 Rychlik, 2001）和 OBCFA 含量（g/100g 脂肪酸；Vlademick 等, 2006）

细菌	发酵产物[d]	Anteiso C13:0	Anteiso C15:0	Anteiso C17:0	Iso C13:0	Iso C15:0	Iso C17:0	Iso C14:0	Iso C16:0	C13:0	C15:0	C17:0	C17:1
白色瘤胃球菌[a]	A	—	9.4	1.3	—	—	0.7	20.6	11.0	—	10.3	1.4	—
溶纤维丁酸弧菌[a]	A, B, F	6.4	16.2	8.6	6.8	10.4	5.7	10.8	11.1	2.9	7.8	4.3	3.5
黄色瘤胃球菌[a]	A, S	—	2.3	2.9	—	37.5	5.2	2.5	7.3	0.1	3.2	0.5	—
解淀粉琥珀酸单胞菌[b] N6	A, P	—	—	—	—	52.6	10.8	1.6	5.3	1.6	5	—	—
解淀粉琥珀酸单胞菌[b] B24	A, P	—	—	—	—	0.1	0.3	—	0.6	1.4	3.3	1.3	0.6
普雷沃氏菌[bc]	A, S	1.2	36.7	4.2	3.0	14.7	2.3	3.3	3.0	1.2	12.1	2.1	—
多对毛螺菌[bc]	A, L, F	—	4.0	2.6	—	1.1	1.1	1.2	1.8	0.3	2.9	0.8	0.1
溶糊精琥珀酸弧菌[c]	A, S	0.8	3.6	1.0	—	0.1	—	0.6	1.5	0.5	4.0	0.7	—
嗜淀粉反刍杆菌[b]	A, S, F	—	1.1	—	—	—	—	—	—	0.5	1.1	0.3	0.1
产琥珀酸丝状杆菌[a]	A, S	3.9	7.7	1.2	—	0.1	0.2	3.6	3.4	9.0	30.2	2.1	—
牛链球菌[b]	L	—	0.9	—	—	—	—	0.4	0.2	0.6	1.7	1.2	0.2
埃氏巨型球菌[c]	A, P, B	—	2.8	—	0.1	0.2	0.2	1.5	0.5	1.5	6.0	4.5	3.0
反刍真杆菌[b] B1C23	B, L, F	—	—	—	—	17.7	1.4	—	—	5.4	49.0	1.5	—
反刍真杆菌[b] GA195	B, L, F	—	30.1	1.7	—	0.4	0.2	6.1	3.7	0.4	6.5	0.4	—
反刍月形单胞菌[b]	A, P, L	—	0.1	—	—	0.2	—	0.3	0.1	1.3	6.0	2.9	2.6

注：[a] 发酵纤维素和半纤维素的细菌；
[b] 发酵淀粉的细菌；
[c] 发酵糖和果胶的细菌；
[d] 发酵产物：A: 乙酸盐；S: 琥珀酸；B: 丁酸盐；F: 甲酸盐；P: 丙酸盐；L: 乳酸。
主要的 OBCFA 带有下划线。主要基材的指示由上标字母给出。

物质还具有许多肠道细菌内毒素所共有的生化特性（Nagaraja 等，1979）。虽然它的内毒素效力较低（Nagaraja 等，1979），但随着埃氏巨型球菌丰度的增加可能诱导部分免疫反应（Plaizier 等，2008）。然而，这一假设还需要进一步的研究来证实。

8.4 通过乳脂中的反式和支链脂肪酸来鉴别亚急性瘤胃酸中毒（风险）

传统上认为当瘤胃内环境每天数小时处于低 pH 值状态即为亚急性瘤胃酸中毒（Plaizier 等，2008）。因此，可使用低于某一 pH 值阈值的观察频率，例如，pH 值<5.6 或 5.8 和 pH 值<6.0 的时间（使用留置在网胃中的瘤胃 pH 值探针），或者曲线下的面积（低于 pH 值阈值的降幅与下降的持续时间之积），或不归一化的采食量水平，以获得酸中毒指数（Gao 和 Oba，2014），这也被用作亚急性瘤胃酸中毒（SARA）严重程度的指标。针对监测偏差可以应用一些信号处理每日原始的 pH 值动力学数据，纠正个体间变异性、传感器漂移和传感器噪声，计算相对 pH 指标，以监测个体内 pH 值的变化（Villot 等，2018）。Denwood 等（2018）也提出了监测偏差，他们评估了偏离可预测模式的偏差，描述了瘤胃 pH 值的时间动态变化。Colman 等（2012）建议通过拟合 logistic 曲线来描述每日 pH 值变化。该曲线由 β_1（代表瘤胃平均 pH 值）和 β_0（代表瘤胃 pH 值范围）来描述。乳脂中的反式脂肪酸和奇数和支链脂肪酸（奇数和支链脂肪酸）可被作为诊断 SARA 的生物标志物用以识别有发生亚急性瘤胃酸中毒风险的奶牛（早期预警和动物间易感性）。因此，亚急性瘤胃酸中毒病例的识别通常是基于每日低于 pH 值阈值的时间，在较小程度上是基于酸中毒指数和 logistic 曲线的参数。目前还没有发现乳脂肪酸与瘤胃 pH 值偏离动力学模式有关，也没有发现与亚急性瘤胃酸中毒的炎症反应有关。

Vlaeminck 等（2006）认为牛奶中异构脂肪酸的比例越高，反映出亚急性瘤胃酸中毒的风险越低，而反式异构 C15:0、C15:0 和 C17:0 的比例越高，则表明亚急性瘤胃酸中毒风险越高。事实上，有报道称，在亚急性瘤胃酸中毒发生期间，乳脂 C15:0 增加（Enjalbert 等，2008；Colman 等，2010）。这与 Prado 等（2019）最近的 meta 分析得出的研究结果一致，他们指出在通过 C15:0（g/d）和 C17:0（g/d）预测产奶量的方程中，日粮淀粉和瘤胃 pH 值分别与正斜率和负斜率相关。反式异构 C17:0（g/d）与瘤胃 pH 值呈负相关，但该方程的决定系数/拟合优度较小。在这项 meta 分析中未报道异构偶链脂肪酸。Colman 等（2015）通过 6 头瘤胃瘘管奶牛酸中毒诱导试验的数据指出，降低异式支链脂肪酸浓度似乎比增加 C15:0 和 C17:0 浓度更能区分亚急性瘤胃酸中毒病例，而 Jing 等（2018）的研究则得出相反的结论。然而，这两篇论文均报道了反式-10 C18:1 和/或反式-10 C18:1 与反式-11 C18:1 的比值与亚急性瘤胃酸中毒有关，尽管在 Colman 等（2015）的一些模型中，反式-10 C18:1 被顺式-9 反式-11 C18:2 取代为主要的判别变量。此外，Jing 等（2018）在泌乳早期的 4 周精料积累期间监测了亚急性瘤胃酸中毒指示性乳脂肪酸（异构偶数和奇数链脂肪酸、反异构奇数链脂肪酸以及反式脂肪酸）。除了反式-10 C18:1 和 C15:0 之外，Δ 反式-11 C18:1 被发现在 4 周期间下降比例最大，因此可作为确定亚急性瘤胃酸中毒易感性动物间变异性的附加参数。在 Jing 等（2018）的工作中，牛奶奇数和支链脂肪酸和反式脂肪酸不仅显示了在亚急性瘤胃酸中毒诱导试验中诊断发病的潜力，而且还可用于在同一牧群中区分对亚急性瘤胃酸中毒敏感性相对较高或较低的奶牛。

此外，Colman 等（2012）试图通过 logistic 曲线参数 β1 和 β0 分别区分与瘤胃 pH 值水平相关的和与瘤胃 pH 值波动相关的乳脂肪酸。指出了反式-10 C18:1 的乳脂比例与 pH 值低且波动较大的情况有关，而稳定的低 pH 值并没有引起从主要形成反式-11 中间物向增加形成反式-10 中间物的转变。反式-11 C18:1、顺式-9、反式-11 C18:2 只受 pH 值波动的影响，不受平均 pH 值水平的影响，而异构脂肪酸受平均 pH 值水平的影响，不受 pH 值日变化的影响。异构脂肪酸（正相关）和反式-10 C18:1（负相关）与平均瘤胃 pH 值之间的关系在 Colman（2012）开展的 6 个 SARA 诱导试验的扩展数据库中得到证实。在所有试验中，瘤胃 pH 值范围均与特定乳奇数和支链脂肪酸和反式脂肪酸有关，但两者之间的关系并不明确，这可能与 6 个试验 pH 值范围的幅度变化较大有关。

未来应开发一个多变量和稳健的模型来识别有亚急性瘤胃酸中毒风险的奶牛。此外，虽然目前并不现实，但为了实际应用，这些诊断性牛奶脂肪酸应纳入常规检测指标。研究团队早期试验表明，拉曼光谱具有测定乳脂中个体和分组反式（单）不饱和脂肪酸的潜力（Stefanov 等，2011），但仍需对这种常规技术进一步投资开发。

9　结论

在瘤胃代谢过程中产生的反式脂肪酸、奇链脂肪酸和支链脂肪酸是诊断瘤胃功能受损、早期识别亚急性瘤胃酸中毒风险的潜在生物标志物，还可用以区分对瘤胃疾病易感性相对较高或较低的动物。另外，预防瘤胃代谢可能有助于生产富含不饱和脂肪酸的乳制品，以促进人体健康。截至目前，无论是利用甲醛蛋白、凝胶蛋白还是固体脂肪基质，包埋技术都能提高脂肪酸从日粮到牛奶的转移效率。

10　参考文献

Anadón, A., Martínez-Larrañaga, M. R., Martínez, M. A., et al., 2010. Acute oral safety study of dairy fat rich in trans-10 C18:1 versus vaccenic plus conjugated linoleic acid in rats. Food Chem. Toxicol. 48:591-8.

Annison, E. F., Linzell, J. L., Fazakerley, S., et al., 1967. The oxidation and utilization of palmitate, stearate, oleate and acetate by the mammary gland of the fed goat in relation to their overall metabolism, and the role of plasma phospholipids and neutral lipids in milk-fat synthesis. Biochem. J. 102:637-47.

Annous, B. A., Becker, L. A., Bayles, D. O., et al., 1997. Critical role of anteiso-C15:0 fatty acid in the growth of Listeria monocytogenes at low temperatures. Appl. Environ. Microbiol. 63:3387-894.

Astorg, P., Arnault, N., Czernichow, S., et al., 2004. Dietary intakes and food sources of n-6 and n-3 PUFA in French adult men and women. Lipids 39:527-35.

Bainbridge, M. L., Lock, A. L. and Kraft, J. 2015. Lipid-encapsulated echium oil (Echium plantagineum) increases the content of stearidonic acid in plasma lipid fractions and milk fat of dairy cows. J. Agric. Food Chem. 63:4827-35.

Balch, D. A. and Rowland, S. J. 1957. Volatile fatty acids and lactic acid in the rumen of dairy cows receiving a variety of diets. Br. J. Nutr. 11:288-98.

Bauman, D. E. and Lock, A. L. 2006. Concepts in lipid digestion and metabolism in dairy cows. Proc. Tri-State Dairy Nutr. Conf., pp. 1-14.

Bernard, L., Toral, P. G. and Chilliard, Y. 2017. Comparison of mammary lipid metabolism in dairy cows and goats fed diets supplemented with starch, plant oil, or fish oil. J. Dairy Sci. 100:9338-51.

Bernard, L., Bonnet, M., Delavaud, C., et al., 2018. Milk fat globule in ruminant: major and minor compounds, nutritional regulation and differences among species. Eur. J. Lipid Sci. Technol. 120:201700039.

Bessa, R. J. B., Maia, M. R. G., Jeronimo, E., et al., 2009. Using microbial fatty acids to improve understanding of the contriubution of solid associated bacteria to microbial mass in the rumen. Anim. Feed Sci. Technol. 150:197-206.

Boerman, P., Firkins, J. L., St-Pierre, N. R., et al., 2015. Intestinal digestibility of long-chain fatty acids in lactating dairy cows: A meta-analysis and meta-regression. J. Dairy Sci. 98:8889-903.

Boerman, J. P., de Souza, J. and Lock, A. L. 2017. Milk production and nutrient digestibility responses to increasing levels of stearic acid supplementation of dairy cows. J. Dairy Sci. 100:2729-38.

Bruss, M. L. 1997. Lipids and ketones. In: Kaneko, J. J., Harvey, J. W. and Bruss, M. L. (Eds), Clinical Biochemistry of Domestic Animals (5th edn.). Academic Press, San Diego, CA, pp. 83-115. Chapter 4.

Bryant, M. P. 1959. Bacterial species of the rumen. Bacterial. Rev. 23:125-53.

Buccioni, A., Decandia, M., Minieri, S., et al., 2012. Lipid metabolism in the rumen: new insights on lipolysis and biohydrogenation with an emphasis on the role of endogenous plant factors. Anim. Feed Sci. Tech. 174:1-25.

Carroll, S. M., DePeters, E. J. and Rosenberg, M. 2006. Efficacy of a novel whey protein gel complex to increase the unsaturated fatty acid composition of bovine milk fat. J. Dairy Sci. 89:640-50.

Castaneda-Gutierrez, E., Benefield, B. C., de Veth, M. J., et al., 2007a. Evaluation of the mechanism of action of conjugated linoleic acid isomers on reproduction in dairy cows. J. Dairy Sci. 90:4253-64.

Castaneda-Gutierrez, E., de Veth, M. J., Lock, A. L., et al., 2007b. Effect of supplementation with calcium salts of fish oil on n-3 fatty acids in milk fat. J. Dairy Sci. 90:4149-56.

Castro-Gómez, M. P., Rodriguez-Alcalá, L. M., Calvo, M. V., et al., 2014. Total milk fat extraction and quantification of polar and neutral lipids of cow, goat, and ewe milk by using a pressurized liquid system and chromatographic techniques. J. Dairy Sci. 97:6719-28.

Chalupa, W., Vecchiarelli, B., Elser, A. E., et al., 1986. Ruminal fermentation *in vivo* as influenced by long-chain fatty-acids. J. Dairy Sci. 69:1293-301.

Chilliard, Y., Ferlay, A., Mansbridge, R. M., et al., 2000. Ruminant milk fat plasticity: nutritional control of saturated, polyunsaturated, trans and conjugated fatty acids. Ann. Zootech. 49:181-205.

Chouinard, P. Y., Girard, V. and Brisson, G. J. 1998. Fatty acid profile and physical properties of milk fat from cows fed calcium salts of fatty acids with varying unsaturation. J. Dairy Sci. 81:471-81.

Colman, E. 2012. Milk fatty acids as biomarkers of subacute ruminal acidosis in dairy cows. PhD dissertation, Ghent University, Belgium, 278p. ISBN 978-90-5989-553-9.

Colman, E., Fokkink, W. B., Craninx, M., et al., 2010. Effect of induction of sub-acute ruminal acidosis (SARA) on milk fat profile and rumen parameters. J. Dairy Sci. 93:4759-73.

Colman, E., Tas, B. M., Waegeman, W., et al., 2012. The logistic curve as a tool to describe the daily ruminal pH pattern and its link with milk fatty acids. J. Dairy Sci. 95:5845-65.

Colman, E., Waegeman, W., De Baets, B., et al., 2015. Prediction of subacute ruminal acidosis based on milk fatty acids: a comparison of linear discriminant and support vector machine approaches for model development. Comp. Electron. Agric. 111:179-85.

Conte, G., Dimauro, C., Serra, A., et al., 2018. A canonical discriminant analysis to study the association between milk fatty acids of ruminal origin and milk fat depression in dairy cows. J. Dairy Sci. 101: 6497-510.

Cortes, C., da Silva-Kazama, D. C., Kazama, R., et al., 2010. Milk composition, milk fatty acid profile, digestion, and ruminal fermentation in dairy cows fed whole flaxseed and calcium salts of flaxseed oil. J. Dairy Sci. 93:3146-57.

Craninx, M., Steen, A., Van Laar, H., et al., 2008. Effect of lactation stage on the odd-and branched-chain milk fatty acids of dairy cattle under grazing and indoor conditions. J. Dairy Sci. 91:2662-77.

Cummings, K. R. and Forrest, R. L. 1997. Aliphatic amide feed supplement for ruminants. US patent 5670191 A.

Demeyer, D. and Doreau, M. 1999. Target and procedures for altering ruminant meat and milk lipids. P. Nutr. Soc. 58:593-607.

Denwood, M. J., Kleen, J. L., Jensen, D. B., et al., 2018. Describing temporal variation in reticulo-ruminal pH using continuous monitoring data. J. Dairy Sci. 101:233-45.

Desai, K. G. and Jin Park, H. 2005. Recent developments in microencapsulation of food ingredients. Drying Technol. 23:1361-94.

de Veth, M. J., Griinari, J. M., Pfeiffer, A. M., et al., 2004. Effect of CLA on milk fat synthesis in dairy cows: comparison of inhibition by methyl esters and free fatty acids, and relationships among studies. Lipids 39:365-72.

de Veth, M. J., Gulati, S. K., Luchini, N. D., et al., 2005. Comparison of calcium salts and formaldehyde-protected conjugated linoleic acid in inducing milk fat depression. J. Dairy Sci. 88:1685-93.

Devillard, E., McIntosh, F. M., Duncan, S. H., et al., 2007. Metabolism of linoleic acid by human gut bacteria: different routes for biosynthesis of conjugated linoleic acid. J. Bacteriol. 189:2566-70.

Dewanckele, L., Vlaeminck, B., Jeyanathan, J., et al., 2017. Effect of pH and 22:6n-3 on in vitro biohydrogenation of 18:2n-6 by different ratios of *Butyrivibrio fibrisolvens* to *Propionibacterium acnes*. J. Dairy Sci. 100 (Suppl. 2):221 (Abstract).

Dewanckele, L., Vlaeminck, B., Hernandez-Sanabria, E., et al., 2018. Rumen biohydrogenation and microbial community changes upon early life supplementation of 22:6n-3 enriched microalgae to goats. Front. Microbiol. 9:573.

Dewanckele, L., Jing, L., Stefańska, B., et al., 2019. Distinct blood and milk 18-carbon fatty acid proportions and buccal bacterial populations in dairy cows differing in reticulorumen pH response to dietary supplementation to rapidly fermentable carbohydrates. J. Dairy Sci. 102:4025-40.

Dewhurst, R. J., Moorby, J. M., Vlaeminck, B., et al., 2007. Apparent recovery of duodenal odd-and branched-chain fatty acids in milk. J. Dairy Sci. 90:1775-80.

Dhiman, T. R., Nam, S. H. and Ure, A. L. 2005. Factors affecting conjugated linoleic acid content in milk and meat. Crit. Rev. Food Sci. Nutr. 45:463-82.

Diedrich, M. and Henschel, K. P. 1990. The natural occurrence of unusual fatty acids. Part 1. Odd numbered fatty acids. Nahrung 34:935-43.

Dodds, P. F., Guadalupe, M., Guzman, F., et al., 1981. Acetoacetyl – CoA reductase activity of lactating bovine mammary fatty acid synthase. J. Biol. Chem. 256:6282-90.

Doreau, M., Troegeler-Meynadier, A., Fievez, V., et al., 2015. Ruminal metabolism of fatty acids: modulation of polyunsaturated, conjugated, and trans fatty acids in meat and milk. In: Watson, R. R. and De Meester, F. (Eds), Handbook of Lipids in Human Function: Fatty Acids. AOCS Press, San Diego, CA, pp. 521-42. Chapter 19.

Elgersma, A. 2015. Grazing increases the unsaturated fatty acid concentration of milk from grass-fed cows: a review of the contributing factors, challenges and future perspectives. Eur. J. Lipid Sci. Technol. 117: 1345-69.

Enjalbert, F., Videau, Y., Nicot, M. C., et al., 2008. Effects of induced subacute ruminal acidosis on milk fat content and milk fatty acid profile. J. Anim. Physiol. An. N. 92, 284-91.

Ferlay, A., Bernard, L., Meynadier, A., et al., 2017. Production of trans and conjugated fatty acids in dairy ruminants and their putative effects on human health: a review. Biochimie 141:107-20.

Fernando, S. C., Purvis, H. T., Najar, F. Z., et al., 2010. Rumen microbial population dynamics during adaptation to a high-grain diet. Appl. Environ. Microbiol. 76:7482-90.

Field, C. J., Blewett, H. H., Proctor, S., et al., 2009. Human health benefits of vaccenic acid. Appl. Physiol. Nutr. Metabol. 34:979-91.

Fievez, V., Vlaeminck, B., Dhanoa, M. S., et al., 2003. Use of principal component analysis to investigate the origin of heptadecenoic and conjugated linoleic acids in milk. J. Dairy Sci. 86:4047-53.

Fievez, V., Colman, E., Castro-Montoya, J. M., et al., 2012. Milk odd and branched chain fatty acids as biomarkers of rumen function-an update. Anim. Feed Sci. Technol. 172, 51-65.

Fotouhi, N. and Jenkins, T. C. 1992. Resistance of fatty acyl amides to degradation and hydrogenation by ruminal microorganisms. J. Dairy Sci. 75:1527-32.

French, E. A., Bertics, S. J. and Armentano, L. E. 2012. Rumen and milk odd-and branched-chain fatty acid proportions are minimally influenced by ruminal volatile fatty acid infusions. J. Dairy Sci. 95: 2015-26.

Fukuda, S., Nakanishi, Y., Chikayama, E., et al., 2009. Evaluation and characterization of bacterial metabolic dynamics with a novel profiling technique, real-time metabolotyping. PLoS ONE 4:e4893.

Fulco, A. J. 1983. Fatty acid metabolism in bacteria. Prog. Lipid Res. 22:133-60.

Gadeyne, F., Van Ranst, G., Vlaeminck, B., et al., 2015. Protection of polyunsaturated oils against ruminal biohydrogenation and oxidation during storage using a polyphenol oxidase containing extract from red clover. Food Chem. 171:241-50.

Gadeyne, F., De Neve, N., Vlaeminck, B., et al., 2016. Transfer to the milk of rumen bypass CLA e-mulsions created by potato tuber peel polyphenol oxidase. Book of abstracts 14th Euro Fed Lipid Congress, 18-21 September, Ghent, Belgium.

Gadeyne, F., De Neve, N., Vlaeminck, B., et al., 2017. State of the art in rumen lipid protection technologies and emerging interfacial protein cross-linking methods. Eur. J. Lipid Sci. Technol. 119:1600345.

Gao, X. and Oba, M. 2014. Relationship of severity of subacute ruminal acidosis to rumen fermentation, chewing activities, sorting behavior, and milk production in lactating dairy cows fed a high-grain diet. J. Dairy Sci. 97:3006-16.

Gebreyowhans, S., Lu, J., Zhang, S., et al., 2019. Dietary enrichment of milk and dairy products with n-3 fatty acids: a review. Int. Dairy J. 97:158-66.

Ghaffari, M. H., Khafipour, E. and Steele, M. A. 2017. Systems biology and ruminal acidosis. In: Ametaj, B. (Ed.), Periparturient Diseases of Dairy Cows. Springer, Cham, Switzerland, pp. 51–69.

Glasser, F., Doreau, M., Ferlay, A., et al., 2007. Technical Note: Estimation of milk fatty acid yield from milk fat data. J. Dairy Sci. 90:2302–4.

Gómez-Cortés, P., Juárez, M. and de la Fuente, M. A. 2018. Milk fatty acids and potential health benefits: an updated vision. Trends Food Sci. Tech. 81:1–9.

Gozho, G. N., Plaizier, J. C., Krause, D. O., et al., 2005. Subacute ruminal acidosis induces ruminal lipopolysaccharide endotoxin release and triggers an inflammatory response. J. Dairy Sci. 88:1399–403.

Gozho, G. N., Krause, D. O. and Plaizier, J. C. 2007. Ruminal lipopolysaccharide concentration and inflammatory response during grain induced subacute ruminal acidosis in dairy cows. J. Dairy Sci. 90:856–66.

Griinari, J. M., Corl, B. A., Lacy, S. H., et al., 2000. Conjugated linoleic acid is synthesized endogenously in lactating dairy cows by $\Delta 9$-desaturase. J. Nutr. 130:2285–91.

Gulati, S. K., Garg, M. R. and Scott, T. W. 2005. Rumen protected protein and fat produced from oilseeds and/or meals by formaldehyde treatment: their role in ruminant production and product quality: a review. Aust. J. Exp. Agr. 45:1189–203.

Gulati, S. K., McGrath, S., Wynn, P. C., et al., 2006. Rumen protected fat reverses the conjugated linoleic acid induced low milk fat content in dairy cows. Canadian J. Anim. Sci. 86:63–70.

Harfoot, C. G. and Hazlewood, G. P. 1997. Lipid metabolism in the rumen. In: Hobson, P. N. and Stewart, C. S. (Eds), The Rumen Microbial Ecosystem. Springer, Dordrecht, the Netherlands, pp. 382–426. Chapter 9.

Heguy, J. M., Juchem, S. O., DePeters, E. J., et al., 2006. Whey protein gel composites of soybean and linseed oils as a dietary method to modify the unsaturated fatty acid composition of milk lipids. Anim. Feed Sci. Tech. 131:370–88.

Honkanen, A. M., Leskinen, H., Toivonen, V., et al., 2016. Metabolism of α-linolenic acid during incubations with strained bovine rumen contents: products and mechanisms. Br. J. Nutr. 115:2093–105.

Hulshof, K. F. A. M., van Erp-Baart, M. A., Anttolainen, M., et al., 1999. Intake of fatty acids in Western Europe with emphasis on trans fatty acids: the TRANSFAIR study. Eur. J. Clin. Nutr. 53:143–57.

Hurley, J. C. 1995. Endotoxemia: methods of detection and clinical correlates. Clin. Microbial. Rev. 8:268–92.

Huws, S. A., Kim, E. J., Lee, M. R. F., et al., 2011. As yet uncultured bacteria phylogenetically classified as *Prevotella*, Lachnospiraceae incertae sedis and unclassified Bacteroidales, Clostridiales and Ruminococcaceae may play a predominant role in ruminal biohydrogenation. Environ. Microbiol. 13:1500–12.

Jaudszus, A., Jahreis, G., Schlörmann W., et al., 2012. Vaccenic acid-mediated reduction in cytokine production is independent of c9, t11 – CLA in human peripheral blood mononuclear cells. Biochim. Biophys. Acta. 1821:1316–22.

Jenkins, T. C. 1996. Feed supplements for ruminants and method for using same. US patent 5547686 A.

Jenkins, T. C. and Bridges, W. C. 2007. Protection of fatty acids against ruminal biohydrogenation in cattle. Eur. J. Lipid Sci. Tech. 109:778–89.

Jenkins, T. C., Wallace, R. J., Moate, P. J., et al., 2008. Board-invited review: Recent advances in biohydrogenation of unsaturated fatty acids within the rumen microbial ecosystem. J. Anim. Sci. 86:397–412.

Jensen, R. G. 2002. The composition of bovine milk lipids: January 1995 to December 2000. J. Dairy Sci. 85:295-350.

Jing, L., Dewanckele, L., Vlaeminck, B., et al., 2018. Susceptibility of dairy cows to subacute ruminal acidosis is reflected in milk fatty acid proportions, with C18:1 trans-10 as primary and C15:0 and C18:1 trans-11 as secondary indicators. J. Dairy Sci. 101:1-14.

Jorjong, S., van Knegsel, A. T. M., Verwaeren, J., et al., 2014. Milk fatty acids as possible biomarkers to early diagnose elevated concentrations of blood plasma nonesterified fatty acids in dairy cows. J. Dairy Sci. 97:7054-64.

Kairenius, P., Leskinen, H., Toivonen, V., et al., 2017. Effect of dietary fish oil supplements alone or in combination with sunflower and linseed oil on ruminal lipid metabolism and bacterial populations in lactating cows. J. Dairy Sci. 101:3021-35.

Kamke, J., Kittelmann, S., Soni, P., et al., 2016. Rumen metagenome and metatranscriptome analysis of low methane yield sheep reveals a Sharpea-enriched microbiome characterised by lactic acid formation and utilisation. Microbiome 4:56.

Kaneda, T. 1977. Fatty acids of the genus *Bacillus*: an example of branched-chain preference. Bacteriol. Rev. 41:391-418.

Kaneda, T. 1991. Iso - and anteiso - fatty acids in bacteria: biosynthesis, function, and taxonomic significance. Microbiol. Rev. 55:288-302.

Keeney, M., Katz, I. and Allison, M. J. 1962. On the probable origin of some milk fat acids in rumen microbial lipids. J. Am. Oil Chem. Soc. 39:198-201.

Kepler, C. R., Hirons, K. P., McNeill, J. J., et al., 1966. Intermediates and products of the biohydrogenation of linoleic acid by *Butyrivibrio fibrisolvens*. J. Biol. Chem. 241:1350-4.

Keweloh, H. and Heipieper, H. J. 1996. Trans unsaturated fatty acids in bacteria. Lipids 31:129-37.

Khafipour, E., Li, S., Plaizier, J. C., et al., 2009. Rumen microbiome composition determined using two nutritional models of subacute ruminal acidosis. Appl. Environ. Microbiol. 75:7115-24.

Kim, Y. J., Liu, R. H., Rychlik, J. L., et al., 2002. The enrichment of a ruminal bacterium (*Megasphaera elsdenii* YJ-4) that produces the trans-10, cis-12 isomer of conjugated linoleic acid. J. Appl. Microbiol. 92:976-82.

Kim, J. H., Kim, Y., Kim, Y. J., et al., 2016. Conjugated linoleic acid: potential health benefits as a functional food ingredient. Annu. Rev. Food Sci. Technol. 7:221-44.

Kliem, K. E. and Shingfield, K. J. 2016. Manipulation of milk fatty acid composition in lactating cows: Opportunities and challenges. Eur. J. Lipid Sci. Technol. 118:1661-83.

Kliem, K. E., Humphries, D. J., Grandison, A. S., et al., 2016. Effect of whey protein and rapeseed oil gel feed supplement on milk fatty acid composition of Holstein cows. J. Dairy Sci. 102:288-300.

Kuhnt, K., Kraft, J., Moeckel, P., et al., 2006. Trans - 11 - 18:1 is effectively $\Delta 9$ - desaturated compared with trans-12-18:1 in humans. Br. J. Nutr. 95:752-61.

Kuhnt, K., Degen, C. and Jahreis, G. 2016. Evaluation of the impact of ruminant transfatty acids on human health: Important aspects to consider. Crit. Rev. Food Sci. Nutr. 56:1964-80.

Lock, A. L. and Bauman, D. E. 2004. Modifying milk fat composition of dairy cows to enhance fatty acids beneficial to human health. Lipids 39:12.

Loor, J. J., Herbein, J. H. and Jenkins, T. C. 2002a. Nutrient digestion, biohydrogenation, and fatty acid profiles in blood plasma and milk fat from lactating Holstein cows fed canola oil or canolamide. Anim.

Feed Sci. Tech. 97:65-82.

Loor, J. J., Quinlan, L. E. and Herbein, J. H. 2002b. Distribution of trans-vaccenic acid and cis9, trans11-conjugated linoleic acid (rumenic acid) in blood plasma lipid fractions and secretion into milk of Jersey cows fed canola or soybean oil. Anim. Res. 51:119-34.

Lorenzon, M. 2015. Product based on conjugated linoleic acid and a method for the manufacture thereof. US patent 9034385 B2.

Lourenço, M., Ramos-Morales, E. and Wallace, R. J. 2010. The role of microbes in rumen lipolysis and biohydrogenation and their manipulation. Animal 4:1008-23.

Lundy, F. P., Block, E., Bridges, W. C., et al., 2004. Ruminal biohydrogenation in Holstein cows fed soybean fatty acids as amides or calcium salts. J. Dairy Sci. 87:1038-46.

Maia, M. R. G., Chaudhary, L. C., Figueres, L., et al., 2007. Metabolism of polyunsaturated fatty acids and their toxicity to the microflora of the rumen. Anton. Leeuw. 91:303-14.

Maia, M. R. G., Chaudhary, L. C., Bestwick, C. S., et al., 2010. Toxicity of unsaturated fatty acids to the biohydrogenating ruminal bacterium, *Butyrivibrio fibrisolvens*. BMC Microbiol. 10:52.

Mao, S. Y., Zhang, R. Y., Wang, D. S., et al., 2013. Impact of subacute ruminal acidosis (SARA) adaptation on rumen microbiota in dairy cattle using pyrosequencing. Anaerobe 24:12-19.

Massart-Leën, A. M., Roets, E., Peeters, G., et al., 1983. Propionate for fatty acid synthesis by the mammary gland of the lactating goat. J. Dairy Sci. 66:1445-54.

Maxin, G., Glasser, F., Hurtaud, C., et al., 2011. Combined effects of trans-10, cis-12 conjugated linoleic acid, propionate, and acetate on milk fat yield and composition in dairy cows. J. Dairy Sci. 94:2051-9.

McIntosh, F. M., Shingfield, K. J., Devillard, E., et al., 2009. Mechanism of conjugated linoleic acid and vaccenic acid formation in human faecal suspensions and pure cultures of intestinal bacteria. Microbiol. 155:285-94.

McKain, N., Shingfield, K. J. and Wallace, R. J. 2010. Metabolism of conjugated linoleic acids and 18:1 fatty acids by ruminal bacteria:products and mechanisms. Microbiol. 156:579-88.

Meade, T. L., Plakias, R. C. and Auth, J. C. 1999. Rumen by-pass feed supplement. US patent 5928687 A.

Meyer, B. J., Mann, N. J., Lewis, J. L., et al., 2003. Dietary intakes and food sources of omega-6 and omega-3 polyunsaturated fatty acids. Lipids 38:391-8.

Mickdam, E., Khiaosa-ard, R., Metzler-Zebeli, B. U., et al., 2016. Rumen microbial abundance and fermentation profile during severe subacute ruminal acidosis and its modulation by plant derived alkaloids in vitro. Anaerobe 39:4-13.

Moallem, U., Lehrer, H., Zachut, M., et al., 2010. Production performance and pattern of milk fat depression of high-yielding dairy cows supplemented with encapsulated conjugated linoleic acid. Animal 4:641-52.

Moore, J. H. and Christie, W. W. 1981. Lipid metabolism in the mammary gland of ruminant animals. In:Christie, W. W. (Ed.), Lipid Metabolism in Ruminant Animals. Pergamon Press, Oxford, UK, pp. 227-77.

Nagaraja, T. G. and Lechtenberg, K. F. 2007. Acidosis in feedlot cattle. Vet. Clin. Food Anim. 23:333-50.

Nagaraja, T. G., Bartley, E. E., Fina, L. R., et al., 1978a. Evidence of endotoxins in the rumen bacteria of cattle fed hay or grain. J. Anim. Sci. 47:226-34.

Nagaraja, T. G., Bartley, E. E., Fina, L. R., et al., 1978b. Quantitation of endotoxin in cell-free rumen fluid of cattle. J. Anim. Sci. 46:759-1767.

Nagaraja, T. G., Fina, L. R., Lassman, B. A., et al., 1979. Characterization of endotoxin from the rumen bacterium *Megasphaera elsdenii*. Am. J. Vet. Res. 40:35-9.

Noble, R. C. 1978. Digestion, absorption and transport of lipids in ruminant animals. Prog. Lipid Res. 17: 55-91.

Odens, L. J., Burgos, R., Innocenti, M., et al., 2007. Effects of varying doses of supplemental conjugated linoleic acid on production and energetic variables during the transition period. J. Dairy Sci. 90: 293-305.

Palmquist, D. L. and Jenkins, T. C. 1987. Process for feeding ruminant animals and composition for use therein. US patent 4642317 A.

Palmquist, D. L. and Jenkins, T. C. 2017. A 100-year review:fat feeding of dairy cows. J. Dairy Sci. 100: 10061-77.

Prado, L. A., Schmidely, Ph., Nozière, P., et al., 2019. Milk saturated fatty acids, odd- and branched-chain fatty acids, and isomers of C18:1, C18:2, and C18:3n-3 according to their duodenal flows in dairy cows:a meta-analysis approach. J. Dairy Sci. 102:3053-70.

Pappritz, J., Lebzien, P., Meyer, U., et al., 2011. Duodenal availability of conjugated linoleic acids after supplementation to dairy cow diets. Eur. J. Lipid Sci. Technol. 113:1443-55.

Perfield, J. W., Lock, A. L., Pfeiffer, A. M., et al., 2004. Effects of amide-protected and lipid-encapsulated conjugated linoleic acid (CLA) supplements on milk fat synthesis. J. Dairy Sci. 87:3010-16.

Petri, R. M., Schwaiger, T., Penner, G. B., et al., 2013. Changes in the rumen epimural bacterial diversity of beef cattle as affected by diet and induced ruminal acidosis. Appl. Environ. Microbiol. 79: 3744-55.

Pitta, D. W., Indugu, N., Vecchiarelli, B., et al., 2018. Alterations in ruminal bacterial populations at induction and recovery from diet-induced milk fat depression in dairy cows. J. Dairy Sci. 101:295-309.

Plaizier, J. C., Krause, D. O., Gozho, G. N., et al., 2008. Subacute ruminal acidosis in dairy cows:the physiological causes, incidence and consequences. Vet. J. 176:21-31.

Plaizier, J. C., Khafipour, E., Li, S., et al., 2012. Subacute ruminal acidosis (SARA), endotoxins and health consequences. Anim. Feed Sci. Tech. 172:9-21.

Plaizier, J. C., Li, S., Tun, H. M., et al., 2017. Nutritional models of experimentally-induced subacute ruminal acidosis (SARA) differ in their impact on rumen and hindgut bacterial communities in dairy cows. Front. Microbiol. 7:2128.

Prom, C. M. and Lock, A. L. 2019. Abomasal infusion of different exogenous emulsifiers alters fatty acid digestibility and milk fat yield of lactating dairy cows. J. Dairy Sci. 102 (Suppl. 1):318 (Abstract).

Rico, J. E., de Souza, J., Allen, M. S., et al., 2017. Nutrient digestibility and milk production responses to increasing levels of palmitic acid supplementation vary in cows receiving diets with or without whole cottonseed. J. Anim. Sci. 95:436-46.

Rigout, S., Hurtaud, C., Lemosquet, S., et al., 2003. Lactational effect of propionic acid and duodenal glucose in cows. J. Dairy Sci. 86:243-53.

Rosenberg, M. and DePeters, E. J. 2010. Method and compositions for preparing and delivering rumen protected lipids, other nutrients and medicaments. US patent 7700127 B2.

Roy, A., Chardigny, J. M., Bauchart, D., et al., 2007. Butters rich either in trans-10-C18:1 or in trans-11-C18:1 plus cis-9, trans-11 CLA differentially affect plasma lipids and aortic fatty streak in experimental atherosclerosis in rabbits. Animal 1:467-76.

Russell, J. B. and Rychlik, J. L. 2001. Factors That Alter Rumen Microbial Ecology. Science 292: 1119-22.

Schmidely, P., Glasser, F., Doreau, M., et al., 2008. Digestion of fatty acids in ruminants: a meta-analysis of flows and variation factors. 1. Total fatty acids. Animal 2:677-90.

Schwarz, F. J., Lierman, T., Möcker, P., et al., 2009. Performance, metabolic parameters and fatty acid composition of milk fat due to dietary CLA and rumen protected fat of dairy cows. Book of abstracts No. 15 of the 60th Annual Meeting of the European Association for Animal Production, 24-27 August 2009, Wageningen Academic Pubishers, Barcelona, Spain, p. 351.

Scott, T. W. and Hills, G. D. L. 1975. Feed supplements for ruminants comprising lipid encapsulated with protein-aldehyde reaction product. US patent 3925560 A.

Sharp, M. E., Latham, M. J., Garvie, E. I., et al., 1973. Two new species of *Lactobacillus* isolated from the bovine rumen, *Lactobacillus ruminis* sp. nov. and *Lactobacillus vitulinus* sp. nov. J. Gen. Microbiol. 77:37-49.

Shennan, D. B. and Peaker, M. 2000. Transport of milk constituents by the mammary gland. Physiol. Rev. 80:925-51.

Shingfield, K. J. and Griinari, J. M. 2007. Role of biohydrogenation intermediates in milk fat depression. Eur. J. Lipid Sci. Technol. 109:799-816.

Shingfield, K. J. and Wallace, R. J. 2014. Synthesis of conjugated linoleic acid in ruminants and humans. In: Sels, B. and Philippaerts, A. (Eds), Conjugated Linoleic Acids and Conjugated Vegetable Oils. Royal Society of Chemistry (RSC) Publishing, pp. 1-65. Chapter 1.

Shingfield, K. J., Kairenius, P., Ärölä, A., et al., 2012. Dietary fish oil supplements modify ruminal biohydrogenation, alter the flow of fatty acids at the omasum, and induce changes in the ruminal *Butyrivibrio population* in lactating cows. J. Nutr. 142:1437-48.

Shingfield, K. J., Bonnet, M. and Scollan, N. D. 2013. Recent developments in altering the fatty acid composition of ruminant-derived foods. Animal 7:132-62.

Simopoulos, A. P. 2001. n-3 fatty acids and human health: Defining strategies for public policy. Lipids 36: S83-9.

Sioen, I. A., Pynaert, H., Matthys, C., et al., 2006. Dietary intakes and food sources of fatty acids for Belgian women, focused on n-6 and n-3 polyunsaturated fatty acids. Lipids 41:415.

Stamey, J. A., Shepherd, D. M., de Veth, M. J., et al., 2012. Use of algae or algal oil rich in n-3 fatty acids as a feed supplement for dairy cattle. J. Dairy Sci. 95:5269-75.

Stefanov, I., Baeten, V., Abbas, O., et al., 2011. Determining milk isolated and conjugated trans unsaturated fatty acids using Fourier transform Raman spectroscopy. J. Agric. Food Chem. 59:12771-83.

Sterk, A., Vlaeminck, B., van Vuuren, A. M., et al., 2012. Effects of feeding different linseed sources on omasal fatty acid flows and fatty acid profiles of plasma and milk fat in lactating dairy cows. J. Dairy Sci. 95:3149-65.

Sukhija, P. S. and Palmquist, D. L. 1990. Dissociation of calcium soaps of long-chain fatty-acids in rumen fluid. J. Dairy Sci. 73:1784-7.

Sultana, H., Ishida, T., Shintaku, T., et al., 2008 Effect of feeding Ca-salts of fatty acids from soybean oil and linseed oil on c9, t11-CLA production in ruminal fluid and milk of Holstein dairy cows. Asian

Austral. J. Anim. 21:1262-70.

Theurer, M. L., Block, E., Sanchez, W. K., et al., 2009. Calcium salts of polyunsaturated fatty acids deliver more essential fatty acids to the lactating dairy cow. J. Dairy Sci. 92:2051-6.

Toral, P. G., Bernard, L., Belenguer, A., et al., 2016. Comparison of ruminal lipid metabolism in dairy cows and goats fed diets supplemented with starch, plant oil, or fish oil. J. Dairy Sci. 99:301-16.

Toral, P. G., Hervás, G., Peiró, V., et al., 2018a. Conditions associated with marine lipid-induced milk fat depression in sheep cause shifts in the *in vitro* ruminal metabolism of 1-13C oleic acid. Animals 8:196.

Toral, P. G., Monahan, F. J., Hervas, G., et al., 2018b. Review:Modulating ruminal lipid metabolism to improve the fatty acid composition of meat and milk. challenges and opportunities. Animal 12:S272-81.

Toral, P. G., Hervás, G. and Frutos, P. 2019. In vitro biohydrogenation of 13C-labeled α-linolenic acid in response to ruminal alterations associated with diet-induced milk fat depression in ewes. J. Dairy Sci. 102:1213-23.

Tyburczy, C., Major, C., Lock, A. L., et al., 2009. Individual trans octadecenoic acids and partially hydrogenated vegetable oil differentially affect hepatic lipid and lipoprotein metabolism in golden Syrian hamsters. J. Nutr. 139:257-63.

Van Nevel, C. J. and Demeyer, D. I. 1996. Effect of pH on biohydrogenation of polyunsaturated fatty acids and their Ca-salts by rumen microorganisms in vitro. Arch. Tierernahr. 49:151-7.

Van Valenberg, H. J. F., Hettinga, K. A., Dijkstra, J., et al., 2013. Concentrations of n-3 and n-6 fatty acids in Dutch bovine milk fat and their contribution to human dietary intake. J. Dairy Sci. 96:4173-81.

Van Vuuren, A. M., van Wikselaar, P. G., van Riel, J. W., et al., 2010. Persistency of the effect of long-term administration of a whey protein gel composite of soybean and linseed oils on performance and milk fatty acid composition of dairy cows. Livest. Sci. 129:213-22.

Van Wijlen, R. P. J. and Colombani, P. C. 2010. Grass-based ruminant production methods and human bioconversion of vaccenic acid with estimations of maximal dietary intake of conjugated linoleic acids. Int. Dairy J. 20:433-48.

Verbeke, R., Lauryssens, M., Peeters, G., et al., 1959. Incorporation of DL-[1-14C] leucine and [1-14C] isovaleric acid into milk constituents by the perfused cow's udder. Biochem. J. 73:24-9.

Villot, C., Meunier, B., Bodin, J., et al., 2018. Relative reticulo rumen pH indicators for subacute ruminal acidosis detection in dairy cows. Animal 12:481-90.

Vlaeminck, B., Fievez, V., Cabrita, A. R. J., et al., 2006. Factors affecting odd-and branched-chain fatty acids in milk:a review. Anim. Feed Sci. Technol. 131:389-417.

Vlaeminck, B., Gervais, R., Rahman, M. M., et al., 2015. Postruminal synthesis modifies the odd-and branched chain fatty acid profile from the duodenum to milk. J. Dairy Sci. 98:4829-40.

Wall, R., Ross, R. P., Fitzgerald, G. F., et al., 2010. Fatty acids from fish:the anti-inflammatory potential of long-chain omega-3 fatty acids. Nutr. Rev. 68:280-9.

Wallace, R. J., McKain, N., Shingfield, K. J., et al., 2007. Isomers of conjugated linoleic acids are synthesized via different mechanisms in ruminal digesta and bacteria. J. Lipid Res. 48:2247-54.

Weimer, P. J., Stevenson, D. M. and Mertens, D. R. 2010. Shifts in bacterial community composition in the rumen of lactating dairy cows under milk fat-depressing conditions. J. Dairy Sci. 93:265-78.

Weinstein, J. A., Taylor, S. J., Rosenberg, M., et al., 2015. Whey protein gel composites in the diet of goats increased the omega-3 and omega-6 content of milk fat. J. Anim. Physiol. An. N. 100:789-800.

Wu, S. H. W. and Papas, A. 1997. Rumen-stable delivery systems. Adv. Drug Deliver. Rev. 28:323-34.

Zened, A., Combes, S., Cauquil, L., et al., 2013. Microbial ecology of the rumen evaluated by 454 GS FLX pyrosequencing is affected by starch and oil supplementation of diets. FEMS Microbiol. Ecol. 83: 504-14.

Zened, A., Meynadier, A., Cauquil, L., et al., 2016. Trans-11 to trans-10 shift of ruminal biohydrogenation of fatty acids is linked to changes in rumen microbiota. Abstr. P-202 (p. 150). Gut Microbiology 2016:Proceedings of the 10th Joint Symposium on Gut Microbiology, 20-23 June 2016, Clermont-Ferrand, France.

Zhu, H., Fievez, V., Mao, S., et al., 2016. Dose and time response of ruminally infused algae on rumen fermentation characteristics, biohydrogenation and *Butyrivibrio* group bacteria in goats. J. Anim. Sci. Biotechnol. 7:22.

第15章 反刍动物温室气体产生的营养因素：对肠道和粪便甲烷排放的影响

Stephanie A. Terry，加拿大农业及农业食品部，澳大利亚悉尼大学；
Carlos M. Romero，加拿大农业及农业食品部，加拿大莱斯布里奇大学；
Alex V. Chaves、Tim A. McAllister，加拿大农业及农业食品部
（谭翠译）

1 前言

畜牧业是温室气体（GHGs）的主要来源之一，约占农业温室气体排放总量的40%（IPCC，2006）（图1）。据估计，畜牧业每年产生7.1 Gt 的 CO_2 当量温室气体，其中动物生产和粪便处理分别占 26.8% 和 31.0%（Gerber 等，2013）。甲烷（CH_4）和氧化亚氮（N_2O）是畜牧业直接排放的两种主要温室气体，其全球变暖潜力分别是 CO_2 的 28 倍和 298 倍（Gerber 等，2013）。据估计，畜牧业 CH_4 和 N_2O 排放分别占农业排放量的 40% 和 48%，而反刍动物占畜牧业排放总量的 80%（Opio 等，2013）。动物肠道发酵和畜禽粪便处理产生的 CH_4 分别占畜禽生产相关 CH_4 排放量的 82% 和 18%。N_2O 的排放主要源自化学肥料、施肥以及来自圈养动物和粪便储存的氮沉积（Adler 等，2015）。

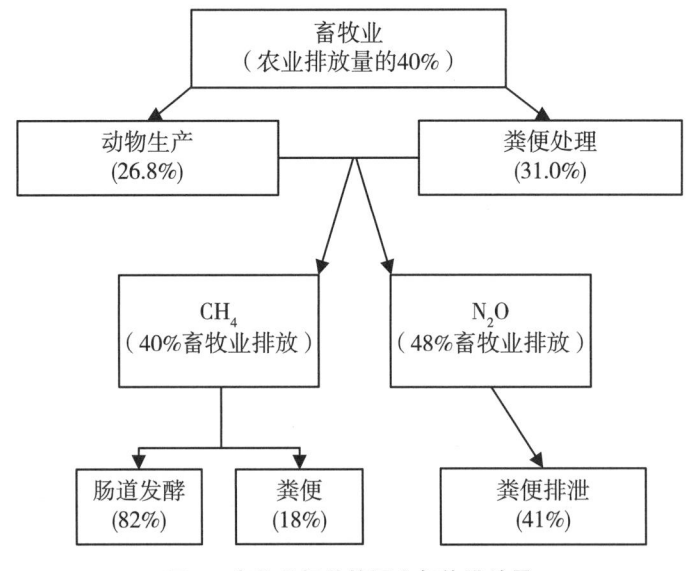

图1　畜牧业相关的温室气体排放量

第15章 反刍动物温室气体产生的营养因素：对肠道和粪便甲烷排放的影响

1.1 温室气体生成

反刍动物产生的 CH_4 是瘤胃微生物发酵产生的副产物，其生化途径有据可查（Huws 等，2018）。瘤胃微生物菌群将淀粉、细胞壁聚合物和蛋白质发酵成单糖和碳骨架，这些物质在厌氧条件下通过初级和次级发酵被转化为挥发性脂肪酸（VFA）、CO_2 和代谢氢 ［H］。瘤胃纤毛虫和厌氧真菌是两类真核微生物，它们产生大量的代谢氢 ［H］，并与古细菌共生（Guyader 等，2014）。原虫和真菌都含有将单糖发酵中间体转化为氢 ［H］ 的特殊细胞器——氢化酶体。产甲烷菌在维持氢 ［H］ 的低分压方面起着重要作用，有利于保持氢化酶体内的氢化酶活性。在发酵过程中，还原型辅助因子 NADH、NADPH 和 FADH 被氧化，释放代谢氢 ［H］，CO_2 通过一系列生化途径被产甲烷菌中的代谢氢 ［H］ 还原为 CH_4（Ungerfeld，2015b）。

粪便 CH_4 与肠道 CH_4 一样，都是有机物（OM）厌氧分解产生的。粪便也是 N_2O、氨（NH_3）和氮氧化物（NOx）的重要排放源，这些气体可能是温室气体和环境污染物的直接或间接来源，影响 N_2O、NH_3 和 NOx 浓度的因素包括饲料类型、粪便养分和粪便处理与存储。N 转化为气体是通过同步的硝化和反硝化过程（图2），NH_3 氧化细菌（如 β-变形菌或 NH_3-氧化古细菌 Thaumarchaeota）和亚硝酸盐（NO_2^-）氧化细菌（即 α-变形菌硝基杆菌和亚硝基螺旋体）都能发生硝化作用，反硝化细菌在系统发育上具有多样性，且具有特定基因编码其催化酶（Maeda 等，2011）。

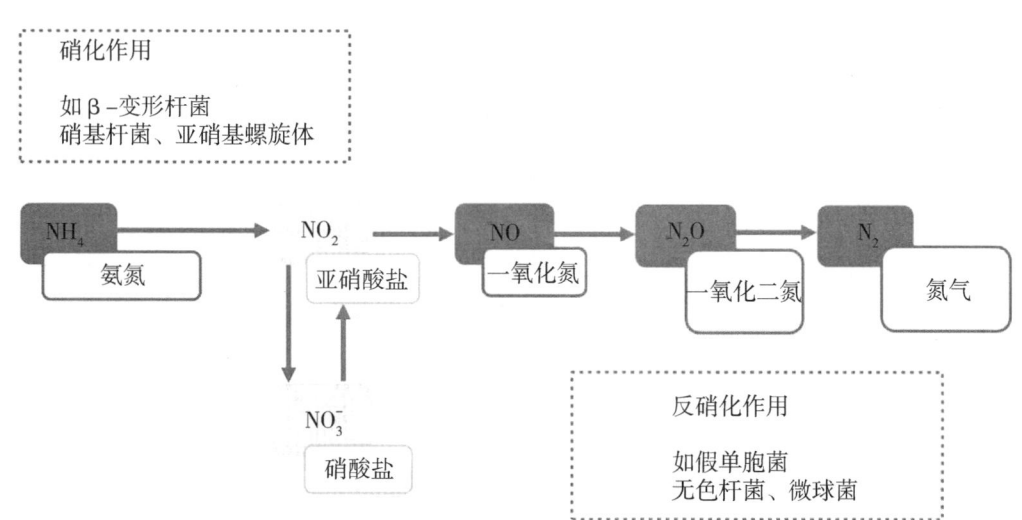

图2 亚硝酸盐途径的硝化和反硝化过程。
N_2O 是一种温室气体，其全球变暖潜势是 CO_2 的 298 倍。

虽然粪便中排放的 NH_3 不是 GHG 的直接来源，但在评估饲料添加剂对空气质量的影响时仍需考虑。氨来自尿液中尿素的快速水解，也可能是 N_2O 的前体物。氨极易挥发，当浓度超过 25~35 mg/kg 的阈值上限时，会对人体健康产生严重影响（NRC，2008）。此外，干 NH_3 或湿 NH_3 沉积可能会导致土壤酸化和地表水富营养化（Hünerberg 等，2013a）。

由于粪 N 是一种缓释形式的 N，更容易被土壤菌群利用，那么将尿 N 转为粪 N 可能更有益于环境。

1.2 平衡肠道甲烷产生和粪便甲烷排放

肠道 CH_4 产量与粪便 CH_4 和 N_2O 排放量之间存在显著的相关关系（Knapp 等，2014）。平衡瘤胃直接产生的甲烷或粪便间接产生的温室气体之间的 GHG 净排放量是具有挑战性的。

在评估日粮调控措施时，需要考虑的另外一个重要因素是 GHG 产量和 GHG 排放强度的差异。尽管全球反刍动物的 GHG 排放量已经减少，但 GHG 的总产量在增加并将继续增加，因为全世界家养反刍动物数量预计将从 32 亿只（头）增加到 2050 年的 53 亿只（头）（Turk，2016）。

与牧场放牧的牛（低 CH_4 产量和高 CH_4 排放强度）相比，养殖场中饲喂标准日粮的牛的 CH_4 产量（g/d）可能更高，但 CH_4 排放强度降低（g/kg 消耗物）。然而，由于日粮能量浓度低，放牧饲养的反刍动物产生的粪便 CH_4 产量只有养殖场反刍动物的一半（Koneswaran 和 Nierenberg，2008），因为放牧动物粪便中的淀粉含量更低（Hales 等，2013）。此外，使微生物发酵从瘤胃转移到后肠的日粮改变可能会降低肠道的 CH_4 产量，但不会改变总的 GHG 净排放量。这个概念被称为污染交换，指一种 GHG 产量的改变导致相同的 GHG 或另外一种 GHG 排放量在其上游或下游的变化（Hristov 等，2013）。通过增加日粮可发酵碳水化合物、氮和脂肪含量来改变日粮消化率的营养调控措施都可能导致污染交换（图3）。

日粮变化	肠道							粪便					总体	
	发酵	纤维降解	N降解	淀粉降解	pH值	CH_4日产量	CH_4强度	淀粉	尿N	粪N	CH_4排放量	N_2O排放量	GHG产量降低	GHG强度降低
精粗比														
精粗比增加	↑	↓	NA	↑	↓	~	↓	↑	NA	NA	↑	~	~	√
酸中毒	↑	↓	NA	↑	↓	~	-	~	NA	NA	~	~	~	~
高粗料	↓	↑	NA	↓	↑	↑	↑	↓	NA	NA	↓	~	~	×
N														
DDGS	~	↑	↑	~	~	↓	↓	↑	↑	↑	↑	↑	×	×
脂肪														
<6%	~	-	NA	-	-	↓	↓	-	NA	NA	-	~	√	√
>6%	↓	↓	NA	~	↑	↓	↓	↑	NA	NA	↑	~	√	×
抑制剂														
硝酸盐	~	-	-	~	-	↓	↓	~	~	~	~	~	√	√
3NOP	-	-	-	-	-	↓	↓	~	~	~	~	~	√	√
植物次级代谢物														
单宁	↓	↓	↓	NA	NA	↓	↓	~	↓	↑	~	↑	√	√

图3 日粮调控对肠道 CH_4 产量和温室气体排放量的影响

符号含义：↑=增加，↓=减少，-=没变化，NA=不适用，~=可变的/未知的。

1.2.1 日粮消化率和可发酵碳水化合物

日粮消化率与肠道 GHG 和粪便 GHG 的产量是内在相联的。日粮越容易发酵，营养成分浪费和 GHG 排放强度就越低。粗料品质、精粗料比例以及精粗料饲料类型等因素都会影响微生物对饲料的降解效率。例如，增加日粮中精料浓度可以通过增加易发酵碳水化合物的比例来减少肠道 CH_4 产量,这是将发酵模式从产生氢 [H] 的乙酸发酵转变为利用氢 [H] 的丙酸发酵,从而降低甲烷生成所需的代谢氢 [H] 含量。在日粮中增加可发酵碳水化合物含量也可以增加饲料的消化率和通过率,这既能提高生产性能,又能减少粪便排泄物中的有机物含量。粪便中的 OM 减少会降低可用于分解的底物数量,从而减少产甲烷菌的氢 [H] 供应,最终降低粪便中的 CH_4 产量。日粮的消化率增加,伴随着更多底物在瘤胃中发酵和产生的还原当量增加,动物每天的肠道 CH_4 产量(g/d)也随之增加。例如,给放牧饲养的泌乳奶牛补充一定水平(0~8 kg/d)的精料,奶牛的 CH_4 产量(g/d)和产奶量都有所增加(Wyngaard 等,2018;Muñoz 等,2015)。

随着日粮中精料水平的增加,瘤胃中降解非纤维碳水化合物的内毛属纤毛原虫(*Entodinium*)数量增加,而降解纤维素的微生物:纤维杆菌属(*Fibrobacter*)、多甲属(*Polyplastron*)和硬甲属(*Ostracodinium*)纤毛原虫数量减少(Zhang 等,2017)。高粗料与高精料日粮相比,真菌菌群的多样性和丰度相似;然而,随着饲料中精料比例的增加,真菌 *Ascomycota*、*Basidiomycota*、*Cercozoa* 和 *Chytridiomycota* 的相对丰度增加,而 *Neocallimastigomycota* 的相对丰度降低(Zhao 等,2018)。其他研究发现日粮中粗料比例增加,瘤胃真菌菌群丰度也随之增加,表明真菌是负责降解瘤胃中的复合纤维(Kumar 等,2015)。由于真核生物会产生甲烷生成所需的氢 [H],真核生物丰度和多样性的变化可能会影响产甲烷菌的丰度和多样性。

高精料日粮降低了古菌群的整体丰度,但没有改变菌群种类的范围(多样性)(Zhang 等,2017;Mao 等,2016)。据报道,反刍动物适应了高粗料日粮转变为高精料日粮后,古菌菌群组成和多样性的变化较小(Henderson 等,2015;Kumar 等,2015),这可能是由于产甲烷菌的低丰度和较稳定的代谢能力对日粮变化不敏感(Kumar 等,2015;Henderson 等,2015)。然而,据报道甲烷微菌目(Methanomicrobium)和甲烷八叠球菌目(Methanocerococcus)对日粮变化敏感,这两类产甲烷菌在高谷物日粮中都不存在,仅在粗饲料日粮中检测到(Friedman 等,2017a)。高谷物日粮中不存在 Methanomicrococcus 可能与瘤胃低 pH 值导致的氧化还原电位增加有关(Friedman 等,2017a)。

含有易发酵底物组成的日粮会导致微生物产生的有机酸超过瘤胃的缓冲能力,导致瘤胃 pH 值长期降低,出现瘤胃酸中毒,其特征是微生物多样性降低和瘤胃功能障碍,包括采食量和饲料消化率下降。出现亚急性瘤胃酸中毒后,羊瘤胃液(Li 等,2017)、奶牛瘤胃液和粪便(Plaizier 等,2017)中细菌的丰度和多样性均降低。高谷物日粮通常会导致淀粉利用菌、乳酸利用菌和丙酸生成菌(普雷沃氏菌、月形单胞菌、链球菌)增加(Plaizier 等,2017;Zhu 等,2018);Henderson 等(2015)发现在饲喂高谷物饲料的反刍动物瘤胃中,普雷沃氏菌属和琥珀弧菌科占主导地位。相反,丁酸弧菌、瘤胃球菌和纤维杆菌等纤维降解菌容易受到低 pH 值影响,高谷物日粮会使这些纤维降解菌的丰度降低(Zhu 等,2018)。Ishaq 等(2017)发现日粮诱导的亚急性酸中毒使稀有真菌类群的丰度和多样性增加,包括与乳酸利用相关的毕赤酵母 *Pichia* 和念珠菌 *Candida*,但也有研究发现高谷物日粮导致真菌的多样性下降(Kumar 等,2015;Tapio 等,2017a)。Hook 等(2011)研究发现瘤胃古菌的丰度不会随着日粮精料的增加和瘤胃 pH 值的降低而改变,这表明古菌能够适应 pH 值的变化,低 pH 值条件仅抑制其功能活性,对古菌的丰度影响不大，这可以解释产甲烷菌丰度与 CH_4 产量之间的相关关系差的原因（Fikins 和 Yu，2015）。

改变瘤胃微生物降解饲料的能力会增加粪便中的养分损失，增加粪便中可用于生成 CH_4 的 OM 含量。尽管增加日粮中过瘤胃淀粉浓度可以调控瘤胃发酵，且可能减少甲烷生成，但后消化道中的淀粉消化也可能受到限制（<60%）（Haque，2018），这就导致粪便中含有更多淀粉，进而可能增加粪便分解过程中的 CH_4 排放量。在饲喂加工谷物和饲草饲料的肉牛粪便中，瘤胃球菌科含量更高；而在饲喂未加工谷物的肉牛粪便中，普雷沃氏菌属占主导地位（Shanks 等，2011）。随着粪便中淀粉含量的增加，拟杆菌增加，而厚壁菌减少，拟杆菌是消化复杂碳水化合物的主要代表菌。饲喂未加工谷物和加工谷物的肉牛粪便中的淀粉浓度分别比饲喂草料的肉牛高 98.4% 和 66.9%，这表明谷物加工不足会增加粪便中的 CH_4 排放量（Shanks 等，2011）。

提高日粮消化率，即提高反刍动物从日粮中获取营养物质的潜力，从而提高动物生长效率，这可能会导致 GHG 总排放量的增加或减少，具体取决于产生的气体类型以及肠道排放和粪便排放之间的平衡。无论如何，当动物生产性能提高（即饲喂天数减少）时，GHG 实际排放量减少，GHG 排放强度会降低，即生产单位产品（肉、牛奶、羊毛等）的 GHG 排放量降低（Hristovet 等，2013）。

1.2.2　氮含量

反刍动物是全球氮排放的重要贡献者。在反刍动物中，氮循环是指饲料、粪便和土壤中的无机氮和有机氮发生的一系列复杂的生物地球化学反应（图 4；Robertson 和 Vitousek，2009）。植物和动物在整个循环中利用氮，但它们以产品形式沉积氮的能力有限。如反刍动物将日粮氮转化为可食用蛋白质（牛奶、肉）的转化率非常低（20%～30%），谷类作物对肥料中氮的利用率很少超过 50%（Fageria 和 Baligar，2005）。来自畜牧业的过量氮则以大量过剩氮释放到环境中，主要通过 NH_3 和 N_2O 排放和/或硝态氮（NO_3-N）的排放（Galloway 等，2004；Powell 等，2011）。NO_3-N 对地下水的污染、广泛的富营养化和 N_2O 的全球变暖效应都是农业中人类活动产生的氮的贡献（Erisman 等，2013）。

反刍动物系统中的氮循环主要是由饲料氮转化为牛奶或肉制品，剩余的氮通过尿液或粪便排出。尿液和粪便中的 N 浓度取决于日粮中的粗蛋白（CP）含量（Dijkstra 等，2013）。根据反刍动物代谢蛋白质（MP）需要来提供相应水平的氮可确保最佳利用和最少的营养损失（Broderick，2003）。反刍动物饲料中的氮含量和类型也对反刍动物如何利用和排泄氮有多种影响。提供的日粮蛋白质包括瘤胃可降解蛋白（RDP）和不可降解蛋白质（RUP），瘤胃可降解蛋白由真蛋白 N 和非蛋白 N 组成，它们被降解后用于微生物蛋白（MCP）的合成和微生物的生长（Bach 等，2005）。反刍动物对日粮蛋白和能量的需求是相互关联的，因为高能量日粮会促进微生物合成，增加动物对瘤胃可降解蛋白的需求（Broderick，2003）。虽然改变日粮中的 CP 含量对肠道 CH_4 排放无明显效应，但用碳水化合物替代 CP 则会影响 CH_4 排放。

用可发酵碳水化合物替代日粮的蛋白质供应是减少尿氮排泄、增加微生物氮捕获和减少 NH_3 排放量的有效方法（Dijkstra 等，2013）。然而，用可发酵碳水化合物替代日粮 CP

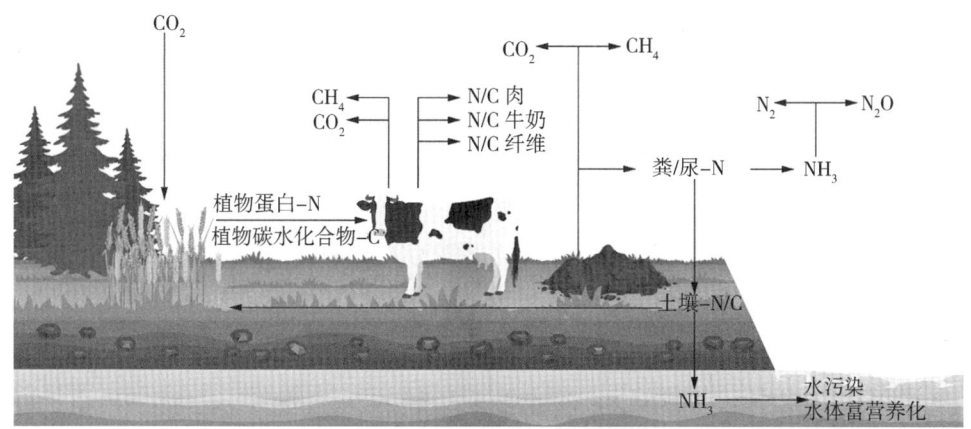

图 4　反刍动物农业中的碳氮循环

会增加肠道 CH_4 产量（Sauvant 等，2011），因为瘤胃中可用于生成甲烷的底物增加。Dijkstra 等（2013）估计，增加日粮中碳水化合物导致的肠道 CH_4 排放量的增加通常被粪便中 N_2O 排放量的减少所抵消。

普雷沃氏菌属（*Prevotella*）是全球反刍动物瘤胃中的主要菌属（Henderson 等，2015），主要参与碳水化合物和 N 的代谢。具体而言，栖瘤胃普雷沃氏菌菌株 23（*Prevotella ruminicola*）可以有效地降解半纤维素和果胶，利用 NH_3-N 和肽作为其生长的氮源，而不是氨基酸（Kim 等，2017）。Belanche 等（2012）研究发现饲喂低蛋白日粮的奶牛瘤胃中布氏普雷沃氏菌（*Prevotella bryantii*）的丰度下降，这表明布氏普雷沃氏菌属在氮代谢中起着重要作用；白色瘤胃球菌、黄色瘤胃球菌、产琥珀酸丝状杆菌和溶纤维丁酸弧菌的相对丰度都下降，表明这些纤维降解菌易受 N 缺乏的影响；相比之下，栖瘤胃普雷沃氏菌、反刍月形单胞菌、牛链球菌、埃氏巨球型菌和 *Aliiglaciecola lipolytica* 等非纤维降解菌不受 N 缺乏的影响，表明低浓度的 NH_3-N 就能满足它们的生长需要（Belanche 等，2012）。Niu 等（2016）评估了两个不同 CP 浓度（15.2% 与 18.5%）日粮对奶牛的影响，发现与高蛋白日粮（18.5%）相比，低蛋白日粮降低了 OM、N 和淀粉的总消化率。Belanche 等（2012）发现低蛋白日粮使瘤胃中原虫和产甲烷菌浓度下降，而 Niu 等（2016）发现不同蛋白含量的日粮并未影响动物的 CH_4 排放量。

日粮蛋白质的降解及微生物蛋白质的合成会降低甲烷生成所需的可用氢［H］，因为该过程既能利用也能产生还原当量（Knapp 等，2014）。如氨基酸的合成增加可以降低甲烷生成，因为氨基酸可充当氢［H］池（Ungerfeld，2015b）。氨基酸产量的增加可能与拟杆菌和普雷沃氏菌的相对丰度增加有关，这两类菌都与蛋白水解活性增加有关（Martinez-Fernandez 等，2016）。增加日粮中可溶性碳水化合物的比例与支链脂肪酸的减少有关，支链脂肪酸是瘤胃微生物从头合成氨基酸所必需的。降低支链脂肪酸的产量可能会减少微生物蛋白质合成和微生物生长（Hall 和 Huntington，2008），从而降低充当甲烷生成替代［H］汇的储量。

日粮蛋白质含量与瘤胃甲烷生成之间的潜在作用尚不清楚。在高纤维日粮中添加蛋白质可以提高微生物蛋白质的合成效率并降低肠道 CH_4 的排放强度,同时增加 N 排泄和粪便中的 N_2O 排放。提高微生物蛋白质合成效率可以将氢 [H] 从甲烷生成转向微生物细胞形成,并提高反刍动物的生产力。

1.2.3 日粮脂肪

以参与甲烷生成的微生物作为调控目标的添加剂可能具有减少肠道 CH_4 产量而又不影响粪便甲烷排放的优势。例如,日粮脂肪可能会通过以下几种方式降低肠道 CH_4 产量:①对产甲烷菌和原虫的毒性作用,②替代可发酵碳水化合物或③通过生物氢化提供可替代的氢 [H] 池(Beauchemin 等,2008;Knapp 等,2014)。

日粮脂质对瘤胃的调控作用很大程度上取决于脂肪成分、浓度和来源,因此,日粮脂肪对瘤胃微生物种群的影响因脂类的性质而异。反刍兽甲烷短杆菌(*Methanobrevibacter ruminantium*)是瘤胃最丰富的产甲烷菌种类,研究发现饱和脂肪酸和油酸可以降低其丰度(Henderson 等,2015;Enjalbert 等,2017)。研究发现添加亚麻籽和椰子油使 CH_4 产量降低,但这与古菌群丰度或多样性的变化不相关(Patra 和 Yu,2013;Martin 等,2016)。同样,给羔羊饲喂亚麻籽油可以增加琥珀酸弧菌科(琥珀酸生产者)和韦荣氏球菌科(丙酸生产者)的相对丰度,降低瘤胃球菌科(纤维降解菌)的丰度(Lyons 等,2017)。琥珀酸弧菌科的丰度增加与低 CH_4 排放相关,而瘤胃球菌科的丰度增加与高 CH_4 排放相关(Wallace 等,2015)。琥珀酸弧菌科通过利用代谢氢 [H] 产生琥珀酸,而瘤胃球菌科则是已知的产氢菌(Wallace 等,2015)。给羔羊饲喂亚麻籽油使甲烷短杆菌的相对丰度降低 19.5%,甲烷球菌的相对丰度增加 34.7%,但该研究中没有测定 CH_4 产量(Lyons 等,2017)。

豆油可以抑制瘤胃真菌 *Neocallimastix frontalis* 的生长(Boots 等,2013),脂类也会抑制纤维降解菌(纤维杆菌和瘤胃球菌)的生长(Enjalbert 等,2017)。增加日粮中不饱和脂肪酸的含量可能与原虫数量降低相关(Oldick 和 Firkins,2000),由于不饱和脂肪酸破坏了原虫和古菌之间的密切关系,进而改变古菌群的多样性(Hristov 等,2012)。

由于日粮脂肪变化导致的瘤胃微生物组成改变可能会引起动物生理反应的高度变化,包括纤维降解受到抑制,进而降低 CH_4 产量。研究证实,反刍动物日粮中脂肪含量增加到 6%~7% 以上则会降低纤维的消化率(Johnson 和 Johnson,1995)。在连续培养试验中,油酸含量高的一种油(Tucumã)通过抑制纤维杆菌来降低 CH_4 产量,而没有改变产甲烷菌丰度(Ramos 等,2018)。降低纤维消化率会增加粪 C,粪 C 是粪便 CH_4 生成的底物。但 Gautam 等(2016)发现,不同来源的干酒糟可溶物(DDGS)和玉米油(日粮脂肪为 3%~5.5% DM)对粪中营养成分或温室气体排放没有影响。

添加日粮脂肪是一种有效的肠道 CH_4 减排措施,日粮脂肪可能会减少每日 CH_4 排放量和 CH_4 的排放强度,这取决于它们的脂肪酸组成。然而,决定日粮脂肪添加水平的主要限制条件是以不降低日粮纤维的降解为前提,一般控制在 6% DM 左右,这样 CH_4 减排潜力可达到 10%~15%。

2 干酒糟及其可溶物(DDGS)

通过评估 DDGS 的研究实例可以体现采用整体评分法研究日粮添加剂效应的重要性。

DDGS是生产乙醇的副产物,由于其脂肪和能量含量高,常用作替代反刍动物日粮中的谷物饲料。玉米酒糟和小麦酒糟由于含油量高,已被证实能降低CH_4产量。根据市场需求,DDGS在低成本日粮中有时可能比谷物饲料更经济。

Griffin等(2012)的荟萃分析发现当用DDGS替代草料时,试验牛的平均日增重和最终体重均增加。黑小麦DDGS可以替代育肥日粮中的大麦(Hordeum vulgare L.)青贮饲料和大麦谷物,对育肥牛的生长性能或胴体质量没有影响。

由于DGGS脂肪含量高,可以有效降低CH_4排放。研究证实用玉米DDGS替代大麦青贮饲料基础日粮中35%(DM基础)的大麦谷物时,试验肉牛的肠道CH_4产量减少16.4%(%DM摄入量)(McGinn等,2009)。类似研究发现用玉米DDGS替代日粮中35%的大麦谷物和5%的双低油菜粕(Brassica napus L.)后,CH_4产量降低15.1%(Hünerberg等,2013a)。Hünerberg等(2013b)在用玉米DDGS替代高谷物育肥日粮中40%的大麦谷物时,肉牛的甲烷排放量也减少了18.0%(%DM摄入量)。尽管这些研究观察到CH_4排放量减少,但用DDGS替代日粮中的谷物饲料显著增加了N的摄入量和排泄量。此外,还有研究证实玉米DDGS会降低淀粉的消化率(Castillo-Lopez等,2014),促使粪便中CH_4的排放量增加。

Castillo-Lopez等(2014,2017)发现日粮中添加DDGS降低了纤维杆菌和拟杆菌的丰度。Castillo-Lopez等(2017)研究发现添加20%的DDGS(DM基础)降低了拟杆菌的丰度,增加了软壁菌门的丰度,值得一提的是,DDGS还增加了瘤胃球菌科的丰度,但Wallace等(2015)认为瘤胃球菌科与CH_4产量增加有关。还有研究发现50%的DDGS(DM基础)使琥珀酸弧菌降低75.4%,拟杆菌和普雷沃氏菌分别增加61.0%和34.6%。

尽管日粮中添加DDGS最初被视为一种有效的GHG减排措施,但使用生命周期方法却发现饲喂玉米或小麦为基础的DDGS日粮的总GHG排放量分别比饲喂大麦对照日粮增加了6.2%和9.3%(Hünerberg等,2014)。尽管粪中CH_4排放量有所减少,但粪中N_2O排放量增加是GHG排放总量增加的主要因素。N_2O排放量的增加是由于日粮中CP含量增加,以及尿N和粪N排泄量增加的结果。

要使DDGS作为一种合适的温室气体减排策略,它的饲喂水平不能超过反刍动物的蛋白质需要量。然而,日粮中较低水平的DDGS不足以达到降低肠道CH_4排放所需要的脂肪浓度(Castillo-Lopez等,2017;Judy等,2016;Hales等,2013)。

3 硝基化合物

硝基化合物具有降低反刍动物肠道CH_4排放的潜力,这些化合物可作为替代电子池利用代谢氢[H]或可能直接抑制产甲烷菌来减少CH_4产生,硝酸盐和3-硝基氧丙醇(3-NOP)是关注较多的两种饲料添加剂,它们能够持续减少肠道CH_4排放。

3.1 硝酸盐

硝酸盐是饲料中存在的一种天然氮,不同类型粗料中的硝酸盐浓度不同。硝酸盐降低肠道CH_4排放的潜力是源于硝酸盐在瘤胃内可充当替代电子池利用代谢氢[H]的能力,

NO_3^- 和 CO_2 均可用作瘤胃中的替代电子受体，它们被还原所释放的能量（ΔG）分别是 371 kJ 和 67 kJ。与产甲烷菌利用氢[H]将 CO_2 还原为 CH_4 相比，硝酸盐利用氢[H]将 NO_2^- 还原为 NH_3 的反应在热力学上更易进行（Ungerfeld 和 Kohn，2006）。如果 NO_3^- 还原为 NO_2^- 的速度快于 NO_2^- 转化为 NH_3 的速度，瘤胃内则可能会发生 NO_2^- 积累（Latham 等，2016），这种积累会导致 NO_2^- 中毒，降低红细胞携带氧气的能力（Lee 和 Beauchemin，2014）。然而，各种研究得出的结论是，如果瘤胃微生物逐渐适应 NO_2^-，可以忽略其毒性风险（Cottle 等，2011；Olijhoek 等，2016）。

3.1.1 反刍动物研究

系列研究发现，在反刍动物日粮中补充 NO_3^- 可使单位 DM 摄入量的 CH_4 排放量减少 16%~35%，还发现补充 NO_3^- 使总 VFA 增加（El-Zaiat 等，2014；Nolan 等，2010；van Zijderveld 等，2010）和 NH_3 浓度增加（Olijhoek 等，2016）。尽管 CH_4 产量下降，但牛奶或牛肉产量均未增加，CH_4 排放量降低会增加可利用的代谢能（Johnson 和 Johnson，1995）。几项研究发现瘤胃内出现 H_2 积累（Lee 等，2015c；Olijhoek 等，2016），可能原因是 H_2 未用于还原 CO_2 或 NO_3^-，H_2 的排放也是代谢能的损失。

决定 NO_3^- 还原的主要因素是在热力学上比 CO_2 还原为 CH_4 更易进行，但 NO_3^- 对产甲烷菌丰度有不同的影响（Zhou 等，2014；Liu 等，2017）。Zhao 等（2018）发现，虽然 NO_3^- 降低了体外发酵 CH_4 产量，但饲喂 NO_3^- 的供体牛的瘤胃液中产甲烷菌总数没有变化。然而，随着 NO_3^- 浓度（0% DM 基础、1% DM 基础、2% DM 基础）增加，甲烷微菌目（*Methanomicrobiales*）的相对丰度减少，而甲烷八叠球菌目（*Methanosarcinales*）的相对丰度增加。日粮中添加 NO_3^- 增加了甲烷球形菌属（*Methanosphaera*）和甲烷微球菌属（*Methanimicrococcus*）的丰度，降低了甲烷盘菌属（*Methanoplanus*）的丰度。相比之下，Asanuma 等（2015）发现，山羊每天摄入 6 g 或 9 g NO_3^- 时，产甲烷菌、原虫和真菌的数量都急剧减少，牛链球菌和反刍月形单胞菌的丰度增加，提示牛链球菌和反刍月形单胞菌可能在 NO_3^- 代谢中发挥作用。

虽然肠道 CH_4 排放减少与 NO_3^- 的添加一致，但这通常并不伴随着瘤胃产甲烷菌数量的减少，由此可以推断 NO_3^- 是抑制产甲烷菌的活性而不是抑制产甲烷菌的生长来降低 CH_4 产量。大量的细菌种类可能参与 NO_3^- 代谢，如铜绿假单胞菌（*Pseudomonas aeruginosa*）和丙酸杆菌属（*Propionibacterium*）、丁酸弧菌属（*Butyrivibrio*）、梭状芽胞杆菌（*Clostridium*）、消化链球菌（*Peptostreptococcus*）、亚硝化单胞菌（*Nitrosomona*）、脱硫弧菌（*Desulfovibrio*）和肠杆菌科（*Enterobacteriaceae*）等菌群是潜在的反硝化菌（Latham 等，2016）。在培养基中，反刍月形单胞菌对 NO_3^- 耐受，一些菌株能将 NO_3^- 还原为 NO_2^-。在无 NO_3^- 的培养基中，韦荣氏球菌（*Veillonella parvula*）、产琥珀酸沃廉氏菌（*Wolinella succinogenes*）的丰度减少，添加 5 mmol/L 硝酸盐后，丰度增加（Iwamoto 等，2002）。

亚硝酸盐是 NO_3^- 还原为 NH_3 的中间体，研究证实亚硝酸盐会降低纤维素降解菌的丰度（Iwamoto 等，2002；Asanuma 等，2015）。但也有报道发现添加 NO_3^- 后，琥珀酸纤维杆菌、黄色瘤胃球菌、白色瘤胃球菌和总原虫的数量没有发生变化或增加（Patra 和 Yu，2014；Zhao 等，2015）。还有报道发现添加 NO_3^- 使原虫数量降低（Nolan 等，2010；El-

Zaiat 等，2014），这可能通过抑制与原虫共生的产甲烷菌而间接降低 CH_4 产量。

3.1.2 粪便

在土壤中，硝化作用是由 NH_3 氧化细菌和古细菌、异养硝化菌和真菌执行的（Norton，2008），而反硝化作用是由异养反硝化菌、反硝化真菌和自养/异养硝化菌执行（Coyne，2008）。β-变形杆菌 NH_3-氧化细菌、奇古菌门 NH_3-氧化古菌、镰刀菌属和木霉属真菌（Maeda 等，2015）能够将 NO_3^- 反硝化生成 N_2O（Maeda 等，2011）。

尽管两种日粮都是等氮日粮，与在日粮（精粗比 45∶55）中添加缓释包膜尿素相比，添加高达 3% DM 的缓释包膜 NO_3^- 可线性降低总的尿氮排泄量（Lee 等，2015a），然而，尿液和粪中的 NO_3-N 排泄量增加，尿中的尿素-N 减少。减少尿中尿素-N 和增加尿中 NO_3-N 排泄可能会减少 NH_3 排放，由于尿素是挥发性 NH_3 的主要来源，但这可能增加 N_2O 的直接排放量，由于土壤微生物对 NO_3-N 的反硝化作用。Zijderveld 等（2011）和 Li 等（2012）发现，使用 NO_3^- 代替尿素对奶牛或绵羊的 N 排泄没有影响，Lee 等（2015b）发现 DM、OM 和淀粉消化率随着 NO_3^- 的补充而增加，进而降低粪中可用于生成 CH_4 的底物含量。

3.2 3-硝基氧丙醇

已有研究证实 3-硝基氧丙醇（3-NOP）可使饲喂高谷物日粮肉牛的肠道 CH_4 排放量减少高达 80%（DM 摄入量基础）（Vyas 等，2016）。3-NOP 通过抑制甲基辅酶来降低甲烷生成，甲基辅酶参与甲基向甲基辅酶 M 还原酶的转移，这是甲烷生成途径的终末反应（Duin 等，2016）。

3.2.1 反刍动物研究

研究发现 3-NOP 可使绵羊（Martínez-Fernández 等，2014）、奶牛（Haisan 等，2014，2017；Hristov 等，2015；Lopes 等，2016；Reynolds 等，2014）和肉牛（Romero-Perez 等，2014，2015；Vyas 等，2016，2018b）的肠道 CH_4 排放量减少 7.7%~80.7%；增加 3-NOP 浓度（0.75~4.5 mg NOP/kg BW）可以线性降低 CH_4 产量（从 6.49% 总能摄入降到 4.34% 总能摄入）（Romero-Perez 等，2014）。

据报道，因降低 CH_4 产量而节省的饲料能会增加奶牛（Haisan 等，2014；Hristov 等，2015）和肉牛（Martinez-Fernandez 等，2018）的体重。其他研究发现瘤胃液中乙酸丙酸比降低（Haisan 等，2014；Romero-Perez 等，2014，2015；Martínez-Fernández 等，2014；Lopes 等，2016；Haisan 等，2017），以及 DM 消化率（Haisan 等，2017）和乳蛋白含量（Reynolds 等，2014）增加。虽然 Vyas 等（2016）发现 3-NOP 可使饲喂大麦青贮饲料肉牛的 CH_4 产量降低 37.6%，料肉比有增加趋势，但给这些牛饲喂高谷物育肥料时，平均日增重呈下降趋势。Vyas 等（2018a）发现，在高粗料日粮或高谷物日粮中添加 3-NOP，都能使肉牛 CH_4 产量降低，饲料转化率提高。

在纯培养中，发现 3-NOP 可以抑制产产甲烷菌的生长，而不影响瘤胃内的其他细菌（Duin 等，2016）。1μM 和 10μM 的 3-NOP 均能抑制 *Methanothermobacter marburgensis* 的生长和 CH_4 产量，但在给药 5 h 后，培养物中的甲烷生成作用恢复，这表明 3-NOP 浓度可能不足以完全抑制产甲烷菌的生长。同样，在纯培养物中，0.25~10μM 浓度的 3-NOP 能够抑制这些产甲烷菌的生长和甲烷产生：反刍兽甲烷短杆菌（*Methanobrevibacter ruminantium*）、史氏甲烷短杆菌（*Methanobrevibacter smithii*）、*Methanobrevibacter millerae*、布氏甲烷杆菌（*Methanobacterium bryantii*）、沃氏甲烷嗜热杆菌（*Methanothermobacter wolfeii*）、*Methanosphaera stadtmanae*、运动甲烷微菌（*Methanomicrobium mobile*）和巴氏甲烷八叠球菌（*Methanosarcina barkeri*）（Duin 等，2016）。动物体内试验研究发现 3-NOP 使甲产烷菌丰度下降了 56.6%~64.7%（Haisan 等，2014；Romero-Perez 等，2015；Martinez-Fernandez 等，2018），但并非所有的体内试验都发现 3-NOP 改变了产甲烷菌的拷贝数（Romero-Perez 等，2014；Lopes 等，2016）。

给婆罗门阉牛每天饲喂 2.5 g 3-NOP，甲烷短杆菌属和 *Methanomassilii coccaceae* 分别降低了 5.6 倍和 4.0 倍，CH_4（g/kg DMI）降低了 38%。*Methanomassilii coccaceae* 的减少可能导致瘤胃中三甲胺和二甲胺的增加，因为该家族产甲烷菌主要利用这些化合物作为生成甲烷的底物（Martinez-Fernandez 等，2018）。与占比 77% 的氢营养型瘤胃产甲烷菌（主要是甲烷杆菌目）相比，占比 20% 的甲基营养型产甲烷菌（*Methanoplasmatales* 和 *Methanosphaera* spp.）在瘤胃甲烷生成中只起次要作用（Henderson 等，2015）。在肉牛中，3-NOP 降低了氢营养型产甲烷菌与甲基营养型产甲烷菌的比例，表明这种添加剂对氢营养型产甲烷菌的抑制作用更大（Martinez-Fernandez 等，2018）。

Martinez-Fernandez 等（2018）利用瘤胃尼龙布袋法研究发现添加 3-NOP 使饲料 24 h DM 降解率降低了 7.1%，但 Haisan 等（2017）发现 3-NOP 使荷斯坦泌乳奶牛的 DM、OM 和中性洗涤纤维（NDF）的消化率增加，这意味着发酵过程中产生的电子载体被用来还原瘤胃中可替代的电子受体。丙酸产量增加（4.9%~26.0%）可以一定程度上解释这种氢[H]的利用，但并未在所有 3-NOP 研究中观察到丙酸盐的增加（Guyader 等，2017）。给架子肉牛提供 200 mg/kg 体重的 3-NOP 后，CH_4（g/d）减少了 37.6%，而 H_2 排放量增加了 89.1%；同样，育肥期 CH_4 减少 84.3%，H_2 增加 99.8%（Vyas 等，2016）。Hristov 等（2015）发现补充 3-NOP 的奶牛的 H_2 排放量增加了 64 倍，但这仅占未用于甲烷生成的代谢氢[H]的 3%（Latham 等，2016）。另外，给饲喂高粗料日粮的公牛补充 3-NOP 可使 CH_4 产量降低 30.2%，而 H_2 排放量无变化；同样使丁酸浓度降低，却使乙酸、异丁酸、异戊酸的浓度增加，但丙酸浓度没有变化（Martinez-Fernandez 等，2018），这表明一些氢[H]可能被转移到其他微生物细胞，反映在 NH_3 和支链 VFA 的浓度增加。

3.2.2 粪便

添加 3-NOP 使粪中 DM、NDF 和 ADF 含量增加（Reynolds 等，2014），尼龙布袋中的 DM 消化率降低（Martinez-Fernandez 等，2018），但也有研究发现 3-NOP 可以提高 DM 和 NDF 的消化率（Haisan 等，2017）。研究未发现 3-NOP 影响 CP 消化率或 N 排泄，这表明 3-NOP 中相对少量的 N 对反刍动物 N 代谢没有明显影响。据报道，3-NOP 在环境中会迅速分解，因此它在粪便中可能没有生物活性；调查发现给牛饲喂 3-NOP 对堆肥和储存粪便中 CO_2、CH_4、N_2O 或 NH_3 排放无影响（Owens 等，未发表）。

硝酸盐和 3-NOP 均被证明可抑制肠道 CH_4 排放，但从生命周期角度评估硝酸盐和 3-NOP 对温室气体排放的影响还非常有限。尽管有研究表明可以通过微生物适应来加速瘤

胃中 NO_2^- 还原为 NH_3 的反应,但考虑到添加 NO_3^- 引起的 NO_2^- 中毒效应限制了其在产业中的应用。多项研究表明,3-NOP 可以显著降低肠道 CH_4 排放,但作用的幅度和效果一致性似乎取决于日粮的精粗比。为了获得在反刍动物中适当使用 3-NOP 的监管批准,还需要确定饲喂不同日粮的反刍动物所需要的适当剂量。此外,为了鼓励生产者采用添加 NO_3^- 和 3-NOP 这些调控技术,还需测定 NO_3^- 和 3-NOP 能否持续提高反刍动物的饲料效率。为了能够在放牧反刍动物的饲养系统中使用 NO_3^- 和 3-NOP 两种添加剂,还需要结合方法学,且这种放牧系统的 GHG 排放强度通常高于集约化养殖系统。

4 植物次生化合物

研究发现多种植物的大量植物次生化合物具有降低肠道 CH_4 排放的潜力。虽然已经确定 200 000 多种植物化学物质,但用作饲料添加剂的次生代谢物主要是精油、皂苷和单宁(Hartmann,2007)。单宁是一种结构稳定的芳香族化合物,能够与碳水化合物、蛋白质和矿物质形成复合物,单宁可以分为水解单宁和缩合单宁,其主要差异是一些水解单宁可以被瘤胃微生物代谢,而缩合单宁(CT)可以抵抗生物降解(Aboagye 等,2018)。

4.1 单宁

大量研究表明,许多富含单宁的粗料或单宁提取物可以降低牛的肠道 CH_4 排放(Waghorn,2008;Hess 等,2006;Grainger 等,2009;Alves 等,2017),但单宁降低 CH_4 生成的实际作用方式尚不清楚,研究者们对此提出了以下一些假设:抑制产甲烷菌、干扰产甲烷菌和原虫之间的共生关系、抑制瘤胃纤毛虫(Bhatta 等,2015),减少可发酵底物,与微生物酶结合(Gonçalves 等,2011)或充当氢[H]池消耗瘤胃氢(Naumann 等,2017)。缩合单宁对日粮蛋白具有较高的结合能力,并且可以增加流向小肠的可代谢蛋白量,这可能会降低微生物蛋白的合成。相比之下,水解单宁对蛋白质的亲和力较低,细菌降解这些单宁物质可产生低分子量的代谢物,包括可能对反刍动物有毒的酚类物质(Patra 和 Saxena,2011)。

4.1.1 反刍动物研究

单宁对瘤胃 CH_4 排放的影响结果并不一致。Grainger 等(2009)发现给泌乳奶牛每天饲喂 266 g 黑荆(*Acacia mearnsii*)后,CH_4 产量最高可降低 30%。Alves 等(2017)也发现黑荆可以使热带牧场放牧的荷斯坦奶牛的 CH_4 排放量(g/kg 产奶量)降低 32%,对牛奶产量(产量、乳脂含量、乳蛋白含量)无不利影响。Koenig 等(2018)发现,在 40% 玉米 DDGS 中添加黑荆缩合单宁(CT)会降低 OM、NDF、ADF、N 和总能的消化率,粪 N 增加 32.4%,尿 N 减少 17.5%,瘤胃 NH_3-N、VFA 浓度和乙酸/丙酸比例也降低,但本研究未测定 CH_4 生成量。Carulla 等(2005)发现给瑞士白山绵羊饲喂 41 g/kg 提取自黑荆的缩合单宁后,OM、CP、NDF 和 ADF 的表观消化率降低(1.89%~10.6%),氨氮减少,随着丙酸盐增加 6.3%,乙酸丙酸比降低,甲烷排放量(DM、OM 摄入量的百分比)下降,很可能是由于抑制了纤维的消化。

当给肉牛和绵羊饲喂含有黑荆 CT 的日粮后，对动物生产性能（饲料转化效率）无影响（Koenig 等，2018；Carulla 等，2005）。Woodward 等（2004）发现饲喂含有 CT 的百脉根粗料时，CH_4 产量（DMI 基础）降低 16%；白坚木树（*Schinopsis quebracho*）的 CT 提取物对 CH_4 排放量无影响，CP 消化率降低（Beauchemin 等，2007）。Aboagye 等（2018）发现，添加具有或缺乏 CT（quebracho）的水解栗子单宁 12 周后，对肉牛的 ADG 或 G：F 无影响，CH_4 产量（g/kg DMI）有下降趋势。日粮中添加单宁对肠道 CH_4 排放调控不一致的原因可能是由于单宁种类、植物化学成分、补充方法和日粮中单宁浓度差异导致。

单宁与瘤胃微生物形成复合物的能力需要微生物细胞壁和分泌的细胞外酶的多酚反应活性（McSweeney 等，2001）。产琥珀酸丝状杆菌、溶纤维丁酸弧菌、嗜淀粉瘤胃杆菌和牛链球菌等细菌对多种单宁具有高亲和力。McSweeney 等（1998）和 McAllister 等（1994）推测蛋白降解菌和真菌对单宁的敏感性低于纤维降解菌。Tan 等（2011）发现饲喂银合欢时，单宁改变了瘤胃内产甲烷菌的多样性，即降低了 *Methanomicrobiales*（15.1%）和 *Methanobacteriales*（6.8%）的比例，增加了 *Thermoplasmatales* 比例（21.9%），使古菌的整体多样性降低。

饲喂单宁最常见的反应是瘤胃 CP 消化率降低，CT 通过 H 键和疏水作用与蛋白质形成复合物（Koenig 和 Beauchemin，2018），这使瘤胃内可用于微生物降解的日粮蛋白降低。也有人提出，虽然蛋白质在瘤胃内被结合，但这些复合物在皱胃中解离，蛋白质进一步被消化，氨基酸由后消化道吸收。然而研究结果并不一致，因为并非所有复合物都会解离，通常导致 CP 消化率降低，以及氮排泄从尿 N 到粪 N 的转变（Patra 和 Saxena，2011）。

4.1.2 粪便

与其他日粮添加剂不同，反刍动物日粮中添加单宁酸对粪便的影响已得到广泛研究（Koenig 和 Beauchemin，2018；Powell 等，2011；Halvorson 等，2017；Jordan 等，2015）。在考虑粪便 GHG 排放方面，饲喂日粮单宁的一个优势是将 N 的排泄部位从尿液转移到粪便。与尿中的尿素相比，粪中以 CT-蛋白复合物形式存在的粪 N 更加稳定，也不太可能导致较高的 NH_3 排放。释放较慢的粪 N 也更容易被植物捕获并用于合成植物蛋白。

单宁对脲酶活性的抑制作用也与底物-单宁复合物的形成有关，单宁还可能降低 N_2O 排放，因为单宁可与蛋白质形成复合物，从而产生不可溶和不可利用的 N（Powell 等，2011）。Powell 等（2011）利用通风室研究日粮中添加单宁对泌乳期荷斯坦奶牛粪便的影响，给奶牛饲喂四个递增浓度（0、4.5 g/kg、9.0 g/kg 和 18.0 g/kg DMI）富含单宁的红坚木树和板栗树混合物，单宁组的 NH_3 累积排放量比对照组低 27%。此外，有 54% 和 66% 的饲用尿素以 NH_3 的形式分别从单宁组和对照组中释放。

作为外用，将 4%（w/w）缩合单宁分别添加到堆肥的山羊粪和氮磷贫乏的土壤中，单宁使堆肥的累积 C 排放量减少 40%，N 排放量减少 36%。与未添加单宁的土壤相比，土壤添加单宁后，N_2O 排放量减少 17%，NH_3 释放量减少 51%（Jordan 等，2015）。众所周知，单宁会隔离有机 N 源，降低无机 N 的可用性。同样，单宁可能作为不稳定的碳源，促进固氮（Kraus 等，2004）。Koenig 等（2018）在育肥牛的 40% 酒糟日粮中添加 2.5% 的澳洲黑荆，使用带有被动 NH_3 采样器的综合水平通量技术来测定围栏中的 NH_3 排放，尽管测量值有限，但饲喂含有单宁的饲料使肉牛的 NH_3-N 排放量降低了 23%。

香茅醇假单胞菌（*Pseudomonas citronellolis*）和香鱼假单胞菌（*Pseudomonas plecoglossicidda*）是两种能够分别利用单宁土壤中单宁酸（可水解）和没食子酸（单宁酸的酚水解产物）的细菌。Bending 和 Read（1996）发现一种可水解的多酚-蛋白可以被外生菌根真菌（*Hysterangium setchellii*、*Lactarius affinis*、*Lactarius controversus*）、欧石楠类菌根（*Hymenoscyphus ericae*）和木材分解真菌（*Hypholoma fasciculare* 和 *Phanerochaete velutina*）降解（Mutabaruka 等，2007）。Mutabaruka 等（2007）发现在含有大量 CT 复合物的酸性土壤系统中，真菌与细菌的比例增加。尽管化学性质不同，但相似的微生物可能栖息在单宁堆肥粪中，在堆肥过程中或施用到土地后分解单宁复合物。

在反刍动物日粮中添加单宁对反刍动物的代谢和 CH_4 排放产生了不一致的结果。似乎从澳洲黑荆（*Acacia mearnsii*）中提取的 CT 对肠道 CH_4 的减排结果最为一致。从粪便 GHG 减排的角度看，大多数 CT 对 NH_3 和 N_2O 的减排结果一致。然而，单宁也会降低 CP 消化率，导致动物生产的 N 损失。虽然这会对反刍动物生产产生负面影响，但它可以改善粪便营养成分，用作肥料和土壤改良剂。

5 碳衍生材料

C 循环是指通过光合作用将大气 CO_2 转化为植物的生物质 C。在反刍动物中，消耗的 C 被分配到代谢的动物副产品和代谢废物中，包括呼吸作用和粪便分解产生的 CO_2（图 4）。沉积的粪便可能会被分解并释放不稳定的 C，从而增加表层土壤中的有机物含量（Sharma 等，2017）。腐殖质（HS）和生物炭最初因具有增加和隔离土壤碳的能力而受到关注。然而，最近研究发现在反刍动物日粮中直接添加 HS 和生物炭，具有降低肠道 CH_4 排放的潜力。

5.1 腐殖质（HS）

土壤中的 OM 是处于不同氧化和腐烂阶段的植物源和动物源前体物的复杂异质混合物（Masoom 等，2016）。HS 主要是由强力抵抗生物降解的聚合分子组成（Stevenson，1995）。根据 HS 在碱性或酸性溶液中的溶解度分为 3 个操作级分。因此，黄腐酸是可溶于碱和酸的小脂肪族化合物，而腐植酸是通过稀碱提取的高分子物质，pH 值为 2 时的腐质沉淀代表 HS 的不溶比例（Stevenson，1995；Lamar 等，2014）。因官能团结构、组成和反应活性差异，HS 的化学性质可能有很大差异，从而差异性调控瘤胃功能（Stevenson，1995）。

HS 能够降低肠道 CH_4 的生成量和影响人畜共患病原体。现有文献研究了各种 HS 衍生物对体外发酵（Sheng 等，2017；Terry 等，2018a；Varadyova 等，2009）和体内代谢的影响（Terry 等，2018b；El-Zaiat 等，2018；Ponce 等，2016），然而，研究之间几乎没有一致性，可能是因为 HS 的浓度和类型尚未得到充分描述。

5.1.1 反刍动物研究

补充 HS 的牛与饲喂莫能菌素的牛的平均日增重、干物质摄入量和饲料效率没有区别，这表明 HS 与离子载体具有相似的能力（McMurphy 等，2009）。然而，本研究中未使用负对照，结果需要谨慎解释。给萨能山羊饲喂高达 3% 腐植酸（DM 基础）的日粮，使山羊产奶量提高，血液胆固醇水平降低（Degirmenci，2012）。Agazzi 等（2007）发现给新生牛喂食 HS，其牛奶消耗量和 ADG 高于对照组，该研究提出假设：HS 的抗生素特性提高了细胞介导的免疫力，降低幼牛消化紊乱和腹泻的风险。

最近研究测试了 HS 抑制瘤胃产甲烷的能力。虽然体外分批培养研究发现 HS 能够持续降低 CH_4 产量（Sheng 等，2017），但在瘤胃模拟技术（Rusitec）连续培养上并未得到证实（Terry 等，2018a）。此外，Terry 等（2018b）发现在大麦青贮饲料中添加 300 mg/kg 动物活体重的 HS 时，对小母牛肠道 CH_4 产量没有影响，总氮留存量增加，表明蛋白质利用率有所提高。

Terry 等（2018a，b）利用 16s rRNA 测序检测了瘤胃微生物种群数量，发现 HS 降低了 Rusitec 固相样品中的纤维杆菌属（*Fibrobacter*）和克里斯滕森菌 R-7（*Christenseneuaceae*）丰度，没有改变总的产甲烷菌数量，但增加了甲烷杆菌数量，降低了甲烷短杆菌和甲烷球菌数量。体内研究发现，HS 降低了变形菌门（Proteobacteria）、互养菌门（Synergistetes）和广古菌门（Euryarchaeota）的相对丰度。

5.1.2 粪便

关于将 HS 掺入粪肥混合物的研究信息很少，有研究发现 HS 降低了小母牛的总氮排泄量，这意味着 HS 有潜力降低粪便中 NH_3 排放。Shi 等（2001）在土壤-粪-尿混合物中添加黑色和棕色腐植酸盐（混合物总质量的 1.7%），与不含腐植酸的粪便-土壤混合物相比，NH_3 排放量减少 39.8%。但在本研究中，HS 是在排泄后添加到粪中的，其结果可能与直接添加 HS 到日粮中结果不同。

多个研究发现 HS 降低了缺氧环境下土壤 CH_4 的排放（Miller 等，2015；Blodau 和 Deppe，2012；Tan 等，2018），研究者认为可能是由于 HS 内的功能结构充当电子受体（即甲烷生成过程中的氢），诱导降低甲烷生成（Martinez 等，2013；Terry 等，2018b）。然而，相矛盾的研究表明 HS 促进了缺氧稻田中 CH_4 产生（Zhou 等，2014）。

5.2 生物炭

生物炭是一种热降解形式的黑碳，通过在限氧条件下加热（350~600℃）植物生物质残留物获得（Cha 等，2016）。生物炭主要由顽固性 C 组成，但也包含一系列无机营养素（Joseph 等，2018）。生物炭的特点是多孔结构、大表面积和高矿物质含量，这些特性取决于原始生物质或原料来源（Cha 等，2016）。同样，其他研究发现生物炭具有高的离子交换和吸附特性（Yuan 等，2017），使其适合用作土壤改良剂、水和空气洗涤器以及解毒剂（Tawheed 和 Baowei，2017）。生物炭可降低耕地中 N_2O 和 CH_4 的排放（Karhu 等 2011；Cayuela 等，2014），其解毒能力为其用作反刍动物饲料添加剂提供可能。

5.2.1 反刍动物研究

Leng 等（2012）首次提出使用生物炭作为日粮的 CH_4 减排策略，假设生物炭的多孔性质可能会促进生物膜的形成或诱导瘤胃内的种间电子转移。此外还假设生物炭会增加瘤胃内甲烷氧化菌（*Methanotrophs*）的数量，尽管体外研究发现甲烷氧化菌只占微生物活性的很小一部分（Kajikawa 等，2003）。生物炭对 CH_4 的吸收可能在降低 CH_4 产量方面发挥着重要作用。然而 Saleem 等（2018）发现生物炭（添加 0.5%~2% DM）似乎不太可能吸收由瘤胃产甲烷菌产生的大量 CH_4。

Leng 等（2012）使用 12 头来自老挝的黄色小牛研究硝酸盐和稻壳产生的生物炭对牛生长性能和 CH_4 排放量的影响，发现生物炭使甲烷产量降低了 22%，动物活体增重增加了令人难以置信的 25%；然而，由于没有使用连续量热法测定 CH_4，研究结果需要谨慎解释。Saleem 等（2018）利用 Rusitec 技术研究发现添加 2.0% DM 松树生物炭，使 CH_4（g/g 消化 DM）减少 22.4%，总 VFA、NH_3-N 和养分消化率（DM、OM、CP、NDF 和 ADF）均得到提高。然而，使用相同的产品，Terry 等（2019）在大麦青贮饲料日粮中添加 2.0% DM 的生物炭，发现小母牛瘤胃中 VFA 浓度、CH_4 产量和营养物质表观消化率不受生物炭的影响。

5.2.2 粪便

Joseph 等（2015）发现给奶牛饲喂糖蜜（0.1 kg/d）和高温贾拉木生物炭（0.33 kg/d）混合物，奶牛的粪便改善了澳大利亚艳色土的土壤特性并增加了 OM 的封存量（0~40cm），其原因是增强了奶牛肠道对生物炭中 N 和 P 的吸收，以及限制了消化后顽固碳的转化，最后增加的稳定碳使土壤肥力提高。

Yuan 等（2017）利用水稻壳制作的生物炭作为鸡粪的复合成分。与堆肥相比，生物炭改良粪中 CO_2 和 N_2O 排放量分别减少 35% 和 27%，降低了细菌 16S rRNA 的丰度。假设生物炭通过土壤剖面增加 OM 的稳定性，以及影响反硝化菌数量，体现在增加了细菌编码硝化相关酶的基因表达。同样，生物炭可以有效地保留混合堆肥家禽垫料中的 NH_3 和 N_2O（Steiner 等，2010）。Jia 等（2016）发现，与纯堆肥相比，稻壳生物炭用作填充剂可使 N_2O 排放的峰值速率降低 60%。

从现有研究来看（Atkinson 等，2010；Jia 等，2016；Steiner 等，2010；Yuan 等，2017），生物炭改良堆肥似乎是一种有效的鸡粪堆肥减排措施。需要更多的研究评估生物炭对牛粪堆肥是否会产生相同的效果，以及给反刍动物饲喂生物炭是否会改变粪便排泄后的温室气体排放。

根据目前的研究结果可知，HS 和生物炭在降低瘤胃 CH_4 排放方面基本上无效。尽管 HS 没有降低反刍动物的 CH_4 排放，但仍然存在相关健康或代谢益处，即 HS 使山羊的平均日增重增加（Agazzi 等，2007）和肉牛的氮沉积增加（Terry 等，2018b）。然而，还要使用性能试验进行评估检测。尽管体外试验发现生物炭可以降低肠道 CH_4 排放，但在体内试验的作用效果尚未得到证实；而体外试验目前提出的生物炭降低肠道 CH_4 排放的机制不能得到当前体内研究结果的支持，需要进一步评估生物炭如何改变瘤胃微生物菌群。有限的研究表明，HS 和生物炭均可有效降低粪便温室气体排放，也许在规定水平上饲喂这些添加剂将改善粪肥的营养成分，而不会对反刍动物的性能产生负面影响。

6 微生物的氢利用

人们普遍认为，瘤胃内氢 [H] 分压的增加将通过降低辅因子的还原氧化来抑制瘤胃发酵（Ungerfeld，2015b）。此外，化学计量表明反刍动物肠道 CH_4 产量的降低会导致更多的能量用于维持和生产（Johnson 和 Johnson，1995）。然而，这两个观念并不总是从肠道 CH_4 减排研究中观察得到。

氢营养型古菌（*Methanobrevibacter*、*Methanobacterium*）是瘤胃中的主要优势产甲烷菌（Henderson 等，2015），可以利用氢［H］或少量甲酸盐作为电子供体将 CO_2 还原为 CH_4（Richards 等，2016）。甲酸对 CH_4 生成的贡献估计为瘤胃中 CH_4 产量的 18%（Tapio 等，2017b）。甲基营养型产甲烷菌（即 *Methanosarcinales*、*Methanosphaera* 和 *Methanomassiliicoccaceae*）在瘤胃中的数量较少，可以利用甲醇和甲胺生成 CH_4（Huws 等，2018）。*Methanosarcinales* spp. 利用乙酸分解途径生成 CH_4，但该途径的产甲烷菌生长速度较慢，且在成熟瘤胃中并不突出（Friedman 等，2017b）。虽然大部分氢［H］被古菌利用，但瘤胃中还有其他几种方式利用氢［H］，包括利用还原当量还原硫酸盐和硝酸盐（NO_3^-），还原生成乙酸和丙酸，以及微生物生物质合成。

硫酸盐和硝酸盐的还原反应在热力学上比 CO_2 还原反应更易进行（Morgavi 等，2010；Haque，2018），并且发现缺氧环境中硫酸盐和 NO_3^- 还原菌胜出产甲烷菌（Scheid 等，2003）。还原生成乙酸的反应在热力学上比甲烷生成反应更难进行，所以瘤胃不太可能建立起持续还原乙酸的优势微生物种群（Fonty 等，2007；Friedman 等，2017b）。然而，如果开发出一种可以自然存在于瘤胃中，并且只利用氢［H］而不利用糖（专性氢营养）的还原性产乙酸菌，它可能会成功地抑制甲烷生成（Ungerfeld，2015a）。

将氢［H］转向到其他代谢池的饲料添加剂是一种新的研究途径代表。Martinez Fernandez 等（2017）在婆罗门牛日粮中添加抗甲烷生成的化合物——氯仿，一半的阉牛还服用了间苯三酚（类黄酮降解的中间产物），它可以利用氢［H］和甲酸盐形成乙酸盐；间苯三酚增加了乙酸浓度、普雷沃氏菌属、瘤胃球菌属和纤维杆菌属的丰度，降低了 H_2 浓度和甲酸盐产量，这是第一个体内试验证明氢［H］可以向还原间苯三酚方向转变，从而抑制甲烷生成。在抑制甲烷生成过程中测定氢［H］的重定向，以及如何改变瘤胃微生物菌群方面还要进一步研究。例如，3-NOP 可以有效减少甲烷生成，然而过量的氢［H］不会被其他还原过程捕获，导致 H_2 排放量增加。将氢［H］重新定向到可用的代谢物上，将进一步降低 CH_4 产量和 CH_4 排放强度，可能不会以 H_2 形式损失太多能量。

有人提出，与其他厌氧环境一样，甲烷生成菌和甲烷利用菌之间存在平衡。甲烷氧化菌是特定的古菌或细菌，它们在有氧条件下可以代谢 CH_4（Leng，2014）。然而，也可以利用硫酸盐、金属氧化物和硝酸盐中的现有氧让 CH_4 发生厌氧氧化（Joye，2012），但它们在瘤胃环境中的存在或重要性还存在争议。

在人工瘤胃系统中，发现只有 0.2%~0.5% 的 CH_4 被硫酸盐还原氧化（Kajikawa 等，2003）。从理论上讲，由于氧气从血流中扩散，甲烷氧化菌更容易在瘤胃壁定殖，但 Wallace 等（2015）利用宏基因组分析肉牛瘤胃微生物菌群，并未检测到甲烷氧化菌。Mitsumori 等（2002）使用甲烷氧化菌特异性引物检测，发现在瘤胃液和瘤胃壁生物膜中均检测到甲烷氧化菌，仅是 I 型甲烷氧化菌。I 型甲烷氧化菌包括甲基单胞菌、甲基杆菌、甲基微菌和甲基球菌，它们利用核酮糖单磷酸途径吸收碳。II 型甲烷氧化菌包括甲基孢囊菌和甲基弯曲菌，它们利用丝氨酸途径吸收碳（Mitsumori 等，2002）。Jin 等（2017）发现甲基球菌科在固体、液体和瘤胃壁相关种群中占优势地位。在瘤胃液分批培养中，Liu

等（2017）发现添加 NO_3^- 降低甲烷生成，增加了 NC10 门细菌的丰度，NC10 细菌是唯一能够厌氧氧化 CH_4 的已知细菌。细菌 Methoxymirabilisoxyfera 可将亚硝酸盐（NO_2^-）转变成一氧化氮（NO），接着将 NO 转变为氮气和氧气，进一步利用产生的氧气氧化甲烷（He 等，2016；Joye，2012）。所有其他具有厌氧氧化 CH_4 能力的微生物都是古细菌。Klieve 等（2012）发现牛的瘤胃内容物中具有与 CH_4 氧化古菌相关的 mcrA 基因序列。这些古菌可利用来自墨西哥湾沉积物中的硫酸盐还原进行厌氧氧化甲烷（Lloyd 等，2006）。

Parmar 等（2015）研究发现 I 型甲烷氧化菌在 50∶50 精粗比日粮中更丰富，而 II 型甲烷氧化菌丰度则在全草料日粮中增加。氧化甲酸的甲酸脱氢酶含量在高粗料日粮中增加，这与 II 型甲烷氧化菌的丰度增加一致。Auffret 等（2018）确定肉牛瘤胃中存在 3 种甲烷氧化菌，包括甲基杆菌、甲基单胞菌和甲基微菌，它们的丰度很低 [（0.1 ± 0.01）%]，其中甲基单胞菌的丰度较高，且与 CH_4 排放呈负相关。与 CH_4 排放低的动物瘤胃相比，CH_4 排放高的动物瘤胃中甲烷氧化菌的整体多样性更大。针对瘤胃甲烷氧化菌的不一致研究结果可能是由于这些稀有种群缺乏测序的深度和广度。

像瘤胃一样营养丰富的环境中，甲烷氧化菌的重要性值得怀疑。然而，当反刍动物饲料中添加 CH_4 抑制剂（如 NO_3^- 和 3-NOP）时，甲烷氧化菌对 CH_4 的消耗可能贡献了部分氢 [H]。甲烷氧化菌在其他厌氧环境中非常重要，如沉积物、海洋海底以及淡水和咸水系统（He 等，2016），在到达大气之前，超过 80% 的 CH_4 已被氧化（Cai 等，2016）。然而，这些都是相对稳定的环境，不会像瘤胃内那样经历一定的通过率或可用性底物的每天变化。随着测序技术进步，我们有能力深入研究瘤胃微生物，对甲烷氧化菌的详细鉴定将会增强我们对瘤胃氢 [H] 平衡的理解。

7 未来趋势及结论

目前，3-NOP 和 NO_3^- 可以降低肠道 CH_4 排放，而对粪中的 GHG 排放几乎没有影响；然而，过量的 [H] 并未以还原型底物形式被完全捕获（表1）。同样，单宁酸可以降低粪中的 GHG 排放，而它们对反刍动物肠道 CH_4 排放的效果差异很大。有机 C 可能具有减少粪便 GHG 排放的潜力，但支持降低肠道 CH_4 排放的研究还很有限。

表1 日粮添加剂及其对反刍动物生产 GHG 减排的影响

日粮添加剂	肠道排放	生产性能提升	粪便排放	性能改善	相互作用	推荐
硝基化合物						
硝酸盐	$\downarrow CH_4 \uparrow H_2$	无	N/A^a，可能 $\uparrow N_2O \downarrow NH_3$	N/A	N/A，可能	是[a]
3-NOP	$\downarrow CH_4 \uparrow H_2$	不一致	N/A	N/A	N/A，不太可能	是[a]
次生化合物						
单宁	变动的，可能 $\downarrow CH_4$	不一致，可能 $\downarrow DMI$	$\downarrow N_2O \downarrow NH_3$	有	有	变化很大

(续表)

日粮添加剂	肠道排放	生产性能提升	粪便排放	性能改善	相互作用	推荐
有机碳						
腐殖质	无	N/A	N/A[a]，可能↓NH_3	N/A[a]，可能↑稳态 C	无	N/A
生物炭	无	N/A	N/A[a]，可能↓N_2O，↓NH_3	N/A[a]，可能↑稳态 C	N/A	N/A

注：[a] 基于有限研究。N/A=信息不可用。

日粮调控被认为是降低反刍动物 GHG 排放最可行的一种减排措施，然而，需确保不会干扰瘤胃代谢和增加粪中 GHG 排放。另外，还需考虑日粮变化如何改变瘤胃微生物组成以及这些变化能够持续的时间。关于日粮添加剂对肠道 CH_4 和粪便 CH_4 排放影响的研究强化了肠道 CH_4 和粪便 CH_4 之间的动态复杂性。尽管肠道 CH_4 产量对 GHG 总排放量（CO_2 当量）贡献要大得多，但粪便产生的 N_2O 却是一种更强效的温室气体。因此，在推荐反刍动物 GHG 减排策略时，重要的是要在整个农场层面验证其效率。

8 更多信息

8.1 进一步阅读

- Extensive review of GHG mitigation strategies from livestock production：'Mitigation of Greenhouse gas emissions in livestock production-FAO Animal production and Health'. 见：http://www.fao.org/docrep/018/i3288e/i3288e00.htm.
- Extensive characterization of livestock production by region and associated GHG production：'Assessment of greenhouse gas emissions and mitigation potential'-FAO Global Livestock Environmental Assessment Model（GLEAM）. 见：http://www.fao.org/gleam/results/en/.
- A review on agricultural Ncycle：Robertson 和 Vitousek（2009）.

8.2 重点期刊/会议

温室气体和畜牧业会议（Greenhouse Gas & Animal Agriculture，GGAA）是每 3 年举行一次的国际会议。

8.3 重大国际研究项目

生物炭项目：在肉牛日粮中添加生物炭降低农业温室气体排放的潜力：http://www.agr.gc.ca/eng/programs-and-services/agriculture-greenhouse-gases program/approved-projects/? id=1508423883267。

9 参考文献

Aboagye, I. A., Oba, M., Castillo, A. R., et al., 2018. Effects of hydrolyzable tannin with or without condensed tannin on methane emissions, nitrogen use, and performance of beef cattle fed a high-forage diet. J. Anim. Sci. 96 (12), 5276-86.

Adler, A. A., Doole, G. J., Romera, A. J., et al., 2015. Managing greenhouse gas emissions in two major dairy regions of New Zealand: a system-level evaluation. Agric. Sys. 135, 1-9.

Agazzi, A., Cigalino, G., Mancin, G., et al., 2007. Effects of dietary humates on growth and an aspect of cell-mediated immune response in newborn kids. Small Rumin. Res. 72 (2-3), 242-5.

Alves, T. P., Dall-Orsoletta, A. C. and Ribeiro-Filho, H. M. N. 2017. The effects of supplementing *Acacia mearnsii* tannin extract on dairy cow dry matter intake, milk production, and methane emission in a tropical pasture. Trop. Anim. Health Prod. 49 (8), 1663-8.

Asanuma, N., Yokoyama, S. and Hino, T. 2015. Effects of nitrate addition to a diet on fermentation and microbial populations in the rumen of goats, with special reference to *Selenomonas ruminantium* having the ability to reduce nitrate and nitrite. Anim. Sci. J. 86 (4), 378-84.

Atkinson, C. J., Fitzgerald, J. D. and Hipps, N. A. 2010. Potential mechanisms for achieving agricultural benefits from biochar application to temperate soils: a review. Plant Soil 337 (1-2), 1-18.

Auffret, M. D., Stewart, R., Dewhurst, R. J., et al., 2018. Identification, comparison, and validation of robust rumen microbial biomarkers for methane emissions using diverse Bos taurus breeds and basal diets. Front. Microbiol. 8, 2642.

Bach, A., Calsamiglia, S. and Stern, M. D. 2005. Nitrogen metabolism in the rumen. J. Dairy Sci. 88 (Suppl. 1), E9-21.

Beauchemin, K. A., McGinn, S. M., Martinez, T. F., et al., 2007. Use of condensed tannin extract from quebracho trees to reduce methane emissions from cattle. J. Anim. Sci. 85 (8), 1990-6.

Beauchemin, K. A., Kreuzer, M., O'Mara, F., et al., 2008. Nutritional management for enteric methane abatement: a review. Aust. J. Exp. Agric. 48 (2), 21-7.

Belanche, A., Doreau, M., Edwards, J. E., et al., 2012. Shifts in the rumen microbiota due to the type of carbohydrate and level of protein ingested by dairy cattle are associated with changes in rumen fermentation. J. Nutr. 142 (9), 1684-92.

Bending, G. D. and Read, D. J. 1996. Nitrogen mobilization from protein-polyphenol complex by ericoid and ectomycorrhizal fungi. Soil Biol. Biochem. 28 (12), 1603-12.

Bhatta, R., Saravanan, M., Baruah, L., et al., 2015. Effects of graded levels of tannin-containing tropical tree leaves on *in vitro* rumen fermentation, total protozoa and methane production. J. Appl. Microbiol. 118 (3), 557-64.

Blodau, C. and Deppe, M. 2012. Humic acid addition lowers methane release in peats of the Mer Bleue bog, Canada. Soil Biol. Biochem. 52, 96-8.

Boots, B., Lillis, L., Clipson, N., et al., 2013. Responses of anaerobic rumen fungaldiversity (phylum Neocallimastigomycota) to changes in bovine diet. J. Appl. Microbiol. 114 (3), 626-35.

Broderick, G. A. 2003. Effects of varying dietary protein and energy levels on the production of lactating dairy cows. J. Dairy Sci. 86 (4), 1370-81.

Cai, Y., Zheng, Y., Bodelier, P. L. E., et al., 2016. Conventional methanotrophs are responsible for atmospheric methane oxidation in paddy soils. Nat. Commun. 7, 11728.

Carulla, J. E., Kreuzer, M., Machmüller, A., et al., 2005. Supplementation of *Acacia mearnsii* tannins decreases methanogenesis and urinary nitrogen in forage-fed sheep. Aust. J. Agric. Res. 56 (9), 961-70.

Castillo-Lopez, E., Ramirez Ramirez, H. A., Klopfenstein, T. J., et al., 2014. Effect of feeding dried distillers grains with solubles on ruminal biohydrogenation, intestinal fatty acid profile, and gut microbial diversity evaluated through DNA pyro-sequencing. J. Anim. Sci. 92 (2), 733-43.

Castillo-Lopez, E., Jenkins, C. J. R., Aluthge, N. D., et al., 2017. The effect of regular or reduced-fat distillers grains with solubles on rumen methanogenesis and the rumen bacterial community. J. Appl. Microbiol. 123 (6), 1381-95.

Cayuela, M. L., van Zwieten, L., Singh, B. P., et al., 2014. Biochar's role in mitigating soil nitrous oxide emissions: a review and meta-analysis. Agric. Ecosyst. Environ. 191, 5-16.

Cha, J. S., Park, S. H., Jung, S.-C., et al., 2016. Production and utilization of biochar: a review. J. Ind. Eng. Chem. 40, 1-15.

Cottle, D. J., Nolan, J. V. and Wiedemann, S. G. 2011. Ruminant enteric methane mitigation: a review. Anim. Prod. Sci. 51 (6), 491-514.

Coyne, M. S. 2008. Biological denitrification. In: Schepers, J. S. and Raun, W. R. (Eds), Nitrogen in Agricultural Systems. American Society of Agronomy, Crop Science Society of America, Soil Science Society of America, Madison, WI, pp. 201-53.

Degirmenci, T. 2012. Effects of diets containing humic acid on the milk yield, milk composition and blood metabolites in Saanen goats. Res. J. Anim. Sci. 6 (1), 4-7.

Dijkstra, J., Oenema, O., van Groenigen, J. W., et al., 2013. Diet effects on urine composition of cattle and N2O emissions. Animal 7, 292-302.

Duin, E. C., Wagner, T., Shima, S., et al., 2016. Mode of action uncovered for the specific reduction of methane emissions from ruminants by the small molecule 3-nitrooxypropanol. Proc. Natl. Acad. Sci. U. S. A. 113 (22), 6172-7.

El-Zaiat, H. M., Araujo, R. C., Soltan, Y. A., et al., 2014. Encapsulated nitrate and cashew nut shell liquid on blood and rumen constituents, methane emission, and growth performance of lambs. J. Anim. Sci. 92 (5), 2214-24.

El-Zaiat, H. M., Morsy, A. S., El-Wakeel, E. A., et al., 2018. Impact of humic acid as an organic additive on ruminal fermentation constituents, blood parameters and milk production in goats and their kids growth rate. J. Anim. Feed Sci. 27 (2), 105-13.

Enjalbert, F., Combes, S., Zened, A., et al., 2017. Rumen microbiota and dietary fat: a mutual shaping. J. Appl. Microbiol. 123 (4), 782-97.

Erisman, J. W., Galloway, J. N., Seitzinger, S., et al., 2013. Consequences of human modification of the global nitrogen cycle. Philos. Trans. R. Soc. Lond., B, Biol. Sci. 368 (1621), 20130116.

Fageria, N. K. and Baligar, V. C. 2005. Enhancing nitrogen use efficiency in crop plants. In: Advances in Agronomy. Academic Press, pp. 97-185.

Firkins, J. L. and Yu, Z. 2015. RUMINANT NUTRITION SYMPOSIUM: how to use data on the rumen microbiome to improve our understanding of ruminant nutrition. J. Anim. Sci. 93 (4), 1450-70.

Fonty, G., Joblin, K., Chavarot, M., et al., 2007. Establishment and development of ruminal hydrogenotrophs in methanogen-free lambs. Appl. Environ. Microbiol. 73 (20), 6391-403.

Friedman, N., Shriker, E., Gold, B., et al., 2017a. Diet-induced changes of redox potential underlie compositional shifts in the rumen archaeal community. Environ. Microbiol. 19 (1), 174-84.

Friedman, N., Jami, E. and Mizrahi, I. 2017b. Compositional and functional dynamics of the bovine rumen methanogenic community across difrerent developmental stages. Environ. Microbiol. 19 (8), 3365-73.

Galloway, J. N., Dentener, F. J., Capone, D. G., et al., 2004. Nitrogen cycles:past, present, and future. Biogeochemistry 70 (2), 153-226.

Gautam, D. P., Rahman, S., Borhan, M. S., et al., 2016. The effect of feeding high fat diet to beef cattle on manure composition and gaseous emission from a feedlot pen surface. J. Anim. Sci. Technol. 58, 22.

Gerber, P. J., Steinfeld, H., Henderson, B., et al., 2013. Tackling climate change through livestock:a global assessment of emissions and mitigation opportunities. Food and Agriculture Organization of the United Nations, Rome.

Gonçalves, R., Mateus, N. and de Freitas, V. 2011. Inhibition of α-amylase activity by condensed tannins. Food Chem. 125 (2), 665-72.

Grainger, C., Clarke, T., Auldist, M. J., et al., 2009. Potential use of *Acacia mearnsii* condensed tannins to reduce methane emissions and nitrogen excretion from grazing dairy cows. Can. J. Anim. Sci. 89 (2), 241-51.

Griffin, W. A., Bremer, V. R., Klopfenstein, T. J., et al., 2012. A meta-analysis evaluation of supplementing dried distillersgrains plus solubles to cattle consuming forage-based diets 1. Prof. Anim. Sci. 28 (3), 306-12.

Guyader, J., Eugene, M., Noziere, P., et al., 2014. Influence of rumen protozoa on methane emission in ruminants:a meta-analysis approach. Animal 8 (11), 1816-25.

Guyader, J., Ungerfeld, E. M. and Beauchemin, K. A. 2017. Redirection of metabolic hydrogen by inhibiting methanogenesis in the rumen simulation technique (RUSITEC). Front. Microbiol. 8, 393.

Haisan, J., Sun, Y., Guan, L. L., et al., 2014. The effects of feeding 3-nitrooxypropanol on methane emissions and productivity of Holstein cows in mid lactation. J. Dairy Sci. 97 (5), 3110-9.

Haisan, J., Sun, Y., Guan, L., et al., 2017. The effects of feeding 3-nitrooxypropanol at two doses on milk production, rumen fermentation, plasma metabolites, nutrient digestibility, and methane emissions in lactating Holstein cows. Anim. Prod. Sci. 57 (2), 282-9.

Hales, K. E., Cole, N. A. and Macdonald, J. C. 2013. Effects of increasing concentrations of wet distillers grains with solubles in steam-flaked, corn-based diets on energy metabolism, carbon-nitrogen balance, and methane emissions of cattle. J. Anim. Sci. 91 (2), 819-28.

Hall, M. B. and Huntington, G. B. 2008. Nutrient synchrony:sound in theory, elusive in practice. J. Anim. Sci. 86 (14 Suppl), E287-92.

Halvorson, J. J., Kronberg, S. L. and Hagerman, A. E. 2017. Effects of dietary tannins on total and extractable nutrients from manure. J. Anim. Sci. 95 (8), 3654-65.

Haque, M. N. 2018. Dietary manipulation:a sustainable way to mitigate methane emissions from ruminants. J. Anim. Sci. Technol. 60, 15.

Hartmann, T. 2007. From waste products to ecochemicals:fifty years research of plant secondary metabolism. Phytochemistry 68 (22-24), 2831-46.

He, Z., Cai, C., Wang, J., et al., 2016. A novel denitrifying methanotroph of the NC10 phylum and its microcolony. Sci. Rep. 6, 32241.

Henderson, G., Cox, F., Ganesh, S., et al., 2015. Rumen microbial community composition varies with diet and host, but a core microbiome is found across a wide geographical range. Sci. Rep. 5, 14567.

Hess, H. D., Tiemann, T. T., Noto, F., et al., 2006. Strategic use of tannins as means to limit methane emission from ruminant livestock. Int. Congr. Ser. 1293, 164-7.

Hook, S. E., Steele, M. A., Northwood, K. S., et al., 2011. Impact of high-concentrate feeding and low ruminal pH on methanogens and protozoa in the rumen of dairy cows. Microb. Ecol. 62 (1), 94-105.

Hristov, A. N., Callaway, T. R., Lee, C., et al., 2012. Rumen bacterial, archaeal, and fungal diversity of dairy cows in response to ingestion of lauric or myristic acid. J. Anim. Sci. 90 (12), 4449-57.

Hristov, A. N., Oh, J., Lee, C., et al., 2013. Mitigation of Greenhouse Gas Emissions in Livestock Production-A Review of Technical Options for non-CO_2 Emissions. Gerber, P. J., Henderson, B. and Makkar, H. P. S. (Eds). Food and Agriculture Organization, Rome, Italy.

Hristov, A. N., Oh, J., Giallongo, F., et al., 2015. An inhibitor persistently decreased enteric methane emission from dairy cows with no negative effect on milk production. Proc. Natl. Acad. Sci. U. S. A. 112 (34), 10663-8.

Hünerberg, M., McGinn, S. M., Beauchemin, K. A., et al., 2013a. Effect of dried distillers grains plus solubles on enteric methane emissions and nitrogen excretion from growing beef cattle1. J. Anim. Sci. 91 (6), 2846-57.

Hünerberg, M., McGinn, S. M., Beauchemin, K. A., et al., 2013b. Effect of dried distillers' grains with solubles on enteric methane emissions and nitrogen excretion from finishing beef cattle. Can. J. Anim. Sci. 93 (3), 373-85.

Hünerberg, M., Little, S. M., Beauchemin, K. A., et al., 2014. Feeding high concentrations of corn dried distillers' grains decreases methane, but increases nitrous oxide emissions from beef cattle production. Agric. Sys. 127, 19-27.

Huws, S. A., Creevey, C. J., Oyama, L. B., et al., 2018. Addressing global ruminant agricultural challenges through understanding the rumen microbiome: past, present, and future. Front. Microbiol. 9, 2161.

IPCC. 2006. IPCC Guidelines for National Greenhouse Gas Inventories-A Primer. Institute for Global Environmental Strategies, Japan.

Ishaq, S. L., AlZahal, O., Walker, N., et al., 2017. An investigation into rumen fungal and protozoal diversity in three rumen fractions, during high-fiber or grain-induced sub-acute ruminal acidosis conditions, with or without active dry yeast supplementation. Front. Microbiol. 8, 1943.

Iwamoto, M., Asanuma, N. and Hino, T. 2002. Ability of *Selenomonas ruminantium*, *Veillonella parvula*, and *Wolinella succinogenes* to reduce nitrate and nitrite with special reference to the suppression of ruminal methanogenesis. Anaerobe 8 (4), 209-15.

Jia, X., Wang, M., Yuan, W., et al., 2016. N2O emission and nitrogen transformation in chicken manure and biochar co-composting. Trans. ASABE 59 (5), 1277-83.

Jin, D., Zhao, S., Zheng, N., et al., 2017. Differences in ureolytic bacterial composition between the rumen digesta andrumen wall based on *ureC* gene classification. Front. Microbiol. 8, 385.

Johnson, K. A. and Johnson, D. E. 1995. Methane emissions from cattle. J. Anim. Sci. 73 (8), 2483-92.

Jordan, G., Predotova, M., Ingold, M., et al., 2015. Effects of activated charcoal and tannin added to compost and to soil on carbon dioxide, nitrous oxide and ammonia volatilization. J. Plant Nutr. Soil Sci. 178 (2), 218-28.

Joseph, S., Pow, D., Dawson, K., et al., 2015. Feeding biochar to cows:an innovative solution for improving soil fertility and farm productivity. Pedosphere 25 (5), 666-79.

Joseph, S., Kammann, C. I., Shepherd, J. G., et al., 2018. Microstructural and associated chemical changes during the composting of a high temperature biochar:mechanisms for nitrate, phosphate and other nutrient retention and release. Sci. Total Environ. 618, 1210-23.

Joye, S. B. 2012. Microbiology:a piece of the methane puzzle. Nature 491 (7425), 538-9.

Judy, J. V., Brown-Brandl, T. M., Fernando, S. C., et al., 2016. 1454 Manipulation of lactating dairy cows diets using reduced-fat distillers' grains, corn oil, and calcium sulfate to reduce methane production measured by indirect calorimetry. J. Anim. Sci. 94 (suppl_5), 706.

Kajikawa, H., Valdes, C., Hillman, K., et al., 2003. Methane oxidation and its coupled electron-sink reactions in ruminal fluid. Lett. Appl. Microbiol. 36 (6), 354-7.

Karhu, K., Mattila, T., Bergström, I., et al., 2011. Biochar addition to agricultural soil increased CH_4 uptake and water holding capacity-results from a short-term pilot field study. Agric. Ecosyst. Environ. 140 (1-2), 309-13.

Kim, J. N., Méndez-García, C., Geier, R. R., et al., 2017. Metabolic networks for nitrogen utilization in *Prevotella ruminicola* 23. Sci. Rep. 7 (1), 7851.

Klieve, A. V., Ouwerkerk, D. and Maguire, A. J. 2012. Archaea in the foregut of macropod marsupials: PCR and amplicon sequence-based observations. J. Appl. Microbiol. 113 (5), 1065-75.

Knapp, J. R., Laur, G. L., Vadas, P. A., et al., 2014. Invited review:enteric methane in dairy cattle production:quantifying the opportunities and impact of reducing emissions. J. Dairy Sci. 97 (6), 3231-61.

Koenig, K. M. and Beauchemin, K. A. 2018. Effect of feeding condensed tannins in high protein finishing diets containing corn distillers grains onruminal fermentation, nutrient digestibility, and route of nitrogen excretion in beef cattle. J. Anim. Sci. 96 (10), 4398-413.

Koenig, K. M., Beauchemin, K. A. and McGinn, S. M. 2018. Feeding condensed tannins to mitigate ammonia emissions from beef feedlot cattle fed high-protein finishing diets containing distillers grains. J. Anim. Sci. 96 (10), 4414-30.

Koneswaran, G. and Nierenberg, D. 2008. Global farm animal production and global warming:impacting and mitigating climate change. Environ. Health Perspect. 116 (5), 578-82.

Kraus, T. E. C., Zasoski, R. J., Dahlgren, R. A., et al., 2004. Carbon and nitrogen dynamics in a forest soil amended with purified tannins from different plant species. Soil Biol. Biochem. 36 (2), 309-21.

Kumar, S., Indugu, N., Vecchiarelli, B., et al., 2015. Associative patterns among anaerobic fungi, methanogenic archaea, and bacterial communities in response to changes in diet and age in the rumen of dairy cows. Front. Microbiol. 6, 781.

Lamar, R. T., Olk, D. C., Mayhew, L., et al., 2014. A new standardized method for quantification of humic and fulvic acids in humic ores and commercial products. J. AOAC Int. 97 (3), 721-30.

Latham, E. A., Anderson, R. C., Pinchak, W. E., et al., 2016. Insights on alterations to the rumen ecosystem by nitrate and nitrocompounds. Front. Microbiol. 7, 228.

Lee, C. and Beauchemin, K. A. 2014. A review of feeding supplementary nitrate to ruminant animals:nitrate toxicity, methane emissions, and production performance. Can. J. Anim. Sci. 94 (4), 557-70.

Lee, C., Araujo, R. C., Koenig, K. M., et al., 2015a. Effects of encapsulated nitrate on eating behavior, rumen fermentation, and blood profile of beef heifers fed restrictively or ad libitum. J. Anim. Sci. 93 (5), 2405-18.

Lee, C., Araujo, R. C., Koenig, K. M., et al., 2015b. Effects of encapsulatednitrate on enteric methane production and nitrogen and energy utilization in beef heifers. J. Anim. Sci. 93 (5), 2391-404.

Leng, R. A. 2014. Interactions between microbial consortia in biofilms: a paradigm shift in rumen microbial ecology and enteric methane mitigation. Anim. Prod. Sci. 54 (5), 519-43.

Leng, R., Preston, T. and Inthapanya, S. 2012. Biochar reduces enteric methane and improves growth and feed conversion in local "Yellow" cattle fed cassava root chips and fresh cassava foliage. Livest. Res. Rural Dev. 24.

Li, L., Davis, J., Nolan, J., et al., 2012. An initial investigation on rumen fermentation pattern and methane emission of sheep offered diets containing urea or nitrate as the nitrogen source. Anim. Prod. Sci. 52 (7), 653-8.

Li, F., Wang, Z., Dong, C., et al., 2017. Rumen bacteria communities and performances of fattening lambs with a lower or greater subacute ruminal acidosis risk. Front. Microbiol. 8, 2506.

Liu, L., Xu, X., Cao, Y., et al., 2017. Nitrate decreases methane production also by increasing methane oxidation through stimulating NC10 population in ruminal culture. AMB Express 7 (1), 76.

Lloyd, K. G., Lapham, L. and Teske, A. 2006. An anaerobic methane-oxidizing community of ANME-1b archaea in hypersaline Gulf of Mexico sediments. Appl. Environ. Microbiol. 72 (11), 7218-30.

Lopes, J. C., de Matos, L. F., Harper, M. T., et al., 2016. Effect of 3-nitrooxypropanol on methane and hydrogen emissions, methane isotopic signature, and ruminal fermentation in dairy cows. J. Dairy Sci. 99 (7), 5335-44.

Lyons, T., Boland, T., Storey, S., et al., 2017. Linseed oil supplementation of lambs' diet in early life leads to persistent changes in rumen microbiome structure. Front. Microbiol. 8, 1656.

Maeda, K., Hanajima, D., Toyoda, S., et al., 2011. Microbiology of nitrogen cycle in animal manure compost. Microb. Biotechnol. 4 (6), 700-9.

Maeda, K., Spor, A., Edel-Hermann, V., et al., 2015. N_2O production, a widespread trait in fungi. Sci. Rep. 5, 9697.

Mao, S. Y., Huo, W. J. and Zhu, W. Y. 2016. Microbiome-metabolome analysis reveals unhealthy alterations in the composition and metabolism of ruminal microbiota with increasing dietary grain in a goat model. Environ. Microbiol. 18 (2), 525-41.

Martin, C., Ferlay, A., Mosoni, P., et al., 2016. Increasing linseed supply in dairy cow diets based on hay or corn silage: effect on enteric methane emission, rumen microbial fermentation, and digestion. J. Dairy Sci. 99 (5), 3445-56.

Martinez, C. M., Alvarez, L. H., Celis, L. B., et al., 2013. Humus-reducing microorganisms and their valuable contribution in environmental processes. Appl. Microbiol. Biotechnol. 97 (24), 10293-308.

Martínez-Fernández, G., Abecia, L., Arco, A., et al., 2014. Effects of ethyl-3-nitrooxy propionate and 3-nitrooxypropanol on ruminal fermentation, microbial abundance, and methane emissions in sheep. J. Dairy Sci. 97 (6), 3790-9.

Martinez-Fernandez, G., Denman, S. E., Yang, C., et al., 2016. Methane inhibition alters the microbial community, hydrogen flow, and fermentation response in the rumen of cattle. Front. Microbiol.

7, 1122.

Martinez-Fernandez, G., Denman, S. E., Cheung, J., et al., 2017. Phloroglucinol degradation in the rumen promotes the capture of excess hydrogen generated from methanogenesis inhibition. Front. Microbiol. 8, 1871.

Martinez-Fernandez, G., Duval, S., Kindermann, M., et al., 2018. 3-NOP vs. halogenated compound: methane production, ruminal fermentation and microbial community response in forage fed cattle. Front. Microbiol. 9, 1582.

Masoom, H., Courtier-Murias, D., Farooq, H., et al., 2016. Soil organic matter in its native state: unravelling the most complex biomaterial on earth. Environ. Sci. Technol. 50 (4), 1670-80.

McAllister, T. A., Bae, H. D., Yanke, L. J., et al., 1994. Effect of condensed tannins from birdsfoot trefoil on endoglucanase activity and the digestion of cellulose filter paper by ruminal fungi. Can. J. Microbiol. 40 (4), 298-305.

McGinn, S. M., Chung, Y.-H., Beauchemin, K. A., et al., 2009. Use of corn distillers' dried grains to reduce enteric methane loss from beef cattle. Can. J. Anim. Sci. 89 (3), 409-13.

McMurphy, C. P., Duff, G. C., Harris, M. A., et al., 2009. Effect of humic/fulvic acid in beef cattle finishing diets on animal performance, ruminal ammonia and serum urea nitrogen concentration. J. Appl. Anim. Res. 35 (2), 97-100.

McSweeney, C., Palmer, B., Kennedy, P., et al., 1998. Effect of Calliandra tannins on rumen microbial function. Anim. Prod. Aust 22, 289.

McSweeney, C. S., Palmer, B., McNeill, D. M., et al., 2001. Microbial interactions with tannins: nutritional consequences for ruminants. Anim. Feed Sci. Technol. 91 (1-2), 83-93.

Miller, K. E., Lai, C.-T., Friedman, E. S., et al., 2015. Methane suppression by iron and humic acids in soils of the Arctic Coastal Plain. Soil Biol. Biochem. 83, 176-83.

Mitsumori, M., Ajisaka, N., Tajima, K., et al., 2002. Detection of Proteobacteria from the rumen by PCR using methanotroph-specific primers. Lett. Appl. Microbiol. 35 (3), 251-5.

Morgavi, D. P., Forano, E., Martin, C., et al., 2010. Microbial ecosystem and methanogenesis in ruminants. Animal 4 (7), 1024-36.

Muñoz, C., Hube, S., Morales, J. M., et al., 2015. Effects of concentrate supplementation on enteric methane emissions and milk production of grazing dairy cows. Livest. Sci. 175, 37-46.

Mutabaruka, R., Hairiah, K. and Cadisch, G. 2007. Microbial degradation of hydrolysable and condensed tannin polyphenol-protein complexes in soils from different land-use histories. Soil Biol. Biochem. 39 (7), 1479-92.

National Research Council. 2008. Acute Exposure Guideline Levels for Selected Airborne Chemicals: Volume 6. Available at: https://www.ncbi.nlm.nih.gov/books/NBK207883/.

Naumann, H. D., Tedeschi, L. O., Zeller, W. E., et al., 2017. The role of condensed tannins in ruminant animal production: advances, limitations and future directions. Rev. Bras. Zootec. 46 (12), 929-49.

Niu, M., Appuhamy, J. A. D. R. N., Leytem, A. B., et al., 2016. Effect of dietary crude protein and forage contents on enteric methane emissions and nitrogen excretion from dairy cows simultaneously. Anim. Prod. Sci. 56 (3).

Nolan, J. V., Hegarty, R. S., Hegarty, J., et al., 2010. Effects of dietary nitrate on fermentation, methane production and digesta kinetics in sheep. Anim. Prod. Sci. 50 (8), 801-6.

Norton, J. M. 2008. Nitrification in agricultural soils. In: Schepers, J. S. and Raun, W. R. (Eds), Nitrogen in Agricultural Systems. American Society of Agronomy, Crop Science Society of America, Soil Science Society of America, Madison, WI, pp. 173-99.

Oldick, B. S. and Firkins, J. L. 2000. Effects of degree of fat saturation on fiber digestion and microbial protein synthesis when diets are fed twelve times daily. J. Anim. Sci. 78 (9), 2412-20.

Olijhoek, D. W., Hellwing, A. L. F., Brask, M., et al., 2016. Effect of dietary nitrate level on enteric methane production, hydrogen emission, rumen fermentation, and nutrient digestibility in dairy cows. J. Dairy Sci. 99 (8), 6191-205.

Opio, C., Gerber, P., Mottet, A., et al., 2013. Greenhouse gas emissions from ruminant supply chains—A global life cycle assessment. Food and Agriculture Organization of the United Nations (FAO), Rome.

Parmar, N. R., Nirmal Kumar, J. I. and Joshi, C. G. 2015. Exploring diet-dependent shifts in methanogen and methanotroph diversity in the rumen of Mehsani buffalo by a metagenomics approach. Front. Life Sci. 8 (4), 371-8.

Patra, A. K. and Saxena, J. 2011. Exploitation of dietary tannins to improve rumen metabolism and ruminant nutrition. J. Sci. Food Agric. 91 (1), 24-37.

Patra, A. K. and Yu, Z. 2013. Effects of coconut and fish oils on ruminal methanogenesis, fermentation, and abundance and diversity of microbial populations *in vitro*. J. Dairy Sci. 96 (3), 1782-92.

Patra, A. K. and Yu, Z. 2014. Combinations of nitrate, saponin, and sulfate additively reduce methane production by rumen cultures in vitrowhile not adversely affecting feed digestion, fermentation or microbial communities. Bioresour. Technol. 155, 129-35.

Plaizier, J. C., Li, S., Tun, H. M., et al., 2017. Nutritional models of experimentally-induced subacute ruminal acidosis (SARA) differ in their impact on rumen and hindgut bacterial communities in dairy cows. Front. Microbiol. 7, 2128.

Ponce, C. H., Arteaga, C. and Flores, A. 2016. Effects of humic acid supplementation on pig growth performance, nitrogen digestibility, odor, and ammonia emission. J. Anim. Sci. 94 (suppl_5), 486.

Powell, J. M., Aguerre, M. J. and Wattiaux, M. A. 2011. Dietary crude protein and tannin impact dairy manure chemistry and ammonia emissions from incubated soils. J. Environ. Qual. 40 (6), 1767-74.

Ramos, A. F. O., Terry, S. A., Holman, D. B., et al., 2018. Tucumã oil shifted ruminal fermentation, reducing methane production and altering the microbiome but decreased substrate digestibility within a RUSITEC fed a mixed hay-concentrate diet. Front. Microbiol. 9, 1647.

Reynolds, C. K., Humphries, D. J., Kirton, P., et al., 2014. Effects of 3-nitrooxypropanol on methane emission, digestion, and energy and nitrogen balance of lactating dairy cows. J. Dairy Sci. 97 (6), 3777-89.

Richards, M. A., Lie, T. J., Zhang, J., et al., 2016. Exploring hydrogenotrophic methanogenesis: a genome scale metabolic reconstruction of *Methanococcus maripaludis*. J. Bacteriol. 198, 3379-90.

Robertson, G. P. and Vitousek, P. M. 2009. Nitrogen in agriculture: balancing the cost of an essential resource. Annu. Rev. Environ. Resour. 34 (1), 97-125.

Romero-Perez, A., Okine, E. K., McGinn, S. M., et al., 2014. The potential of 3-nitrooxypropanol to lower enteric methane emissions from beef cattle1. J. Anim. Sci. 92 (10), 4682-93.

Romero-Perez, A., Okine, E. K., McGinn, S. M., et al., 2015. Sustained reduction in methane production from long-term addition of 3-nitrooxypropanol to a beef cattle diet. J. Anim. Sci. 93 (4),

1780-91.

Saleem, A. M., Ribeiro, G. O., Yang, W. Z., et al., 2018. Effect of engineered biocarbon on rumen fermentation, microbial protein synthesis, and methane production in an artificial rumen (RUSITEC) fed a high forage diet. J. Anim. Sci. 96 (8), 3121-30.

Sauvant, D., Giger-Reverdin, S., Serment, A., et al., 2011. Influences des régimes et de leur fermentation dans le rumen sur la production de méthane par les ruminants. INRA Product. Anim. 24, 433-46.

Scheid, D., Stubner, S. and Conrad, R. 2003. Effects of nitrate-and sulfate-amendment on the methanogenic populations in rice root incubations. FEMS Microbiol. Ecol. 43 (3), 309-15.

Shanks, O. C., Kelty, C. A., Archibeque, S., et al., 2011. Community structures of fecal bacteria in cattle from different animal feeding operations. Appl. Environ. Microbiol. 77 (9), 2992-3001.

Sharma, P., Laor, Y., Raviv, M., et al., 2017. Compositional characteristics of organic matter and its water-extractable components across a profile of organically managed soil. Geoderma 286, 73-82.

Sheng, P., Ribeiro, G. O., Wang, Y., et al., 2017. Humic substances supplementation reduces ruminal methane production and increases the efficiency of microbial protein synthesis *in vitro*. J. Anim. Sci. 95 (suppl_4), 300.

Shi, Y., Parker, D., Cole, N., et al., 2001. Surface amendments to minimize ammonia emissions from beef cattle feedlots. Transactions of the ASAE 44 (3), 677-82.

Steiner, C., Das, K. C., Melear, N., et al., 2010. Reducing nitrogen loss during poultry litter composting using biochar. J. Environ. Qual. 39 (4), 1236-42.

Stevenson, F. J. 1995. Humus Chemistry: Genesis, Composition, Reactions, Second Edition. J. Chem. Educ. 72, A93.

Tan, H. Y., Sieo, C. C., Lee, C. M., et al., 2011. Diversity of bovine rumen methanogens *in vitro* in the presence of condensed tannins, as determined by sequence analysis of 16S rRNA gene library. J. Microbiol. 49 (3), 492-8.

Tan, W., Jia, Y., Huang, C., et al., 2018. Increased suppression of methane production by humic substances in response to warming in anoxic environments. J. Environ. Manage. 206, 602-6.

Tapio, I., Fischer, D., Blasco, L., et al., 2017a. Taxon abundance, diversity, co-occurrence and network analysis of the ruminal microbiota in response to dietary changes in dairy cows. PLoS ONE 12 (7), e0180260.

Tapio, I., Snelling, T. J., Strozzi, F., et al., 2017b. The ruminal microbiome associated with methane emissions from ruminant livestock. J. Anim. Sci. Biotechnol. 8, 7.

Tawheed, M. E. S. and Baowei, Z. 2017. Review paper: the fundamentals of biochar as a soil amendment tool and management in agriculture scope: an overview for farmers and gardeners. J. Agric. Chem. Environ. 6, 38-61.

Terry, S. A., Ramos, A. F. O., Holman, D. B., et al., 2018a. Humic substances alter ammonia production and the microbial populations within a RUSITEC fed a mixed hay-concentrate diet. Front. Microbiol. 9, 1410.

Terry, S. A., Ribeiro, G. O., Gruninger, R. J., et al., 2018b. Effect of humic substances on rumenfermentation, nutrient digestibility, methane emissions, and rumen microbiota in beef heifers. J. Anim. Sci. 96, 3863-77.

Terry, S. A., Ribeiro, G. O., Gruninger, R. J., et al., 2019. A pine enhanced biochar does not decrease enteric CH_4 emissions, but alters the rumen microbiota. Front. Vet. Sci. 6, 308.

Turk, J. 2016. Meeting projected food demands by 2050: understanding and enhancing the role of grazing ruminants. J. Anim. Sci. 94 (suppl_6), 53-62.

Ungerfeld, E. M. 2015a. Limits to dihydrogen incorporation into electron sinks alternative to methanogenesis in ruminal fermentation. Front. Microbiol. 6, 1272.

Ungerfeld, E. M. 2015b. Shifts in metabolic hydrogen sinks in the methanogenesis-inhibited ruminal fermentation: a meta-analysis. Front. Microbiol. 6, 37.

Ungerfeld, E. and Kohn, R. 2006. The role of thermodynamics in the control of ruminal fermentation. In: Sejrsen, K., Hvelplund, T. and Nielsen, M. O. (Eds), RuminantPhysiology: Digestion, Metabolism and Impact of Nutrition on Gene Expression, Immunology and Stress. Wageningen Academic Publishers, Wageningen, Netherlands, pp. 55-85.

Van Wyngaard, J. D. V., Meeske, R. and Erasmus, L. J. 2018. Effect of concentrate feeding level on methane emissions, production performance and rumen fermentation of Jersey cows grazing ryegrass pasture during spring. Anim. Feed Sci. Technol. 241, 121-32.

Van Zijderveld, S. M., Gerrits, W. J. J., Apajalahti, J. A., et al., 2010. Nitrate and sulfate: effective alternative hydrogen sinks for mitigation of ruminal methane production in sheep. J. Dairy Sci. 93 (12), 5856-66.

Van Zijderveld, S. M., Gerrits, W. J. J., Dijkstra, J., et al., 2011. Persistency of methane mitigation by dietary nitrate supplementation in dairy cows. J. Dairy Sci. 94 (8), 4028-38.

Varadyova, Z., Kisidayová, S. and Jalc, D. 2009. Effect of humic acid on fermentation and ciliate protozoan population in rumen fluid of sheep in vitro. J. Sci. Food Agric. 89 (11), 1936-41.

Vyas, D., McGinn, S. M., Duval, S. M., et al., 2016. Effects of sustained reduction of enteric methane emissions with dietary supplementation of 3-nitrooxypropanol on growth performance of growing and finishing beef cattle1. J. Anim. Sci. 94 (5), 2024-34.

Vyas, D., Alemu, A. W., McGinn, S. M., et al., 2018a. The combined effects of supplementing monensin and 3-nitrooxypropanol on methane emissions, growth rate, and feed conversion efficiency in beef cattle fed high-forage and high-grain diets. J. Anim. Sci. 96 (7), 2923-38.

Vyas, D., McGinn, S. M., Duval, S. M., et al., 2018b. Optimal dose of 3-nitrooxypropanol for decreasing enteric methane emissions from beef cattle fed high-forage and high-grain diets. Anim. Prod. Sci. 58 (6), 1049-55.

Waghorn, G. 2008. Beneficial and detrimental effects of dietary condensed tannins for sustainable sheep and goat production—progress and challenges. Anim. Feed Sci. Technol. 147 (1-3), 116-39.

Wallace, R. J., Rooke, J. A., McKain, N., et al., 2015. The rumen microbial metagenome associated with high methane production in cattle. BMC Genomics 16, 839.

Woodward, S. L., Waghorn, G. C. and Laboyrie, P. G. 2004. Condensed tannins in birdsfoot trefoil (Lotus corniculatus) reduce methane emissions from dairy cows. In: Proceedings of the New Zealand Society of Animal Production. New Zealand Society of Animal Production, Hamilton, New Zealand, pp. 160-4.

Yuan, Y., Chen, H., Yuan, W., et al., 2017. Is biochar-manure co-compost a bettersolution for soil health improvement and N_2O emissions mitigation? Soil Biol. Biochem. 113, 14-25.

Zhang, J., Shi, H., Wang, Y., et al., 2017. Effect of dietary forage to concentrate ratios on dynamic profile changes and interactions of ruminal microbiota and metabolites in Holstein heifers. Front. Microbiol. 8, 2206.

Zhao, L., Meng, Q., Ren, L., et al., 2015. Effects of nitrate addition on rumen fermentation, bacterial biodiversity andabundance. Asian-Australas. J. Anim. Sci. 28 (10), 1433-41.

Zhao, L., Meng, Q., Li, Y., et al., 2018. Nitrate decreases ruminal methane production with slight changes to ruminal methanogen composition of nitrate-adapted steers. BMC Microbiol. 18 (1), 21.

Zhou, S., Xu, J., Yang, G. and Zhuang, L. 2014. Methanogenesis affected by the co-occurrence of iron (Ⅲ) oxides and humic substances. FEMSMicrobiol. Ecol. 88 (1), 107-20.

Zhu, Z., Kristensen, L., Difford, G. F., et al., 2018. Changes in rumen bacterial and archaeal communities over the transition period in primiparous Holstein dairy cows. J. Dairy Sci. 101 (11), 9847-62.

第 16 章 瘤胃微生物与宿主间的相互作用及对饲料转化效率、甲烷产量和其他生产性能的影响

Elie Jami，以色列农业研究机构 Volcani 中心；
Itzhak Mizrahi，以色列本-古里安大学

（谭翠译）

1 前言

瘤胃微生物的重要任务是为反刍动物提供大部分的营养需要，即为反刍动物提供高达 70% 的代谢能需要和蛋白需要（Siciliano-Jones 和 Murphy，1989；Bergman，1990），这些都是由于瘤胃中微生物种类繁多，且在不同营养水平上发挥作用（Flint 等，2008；Moraïs 和 Mizrahi，2019a，b）。因此，宿主—微生物相互作用是反刍动物的显著特点。在过去 70 年对反刍动物的研究中，研究者就提出微生物组成差异可能会影响动物生理机能、生产效率和废物排放，这些观点远早于高通量技术的发展，近十年研究人员利用高通量测序技术研究微生物组成（Krause 等，2003）。Hungate（2013）在其开创性著作《瘤胃及其微生物》中提出，通过调控微生物菌群组成来提高日粮纤维消化率，进而提高动物生产力。事实上，关于早期瘤胃微生物组的研究主要集中在可培养细菌方面，研究者认为可通过改善可培养细菌的功能来提高动物生产力（Krause 等，2003），然而，这些早期通过改变瘤胃微生物组成来提高动物生产效率的尝试大多未能达到预期目标（Attwood 等，1988；Flint 等，1989；Wallace 和 Walker，1993；Miyagi 等，1995；Krause 等，1999，2003）。为了实现这一目标，必须首先了解每个微生物组成部分的功能，以及对整个微生物菌群和宿主的影响。

如今，随着人们对瘤胃微生物群落组成的整体研究，以及微生物组整体新观点的涌现，可以利用基本生态学原理对微生物组整体结构和宿主生理反应进行研究，以便人们更加了解微生物组的作用，及其各部分微生物对动物生产效率、动物健康和废物排放（如甲烷）的作用。本章重点介绍瘤胃微生物组对能量利用、甲烷排放和决定微生物组成和选择的潜在遗传因素等方面的作用。

2 瘤胃核心菌群、微生物组成的恢复和变化

2.1 核心菌群

鉴定常见微生物的特征可以了解瘤胃生态系统的更多基本需求，因为它们可能在瘤胃

代谢中发挥关键作用。几项研究发现同一谱系和不同谱系的反刍动物瘤胃中存在共享的核心微生物菌群（Jami 和 Mizrahi，2012a；Henderson 等，2015）。Henderson 等（2015）对32 种反刍动物和非反刍动物的微生物组进行综合分析，发现在不同地理范围、动物谱系和饲养管理水平下瘤胃微生物区系组成存在共享性和分散特性，且发现不同谱系的反刍动物之间存在一个属和种级别的核心细菌菌群，包括普雷沃氏菌属（瘤胃中最主要的菌属）、丁酸弧菌属和瘤胃球菌属（瘤胃中主要的纤维降解菌），以及未分类的毛螺菌科、瘤胃球菌科、梭菌目（所有厚壁菌门）和拟杆菌目（拟杆菌门）。牛的瘤胃中还富含纤维杆菌，这是一种重要的纤维降解菌，在饲喂高粗料日粮牛的瘤胃中含量最高；*Methanobrevibacter gottschalkii* 和 *Methanobrevibacter ruminantium* 两个簇的产甲烷菌是瘤胃中的核心产甲烷菌群代表（Henderson 等，2015）。不同谱系动物的前肠道中存在一定范围的共享菌群，这表明核心菌群在瘤胃功能和代谢中起着关键作用（Shade 和 Handelsman，2012）。最近的研究证实奶牛瘤胃的核心菌群与奶牛的生理特征之间相关性更高（Li 等，2019；Wallace 等，2019）。

2.2 菌群变化

在所有测试的模型动物中，包括反刍动物，尽管严格考虑了饲养环境、管理水平和日粮差异等外部因素，但不同动物个体的微生物组成和丰度仍然存在内部个体差异，（Brulc 等，2009；Jami 和 Mizrahi，2012a；Henderson 等，2015）。Brulc 等（2009）开创性地利用鸟枪法宏基因组测序技术测定瘤胃微生物组成，发现 3 头阉牛的微生物组成存在显著差异，尽管饲喂相同的日粮，但其中 1 头公牛的微生物组成与其他两头公牛明显不同。同样，在相同日粮和管理水平下，利用 Bray-Curtis 指数对 16 头奶牛的瘤胃微生物菌群进行分析，其 Bray-Curtis 相异系数为 0.51，Bray-Curtis 指数不仅考虑样本中菌群类别的有无，还考虑不同菌种的相对丰度；一些菌属的存在和丰度相对稳定，而另一些菌属在丰度上可能表现出两个数量级的差异（Jami 和 Mizrahi，2012a,b）。

2.3 菌群韧性

虽然多个研究一致指出，不同宿主动物的微生物菌群组成存在很大差异（Brulc 等，2009；Li 等，2009；Jami 和 Mizrahi，2012a），但同一头奶牛在不同采样时间点的微生物组成的稳定性非常高（Li 等，2009；Welkie 等，2009）。Welkie 等（2009）利用自动核糖体间隔基因分析（Automated ribosomal intergenic spacer analysis，ARISA）检测一个饲养周期中瘤胃细菌菌群变化，也证实同一头奶牛在不同阶段的瘤胃微生物菌群的稳定性极高，而不同奶牛的微生物菌群组成存在巨大差异。改变动物日粮后，对公牛的微生物区系组成变化进行长期跟踪检测，发现在饲喂新日粮 25 d 后，微生物菌群组成基本趋于稳定（Snelling 等，2019）。微生物区系组成具有一定弹性，即使是很大的干扰，如转变宿主，即用一头奶牛的瘤胃液几乎完全替换另一头奶牛的瘤胃液，但在短短几周内，瘤胃微生物区系组成将恢复到接近之前的微生物组成水平（Weimer 等，2017），这表明宿主可能是决定微生物区系重建的主要因素。最近的研究表明，动物个体的遗传与其瘤胃微生物组和某部分微生物组成的遗传之间存在联系（Roehe 等，2016；Li 等，2016；Sasson 等，2017；

Wallace 等，2019）。Zhou 等（2018）在一个规模较大的研究中发现，在转变宿主后，每头奶牛都表现出独特的微生物菌群重建模式，进一步强调了宿主对微生物区系组成的强有力调控。

3 依赖微生物组的特性

瘤胃微生物活动的发酵产物是反刍动物的主要能量来源，占其70%的代谢需求（Wolin，1979；Bergman，1990）。摄入的植物饲料经过一系列降解，从复杂的聚合物到中间分子，再到终产物，终产物要么被动物吸收，要么被排放到环境中。研究者们利用了多种方法和指标来评定动物将饲料转化为可用于生长和生产产品的效率水平，包括饲料转化率（FCR）、剩余采食量（RFI）、能量校正乳/干物质摄入量（ECM/DMI），以及最近的残留摄入量和增重（RGI）（NRC，2001；Berry 和 Crowley，2012）。尽管每个效率指标的计算方法不同，且要求观察的生理参数不同，但它们的总体基本原理是相似的，即测定摄入的饲料能值与反刍动物用于维持和生产所用能值之间的总体比率（Mizrahi，2012）。根据研究问题，需要更具体的测定来补充全局特征，如牛奶中的蛋白质、碳水化合物和脂质含量（Jami 等，2014）、宿主的健康参数（Jewell 等，2015）、或与能量摄入相关的瘤胃特异性代谢产物（VFA 组成和含量）、能量损失（甲烷）（Hernandez-Sanabria 等，2010；Shabat 等，2016；Tapio 等，2017）。瘤胃微生物区系组成和微生物的基因表达对动物生理可能产生影响，主要是生产性能和甲烷排放，这是过去十年的许多研究重点，研究结果总结见表1。如上所述，不同宿主动物的微生物组中降解和发酵植物纤维的微生物的基本功能是相似的（Moraïs 和 Mizrahi，2019a）。然而，特定微生物类群的组成和丰度及其基因表达的变化与甲烷排放以及特定的瘤胃代谢产物相关，这些代谢产物对宿主生产性状有较大影响，如乳成分和饲料的能量利用效率（Shi 等，2014；Shabat 等，2016；Kamke 等，2016；Li 和 Guan，2017）。在肉牛中，普雷沃氏菌属菌种与动物生产力和乳成分参数的变化有关（Carberry 等，2012；Jami 等，2014；McCann 等，2014；Jewell 等，2015；Shabat 等，2016；Indugu 等，2017）。Jami 等（2014）研究了15头奶牛的微生物区系组成的丰度与奶牛生理参数之间的关系，发现厚壁菌门/拟杆菌门的比值与每日乳脂产量呈正相关关系，还发现厚壁菌门/拟杆菌门的比值差异主要是因为普雷沃氏菌属（拟杆菌门）丰度的巨大变化，普雷沃氏菌属与乳脂产量呈负相关关系。小鼠和人类上的微生物组研究结果也可以反映这些关系，其中拟杆菌门丰度的降低与小鼠血液和组织中脂肪增加相关（Turnbaugh 等，2006）。Jewell 等（2015）调查了奶牛瘤胃中微生物种群在两个泌乳期的动态变化，发现普雷沃氏菌属及其菌种与饲料利用效率相关，且与普雷沃氏菌属相关的特定 OTUs（物种相似性为97%）与生产效率呈负相关，但该研究还发现各种普雷沃氏菌种与高的饲料效率相关，这与另一研究发现普雷沃氏菌属与奶牛产奶量增加有关一致（Indugu 等，2017）。由于该菌属在瘤胃中的高度多样性，不同的普雷沃氏菌种可能对瘤胃生理产生不同的影响（Ley，2016）。我们对该菌属的功能研究仅局限于一些可培养的菌种，这限制了人们对其完整功能范围的探索。Stewart 等（2019）使用深度宏基因组测序从瘤胃中获得了约5 000个基因组，阐明普雷沃氏菌种可能具有多种功能。一些研究根据 RFI 值（低

RFI、高 RFI）对阉牛进行分组，再利用 PCR-DGGE 指纹图谱技术研究瘤胃细菌菌群和古菌菌群，发现特定细菌菌群的丰度高低取决于各类微生物菌群及 VFA 发酵模式（Guan 等，2008；Hernandez-Sanabria 等，2010）。此外，使用 16S rDNA 克隆文库比较研究瘤胃产甲烷菌群，发现低 RFI 阉牛的产甲烷菌群多样性较低，而且产甲烷菌的组成与宿主效率之间可能存在相关性（Zhou 和 Hernandez-Sanabria，2009；Zhou 等，2010）。随后对不同饲料效率和饲喂不同日粮的 58 头公牛的瘤胃产甲烷菌和细菌种群进行分析，发现琥珀酸弧菌属（*Succinivibrio* spp.）和优杆菌属（*Eubacterium* spp.）与阉牛饲料效率增加相关，作者认为低 RFI 公牛瘤胃中琥珀酸弧菌的丰度较高，且有较高的乙酸盐和较低的异戊酸盐，并使瘤胃代谢从甲烷生成转移到丙酸生成（Hernandez-Sanabria 等，2012）。有趣的是，琥珀酸弧菌科及其菌种经常与绵羊和沙袋鼠的低甲烷表型有关（Pope 等，2011；Wallace 等，2015b）。奶牛上也证实了依赖微生物组的代谢调控：从甲烷生成转向 VFA 生成（Shabat 等，2016）。

表 1　评估瘤胃微生物区系组成与动物生理学之间联系的研究列表总结

动物	方法	主要发现	参考文献
肉牛	PCR-DGGE	发现与短链脂肪酸（SCFA）组成相关的饲料效率表型和特定细菌	Guan 等，2008
肉牛	PCR-DGGE	高 RFI 和低 RFI 之间 *M. stadtmanae* 和特定甲烷短杆菌菌株的丰度差异，高 RFI 动物的产甲烷菌总体分类多样性越高	Zhou 等，2009
肉牛	PCR-DGGE，qPCR	在低能量日粮下与特定 VFA 和不同 RFI 表型相关的特定细菌和古菌 OTU	Hernandez-Sanabria 等，2010
肉牛	PCR-DGGE，qPCR	琥珀酸弧菌属与甲烷排放低相关，且饲料利用效率高的动物瘤胃乙酸含量更高	Hernandez-Sanabria 等，2012
肉牛	PCR-DGGE，qPCR	低效率动物中普雷沃氏菌属的丰度高，细菌分布和饲料效率之间的联系因日粮不同而不同	Carberry 等，2012
肉牛	克隆文库和 16S 扩增子测序	高 RFI 动物和低 RFI 动物之间的甲烷短杆菌的丰度不同	Carberry 等，2014
绵羊	鸟枪法宏基因组学和宏转录组测序	与氢营养产甲烷途径相关的基因在甲烷排放高的绵羊中表达更高	Shi 等，2014
奶牛	16S 扩增子测序	发现了细菌菌属与生产参数之间的相关性，其中普雷沃氏菌属与乳脂产量呈负相关	Jami 等，2014
肉牛	16S 扩增子测序	高丰度的普雷沃氏菌属与动物的低效表型有关	McCann 等，2014
绵羊	16S 扩增子测序	鉴定了与高甲烷排放和低甲烷排放相关的三种不同"瘤胃型"的菌群	Kittelmann 等，2014

(续表)

动物	方法	主要发现	参考文献
奶牛	16S 扩增子测序	与高效奶牛相关的核心 OTUs（普雷沃氏菌属），与低效奶牛相关的核心 OTUs（普雷沃氏菌属，丁酸弧菌属）	Jewell 等，2015
奶牛	16S 扩增子测序，鸟枪法宏基因组学，代谢组学	埃氏巨型球菌和丙烯酸酯途径参与提高饲料效率，高效奶牛的细菌多样性较低	Shabat 等，2016
绵羊	宏转录组测序	低甲烷排放表型的绵羊富含阿扎夏普氏菌（*Sharpea azabuensis*）和巨球型菌（*Megasphaera* spp.）及其丙烯酸酯途径	Kamke 等，2016
奶牛	16S 扩增子测序	普雷沃氏菌属、S24-7 和琥珀酸弧菌谱系与产奶量呈正相关	Indugu 等，2017
奶牛	16S 扩增子测序	*M. Ruminantium* 和 *M. gottschalkii* 与甲烷排放低有关	Danielsson 等，2017
肉牛	宏转录组测序	低饲料效率动物的代谢途径多样性更高	Li & Guan，2017
奶牛	鸟枪法宏基因组学	高效奶牛中具有较高丰度的拟杆菌和较低丰度的产甲烷菌	Delgado 等，2019
肉牛	鸟枪法宏基因组学	确定了与 FCR 相关的微生物基因编码的蛋白参与细胞壁生物合成、半纤维素和纤维素降解、宿主-微生物组的交联；与 RFI 相关的微生物基因编码的蛋白参与维生素 B_{12} 生物合成、环境信息处理和细菌迁移	Lima 等，2019
奶牛	16S 扩增子测序，代谢组学，宿主基因分型	核心微生物更能关联和预测多种宿主性状，包括饲料效率、甲烷排放和 VFAs	Wallace 等，2019

注：DGGE=变性梯度凝胶电泳；SCFA=短链脂肪酸；RFI=剩余采食量；qPCR=定量 PCR；OTUs=操作分类单元；VFA=挥发性脂肪酸；FCR=饲料转化率。

  Shabat 等（2016）利用 78 头具有极端 RFI 表型（38 头低效，40 头高效）的奶牛进行研究，发现奶牛细菌菌群分类组成、细菌基因组成和生态特征与饲料效率表型具有显著相关性，且可用于预测饲料效率表型；也发现高效奶牛的微生物组多样性低于低效奶牛，其基因组成更倾向于为奶牛提供可用的终产物，如丙酸盐；作者还发现埃氏巨球型菌（*Megasphaera elsdenii*）和尖锐粪球菌（*Coprococcus catus*），以及其编码的丙烯酸酯途径，该中心途径与动物生产高效率相关，即利用乳酸盐中间产物生成丙酸盐；研究还提出丙烯酸酯途径与 VFA 生成和甲烷生成之间相互作用模型，认为两种表型（低效—高效）虽然不一定结合，但与饲料效率有关。这些发现进一步支持了之前的研究结果，即琥珀酸弧菌科菌群增加与不同 VFA 模式、较低甲烷排放和较高效率相关（Wallace 等，2015b）。溶糊精琥珀酸弧菌（*Succinivibrio dextrinosolvens*）是瘤胃内的一种已知常驻菌，发现在特定生长条件下可以增加其乳酸产量（O'Herrin 和 Kenealy，1993）。综合这些研究结果可知，乳酸作为特定级联发酵的中间体，是决定反刍动物发酵效率的核心。需要注意的是生理参数和微生物组之间的联系可能因特定日粮而定，即在不同日粮中可能观察到不同的联系（Carberry 等，2012）。

4 甲烷产生

甲烷生成仅由栖息在瘤胃中的产甲烷古菌（产甲烷菌）负责。产甲烷菌作为整个瘤胃生态系统的电子汇，驱动发酵过程的方向，否则发酵产生的 H_2 积累会抑制瘤胃发酵（McAllister 和 Newbold，2008；vanLingen 等，2016）。瘤胃中的主要产甲烷途径是氢营养型产甲烷途径，即利用发酵产物 H_2 还原 CO_2 生成甲烷。瘤胃甲烷生成尽管是必要的，但对环境有害，也造成2%~12%的饲料能量损失（Johnson 和 Johnson，1995）。因此，研究微生物组对甲烷产量的影响与研究其对生产效率的影响密切相关（Shabat 等，2016）。反刍动物的甲烷生成是瘤胃生态系统的一个重要组成部分，且甲烷排放量的变化幅度很大。几项研究表明，高的甲烷排放量与牛羊瘤胃中产甲烷菌的绝对丰度间的相关关系很弱或完全无关（Shi 等，2014；Kittelmann 等，2014；Tapio 等，2017）。然而，最近利用全基因组测序研究阉牛的瘤胃微生物，发现古菌的比例可以预测甲烷排放，其相关性系数为0.49（Wallace 等，2015a）。

产甲烷菌的菌群组成差异和甲烷生成的相关基因表达水平差异共同决定了牛羊甲烷排放高低表型，其中几种产甲烷菌类群与甲烷排放增加相关。一般情况下，成熟瘤胃中最丰富的产甲烷菌属是甲烷短杆菌属（*Methanobrevibacter*）（Janssen 和 Kirs，2008；Friedman 等，2017；Tapio 等，2017）。研究者将甲烷短杆菌属分为两个进化簇，即 SGMT 簇（*smithii-gottschalkii-millerae-thaurei*）和 RO 簇（*ruminantium-olleyae*）（King 等，2011），高丰度的 SMGT 簇产甲烷菌具有更高的产甲烷潜力（Danielsson 等，2012，2017；Shi 等，2014）。Shi 等（2014）利用元转录组学，观察到产甲烷高的绵羊瘤胃中 *Methanobrevibacter gottschalkii* 丰度增加，且编码 CO_2/H_2 产甲烷途径的基因转录子丰度也显著增加，但产甲烷菌丰度没有明显差异。

尽管产甲烷菌是瘤胃甲烷的唯一生产者，但它们要依赖细菌、原虫和真菌的上游发酵过程及其发酵产物，这些都会影响瘤胃的甲烷产量。Kittelmann 等（2014）确定了三种不同的细菌群落类型与绵羊的甲烷排放高低表型相关，称为"瘤胃型"（ruminotypes），其中高甲烷排放瘤胃型是富集产氢细菌，如瘤胃球菌、毛螺旋菌、Catabacteriaceae、粪球菌、其他梭菌、普雷沃氏菌、拟杆菌和 α-变形菌。低甲烷排放瘤胃型是富集产丙酸盐的卵状奎因氏菌（*Quinella ovalis*）、产乳酸盐和琥珀酸盐的细菌，如纤维杆菌属、*Kandleria vitulina*、欧陆森氏菌属（*Olsenella*）、布氏普雷沃氏菌（*Prevotella bryantii*）和 *Sharpea azabuensis*（Kittelmann 等，2014）。正如在饲料效率部分提到，产乳酸和琥珀酸的细菌，如溶糊精琥珀酸弧菌（*Succinivibrio dextrinosolvens*）可能是沙袋鼠甲烷排放量低的原因，且与牛的高饲料效率相关（Pope 等，2011；Wallace 等，2015b）。另一研究同样发现乳酸生成菌 *Sharpeae albenenzis* 在甲烷排放低的羊瘤胃中富集（Kamke 等，2016），作者认为，该菌种将原本用于甲烷生成的 H_2 转换到用于生成乳酸,乳酸又通过埃氏巨球型菌(*Megasphaera elsdenii*)的丙烯酸酯途径转化为丙酸。在高效、低 CH_4 排放奶牛中观察到的高丰度乳酸利用菌及其基因也证实了相关结论（Shabat 等，2016）。这些研究发现总体表明,瘤胃营养系统中微生物之间的电子转移差异导致了菌群组成状态的差异（Moraïs 和 Mizrahi,2019a）,进而导致瘤胃发酵终产物产量不同。可替代的菌群状态模型表明,一个瘤胃生态系统可能转向利于 H_2 生成的代谢,从而导致更高的甲烷产量,而另一个则转向利于生成 VFA 方向所需的乳酸生成和利用（Moraïs 和 Mizrahi,2019a）。

除了观察到的细菌区系和甲烷生成之间的联系外,瘤胃纤毛原虫还通过与产甲烷菌的互利共生在增加甲烷排放方面发挥了许多作用（Newbold 等,2015）,该研究主题是过去 30 年里许多研究的焦点。与细菌相反,纤毛原虫不是瘤胃正常功能必不可少的,利用驱除原虫的方法,即通过各种方式从瘤胃中移除瘤胃纤毛原虫,这为评估原虫对瘤胃生态系统和宿主生理功能的影响提供了研究基础。最近一项 Meta-分析总结了近 30 年关于试验动物驱除原虫的研究结果,揭示驱除原虫和具有原虫的反刍动物（绵羊和牛）之间存在显著的生理差异,包括消除纤毛原虫的微生物蛋白质供应和甲烷排放变化（Guyader 等,2014；Newbold 等,2015）。具体而言,Meta-分析揭示驱除原虫可一致降低动物体的甲烷排放量,降低幅度高达 11%。体外研究进一步表明,在原虫区系内,全毛目原虫更多地参与甲烷产生（Belanche 等,2015）。此外,另外一项 Meta-分析比较了 79 个研究中的甲烷排放和原虫细胞丰度,发现动物体内原虫细胞数量与甲烷排放之间具有很强的线性相关关系（Guyader 等,2014）。由于大多瘤胃原虫会产生大量 H_2,研究者认为原虫与产甲烷菌之间的互利共生是源于原虫提供的富氢环境受到氢营养型产甲烷菌的青睐（Newbold 等,2015）。在多泡多甲虫（*Poly. multivesiculatum*）和巴氏甲烷八叠球菌（*Methanosarcina barkeri*）的共培养试验中也观察到了这种互利共生现象,即培养物中的 H_2 减少与甲烷的增加同时发生（Ushida 等,1997）。

使用原位杂交技术研究发现原虫在很大程度上被产甲烷菌定殖,再次证实原虫和产甲烷菌之间存在很强的互利共生关系（Finlay 等,1994；Lloyd 等,1996）。Lloyd 等（1996）发现产甲烷菌附着在原虫的外膜和原虫内部,而且是在原虫食物液泡的外部,这表明产甲烷菌不是原虫捕食的食物,而是作为原虫的共生体。此外,与瘤胃中自由活动的细菌相比,原虫携带了更高的产甲烷菌/细菌比率,这表明产甲烷菌和原虫之间存在特定的趋向性（Levy 和 Jami,2018）。然而,在校正原虫大小时,Belanche 等（2014）没有观察到这种富集,问题仍是产甲烷菌在原虫体内和周围的积累比例是否高于细菌。虽然这个问题仍然悬而未决,但一些研究发现,与游离原核种群相比,与原虫相关的产甲烷菌的组成不同（Tymensen 等,2012；Tymensen 和 McAllister,2012；Belanche 等,2014；Levy 和 Jami,2018）。与原虫相关的产甲烷菌绝大多数是甲烷短杆菌属（Görtz,2006；Tymensen 等,2012；Belanche 等,2014；Levy 和 Jami,2018）。甲烷短杆菌属产甲烷菌通常是氢营养型,这进一步证实产甲烷菌-原虫之间的关系是基于 H_2 的种间电子转移（Janssen 和 Kirs,2008）。此外,观察发现驱除原虫后产甲烷菌丰度可能没有差异,但甲烷排放量降低,推测与原虫相关的产甲烷菌在甲烷生成方面更加活跃,即参与甲烷生成的贡献更大。Levy 和 Jami（2018）发现,当把原虫按大小分开时,大型原虫具有较高丰度的 SGMT 簇甲烷短杆菌相关的OTUs，Danielsson等（2017）研究发现SGMT簇甲烷短杆菌与牛的甲烷排放量高相关，这进一步支持与原虫相关的产甲烷菌在甲烷生成方面更加活跃这一假设。

5 氮化合物：利用和排放

瘤胃系统中的氮循环可能会影响瘤胃菌群组成和宿主属性。在研究饲料效率与瘤胃微生物菌群及其功能之间的联系时，发现饲料效率低的奶牛富含这些与蛋白质消化和氨基酸生物合成相关功能的菌群（Shabat等，2016）。此外，纤毛原虫是通过增加对细菌的捕食来降低动物对氮的利用（Newbold等，2015）。具体来说，内毛属（*Entodinium*）纤毛原虫会降低宿主动物对微生物蛋白质利用，通过驱除原虫消除内毛属纤毛原虫，可以增加30%的蛋白质供应（Newbold等，2015）。瘤胃中微生物的蛋白水解活性因动物而异，过高的蛋白水解活性，如脱氨基，并不利于氮的有效利用（Hartinger等，2018）。因此，氮利用效率低会对动物生产和环境产生负面影响，导致大量N_2O释放到大气中（Huws等，2018）。特定菌群（如高产氨细菌）的丰度差异可能导致氮的利用效率不同，因此调控这些菌群可以提高动物的氮利用效率（Firkins等，2007；Hartinger等，2018），这可能是今后研究的重要内容，以便更好地了解菌群组成与N循环效率之间的潜在关系。

6 微生物组和宿主遗传

虽然在微生物组与宿主生理和生产力之间的关系研究方面取得了重要进展，但关于宿主遗传对微生物组成和选择性调控问题方面仍未得到充分研究，特别是与人类微生物组的大量研究相比（Rothschild等，2018）。宿主遗传是否会影响微生物组成，进而影响与能量获取和甲烷排放相关的微生物特征，这些会指导我们通过选择性育种合理选择有利的基因表型（Myer，2019）。在人类和小鼠上的类似研究数据更多，一些研究指出微生物菌群组成取决于宿主的基因组成（Benson等，2010；Goodrich等，2014，2016；Bonder等，2016；Turpin等，2016）。然而，最近的一项人类研究质疑宿主遗传对微生物组成和丰度的影响程度（Rothschild等，2018）。通过比较不同品种及其杂交品种肉牛的微生物组成，指示宿主遗传会影响宿主的微生物菌群组成和生理参数（Guan等，2008；Hernandez-Sanabria等，2013；Paz等，2016）。Roehe等（2016）研究了两个不同公牛品种的宿主遗传、微生物组成（特别是古菌：细菌比值）和甲烷之间的联系，进一步支持动物选择是基于遗传学进行。Roehe等（2016）提出宿主动物可以通过一些机制控制微生物组成及其后续影响，如唾液对pH值缓冲能力或瘤胃内饲料滞留时间，后者是一种遗传性状。Sasson等（2017）研究了47头不同饲料效率奶牛的微生物丰度数据与基因组图谱，确定了22个与瘤胃代谢和宿主生理性状相关的OTUs，这些遗传性状是可测量的。研究表明，与其他瘤胃微生物相比，这些可遗传的OTUs与宿主生理和瘤胃代谢物的联系更紧密（Sasson等，2017）。最近，三项规模较大的研究是使用全基因组关联分析（GWAS）检测了数百头动物，以建立宿主遗传与微生物组成之间的联系，三项研究的结果存在一些相似之处（Difford等，2018；Li等，2019；Wallace等，2019）。Difford等（2018）利用750

头奶牛的数据，确定了物种水平上6%的细菌菌群和12%的古菌菌群的丰度是可遗传的（$h^2>0.15$），即受宿主基因组控制，并且甲烷排放与微生物组特征和宿主遗传相关，但这两种关联在很大程度上是独立的。宿主遗传控制甲烷排放和微生物组的独立性表明，动物的甲烷排放差异可能不是宿主遗传对微生物组影响的结果。因此，作者建议应考虑两条平行的甲烷减排途径：一是育种选择动物基因组中低甲烷排放相关的性状，二是调控能够降低甲烷排放的微生物组（Difford等，2018）。相比之下，Li等（2019）利用669头公牛进行研究，发现59种微生物菌群（56种细菌和3种古菌）的丰度与动物遗传相关，这些菌群也与宿主的饲料效率性状和瘤胃代谢产物有关，研究还揭示了19个单核苷酸多态性（SNP），其中5个饲料效率相关的位点与12种微生物菌群类别相关。因此，这项研究表明，宿主对瘤胃微生物亚群的控制可能直接影响动物生产效率。此外，利用交互网络推理发现4个可遗传的菌群类别与大量菌群类别进行相互作用。Wallace等（2019）对欧洲4个国家的1 000头奶牛进行了大规模的全基因组关联分析（GWAS），通过瘤胃微生物组测序，以及大量代谢和生理参数的测定，发现"核心"菌群类别与宿主遗传的变异有关，"核心"菌群类别是指特定农场内至少50%的奶牛反复出现的菌群类别，因此宿主遗传可以在很大程度上解释"核心"菌群类别的丰度。在这个"核心"微生物菌群中，发现39个菌群类别的遗传力高达$h^2=0.6$。此外，使用表型特征的网络推理表明，与非核心菌群类别相比，这些可遗传的核心微生物与动物生产力参数的联系更紧密。此外，Li等（2019）也得到一致的研究结果，发现这些可遗传的微生物是微生物菌群相互作用网络的核心，表明它们是关键的微生物菌种。这些研究结果表明，通过控制核心微生物的一个子集，宿主就可以有效地间接调控更广泛的微生物，进而对瘤胃生理进行间接调控。这些结果支撑了选择核心微生物菌群的观点，确定核心微生物菌群对动物生理特征方面发挥的重大作用。尽管与最新的人类研究看似矛盾，但结果中仍有一些相似之处，而实验设计的差异可能部分解释了这些差异。首先，与人类一样，只有相对较小的菌群类别子集可能与宿主遗传相关（1.9%）（Rothschild等，2018），这对于牛也是如此（0.25%的OTUs）（Wallace等，2019）。然而，就丰度而言，这些菌群类别占肉牛微生物菌群的60%，而人类的可遗传菌群类别占微生物菌群的6%左右。此外，与人类研究相反，动物研究对许多生理和管理参数进行了更严格的控制，而人类研究不能绝对确定和控制人与人之间的食物差异。实现环境因素同质化可能能够揭示宿主遗传对微生物组成结构的强力影响。然而，很明显，日粮等环境因素可能对微生物组组成产生更大影响，但甲烷排放等产出参数的遗传性表明，宿主遗传可能影响每个动物中可替代的稳定的微生物菌群状态的去留。因此需要根据反刍动物的遗传来设计新的育种选择标准，进而引导微生物菌群向农业有利的表型方向发展。

7 参考文献

Attwood, G. T., Lockington, R. A., Xue, G. P., et al., 1988. Use of a unique gene sequence as a probe to enumerate a strain of *Bacteroides ruminicola* introduced into the rumen. Applied and Environmental Microbiology 54, 534-9.

Belanche, A., de la Fuente, G. and Newbold, C. J. 2014. Study of methanogen communities associated with different rumen protozoal populations. FEMS Microbiology Ecology 90, 663-77.

Belanche, A., de la Fuente, G. and Newbold, C. J. 2015. Effect of progressive inoculation of fauna-free sheep with holotrich protozoa and total-fauna on rumen fermentation, microbial diversity and methane emissions. FEMS Microbiology Ecology 91 (3).

Benson, A. K., Kelly, S. A., Legge, R., et al., 2010. Individuality in gut microbiota composition is a complex polygenic trait shaped by multiple environmental and host genetic factors. Proceedings of the National Academy of Sciences of the United States of America 107, 18933-8.

Bergman, E. N. 1990. Energy contributions of volatile fatty acids from the gastrointestinal tract in various species. Physiological Reviews 70, 567-90.

Berry, D. P. and Crowley, J. J. 2012. Residual intake and body weight gain: a new measure of efficiency in growing cattle. Journal of Animal Science 90, 109-15.

Bonder, M. J., Kurilshikov, A., Tigchelaar, E. F., et al., 2016. The effect of host genetics on the gut microbiome. Nature Genetics 48, 1407-12.

Brulc, J. M., Antonopoulos, D. A., Miller, M. E., et al., 2009. Gene-centric metagenomics of the fiber-adherent bovine rumen microbiome reveals forage specific glycoside hydrolases. Proceedings of the National Academy of Sciences of the United States of America 106, 1948-53.

Carberry, C. A., Kenny, D. A., Han, S., et al., 2012. Effect of phenotypic residual feed intake and dietary forage content on the rumen microbial community of beef cattle. Applied and Environmental Microbiology 78, 4949-58.

Carberry, C. A., Waters, S. M., Waters, S. M., et al., 2014. Rumen methanogenic genotypes differ in abundance according to host residual feed intake phenotype and diet type. Applied and Environmental Microbiology 80, 586-94.

Danielsson, R., Schnürer, A. and Arthurson, V. 2012. Methanogenic population and CH_4 production in Swedish dairy cows fed different levels of forage. Applied and Environmental Microbiology 78 (17), 6172-9.

Danielsson, R., Dicksved, J., Sun, L., et al., 2017. Methane production in dairy cows correlates with rumen methanogenic and bacterial community structure. Frontiers in Microbiology 8, 226.

Delgado, B., Bach, A., Guasch, I., et al., 2019. Whole rumen metagenome sequencing allows classifying and predicting feed efficiency and intake levels in cattle. Scientific Reports 9, 11.

Difford, G. F., Plichta, D. R., Løvendahl, P., et al., 2018. Host genetics and the rumen microbiome jointly associate with methane emissions in dairy cows. PLoS Genetics 14, e1007580.

Finlay, B. J., Esteban, G., Clarke, K. J., et al., 1994. Some rumen ciliates have endosymbiotic methanogens. FEMS Microbiology Letters 117, 157-61.

Firkins, J. L., Yu, Z. and Morrison, M. 2007. Ruminal nitrogen metabolism: perspectives for integration of microbiology and nutrition for dairy. Journal of Dairy Science 90 Suppl. 1, E1-16.

Flint, H. J., Bisset, J. and Webb, J. 1989. Use of antibiotic resistance mutations to track strains of obligately anaerobic bacteria introduced into the rumen of sheep. The Journal of Applied Bacteriology 67, 177-83.

Flint, H. J., Bayer, E. A., Rincon, M. T., et al., 2008. Polysaccharide utilization by gut bacteria: potential for new insights from genomic analysis. Nature Reviews. Microbiology 6, 121-31.

Friedman, N., Jami, E. and Mizrahi, I. 2017. Compositional and functional dynamics of the bovine rumen methanogenic community across different developmental stages. Environmental Microbiology 19（8），3365-73.

Goodrich, J. K., Waters, J. L., Poole, A. C., et al., 2014. Human genetics shape the gut microbiome. Cell 159, 789-99.

Goodrich, J. K., Davenport, E. R., Beaumont, M., et al., 2016. Genetic determinants of the gut microbiome in UK twins. Cell Host & Microbe 19, 731-43.

Görtz, H. -D. 2006. Symbioticassociations between ciliates and prokaryotes. In: Dworkin, M., Falkow, S., Rosenberg, E., Schleifer, K. -H. and Stackebrandt, E. (Eds), The Prokaryotes: Volume 1: Symbiotic Associations, Biotechnology, Applied Microbiology. Springer New York, New York, NY, pp. 364-402.

Guan, L. L., Nkrumah, J. D., Basarab, J. A., et al., 2008. Linkage of microbial ecology to phenotype: correlation of rumen microbial ecology to cattle's feed efficiency. FEMS Microbiology Letters 288, 85-91.

Guyader, J., Eugène, M., Nozière, P., et al., 2014. Influence of rumen protozoa on methane emission in ruminants: a meta-analysis approach. Animal 8, 1816-25.

Hartinger, T., Gresner, N. and Südekum, K. -H. 2018. Does intra-ruminal nitrogen recycling waste valuable resources? A review of major players and their manipulation. Journal of Animal Science and Biotechnology 9, 33.

Henderson, G., Cox, F., Ganesh, S., et al., 2015. Rumen microbial community composition varies with diet and host, but a core microbiome is found across a wide geographical range. Scientific Reports 5.

Hernandez-Sanabria, E., Guan, L. L., Goonewardene, L. A., et al., 2010. Correlation of particular bacterial PCR-denaturing gradient gel electrophoresis patterns with bovine ruminal fermentation parameters and feed efficiency traits. Applied and Environmental Microbiology 76, 6338-50.

Hernandez-Sanabria, E., Goonewardene, L. A., Wang, Z., et al., 2012. Impact of feed efficiency and diet on adaptive variations in the bacterial community in the rumen fluid of cattle. Applied and Environmental Microbiology 78, 1203-14.

Hernandez-Sanabria, E., Goonewardene, L. A., Wang, Z., et al., 2013. Influence of sire breed on the interplay among rumen microbial populations inhabiting the rumen liquid of the progeny in beef cattle. PLoS ONE 8, e58461.

Hungate, R. E. 2013. *The Rumen and Its Microbes*. Elsevier.

Huws, S. A., Creevey, C. J., Oyama, L. B., et al., 2018. Addressing global ruminant agricultural challenges through understanding the rumen microbiome: past, present, and future. Frontiers in Microbiology 9, 2161.

Indugu, N., Vecchiarelli, B., Baker, L. D., et al., 2017. Comparison of rumen bacterial communities in dairy herds of different production. BMC Microbiology 17, 190.

Jami, E. and Mizrahi, I. 2012a. Composition and similarity of bovine rumen microbiota across individual animals. PLoS ONE 7, e33306.

Jami, E. and Mizrahi, I. 2012b. Similarity of the ruminal bacteria across individual lactating cows. Anaerobe 18, 338-43.

Jami, E., White, B. A. and Mizrahi, I. 2014. Potential role of the bovine rumen microbiome in modulating milk composition and feed efficiency. PLoS ONE 9, e85423.

Janssen, P. H. and Kirs, M. 2008. Structure of the archaeal community of the rumen. Applied and Environmental Microbiology 74, 3619-25.

Jewell, K. A., McCormick, C. A., Odt, C. L., et al., 2015. Ruminal bacterial community composition in dairy cows is dynamic over the course of two lactations and correlates with feed efficiency. Applied and Environmental Microbiology 81, 4697-710.

Johnson, K. A. and Johnson, D. E. 1995. Methane emissions from cattle. Journal of Animal Science 73, 2483-92.

Kamke, J., Kittelmann, S., Soni, P., et al., 2016. Rumen metagenome and metatranscriptome analyses of low methane yield sheep reveals a Sharpea-enriched microbiome characterised by lactic acid formation and utilisation. Microbiome 4, 56.

King, E. E., Smith, R. P., St-Pierre, B., et al., 2011. Differences in the rumen methanogen populations of lactating Jersey and Holstein dairy cows under the same diet regimen. Applied and Environmental Microbiology 77, 5682-7.

Kittelmann, S., Pinares-Patiño, C. S., Seedorf, H., et al., 2014. Two different bacterial community types are linked with the low-methane emission trait in sheep. PLoS ONE 9, e103171.

Krause, D. O., Smith, W. J., Ryan, F. M., et al., 1999. Use of 16S-rRNA based techniques to investigate the ecological succession of microbial populations in the immature lamb rumen: tracking of a specific strain of inoculated ruminococcus and interactions with other microbial populations in vivo. Microbial Ecology 38, 365-76.

Krause, D. O., Denman, S. E., Mackie, R. I., et al., 2003. Opportunities to improve fiber degradation in the rumen: microbiology, ecology, and genomics. FEMS Microbiology Reviews 27, 663-93.

Levy, B. and Jami, E. 2018. Exploring the prokaryotic community associated within the rumen ciliate protozoa population. Frontiers in Microbiology 9, 2526.

Ley, R. E. 2016. Gut microbiota in 2015: prevotella in the gut: choose carefully. Nature Reviews. Gastroenterology and Hepatology 13, 69-70.

Li, F. and Guan, L. L. 2017. Metatranscriptomic profiling reveals linkages between the active rumen microbiome and feed efficiency in beef cattle. Applied and Environmental Microbiology 83 (9).

Li, M., Penner, G. B., Hernandez-Sanabria, E., et al., 2009. Effects of sampling location and time, and host animal on assessment of bacterial diversity and fermentation parameters in the bovine rumen. Journal of Applied Microbiology 107, 1924-34.

Li, Z., Wright, A.-D. G., Si, H., et al., 2016. Changes in the rumen microbiome and metabolites reveal the effect of host genetics on hybrid crosses. Environmental Microbiology Reports 8, 1016-23.

Li, F., Li, C., Chen, Y., et al., 2019. Host genetics influence the rumen microbiota and heritable rumen microbial features associate with feed efficiency in cattle. Microbiome 7, 92.

Lima, J., Auffret, M. D., Stewart, R. D., et al., 2019. Identification of rumen microbial genes involved in pathways linked to appetite, growth, and feed conversion efficiency in cattle. Frontiers in Genetics 10, 701.

Lloyd, D., Williams, A. G., Amann, R., et al., 1996. Intracellular prokaryotes in rumen ciliate protozoa: detection by confocal laser scanning microscopy after in situ hybridization with fluorescent 16S rRNA probes. European Journal of Protistology 32, 523-31.

McAllister, T. A. and Newbold, C. J. 2008. Redirecting rumen fermentation to reduce methanogenesis. Australian Journal of Experimental Agriculture 48, 7-13.

McCann, J. C., Wiley, L. M., Forbes, T. D., et al., 2014. Relationship between the rumen microbiome and residual feed intake-efficiency of brahman bulls stocked on bermudagrass pastures. PLoS ONE 9, e91864.

Miyagi, T., Kaneichi, K., Aminov, R. I., et al., 1995. Enumeration of transconjugated *Ruminococcus albus* and its survival in the goat rumen microcosm. Applied and Environmental Microbiology 61, 2030-2.

Mizrahi, I. 2012. The role of the rumen microbiota in determining the feed efficiency of dairy cows. In: Beneficial Microorganisms in Multicellular Life Forms, pp. 203-10. Springer.

Moraïs, S. and Mizrahi, I. 2019a. The road not taken: the rumen microbiome, functional groups, and community states. Trends in Microbiology 27, 538-49.

Moraïs, S. and Mizrahi, I. 2019b. Islands in the stream: from individual to communal fiber degradation in the rumen ecosystem. FEMS Microbiology Reviews 43, 362-79.

Myer, P. R. 2019. Bovine genome-microbiome interactions: metagenomic frontier for the selection of efficient productivity in cattle systems. mSystems 4 (3).

National Research Council. 2001. Nutrient Requirements of Dairy Cattle. National Academy Press, Washington DC.

Newbold, C. J., de la, Fuente, G., Belanche, A., et al., 2015. The role of ciliate protozoa in the rumen. Frontiers in Microbiology 6.

O'Herrin, S. M. and Kenealy, W. R. 1993. Glucose and carbon dioxide metabolism by *Succinivibrio dextrinosolvens*. Applied and Environmental Microbiology 59, 748-55.

Paz, H. A., Anderson, C. L., Muller, M. J., et al., 2016. Rumen bacterial community composition in holstein and jersey cows is different under same dietary condition and is not affected by sampling method. Frontiers in Microbiology 7, 1206.

Pope, P. B., Smith, W., Denman, S. E., et al., 2011. Isolation of Succinivibrionaceae implicated in low methane emissions from tammar wallabies. Science 333, 646-8.

Roehe, R., Dewhurst, R J, Duthie, C-A, et al., 2016. Bovine host genetic variation influences rumen microbial methane production with best selection criterion for low methane emitting and efficiently feed converting hosts based on metagenomic gene abundance. PLoS Genetics 12, e1005846.

Rothschild, D., Weissbrod, O., Barkan, E., et al., 2018. Environment dominates over host genetics in shaping human gut microbiota. Nature 555, 210-15.

Sasson, G., Kruger, B-S. S., Seroussi, E., et al., 2017. Heritable bovine rumen bacteria are phylogenetically related and correlated with the cow's capacity to harvest energy from its feed. mBio 8.

Shabat, S. K. B., Sasson, G., Doron-Faigenboim, A., et al., 2016. Specific microbiome-dependent mechanisms underlie the energy harvest efficiency of ruminants. The ISME Journal 10, 2958.

Shade, A. and Handelsman, J. 2012. Beyond the Venn diagram: the hunt for a core microbiome. Environmental Microbiology 14, 4-12.

Shi, W., Moon, C. D., Leahy, S. C., et al., 2014. Methane yield phenotypes linked to differential gene expression in the sheep rumen microbiome. Genome Research 24, 1517-25.

Siciliano-Jones, J. and Murphy, M. R. 1989. Production of volatile fatty acids in the rumen and cecum-colon of steers as affected by forage: concentrate and forage physical form. Journal of Dairy Science 72, 485-92.

Snelling, T. J., Auffret, M. D., Duthie, C-A., et al., 2019. Temporal stability of the rumen microbiota in beef cattle, and response to diet and supplements. Animal Microbiome 1, 16.

Stewart, R. D., Auffret, M. D., Warr, A., et al., 2019. Compendium of 4,941 rumen metagenome-assembled genomes for rumen microbiome biology and enzyme discovery. Nature Biotechnology 37, 953-61.

Tapio, I., Snelling, T. J., Strozzi, F., et al., 2017. The ruminal microbiome associated with methane emissions from ruminant livestock. Journal of Animal Science and Biotechnology 8, 7.

Turnbaugh, P. J., Ley, R. E., Mahowald, M. A., et al., 2006. An obesity-associated gut microbiome with increased capacity for energy harvest. Nature 444, 1027-31.

Turpin, W., Espin-Garcia, O., Xu, W., et al., 2016. Association of host genome with intestinal microbial composition in a large healthy cohort. Nature Genetics 48, 1413-17.

Tymensen, L. D. and McAllister, T. A. 2012. Community structure analysis of methanogens associated with rumen protozoa reveals bias in universal archaeal primers. Applied and Environmental Microbiology 78, 4051-6.

Tymensen, L. D., Beauchemin, K. A. and McAllister, T. A. 2012. Structures of free-living and protozoa-associated methanogen communities in the bovine rumen differ according to comparative analysis of 16S rRNA and mcrA genes. Microbiology 158, 1808-17.

Ushida, K., Newbold, C. J. and Jouany, J.-P. 1997. Interspecies hydrogen transfer between the rumen ciliate *Polyplastron multivesiculatum* and, *Methanosarcina barkeri*. The Journal of General and Applied Microbiology 43, 129-31.

van Lingen, H. J., Plugge, C. M., Fadel, J. G., et al., 2016. Thermodynamic driving force of hydrogen on rumen microbial metabolism: a theoretical investigation. PLoS ONE 11, e0161362.

Wallace, R. J. and Walker, N. D. 1993. Isolation and attempted introduction of sugar alcohol-utilizing bacteria in the sheep rumen. The Journal of Applied Bacteriology 74, 353-9.

Wallace, R. J., John, Wallace, R., Rooke, J. A., et al., 2015a. Archaeal abundance in post-mortem ruminal digesta may help predict methane emissions from beef cattle. Scientific Reports 4.

Wallace, R. J., Rooke, J. A., McKain, N., et al., 2015b. The rumen microbial metagenome associated with high methane production in cattle. BMC Genomics 16, 839.

Wallace, R. J., Sasson, G., Garnsworthy, P. C., et al., 2019. A heritable subset of the core rumen microbiome dictates dairy cow productivity and emissions. Science Advances 5, eaav8391.

Weimer, P. J., Cox, M. S., de, Paula, T. V., et al., 2017. Transient changes in milk production efficiency and bacterial community composition resulting from near-total exchange of ruminal contents between high-and low-efficiency Holstein cows. Journal of Dairy Science 100, 7165-82.

Welkie, D. G., Stevenson, D. M. and Weimer, P. J. 2009. ARISA analysis of ruminal bacterial community dynamics in lactating dairy cows during the feeding cycle. Anaerobe 16, 94-100.

Wolin, M. J. 1979. The rumen fermentation: a model for microbial interactions in anaerobic ecosystems. In: Alexander, M. (Ed.), Advances in Microbial Ecology (vol. 3). Springer US, Boston, MA, pp. 49-77.

Zhou, M. I. and Hernandez-Sanabria, E. 2009. Assessment of the microbial ecology of ruminal methanogens in cattle with different feed efficiencies. Applied and Environmental Microbiology 75, 6524-33.

Zhou, M., Hernandez Sanabria, E. and Guan, L. 2009. Assessment of the microbial ecology of ruminal methanogens in cattle with different feed efficiencies. Applied and Environmental Microbiology 75, 6524-33.

Zhou, M., Hernandez-Sanabria, E. and Guan, L. L. 2010. Characterization of variation in rumen methanogenic communities under different dietary and host feed efficiency conditions, as determined by PCR-denaturing gradient gel electrophoresis analysis. Applied and Environmental Microbiology 76, 3776-86.

Zhou, M., Peng, Y.-J., Chen, Y., et al., 2018. Assessment of microbiome changes after rumen transfaunation: implications on improving feed efficiency in beef cattle. Microbiome 6, 62.

第17章 瘤胃对奶牛免疫功能的调节

S. Aditya,奥地利维也纳兽医大学,印度尼西亚布拉维贾瓦大学;

E. Humer、Q. Zebeli,奥地利维也纳兽医大学

(金巍译)

1 前言

现代奶牛的能量需要一般通过饲喂高谷物日粮来满足。但是这种饲养模式降低了反刍活动,减少了唾液的分泌,还导致瘤胃中挥发性脂肪酸(VFA)浓度升高,最终引发奶牛 SARA(亚急性瘤胃酸中毒)(Plaizier 等,2008;Kleen 和 Cannizzo,2012)。SARA 是指瘤胃 pH 值间歇性低于 5.6 的时间超过 3 h/d(gozh 等,2005;Plaizier 等,2008)或 pH 值低于 5.8 的时间超过 5~6 h/d(Zebeli 等,2008)。虽然牛瘤胃可通过增加瘤胃乳头上皮细胞的表面积以适应高谷物日粮,但持续饲喂高谷物日粮可导致细胞老化速度降低到引起瘤胃上皮角化不全和角化过度的水平,阻碍 VFA 的吸收、降低瘤胃 pH 值、提高 SARA 的患病风险(Zebeli 和 Metzler-Zebeli,2012)。

健康瘤胃的鳞状多层上皮是重要营养素(如 VFA 和电解质)的主要吸收部位,其高度的选择性可阻止微生物和腔内毒素进入血液循环(Plaizier 等,2018;Aschenbach 等,2019)。然而,SARA 可使选择性瘤胃上皮屏障功能受损,使瘤胃中的抗原转移至血液和淋巴系统中(Wu 等,2016)。具体地说,多种腔内毒素(如内毒素和生物胺)可能通过改变上皮细胞紧密连接屏障的结构和功能干扰上皮的防御功能,从而破坏上皮细胞的完整性,同时通过改变细胞通路使腔内毒素异位(Berkes 等,2003)。

越来越多的证据表明,SARA 可促进革兰氏阴性菌(GNB)的生长和裂解,并释放大量游离的脂多糖(LPS)(Nagaraja 等,1978;Beutler 和 Rietschel,2003;Plaizier 等,2012)和生物胺(BA),如乙醇胺和组胺(Dong 等,2011)。虽然一些生物胺被看作是局部炎性因子(Gozho 等,2005),但胃肠道中的 LPS 是一种含量丰富的潜在促炎因子。LPS 是 GNB 细胞壁外膜的一部分,它与免疫和疾病的关系已得到广泛的关注和研究(Emmanuel 等,2008;Li 等,2012a;Plaizier 等,2012)。一旦胃肠道中的 LPS 进入循环系统,就会触发促炎级联反应,也就是常说的低度急性时相反应(APR),其特点是血清急性时相蛋白(APP)中度升高。APP 与脂多糖结合蛋白(LBP)、血清淀粉样蛋白 A(SAA)以及结合珠蛋白(Hp;Ceciliani 等,2012)均是牛瘤胃 LPS 异位的主要标志物。在肝细胞和胆汁中,SAA 通过形成 LPS-SAA-脂蛋白复合物清除进入循环系统的 LPS(Ametaj 等,2010a;Zebeli 和 Metzler-Zebeli,2012)。

另外,LBP 可将 LPS 转移到免疫细胞(如巨噬细胞、单核细胞和中性粒细胞)中脱

毒（Gallay 等，1994；Schumann 等，1994）。更具体地说，当 LBP 将 LPS 转运到巨噬细胞时，巨噬细胞的膜表面受体 CD-14 与 TLR4 以及髓样分化因子 2（MD-2）相互作用，活化巨噬细胞（Chow 等，1999；da Silva Correia 等，2001）。这些信号可被髓样分化初级应答基因 88（MyD88）识别，激活核因子 κ-B（NFκ-B）和促炎细胞因子，如肿瘤坏死因子（TNF）-α 和白介素（IL）-1β、IL-6 和 IL-8（Erridge 等，2002；Emmanuel 等，2008；Ceciliani 等，2012；Plaizier 等，2012）。这些促炎细胞因子激活宿主的炎症反应，引起发烧、应激、采食量降低、脂肪动员以及其他代谢变化（Zebeli 和 MetzlerZebeli，2012；Abaker 等，2017）。

LPS 除了引起全身性炎症反应，也会增加支持免疫应答的营养需要量，因此用于乳成分合成的营养物质随之减少（Dong 等，2011）。此外，当 LPS 通过体循环进入乳腺，可能对乳腺上皮细胞的功能产生损伤（Dong 等，2011）。有报道显示，高谷物日粮饲喂奶牛会导致乳腺血液中 LPS 浓度（用鲎试剂法测定）升高、乳腺组织的表观遗传改变（Dong 等，2014）以及牛奶中的 LBP 浓度升高（Khafipour 等，2009a）。大量文献综述都对 SARA 和胃肠道中 LPS 对奶牛免疫功能和代谢的影响给予了极大的关注（Plaizier 等，2008，2012；Dong 等，2011；Kleen 和 Cannizzo，2012）。然而，对 SARA 和 LPS 的关系及其对炎症、全身代谢、乳腺免疫系统影响的知识还需要系统的总结和回顾。本章旨在更新并总结最近关于 SARA 及其导致的瘤胃游离 LPS 浓度增加对牛的代谢和健康影响的研究，特别是对奶牛乳腺影响的研究。

2 亚急性瘤胃酸中毒（SARA）在牛群中的发病率

SARA 是一种严重且高发的奶牛疾病，特别容易在泌乳前期和中期发生，给乳业带来重大经济损失（Garrett 等，1997；Plaizier 等，2008）。研究显示，泌乳早期 SARA 的发生率为 11%~27%，中期为 18%~27%（Garrett 等，1997；Kleen 等，2004；Tajik 等，2009）。将奶牛围产期日粮从富含牧草的干奶牛日粮替换为富含淀粉的泌乳奶牛日粮，这个过程需要至少 4~5 周（Enemark，2008；Humer 等，2018a），过于突然的转换会增加奶牛的患病风险。短期适应高谷物日粮也会增加奶牛患 SARA 的风险（Pourazad 等，2016）。泌乳中期奶牛采食量过高，尤其是当日粮中缺少中性洗涤纤维（NDF）时也会增加患病风险（Nordlund 等，1995；Stone，2004）。奶牛的胎次起着重要作用，与经产牛相比初产牛患病率更高（Humer 等，2015）。这可能与两者瘤胃上皮以及菌群的发育完全度有关，两者采食行为和体重也存在差异，同时瘤胃 pH 值的自我调节能力也存在差异（Humer 等，2018a）。

一些国家的实地调研显示，奶牛场 SARA 整体发病率在 8%~33%（表 1）。引起差异的因素很多，如爱尔兰和澳大利亚以放牧为主的饲养模式，在饲喂和牛群管理上也存在差异。大型牛群的高发病率可能是由于对牛群的观察不足导致的（Kleen 等，2013）。通常 SARS 会通过降低牛的体况、采食量、产奶量、泌乳能量效率（Yang 和 Beauchemin，2006）降低泌乳性能（Nagaraja 和 Lechtenberg，2007；Plaizier 等，2008）。总的来说，每头 SARA 牛带来的经济损失可达每天 1.12 美元（Stone，1999）。

表1 一些国家奶牛亚急性瘤胃酸中毒（SARA）的患病率

发病率（%）	国家	参考文献
33	意大利	Morgante 等，2007
28	伊朗	Tajik 等，2009
26	英国	Atkinson，2013
22	丹麦	Enemark 和 Jørgensen，2001
20	德国	Kleen 等，2013
16	希腊	Kitkas 等，2013
14	荷兰	Kleen 等，2009
11	爱尔兰	O'Grady 等，2008
8	澳大利亚	Bramley 等，2008

为了把损失降到最低，对 SARA 的早期发现就显得至关重要。尽管 SARA 被认为是一种亚临床疾病，但通过亚临床体征的辅助也可查出患病牛，例如反刍次数降低（Zebeli 等，2010）、乳脂含量下降（Zebeli 和 Ametaj，2009）、腹泻、泡沫粪便（Nordlund 和 Garrett，1994；Kleen 等，2003），粪便中有未消化的谷物（Enemark，2008）。另外，SARA 发病可能与其他代谢紊乱有关（Aditya 等，2017），如猝死症、脂肪肝（Ametaj 等，2005）和蹄叶炎（Nocek，1997）。关于 SARA 增加奶牛代谢紊乱风险的原因尚未完全确定。然而，瘤胃稳态的改变（如革兰氏阴性菌含量升高）和免疫功能紊乱等都起着一定作用（Plaizier 等，2008，2012；Dong 等，2011；Kleen 和 Cannizzo，2012；Zebeli 和 MetzlerZebeli，2012）。

3 瘤胃健康、代谢活动和紊乱

虽然 APR（急性时相反应）激活对消除引起炎症的因子和重建机体稳态必不可少，但长期的炎症反应对宿主存在负面影响（Morris 和 Li，2012；Lacetera，2016）。例如，与 APR 激活对应的高能量需求可导致饲料利用率下降，并可能加剧奶牛能量负平衡（NEB），尤其在泌乳早期（Zebeli 和 Metzler-Zebeli，2012；Lacetera，2016）。APR 可引起包括乳腺在内的机体不同组织的能量和脂类代谢的改变（Kushibiki 等，2002；Khovidhunkit 等，2004）。此外，有研究表明，LPS 和激活的 APR 可能与真胃异位、脂肪肝、肝脓肿、蹄叶炎和奶牛卧倒不起综合征等多种代谢疾病的发生有关（Nocek，1997；Kleen 等，2003；Plaizier 等，2008；Zebeli 等，2015）。然而，目前关于奶牛瘤胃健康与全身代谢相互作用的机理的研究有限。对于瘤胃中潜在有毒化合物（如 LPS 和 BA）浓度增加的后果，以及奶牛全身循环代谢途径可能的变化研究也很少。

关于瘤胃发酵失调对全身代谢影响的大多数研究都集中在单一代谢产物上，比如 NEFA、BHBA（β-羟基丁酸）、胆固醇和肝酶（显示肝组织损伤）（Zebeli 等，2011；Marchesini 等，2013）。一些研究报道了瘤胃发酵紊乱与肝酶活性增加之间的关联（Humer 等，2018b；Kröger 等，2019），这与高精料日粮饲喂期间瘤胃内 LPS 和 BA（生物胺）水平增加相关（Humer 等，2018b）。这一结果暗示这些变化可能是由在肝脏库普弗细胞中 LPS 和其他循环毒素清除率提高引起的（Lechowski，1997；Marchesini 等，2013）。有研究表明，奶牛采食高谷物日粮 2 周，即使瘤胃 pH 值升高了，肝酶的水平也会继续升高（图1），由此可推测，高精料日粮饲喂后，瘤胃 pH 值几天内即可调节恢复正常水平，而肝脏则需更长的时间才能恢复健康（Kröger 等，2017；Khiaosa-ard 等，2018）。瘤胃发酵异常引起肝酶活性增强，暗示饲喂高谷物日粮对于 LPS 含量和肝脏健康有着负面和潜在的累积效应，这可能会导致奶牛整体的健康损害。

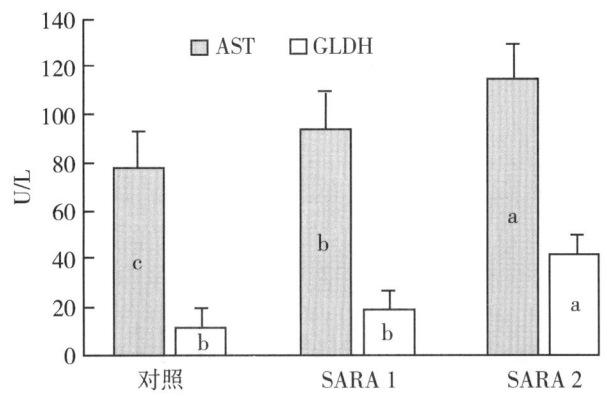

图 1 连续 2 周饲喂纯草料日粮（对照）或 65% 饲喂高精料日粮［诱发亚急性瘤胃酸中毒（SARA，SARA1 和 SARA2）］的奶牛血液中的肝酶酶活［天冬氨酸氨基转移酶（AST）；谷氨酸脱氢酶（GLDH）］。不同字母表示基线、SARA 1 和 SARA 2 之间的差异，$P<0.05$。资料来源：Humer 等（2018b）。

近年来，代谢组学技术已能够用于检测多种代谢物质，可反映关键代谢途径的变化。这有助于我们更好地理解营养、代谢及健康之间的相互关系。代谢组学是指在特定条件下对生物体中所有代谢物进行定量分析。代谢物是代谢途径的中间和最终产物。与现有的生物标志物（如 APP 和肝酶）相比，它们能更快速地反映机体生理功能失调。代谢组学技术可以在代谢失调的早期阶段发现病变，帮助识别重要的奶牛疾病标志物，如瘤胃发酵紊乱（Ametaj，2010b）。

代谢组学技术已应用于饲喂不同水平谷物饲料的奶牛瘤胃液样品的检测（Ametaj 等，2010b；Saleem 等，2012）。最近，一种代谢组学方法被用于分析奶牛的血液样本，这些奶牛一组饲喂纯牧草日粮，一组饲喂高精料日粮（51% 精料，可以诱发 SARA）（Humer 等，2018b）。采用不同的数据分析方法发现，瘤胃内容物中检测出的潜在的有毒物质（如 BA 和 LPS）与血液代谢谱变化有关联（Humer 等，2018b）。多元统计分析显示，SARA 奶牛瘤胃内 LPS 和 BA（组胺、乙醇胺、异丙胺、吡咯烷胺、腐胺、尸胺和亚精胺）浓度升高，并伴有血液代谢组显著变化，特别是磷脂酰胆碱（PC；图 2）、溶血磷脂酰胆碱（lysoPC）和鞘磷脂浓度降低。有研究发现在饲喂高精料日粮（含高达 45% 的大麦谷物）的奶牛瘤胃液中 PC 浓度下降（Saleem 等，2012），这可能是由于瘤胃 pH 值低期间，原虫数量下降，而原虫是瘤胃液中 PC 的主要来源（Jouany 等，1988；Goad 等，1998；Khafipour 等，2009）。因为 lysoPC 是 PC 的水解产物，所以 PC 浓度的降低也可能引起 lysoPC 浓度的降低（Hailemariam 等，2014a）。研究还发现，肌内注射大肠杆菌（O26:B6）源 LPS 的牛或者患乳房炎、子宫炎、胎衣不下、蹄叶炎等疾病的牛血浆中的 lysoPC 浓度下降（Hailemariam 等，2014a；Humer 等，2018c）。

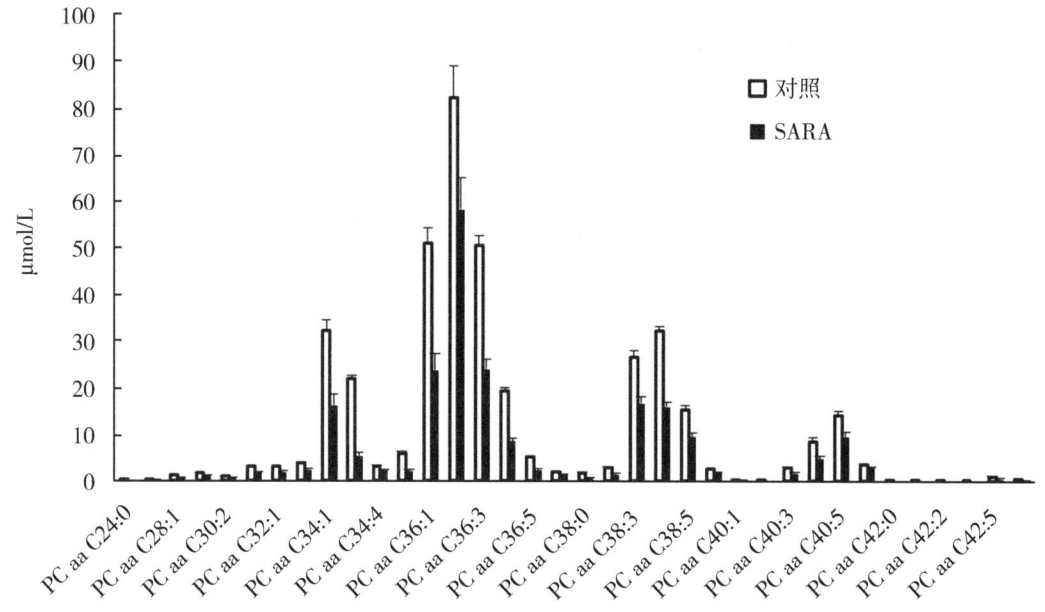

图 2　饲喂纯草料日粮（对照）或 65% 高精料日粮诱发亚急性瘤胃酸中毒（SARA）的奶牛血液中磷脂酰胆碱（PC）和二酰基残基（aa）的浓度

　　PC 浓度降低的进一步解释可能是由胆固醇含量降低引起的，因为 PC 通常与胆固醇和三酰甘油有关（Gruffat 等，1996）。反刍动物日粮中高水平的易发酵碳水化合物通常与胆固醇合成前体的产量下降有关（例如乙酸盐；Liepa 等，1978；Neubauer 等，2018）。PC 和 lysoPC 浓度降低可能是饲喂高谷物日粮奶牛的标志。鞘磷脂则属于牛磷脂一类（Nilsson 和 Duan，2006）。多元分析和相关分析显示 PC、鞘磷脂和胆固醇属于一类，它们之间存在强正相关。

　　图 3 显示了 SARA 对奶牛血液中氨基酸（AA）浓度的影响。SARA 奶牛血液中精氨酸、瓜氨酸、异亮氨酸、蛋氨酸、苯丙氨酸、色氨酸和酪氨酸含量下降，而甘氨酸和丝氨酸含量升高。SARA 对血液中氨基酸浓度降低的影响与先前的研究结果一致。这些研究发现，牛在外部大肠杆菌 LPS 刺激下（Humer 等，2018c）或患有一种或几种围产期疾病时，血液氨基酸浓度显著降低（Hailemariam 等，2014b）。其内在的机制可能是免疫细胞对氨基酸的需求增加了蛋白质分解代谢，导致精氨酸或色氨酸等氨基酸的消耗增加（Le Floch 等，2004；Hailemariam 等，2014b）。SARA 奶牛仅甘氨酸和丝氨酸消耗增加。甘氨酸及其前体丝氨酸的生成量增加，可能是奶牛平衡 SARA 引发的炎症和氧化应激的一种

途径，因为研究发现甘氨酸具有抗氧化、抗炎、细胞保护和免疫调节的特性，以及抵抗损伤和疾病的作用（Razak等，2017）。

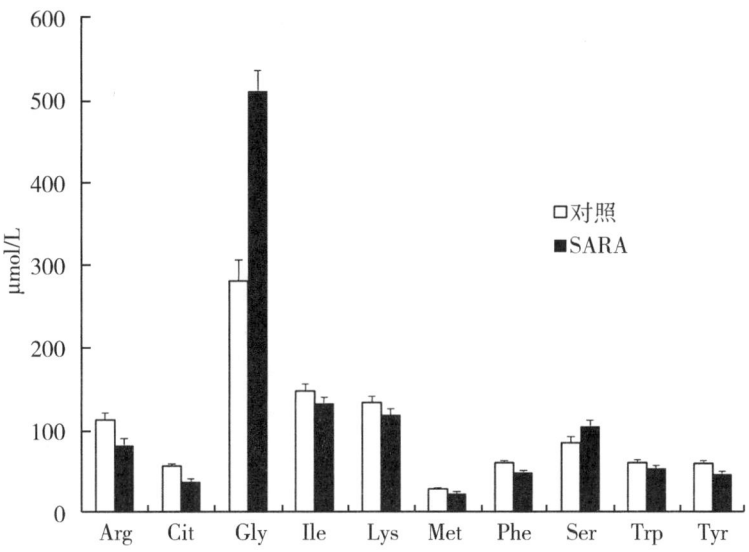

图 3 饲喂纯草料日粮（对照）或 65%高精料日粮诱发亚急性瘤胃酸中毒（SARA）的奶牛血液中的氨基酸浓度

多元分析发现，瘤胃液中几种有害化合物（LPS、组胺、乙醇胺、吡咯烷和亚精胺）的浓度与血中氨基酸、PC、lysoPC 和鞘磷脂的浓度呈负相关，这暗示高谷物日粮诱发的 SARA 奶牛瘤胃中有毒物质释放增多影响了氨基酸和脂类的代谢途径。

4 瘤胃健康和乳腺免疫系统

SARA 奶牛的常见症状就是乳脂率下降。奶牛科学家早就认识到瘤胃发酵失调（即 SARA）对乳腺代谢的影响。最新数据显示，来自胃肠道的内源性 LPS 穿过血乳屏障侵入乳腺，引发局部免疫反应。文献报道，SARA 奶牛乳中 LBP 含量（Khafipour 等，2009）和乳腺血液中促炎因子含量升高（Zhou 等，2014）。这些结果与外源性 LPS 诱发的奶牛乳腺炎的作用相似，也会导致乳中 LBP（Bannerman 等，2003）和乳腺血液中促炎因子含量升高（Lee 等，2003；Wellnitz 等，2011）。似乎内源性 LPS（来自胃肠道细菌）和外源性 LPS（来自环境）均能破坏血乳屏障。Humer 等（2018b）发现，奶牛肌内注射外源性 LPS 刺激时，与非 SARA 牛相比，SARA 牛乳中牛奶淀粉样蛋白 A（MAA）浓度升高。似乎 SARA 会放大奶牛机体对外源 LPS 等感染性外源刺激的炎症反应。

Gott 等（2015）的研究观察到了相反的结果，他们发现乳房灌注 LPS 攻毒后，饲喂对照组日粮奶牛的乳 MAA 浓度和体细胞数比饲喂高淀粉日粮（旨在诱导慢性瘤胃低 pH 值）奶牛的高，暗示奶牛对 LPS 有一定的耐受性。在 Gott 等（2015）的研究中观察到不同结果的一个可能的原因是奶牛遭受较强的 SARA 和长期接受较高剂量的 LPS 攻毒。在 Humer 等（2018b）的研究中，SARA 奶牛较强反应的一个可能的解释是患轻度 SARA 的

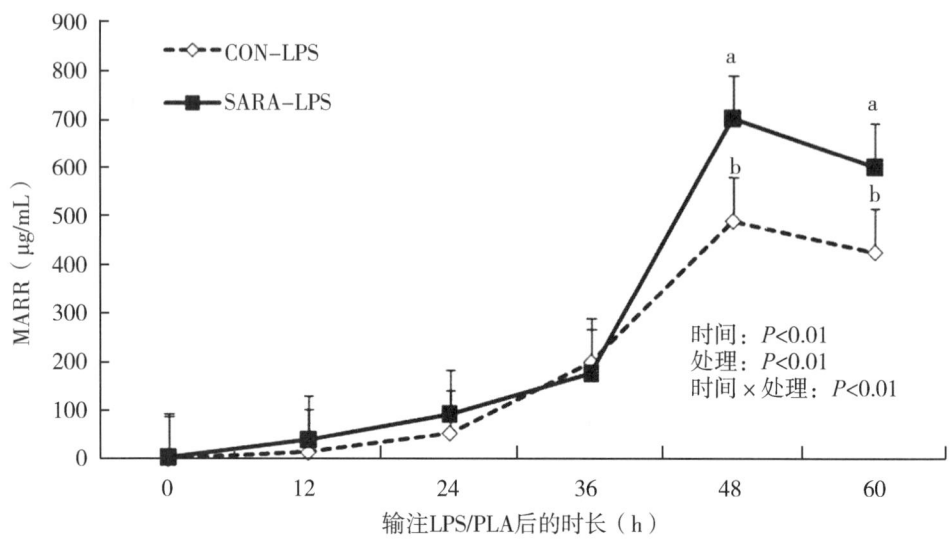

图 4　SARA（SARA-LPS）和非 SARA（CON-LPS）奶牛接受 LPS 灌注后乳中
牛奶淀粉样蛋白 A（MAA）浓度

奶牛可能长期受到低剂量的 LPS 刺激，因此产生低浓度的 APP（例如 LBP）。在人类上的研究表明，低剂量的 LPS 诱导机体出现对后续毒性剂量的 LPS 的耐受状态，非常低的剂量甚至可能产生相反的效果（Morris 和 Li，2012）。然而，由于 Gott 等（2015）未报道瘤胃 pH 值动态变化和 LPS 或者 LBP 浓度的数据，这一种解释尚在假设层面。显然，有必要通过进一步研究阐明奶牛瘤胃发酵失调后对外部感染因子的反应。

近期报道称，LPS 可以通过改变奶牛乳腺上皮细胞间紧密连接（TJ）的闭合蛋白（claudins）破坏血乳屏障。闭合蛋白是控制紧密连接屏障功能的重要蛋白（Beeman 等，2012；Schlingmann 等，2016）。其成分的改变可能导致 LPS 破坏血乳屏障。有趣的是，通过 LPS/TLR-4 信号激活 NF-κB 通路被推测可引起紧密连接通透性的变化（Kobayashi 等，2013）。一旦 LPS 破坏血乳屏障，牛的乳腺上皮细胞将变成下一道防线（Strandberg 等，2005；Zbinden 等，2014）。

乳腺的大部分防御机制都是通过白细胞和可溶性免疫分子（如炎性标志物和抗菌因子等）运行的非特异性免疫。在急性乳房炎期间，牛奶中的抗菌蛋白（如乳铁蛋白和溶菌酶）通常会增加。Jin 等（2016）在 SARA 奶牛乳腺上皮细胞中观察到防御素（如舌抗菌肽）的表达增加。这些作者猜测潜在的机制之一是 NF-κB 因子的激活。在 SARS 牛中，反映炎症和氧化应激的其他杀菌蛋白（如 β-N-乙酰氨基葡萄糖苷酶和髓过氧化物酶）的活性显著提高，表明牛的乳腺发生了感染。

当 LPS 侵入乳腺时，会产生过多的促炎细胞因子，如 IL-1β、IL-6 和 TNF-α，它们通过 LPS/TLR4 信号通路诱导局部炎症（Akira 等，2006；Ingman 等，2014）。一般情况下，牛乳腺上皮细胞通过特定的模式识别受体（如 TLR4）来识别病原体，比如 TLR4 与 LPS-LBP 复合物结合并被激活，从而促进机体对乳腺内感染的先天性免疫反应（Ibeagha-Awemu 等，2008）。总的来说，约 10 个牛 TLRs 已经被鉴定出来，每个都能识别特定配体或病原体的相关分子模式（McGuire 等，2006）。然而，有报道称一些病原体可以激活不止一个 TLR（Swanson 等，2004）。在这一方面，TLR4 主要识别 LPS，而 TLR2 通常由其他细胞壁成分触发，包括革兰氏阳性细菌（GPB）上发现的肽聚糖、脂磷壁酸等物质（Eckel 和 Ametaj，2016）。随着培养液中 LPS 浓度的增加，牛乳腺上皮细胞 TLR4 及其下游信号分子 TLR2 表达上调，同时这些受体的特异性抗体表面表达增加（Ibeagha-Awemu 等，2008）。TLR2 的上调可能是由于 TLR2 和 TLR4 的互作，因为 TLR2 是 TLR4 激活和 NF-κB 产生的刺激反应（Faure 等，2001；Fan 等，2003）。

依赖于 TLR4 的 TLR2 表达上调似乎依赖于位于 TLR2 上的 NF-κB 位点（Ibeagha-Awemu 等，2008）。然而，一些研究发现 TLR2 并未参与 LPS 信号通路（Heine 等，1999；Takeuchi 等，1999）。与其他细胞类型相比，乳腺上皮细胞对某些致病化合物的反应可能不同（Ibeagha-Awemu 等，2008）。虽然人们一般认为金黄色葡萄球菌只上调 TLR2，但 Goldammer 等（2004）在金黄色葡萄球菌诱发奶牛乳房炎的试验中，观察到 TLR2 和 TLR4 的表达协同上调。乳腺上皮细胞似乎不仅有对大肠杆菌进行强大防御所需的免疫储备，而且能够对不同类型乳腺炎病原菌做出有效应答（Ibeagha-Awemu 等，2008）。

值得注意的是，机体出现感染后，促炎因子过量产生可能导致乳腺上皮细胞损伤，也可诱发严重的全身性病变，如慢性肠炎、动脉粥样硬化，甚至发生感染性休克（Takeda 和 Akira，2005）。Kobayashi 等（2013）的研究清楚地证明了乳腺血液中炎性细胞因子（IL-1β 和 IL-8）的浓度与泌乳性能参数呈负相关。SARA 奶牛产奶量下降可能部分归因于局部炎症反应。炎症反应导致用于合成牛奶成分的可利用营养素或乳前体物质减少。

LPS 进入乳腺组织也可能激活中性粒细胞，反过来，中性粒细胞也能产生大量杀菌分子，包括蛋白质、多肽和活性氧（Dong 等，2011）。因为活性氧是不稳定的含氧分子，会对其他分子做出反应，如蛋白质、脂质、细胞中的 DNA 和 RNA，促进氧化，导致组织损伤（Abuelo 等，2015；Zebeli 等，2015）。这会引起功能失调性炎症，导致代谢应激，增加奶牛对健康失调的易感性（Sordillo 和 Aitken，2009）。有报道称，乳腺分泌组织对 LPS 高度敏感（Schmitz 等，2004；Blum，2000），这是由于乳腺中存在高水平的氧自由基和脂质过氧化代谢物，高产奶牛尤其敏感（Shi 等，2016）。因此，SARA 是危害乳房健康的一个因素。

长期的 SARA 会损害泌乳奶牛肝脏和乳腺的抗氧化机制（Abaker 等，2017；Memon 等，2019）。这一机制涉及丙二醛浓度的增加（MDA）和丝裂原活化蛋白激酶（MAPK）促炎基因表达上调，还涉及核因子红细胞 2 相关因子（Nrf2）水平的下降，该因子与 SARA 奶牛乳腺组织中的蛋白质表达和抗氧化基因有关（Memon 等，2019）。Nrf2 蛋白在抵抗炎症触发的 ROS 过程中的抗氧化防御蛋白表达中起关键作用（Kansanen 等，2013）。Nrf2 表达的减少表明机体抗氧化系统遭到抑制，从而导致乳腺组织发生氧化应激。

LPS 或其他免疫原性化合物易位进入乳腺引起的乳腺局部免疫反应也可能损害乳头的免疫防御机制，甚至破坏其屏障功能（Pareek 等，2005；Dong 等，2011）。与健康奶牛相比，患有 SARA 的奶牛可能更容易受到细菌入侵并在乳腺组织中定植（Sordillo 和 Streicher，2002）。当细菌穿过第一道解剖学防线——乳头和乳头管屏障后，它们可能会避开乳腺细胞和体液免疫防御系统（Sordillo 和 Streicher，2002；Zhao 和 Lacasse，2008）。除了已知的 SARA 会引起瘤胃微生物组变化外，高谷物日粮饲喂也可能影响乳微生物的组成。最近的一项研究发现，在 SARA 奶牛中，几种引起乳房炎的病原菌，如嗜麦芽窄养单胞菌、副乳房链球菌和短波单胞菌的比例升高（Zhang 等，2015），与健康奶牛相比，SARA 奶牛乳中的几种嗜冷细菌（短波单胞菌、鞘氨醇杆菌、产碱杆菌、肠杆菌和乳酸杆菌）丰度升高。因此，高谷物日粮饲养可能会增加奶牛患革兰氏阴性菌感染的乳腺炎的风险，也会降低原奶的品质和安全性，并缩短加工牛奶的保质期（Zhang 等，2015）。

5 结论

本章将 SARA 以及由此产生的内源 LPS 从胃肠道转移到循环系统对奶牛肝脏健康、全身代谢、乳房代谢和健康的影响的最新研究成果进行了总结。目前的证据还表明，内源性 LPS 可以在破坏血乳屏障后侵入乳腺，从而引发局部免疫反应，与外源 LPS 造成的血乳屏障损伤的原理类似。因长期的 SARA 发作而产生的内源 LPS 可能会改变由外源 LPS 引起的乳腺炎症，并可能降低受感染奶牛的乳房防护能力。

6 致谢

首先要感谢奥地利教育和研究国际合作署（OeAD），东盟欧洲学术大学网络为在奥地利学习期间提供资金支持，也感谢维也纳科学与技术大学（维也纳科技基金）对合作项目（INFLACOW ls12-010）的支持。

7 补充信息

本章综述的基础知识来自 Zebeli 和 Metzler-Zebeli 的文章《奶牛瘤胃消化紊乱与日粮诱导炎症之间的相互作用》（详见参考文献 Zebeli and Metzler-Zebeli，2012）。

此外，目前有一些支持研究 SARA 发病率与瘤胃健康的综述/研究项目，包括：

- 特约综述：降低奶牛 SARA 中毒风险的实用饲养管理建议（https://www.ncbi.nlm.nih.gov/pubmed/29153519）。
- 奶牛 SARA 的诊断与处理（https://www.sciencedirect.com/science/article/abs/pii/S0749072017300579）。
- 补充植物性化合物或自溶酵母可调节亚急性瘤胃酸中毒干奶牛瘤胃生物胺和血浆代谢组（https://www.sciencedirect.com/science/article/pii/S0022030218306672）。

8 参考文献

Abaker, J. A., Xu, T. L., Jin, D., et al., 2017. Lipopolysaccharide derived from the digestive tract provokes oxidative stress in the liver of dairy cows fed a high-grain diet. J. Dairy Sci. 100 (1), 666-78.

Abuelo, A., Hernandez, J., Benedito, J. L., et al., 2015. The importance of theoxidative status of dairy cattle in the periparturient period: revisiting antioxidant supplementation. J. Anim. Physiol. Anim. Nutr. 99 (6), 1003-16.

Aditya, S., Humer, E., Pourazad, P., et al., 2017. Intramammary infusion of *Escherichia coli* lipopolysaccharide negatively affects feed intake, chewing, and clinical variables, but some effects are stronger in cows experiencing subacute rumen acidosis. J. Dairy Sci. 100, 1363-77.

Akira, S., Uematsu, S. and Takeuchi, O. 2006. Pathogen recognition and innate immunity. Cell 124 (4), 783-801.

Ametaj, B. N., Bradford, B. J., Bobe, G., et al., 2005. Strong relationships between mediators of the acute phase response and fatty liver in dairy cows. Can. J. Anim. Sci. 85 (2), 165-75.

Ametaj, B. N., Zebeli, Q. and Iqbal, S. 2010a. Nutrition, microbiota, and endotoxinrelated diseases in dairy cows. R. Bras. Zootec. 39 (suppl. spe), 433-44.

Ametaj, B. N., Zebeli, Q., Saleem, F., et al., 2010b. Metabolomics reveals unhealthy alterations in rumen metabolism with increased proportion of cereal grain in the diet of dairy cows. Metabolomics 6 (4), 583-94.

Aschenbach, J. R., Zebeli, Q., Patra, A. K., et al., 2019. Symposium review: the importance of the ruminal epithelial barrier for a healthy and productive cow. J. Dairy Sci. 102 (2), 1866-82.

Atkinson, O. 2013. A cross-sectional survey to investigate prevalence of and clinical indicators for Subacute Ruminal Acidosis (SARA) in lactating cows on UK dairy farms. Diploma in Cattle Health and Production. Dissertation. Royal College of Veterinary Surgeons, London.

Bannerman, D. D., Paape, M. J., Hare, W. R., et al., 2003. Increased levels of LPS binding protein in bovine blood and milk following bacterial lipopolysaccharide challenge. J. Dairy Sci. 86 (10), 3128-37.

Beeman, N., Webb, P. G. and Baumgartner, H. K. 2012. Occludin is required for apoptosis when claudin-claudin interactions are disrupted. Cell Death Dis. 3 (2), e273.

Berkes, J., Visvanathan, V. K., Savkovic, S. D., et al., 2003. Intestinal epithelial responses to enteric pathogens: effects on the tight junction barrier, ion transport, and inflammation. Gut 52 (3), 439-51.

Beutler, B. and Rietschel, E. T. 2003. Innate immune sensing and itsroots: the story of endotoxin. Nat. Rev. Immunol. 3 (2), 169-76.

Blum, J. W., Dosogne, H., Hoeben, D., et al., 2000. Tumor necrosis factor-α and nitrite/nitrate responses during acute mastitis induced by *Escherichia coli* infection and endotoxin in dairy cows. Domest. Anim. Endocrinol. 19 (4), 223-35.

Bramley, E., Lean, I. J., Fulkerson, W. J., et al., 2008. The definition of acidosis in dairy herds predominantly fed on pasture and concentrates. J. Dairy Sci. 91 (1), 308-21.

Carlsson, A., Björck, L. and Persson, K. 1989. Lactoferrin and lysozyme in milk during acute mastitis and their inhibitory effect in Delvotest P. J. Dairy Sci. 72 (12), 3166-75.

Ceciliani, F., Ceron, J. J., Eckersall, P. D., et al., 2012. Acute phase protein in ruminants. J. Proteomics 75 (14), 4207-31.

Chow, J. C., Young, D. W., Golenbock, D. T., et al., 1999. Toll-like receptor-4 mediates lipopolysaccharide-induced signal transduction. J. Biol. Chem. 274 (16), 10689-92.

da Silva Correia, J., Soldau, K., Christen, U., et al., 2001. Lipopolysaccharide is in close proximity to each of the proteins in its membrane receptor complex transfer from CD14 to TLR4 and MD-2. J. Biol. Chem. 276 (24), 21129-35.

Dong, G., Liu, S., Wu, Y., et al., 2011. Diet-induced bacterial immunogens in the gastrointestinal tract of dairy cows: impacts on immunity and metabolism. Acta. Vet. Scand. 53 (1), 48.

Dong, G., Qiu, M., Ao, C. Z., et al., 2014. Feeding a high-concentrate corn straw diet induced epigenetic alterations in the mammary tissue of dairy cows. PLoS ONE 9 (9), e107659.

Eckel, E. F. and Ametaj, B. N. 2016. Invited review: role of bacterial endotoxins in the etiopathogenesis of periparturient diseases of transition dairy cows. J. Dairy Sci. 99 (8), 5967-90.

Emmanuel, D. G., Dunn, S. M. and Ametaj, B. N. 2008. Feeding high proportions of barley grain stimulates an inflammatory response in dairy cows. J. Dairy Sci. 91 (2), 606-14.

Enemark, J. M. D. 2008. The monitoring, prevention and treatment of sub-acute ruminal acidosis (SARA): a review. Vet. J. 176 (1), 32-43.

Enemark, J. M. D. and Jørgensen, R. J. 2001. Subclinical rumen acidosis as a cause of reduced appetite in newly calved dairy cows in Denmark: results of a poll among Danish dairy practitioners. Vet. Quart. 23 (4), 206-10.

Erridge, C., Bennett-Guerrero, E. and Poxton, I. R. 2002. Structure and function of lipopolysaccharides. Microbes Infect. 4 (8), 837-51.

Fan, J., Frey, R. S. and Malik, A. B. 2003. TLR4 signaling induces TLR2 expression in endothelial cells via neutrophil NADPH oxidase. J. Clin. Invest. 112 (8), 1234-43.

Faure, E., Thomas, L., Xu, H., et al., 2001. Bacterial lipopolysaccharide and IFN-gamma induce toll-like receptor 2 and toll-like receptor 4 expression in human endothelial cells: role of NF-kappa B activation. J. Immunol. 166 (3), 2018-24.

Gallay, P., Heumann, D., Le, D., et al., 1994. Mode of action of antilipopolysaccharide-binding protein antibodies for prevention of endotoxemic shock in mice. Proc. Natl. Acad. Sci. U. S. A. 91 (17), 7922-6.

Garrett, E. F., Nordlund, K. V., Goodger, W. J., et al., 1997. A cross-sectional field study investigating the effect of periparturient dietary management on ruminal pH in early lactation dairy cows. J. Dairy Sci. 80 (1), 169.

Goad, D. W, Goad, C. L. and Nagaraja, T. G. 1998. Ruminal microbial and fermentative changes associated with experimentally induced subacute acidosis in steers. J. Anim. Sci. 76, 234-41.

Goldammer, T., Zerbe, H., Molenaar, A., et al., 2004. Mastitis increases mammary mRNA abundance of beta-defensin 5, toll-like-receptor 2 (TLR2), and TLR4 but not TLR9 in cattle. Clin. Diagn. Lab. Immunol. 11 (1), 174-85.

Gott, P. N., Hogan, J. S. and Weiss, W. P. 2015. Effects of various starch feeding regimens on responses of dairy cows to intramammary lipopolysaccharide infusion. J. Dairy Sci. 98 (3), 1786-96.

Gozho, G. N., Plaizier, J. C., Krause, D. O., et al., 2005. Subacute ruminal acidosis induces ruminal lipopolysaccharide endotoxin release and triggers an inflammatory response. J. Dairy Sci. 88 (4), 1399-403.

Gruffat, D., Durand, D., Graulet, B., et al., 1996. Regulation of VLDL synthesis and secretion in the liver. Reprod. Nutr. Dev. 36 (4), 375-89.

Hailemariam, D., Mandal, R., Saleem, F., et al., 2014a. Identification of predictive biomarkers of disease state in transition dairy cows. J. Dairy Sci. 97 (5), 2680-93.

Hailemariam, D., Mandal, R., Saleem, F., et al., 2014b. Metabolomics approach reveals altered plasma amino acid and sphingolipid profiles associated with pathological state in transition dairy cows. Curr.

Metab. 2, 184-95.

Heine, H., Kirschning, C. J., Lien, E., et al., 1999. Cutting edge: cells that carry a null allele for toll-like receptor 2 are capable of responding to endotoxin. J. Immunol. 162 (12), 6971-5.

Humer, E., Khol-Parisini, A., Gruber, L., et al., 2015. Long-term reticuloruminal pH dynamics and markers of liver health in earlylactating cows of various parities fed diets differing in grain processing. J. Dairy Sci. 98 (9), 6433-48.

Humer, E., Petri, R. M., Aschenbach, J. R., et al., 2018a. Invited review: practical feeding management recommendations to mitigate the risk of subacute ruminal acidosis in dairy cattle. J. Dairy Sci. 101 (2), 872-88.

Humer, E., Kröger, I., Neubauer, V., et al., 2018b. Supplementing phytogenic compounds or autolyzed yeast modulates ruminal biogenic amines and plasma metabolome in cows experiencing subacute rumen acidosis. J. Dairy Sci. 101 (10), 9559-74.

Humer, E., Aditya, S. and Zebeli, Q. 2018c. Innate immunity and metabolomic responsesin dairy cows challenged intramammarily with lipopolysaccharide after subacute ruminal acidosis. Animal 12 (12), 2551-60.

Ibeagha-Awemu, E. M., Lee, J. W., Ibeagha, A. E., et al., 2008. Bacterial lipopolysaccharide induces increased expression of tolllike receptor (TLR) 4 and downstream TLR signaling molecules in bovine mammary epithelial cells. Vet. Res. 39 (2), 11.

Ingman, W. V., Glynn, D. J. and Hutchinson, M. R. 2014. Inflammatory mediators in mastitis and lactation insufficiency. J. Mammary. Gland. Biol. Neoplasia 19 (2), 161-7.

Iqbal, S. 2013. Novel mucosal vaccines to improve immune and health status of periparturient dairy cows and increase their productive potentials. PhD Thesis. University of Alberta.

Jin, D., Chang, G., Zhang, K., et al., 2016. Rumen-derived lipopolysaccharide enhances the expression of lingual antimicrobial peptide in mammary glands of dairy cows fed a high-concentrate diet. BMC. Vet. Res. 12 (1), 128.

Jouany, J. P., Demeyer, D. I. and Grain, J. 1988. Effect of defaunating the rumen. Anim. Feed Sci. Technol. 21 (2-4), 229-65.

Kansanen, E., Kuosmanen, S. M., Leinonen, H., et al., 2013. The Keap1-Nrf2 pathway: mechanisms of activation and dysregulation in cancer. Redox Biol. 1 (1), 45-9.

Khafipour, E., Krause, D. O. and Plaizier, J. C. 2009. A grain-based subacute ruminal acidosis challenge causes translocation of lipopolysaccharide and triggers inflammation. J. Dairy Sci. 92 (3), 1060-70.

Khiaosa-Ard, R., Pourazad, P., Aditya, S., et al., 2018. Factors related to variation in the susceptibility to subacute ruminal acidosis in early lactating Simmental cows fed the same grain-rich diet. Anim. Feed Sci. Technol. 238, 111-22.

Khovidhunkit, W., Kim, M. S., Memon, R. A., et al., 2004. Effects of infection and inflammation on lipid and lipoprotein metabolism: mechanisms and consequences to the host. J. Lipid Res. 45 (7), 1169-96.

Kim, K. N., Ko, Y. J., Yang, H. M., et al., 2013. Anti-inflammatory effect of essential oil and its constituents from fingered citron (*Citrus medica* L. var. sarcodactylis) through blocking JNK, ERK and NF-κB signaling pathways in LPS-activated RAW 264.7 cells. Food Chem. Toxicol. 57, 126-31.

Kitkas, G. C., Valergakis, G. E., Karatzias, H., et al., 2013. Subacute ruminal acidosis: prevalence and risk factors in Greek dairy herds. Iran. J. Vet. Res. 14, 183-9.

Kleen, J. L. and Cannizzo, C. 2012. Incidence, prevalence and impact of SARA in dairy herds. Anim. Feed. Sci. Technol. 172 (1-2), 4-8.

Kleen, J. L., Hooijer, G. A., Rehage, J., et al., 2003. Subacute ruminal acidosis (SARA): a review. J. Vet. Med. A 50 (8), 406-14.

Kleen, J. L., Hooijer, G. A., Rehage, J., et al., 2004. Rumenocentesis (rumen puncture): a viable instrument in herd health diagnosis. DTW. Dtsch. Tierarztl. Wochenschr. 111 (12), 458-62.

Kleen, J. L., Hooijer, G. A., Rehage, J., et al., 2009. Subacute ruminal acidosis in Dutch dairy herds. Vet. Rec. 164 (22), 681-3.

Kleen, J. L., Upgang, L. and Rehage, J. 2013. Prevalence and consequences of subacute ruminal acidosis in German dairy herds. Acta Vet. Scand. 55, 48.

Kobayashi, K., Oyama, S., Numata, A., et al., 2013. Lipopolysaccharide disrupts the milk-blood barrier by modulating claudins in mammary alveolar tight junctions. PLoS ONE 8 (4), e62187.

Kröger, I., Humer, E., Neubauer, V., et al., 2017. Modulation of chewing behavior and reticular pH in nonlactating cows challenged with concentrate-rich diets supplemented with phytogenic compounds and autolyzed yeast. J. Dairy Sci. 100 (12), 9702-14.

Kröger, I., Humer, E., Neubauer, V., et al., 2019. Feeding diets moderate in physically effective fibre alters eating and feed sorting patterns without improving ruminal pH, but impaired liver health in dairy cows. Animals (Basel) 9 (4).

Kushibiki, S., Hodate, K., Shingu, H., et al., 2002. Alterations in lipid metabolism induced by recombinant bovine tumornecrosis factor-alpha administration to dairy heifers. J. Anim. Sci. 80 (8), 2151-7.

Lacetera, N. 2016. Metabolic stress, heat shock proteins, and innate immune response. In: Amadori, M. (Ed.), The Innate Immune Response to Noninfectious Stressors. Human and Animal Models. Elsevier, Oxford, pp. 107-32.

Lechowski, R. 1997. The influence of metabolic acidosis in new-born calves on biochemical profile of the liver. Comp. Haem. Int. 7 (3), 172-6.

Lee, J. W., Paape, M. J., Elsasser, T. H., et al., 2003. Elevated milk soluble CD14 in bovine mammary glands challenged with *Escherichia coli* lipopolysaccharide. J. Dairy Sci. 86 (7), 2382-9.

Le Floćh, N., Melchior, D. and Obled, C. 2004. Modifications of protein and amino acid metabolism during inflammation and immune system activation. Livest. Prod. Sci. 87 (1), 37-45.

Li, S., Khafipour, E., Krause, D. O., et al., 2012. Effects of subacute ruminal acidosis challenges on fermentation and endotoxins in the rumen and hindgut in dairy cows. J. Dairy Sci. 95 (1), 294-303.

Liepa, G. U., Beitz, D. C. and Linder, J. R. 1978. Cholesterol synthesis in ruminating and nonruminating goats. J. Nutr. 108 (3), 535-43.

Marchesini, G., De Nardi, R., Gianesella, M., et al., 2013. Effect of induced ruminal acidosis on blood variables in heifers. BMC Vet. Res. 9, 98.

McGuire, K., Jones, M., Werling, D., et al., 2006. Radiation hybrid mapping of all 10 characterized bovine Toll-like receptors. Anim. Genet. 37 (1), 47-50.

Memon, M. A., Wang, Y., Xu, T., et al., 2019. Lipopolysaccharide induces oxidative stress by triggering MAPK and Nrf2 signalling pathways in mammary glands of dairy cows fed a high-concentrate diet. Microb. Pathog. 128, 268-75.

Morgante, M., Stelleta, C., Berzaghi, P., et al., 2007. Subacute rumen acidosis in lactating cows: an investigation in intensive Italian dairy herds. J. Anim. Physiol. Anim. Nutr. 91 (5-6), 226-34.

Morris, M. and Li, L. 2012. Molecular mechanisms and pathological consequences of endotoxin tolerance and priming. Arch. Immunol. Ther. Exp. (Warsz.) 60 (1), 13-8.

Nagaraja, T. G. and Lechtenberg, K. F. 2007. Acidosis in feedlot cattle. Vet. Clin. North Am. Food Anim. Pract. 23, 333-50.

Nagaraja, T. G., Bartley, E. E., Fina, L. R., et al., 1978. Relationship of rumen gram-negative bacteria and free endotoxin to lactic acidosis in cattle. J. Anim. Sci. 47 (6), 1329-37.

Neubauer, V., Petri, R., Humer, E., et al., 2018. High-grain diets supplemented with phytogenic compounds or autolyzed yeast modulate ruminal bacterial community and fermentation in dry cows. J. Dairy Sci. 101 (3), 2335-49.

Nilsson, A. and Duan, R. D. 2006. Absorption and lipoprotein transport of sphingomyelin. J. Lipid Res. 47 (1), 154-71.

Nocek, J. E. 1997. Bovine acidosis: implications on laminitis. J. Dairy Sci. 80 (5), 1005-28.

Nordlund, K. V. and Garrett, E. F. 1994. Rumenocentesis: a technique for collecting rumen fluid for the diagnosis of subacute rumen acidosis in dairy herds. Bov. Pract. 28, 109-12.

Nordlund, K. V., Garrett, E. F. and Oetzel, G. R. 1995. Herd-based rumenocentesis: a clinical approach to the diagnosis of subacute rumen acidosis. Compend. Contin. Educ. Vet. 17, S48-56.

O'Grady, L., Doherty, M. L. and Mulligan, F. J. 2008. Subacute ruminal acidosis (SARA) in grazing Irish dairy cows. Vet. J. 176 (1), 44-9.

Pareek, R., Wellnitz, O., Van Dorp, R., et al., 2005. Immunorelevant gene expression in LPS-challenged bovine mammary epithelial cells. J. Appl. Genet. 46 (2), 171-7.

Plaizier, J. C., Krause, D. O., Gozho, G. N., et al., 2008. Subacute ruminal acidosis in dairy cows: the physiological causes, incidence and consequences. Vet. J. 176 (1), 21-31.

Plaizier, J. C., Khafipour, E., Li, S., et al., 2012. Subacute ruminal acidosis (SARA), endotoxins, and health consequences. Anim. Feed. Sci. Technol. 172 (1-2), 9-21.

Plaizier, J. C., Danesh Mesgaran, M., Derakhshani, H., et al., 2018. Review: enhancing gastrointestinal health in dairy cows. Animal 12 (s2), s399-418.

Pourazad, P., Khiaosa-Ard, R., Qumar, M., et al., 2016. Transient feeding of a concentrate-rich diet increases the severity of subacute ruminal acidosis in dairy cattle. J. Anim. Sci. 94 (2), 726-38.

Razak, M. A., Begum, P. S., Viswanath, B., et al., 2017. Multifarious beneficial effect of nonessential amino acid, glycine: a review. Ox. Med. Cell. Long.

Saleem, F., Ametaj, B. N., Bouatra, S., et al., 2012. A metabolomics approach to uncover the effects of grain diets on rumen health in dairy cows. J. Dairy Sci. 95 (11), 6606-23.

Schlingmann, B., Overgaard, C. E., Molina, S. A., et al., 2016. Regulation of claudin/zonula occludens-1 complexes by hetero-claudin interactions. Nat. Commun. 7, 12276.

Schmitz, S., Pfaffl, M. W., Meyer, H. H. D., et al., 2004. Short-term changes of mRNA expression of various inflammatory factors and milk proteins in mammary tissue during LPS-induced mastitis. Domest. Anim. Endocrinol. 26 (2), 111-26.

Schumann, R. R., Rietschel, E. T. and Loppnow, H. 1994. The role of CD14 and lipopolysaccharide-binding protein (LBP) in the activation of different cell types by endotoxin. Med. Microbiol. Immunol. 183 (6), 279-97.

Shi, H., Guo, Y., Liu, Y., et al., 2016. The in vitro effect of lipopolysaccharide on proliferation, inflammatory factors and antioxidant enzyme activity in bovine mammary epithelial cells. Anim. Nutr. 2 (2),

99-104.

Sordillo, L. M. and Aitken, S. L. 2009. Impact of oxidative stress on the health and immune function of dairy cattle. Vet. Immunol. Immunopathol. 128 (1-3), 104-9.

Sordillo, L. M. and Streicher, K. L. 2002. Mammary gland immunity and mastitis susceptibility. J. Mammary Gland. Biol. Neoplasia 7 (2), 135-46.

Stone, W. C. 1999. The effect of subclinical rumen acidosis on milk components. In: Proceeding of the Cornell Nutrition Conference of Feed Manufacturers, Syracuse, NY. Cornell University, Ithaca, NY, pp. 40-6.

Stone, W. C. 2004. Nutritional approaches to minimize subacute ruminal acidosis and laminitis in dairy cattle. J. Dairy Sci. 87 (E. Suppl.), E13-26.

Strandberg, Y., Gray, C., Vuocolo, T., et al., 2005. Lipopolysaccharide and lipoteichoic acid induce different innate immune immune responses in bovine mammary epithelial cells. Cytokine 31 (1), 72-86.

Swanson, K., Gorodetsky, S., Good, L., et al., 2004. Expression of a beta-defensin mRNA, lingual antimicrobialpeptide, in bovine mammary epithelial tissue is induced by mastitis. Infect. Immun. 72 (12), 7311-4.

Tajik, J., Nadalian, M. G., Raoofi, A., et al., 2009. Prevalence of subacute ruminal acidosis in some dairy herds of Khorasan Razavi province, Northeast of Iran. Iran. J. Vet. Res. 10, 28-32.

Takeda, K. and Akira, S. 2005. Toll-like receptors in innate immunity. Int. Immunol. 17 (1), 1-14.

Takeuchi, O., Hoshino, K., Kawai, T., et al., 1999. Differential roles of TLR2 and TLR4 in recognition of gram-negative and gram-positive bacterial cell wall components. Immunity 11 (4), 443-51.

Wall, S. K., Hernandez-Castellano, L. E., Ahmadpour, A., et al., 2016. Differential glucocorticoid-induced closure of the blood-milk barrier during lipopolysaccharide- and lipoteichoicacid-induced mastitis in dairy cows. J. Dairy Sci. 99 (9), 7544-53.

Wellnitz, O., Arnold, E. T. and Bruckmaier, R. M. 2011. Lipopolysaccharide and lipoteichoic acid induce different immune responses in the bovine mammary gland. J. Dairy Sci. 94 (11), 5405-12.

Wu, T., Wang, C., Ding, L., et al., 2016. Arginine relieves the inflammatory response and enhances the casein expression in bovine mammary epithelial cells induced by lipopolysaccharide. Mediators Inflamm. 2016, 9618795.

Yang, W. Z. and Beauchemin, K. A. 2006. Increasing the physically effective fiber content of dairy cow diets may lower efficiency of feed use. J. Dairy Sci. 89 (7), 2694-704.

Zbinden, C., Stephan, R., Johler, S., et al., 2014. The inflammatory response of primary bovine mammary epithelial cells to *Staphylococcus aureus* strains is linked to the bacterial phenotype. PLoS ONE 9 (1), e87374.

Zebeli, Q. and Ametaj, B. N. 2009. Relationships between rumen lipopolysaccharide and mediators of inflammatory response with milk fat production and efficiency in dairy cows. J. Dairy Sci. 92 (8), 3800-9.

Zebeli, Q. and Metzler-Zebeli, B. U. 2012. Interplay between rumen digestive disorders and diet-induced inflammation in dairy cattle. Res. Vet. Sci. 93 (3), 1099-108.

Zebeli, Q., Dijkstra, J., Tafaj, M., et al., 2008. Modeling the adequacy of dietary fiber in dairy cows based on the responses of ruminal pH and milk fat production to composition of the diet. J. Dairy Sci. 91 (5), 2046-66.

Zebeli, Q., Mansmann, D., Steingass, H., et al., 2010. Balancing diets for physically effective fibre and ruminally degradable starch: a key to lower the risk of sub-acute rumen acidosis and improve productivity of dairy cattle. Livest. Sci. 127 (1), 1-10.

Zebeli, Q., Dunn, S. M. and Ametaj, B. N. 2011. Perturbations of plasma metabolites correlated with the rise of rumen endotoxin in dairy cows fed diets rich in easily degradable carbohydrates. J. Dairy Sci. 94 (5), 2374-82.

Zebeli, Q., Ghareeb, K., Humer, E., et al., 2015. Nutrition, rumen health and inflammation in the transition period and their role onoverall health and fertility in dairy cows. Res. Vet. Sci. 103, 126-36.

Zhang, R., Huo, W., Zhu, W., et al., 2015. Characterization of bacterial community of raw milk from dairy cows during subacute ruminal acidosis challenge by highthroughput sequencing. J. Sci. Food Agric. 95 (5), 1072-9.

Zhang, K., Chang, G., Xu, T., et al., 2016. Lipopolysaccharide derived from the digestive tract activates inflammatory gene expression and inhibits casein synthesis in the mammary glands of lactating dairy cows. Oncotarget 7 (9), 9652-65.

Zhao, X. and Lacasse, P. 2008. Mammary tissue damage during bovine mastitis: causes and control. J. Anim. Sci. 86 (13Suppl.), 57-65.

Zhou, J., Dong, G., Ao, C., et al., 2014. Feeding a high-concentrate corn straw diet increased the release of endotoxin in the rumen and pro-inflammatory cytokines in the mammary gland of dairy cows. BMC. Vet. Res. 10, 172.

第四部分

优化瘤胃微生物功能的营养策略

第18章 瘤胃微生物在反刍动物生产系统中的作用

Sinéad M. Waters、David A. Kenny，爱尔兰Teagasc动物与生物科学研究部；
Paul E. Smith、Teagasc，动物与生物科学研究部，
爱尔兰都柏林大学健康与农业科学都柏林大学学院

(罗玉衡译)

1 前言

反刍动物自身不具有降解以结构性多糖为主的植物组分所必需的消化酶（Flint等，2012）。尽管如此，它们还是能有效利用储存于多种牧草中的能量，而绝大多数哺乳动物都不具备直接利用这些牧草的能力。到2050年，全球人口预计将超过90亿人（UN，2017），而反刍动物在为全球人口提供食物方面发挥着至关重要的作用。由于反刍动物能够将植物中储存的难以获取的能量转化为优质蛋白质和能量源，人类对肉牛、水牛、绵羊和山羊等少数家畜物种的食物依赖性增加。据估计，到2050年，全球农业产出需要实现近50%的增长（FAO，2017）。实际上，由于反刍动物具备将饲草转化为优质食物来源的独特能力，进一步发展反刍动物养殖可能实现农业产出的增长从而满足全球粮食需求。

在长达5 000万年的进化过程中，瘤胃微生物种群与宿主之间已形成稳定的共生关系，使反刍动物具备了从饲草型日粮中获取所需营养的能力（Sasson等，2017）。瘤胃中的微生物，包括细菌、真菌、原生动物和古菌，都具有各自独特的获取营养的方式。一些细菌、真菌和原生动物能够直接降解动物摄入的食物。而其他一些微生物，主要是古菌和部分细菌，则以其他微生物的发酵产物为利用底物。反刍动物本身与瘤胃生态系统中营养层级较低的微生物类似，因为宿主对瘤胃微生物产生的挥发性脂肪酸（VFAs）具有高度依赖性。据估计，反刍动物瘤胃产生的VFAs约占反刍动物能量需求的63%（Bergman，1990）。瘤胃微生物与宿主的这种互惠共生关系可以为前者提供两方面的益处：来自摄入饲料的大量底物，以及对瘤胃专性厌氧菌至关重要的厌氧环境。

在畜牧生产系统中，饲料成本占比高，对农场盈利能力影响较大（Kenny等，2018）。自然条件下，反刍动物的日粮以饲草为基础。鉴于其成本竞争力大于谷物，牧草长期以来被用作放牧反刍家畜的主要饲料来源。然而，谷物也被用作反刍动物日粮组分之一，尤其是在西方国家，其目的在于提高动物的生产性能，和/或在牧草不足期间作为营养补充。饲喂禾本科牧草以及其他牧草（如豆科）不仅有利于农场盈利，而且在一定程度上减缓了人类和家畜之间对稀缺耕地资源的竞争。

根据生产目的来选育有效利用饲料的动物可以提高农场盈利能力。饲料效率通常被认

为是一个多因素的复杂性状，可以用多种方式进行描述，包括采食量与体增重的比值或动物预期采食量与实际采食量的回归关系（Kenny 等，2018）。但目前仍然缺乏根据饲料效率来区分动物个体的精准生物学方法。虽然在许多研究中，日粮已被承认是导致瘤胃微生物组成和功能发生重大变化的原因，但越来越多的证据表明，瘤胃微生物区系可能随宿主动物的饲料效率而变化（Guan 等，2008；Carberry 等，2012；Hernandez-Sanabria 等，2012；Shabat 等，2016；Ellison 等，2017；McGovern 等，2018；Delgado 等，2019）。事实上，某些瘤胃微生物丰度的变化可能影响反刍动物对饲料能量的利用（Cantalapiedra Hijar 等，2018 年）。

值得注意的是，不能将饲料效率作为一种独立于动物所处环境的性状，因为动物所接受的营养管理可能会在很大程度上影响这一性状（Kenny 等，2018）。事实上，有证据表明，不同日粮能够改变反刍动物的饲料效率，也可以根据动物的饲料效率差异来调整日粮。饲料效率，特别是剩余采食量（RFI）在不同生产阶段具有相对较高的重复性（Kelly 等，2010）。然而有报道也表明，日粮变换也会导致动物饲料效率在原有基础上的改变（Durunna 等，2012；Thompson，2015），这意味着动物基因型可能会影响瘤胃微生物区系发酵不同类型饲料的能力。

动物的饲料效率及其相关的温室气体（greenhouse gas，GHG）产量之间可能存在关联。大量研究显示，饲料效率更高的动物甲烷排放量减少（表4），日粮氮存留量增加（Sharma 等，2018）。这一结果是可以预期的，因为甲烷的产生可能导致 6%~12% 动物总能摄入量（gross energy intake，GEI）的损失（Johnson 和 Johnson，1995；Pacheco 等，2014）。考虑到全球畜牧业面临着越来越大的减少碳足迹的压力，培育更为高效的动物品种既有利于节能减排，又能永久提高农场的盈利能力。

日粮除了改变瘤胃微生物组成也会影响动物的生产性能。例如，放牧于高质量牧草场可以减少瘤胃甲烷的产量（Jonker 等，2018b）。类似研究也表明饲喂豆科牧草与甲烷产量的降低有关（Enriquez-Hidalgo 等，2014a；Niderkorn 等，2015）。其实，针对放牧前草量较低的高质量草地的研究显示，良好的草场管理可以同时提高草地和牛的性能，也可降低甲烷排放量（Hart 等，2009；Wims 等，2010）并提高动物的平均日增重（average daily gain，ADG；Boland 等，2013）。

本章目的是在放牧反刍动物生产系统的背景下，更好地理解瘤胃微生物区系与饲料效率之间的关系。此外，我们将聚焦于牧草（主要是禾本科和豆科）的作用，及其改变瘤胃微生物组的能力和由此引发的对动物生产性能的影响。我们首先描述了与饲料效率和甲烷生成有关的主要瘤胃特征，其次讨论了牧场组成和管理对瘤胃功能和生产性能的潜在影响。我们将主要以肉牛的生产系统为例，辅以其他反刍动物种类和一定的体外研究结果加以说明。

2 日粮和瘤胃微生物组

如前所述，反刍动物本身缺乏降解复杂植物多糖所必需的酶。因此，反刍动物依靠与其前胃内厌氧微生态系统的共生关系从饲草中获取必需营养素（Henderson 等，2015）。

反刍动物及其瘤胃微生物菌群的这种独特的协同进化，使其可以将人类无法消化的植物基质转化为优质乳制品和肉制品，造福全球。

在爱尔兰或新西兰等全年气候较为温和的国家，牧草是反刍动物的主要饲料来源。然而，无论其来源是牧草还是谷物，目前为止植物源碳水化合物仍然是瘤胃微生物种群最重要的营养来源，必须持续供应以维持瘤胃微生物生长（Hungate，1966）。众所周知，不同来源的饲料可以改变家畜的生产性能。长期以来，日粮组成的差异被认为是调节瘤胃微生物组成和潜在功能的主要因素（Hungate，1966；Carberry 等，2012、2014a；Henderson 等，2015；Ellison 等，2017）。因此，日粮对瘤胃微生物群的影响决定了后者利用饲料营养素的效率，这很可能导致宿主与日粮相关的生产性能的差异（McCann 等，2014a）。

许多研究表明瘤胃微生物群落对不同日粮的反应存在差异。例如，与精料和粗料比例接近1∶1的日粮相比，当日粮中的精料浓度超过90%时，杂交牛瘤胃微生物中古菌和原虫的丰度显著降低（Rooke 等，2014）。同样，随着日粮精料水平增加，瘤胃中产甲烷菌的比例也发生改变。与高粗料日粮相比，采食高精料日粮可使瘤胃中的产甲烷菌总数，以及斯氏甲烷球形菌（*Methanosphaera stadtmanae*）、反刍兽甲烷短杆菌（*Methanobrevibacter ruminantium*）和史氏甲烷短杆菌（*Methanobrevibacter smithii*）等产甲烷菌的丰度增加（Carberry 等，2014b）。当肉牛日粮类型逐渐由高粗料转变为高精料时，其瘤胃内埃氏巨球型菌（*Megasphaera elsdenii*）、牛链球菌（*Streptococcus bovis*）、反刍月形单胞菌（*Selenomonas ruminantium*）、短普雷沃氏菌（*Prevotella bryantiielsdenii*）等细菌数量随精料比例上升而增加，而溶纤维丁酸弧菌（*Butyrivibrio fibrisolvens*）和产琥珀酸丝状杆菌（*Fibrobacter succinogenes*）的数量相应减少（Fernando 等，2010）。

现代瘤胃微生物研究特别关注生态学，不同的微生物群各自占据瘤胃生态系统中的特定生态位，这有助于解释为什么饲喂不同日粮时瘤胃微生物种群总是发生变化。例如，一种专门降解纤维素的瘤胃细菌——产琥珀酸丝状杆菌（*F. succinogenes*）（Stewart 等，1997）已被证明能产生胞外纤维素酶（Forsberg 等，1981），因而当牛采食饲草型日粮时，该菌在瘤胃中的丰度增加（Henderson 等，2015）。

随着微生物种群的变化，瘤胃中特定微生物的作用也发生改变，故可以预期瘤胃发酵的终产物（如VFAs、乳酸、CO_2、H_2和CH_4）也会发生波动。瘤胃内特定的纤维素降解细菌、真菌和原虫等微生物可以产生纤维素酶，它们的终产物可以被其他不能直接降解纤维素的瘤胃微生物利用（Wolin 等，1997 年）。因此，根据微生物摄入的营养水平，其终产物产量的波动可能会带来不同影响。当产琥珀酸丝状杆菌与反刍月形单胞菌共培养时，前者产生的琥珀酸可以支持后者在纤维素培养基上生长良好（Scheifinger 和 Wolin，1973）。虽然我们很容易理解日粮组成对初级微生物的直接影响，但也不可忽视在其他营养条件下瘤胃发酵产生的间接和连锁效应。

3 瘤胃纤维素降解

反刍动物能够利用更简单的哺乳动物（如人类）无法降解的纤维物质（Morgavi 等，2010）。结构多糖、纤维素、半纤维素和果胶是瘤胃微生物从牧草中获得的主要能量来源

(Dehority，1991）。细胞壁的降解通过瘤胃细菌和真菌的合作完成，瘤胃原虫的贡献较小（Chesson 和 Forsberg，1997）。

不同的微生物群拥有特定的利用纤维素的策略（适应性），其中一些会在本书后面的章节讨论。虽然有许多植物来源的多糖都可以被瘤胃微生物降解，本节重点介绍纤维素的降解，同时也可以参考本章中其他植物性产品的利用情况。纤维素的降解由一系列酶催化，包括糖基水解酶和内切葡聚糖酶。植物多糖在酶的作用下转化为可溶性低聚糖，被细胞吸收用于进一步加工（Arntzen 等，2017）。植物细胞壁的酶解仅限于摄入的植物颗粒表面的多糖（Chesson 和 Forsberg，1997）。根据 Chesson（1993）提出的植物细胞壁降解机理模型，细胞壁的降解不是选择性的，相反，酶的作用针对暴露于摄入的植物基质表面的所有多糖。图1展示的是半纤维素中嵌入纤维素的植物颗粒酶解过程的简化版本。

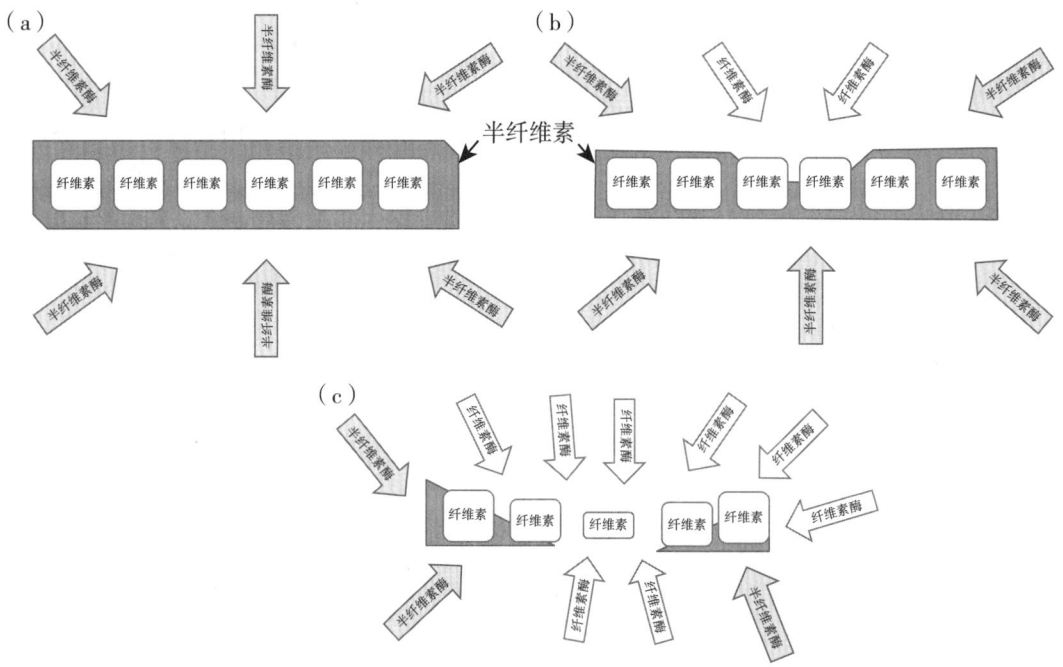

图1 基于 Chesson（1993）提出的模型，瘤胃微生物对植物物质进行酶降解的示意图，其中纤维素嵌入半纤维素中。箭头表示活性酶。酶的颜色对应于目标底物；灰色箭头代表以半纤维素为目标的半纤维素酶，而白色箭头代表以纤维素为目标的纤维素酶。（a）植物颗粒的半纤维素酶靶向半纤维素成分；（b）随着纤维素的暴露和纤维素酶的产生，半纤维素开始降解；（c）随着纤维素的进一步暴露，纤维素酶的产生增加。

3.1 瘤胃细菌对纤维素的降解

产琥珀酸丝状杆菌、黄色瘤胃球菌（*Ruminococcus flavefaciens*）和白色瘤胃球菌（*Ruminococcus albus*）是瘤胃内主要的纤维素降解菌（Weimer，1996）。虽然纤维杆菌属（*Fi-*

brobacter) 和瘤胃球菌属（*Ruminococcus*）的成员是主要的纤维素分解细菌，但二者降解木质纤维素化合物的方式有所不同。据鉴定，瘤胃内半数以上的纤维素糖基水解酶，以及超过 1/3 的半纤维素糖基水解酶均来自上述两个菌属（Dai 等，2015）。

黄色瘤胃球菌和白色瘤胃球菌都是梭菌目（Clostridiales）的成员，它们通过产生纤维小体（cellulosomes）来降解植物基质，这些纤维小体是由锚定蛋白（dockerins）组成的多酶复合物（用于酶的附着）和被称为"支架"（scaffoldin）的结构蛋白构成（Artzi 等，2017）。纤维小体复合物协同促进纤维素酶和半纤维素酶的产生，从而促进植物细胞壁多糖的有效代谢（Bayer 等，2004；Han 等，2004；Devendran 等，2016）。当发生某些偏差时，纤维小体复合物可以允许各种酶附着在纤维素分解细菌的外表面，以促进其有序降解多种复杂植物多糖。黄色瘤胃球菌和白色瘤胃球菌的纤维素小体的复杂性不同，前者能够合成大量锚定蛋白和多种支架蛋白，从而提高酶活性（Artzi 等，2017；Seshadri 等，2018）。

多糖的来源和组成多糖的子单位（如纤维素、果胶、木聚糖、纤维二糖等）已被证明可以影响纤维小体的基因表达和酶谱，进而影响纤维小体复合物的组成（Han 等，2004）。例如，当生长于含有替代碳源的培养基时，*Clostridium cellulovorans* 胞内的一些底物特异性酶的基因表达量随相应底物的增加而升高；当生长于含有纤维素的培养基时，cbpA（一种纤维素结合蛋白）的表达量升高，而当培养基中含有木聚糖时，编码木聚糖酶的 *xynA* 基因表达量也相应升高。此前已有关于细菌纤维小体的全面综述（Artzi 等，2017 年）。

产琥珀酸丝状杆菌需要与其底物紧密接触才能有效降解纤维素，但不通过产生降解纤维素的纤维小体，而是依赖外膜蛋白的组合（Jun 等，2007；Suen 等，2011）。在产琥珀酸丝状杆菌 S85 的基因组中已鉴定出 10 个已知的与黏菌类微生物（slime moulds，真菌的一个类群）相关的基因，推测它们可能在与纤维素结合的过程中发挥作用（Suen 等，2011）。此外，外膜菌毛可能促进产琥珀酸丝状杆菌对纤维素的黏附。Jun 等（2007）采用蛋白质组学分析比较了产琥珀酸丝状杆菌及其黏附素突变株 Ad1 和 Ad4 的结合能力，发现野生型产琥珀酸丝状杆菌只产生Ⅳ型菌毛蛋白，而上述两种突变菌株都无法产生这种蛋白。同时，由于突变株既缺乏合成菌毛的能力，也无法牢固附着于晶体纤维素，说明菌毛蛋白对产琥珀酸丝状杆菌附着于纤维素至关重要。

产琥珀酸丝状杆菌具有水解多种植物多糖的能力，但它只能利用纤维素进行生长，因此可以推测它通过水解各种复杂碳水化合物来获取纤维素（Suen 等，2011；Dai 等，2015）。外膜囊泡为蛋白质提供了一种保护性的运输载体，使其能够集中作用于距离稍远的靶标（Kulp 和 Kuehn，2010），这被认为是纤维杆菌降解植物细胞的主要方式（Arntzen 等，2017）。产琥珀酸丝状杆菌也会通过基因表达差异体现对底物的响应（Neumann 等，2018）。

发酵终产物会由于纤维素降解方式不同而变化。琥珀酸是产琥珀酸丝状杆菌发酵的主要终产物，其次是乙酸（Gokarn 等，1997），而氢（H_2）的产生与该物种无关（Joblin 等，2002）。反之，可以利用纤维素的瘤胃球菌则产生 H_2、乙酸、甲酸和 CO_2（Zheng 等，2014；Rooke 等，2014）。黄色瘤胃球菌也产生琥珀酸，但产量远低于产琥珀酸丝状杆菌

（Gokarn 等，1997），而在与产甲烷菌共培养时，黄色瘤胃球菌的琥珀酸产量进一步降低，但乙酸产量增加（Latham 和 Wolin，1977；Wolin 等，1997）。

普雷沃氏菌（*Prevotella*）已被确定为瘤胃微生物区系中最常见的一类细菌（Stewart 等，1997；Stevenson 和 Weimer，2007；Henderson 等，2015），并具有降解半纤维素（Rubino 等，2017）、淀粉、木聚糖、果胶的能力（Matsui 等，2000）和利用氮的功能（Kim 等，2017）。虽然该属成员在瘤胃内的作用存在相似性，但在种水平上可能具有独特性。例如，Matsui 等（2000）的体外研究发现，*P. ruminicola*、*P. bryantii*、*P. albensis* 和 *P. brevis* 对不同生长介质的响应表现为生长速率和酶产量的差异。

尽管普雷沃氏菌在纤维降解过程中起作用，但可能并不是一种关键的纤维素分解细菌。虽然已经被证明可以在纤维二糖培养基上生长（Matsui 等，2000），但普雷沃氏菌的作用更倾向于将纤维素暴露于其他瘤胃微生物（Huws 等，2016；Rubino 等，2017）。普雷沃氏菌是瘤胃中植物基质的初级定植者，体外试验也表明含有牧草的培养基可以提高其丰度（Mayorga 等，2016；Elliott 等，2018）。此外，普雷沃氏菌基因组编码了瘤胃内 14% 的低聚糖降解酶和 13.5% 的半纤维素酶（Dai 等，2015）。

梭菌目的其他成员也在纤维的降解过程中起类似作用。梭菌属（*Clostridium*）的成员已被证明能产生纤维小体，而真杆菌属（*Eubacterium*）的一些菌种已被证明可以编码纤维素酶和半纤维素酶（Dai 等，2015）。其他微生物，如毛螺菌科（*Lachnospiraceae*）的某些菌属，包括 *Butteriviribrio* 和 *Pseudobutiviribrio*，则可以发酵半纤维素和木聚糖（Krause 等，2003），因而也对纤维素的暴露有贡献，但某些梭菌目微生物在效率较低的瘤胃中呈现更高丰度（Shabat 等，2016）。

那些利用易发酵可溶性碳水化合物的微生物是瘤胃内纤维物质的初级定植者（Brulc 等，2009），而纤维降解的主要产物形成于次级定植过程中。在瘤胃内孵育 4 h 后，植物中的干物质（DM）消失率显著增加（1～2 h 为 2.8%，4～8 h 为 31.7%）（Huws 等，2016），瘤胃内毛螺菌科、*Pseudobutyrivibrio* 和 *Butyrivibrio* 的比例也呈同步上升趋势（Mayorga 等，2016；Huws 等，2016）。因此，鉴于毛螺菌科细菌比例的增加与植物干物质的消失趋同，该科成员很可能积极参与纤维的分解，并极有可能将纤维素暴露于其他次级微生物。表 1 概述了瘤胃内的优势纤维水解细菌的主要发酵终产物。

3.2 真菌和纤维素

瘤胃中的厌氧真菌属于 Neocallimastigomycota 菌门（Gruninger 等，2014），能够合成多种高效酶并通过假根机械破坏植物结构，在植物基质的降解过程中起关键作用（Orpin，1977；Choudhury 等，2015；Huws 等，2018）。现有研究已报道了该门的 11 个菌属，包括 *Orpinomyces*、*Neocallimastix*、*Cyllamyces*、*Piromyces*、*Anaeromyces* 和 *Caecomyces*（Choudhury 等，2015；Gruninger 等 2014），以及通过分子生物学手段新发现的 5 个菌属：*Buwchfawromyces*（Callaghan 等，2015）、*Oontomyces*（发现于骆驼前胃）（Dagar 等，2015）、*Pecoramyces*（Hanafy 等，2017）、*Liebetanzomycespolymorphus*（Joshi 等，2018）和 *Feramyces*（Hanafy 等，2018）。下文提及的瘤胃真菌将仅限于厌氧真菌。

表 1 主要溶纤菌发酵终产物

属	种	主要发酵终产物	H_2	CO_2	研究
纤维杆菌属	产琥珀酸丝状杆菌	乙酸盐、琥珀酸盐	否	否	Gokarn 等，1997；Joblin 等，2002
瘤胃球菌属	黄色瘤胃球菌	乙酸盐、甲酸盐、琥珀酸盐	是	是	Latham 和 Wolin，1977
	白色瘤胃球菌	乙酸盐、甲酸盐	是	是	Miller 和 Wolin，1973；Zheng 等，2014
丁酸弧菌属	溶纤维丁酸弧菌	丁酸盐、甲酸盐、乳酸（乙酸盐）	是	是*	Marounek 和 Dušková，1999；Emerson 和 Weimer，2017
假丁酸弧菌	瘤胃假丁酸弧菌	丁酸盐、甲酸盐、乳酸（乙酸盐）	—	—	Van Gylswyk 等，1996
毛螺旋菌属	毛螺菌	乙酸盐、甲酸盐（乳酸）	是	是	Dušková 和 Marounek，2001
普雷沃氏菌属	栖瘤胃普雷沃氏菌	乙酸盐、甲酸盐、丙酸盐、琥珀酸盐	次要	是*	Marounek 和 Dušková，1999；Emerson 和 Weimer，2017
	阿尔本斯普雷沃氏菌	乙酸盐、甲酸盐、丙酸盐、琥珀酸盐	是	是*	Emerson 和 Weimer，2017
	revis	乙酸盐、甲酸盐、丙酸盐、琥珀酸盐	次要	否*	Emerson 和 Weimer，2017
	byranti	乙酸盐、甲酸盐、丙酸盐、琥珀酸盐	次要	否*	Emerson 和 Weimer，2017

注：1 终产物的顺序与最大产量的产物顺序并不一致。
2 括号中的产物为少量。

* 计算过程基于 Emerson 和 Weimer（2017）的方法，其中 CO_2 的产量根据其他已知瘤胃发酵终产物途径的 CO_2 产量的化学计量法来估计，使用方程式 [CO_2 = 乙酸盐 + 2（丁酸盐）- 琥珀酸盐] 进行计算。

Gordon 和 Phillips（1993）的研究发现，去除瘤胃真菌可导致绵羊采食量（以稻草为主的日粮）下降 40%，这为真菌在瘤胃发酵中的关键作用提供了直接证据。一些瘤胃真菌，特别是 *Piromyces* 和 *Neocallimastix* 属，对底物几乎没有偏好性和选择性，但可能具有较强的降解木质纤维素的能力，甚至与其他底物相比，在以木质纤维素为底物时，这些真菌的生长率可以提高 20%（Solomon 等，2016）。然而，最近对 4 种常见瘤胃真菌的转录组分析表明，瘤胃内可能存在独特的真菌生态位（Gruninger 等，2018）。早前研究发现，在 *Piromyces rhizinflata* 的基因组中，编码果胶酶的碳水化合物活性酶（CAZymes）转录本数量几乎是其他种类真菌的两倍，这可能反映了该菌种对果胶的利用率较高。瘤胃中超过 8.5% 的纤维素酶已被证明存在于 *Neocallimastix* 和 *Piromyces* 属真菌的基因组中（Dai 等，2015）。与纤维素分解细菌类似，瘤胃真菌也可以产生纤维小体（Haitjema 等，2017）。

近年来，真菌的纤维分解特性受到生物技术产业的关注（Ribeiro 等，2016；Edwards 等，2017），进一步肯定了真菌源纤维素水解酶的有效性。目前认为瘤胃真菌纯培养的主要终产物包括甲酸、乙酸、乳酸、乙醇、CO_2 和 H_2，以及微量或少量琥珀酸，不产生丙酸和丁酸（Borneman 等，1989；Edwards 等，2017）。

3.3 瘤胃原虫和纤维素的降解

据推测，原虫在瘤胃纤维物质的消化过程中发挥积极作用，并产生类似于细菌发酵的终产物，包括乙酸、丁酸和 H_2（Choudhury 等，2015）。瘤胃原虫可以产生纤维素酶和半纤维素酶（Williams 和 Coleman，1997）。对瘘管绵羊纤毛虫（ciliate protozoa）的 cDNA 文库分析表明，内毛目（Entodiniomorphida）原虫可以产生木聚糖酶、纤维素酶、果胶裂解酶和果胶降解蛋白（Ricard 等，2006）。*Epidinium* 和 *Polyplastron* 属原虫的基因组编码了超过 9% 的瘤胃纤维素酶和 3% 的半纤维素酶（Dai 等，2015）。

瘤胃原虫具有捕食细菌和真菌的行为（Williams 和 Coleman，1997）。体外研究发现，当原虫与真菌共培养时可以减少纤维素的降解（Morgavi 等，1994），与此一致的是一个公认的事实，即去除原生动物有利于瘤胃微生物功能的发挥。图 2 简单阐释了具原虫和去原虫瘤胃功能的差异；图 3 显示了瘤胃纤毛原虫的显微结构。

Newbold 等（2015）对 23 项瘤胃去原虫的相关研究进行了荟萃分析，并推断去除瘤胃原虫可使有机物（OM）、中性洗涤纤维（NDF）和酸性洗涤纤维（ADF）的消化率分别降低 7%、20% 和 16%。对来自一头泌乳泽西奶牛的原虫、细菌和真菌的混合体外培养研究表明，与真菌和细菌的单一培养物相比，原虫对鸭茅干草的降解能力显著降低，前者效率几乎是后者的两倍（Lee 等，2000）。同一研究的结果还表明，原虫与细菌和真菌的共培养也会降低底物的降解率。在羧甲基纤维素测定过程中，也观察到狗牙根降解产生内切葡聚糖酶的类似情况。

由于体外纯培养难以维持原虫群落，目前很难分析瘤胃原虫对纤维素分解的贡献，相关研究也难以开展（Newbold 等，2015）。然而据估计，去除瘤胃原虫会减少真菌（92%）、白色瘤胃球菌和黄色瘤胃球菌（22%）的数量（Newbold 等，2015），因此原虫可能影响瘤胃内植物组织分解相关酶的释放（Williams 和 Coleman，1997）。

此外，基于组学的研究有助于更好地了解纤毛虫对瘤胃功能的作用，但这需要扩大瘤胃原虫基因组检测的样本数。最新研究报道了有尾内毛虫（*Entodinium caudatum*）大核基因组的初稿草图（Park 等，2018），这可能有助于我们进一步理解这些难以培养的微生物。

因此，越来越多的证据表明瘤胃原虫能够产生纤维素酶，随着更多原虫基因组的发布，它们作为瘤胃微生物组中关键纤维素分解成员的作用可能会进一步被揭示。同样值得期待的是，新方法的相继发现将有望实现瘤胃原虫的纯培养。

4 瘤胃微生物组和饲料效率

提高反刍动物饲料效率被视为降低饲料成本的一种方法。作为生长性能之一，饲料效

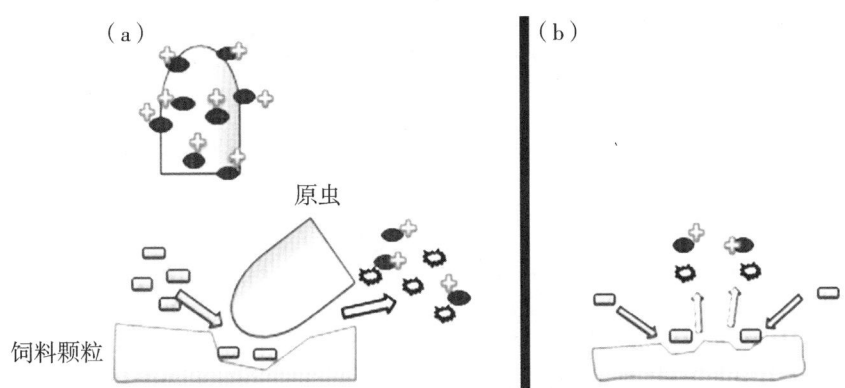

图 2 具原虫和去原虫瘤胃之间差异的详细示意图。(a) 具原虫瘤胃。原虫（蓝色）降解饲料颗粒（绿色）。在降解过程中，细菌（红色）进入饲料颗粒的可消化部分，细菌和原虫产生的发酵终产物（黑色）被产甲烷菌（紫色）利用以产生甲烷（橙色）。产甲烷菌附着于原虫以利用其氢小体产生的过量氢。(b) 去原虫瘤胃。在没有原生动物的情况下，摄入的饲料颗粒的降解导致发酵产物和甲烷产量减少（彩图8）。

率既可以用比率来描述（如饲料转化率，FCR），也可以用基于回归方程的残差来描述（如剩余采食量，RFI）（Berry 和 Crowley，2013），其中 FCR（采食量与增重之比）是衡量饲料效率的传统标准，但 RFI（实际采食量与预期的维持和生长所需采食量之差）已逐渐成为衡量饲料效率的首选标准（Kenny 等，2018）。Berry 和 Crowley（2013）针对不同国家多个品种奶牛和肉牛的 39 项研究进行了荟萃分析，估测生长期动物的 RFI 和 FCR 遗传力分别为 0.33 和 0.23，而此前的研究估测 RFI 的遗传力介于 0.07~0.62。针对 2 000 头澳大利亚安格斯牛群体的最新研究估测其 RFI 和 FCR 的遗传力分别为 0.40 和 0.20（Torres-Vázquez 等，2018）。

如前所述，瘤胃是反刍动物消化饲料的主要场所，日粮可被视为引起瘤胃微生物组成变化的主要贡献者之一。因此，在观察饲料效率时不能独立于动物所处的饲养管理环境（Kenny 等，2018）。由于饲料效率被认为是一个重要的经济性状，本节将重点总结相关文献，以描述与饲料效率相关的瘤胃微生物组特征。

基于宏基因组学和宏转录组学的大量研究突出了宿主饲料效率对瘤胃微生物组成和假定功能的影响。事实上，宏基因组学相关研究对这一领域的贡献越来越多。早期研究的重点主要是某些微生物类群是否存在及其在瘤胃中的功能，这些功能可以根据其与动物性状、发酵产物、体外试验和其他实验室指标的相关性来考虑。因此，随着宏转录组学和宏基因组学研究数量的增加，人们对瘤胃及瘤胃微生物功能的理解也在加深。大多数研究表明，饲料效率不同的动物之间，瘤胃微生物组存在少数显著的差异（Cantalapiedra-Hijar 等，2018）。然而正如前人研究所示，与饲料效率有关的瘤胃微生物种群的变化也可能是采食量或瘤胃滞留时间改变的结果。

除了产甲烷菌可以通过产甲烷过程消耗日粮能量以外，确定特定的瘤胃微生物与宿主饲料效率的稳定相关性是比较困难的。其实产甲烷菌的种群大小也不总是与宿主的 RFI 差

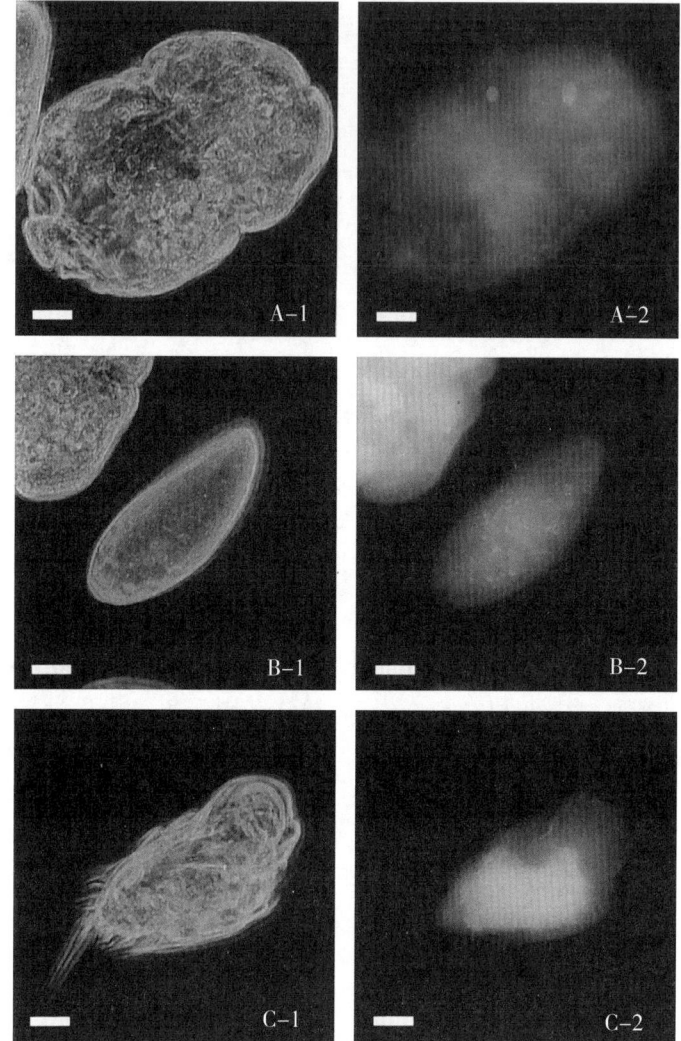

1：相差显微镜，2：荧光显微镜。A：*Polyplastron multivesiculatum*，B：*Isotricha prostoma*，C：*Ophryoscolex caudatus*。标尺表示 20 μm。

图 3　瘤胃纤毛虫（改编自 Tokura 等，1999）

异有关（Carberry 等，2014a；Dini 等，2019）。与总体数量相比，瘤胃产甲烷菌种类的差异更可能与 RFI 有关。基于古菌 16S rRNA 基因的克隆库分析表明，不同饲料效率的阉公牛其瘤胃产甲烷菌群体丰度并无差异，但在高 RFI 阉公牛瘤胃中，属于 *Methanosphaera stadtmanae* 和甲烷短杆菌属 AbM4 菌株的可操作分类单元（operational taxonomic units，OTU）丰度分别较低 RFI 公牛高 1.92 倍和 2.26 倍（Zhou 等，2009）。与低 RFI 肉母牛相比，高 RFI 母牛瘤胃中属于 *Methanobrevibacter smithii* 的 OTU 略有增加（Carberry 等，2014b），而属于 *Methanobrevibacter millerae* YE315（97%）和甲烷短杆菌属 AbM4 菌株（99%）的 OTU 则被确定与 RFI 呈负相关（McGovern 等，2018）。最新的宏基因组分析结果表明，甲烷短杆菌属的丰度在饲料效率较低的奶牛（饲喂基于羊茅、黑麦草和浓缩料的混合日粮）瘤胃中更高（Delgado 等，2019）。如果产甲烷菌物种和种群结构的变化有利于提高饲料效率，可能导致甲烷生成效率降低。此外，为产甲烷菌提供底物的其他瘤胃微生物的变化，或可供选择的氢汇（alternative hydrogen sinks）的增加也可能起类似作用（Delgado 等，2019）。从这个角度而言，产甲烷菌群的组成比其丰度更倾向于与甲烷生成有关（Tapio 等，2017）。因此，与甲烷产生呈负相关的产甲烷菌可能有利于提高饲料效率，但相关研究仍需更加深入。相关研究也为这一假设提供了支持，即甲烷短杆菌属 AbM4 菌株的丰度在饲料效率较高但甲烷产量较低的泌乳奶牛瘤胃中较低（Arndt 等，2015）。并且据报道，饲料效率较低的奶牛瘤胃微生物组产甲烷途径显著富集，同时 *Methanobrevibacter ruminantium* 的丰度明显增加（Shabat 等，2016），进一步将特定种类的产甲烷菌与饲料效率和潜在的甲烷产量联系起来。

现有研究的焦点主要集中于宿主饲料效率与瘤胃原核微生物的关系。尽管针对瘤胃原虫和真菌也进行了一些定量 PCR（采用通用引物）研究，但现有研究结果几乎没有发现这两种瘤胃微生物与宿主饲料效率的联系（Carberry 等，2012）。因此，选择能够有效识别瘤胃真菌和原虫群落差异的特异性引物是更具针对性的方法，可以更为深入地研究真核微生物与宿主饲料效率的关系。一项针对瘤胃去原虫研究的荟萃分析结果表明，去除瘤胃原虫能够提高宿主的 FCR 和 ADG（9%）并改善育肥能量的利用率（11%），同时减少热量（5%）产生（Newbold 等，2015）。考虑到饲料效率与甲烷产量之间可能存在的间接联系，在本书的其他部分会具体介绍瘤胃真菌和原虫对甲烷生成的影响。

如前所述，普雷沃氏菌属通常被认为是瘤胃中丰度最高的菌属之一，在多种底物的降解过程中起不同作用。普雷沃氏菌属与宿主饲料效率的关系加深了我们对其复杂自然特性的认知。例如，属于普雷沃氏菌属的多个 OTU 丰度在高、低 RFI 奶牛瘤胃中存在差异（Jewell 等，2015）。公牛 RFI 与其瘤胃普雷沃氏菌 OTU 的丰度之间也存在类似关系（McGovern 等，2018）。但不一致的结果也有发现，例如饲料效率较低的放牧婆罗门公牛瘤胃普雷沃氏菌 OTU 丰度增加（McCann 等，2014b），但在饲喂混合饲料的荷斯坦奶牛瘤胃中却发现了相反结果（Delgado 等，2019）。此外，Myer 等（2015）的研究并未发现普雷沃氏菌丰度与阉公牛较低的饲料效率有关。

有趣的是，日粮可能会影响不同种类的普雷沃氏菌与宿主饲料效率的关系。例如，在采食高精料日粮时，拥有较高饲料效率的阉公羊瘤胃 *P. ruminicola* 丰度较高，而当日粮以粗料为主时，该菌丰度降低（Ellison 等，2017）。类似的影响也被发现存在于普雷沃氏菌属的其他菌种。正如我们之前讨论的，普雷沃氏菌属的多个菌种在不同的体外培养基上呈现不同的生长速率。因此在种水平上，日粮和宿主饲料效率都可能影响普雷沃氏菌的丰度。

根据前面提到的一些研究结果，对饲喂以牧草/粗料为主的日粮的宿主而言，较高的普雷沃氏菌属丰度似乎与较低的饲料效率有关，而对饲喂高精料日粮的宿主而言，瘤胃普雷沃氏菌属的丰度与饲料效率之间可能存在相反关系。未来可能借助于宏基因组学相关方法来比较和确认瘤胃微生物组与不同日粮下宿主饲料效率的关系。

需要注意的是，虽然本章不会深入讨论测序数据的分析，但基于预定义的序列相似性来识别同一属的不同 OTU 是存在缺陷的。读者可从其他文献参考序列 OTU 分组相关的生物信息学方法的缺点（Callahan 等，2017）。

瘤胃球菌属（*Ruminococcus*）是另一类与 RFI 有关的瘤胃微生物。研究表明，在饲喂高粗料日粮时，低 RFI 肉牛瘤胃白色瘤胃球菌丰度相比高 RFI 肉牛增加了 1.7 倍（Carberry 等，2012），且无论采食何种日粮，前者瘤胃白色瘤胃球菌的相对丰度总是略高于后者（$P=0.08$）。对西门塔尔牛瘤胃液相和固相样本的 16S rRNA 测序分析也获得了类似结果，在采食高精料日粮的情况下，瘤胃球菌属的丰度与西门塔尔牛的 RFI 呈负相关，进而与更高的瘤胃发酵效率有关（McGovern 等，2018）。但也有研究表明，当饲喂高粗料日粮时，较高的白色瘤胃球菌丰度与绵羊的高 RFI 相关，而低 RFI 绵羊瘤胃中的布氏瘤胃球菌丰度更高（Ellison 等，2017）。有趣的是，在这项研究中，当喂饲高精料日粮时，白色瘤胃球菌的丰度在低 RFI 绵羊瘤胃中却更高。这些瘤胃球菌与宿主 RFI 相关性的变化是可以解释的，因为并非所有的白色瘤胃球菌菌株都能高效利用纤维素（Morris 和 Cole，1987）。这项研究还表明，无论饲喂何种日粮，黄色瘤胃球菌的丰度在饲料效率较低的绵羊瘤胃中都较高（Ellison 等，2017），但其他研究结果并不支持这种相关性（Carberry 等，2012；Shabat 等，2016）。正是由于普雷沃氏菌和瘤胃球菌不同菌种的差异，基于菌种和菌株水平的体内研究才真正有助于了解这些微生物与宿主饲料效率的关系。

其他与较高的饲料效率有关的微生物群包括产琥珀酸丝状杆菌（Elolimy 等，2018；McGovern 等，2018）、*Shuttleworthia* 属（Jewell 等，2015）、埃氏巨球型菌、灵巧粪球菌（*Coprococcus catus*）（Shabat 等，2016），以及能对 *Psuedobutyivibrio ruminis* 和溶糊精琥珀酸弧菌（*Succinivibrio dextrinsolven*）相关的饲料效率产生负面影响的反刍真杆菌（*Eubacterium ruminantium*）（Elolimy 等，2018），但在所有这些研究中，精料在日粮中的占比很大。已知 Succinivibrionaceae 科细菌产生的琥珀酸可以被其他微生物利用产生丙酸，并且已被证明与反刍动物产奶量呈正相关（Indugu 等，2017）。然而与通过丙烯酸盐途径生成丙酸相比，由琥珀酸合成丙酸的效率较低（Shabat 等，2016），这也可用于解释 Succinivibrionaceae 科细菌与宿主饲料效率之间的负相关关系。此外据报道，一种未分类的 Succinivibrionaceae 科细菌与生长期羔羊的瘤胃乳头宽度呈中度负相关（Yang 等，2018）。

宏基因组学结合宏转录组学分析可能是理解瘤胃微生物组与饲料效率之间关系的更好方法，然而日粮似乎是影响二者关系的一个重要的协变量。作者的研究小组与其他团队的研究结果均表明，日粮的确可以影响饲料效率，但饲料效率与特定的微生物群有关，导致宿主的饲料效率又随日粮而异（Carberry 等，2012；Ellison 等，2017）。有人提出，更为有效的瘤胃微生物群的多样性应该更低，以便其更有针对性地从饲料中获取能量。对不同 RFI 的泌乳奶牛的瘤胃内容物全基因组测序分析表明，当日粮精料与粗料的比例为 70∶30 时，饲料效率更高的奶牛瘤胃微生物种类和基因的多样性更低（Shabat 等，2016），但笔者认为这一现象可能与日粮有关。Patil 等（2018）采用基于鸟枪法的宏基因组学分析发现，饲喂高粗料日粮的低 RFI 塔尔基母羔羊瘤胃微生物的多样性较高（$P=0.073$），但在比较高、低 RFI 个体的瘤胃代谢网络时，拥有较高饲料效率的羔羊瘤胃微生物组代谢网络和酶的多样性更高。

虽然上述两项研究结果的差异可能与物种有关，但我们不应忽视日粮对饲料效率的影响,因为以粗料为主的日粮可以为瘤胃微生物提供比精料更为复杂的多糖。随着动物日粮从粗料转向精料,瘤胃微生物代谢网络的复杂性降低(Wolff 等,2017)。上述两项研究(Shabat 等 2016;Patil 等,2018)中描绘的微生物及其代谢网络的变化进一步说明,高效瘤胃的特征很可能取决于日粮种类。综上所述,当宿主采食粗料时,瘤胃微生物多样性倾向于更高以促进复杂多糖的降解。相比之下,对于更易消化的谷物类日粮而言,瘤胃微生物多样性较低可使发酵更具有针对性,但是这一理论假设与宿主饲料效率的关系有必要进一步研究。

4.1 挥发性脂肪酸(volatile fatty acids,VFAs)组成与饲料效率的关系

饲料效率与 VFA 产量之间的关系似乎也受到日粮的影响,因为饲料效率会根据动物采食粗料或精料型日粮而呈现相反的结果。人们更倾向于使用能够产生丙酸生成前体物的日粮,因为丙酸合成过程与甲烷产生途径是底物竞争关系(McAllister 和 Newbold,2008),从而可以为动物提供更多有效能量。在饲喂高精料日粮时,饲料效率较高的奶牛瘤胃 VFA 浓度更高,同时丙酸和丁酸的浓度也较高(Shabat 等,2016)。与高 RFI 羔羊相比,当精料比例为 75% 时,低 RFI 羔羊瘤胃丁酸浓度增加,但丙酸产量较低,二者瘤胃中的总 VFA 浓度无差异(Liang 等,2017)。而采食高粗料日粮时并未发现瘤胃 VFA 浓度的改变(Fitzsimons 等,2013,2014a,b;McDonnell 等,2016)。然而 McDonnell 等(2016)的研究发现,饲喂青贮饲料可以减少低 RFI 小母牛瘤胃中的丙酸含量,且使低 RFI 妊娠肉牛瘤胃氨(虽然不属于 VFA)浓度降低(Fitzsimons 等,2014a)。表 2 列举了不同研究中具有 RFI 差异的动物瘤胃 VFA 概况。表中主要 VFA(乙酸、丙酸和丁酸)的平均比例说明 VFA 与 RFI 没有显著联系。

4.2 消化率与饲料效率的关系

表 2 重点关注了瘤胃氨(NH_3)浓度和以牧草为主的日粮,平均而言,低 RFI 动物瘤胃 NH_3 浓度(mg/L)较高 RFI 动物高 14.5%。考虑到关于瘤胃 NH_3 浓度的相关报道较少,这一结论可能不够准确,但至少表明低 RFI 动物对氮的消化率增加。尽管由于瘤胃氨浓度相关数据缺乏代表性导致高粗料日粮与高精料日粮之间的比较受到限制,但我们推测两种日粮之间的差异可能并不明显。高精料日粮通常具有更高的能量密度,因而可能促进蛋白质和能量需求更为平衡的瘤胃发酵,从而促使瘤胃细菌利用过剩的 NH_3,这一点会在之后的章节进一步讨论。

通常认为饲料效率较高的个体瘤胃具有更强的摄取和吸收日粮营养物质的能力,尤其是氮。已有研究证明饲料效率更高的动物个体氮代谢增加,可能是对 DMI 降低的一种补偿(Rius 等,2012)。研究还发现,饲料效率较高的公牛和阉公牛瘤胃乳头上皮较厚,说明其拥有较高的代谢活性(Lam 等,2018)。

饲料效率较高的肉牛瘤胃转录组学特征表明,与消化、细胞增殖和生存能力相关的通路活性增强(Guan 等,2017)。其他转录组学研究表明,饲料效率较高和较低的牛瘤胃上皮细胞转录本聚类为不同集群。例如,饲料效率高的杂交阉牛瘤胃上皮与细胞旁渗透性相关的基因表达量明显增加,从而可能帮助增加营养物质的吸收(Kong 等,2016)。此外,

饲料效率更高的阉牛和小母牛瘤胃上皮溶质载体基因（如 SLC16A3）的表达量也有增加趋势（Elolimy 等，2018）。Patil 等（2018）发现，饲料效率较高的母羊瘤胃微生物组倾向于合成更多的代谢产物，这些代谢物可能转移至宿主，从而使微生物和宿主代谢网络之间的联系更紧密。因此，高效的瘤胃微生物群应该包含与宿主代谢网络更为一致的微生物代谢网络，从而提高有益的瘤胃发酵终产物的利用率（表 2）。

表 2 不同日粮的研究中，高（H）、低（L）剩余采食量（RFI）动物产生的 VFA 浓度差异

动物	性别	主要日粮成分	pH 值	RFI	总 VFA (mM/L)	乙酸 (%)	丙酸 (%)	丁酸 (%)	A：P	戊酸 (%)	NH_3 (mg/L)
奶牛[1]	F	精料	—	H	79.3	41.94	24.53	22.72	—	3.97	—
肉牛	M	精料	5.72	H	95.1	51.9	30.5	12.5	1.97	5.1	43.6
绵羊	M	精料	6.65	H	131.02	53.5	30.51[a]	13.44[a]	2.04[b]	1.6	—
肉牛	M	精料	—	H	55.35	56.4	32.6	6.1[a*]	1.73[a*]	1.2	—
肉牛	M	精料	—	H	58.55	54.48	31.45	9.51	1.87	1	(0.096mM)
肉牛[2]	M	精料	—	H	86.025	52.285	33.98	8.695	1.695	1.735	(0.15mM)
肉牛[3]	M	TMR	—	H	73.8	49.7	32.4	10.2	—	3.12	—
肉牛[2]	F	禾本科牧草/青贮牧草/全混合日粮	6.8	H	119	66.2	20	10.8	3.42	1.3	—
奶牛	F	禾本科牧草	6.09	H	145.2	61.5	21.7	11.5	2.9	2.1	(12.9mM)[a]
肉牛	F	禾本科牧草	6.57	H	87.6	66.9	24.1	10.6	3.27	2.1	110.6
肉牛	F	青贮牧草	6.88	H	71.3	61.3	24.1	10.9	2.57	3.9	105.5
肉牛	F	青贮牧草	6.85	H	80	62.3	14.1	20.4	4.6	3.2	38.2
肉牛[3]	F	青贮牧草	6.77	H	54.3	68	16.6	10.3	4.1	4.9	20.7[a]
肉牛[4]	F	青贮牧草	6.84	H	85.2	68.3	18.9[a]	10.4	3.66[c]	2.4	73.5
绵羊	M	苜蓿颗粒饲料	—	H	—	65.2	17.3	14.1	—	1.5	—
		均值	6.57		87.88	59.85	25.33	11.39	2.74	2.51	65.35

（续表）

乳酸(mg/L)	RFI	pH	总VFA(mM/L)	乙酸%	丙酸%	丁酸%	A：P	戊酸%	NH₃(mg/L)	乳酸(mg/L)	文献
—	L	—	88.14	40.4	25.32	25	—	4.25	—	—	Shabt 等（2016）
121.4	L	5.76	9.13	54	27.5	13.9	2.27	4.6	52.2	127.2	Fitzsimons 等（2014b）
—	L	6.33	142.04	49.23	39.89[b]	8.23[b]	1.32[b]	1.82	—	—	Liang 等（2017）
—	L	—	96.74	54.4	25.9	15[b*]	2.11[b*]	1.7	—	—	Guan 等（2008）
—	L	—	64.17	54.92	33.41	7.26	1.69	1.04	(0.11 mM)	—	Hernandez-Sanabria 等（2010）
—	L	—	81.205	53.665	33.615	8.255	1.745	1.505	(0.14 mM)	—	Hernandez-Sanabria 等（2012）
—	L	—	74.9	49.1	33.1	10	—	3.33	—	—	Lam 等（2018）
—	L	6.9	116	66.7	19.6	10.5	3.46	1.3	—	—	McDonnell 等（2016）
—	L	6.11	142	62	21.41	11.7	2.9	1.8	(15.5 mM)[b]	—	Rius 等（2012）
40.6	L	6.42	93.4	65.7	24.6	11.3	3.14	2.1	134.8	61	Lawrence 等（2013）
33.1	L	6.94	68.7	60.9	24.6	10.5	2.45	3.6	118.6	40.5	Lawrence 等（2013）
29.4	L	6.81	75.2	62.5	18.2	16.1	3.5	3.2	53.7	16.6	Fitzsimons 等（2013b）
24	L	6.98	54	67.4	16.5	10.4	4.1	5.5	10.2[b]	18.9	Fitzsimons 等（2014a）
20.2	L	6.82	79.9	67.1	20.2[b]	10.3	3.33[d]	2.7	90.5	24.0	Lawrence 等（2011）
—	L	—	—	66.2	18.1	12.5	—	1.5	—	—	Ellison 等（2017）
44.78		6.56	90.74	59.56	25.88	11.14	2.61	2.55	76.67	48.03	

注：¹ 根据所有报道结果换算的 VFA 浓度以 mM 计。A：P 也以 mM 计。

² VFA 浓度取各种日粮的平均值。

³ 在农场记录。

⁴ 怀孕的动物。

[a,b] 根据原始论文，用具有不同下标的平均值推断 H 组和 L 组之间的显著差异。

[a*,b*] 根据报道的数据将 VFA 的单位由 mM 数据转换为%。带有星号的不同下标表示以 mM 计的显著差异。

5 瘤胃微生物群和甲烷产生

产甲烷菌属于古细菌，甲烷是其代谢终产物（Deppenmeier 和 Müller，2008）（表3）。瘤胃独特内环境的特征是食糜通过速率相对较快，并且 CO_2 和 H_2 供应充足，从而形成了与其他厌氧系统完全不同的古菌群落（Patra 等，2017）。存在于瘤胃液且附着在饲料颗粒或原虫上的产甲烷菌被认为对甲烷生成的贡献最大（Morgavi 等，2010）。

由于产甲烷菌是瘤胃内唯一的甲烷来源，产甲烷菌丰度的增加似乎应该与甲烷排放量的增加有关，这一假设看似合理。然而研究表明，与瘤胃甲烷产量密切相关的可能是产甲烷菌的组成而非其种群数量（Tapio 等，2017）。高、低甲烷排放量奶牛瘤胃古菌的相对丰度并无差异（Danielsson 等，2017），阉公牛瘤胃古菌丰度和甲烷产量（CH_4 g/kg DMI）的相关性同样较弱（Wallace 等，2014）。对绵羊的研究也表明，高（Kittelmann 等，2014）、低甲烷排放量（Shi 等，2014）个体瘤胃古菌的相对丰度也没有显著差异。但宏基因组分析呈现了不同结果，即高甲烷排放量杂交牛瘤胃产甲烷菌相对丰度是低甲烷排放量个体的两倍（Auffret 等，2017）。

Danielsson 等（2017）报道了高、低甲烷排放量奶牛瘤胃古菌特定种群丰度的差异。奶牛瘤胃中 *M. gottcshalkii* 相对丰度的增加与较高的甲烷排放量有关，而反刍兽甲烷短杆菌（*M. ruminantium*）的丰度则在低甲烷排放量奶牛的瘤胃中更高。在绵羊瘤胃中也发现了类似的 *M. gottchalkii* 丰度与甲烷排放量的关联（Shi 等，2014）。反刍兽甲烷短杆菌和 *M. gottcshalkii* 丰度的这种变化可能是不同形式的甲基辅酶 M 还原酶（*mcr*）差异表达的结果（Tapio 等，2017）。甲烷短杆菌这一进化分支可进一步分成两个亚组，即 SGMT（*M. smithii*、*M. millerae*、*M. thaueri* 和 *M. gottschalkii*）和 RO 簇（反刍兽甲烷短杆菌和 *M. olleyae*），SGMT 簇产甲烷菌可以同时表达 *mcrI* 和 *mcr*II，而 RO 簇产甲烷菌仅能表达 *mcrI*（Leahy 等，2010；Tapio 等，2017）。

表3 瘤胃中3种主要的产甲烷途径以及反应自由能的估计值

途径	公式	$\Delta G°'a$（kJ/mol CH_4）
氢营养型产甲烷	$CO_2+4H_2 \rightarrow CH_4+2H_2O$	−135 到−130.4
甲基营养型产甲烷	$CH_3OH+H_2 \rightarrow CH_4+H_2O$	−113 到−112.5
乙酸营养型产甲烷	$CH_3COOH \rightarrow CH_4+CO_2$	−36 到−33

注：表格改编自 Liu 和 Whitman（2008），Ferry（2012）。建议读者阅读原始文献，以更深入讨论甲烷产生的生物化学原理。

[1] 甲酸盐也可以被利用，但由于甲酸在被利用前会被转化为 CO_2，因此未将其单独列出。

[2] 热力学上最有利的甲基型反应。

mcrI 和 *mcr*II 的表达受到瘤胃中可利用 H_2 浓度的调节，前者通常在 H_2 浓度较低时表达，后者则在 H_2 浓度较高时表达（Reeve 等，1997）。因此，拥有较高甲烷排放量的个体（牛、绵羊）瘤胃内 *M. gottschalkii* 丰度增加说明 H_2 浓度增加，同时，瘤胃中 SGMT 簇产

甲烷菌的大量存在可能暗示能够为甲烷生成提供底物的微生物数量增加（Danielsson等，2017）。同样，由于 *mcrI* 的表达发生于低 H_2 环境，RO 簇产甲烷菌的大量存在可能暗示瘤胃微生物群中含有较低丰度的产 H_2 微生物，和/或较高丰度的耗 H_2 微生物，从而使该簇产甲烷菌丰度与较低的甲烷产量相关。Kittelmann 等（2013）的研究表明，SGMT 和 RO 簇产甲烷菌与高、低丰度的产 H_2 微生物之间存在明显关联。他们观察到 *M. gottschalkii* 与瘤胃球菌科细菌的丰度之间（R=0.90），及反刍兽甲烷短杆菌与纤维杆菌科（*Fibrobacteraceae*）细菌的丰度之间（R=0.72）呈明显正相关。有趣的是，同一项研究也显示 *M. gottschalkii* 与反刍兽甲烷短杆菌丰度呈现负相关（R=-0.51），作者认为这是产甲烷底物竞争的结果。因此，通过增加瘤胃内 RO 簇产甲烷菌的丰度可能有利于降低甲烷产量。

与瘤胃中的产甲烷菌组成类似，特定瘤胃细菌的丰度，特别是产 H_2 细菌的丰度与甲烷的产生有关（Tapio 等，2017）。瘤胃细菌的底物特异性以及产 H_2 和耗 H_2 的菌群差异很大（Stewart 等，1997）。Kittelmann 等（2014）在绵羊上发现了 3 种不同的与甲烷排放相关的瘤胃细菌型（ruminotypes）。瘤胃细菌型 Q 和 S 与绵羊较低的甲烷排放量有关，且与丙酸的产生和乳酸与琥珀酸的组合相关细菌群落丰度较高。反之，瘤胃细菌型 H 存在于高甲烷排放量的绵羊瘤胃中，其特征是产 H_2 微生物的丰度较高。需要注意的是，丙酸可以替代 H_2 作为电子受体且与较低的甲烷产量有关（Janssen，2010）。同时，琥珀酸和乳酸也是产丙酸的前体物（Wolin 等，1997），一般情况下它们与瘤胃中较低的 H_2 浓度或 H_2 的零生成有关（Kittelmann 等，2014），从而导致可用于产甲烷的 H_2 减少。据 Wallace 等（2015）报道，与甲烷排放量较低的阉公牛相比，甲烷排放量较高的牛瘤胃中的 Succinivibrionaceae 科产琥珀酸细菌的丰度降低了 4 倍。其他研究也表明，与甲烷排放量较高的个体相比，甲烷排放量较低的个体瘤胃内产丙酸菌的丰度较高（Kamke 等，2016；Edwards 等，2017）。体外纯培养以及与产甲烷菌的共培养研究显示，产乳酸菌（如 *Sharpea* 和 *Kandleria* 属）纯培养及其与产甲烷菌的共培养相比，培养体系中的乳酸、甲酸和乙酸浓度并无显著变化（Kumar 等，2018）。因此，我们可以据此假设瘤胃中的这些细菌会降低 H_2 的可利用性，从而降低产甲烷底物的可利用性（Kittelmann 等，2014；Kamke 等，2016）。

除产甲烷菌和细菌以外，瘤胃原虫作为关键的产 H_2 微生物在产甲烷过程中发挥着重要作用（Morgavi 等，2012）。Guyader 等（2014）对瘤胃原虫相关研究的荟萃分析表明，甲烷产量与 $\log_{10}$（原虫数量）之间存在强正相关。根据其他的大数据分析结果，去除瘤胃原虫可减少 11% 的甲烷排放，但不会显著降低产甲烷菌的数量（Newbold 等，2015）。这种甲烷产量与产甲烷菌数量的矛盾或许是可以解释的，因为有报道发现反刍兽甲烷短杆菌 M_1 的黏附样蛋白（MRU1499）能够结合在多种瘤胃原虫表面，同时也能与可以产生 H_2 的细菌 *Butterivibrio proteolaticus* 结合，说明这种产甲烷菌或许能够在去除瘤胃原虫的情况下继续产生甲烷（Ng 等，2016）。

Belanche 等（2015）曾经采用绵羊进行了瘤胃原虫的重组研究，他们发现将特定的 3 种原虫和瘤胃液（全种类原虫）引入去原虫的绵羊瘤胃后，绵羊的甲烷排放量增加。与细菌不同，目前与产甲烷相关的特定原虫的报道并不多见。上述 Belanche 等（2015）的研究表明，与引入特定种类原虫相比，引入瘤胃液可使绵羊瘤胃原虫的多样性更大、总丰度更高（14.4 倍），但并没有发现两组动物甲烷排放量或产甲烷菌浓度的统计学差异。此外，Kittelmann 等（2014）的扩增子测序分析也并未发现低和高甲烷排放量的绵羊瘤胃中存在任何独特的纤毛虫聚类。

与瘤胃原虫类似，特定瘤胃真菌种群与产甲烷之间的关系也未见详细报道。对绵羊的研究未能确定高甲烷、低甲烷排放量个体真菌群落的明显差异（Kittelmann 等，2014），而对奶牛的研究也未能明确瘤胃真菌种群与甲烷排放量之间的相关性（Cunha 等，2017）。根据 Cunha 等（2017）的研究，约 73.19% 的瘤胃真菌未能被明确分类，因此，对真菌类群的鉴定不力可能导致相关报道的缺乏，作者同时也承认取样方法可能会影响真菌种类的鉴定，因为厌氧真菌通常附着于饲料颗粒。针对瘘管牛瘤胃真菌的体外厌氧培养显示，真菌数量与甲烷排放量之间存在正相关关系（Aydin 等，2017），这可能说明瘤胃真菌的种群大小与甲烷排放量相关。

6 甲烷产生与剩余采食量

越来越多的研究为饲料效率和甲烷产量之间的关系提供了证据。饲料效率较高的动物甲烷生成量较低，这种趋势在各项研究中越来越明显（Nkrumah 等，2006；Hegarty 等，2007；Fitzsimons 等，2013；Alemu 等，2017；Sharma 等，2018）。饲料效率与甲烷产量之间的这种关系似乎是合理的，因为瘤胃甲烷的生成过程需要消耗高达 12% 的 GEI（Johnson 和 Johnson，1995），这与动物生产的本质是背道而驰的。

然而，饲料效率与甲烷产量关系的这种一致性受到挑战。有研究发现，被认为效率更高的动物甲烷生成量反而增加（McDonnell 等，2016；Flay 等，2019）。有人推测其原因可能是被认定为饲料效率更高的动物消化能力增加，导致产甲烷菌可利用的底物增加（Flay 等，2019）。实际上效率更高的动物瘤胃拥有更强的发酵能力，可以从摄取的饲料中释放更多能量。因此，DMI 的降低与较低的 RFI 有关，再加上更有效的瘤胃发酵，很可能使得这些动物能够从更少的饲料中获取足够的能量。然而 DMI 的降低，以及微生物对饲料中不易消化成分降解能力的增强也可能延长瘤胃滞留时间，从而可能导致甲烷产量的增加（Goopy 等，2014）。例如，随着绵羊采食量的增加，其瘤胃通过速率呈显著的线性增加，而甲烷产量则线性减少（Hammond 等，2014）。此外，尽管检测技术可能存在一些偏差，但总体而言随着采食量的增加，消化道和瘤胃的总平均滞留时间同时呈现线性减少。

近期报道表明低 RFI 羔羊的瘤胃较小且小肠较长（Zhang 等，2017），同时与高 RFI 羔羊相比，低 RFI 羔羊的 DMI 约有 18% 的降低。因此，与低 RFI 相关的 DMI 的降低可能是由于瘤胃容积的减少，同时消化率的增加可能是小肠吸收能力增强的结果。虽然进一步的研究仍有必要，但目前的研究结果说明，甲烷产量和排放量的降低可能与高可消化日粮较高的饲料效率呈正相关，因为高可消化日粮最有可能缩短瘤胃滞留时间，并且减少 DMI 的限制。

Flay 等（2019）假设，在这些研究中观察到的 RFI 和甲烷产量之间的不同联系可能是由日粮或性别差异造成的。表 4 中列举了各项研究中饲料效率（主要是 RFI）与甲烷产

表 4 饲喂不同日粮下剩余采食量（RFI）对甲烷排放的影响

物种	主要日粮组成	高 RFI	DMI	ADG	MP	MY	低 RFI	DMI	ADG	MP	MY	显著性	参考文献	备注
肉牛[1]GF	大麦青贮饲料	0.292	7.9	1.03	222.2a	28.5	−0.25	7.4	1.14	202.6b	27.7	MP	Alemu 等（2017）	
肉牛[1]RC	大麦青贮饲料	0.292	6.3	1.03	164.5	26.5	−0.25	6	1.14	156.3	26.5	ND	Alemu 等（2017）	
肉牛	大麦精料	—	14.13	1.229	190.2a	14.7	—	8.38	1.126	142.3b	16.3	MP	Hegarty 等（2007）	
奶牛	苜蓿草粉	—	12.4	—	256	20.7b	—	11.3	—	253	22.7a	MY	Flay 等（2019）	
肉牛[3]	优质禾本科牧草	0.68	14.0	—	227.24	—	−0.69	13.1	—	181	—	NA	Jones 等（2011）	
肉牛[3]	劣质禾本科牧草	0.68	10.7	—	125.06	—	−0.69	10.2	—	132.6	—	NA	Jones 等（2011）	
肉牛	青贮牧草	0.54	7.96	0.6	297a	36	−0.49	6.95	0.59	260b	38	MP	Fitzsimons 等（2013）	
肉牛	青贮牧草/放牧/全混合日粮	0.66	7.54	1.52	146	20.2	−0.74	7.18	1.55	156	22.4	ND	McDonnell（2016）	g/d 是根据报告的 g/LW 的估计值计算的，也对 EBV DMI 进行了计算，也按每 500 kg BW 计算
肉牛	全混合日粮	0.83	10.6	0.8	265a	28.1a	−0.78	9.33	0.83	194b	20.3b	MP, MY	Dini 等（2019）	
内洛尔	全混合日粮	0.362	8.12	0.547	163a	25.7	−0.506	7.09	0.467	144b	24.4	MP	Mercadante 等（2015）	
水牛	全混合日粮	0.05	9.2	0.74	222.2a	24.8	−0.05	9.5	0.76	163.4a	17.2	MP	Sharma（2017）	
内洛尔	全混合日粮	0.787	7.86	1.12	107	11.4	−0.683	6.31	1.11	101	11.9	ND	Oliveira 等（2018）	
奶牛[4]	低精料	0.44	20.9	—	635	30.7	−0.62	18.6	—	595	32.4	ND	Oliveira 等（2018）	
奶牛[4]	低精料	0.98	15	—	499	32.6	−0.74	14.9	—	491	32.5	ND	Oliveira 等（2018）	
奶牛[4]	高精料	0.44	23.7	—	493	28.2	−0.62	21.6	—	467	27.9	ND	Oliveira 等（2018）	
奶牛[4]	高精料	0.98	17.8	—	504	21.4	−0.74	17	—	527	24.5	ND	Oliveira 等（2018）	

注：RFI=剩余采食量（kg DMI/d）。
MP=甲烷日排放量（g/d），MY=甲烷产量（CH_4 g/kg DMI）。
ND=无差异，NA=未获得数据。
[1] 使用 GreenFeed 系统估算的甲烷排放量。
[2] 使用呼吸测热室估算的甲烷排放量
[3] 本表的日排放量是将原始值转换为每组平均活重的值。本表的日排放量修正为 500 kg 活重。
[4] 排放量为 L/d。

量的关系。总体而言,饲料效率更高的动物甲烷每日排放量和甲烷产量均较低,且与日粮无关。然而当饲喂牧草为主的日粮时,类似关系似乎并不存在,并且(尽管仅有少数研究报道)低 RFI 动物甲烷每日排放量减少约 6.5%,但甲烷产量却增加了 7.5%。

如果测定 RFI 期间的日粮与测定甲烷排放期间的日粮不同,则可能导致"低 RFI 动物甲烷产量较低"的结论无法成立。例如,根据父母的估计育种值(EBV)选择后代的 RFI,如果父母代日粮以谷物为主,当其低 RFI 后代饲喂牧草型日粮时似乎并不能减少甲烷的产量(Velazco 等,2017)。因此,虽然减少每日甲烷排放量的饲料和饲养管理可能对饲料效率和生产性能有促进作用,但饲料中牧草的比例对甲烷产量和饲料效率的影响仍需进一步研究。

我们可以假设,瘤胃内某些微生物类群可能对饲料效率和甲烷产量产生影响,尤其是当动物采食牧草型日粮时。例如,降解纤维素的瘤胃球菌可以产生 H_2。当白色瘤胃球菌与产甲烷菌共培养时,乙酸和 H_2 的产量增加,为甲烷生成提供更多的底物(Wolin 等,1997)。白色瘤胃球菌已被证明在纤维降解中发挥积极作用,且与饲料效率有关。因此,当采食高纤维日粮时,瘤胃球菌的存在很可能利于纤维降解,从而可能为动物提供更多能量。然而,瘤胃发酵的增强也可能为甲烷排放提供更多可利用底物。此外,由于琥珀酸是丙酸产生的前体,Succinivibrionaceae 丰度的增加与甲烷产量的减少有关,但该科细菌与饲料效率却呈负相关。因此有必要进一步研究这些细菌与甲烷生产和 RFI 的关系,并在未来研究中更多地关注饲草型日粮的影响。

7 饲用植物对动物生产性能的影响

如前所述,全球大量植物基质无法直接作为人类的食物来源。因此我们依靠反刍动物与瘤胃微生物群的共生关系,将不可消化的植物物质转化为可食用的优质乳制品和肉制品。牧草是全球反刍家畜的传统食物来源,特别是在爱尔兰或新西兰等全年气候较为温和的国家。由于其畜牧生产主要以牧草为基础,上述国家通常以在日粮中添加最大比例的牧草为饲养目标。目前,在日粮中添加大量谷物(精料)也成为肉牛育肥和高产奶牛养殖的一种趋势,但长期以来牧草一直被认为是反刍动物最经济的饲料来源,有助于提高农场盈利能力。由于牧草能量浓度较低,谷物通常被用于提高动物的生产性能。

不同种类的植物以及同种植物之间的营养组成都存在差异。一项研究考察了来自 30 个不同国家的大约 136 种饲料植物的蛋白质、碳水化合物、纤维、木质素和矿物质含量的差异(Lee,2018)。平均而言,禾本科植物比豆科植物的纤维(NDF)含量更高(59% vs 42%),而蛋白质含量则相反(15% vs 21%)。然而同一种植物的营养成分也存在很大差异。例如,多年生黑麦草的 NDF、干物质消化率和粗蛋白质(CP)含量跨度均较大,分别为 34%~62%、56%~86% 和 6%~34%(Lee,2018)。

简言之,植物中主要含有两种类型多糖:贮存多糖如淀粉,以及构成大部分植物细胞壁的结构多糖(Chesson 和 Forsberg,1997)。大多数植物细胞壁可分为初生细胞壁和次生细胞壁,初生细胞壁主要存在于幼嫩植物,次生细胞壁存在于更成熟的植物体(图 4)。

生长中的细胞壁主要由嵌入半纤维素复杂基质中的纤维素微纤维(Cosgrove,

2005）组成，这些微纤维主要由木聚糖（木葡聚糖和阿拉伯木聚糖）和果胶多糖构成，有助于提高细胞壁强度（Harholt 等，2010；Zamil 和 Geitmann，2017）。初生细胞壁的纤维素成分估计为 15%~40%（Cosgrove 和 Jarvis，2012）。由于纤维素含量高而木聚糖含量低，大多数初生细胞壁的降解预计需要 8~12 h，剩余材料主要由降解速率较慢的木聚糖组成（Chesson 和 Forsberg，1997）。相邻的植物细胞壁通过富含果胶的胞间层连接在一起（Zamil 和 Geitmann，2017）。木质素主要与植物的次生细胞壁和成熟有关。次生细胞壁主要由半纤维素形成的微纤维组成，胞间层的果胶被木质素取代（Cosgrove 和 Jarvis，2012；Zamil 和 Geitmann，2017）。

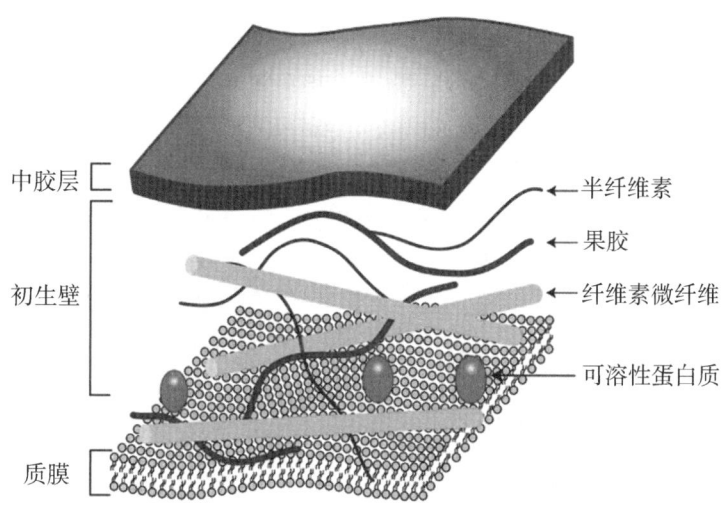

图 4　植物细胞壁示意图（改编自 Flint 等，2012）

7.1　草地管理和动物生产性能

我们应根据不同种类牧草营养价值的差异选择适宜的饲草作物来提高动物的生产性能。如前所述，即使同一物种的牧草其营养价值的差异也较大（Lee，2018）。虽然本章无意深入探讨牧场管理的相关知识，但我们必须考虑到牧场管理对优化畜牧生产效率的重要性。

由于纤维的消化速度较慢甚至在某些情况下难以被消化，植物中纤维浓度过高会对瘤胃消化率产生负面影响（Buxton 和 Redfearn，1997）。据推测，瘤胃中的纤维消化率不受纤维降解微生物的限制，而是受到可降解纤维素的限制（Weimer，1996）。本质上，改变瘤胃发酵的是微生物是否进入植物细胞可发酵部分，而不是微生物存在与否。木质素作为一种结构性化合物，随植物生长沉积在植物细胞壁中，为细胞壁多糖提供刚性保护，阻碍微生物的分解（Vanholme 等，2010）。虽然木质素的沉积有利于维持植物结构，但由于其能抵抗真菌和细菌的降解（Raffrenato 等，2017），其对牧草的消化率（Jung 和 Allen，1995）和动物生产性能（Duble 等，1971）具有负面影响。在 Chesson（1993）构建的模型中，虽然木质素在瘤胃中的可降解（实际上木质素几乎不能被降解）程度最小，但微生物似乎能够降解木质素-碳水化合物复合体中的非木质素部分。因此，不可消化木质素的比例及其在细胞壁中的位置（包括是否存在于饲料颗粒表面）都是瘤胃降解的限制因素。

在牧草质量和消化率最佳且次生细胞壁不太明确的生长阶段，放牧草地的管理一般以动物对牧草的消耗量为目标。然而对于放牧系统而言，实现放牧数量和质量之间的平衡至关重要（Curran 等，2010），因此必须允许植物有足够的生长时间以确保最佳牧草供应。当草地管理良好时，低矮牧草（low herbage swards）（植物处于再生阶段）茎的比例更低，消化率更高（Holmes 等，1992）。在低矮牧草的草场放牧时，动物对牧草的粗蛋白质和有机物消化率增加，同时 NDF 的消化率降低（Curran 等，2010；Wims 等，2010；Boland 等，2013）。与此同时，植物叶和茎比例的增加与低矮草层数量有关，而叶和茎比例的减少则与放牧前较高的草层高度有关（Wims 等，2010）。据报道，放牧于高质量草场的奶牛或牛肉产奶量、平均日增重（ADG）和甲烷排放量均较高（Hart 等，2009；Wims 等，2010；Curran 等，2010；Boland 等，2013；Muñoz 等，2016）。动物生产性能因草场质量的改善而获益很可能与瘤胃微生物组的有利变化有关。

目前，二代测序技术（next-generation sequecing，NGS）已经被用于观察草场质量对瘤胃微生物组的影响，但相关研究并不详尽。虽然前人研究考察了草地消化率（sward digestibility）对瘤胃微生物组的影响，但仅限于细菌和真菌的数量而非种类，结果发现两种微生物数量均不受影响（Hart 等，2009）。如果我们采用更为先进的微生物生态学技术研究草地质量，以及草地管理引起的消化率的变化，可能会发现与草地质量有关的动物生产性能差异导致的瘤胃微生物组变化。

7.2 豆科牧草和动物生产性能

与禾本科植物相比，豆科植物的纤维比例较少且粗蛋白比例较高，由于其可以固定大气中的游离氮为己所用，因而是放牧草地的一种常见牧草。豆科牧草也被认为更易消化，含有豆科牧草的饲料通常可以提高 DMI（表5）。

已有研究表明，与其他植物结构组分相比，瘤胃微生物可以更快降解红三叶草和苜蓿细胞壁内的果胶（Chesson 和 Monro，1982）。此外，可以利用果胶的多对毛螺菌（*L. multiparus*）对三叶草的降解速率似乎快于禾本科牧草（Cheng 等，1979），这可能是由于三叶草中的果胶含量较高。在接种瘤胃液的情况下，豆科植物细胞壁的破裂与禾本科植物不同，前者更倾向于细胞解体而非凹陷变形（Cheng 等，1980）。因此与禾本科植物相比，饲喂豆科植物可以提高动物消化率，其原因可能在于豆科植物的降解率较高，且细胞壁破坏程度更大。豆科植物在降解过程中的这种细胞破裂可能使动物摄入的饲料颗粒变小，继而增加其瘤胃通过性，从而可以解释为什么豆科牧草能被快速消化并能提高动物的 DMI（表5）。

通过总结35项研究结果发现，放牧草地中增加白三叶草（WC）（*Trifolium repens*）可将奶牛的产奶量提高1.4 kg，且乳固形物含量增加0.12 kg（Dineen 等，2018）。同时，与仅含有多年生黑麦草（PRG）的草地相比，放牧于含有 PRG 和 WC 的混合草地可提高绵羊 ADG，缩短出栏时间，增加胴体重，并减少羔羊的寄生虫疾病负担（Grace 等，2018）。

放牧于含有 WC 的混合草地也会降低奶牛的日甲烷产量（Enriquez-Hidalgo 等，

2014a），在室内饲养试验中，增加日粮中的 WC 水平也呈现类似效果（Lee 等，2004a）。然而，在以 PRG 为主的日粮中添加 WC 未能降低绵羊的甲烷产量（Niderkorn 等，2017）。但关于豆科牧草对瘤胃微生物组的影响仍少有报道，仅限于放牧奶牛（Smith 等，2019）。尽管如此，Niderkorn 等（2017）进行的绵羊和放牧瘘管牛的试验，以及 Enriquez Hidalgo 等（2014a,b）进行的奶牛甲烷排放量测定试验均涉及微生物的代谢产物（VFA）分析。Enriquez Hidalgo 等（2014b）和 Niderkorn 等（2017）的研究结果均显示，奶牛或绵羊的日粮中添加 WC 可略微降低瘤胃丁酸浓度，并且氨浓度有相似的增加趋势。

表 5 与牧草（主要是多年生黑麦草）相比，饲喂各种豆类对动物生产性能的影响

饲料	物种	DMI	MP	MY	ADG	牛奶产量	研究
白三叶草 vs 禾本科牧草	绵羊	-	-	-	↑	-	Grace 等，2018
	绵羊	↑	↑	ND	-	-	Niderkorn 等，2017
	绵羊[1]	ND	ND	ND	-	-	Hammond 等，2011
	奶牛	↑	-	-	-	↑	Dewhurst 等，2003
	奶牛	-	ND	↓	-	ND	Enriquez-Hidalgo 等，2014a
	奶牛	↑	↑	↓	-	↑	Lee 等，2004a
红三叶草 vs 禾本科牧草	绵羊	↑	ND	↓	-	-	Niderkorn 等，2015
	绵羊	↑	-	-	↑	-	Fraser 等，2004
	奶牛	↑	-	-	-	↑	Dewhurst 等，2003
紫花苜蓿 vs 禾本科牧草	肉牛[2]	↓	ND	↑	-	-	Chaves 等，2006
	肉牛[3]	ND	ND	ND	-	-	Chaves 等，2006
	奶牛	↑	-	↑	-	↑	Dewhurst 等，2003
	绵羊	↑	-	-	↑	-	Fraser 等，2004
	山羊	ND	ND	ND	-	-	Puchala 等，2012

注：DMI=干物质摄入量，ADG=平均日增重，MP=甲烷日排放量，MY=甲烷产量。
ND=无差异。
[1] 用于分析的 3 项研究的平均值。研究考查了不同的饲喂水平。
[2] 使用烷烃方法估算的摄入量。
[3] 使用康奈尔净碳水化合物和蛋白质系统模型估算的摄入量。
[2,3] 3 个地点的平均值。
ND=无差异。
-=无报道。

Niderkorn 等（2017）发现随着日粮中 WC 浓度的相应增加，粗蛋白比例显著增加，Enriquez-Hidalgo 等（2014a,b）的研究也发现类似结果（数值增加幅度较小）。同时这 3 项研究都发现日粮中添加 WC 可以相应减少 NDF 含量。

饲喂 WC 降低甲烷产量的原因尚不完全清楚，有人认为可能与 DMI 增加（$P <$

0.07）有关（Enriquez-Hidalgo 等，2014a），因为 DMI 升高表明瘤胃滞留时间缩短，这两者在之前的研究中都被发现与甲烷产量的降低有关（Hammond 等，2014；Goopy 等，2014）。然而我们也注意到，在上述两项（Hammond 等，2014；Goopy 等，2014）研究中，与只饲喂 PRG 的动物相比，采食含有 WC 日粮的奶牛和绵羊 DMI 更高，因此我们不能用更快的瘤胃通过率来解释 WC 为什么能降低甲烷产量。

除了物种差异带来的相关影响，Niderkorn 等（2017）和 Enriquez Hidalgo 等（2014a，b）的研究在试验方法上也存在差异，后者的试验动物所采食的 WC 是混合放牧系统的一部分，而前者的试验动物采食的是单一品种 WC 和 PRG 的混合物。因此，上述 WC 的甲烷减排潜力只能在混合放牧系统中得以体现。有人曾在天然的混合草场观察到 WC 形态的变化（Guy 等，2018），尚不清楚这些形态是否属于 WC 的不确定种。因此，在 Niderkorn 等（2017）的试验中，单一品种 WC 和 PRG 的混播不太可能发生未确定的种间相互作用，这可能是上述研究结果存在差异的原因。

此外，上述两项研究采用了不同的施肥措施。用于放牧奶牛的草地施肥量为 260 kg N/hm^2（60 kg/hm^2 尿素和 200 kg/hm^2 硝酸铵钙），而绵羊试验中的 PRG 草地施肥量为 100 kg/hm^2（作者并未明确肥料种类），WC 的单一品种草地则未施用化肥。此前研究已证明，对 PRG/WC 混合草地施用化肥会减少其固氮量（Harris 和 Clark，1996；Ledgard 等，2001）。与此相关的是，WC 的固氮能力与土壤硝酸盐的吸收量呈负相关关系（Griffith 等，2000）。这种因施肥引起的草地表型改变可能对减少甲烷排放量有某些未知益处。Smith 等（2019）对 Enriquez Hidalgo 等（2014a）试验中的奶牛进行了瘤胃微生物分析，他们发现这些奶牛瘤胃古菌的相对丰度降低了 21%，同时对 PRG/WC 组奶牛瘤胃甲烷短杆菌属的丰度产生了中度负面影响。该组奶牛的乳脂率也有所增加，这可能表明 PRG/WC 组奶牛瘤胃中的乙酸产量较高（Urrutia 和 Harvatine，2017），与此同时，放牧于相同草地的瘘管奶牛瘤胃中氨浓度高于仅放牧于 PRG 草地的奶牛（Enriquez-Hidalgo 等，2014b）。上述与 PRG/WC 草地放牧相关的瘤胃特征与饲料中含有硝酸盐导致的瘤胃特征类似。体外研究表明，硝酸盐能够降低产甲烷菌的丰度，同时降低甲烷和丁酸的产量，并增加乙酸比例（Zhou 等，2012；Liu 等，2017）。通常认为硝酸盐对 H_2 的亲和力强于 CO_2（Zhou 等，2012），可以减少甲烷生成过程中的 H_2。此外，向阉公牛的日粮中补充硝酸盐可以减少产甲烷菌数量（Zhao 等，2018），而在泌乳奶牛日粮中补充硝酸盐也会增加瘤胃乙酸和氨的产量，虽然乳脂率并没有显著增加（Farra 和 Satter，1971）。因此我们假设，结合瘤胃微生物分析和甲烷产量估算，有望观察相同施肥措施下 PRG/WC 和 PRG 草地的无机氮含量，以探寻可以解释甲烷产量降低的差异。

迄今为止，关于 WC 影响瘤胃微生物谱的体内研究并不多见。最近有一项研究采用交叉试验设计研究了 4 头荷斯坦（Holstein Friesian）泌乳奶牛的瘤胃微生物群，这些动物均放牧于纯种 PRG 和 WC 草场（Bowen 等，2018）。虽然该项研究的重点是研究瘤胃液相或固相样品微生物群落的差异，但通过考察微生物平均丰度发现，微生物群落组成的变化与日粮有关。笔者实验室通过分析 Enriquez Hidalgo 等（2014a）研究中使用的奶牛瘤胃微生物组也发现了类似的显著差异（Smith 等，2019），这或许可以解释上文提到的甲烷产量的减少。

早前研究表明，RFI 较低的泌乳奶牛瘤胃微生物产生的氨浓度较高，同时其粪便中的氮含量减少（Rius 等，2012）。虽然在该项研究中未检测到尿氮的增加，但作者认为瘤胃氨浓度的增加可能表明瘤胃蛋白质的水解增加，随后提高了尿氮损失。如前所述，虽然体外试验表明氨浓度的增加与甲烷产量的减少有关，但其对环境的积极影响可能被动物体尿氮输出量的增加所抵消，从而增加氧化亚氮的生成量。有趣的是，针对低 RFI 水牛小母牛（Sharma 等，2018）和瘤胃氨浓度较低的怀孕小母牛（Fitzsimons 等，2014a）的研究表明，二者的氮输出量也较低。近期一项关于 Romeny 阉公羊的研究表明，在摄入 3 种 PRG 的情况下，瘤胃氨浓度与每日尿氮（单位：g）和进入尿液中的食入氮密切相关（相关系数分别为 0.72 和 0.64），而与每日沉积氮（单位：g）和沉积氮占食入氮的比例呈负相关（相关系数分别为-0.54 和-0.61）（Jonker 等，2018a）。因此，虽然提高瘤胃中的氨浓度可能有利于甲烷减排策略，但其可能对 RFI 产生负面影响并增加氮损失。

红三叶草（RC，*Trifolium pratense*）也是一种具有减排潜力的植物。体外试验已证明，与禾本科牧草相比，RC 可以降低甲烷产量（Navarro Villa 等，2011；Belanche 等，2013），这与 WC 类似。采用荷兰泰克斯谢尔（Texel）绵羊进行的在体研究也发现 RC 可以降低瘤胃甲烷产量（Niderkorn 等，2015）。同时 RC 也与 WC 一样可以提高动物产奶量和乳中的固形物含量。但在比较含有 RC 或 WC 的混合草场的放牧效果时发现，虽然含有 WC 的混合草场可以略微增加牛奶中的蛋白质，但动物的产奶量没有明显差异（Steinshamn 和 Thuen，2008）。当青贮牧草中含有 RC 或 WC 时可以提高奶牛的产奶量，但也有报道认为添加三叶草会降低食入氮转化为乳中氮的效率（Dewhurst 等，2003）。与 PRG 相比，采用 RC 饲养可以缩短羔羊屠宰天数（38 d vs 66 d），增加活重（305 g/d vs 184 g/d）和屠宰率（48% vs 46%）（Fraser 等，2004）。

与禾本科牧草相比，日粮添加 RC 可以显著改变瘤胃微生物组成。与普通青贮牧草相比，饲喂青贮 RC 可以增加杂交阉牛瘤胃中 *B. proteoclasticus* 和 *R. albus* 的 DNA 浓度，降低黄化瘤胃球菌和产琥珀酸丝状杆菌的 DNA 浓度（Huws 等，2010）。体外研究也发现类似结果，即 *Butyrivibrio* 属细菌的丰度与 RC 的相关性高于 PRG（Elliott 等，2018）。另一项体外研究也发现，RC 在降低上述瘤胃球菌和产琥珀酸丝状杆菌数量的同时，还会减少 *B. fibrisolvens*、*P. ruminicola*、*S. ruminantium*、真菌和产甲烷菌的数量（Belanche 等，2013）。有趣的是，RC 对瘤胃原虫的影响存在不一致的报道。一些数据显示 RC 对原虫整体没有影响（Belanche 等，2013），但某些研究发现 RC 对不同种类原虫的影响存在差异（Niderkorn 等，2015）。扩增子测序表明，在使用相同精料和饲料添加剂（油籽和异油酸）的情况下，与饲喂干草相比，以 70∶30 的比例饲喂含有青贮 RC 的日粮可以降低阉公牛瘤胃 *Butyrivibrio* 的数量（Petri 等，2014）。此外，RC 可以提高纤维杆菌（Fibrobacter）的丰度。体外发酵试验也已证实，*Psuedobutyrivibrio* 丰度的降低与 RC 的相关性高于 PRG（Elliott 等，2018）。瘤胃中 *Psuedobutyrivibrio* 丰度的降低可以被视为 RC 的一种积极作用，因为该属细菌被认为对 RFI 有负面影响，并且体外研究发现当甲烷产量较低时该属细菌丰度也降低（Mi 等，2017）。

豆科植物（尤其是 WC）中的果胶浓度比禾本科植物更高（Thomson，1984），而 *Butterivibrio* 属细菌是已知的果胶降解菌。因此，与 RC 相关的 *Butterivibrio* 丰度或数量的增加可能仅反映了其可发酵底物的增加。而在 WC 相关研究中发现的 VFA 组成的变化，以及与甲烷产量降低的相关变化可能是由于底物（果胶）发酵的改变。与葡萄糖相比，当以果胶为体外碳源底物时，*B. fibrisolvens* 和 *P. ruminicola* 产生的丁酸比例都有所降低，同时乙酸浓度相应增加（Marounek 和 Dušková，1999）。这些高丰度微生物终产物的变化可以解释饲喂豆科牧草时瘤胃 VFA 谱的相应变化。此外据报道，上述情况下 *B. fibrisolvens* 终产物中 H_2 和甲酸的浓度显著降低，这可能导致产甲烷菌可利用底物的减少。在仅含有 WC 的单一品种草场放牧不会导致肉牛瘤胃 *Butterivibrio* 比例的变化，但有增加 Lachnospiraceae 科细菌的趋势，且 *Lachnospira* 属的比例显著上升（Bowen 等，2018）。与此类似，相较于葡萄糖，以果胶作为体外培养碳源时，*Lachnospira multiparus* 产生的乙酸浓度增加，而甲酸和 H_2 浓度减少（Dušková 和 Marounek，2001），从而可能有助于降低甲烷产量。反刍动物饲喂三叶草对甲烷生成和瘤胃微生物群的影响值得进一步研究。

由于植物中的多酚氧化酶（PPO）可以保护其蛋白质免受瘤胃降解，因此饲喂 RC 通常与更高的氮效率相关（Huws 等，2018）。在暴露于氧气时 PPO 被激活，将酚类转化为醌类，而醌类与蛋白质、胺和酰胺等细胞成分交联，从而减少蛋白质水解（Lee 等，2004b）。因此虽然 RC 中的 PPO 被认为有利于减少瘤胃蛋白质降解，但这种酶的激活必须发生在瘤胃以外，例如动物咀嚼或青贮过程（Lee，2014）。以往研究已发现高纤维、低蛋白日粮可以降低黄化瘤胃球菌种群，同时也趋于降低产琥珀酸丝状杆菌和真菌数量（Belanche 等，2012）。因日粮 RC 导致的上述微生物的减少表明，PPO 对这些微生物的氮供应做出的贡献是有限的。

无论是在饲喂牧草或精料的情况下添加 RC 和 WC 都被证明可以改善牛奶中的氮沉积量，但当 RC 和 WC 被用作唯一的饲料时却会降低氮沉积量（Dewhurst 等，2003）。单一品种草场与混合草场的放牧效果与此相似。与单一品种草地放牧相比，RC 包合率为 25% 的草地放牧可增加沉积氮（单位：g）与食入氮的比例，从 0.218 增加到 0.273（Niderkorn 等，2015）。PPO 的有利作用只有在日粮有效供应可发酵能量的情况下，允许微生物蛋白质生产时才能实现，否则当日粮氮含量远超动物需求量时，过剩的氮被直接转化为微生物蛋白质（Lee，2014）。总之，当氮摄入量高于动物的需要量时，PPO 的积极作用无法体现。

有趣的是，在我们提到的大多数研究中，日粮 WC 和 RC 都可以增加 DMI（Dewhurst 等，2003；Fraser 等，2004；Niderkorn 等，2015、2017）。以往研究已表明 DMI 的增加也与食糜过瘤胃速率和瘤胃滞留时间有关，这有利于减少甲烷排放（Pinares Patiño 等，2003；Goopy 等，2014）。因此，上述两种三叶草对动物产奶量、生长速度和甲烷减排等方面表现出的有益作用可能是 DMI 的增加减少了瘤胃滞留时间的结果。然而，如前所述，这种与日粮三叶草相关的 DMI 的增加，及其对甲烷排放的积极作用仍然存疑。

紫花苜蓿对动物生产性能也有促进作用。其中一个明显的例子是，饲喂苜蓿可以提高母羊体重及其后代的哺乳期体重（Corner Thomas 等，2014）。放牧于苜蓿或 PRG 草场的肥育期肉羊也有类似结果，苜蓿可以增加肉羊活体重，并缩短其出栏时间（Fraser 等，2004）。研究还表明，在代乳料和开食料中添加苜蓿可以提高断奶湖羊体内瘤胃球菌的数量（Yang 等，2018），并显著增加瘤胃乳头长度（Yang 等，2015）。饲喂苜蓿能提高羔羊体重可能是其促进瘤胃乳头发育，或减少瘤胃纤毛虫的结果（Puchala 等，2012）。与此相反，当奶牛饲喂含青贮饲料的精粗比为 60 : 40 的全混合日粮（TMR）时，瘤胃原虫总数随着紫花苜蓿（作为日粮中的主要牧草）比例的增加呈现线性增加（Hassanat 等，2014）。苜蓿含有次生植物化合物皂苷，能够与原生动物细胞膜中的胆固醇相互作用，从而导致其细胞破裂（Wina 等，2005）。此外，与饲料中添加羊草（*Leymus chinensis*）相比，添加紫花苜蓿干草可以提高奶牛瘤胃 *Selenomonas* 和 *Prevotella* 属细菌的丰度，但并未发现瘤胃 VFA 浓度或组成的显著差异（Zhang 等，2014）。

紫花苜蓿究竟是否具有甲烷减排潜力仍无定论。表 5 展示了上述豆科牧草对动物生产性能的影响。大多数研究表明紫花苜蓿比禾本科牧草更有利于提高 DMI。在一些研究中，WC 和 RC 已被证明可以通过提高 DMI 来减少甲烷产生，但并无有力证据表明紫花苜蓿也具有类似作用，需要进一步研究。同样，与禾本科植物相比，在提高氮利用率方面，紫花苜蓿似乎也不是最佳选择。据 Dewhurst 等（2003 年）报道，与三叶草青贮饲料和三叶草/禾本科牧草混合青贮饲料相比，紫花苜蓿青贮饲料的氮效率最低，并且使用康奈尔净碳水化合物和蛋白质系统模型（Cornell Net Carbohydrate and Protein System model）预测其会导致氮排泄过量（Chaves 等，2006）。

豆科牧草可能通过降低甲烷浓度从根本上实现其减排作用。我们在前面的章节已经提到，豆科牧草可以提高奶产量相关参数和胴体重，因此可以在不增加甲烷排放量的情况下促进动物生产，这在某种程度上增加了成本效益。实际上，尽管有报道称饲喂豆科牧草会减少氮沉积，但针对 PRG 草场的相关研究已表明，施用合成氮肥或放牧于含有 WC 的草场可以抵消这种负面影响（Enriquez-Hidalgo 等，2018）。因此，将 WC 纳入放牧草场可能有助于动物产品的可持续增长，并可能降低每千克畜产品的温室气体排放量。

读者在评估一些单一品种豆科牧草的相关研究结果时应更为谨慎。正常的生产流程不包括长时间放牧于纯种豆科牧草场地。相反，豆科与禾本科牧草的混合草场才是常用的。因此，我们应该有条件地接受纯种草场的相关研究结果，并且未来的研究可能需要在更贴近实际生产的条件下（如农场）进行。

7.3 禾本科牧草和动物生产性能

由于产量较大且有利于永久性草地的形成，PRG 是放牧系统中最常见的播种草种之一。筛选具有减排潜力的禾本科牧草品种可能是减少甲烷排放的实用手段（Jonker 等，2018b）。目前新西兰已经育成一种经过基因改造的、具有减排潜力的高脂黑麦草，但相关数据尚未公布（Eckard 和 Clark，2018）。PRG 的甲烷减排潜力因品种而异。

有学者近期以 Romeny 阉公羊为对象，研究了放牧于常规二倍体、高糖分二倍体和四倍体品种 PRG 草地对甲烷排放和氮分配的影响（Jonker 等，2018a,b）。与传统二倍体品种相比，放牧于高糖分二倍体和四倍体品种草场可以降低每日甲烷排放量和甲烷产量，且后者的降低幅度最大。

然而在氮摄入量相似的同一组绵羊中，放牧于高糖分二倍体和四倍体品种草地增加了每日尿氮量，且后者提高了绵羊尿氮分配量，作者认为其原因之一可能是以干物质为基础的 OM 表观消化率（DOMD）降低，这再次强调了能量的可用性对瘤胃氮利用效率的重要性。同样，由于 DOMD 的减少（伴随着 DMI 的降低）降低了产甲烷底物的可用性，从而减少了甲烷排放量（日排放量和产量）。对牦羊和处于 3 月龄和 6 月龄的羔羊而言，季节可能干扰牧草对动物生产性能的影响，例如春季放牧于高糖分品种草地的 ADG 最高，而秋季放牧于四倍体品种草地的 ADG 最高，但无论放牧于哪种草地，动物的每公顷活重增加量均相似（Cosgrove 等，2015）。

尽管绵羊的甲烷总产量在放牧于高糖品种草地时高于四倍体品种草地（16.2 g/d vs 14.6 g/d；19.4 g/kg DMI vs 18.4 g/kg DMI），与后者相比，前者尿液中的氮分配在数量上较低，这凸显了高糖品种 PRG 相比于常规二倍体品种的减排优势，因为其在减少甲烷排放量的同时还优化了氮分配。饲喂高糖品种（AberMagic）干草可减少泌乳奶牛尿氮分配，并增加其粪氮分配（占摄入氮的比例）（Staerfl 等，2012）。虽然同一项研究也显示饲喂低糖品种 PRG 会降低奶牛产奶量，但并未发现甲烷排放量的显著降低以及 CP 的显著升高，这也许能解释动物生产性能的差异。

Edwards 等（2007）对高糖品种禾本科牧草的相关研究进行了综述。在 7 项相关研究中有 3 项表明高糖品种草可以降低动物尿氮分配；在 14 项研究中有 3 项表明其可以增加产奶量；在 5 项研究中有 2 项发现其可以增加肉牛采食量，且提高羔羊增重。因此，需要进一步研究高糖牧草品种在动物生产中的作用，以及动物甲烷排放响应的变化。

关于新鲜高糖分禾本科牧草影响瘤胃微生物群的相关报道很少。但是一些研究检测了不同精粗比饲料中添加低质量高糖分干草对瘤胃微生物的影响。虽然没有直接可比性，但这些研究结果可能为高糖分牧草对瘤胃的潜在影响提供证据。研究发现高糖分干草显著改变非泌乳期奶牛的瘤胃微生物组成，且随着高质量干草浓度的增加微生物组成进一步改变（Klevenhusen 等，2017）。相反，无论添加浓度是否改变，采食低质量干草后，瘤胃液相和固相样品中的微生物聚类无明显分离，说明干草类型能够改变瘤胃微生物结构。在科水平上，瘤胃液相样本中 *Succinivibrionaceae* 的丰度随着高糖分干草浓度的增加呈线性下降。在属水平上，*Ruminobacter* 丰度的变化也遵循类似规律。有趣的是，当饲料中分别含有低糖分和高糖分干草且精料比例相同时，瘤胃固相样本中纤维杆菌（*Fibrobacter*）属的丰度减少，且随着精料水平降低，纤维杆菌数量逐渐增加，其原因据推测是由于低质量干草中的纤维含量较高。此外，随着日粮中高糖分干草浓度的增加，瘤胃丙酸和丁酸浓度呈线性下降，而乙酸和异丁酸呈线性上升。但遗憾的是这些研究并没有记录动物的产奶量和氮沉积相关参数。在饲喂上述两种极端日粮的动物瘤胃中，液相和固相样本中的氨浓度似乎都较高，这很可能是因为高糖分干草中的 CP 浓度较高。

动物的生产性能很可能受高糖分牧草营养成分的影响，因为不同品种高糖分禾本科牧草 CP 水平似乎存在差异。Edwards 等（2007）的研究表明，增加高糖分禾本科牧草水溶性碳水化合物（WSC）与 CP 的比值有可能促进氮沉积。有人假设可发酵能量的增加可能导致微生物对氨的利用率增加。在各种模拟条件下对高糖分禾本科牧草的分析（Ellis 等，2011）表明，氮利用率的改善取决于多种条件。以牺牲 CP 为代价增加 WSC 含量和 WSC：CP 与相对于氮摄入量的尿氮排出量的减少有关。禾本科牧草 WSC：CP 与动物的尿氮排泄之间也存在强负相关（-0.90）。有人预测，以牺牲 CP 为代价增加 WSC 可能降低产奶量（-1.5 kg/d），然而，以牺牲 NDF 为代价增加牧草中的 WSC 含量则可能提高产奶量（+2.4 kg/d）。碳水化合物是微生物生长所必需的，如果缺乏稳定的碳水化合物供应，瘤胃微生物用于生长的氨摄取量会低于最佳水平（Hristov 和 Jouany，2005）。为了更好地理解高糖分禾本科牧草与饲料效率、产奶量和环境输出之间的关系，有必要更全面（包括对瘤胃微生物组成的考察）地评估高糖分牧草的饲喂效果。

8 结论

反刍动物利用人类无法消化的植物基质生产高质量肉类和乳品的能力，已经并且仍会在未来几十年内支撑全球人口的空前增长。然而，亟须解决为满足全球动物产品需求而增加的牲畜数量对环境的影响。

不同种类牧草对家养反刍动物生产性能的影响存在差异。由于牧草成本较低，基于放牧的生产体系在反刍动物生产中占据较大比重。供给动物的饲料营养成分取决于饲料种类，这可能影响瘤胃生态系统的组成和功能。此外，牧场管理可能影响植物的解剖结构，从而改变其营养成分，进而影响动物的生产性能，但这是否影响瘤胃微生物组成尚需进一步研究。一般而言，在植物处于再生阶段时进行针对性地放牧更可行，因为此阶段的植物更易被消化，对动物更有利。与此同时，更易被消化的植物基质瘤胃滞留时间更短，从而可能减少甲烷的产生。

在改善全球家畜生产的可持续性方面，策略性的日粮管理和动物育种对瘤胃微生物群的影响都是潜在的有效方法（Tapio 等，2017）。选育拥有更高效瘤胃微生态系统的动物品种是温室气体减排的长效手段。然而，中短期的日粮干预，特别是生长早期的日粮干预（Yanez-Ruiz 等，2010），是改变瘤胃发酵效率的替代方法。

反刍动物饲料效率的测定通常是在特定的饲养环境中进行的，谷物是其大宗饲料原料。因此，目前已知的关于饲料效率、瘤胃微生物组成和甲烷排放量关系的数据，主要是基于以谷物为主要日粮成分的饲养试验。然而，尽管如此，鉴于反刍家畜年度饲料预算的大部分来自新鲜和/或青贮牧草，我们迫切需要了解饲草型日粮对瘤胃及瘤胃微生物组成和功能的影响。

随着下一代测序（NGS）技术的发展，我们进一步了解了特定种类瘤胃原虫对宿主饲料效率的作用及其对环境的影响。然而，由于现有微生物数据库中缺乏相关的代表性数据，瘤胃真菌和原虫的种类鉴定和功能尚不清楚。瘤胃中的真核生物，特别是厌氧真菌具有降解牧草的强大能力。因此，我们需要同时在体内和体外进一步了解这些真核生物的作用。

最后，我们需要更全面地考虑具有减排潜力的措施。某些方法可能会对反刍动物的甲烷和氮排放产生不利影响，这在验证管理策略的总体减排潜力时需要重点考虑。

9 展望

该领域未来的研究重点是提高瘤胃微生物群消化饲料的能力，以提高饲料效率并最终降低饲料成本。这将促使我们通过培养、深度测序和改进参考数据库来加深对瘤胃微生物组成和功能的了解。未来的另一个目标是应用碳水化合物活性酶（CAZymes）促进肉和奶的产出，这些酶产品需要实现大规模工业化生产以作为饲料添加剂。如本章所述，目前国际上正在为减少农业碳足迹作出重大努力，特别是减少不同饲养系统中反刍动物的甲烷排放。通过改变营养策略（包括补充饲料添加剂）来减少甲烷排放是目前正在进行的工作。同时，部分工作也集中在通过合理的育种策略来培育饲料利用效率较高，且甲烷排放量较低的牛肉和奶牛新品种。爱尔兰 Teagasc、法国 INRA、新西兰 AgResearch、加拿大阿尔伯塔大学、加拿大农业及农业食品部、美国农业部和世界各地的其他机构都在进行此项育种工作。所有这些机构的研究都由国际组织（如全球气候变化研究联盟牲畜研究小组）进行联系和支持。

10　参考文献

Alemu, A. W., Vyas, D., Manafiazar, G., et al., 2017. Enteric methane emissions from low-and high-residual feed intake beef heifers measured using GreenFeed and respiration chamber techniques. J. Anim. Sci. 95 (8), 3727-37.

Arndt, C., Powell, J. M., Aguerre, M. J., et al., 2015. Feed conversion efficiency in dairy cows: repeatability, variation in digestion and metabolism of energy and nitrogen, and ruminal methanogens. J. Dairy. Sci. 98 (6), 3938-50.

Arntzen, M. Ø, Varnai, A., Mackie, R. I., Eijsink, V. G. H., et al., 2017. Outer membrane vesicles from *Fibrobacter succinogenes* S85 contain an array of carbohydrate-active enzymes with versatile polysaccharide-degrading capacity. Environ. Microbiol. 19 (7), 2701-14.

Artzi, L., Bayer, E. A. and Morais, S. 2017. Cellulosomes: bacterial nanomachines for dismantling plant polysaccharides. Nat. Rev. Microbiol. 15 (2), 83-95.

Auffret, M. D., Stewart, R., Dewhurst, R. J., et al., 2017. Identification, comparison, and validation of robust rumen microbial biomarkers for methane emissions using diverse Bos taurus breeds and basal diets. Front. Microbiol. 8, 2642.

Aydin, S., Yildirim, E., Ince, O., et al., 2017. Rumen anaerobic fungi create new opportunities for enhanced methane production from microalgae biomass. Algal Res. 23, 150-60.

Bayer, E. A., Belaich, J. P., Shoham, Y., et al., 2004. The cellulosomes: multienzyme machines for degradation of plant cell wall polysaccharides. Annu. Rev. Microbiol. 58, 521-54.

Belanche, A., Doreau, M., Edwards, J. E., et al., 2012. Shifts in the rumen microbiota due to the type of carbohydrate and level of protein ingested by dairy cattle are associated with changes in rumen fermentation. J. Nutr. 142 (9), 1684-92.

Belanche, A., Lee, M. R. F., Moorby, J. M., et al., 2013. Comparison of ryegrass and red clover on the fermentation pattern, microbial community and efficiency of diet utilisation in the rumen simulation technique (Rusitec). Anim. Prod. Sci. 53 (10), 1052-64.

Belanche, A., De La Fuente, G. and Newbold, C. J. 2015. Effect of progressive inoculation of fauna-free sheep with holotrich protozoa and total-fauna on rumen fermentation, microbial diversity and methane emissions. FEMS Microbiol. Ecol. 91 (3), fiu026.

Bergman, E. N. 1990. Energy contributions of volatile fatty acids from the gastrointestinal tract in various species. Physiol. Rev. 70 (2), 567-90.

Berry, D. P. and Crowley, J. J. 2013. CELL BIOLOGY SYMPOSIUM: genetics of feed efficiency in dairy and beef cattle. J. Anim. Sci. 91 (4), 1594-613.

Boland, T. M., Quinlan, C., Pierce, K. M., et al., 2013. The effect of pasture pregrazing herbage mass on methane emissions, ruminal fermentation, and average daily gain of grazing beef heifers. J. Anim. Sci. 91 (8), 3867-74.

Borneman, W. S., Akin, D. E. and Ljungdahl, L. G. 1989. Fermentation products and plant cell wall-degrading enzymes produced by monocentric and polycentric anaerobic ruminal fungi. Appl. Environ. Microbiol. 55 (5), 1066-73.

Bowen, J. M., McCabe, M. S., Lister, S. J., et al., 2018. Evaluation of microbial communities associated with the liquid and solid phases of the rumen of cattleoffered a diet of perennial ryegrass or white clover. Front. Microbiol. 9, 2389.

Brulc, J. M., Antonopoulos, D. A., Miller, M. E., et al., 2009. Gene-centric metagenomics of the fiber-adherent bovine rumen microbiome reveals forage specific glycoside hydrolases. PNAS 106 (6), 1948-53.

Buxton, D. R. and Redfearn, D. D. 1997. Plant limitations to fiber digestion and utilization. J. Nutr. 127 (5 Suppl.), 814S-8S.

Callaghan, T. M., Podmirseg, S. M., Hohlweck, D., et al., 2015. *Buwchfawromyces eastonii* gen. nov., sp. nov.: a new anaerobic fungus (Neocallimastigomycota) isolated from buffalo faeces. MycoKeys 9, 11-28.

Callahan, B. J., McMurdie, P. J. and Holmes, S. P. 2017. Exact sequence variants should replace operational taxonomic units in marker-gene data analysis. ISME J. 11 (12), 2639-43.

Cantalapiedra-Hijar, G., Abo-Ismail, M., Carstens, G. E., et al., 2018. Review: biological determinants of between-animal variation in feed efficiency of growing beef cattle. Animal 12 (S2), s321-35.

Carberry, C. A., Kenny, D. A., Han, S., et al., 2012. Effect of phenotypic residual feed intake and dietary forage content on the rumen microbial community of beef cattle. Appl. Environ. Microbiol. 78 (14), 4949-58.

Carberry, C. A., Kenny, D. A., Kelly, A. K., et al., 2014a. Quantitative analysis of ruminal methanogenic microbial populations in beef cattle divergent in phenotypic residual feed intake (RFI) offered contrasting diets. J. Anim. Sci. Biotechnol. 5 (1), 41.

Carberry, C. A., Waters, S. M., Kenny, D. A., et al., 2014b. Rumen methanogenic genotypes differ in abundance according to host residual feed intake phenotype and diet type. Appl. Environ. Microbiol. 80 (2), 586-94.

Chaves, A. V., Thompson, L. C., Iwaasa, A. D., et al., 2006. Effect of pasture type (alfalfa vs. grass) on methane and carbon dioxide production by yearling beef heifers. Can. J. Anim. Sci. 86 (3), 409-18.

Cheng, K. J., Dinsdale, D. and Stewart, C. S. 1979. Maceration of clover and grass leaves by Lachnospira multiparus. Appl. Environ. Microbiol. 38 (4), 723-9.

Cheng, K. J., Fay, J. P., Howarth, R. E., et al., 1980. Sequence of events in the digestion of fresh legume leaves by rumen bacteria. Appl. Environ. Microbiol. 40 (3), 613-25.

Chesson, A. 1993. Mechanistic models of forage cell wall degradation. In: Jung, H. G., Buxton, D. R.,

Hatfield, R. D. and Ralph, J. (Eds), Forage Cell Wall Structure and Digestibility. ASA-CSSA-SSSA, Madison, WI, pp. 347-76.

Chesson, A. and Forsberg, C. 1997. Polysaccharide degradation by rumen microorganisms. In: Hobson, P. N. and Stewart, C. S. (Eds), The Rumen Microbial Ecosystem. Blackie Academic and Professional, London, pp. 329-81.

Chesson, A. and Monro, J. A. 1982. Legume pectic substances and their degradation in the ovine rumen. J. Sci. Food Agric. 33 (9), 852-9.

Choudhury, P. K., Salem, A. Z. M., Jena, R., et al., 2015. Rumen microbiology: an overview. In: Puniya, A., Singh, R. and Kamra, D. (Eds), Rumen Microbiology: From Evolution to Revolution. Springer, New Delhi.

Corner-Thomas, R. A., Kemp, P. D., Morris, S. T., et al., 2014. Grazing alternative herbages in lactation increases the liveweight of both ewe lambs and their progeny at weaning. Anim. Prod. Sci. 54 (10), 1741-6.

Cosgrove, D. J. 2005. Growth of the plant cell wall. Nat. Rev. Mol. Cell. Biol. 6 (11), 850-61.

Cosgrove, D. J. and Jarvis, M. C. 2012. Comparative structure and biomechanics of plant primary and secondary cell walls. Front. Plant Sci. 3 (204).

Cosgrove, G., Taylor, P. and Jonker, A. 2015. Sheep performance on perennial ryegrass cultivars differing in concentration of water-soluble carbohydrate. J. N. Z. Grassl. Assoc. 77, 123-30.

Cunha, C. S., Veloso, C. M., Marcondes, M. I., et al., 2017. Assessing the impact of rumen microbial communities on methane emissions and production traits in Holstein cows in a tropical climate. Syst. Appl. Microbiol. 40 (8), 492-9.

Curran, J., Delaby, L., Kennedy, E., et al., 2010. Sward characteristics, grass dry matter intake and milk production performance are affected by pre-grazing herbage mass and pasture allowance. Livest. Sci. 127 (2-3), 144-54.

Dagar, S. S., Kumar, S., Griffith, G. W., et al., 2015. A new anaerobic fungus (*Oontomyces anksri* gen. nov., sp. nov.) from the digestive tract of the Indian camel (*Camelus dromedarius*). Fungal Biol. 119 (8), 731-7.

Dai, X., Tian, Y., Li, J., et al., 2015. Metatranscriptomic analyses of plant cell wall polysaccharide degradation by microorganisms in the cow rumen. Appl. Environ. Microbiol. 81 (4), 1375-86.

Danielsson, R., Dicksved, J., Sun, L., et al., 2017. Methane production in dairy cows correlates with rumen methanogenic and bacterial community structure. Front. Microbiol. 8, 226.

Dehority, B. A. 1991. Effects of microbial synergism on fibre digestion in the rumen. Proc. Nutr. Soc. 50 (2), 149-59.

Delgado, B., Bach, A., Guasch, I., et al., 2019. Whole rumen metagenome sequencing allows classifying and predicting feed efficiency and intake levels in cattle. Sci. Rep. 9 (1), 11.

Deppenmeier, U. and Müller, V. 2008. Life close to the thermodynamic limit: how methanogenic archaea conserve energy. Results Probl. Cell Differ. 45, 123-52.

Devendran, S., Abdel-Hamid, A. M., Evans, A. F., et al., 2016. Multiple cellobiohydrolases and cellobiose phosphorylases cooperate in the ruminal bacterium *Ruminococcus albus* 8 to degrade cellooligosaccharides. Sci. Rep. 6, 35342.

Dewhurst, R. J., Fisher, W. J., Tweed, J. K. S., et al., 2003. Comparison of grass and legume silages for milk production. 1. Production responses with different levels of concentrate. J. Dairy Sci. 86 (8),

2598-611.

Dineen, M., Delaby, L., Gilliland, T., et al., 2018. Meta-analysis of the effect of white clover inclusion in perennial ryegrass swards on milk production. J. Dairy Sci. 101 (2), 1804-16.

Dini, Y., Cajarville, C., Gere, J. I., et al., 2019. Association between residual feed intake and enteric methane emissions in Hereford steers. Transl. Anim. Sci. 3 (1), 239-46.

Duble, R. L., Lancaster, J. A. and Holt, E. C. 1971. Forage characteristics limiting animal performance on warm-season perennial grasses. Agron. J. 63 (5), 795-8.

Durunna, O. N., Colazo, M. G., Ambrose, D. J., et al., 2012. Evidence of residual feed intake reranking in crossbred replacement heifers. J. Anim. Sci. 90 (3), 734-41.

Dušková, D. and Marounek, M. 2001. Fermentation of pectin and glucose, and activity of pectin-degrading enzymes in the rumen bacterium *Lachnospira multiparus*. Lett. Appl. Microbiol. 33 (2), 159-63.

Eckard, R. J. and Clark, H. 2018. Potential solutions to the major greenhouse-gas issues facing Australasian dairy farming. Anim. Prod. Sci.

Edwards, G. R., Parsons, A., Rasmussen, S., et al., 2007. High sugar ryegrasses for livestock systems in New Zealand. Proc. N. Z. Grassland Assoc. 66, 161-71.

Edwards, J. E., Forster, R. J., Callaghan, T. M., et al., 2017. PCR and omics based techniques to study the diversity, ecology and biology of anaerobic fungi: insights, challenges and opportunities. Front. Microbiol. 8, 1657.

Elliott, C. L., Edwards, J. E., Wilkinson, T. J., et al., 2018. Using 'omic approaches to compare temporal bacterial colonization of *Lolium perenne*, *Lotus corniculatus*, and *Trifolium pratense* in the rumen. Front. Microbiol. 9, 2184.

Ellis, J. L., Dijkstra, J., Bannink, A., et al., 2011. The effect of high-sugar grass on predicted nitrogen excretion and milk yield simulated using a dynamic model. J. Dairy Sci. 94 (6), 3105-18.

Ellison, M. J., Conant, G. C., Lamberson, W. R., et al., 2017. Diet and feed efficiency status affect rumen microbial profiles of sheep. Small Rumin. Res. 156, 12-9.

Elolimy, A. A., Abdelmegeid, M. K., McCann, J. C., et al., 2018. Residual feed intake in beef cattle and its association with carcass traits, ruminal solid-fraction bacteria, and epithelium gene expression. J. Anim. Sci. Biotechnol. 9, 67.

Emerson, E. L. and Weimer, P. J. 2017. Fermentation of model hemicelluloses by *Prevotella* strains and *Butyrivibrio fibrisolvens* in pure culture and in ruminal enrichment cultures. Appl. Microbiol. Biotechnol. 101 (10), 4269-78.

Enriquez-Hidalgo, D., Gilliland, T., Deighton, M. H., et al., 2014a. Milk production and enteric methane emissions by dairy cows grazing fertilized perennial ryegrass pasture with or without inclusion of white clover. J. Dairy Sci. 97 (3), 1400-12.

Enriquez-Hidalgo, D., Hennessy, D., Gilliland, T., et al., 2014b. Effect of rotationally grazing perennial ryegrass white clover or perennial ryegrass only swards on dairy cow feeding behaviour, rumen characteristics and sward depletion patterns. Livest. Sci. 169, 48-62.

Enriquez-Hidalgo, D., Gilliland, T. J., Egan, M., et al., 2018. Production and quality benefits of white clover inclusion into ryegrass swards at different nitrogen fertilizer rates. J. Agric. Sci. 156 (3), 378-86.

FAO. 2017. The Future of Food and Agriculture-Trends and Challenges. FAO, Rome. Farra, P. A. and Satter, L. D. 1971. Manipulation of the ruminal fermentation. III. Effect of nitrate on ruminal volatile fatty acid production and milk composition. J. Dairy Sci. 54 (7), 1018-24.

Fernando, S. C., Purvis, H. T., Najar, F. Z., et al., 2010. Rumen microbial population dynamics during adaptation to a high-grain diet. Appl. Environ. Microbiol. 76 (22), 7482-90.

Ferry, J. G. (Ed.). 2012. Methanogenesis: Ecology, Physiology, Biochemistry & Genetics. Springer Science and Business Media.

Fitzsimons, C., Kenny, D. A., Deighton, M. H., et al., 2013. Methane emissions, body composition, and rumen fermentation traits of beef heifers differing in residual feed intake. J. Anim. Sci. 91 (12), 5789-800.

Fitzsimons, C., Kenny, D. A., Fahey, A. G., et al., 2014a. Feeding behavior, ruminal fermentation, and performance of pregnant beef cows differing in phenotypic residual feed intake offered grass silage. J. Anim. Sci. 92 (5), 2170-81.

Fitzsimons, C., Kenny, D. A. and McGee, M. 2014b. Visceral organ weights, digestion and carcass characteristics of beef bulls differing in residual feed intake offered a high concentrate diet. Animal 8 (6), 949-59.

Flay, H. E., Kuhn-Sherlock, B., Macdonald, K. A., et al., 2019. Hot topic: selecting cattle for low residual feed intake did not affect daily methane production but increased methane yield. J. Dairy Sci. 102 (3), 2708-13.

Flint, H. J., Scott, K. P., Duncan, S. H., et al., 2012. Microbial degradation of complex carbohydrates in the gut. Gut Microbes 3 (4), 289-306.

Forsberg, C. W., Beveridge, T. J. and Hellstrom, A. 1981. Cellulase and xylanase release from *Bacteroides succinogenes* and its importance in the rumen environment. Appl. Environ. Microbiol. 42 (5), 886-96.

Fraser, M. D., Speijers, M. H. M., Theobald, V. J., et al., 2004. Production performance and meat quality of grazing lambs finished on red clover, lucerne or perennial ryegrass swards. Grass Forage Sci. 59 (4), 345-56.

Gokarn, R. R., Eiteman, M. A., Martin, S. A., et al., 1997. Production of succinate from glucose, cellobiose, and various cellulosic materials by the ruminal anaerobic bacteria *Fibrobacter succinogenes* and *Ruminococcus flavefaciens*. Appl. Biochem. Biotechnol. 68 (1-2), 69-80.

Goopy, J. P., Donaldson, A., Hegarty, R., et al., 2014. Low-methane yield sheep have smaller rumens and shorter rumen retention time. Br. J. Nutr. 111 (4), 578-85.

Gordon, G. L. R. and Phillips, M. W. 1993. Removal of anaerobic fungi from the rumen of sheep by chemical treatment and the effect on feed consumption and *in vivo* fibre digestion. Lett. Appl. Microbiol. 17 (5), 220-3.

Grace, C., Lynch, M. B., Sheridan, H., et al., 2018. Grazing multispecies swards improves ewe and lamb performance. Animal 17, 1-9.

Griffith, G. S., Cresswell, A., Jones, S., et al., 2000. The nitrogen handling characteristics of white clover (*Trifolium repens* L.) cultivars and a perennial ryegrass (*Lolium perenne* L.) cultivar. J. Exp. Bot. 51 (352), 1879-92.

Gruninger, R. J., Puniya, A. K., Callaghan, T. M., et al., 2014. Anaerobic fungi (phylum Neocallimastigomycota): advances in understanding their taxonomy, life cycle, ecology, role and biotechnological potential. FEMS Microbiol. Ecol. 90 (1), 1-17.

Gruninger, R. J., Nguyen, T. T. M., Reid, I. D., et al., 2018. Application of transcriptomics to compare the carbohydrate active enzymes that are expressed by diverse genera of anaerobic fungi to degrade plant cell wall carbohydrates. Front. Microbiol. 9, 1581-.

Guan, L. L., Nkrumah, J. D., Basarab, J. A., et al., 2008. Linkage of microbial ecology to phenotype: correlation of rumen microbial ecology to cattle's feed efficiency. FEMS Microbiol. Lett. 288 (1), 85-91.

Guan, L., Li, F., Bulumulla, A., et al., 2017. The role of rumen microbiome on feed efficiency of grazing cattle. 28th Annual Florida Ruminant Nutrition Symposium, 770, 137.

Guy, C., Hennessy, D., Gilliland, T. J., et al., 2018. Growth, morphology and biological nitrogen fixation potential of perennial ryegrass-white clover swards throughout the grazing season. J. Agric. Sci. 156 (2), 188-99.

Guyader, J., Eugene, M., Noziere, P., et al., 2014. Influence of rumen protozoa on methane emission in ruminants: a meta-analysis approach. Animal 8 (11), 1816-25.

Haitjema, C. H., Gilmore, S. P., Henske, J. K., et al., 2017. A parts list for fungal cellulosomes revealed by comparative genomics. Nat. Microbiol. 2 (8), 17087.

Hammond, K. J., Hoskin, S. O., Burke, J. L., et al., 2011. Effects of feeding fresh white clover (Trifolium repens) or perennial ryegrass (*Lolium perenne*) on enteric methane emissions from sheep. Anim. Feed Sci. Tech. 166-167, 398-404.

Hammond, K. J., Pacheco, D., Burke, J. L., et al., 2014. The effects of fresh forages and feed intake level on digesta kinetics and enteric methane emissions from sheep. Anim. Feed Sci. Tech. 193, 32-43.

Han, S. O., Cho, H. Y., Yukawa, H., et al., 2004. Regulation of expression of cellulosomes and non-cellulosomal (hemi) cellulolytic enzymes in *Clostridium cellulovorans* during growth on different carbon sources. J. Bacteriol. 186 (13), 4218-27.

Hanafy, R. A., Elshahed, M. S., Liggenstoffer, A. S., et al., 2017. *Pecoramyces ruminantium*, gen. nov., sp. nov., an anaerobic gut fungus from the feces of cattle and sheep. Mycologia 109 (2), 231-43.

Hanafy, R. A., Elshahed, M. S. and Youssef, N. H. 2018. *Feramyces austinii*, gen. nov., sp. nov., an anaerobic gut fungus from rumen and fecal samples of wild Barbary sheep and fallow deer. Mycologia 110 (3), 513-25.

Harholt, J., Suttangkakul, A. and Vibe Scheller, H. 2010. Biosynthesis of pectin. Plant Physiol. 153 (2), 384-95.

Harris, S. L. and Clark, D. A. 1996. Effect of high rates of nitrogen fertiliser on white clover growth, morphology, and nitrogen fixation activity in grazed dairy pasture in northern New Zealand. N. Z. J. Agr. Res. 39 (1), 149-58.

Hart, K. J., Martin, P. G., Foley, P. A., et al., 2009. Effect of sward dry matter digestibility on methane production, ruminal fermentation, and microbial populations of zero-grazed beef cattle. J. Anim. Sci. 87 (10), 3342-50.

Hassanat, F., Gervais, R., Massé, D. I., et al., 2014. Methane production, nutrient digestion, ruminal fermentation, N balance, and milk production of cows fed timothy silage-or alfalfa silage-based diets. J. Dairy Sci. 97 (10), 6463-74.

Hegarty, R. S., Goopy, J. P., Herd, R. M., et al., 2007. Cattle selected for lower residual feed intake have reduced daily methane production. J. Anim. Sci. 85 (6), 1479-86.

Henderson, G., Cox, F., Ganesh, S., et al., 2015. Rumen microbial community composition varies with diet and host, but a core microbiome is found across a wide geographical range. Sci. Rep. 5, 14567.

Hernandez-Sanabria, E., Guan, L. L., Goonewardene, L. A., et al., 2010. Correlation of particular bacterial PCR-denaturing gradient gel electrophoresis patterns with bovine ruminal fermentation parameters and feed efficiency traits. Appl. Environ. Microbiol. 76 (19), 6338-50.

Hernandez-Sanabria, E., Goonewardene, L. A., Wang, Z., et al., 2012. Impact of feed efficiency and diet on adaptive variations in the bacterial community in the rumen fluid of cattle. Appl. Environ. Microbiol. 78 (4), 1203-14.

Holmes, C. W., Hoogendoorn, C. J., Ryan, M. P. and Chu, A. C. P. 1992. Some effects of herbage composition, as influenced by previous grazing management, on milk production by cows grazing on ryegrass/white clover pastures. 1. Milk production in early spring: effects of different regrowth intervals during the preceding winter period. Grass Sci. 47 (4), 309-15.

Hristov, A. and Jouany, J. 2005. Factors affecting the efficiency of nitrogen utilization in the rumen. In: Hristov, A. N. and Pfeffer, E. (Eds), Nitrogen and Phosphorus Nutrition of Cattle: Reducing the Environmental Impact of Cattle Operations. CAB International, Wallingford, UK, pp. 117-66.

Hungate, R. E. 1966. Variations in the rumen. In: The Rumen and Its Microbes. Academic Press, New York and London, pp. 376-418.

Huws, S. A., Lee, M. R., Muetzel, S. M., et al., 2010. Forage type and fish oil cause shifts in rumen bacterial diversity. FEMS Microbiol. Ecol. 73 (2), 396-407.

Huws, S. A., Edwards, J. E., Creevey, C. J., et al., 2016. Temporal dynamics of the metabolically active rumen bacteria colonizing fresh perennial ryegrass. FEMS Microbiol. Ecol. 92 (1), fiv137.

Huws, S. A., Creevey, C. J., Oyama, L. B., et al., 2018. Addressing global ruminant agricultural challenges through understanding the rumen microbiome: past, present, and future. Front. Microbiol. 9, 2161.

Indugu, N., Vecchiarelli, B., Baker, L. D., et al., 2017. Comparison of rumen bacterial communities in dairy herds of different production. BMC Microbiol. 17 (1), 190.

Janssen, P. H. 2010. Influence of hydrogen on rumen methane formation and fermentation balances through microbial growth kinetics and fermentation thermodynamics. Anim. Feed Sci. Tech. 160 (1-2), 1-22.

Jewell, K. A., McCormick, C. A., Odt, C. L., et al., 2015. Ruminal bacterial community composition in dairy cows is dynamic over the course of two lactations and correlates with feed efficiency. Appl. Environ. Microbiol. 81 (14), 4697-710.

Joblin, K. N., Matsui, H., Naylor, G. E., et al., 2002. Degradation of fresh ryegrass by methanogenic co-cultures of ruminal fungi grown in the presence or absence of *Fibrobacter succinogenes*. Curr. Microbiol. 45 (1), 46-53.

Johnson, K. A. and Johnson, D. E. 1995. Methane emissions from cattle. J. Anim. Sci. 73 (8), 2483-92.

Jones, F. M., Phillips, F. A., Naylor, T., et al., 2011. Methane emissions from grazing Angus beef cows selected for divergent residual feed intake. Anim. Feed Sci. Tech. 166-167, 302-7.

Jonker, A., Cheng, L., Edwards, G. R., et al., 2018a. Nitrogen partitioning differs in sheep offered a conventional diploid, a high sugar diploid or a tetraploid perennial ryegrass cultivar at two feed allowances. Anim. Feed Sci. Tech. 245, 32-40.

Jonker, A., Molano, G., Sandoval, E., et al., 2018b. Methane emissions differ between sheep offered a conventional diploid, a high-sugar diploid or a tetraploid perennial ryegrass cultivar at two allowances at three times of the year. Anim. Prod. Sci. 58 (6), 1043-8.

Joshi, A., Lanjekar, V. B., Dhakephalkar, P. K., et al., 2018. *Liebetanzomyces polymorphus* gen. et sp. nov., a new anaerobic fungus (Neocallimastigomycota) isolated from the rumen of a goat. MycoKeys 40

(40), 89-110.

Jun, H. S., Qi, M., Gong, J., et al., 2007. Outer membrane proteins of *Fibrobacter succinogenes* with potential roles in adhesion to cellulose and in cellulose digestion. J. Bacteriol. 189 (19), 6806-15.

Jung, H. G. and Allen, M. S. 1995. Characteristics of plant cell walls affecting intake and digestibility of forages by ruminants. J. Anim. Sci. 73 (9), 2774-90.

Kamke, J., Kittelmann, S., Soni, P., et al., 2016. Rumen metagenome and metatranscriptome analyses of low methane yield sheep reveals a Sharpea-enriched microbiome characterised by lactic acid formation and utilisation. Microbiome 4 (1), 56.

Kelly, A. K., McGee, M., Crews Jr., et al., 2010. Repeatability of feed efficiency, carcass ultrasound, feeding behavior, and blood metabolic variables in finishing heifers divergently selected for residual feed intake. J. Anim. Sci. 88 (10), 3214-25.

Kenny, D. A., Fitzsimons, C., Waters, S. M., et al., 2018. Invited review: improving feed efficiency of beef cattle-the current state of the art and future challenges. Animal 12 (9), 1815-26.

Kim, J. N., Méndez-García, C., Geier, R. R., et al., 2017. Metabolic networks for nitrogen utilization in *Prevotella ruminicola* 23. Sci. Rep. 7 (1), 7851.

Kittelmann, S., Seedorf, H., Walters, W. A., et al., 2013. Simultaneous amplicon sequencing to explore co-occurrence patterns of bacterial, archaeal and eukaryotic microorganisms in rumen microbial communities. PLoS ONE 8 (2), e47879.

Kittelmann, S., Pinares-Patiño, C. S., Seedorf, H., et al., 2014. Two different bacterial community types are linked with the low-methane emission trait in sheep. PLoS ONE 9 (7), e103171.

Klevenhusen, F., Petri, R. M., Kleefisch, M. T., et al., 2017. Changes in fibre-adherent and fluid-associated microbial communities and fermentation profiles in the rumen of cattle fed diets differing in hay quality and concentrate amount. FEMS Microbiol. Ecol. 93 (9), fix100.

Kong, R. S. G., Liang, G., Chen, Y., et al., 2016. Transcriptome profiling of the rumen epithelium of beef cattle differing in residual feed intake. BMC Genomics 17, 592.

Krause, D. O., Denman, S. E., Mackie, R. I., et al., 2003. Opportunities to improve fiber degradation in the rumen: microbiology, ecology, and genomics. FEMS Microbiol. Rev. 27 (5), 663-93.

Kulp, A. and Kuehn, M. J. 2010. Biological functions and biogenesis of secreted bacterial outer membrane vesicles. Annu. Rev. Microbiol. 64, 163-84.

Kumar, S., Treloar, B. P., Teh, K. H., et al., 2018. Sharpea and Kandleria are lactic acid producing rumen bacteria that do not change their fermentation products when co-cultured with a methanogen. Anaerobe 54, 31-8.

Lam, S., Munro, J. C., Zhou, M., et al., 2018. Associations of rumen parameters with feed efficiency and sampling routine in beef cattle. Animal 12 (7), 1442-50.

Latham, M. J. and Wolin, M. J. 1977. Fermentation of cellulose by *Ruminococcus flavefaciens* in the presence and absence of *Methanobacterium ruminantium*. Appl. Environ. Microbiol. 34 (3), 297-301.

Lawrence, P., Kenny, D. A., Earley, B., et al., 2011. Grass silage intake, rumen and blood variables, ultrasonic and body measurements, feeding behavior and activity in pregnant beef heifers differing in phenotypic residual feed intake. Journal of Animal Science 89, 3248-61.

Lawrence, P., Kenny, D. A., Earley, B., et al., 2013. Intake of conserved and grazed grass and performance traits in beef suckler cows differing in phenotypic residual feed intake. Livestock Science 152, 154-66.

Leahy, S. C., Kelly, W. J., Altermann, E., et al., 2010. The genome sequence of the rumen methanogen *Methanobrevibacter ruminantium* reveals new possibilities for controlling ruminant methane emissions. PLoS ONE 5 (1), e8926.

Ledgard, S. F., Sprosen, M. S., Penno, J. W., et al., 2001. Nitrogen fixation by white clover in pastures grazed by dairy cows: temporal variation and effects of nitrogen fertilization. Plant Soil 229 (2), 177-87.

Lee, M. R. F. 2014. Forage polyphenol oxidase and ruminant livestock nutrition. Front. Plant Sci. 5, 694.

Lee, M. A. 2018. A global comparison of the nutritive values of forage plants grown in contrasting environments. J. Plant Res. 131 (4), 641-54.

Lee, S. S., Ha, J. K. and Cheng, K. 2000. Relative contributions of bacteria, protozoa, and fungi to in vitro degradation of orchard grass cell walls and their interactions. Appl. Environ. Microbiol. 66 (9), 3807-13.

Lee, J., Woodward, S., Waghorn, G., et al., 2004a. Methane emissions by dairy cows fed increasing proportions of white clover (*Trifolium repens*) in pasture. Proc. N. Z. GrasslandAssess. 66, 151-5.

Lee, M. R., Winters, A. L., Scollan, N. D., et al., 2004b. Plant-mediated lipolysis and proteolysis in red clover with different polyphenol oxidase activities. J. Sci. Food Agric. 84 (13), 1639-45.

Liang, Y. S., Li, G. Z., Li, X. Y., et al., 2017. Growth performance, rumen fermentation, bacteria composition, and gene expressions involved in intracellular pH regulation of rumen epithelium in finishing Hu lambs differing in residual feed intake phenotype. J. Anim. Sci. 95 (4), 1727-38.

Liu, Y. and Whitman, W. B. 2008. Metabolic, phylogenetic, and ecological diversity of the methanogenic archaea. Ann. N. Y. Acad. Sci. 1125, 171-89.

Liu, L., Xu, X., Cao, Y., et al., 2017. Nitrate decreases methane production also by increasing methane oxidation through stimulating NC10 population in ruminal culture. AMB Express 7 (1), 76.

Marounek, M. and Dušková, D. 1999. Metabolism of pectin in rumen bacteria *Butyrivibrio fibrisolvens and Prevotella ruminicola*. Lett. Appl. Microbiol. 29 (6), 429-33.

Matsui, H., Ogata, K., Tajima, K., et al., 2000. Phenotypic characterization of polysaccharidases produced by four Prevotella type strains. Curr. Microbiol. 41 (1), 45-9.

Mayorga, O. L., Kingston-Smith, A. H., Kim, E. J., et al., 2016. Temporal metagenomic and metabolomic characterization of fresh perennial ryegrass degradation by rumen bacteria. Front. Microbiol. 7, 1854.

McAllister, T. A. and Newbold, C. J. 2008. Redirecting rumen fermentation to reduce methanogenesis. Aust. J. Exp. Agric. 48 (2), 7-13.

McCann, J. C., Wickersham, T. A. and Loor, J. J. 2014a. High-throughput methods redefine the rumen microbiome and its relationship with nutrition and metabolism. Bioinform. Biol. Insights 8, 109-25.

McCann, J. C., Wiley, L. M., Forbes, T. D., et al., 2014b. Relationship between the rumen microbiome and residual feed intake-efficiency of Brahman bulls stocked on bermudagrass pastures. PLoS ONE 9 (3), e91864.

McDonnell, R. P., Hart, K. J., Boland, T. M., et al., 2016. Effect of divergence in phenotypic residual feed intake on methane emissions, ruminal fermentation, and apparent whole-tract digestibility of beef heifers across three contrasting diets. J. Anim. Sci. 94 (3), 1179-93.

McGovern, E., Kenny, D. A., McCabe, M. S., et al., 2018. 16S rRNA sequencing reveals relationship between potent cellulolytic genera and feed efficiency in the rumen of bulls. Front. Microbiol. 9, 1842-.

Mercadante, M. E. Z., Caliman, A. P. D. M., Canesin, R. C., et al., 2015. Relationship between residual feed intake and enteric methane emission in Nellore cattle. R. Bras. Zootec. 44 (7), 255-62.

Mi, J., Zhou, J., Huang, X., et al., 2017. Lower methane emissions from yak compared with cattle in Rusitec fermenters. PLoS ONE 12 (1), e0170044.

Miller, T. L. and Wolin, M. J. 1973. Formation of hydrogen and formate by *Ruminococcus albus*. J. Bacteriol. 116 (2), 836-46.

Morgavi, D. P., Sakurada, M., Mizokami, M., et al., 1994. Effects of ruminal protozoa on cellulose degradation and the growth of an anaerobic ruminal fungus, Piromyces sp. strain OTS1, in vitro. Appl. Environ. Microbiol. 60 (10), 3718-23.

Morgavi, D. P., Forano, E., Martin, C., et al., 2010. Microbial ecosystem and methanogenesis in ruminants. Animal 4 (7), 1024-36.

Morgavi, D. P., Martin, C., Jouany, J. P., et al., 2012. Rumen protozoa and methanogenesis: not a simple cause-effect relationship. Br. J. Nutr. 107 (3), 388-97.

Morris, E. J. and Cole, O. J. 1987. Relationship between cellulolytic activity and adhesion to cellulose in Ruminococcus albus. Microbiology 133 (4), 1023-32.

Muñoz, C., Letelier, P. A., Ungerfeld, E. M., et al., 2016. Effects of pregrazing herbage mass in late spring on enteric methane emissions, dry matter intake, and milk production of dairy cows. J. Dairy Sci. 99 (10), 7945-55.

Myer, P. R., Smith, T. P. L., Wells, J. E., et al., 2015. Rumen microbiome from steers differing in feed efficiency. PLoS ONE 10 (6), e0129174.

Navarro-Villa, A., O'Brien, M., Lopez, S., et al., 2011. *In vitro* rumen methane output of red clover and perennial ryegrass assayed using the gas production technique (GPT). Anim. Feed Sci. Tech. 168 (3-4), 152-64.

Neumann, A. P., Weimer, P. J. and Suen, G. 2018. A global analysis of gene expression in *Fibrobacter succinogenes* S85 grown on cellulose and soluble sugars at different growth rates. Biotechnol. Biofuels 11, 295.

Newbold, C. J., De La Fuente, G., Belanche, A., et al., 2015. The role of ciliate protozoa in the rumen. Front. Microbiol. 6, 1313.

Ng, F., Kittelmann, S., Patchett, M. L., et al., 2016. An adhesin from hydrogen-utilizing rumen methanogen *Methanobrevibacter ruminantium* M1 binds a broad range of hydrogen-producing microorganisms. Environ. Microbiol. 18 (9), 3010-21.

Niderkorn, V., Martin, C., Rochette, Y., et al., 2015. Associative effects between orchardgrass and red clover silages on voluntary intake and digestion in sheep: evidence of a synergy on digestible dry matter intake. J. Anim. Sci. 93 (10), 4967-76.

Niderkorn, V., Martin, C., Le Morvan, A., et al., 2017. Associative effects between fresh perennial ryegrass and white clover on dynamics of intake and digestion in sheep. Grass For. Sci. 72 (4), 691-9.

Nkrumah, J. D., Okine, E. K., Mathison, G. W., et al., 2006. Relationships of feedlot feed efficiency, performance, andfeeding behavior with metabolic rate, methane production, and energy partitioning in beef cattle. J. Anim. Sci. 84 (1), 145-53.

Olijhoek, D. W., Løvendahl, P., Lassen, J., et al., 2018. Methane production, rumen fermentation, and diet digestibility of Holstein and Jersey dairy cows being divergent in residual feed intake and fed at 2 forage-to-concentrate ratios. J. Dairy Sci. 101 (11), 9926-40.

Oliveira, L. F., Ruggieri, A. C., Branco, R. H. et al., 2018. Feed efficiency and enteric methane production of Nellore cattle in the feedlot and on pasture. Anim. Prod. Sci. 58 (5), 886-93.

Orpin, C. G. 1977. Invasion of plant tissue in the rumen by the flagellate *Neocallimastix frontalis*. J. Gen. Microbiol. 98 (2), 423-30.

Pacheco, D., Waghorn, G. and Janssen, P. H. 2014. Decreasing methane emissions from ruminants grazing forages: a fit with productive and financial realities? Anim. Prod. Sci. 54 (9), 1141-54.

Park, T., Wijeratne, S., Meulia, T., et al., 2018. Draft macronuclear genome sequence of the ruminal ciliate *Entodinium caudatum*. Microbiol. Resour. Announc. 7 (1), e00826-18.

Patil, R. D., Ellison, M. J., Wolff, S. M., et al., 2018. Poor feed efficiency in sheep is associated with several structural abnormalities in the community metabolic network of their ruminal microbes. J. Anim. Sci. 96 (6), 2113-24.

Patra, A., Park, T., Kim, M., et al., 2017. Rumen methanogens and mitigation of methane emission by anti-methanogenic compounds and substances. J. Anim. Sci. Biotech. 8, 13.

Petri, R. M., Mapiye, C., Dugan, M. E. R., et al., 2014. Subcutaneous adipose fatty acid profiles and related rumen bacterial populations of steers fed red clover or grass hay diets containing flax or sunflowerseed. PLoS ONE 9 (8), e104167.

Pinares-Patiño, C., Ulyatt, M. J., Lassey, K. R., et al., 2003. Rumen function and digestion parameters associated with differences between sheep in methane emissions when fed chaffed lucerne hay. J. Agr. Sci. 140 (2), 205-14.

Puchala, R., Animut, G., Patra, A. K., et al., 2012. Effects of different fresh-cut forages and their hays on feed intake, digestibility, heat production, and ruminal methane emission by Boer x Spanish goats. J. Anim. Sci. 90 (8), 2754-62.

Raffrenato, E., Fievisohn, R., Cotanch, K. W., et al., 2017. Effect of lignin linkages with other plant cell wall components on *in vitro* and *in vivo* neutral detergent fiber digestibility and rate of digestion of grass forages. J. Dairy Sci. 100 (10), 8119-31.

Reeve, J. N., Nolling, J., Morgan, R. M., et al., 1997. Methanogenesis: genes, genomes, and who's on first? J. Bacteriol. 179 (19), 5975-86.

Ribeiro, G. O., Gruninger, R. J., Badhan, A., et al., 2016. Mining the rumen for fibrolytic feed enzymes. Anim. Front. 6 (2), 20-6.

Ricard, G., McEwan, N. R., Dutilh, B. E., et al., 2006. Horizontal gene transfer from Bacteria to rumen Ciliates indicates adaptation to their anaerobic, carbohydrates-rich environment. BMC Genomics 7, 22.

Rius, A. G., Kittelmann, S., Macdonald, K. A., et al., 2012. Nitrogen metabolism and rumen microbial enumeration in lactating cows with divergent residual feed intake fed high-digestibility pasture. J. Dairy Sci. 95 (9), 5024-34.

Rooke, J. A., Wallace, R. J., Duthie, C. A., et al., 2014. Hydrogen and methane emissions from beef cattle and their rumen microbial community vary with diet, time after feeding and genotype. Br. J. Nutr. 112 (3), 398-407.

Rubino, F., Carberry, C., Waters, S. M., et al., 2017. Divergent functional isoforms drive niche specialisation for nutrient acquisition and use in rumen microbiome. ISME J. 11 (4), 932-44.

Sasson, G., Kruger Ben-Shabat, S., Seroussi, E., et al., 2017. Heritable bovine rumen bacteria are phylogenetically related and correlated with the cow's capacity to harvest energy from its feed. mBio 8 (4), 12.

Scheifinger, C. C. and Wolin, M. J. 1973. Propionate formation from cellulose and soluble sugars by combined cultures of *Bacteroides succinogenes* and *Selenomonas ruminantium*. Appl. Microbiol. 26 (5), 789-95.

Seshadri, R., Leahy, S. C., Attwood, G. T., et al., 2018. Cultivation and sequencing of rumen microbiome members from the Hungate1000 Collection. Nat. Biotech. 36, 359-67.

Shabat, S. K., Sasson, G., Doron-Faigenboim, A., et al., 2016. Specific microbiome-dependent mechanisms underlie the energy harvest efficiency of ruminants. ISME J. 10 (12), 2958-72.

Sharma, V. K., Kundu, S. S., Datt, C., et al., 2018. Buffalo heifers selected for lower residual feed intake have lower feed intake, better dietary nitrogen utilisation and reduced enteric methane production. J. Anim. Physiol. Anim. Nutr. 102 (2), e607-14.

Shi, W. B., Moon, C. D., Leahy, S. C., et al., 2014. Methane yield phenotypes linked to differential gene expression in the sheep rumen microbiome. Genome Res. 24 (9), 1517-25.

Smith, P., Enriquez-Hidalgo, D., Hennessy, D., et al., 2019. Archaea and members of the family, Lachnospiraceae, are altered in the rumenmicrobiome of cattle divergent in methane yield grazing perennial ryegrass swards with and without white clover. Proc. Br. Soc. Anim. Sci. 86.

Solomon, K. V., Haitjema, C. H., Henske, J. K., et al., 2016. Early-branching gut fungi possess a large, comprehensive array of biomass-degrading enzymes. Science 351 (6278), 1192-5.

Staerfl, S. M., Amelchanka, S. L., Kälber, T., et al., 2012. Effect of feeding dried high-sugar ryegrass ('AberMagic') on methane and urinary nitrogen emissions of primiparous cows. Livest. Sci. 150 (1-3), 293-301.

Steinshamn, H. and Thuen, E. 2008. White or red clover-grass silage in organic dairy milk production: grassland productivity and milk production responses with different levels of concentrate. Livest. Sci. 119 (1-3), 202-15.

Stevenson, D. M. and Weimer, P. J. 2007. Dominance of *Prevotella* and low abundance of classical ruminal bacterial species in the bovine rumen revealed by relative quantification real-time PCR. Appl. Microbiol. Biotechnol. 75 (1), 165-74.

Stewart, C., Flint, H. and Bryant, M. 1997. The rumen bacteria. In: Hobson, P. N. and Stewart, C. S. (Eds), The Rumen Microbial Ecosystem. Springer Netherlands, pp. 10-72.

Suen, G., Weimer, P. J., Stevenson, D. M., et al., 2011. The complete genome sequence of *Fibrobacter succinogenes* S85 reveals a cellulolytic and metabolic specialist. PLoS ONE 6 (4), e18814.

Tapio, I., Snelling, T. J., Strozzi, F., et al., 2017. The ruminal microbiome associated with methane emissions from ruminant livestock. J. Anim. Sci. Biotech. 8, 7.

Thompson, S. C. 2015. The effect of diet type on residual feed intake and the use of infrared thermography as a method to predict efficiency in beef bulls. MSc Thesis. University of Manitoba.

Thomson, D. 1984. The nutritive value of white clover. British Grassland Society Occasional Symposium 16, 78-92.

Tokura, M., Tajima, K. and Ushida, K. 1999. Isolation of *Methanobrevibacter* sp. as a ciliate-associated ruminal methanogen. J. Gen. Appl. Microbiol. 45, 43-7.

Torres-Vázquez, J. A., Van Der Werf, J. H. J. and Clark, S. A. 2018. Genetic and phenotypic associations of feed efficiency with growth and carcass traits in Australian Angus cattle. J. Anim. Sci. 96 (11), 4521-31.

United Nations, Department of Economic and Social Affairs, Population Division 2017. World population prospects: the 2017 revision, key findings and advance tables. Working Paper No. ESA/P/WP/248.

Urrutia, N. L. and Harvatine, K. J. 2017. Acetate dose-dependently stimulates milk fat synthesis in lactating dairy cows. J. Nutr. 147 (5), 763-9.

Van Glyswyk, N. O., Hippe, H. and Rainey, F. A. 1996. *Pseudobutyrivibrio ruminis* gen. nov., sp. nov., a butyrate-producing bacterium from the rumen that closely resembles *Butyrivibrio fibrisolvens* in phenotype. Int. J. Syst. Bacteriol. 46 (2), 559-63.

Vanholme, R., Demedts, B., Morreel, K., et al., 2010. Lignin biosynthesis and structure. Plant Physiol. 153 (3), 895-905.

Velazco, J. I., Herd, R. M., Cottle, D. J., et al., 2017. Daily methane emissions and emission intensity of grazing beef cattle genetically divergent for residual feed intake. Anim. Prod. Sci. 57 (4), 627-35.

Wallace, R. J., Rooke, J. A., Duthie, C. A., et al., 2014. Archaeal abundance in post-mortem ruminal digesta may help predict methane emissions from beef cattle. Sci. Rep. 4, 5892.

Wallace, R. J., Rooke, J. A., McKain, N., et al., 2015. The rumen microbial metagenome associated with high methane production in cattle. BMC Genomics 16, 839.

Weimer, P. J. 1996. Why don't ruminal bacteria digest cellulose faster? J. Dairy Sci. 79 (8), 1496-502.

Williams, A. G. and Coleman, G. S. 1997. The rumen protozoa. In: Hobson, P. N. and Stewart, C. S. (Eds), The Rumen Microbial Ecosystem, Springer, pp. 73-139.

Wims, C. M., Deighton, M. H., Lewis, E., et al., 2010. Effect of pregrazing herbage mass on methane production, dry matter intake, and milk production of grazing dairy cows during the mid-season period. J. Dairy Sci. 93 (10), 4976-85.

Wina, E., Muetzel, S. and Becker, K. 2005. The impact of saponins or saponin-containing plant materials on ruminant production-a review. J. Agric. Food Chem. 53 (21), 8093-105.

Wolff, S. M., Ellison, M. J., Hao, Y., et al., 2017. Diet shifts provoke complex and variable changes in the metabolic networks of the ruminal microbiome. Microbiome 5 (1), 60.

Wolin, M. J., Miller, T. L. and Stewart, C. S. 1997. Microbe-microbe interactions. Hobson, P. N. and Stewart, C. S. (Eds), The Rumen Microbial Ecosystem. Springer, pp. 467-91.

Yanez-Ruiz, D. R., Macias, B., Pinloche, E., et al., 2010. The persistence of bacterial and methanogenic archaeal communities residing in the rumen of young lambs. FEMS Microbiol. Ecol. 72 (2), 272-8.

Yang, B., He, B., Wang, S. S., et al., 2015. Early supplementation of starter pellets with alfalfa improves the performance of pre-and postweaning Hu lambs. J. Anim. Sci. 93 (10), 4984-94.

Yang, B., Le, J., Wu, P., et al., 2018. Alfalfa intervention alters rumen microbial community development in Hu lambs during early life. Front. Microbiol. 9, 574-.

Zamil, M. S. and Geitmann, A. 2017. The middle lamella—more than a glue. Phys. Biol. 14 (1), 015004.

Zhang, R., Zhu, W., Zhu, W., et al., 2014. Effect of dietary forage sources on rumen microbiota, rumen fermentation and biogenic amines in dairy cows. J. Sci. Food Agric. 94 (9), 1886-95.

Zhang, X., Wang, W., Mo, F., et al., 2017. Association of residual feed intake with growth and slaughtering performance, blood metabolism, and body composition in growing lambs. Sci. Rep. 7 (1), 12681.

Zhao, L., Meng, Q., Li, Y., et al., 2018. Nitrate decreases ruminal methane production with slight changes to ruminal methanogen composition of nitrate-adapted steers. BMC Microbiol. 18 (1), 21.

Zheng, Y., Kahnt, J., Kwon, I. H., et al., 2014. Hydrogen formation and its regulation in

Ruminococcus albus: involvement of an electron-bifurcating [FeFe] -hydrogenase, of a non-electron-bifurcating [FeFe] - hydrogenase, and of a putative hydrogen - sensing [FeFe] - hydrogenase. J. Bacteriol. 196 (22), 3840-52.

Zhou, M., Hernandez-Sanabria, E. and Guan, L. L. 2009. Assessment of the microbial ecology of ruminal methanogens in cattle with different feed efficiencies. Appl. Environ. Microbiol. 75 (20), 6524-33.

Zhou, Z., Yu, Z. and Meng, Q. 2012. Effects of nitrate on methane production, fermentation, and microbial populations in *in vitro* ruminal cultures. Bioresour. Technol. 103 (1), 173-9.

第 19 章 优化瘤胃功能：青贮饲料和精料提高奶牛饲料转化率、减少甲烷和氮排放的作用

Aila Vanhatalo、Anni Halmemies-Beauchet-Filleau

赫尔辛基大学，瑞典

（米见对译）

1 前言

在全球温室气体排放中，反刍动物贡献了包括甲烷和一氧化二氮在内的大量温室气体。因此，近年来许多研究在寻求通过营养途径减少温室气体排放的方法，特别是针对奶牛养殖过程。多种日粮调控方案正在研发中，包括日粮、日粮管理和瘤胃改良剂（即饲喂直接或者间接抑制甲烷产生的特定物质），或者通过生物控制直接减少产甲烷菌的数量（Knapp 等，2014）。一些饲料添加剂，如甲烷抑制剂 3-硝基氧基丙醇（3-NOP），不仅在密集型奶牛日粮中非常有应用前景（Dijkstra 等，2018；Van Gastelen 等，2019），而且在全草料肉牛饲养中也很有前景（Martinez-Fernandez 等，2018）。这些方案还没有在实践中得到广泛运用。但是，基于改变日粮精粗比的营养调控方案往往比瘤胃改良剂更具适应性和实用性。据估计，通过改善饲养和营养（包括提高饲草品质、饲喂谷物和脂肪）降低温室气体排放的潜力通常为 10%~30%（Hristov 等，2013；Knapp 等，2014）。然而，Hristov 等（2013）认为提高饲草品质和营养利用效率是降低每单位动物产品甲烷排放强度的有效方法。

无论是基于粗饲料或精料的日粮策略，其减少温室气体排放的有效性在很大程度上取决于对瘤胃挥发性脂肪酸（VFA）发酵的影响。根据 Knapp 等（2014）的综述，任何有利于生成丙酸的日粮组成的变化都可以减少等量的甲烷产生，而有利于生成乙酸和丁酸的日粮则会产生氢气用于生成甲烷，从而增加甲烷的产量。其他的发酵过程，如瘤胃蛋白质降解并同化为微生物蛋白，以及瘤胃内发生的脂肪酸生物氢化作用，都有助于氢气的平衡，前者导致氢气的净消耗或净产生，而后者导致 H_2 的净消耗。

由于牧草是奶牛日粮的主要成分，在当地生产高质量的牧草对奶农维持和确保盈利至关重要。在农场层面选择适当的减少温室气体的营养策略在很大程度上取决于农场的地理位置，其决定了当地的气候条件和用于青贮的饲料植物种类。例如，依据 Bernardes 等（2018）的综述，高温会影响青贮作物的产量和营养价值，而在寒冷地区，短而凉爽的生长季节可能会限制玉米等对低温敏感的作物的使用。这在北欧等北部地区经常发生，在这些地区，青贮饲料的生产主要是使用牧草而不是玉米。Bernardes 等（2018）认为，气候条件影响青贮生产的各个阶段，其中温度是最大的限制因素。青贮饲料生产既取决于可控

因素，例如植物种类和成熟阶段、收获和青贮方法以及添加剂的使用，也取决于不可控的气候相关因素，这使得必须考虑青贮发酵的品质和营养价值的年变化。

以精料为基础的方案包括了增加日粮中精料的比例和/或改变碳水化合物的类型（如纤维 vs 淀粉）或脂质类型（如脂肪酸的组成）。尽管日粮中的精料成分，如谷物和油籽，相对于牧草，对营养价值的年度变化不那么敏感，它们对奶牛日粮的贡献很大程度上取决于牧草的干物质（DM）摄入量和营养成分的消化率，例如，粗纤维。Hristov 等（2013）研究发现，日粮中添加精料可以降低甲烷排放强度，特别是当添加量超过日粮干物质的 40%，瘤胃功能未受到损害。另外，增加奶牛日粮中的精粗比与减少使用人类可食用日粮（如谷物）的举措相矛盾。可持续的奶牛饲养策略应该充分利用反刍动物的独特能力，将人类不可食用的生物质转化为高质量的动物源性蛋白质食品，即牛奶和肉类。

在这一章中，作者重点回顾了最近使用青贮和/或精料进行的奶牛生理或牛奶生产相关的研究文献，包括对甲烷产生的监测。本章以欧洲和北美为重点，综述了温带地区奶牛日粮对生产参数、饲料转化率（FE）、氮素利用率（NUE）和甲烷排放强度的影响。作者评估了植物种类、青贮作物的成熟阶段以及日粮精粗比和精料组成等因素在不影响动物生产性能的前提下减少奶牛生产对环境的影响。

2 青贮的作用：禾本牧草、豆科牧草和玉米

2.1 青贮植物的种类

温带地区的气候条件差别很大，因此用于制作青贮饲料的牧草品种也有很大差异，主要的植物种类包括牧草、豆科牧草和玉米。在北部地区，短而凉爽的生长季节加上寒冷的冬季，限制了用于青贮的多年生牧草和豆科植物的选择（Bernardes 等，2018）。使用最广泛的多年生牧草品种有梯牧草（*Phelum pratense*）、多年生黑麦草（*Lolium perenne* L.）和各种羊茅属牧草，如草甸羊茅（*Festuca pratensis*）和高羊茅（*Festuca arundinacae* L.），以及豆科植物，如红三叶草（*Trifolium pratense*）和苜蓿（*Medicago sativa* L.）（Wilkinson 和 Rinne，2018；Bernardes 等，2018）。尽管紫花苜蓿的生产仅限于该地区的南部，但气候变化可能会使其在未来可以在更北的区域生产，例如斯堪的纳维亚半岛（Järvenranta 等，2016）。

玉米（*Zea mays* L.）是一种原产热带的作物，但作为一种有价值的饲料作物，从温带地区到热带地区，它被用于任何玉米可以生长的地方。玉米的特点是每公顷可以得到高产量的低成本淀粉和相对高浓度的代谢能，这使得玉米对农民非常有吸引力（Wilkinson 和 Rinne，2018）。植物育种家已经培育出早熟的玉米品种，这些品种可用于在北方条件下制作全株青贮。因此，用于制作青贮饲料的玉米种植的面积正在逐渐向北方扩展，气候变化可能也促成了这一转变。在北欧的边缘地区，如斯堪的纳维亚国家，在耕作中使用可生物降解的薄膜也可以提早播种和收获玉米。

冷季牧草的特点是由于低温和长日照延迟了细胞壁的木质化，使其具有高消化率（Huhtanen 等，2013；Bernardes 等，2018）。由于豆科植物因早熟而导致的消化率下降速

度比其他牧草慢，混合这些植物品种用于制作青贮饲料能够延长这些牧草的最佳收获期（Kuoppala，2010）。然而，这些牧草的非结构性碳水化合物（NSC），包括水溶性碳水化合物（WSC）和淀粉的含量往往较低，且随气候条件变化很大，会影响牧草和豆科牧草的青贮潜力和饲料价值，但可以通过改变收获时间和使用添加剂等青贮方法进行调控（Vanhatalo 和 Jaakkola，2016）。也有富集水溶性碳水化合物的高糖黑麦草品种可用于青贮（Moorby 等，2006）。

2.2 牧草青贮

在牧草青贮管理因素中，改变牧草种类和收获时间是降低甲烷排放最具潜力的措施，而施氮量、使用添加剂或种植高糖黑麦草对甲烷排放没有影响（表1）。

在高比例奶牛粗日粮中饲喂混合青贮黑麦草与梯牧草（Warner 等，2016，2017）、青贮黑麦草（Brask 等，2013a）或青贮梯牧草（Pang 等，2018），显著提高了干物质摄入量、能量校正乳（ECM）和饲料转化率（FE），并降低了高达20%的甲烷排放强度（以每千克能量校正乳排放的每克甲烷计）（Warner 等，2016，2017）。而且，通过改善青贮饲料品质而降低甲烷的排放量与干物质采食量无关，且在泌乳后期比早期更小（Warner 等，2017）。这不是由于乙酸与丙酸的比例造成的，因为牧草的成熟度没有变化（Brask 等，2013a；Warner 等，2016）。然而，这些积极的结果是以降低氮利用效率为代价取得的，当使用极早刈割的叶期牧草青贮时，氮素利用效率（NUE）显著降低（-35%）（Warner 等，2017），反映出叶期粗蛋白质（CP）含量远高于抽穗期牧草。高施氮量的牧草（150 kg N/hm² vs 65 kg N/hm²）也与降低氮素利用率有关，这是因为高施氮量的牧草青贮中粗蛋白质含量提高了5%（Warner 等，2016）。

早熟青贮饲草对奶牛干物质摄入量和产奶量的积极作用已被证实（Rinne，2000；Harrison 等，2003），早熟青贮饲料中的高氮含量也会导致动物氮的损失（如 Rinne 等，1997）。因此，不推荐使用如 Warner 等（2017）研究中使用的过早收获的牧草。然而，使用由早期刈割的初级生长的牧草制成的青贮饲料可以提高饲料转化率，降低甲烷排放强度，并确保高产奶牛良好的产奶量，尽管在氮利用率方面有所降低。在早熟阶段收获的牧草会导致甲烷排放的减少似乎与瘤胃发酵模式无关，因为早熟牧草青贮对丙酸浓度的影响很小且不一致（Harrison 等，2003；Warner 等，2016）。早熟牧草青贮中高浓度的硝酸盐含量或高能量有利于微生物的生长（Knapp 等，2014；Warner 等，2016），从而作为丙酸产生的氢替代库，进而降低甲烷排放。高糖黑麦草青贮提高了全青贮日粮的氮素利用率（表1；Staerfl 等，2012），这与在肉牛生产中研究结果一致（Merry 等，2006），但降低了产奶参数，对甲烷排放强度的影响较小。Bertilsson 等（2017）认为，高糖黑麦草青贮中水溶性碳水化合物（WSC）水平的提高是以粗蛋白质和纤维的损失为代价的，但它们对奶牛生产性能的影响很小。然而，积极的结果归因于更有利的氮分配，即更多的氮进入牛奶和粪便，而形成尿液的氮较少。综上所述，改善牧草青贮氮素利用率低的问题值得进一步研究。尽管早期青贮对产奶量产生了积极的影响（Muck 等，2018），但 Ellis 等（2016）的一项研究并未发现这种结果，也没有发现牧草青贮可以降低甲烷排放强度（表1）。

第19章 优化瘤胃功能：青贮饲料和精料提高奶牛饲料转化率、减少甲烷和氮排放的作用

表1 不同成熟度、青贮方式或不同植物种类的日粮替代基础日粮对奶牛生产性能的影响

基础日粮[2]	植物种类	替代日粮	F:C[3]	相对于对照组的变化（%） DMI	ECM	FE	NUE	瘤胃 C2/C3	甲烷排放强度（g CH₄/kg ECM） 对照组	试验组	变化（%）	参考文献
牧草青贮												
后期刈割	黑麦草[4]	早期刈割，一茬	65:35	9	11	2	-7	-2	16.9	14.7	-13	Brask 等（2013a）
后期刈割	梯牧草	早期刈割，一茬	60:40	10	13	3	-17	NR	14.0	13.1	-7	Pang 等（2018）
后期刈割，低氮	黑麦草：梯牧草	早期刈割，二茬	80:20	6	32	25	-13	18	15.9	12.5	-21	Warner 等（2016）
后期刈割，高氮		早期刈割，二茬		20	34	12	-8	-1	16.3	13.1	-20	
低氮		高氮		-4	-2	2	-19	-4	15.2	14.9	-2	
后期刈割[5]	黑麦草：梯牧草	极早刈割叶	80:20	9	12	6	-35	NR	14.0	11.2	-20	Warner 等（2017）
后期刈割[6]		一茬[7]		4	11	8	-35	NR	12.9	10.2	-21	
无添加	黑麦草：梯牧草	重复播种的牧草	75:25	3	2	-1	4	NR	16.0	16.0	0	Ellis 等（2016）
低糖	黑麦草	高糖	100:0	-8	-15	-8	46	NR	16.5	17.2	4	Staerfl 等（2012）
豆科青贮												
禾本科牧草	梯牧草	紫花苜蓿	60:40	9	-2	-10	-15	-1	12.8	13.4	5	Hassanat 等（2014）

（续表）

基础日粮[2]	植物种类	替代日粮	相对于对照组的变化（%）					甲烷排放强度（g CH$_4$/kg ECM）			参考文献
			F:C[3]	DMI	ECM	FE	NUE	瘤胃 C2/C3	对照组	试验组	变化(%)	
红三叶草:牧草 30:70	NR:梯牧草	红三叶草:牧草 70:30	60:40	-1	0	1	-10	NR	15.2	15.5	5	Gidlund 等（2017）
牧草:红豆草:玉米 86:0:14	NR	牧草:红豆草:玉米 42:42:16	70:30	5	7	2	-5	NR	15.0	13.9	-7	Huyen 等（2016）
玉米青贮	LG30218	晚期刈割玉米	80:20	0	0	0	-3	13	12.8	11.9	-7	Hatew 等（2016）
早期刈割玉米	NR	褐色中脉玉米	65:35	6	8	1	5	-4	14.0	12.6	-10	Hassanat 等（2017）
常规玉米	未成熟时刈割	玉米乳线处于2/3的玉米	60:40	20	15	-4	6	-13	15.0	14.9	-1	Benchaar 等（2014）
大麦	NR	玉米	80:20	8	7	-1	18	6	16.6	15.0	-10	Van Gastelen 等（2015）
禾本科牧草	黑麦草[4]	玉米	65:35	-1	0	2	24	-22	14.7	13.8	-6	Brask 等（2013a）
早期刈割牧草				7	11	4	15	-23	16.9	13.8	-19	
晚期刈割牧草	NR	牧草:玉米 25:75	50:50	11	5	-5	-7	NR	15.0	14.3	-5	Reynolds 等（2010）
牧草:玉米 75:25												

第19章 优化瘤胃功能：青贮饲料和精料提高奶牛饲料转化率、减少甲烷和氮排放的作用

（续表）

基础日粮[2]	植物种类	替代日粮	F∶C[3]	相对于对照组的变化（%）					甲烷排放强度（g CH$_4$/kg ECM）			参考文献
				DMI	ECM	FE	NUE	瘤胃 C2/C3	对照组	试验组	变化（%）	
牧草∶玉米 75∶25[8]	第三茬黑麦草	牧草∶玉米 25∶75	50∶50	28	9	-15	-10	NR	16.3	14.2	-13	Hammond 等 (2016)
与上述日粮相同				19	24	4	15	NR	16.9	16.2	-4	
牧草∶玉米 75∶25	NR	牧草∶玉米 25∶75	50∶50	7	-2	-8	1	NR	12.9	12.0	7	Livingstone 等 (2015)
禾本科牧草	黑麦草∶鸭茅	处于玻璃状阶段的玉米	45∶55	0	-2	-2	6	-5	14.9	13.4	-10	Doreau 等 (2014)
红三叶草	NR	玉米	60∶40	0	2	2	7	-16	14.6	14.1	-4	Benchaar 等 (2015)
紫花苜蓿	NR	玉米	60∶40	5	-1	-6	14	-26	13.9	14.4	3	Hassanat 等 (2013)
紫花苜蓿 80∶20	NR	紫花苜蓿∶玉米 55∶45	55∶45	-1	0	1	15	-18	17.8	18.1	2	Arndt 等 (2015)

注：[1]干物质采食量（DMI）和能量校正乳（ECM）的计算参考 Sjaunja 等（1991）的报道。饲料转化率（FE）根据能量校正乳/干物质采食量计算，氮素利用效率根据奶中氮的总含量/总氮摄入量计算，瘤胃乙酸/丙酸比根据其在瘤胃液中的摩尔含量计算，NR一代表未报道目不可估算。[2]混合日粮中各成分的比例为干物质基础。[3]精粗比（F∶C）为干物质基础。[4]三叶草含量<10%。[5]泌乳96 d。[6]泌乳218 d。[7]处理组由处于叶期中期的牧草和5%切碎的小麦秸秆组成。[8]甲烷排放用 GreedFeed 系统测量。[9]甲烷排放用呼吸室测量。

2.3 豆科牧草青贮

关于豆科牧草青贮对奶牛生产中甲烷排放强度影响的研究较少（表1）。在奶牛日粮中，用紫花苜蓿替代梯牧草在减少甲烷排放方面没有效果，但会增加干物质摄入量，饲料转化率下降，尤其是氮素利用效率降低（Hassanat 等，2014；表1）。在以牧草青贮为基础的青贮饲料中添加含缩合单宁的红豆苷（*Onobrychis viciciifolia*）增加了干物质采食量和能量校正乳的产量，引起甲烷排放强度轻微下降（Huyen 等，2016；表1）。用 70:30 的红三叶草和梯牧草混合青贮饲料替代 70:30 的梯牧草和红三叶草混合青贮饲料，没有影响奶牛的干物质摄入量、能量校正乳的产量、饲料转化率或甲烷排放强度，但导致了氮素利用效率的降低（Gidlund 等，2017；表1），这与 Van Dorland 等（2007）的研究结果一致，用红三叶草或白三叶草青贮替代部分黑麦草青贮并不影响甲烷排放，但略微增加了氮素向环境中的排放。即便如此，根据 Phelan 等（2015）的综述，与牧草相比，豆科牧草通常会降低动物生产每千克牛奶或肉类的甲烷排放量。然而，这只发生在豆科牧草有更高的采食量和过瘤胃率或豆科含有复合单宁时。

比较豆科牧草和其他牧草青贮日粮的结果表明，豆科牧草的干物质采食量和产奶潜力优于普通牧草（Vanhatalo 和 Jaakkola，2016）。而且，豆科牧草通常被认为是一种具有一定经济效益，并可以替代牧草和/或玉米为基础的日粮，因为它们具有生物固氮能力，是减少对合成氮肥和化石能源依赖的有效手段（Vanhatalo 和 Jaakkola，2016）。尽管豆科牧草的消化率较低，但豆科牧草的采食量高于牧草青贮，这是因为豆科牧草的纤维含量较低，在瘤胃内可以快速发酵和降解，并且过瘤胃速率较高（Kuoppala 等，2009；Kuoppala，2010；Dewhurst，2013）。

豆科牧草减缓甲烷排放的不同结果可能与青贮发酵质量和青贮饲料中豆科牧草的使用比例有关。需要指出的是，豆科牧草通常与其他牧草或植物混合生长，而不是单独生长，因为混合生长的牧草年产量更高（Phelan 等，2015）。以豆科牧草为基础的日粮与以牧草青贮为基础的日粮相比，降低了氮素利用率，这是由于其固有的高粗蛋白质含量，特别是紫花苜蓿。然而，Dewhurst（2013）指出，红三叶草和紫花苜蓿等豆科牧草的氮组分也存在差异，这可能会不同程度地影响氮素利用率。显然，饲喂豆科牧草降低甲烷排放强度的潜力及其对氮素利用率的影响还有待进一步研究。

2.4 青贮玉米

Hatew 等（2016）在收获时将玉米的成熟度从早期（20% DM）推迟到后期（40% DM），有效地降低了高粗日粮中甲烷的排放强度，但不影响干物质摄入量、能量校正乳的产量、饲料转化率或氮素利用率（表1）。这是由于随着玉米的成熟，淀粉含量显著增加，瘤胃淀粉降解率和中性洗涤纤维（NDF）含量降低所致。然而，尽管淀粉采食量增加，但这并非如预期的那样可归因于瘤胃 pH 值的降低和丙酸的增加。相反，随着玉米成熟度的增加，乙酸与丙酸的比例有增加的趋势。这些结果表明，在较高成熟度时收获全株玉米，而不是采用目前推荐的做法（30%~35%，Khan 等，2015）具有减少肠道甲烷排放的潜力。

Jung 等（2011）比较了具有更高细胞壁消化率和采食特性的玉米品种与传统玉米品种的差异（Hassanat 等，2017；表 1）。结果表明，用较易消化的褐色中脉玉米青贮（BMCS，DM：34%，淀粉：283 g/kg DM）替代常规玉米青贮（DM：40%，淀粉：269 g/kg DM），不仅增加了干物质采食量和能量校正乳的产量，而且提高了氮素利用率，降低了甲烷的排放强度。同样，甲烷的减少与瘤胃发酵模式无关，因为各处理之间没有变化。此外，通过使用褐色中脉玉米青贮（BMCS），粪氮排泄量减少，通过将氮从尿液转移到粪氮减少了氮的挥发。然而，粪便中挥发性固体含量（如可降解有机排泄物）的增加会导致粪便储存过程中甲烷排放的增加。然而，玉米青贮类型（Falkone vs LG30224）对奶牛生产性能和甲烷排放量的影响不大，尽管在 Falkone 的玉米青贮日粮中玉米占比为 65%，其中的瘤胃中性洗涤纤维（NDF）消化率较低，但淀粉含量较高（De Boever 等，2017）。

用玉米青贮（DM：31%，淀粉：322 g/kg DM）完全替代大麦青贮（DM：32%，淀粉：139 g/kg DM）显著提高了干物质采食量、产奶量和氮素利用率，但对饲料转化率和甲烷排放强度没有影响（Benchaar 等，2014；表 1）。然而，以玉米青贮代替大麦青贮降低了甲烷能量的损失，并降低了瘤胃乙酸/丙酸比。随着日粮中玉米青贮量的增加，氮素利用率的提高是由于尿氮损失的减少，表明粪便中一氧化二氮（N_2O）和氨的排放较低。

2.5 玉米青贮与牧草和豆科牧草青贮的比较

由于玉米固有的高能量和低粗蛋白质的特征，研究在日粮中用玉米青贮饲料混合或替代粗蛋白质含量高的牧草或豆科牧草是否会增加氮素利用率和降低甲烷排放强度是非常有意义的。用玉米青贮饲料（DM：32%，淀粉：322 g/kg DM）代替牧草青贮饲料在限制性高粗日粮中改善了氮素利用率并降低了甲烷排放强度，但不影响饲料转化率或产奶量，增加了牛奶蛋白的产量（Van Gastelen 等，2015；表 1）。甲烷的减少与乙酸与丙酸的比例无关，因为其在不同处理之间没有变化。用玉米青贮饲料（DM：31%，淀粉：150g/kg DM）代替高粗日粮中早割或晚割的青贮牧草对干物质摄入量或产奶量没有影响，但氮素利用率得到改善，特别在早割牧草青贮中结果更加显著。甲烷排放强度的降低与晚割牧草青贮有关（Brask 等，2013a；表 1）。在这些日粮条件下，甲烷的减少归因于瘤胃中乙酸/丙酸比的显著降低以及玉米青贮饲料的低瘤胃纤维消化率。

在 50：50 精粗比的日粮中，将牧草和玉米混合青贮饲料中玉米的比例从 25% 提高到 75%，可增加干物质采食量和产奶量，降低甲烷排放强度，但对饲料转化率和氮素利用率的影响在不同试验中是不一致的（Reynolds 等，2010；Hammond 等，2016；表 1）。然而，在类似的试验设置下，Livingstone 等（2015，表 1）没有发现这些青贮处理在上述任何参数上的差异，因为牧草的中性洗涤纤维含量极低。在低粗日粮条件下，Doreau 等（2014）（表 1）除了在玉米青贮日粮发现较低的甲烷排放强度外，其他参数均无差异。但各处理间瘤胃乙酸/丙酸比例仍然没有变化。

用玉米青贮替代红三叶草和紫花苜蓿青贮后对干物质摄入量、生产性能和甲烷排放强度的影响较小，但明显改善了氮素利用率（Hassanat 等，2013；Benchaar 等，2015；Arndt 等，2015；表 1）。在这些研究中，玉米青贮饲料的成熟度范围为 36%~38%DM 和淀粉含量的范围为 290~339 g/kg DM，豆科牧草的淀粉含量低于 18 g/kg DM（Hassanat 等，2013；Benchaar 等，2015）。在日粮中用玉米青贮代替红三叶草和紫花苜蓿后，尽管不影响甲烷排放强度，但瘤胃 pH 值和瘤胃乙酸/丙酸比显著降低（Hassanat 等，2013；Benchaar 等，2015；Arndt 等，2015）。青贮玉米替代豆科牧草对降低尿氮和粪氮的积极变化可能导致氨和 N_2O 排放的降低。但是，淀粉含量较高的玉米青贮减少了瘤胃中的纤维消化率，可能导致粪便储存过程中甲烷排放增加（Hassanat 等，2013）。

在高牧草青贮为基础的日粮中用玉米青贮替代牧草或豆科牧草青贮能持续带来环境效益，如降低甲烷排放强度，并改善氮素利用率，特别是在粗蛋白质含量较高的豆科牧草青贮中，且不会影响牛奶产量。然而，尽管高牧草青贮为基础的日粮中淀粉含量随玉米青贮比例的增加而增加，但甲烷排放强度的降低并不如预期的那样与 pH 值和乙酸/丙酸比降低的瘤胃发酵模式有关。例如，在高粗日粮中，用玉米青贮替代牧草青贮维持了较高的瘤胃 pH 值，并提高了瘤胃丁酸产量（Van Gastelen 等，2015）。因此，在高粗日粮中，降低瘤胃 pH 值并不一定会提高丙酸产量（Dijkstra 等，2011）。由于玉米淀粉对瘤胃发酵具有相当的抗性，而易在小肠中被酶消化（Owens 等，1986）。因此，有一种可能是从乙酸转变为丁酸，有利于减少瘤胃中氢气的产量（Moss 等，2000），氢气产量的减少归因于淀粉的消化转移到了后肠道，从而减少了甲烷的排放。然而，改变瘤胃甲烷生成和减少甲烷产量可能需要特定的日粮淀粉浓度（Hassanat 等，2013；Van Gastelen 等，2015）。

有趣的是，玉米青贮日粮代替豆科牧草日粮，明显降低了乙酸/丙酸的比例（Hassanat 等，2013；Benchaar 等，2015；Arndt 等，2015），但没有导致甲烷排放强度的降低。这可能与日粮中性洗涤纤维中碳水化合物的类型有关，其会影响甲烷的排放（Arndt 等，2015）。玉米中性洗涤纤维发酵产生的甲烷比紫花苜蓿中性洗涤纤维发酵产生的甲烷要多。因此，每克紫花苜蓿中性洗涤纤维发酵产生的甲烷排放量减少，抵消了由于紫花苜蓿中性洗涤纤维含量高而导致的高甲烷排放。Hassanat 等（2013）也发现了玉米和紫花苜蓿在中性洗涤纤维发酵方面也存在类似的差异。此外，Brask 等（2013a）的研究表明，尽管日粮中性洗涤纤维含量相当，但玉米青贮日粮的瘤胃中性洗涤纤维消化率较牧草青贮日粮低。日粮碳水化合物类型对甲烷排放和瘤胃消化动力学的影响有待进一步研究。研究应同时检测瘤胃发酵、消化动力学和微生物群，并测定温室气体和产奶量，以深入探讨影响瘤胃功能和温室气体形成的机制。随着日粮中玉米比例的增加，氮损失减少，这表明日粮从豆科牧草转向玉米青贮，导致粪便氨气和 N_2O 排放量的降低（Hassanat 等，2013；Arndt 等，2015）。

然而，应该强调的是，关于饲养策略减少温室气体排放潜力的结论还取决于该策略在农场是否可行，而不仅是在单个动物的水平上（Van Middelaar 等，2013）。在单个动物水平上，以玉米青贮代替奶牛日粮中的牧草青贮是一种快速高效减少温室气体排放的策略。然而，将这一策略应用于集约化的荷兰农场，需要减少草地面积，会面临与欧盟法规相冲突的问题。另外，将该策略应用于集约化农业以减少草地面积，由于土地利用方式的变化，即将草地改为玉米地，可能会导致更高的温室气体排放。

3 精料的作用：脂质、碳水化合物和蛋白质

3.1 脂质

脂质补充水平。脂质添加被广泛用于增加日粮能量密度，以满足高产奶牛在泌乳早期和中期的能量需要，并提高产奶的能量利用率。然而，日粮脂肪含量不应超过干物质的6%~7%（Beauchemin 等，2008），否则可能会抑制干物质采食量和瘤胃纤维消化率，甚至进一步影响产奶量，从而抵消增加日粮能量密度的优势（Bayat 等，2017；Halmemies-Beauchet-Filleau 等，2017）。脂质也是工业化国家改善牛奶或肉类脂肪酸组成、减少反刍动物温室气体排放的最有效和切实可行的手段之一，但其有效性取决于多种因素，包括脂质添加水平、脂肪酸组成（如链的长度和不饱和水平）、脂质供给的形式（如油 vs 种子）和基础日粮的类型（Eugène 等，2008；Beauchemin 等，2008；Shingfield 等，2013；表2）。为了保障人类的长期健康，其目标是减少饱和脂肪酸的比例，增加顺式单不饱和脂肪酸和 ω-3 脂肪酸的比例，改善反刍动物产品中 ω-6 和 ω-3 脂肪酸的平衡（Shingfield 等，2013）。

在广泛的饮食条件下，Martin 等（2010）报告称，在日粮干物质上每添加 1% 的脂质，甲烷排放量平均减少 3.8%。一种或多种机制可能有助于提高不同脂质在瘤胃中降低甲烷的潜力。这些因素包括：瘤胃中发酵的有机物（OM）含量较低（脂质降低了干物质采食量和/或替代了反刍动物日粮中瘤胃可发酵的组分），对瘤胃纤维素降解菌、产甲烷菌和/或附着原虫具有直接毒性或抑制作用。在不饱和脂肪补充的情况下，瘤胃发酵从乙酸转向丙酸（消耗而不是产生氢气）和脂肪酸的生物氢化过程（Martin 等，2010）。

中链饱和脂肪酸。中链饱和脂肪酸如肉豆蔻酸（C14:0）或富含月桂酸（C12:0）的椰子油，在日粮干物质基础上添加量为 3.3%~5% 时，可使奶牛瘤胃甲烷排放强度最高降低 30%。其主要机制可能是降低了日粮干物质采食量（表2）和瘤胃可发酵有机物（Bayat 等，2018）。中链饱和脂肪酸也可能对产甲烷菌（Beauchemin 等，2008）或原虫（Hristov 等，2011b）表现出毒性作用，并损害纤维消化能力（Hollmann 等，2012），但这些作用在不同研究之间并不一致。Bayat 等（2018）报道，尽管瘤胃甲烷产量显著下降，但特定微生物类群的多样性只有轻微改变，对细菌总数、产甲烷菌或纤毛虫原生动物或纤维消化没有影响。在日粮中加入中链饱和脂肪酸通常会改善奶牛生产中的饲料转化率和氮素利用率，但伴随着能量校正乳的产量急剧下降（表2），另外，这些脂质补充剂相对较高的价格使得在商业奶牛场中应用的可能性不大。此外，日粮中添加 C12:0 和 C14:0 增加了其在乳脂中的含量（Odongo 等，2007；Hollmann 等，2012；Bayat 等，2018），这对消费者来说是一种不受欢迎的营养变化。

不饱和脂肪酸。植物不饱和脂肪酸，如油酸（C18:1 n-9）和必需脂肪酸亚油酸（C18:2 n-6）和 α-亚麻酸（C18:3 n-3）被认为有益人体健康。因此，它们在奶牛日粮中的使用通常会使反刍动物牛奶和肉类中 ω-脂肪酸增加和饱和脂肪酸减少，这可能是一种缓解甲烷排放的可行方法。

表 2 添加脂肪对奶牛生产性能的影响

脂质来源	形式	日粮 DM 中脂质的含量	基础日粮组成	F:C³	相对于未添加的对照组日粮的变化（%） DMI	ECM	FE	NUE	瘤胃 C2/C3	甲烷排放强度 g CH₄/kg ECM 对照组日粮	脂质组日粮	变化	来源
中链饱和脂肪酸													
豆蔻酸	油	5%	玉米青贮:半干草料 55:35:10	60:40	-7	-10	-4	-1	NR	28.4	20.4	-28	Odongo 等 (2007)
豆蔻酸	甲酯	5%	禾本科牧草青贮	60:40	-31	-20	17	2	-6	22.7	18.8	-17	Bayat 等 (2018)
椰子	油	1.3%	玉米:紫花苜蓿:禾本科牧草青贮 75:15:10	50:50	-7	4	12	9	NR	13.9	12.9	-7	Hollmann 等 (2012)
		2.7%			-22	-18	4	11	NR	13.9	9.9	-29	
		3.3%			-29	-24	8	22	NR	13.9	9.9	-29	
单一不饱和脂肪酸													
油菜籽	饼粕	2%~3%	牧草:玉米青贮 55:45	50:50	3	11	8	8	-1	14.6	13.6	-7	Brask 等 (2013b)
	粉碎				-2	-8	-6	4	-1	14.6	12.1	-17	
	油				-14	4	21	6	7	14.6	12.0	-18	
油菜籽	粉碎	3%	玉米青贮	65:35	-5	-10	-6	7	-2	13.7	14.0	3	Brask 等 (2013a)
			早割牧草青贮		-2	-1	1	0	-1	15.6	13.9	-11	
			晚割牧草青贮		1	3	3	3	1	17.8	16.1	-10	
油菜籽	粉碎	2%	玉米:牧草青贮 75:25	50:50	2	4	1	4	NR	14.7	14.2	-3	Kliem 等 (2019)
油菜籽	油	5%	牧草青贮	60:40	-12	3	17	11	3	22.7	17.5	-23	Bayat 等 (2018)

第 19 章 优化瘤胃功能：青贮饲料和精料提高奶牛饲料转化率、减少甲烷和氮排放的作用

（续表）

脂质来源	形式	日粮 DM 中脂质的含量	基础日粮组成	F : C³	相对于未添加的对照组日粮的变化（%）					甲烷排放强度 g CH₄/kg ECM			来源
					DMI	ECM	FE	NUE	瘤胃 C2/C3	对照组日粮	脂质组日粮	变化	
多不饱和脂肪酸													
红花籽	油	5%	牧草青贮	60 : 40	-6	2	9	6	0	22.7	17.5	-23	Bayat 等 (2018)
大豆	油	3.5%	玉米青贮：半干紫花苜蓿干草 45 : 40 : 15	65 : 35	-3	0	4	NR	NR	18.3	18.2	0	Sauer 等 (1998)
葵花籽	油	5%	牧草青贮	65 : 35	-2	-2	0	-6	-5	18.9	14.5	-23	Bayat 等 (2017)
				35 : 65	-11	-16	-6	13	3	14.2	14.5	2	
亚麻籽	油	5%	牧草青贮	50 : 50	-12	-16	-5	6	-5	15.4	13.0	-16	Bayat 等 (2015)
亚麻籽	完整籽实	5%	玉米青贮：干草 90 : 10	65 : 35	-2	-1	0	-6	NR	17.7	15.9	-10	Martin 等 (2008)
	压片				-16	-16	-1	4	NR	17.7	13.1	-26	
	油				-26	-26	0	12	NR	17.7	8.5	-52	
亚麻籽	压片	1.8%	玉米青贮：干草 90 : 10	60 : 40	-2	-15	-14	-2	-9	15.4	17.1	11	Ferlay 等 (2013)
		3.6%			-5	-16	-12	-3	-14	15.4	15.9	4	
		5.4%			-11	-5	7	12	-27	15.4	9.4	-39	
		1.8%	干草	50 : 50	-9	-4	6	5	-6	19.8	17.4	-12	Martin 等 (2016)
		3.6%			-4	5	10	8	-6	19.8	15.3	-23	
		5.4%			-4	-3	1	12	-12	19.8	12.2	-39	

（续表）

脂质来源	形式	日粮DM中脂质的含量	基础日粮组成	F：C³	相对于未添加的对照组日粮的变化（%）					甲烷排放强度 g CH₄/kg ECM			来源
					DMI	ECM	FE	NUE	瘤胃C2/C3	对照组日粮	脂质组日粮	变化	
亚麻籽	压片	2%	玉米：牧草青贮 75：25	50：50	0	2	2	4	NR	14.7	113.4	-9	Kliem 等（2019）
亚麻籽和棕榈混合	油混钙盐	2%		50：50	-2	3	6	1	NR	14.7	12.8	-13	
亚麻籽	油	4%	玉米青贮	60：40	-9	-14	-6	1	-22	14.1	12.1	-14	Benchaar 等（2015）
			红三叶草青贮	60：40	-2	2	4	5	-4	14.6	13.0	-11	
亚麻籽	油	5%	牧草青贮	60：40	-8	3	12	5	2	22.7	17.5	-23	Bayat 等（2018）
鱼	油	0.8%	玉米青贮：干紫花苜蓿：干草 55：25：20	52：48⁵	4	8	4	-7	-1	13.5	12.6	-7	Pirondini 等（2015）
		0.8%		52：48⁶	-2	-3	-1	1	0	12.4	13.2	6	
		0.3%	干紫花苜蓿	74：26	0	-6	-6	3	2	21.8	23.9	10	Moate 等（2013）
富含二十二碳六烯酸的海藻	粉	0.6%			-6	-14	-8	5	-1	21.8	25.7	18	
		1%			-11	-15	-5	0	4	21.8	24.1	11	
		0.3%⁷	玉米：牧草青贮 70：30	70：30	0	-12	-12	0	NR	9.5	11.1	16	Klop 等（2016）

注：¹干物质采食量（DMI），能量校正乳（ECM）根据 Sjaunja 等（1991）计算，饲料转化率（FE）根据能量校正乳（ECM）/干物质（DM）采食量计算，氮素利用率（NUE）根据奶中氮的总含量/总氮摄入量计算。²瘤胃乙酸与丙酸比（瘤胃C2/C3）根据其在瘤胃液中的摩尔比计算。³精粗比（F：C）。⁴干物质基础。⁵干物质基础上含有<10%的三叶草。⁶精料中淀粉含量低。⁷精料中淀粉含量高。⁸日粮干物质基础上补充DHA（%）。

第 19 章　优化瘤胃功能：青贮饲料和精料提高奶牛饲料转化率、减少甲烷和氮排放的作用

针对泌乳奶牛的试验表明，脂类对动物生产性能和瘤胃甲烷生成的影响与脂类添加水平和不饱和程度成正比（表 2）。至于饱和脂肪酸，减少甲烷排放的主要机制可能是降低了干物质（DM）采食量（表 2）。Martin 等（2016）开展的一项剂量效应试验表明，在 3 个水平中添加富含 C18:3 n-3 的亚麻籽，使日粮中脂肪的最高含量为干物质基础的 5.4%，最高可降低 39% 的瘤胃甲烷排放强度，但由于基础日粮组成的差异，较低水平脂肪添加的作用效果存在差异。除干物质采食量降低外，瘤胃乙酸/丙酸比例和原虫数量均降低，而瘤胃产甲烷菌数量和纤维消化率未发生变化。适量的脂质（日粮干物质的 1%~2%）并不会抑制采食量，但会改变乳中脂肪酸组成（Halmemies-Beauchet-Filleau 等，2011）。与富含 C18:3 n-3 的亚麻籽可降低 39%~52% 的甲烷排放相比，在日粮干物质中脂质含量为 5% 的情况下，富含 C18:1 n-9 的油菜籽和富含 C18:2n-6 的红花籽、葵花籽和亚麻荠籽导致甲烷排放强度降低幅度较低（最高 23%）（表 2）。在表 2 所示的大多数研究中，试验期为 4~6 周，但最近 Alstrup 等（2015）证实，植物脂类在整个泌乳过程中抑制了瘤胃甲烷生成。然而，需要更多的全哺乳期研究来证实脂质减少瘤胃甲烷排放的持久性。

一般认为，完整的油菜籽在一定程度上保护了油脂不受微生物代谢的影响，并且可能会限制油脂对瘤胃微生物和营养物质消化率的影响。然而，Martin 等（2008）报告称，完整亚麻籽、挤压亚麻籽和亚麻籽油日粮在有机物和纤维消化率方面没有差异。虽然纯油通常能更有效地减少了瘤胃中甲烷的生成，但加工油籽（如辗轧或压饼）更受青睐，因为其对干物质采食量的不利影响较小和通常具有更低的价格（Beauchemin 等，2008；表 2）。此外，作为全混合日粮（TMR）的一部分，与单独饲喂草料的精料相比，给予高水平的不饱和脂肪酸会导致干物质采食量的下降（Bayat 等，2015；Halmemies-Beauchet-Filleau 等，2017）。这可能是由于相对于全混合日粮（TMR），在单独日粮中瘤胃中不饱和脂肪酸的释放更为迅速，导致大量的游离不饱和脂肪酸对纤维素分解细菌产生毒性（Maia 等，2007）。

普通日粮对于生产性能和甲烷排放的影响与脂质密切相关。在纤维含量较高的 50% 的粗料日粮或干物质含量更高的牧草青贮日粮中，如红三叶草青贮饲料或牧草干草，干物质基础上高达 5% 的不饱和脂肪酸，维持了能量校正乳的产量和提高了饲料转化率。此外，瘤胃甲烷排放的降低呈剂量依赖性（表 2）。相比之下，在淀粉日粮（以玉米青贮或富含淀粉为基础的日粮）中，低脂水平（占日粮干物质的 2%~4%）往往会影响能量校正乳的产量和饲料转化率，但是不同研究对瘤胃甲烷排放影响的报道各异（表 2）。在高脂水平（占日粮干物质的 4%~5.5%）下，瘤胃甲烷排放强度通常是降低的，但通常也伴随着能量校正乳的产量降低（表 2）。因此，以不饱和脂肪补充剂作为添加剂来减少甲烷排放更适合于富含纤维的日粮，但由于对能量校正乳产量的负面影响，在淀粉日粮中使用不饱和脂肪添加剂的效果有限。这可能与相对于高纤维日粮而言，高淀粉日粮中不饱和脂肪对瘤胃纤维消化的不利影响有关，导致瘤胃中乙酸的形成减少，进而降低乳脂含量（Benchaar 等，2015；Bayat 等，2017）。此外，在淀粉日粮中添加多不饱和脂肪酸可直接导致脂肪酸在瘤胃中由反式 11 向反式 10 转化，部分反式 10 异构体在牛乳腺中具有潜在的抗脂肪生成作用，从而抑制了乳脂的合成

（MFD；Shingfield 等，2010；Ventto 等，2017）。

在不饱和脂质添加的情况下，日粮总氮在牛奶蛋白质合成的效率一般是不变的或略有提高（表2）。由于日粮中的脂质，乳蛋白含量或产量、日粮和粗蛋白质摄入量可能在某些情况下减少（Benchaar 等，2015；HalmemiesBeauchet-Filleau 等，2017；Bayat 等，2018），这解释了为什么对于氮素利用率会不受影响或略有提高。由于补充脂质，牛奶蛋白的合成可能降低，这归因于对能量摄入的负面影响、葡萄糖供应和微生物蛋白合成的不足（Lock 和 Shingfield，2004；Halmemies-Beauchet-Filleau 等，2017）。

在体外试验中，鱼油或特定藻类产品中存在的二十碳五烯酸（$C20:5\ n-3$）和二十二碳六烯酸（$C22:6\ n-3$）对甲烷的产生有很强的抑制作用（Martin 等，2010）。然而，当泌乳奶牛饲喂低水平脂质（最高为日粮 DM 的 1%）时，能量校正乳的产量和饲料转化率降低，但对瘤胃甲烷排放没有任何影响。这表明，富含 20 碳和 22 碳多不饱和脂肪酸的脂类添加剂在实际生产中并不是减少奶牛温室气体排放的有效物质。

3.2 碳水化合物

精料补充水平。通常降低奶牛日粮中的精粗比（即增加日粮中的淀粉）会提高采食量，因为降低了高体积密度日粮纤维的贡献（表3，Allen，2000）。此外，日粮中精料比例越高，有机物在瘤胃内的消化越强，这反映了精料中非结构性碳水化合物（NSC，如淀粉和糖）的固有消化率相对于结构性碳水化合物（半纤维素和纤维素）更高（Bayat 等，2017）。与饲料中该精料水平相一致的是淀粉的全肠道消化率的增加，但纤维消化率也往往同时受到不利影响（Niu 等，2016；Bayat 等，2017），这可能会抵消淀粉对有机物消化率的整体影响。采食量的变化反映在能量校正乳（ECM）的产量上，除了富含脂质的日粮外，其他日粮的饲料利用效率不受影响（表3）。补充精料后，乳蛋白有增加的趋势，乳脂则呈现下降的趋势（Niu 等，2016；Bayat 等，2017）。乳蛋白含量的增加可能是由于日粮中含有更多具有高代谢能量密度的谷物导致的。随着日粮中淀粉水平的增加，粗饲料和纤维摄入量也随之减少。中性洗涤纤维可在瘤胃发酵产生可生成脂肪的挥发性脂肪酸，这可能是饲喂粗饲料水平低的日粮导致乳脂降低的原因。

除了 Olijhoek 等（2018）的研究，表3 中报告的低精料和高精料日粮均为等氮日粮（干物质中粗纤维的含量在 15%~18%），其中高精料日粮的粗蛋白质含量比粗料日粮高 20%，导致氮素利用率降低。高精料日粮中较低的瘤胃氨浓度表明：高精料日粮中氮素利用率普遍提高（表3）可能是由于瘤胃微生物可降解蛋白质和可用能量的平衡实现的（Bayat 等，2017），同时归因于优质的日粮蛋白。

随着日粮精料水平的提高，瘤胃甲烷产量的降低已得到证实（表3；Martin 等，2010）。在干物质精料含量高达 30%~40% 的日粮中，牛的甲烷排放量相对稳定，而在某些肉牛生产系统中常见的精料含量为 80%~90% 的日粮中，甲烷排放量迅速下降至较低水平（Martin 等，2010）。在精料中用非结构性碳水化合物（NSC，如淀粉和糖）替代结构性碳水化合物（纤维）会改变瘤胃理化环境和微生物种群，有利于淀粉发酵的

第19章 优化瘤胃功能：青贮饲料和精料提高奶牛饲料转化率、减少甲烷和氮排放的作用

微生物和丙酸的形成（Martin 等，2010）。然而，即使瘤胃甲烷生成显著减少，也并不总是伴随着瘤胃挥发性脂肪酸向丙酸的转变（Aguerre 等，2011）。研究中使用的粗日粮为玉米和牧草的混合青贮饲料，以青贮为基础的牛瘤胃发酵更能抵抗精料的添加（Huhtanen 等，2013）。另外，Bayat 等（2017）报道称，在青贮饲草基础日粮中，随着精料添加量从35%增加到65%，乙酸/丙酸比（-28%）和甲烷排放强度（-25%）显著降低。然而，当日粮中添加脂类时，尽管乙酸/丙酸比显著降低（-22%），但甲烷含量没有降低，与未添加脂类的日粮相似。需要注意的是，瘤胃液中VFA的浓度并不直接反映VFA的产量，但可以反映瘤胃中VFA的生产和吸收的平衡。正如 Aguerre 等（2011）指出的，这可能导致在某些情况下瘤胃液中的VFA组成与瘤胃甲烷产量之间存在明显的不一致。

除了瘤胃发酵的模式，高精料下瘤胃 pH 值的降低可导致原虫数量的减少或在 pH 值低于6时直接抑制甲烷的生成，进而有助于减少瘤胃甲烷的产量（Martin 等，2010；Van Kessel 和 Russell，1996）。虽然在奶牛日粮中加入高水平的精料是一种有效的调控甲烷的策略（表3），但它的缺点是增加了亚急性瘤胃酸中毒（SARA）的风险（Krause 和 Oetzel，2006），并且会与人类竞争食物，谷物生产过程中也伴随着温室气体排放和饲料成本的增加。此外，纤维饲料而非淀粉精料，是牛日粮中适合利用的最主要的天然成分。

碳水化合物的来源。来自食品和生物能源工业的纤维状、人类不可食用的副产品可以部分或全部用于替代高产奶牛日粮中的含淀粉谷物，并可以提供一种具有经济效益的和可持续的饲养策略，以促进循环经济。此外，与大量淀粉和其他容易发酵的碳水化合物会导致瘤胃 pH 值降低、改变瘤胃环境和易引发 SARA 相反，富含纤维的日粮能促进瘤胃和动物健康（Krause 和 Oetzel，2006）。

在不同的等氮日粮条件下，在精粗比和日粮类型方面，以果肉、大豆壳或谷物麸皮为主要碳水化合物来源的富含纤维的精料饲喂奶牛，与富含谷物淀粉的精料相比，干物质采食量、能量校正乳（ECM）的产量、饲料转化率（FE）和氮素利用率（NUE）相似（表4）。但需要注意的是，所有这些试验均在泌乳中后期进行，能量校正乳（ECM）的产量约为 30 kg/d。因此，在泌乳早期和泌乳高峰泌乳量和营养需求水平较高时，对动物生产性能的影响可能不同。Piccioli-Cappelli 等（2014）报道，在泌乳早期（泌乳 30 d），饲喂低或高易发酵碳水化合物日粮的奶牛干物质采食量和泌乳性能无差异（日粮干物质中淀粉和糖的比例分别为 18% 和 25%，产奶量为 37 kg/d）。然而，血液中能量代谢物和激素浓度的变化以及体重的下降，表明低淀粉日粮下机体调动了身体储备。这与高淀粉日粮相反，高淀粉日粮会导致能量的留存和体重增加。由此可见，大部分而不是所有的谷物淀粉可以被高消化率的纤维副产品如甜菜粕和大豆皮所替代，产奶量也会高达 30 kg/d，而不会显著降低奶牛的泌乳性能。此外，Cabezas-Garcia 等（2017）证明，用高能量值的早刈青贮替代大麦和晚刈青贮是可行的，而不影响能量校正乳（ECM）的产量、氮素利用率（NUE）和甲烷排放强度。在这些研究中，精料在日粮干物质中的比例由 60% 逐渐降低至 45%，淀粉由 25% 逐渐降低至 17%，而中性洗涤纤维由 36% 逐渐升高至 42%。

表 3 精料水平和日粮淀粉对奶牛生产性能的影响

对照组中精料的主要组成（CC）	CC在干物质基础的含量（%）	替换组中精料的主要组成（SC）	SC在干物质基础的含量（%）	淀粉在SC和CC干物质基础上的含量（%）	基础日粮的组成	对比空白组的变化率（%）[1]				瘤胃 C_2/C_3	甲烷排放强度（g CH_4/kg ECM）			参考文献
						DMI	ECM	FE	NUE		CC日粮	SC日粮	变化率（%）	
玉米、豆粕和豆皮	32	更多的玉米	39	23 vs 20	玉米：青贮苜蓿	0	-1	0	3	-1	18.9	17.2	-9	Aguerre等（2011）
			46	26 vs 20	青贮50:50[2]	3	3	0	3	-4	18.9	16.8	-11	
			53	29 vs 20		4	3	-1	7	4	18.9	15.2	-20	
玉米、豆粕干酒糟	47	更多的玉米和豆粕	63	32 vs 21	紫花苜蓿干草	3	1	-2	10	NR	14.8	13.7	-8	Niu等（2016）
大麦、小麦、菜籽粕	35	更多的谷物	65	32 vs 14	青贮牧草	23	15	-6	-4	-28	18.9	14.2	-25	Bayat等（2017）
上述日粮基础上加植物油		上述日粮基础上加植物油		29 vs 11		11	-2	-12	15	-22	14.5	14.5	0	
大麦、菜籽饼、豆粕	32	更多大麦	61	22 vs 11	牧草-青贮三叶草	15	10	-4	-15	-31	15.3	12.7	-17	Olijhoek等

注：[1]干物质采食量（DMI）和能量校正乳（ECM）的计算参考Sjaunja等（1991）的报道。饲料转化率（FE）根据能量校正乳/干物质采食量计算，氮素利用效率（NUE）根据奶中氮的总含量/总氮摄入量计算，瘤胃乙酸/丙酸比根据其在瘤胃液中的摩尔含量计算。[2]青贮饲料在干物质基础上含有50%的玉米和50%的青贮紫花苜蓿，NR-代表未报道日不可估算。

第19章 优化瘤胃功能:青贮饲料和精料提高奶牛饲料转化率、减少甲烷和氮排放的作用

表4 精料中碳水化合物和蛋白来源及水平对奶牛生产性能的影响

对照组中精料的主要组分(CC)	在替代日粮中不同的精料(SC)	在SC和CC日粮中干物质基础中的占比(%)	基础日粮的组成[2]	F:C[3]	DMI	ECM	FE	NUE	瘤胃C2/C3	CC日粮	SC日粮	变化率(%)	参考文献
碳水化合物来源		淀粉											
柑橘和甜菜浆,大豆壳	小麦,小麦,饲料	15 vs 10	GS:MS 70:30	70:30	3	0	-3	-4	NR	13.0	13.3	2	Hart等(2015)
棕榈仁提取物		19 vs 14	GS:MS 30:70		2	-2	-4	3	NR	12.6	12.1	-4	
玉米粉,大豆壳	减少玉米粉,增加大豆壳	28 vs 24[4]	MS:LH:GH 55:25:20	50:50	0	3	3	1	0	13.5	12.6	-8	Pirondini等(2015)
		28 vs 24[5]	MS:LH:GH 55:25:20		-6	-8	-2	10	1	12.4	13.2	5	
甜菜浆,麦麸,小麦	燕麦,大麦,棕桐仁饼	15 vs 3	Early GS	66:34	5	1	-4	-2	NR	13.1	13.0	1	Pang等(2018)
		15 vs 3	Late GS		2	7	2	1	NR	14.0	13.9	1	
甜菜浆,大豆壳,干酒糟,玉米粒	小麦,玉米,中等粒,小麦,小麦淀粉	23 vs 6[6]	GS:GH 85:15	50:50	-3	-3	0	9	-13	14.9	13.3	-11	Bougouin等(2018)
		23 vs 6[7]	GS:GH 85:15		-4	-1	3	1	-14	15.3	11.9	-22	
蛋白水平		蛋白											
非蛋白日粮	豆粕	17 vs 15	GS	60:40	0	1	1	-9	NR	17.5	16.9	-3	Gidlund等(2015)
		19 vs 15	GS		1	5	3	-18	NR	17.5	15.9	-9	
		21 vs 15	GS		0	3	3	-25	NR	17.5	17.8	2	

（续表）

对照组中精料的主要组分（CC）	在替代日粮中不同的精料（SC）	在SC和CC日粮中干物质基础中的占比（%）	基础日粮的组成[2]	对比空白组的变化率（%）[1]					瘤胃C2/C3	甲烷排放强度，（g CH$_4$/kg ECM）			参考文献
				F:C[3]	DMI	ECM	FE	NUE		CC日粮	SC日粮	变化率（%）	
非蛋白日粮	菜籽粕	17 vs 15	GS		1	3	1	-8	NR	17.5	16.9	-3	Gidlund 等（2017）
		18 vs 15	GS		4	7	2	-14	NR	17.5	16.1	-8	
		20 vs 15	GS		2	7	5	-19	NR	17.5	15.8	-10	
		16 vs 15	GS:RCS 70:30	60:40	5	8	3	-1	NR	16.7	15.6	-7	
		17 vs 15	GS:RCS 70:30		10	10	0	-14	NR	16.7	15.2	-9	
		19 vs 15	GS:RCS 70:30		10	6	-4	-22	NR	16.7	14.9	-11	
非蛋白日粮	菜籽粕	17 vs 16	GS:RCS 30:70		4	3	-1	-4	NR	17.8	15.0	-16	
		19 vs 16	GS:RCS 30:70		9	4	-5	-14	NR	17.8	16.4	-8	
		19 vs 16	GS:RCS 30:70		7	7	0	-20	NR	17.8	16.3	-8	
减少大豆蛋白	更多的大豆蛋白	16 vs 14	GS:MS 75:25	50:50	8	1	-6	-10	NR	16.7	15.6	-7	Reynolds 等（2010）[8]
		18 vs 14	GS:MS 75:25		0	6	6	-22	NR	16.7	15.2	-9	
减少大豆蛋白、菜籽粕	更多大豆蛋白、菜籽粕	16 vs 14	GS:MS 25:75		2	0	-1	-13	NR	17.8	15.0	-16	
		18 vs 14	GS:MS 25:75		5	2	-3	-24	NR	17.8	16.4	-8	

第19章　优化瘤胃功能：青贮饲料和精料提高奶牛饲料转化率、减少甲烷和氮排放的作用

（续表）

对照组中精料的主要组分（CC）	在替代日粮中不同的精料（SC）	在 SC 和 CC 日粮中干物质基础的占比（%）	基础日粮的组成[2]	F：C[3]	对比空白组的变化率（%）[1]				瘤胃 C2/C3	甲烷排放强度，(g CH$_4$/kg ECM)			参考文献
					DMI	ECM	FE	NUE		CC 日粮	SC 日粮	变化率（%）	
减少豆粕	更多豆粕	19 vs 15[9]	LH	45：55	0	3	3	−18	NR	14.4	14.0	−3	Niu 等 (2016)
蛋白来源													
豆粕	菜籽粕	19 vs 19	GS	60：40	2	2	3	5	NR	16.9	16.3	−4	Gidlund 等 (2015)
豆粕	粉碎的蚕豆	16 vs 16	LS：MS 65：35	55：45	0	−2	−2	−3	1	15.0	15.2	1	Cherif 等 (2018)
	粉碎的蚕豆	16 vs 16	LS：MS 65：35		1	−2	−3	−3	9	15.0	15.4	3	
豆粕：菜籽粕 65：35[10]	蚕豆 16%	16 vs 16	GS	60：40	0	5	5	5	NR	17.6	15.9	−9	Johnston 等 (2019)
	33%	16 vs 16	GS		0	0	−1	2	NR	17.6	16.2	−8	
	47%	16 vs 16	GS		−1	1	2	−8	NR	17.6	16.9	−4	
菜籽粕	蚕豆	19 vs 19	GS	60：40	−2	−4	−3	4	NR	15.2	16.1	6	Ramin 等 (2017)
	豌豆	18 vs 19	GS		0	−6	−6	−6	NR	15.2	16.7	10	

注：[1]干物质采食量（DMI）和能量校正乳（ECM）的计算参考 Sjaunja 等（1991）的报道。饲料转化率（FE）根据能量校正乳/干物质采食量计算。氮素利用效率（NUE）根据奶中氮的总含量/总氮摄入量计算。瘤胃乙酸丙酸比根据其在瘤胃液中的摩尔含量计算，NR-代表未有报道且不可估计。[2]牧草（G）、干草（H）、紫花苜蓿（L）、玉米（M）、红三叶草（RC）、青贮饲料（S）。[3]精粗比（F：C）为干物质基础。[4]日粮中未添加鱼油。[5]日粮中未添加碳酸氢盐。[6]日粮中添加碳酸氢盐。[7]日粮中添加碳酸氢盐。[8]Reynolds 关于日粮和牛奶组分的个人见解。[9]两种粗精比 53：47 和 38：62 的平均值。[10]蛋白混合日粮包括干物质基础上 65%的豆粕和 35%的菜籽粕。

虽然 Benchaar 等（2001）构建的模型表明：用淀粉替代纤维类精料可以减少甲烷的排放，但日粮中淀粉的临界浓度为干物质的 20%~22%，这是减少瘤胃甲烷生成的前提（表 3 和表 4；Hassanat 等，2013）。然而，在添加脂肪的日粮中，即使干物质中淀粉的含量为 28%，也不会改变瘤胃甲烷的生成（Pirondini 等，2015，表 4；Bayat 等，2017，表 3）。瘤胃原虫数量的减少和瘤胃发酵转向丙酸似乎是高淀粉日粮中甲烷排放强度降低的主要因素（Pirondini 等，2015；Bougouin 等，2018）。虽然高淀粉日粮在降低奶牛甲烷排放强度方面有很好的潜力，但要显著降低甲烷排放，谷物淀粉的含量要非常高。这与反刍动物可将不可食用的纤维生物质转化为高质量的人类食物的能力相矛盾。

3.3 蛋白质

蛋白质添加水平。以传统的优质蛋白质饲料油菜籽和豆粕为主的日粮通常会增加奶牛的干物质采食量（Pirondini 等，2015；Bougouin 等，2018），但是在某些情况下的影响可以忽略不计（表 4）。虽然传统蛋白质饲料中更好的氨基酸平衡或更高的产奶量需要增加营养需求也可能会导致采食量增加（Gidlund 等，2017），但是采食量的增加可能部分也归因于日粮纤维和粗蛋白质消化率的提高（Broderick，2003；Jaakkola 等，2009）。优质粗蛋白质添加改善了能量校正乳（ECM）和乳蛋白的产量，这可能是由于为乳腺提供了更多的必需氨基酸（Gidlund 等，2017）。

但当日粮粗蛋白质含量高于 14%~15% 以上时，能量校正乳（ECM）的产量总体上增长相当缓慢，似乎趋于平稳，甚至当饲料粗蛋白质浓度高于 18%~20% 时出现下降（表 4；Broderick，2003）。有趣的是，不管饲料粗蛋白质浓度范围有多大，菜籽粕添加对奶牛的生产性能的影响类似（Jaakkola 等，2009；Gidlund 等，2017；表 4）。用脱水紫花苜蓿替代豆粕也降低了 7% 的产奶量（Doreau 等，2014）。由此可见，油菜籽和大豆的蛋白质较粗料中的蛋白质在提高奶牛的产奶量方面具有明显优势。

日粮粗蛋白质含量是乳品生产中氮素利用率（NUE）的最佳预测指标（Huhtanen 和 Hristov，2009）。确实，氮素利用率（NUE）随日粮粗蛋白质供应的增加呈线性下降趋势（表 4），效率通常在 20%~40%（Dijkstra 等，2011）。在低蛋白质日粮中，粪氮排泄量占氮摄入量的比例（高达 50%）高于尿氮（低至 25%），但随着日粮蛋白增加，粪便氮的贡献减少，而尿氮呈指数增长，最高可达氮摄入量的 60%（Dijkstra 等，2011）。这在一定程度上与瘤胃微生物无法利用多余的可降解蛋白质有关，因为从瘤胃清除多余氨氮的主要途径是在肝脏转化为尿素，然后通过尿液排出（Castillo 等，2000）。

从理论上讲，日粮粗蛋白水平的提高可以降低瘤胃甲烷排放，最明显的原因是蛋白质发酵产生的甲烷少于碳水化合物发酵产生的甲烷（Bannink 等，2006）。在实践中，通过增加蛋白质水平来降低瘤胃甲烷排放强度的潜力很小；从低到中等水平的蛋白质饲料可获得最小的瘤胃甲烷排放强度（最佳可减少 15% 左右）和泌乳性能最大的改善（表 4）。由于蛋白质饲料价格昂贵，奶牛日粮中粗蛋白含量过高是不合适的，而在高粗蛋白水平下，牛奶和甲烷排放的改善会减少甚至逆转（表 4），导致更显著的氮负荷通过粪便和尿液释放到环境中。

蛋白质的来源。相比豆粕和其他蛋白质来源，在牛奶生产中添加菜籽粕一般会增加干物质采食量、能量校正乳（ECM）和乳蛋白的产量（Huhtanen 等，2011；Martineau 等，2013；表4）。Huhtanen 等（2011）认为油菜籽粕对牛奶产量的影响更大是因为增加的或更平衡的氨基酸供应（尤其是组氨酸），产奶所需能量的增加也会影响干物质采食量。相对于大豆，菜籽添加日粮的氮素利用率略有提高（表4）。比较菜籽粕和大豆对肠道甲烷产量影响的报道很少。Gidlund 等（2015；表4）报告了在不同的日粮粗蛋白质水平下，使用菜籽相对于豆粕的甲烷排放强度略有下降。

豆类中蚕豆（*Vicia faba*）和豌豆（*Pisum sativum*）种子相对富含蛋白质（23%~30% DM）和淀粉（45%~50% DM），使它们成为温带地区奶牛自产蛋白质和能量的主要来源。与菜籽或大豆相比，这些替代谷物豆类的蛋白质在瘤胃中可降解性更强，蛋氨酸含量更低，这可能会限制奶牛的泌乳性能（Halmemies-Beauchet-Filleau 等，2018）。然而，用蚕豆或豌豆部分或全部替代豆粕中的蛋白质，奶牛的泌乳性能相似（Halmemes-Beauchetfilleau 等，2018，表4）。Puhakka 等（2016）报道了添加蚕豆蛋白导致牛奶蛋白产量下降和尿素浓度增加，从尿中排出氮的比例来看，蚕豆蛋白质的利用效率低于菜籽，导致氮向环境排放增加。然而，在大多数研究中，替代谷物豆类的氮素利用率似乎与大豆和菜籽粕相当（表4）。在奶牛日粮中添加蚕豆或豌豆可以增加淀粉摄入量，使瘤胃发酵向丙酸方向转变，从而减少瘤胃甲烷的产生。然而，在最近的研究中表明对甲烷排放强度的影响可以忽略不计（Ramin 等，2017；Cherif 等，2018；Johnston 等，2019；表4）。

4 案例研究：粉碎油菜籽对奶牛产奶量、乳脂组成和瘤胃甲烷排放的影响

前言。富含 C18:1 n-9 的油菜籽等不饱和脂肪酸除了可以减少瘤胃甲烷的产生外，还可以通过降低饱和脂肪酸的比例和增加不饱和脂肪酸的比例来改变反刍动物肉和奶的脂质组成。这是非常重要的，因为牛奶和乳制品对人类 C12:0、C14:0 和软脂酸（C16:0）的摄入量有很大的贡献，这些饱和脂肪酸的过量摄入会增加心血管疾病的风险，并降低胰岛素敏感性（Shingfield 等，2013）。奶牛日粮中脂质的形式影响脂质的生物利用率和最终产品中的脂质组成。添加经过粉碎的油菜籽以释放种子中的脂质提高吸收效率是必要的（Kairenius 等，2009）。除反式脂肪酸增加较低外，日粮中添加粉碎的菜籽与添加菜籽油牛乳中的脂肪酸组成相似。与油脂相比，在储存饲料期间，根据需要粉碎油菜籽也能最大限度地降低不饱和脂肪酸氧化变质的风险。本研究旨在研究实际农场条件下，粉碎油菜籽对奶牛乳脂组成和瘤胃甲烷排放的影响。

材料和方法。这项研究是在芬兰赫尔辛基大学 Viikki 研究农场进行（详情见 Halmemies-BeauchetFilleau 等，2019）。整个芬兰艾尔郡奶牛群饲喂对照日粮3周（阶段1），随后饲喂油菜籽富脂日粮4周（阶段2）。在此之后，所有奶牛转回对照日粮饲养3周（阶段3）。自由采食以优质牧草青贮（可消化 OM 为 696 g/kg DM，在混合日粮中 DM 含量为 60%）为基础的全混合日粮（TMR）。在第一次刈割后的预枯牧草（主要是梯牧草和草甸羊茅），用甲酸基添加剂进行大捆青贮。混合日粮中的精料（在混合日粮中的干物质含量为 40%）包括本地种植的谷物、作为蛋白质补充的油菜籽饲料、糖化甜菜粕以及维生素和矿物

质。在混合日粮制备过程中，使用普通锤式研磨机（筛孔大小为6~8 mm）将油菜籽蛋白作为脂质提取粉（对照日粮）或全脂种子（试验日粮）进行等氮供给。

试验日粮中油菜籽脂的添加量约为 50 g/kg DM，对照组日粮中以大麦为主，试验日粮中以燕麦为主。当参观挤奶机器人时（Lely Astronaut A3, Lely, Maassluis, the Netherlands），在试验开始时，每天产奶量小于 30 kg、介于 30~40 kg 和超过 40 kg 的奶牛在整个研究期间每天分别饲喂 3 kg、4 kg 或 5 kg 的标准精料。挤奶机器人配备 GreenFeed 系统（C-Lock Inc., Rapid City, SD, USA），测量瘤胃甲烷、二氧化碳和氢气的排放。

结果和讨论。饲喂试验日粮的奶牛没有健康问题，但干物质采食量比对照日粮平均减少 4%（详情见 Halmemies-Beauchet-Filleau 等，2019）。这并不出人意料，因为高脂肪添加通常会抑制干物质采食量（Huhtanen 等，2008；Halmemies-Beauchet-Filleau 等，2017）。与对照日粮相比，试验日粮未影响能量校正乳（ECM）的产量，饲料转化率（FE）略有提高，从 1.34 提高到 1.40。蛋白质产量和牛奶尿素含量也不受日粮中菜籽蛋白质类型的影响。虽然试验日粮对乳脂产量没有影响，但改变了乳脂组成（表5）。试验日粮乳脂中总饱和脂肪酸含量比对照日粮低 17%（表5）。此外，C10 到 C16 饱和脂肪酸被认为是人体血液中增加胆固醇的关键脂肪酸，在试验日粮牛奶中的含量明显低于对照组的含量。事实上，众所周知，增加长链脂肪酸的供应会抑制乳腺中饱和脂肪酸的从头合成（Shingfield 等，2010）。试验组乳脂中全部单不饱和脂肪酸含量高于对照日粮 58%，主要为 C18:1 n-9。粉碎油菜籽对牛奶中多不饱和脂肪酸的影响不大。此外，与富含多不饱和脂肪酸的乳脂相比（Havemose 等，2006），乳脂肪和单不饱和脂肪酸含量高的富含脂肪的乳制品更不易发生氧化变质（Lin 等，1996）。

表5 灌装牛奶中脂肪酸的组成

脂肪酸（g/100 g 总脂肪酸）	对照日粮	试验日粮	变化率%
10:0	3.9	2.0	-49
12:0	4.6	2.2	-52
14:0	13	8.5	-35
16:0	31	21	-31
18:0	9.7	18	+82
18:1 $n-9$	16	28	+70
18:2 $n-6$	1.3	1.1	
18:3 $n-3$	0.4	0.4	
总饱和脂肪酸	74	61	-17
总单不饱和脂肪酸	23	36	+58
总多不饱和脂肪酸	2.6	2.3	
总反式脂肪酸	3.6	5.0	

资料来源：Halmemies-Beauchet-Filleau 等（2019）。

第19章 优化瘤胃功能：青贮饲料和精料提高奶牛饲料转化率、减少甲烷和氮排放的作用

与对照组相比，试验日粮的瘤胃甲烷、二氧化碳和氢气排放量分别降低18%、5%和36%。在饲喂富含高可消化牧草青贮饲料的日粮中，粉碎的油菜籽显著降低了奶牛瘤胃中氢的负荷和甲烷的生成（图1）。干物质采食量的小幅下降不能完全解释试验日粮瘤胃中氢和甲烷排放的减少。可能是瘤胃发酵模式转向了丙酸，增加了氢的利用率。在某些研究中表明，在奶牛日粮中添加菜籽或其他不饱和脂类会降低瘤胃乙酸/丙酸比例，但并非所有研究都是如此（Hristov等，2011a；表2）。虽然瘤胃不饱和脂肪酸的生物氢化也是一种氢的利用方式，但一般认为其对瘤胃甲烷生成的贡献非常低（Martin等，2010）。试验日粮和对照日粮的甲烷排放强度分别为12.1 g/kg ECM和15.1 g/kg ECM。奶牛在芬兰每年平均生产10 300 ECM。45头奶牛，从对照日粮转换为试验日粮，每年可减少1 390 t的瘤胃甲烷排放，这相当于整个牧群一年中2个月的甲烷排放量。

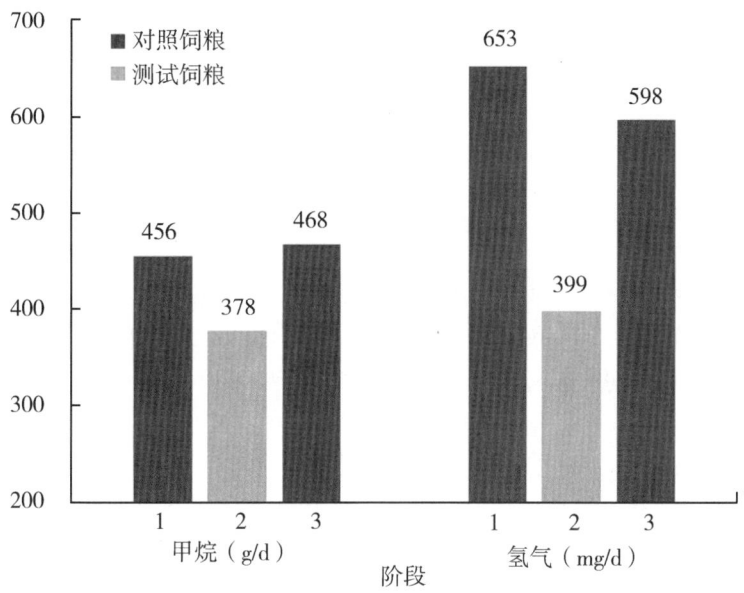

图1 奶牛在饲喂对照日粮的第1阶段、测试日粮的第2阶段和对照日粮的第3阶段瘤胃气体排放特征（Halmemies-Beauchet-Filleau等，2019）

结论。在以高度可消化的青贮饲料为基础的奶牛日粮中，用研磨的菜籽替代菜籽粕（提供日粮DM中5%的脂肪）对产奶量、饲料转化率或动物健康均无不良影响。在整个牧群水平上，粉碎的油菜籽通过降低中链饱和脂肪酸和提高C18:1 n-9的比例来改善乳脂组成。此外，在商业实践中饲喂粉碎的油菜籽显著抑制了瘤胃甲烷的产生。

致谢。这项研究得到了欧洲创新技术研究所（EIT）的部分资助（EIT Food Project 18095：乳制品中饱和脂肪酸的减少），并与Valio Ltd和雷丁大学合作。

5 总结和未来趋势

可用于减少奶牛生产过程中温室气体排放的营养策略，包括各种瘤胃改良剂（目前正在开发中）和目前可实际使用的牧草和/或精饲料为基础的策略。考虑到减少人类可食用的粮食（如动物饲料中的谷物），应强调以牧草为基础的日粮策略，特别是反刍动物具有以纤维消化为主的特征。在温带地区，可用于制作青贮饲料的主要植物品种包括牧草、豆科牧草和玉米，但它们在不同地区的可用性取决于当地的气候条件。

青贮饲料。依据牧草成熟度进行收割，是减少冷季牧草青贮在奶牛生产中"碳足迹"最有潜力的方式。早期收割牧草会增加了奶牛的干物质采食量、能量校正乳（ECM）和饲料转化率（FE），降低甲烷排放强度，但会导致氮素利用效率（NUE）降低。减少甲烷排放和氮素利用率降低（NUE）之间的权衡是复杂的，显然需要进一步研究。其他管理因素，如施氮量、在青贮饲料中使用添加剂或培育高糖牧草品种的作用有限。

豆科牧草青贮。与普通牧草相比，豆科牧草青贮对甲烷排放强度影响的数据有限，且会导致氮素利用率（NUE）降低，表明豆科牧草在降低奶牛生产中环境足迹方面的潜力不大。与之相反的研究表明：如果饲用豆科牧草具有较高的干物质采食量和瘤胃通过率，其甲烷排放强度要低于其他牧草。如此相互矛盾的结果可能是因为在实际生产中豆科牧草与其他牧草混合种植和饲喂，或者其他植物和青贮饲料的营养和发酵质量不同年份之间差异巨大。为了充分发挥豆科牧草对牧草生产、饲料干物质采食量和动物生产性能的有益影响，还需要进一步研究饲用豆科牧草在减少奶牛生产中环境足迹的潜力。

玉米青贮饲料。含淀粉的玉米青贮具有代谢能高和粗蛋白质含量低的特点，是一种极具价值的饲料作物，适合与粗蛋白质含量较高的牧草和豆科植物混合使用。可用于减少玉米青贮"碳足迹"的方法包括在收获时将玉米作物的成熟度推迟到后期（40% DM），以及使用具有较高细胞壁消化率和易被采食特性的玉米品种，如褐色叶脉的玉米。这些方法有可能降低高粗日粮中10%的甲烷排放。

用玉米青贮替代牧草或豆科牧草青贮，可降低以高牧草青贮为基础的日粮中甲烷排放强度和提高氮素利用率降低（NUE）（尤其是粗蛋白含量高的豆科牧草青贮）而不影响牛奶产量，均能带来有益的环境效益。然而，还需要进一步的研究从而优化这些牧草在奶牛生产中的应用。豆科牧草氮组分对氮素利用率降低（NUE）的影响以及牧草碳水化合物类型对甲烷排放的影响还有待进一步研究。还需要应用生命周期分析法，比较使用玉米和多年生青贮作物对环境的影响。

精料中的脂质。日粮干物质中高达5%的不饱和植物脂有可能以剂量依赖性的方式减少20%~40%的瘤胃甲烷排放，且不会对动物的能量校正乳的产量和饲料转化率产生负面影响。脂质的影响似乎在整个哺乳期都存在，但还需要更多的长期研究来证实这一点。在高脂日粮中，将脂类作为混合日粮的一部分饲喂，优于单独饲喂。相比之下，在淀粉型日粮（以玉米青贮或富含淀粉的精料）中，由于脂肪对能量校正乳的产量有负面影响，所以补充脂肪的动力有限。这可能与基础日粮中含有大量淀粉时不饱和脂肪对瘤胃纤维消化的不利影响有关。

第 19 章 优化瘤胃功能：青贮饲料和精料提高奶牛饲料转化率、减少甲烷和氮排放的作用

精料中的碳水化合物。提高谷物淀粉在奶牛日粮中的比例可提高采食量、能量校正乳（ECM）的产量和氮素利用率（NUE）。日粮干物质（DM）中淀粉的临界浓度为 20%~22%，可调控瘤胃甲烷的生成。报道称，当日粮干物质中淀粉含量达到 20%~32% 时，可降低 20%~25% 的甲烷排放。然而，谷物中含有大量易发酵的碳水化合物，会使牛易患亚急性瘤胃酸中毒（SARA），并与人类竞争食物。食品和生物能源工业中纤维性的、人类不可食用的副产品提供了一种具有成本效益和理论上可行的饲养策略，以实现循环经济。大豆皮、甜菜粕和麸皮部分或全部替代奶牛日粮中富含淀粉的谷物，对奶牛生产性能没有影响，也不会增加瘤胃甲烷的产量。然而，在这些研究中，泌乳中期奶牛的产奶量并未超过 30 kg/d，因此，需要在具有更高泌乳水平和泌乳早期的奶牛中进行更多的研究，以证实这些有前景的发现。

精料中的蛋白质。优质蛋白质的来源，如油菜籽和豆粕通常会增加奶牛的干物质（DM）采食量，尽管在某些情况下的影响可以忽略不计。低至中水平的蛋白质饲料（日粮粗蛋白质含量为 15%~18%，取决于基础日粮中粗蛋白质的水平）具有最小的瘤胃甲烷排放强度（最多可降低 15% 左右），且泌乳性能的提高最大。奶牛日粮中饲喂高比例的粗蛋白质（CP）（CP 高于 18%~20% DM）是没必要的，因为蛋白饲料价格昂贵，且在高粗蛋白水平下，对牛奶和甲烷产量的影响会减少甚至逆转，只会导致粪尿对环境产生更显著的氮负荷。有趣的是，在提高奶牛产奶量方面，传统奶牛蛋白饲料中的油菜籽和大豆的蛋白质优于牧草。在泌乳性能方面，菜籽蛋白略优于大豆、蚕豆和豌豆，但在瘤胃甲烷排放方面，这些蛋白质来源之间的差异可以忽略不计。在温带地区，由于相对高含量的蛋白质和淀粉，蚕豆和豌豆很有希望用于奶牛日粮的本土蛋白和能源饲料。需要更多的研究来寻找提高牧草蛋白利用效率和替代豆科谷物的方法，以提高其在奶牛生产中的氮素利用效率（NUE）。

6 补充材料

以下的综述文章或荟萃分析可以提供更多的信息：

Dewhurst, R. J. 2013. Milk production from silage: comparison of grass, legume and maize silages and their mixtures. Agric. Food Sci. 22, 57–69.

- Huhtanen, P. and Hristov, A. N. 2009. A meta-analysis of the effects of dietary protein concentration and degradability on milk protein yield and milk N efficiency in dairy cows. J. Dairy Sci. 92, 3222–32.
- Hristov, A. N., Oh, J., Firkins, J. L., Dijkstra, J., Kebreab, E., Waghorn, G., Makkar, H. P. S., Adesogan, A. T., Yang, W., Lee, C., Gerber, P. J., Henderson, B. and Tricarico, J. M. 2013. Special topics-mitigation of methane and nitrous oxide emissions from animal operations: I. A review of entericmethane mitigation options. J. Anim. Sci. 91, 5045–69.
- Ramin, M. and Huhtanen, P. 2013. Development of equations for predicting methane emissions from ruminants. J. Dairy Sci. 96, 2476–93.

7　参考文献

Aguerre, M. J., Wattiaux, M. A., Powell, J. M., et al., 2011. Effect of forage-to-concentrate ratio in dairy cow diets on emission of methane, carbon dioxide, and ammonia, lactation performance, and manure excretion. J. Dairy Sci. 94 (6), 3081-93.

Allen, M. S. 2000. Effects of diet on short-term regulation of feed intake by lactating dairy cattle. J. Dairy Sci. 83 (7), 1598-624.

Alstrup, L., Hellwing, A. L. F., Lund, P., et al., 2015. Effect of fat supplementation and stage of lactation on methane production in dairy cows. Anim. Feed Sci. Technol. 207, 10-9.

Arndt, C., Powell, J. M., Aguerre, M. J., et al., 2015. Performance, digestion, nitrogen balance, and emission of manure ammonia, enteric methane, and carbon dioxide in lactating cows fed diets with varying alfalfa silage-to-corn silage ratios. J. Dairy Sci. 98 (1), 418-30.

Bannink, A., Kogut, J., Dijkstra, J., et al., 2006. Estimation of the stoichiometry of volatile fatty acid production in the rumen of lactating cows. J. Theor. Biol. 238 (1), 36-51.

Bayat, A. R., Kairenius, P., Stefánski, T., et al., 2015. Effect of Camelina oil or live yeasts (*Saccharomyces cerevisiae*) on ruminal methane production, rumen fermentation, and milk fatty acid composition in lactating cows fed grass silage diets. J. Dairy Sci. 98 (5), 3166-81.

Bayat, A. R., Ventto, L., Kairenius, P., et al., 2017. Dietary forage to concentrate ratio and sunflower oil supplement alter rumen fermentation, ruminal methane emissions, and nutrient utilization in lactating cows. Transl. Anim. Sci. 1 (3), 277-86.

Bayat, A. R., Tapio, I., Vilkki, J., et al., 2018. Plant oil supplements reduce methane emissions and improve milk fatty acid composition in dairy cows fed grass silage-based diets without affecting milk yield. J. Dairy Sci. 101 (2), 1136-51.

Beauchemin, K. A., Kreuzer, M., O'Mara, F., et al., 2008. Nutritional management for enteric methane abatement: a review. Aust. J. Exp. Agric. 48 (2), 21-7.

Benchaar, C., Pomar, C. and Chiquette, J. 2001. Evaluation of dietary strategies to reduce methane production in ruminants: a modelling approach. Can. J. Anim. Sci. 81 (4), 563-74.

Benchaar, C., Hassanat, F., Gervais, R., et al., 2014. Methane production, digestion, ruminal fermentation, nitrogen balance, and milk production of cows fed corn silage-or barley silage-based diets. J. Dairy Sci. 97 (2), 961-74.

Benchaar, C., Hassanat, F., Martineau, R., et al., 2015. Linseed oil supplementation to dairy cows fed diets based on red clover silage or corn silage: effects on methane production, rumen fermentation, nutrient digestibility, N balance, and milk production. J. Dairy Sci. 98 (11), 7993-8008.

Bernardes, T. F., Daniel, J. L. P., Adesogan, A. T., et al., 2018. Silage review: unique challenges of silages made in hot and cold regions. J. Dairy Sci. 101 (5), 4001-19.

Bertilsson, J., Åkerlind, M. and Eriksson, T. 2017. The effects of high-sugar ryegrass/red clover silage diets on intake, production, digestibility, and N utilization in dairy cows, as measured *in vivo* and predicted by the NorFor model. J. Dairy Sci. 100 (10), 7990-8003.

Bougouin, A., Ferlay, A., Doreau, M., et al., 2018. Effects of carbohydrate type or bicarbonate addition to grass silage-based diets on enteric methane emissions and milk fatty acid composition in dairy cows. J. Dairy Sci. 101 (7), 6085-97.

Brask, M., Lund, P., Hellwing, A. L. F., et al., 2013a. Enteric methane production, digestibility and rumen fermentation in dairy cows fed different forages with and without rapeseed fat supplementation. Anim. Feed Sci. Technol. 184 (1-4), 67-79.

Brask, M., Lund, P., Weisbjerg, M. R., et al., 2013b. Methane production and digestion of different physical forms of rapeseed as fat supplements in dairy cows. J. Dairy Sci. 96 (4), 2356-65.

Broderick, G. A. 2003. Effects of varying dietary protein and energy levels on the production of lactating dairy cows. J. Dairy Sci. 86 (4), 1370-81.

Cabezas-Garcia, E. H., Krizsan, S. J., Shingfield, K. J., et al., 2017. Effects of replacement of late-harvested grass silage and barley with early-harvested silage on milk production and methane emissions. J. Dairy Sci. 100 (7), 5228-40.

Castillo, A. R., Kebreab, E., Beever, D. E., et al., 2000. A review of efficiency of nitrogen utilisation in lactating dairy cows and its relationship with environmental pollution. J. Anim. Feed Sci. 9 (1), 1-32.

Cherif, C., Hassanat, F., Claveau, S., et al., 2018. Faba bean (Vicia faba) inclusion in dairy cow diets: effect on nutrient digestion, rumen fermentation, nitrogen utilization, methane production, and milk performance. J. Dairy Sci. 101 (10), 8916-28.

De Boever, J. L., Goossens, K., Peiren, N., et al., 2017. The effect of maize silage type on the performances and methane emission of dairy cattle. J. Anim. Physiol. Anim. Nutr. 101 (5), e246-56.

Dewhurst, R. J. 2013. Milk production from silage: comparison of grass, legume and maize silages and their mixtures. AFSci 22 (1), 57-69.

Dijkstra, J., Oenema, O. and Bannink, A. 2011. Dietary strategies to reducing N excretion from cattle: implications for methane emissions. Curr. Opin. Env. Sust. 3 (5), 414-22.

Dijkstra, J., Bannik, A., France, J., et al., 2018. Short communication: antimethanogenic effects of 3-nitroxypropanol depend on supplementation dose, dietary fiber content, and cattle type. J. Dairy Sci. 101 (10), 9041-7.

Doreau, M., Ferlay, A., Rochette, Y., et al., 2014. Effects of dehydrated Lucerne and soya bean meal on milk production and composition, nutrient digestion, and methane and nitrogen losses in dairy cows receiving two different forages. Animal 8 (3), 420-30.

Ellis, J. L., Hindrichsen, I. K., Klop, G., et al., 2016. Effects of lactic acid bacteria silage inoculation on methane emission and productivity of Holstein Friesian dairy cattle. J. Dairy Sci. 99 (9), 7159-74.

Eugène, M., Massé, D., Chiquette, J., et al., 2008. Meta-analysis on the effects of lipid supplementation on methane production in lactating dairy cows. Can. J. Anim. Sci. 88 (2), 331-7.

Ferlay, A., Doreau, M., Martin, C., et al., 2013. Effects of incremental amounts of extruded linseed on the milk fatty acid composition of dairy cows receiving hay or corn silage. J. Dairy Sci. 96 (10), 6577-95.

Gidlund, H., Hetta, M., Krizsan, S. J., et al., 2015. Effects of soybean meal or canola meal on milk production and methane emissions in lactating dairy cows fed grass silage-based diets. J. Dairy Sci. 98 (11), 8093-106.

Gidlund, H., Hetta, M. and Huhtanen, P. 2017. Milk production and methane emissions from dairy cows fed a low or high proportion of red clover silage and an incremental level of rapeseed expeller. Liv. Sci. 197, 73-81.

Halmemies-Beauchet-Filleau, A., Kokkonen, T., Lampi, A. M., et al., 2011. Effect of plant oils and

Camelina expeller on milk fatty acid composition in lactating cows fed diets based on red clover silage. J. Dairy Sci. 94 (9), 4413-30.

Halmemies-Beauchet-Filleau, A., Shingfield, K. J., Simpura, I., et al., 2017. Effect of incremental amounts of Camelina oil on milk fatty acid composition in lactating cows fed diets based on a mixture of grass and red clover silage and concentrates containing Camelina expeller. J. Dairy Sci. 100 (1), 305-24.

Halmemies-Beauchet-Filleau, A., Rinne, M., Lamminen, M., et al., 2018. Alternative and novel feeds for ruminants: nutritive value, product quality and environmental aspects. Animal 12 (s2), s295-309.

Halmemies-Beauchet-Filleau, A., Jaakkola, S., Kokkonen, T., et al., 2019. Rapeseed lipids to decrease saturated fatty acids in milk and ruminal methane emissions of dairy cows. In: Proceedings of the 10th Nordic Feed Science Conference, pp. 69-73.

Hammond, K. J., Jones, A. K., Humphries, D. J., et al., 2016. Effects of diet forage source and neutral detergent fiber content on milk production of dairy cattle and methane emissions determined using GreenFeed and respiration chamber techniques. J. Dairy Sci. 99 (10), 7904-17.

Harrison, J., Huhtanen, P. and Collins, M. 2003. Perennial grasses. In: Buxton, D. R., Muck, R. E. and Harrison, J. H. (Eds), Silage Science and Technology. American Society of Agronomy, Crop Science Society of America and Soil Science of Society of America, Madison, WI, pp. 635-747.

Hart, K. J., Huntington, J. A., Wilkinson, R. G., et al., 2015. The influence of grass silage-to-maize silage ratio and concentrate composition on methane emissions, performance and milk composition of dairy cows. Animal 9 (6), 983-91.

Hassanat, F., Gervais, R., Julien, C., et al., 2013. Replacing alfalfa silage with corn silage in dairy cow diets: effects on enteric methane production, ruminal fermentation, digestion, N balance, and milk production. J. Dairy Sci. 96 (7), 4553-67.

Hassanat, F., Gervais, R., Massé, D. I., et al., 2014. Methane production, nutrient digestion, ruminal fermentation, N balance, and milk production of cows fed timothy silage-or alfalfa silage-based diets. J. Dairy Sci. 97 (10), 6463-74.

Hassanat, F., Gervais, R. and Benchaar, C. 2017. Methane production, ruminal fermentation characteristics, nutrient digestibility, nitrogen excretion, and milk production of dairy cows fed conventional or brown midrib corn silage. J. Dairy Sci. 100 (4), 2625-36.

Hatew, B., Bannink, A., Van Laar, H., et al., 2016. Increasing harvest maturity of whole-plant corn silage reduces methane emission of lactating dairy cows. J. Dairy Sci. 99 (1), 354-68.

Havemose, M. S., Weisbjerg, M. R., Bredie, W. L. P., et al., 2006. Oxidative stability of milk influenced by fatty acids, antioxidants, and copper derived from feed. J. Dairy Sci. 89 (6), 1970-80.

Hollmann, M., Powers, W. J., Fogiel, A. C., et al., 2012. Enteric methane emissions and lactational performance of Holstein cows fed different concentrations of coconut oil. J. Dairy Sci. 95 (5), 2602-15.

Hristov, A. N., Domitrovich, C., Wachter, A., et al., 2011a. Effect of replacing solvent-extracted canola meal with high-oil traditional canola, high-oleic acid canola, or high-erucic acid rapeseed meals on rumen fermentation, digestibility, milk production, and milk fatty acid composition in lactating dairy cows. J. Dairy Sci. 94 (8), 4057-74.

Hristov, A. N., Lee, C., Cassidy, T., et al., 2011b. Effects of lauric and myristic acids on ruminal fermentation, production, and milk fatty acid composition in lactating dairy cows. J. Dairy Sci. 94 (1), 382-95.

第19章 优化瘤胃功能：青贮饲料和精料提高奶牛饲料转化率、减少甲烷和氮排放的作用

Hristov, A. N., Oh, J., Firkins, J. L., et al., 2013. Special topics—mitigation of methane and nitrous oxide emissions from animal operations: I. A review of enteric methane mitigation options. J. Anim. Sci. 91 (11), 5045-69.

Huhtanen, P. and Hristov, A. N. 2009. A meta-analysis of the effects of dietary protein concentration and degradability on milk protein yield and milk N efficiency in dairy cows. J. Dairy Sci. 92 (7), 3222-32.

Huhtanen, P., Rinne, M. and Nousiainen, J. 2008. Evaluation of concentrate factors affecting silage intake of dairy cows: a development of the relative total diet intake index. Animal 2 (6), 942-53.

Huhtanen, P., Hetta, M. and Swensson, C. 2011. Evaluation of canola meal as a protein supplement for dairy cows: a review and a meta-analysis. Can. J. Anim. Sci. 91 (4), 529-43.

Huhtanen, P., Jaakkola, S. and Nousiainen, J. 2013. An overview of silage research in Finland: from ensiling innovation to advances in dairy cow feeding. AFSci 22 (1), 35-56.

Huyen, N. T., Desrues, O., Alferink, S. J. J., et al., 2016. Inclusion of sainfoin (*Onobrychis viciifolia*) silage in dairy cow rations affects nutrient digestibility, nitrogen utilization, energy balance, and methane emissions. J. Dairy Sci. 99 (5), 3566-77.

Jaakkola, S., Saarisalo, E. and Heikkilä, T. 2009. Formic acid treated whole crop barley and wheat silages in dairy cow diets: effects of crop maturity, proportion in the diet, and level and type of concentrate supplementation. AFSci 18 (3-4), 234-56.

Järvenranta, K., Kuoppala, K., Rinne, M., et al., 2016. Legumes in ruminant production systems in European cold climates-limitations and opportunities. Legume Perspectives Issue 12, April 2016, pp. 34-5.

Johnston, D. J., Theodoridou, K. and Ferris, C. P. 2019. The impact of field bean inclusion level in dairy cow diets on cow performance and nutrient utilisation. Liv. Sci. 220, 166-72.

Jung, H. G., Mertens, D. R. and Phillips, R. L. 2011. Effect of reduced ferulate-mediated lignin/arabinoxylan cross-linking in corn silage on feed intake, digestibility, and milk production. J. Dairy Sci. 94 (10), 5124-37.

Kairenius, P., Toivonen, V., Ahvenjärvi, S., et al., 2009. Effects of rapeseed lipids in the diet on ruminal lipid metabolism and milk fatty acid composition in cows fed grass silage based diets. In: Ruminant Physiology: Digestion, Metabolism and Effects of Nutrition on Reproduction and Welfare. Proceedings of the XIth Interface Symposium Ruminant Physiology. Wageningen Academic Publishing, the Netherlands, pp. 232-3.

Khan, N. A., Yu, P., Ali, M., et al., 2015. Nutritive value of maize silage in relation to dairy cow performance and milk quality. J. Sci. Food Agric. 95 (2), 238-52.

Kliem, K. E., Humphries, D. J., Kirton, P., et al., 2019. Differential effects of oilseed supplements on methane production and milk fatty acid concentrations in dairy cows. Animal 13 (2), 309-17.

Klop, G., Hatew, B., Bannink, A., et al., 2016. Feeding nitrate and docosahexaenoic acid affects enteric methane production and milk fatty acid composition in lactating dairy cows. J. Dairy Sci. 99 (2), 1161-72.

Knapp, J. R., Laur, G. L., Vadas, P. A., et al., 2014. Invited review: Enteric methane in dairy cattle production: quantifying the opportunities and impact of reducing emissions. J. Dairy Sci. 97 (6), 3231-61.

Krause, K. M. and Oetzel, G. R. 2006. Understanding and preventing subacute ruminal acidosis in dairy herds: a review. Anim. Feed Sci. Technol. 126 (3-4), 215-36.

Kuoppala, K. 2010. Influence of harvesting strategy on nutrient supply and production of dairy cows consuming diets based on grass and red clover silage. Doctoral dissertation. MTT Science, no. 11, 49p.

Kuoppala, K., Ahvenjärvi, S., Rinne, M., et al., 2009. Effects of feeding grass or red clover silage cut at two maturity stages in dairy cows. 2. Dry matter intake and cell wall digestion kinetics. J. Dairy Sci. 92 (11), 5634-44.

Lin, M. P., Sims, C. A., Staples, C. R., et al., 1996. Flavor quality and texture of modified fatty acid high monoene, low saturate butter. Food Res. Int. 29 (3-4), 367-71.

Livingstone, K. M., Humphries, D. J., Kirton, P., et al., 2015. Effects of forage type and extruded linseed supplementation on methane production and milk fatty acid composition of lactating dairy cows. J. Dairy Sci. 98 (6), 4000-11.

Lock, A. L. and Shingfield, K. J. 2004. Optimizing milk composition. In: Kebreab, E., Mills, J. and Beever, D. E. (Eds), Dairying-Using Science to Meet Consumers Needs. Br. Soc. Anim. Sci. Publ. 29. Nottingham University Press, Loughborough, UK, pp. 107-88.

Maia, M. R., Chaudhary, L. C., Figueres, L., et al., 2007. Metabolism of polyunsaturated fatty acids and their toxicity to the microflora of the rumen. Antonie Leeuwenhoek 91 (4), 303-14.

Martin, C., Rouel, J., Jouany, J. P., et al., 2008. Methane output and diet digestibility in response to feeding dairy cows crude linseed, extruded linseed, or linseed oil. J. Anim. Sci. 86 (10), 2642-50.

Martin, C., Morgavi, D. P. and Doreau, M. 2010. Methane mitigation in ruminants: from microbe to the farm scale. Animal 4 (3), 351-65.

Martin, C., Ferlay, A., Mosoni, P., et al., 2016. Increasing linseed supply in dairy cow diets based on hay or corn silage: effect on enteric methane emission, rumen microbial fermentation, and digestion. J. Dairy Sci. 99 (5), 3445-56.

Martineau, R., Ouellet, D. R. and Lapierre, H. 2013. Feeding canola meal to dairy cows: a meta-analysis on lactational responses. J. Dairy Sci. 96 (3), 1701-14.

Martinez-Fernandez, G., Duval, S., Kinderman, M., et al., 2018. 3-NOP vs. halogenated compound: methane production, ruminal fermentation and microbial community response in forage fed cattle. Front. Microbiol. 9, 1-13.

Merry, R. J., Lee, M. R. F., Davies, D. R., et al., 2006. Effects of high-sugar ryegrass silage and mixtures with red clover silage on ruminant digestion. 1. *In vitro* and *in vivo* studies of nitrogen utilization. J. Anim. Sci. 84 (11), 3049-60.

Moate, P. J., Williams, S. R. O., Hannah, M. C., et al., 2013. Effects of feeding algal meal high in docosahexaenoic acid on feed intake, milk production, and methane emissions in dairy cows. J. Dairy Sci. 96 (5), 3177-88.

Moss, A. R., Jouany, J.-P. and Newbold, J. 2000. Methane production by ruminants: its contribution to global warming. Ann. Zootech. 49 (3), 231-53.

Moorby, J. M., Evans, R. T., Scollan, N. D., et al., 2006. Increased concentration of water-soluble carbohydrate in perennial ryegrass (*Lolium perenne* L.). Evaluation in dairy cows in early lactation. Grass Sci. 61 (1), 52-9.

Muck, R. E., Nadeau, E. M. G., McAllister, T. A., et al., 2018. Silage review: recent advances and future uses of silage additives. J. Dairy Sci. 101 (5), 3980-4000.

Niu, M., Appuhamy, J. A. D. R. N., Leytem, A. B., et al., 2016. Effect of dietary crude protein and forage contents on enteric methane emissions and nitrogen excretion from dairy cows simultaneously. Anim.

Prod. Sci. 56 (3), 312-21.

Odongo, N. E., Or-Rashid, M. M., Kebreab, E., et al., 2007. Effect of supplementing myristic acid in dairy cow rations on ruminal methanogenesis and fatty acid profile in milk. J. Dairy Sci. 90 (4), 1851-8.

Olijhoek, D. W., Løvendahl, P., Lassen, J., et al., 2018. Methane production, rumen fermentation, and diet digestibility of Holstein and Jersey dairy cows being divergent in residual feed intake and fed at 2 forage-to-concentrate ratios. J. Dairy Sci. 101 (11), 9926-40.

Owens, F., Zinn, R. A. and Kim, Y. K. 1986. Limits to starch digestion in the ruminant small intestine. J. Anim. Sci. 63, 1634-48.

Pang, D., Yan, T., Trevisi, E., et al., 2018. Effect of grain-or by-product-based concentrate fed with early-or late-harvested first-cut grass silage on dairy cow performance. J. Dairy Sci. 101 (8), 7133-45.

Phelan, P., Moloney, A. P., McGeough, E. J., et al., 2015. Forage legumes for grazing and conserving in ruminant production systems. Crit. Rev. Plant Sci. 34 (1-3), 281-326.

Piccioli-Cappelli, F., Loor, J. J., Seal, C. J., et al., 2014. Effect of dietary starch level and high rumen-undegradable protein on endocrine-metabolic status, milk yield, and milk composition in dairy cows during early and late lactation. J. Dairy Sci. 97 (12), 7788-803.

Pirondini, M., Colombini, S., Mele, M., et al., 2015. Effect of dietary starch concentration and fish oil supplementation on milk yield and composition, diet digestibility, and methane emissions in lactating dairy cows. J. Dairy Sci. 98 (1), 357-72.

Puhakka, L., Jaakkola, S., Simpura, I., et al., 2016. Effects of replacing rapeseed meal with fava bean at 2 concentrate crude protein levels on feed intake, nutrient digestion, and milk production in cows fed grass silage-based diets. J. Dairy Sci. 99 (10), 7993-8006.

Ramin, M., Höjer, A. and Hetta, M. 2017. The effects of legume seeds on the lactation performance of dairy cows fed grass silage-based diets. AFSci 26 (3), 129-37.

Reynolds, C. K., Crompton, L. A., Mills, J. A. N., et al., 2010. Effects of diet protein level and forage source on energy and nitrogen balance and methane and nitrogen excretion in lactating dairy cows. In: Crovetto, G. M. (Ed.), Energy and Protein Metabolism and Nutrition. Proceedings of the 3rd EAAP International Symposium Energy Protein Metabolism and Nutrition. Wageningen Academic Publishing, the Netherlands, pp. 463-4.

Rinne, M., Jaakkola, S. and Huhtanen, P. 1997. Grass maturity effects on cattle fed silage-based diets. 1. Organic matter digestion, rumen fermentation and nitrogen utilization. Anim. Feed Sci. Technol. 67 (1), 1-17.

Rinne, M. 2000. Influence of the timing of the harvest of primary grass growth on herbage quality and subsequent digestion and performance in the ruminant animal. Academic dissertation. University of Helsinki, Department of Animal Science Publications no. 54. Helsinki University Press, Helsinki, 42p.

Sauer, F. D., Fellner, V., Kinsman, R., et al., 1998. Methane output and lactation response in Holstein cattle with monensin or unsaturated fat added to the diet. J. Anim. Sci. 76 (3), 906-14.

Shingfield, K. J., Bernard, L., Leroux, C., et al., 2010. Role of trans fatty acids in the nutritional regulation of mammary lipogenesis in ruminants. Animal 4 (7), 1140-66.

Shingfield, K. J., Bonnet, M. and Scollan, N. D. 2013. Recent developments in altering the fatty acid composition of ruminant-derived foods. Animal 7 (Suppl. 1), 132-62.

Sjaunja, L. O., Bævre, L., Junkkarinen, L., et al., 1991. A Nordic proposal for an energy corrected milk (ECM) formula. In: Proceedings of the 27th Biennial Session of the International Committee for Animal Recording (ICAR), Paris, France. Wageningen Academic Publishing, the Netherlands, pp. 156-7.

Staerfl, S. M., Amelchanka, S. L., Kälber, T., et al., 2012. Effect of feeding dried high-sugar ryegrass ('AberMagic') on methane and urinary nitrogen emission of primiparous cows. Liv. Sci. 150, 203-301.

Van Dorland, H. A., Wettstein, H. -R., Leuenberger, H., et al., 2007. Effect of supplementation of fresh and ensiled clovers to ryegrass on nitrogen loss and methane emission of dairy cows. Liv. Sci. 111 (1-2), 57-69.

Van Gastelen, S., Antunes-Fernandes, E. C., Hettinga, K. A., et al., 2015. Enteric methane production, rumen volatile fatty acid concentrations, and milk fatty acid composition in lactating Holstein-Friesian cows fed grass silage-or corn silage-based diets. J. Dairy Sci. 98 (3), 1915-27.

Van Gastelen, S., Dijkstra, J. and Bannink, A. 2019. Are dietary strategies to mitigate enteric methane emission equally effective across dairy cattle, beef cattle, and sheep? J. Dairy Sci. 102 (7), 6109-30.

Vanhatalo, A. and Jaakkola, S. 2016. Intake and performance with temperate forage legume-based ruminant production systems. Legume Perspect. (12), 17-9.

Van Kessel, J. A. S. and Russell, J. B. 1996. The effect of pH on ruminal methanogenesis. FEMS Microb. Ecol. 20 (4), 205-10.

Van Middelaar, C. E., Berentsen, P. B. M., Dijkstra, J., et al., 2013. Evaluation of a feeding strategy to reduce greenhouse gas emissions from dairy farming: the level of analysis matters. Agric. Syst. 121, 9-22.

Ventto, L., Leskinen, H., Kairenius, P., et al., 2017. Diet-induced milk fat depression is associated with alterations in ruminal biohydrogenation pathways and formation of novel fatty acid intermediates in lactating cows. Br. J. Nutr. 117 (3), 364-76.

Warner, D., Hatew, B., Podesta, S. C., et al., 2016. Effects of nitrogen fertilisation rate and maturity of grass silage on methane emission by lactating dairy cows. Animal 10 (1), 34-43.

Warner, D., Bannink, A., Hatew, B., et al., 2017. Effects of grass silage quality and level of feed intake on enteric methane production in lactating dairy cows. J. Anim. Sci. 95 (8), 3687-700.

Wilkinson, J. M. and Rinne, M. 2018. Highlights of progress in silage conservation and future perspectives. Grass Sci. 73 (1), 40-52.

第20章 应用舍饲与谷物提高反刍动物饲料利用率并减少甲烷产生

Kristian Hales、James E. Wells，美国农业部农业研究局
（USDA-ARS）-美国肉用动物研究中心；
Jeferson Lourenco、Darren S. Seidel、Osman Yasir Koyun、
Dylan Davis、Christina Welch、Todd R. Callaway，美国佐治亚大学
（杨鼎译）

1 前言

肉牛是动物进化的一个奇迹，它将低质饲草转化成可供人类食用的优质蛋白质。反刍动物具有非凡的能力，将太阳能转化成肉、奶和纤维，而这一切又是基于宿主动物与生活在其消化道内的微生物共生关系（Hungate，1966）。动物能量代谢依赖挥发性脂肪酸（VFAs），而VFAs的生成离不开混合微生物对饲料的分步发酵降解（Russell，2002）。可是，由瘤胃微生态产生的每一种VFAs的摩尔比例变化不仅受日粮组分的影响，同时也受消化道常驻微生物种群的影响，这对动物生产效率、胴体品质和食品安全具有深远影响（Russell和Hespell，1981；Depenbusch等，2008；Verdú等，2015；Wilson等，2016）。当反刍动物进化到降解饲草时，不仅具有发酵饲草的能力，还包括降解谷物的能力，包括：小麦、玉米、大麦、高粱和燕麦。

此时瘤胃环境高度被还原（Russell，2002），这意味着还原当量（如NADH、NAD）变得非常有限，进而危害饲料发酵的可持续性。微生物种间氢气传递重新生成NAD，进而生成甲烷，该发现也是研究瘤胃微生物群落共生关系的一个里程碑（Iannotti等，1973；Thiele和Zeikus，1988）。然而，由于没有其他可替代电子源，甲烷作为饲料厌氧降解过程中重要的还原当量库，使瘤胃微生物物种间转移氢气生成甲烷就显得尤为重要。尽管瘤胃微生物种群受益于产生的甲烷，但同时损失了大量的碳和能量，占动物总能量（GE）的2%~8%（Dong et al，2006）；在某些情况下，特别是当牛被喂食牧草时，甲烷可以占DE的12%（Johnson和Johnson，1995）。值得注意的是，甲烷作为一种强效的温室气体，在全球的重要性与日俱增。

排放温室气体（Greenhouse Gases，GHGs）是畜牧业影响全球气候变化的主要原因，分为直接原因（如肠道发酵）或间接原因（如饲料生产）（Steinfeld等，2006；Beauchemin，2009；Haque，2018）。二氧化碳（CO_2）、甲烷（CH_4）和一氧化二氮（N_2O）是家畜生产过程中产生的主要温室气体（Hungate，1966）。反刍动物是家畜排放温室气体的主要贡献者，约占家畜总排放量的80%（Opio等，2012）；而猪和家禽的贡献率分别约为9%和8%（Gerber等，2013）。反刍动物消化了能在光合作用中利用CO_2的植

物,所以动物排放的 CO_2 并不被视为气候变化的净贡献者(Steinfeld 等,2006)。因此,CH_4 和 N_2O 被认为是反刍动物生产系统中消化道发酵和饲料生产中产生的主要温室气体,在全球变暖潜势(GWP)中分别占 25 CO_2 当量和 298 CO_2 当量(Forster 等,2007;Eckard 等,2010)。

自 20 世纪 40 年代中期以来,美国牛肉生产行业越来越依赖谷物饲料来生产健康、优质蛋白质源。通过添加玉米等含淀粉饲料,微生物发酵的终产物发生了改变,对发酵效率和宿主生理产生重大影响。淀粉发酵会产生更大比例的丙酸,这将提高谷饲肉牛肌内大理石纹水平。本书其他章节将讨论瘤胃微生物发酵的终产物和瘤胃固有的低发酵效率,但在本章中,我们介绍通过肉牛谷饲来降低瘤胃微生物有毒发酵产物——甲烷。

2 肉牛谷饲的类型

美国的肉牛养殖业主要集中在占美国版图 1/3 的中部大平原地区。靠近粮食带意味着将谷物运输给肉牛在物流上和经济上都是可行的,因此美国的许多肉牛养殖场都位于中部高原地区。谷物是美国农业经济的重要组成部分,2016 年谷物产量超过 4.75 亿 t,包括小麦、玉米、大麦、高粱、大米和燕麦。由于谷物能满足现代生产高能量水平需求,谷物生产一部分供人类消费,但大部分用于饲料工业。淀粉通常是反刍动物日粮中用于提高产量的主要营养物质,而谷物是淀粉的有效来源(Theurer,1986)。

谷粒的果仁由 3 个主要部分组成,即麸皮(或外壳)、胚乳(淀粉的主要储存部位)和胚芽(或胚)。虽然谷物是优质淀粉源,但实际上谷粒细胞壁(外层)是无法消化的,需要进行饲喂前深度处理。一些常见的加工程序包括干法辗轧、研磨、破裂、蒸汽辗轧、蒸汽剥落、制粒和添加酶制剂。干法辗轧、破裂和研磨通常被归类在一起,因为每一种机械过程都通过打破种皮和减小颗粒尺寸来增加用于消化的谷物的表面积(Rowe 等,1999)。谷物蒸汽辗轧前处理是将谷物送入蒸箱加热来软化谷仁。还有一种是在蒸汽压片前将谷物置于高蒸汽压环境下。后一种方法使淀粉部分糊化,并破坏种皮和胚乳,使籽粒更容易被淀粉分解细菌降解(Rowe 等,1999)。制粒将小颗粒结合在一起,形成大颗粒,通过改变最终颗粒的密度和大小,可以控制消化的部位和速度(Rowe 等,1999)。在美国每年生产的谷物清单中,用于牲畜口粮的主要谷物是大麦、玉米和小麦。

2.1 大麦

大麦(*Hordeum vulgare* L.)产量仅次于玉米、水稻、小麦,居于第四位。2014 年,大麦总产量超过 1.44 亿 t,俄罗斯、法国、德国、澳大利亚和乌克兰是世界前五大大麦生产国。过去 5 年中,尽管大麦播种面积逐年下降,但通过高产品种和高效耕作方式使大麦总产量得以增加。大麦以其优质的营养价值成为肉牛高产日粮中最合适的补充。淀粉一般占大麦籽粒的 52%~60%,蛋白质和能量分别占大麦籽粒的 13% 和 85% 左右。虽然谷物通常钙含量低,磷含量高,但与玉米、小麦和高粱相比,大麦提供了更多的钙和 ß-葡聚糖。在大麦籽粒内,存在多种淀粉类型,包括蜡质淀粉、普通淀粉、高直链淀粉(Zhu,2017)。虽然大麦的营养价值对饲养牛是有益的,但大麦籽粒被纤维壳包围,在瘤胃中很难消化，需要一些如干碾等前加工，以最大限度地利用其营养（Beauchemin 等，1994）。有研究表明，干辗大麦可使整个大麦的消化率提高 32.7%（Toland，1976）。

2.2 玉米

由于其营养价值、高产和适口性等优点，玉米（*Zea Mays* L.）是家畜养殖业最常用的谷物。玉米不仅可以用来喂牛（加工或全株），还可以从玉米生产中获得多种其他形式的副产品，包括白酒糟（湿的和干的）、玉米青贮和玉米麸质（Firkins 等，1985；Fron 等，1996）。畜牧饲料工业受益于乙醇工业生产中多种玉米副产品，如浓缩可溶蒸馏液、湿式蒸馏谷物和干式蒸馏谷物。釜馏物是一种液体产品，由麦芽浆蒸馏而成，可用于替代牛日粮中的水，以减少干物质摄入而不影响牛的生产性能。釜馏物也可以脱水，制成浓缩的釜馏物或糖浆。从乙醇生产中产生的固体部分是常见的酒糟。酒糟具有易消化性，可替代全株玉米，并且由于其高脂肪，使它具有更高的能量值。通过机械加工，玉米的消化率提高了 5%~10%，并通过裂开果皮（光滑外表面），使微生物能够接触到籽粒内的淀粉，从而提高了营养物质的利用率。玉米副产品含 20%~35% 的粗蛋白质、70%~100% 的 TDN；而全株玉米含 70%~72% 的淀粉，10% 的蛋白质和 87% 的 TDN。（Beauchemin 等，1994）。

2.3 小麦

小麦（*Triticum aestivum* L.）由于其用途广泛，是畜牧饲料工业中重要的组成部分。虽然大多数谷物在春季种植，夏季生长，但小麦可在秋季种植，为牲畜在冬季提供良好的牧草来源。除了放牧，小麦还可以用来制作干草、青贮饲料，或者作为谷物喂养。小麦种子大约含有 60% 的淀粉，16% 的蛋白质、80% 的 TDN 或能量。麦仁由 3 部分组成：麸皮（外层）、胚乳（85%~86% 淀粉）和胚芽（胚）。麦麸或麦仁的外层能够逃避瘤胃发酵，因此需要做进一步的处理，如蒸汽剥落或干辗（Kreikemeier 等，1990）。如果饲喂全麦会增加咀嚼时间，同时干物质消失会降低消化率。就像其他谷物一样，小麦可以作为定量配给的有效能量来源，但其低钙高磷的特点，使得在饲喂小麦的同时要注意补钙。动物试验表明，与其他谷物相比，进一步加工后的小麦，其干物质消失的最多（Herrera-Saldana 等，1990）。

2.4 其他谷物

燕麦（*Avena sativa*）在全球谷物产量中排名第六，仅次于小麦、玉米、水稻、大麦和高粱（Stevens 等，2004）。与玉米类似，燕麦在畜牧业中有多种用途，包括牧草、饲料、青贮、半干青贮料、麦秆和干草。尽管燕麦还有其他用途，但 74% 的燕麦用于牲畜饲料（Stevens 等，2004）。与大麦相似，燕麦是有特别营养价值的 ß-葡聚糖和矿物质的重要来源，同时提供了约 13% 的粗蛋白质。虽然燕麦籽粒比其他谷物更软，给研磨过程造成了困难，但其籽粒含有 60% 的淀粉，使它成为饲料中有效的能量来源。高粱（*Sorghum bicolor*）不像其他谷物那样被广泛种植和应用（Etuk 等，2012）。高粱淀粉含量比燕麦高（74% vs 60%），蛋白质含量略低（12.3% vs 13%）。作为饲料，高粱与玉米相似，但仅有51%的高粱产量作为饲草、谷物或青贮。植物成熟度对高粱饲料至关重要，因为青绿高粱含有一种化合物——蜀黍苷。蜀黍苷是一种氰苷，在水解过程中产生氰化氢（HCN），对动物是致命的毒素（Etuk等，2012）。

3　谷物生产

谷物生产是世界经济的重要组成部分，2010年谷物产量超过6亿t。然而，有几个因素决定了并不是世界上每个地方都适合生产所有的谷物，这些因素包括：作物的产地，环境、文化和经济（Awika，2011）。决定谷物产地的主要因素是当地的环境条件。

3.1　玉米

就产量而言，玉米是所有谷物中最重要的，2010年玉米产量超过8亿t，其中40%由美国生产（Awika，2011）。由于玉米易受到干旱和霜冻的影响，所以它广泛种植于雨水充足或灌溉能力强的地区，在春天种植，夏季生长。美国多州都种植玉米，最集中的是中西部各州，俗称"玉米带"。组成"玉米带"的州主要是印第安纳州、伊利诺斯州、爱荷华州、明尼苏达州、密苏里州、内布拉斯加州和堪萨斯州，以及周边的一些州。

美国农业部（USDA）将产品附加值定义为通过改变产品的物理状态或形式来提高其价值。从这个意义上说，肉牛是一种巨大的玉米增值产品。肉牛进化已经和瘤胃微生物形成一种共生关系，使肉牛能够有效消化其他动物无法利用的饲料。瘤胃微生物降解玉米（和所有谷物）中的淀粉，产生VFA，为动物提供能量，促进动物生长，转化率为每磅（1磅≈0.454 kg）增重消耗5~7磅饲料。饲料玉米通常不适合人类直接食用，但饲养肉牛可将其转化为可消化、高营养价值的动物蛋白。例如，每蒲式耳（bushel，1蒲式耳=27.216 kg）玉米平均重56磅，目前每蒲式耳价值3.61美元，约合每磅0.065美元。目前的活牛价格为每磅1.26美元，而在美国一家知名联锁超市买的谷物喂养的精选肋眼牛排价格为每磅9.99美元。根据这些信息，可以计算出每1 000磅玉米价值65美元，每1 000磅活牛价值1 260美元，尽管这种简化的计算忽略了一些其他成本（例如，不同的饲料转换效率、维持饲养场的成本、运输、机会成本、债务偿还等），把肉牛产品运到市场，再到消费者的餐桌上。玉米是肉牛日粮的重要组成部分，因为玉米能够满足肉牛高耗能生产系统的能量需求。由于玉米和肉牛巨大的价格差异，作为玉米的增值产品，养殖者必将会从肉牛养殖中获益。

3.2　小麦

小麦可以说是用途最广的谷物。由于其抗干旱和抗温差特性，小麦能在冬季和夏季，以及更广泛的环境中种植（Awika，2011）。由于其在不同气候和温度下的多样性，小麦的生产分布在美国和加拿大的部分地区。随着气候和温度的不同，小麦产量在美国和加拿大部分地区也显示出多样化。"小麦带"从得克萨斯州中部延伸到阿尔伯塔省中部，包括俄克拉荷马州、堪萨斯州、内布拉斯加州、蒙大拿州和达科他州。

3.3 大麦

大麦比其他谷物更具有耐寒性,能在更寒冷的气候中生存。美国主要的大麦产地在北部和西北部的蒙大拿州、北达科他州、爱达荷州和华盛顿州,2008 年至 2012 年平均年产量为 2.05 亿蒲式耳。在北美,加拿大的大麦产量最多,比美国多出 40%左右。

3.4 其他谷物

在美国,高粱和燕麦的产量远低于玉米、小麦、大麦甚至大米(尽管大米一般不用作牲畜饲料)。北美的大部分燕麦和大麦都是由加拿大生产的。在美国,燕麦产量最集中的地区是爱荷华州、明尼苏达州、南达科他州、北达科他州和威斯康星州。与其他谷物相比,高粱更耐高温和干旱,这使得它在非洲国家很受欢迎(Awika,2011)。在美国,它主要生长在从得克萨斯州南部到南达科他州的"高粱带"的干旱地带,包括俄克拉荷马州、新墨西哥州、科罗拉多州、堪萨斯州和内布拉斯加州。

4 影响反刍动物产甲烷的日粮因素

由反刍动物产生的大多数消化道 CH_4 是在厌氧条件下,通过瘤胃微生物(包括细菌、原虫和真菌)发酵碳水化合物、蛋白质和脂类生成,接下来生成 VFA,主要是乙酸、丙酸、丁酸作为动物能量的来源(Martin 等,2010)。该发酵过程的副产物(CO_2、H_2 和 CH_4)则通过打嗝的方式排出体外(Boadi 等,2004;Kebreab 等,2006)。生成甲烷的发酵过程是一个氧化反应,还原辅助因子(NADH、NADPH、FADH)被重新氧化成(NAD^+、$NADP^+$、FAD^+),通过脱氢反应在瘤胃内产生 H_2。产生的 H_2 被产甲烷菌(一种有别于真细菌的微生物)利用,将 CO_2 或甲酸还原为 CH_4,(McAllister 和 Newbold,2008;Martin 等,2010)。然而,还有一些瘤胃产甲烷菌,如 Methanomassiliicoccaceae,通过从其他代谢物(如甲胺)中还原甲基来生成甲烷(Martinez-Fernandez 等,2018)。这一代谢途径对瘤胃产生甲烷具有重要意义,因为果胶含有较高比例的甲基化合物。此外,乙酸产生的同时释放了 H_2,而丙酸成为了净 H_2 库。因此,考虑到可供产甲烷菌还原 CO_2 生成 CH_4 反应的 H_2 较少,调整日粮结构使瘤胃中丙酸提高,乙酸减少,能够减少瘤胃中 CH_4 产量(Beauchemin,2009)。

反刍动物释放 CH_4 的多少取决于多种因素,包括:碳水化合物摄入量、碳水化合物来源类型、瘤胃 pH 值、过瘤胃时间、瘤胃发酵速率和甲烷生成速率。在不影响动物生长性能的前提下,通过降低采食量和(或)瘤胃发酵速率可抑制消化道 CH_4 的日排放。由于脂肪可替代碳水化合物,添加脂肪的日粮可降低瘤胃内碳水化合物的发酵;但会影响纤维消化率,导致反刍动物饲料转化率降低(Martin 等,2008;Beauchemin,2009)。干物质采食量(DMI)是影响反刍动物甲烷产量的另一个主要因素,DMI 与甲烷产量呈正相关关系。然而,单位采食量的甲烷产量(g CH_4/kg DMI)随着 DMI 的增加而降低,这表明瘤胃的高运转会导致低饲料消化率(Buddle 等,2011)。

碳水化合物类型在 CH_4 生产中起着关键作用,因为它可以影响瘤胃 pH 值,并随后改变微生物群。与可溶性碳水化合物相比,纤维素和半纤维素的消化率与甲烷产量高度相关(Hook 等,2010)。非泌乳期奶牛饲草中半纤维素消化率与甲烷排放呈正相关;然而,也有研究表明纤维素消化率与甲烷排放之间存在负相关关系(Holter 和 Young,1992)。另一项研究发现,甲烷产量似乎与日粮中精料的比例存在曲线关系,当精料占 30%~40%时,生产甲烷损失总能量(GE)的 6%~7%,而当精料水平达到 80%~90%时,生产甲烷损失总能也降低到 2%~3%(Sauvant 和 Giger-Reverdin,2007)。另外,在奶牛饲养中提高精料水平以抑制甲烷排放的前景却并不乐观,因为精料水平超过日粮的 50%会对牛奶质量产生不利影响(Beauchemin 等,2008)。

重要的是,当提升快速发酵性碳水化合物比例后,会加快瘤胃转化速率,降低饲料过瘤胃时间和瘤胃 pH 值,以致碳水化合物的降解(生成甲烷)由瘤胃转移向后消化道和粪便(Hindrichsen 等,2006;Hook 等,2010)。饲喂反刍动物前,对饲料进行研磨似乎会减少甲烷的产生,其原因最可能是提高了瘤胃消化和胃肠道流速(GIT);缩短了瘤胃生产甲烷的时间(Johnson 和 Johnson,1995)。此外,瘤胃内碳水化合物的快速发酵提高了 VFAs 的产量,但如果 VFAs 生成的量大于瘤胃吸收量,pH 值还会下降,这样会增加瘤胃酸中毒的风险并导致瘤胃微生物失衡(Plaizier 等,2008)。

5 淀粉和饲草在甲烷形成中的作用

当需要动物处于高水平生产性能时,淀粉是肉牛饲料增加消化能水平的主要营养物质(Theurer,1986)。淀粉分为两种:直链淀粉和支链淀粉。直链淀粉是一种线性分子,只占淀粉的 20%~30%,主要由 α-1,4 糖苷键和很少的 α-1,6 分枝点组成(图 1)。

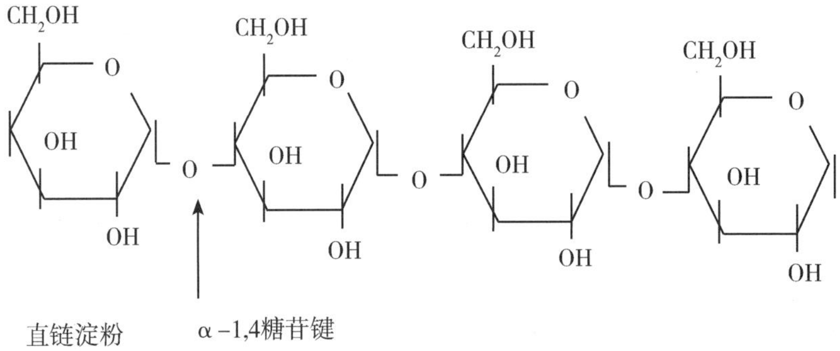

图 1 直链淀粉(淀粉),以 α-1,4 糖苷键相连

另外,支链淀粉是一种更大的分子(尽管密度较小),它被描述为"蓬松的",因为它有大量的 α-1,6 键和 α-1,4 键。一般来说,谷物由 50%以上的淀粉组成,淀粉颗粒的大小取决于直链淀粉和支链淀粉的比例。淀粉颗粒的大小与直链淀粉的含量呈负相关。直链淀粉被淀粉酶(α 和 β)等酶分解成糖,并迅速发酵产生 VFA(乙酸和丙酸)和乳酸。支链淀粉也可被淀粉酶(α 和 β)分解,但由于 α-1,6 分枝点的增加,产物包括糖和极限糊精(图 2)。

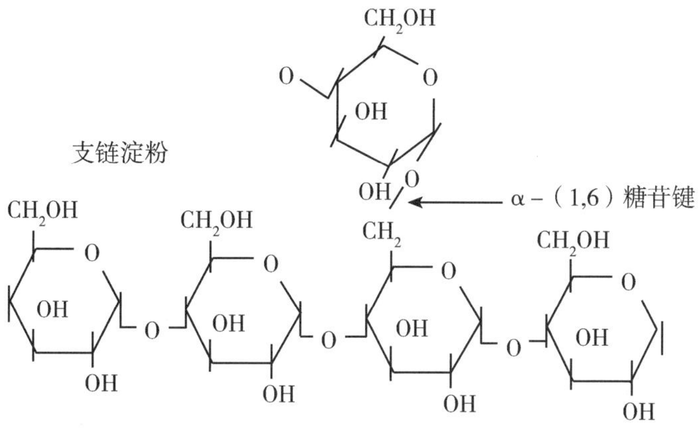

图 2 支链淀粉结构，以 α-1,4 糖苷键和 α-1,6 糖苷键相连

糖在瘤胃内快速发酵，而极限糊精需要额外的酶才能完全降解，如极限糊精酶。淀粉在瘤胃中由淀粉分解细菌发酵［即牛链球菌、瘤胃普雷沃氏菌、反刍杆菌属（原拟杆菌）嗜淀粉菌、反刍月形单胞菌、淀粉分解琥珀单胞菌］产生乙酸盐、丙酸盐和乳酸盐。随着瘤胃内淀粉发酵，丙酸产量增加，乙酸与丙酸的比值降低，为动物提供更多的有效能量。虽然乳酸对动物是有害的，比如酸中毒；但其能平衡过量的饲草，可以预防许多其他负面影响。

淀粉是高产奶牛产葡萄糖能的关键来源，也是瘤胃微生物发酵的重要能源。(Koenig 等，2003)。淀粉一旦进入瘤胃，主要由淀粉分解菌来降解淀粉，部分由原虫和真菌降解。瘤胃微生物产生的酶能够水解直链淀粉和支链淀粉的糖苷键，释放各种低聚糖。通过胰腺分泌 α-淀粉酶，将直链淀粉和支链淀粉水解成糊精和更小的低聚糖，触发了瘤胃分解淀粉后半程。这一过程以小肠分泌的麦芽糖酶和异麦芽糖酶的作用而结束，但有研究表明，由于进化上很少有淀粉到达反刍动物小肠，所以限制了淀粉在反刍动物小肠内的消化（Ortega Cerrilla 和 Mendoza-Martínez，2003；Gómez 等，2016）。淀粉分解位点影响反刍动物对底物的吸收。瘤胃淀粉分解降低了肠内 CH_4 的形成，生成可替代 H_2 库用于形成甲烷，并产生 VFAs 可被宿主吸收和为微生物蛋白质合成提供能量。然而，淀粉消化率的降低有益于预防酸中毒和增加产糖底物。另外，反刍动物小肠淀粉降解比微生物发酵降解具有更高的能量效率，主要是因为微生物还原 CH_4、发酵热损失，以及更高的代谢效率。值得注意的是，更低的瘤胃淀粉消化率与较高的小肠淀粉消化率并不相关；但它与较高的后肠消化率和较低的全肠道消化率有关（Larsen 等，2009）。

日粮中谷物的选择对甲烷的生产水平以及牛奶和肉类的生产效率也至关重要，这为通过当前的技术和饲养系统降低甲烷排放提供了潜在的方法（Martin 等，2010；Moate 等，2016）。将澳大利亚肉牛日粮由玉米改为小麦，将使甲烷产量（g CH_4/kg DMI）显著下降（Moate 等，2018）。有研究证明，日粮主要成分为小麦的和主要成分为玉米的相比，小麦比玉米所生成甲烷产量要低（11.1<19.5 g CH_4/kg DMI），甲烷效率也低（7.6<15.7 g CH_4/kg 牛奶）（Moate 等，2019）。然而，有人指出在牛奶产量和甲烷生成方面，一些牛对随时间变化的日粮具有"适应性"，但并非所有牛的反应都是相同的（Moate 等，

2018）。因此，谷物可消化淀粉在瘤胃甲烷生成中发挥了重要的作用。但目前还不清楚还原甲烷的作用机理，这为通过分析微生物组的变化以及该变化是如何影响甲烷形成提出了问题导向。

饲草质量也对瘤胃生成 CH_4 有影响。优质牧草（即幼嫩植物）由于含有较多的易发酵碳水化合物和较少的中性洗涤纤维（NDF），可通过改变发酵路径降低 CH_4 产量，从而提高消化率和通过率（Beever 等，1986）。而成熟饲草由于 C：N 比较高，使 CH_4 产量较高（如甲烷产量/kg DMI），导致消化率降低（Milich，1999）。这与人们普遍接受的增加成熟牧草的有效氮能提高牧草消化率的认识形成了对比。

不同类型的牧草由于所含化学成分的不同也会影响 CH_4 的排放。谷类饲料由于含有大量淀粉，更有利于形成丙酸，而不是乙酸，这样可以减少瘤胃 CH_4 的形成（Beauchemin 等，2008）。豆科牧草由于含有缩合单宁，低纤维，高 DMI，以及高流通速度，导致 CH_4 产量较低（Beauchemin 等，2008）。然而，其他关于牧草类型对甲烷产量影响的研究结果并不一致（Benchaar 等，2001；Hammond 等，2013）。有研究表明，C4 植物比 C3 植物产生更多的 CH_4（Archimède 等，2011）。此外，饲料加工和保存也会影响 CH_4 的排放。例如，当饲草被切碎或颗粒化时，较小的颗粒在瘤胃中分解得更少，所以可以减少每千克 DMI 的 CH_4 产量（Boadi 等，2004；Martin 等，2010）。青贮饲料可消化的低聚糖在发酵过程中会产生少量的 CH_4（Boadi 等，2004）。

6 瘤胃 H_2 库和甲烷的产生

甲烷对瘤胃正常生理功能至关重要，因为它作为一个还原物库阻止了 H_2 的积累。消除甲烷将抑制参与还原辅助因子氧化反应中的脱氢酶活性（Wolin，1975；McAllister 和 Newbold，2008）。微生物发酵底物产生各种不同的最终产物中，在 H_2 产量方面是不相等的。因此，积累的 H_2 会抑制瘤胃微生物氧化具有电子传递作用的辅助因子，从而减少发酵过程产生的能量（Beauchemin，2009；Martin 等，2010）。瘤胃中形成乙酸和丁酸的过程会释放 H_2，这有利于形成 CH_4，但与形成丙酸成竞争关系，因为丙酸的形成也需要 H_2 的参与（Boadi 等，2004）。效率是一个需要定义的复杂生产性状，但在特定的甲烷产量或产甲烷细菌数量上，被定义为生产效率高的动物与生产效率低的动物之间没有区别（Freetly 等，2015）。

瘤胃是一个厌氧发酵室，生活其中的微生物种群处于共生关系，互相交换代谢物以促进彼此的生长；因此这种相互作用被称为"交叉饲养"（Schultz 和 Breznak，1979）。甲烷的形成被认为是产 H_2 微生物种群（如溶纤维真菌和细菌）和噬 H_2 产甲烷菌之间的交叉饲养，以消除 H_2 并改善纤维发酵（Kobayashi，2010）。目前减少甲烷策略主要考虑的是找到与生成甲烷竞争 H_2 库的可替代代谢途径。

在瘤胃中有几种消耗 H_2 的代谢途径。以产甲烷为主，其次是产丙酸（延胡索酸的还原反应）。其他途径如硝酸盐和亚硝酸盐还原、还原生成乙酸，和不饱和脂肪酸的生物氢化作用对瘤胃中 H_2 的消耗较小（Kobayashi，2010）。

由于莫能菌素对原虫和革兰氏阳性细菌（包括为甲烷生成提供底物的瘤胃球菌、链球菌和乳酸菌)有抑制作用,作为缓解甲烷生成的策略,莫能菌素引起了人们的兴趣(Russell 和 Strobel,1989)。莫能菌素放过了革兰氏阴性菌,使瘤胃转向生产丙酸,所以可以推论莫能菌素并没有限制产甲烷菌来影响 CH_4 的生成(Martin 等,2010),但却抑制了为形成甲烷提供底物的细菌和原虫(Bergen 和 Bates 1984;Russell 和 Strobel,1989)。该假设得到了以下试验结果的支持:在体外瘤胃液中添加莫能菌素后,CH_4 的产量会持续下降,直到提供 H_2 时,CH_4 的形成才会恢复(Russell 和 Strobel,1989)。

莫能菌素通常作为预混料或通过缓释胶囊补充到日粮中(Beauchemin 等,2008)。有研究表明,莫能菌素降低 CH_4 产量的作用似乎具有剂量依赖性,较低剂量(10~15 mg/kg)可引起奶牛的有效产奶反应,但对 CH_4 无影响;而较高剂量(24~35 mg/kg)可降低 CH_4 产量(Eckard 等,2010)。然而,对于莫能菌素抑制作用的可持续性,不同的研究结果却存在着矛盾。有研究表明,长期给奶牛服用莫能菌素可以在不影响产奶量下,6 个月内减少了 7% 的甲烷(Odongo 等,2007;Ellis 等,2008)。还有研究称,莫能菌素(33 mg/kg)可使肉牛的 CH_4 排放减少高达 30%,但停药后在 2 个月内 CH_4 水平又恢复到处理前的水平(Guan 等,2006)。

在瘤胃中,二羧酸如延胡索酸和苹果酸是生成丙酸的前体,替代 H_2 库以减缓甲烷生成(Nisbet 和 Martin,1993;Martin 和 Streeter,1995;McAllister 和 Newbold,2008)。已有研究报道,补充延胡索酸或苹果酸可降低 CH_4 的生成(Callaway 和 Martin,1996;McGinn 等,2004;Foley 等,2009;Wood 等,2009)。然而,作者们却认为补充有机酸的效果似乎是受到日粮的影响。当日粮中添加高精料时,甲烷形成获得了更强的还原反应,这很可能是受瘤胃中乙酸:丙酸(A:P)比的影响,此外,它还具有作为 H_2 库的功能。添加高水平有机酸降低了 DMI 和瘤胃 pH 值,不利于瘤胃纤维发酵(Beauchemin 等,2009)。为了克服以上不足,通过将有机酸与脂肪包裹在一起,以减缓它们在瘤胃的释放(Wallace 等,2006;Martin 等,2010),但其成本太高不适宜广泛推广。

7 利用谷物提高饲料转化率并降低甲烷产量

一般情况下,反刍动物会以甲烷的形式损失其摄入能量的 2%~12%(Ferrell,1988;Harper 等,1999)。因此,减少肉牛甲烷排放的策略不仅对环境有益,也对动物有益,因为如果碳和氢仅以可代谢能量形式被动物利用,那也意味着优秀的动物生产性能。如果 H_2 仅是简单的以另一种形式被固定或作为 H_2 气体排放,那么它对动物来说仍然是一种损失。简单地减少产甲烷菌的数量可以减少甲烷的产量(至少是暂时的),但这种甲烷还原并不影响动物的生产性能(Patra 和 Saxena,2009;Hook 等,2010;Wright 和 Klieve,2011)。因此,必须将甲烷损失的能量转换为动物可用的其他形式(如丙酸),减少可代谢能量的浪费。综上所述,肉牛甲烷产量受多种因素影响,如日粮中离子载体和脂质的添加、采食量的水平,甚至还受日粮类型的影响(Johnson 和 Johnson,1995;Grainger 和 Beauchemin,2011),在本节中,作者着重讨论了后一个因素:提供给牛的日粮类型。更具体地说,作者关注的是谷物的使用,以及增加它们在饲料中所占比例(以牺牲粗饲料来源为代价)如何影响饲料效率和甲烷的生成。

快速可降解碳水化合物（如淀粉）在瘤胃的发酵促进了丙酸的生成，为甲烷生成创造一个替代的氢库（Grainger 和 Beauchemin，2011）。此外，这种发酵方式降低了瘤胃 pH 值，抑制了瘤胃产甲烷菌的生长，并降低了瘤胃原虫的数量。有人认为，瘤胃原虫在吞食和发酵淀粉颗粒的过程中起到了隔离淀粉的作用。玉米日粮的瘤胃内毛虫的含量高于大麦日粮，进一步证明了原虫在瘤胃淀粉发酵中发挥了重要作用（Xia 等，2015）。原虫数量的减少限制了氢气从原虫向产甲烷菌的转移，有助于减少甲烷的生成（Grainger 和 Beauchemin，2011）。

Harper 等测量了放牧和舍饲条件下肉牛甲烷的排放。他们发现牛在牧场上吃草时，会浪费摄入的 7.7%~8.4% 的 GE 用于生成甲烷。然而，当同一组小母牛饲喂高度可消化的高谷物日粮时，甲烷的能量损失降至仅为 GE 摄入量的 1.9%~2.2%。这种几乎 4 倍大小的差异表明，饲喂低品质（高纤维）日粮的牛比饲喂高品质、高谷物日粮的牛会产生更多的甲烷（Harper 等，1999）。同样，为了量化不同精料比下产生的甲烷量，Hales 等（2014）给小公牛饲喂了 2%~14% 的苜蓿干草日粮。他们发现，随着日粮中饲草水平的增加（或干-碎玉米水平的降低），肉牛的干物质消化率、消化能和代谢能呈线性下降。这导致了增重时的低能量保留。这种较低的能量保留归因于几个因素，包括生成甲烷造成的较高能量损失，因为牛的日粮中有更多的饲草。例如，当日粮中饲草含量从 2% 增加到 14% 时，甲烷引起的能量摄入损失百分比从 3.07% 增加到 4.18%。作者得出结论，甲烷损失随饲草水平的增加而线性增加，这与动物保留的能量减少相一致。

从本质上讲，与发酵可溶性碳水化合物相比，发酵细胞壁成分（纤维素和半纤维素）会导致更高的乙酸与丙酸的摩尔比，从而比发酵碳水化合物产生更多的甲烷（Johnson 和 Johnson，1995）。由于甲烷的产量与瘤胃中碳水化合物发酵产生的最终产物有关（Fahey 和 Berger，1988），以丙酸为代价生成更多的乙酸会生成更多的甲烷，见图 3。

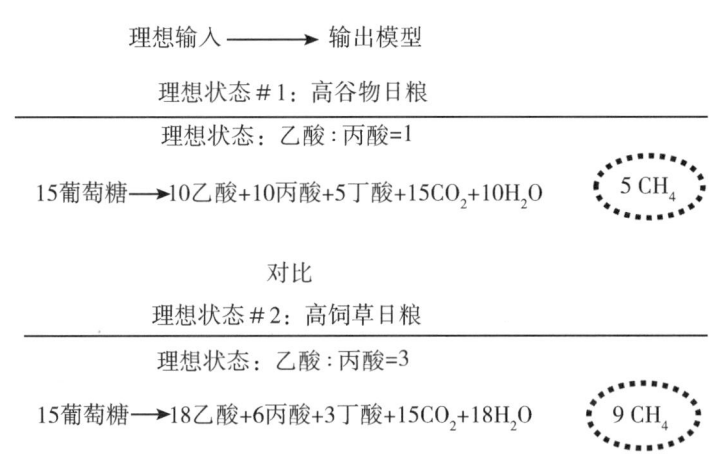

图 3　在高谷物和高饲料两种假想情况下，15 个葡萄糖分子的瘤胃发酵理论模型

如果对肉牛饲喂高谷物或高饲草日粮，会产生不同的乙酸/丙酸比，即发酵同等数量葡萄糖可产生截然不同数量的甲烷。事实上如图 1 所示，饲喂高饲草料日粮所产生的甲烷是高谷物日粮的 1.8 倍。由此可见，乙酸产量与甲烷产量呈正相关，而丙酸产量与甲烷产量呈负相关。根据 Fahey 和 Berger（1988）的研究，这些关系产生于丙酸形成时，葡萄糖中存在更多的 C 原子和 H 原子（Fahey 和 Berger，1988）。而当葡萄糖转化为乙酸时，会释放更多的 H 原子，从而生成更多的甲烷。

自然界中大约 2/3 的甲烷来自乙酸盐，更确切地说，来自乙酸的甲基（Ferry，1992；Wolfe，1993）。如前所述，从生产力的角度来看，甲烷被认为是一种能源的浪费。因此，当生成更多丙酸（而不是乙酸）时，动物生产性能通常会提高，其原因之一是更多的 C 原子和 H 原子变成丙酸而不是甲烷，增加了日粮的代谢能水平（Fahey 和 Berger，1988）。在实际应用中，乙酸/丙酸比通常在 0.9~4.0 范围内变化，因此相应的甲烷损失也很大。然而，正如本节所强调的，文献清楚地表明，给反刍动物喂食更多的谷物可以大大减少它们的甲烷排放。它不仅可以降低消耗每千克 DM 的甲烷排放量，而且可以减少每千克最终产品（如牛奶、肉类）的甲烷总排放量，因为饲喂更多的谷物通常会提高动物的生产性能。

因此，制定减少肉牛甲烷排放策略至关重要，在不减少甚至改善动物生产性能的同时实现这一目标，实现提高肉牛饲喂谷物水平和减少甲烷排放"一举两得"。

8 谷物发酵微生物学

瘤胃微生物区系是由古生菌、细菌、真菌和原虫组成的密集而多样的联合体，它们都通过反刍动物宿主的饮食来争夺资源。独特的是，这些微生物为反刍动物提供了一个进化生态位，在这个生态位中，微生物能够利用复杂的碳水化合物，如纤维素和半纤维素（分解纤维微生物），通过特定的代谢途径产生可被动物吸收和利用的发酵副产物。反刍动物宿主反过来为微生物提供适宜的温度（39℃）、厌氧环境和富营养的培养基。淀粉发酵微生物含有 α-淀粉酶和 β-淀粉酶，这使它们能够降解直链淀粉和支链淀粉形式的葡萄糖多糖。一般来说，大多数淀粉和饲草发酵会产生短链脂肪酸（甲酸、乙酸、丙酸、丁酸和支链 SCFA），产生三磷酸腺苷用于微生物细胞能量，微生物粗蛋白质合成，热+气体（甲烷和二氧化碳）作为动物和微生物的一种能量损失。

Robert Hungate 和 Marvin Bryant 是现代瘤胃微生物学和微生物生态学的先驱（Krause 等，2013），同时也是两位首先研究瘤胃微生物分离、纯化培养、反刍动物共生微生物相互作用的微生物学家（Hungate，1944；Hungate，1947；Bryant，1959）。最后，反刍动物依靠瘤胃中大多数前消化道发酵终产物，同时还有第二个后消化道发酵位点，产生占反刍动物总消耗能 10%~20%，这与单胃动物一样（Russell，2002）。体外研究表明，盲肠食糜产甲烷量是瘤胃食糜的 6.8%（Freetly 等，2015）。此外，在瘤胃和盲肠两个消化点之间没有观察到产甲烷细菌种群的差异。然而，本书要对瘤胃中主要利用淀粉微生物和产生甲烷的古菌进行全面讨论。

Hungate 和 Bryant 的早期工作是微生物的基本生态位分类：纤维溶解性、简单碳水化合物、专性氨基酸和产甲烷发酵微生物。虽然可以很容易将这些瘤胃微生物分类，但值得注意的是，目前的技术已经表明，微生物的多样性可能比科学家之前认识到的要大。瘤胃的微生物生态非常丰富，通过在纯培养实验中培养和研究，认识到多种发酵淀粉的细菌和产甲烷的古菌。表1说明了从罗素（2002）改良的淀粉降解和产甲烷微生物的生态位和副产物。

表1 瘤胃微生物菌种、生态位和副产物

物种	初级生态位	产物
溶纤维丁酸弧菌	纤维素、半纤维素、淀粉、果胶和糖	丁酸盐、甲酸盐、乳酸盐和乙酸盐
嗜淀粉瘤胃杆菌	淀粉	琥珀酸盐、甲酸盐和乙酸盐
反刍月形单胞菌	糖、淀粉和乳酸	乳酸、乙酸、丙酸、丁酸和 H_2
瘤胃普雷沃氏菌、阿尔伯普雷沃氏菌、布氏普雷沃氏菌、短普雷沃氏菌	淀粉、半纤维素、果胶、β-葡聚糖、蛋白质	琥珀酸、乙酸、甲酸、丙酸
溶淀粉琥珀酸单胞菌	淀粉	琥珀酸、乙酸、丙酸
牛链球菌	淀粉、糖	乳酸盐、乙酸盐、甲酸盐、乙醇
瘤胃甲烷短杆菌	氢气、二氧化碳、甲酸盐	甲烷
斯氏甲烷球形菌	氢气、甲醇	

资料来源：Liu 和 Whitman，2008；Russell，2002。

随着研究从基于培养的方法转移到二代测序和 Hungate1000 项目数据，发现了关于微生物种群变化的进一步信息（McAllister 等，2015），（Seshadri 等，2018）。使用淀粉作为选择剂进行瘤胃液体外培养，增加了与淀粉饲养相关的普雷沃氏菌种群（Bandarupalli，2017）。然而体外研究表明，过去认为在淀粉发酵中起重要作用的许多细菌（如牛链球菌、嗜淀粉瘤胃杆菌、淀粉琥珀单胞菌、纤纤维溶丁酸弧菌），在饲喂玉米或大麦日粮的牛瘤胃液中却没有检测到高细菌浓度（Xia 等，2015）。由于高谷物日粮而产生泡沫性臌胀症肉牛，增加了瘤胃中梭状芽胞杆菌、真杆菌和丁酸弧菌的数量，而降低了普雷沃氏菌的数量（Pitta 等，2016）。

9 参与发酵的细菌与古细菌

9.1 溶纤维丁酸弧菌

丁酸弧菌属是一种多功能革兰氏阴性杆菌，具有降解戊糖、己糖以及淀粉和半纤维素的能力。Bryant首先分离了丁酸弧菌，并根据该菌代谢产生丁酸的代谢产物特点进行分类，包括：己糖、丙酮酸、乙酰辅酶A、丁酸（$NADP^+$再生）和乙酸（生成ATP）。有些丁酸弧菌具有丁酸激酶，且不会产生乳酸；然而没有丁酸激酶的丁酸弧菌会在没有乙酸的情况下，将碳骨架还原为乳酸。

9.2 嗜淀粉瘤胃杆菌

嗜淀粉瘤胃杆菌是一个很好的降解淀粉的细菌模型，因为它仅利用葡萄糖α 1-4 聚合物作为能量，而且它是第一个被发现在细胞表面有类似于肠道拟杆菌的淀粉结合位点的瘤胃细菌（Anderson，1995）。嗜淀粉瘤胃杆菌（*R. amylophilus*）是一种革兰氏阴性、厌氧和水解蛋白质（仅与氨氮）的微生物，依赖麦芽糖、麦芽糊精和淀粉降解多糖作为能量来源。

9.3 反刍月形单胞菌

反刍月形单胞菌（*Selenomonas ruminantium*）是一种革兰氏阴性、严格厌氧菌，具有杆状结构和独特的鞭毛，鞭毛允许微生物在两个轴上旋转。反刍月形单胞菌在有糖条件下生长非常迅速，几乎严格以同乳酸方式（丙酮酸转化为乳酸）发酵。尽管反刍月形单胞菌的生态位与单糖或双糖有关，但在单糖缺失的条件下，反刍月形单胞菌可利用来自其他淀粉发酵微生物的乳酸和糊精，作为产生 ATP 的主要来源。

9.4 普雷沃氏菌

瘤胃普雷沃氏菌（*Prevotella ruminicola*）因其对胆盐和磷酸己糖途径的敏感性，成为第一个从瘤胃拟杆菌（*Bacteroides ruminicola*）大类中重新分类的微生物（Russell，2002）。20 世纪 90 年代中期，根据酶活性和木糖利用情况（羧甲基纤维素酶-阴性、生产脱氧核糖核酸酶、不利用木糖），瘤胃普雷沃氏菌被进一步划分为以下不同种类（栖瘤胃拟杆菌、阿尔伯弧菌、布氏普雷沃氏菌和短普雷沃氏菌）；短普雷沃氏菌，羧甲基纤维素酶-阳性，生产脱氧核糖核酸酶，利用木糖；布氏普雷沃氏菌，羧甲基纤维素酶-阳性，但不产生脱氧核糖核酸酶，均属于瘤胃普雷沃氏菌（Avgustin 等，1997）。普雷沃氏菌属是多形性杆状细菌，可产生大量的琥珀酸，利用除纤维素外的多种底物获取能量，是瘤胃内最丰富的典型菌属。

9.5 溶淀粉琥珀酸单胞菌

溶淀粉琥珀单胞菌（*Succinomonas amylolytica*）是一种严格厌氧菌，呈杆状到球状的细胞，在瘤胃中位居淀粉降解生态位（Bryant 等，1957）。溶淀粉琥珀单胞菌在去除瘤胃液后，在富含胰蛋白酶和酵母的培养基中生长良好，解淀粉菌生长良好，但在无碳酸氢盐培养基中则完全不生长。溶淀粉琥珀单胞菌主要副产物是琥珀酸和乙酸，仅产生少量丙酸。

9.6 牛链球菌

牛链球菌（*Streptococcus bovis*）是一种卵形的、革兰氏阳性、兼性厌氧菌，最佳培养条件为缺氧的还原环境。牛链球菌是瘤胃中生长最快的细菌之一，已公布的传代时间短至 24min（Russell 和 Robinson，1984）。然而，这种快速传代也带来不利的后果，因为当环境中含有充足的糖和淀粉时，牛链球菌会同时发酵产生过量的乳酸，这导致反刍动物瘤胃的 pH 值下降和酸性条件的增加。

9.7 参与发酵的古细菌

Methanobrevibacter ruminantium 和 *Methanosphaera stadtmanae* 都属于古细菌甲烷杆菌科。

甲烷是瘤胃发酵的主要副产品，成年牛一天可产生约400L甲烷。古细菌在牛体内广泛分布，其中 *Methanobrevibacter gottschalkii* 和 *Methanobrevibacter ruminantium* 占古细菌总数的74%（Henderson 等，2015）。5个主要的产甲烷菌群组成了89%以上的古细菌群落（Henderson 等，2015）。

反刍动物微生物学家和营养学家研究了减少甲烷产量的替代方案，因为多达10%的日粮 GE 会因为甲烷的生成而被损失（Blaxter，1962）。然而，产甲烷是瘤胃的一个自然过程，产甲烷菌需要利用 H^+ 和 CO_2，来维持较低的氢离子浓度，而甲烷作为一种副产物被生产出来。与淀粉分解菌相比，产甲烷菌往往与纤维溶解菌有更强的共生关系和生产效率。人们早就知道，当添加谷物到瘤胃日粮中时，丙酸产量增加，甲烷产量却减少了（Van Kessel，1996；Kowalczyk，1986）。

泡沫性臌胀症是许多牧场牛群中遇到的一种疾病，其原因是饲喂较高比例谷物饲料后，黏性瘤胃液捕获了气体并产生气泡，与对照组相比，泡沫性鼓胀症牛的古细菌-产甲烷短杆菌较高（Pitta 等，2016）。见图4。

10 过瘤胃时间

过瘤胃时间指的是饲料在瘤胃中保留的时间，是最有助于饲料降解和营养物质吸收的因素之一。饲料经过瘤胃后，通过网胃-瓣胃孔进入消化道其余部分。饲料的过瘤胃时间受多种因素影响，动物日粮是其中之一。由于分解速度快，富含易发酵淀粉的精饲料减少了过瘤胃时间。在接种稀释后绵羊瘤胃液 RUSITEC（瘤胃模拟技术）发酵罐试验中，DMI、VFA 和氨氮产量均得到不同程度提高（Martínez 等，2009）。

当饲料的过瘤胃时间减少时，产甲烷菌在瘤胃中甲烷产率也会降低，因为产甲烷菌的生长速度相对较慢，无法维持足够数量来满足较高的产甲烷量。提高瘤胃液稀释率可以抑制产甲烷菌的生长，限制它们在瘤胃中的丰度（Eun 等，2004），最终减少产生的甲烷。通过改变饮食或其他因素可以实现上述过程。Okine 等发现用惰性物质取代瘤胃的自由容积，甲烷生成量下降29%，瘤胃液稀释率增加43%。1/4 的瘤胃甲烷产量变化是由进食期间和进食后，瘤胃液流速引起的（Okine 等，1989）。

通过遗传育种可以选择具有特定甲烷生成量水平表型的动物，这些动物具有相似的甲烷产量和过瘤胃时间。产甲烷生成量较少的绵羊在固体稀释率和此液体稀释率下的过瘤胃时间上都有所减少，分别占甲烷量变化的59%和70%（Goopy 等，2014）。甲烷产出量较少的绵羊瘤胃中的颗粒量（g DM）也比甲烷产出量多的绵羊少（Goopy 等，2014）。甲烷产量与瘤胃液以及颗粒平均过瘤胃时间呈正相关。如前所述，甲烷产量与过瘤胃时间相关。

11 饲料引起的酸中毒和其他负面影响

如果过快地将过多的谷物添加入饲料中，瘤胃 pH 值会迅速下降，引起瘤胃酸中毒。酸中毒可分为急性或亚急性（慢性）。当牛食用大量易发酵淀粉后，瘤胃 pH 值急剧下降

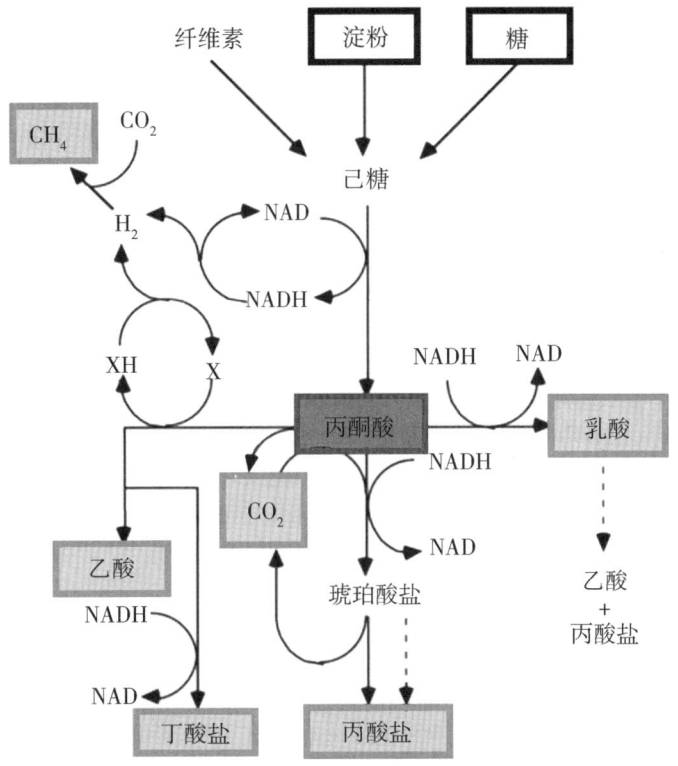

图 4 白色方框表示反刍动物日粮单个碳水化合物底物。浅灰色方框表示微生物发酵副产物。深灰色方框表示丙酮酸,是多个微生物代谢途径的枢纽。资料引自 Russell 和 Rychilk (2001)。

时,就会引发急性酸中毒。当瘤胃 pH 值持续过低(pH 值在 5.1~5.6)时,就会发生亚急性酸中毒,这种情况可能发生在舍饲环境中,当饲料中的精料比达到极限时,就会发生亚急性酸中毒(Brown 等,2006)。亚急性酸中毒不会表现出明显的症状,牛甚至在死亡之前都不会表现出病症(Owens 等,1998)。日粮中添加谷物并不能影响所有的动物。在相同的饮食条件下,有些牛出现了严重副作用,而有些牛只出现了如腹泻等轻微副作用(Brown 等,2006)。无论严重程度如何,都必须及时治疗,因为它会导致动物 DMI 减少或完全"不进食"。为了避免瘤胃环境微生物种群急剧被扰乱,使用比例递增日粮可以逐渐使瘤胃微生物种群适应并减少牛链球菌的丰度,牛链球菌会过度生长,快速产生乳酸,从而降低瘤胃的 pH 值(Wells 等,1997)。这有助于瘤胃微生物适应新环境。

瘤胃 pH 值下降会增加酸中毒风险,同时也会破坏瘤胃上皮内壁,导致角化或角质化不全,使用于营养吸收的瘤胃上皮细胞变硬。与 0% 和 30% 谷物日粮相比,60% 谷物日粮增厚了山羊瘤胃上皮细胞。这种损害最终导致从瘤胃吸收的营养物质和 VFAs 的减少,因为它们不能通过上皮细胞进入血液被动物吸收(Hinders 和 Owen,1965)。营养吸收不良导致动物的可用能量下降,整体生产性能下降。

高谷物日粮不仅降低了 pH 值,损害了瘤胃内壁,还增加了肝脓肿的患病率。当饲喂淀粉导致的低 pH 值，损害瘤胃上皮时，病原菌就有机会在瘤胃定植。例如厌氧坏死性梭杆菌（*Fusobacterium necrophorum*），是肝脓肿的主要病因（Nagaraja 和 Chengappa，1998）。在瘤胃定植后，细菌可以通过受损的上皮细胞进入血液，最终被肝脏过滤，导致脓肿（Nagaraja 和 Chengappa，1998）。这一过程可以发生在泌乳和产肉的各个阶段，因此当对牛喂食高浓缩饲料时，这种情况在饲养场最常见。

肝脓肿是舍饲生产系统的主要问题，12%~32% 的舍饲幼牛在屠宰时表现出严重的肝损伤（Brink 等，1990）。这给肉联厂商造成了经济损失，因为他们不能出售肝脏，还必须花更多时间修剪胴体。肝脓肿不仅会导致肝脏受损，而且动物的前期生产性能也会受到影响。肝脓肿牛的采食量和增重都会受到影响（Brink 等，1990）作为过去的一项预防措施，饲料中加入抗生素磷酸泰乐菌素，可以减少瘤胃坏死性梭杆菌，预防肝脓肿并提高增重（Brown 等，1975）。

12 小结

使用含淀粉的谷物作为饲料原料可以提高生产效率，因为这些能量密集的饲料比饲草更易于微生物发酵，而淀粉的发酵通常会导致 VFAs 的转变以及更少的温室气体。对还原当量库 CH_4 的还原，积累了代谢反应中大量 NADH，进一步确保提高丙酸生成量，而丙酸是一种被还原的挥发性脂肪酸。并且丙酸是一种生糖 VFA，它在反刍动物代谢时会产生更多的 ATP，生成较高水平的肌内大理石纹。

此外，由于添加谷物使发酵谱偏离了生成乙酸盐的路径，使瘤胃碳水化合物发酵产生的二氧化碳也随之减少，这确保了动物生产的可持续性。因此，饲喂谷物对反刍动物生理和生产效率有着深远的影响，同时还降低了反刍动物生产的碳足迹。

高淀粉日粮不仅对牛的胴体品质和生长效率有许多好处，还能降低甲烷产量。然而，饲喂谷物仍有负面影响，包括降低 pH 值和诱导牛的酸中毒，以及肝脓肿。高淀粉日粮对养殖企业和环境可持续性都有好处，这对维持环境可持续性很重要，但这些好处的生化机制必须进一步明确。饲喂淀粉对瘤胃和后消化道微生物种群组成有深远的影响，并与牛肉生产效率和可持续性有关。从经济和环境的角度来看，高谷物日粮喂养还是有利的，除非找到其他方法能够替代淀粉喂养，并取得与淀粉喂养相同的功效。

13 更多信息

肉牛谷物饲喂的话题与当前有关环境影响和长期生产可持续性的讨论密不可分。然而，本章主要集中在动物内部的影响，以及这些饮食对微生物种群的影响。我们必须了解当牛饲喂高淀粉日粮时，瘤胃微生物种群的演替。如果能充分了解瘤胃微生物和代谢物的变化，我们就可以开始准备特定的干预或补救措施来改善负面影响（如急性的和亚急性的酸中毒）、减少发酵生产终产物（如甲烷和 CO_2）的浪费。此外，我们还必须了解其他最低成本饲料成分对微生物种群的影响，以及它们对动物健康、免疫和食品安全的影响。

为了进一步了解高谷物日粮对瘤胃生态系统和生理机能的影响，有许多研究机构参与了该项研究。它们是美国农业部农业研究局，尤其以肉类动物研究中心最为著名，接下来是位于阿伯丁的罗维特研究所（现在是阿伯丁大学的一部分）、罗斯林研究所，法国农业科学研究院（INRA），当然还有澳大利亚的联邦科学与工业研究组织（CSIRO）。世界各地的许多大学的研究人员都参与了扩大我们的知识和视野，坦率地说，太多了，以至不能在这里一一列出，因为高质量的科学家总会不可避免地会在不经意间被排除在名单之外。这个领域的研究正在进行而且非常活跃，每年都有新的发现。许多该领域最活跃的研究人员参加了胃肠功能大会（https://www.congressgastrofunction.org/），以及"法国农业科学研究院-阿伯丁"胃肠道微生物学联合研讨会。为了研究本章节所描述的许多问题的最佳新资源之一是以下书籍：*Rumen Health Compendium*（Tedeschi 和 Nagaraja，2020）。

14 参考文献

Anderson, K L. 1995. "Biochemical Analysis of Starch Degradation by *Ruminobacter Amylophilus* 70." Applied and Environmental Microbiology 61（4）：1488-91.

Archimède, H., M. Eugène, C., Marie Magdeleine, M., et al., 2011. "Comparison of Methane Production between C3 and C4 Grasses and Legumes." Animal Feed Science and Technology 166-167（June）：59-64.

Avgustin, G., R. J. Wallace, and H. J. Flint. 1997. "Phenotypic Diversity among Ruminal Isolates of Prevotella Ruminicola：Proposal of *Prevotella Brevis* Sp. Nov., *Prevotella Bryantii* Sp. Nov., and *Prevotella Albensis* Sp. Nov. and Redefinition of *Prevotella Ruminicola*." International Journal of Systematic Bacteriology 47（2）：284-88.

Awika, Joseph M. 2011a. "Major Cereal Grains Production and Use around the World." InACS Symposium Series, edited by Joseph M. Awika, Vieno Piironen, and Scott Bean, 1089：1-13. Washington, DC：American Chemical Society.

Bandarupalli, V. 2017. "Identification of Novel Rumen Bacteria Using Starch as a Selective Nutrient in Batch Cultures." Journal of Animal Science 95（suppl_ 4）：305-305.

Beauchemin, K. A. 2009. "Dietary Mitigation of Enteric Methane from Cattle." CAB Reviews：Perspectives in Agriculture, Veterinary Science, Nutrition and Natural Resources 4（035）.

Beauchemin, K. A., M. Kreuzer, F. O'Mara, and T. A. McAllister. 2008. "Nutritional Management for Enteric Methane Abatement：A Review." Australian Journal of Experimental Agriculture 48（2）：21.

Beauchemin, K. A., T. A. McAllister, Y. Dong, et al., 1994. "Effects of Mastication on Digestion of Whole Cereal Grains by Cattle2." Journal of Animal Science 72（1）：236-46.

Beever, D. E., M. S. Dhanoa, H. R. Losada, et al., 1986. "The Effect of Forage Species and Stage of Harvest on the Processes of Digestion Occurring in the Rumen of Cattle." British Journal of Nutrition 56（2）：439-54.

Benchaar, C., C. Pomar, and J. Chiquette. 2001. "Evaluation of Dietary Strategies to Reduce Methane Production in Ruminants：A Modelling Approach." Canadian Journal of Animal Science 81（4）：563-74.

Bergen, Werner G., and Douglas B. Bates. 1984. "Ionophores：Their Effect on Production Efficiency and Mode of Action." Journal of Animal Science 58（6）：1465-83.

Blaxter, K. L. 1962. "The Energy Metabolism of Ruminants." Hutchinson Scientific and Technical,.

http://www.researchgate.net/publication/271201780_The_Energy_Metabolism_of_Ruminants.

Boadi, D., C. Benchaar, J. Chiquette, et al., 2004. "Mitigation Strategies to Reduce Enteric Methane Emissions from Dairy Cows: Update Review." Canadian Journal of Animal Science 84 (3): 319-35.

Brink, D R, S R Lowry, R A Stock, et al., 1990. "Severity of Liver Abscesses and Efficiency of Feed Utilization of Feedlot Cattle." Journal of Animal Science 68 (5): 1201.

Brown, H., R. F. Bing, H. P. Grueter, et al., 1975. "Tylosin and Chlortetracycline for the Prevention of Liver Abscesses, Improved Weight Gains and Feed Efficiency in Feedlot Cattle." Journal of Animal Science 40 (2): 207-13.

Brown, M. S., C. H. Ponce, and R. Pulikanti. 2006. "Adaptation of Beef Cattle to High-Concentrate Diets: Performance and Ruminal Metabolism1." Journal of Animal Science 84 (suppl_13): E25-33.

Bryant, M P, Nola Small, Cecelia Bouma, et al., 1957. "Bacteroides *Ruminicola N.* sp. and *Succinimonas Amylolytica* the New Genus and Species Species of Succinic Acid-Producing Anaerobic Bacteria of the Bovine Rumen," 9.

Bryant, Marvin P. 1959. "Bacterial Species of the Rumen." Bacteriological Reviews 23 (3): 125-53.

Buddle, Bryce M., Michel Denis, Graeme T. Attwood, et al., 2011. "Strategies to Reduce Methane Emissions from Farmed Ruminants Grazing on Pasture." The Veterinary Journal 188 (1): 11-17.

Callaway, T R, and S A Martin. 1996. "Effects of Organic Acid and Monensin Treatment on *in Vitro* Mixed Ruminal Microorganism Fermentation of Cracked Corn." Journal of Animal Science 74 (8): 1982.

Depenbusch, B. E., T. G. Nagaraja, J. M. Sargeant, et al., 2008. "Influence of Processed Grains on Fecal pH, Starch Concentration, and Shedding of Escherichia Coli O157 in Feedlot Cattle1." Journal of Animal Science 86 (3): 632-39.

Eckard, R. J., C. Grainger, and C. A. M. de Klein. 2010. "Options for the Abatement of Methane and Nitrous Oxide from Ruminant Production: A Review." Livestock Science 130 (1-3): 47-56.

Ellis, J. L., J. Dijkstra, E. Kebreab, et al., 2008. "Aspects of Rumen Microbiology Central to Mechanistic Modelling of Methane Production in Cattle." The Journal of Agricultural Science 146 (2): 213-33.

Esther, M., O. Cerrilla, and GM Martínez. 2003. "Starch Digestion and Glucose Metabolism in the Ruminant: A Review." Interciencia 28.

Etuk, Edeheudim. 2012. "Nutrient Composition and Feeding Value of Sorghum for Livestock and Poultry: A Review."

Eun, J-S., V. Fellner, and M. L. Gumpertz. 2004. "Methane Production by Mixed Ruminal Cultures Incubated in Dual-Flow Fermentors." Journal of Dairy Science 87 (1): 112-21.

Fahey, G. C., and Berger, L. L. 1988. "The Ruminant Animal: Digestive Physiology and Nutrition. Prentice Hall,." In , 269-97. Englewood Cliffs, NJ,: Prentice Hall,.

Ferrell, C. L. 1988. "The Ruminant Animal Digestive Physiology and Nutrition." Carbohydrate Nutrition of Ruminants. In: Church, D. C. (Ed.), 269-97.

Ferry, J G. 1992. "Methane from Acetate." Journal of Bacteriology 174 (17): 5489-95.

Firkins, J. L., L. L. Berger, and G. C. Fahey. 1985. "Evaluation of Wet and Dry Distillers Grains and Wet and Dry Corn Gluten Feeds for Ruminants." Journal of Animal Science 60 (3): 847-60.

Foley, P. A., D. A. Kenny, J. J. Callan, et al., 2009. "Effect of DL-Malic Acid Supplementation on Feed Intake, Methane Emission, and Rumen Fermentation in Beef Cattle." Journal of Animal Science 87 (3): 1048-57.

Forster, P., Ramaswamy, V., Artaxo, P., et al., 2007 "Changes in Atmospheric Constituents and in

Radiative Forcing." Cambridge, UK and New York, NY.:Cambridge University Press.

Freetly, H. C., A. K. Lindholm-Perry, K. E. Hales, et al., 2015. "Methane Production and Methanogen Levels in Steers That Differ in Residual Gain123." Journal of Animal Science 93 (5):2375-81.

Fron, Melanie, Humberto Madeira, Chris Richards, et al., 1996. "The Impact of Feeding Condensed Distillers Byproducts on Rumen Microbiology and Metabolism." Animal Feed Science and Technology 61 (1-4):235-45.

Gavrilova, O., A. Leip, H. Dong, et al., 2019. Emmisions from Livestock and Manure Management. In: CALVO BUENDIA, E.; TANABE, K.; KRANJC, A.; BAASANSUREN, J.; FUKUDA, M.; NGARIZE, S.; OSAKO, A.; PYROSHENKO, Y. SHERMANAU, P.; FEDERICI, S. (Ed.). 2019 Refinement to the 2006 guidelines for National Greenhouse Gas Inventories. Agriculture, forestry and other land use. Geneve: IPCC, 2019. v. 4. cap. 10. http://www.alice.cnptia.embrapa.br/handle/doc/1119362.

Gerber, P. J., H. Steinfeld, B. Henderson, et al., 2013. "Tackling Climate Change through Livestock A Global Assessment of Emissions and Mitigation Opportunities."

Goopy, John P., Alastair Donaldson, Roger Hegarty, et al., 2014. "Low-Methane Yield Sheep Have Smaller Rumens and Shorter Rumen Retention Time." British Journal of Nutrition 111 (4):578-85.

Grainger, C., and K. A. Beauchemin. 2011. "Can Enteric Methane Emissions from Ruminants Be Lowered without Lowering Their Production?" Animal Feed Science and Technology 166-167 (June):308-20.

Guan, H., K. M. Wittenberg, K. H. Ominski, et al., 2006. "Efficacy of Ionophores in Cattle Diets for Mitigation of Enteric Methane1." Journal of Animal Science 84 (7):1896-1906.

Hammond, K. J., J. L. Burke, J. P. Koolaard, et al., 2013. "Effects of Feed Intake on Enteric Methane Emissions from Sheep Fed Fresh White Clover (*Trifolium Repens*) and Perennial Ryegrass (*Lolium Perenne*) Forages." Animal Feed Science and Technology 179 (1-4):121-32.

Haque, Md Najmul. 2018. "Dietary Manipulation: A Sustainable Way to Mitigate Methane Emissions from Ruminants." Journal of Animal Science and Technology 60 (1):15.

Harper, L A, O T Denmead, J R Freney, et al., 1999. "Direct Measurements of Methane Emissions from Grazing and Feedlot Cattle." Journal of Animal Science 77 (6):1392.

Henderson, Gemma, Faith Cox, et al., 2015. "Rumen Microbial Community Composition Varies with Diet and Host, but a Core Microbiome Is Found across a Wide Geographical Range." Scientific Reports 5 (1):14567.

Herrera-Saldana, R. E., J. T Huber, and M. H. Poore. 1990. "Dry Matter, Crude Protein, and Starch Degradability of Five Cereal Grains." Journal of Dairy Science 73 (9):2386-93.

Hinders, R. G., and F. G. Owen. 1965. "Relation of Ruminal Parakeratosis Development to Volatile Fatty Acid Absorption." Journal of Dairy Science 48 (8):1069-73.

Hindrichsen, I. K., H.-R. Wettstein, A. Machmüller, et al., 2006. "Methane Emission, Nutrient Degradation and Nitrogen Turnover in Dairy Cows and Their Slurry at Different Milk Production Scenarios with and without Concentrate Supplementation." Agriculture, Ecosystems & Environment 113 (1-4):150-61.

Holter, J. B., and A. J. Young. 1992. "Methane Prediction in Dry and Lactating Holstein Cows." Journal of Dairy Science 75 (8):2165-75.

Hook, Sarah E., André-Denis G. Wright, and Brian W. McBride. 2010. "Methanogens: Methane Producers of the Rumen and Mitigation Strategies." Archaea 2010:1-11.

Hungate, R. E. 1944. "Studies on Cellulose Fermentation: I. The Culture and Physiology of an Anaerobic Cellulose-Digesting Bacterium." Journal of Bacteriology 48 (5): 499-513.

Hungate, R. E. 1947. "Studies on Cellulose Fermentation: III. The Culture and Isolation for Cellulose-Decomposing Bacteria from the Rumen of Cattle." Journal of Bacteriology 53 (5): 631-45.

Hungate, R. E. 1966. The Rumen and Its Microbes. New York: Academic Press.

Hungate1000 project collaborators, Rekha Seshadri, Sinead C Leahy, et al., 2018. "Cultivation and Sequencing of Rumen Microbiome Members from the Hungate1000 Collection." Nature Biotechnology 36 (4): 359-67.

Huntington, G B. 1997. "Starch Utilization by Ruminants: From Basics to the Bunk." Journal of Animal Science 75 (3): 852.

Iannotti, E. L., D. Kafkewitz, M. J. Wolin, and M. P. Bryant. 1973. "Glucose Fermentation Products of Ruminococcus Albus Grown in Continuous Culture with *Vibrio Succinogenes*: Changes Caused by Interspecies Transfer of H_2." Journal of Bacteriology 114 (3): 1231-40.

Johnson, K. A., and D. E. Johnson. 1995. "Methane Emissions from Cattle." Journal of Animal Science 73 (8): 2483-92.

Kebreab, E., K. Clark, C. Wagner-Riddle, et al., 2006. "Methane and Nitrous Oxide Emissions from Canadian Animal Agriculture: A Review." Canadian Journal of Animal Science 86 (2): 135-57.

Kobayashi, Yasuo. 2010. "Abatement of Methane Production from Ruminants: Trends in the Manipulation of Rumen Fermentation." Asian-Australasian Journal of Animal Sciences 23 (3): 410-16.

Koenig, K. M., K. A. Beauchemin, and L. M. Rode. 2003. "Effect of Grain Processing and Silage on Microbial Protein Synthesis and Nutrient Digestibility in Beef Cattle Fed Barley-Based Diets1, 2." Journal of Animal Science 81 (4): 1057-67.

Kowalczyk, J. 1986. An Introduction to Rumen Studies: By J. W. Czerkawski. Vol. 22. Oxford, UK: Elsevier Press.

Krause, D O., Nagaraja, T G., Wright, A D G., et al., 2013. "Board-Invited Review: Rumen Microbiology: Leading the Way in Microbial Ecology." J ANIM SCI 2013, 91 (1): 331-41.

Kreikemeier, K K., D L Harmon, R T., Brandt., et al., 1990. "Steam-Rolled Wheat Diets for Finishing Cattle: Effects of Dietary Roughage and Feed Intake on Finishing Steer Performance and Ruminal Metabolism." Journal of Animal Science 68 (7): 2130.

Larsen, M., P. Lund, M. R. Weisbjerg, et al., 2009. "Digestion Site of Starch from Cereals and Legumes in Lactating Dairy Cows." Animal Feed Science and Technology 153 (3-4): 236-48.

Liu, Yuchen, and William B. Whitman. 2008. "Metabolic, Phylogenetic, and Ecological Diversity of the Methanogenic Archaea." Annals of the New York Academy of Sciences 1125 (1): 171-89.

M Gómez, Luis, Sandra L Posada, Universidad de Antioquia, et al., 2016. "Starch in Ruminant Diets: A Review." Revista Colombiana de Ciencias Pecuarias 29 (2).

Martin, C., D. P. Morgavi, and M. Doreau. 2010. "Methane Mitigation in Ruminants: From Microbe to the Farm Scale." Animal 4 (3): 351-65.

Martin, C., J. Rouel, J. P. Jouany, et al., 2008. "Methane Output and Diet Digestibility in Response to Feeding Dairy Cows Crude Linseed, Extruded Linseed, or Linseed Oil1." Journal of Animal Science 86 (10): 2642-50.

Martin, S. A., and M. N. Streeter. 1995. "Effect of Malate on in Vitro Mixed Ruminal Microorganism Fermentation." Journal of Animal Science 73 (7): 2141-45.

Martinez-Fernandez, Gonzalo, Stephane Duval, et al., 2018. "3-NOP vs. Halogenated Compound: Methane Production, Ruminal Fermentation and Microbial Community Response in Forage Fed Cattle." Frontiers in Microbiology 9 (August): 1582.

McAllister, T. A., S. J. Meale, E. Valle, et al., 2015. "RUMINANT NUTRITION SYMPOSIUM: Use of Genomics and Transcriptomics to Identify Strategies to Lower Ruminal Methanogenesis1, 2, 3." Journal of Animal Science 93 (4): 1431-49.

McAllister, T. A., and C. J. Newbold. 2008. "Redirecting Rumen Fermentation to Reduce Methanogenesis." Australian Journal of Experimental Agriculture 48 (2): 7.

McGinn, S. M., K. A. Beauchemin, T. Coates, and D. Colombatto. 2004. "Methane Emissions from Beef Cattle: Effects of Monensin, Sunflower Oil, Enzymes, Yeast, and Fumaric Acid1." Journal of Animal Science 82 (11): 3346-56.

Milich, L. 1999. "The Role of Methane in Global Warming: Where Might Mitigation Strategies Be Focused?" Global Environmental Change 9 (3): 179-201.

Moate, P. J., S. R. O. Williams, M. H. Deighton, et al., 2019. "Effects of Feeding Wheat or Corn and of Rumen Fistulation on Milk Production and Methane Emissions of Dairy Cows." Animal Production Science 59 (5): 891.

Moate, Peter J., Matthew H. Deighton, S. Richard O. Williams, et al., 2016. "Reducing the Carbon Footprint of Australian Milk Production by Mitigation of Enteric Methane Emissions." Animal Production Science 56 (7): 1017.

Moate, P. J., J. L. Jacobs, M. C. Hannah, et al., 2018. "Adaptation Responses in Milk Fat Yield and Methane Emissions of Dairy Cows When Wheat Was Included in Their Diet for 16 Weeks." Journal of Dairy Science 101 (8): 7117-32.

Nagaraja, T G, and M M Chengappa. 1998. "Liver Abscesses in Feedlot Cattle: A Review." Journal of Animal Science 76 (1): 287.

Nisbet, D. J., and S. A. Martin. 1993. "Effects of Fumarate, l-Malate, and An Aspergillus Oryzae Fermentation Extract Ond-Lactate Utilization by the Ruminal Bacterium *Selenomonas Ruminantium*." Current Microbiology 26 (3): 133-36.

Odongo, N. E., R. Bagg, G. Vessie, et al., 2007. "Long-Term Effects of Feeding Monensin on Methane Production in Lactating Dairy Cows." Journal of Dairy Science 90 (4): 1781-88.

Okine, E. K., G. W. Mathison, and R. T. Hardin. 1989. "Effects of Changes in Frequency of Reticular Contractions on Fluid and Particulate Passage Rates in Cattle." Journal of Animal Science 67 (12): 3388.

Opio, C, Gerber, P., and Mottet,. 2012. "Greenhouse Gas Emission from Ruminant Supply Chains." Food and Agriculture Organization/World Health Organization.

Owens, F N, D S Secrist, W J Hill, et al., 1998. "Acidosis in Cattle: A Review." Journal of Animal Science 76 (1): 275.

Patra, Amlan K., and Jyotisna Saxena. 2009. "Dietary Phytochemicals as Rumen Modifiers: A Review of the Effects on Microbial Populations." Antonie van Leeuwenhoek 96 (4): 363-75.

Pitta, D. W., W. E. Pinchak, N. Indugu, et al., 2016. "Metagenomic Analysis of the Rumen Microbiome of Steers with Wheat-Induced Frothy Bloat." Frontiers in Microbiology 7 (May).

Plaizier, J. C., D. O. Krause, G. N. Gozho, et al., 2008. "Subacute Ruminal Acidosis in Dairy Cows: The Physiological Causes, Incidence and Consequences." The Veterinary Journal 176 (1): 21-31.

Rowe, J. B., M. Choct, and D. W. Pethick. 1999. "Processing Cereal Grains for Animal Feeding." Australian Journal of Agricultural Research 50 (5): 721.

Russell, J. B. 2002. Rumen Microbiology and Its Role in Ruminant Nutrition. Ithaca, NY.: Cornell University Press.

Russell, J B, and H J Strobel. 1989. "Effect of Ionophores on Ruminal Fermentation." Applied and Environmental Microbiology 55 (1): 1-6.

Russell, James B. 2002. Rumen Microbiology and Its Role in Ruminant Nutrition. Department of Microbiology, Cornell University.

Russell, James B., and Robert B. Hespell. 1981. "Microbial Rumen Fermentation." Journal of Dairy Science 64 (6): 1153-69.

Russell, J. B., and P. H. Robinson. 1984. "Compositions and Characteristics of Strains of *Streptococcus Bovis*." Journal of Dairy Science 67 (7): 1525-31.

Sauvant, Daniel, and Sylvie Giger-Reverdin. 2007. "Empirical Modelling by Meta-Analysis of Digestive Interactions and CH_4 Production in Ruminants." In, 124: np. Wageningen Academic Publisher. https://hal.archives-ouvertes.fr/hal-01173426.

Schultz, J. E., and John A. Breznak. 1979. "Cross-Feeding of Lactate Between *Streptococcus Lactis* and *Bacteroides* Sp. Isolated from Termite Hindguts." Applied and Environmental Microbiology 37 (6): 1206-10.

Steinfeld, H., Gerber, P., Wassenaar, T., et al., 2006. "Livestock's Long Shadow: Environmental Issues and Options." Rome, Italy.: Food and Agriculture Organization,.

Stevens, E., Armstrong, K., Bezar, H., et al., 2004. "Fodder Oats an Overview. In: Suttie, J. M. and Reynolds, S. G. (Eds), Fodder Oats: A World Overview. Plant Production and Protection Series." No. 33. Rome: Food and Agriculture Organization.

Tedeschi, Luis O., and T. G. Nagaraja. 2020. Rumen Health Compendium. XanEdu Publishing Inc.

Theurer, C. Brent. 1986. "Grain Processing Effects on Starch Utilization by Ruminants." Journal of Animal Science 63 (5): 1649-62.

Thiele, Jurgen H., and J. Gregory Zeikus. 1988. "Control of Interspecies Electron Flow during Anaerobic Digestion: Significance of Formate Transfer versus Hydrogen Transfer during Syntrophic Methanogenesis in Flocs." Applied and Environmental Microbiology 54 (1): 20-29.

Toland, Pc. 1976. "The Digestibility of Wheat, Barley or Oat Grain Fed Either Whole or Rolled at Restricted Levels with Hay to Steers." Australian Journal of Experimental Agriculture 16 (78): 71.

Van Kessel, J. 1996. "The Effect of PH on Ruminal Methanogenesis." FEMS Microbiology Ecology 20 (4): 205-10.

Verdú, M., A. Bach, and M. Devant. 2015. "Effect of Concentrate Feeder Design on Performance, Eating and Animal Behavior, Welfare, Ruminal Health, and Carcass Quality in Holstein Bulls Fed High-Concentrate Diets1." Journal of Animal Science 93 (6): 3018-33.

Wallace, R. J., T. A. Wood, A. Rowe, et al., 2006. "Encapsulated Fumaric Acid as a Means of Decreasing Ruminal Methane Emissions." International Congress Series 1293 (July): 148-51.

Wells, James E, Denis O Krause, Todd R Callaway, et al., 1997. "A Bacteriocin-Mediated Antagonism by Ruminal *Lactobacilli* against *Streptococcus Bovis*." FEMS Microbiology Ecology 22 (3): 237-43.

Wilson, B. K., B. P. Holland, D. L. Step, et al., 2016. "Feeding Wet Distillers Grains plus Solubles with and without a Direct-Fed Microbial to Determine Performance, Carcass Characteristics, and Fecal Shedding of Escherichia Coli O157: H7 in Feedlot Heifers1." Journal of Animal Science 94 (1):

297-305.

Wolfe, R. S. 1993. "An Historical Overview of Methanogenesis." Springer US.

Wolin, M. J. 1975. "Interactions between the Bacterial Species in the Rumen. In: McDonald, W. and Warner, A. C. I. (Eds), Digestion and Metabolism in the Ruminant." Armidale, Australia,: Univ. New Eng. Pub. Unit,.

Wood, T. A., R. J. Wallace, A. Rowe, et al., 2009. "Encapsulated Fumaric Acid as a Feed Ingredient to Decrease Ruminal Methane Emissions." Animal Feed Science and Technology 152 (1-2):62-71.

Wright, André-Denis G., and Athol V. Klieve. 2011. "Does the Complexity of the Rumen Microbial Ecology Preclude Methane Mitigation?" Animal Feed Science and Technology 166-167 (June):248-53.

Xia, Yun, Yunhong Kong, Robert Seviour, et al., 2015. "In Situ Identification and Quantification of Starch-Hydrolyzing Bacteria Attached to Barley and Corn Grain in the Rumen of Cows Fed Barley-Based Diets." Edited by Cindy Nakatsu. FEMS Microbiology Ecology 91 (8):fiv077.

Zhu, Fan. 2017. "Barley Starch:Composition, Structure, Properties, and Modifications:Barley Starch Review…." Comprehensive Reviews in Food Science and Food Safety 16 (4):558-79.

第21章 植物次生产物：在反刍动物可持续生产及营养中的有益作用

David R. Yáñez-Ruiz、Alejandro Belanche

扎伊丁实验站（Estación Experimental del Zaidín），

西班牙高等科研理事会（CSIC），西班牙

（裴彩霞译）

1 前言

在满足高产动物的营养需要、提高畜牧生产效率的反刍动物集约化饲养过程中往往会引起瘤胃微生物的生态系统失调。满足反刍动物的能量和营养需求，同时避免消化和系统代谢失调，对于确保动物奶和肉高效产出至关重要。当前的集约化管理系统为了高产往往在日粮中加入大量谷物或易降解的副产品（Zebeli等，2010）。虽然这些饲喂方式在短期内提高了产量，但不符合反刍动物消化生理（Zebeli和Ametaj，2009），最严重的后果是瘤胃生态系统受损，导致乳酸酸中毒、肠道炎症和/或腹泻。过去常用低于治疗水平的抗生素来提高饲料转化效率（如牛奶和肉类）和/或预防代谢紊乱和健康问题。然而，一些国家对抗生素使用的限制和禁令（如欧盟）已经促使科学家和饲料行业寻找替代产品。植物次生产物（Plant secondary compounds），也称为植原质（Phytogenics）或植物化学物质（Phyotochemicals）（Patra，2010），是经过提取和浓缩后的植物产生的次生产物，由于其可调节肠道微生物的活性而受到广泛关注。此外，目前消费者越来越拒绝在食品生产中使用合成物质，故具有促生长活性的植物源性化合物正在动物饲料添加剂市场上占有一席之地（Dhama等，2015）。在过去20年，植物次生产物在反刍动物营养中使用的研究逐渐增多。

植物次生产物饲料添加剂是一类极为复杂的化合物，其化学结构和生物活性差异极大（Kroon和Williamson，1999）。植物中的植物次生产物的含量由于植物的部位、收获季节和产地等内在因素和添加剂生产过程等外在因素而差异很大（Ganguly，2013）。植物次生产物是植物进化过程中为了抵御微生物、昆虫和食草动物而逐渐形成的差异很大的物质（Cowan，1999），由于其种类多样，使其能成为提高动物生产性能而具有多种不同生物活性添加剂的宝贵资源。动物生产实验中使用的各种植物次生产物饲料添加剂很难分类，部分原因是没有简明的分类方法（Valenzuela Grijalva等，2017）。根据不同的标准有不同的分类，主要有：①植物来源；②化学成分；③加工方法；④作用机制。然而，一些次生产物可能通过不同的机制发挥作用，从而使其分类更为复杂。根据植物次生产物的不同生物活性，提出了4种主要机制以解释研究中动物生理和生产性能的变化（Valenzuela Grijalva

第21章 植物次生产物：在反刍动物可持续生产及营养中的有益作用

等，2017）。

（1）改善饲料特性和动物饲料采食量；
（2）调节微生物活性；
（3）促进营养物质的消化和吸收；
（4）影响靶组织的合成代谢活性。

虽然植物次生产物的作用可能是上述机制的组合，但本章将主要关注其对瘤胃微生物群的调节作用。由于植物次生产物中存在多种成分，它们的抗菌活性很可能不是单一的作用模式，而是涉及细胞的多个靶点（Acamovic和Brooker，2005）。有4种公认的作用模式来解释植物次生产物饲料添加剂如何发挥其抗菌作用以及对瘤胃微生物群的改变：①抑制细胞壁合成；②破坏细胞壁结构和功能（例如改变细胞质膜的通透性）；③抑制核酸合成；④抑制蛋白质合成或独特的细菌代谢途径。这些作用导致核心菌群活性的丧失，从而导致细菌死亡或生长迟缓（Patra和Saxena，2009a）。根据文献综述中使用的化学成分和分类（Calsamiglia等，2007；Benchaar等，2008；Patra和Saxena，2009a），植物次生产物可分成①精油；②单宁；③皂苷（图1）3类，本章将分别讲述各类的具体作用模式。

上述机制既对反刍动物有抗营养/毒性作用，又具有益处。虽然有大量关于牧草中有毒的化合物以及反刍动物如何对抗这些化合物的文献（Estelle，2010）。但是，本章的重点集中在探讨植物次生产物的积极影响。

大多数关于调节瘤胃发酵的植物次生产物筛选的研究都是基于实验室（即体外）的，并且已经鉴定了大量植物次生产物及其主要成分对瘤胃微生物发酵的影响（Patra和Saxena，2011）。

体外产气（*In vitro* gas production，IVGP）系统已广泛用于快速筛选影响瘤胃发酵的化学物质、植物种类、植物提取物和日粮成分。此方法已研究了一系列化学药品（Busquet等，2005a；Bodas等，2008；Garcia Gonz等，2008；Durmic等，2014）和日粮成分（Patra和Yu，2013；Hatew等，2015）的作用模式。IVGP系统为不同浓度化学成分单独或不同组合的广泛评估提供了可能（Busquet等，2005a；Garcia Gonz等，2008）。例如，Bodas等（2008）研究了450种植物提取物对甲烷生成能力的影响，发现其中35种减少甲烷生成量15%以上，其中6种甲烷抑制率大于25%而对消化或发酵无不良影响。然而，体外方法无法为化学药品长期效果提供可靠信息（Castro-Montoya等，2015），后面部分有同一种植物提取物体外研究结果不一致的进一步论述。此外，筛选研究的结果（Bodas等，2008；Durmic等，2014）通常还尚无定论，并且可能由于剂量、测试物的化学结构、日粮、应用处理的组合、供体动物瘤胃微生物的适应性或将制剂加入系统的形式等方面的不同而引起相互冲突的结果（Cardozo等，2004；Cardozo等，2005）。体外和体内方法之间的直接比较（如抗甲烷化合物）较少，并且显示出很大的差异。如Yáñez-Ruiz等（2016）所述，使用相同化合物和类似剂量时，其效果差异的研究结果可能由以下因素进行解释。①被测化合物通常通过瘤胃瘘管添加，添加时间，通常与喂食时间一致，每天1~2次，因此，可能无法迅速且充分地与瘤胃内容物混合；②活性化合物在体外和体内降解率的差异；③体外研究时，由于过滤等过程暴露于氧气中以及瘤胃固体食糜的去除，使瘤胃内容物接种物微生物的浓度降低和细菌群落结构变化（Soto等，2013）；

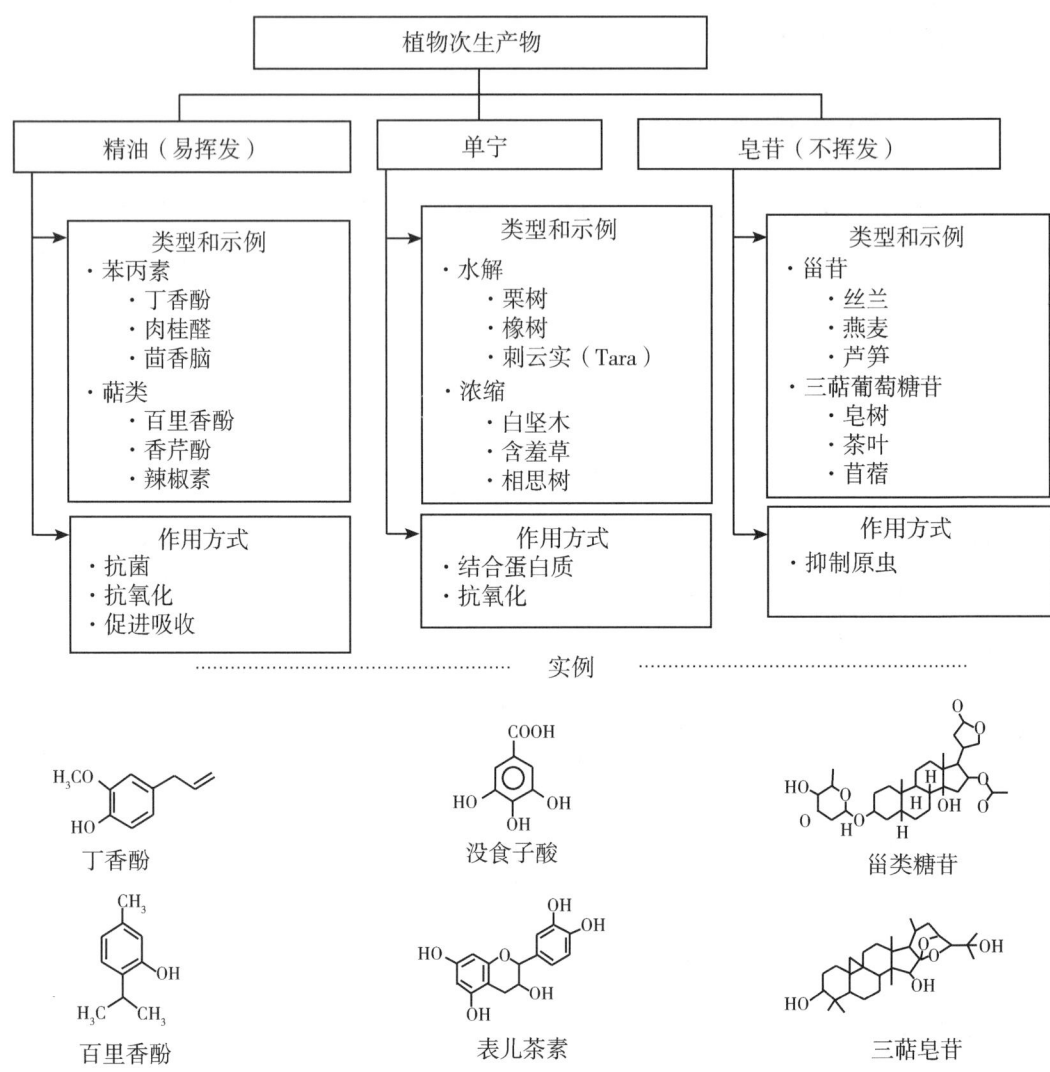

图 1 植物次级产物的分类和作用方式

④一些化合物可能从瘤胃中流出或通过瘤胃壁被吸收；⑤瘤胃生态系统在体内可对受试化合物产生适应性，而体外接种的微生物群是无法模拟的。

已有大量关于多种植物次生产物对瘤胃发酵影响的体外研究，并进行了广泛的综述（Frutos 等，2004；Calsamiglia 等，2007；Benchaar 等，2008；Hart 等，2008；Patra 和 Saxena，2009a，2011；Wang 等，2012；Vasta 等，2019）。Cieslak 等（2013）研究了植物提取物的作用模式，指出精油、单宁和皂苷在瘤胃内的作用模式不同，并得出结论：虽然有大量的体外研究工作，仍需要更多的体内和生产基础的研究来真实反映植物提取物通过调节瘤胃微生物对整体饲料效率（氮和能量）和动物生产性能（如牛奶和肉）的影响。本章介绍在反刍动物营养中使用植物次生产物饲料添加剂的最新进展，重点介绍通过体内研究获得的结果，特别是与瘤胃功能和微生物组的相互作用。详述了植物次生产物的抗菌特性、作用机制，以及对瘤胃蛋白质代谢、肠道甲烷产生、动物生产性能的影响和相关的挑战。

2 精油

2.1 组成

精油（Essential oils，EO）（也称为挥发油或乙醚油）是挥发性亲脂性次级代谢产物的复杂混合物，通常是用沸水和水蒸气蒸馏的方法从植物材料中提取，也可用溶剂萃取、超临界 CO_2 萃取和压榨提取（Benchaar 和 Greathead，2011）。EO 是植物特有的味道和香气的来源（Greathead，2003），在保护植物免受非生物和生物胁迫方面也发挥着重要作用，并对授粉和传播种子的生物起着引诱作用（Wink 和 Schimmer，2018）。EO 可以提取自植物的所有部位，包括叶子、根、花、花瓣、种子、果实、茎和树皮。EO 的产量和组成因植物种类（Martinez 等，2006）、地理位置、气候条件、土壤（Vokou 等，1993）、植物部分（例如种子与叶子；Delaquis 等，2002）、植物健康（例如虫害）和提取方法的不同而不同（Anitescu 等，1997）。

化学上，EO 通常由萜烯或苯丙烯次级代谢产物组成（图1）。萜烯包含结构和功能上不同的类型，是由异戊二烯（二磷酸异丙烯酯）合成，并可根据作为其骨架部分的 5-C-基（C5）异戊二烯单元的数量进行分类（Benchaar 等，2007）。萜烯与其他元素（即氧）相连则称为萜类或类异戊二烯（Benchaar 等，2007）。EO 中最常见为单萜烯，占 EO 的 90%（例如芳樟醇、香芹酚、百里香酚、柠檬烯）。另外，苯丙烷是从苯丙氨酸衍生而来的芳香族化合物，由带有 C3 丙酸侧链的 C6 苯环组成（Hart 等，2008），肉桂醛、丁香酚和茴香脑属于这样的化合物。此外，一些 EO 还含有萜烯或苯丙烯以外的其他化合物，如大蒜油含有大量二烯丙基二硫化物、二烯丙基硫化物或阿霍烯（ajoenes）等有机硫化合物。

2.2 活性

EO 通常由高浓度的 2 或 3 种主要成分为特征，其中萜烯类和萜类占 20%~70%，并解释了主要的生物效应（Okoh 等，2010）。值得注意的是，性质的复杂性取决于这些化合物是否具有叠加性、拮抗性或协同效应（Burt，2004）。先前为了测试不同 EO 的抗菌活性而进行的研究表明，不同 EO 对相同微生物抗菌力存在差异（Burt，2004；Pauli 和 Schilcher，2010）。有许多假说可以解释 EO 抗菌的活性。EO 及其组分的一个重要特征是其疏水性，这使它们能够渗透到细菌细胞膜和线粒体的脂质中，干扰结构并使其渗透性增加（Knobloch 等，1986；Sikkema 等，1994），然后引起离子和其他细胞内容物的流失（Carson 等，2002；Ultee 等，2002）。虽然细菌可以耐受的一定物质的流失而不会丧失活力，但细胞内容物的大量或关键分子和离子的损失将导致微生物死亡（Denyer 和 Hugo，1991）。EO 的成分也可作用于嵌入细胞质膜的细胞蛋白质，如 ATP 酶（Knobloch 等，1989）。这种作用的机制可能是亲脂性碳氢化合物分子的积聚和脂质-蛋白质相互作用的干扰，或亲脂性化合物与蛋白质疏水部分的直接相互作用（Sikkema 等，1995）。由于其成分复杂，EO 的抗菌活性可能是由于与大量分子靶点的相互作用，而不是特定的作用模式（Benchaar 等，2007）。尽管有报道称，由于缺乏保护性外膜，革兰氏阳性细菌对 EO 的敏感性更高，但一些作者（Burt，2004）观察到，革兰氏阴性细菌由于外膜中脂多糖的丢失而敏感性增强。由于这些原因，EO 的抗菌特性可以被认为是非特异性的，且由于 EO 成分和不同微生物的敏感性的高度差异使其对瘤胃微生物种群的影响难以预测（Pauli 和 Schilcher，2010）。瘤胃微生物组成的个体差异以及对 EO 抗菌活性模式了解的局限性也增加了上述的困难（Patra 和 Yu，2015）。

此外，EO 可通过与影响细菌的相同生物学机制影响真菌、原虫和病毒（Jouany 和 Morgavi，2007）；然而，对单个物种或菌株影响的信息较少，细菌也是如此（Burt，2004）。

2.3 精油对瘤胃微生物组和功能的影响

在 DNA 分子生物学技术发展之前，EO 对瘤胃微生物种群的影响主要是通过培养方法进行研究。20 世纪 60 年代的初步研究（Nagy 等，1964）表明，添加 EO 可显著降低体外瘤胃细菌的活性。一些细菌，如栖瘤胃普雷沃氏菌（*Prevotella ruminicola*）、斯氏梭菌（*Clostridium sticklandii*）和厌氧消化链球菌（*Peptostreotococcus anaerobius*）似乎比其他细菌（如牛链球菌，*Streptococcus bovis*）对 EO 更敏感（McIntosh 等，2003）。Evans 和 Martin（2000）发现 90 μg/mL 的百里香酚抑制反刍月形单胞菌（*Selenomonas ruminantium*）的生长，但对牛链球菌生长的抑制需要 180 μg/mL。基于分子生物学技术的方法比培养的方法能更好地揭示了 EO 对瘤胃微生物种群的影响。EO 可有效降低产甲烷菌的数量，但也会影响产琥珀酸丝状杆菌（*Fibrobacter succinogenes*）、白色瘤胃球菌（*Ruminococcus albus*）、黄色瘤胃球菌（*Ruminococcus flavefaciens*）等纤维素分解细菌的数量（Cobellis 等，2016）。此外，大部分 EO 的研究均减少了原虫的数量。

一些研究采用分子技术确定 EO 对高氨生产（Hyper Ammonia-Producing，HAP）细菌的影响，这些细菌似乎对 EO 敏感（Wallace 等，2002；McIntosh 等，2003）。Patra 和 Yu（2012）证实了栖瘤胃普雷沃氏菌、*Prevotella bryantii*、嗜氨梭菌（*Clostridium aminophilum*）、斯氏梭菌和溶纤维丁酸弧菌（*Butyrivibrio fibrisolvens*）在体外对测试的大部分 EO 有高度敏感性。Patra 和 Yu（2015）还利用微阵列分析指出，EO 对瘤胃细菌群落组成的影响取决于所用 EO 的化学性质。具有酚类结构的化合物，如牛至（Oregano）精油，在降低厚壁菌门、梭状芽孢杆菌纲和丁酸弧菌属的丰度方面表现出比其他 EO 有更强的抗菌活性（Patra 和 Yu，2015）。然而，牛至精油增加了普雷沃氏菌属细菌的丰度，可能是由于其能抵抗牛至精油的抗菌作用。EO、饲料成分和宿主之间存在复杂的相互作用，因此，瘤胃发酵特性的结果（例如 VFA、CH_4 和 NH_3 产生）之间的相关性、饲料降解率和微生物丰度可以提供更多关于瘤胃微生物组动力学和功能的信息，并促进开发有效的减排技术。为了进一步研究使用 EO 对 CH_4 排放的影响，一些研究还量化了 EO 对产甲烷菌数量的影响，并给出了正面（Khorrami 等，2015）和负面反应（Tomkins 等，2015）。然而，缺乏产甲烷菌数量与 CH_4 产量间关系的证实研究（Morgavi 等，2010），使得该类菌群的量化信息不多。除研究 EO 对一些微生物物种丰度的影响外，对整个微生物组进行更全面的评估在理论上将提供更多的信息。Schären 等（2017）利用 Illumina 测序方法研究了围产期奶牛莫能菌素和 EO 混合物对瘤胃发酵、产奶量和细菌瘤胃数量的影响。有趣的是，虽然莫能菌素导致了一些细菌群的增加或减少，但 EO 并没有引起瘤胃发酵和生产性能的任何变化。值得指出的是，添加 EO 不会引起瘤胃发酵的巨大变化，并在 DMI 或产奶量方面没有作用。Schären 等（2017）的结果也证实，适用于莫能菌素的假说，即其对细菌的影响取决于细胞壁组成和厚度的不同，而不是革兰氏阴性菌和革兰氏阳性菌之间的明确差异，这可能适用于 EO。

已经有大量的体外研究（大部分使用分批培养，一些使用发酵罐）测试 EO 调节瘤胃微生物种群、提高营养利用效率的能力（Calsamiglia 等，2007），其中相当一部分研究表明 EO 对 VFA 产生、蛋白质代谢、纤维消化和 CH_4 生产有积极影响。如前所述，体内研究的结果较少，有时仅报告对瘤胃发酵或生产性能的影响，但对两者共同影响的研究较少。鉴于已发表的大量研究为体外的结果及上述原因，这里仅引用体内和近 10 年（2010—2019 年）关于对瘤胃功能和生产性能影响的多数研究（表1）。

通常，EO 对某些瘤胃微生物的抑制作用体现为抑制脱氨和产甲烷作用，结果降低 NH_3-N、甲烷和乙酸盐浓度，同时提高丙酸盐和丁酸盐浓度（Busqet 等，2005b；Cardozo 等，2006）。此外，抑制瘤胃中的蛋白质降解有可能增加宿主动物的肠道氨基酸供应（Wallace，2004）。然而，所有这些影响很少同时发生（Hart 等，2008）。瘤胃中 NH_3-N 产生的减少与 HAP 细菌的抑制有关（Russell 等，1991）。这类细菌仅占瘤胃细菌总数的 1%（Russell，1991），但其特点是脱氨能力高，碳水化合物利用能力低（Eschenlauer 等，2002）。HAP 细菌具有非常高的脱氨活性，这受日粮类型和物理形态的影响。所有已知的 HAP 细菌均为革兰氏阳性，并且对莫能菌素等离子载体有高度敏感性（Eschenlauer 等，2002）。迄今为止，已鉴定出约 14 种不同的 HAP 细菌，脱氨活性最高的为斯氏梭菌、嗜氨梭菌和栖瘤胃普雷沃氏菌（Szumacher-Strabel 和 Cieslak，2012）。某些 EO 的添加也会改变富含淀粉的基质中定植的细菌。当给动物饲喂高蛋白质含量的日粮时，EO 抑制蛋白质和淀粉的降解以及抑制 HAP 细菌可产生积极影响（Patra 和 Saxena，2009a）。

减少反刍动物 CH_4 的产生有 3 种途径：直接抑制甲烷产生，减少氢气的产生，和为瘤胃氢气的利用提供替代途径（清除氢气的替代池）（Patra，2011）。EO 的作用与前两种途径有关，通过减少饲料降解（如微生物定植）或直接抑制原虫和/或产甲烷菌（Patra 等，2017）。如前所述，使用 EO 减少 CH_4 的研究大多是体外研究的结果，只有少数研究是体内试验（表1）。在大多数情况下，体内研究显示，由于使用 EO 的剂量和类型不同，结果相互矛盾。由于不同 EO 的化学成分、日粮和喂养制度的差异，也使在瘤胃中达到有效 EO 浓度所需的 EO 剂量很难确定（Bodas 等，2012）。Wang 等（2009）表明，将从牛至中提取的 EO 混合物（0.25 g/d）喂给绵羊 15 d 可降低 CH_4 产量，而 Beauchemin 和 McGinn（2006）以及 Tomkins 等（2015）发现商业混合精油（blend of essential oils, BEO）（1 g/d 和 2 g/d；Crina® 反刍动物；阿克苏表面化学有限公司，英国赫特福德郡）没有降低肉牛 CH_4 产量。最近，两项使用商业 BEO（Agolin®）的研究表明其可以降低奶牛 CH_4 产量（Elcoso 等，2019；Hart 等，2019）。从这些研究可以清楚地看出，系统地评估 EO 的作用需要用体内试验进一步研究。

表 1 体内研究香精油对瘤胃发酵和动物性能影响（ns：不显著；+：增加；-：降低）的近年文献总结

文献	研究细节					瘤胃发酵										动物生产性能				
	动物	主要化合物	g/kg DM	处理时间	DMI	pH	NH_3	VFA	Ac	Pr	Bu	CH_4	Bac	Ptz	Meth	MY	Fat	Pr	ADG	FE
Blanch 等 (2016)	奶牛	肉桂醛、大蒜油	200~400 mg/d	4周	ns	ns	ns	ns	ns	+	ns	-				ns	-	-		
Braun 等 (2019)	奶牛	BEO BIX12	1.2 g/(动物·d)	20 d												+				+
Canaes 等 (2017)	奶山羊	柠檬醛油	0.02~0.24 mL/kg BW	14 d	ns	ns	ns	ns	ns	ns	ns					ns	ns	ns		
Castro-Montoya 等 (2015)	奶牛	BEO Agolin®	0.2 g/d	8周								-				ns	ns	ns		
Chaves 等 (2011)	羔羊	肉桂醛	100~400 mg/kg BW	18周	ns	ns	ns	ns	ns	ns	ns								ns	ns
Cobellis 等 (2016)	绵羊	迷迭香油	1.75v%	21 d	ns	ns	ns	ns					ns	ns	ns					
Elcoso 等 (2019)	奶牛	BEO Agolin®	1 g/d	8周	ns	ns	ns	ns	+	-	ns	-				+	+	ns		
de Jesus 等 (2013)	奶牛	腰果油、蓖麻油	500 mg/kg BW	21 d	ns	ns	ns	ns	ns	+	+		+	+		+	+	+		
Flores 等 (2013)	奶牛	BEO XTreat®	200~600 mg/d	6周	ns	ns	ns	ns			ns					ns	ns	ns		
Giannenas 等 (2011)	奶山羊	BEO Crina®	50~150 mg/kg DM	5个月	ns	ns	ns	ns	ns	+	ns		+	+		+	ns	ns	ns	

（续表）

文献	动物	研究细节			瘤胃发酵											动物生产性能				
		主要化合物	g/kg DM	处理时间	DMI	pH	NH_3	VFA	Ac	Pr	Bu	CH_4	Bac	Ptz	Meth	MY	Fat	Pr	ADG	FE
Ghizzi 等 (2018)	奶牛	腰果酚、腰果树油、蓖麻油	500 mg/kg DM	5周	ns	ns	ns	ns	ns	ns	ns					ns	ns	ns		
Hart 等 (2019)	奶牛	BEO Agolin®	1 g/d	22周	+	ns	ns	ns			ns					+	ns	ns		
Kholifa 等 (2017)	奶山羊	迷迭香	10 g/d	12周	ns	ns	ns	+	ns	+	+	-				+	+	+		+
Kholifa 等 (2017)	奶山羊	香茅	10 g/d	12周	ns	ns	-	+	ns	+	+					+	+	+		+
Khorrami 等 (2015)	肉牛	麝香草	0.5 g/d	21 d	ns	ns	ns	ns	ns	+	+			-	-					
Khorrami 等 (2015)	肉牛	肉桂	0.5 g/d	21 d	ns	ns	ns	ns	ns	ns	ns			-	-					
Klop 等 (2017)	奶牛	BEO Agolin®	0.17 g/kg DM	2周	ns	ns	ns	+	ns	ns	ns	-				ns	ns			
Lin 等 (2013)	绵羊	BEO	0.5~1 g/d	21 d	+	ns	ns	-	ns	ns	ns		ns	-	ns					
Ornaghi 等 (2017)	肉牛	丁香, 肉桂	3.5~7 g/d	187 d	+												+		+	+
Santosa 等 (2010)	奶牛	BEO Agolin®	1 g/d	21 d												ns	+	ns		
Schären 等 (2017)	奶牛	BEO Crina®	1 g/d	3周									ns	ns		ns	ns	ns		

（续表）

文献	研究细节				瘤胃发酵										动物生产性能			
	动物	主要化合物	g/kg DM	处理时间	DMI	pH	NH_3	VFA	Ac	Pr	Bu	CH_4	Bac	Ptz	Meth	MY	Fat	Pr	ADG	FE
Silva 等 (2018)	奶牛	BEO Crina®	44 mg/kg DM	21 d	ns	ns	ns	ns	ns	ns	ns					ns	ns	ns		
Silva 等 (2019)	肉牛	腰果油，蓖麻油	30 mg/kg DM	120 d	+															
Soltan 等 (2018)	绵羊	BEO GRASP®	200~400/kg DM	15 d	ns	ns	ns	ns	ns	+	ns	−		−						
Souza 等 (2019)	肉牛	BEO	2~4 g/(动物·d)	73 d															+	+
Spanghero 等 (2009)	奶牛	BEO			ns	ns	ns	ns	ns	ns	ns					ns	ns	ns	ns	
Tager 和 Krause (2011)	奶牛	肉桂醛，丁香酚	0.5~10 g/d	21 d	ns	ns	ns	ns	ns	ns	ns					ns				
Tager 和 Krause (2011)	奶牛	辣椒	0.25 g/d	21 d	ns	ns	ns	ns	ns	ns	ns					ns				
Tekippe 等 (2011)	奶牛	牛至叶	500 g/d	21 d	ns	ns	ns	ns	ns	ns	ns					ns	ns	ns		
Tekippe 等 (2013)	奶牛	BEO XTract®	525 mg/d	28 d	ns	ns	ns	ns	ns	ns	ns					ns	ns	ns		
Tomkins 等 (2015)	肉牛	BEO Crina®	1~2 g/d	40 d	ns	ns	ns	ns	ns	ns	ns	ns		ns	ns					
Wall 等 (2014)	奶牛	BEO XTract	200~600 mg/d	42 d	+	ns	ns	ns	ns	ns	ns					+	ns	ns		
Yanga 等 (2010)	肉牛	丁香酚	400~600 mg/kg DM	21 d	ns	ns	ns	ns	ns	ns	ns						ns	+	ns	

注：BEO，混合精油；DMI，DM 采食量；VFA，挥发性脂肪酸；C2，乙酸盐；C3，丙酸盐；C4，丁酸盐；Ptz，原虫；Bac，细菌；Meth，产甲烷菌；MY，产奶量；Fat，乳脂率%；Pr，乳蛋白含量；ADG，平均日增重；FE，饲料效率。

尽管人们对通过 EO 改善瘤胃功能的兴趣日益增长，但对 EO 的作用方式和活性仍知之甚少。Cobellis 等（2016）最近回顾了一些瘤胃微生物对 EO 的反应，并强调其大多数是体外研究的结果，而体内研究主要通过显微镜进行原虫计数。

2.4 精油对动物生产性能的影响

EO 对（哺乳期或生长期）生产阶段反刍动物的一个积极影响是增加采食量。然而，情况并非总是如此。虽然一些绵羊或奶牛的体内研究表明，补充 EO 不会影响采食量（Yang 等，2007；Wang 等，2009），但另一些研究发现肉桂醛和丁香酚降低肉牛和奶牛采食量（Cardozo 等，2006；Calsamiglia 等，2007）。在本章分析的 24 项体内研究（表1）中，只有 5 项研究增加了采食量，并且它们不一定对畜产品产量（牛奶或肉类）产生积极影响。然而，在大多数报道饲料效率的研究中，补充 EO 在不改变饲料摄入量的情况下提高产量，从而使饲料效率显著提高。最近的一项荟萃分析汇编了长期使用商用 BEO（Agolin）的实验，结果表明产奶量增加（3.8%）但采食量没有变化（Belanche 等，2019）。然而，这种生产性能提高背后的机制仍有待阐明。先前的研究表明，喂食不同 EO 能增强动物对代谢物的消化和吸收（Franz 和 Novak，2009），其作用机制没有单胃动物的清楚（Muthusamy 和 Sankar，2015）。在反刍动物中，最合理的机制是 EO 对瘤胃微生物发酵的调节（Benchaar 等，2012），但仅测定 VFA 和 NH_3 浓度可能无法发现。可能还有其他机制。由于 EO 对心血管疾病、某些肿瘤、炎症过程和自由基增殖失控的疾病有着积极作用，早在 5 000 年前，中国传统医药疗法就开始使用 EO（Calsamiglia 等，2007）。这些特性取决于它们清除自由基、抑制膜脂过氧化、螯合金属和刺激抗氧化酶活性的能力（Lee 等，2003）。此外，改善营养吸收也被认为是另一个因素（Oh 等，1968）。Braun 等（2019）报道，EO 对奶牛产奶量的有益影响与通过刺激瘤胃上皮中表达的阳离子转运蛋白来增强吸收有关，从而增加钙和铵等阳离子的吸收。在某些研究中观察 EO 提高生产性能的原因，除了增强消化外，还需要更多的研究来验证。

添加 EO 对畜产品产量影响的结果存在差异。首要的局限是体内研究的数量和使用的样本量（每组动物数量）较少，因此通常是数值上存在差异，但没有达到统计意义。另外，EO 在全世界范围内广泛用于商业养殖，但由于难以控制和测量关键变量（如动物个体饲料采食量）或缺乏企业数据，因此来自农场研究的科学报告很少。在本章中，21 项奶牛研究中，有 4 项报告了对产奶量的影响，8 项发现了积极的效果，平均改善 2%~4%。5 项生长试验中，2 项报告了积极的效果。在许多情况下，产奶量或日增重的提高并非如前所述的更高 VFA 浓度导致的结果。体内研究存在差异的原因之一是 EO 使用的持续时间从 14 d 到 5 个月不等（表1）。典型的体内研究运用交叉或拉丁方设计，其中所有动物在不同时期接受所有处理（对照和不同水平的 EO）。但是这种设计处理的时间较短，在大多数情况下不会产生有益的效果。要分辨出处理的差异可能需要延长试验期，并且在使用短期交叉设计的试验中可能存在前期试验的残余效应。最近 Belanche 等（2019）报道，商业 BEO（Agolin）的效果至少需要 4 周才能体现其能增加脂肪和蛋白质校正乳产量和饲料利用率以及减少 CH_4 产量。这些发现可以解释文献中关于不同 EO 和短期试验产奶量间高度不一致的结果（Benchaar 等，2008），这可能与作用模式和目标物种（Pauli 和

Schilcher，2010）以及微生物瘤胃生态系统的复杂性和冗余性（Weimer，2015）有关。

3 单宁

3.1 组成

单宁和黄酮类化合物是多酚类物质，是指一类含有酚类成分的植物次生代谢产物，含有至少一个羟基取代基。从化学角度来看，由于其包含多种低聚物和聚合物，因此很难定义单宁（Harborne 和 Baxter，1999；Schofield 等，2001）。尽管它们具有异质性，但都具有高分子量，并且在一定程度上具有与蛋白质、多糖、生物碱、核酸和矿物质等形成可逆和不可逆复合物的能力（Schofield 等，2001；Frutos 等，2004；Mueller-Harvey，2006）。

传统上，单宁分为两大类：缩合单宁（Condensed tannins，CT）和水解单宁（Hydrolysable tannins，HT）。CT 或原花青素（Proanthocyanidins）可多达 50 个类黄酮单元（黄烷-3-醇、黄烷-3,4-二醇）的非分支聚合物，通常具有比 HT 更高的分子量（从 1 000~20 000 到 500~3 000 Da）（Mueller-Harvey，2006）。HT 是非类黄酮化合物，分为没食子酸鞣质（gallotannins）、鞣花酸鞣质（ellagitannins），前者，HT 的核心是一种葡萄糖，它被酯化为没食子酸，而后者，葡萄糖被酯化为鞣花酸和没食子酸。这些单宁可通过酸、碱、热水和酶作用水解。

白坚木（*Schinopsis lorentzii*）和栗子（*Castanea sativa*）是研究最多的两种 CT 和 HT 来源（Patra 和 Saxena，2011）。单宁在植物中广泛分布，尤其是在树木、灌木和草本豆科植物中（Rhoades，1985；Perevolotsky，1994），并且随着检测技术的发展，含有单宁化合物的物种种类不断增加。一般来说，新叶和新花中的单宁含量更高（更容易被食草动物吃掉）（Rhoades，1985；Terrill 等，1992；Van Soest，1994）。与许多以前的综述中报道一样，环境和季节以及物候发育因素影响植物中单宁的含量（Frutos 等，2004）。一般来说，高温、水分胁迫、极端光照强度和土壤贫瘠会增加植物的单宁含量（Rhoades，1985；Van Soest，1994）。在植物生长期间，植物生长迅速，可用于合成单宁的原料很少，而开花期间，生长速度减缓，过量的碳可用于单宁的合成（Iason 等，1993）。

单宁在反刍动物日粮中普遍存在，尤其是放牧动物。Fraisse 等（2007）测定了同一牧场上种植的物种发现了 170 种不同的酚类化合物，并估计一头放牧的奶牛每天可以消耗高达 500 g 的酚类化合物。草地中的许多豆科牧草如冠状岩黄芪（*Hedysarum coronarium*）、鸡脚草（*Lotus corniculatus*）、大三叶草（*Lotus pedunculatus*）、野豌豆（*Vicia sativa*）、红豆草（*Onobrychis coronarium*）和胡枝子（*Lespedeza cuneate*）等都含有单宁。在许多地区，鉴于人类和畜牧业对土地和水资源利用方面的竞争，食草动物采食的是富含单宁的野生灌木和牧草，如红花（*Carthamus tinctorius*）、菊苣（*Cichorium intybus*）或热带植物（*Calliandra calothyrsus*, *Clitoria fairchildiana*, *Desmodium ovalifolium*, *Flemingia macrophylla*, *Leucaena leucocephala*, *Leucaena macrophylla* and *Leucaena pallida*）。在地中海的干旱地区，绵羊和山羊会采食富含单宁的树叶，如相思树（*Acacia cyanophylla*）、橡树（*Quercus pyrenaica*）、摩洛哥坚果树（*Argania spinosa*）或刺槐（*Ceratonia siliqua*）。近来，一些含有大量

单宁的农业副产品也被视为反刍动物的饲料添加剂。

3.2 活性

减少瘤胃蛋白质降解可能是单宁最显著和最广为人知的作用（Frutos 等，2004）。单宁对蛋白质有很强的亲和力，中性的瘤胃 pH 值有利于单宁-蛋白质复合物的形成。一般来说，蛋白质降解的减少与 NH_3-N 产生的减少和更多的非 NH_3-N 流入十二指肠有关（Frutos 等，2004）。来自不同植物的单宁显示出不同的物理和化学性质（Mangan，1988），因此具有非常不同的生物活性（Zucker，1983）。单宁对蛋白质的高亲和力是由于酚基团的数量较多，这为与肽的羰基发生结合提供了许多可能性（Leinmaller 等，1991）。这种复合物的形成是特定的，涉及单宁和蛋白质，以及参与分子之间的亲和程度（Zucker，1983；Mangan，1988；Hagerman 和 Butler，1991）。促进复合物形成的主要因素包括其相对较高的分子量和结构柔度（Hagerman 和 Butler，1991；Mueller-Harvey，2006）。对单宁最具亲和力的蛋白质相对较大且疏水，具有开放且柔性的结构，通常富含脯氨酸（Mueller Harvey，2006）。单宁还与碳水化合物尤其是半纤维素、纤维素、淀粉和果胶形成复合物。越来越多的证据表明，食用单宁的动物瘤胃中的纤维降解可以显著减少（McSweeney 等，2001b）。单宁抑制饲料成分瘤胃降解的机制尚不清楚，但似乎至少依赖于3种作用模式：①底物缺乏；②酶抑制；③抗微生物活性。关于底物缺乏，假设单宁可防止或至少干扰瘤胃微生物附着到饲料颗粒上（Frutos 等，2004），从而影响植物降解（Chiquette 等，1988；McAllister 等，1994）。另外，与蛋白质和碳水化合物形成复合物使微生物无法获得这些营养素（Frutos 等，2004）。单宁也是螯合剂，这可能会降低微生物代谢所需的某些金属元素的可利用性。关于酶的抑制，单宁可与微生物（细菌和真菌）的酶发生反应，抑制其活性（McSweeney 等，2001b）。最后，单宁具有直接的抗菌作用，例如改变最终形成瘤胃微生物群的微生物细胞膜的渗透性（Frutos 等，2004）。多酚对金属离子的螯合作用可导致缺铁、金属酶活性的降低以及对微生物生长所必需的氧化磷酸化的抑制。因此，已知金属螯合剂具有抗菌活性，并导致间接而重大的生态变化（Smith 等，2005）。由于这些抗微生物特性，瘤胃中不同微生物群（如细菌、产甲烷菌、原虫和真菌）对单宁的耐受程度似乎有很大差异。所有这些作用模式都可能因单宁来源的不同而不同，HT 和 CT 在这方面的差异是众所周知的。

关于日粮单宁（CT 和 HT）对反刍动物营养和产生性能影响的数据已被广泛综述，包括添加纯化物质或饲料和浓缩饲料成分中存在的天然单宁（Makkar，2003；Min 等，2003）。因此，这里我们重点关注表2所示的最新体内研究结果。

3.3 单宁对瘤胃功能的影响

根据日粮组成、动物种类、单宁来源及其在日粮中的含量，单宁既有不利影响也有有利影响（表2）。大量研究表明，与低剂量或中等剂量相比，高剂量的单宁（CT 和 HT）显著降低了自由采食量，并最终降低了动物生产性能（Frutos 等，2004；Jayanegara 等，2012）。这可能是由于其涩的味道降低了日粮的适口性。最近的一项荟萃分析（Min 和 Solaiman，2018）发现绵羊日粮中的 CT 含量与 DM 采食量之间呈负线性相关，但山羊

表 2 体内研究单宁对瘤胃发酵和动物性能影响（ns: 不显著；+: 增加；-: 降低）的近年文献的总结

文献	研究细节					瘤胃发酵											动物生产性能												
	动物	F：C	植物	形式	g/kg DM	DMI	pH	NH_3	VFA	C2	C3	C4	CH_4	Bac	Ptz	Meth	MY	Fat	Pr	MUN	BUN	OMd	NDFd	CPd	MN	UNE	FNE	Nr	ADG	FE
Cieslak 等 (2012)	奶牛	60/40	越桔	CT-E	2	---	---	ns	-	+	-	---		ns	---		+	ns	ns	ns		ns								
Dschaak 等 (2011)	奶牛	59/41	漆树	CT-E	22	---		ns	-	++	++	+					ns	ns	ns		ns	ns	ns	ns				ns		++
Dschaak 等 (2011)	奶牛	41/59	漆树	CT-E	22	---		ns	-	++	++	+					ns	ns	ns	---		ns	ns	ns				ns		+
Aguerre 等 (2016)	奶牛	51/49	漆树	CT-E	4-18	---	ns	++	ns	ns	ns	ns					ns	ns	+	++		ns	ns	ns		-	++	ns		++
Focant 等 (2019)	奶牛	64/36	啤酒花(蛇床草)	CT-E	2.9	ns		ns	ns	ns	ns	ns	ns				ns	ns	ns	---		ns	ns	ns		-	ns	ns		
Focant 等 (2019)	奶牛	64/36	栎树	HT-E	8.7	ns		ns	ns	ns	ns	ns	ns				ns	ns	ns	---		ns	ns	ns		---	+	ns		
Focant 等 (2019)	奶牛	64/36	啤酒花(蛇床草)	HT-E	11	ns		ns	ns	ns	ns	ns	ns				++	ns	ns	---		ns	ns	ns		---	ns	++		
Benchaar 等 (2008)	奶牛	40/60	漆树	CT-E	6.4	ns		ns	ns	ns	ns	ns			ns		ns	ns	ns	---		ns	ns	ns				ns		
Broderick 等 (2004)	奶牛	50/50	鸡脚草	CT-P	6-17	+++	ns	---	ns	ns	ns						ns	ns	ns	---		---	---	---		---	++	++		---
Broderick 等 (2004)	奶牛	48/52	鸡脚草	CT-P	5-15	+	ns	---	ns	ns	ns						ns	ns	ns	---		---	---	---		---	++	ns		ns
Junior 等 (2017)	母牛	50/50	黑荆树	CT-E	2	ns							ns		ns							ns						ns		

第21章 植物次生产物：在反刍动物可持续生产及营养中的有益作用

（续表）

文献	研究细节 动物	F:C	植物	形式	g/kg DM DMI	瘤胃发酵 pH	NH$_3$	VFA	C2	C3	C4	CH$_4$	Bac	Ptz	Meh	动物生产性能 MY	Fat	Pr	MUN	BUN	OMd	NDFd	CPd	MN	UNE	FNE	Nr	ADG	FE
Grainger 等 (2009)	母牛	88/12	黑荆树	CT-E	8-16	-						---				---	---							---	+++	ns		
Piñeiro-Vázquez 等 (2018)	小母牛	96/4	红破斧木	CT-E	7-29	ns	ns	ns	ns	++	ns	---		ns			-		ns		-		ns						
Aboagye 等 (2019)	小母牛	95/5	盐肤木	HT-E	15	ns	-	+	ns	ns	ns	-							---		ns	ns	ns		+	ns	ns		
Aboagye 等 (2019)	小母牛	95/5	单宁酸	CT-E	15	ns	ns	ns	ns	ns	ns	ns		ns					---		ns	ns	---		ns	+++	ns		
Aboagye 等 (2019)	小母牛	95/5	栗子	HT-E	15	ns	ns	ns	ns	ns	ns	ns		ns					---		ns	ns	---		ns	+++	ns		
Dickhoefer 等 (2016)	小母牛	50/50	白坚木	CT-E	1-10	---	ns	+++	---	+	++													---	---				
Koenig 和 Beauchemin (2018)	阉公牛	9/91	黑荆树	CT-E	13	ns	-	---	ns	-	++								---	ns					---	+++	ns		
Doce 等 (2013)	阉公牛	100/0	橡树	CT-L	76-112	ns	-	ns	ns	+	ns	ns													---				
Koenig 等 (2018)	阉公牛	9.0/91	黑荆树	CT-E	13	ns	---	ns	---	ns	-	-							---									ns	ns
Aboagye 等 (2018)	阉公牛	95/5	栗子	HT-E	2.5	ns	ns	ns	ns	ns	-			ns					ns						ns			ns	ns

（续表）

文献	研究细节					瘤胃发酵											动物生产性能													
	动物	F:C	植物	形式	g/kg DM	DMI	pH	NH$_3$	VFA	C2	C3	C4	CH$_4$	Bac	Ptz	Meth	MY	Fat	Pr	MUN	BUN	OMd	NDFd	CPd	MN	UNE	FNE	Nr	ADG	FE
Adoagye等(2018)	阉公牛	95/5	栗子	HT-E	15	ns	ns	--	ns	ns	ns	ns	ns		ns						ns					ns			ns	ns
Adoagye等(2018)	阉公牛	95/5	多种	HT+CT	2.5	ns	ns	--	ns	ns	ns	ns	ns		ns						ns					--			ns	ns
Adoagye等(2018)	阉公牛	95/5	多种	HT+CT	15	ns	ns	--	ns	ns	ns	++	-								ns					--			ns	ns
Krueger等(2010)	阉公牛	9.0/91	栗子	HT-E	15	ns	ns/-	ns	ns	ns	ns	ns	ns/+																	
Krueger等(2010)	阉公牛	9.0/91	黑荆树	CT-E	15	ns	ns	ns	ns	ns	ns	ns/+	ns																	
Bhatta(2012)	山羊	65/35	黑荆树	CT+HT-E	3-6	ns	-	-	-	+	ns	-	-	-	-							-	-	-	-	-	+	-	ns	-
Gunun等(2016)	山羊	60/40	A.thwaitesianum	CT-E	0.8-2	+	ns	ns	ns	ns	+	ns	-	ns	ns					ns	ns	ns	ns	ns	ns	--	ns	++	ns	
Animut等(2008a)	山羊	100/0	胡枝子	CT-P	50-150	-	+	+	-	-	-	ns	--	--	--					-	--	--	--	--	--		++	++	ns	
Animut等(2008b)	山羊	100/0	Schinopsis+Castianea	CT+HT-P	5	ns	ns	ns	ns	ns	ns	ns	ns	ns	ns					ns	ns	ns	ns	ns	ns	ns	++	++	ns	
Abarghuei等(2011)	绵羊	67/33	黎巴嫩栎	HT-L	3.8	ns	ns	ns	ns	ns	ns	ns	ns		-							-	-	-			+	ns	ns	
Dentinho等(2014)	绵羊	60/40	胶蔷树	CT-E	0.1-2	ns	ns	-	-	ns	ns	ns	ns								ns	ns	-	-	ns	ns	ns	ns	ns	
Lima等(2019)	绵羊	60/40	细花含羞草	CT-E	3	ns	-	ns	ns	ns	ns	ns	--	--							ns	ns	ns	ns	ns	ns	ns	ns	ns	

第 21 章 植物次生产物：在反刍动物可持续生产及营养中的有益作用

（续表）

文献	研究细节					瘤胃发酵																动物生产性能								
	动物	F:C	植物	形式	g/kg DM	DMI	pH	NH_3	VFA	C2	C3	C4	CH_4	Bac	Ptz	Meth	MY	Fat	Pr	MUN	BUN	OMd	NDFd	CPd	MN	UNE	FNE	Nr	ADG	FE
Piñeiro-Vázquez 等 (2017)	绵羊	100/0	多种	CT-L	40-66	ns			ns	ns	ns	ns	ns									ns	+	ns						
Rira 等 (2015)	绵羊	100/0	多种	CT-L	17-40	+++	-	ns	+++	--	++	ns	--	ns	ns	-						-								
Theodoridou 等 (2010)	绵羊	100/0	红豆草	CT-P	3-13	ns		---	ns	ns	ns	ns			ns							ns		---		-	+++	ns		
Malik 等 (2017)	绵羊	60/40	孟加拉榕树	CT-L	11	ns		ns	ns	ns	ns	ns	--		--															
Malik 等 (2017)	绵羊	60/40	菠萝蜜	CT-L	7.2			ns	ns	ns	ns	ns	--		--															
Malik 等 (2017)	绵羊	60/40	印楝	CT-L	7.4	-		-	-	ns	ns	ns	--		--															
Liu 等 (2011)	绵羊	52/48	栗子	HT-E	1-3	ns		ns	ns	ns	ns	ns	--		--															
Vasta 等 (2010)	绵羊	30/70	白坚木	CT-E	64									ns												+++				ns	
Utsumi 等 (2013)	绵羊	61/39	漆树	CT-E	75	-	+	++	++	++	+	ns																		
Utsumi 等 (2013)	绵羊	61/39	漆树	CT-E	75	---		ns	ns	---	+	ns																		
Sliwinski 等 (2002)	羔羊	55/45	栗子	HT-E	1-2	ns		ns	ns	ns	ns	ns	ns	-	ns							ns	ns	ns		ns	ns	ns		
Salami 等 (2018)	羔羊	15/85	栗子	HT-E	40	-		ns	+	ns	+	ns	ns		ns	ns														

（续表）

文献	研究细节					瘤胃发酵												动物生产性能										
	动物	F:C	植物	形式	g/kg DM	DMI	pH	NH_3	VFA	C2	C3	C4	CH_4	Bac	Ptz	Meth	MY	Fat	Pr	MUN	BUN	OMd	NDFd	CPd	MN	UNE	FNE	Nr	ADG	FE
Salami 等 (2018)	羔羊	15/85	剩云实	HT-E	40	ns	ns	ns	ns	ns	ns	ns	ns	ns	-	ns													
Salami 等 (2018)	羔羊	15/85	尼格拉相思	CT-E	40	ns	ns	ns	-	ns	ns	+	ns	ns	ns	-													
Salami 等 (2018)	羔羊	15/85	儿茶钩藤	CT-E	40	ns	ns	ns	-	ns	ns	ns	ns	ns	-	-													
Jayanegara 等 (2012)	多种		多种	HT+CT	高达177	+	ns	--	ns	ns	ns	ns	--	ns	ns							--	--	--				ns	
Min 和 Solaiman (2018)	多种		多种	HT+CT	高达100	-		-	-				--			-						-	-					ns	

注：F:C，饲料浓缩比；DMI，DM 采食量；VFA，挥发性脂肪酸；C2，乙酸盐；C3，丙酸盐；C4，丁酸盐；Ptz，原虫；Bac，细菌；Meth，产甲烷菌；MY，产奶量；Fat，乳脂率%；Pr，乳蛋白含量；OMd，OM 消化率；NDFd，NDF 消化率；CPd，CP 消化率；MN，微生物氮流量；UNE，尿氮排泄量；FNE，粪便 N 排泄量；Nr，N 保留率；ADG，平均日增重；FE，饲料效率。

的这种相关是二次的，表明低剂量的单宁可以刺激山羊的采食。尽管不同研究单宁浓度和来源可能不同，但最近的研究中反刍动物饲料添加的浓度在 5~60 mg/kg DM。

在一项基于体外和体内研究的荟萃分析中，Jayanegara 等（2012）报告了单宁添加引起总 VFA 浓度线性下降，并使体外乙酸盐摩尔比例有增加的趋势。然而，同一研究（Jayanegara 等，2012）报告，体内研究单宁对总 VFA 和单个摩尔比例没有显著影响。最近的体内研究与后者的观察结果一致，表明补充单宁可以增加（5 项研究）或减少（7 项研究），但在大多数研究中（28 项）对瘤胃 VFA 浓度没有影响。另外，在某些情况下（9 项研究），补充 CT 导致丙酸的摩尔比例增加。瘤胃发酵中这种细微变化的原因尚不清楚，但可能与单宁蛋白和单宁纤维复合物的形成有关。由于纤维分解菌高度依赖蛋白质和纤维为底物发酵产生乙酸盐和丁酸盐等主要产物，因此这些复合物可能对纤维分解菌产生有害影响（Belanche 等，2012a）。

单宁被认为通过改变膜的通透性对一些瘤胃微生物具有抗菌作用（Frutos 等，2004）。然而，这种抗菌效果高度依赖于单宁的剂量和性质以及微生物种类。CT 对一些瘤胃微生物的半纤维素酶、内切葡聚糖酶和蛋白水解酶有直接抑制作用，如产琥珀酸丝状杆菌、溶纤维丁酸弧菌、嗜淀粉瘤胃杆菌和牛链球菌（Vasta 等，2019）。相反，栖瘤胃普雷沃菌能够通过产生保护性细胞外物质来抵消单宁的负面影响（Jones 等，1994）。一些瘤胃细菌和厌氧真菌能够将没食子酸（HT 的主要成分）完全降解为乙酸盐和丁酸盐（Bhat 等，1998），然而最近的研究并未发现日粮中 HT 的含量与瘤胃中这些发酵产物积累之间存在明确的相关性（表 2）。当山羊摄入栗子和白坚木单宁时，瘤胃中硬壁菌门的浓度显著下降，而拟杆菌门（一般为革兰氏阴性）丰度增加，后肠道中的结果相反，这可能是由于后消化道中单宁-蛋白质复合物的分解造成的（Min 和 Solaiman，2018）。使用两种不同品种的绵羊（Texel 和 Blackbelly）补充来自 3 种不同植物物种（银合欢 *Leucaena leucocephala*、南洋樱 *Gliricidia sepium* 和木薯 *Manihot esculenta*）的 CT（17~40 mg/kg DM），发现瘤胃中黄色瘤胃球菌和产甲烷菌的浓度下降；而在细菌总数、白色瘤胃球菌、产琥珀酸丝状杆菌浓度以及细菌和产甲烷菌多样性方面均未发现差异（Rira 等，2015）。McSweeney 等（2001b）报道，从 *Caliandra calothyrsus* 中提取的缩合单宁减少了产琥珀酸丝状杆菌和瘤胃球菌等纤维素分解细菌的数量，相比之下，原虫和真菌以及拟杆菌属、卟啉单胞菌属、普雷沃菌属等属的微生物群似乎受影响较小。另一项研究（Vasta 等，2010）表明，绵羊补充白坚木（64mg/kg DM）后瘤胃中溶纤维丁酸弧菌浓度较高。此外，最近的一项研究表明，羔羊日粮中添加不同来源的 HT（栗子和刺云实）和 CT（含羞草和钩藤儿茶）40 mg/kg DM 时，瘤胃主要细菌浓度和产甲烷菌种类相似，但栗子和含羞草能引起瘤胃细菌群落结构的巨大变化，但刺云实和钩藤儿茶不会（Salami 等，2018）。不同来源的单宁之间的差异似乎支持了单宁对瘤胃微生物的抗菌活性与其分子量相关的假设。另一项使用分子指纹技术的研究（Smith 和 Mackie，2004）报告，浓缩单宁能显著增加大鼠胃肠道中的抗单宁细菌的比例。DGGE 条带测序分析和单宁抗性菌株的特征表明，单宁对肠杆菌科和拟杆菌属细菌具有选择性。总的来说，这些研究表明，含单宁或其提取物的日粮对瘤胃和小肠及大肠微生物群的组成和多样性有直接影响（Min 和 Solaiman，2018）。然而，这些发现表明单宁对瘤胃细菌群落的影响是一种多因素效应，需要利用新一代分子技术进行进一步研究，以充分阐明其作用模式。

体外研究表明，单宁可能对厌氧真菌具有抗菌活性（Patra 和 Saxena，2009b）。Liu 等（2011）指出，补充栗子 HT 的绵羊瘤胃中厌氧真菌水平较低；然而，更广泛研究表明，不同来源的 HT 和 CT 不影响瘤胃中真菌的浓度（Salami 等，2018）。因此，与纤维素分解细菌相比，瘤胃真菌的纤维降解能力可能对 CT 的抑制作用不太敏感（McSweeney 等，2001b）。

最近的各种研究表明，补充单宁能减少瘤胃产甲烷作用（表 2）。Min 和 Solaiman（2018）报道，日粮中的 CT 含量与 CH_4 排放量呈负相关。这种抑制作用似乎依赖于对产甲烷菌的抗菌作用，以及由于瘤胃中纤维消化和原虫数量减少而导致的 H_2 产量的间接减少（Patra 等，2017）。大量研究表明，补充单宁的绵羊瘤胃中的产甲烷菌浓度较低（Liu 等，2011；Rira 等，2015；Min 和 Solaiman，2018），而不影响总体产甲烷菌群结构（Salami 等，2018）。据报道，随着日粮中 CT 含量的增加，只有甲烷杆菌属（*Methanobrevibacter* spp）菌呈线性减少（Min 和 Solaiman，2018）。研究表明，单宁的类型和分子量对其降低 CH_4 产量和瘤胃产甲烷菌丰度的潜力非常重要，而高分子量 CT 更有效（Patra 等，2017）。Min 等（2003）报道，为了降低瘤胃甲烷产量和优化绵羊生产，饲料中 CT 的最佳水平为 22~40 mg/kg DM。虽然 HT 也会影响产甲烷菌，但它们通常对动物的毒性更大，因此其使用率往往低于 CT（Patra 和 Saxena，2011）。然而，单宁在减少 CH_4 排放方面存在很大差异，有时会与含单宁植物的营养特征相混淆，关于实际作用方式的科学证据很少。还需要在此方面进行进一步研究。

单宁对瘤胃原虫影响的报道相互矛盾。一小部分研究认为，单宁能增加原虫数量，而有的研究表明，单宁可减少原虫数量或没有影响（Patra 和 Saxena，2009a）。研究中的这种差异表明，不同植物中的单宁对原虫的作用并不相同。最新的文献（表 2）可以看出这种抗原虫作用似乎依赖于日粮。事实上，大多数关于单宁导致原虫数量显著下降的研究都是在喂食大量精料（33%~82%）；相反，大多数使用基于牧场日粮的研究均显示没有显著的效果。众所周知，适量的精料会增加瘤胃原虫水平（Belanche 等，2011），这可能会放大单宁的抗原虫活性。先前的研究表明，这种抗原虫的活性可能因原虫物种而异，特别是，全毛虫（holotrichs）似乎比最丰富的原虫类群更容易受到单宁的影响（Makkar，2003；Patra 和 Saxena，2009b）。然而，最近的 3 项研究表明，食用橡树叶（Abarghuei 等，2011）、富含 CT 的各种植物叶（Rira 等，2015）和新鲜红豆草（Theodoridou 等，2010）能降低绵羊瘤胃中内毛虫（Entodiniomorphs）的比例，导致某些种类被完全清除（如双毛属 *Diplodinium* 和真双毛属 *Eudiplodiniium*）。这些发现表明，单宁也具有抗原虫的活性，由于内毛虫分解细菌氮的能力比全毛虫大，这方面间接有利于瘤胃氮效率（Belanche 等，2012a）。尽管有这些新的发现，但需要进一步的研究才能真正理解单宁作为瘤胃微生物调节剂的作用。

3.4 单宁对动物生产性能的影响

普遍认为单宁能降低日粮的消化率，其主要是在 pH 值 3.5~8 能与蛋白质形成稳定的氢键而影响其消化率。这些复合物在瘤胃 pH 下是稳定的，但当 pH 低于 3.5 ［如皱胃

（pH 2.5~3）]或高于8[如十二指肠（pH 8）]时,这些复合物会离解（Frutos等,2004）。因此,补充单宁可以使蛋白质的降解从瘤胃转移到肠道。饲料在肠道消化率的增加通常无法完全补偿在瘤胃中降解的减少,这可能是由于单宁-蛋白质复合物不能在皱胃中完全分解,也可能是单宁在肠道中仍然存在,并与消化酶和饲料蛋白质形成新的复合物,或者是由于在肠道内通常能被消化的黏膜蛋白与单宁相互作用,而一旦与CT复合就会变得难以消化,从而影响肠道的吸收。因此,单宁会降低饲料的总消化率的最清楚证据之一是随着单宁含量的增加粪N排泄量（FNE）增加。最新研究一致认为,FNE的增加与补充单宁的动物尿N排泄量的减少有关,这可以最大限度地减少NH_3挥发和/或地下水污染,并最终减少对环境的负面影响（Gill等,2010）。日粮中补充单宁能使动物十二指肠蛋白质流量的增加量超过肠道氨基酸吸收量的减少程度时,反刍动物可以从中获益。

由于单宁影响动物采食量和饲料的消化利用,因此可能影响动物生产性能。总的来说,高单宁摄入量（高于50 g/kg DM）由于能降低自由采食量和饲料消化率以及改变动物生理（较高的流涎率、黏膜干扰等）,对生产性能有明显的负面影响（Serrano等,2009）。早期研究观察到喂高水平CT（>76 g/kg DM）的羔羊的增重减少;而一些学者指出,继续摄入单宁可能会导致部分适应,使其有害影响减弱甚至消失（Barry和Manley,1984）。摄入低于50 g/kg DM的单宁可提高反刍动物对饲料的消化利用率,这主要是因为反刍动物的瘤胃蛋白质降解减少,小肠对氨基酸的吸收增加（Min和Solaiman,2018）。最近的研究表明,在10~30 g/kg DM的水平下,补充各种CT源的羔羊平均日增重分别增加了8%（Sharifi和Chaji,2019）和35%（Pathak等,2017）。产奶量方面,大多最近的研究表明单宁对产奶量或牛奶成分均无显著影响,尽管许多研究显示牛奶和血浆尿素氮会持续下降（Tiemann等,2008;Aboagye等,2019）。低水平的尿素氮和生产性能的提高表明动物的氮效率提高。有研究报道,越橘（*Vaccinium vitis*）（Cieslak等,2012）和橡树CT（Focant等,2019）可以提高奶牛产奶量（+2.3%~+5%）,白坚木单宁可以提高奶牛饲料效率（+2.3%~+7%）（Dschaak等,2011;Aguerre等,2016）。包括单宁在内的植物提取物改变反刍动物畜产品（即牛奶和肉类）脂肪酸组成的能力目前受到极大关注。最近的研究表明,单宁可以抑制一些参与瘤胃脂肪酸生物氢化反应的细菌（如*Butiryvibrio proteoclasticus*）,该反应将十八（碳）烯酸（vaccenic acid）转化为硬脂酸（stearic acid）（Vasta等,2010）。因此,有人假设,单宁可以通过增加不饱和脂肪酸的含量来提高产品质量。

4 皂苷

4.1 组成

皂苷主要是由植物产生的天然次生产物,也可以由低等海洋动物和一些细菌产生（Yoshiki等,1998）。这个名字来源于其在水溶液中能形成稳定的泡沫,这与肥皂类似。如Hart等（2008）所述,皂苷是高分子量糖苷,其中糖与疏水性苷元（皂苷元）相连,该苷元在性质上可能是三萜或甾体。苷元可以包含一个或多个不饱和C-C键。在C3位置连接有一个糖分子的皂贰为单桥糖皂苷，在 C26 或 C28 位置具有额外糖部分的皂贰为双桥糖皂苷。皂苷结构的多样性和复杂性源于苷元结构的变异性、侧链的性质以及这些部分在苷元上的附着位置。研究表明，皂苷的连接类型和糖组成与其生物活性直接相关（Teferedenge，2000；Ramos Morales 等，2017）。

皂苷存在于植物的不同部位，如根、块茎、树皮、叶、种子和果实。三萜皂苷主要存在于栽培作物（主要是双子叶植物）中，而甾体皂苷存在于单子叶植物中，但有些植物同时含有三萜和甾体。幼叶比成熟叶含有更多的皂苷。因此，它们通常存在于最容易受到真菌或细菌攻击或昆虫掠食的组织中（Morrissey 和 Osbourn，1999；Francis 等，2002）。

虽然在许多植物中都含有皂苷，但本节仅关注与反刍动物生产性能相关的皂苷（表3）。通常富含皂苷的植物本身不被用作动物饲料；但从特定的植物部位提取的皂苷可用作饲料添加剂。三萜皂苷已在许多豆类植物中检测到，如皂树、紫花苜蓿、豌豆、豆类和大豆，以及栗子、茶叶、藜麦、葱、菠菜、甜菜、向日葵或人参。丝兰、燕麦、芦笋、葱、辣椒、茄子、番茄籽或山药中含有甾体皂苷。就重要性而言，丝兰和皂树是最常见的皂苷来源，它们的提取物可在市场上买到，不仅可以作为反刍动物的饲料添加剂，还可以作为饮料中的发泡剂和化妆品中的乳化剂。丝兰是原产于北美的沙漠植物，而皂树（*Quillaja saponaria*）（皂角树）则起源于智利。丝兰提取物含有 4.4% 的甾体皂苷，而皂树提取物含有 10% 的总皂苷，两种提取物都含有超过 20 种不同的活性皂苷异构体（Wina 等，2005）。此外，在过去几十年中，从茶籽（*Camellia sinensis*）、苜蓿（*Medicago sativa*）和无患子（*Sapindus*）果实（如植物无患子 *S. saponaria* 和毛瓣无患子 *S. rarak*）中分离出的新型商业皂苷具有在动物生产中应用的潜力（Patra 和 Saxena，2009a）。

4.2 活性

纤毛虫栖息于瘤胃生态系统，但不是必需的，并且由于其捕食瘤胃细菌是细菌蛋白质大量周转的主要原因（Belanche 等，2012a）。尽管对纤维降解有负面影响，但通过消除瘤胃中的原虫（灭绝原虫），反刍动物氮的利用率可能会得到改善（Newbold 等，2015）。因此，人们普遍认为，去除或降低瘤胃原虫水平可提高反刍动物的生产性能，特别是使用低蛋白日粮时。此外，由于原虫是瘤胃产氢的关键微生物，因此已建议将灭绝原虫作为有效的减少甲烷排放的策略（Belanche 等，2015），平均减少 11%（Newbold 等，2015）。但在商业化养殖中从瘤胃中去除原虫的实用性仍然是一个挑战。皂苷通过与原虫膜表面的甾醇形成复合物来杀死或破坏原虫，膜受损的原虫最终也会解体死亡（Wallace 等，2002）。因此，抑制原虫一直是皂苷研究的目标。瘤胃原虫不同物种对皂苷的敏感性可能存在差异，这是因为它们细胞膜的甾醇组成不同，这导致喂食皂苷可能会导致部分灭绝原虫（Patra 和 Saxena，2009b）。体外研究表明，丝兰皂苷（1%）抑制纤毛虫（内毛虫的纤毛和全毛虫的收缩）的运动，并降低它们捕食瘤胃细菌的活性（Wallace 等，1994）。一些文献详细综述了各种皂苷来源对体外原虫浓度和活性持续的负面影响（Wina 等，2005；Patra 和 Saxena，2009b）。然而，这些提取物在体内条件下的作用存在更大差异。

第21章 植物次生产物：在反刍动物可持续生产及营养中的有益作用

表3 体内研究皂苷对瘤胃发酵和动物性能影响（ns：不显著；+：增加；−：降低）的近年文献的总结

文献	研究细节				瘤胃发酵																	动物生产性能					
	动物	F:C	植物	g/kg DM	DMI	pH	NH₃	VFA	C2	C3	C4	CH₄	Bac	Ptz	Meth	MY	Fat	Prot	OMd	NDFd	CPd	MN	UNE	FNE	Nr	ADG	FE
Valdez等(1986)	奶牛	45:55	丝兰	0.077			ns	+	ns	ns	ns	ns	ns	−		ns	ns	ns	ns							+	
Holtshausen等(2009)	奶牛	51:49	丝兰	1	ns	ns	−	ns	ns	ns	ns	ns		ns		ns	ns	ns	ns	ns	ns					ns	−−
Holtshausen等(2009)	奶牛	51:49	皂树	1	ns	ns	ns	ns	ns	ns	ns	ns		ns		ns	ns	ns	ns	ns	ns					ns	ns
Holtshausen等(2009)	奶牛	51:49	丝兰	2.7	ns	ns	ns	ns	ns	ns	ns	ns		ns		ns	ns	ns	ns	ns	ns						
Anantasook等(2015)	奶牛	50:50	雨树	5.3	ns	ns	−	ns	−	++	ns	−−	ns	−	−	+	ns	+	ns	ns	ns						
Poungchompu等(2009)	母牛	多种	多种	6.3	ns	ns	ns	++	−−−	+++	ns	−−−	ns	−−−	ns				−	ns	ns						
Guyader等(2015)/Popova等(2019)	母牛	50:50	茶树	5	−	ns	ns	+	ns	ns	ns	ns	ns	ns	ns				ns	ns	ns	ns	ns	ns	ns		
Molina-Botero等(2019)	小母牛	80:20	多种	3.3	ns	ns	ns	ns	ns	ns	ns	ns	ns	ns	ns				ns	ns	ns					ns	
Molina-Botero等(2019)	小母牛	85:15	多种	6.6	ns	ns	ns	ns	ns	ns	ns	ns	ns	ns	ns				ns	ns	ns					ns	

（续表）

文献	研究细节					瘤胃发酵												动物生产性能										
	动物	F:C	植物	g/kg DM	DMI	pH	NH$_3$	VFA	C2	C3	C4	CH$_4$	Bac	Ptz	Meth	MY	Fat	Prot	OMd	NDFd	CPd	MN	UNE	FNE	Nr	ADG	FE
Molina-Botero等(2019)	小母牛	90:10	多种	10	ns				ns	ns	ns	ns	ns	ns	ns				ns	-	ns					ns	
Ramirez-Restrepo等(2016a)	阉牛	15:85	茶树	1.4~2.2	ns	ns		ns	ns	ns	+	ns	ns	+	ns												
Wang等(2019)	阉牛	35:65	油茶	9 g/d		ns		+	-	ns	ns																
Ramirez-Restrepo等(2016b)	阉牛	15:85	茶树	0.8~3	ns	+		-	+	-	-	ns															
Malik等(2009)	水牛	60:40	紫花苜蓿	6	+	+	ns	ns	-	+	ns		ns	-	+												
Santoso等(2007)	山羊	70:30	无柄威应草	0.3~0.5	ns	++	--	-	-	++	-		-						-	-	--	++	--	++	++		
Zhou(2012)	山羊	50:50	苦丁茶	0.4~0.8	ns	ns		ns	ns	ns	ns	ns			ns				ns	ns	ns						
Klita等(1996)	绵羊	100:0	紫花苜蓿	10~40	ns			ns	ns	ns	ns		---						-	-	-						
Yuan等(2007)	绵羊	60:40	山茶	5		-		ns	ns	ns	ns	---															

第21章 植物次生产物：在反刍动物可持续生产及营养中的有益作用

（续表）

文献	研究细节				瘤胃发酵											动物生产性能							
	动物	F：C	植物	g/kg DM	DMI	pH	NH_3	VFA	C2	C3	C4	CH_4	Bac	Ptz	Meth	MY	Fat	Prot	OMd	NDFd	CPd	MN	UNE	FNE	Nr	ADG	FE
Camul-Solis 等 (2014)	绵羊	94：6	丝兰	1.8~7.4	ns	ns	ns	ns	ns	ns	ns								ns	ns	ns	ns					
Wang 等 (2009)	绵羊	75：25	丝兰	0.1	ns	ns	---	+++	ns	+	ns	--							ns								
Zhou 等 (2011)	绵羊	60：40	山茶	3		-	++					--		-	ns							++					
Wina 等 (2006)	绵羊	65：35	毛瓣无患子	12~18			--	ns	ns	+	--	---		---					ns	ns	ns		ns	ns			
Sliwiński 等 (2002)	羔羊	55：45	丝兰	0.002-0.03	ns	ns		ns	ns	ns	ns			ns	ns				ns	ns	ns		ns	ns			
Mao 等 (2010)	羔羊	60：40	茶树	4	ns	++	++	+	ns	ns	ns	---	ns	-	ns							++				ns	

注：F：C，饲料浓缩比；DMI，DM 采食量；VFA，挥发性脂肪酸；C2，乙酸盐；C3，丙酸盐；C4，丁酸盐；Ptz，原虫；Bac，细菌；Meth，产甲烷菌；MY，产奶量；Fat，乳脂率％；Pr，乳蛋白含量；OMd，OM 消化率；NDFd，NDF 消化率；CPd，CP 消化率；MN，微生物氮流量；FNE，粪便 N 排泄量；Nr，N 保留率；ADG，平均日增重；FE，饲料效率。

特定剂量下特定皂苷对原虫的抑制趋于稳定［Yaccca 原虫的抑制率为53%，皂树 Quillaja 抑制率为29%，无柄感应草（*Biophytum petersianum*）抑制率为40%］，超过该剂量后抑制作用没有变化（Pen 等，2006；Santoso 等，2007）。因此，目前的研究重点在于不同类型的皂苷的组合，以研究其潜在的协同抗原虫活性。并且饲料的组成对原虫群落有很大的影响，因此，皂苷对瘤胃原虫的影响可能依赖于饲料。一些学者证明了此假设，将皂苷添加到高粗料日粮比添加到高精料日粮对瘤胃原虫水平的下降幅度更大（Hess 等，2003；Poungchompu 等，2009）。同样，反刍动物物种同样影响皂苷的效果，最近的一项研究表明，与其他动物不同，牛对高剂量茶皂苷的耐受性不高（Ramírez Restrepo 等，2016a）。

4.3 皂苷对瘤胃功能的影响

各种皂苷均有抗瘤胃原虫的活性，大多数研究表明所有原虫对皂苷都具有相似的敏感性，因为在添加和不添加皂苷的动物间原虫物种分布没有显著的差异（即全毛虫和内毛虫）（Śliwiński 等，2002；Holtshausen 等，2009；Popova 等，2019）。

皂苷对瘤胃细菌群落的影响尚未被详细研究，结果也是矛盾的。例如，当牛添加雨树（*Samanea saman*）皂苷（Anantasook 等，2015）和茶皂苷（2 mg/kg DM）（Ramírez Restrepo 等，2016a）时，瘤胃产琥珀酸丝状杆菌的浓度增加，而在添加茶皂苷（3 mg/kg DM）的绵羊中观察到相反的情况（Zhou 等，2011）。在体内研究中，皂苷对瘤胃细菌群落影响的差异也体现在白色瘤胃球菌和黄色瘤胃球菌上（Ramírez Restrepo 等，2016b；Wang 等，2019）。这些研究清楚地表明，皂苷对瘤胃细菌的影响具有物种依赖性，这可能为瘤胃代谢提供一种选择性操纵方法。例如，有人推测，在高精料饲料中添加皂苷可以通过抑制乳酸产生菌（如牛链球菌）的生长来降低瘤胃酸中毒的发生率（Russell 和 Rychlik，2001）。Belanche 等（2016）利用测序研究瘤胃微生物组，描述了常春藤果实皂苷对瘤胃模拟技术（Rusitec）的作用模式。这项研究得出这样的结论，常春藤果实皂苷对细菌群落影响较小，导致微生物发酵没有差异。但是，常春藤皂苷改变了产甲烷菌群落的结构，降低了其多样性。常春藤皂苷对产甲烷菌的这种特殊抗菌作用被认为是导致甲烷产量大幅下降（-40%）的主要抗产甲烷机制。

最近的体内研究（表3）结果表明，在日粮中添加皂苷可能会减少甲烷的产生，这可能是由于原虫和/或产甲烷古细菌数量的减少。例如，以低剂量（0.1 g/kg DM）补充丝兰皂苷可促使绵羊甲烷产量减少15%（Wang 等，2009）。同样，茶皂苷也能减少瘤胃原虫数量而使绵羊（-9%）和羔羊（-27%）甲烷生成量持续下降（Yuan 等，2007；Mao 等，2010）。由于这种抗原虫的作用，茶皂苷对不灭绝原虫反刍动物的甲烷产生的抑制作用比灭绝原虫的几乎高1倍（-11% vs -5.7%）（Zhou 等，2011）。因此，皂苷对瘤胃古菌的直接抑制活性取决于皂苷的类型、浓度和日粮类型。例如，高剂量（4 mg/mL）毛瓣无患子果实的皂苷会降低产甲烷菌的浓度，而低剂量则没有影响（Wina 等，2005）。类似地，添加印度田菁（*S. sesban*）在精料型日粮条件下降低甲烷产量的程度比粗料型日粮时更为明显，尽管其与产甲烷菌减少的程度没有相关性（Goel 等，2008）。Guo 等（2008）使用体外模型发现茶皂素减少了甲烷的产生和甲基辅酶-M 还原酶亚单位 A (mcrA)的表达水平，说明这种影响是通过减少原虫细胞和附着在原虫上的产甲烷菌引起的。因此，皂苷对瘤胃纤毛虫和产甲烷菌的影响在不同的研究中差异很大。Patra 等（2017）认为皂苷可能对产甲烷菌的直接影响很小，但通过降低原虫上附着的产甲烷菌的丰度间接发挥作用。已经证明，原虫上附着的产甲烷菌非常活跃，可产生占甲烷总产量 37%的甲烷（Finlay 等，1994）。

尽管许多研究表明皂苷抑制了瘤胃原虫，但一些人指出，抗原虫的作用是暂时的，这是将皂苷用作饲料添加剂的主要缺点。原虫"本身"似乎不会对皂苷产生抗药性（Newbold 等，1997），但瘤胃细菌可以通过增强其对皂苷去糖基化而失去活性的能力来适应皂苷的存在（Patra 和 Saxena，2009b）。克服此局限性的策略有①间歇性补充皂苷以防止细菌适应（Newbold 等，1997）；②通过不同类型皂苷的组合（Molina Botero 等，2019）；③使用化学修饰的皂苷使其更耐瘤胃降解（Ramos Morales 等，2017）。然而，这些策略需要进一步的研究来证明在体内的有效性，以克服瘤胃对皂苷的适应性挑战。

4.4 皂苷对动物生产性能的影响

皂苷的抗原虫作用也对反刍动物生产性能有一定影响，且影响取决于使用的饲料和皂苷。尽管众所周知灭绝原虫可以降低瘤胃中的纤维降解，但人们普遍认为灭绝原虫对反刍动物的生产性能有积极影响（Newbold 等，2015）。鉴于瘤胃原虫对氮代谢的负面影响，当动物饲喂低蛋白日粮时，这些积极影响尤其重要（Firkins 等，2007）。因此，对蛋白质需求量高而饲料真蛋白含量低时，补充皂苷更有用，但不影响能量需要（Eugène 等，2004）。最近的研究显示，皂苷对动物生产性能（即产奶量和增重）没有影响。皂苷对高产奶牛的影响缺乏一致性，可能是由于一些研究使用皂苷水平较低，而在其他研究不添加皂苷的饲料满足了营养需求，因此皂苷能提高动物生产性能的空间很小。尽管存在这一限制，但仍有少数文献报道，奶牛补充雨树（*Samanea saman*）皂苷提取物时产奶量增加（+10%）（Anantasook 等，2015），山羊喂食茶皂苷时增重提高（+9%）（Jadhav 等，2017）。虽然很难解释皂苷（或其代谢物）提高动物生产性能的确切作用机制，但一些研究表明微生物蛋白质合成均增加。据报道，山羊补充无柄感应草皂苷可增加高达 37%的微生物氮流量（Santoso 等，2007），不灭绝原虫的绵羊微生物氮流量增加量类似（增加 29%）（Mao 等，2010；Jadhav 等，2017），但灭绝原虫的绵羊仅增加了 8.6%（Zhou 等，2011）。这些结果似乎表明，皂苷对氮代谢的积极影响主要是由其抗原虫作用介导的。

5 未来趋势和结论

植物提取物包括多种化合物，在反刍动物饲养中具有潜在的应用前景，特别是精油、单宁和皂苷已被证明具有改变瘤胃发酵和提高产量的作用。作用机制主要包括更有效地利用日粮蛋白质和能量，减少甲烷排放，这有时会导致产奶量或增重增加。然而，尽管在过去几十年中进行了广泛的研究，植物提取物作为反刍家畜饲料添加剂的使用仍然受到若干因素的限制。除关于高度可变的成分和剂量相关的研究外，缺乏可靠的体内研究来证实体外的结果是主要限制之一。将体外使用的剂量转化为体内条件的剂量具有挑战性，但如果没有此类研究,就无法得到农场中成功的补充策略。另一个需要考虑的重要因素是体内试验的持续时间。虽然许多植物提取物(如皂苷)没有持久的效果,但其他提取物(如精油)需要较长时间的使用才能实现生产性能的持续提高。

未来的研究需要动物饲料公司(包括开发和提供植物提取物添加剂的公司)加强与科学家之间的合作,以利用现有的农场数据,更好地了解植物提取物对动物健康和生产性能的影响。

6 参考文献

Abarghuei, M., Rouzbehan, Y. and Alipour, D. 2011. Effect of oak (*Quercus libani Oliv.*) leave tannin on ruminal fermentation of sheep. Journal of Agricultural Science and Technology 13, 1021–32.

Aboagye, I. A., Oba, M., Castillo, A. R., et al., 2018. Effects of hydrolyzable tannin with or without condensed tannin on methane emissions, nitrogen use, and performance of beef cattle fed a high-forage diet. Journal of Animal Science 96 (12), 5276–86.

Aboagye, I. A., Oba, M., Koenig, K. M., et al., 2019. Use of gallic acid and hydrolyzable tannins to reduce methane emission and nitrogen excretion in beef cattle fed a diet containing alfalfa silage. Journal of Animal Science 97 (5), 2230–44.

Acamovic, T. and Brooker, J. D. 2005. Biochemistry of plant secondary metabolites and their effects in animals. Proceedings of the Nutrition Society 64 (3), 403–12.

Aguerre, M. J., Capozzolo, M. C., Lencioni, P., et al., 2016. Effect of quebracho-chestnut tannin extracts at 2 dietary crude protein levels on performance, rumen fermentation, and nitrogen partitioning in dairy cows. Journal of Dairy Science 99 (6), 4476–86.

Anantasook, N., Wanapat, M., Cherdthong, A., et al., 2015. Effect of tannins and saponins in Samanea saman on rumen environment, milk yield and milk composition in lactating dairy cows. Journal of Animal Physiology and Animal Nutrition 99 (2), 335–44.

Animut, G., Puchala, R., Goetsch, A. L., et al., 2008a. Methane emission by goats consuming diets with different levels of condensed tannins from lespedeza. Animal Feed Science and Technology 144 (3–4), 212–27.

Animut, G., Puchala, R., Goetsch, A. L., et al., 2008b. Methane emission by goats consuming different sources of condensed tannins. Animal Feed Science and Technology 144 (3–4), 228–41.

Anitescu, G., Doneanu, C. and Radulescu, V. 1997. Isolation of coriander oil:comparison between steam distillation and supercritical CO_2 extraction. Flavour and Fragrance Journal 12 (3), 173–6.

Barry, T. N. and Manley, T. R. 1984. The role of condensed tannins in the nutritional value of Lotus pedunculatus for sheep:2. Quantitative digestion of carbohydrates and proteins. British Journal of Nutrition 51 (3), 493–504.

Beauchemin, K. A. and McGinn, S. M. 2006. Methane emissions from beef cattle:effects of fumaric acid, essential oil, and canola oil. Journal of Animal Science 84 (6), 1489–96.

Belanche, A., Abecia, L., Holtrop, G., et al., 2011. Study of the effect of presence or absence of protozoa on rumen fermentation and microbial protein contribution to the chyme. Journal of Animal Science 89 (12), 4163–74.

Belanche, A., De la Fuente, G., Moorby, J. M., et al., 2012a. Bacterial protein degradation by different rumen protozoal groups. Journal of Animal Science 90 (12), 4495-504.

Belanche, A., Doreau, M., Edwards, J. E., et al., 2012b. Shifts in the rumen microbiota due to the type of carbohydrate and level of protein ingested by dairy cattle are associated with changes in rumen fermentation. Journal of Nutrition 142 (9), 1684-92.

Belanche, A., de la Fuente, G. and Newbold, C. J. 2015. Effect of progressive inoculation of fauna-free sheep with holotrich protozoa and total-fauna on rumen fermentation, microbial diversity and methane emissions. FEMS Microbiology Ecology 91 (3), 1-10.

Belanche, A., Pinloche, E., Preskett, D., et al., 2016. Effects and mode of action of chitosan and ivy fruit saponins on the microbiome, fermentation and methanogenesis in the rumen simulation technique. FEMS Microbiology Ecology 92 (1), 1-13.

Belanche, A., Newbold, C., Morgavi, D., et al., 2019. A meta-analysis describing the effects of a commercial essential oils blend on performance, digestion and methane emissions in dairy cows. Animals (in press).

Benchaar, C. and Greathead, H. 2011. Essential oils and opportunities to mitigate enteric methane emissions from ruminants. Animal Feed Science and Technology 166-167, 338-55.

Benchaar, C., Petit, H. V., Berthiaume, R., et al., 2007. Effects of essential oils on digestion, ruminal fermentation, rumen microbial populations, milk production, and milk composition in dairy cows fed alfalfa silage or corn silage. Journal of Dairy Science 90 (2), 886-97.

Benchaar, C., Calsamiglia, S., Chaves, A. V., et al., 2008. A review of plant-derived essential oils in ruminant nutrition and production. Animal Feed Science and Technology 145 (1-4), 209-28.

Benchaar, C., Lettat, A., Hassanat, F., et al., 2012. Eugenol for dairy cows fed low or high concentrate diets: effects on digestion, ruminal fermentation characteristics, rumen microbial populations and milk fatty acid profile. Animal Feed Science and Technology 178 (3-4), 139-50.

Bhat, T. K., Singh, B. and Sharma, O. P. 1998. Microbial degradation of tannins-a current perspective. Biodegradation 9 (5), 343-57.

Bhatta, R., Enishi, O., Yabumoto, Y., et al., 2012. Methane reduction and energy partitioning in goats fed two concentrations of tannin from Mimosa spp. J. Agric. Sci, 84, 409-415.

Blanch, M., Carro, M. D., Ranilla, M. J., et al., 2016. Influence of a mixture of cinnamaldehyde and garlic oil on rumen fermentation, feeding behaviour and performance of lactating dairy cows. Animal Feed Science and Technology 219, 313-23.

Bodas, R., López, S., Fernandez, M., et al., 2008. *In vitro* screening of the potential of numerous plant species as antimethanogenic feed additives for ruminants. Animal Feed Science and Technology 145 (1-4), 245-58.

Bodas, R., Prieto, N., García-González, R., et al., 2012. Manipulation of rumen fermentation and methane production with plant secondary metabolites. Animal Feed Science and Technology 176 (1-4), 78-93.

Braun, H. S., Schrapers, K. T., Mahlkow-Nerge, K., et al., 2019. Dietary supplementation of essential oils in dairy cows: evidence for stimulatory effects on nutrient absorption. Animal: an International Journal of Animal Bioscience 13 (3), 518-23.

Broderick, G. A., Udén, P., Murphy, M. L., et al., 2004. Sources of variation in rates of in vitro ruminal protein degradation. Journal of Dairy Science 87 (5), 1345-59.

Burt, S. 2004. Essential oils: their antibacterial properties and potential applications in foods—a review. International Journal of Food Microbiology 94 (3), 223-53.

Busquet, M., Calsamiglia, S., Ferret, A., et al., 2005a. Screening for the effects of natural plant extracts and secondary plant metabolites on rumen microbial fermentation in continuous culture. Animal Feed Science and Technology 123, 597-613.

Busquet, M., Calsamiglia, S., Ferret, A., et al., 2005b. Effects of cinnamaldehyde and garlic oil on rumen microbial fermentation in a dual flow continuous culture. Journal of Dairy Science 88 (7), 2508-16.

Calsamiglia, S., Busquet, M., Cardozo, P. W., et al., 2007. Invited review: essential oils as modifiers of rumen microbial fermentation. Journal of Dairy Science 90 (6), 2580-95.

Canaes, T. S., Zanferari, F., Maganhe, B. L., et al., 2017. Increasing dietary levels of citral oil on nutrient total tract digestibility, ruminal fermentation, and milk composition in Saanen goats. Animal Feed Science and Technology 229, 47-56.

Canul-Solis, J. R., Piñeiro-Vázquez, A. T., Briceño-Poot, E. G., et al., 2014. Effect of supplementation with saponins from Yucca schidigera on ruminal methane production by Pelibuey sheep fed Pennisetum purpureum grass. Animal Production Science 54 (10), 1834-7.

Cardozo, P. W., Calsamiglia, S., Ferret, A., et al., 2004. Effects of natural plant extracts on ruminal protein degradation and fermentation profiles in continuous culture. Journal of Animal Science 82 (11), 3230-6.

Cardozo, P. W., Calsamiglia, S., Ferret, A., et al., 2005. Screening for the effects of natural plant extracts at different pH on *in vitro* rumen microbial fermentation of a high-concentrate diet for beef cattle. Journal of Animal Science 83 (11), 2572-9.

Cardozo, P. W., Calsamiglia, S., Ferret, A., et al., 2006. Effects of alfalfa extract, anise, capsicum, and a mixture of cinnamaldehyde and eugenol on ruminal fermentation and protein degradation in beef heifers fed a high-concentrate diet. Journal of Animal Science 84 (10), 2801-8.

Carson, C. F., Mee, B. J. and Riley, T. V. 2002. Mechanism of action of Melaleuca alternifolia (tea tree) oil on *Staphylococcus aureus* determined by time-kill, lysis, leakage, and salt tolerance assays and electron microscopy. Antimicrobial Agents and Chemotherapy 46 (6), 1914-20.

Castro-Montoya, J., Peiren, N., Cone, J. W., et al., 2015. *In vivo* and *in vitro* effects of a blend of essential oils on rumen methane mitigation. Livestock Science 180, 134-42.

Chaves, A. V., Dugan, M. E. R., Stanford, K., et al., 2011. A dose-response of cinnamaldehyde supplementation on intake, ruminal fermentation, blood metabolites, growth performance, and carcass characteristics of growing lambs. Livestock Science 141 (2-3), 213-20.

Chiquette, J., Cheng, K.-J., Costerton, J. W., et al., 1988. Effect of tannins on the digestibility of two isosynthetic strains of birdsfoot trefoil (*Lotus corniculatus* L.) using in vitro and in sacco techniques. Canadian Journal of Animal Science 68 (3), 751-60.

Cieslak, A., Zmora, P., Pers-Kamczyc, E., et al., 2012. Effects of tannins source (*Vaccinium vitis idaea* L.) on rumen microbial fermentation *in vivo*. Animal Feed Science and Technology 176 (1-4), 102-6.

Cieslak, A., Szumacher-Strabel, M., Stochmal, A., et al., 2013. Plant components with specific activities against rumen methanogens. Animal: an International Journal of Animal Bioscience 7 (Suppl. 2), 253-65.

Cobellis, G., Trabalza-Marinucci, M. and Yu, Z. 2016. Critical evaluation of essential oils as rumen modifiers in ruminant nutrition: a review. Science of the Total Environment 545-546, 556-68.

Cowan, M. M. 1999. Plant products as antimicrobial agents. Clinical Microbiology Reviews 12 (4), 564–82.

de Jesus, F., Del Valle, E. T. A., Calomeni, G. D., et al., 2013. Influence of a blend of functional oils or monensin onnutrient intake and digestibility, ruminal fermentation andmilk production of dairy cows. Animal Feed Science and Technology 219, 59–67.

Delaquis, P. J., Stanich, K., Girard, B., et al., 2002. Antimicrobial activity of individual and mixed fractions of dill, cilantro, coriander and eucalyptus essentialoils. International Journal of Food Microbiology 74 (1-2), 101–9.

Dentinho, M. T. P., Belo, A. T. and Bessa, R. J. B. 2014. Digestion, ruminal fermentation and microbial nitrogen supply in sheep fed soybean meal treated with *Cistus ladanifer* L. tannins. Small Ruminant Research 119 (1-3), 57–64.

Denyer, S. P. and Hugo, W. B. 1991. Biocide-induced damage to the bacterial cytoplasmic membrane. In: Denyer, S. P. and Hugo, W. B. (Eds), Mechanisms of Action of Chemical Biocides. Blackwell Scientific Publications, Oxford, UK, pp. 171–88.

Dhama, K., Latheef, S. K., Mani, S., et al., 2015. Multiple beneficial applications and modes of action of herbs in poultry health and production: a review. International Journal of Pharmacology 11 (3), 152–76.

Dickhoefer, U., Ahnert, S. and Susenbeth, A. 2016. Effects of quebracho tannin extract on rumen fermentation and yield and composition of microbial mass in heifers. Journal of Animal Science 94 (4), 1561–75.

Doce, R. R., Belenguer, A., Toral, P. G., et al., 2013. Effect of the administration of young leaves of Quercus pyrenaica on rumen fermentation in relation to oak tannin toxicosis in cattle. Journal of Animal Physiology and Animal Nutrition 97 (1), 48–57.

Dschaak, C. M., Williams, C. M., Holt, M. S., et al., 2011. Effects of supplementing condensed tannin extract on intake, digestion, ruminal fermentation, and milk production of lactating dairy cows. Journal of Dairy Science 94 (5), 2508–19.

Durmic, Z., Moate, P. J., Eckard, R., et al., 2014. In vitro screening of selected feed additives, plant essential oils and plant extracts for rumen methane mitigation. Journal of the Science of Food and Agriculture 94 (6), 1191–6.

Elcoso, G., Zweifel, B. and Bach, A. 2019. Effects of a blend of essential oils on milk yield and feed efficiency of lactating dairy cows. Applied Animal Science 35 (3), 304–11.

Eschenlauer, S. C. P., McKain, N., Walker, N. D., et al., 2002. Ammonia production by ruminal microorganisms and enumeration, isolation, and characterization of bacteria capable of growth on peptides and amino acids from the sheep rumen. Applied and Environmental Microbiology 68 (10), 4925–31.

Estelle, R. E. 2010. Coping with shrub secondary metabolites by ruminants. Small Ruminant Research 94 (1-3), 1–9.

Eugène, M., Archimède, H. and Sauvant, D. 2004. Quantitative meta-analysis on the effects of defaunation of the rumen on growth, intake and digestion in ruminants. Livestock Production Science 85 (1), 81–97.

Evans, J. D. and Martin, S. A. 2000. Effects of thymol on ruminal microorganisms. Current Microbiology 41 (5), 336–40.

Finlay, B. J., Esteban, G., Clarke, K. J., et al., 1994. Some rumen ciliates have endosymbiotic methanogens. FEMS Microbiology Letters 117 (2), 157-61.

Firkins, J. L., Yu, Z. and Morrison, M. 2007. Ruminal nitrogen metabolism: perspectives for integration of microbiology and nutrition for dairy. Journal of Dairy Science 90 (Suppl. 1), E1-16.

Flores, A. J., Garciarena, A. D., Vieyra, J. M. H., et al., 2013. Effects of specific essential oil compounds on the ruminal environment, milk production and milk composition of lactating dairy cows at pasture. Animal feed science and technology, 186 (1-2), 20-26.

Focant, M., Froidmont, E., Archambeau, Q., et al., 2019. The effect of oak tannin (Quercus robur) and hops (Humulus lupulus) on dietary nitrogen efficiency, methane emission, and milk fatty acid composition of dairy cows fed a low-protein diet including linseed. Journal of Dairy Science 102 (2), 1144-59.

Fraisse, D., Carnat, A., Viala, D., et al., 2007. Polyphenolic composition of a permanent pasture: variations related to the period of harvesting. Journal of the Science of Food and Agriculture 87 (13), 2427-35.

Francis, G., Kerem, Z., Makkar, H. P., et al., 2002. The biological action of saponins in animal systems: a review. British Journal of Nutrition 88 (6), 587-605.

Franz, C. and Novak, J. 2009. Sources of essential oils. In: Başer, K. H. C. and Buchbauer, G. (Eds), Handbook of Essential Oils: Science, Technology, and Applications. CRC Press, Boca Raton, FL, pp. 48-90.

Frutos, P., Hervas, G., Giraldez, F. J., et al., 2004. Review. Tannins and ruminant nutrition. Spanish Journal of Agricultural Research 2 (2), 191-202.

Ganguly, S. 2013. Herbal and plant derived natural products as growth promoting nutritional supplements for poultry birds: a review. Journal of Pharmaceutical and Scientific Innovation 2 (3), 12-3.

García-González, R., López, S., Fernandez, M., et al., 2008. Screening the activity of plants and spices for decreasing ruminal methane production in vitro. Animal Feed Science and Technology 147 (1-3), 36-52.

Ghizzi, L. G., Del Valle, T. A., Takiya, C. S., et al., 2018. Effects of functional oils on ruminal fermentation, rectal temperature, and performance of dairy cows under high temperature humidity index environment. Animal Feed Science and Technology 246, 158-66.

Giannenas, I., Skoufos, J., Giannakopoulos, C., et al., 2011. Effects of essential oils on milk production, milk composition, and rumen microbiota in Chios dairy ewes. Journal of Dairy Science 94 (11), 5569-77.

Gill, M., Smith, P. and Wilkinson, J. M. 2010. Mitigating climate change: the role of domestic livestock. Animal: an International Journal of Animal Bioscience 4 (3), 323-33.

Goel, G., Makkar, H. P. S. and Becker, K. 2008. Effects of Sesbania sesban and Carduus pycnocephalus leaves and Fenugreek (*Trigonella foenum-graecum* L.) seeds and their extracts on partitioning of nutrients from roughage-and concentrate based feeds to methane. Animal Feed Science and Technology 147 (1-3), 72-89.

Grainger, C., Clarke, T., Auldist, M. J., et al., 2009. Potential use of *Acacia mearnsii* condensed tannins to reduce methane emissions and nitrogen excretion from grazing dairy cows. Canadian Journal of Animal Science 89 (2), 241-51.

Greathead, H. 2003. Plants and plant extracts for improving animal productivity. Proceedings of the Nutrition Society 62 (2), 279-90.

Gunun, P., Wanapat, M., Gunun, N., et al., 2016. Effects of condensed tannins in mao (Antidesma thwaitesianum Muell. Arg.) seed meal on rumen fermentation characteristics and nitrogen utilization in goats. Asian-Australasian Journal of Animal Sciences 29 (8), 1111-9.

Guo, Y. Q., Liu, J. X., Lu, Y., et al., 2008. Effect of tea saponin on methanogenesis, microbial community structure and expression of mcrA gene, in cultures of rumen micro-organisms. Letters in Applied Microbiology, 47 (5), 421-426. .

Guyader, J., Eugène, M., Doreau, M., et al., 2015. Nitrate but not tea saponin feed additives decreased enteric methane emissions in nonlactating cows. Journal of Animal Science 93 (11), 5367-77.

Hagerman, A. and Butler, L. G. 1991. Tannins and lignins. In: Rosenthal, G. A. and Berenbaum, M. R. (Eds), Herbivores: Their Interactions with Secondary Plant Metabolites (vol. 1), pp. 355-88.

Harborne, J. B. and Baxter, H. 1999. The Handbook of Natural Flavonoids (vols. 1 and 2). John Wiley and Sons, Hoboken, NJ.

Hart, K. J., Yanez-Ruiz, D. R., Duval, S. M., et al., 2008. Plant extracts to manipulate rumen fermentation. Animal Feed Science and Technology 147 (1-3), 8-35.

Hart, K. J., Jones, H. G., Waddams, K. E., et al., 2019. An essential oil blend decreases methane emissions and increases milk yield in dairy cows. Open Journal of Animal Sciences 09 (3), 259-67.

Hatew, B., Cone, J. W., Pellikaan, W. F., et al., 2015. Relationship between *in vitro* and *in vivo* methane production measured simultaneously with different dietary starch sources and starch levels in dairy cattle. Animal Feed Science and Technology 202, 20-31.

Hess, H. D., Kreuzer, M., Díaz, T. E., et al., 2003. Saponin rich tropical fruits affect fermentation and methanogenesis in faunated and defaunated rumen fluid. Animal Feed Science and Technology 109 (1-4), 79-94.

Holtshausen, L., Chaves, A. V., Beauchemin, K. A., et al., 2009. Feeding saponin-containing Yucca schidigera and Quillaja saponaria to decrease enteric methane production in dairy cows. Journal of Dairy Science 92 (6), 2809-21.

Iason, G. R., Hartley, S. E. and Duncan, A. J. 1993. Chemical composition of Calluna vulgaris (Ericaceae): do responses to fertilizer vary with phenological stage? Biochemical Systematics and Ecology 21 (3), 315-21.

Jadhav, R. V., Kannan, A., Bhar, R., et al., 2017. Effect of tea (Camellia sinensis) seed saponin supplementation on growth performance, nutrient utilization, microbial protein synthesis and hemato-biochemical attributes of gaddi goats. Animal Nutrition and Feed Technology 17 (2), 255-68.

Jayanegara, A., Leiber, F. and Kreuzer, M. 2012. Meta-analysis of the relationship between dietary tannin level and methane formation in ruminants from *in vivo* and *in vitro* experiments Journal of Animal Physiology and Animal Nutrition 96 (3), 365-75.

Jones, G. A., McAllister, T. A., Muir, A. D., et al., 1994. Effects of saifoin (Onobrychis viciifolia Scop.) condensed tannins on growth and proteolysis by 4 strains of ruminal bacteria. Applied and Environmental Microbiology 60 (4), 1374-8.

Jouany, J. P. and Morgavi, D. P. 2007. Use of 'natural' products as alternatives to antibiotic feed additives in ruminant production. Animal: an International Journal of Animal Bioscience 1 (10), 1443-66.

Junior, F. P., Cassiano, E. C. O., Martins, M. F., et al., 2017. Effect of tannins-rich extract from *Acacia mearnsii* or monensin as feed additives on ruminal fermentation efficiency in cattle. Livestock Science 203, 21-9.

Kholifa, A. E., Matloupa, O. H., Morsy, T. A., et al., 2017. Rosemary and lemongrass herbs as phytogenic feed additives to improve efficient feed utilization, manipulate rumen fermentation and elevate milk production of Damascus goats. Livestock Science 204, 39-46.

Khorrami, B., Vakili, A. R., Mesgaran, M. D., et al., 2015. Thyme and cinnamon essential oils: potential alternatives for monensin as a rumen modifier in beef production systems. Animal Feed Science and Technology 200, 8-16.

Klita, P. T., Mathison, G. W., Fenton, T. W., et al., 1996. Effects of alfalfa root saponins on digestive function in sheep. Journal of Animal Science 74 (5), 1144-56.

Klop, G., Dijkstra, J., Dieho, K., et al., 2017. Enteric methane production in lactating dairy cows with continuous feeding of essential oils or rotational feeding of essential oils and lauric acid. Journal of Dairy Science 100 (5), 3563-75.

Knobloch, K., Weis, N. and Weigand, H. 1986. Mechanism of antimicrobial activity of essential oils. Planta Medica 52 (6), 556-.

Knobloch, K., Pauli, A., Iberl, B., et al., 1989. Antibacterial and antifungal properties of essential oil components. Journal of Essential Oil Research 1 (3), 119-28.

Koenig, K. M. and Beauchemin, K. A. 2018. Effect of feeding condensed tannins in high protein finishing diets containing corn distillers grains on ruminal fermentation, nutrient digestibility, and route of nitrogen excretion in beef cattle. Journal of Animal Science 96 (10), 4398-413.

Koenig, K. M., Beauchemin, K. A. and McGinn, S. M. 2018. Feeding condensed tannins to mitigate ammonia emissions from beef feedlot cattle fed high-protein finishing diets containing distillers grains. Journal of Animal Science 96 (10), 4414-30.

Kroon, P. A. and Williamson, G. 1999. Hydroxycinnamates in plants and food: current and future perspectives. Journal of the Science of Food and Agriculture 79 (3), 355-61.

Krueger, W. K., Gutierrez-Banuelos, H., Carstens, G. E., et al., 2010. Effects of dietary tannin source on performance, feed efficiency, ruminal fermentation, and carcass and non-carcas traits in steers fed a high-grain diet. Animal Feed Science and Technology 159 (1-2), 1-9.

Lee, S. E., Hwang, H. J., Ha, J. S., et al., 2003. Screening of medicinal plant extracts for antioxidant activity. Life Sciences 73 (2), 167-79.

Leinmüller, E., Steingass, H. and Menke, K. 1991. Tannins in ruminant feedstuffs. In: Bittner, A. (Ed.), Animal Research and Development (vol. 33). Institut fur Wissenschaftliche Zusammenarbeit, Germany, pp. 9-62.

Lima, P. R., Apdini, T., Freire, A. S., et al., 2019. Dietary supplementation with tannin and soybean oil on intake, digestibility, feeding behavior, ruminal protozoa and methane emission in sheep. Animal Feed Science and Technology 249, 10-7.

Liu, H., Vaddella, V. and Zhou, D. 2011. Effects of chestnut tannins and coconut oil on growth performance, methane emission, ruminal fermentation, and microbial populations in sheep. Journal of Dairy Science 94 (12), 6069-77.

Makkar, H. P. S. 2003. Effects and fate of tannins in ruminant animals, adaptation to tannins, and strategies to overcome detrimental effects of feeding tannin-rich feeds. Small Ruminant Research 49 (3), 241-56.

Malik, P., Singhal, K. and Deshpande, S. 2009. Effect of saponin rich lucerne fodder supplementation on rumen fermentation, bacterial and protozoal population in buffalo bulls. Indian Journal of Animal Sciences

79, 912-6.

Malik, P. K., Kolte, A. P., Baruah, L., et al., 2017. Enteric methane mitigation in sheep through leaves of selected tanniniferous tropical tree species. Livestock Science 200, 29-34.

Mangan, J. L. 1988. Nutritional effects of tannins in animal feeds. Nutrition Research Reviews 1 (1), 209-31.

Mao, H.-L., Wang, J.-K., Zhou, Y.-Y., et al., 2010. Effects of addition of tea saponins and soybean oil on methane production, fermentation and microbial population in the rumen of growing lambs. Livestock Science 129 (1-3), 56-62.

Martinez, S., Madrid, J., Hernandez, F., et al., 2006. Effect of thyme essential oils (*Thymus hyemalis* and *Thymus zygis*) and monensin on in vitro ruminal degradation and volatile fatty acid production. Journal of Agricultural and Food Chemistry 54 (18), 6598-602.

McAllister, T. A., Bae, H. D., Jones, G. A., et al., 1994. Microbial attachment and feed digestion in the rumen. Journal of Animal Science 72 (11), 3004-18.

McIntosh, F. M., Williams, P., Losa, R., et al., 2003. Effects of essential oils on ruminal microorganisms and their protein metabolism. Applied and Environmental Microbiology 69 (8), 5011-4.

McSweeney, C. S., Palmer, B., Bunch, R., et al., 2001a. Effect of the tropical forage calliandra on microbial protein synthesis and ecology in the rumen. Journal of Applied Microbiology 90 (1), 78-88.

McSweeney, C. S., Palmer, B., McNeill, D. M., et al., 2001b. Microbial interactions with tannins: nutritional consequences for ruminants. Animal Feed Science and Technology 91 (1-2), 83-93.

Min, B. R. and Solaiman, S. 2018. Comparative aspects of plant tannins on digestive physiology, nutrition and microbial community changes in sheep and goats: a review. Journal of Animal Physiology and Animal Nutrition 102 (5), 1181-93.

Min, B. R., Barry, T. N., Attwood, G. T., et al., 2003. The effect of condensed tannins on the nutrition and health of ruminants fed fresh temperate forages: a review. Animal Feed Science and Technology 106 (1-4), 3-19.

Molina-Botero, I. C., Arroyave-Jaramillo, J., Valencia-Salazar, S., et al., 2019. Effects of tannins and saponins contained in foliage of *Gliricidia sepium* and pods of *Enterolobium cyclocarpum* on fermentation, methane emissions and rumen microbial population in crossbred heifers. Animal Feed Science and Technology 251, 1-11.

Morgavi, D. P., Forano, E., Martin, C., et al., 2010. Microbial ecosystem and methanogenesis in ruminants. Animal: an International Journal of Animal Bioscience 4 (7), 1024-36.

Morrissey, J. P. and Osbourn, A. E. 1999. Fungal resistance to plant antibiotics as a mechanism of pathogenesis. Microbiology and Molecular Biology Reviews: MMBR 63 (3), 708-24.

Mueller-Harvey, I. 2006. Unravelling the conundrum of tannins in animal nutrition and health. Journal of the Science of Food and Agriculture 86 (13), 2010-37.

Muthusamy, N. and Sankar, V. 2015. Phytogenic compounds used as a feed additives in poultry production. International Journal of Science Environment and Technology 4, 167-71.

Nagy, J. G., Steinhoff, H. W. and Ward, G. M. 1964. Effects of essential oils of sagebrush on deer rumen microbial function. Journal of Wildlife Management 28 (4), 785-90.

Newbold, C. J., El Hassan, S. M., Wang, J., et al., 1997. Influence of foliage from African multipurpose trees on activity of rumen protozoa and bacteria. British Journal of Nutrition 78 (2), 237-49.

Newbold, C. J., de la Fuente, G., Belanche, A., et al., 2015. The role of ciliate protozoa in the rumen. Frontiers in Microbiology 6, 1313.

Oh, H. K., Jones, M. B. and Longhurst, W. M. 1968. Comparison of rumen microbial inhibition resulting from various essential oils isolated from relatively unpalatable plant species. Applied Environmental Microbiology 16 (1), 39-44.

Okoh, O. O., Sadimenko, A. P. and Afolayan, A. J. 2010. Comparative evaluation of the antibacterial activities of the essential oils of *Rosmarinus officinalis* L. obtained by hydrodistillation and solvent free microwave extraction methods. Food Chemistry 120 (1), 308-12.

Ornaghi, M. G., Passetti, R. A. C., Torrecilhas, J. A., et al., 2017. Essential oils in the diet of young bulls: effect on animal performance, digestibility, temperament, feeding behaviour and carcass characteristics. Animal Feed Science and Technology 234, 274-83.

Pathak, A. K., Dutta, N., Pattanaik, A. K., et al., 2017. Effect of condensed tannins from Ficus infectoria and Psidium guajava leaf meal mixture on nutrient metabolism, methane emission and performance of lambs. Asian-Australasian Journal of Animal Sciences 30 (12), 1702-10.

Patra, A. K. 2010. Meta-analyses of effects of phytochemicals on digestibility and rumen fermentation characteristics associated with methanogenesis. Journal of the Science of Food and Agriculture 90 (15), 2700-8.

Patra, A. K. 2011. Effects of essential oils on rumen fermentation, microbial ecology and ruminant production. Asian Journal of Animal and Veterinary Advances 6 (5), 416-28.

Patra, A. K. and Saxena, J. 2009a. Dietary phytochemicals as rumen modifiers: a review of the effects on microbial populations. Antonie Van Leeuwenhoek 96 (4), 363-75.

Patra, A. K. and Saxena, J. 2009b. The effect and mode of action of saponins on the microbial populations and fermentation in the rumen and ruminant production. Nutrition Research Reviews 22 (2), 204-19.

Patra, A. K. and Saxena, J. 2011. Exploitation of dietary tannins to improve rumen metabolism and ruminant nutrition. Journal of the Science of Food and Agriculture 91 (1), 24-37.

Patra, A. K. and Yu, Z. 2012. Effects of essential oils on methane production and fermentation by, and abundance and diversity of, rumen microbial populations. Applied and Environmental Microbiology 78 (12), 4271-80.

Patra, A. K. and Yu, Z. 2013. Effects of coconut and fish oils on ruminal methanogenesis, fermentation, and abundance and diversity of microbial populations in vitro. Journal of Dairy Science 96 (3), 1782-92.

Patra, A. K. and Yu, Z. 2015. Essential oils affect populations of some rumen bacteria *in vitro* as revealed by microarray (RumenBactArray) analysis. Frontiers in Microbiology 6, 297.

Patra, A., Park, T., Kim, M., et al., 2017. Rumen methanogens and mitigation of methane emission by anti-methanogenic compounds and substances. Journal of Animal Science and Biotechnology 8, 13.

Pauli, A. and Schilcher, H. 2010. *In vitro* antimicrobial activities of essential oilsmonographed in the European pharmacopoeia 6th edition. In: Başer, K. H. C. and Buchbauer, G. (Eds), Handbook of Essential Oils: Science, Technology, and Applications. CRC Press, Boca Raton, FL, pp. 343-548.

Pen, B., Sar, C., Mwenya, B., et al., 2006. Effects of Yucca schidigera and Quillaja saponaria extracts on in vitro ruminal fermentation and methane emission. Animal Feed Science and Technology 129 (3-4), 175-86.

Perevolotsky, A. 1994. Tannins in Mediterranean woodland species: lack of response to browsing and thinning. Oikos 71 (2), 333-40.

第 21 章 植物次生产物：在反刍动物可持续生产及营养中的有益作用

Piñeiro-Vázquez, A. T., Canul-Solis, J. R., Casanova-Lugo, F., et al., 2017. Emisión de metano en ovinos alimentados con Pennisetum purpureum y árboles que contienen taninos condensados. Revista mexicana de ciencias pecuarias 8 (2), 111-9.

Piñeiro-Vázquez, A. T., Canul-Solis, J. R., Jiménez-Ferrer, G. O., et al., 2018. Effect of condensed tannins from Leucaena leucocephala on rumen fermentation, methane production and population of rumen protozoa in heifers fed low-quality forage. Asian-Australasian Journal of Animal Sciences 31 (11), 1738-46.

Popova, M., Guyader, J., Silberberg, M., et al., 2019. Changes in the rumen microbiota of cows in response to dietary supplementation with nitrate, linseed, and saponin alone or in combination. Applied and Environmental Microbiology 85 (4), e02657-02618.

Poungchompu, O., Wanapat, M., Wachirapakorn, C., et al., 2009. Manipulation of ruminal fermentation and methane production by dietary saponins and tannins from mangosteen peel and soapberry fruit. Archives of Animal Nutrition 63 (5), 389-400.

Ramírez-Restrepo, C. A., O'Neill, C. J., López-Villalobos, N., et al., 2016a. Effects of tea seed saponin supplementation on physiological changes associated with blood methane concentration in tropical Brahman cattle. Animal Production Science 56 (3), 457-65.

Ramírez-Restrepo, C. A., Tan, C., O'Neill, C. J., et al., 2016b. Methane production, fermentation characteristics, and microbial profiles in the rumen of tropical cattle fed tea seed saponin supplementation. Animal Feed Science and Technology 216, 58-67.

Ramos-Morales, E., De La Fuente, G., Duval, S., et al., 2017. Antiprotozoal effect of saponins in the rumen can be enhanced by chemical modifications in their structure. Frontiers in Microbiology 8, 399.

Rhoades, D. F. 1985. Offensive-defensive interactions between herbivores and plants: their relevance in herbivore population dynamics and ecological theory. American Naturalist 125 (2), 205-38.

Rira, M., Morgavi, D. P., Archimède, H., et al., 2015. Potential of tannin-rich plants for modulating ruminal microbes and ruminal fermentation in sheep. Journal of Animal Science 93 (1), 334-47.

Russell, J. B. 1991. Intracellular pH of acid-tolerant ruminal bacteria. Applied and Environmental Microbiology 57 (11), 3383-4.

Russell, J. B. and Rychlik, J. L. 2001. Factors that alter rumen microbial ecology. Science 292 (5519), 1119-22.

Russell, J. B., Onodera, R. and Hino, T. 1991. Ruminal protein fermentation: new perspectives on previous contradictions. In: Tsuda, T., Sasaki, Y. and Kawashima, R. (Eds), Physiological Aspects of Digestion and Metabolism in Ruminants. Academic Press Limited, London, UK, pp. 691-7.

Salami, S. A., Valenti, B., Bella, M., et al., 2018. Characterisation of the ruminal fermentation and microbiome in lambs supplemented with hydrolysable and condensed tannins. FEMS Microbiology Ecology 94 (5), fiy061.

Santoso, B., Kilmaskossu, A. and Sambodo, P. 2007. Effects of saponin from Biophytum petersianum Klotzsch on ruminal fermentation, microbial protein synthesis and nitrogen utilization in goats. Animal Feed Science and Technology 137 (1-2), 58-68.

Santosa, M. B., Robinson, P. H., Williams, P., et al., 2010. Effects of addition of an essential oil complex to the diet of lactating dairy cows on whole tract digestion of nutrients and productive performance. Animal Feed Science and Technology 157 (1-2), 64-71.

Schären, M., Drong, C., Kiri, K., et al., 2017. Differential effects of monensin and a blend of essential oils on rumen microbiota composition of transition dairy cows. Journal of Dairy Science 100 (4), 2765-83.

Schofield, P., Mbugua, D. M. and Pell, A. N. 2001. Analysis of condensed tannins: a review. Animal Feed Science and Technology 91 (1-2), 21-40.

Serrano, J., Puupponen-Pimia, R., Dauer, A., et al., 2009. Tannins: current knowledge of food sources, intake, bioavailability and biological effects. Molecular Nutrition and Food Research 53 (Suppl. 2), S310-29.

Sharifi, A. and Chaji, M. 2019. Effects of processed recycled poultry bedding with tannins extracted from pomegranate peel on the nutrient digestibility and growth performance of lambs. South African Journal of Animal Science 49 (2), 290-300.

Sikkema, J., de Bont, J. A. and Poolman, B. 1994. Interactions of cyclic hydrocarbons with biological membranes. Journal of Biological Chemistry 269 (11), 8022-8.

Sikkema, J., de Bont, J. A. and Poolman, B. 1995. Mechanisms of membrane toxicity of hydrocarbons. Microbiological Reviews 59 (2), 201-22.

Silva, G. G., Takiya, C. S., Del Valle, T. A., et al., 2018. Nutrient digestibility, ruminal fermentation, and milk yield in dairy cows fed a blend of essential oils and amylase. Journal of Dairy Science 101 (11), 9815-26.

Silva, R. B. D., Pereira, M. N., Araujo, R. C. D., et al., 2020. A blend of essential oils improved feed efficiency and affected ruminal and systemic variables of dairy cows. Translational Animal Science, 4 (1), 183.

Śliwiński, B. J., Kreuzer, M., Wettstein, H. R., et al., 2002. Rumen fermentation and nitrogen balance of lambs fed diets containing plant extracts rich in tannins and saponins, and associated emissions of nitrogen and methane. Archives of Animal Nutrition 56 (6), 379-92.

Smith, A. H. and Mackie, R. I. 2004. Effect of condensed tannins on bacterial diversity and metabolic activity in the rat gastrointestinal tract. Applied and Environmental Microbiology 70 (2), 1104-15.

Smith, A. H., Zoetendal, E. and Mackie, R. I. 2005. Bacterial mechanisms to overcome inhibitory effects of dietary tannins. Microbial Ecology 50 (2), 197-205.

Soltan, Y. A., Natel, A. S., Araujo, R. C., et al., 2018. Progressive adaptation of sheep to a microencapsulated blend of essential oils: ruminal fermentation, methan emission, nutrient digestibility, and microbial protein synthesis. Animal Feed Science and Technology 237, 8-18.

Soto, E. C., Molina-Alcaide, E., Khelil, H., et al., 2013. Ruminal microbiota developing in different *in vitro* simulation systems inoculated with goats' rumen liquor. Animal Feed Science and Technology 185 (1-2), 9-18.

Souza, K. Ad, Monteschio, JdO., Mottin, C., et al., 2019. Effects of diet supplementation with clove and rosemary essential oils and protected oils (eugenol, thymol and vanillin) on animal performance, carcass characteristics, digestibility, and ingestive behavior activities for Nellore heifers finished in feedlot. Livestock Science 220, 190-5.

Spanghero, M., Robinson, P. H., Zanfi, C., et al., 2009. Effect of increasing doses of a microencapsulated blend of essential oils on performance of lactating primiparous dairy cows. Animal Feed Science and Technology 153 (1-2), 153-7.

Szumacher-Strabel, M. and Cieślak, A. 2012. Dietary possibilities to mitigate rumen methane and ammonia production. In: Liu, G. (Ed.), Greenhouse Gases-Capturing, Utilization and Reduction. IntechOpen.

University of North Dakota, United States of America, pp. 199-218.

Tager, L. R. and Krause, K. M. 2011. Effects of essential oils on rumen fermentation, milk production, and feeding behavior in lactating dairy cows. Journal of Dairy Science 94 (5), 2455-64.

Teferedegne, B. 2000. New perspectives on the use of tropical plants to improve ruminant nutrition. Proceedings of the Nutrition Society 59 (2), 209-14.

Tekippe, J. A., Hristov, A. N., Heyler, K. S., et al., 2011. Rumen fermentation and production effects of *Origanum vulgare* L. leaves in lactating dairy cows. Journal of Dairy Science 94 (10), 5065-79.

Tekippe, J. A., Tacoma, R., Hristov, A. N., et al., 2013. Effect of essential oils on ruminal fermentation and lactation performance of dairy cows. Journal of Dairy Science 96 (12), 7892-903.

Terrill, T. H., Douglas, G. B., Foote, A. G., et al., 1992. Effect of condensed tannins upon body growth, wool growth and rumen metabolism in sheep grazing Sulla (*Hedysarum coronarium*) and perennial pasture. Journal of Agricultural Science 119 (2), 265-73.

Theodoridou, K., Aufrère, J., Andueza, D., et al., 2010. Effects of condensed tannins in fresh sainfoin (Onobrychis viciifolia) on *in vivo* and *in situ* digestion in sheep. Animal Feed Science and Technology 160 (1-2), 23-38.

Tiemann, T. T., Lascano, C. E., Wettstein, H. R., et al., 2008. Effect of the tropical tannin-rich shrub legumes *Calliandra calothyrsus* and *Flemingia macrophylla* on methane emission and nitrogen and energy balance in growing lambs. Animal: an International Journal of Animal Bioscience 2 (5), 790-9.

Tomkins, N. W., Denman, S. E., Pilajun, R., et al., 2015. Manipulating rumen fermentation and methanogenesis using an essential oil and monensin in beef cattle fed a tropical grass hay. Animal Feed Science and Technology 200, 25-34.

Ultee, A., Bennik, M. H. and Moezelaar, R. 2002. The phenolic hydroxyl group of carvacrol is essential for action against the food-borne pathogen *Bacillus cereus*. Applied and Environmental Microbiology 68 (4), 1561-8.

Utsumi, S. A., Cibils, A. F., Estell, R. E., et al., 2013. Effects of adding protein, condensed tannins, and polyethylene glycol to diets of sheep and goats fed one-seed juniper and low quality roughage. Small Ruminant Research 112 (1-3), 56-68.

Valdez, F. R., Bush, L. J., Goetsch, A. L., et al., 1986. Effect of steroidal sapogenins on ruminal fermentation and on production of lactating dairy cows. Journal of Dairy Science 69 (6), 1568-75.

Valenzuela-Grijalva, N. V., Pinelli-Saavedra, A., Muhlia-Almazan, A., et al., 2017. Dietary inclusion effects of phytochemicals as growth promoters in animal production. Journal of Animal Science and Technology 59, 8.

Van Soest, J. P. 1994. Nutritional Ecology of the Ruminant. Comstock Publishing Associates, Cornell University Press, Ithaca, NY and London, UK.

Vasta, V., Yanez-Ruiz, D. R., Mele, M., et al., 2010. Bacterial and protozoal communities and fatty acid profile in the rumen of sheep fed a diet containing added tannins. Applied and Environmental Microbiology 76 (8), 2549-55.

Vasta, V., Daghio, M., Cappucci, A., et al., 2019. Invited review: plant polyphenols and rumen microbiota responsible for fatty acid biohydrogenation, fiber digestion, and methane emission: experimental evidence and methodological approaches. Journal of Dairy Science 102 (5), 3781-804.

Vokou, D., Kokkini, S. and Bessiere, J.-M. 1993. Geographic variation of Greek oregano (*Origanum vulgare* ssp. hirtum) essential oils. Biochemical Systematics and Ecology 21 (2), 287-95.

Wall, E. H., Doane, P. H., Donkin, S. S., et al., 2014. The effects of supplementation with a blend of cinnamaldehyde and eugenol on feed intake and milk production of dairy cows. Journal of Dairy Science 97 (9), 5709-17.

Wallace, R. J. 2004. Antimicrobial properties of plant secondary metabolites. Proceedings of the Nutrition Society 63 (4), 621-9.

Wallace, R. J., Arthaud, L. and Newbold, C. J. 1994. Influence of Yucca shidigera extract on ruminal ammonia concentrations and ruminal microorganisms. Applied and Environmental Microbiology 60 (6), 1762-7.

Wallace, R. J., McEwan, N. R., McIntosh, F. M., et al., 2002. Natural products as manipulators of rumen fermentation. Asian-Australasian Journal of Animal Sciences 15 (10), 1458-68.

Wang, C. J., Wang, S. P. and Zhou, H. 2009. Influences of flavomycin, ropadiar, and saponin on nutrient digestibility, rumen fermentation, and methane emission from sheep. Animal Feed Science and Technology 148 (2-4), 157-66.

Wang, J. K., Ye, J. A. and Liu, J. X. 2012. Effects of tea saponins on rumen microbiota, rumen fermentation, methane production and growth performance—a review. Tropical Animal Health and Production 44 (4), 697-706.

Wang, B., Ma, M. P., Diao, Q. Y., et al., 2019. Saponin-induced shifts in the rumen microbiome and metabolome of Young cattle. Frontiers in Microbiology 10, 356.

Weimer, P. J. 2015. Redundancy, resilience, and host specificity of the ruminal microbiota: implications for engineering improved ruminal fermentations. Frontiers in Microbiology 6, 296.

Wina, E., Muetzel, S. and Becker, K. 2005. The impact of saponins or saponin-containing plant materials on ruminant production-a review. Journal of Agricultural and Food Chemistry 53 (21), 8093-105.

Wina, E., Muetzel, S. and Becker, K. 2006. Effects of daily and interval feeding of Sapindus rarak saponins on protozoa, rumen fermentation parameters and digestibility in sheep. Asian-Australasian Journal of Animal Sciences 19 (11), 1580-7.

Wink, M. and Schimmer, O. 2018. Modes of action of defensive secondary metabolites. In: Wink, M. and Schimmer, O. (Eds), Annual Plant Reviews Book Series, Volume 3: Functions of Plant Secondary Metabolities and their Exploitation in Biotechnology. John Wiley and Sons, Hoboken, NJ, pp. 18-137.

Yáñez-Ruiz, D. R., Bannink, A., Dijkstra, J., et al., 2016. Design, implementation and interpretation of *in vitro* batch culture experiments to assess enteric methane mitigation in ruminants—a review. Animal Feed Science and Technology 216, 1-18.

Yang, W. Z., Benchaar, C., Ametaj, B. N., et al., 2007. Effects of garlic and juniper berry essential oils on ruminal fermentation and on the site and extent of digestion in lactating Cows1. Journal of Dairy Science 90 (12), 5671-81.

Yanga, W. Z., Benchaar, C., Ametaj, B. N., et al., 2010. Dose response to eugenol supplementation in growing beef cattle: ruminal fermentation and intestinal digestion. Animal Feed Science and Technology 158 (1-2), 57-64.

Yoshiki, Y., Kudou, S. and Okubo, K. 1998 Relationship between chemical structures and biological activities of triterpenoid saponins from soybean. Bioscience, Biotechnology, and Biochemistry 62 (12), 2291-9.

Yuan, Z., Zhang, C., Zhou, L., et al., 2007. Inhibition of methanogenesis by tea saponin and tea saponin plus disodium fumarate in sheep. Journal of Animal and Feed Sciences 16 (Suppl. 2), 560-5.

Zebeli, Q. and Ametaj, B. N. 2009. Relationships between rumen lipopolysaccharide and mediators of inflammatory response with milk fat production and efficiency in dairy cows. Journal of Dairy Science 92 (8), 3800-9.

Zebeli, Q., Mansmann, D., Steingass, H., et al., 2010. Balancing diets for physically effective fibre and ruminally degradable starch: a key to lower the risk of sub-acute rumen acidosis and improve productivity of dairy cattle. Livestock Science 127 (1), 1-10.

Zhou, Y. Y., Mao, H. L., Jiang, F., et al., 2011. Inhibition of rumen methanogenesis by tea saponins with reference to fermentation pattern and microbial communities in Hu sheep. Animal Feed Science and Technology 166-167, 93-100.

Zhou, C. S., Xiao, W. J., Tan, Z. L., et al., 2012. Effects of dietary supplementation of tea saponins (Ilex kudingcha CJ Tseng) on ruminal fermentation, digestibility and plasma antioxidant parameters in goats. Animal Feed Science and Technology 176 (1-4), 163-9.

Zucker, W. V. 1983. Tannins: does structure determine function? An ecological perspective. American Naturalist 121 (3), 335-65.

第 22 章 益生菌作为反刍动物添加剂的应用

Frédérique Chaucheyras-Durand、Lysiane Dunière

拉曼动物营养和克莱蒙奥弗涅大学,法国国家农业食品与环境研究院,UMR 454 MEDIS,法国

(杨春蕾译)

1 前言

良好的瘤胃功能是牛最重要的特征之一,因为其可以影响牛的健康和生产力,进而有助于确保为不断增长的人口提供充足的食物。在其生命中,反刍动物面临着几个时期的压力,从妊娠和产犊到高产奶或肉的需求。众说周知,目前的集约化养殖对消化道微生物构成了挑战,使动物处于代谢失调的风险中。近来许多研究致力于调查某些生命周期阶段对瘤胃功能和健康的影响。研究表明瘤胃微生物和瘤胃上皮壁在这些阶段对维持瘤胃最佳功能起着关键作用。

本章首先回顾益生菌对反刍动物生命周期中关键时期的靶向调控。然后关注于益生菌的定义、递送机制和管理。本章其余部分从幼龄反刍动物的调控潜力开始总结评估益生菌的益处和作用方式。然后从以下几个方面总结益生菌对成年反刍动物的影响:饲料效率、甲烷生成、病原体控制和免疫系统支持。

2 益生菌对反刍动物生命周期中关键时期的靶向调控

产犊对于新生反刍动物来说是一个非常应激的事件,特别是从微生物学的角度来看。通过与母体和养殖场环境的直接接触,胃肠道(GIT)的微生物定植始于出生期间和出生后(Yeoman 等,2018)。典型的瘤胃菌群成员,例如产甲烷菌、纤维分解菌或变形杆菌,在犊牛出生后不到 20min 便可在瘤胃中被检测到(Guzman 等,2015),并具有代谢活性(Guzman 等,2016)。初生犊牛的生理机能(即食管沟闭合反射)在断奶前阶段阻止牛奶进入瘤胃室。然而,食管沟的不成熟和少量固体饲料的摄入导致了断奶前瘤胃微生物的接种和瘤胃微生物与生理缓慢且不成熟的发育(Meale 等,2017)。瘤胃的微生物定植始于需氧和兼性厌氧菌群,然后逐渐被严格厌氧的微生物取代(Rey 等,2014)。瘤胃发育受损会影响营养物质的消化以及动物的进一步生长和生产(Steele 等,2016)。

消化问题是奶犊牛和小母牛死于疾病最普遍的原因之一。在北美奶业产业,小母牛断奶前平均死亡率高达 56.4%(UDSA,2014a)。大部分犊牛腹泻问题发生在出生后前 3 周(McGuirk 等,2008)。在爱沙尼亚进行的一项为期两年的研究中,代谢和消化失调被认为

是农场1月龄以下（43%），1~5月龄（29.5%）和6~19月龄（32.3%）犊牛死亡的主要原因（Mõtus等，2018）。在美国，断奶前小母牛被淘汰的案例中，有21.1%被认定为与消化功能紊乱相关的低生产性能相关（UDSA，2014a）。

在出生时，犊牛免疫力的获取依靠初乳和牛奶的摄入，使得免疫球蛋白（Ig）和中性粒细胞，以及分泌免疫相关成分如细胞因子、抗菌肽和蛋白质的巨噬细胞得以转移（Stelwagen等，2009）。IgG是初乳中的主要免疫球蛋白并且其浓度可用于表征初乳质量（Johnsen等，2019）。低质量或限制摄入初乳会增加死亡率并且减缓犊牛生长（McGuirk和Collins，2004），出生后立即饲喂养高质量初乳能够提高犊牛的断奶体重和体重增加（Priestley等，2013）。这些结果强调了饮食和免疫功能的相关性。初乳免疫被认为是奶犊牛胃肠道疾病管理的重要环节（McGuirk，2008）。与喂饲初乳的犊牛相比，缺乏初乳的犊牛小肠组织中黏附的大肠杆菌比例显著较高，而双歧杆菌的比例较低（Malmuthuge等，2015）。在新生羊羔体内，肠道微生物被认为可以提供涉及功能性免疫应答建立的关键信号，例如，在维持回肠潘氏斑淋巴滤泡的发育和功能方面（Reynnolds和Morris，1984）。

断奶通常发生在奶犊牛6~8周龄时并且象征着瘤胃发育中最显著的变化之一，导致肠道质量、免疫和代谢功能发生重要改变（Baldwin等，2004）。管理不善的断奶阶段往往与痛苦、生长迟缓和腹泻相关（Roth等，2009）。在断奶时期，牛奶突然或者逐步从犊牛饮食中去除，替换为固体饲料进入瘤胃。此时，微生物总活性增加，表现为某些瘤胃微生物群落的生长、瘤胃酶产量和发酵终产物的增加（Jiao等，2015）。断奶会导致瘤胃发生强烈且迅速的生理进化（例如乳头发育，体积增大，瘤胃壁厚度增加）（综述于Meale等，2017）。断奶后会立即观察到微生物多样性的下降，这可能是由于过渡到以饲料为基础的饮食的剧烈变化、瘤胃室内的大量发酵和pH值波动引起的（Meale等，2016）。犊牛瘤胃发育也会受到开食料物理形态和化学成分的影响（Khan等，2016）。

黏膜上皮被认为是GIT和宿主之间的屏障。上皮细胞在识别瘤胃微生物组、食糜中存在的病原体和化学物质方面起着关键作用，因此影响着黏膜免疫系统的发育（Malmuthuge等，2012）。瘤胃微生物配体的持续刺激会导致GIT黏膜上皮的Toll样受体（TLRs）下调以限制慢性和非必要的炎症反应。上皮表达分子如β-防御素（抗菌肽）和PGLYRP1（肽聚糖识别蛋白）的基因表达在断奶前受到抑制，但在断奶后增加。Malmuthuge等（2012）提出TLRs在断奶前驱动原发性先天免疫，但其作用随着动物年龄的增长减弱而偏向于其他先天免疫机制。Bush等（2019）对反刍动物从出生到成年的整个GIT进行了全面的转录组学研究，观察到瘤胃随动物年龄演变出强大的免疫转录组学特征。

专注于瘤胃发育的研究表明，在生命早期可通过控制饲养管理来调节微生物菌群的发育，并对动物生命后期产生影响（Abecia等，2013，2014；Yáñez-Ruiz等，2015）。通过改变山羊幼崽饲料以减少甲烷排放的研究表明其对瘤胃细菌和古菌菌群的影响能够持续到处理后的4个月；同时还有母体效应的存在，因为子代细菌群体和代谢产物谱的改变也与接受处理的母体有关（Abecia等，2018）。这些结果也强调了母体印记对瘤胃微生物潜在的长期影响。Soberon和Van Amburgh（2013）的meta分析表明与断奶前限制性母乳喂养的小母牛相比，断奶前自由母乳喂养的小母牛在第一次泌乳期的产奶量更高。在断奶前饲喂高水平的营养物质会影响参与乳腺形态和生理发育的细胞功能，并增加奶产量（Hare

等，2019）。这些数据表明生命早期的饮食干预，尽管不是专门针对瘤胃，也能对动物后期的生产性能产生影响。

在反刍动物一生中，它们需要从出生地被运输并在某些地方如其他的农场、饲养场或拍卖市场与其他动物混合饲养。2013年，美国28.6%的奶牛场引进了新的动物加入它们的畜群（UDSA，2014b）。这一时期会伴随着例如装卸和运输、缺乏饲料或水、与其他动物混合或高温变化等诸多应激。众所周知，应激会对动物免疫系统产生负面影响（Blecha等，1984），当与其他动物由于混合饲养更有可能接触到传染性病原体时，这种负面影响可能是非常有害的。应激会损害牛的生产性能和健康，通常伴随着摄食量的减少（Hutcheson和Cole，1986；Silanikove，2000）。Ashenafi等（2018）综述了与长距离运输相关的应激的影响，并且指出这会使代谢、免疫能力和行为发生改变，降低生育率，增加发病率和死亡率。

育肥期的肉牛或泌乳期的奶牛通常饲喂高谷物饲料以满足其能量需求。众所周知，给动物提供大量的可发酵淀粉或糖类可以改变瘤胃微生物群落和功能（Petri等，2013a，b，2018）。随着瘤胃pH值下降，乳酸生成者的数量可能超过乳酸利用者，从而导致瘤胃微生物群落结构的变化（Russell和Wilson，1996）。从粗饲料到高谷物饲料的转化中，牛瘤胃中纤维分解菌瘤胃球菌（*Ruminococcus* sp.）和产琥珀酸丝状杆菌（*Fibrobacter succinogenes*）的丰度会降低。相反，乳酸利用菌反刍月形单胞菌（*Selenomonas ruminantium*）和埃氏巨型球菌（*Megasphaera elsdenii*）有所增加（Petri等，2013b）。乳酸杆菌和链球菌出现在临床酸中毒的牛瘤胃中，可能反映了这些菌种对低pH值的耐受和它们在过量可发酵碳水化合物中增殖的能力（Petri等，2013b）。瘤胃长期低pH值会对采食量、微生物代谢和营养物质降解产生负面影响，并且导致例如酸中毒、炎症、蹄叶炎、腹泻和乳脂抑制等生理紊乱的发生（Kleen等，2003；Villot等，2018）。

这种微生物失调可能会引起潜在影响动物健康的有害分子的释放。酸中毒与释放到瘤胃和后肠的革兰氏阴性菌的脂多糖（LPS）有关（Khafipour等，2009；Khiaosa-Ard和Zebeli，2018）。LPS从瘤胃转移到内循环会引发炎症反应伴随着外周血急性期反应蛋白浓度的增加（Plaizier等，2012）。低pH值条件下某些瘤胃细菌生成的组胺（和其他生物胺）（Silberberg等，2013；Wang等，2013）会导致瘤胃上皮细胞炎症反应的增加（Sun等，2017）。最后，当大量的营养和能量从维持基本的机体内稳态和生产中被转移时，血液或组织中免疫系统的激活可能引发全身炎症反应，最终导致动物生产性能不佳，并产生重大的经济影响（Zebeli等，2015）。抗酸中毒肉用公牛瘤胃乳头中TLR2和TLR4的表达高于敏感肉用公牛，表明抗酸中毒肉用公牛的宿主先天性免疫应答更强。作者认为TLRs表达的增加可能通过刺激抗性肉用公牛瘤胃的屏障功能来保护瘤胃上皮免受亚急性瘤胃酸中毒（SARA）损伤（Chen等，2012）。研究表明，唾液中的白细胞可以从瘤胃腔内来回迁移，与口腔内的淋巴组织和前胃壁释放的细胞因子或其他介质发生串扰。在128头牛的野外试验中发现，瘤胃液中总Ig和IgM的浓度受瘤胃pH水平和挥发性脂肪酸（VFA）浓度的调节。这些发现表明牛前胃的代谢活动受到先天免疫系统的控制（Trevisi等，2018）。

表层微生物附着于瘤胃上皮并且和参与炎症反应调节的上皮细胞密切接触。表层微生物的多样性不同于瘤胃液相或固相中所观察到的微生物（Sadet 等，2007）。当干草替代精料后，虽然瘤胃上皮细胞内模式识别受体（TLR）、屏障功能、pH 值调节和营养吸收的基因表达保持稳定，瘤胃表层微生物群落的多样性受到强烈影响，伴随着从厚壁菌门到变形菌门的转变（Petri 等，2018）。Chen 等（2011）观察到在高干草或高谷物饲喂下表层微生物存在差异，发现密螺旋体属（*Treponema* sp.），瘤胃杆菌属（*Ruminobacter* sp.）和毛螺菌科（*Lachnospiraceae*）仅在高谷物饲喂下存在。这些结果与先前发表的数据相矛盾，其作者认为宿主在调控微生物多样性和密度以及对瘤胃环境变化的响应中发挥作用。

在泌乳期开始时，尽管日粮由高纤维转变为高能量含量，通常无法达到奶牛能量需求，需要调动身体储备。因此，为了获得最佳的繁殖性能和产奶量，应当保持瘤胃功能处于最佳状态。此时，瘤胃发生剧烈变化，对瘤胃微生物、发酵特性和上皮通透性产生影响（Bach 等，2018；Minuti 等，2015）。Zhu 等（2018）观察到初产奶牛过渡期细菌和古菌群落的变化可能与短链脂肪酸特征相关。产犊后拟杆菌门相对丰度降低，变形菌门相对丰度增加。在过渡期，产甲烷菌群落组成的显著变化主要发生在 *Methanosphaera* 和 *Methanomassiliicoccus* 中，尽管变化模式在不同的属中有所不同。

瘤胃微生物组成似乎与产奶量有关。瘤胃中厚壁菌门与拟杆菌门的比例与荷斯坦牛饲料转化率（FE）和乳脂产量变化呈负相关（Jami 等，2014）。一项研究采用全基因组测序以确定荷斯坦牛瘤胃微生物与饲料转化率之间的关系。在效率较高的奶牛中拟杆菌门和普雷沃氏菌属的相对丰度较高，而厚壁菌门和一些古菌种群的相对丰度较低（Delgado 等，2019）。

过渡期（从产犊前 3 周到产犊后 3 周）与奶牛健康有着重要联系，且与免疫能力降低、能量负平衡、低钙血症、过度全身炎症反应（即使没有微生物感染迹象）和氧化应激状态相关（Trevisi 和 Minuti，2018）。产犊前 2 周到产后 3 周，瘤胃上皮细胞中参与免疫反应调节的基因表达有很大差异（Bach 等，2018）。炎症反应被认为与泌乳期高精料饲喂引起的瘤胃内毒素释放增加直接关联（Abaker 等，2017）。Ingvartsen 等（2003）综述了高产奶牛相关疾病并发现免疫状态和饮食的失衡会导致代谢紊乱和生殖问题、乳腺炎、酮症和跛足风险增加。

Pitta 等（2018）综述了有关基于微生物和饮食互作提高奶牛生产性能的最新研究。他们指出了许多调控瘤胃微生物过程的策略（图 1）。益生菌的潜在作用可以体现在不同的领域，如纤维消化、蛋白质供应和微生物生长、生物氢化和甲烷生成。反刍动物不仅受其微生物菌群和免疫系统的影响，也受饮食干预和环境胁迫的影响。通过了解胃肠道微生物的多样性、稳定性、代谢产物以及与上皮和免疫系统的互作，调控胃肠道微生物来维持动物肠道健康仍需要更加深入的研究（Gaggìa 等，2010）。

3 定义、递送机制和管理

3.1 定义

"益生菌"一词来源于拉丁语"*pro*"（"支持"）和希腊语"*bios*"（"生命"），在

20世纪60年代（1960s）首次被提出，与"抗生素"（"对抗生命"）形成对比，用来定义原虫产生的能够支持其他微生物生长的物质（Morelli 和 Capurso，2012）。在20世纪80年代（1980s）末，Fuller教授提出了益生菌的定义，即"通过改善肠道平衡对宿主产生有益影响的一种活体微生物饲料添加剂"。那时，益生菌仅关注于家畜。随着其在人类中的广泛应用，联合国粮食及农业组织（FAO）/世界卫生组织（WHO）在2001年提出了一个略有不同的定义："当给予足够量时给宿主带来健康益处的活体微生物"。2013年，国际益生菌和益生元科学协会的科学专家小组对这一定义进行了审查，一致认为FAO/WHO对益生菌的定义仍然是切题的。在本章中，我们仅考虑活体微生物而不是死亡或灭活的微生物，例如酵母培养物或微生物组分，不是上面所定义的益生菌。

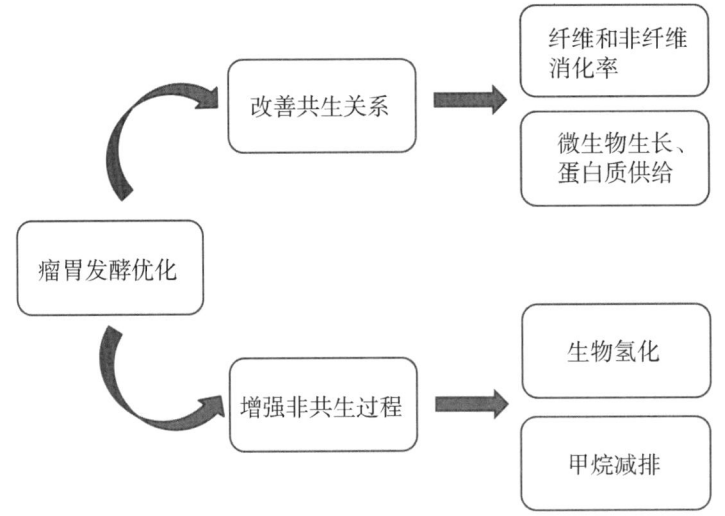

图1　调控瘤胃微生物过程以增强瘤胃功能的策略（改编自Pitta等，2018）

我们需要强调的是，对于反刍动物，"益生菌"这个词实际上不常用。事实上，由于以下几个原因这个严格的定义并不完全适合：

- 主要的消化部位是瘤胃并不是肠道。
- 关注点主要是营养作用，而不是健康益处。
- 活体微生物通常被加入饲料中，因此被认为是饲料添加剂。
- 在欧盟（EU）和其他国家，这些微生物属于饲料添加剂类别（动物技术添加剂）（EC Regulation 1831/2003）。

在大多数国家，包含活体微生物的商业化产品被认为是饲料添加剂（FA）或直接饲喂的微生物（DFM），本章将使用这些术语，而不是"益生菌"。

3.2　递送机制

在反刍动物中，微生物FA通过饲料进行递送，或与矿物质/维生素预混料一起包含在奶粉中或包含在颗粒精料中。由于每克纯添加剂的活性细胞浓度较高，其在饲料中的比例通常较低以确保每只动物的添加量精确。评估微生物FA与某些矿物质（也就是铜）的兼容性是很重要的，这些矿物质可能对活细胞具有一定的毒性。同样重要的是，要确保涉及到高温、高压和湿度等条件的制粒过程不会破坏微生物细胞。Sullivan 和 Bradford（2011）已经对此提出了担忧并比较了喂给奶牛的包含高浓度活酵母细胞的不同商业活性干酵母（ADY）。测试产品不能持续满足产品要求，在40℃下储存3周后活力显著下降。由于维生素A和E以及微量矿物质如硒的抗氧化作用，当ADY产品用包含维生素微量矿物质的预混料稀释时，可以减少高温下活力的损失。FA生产商需要确保正确的配方来解决这些问题。在将其加入颗粒精料的情况下，一些公司提出了如微胶囊球等技术来保护微生物FA免受恶劣饲料生产条件影响，确保活性微生物在消费前的最佳稳定性。如果与精油、植物提取物或抗生素等其他饲料添加剂结合使用，确保活性微生物不受这些已知具有抗菌作用的化合物的负面影响是非常重要的。一种特殊的情况是牛的饲养场补充。在大规模育肥操作中，可以用包含DFM的液体添加剂饲喂动物，主要是为了控制肠道中的病原体（将在本章后面进行讨论）。在这种情况下，DFM（细菌）通过冷藏容器运送并每天喂给畜群，避免了饲料稳定性问题。

在包括人类在内的许多物种中，益生菌的概念是基于活性微生物的日常分布，因为这些微生物不需要在消化道定植就能起作用。在反刍动物中，每天也要通过饲料给予微生物FA。研究表明，重复分配活酵母产品可以确保羔羊瘤胃中细胞的稳定浓度（Durand-Chaucheyras等，1998）。然而，当停止添加时，24小时后检测到的活细胞浓度下降且几天后在瘤胃内全部消失，在最后一次添加后几天，动物粪便中有一些回收的酵母产品。

一些酵母，如布拉迪酵母（*Saccharomyces boulardii*）被认为有能力在肠道中存活，并能活着到达肠道下腔室，证明了其目前应用于北美断奶前反刍动物以促进肠道健康的合理性。益生菌主要针对下消化道健康。然而，关于它们在肠道这部分的潜在存活性信息仍然有限。这在一定程度上是由于缺乏检测方法，无法对肠道共生菌群中的活/能独立生存的益生菌进行特定的检测/量化，因为后者通常包含与使用的益生菌相同的菌种（Fomenky等，2017）。然而，很明显反刍动物的消化环境为益生菌的生存提供了不利的条件。需要研究创新配方来帮助往下消化道递送活性微生物。微胶囊化是一种新兴技术，可用于保护益生菌免受不利环境条件的影响。海藻酸盐、壳聚糖、卡拉胶、树胶、明胶、乳清蛋白、淀粉等材料和压制包衣法的使用可以用于不同的微胶囊化技术（Riaz 和 Masud，2013）。在胶囊特性、益生菌保护、添加成本和易用性之间找到最佳平衡是非常重要的。表1总结了在反刍动物中作为益生菌使用的微生物种类和商业可用性。它们对目标物种的影响和作用方式将稍后在本章介绍。

表1 在反刍动物中被用作益生菌的酵母和细菌菌种

酵母	细菌
酿酒酵母	嗜酸乳杆菌/卷曲乳杆菌
布拉迪酵母	鼠李糖乳杆菌
	屎肠球菌
	枯草芽孢杆菌

(续表)

酵母	细菌
	地衣芽孢杆菌
	费氏丙酸杆菌
	产丙酸丙酸杆菌
	埃氏巨型球菌

3.3 管理

在欧盟，饲料添加剂的 EC 1831/2003 号条例涵盖了包括反刍动物在内的动物营养应用的微生物 FA。该条例适用于所有 FA 和预混料，但不适用于 2001/82/EC 指令中定义的加工助剂或兽药产品。只有经过授权的添加剂才能上市和使用。授权颁发给用于特定动物物种或种类以及特定使用条件的饲料。微生物 FA 被分类为动物技术添加剂，作为消化增强剂或肠道菌群稳定剂。授权在整个欧洲经济区有效期为 10 年，可续期 10 年。授权程序是一个漫长而严格的过程（图 2），这导致欧盟实际授权用于反刍动物（奶牛、肉牛、小型反刍动物和生长中反刍动物）的微生物添加剂数量非常少。

其他国家拥有自己的监管流程，也可能是复杂且苛刻的。在美国，当专家认为添加到食品中的化学品或物质是安全的，食品药品监督管理局（FDA）才会授权指定"公认安全"（GRAS），因此免除了通常的联邦食品、药品和化妆品法案（FFDCA）食品添加剂耐受性要求的约束。上市益生菌被认为是 GRAS。在其他没有明确规定的国家，微生物 FA 通常被认为是安全的，但不需要具体的功效证明。

人们越来越担心的是，肠道细菌在反复接触抗生素时可能产生抗生素耐药性（AMR）。在这种情况下，携带 AMR 基因的益生菌可能对动物宿主是一个问题。Amachawadi 等（2018）研究了应用于牛和猪的含有屎肠球菌（*Enterococcus faecium*）的商业益生菌，特别是它们对抗生素的敏感性或耐药性以及毒力基因的存在。他们发现这些商业益生菌都不携带毒力基因。但是，某些益生菌对医学上重要的抗生素表现出 AMR，并对氯霉素、红霉素、青霉素、卡那霉素、林可霉素和四环素具有多重耐药性，这表明它们可能是动物肠道中 AMR 的来源。改善反刍动物饲料中使用的微生物菌株的特性，特别是通过使用全基因组测序（WGS）获得基因组信息，以确保这些微生物的绝对安全是很重要的。

4 益生菌的益处和作用方式：幼龄反刍动物

如前所述，断奶前是反刍动物生命周期中最关键的时期，死亡率和发病率最高。优化断奶前阶段对于确保成功的畜群管理至关重要。微生物 FA 对于乳制品行业来说是一个很有前途的机会，特别是在减少抗生素使用的情况下。

目前已发表的试验对于微生物 FA 在生产性能上有益作用的报道并不一致（Alugongo

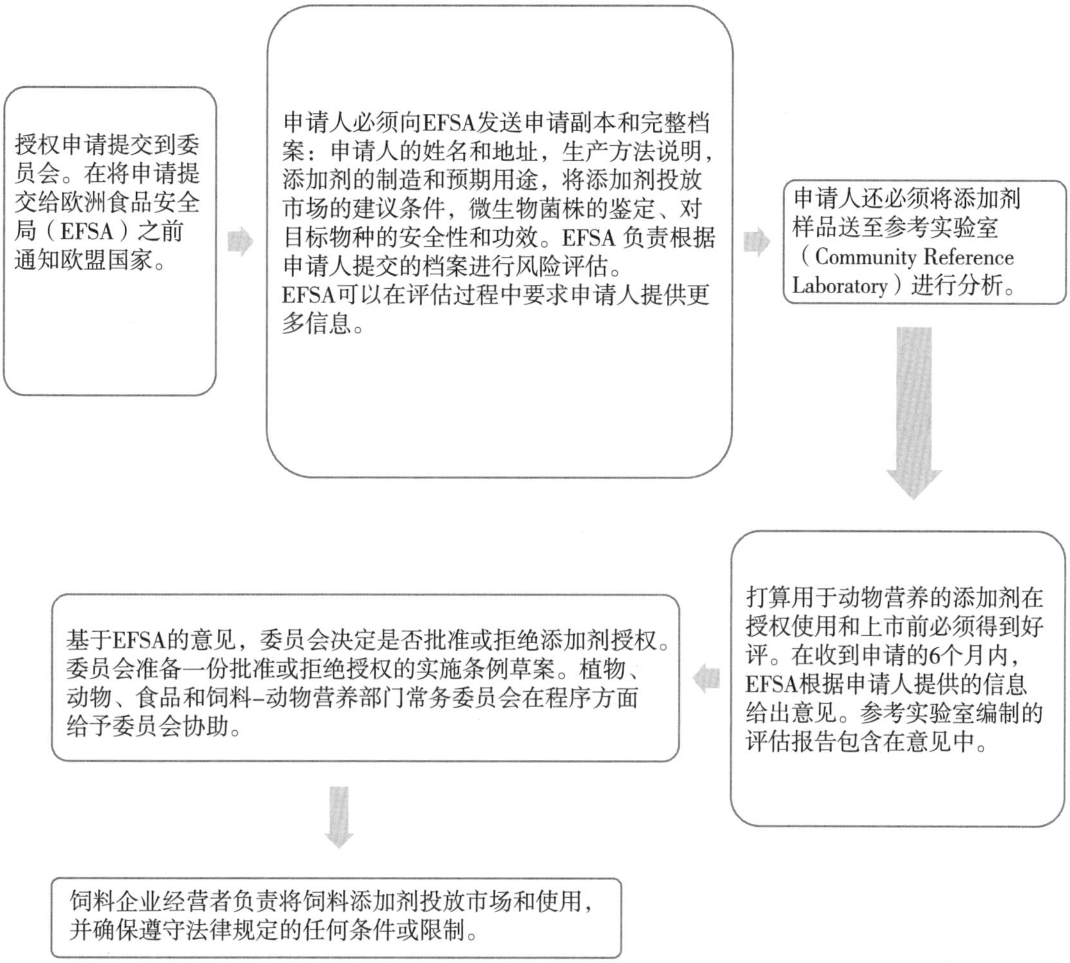

```
┌─────────────┐    ┌──────────────────────────┐    ┌─────────────┐
│ 授权申请提交 │    │ 申请人必须向EFSA发送申请  │    │ 申请人还必须 │
│ 到委员会。在 │    │ 副本和完整档案：申请人的  │    │ 将添加剂样品 │
│ 将申请提交给 │ ⇒  │ 姓名和地址，生产方法说明， │ ⇒ │ 送至参考实验 │
│ 欧洲食品安全 │    │ 添加剂的制造和预期用途，  │    │ 室（Community│
│ 局（EFSA）之 │    │ 将添加剂投放市场的建议条  │    │ Reference   │
│ 前通知欧盟国 │    │ 件，微生物菌株的鉴定、对  │    │ Laboratory）│
│ 家。         │    │ 目标物种的安全性和功效。  │    │ 进行分析。   │
│             │    │ EFSA负责根据申请人提交的  │    │             │
│             │    │ 档案进行风险评估。        │    │             │
│             │    │ EFSA可以在评估过程中要求  │    │             │
│             │    │ 申请人提供更多信息。      │    │             │
└─────────────┘    └──────────────────────────┘    └─────────────┘
                                                          ⇓
┌─────────────────────────────────────────┐    ┌─────────────────┐
│ 基于EFSA的意见，委员会决定是否批准或拒  │    │ 打算用于动物营养 │
│ 绝添加剂授权。委员会准备一份批准或拒绝  │    │ 的添加剂在授权使 │
│ 授权的实施条例草案。植物、动物、食品和  │ ⇐  │ 用和上市前必须得 │
│ 饲料-动物营养部门常务委员会在程序方面   │    │ 到好评。在收到申 │
│ 给予委员会协助。                        │    │ 请的6个月内，EFSA│
│                                         │    │ 根据申请人提供的 │
│                                         │    │ 信息给出意见。参 │
│                                         │    │ 考实验室编制的评 │
│                                         │    │ 估报告包含在意见 │
│                                         │    │ 中。             │
└─────────────────────────────────────────┘    └─────────────────┘
              ⇓
┌─────────────────────────────────────────┐
│ 饲料企业经营者负责将饲料添加剂投放市场  │
│ 和使用，并确保遵守法律规定的任何条件或  │
│ 限制。                                  │
└─────────────────────────────────────────┘
```

图 2　欧盟动物营养微生物饲料添加剂授权的现行程序
[来自动物营养使用添加剂（EC）N°1831/2003 号条例]。

等，2017；Fomenky 等，2017；Galvão 等，2005；He 等，2017）。这可能是由于研究设计、日粮类型、FA 菌株和剂量、持续时间和施用模式（在牛奶中，在开食料中，或两者均有，以及与其他添加剂结合与否）差异，而且在来自同一畜群的犊牛中也观察到较大的个体差异。当通过开食料给予微生物 FA 时，这种固体饲料在出生后前 2 周的消耗量通常较低，使犊牛无法充分接种活的微生物细胞，从而影响了添加剂的功效。因此，通过牛奶给予微生物 FA 受到了关注，因为其可以更好地控制给予剂量。当活益生菌分布在牛奶中时，其可以到达肠道并对肠道健康和犊牛免疫系统产生有益影响。

正如已经指出的那样，瘤胃微生物接种在动物出生时就开始了（Yeoman 等，2018），并且在生命的最初几天，瘤胃细菌群落组成会发生快速改变（Jami 等，2013；Rey 等，2012）。在一项与母羊一起喂养的羔羊试验中，每日给予活酵母添加剂（*Saccharomyces cerevisiae* CNCM I-1077）加速了幼小羔羊瘤胃微生物功能群落的建立（Chaucheyras-

Durand 和 Fonty，2002）。在无菌隔离器中饲养的无菌羔羊中，添加剂也会刺激纤维分解菌群的建立（Chaucheyras-Durand 和 Fonty，2001）。通过专用 DNA 微阵列评估，研究发现相同的菌株可以改善人工喂养新生羔羊瘤胃成熟中的微生物定植，尤其是纤毛原虫、真菌和纤维分解细菌产琥珀酸丝状杆菌（Chaucheyras-Durand，未发表数据），使其成为潜在的趋向于高效纤维降解的特有生态系统（Comtet-Marre 等，2018）。这表明其可能对纤维消化效率产生积极影响（Chaucheyras-Durand，未发表数据）。这种酵母添加剂（*S. cerevisiae* CNCM I-1077）在断奶后 3 周的犊牛性能试验中也具有有益效果，伴随着生长和最终体重均显著增加，同时增重与饲料比更好，这可能是因为改善了瘤胃微生物成熟（Terré 等，2015）。酵母添加对瘤胃的作用方式可能包括改善厌氧环境，因为添加酵母后，由于活酵母细胞具有清除氧气的能力，幼龄羔羊瘤胃内容物的氧化还原电位降低（Chaucheyras-Durand 和 Fonty，2002）。酵母 FA 也可能提供了关键营养、B 族维生素或辅因子，能够促进纤维分解微生物菌群的生长和在饲料颗粒的定植（Chaucheyras-Durand 等，2016）。

值得注意的是，这种有益效果可能取决于幼龄动物最初的健康状况。非常好的健康状况加上最佳的农场清洁度和良好的管理，可能是有助于解释为什么在缺乏任何攻击源或致病剂的情况下，不能观察到微生物 FA 对动物益处的原因之一（Alugongo 等，2017）。在生长迟缓的犊牛中，芽孢杆菌益生菌的补充对体重、采食量、饲料转化和血清中生长因子水平具有有益作用。相比于未补充组，补充组中参与生成能量和短链脂肪酸的细菌（如 *Proteobacteria*、*Rhodospirillaceae*、*Campylobacterales* 和 *Butyricimonas*）数量增加，而不良支原体减少（Du 等，2018）。给奶犊牛添加 *Bacillus amyloliquefaciens* H57 可提高增重，尽管瘤胃群落结构无显著差异。*Bacillus amyloliquefaciens* H57 在瘤胃中的低丰度表明该益生菌并不直接负责增重而是影响动物行为（饲料消耗）或改变瘤胃群落功能（Schofield 等，2018）。

Calvão 等（2005）研究了犊牛由于缺乏初乳血清中 IgG 浓度较低。给犊牛饲喂含有 *Saccharomyces boulardii* strain CNCM I-1079 的代乳粉显著减少了腹泻和相关兽医费用。大肠杆菌是引起犊牛腹泻的主要病原体之一，即使在没有发现这种病原菌的情况下，DFM 也通常被用于缓解腹泻。在一项 meta 分析中，Signorini 等（2012）得出结论，给断奶前奶犊牛补充乳酸菌（LAB）益生菌产生了保护作用并减少了腹泻发生率。其他研究也表明用 *E. coli* Nissle 1917（von Buenau 等，2005）或混合微生物饲喂奶犊牛可以预防腹泻（Mokhber-Dezfouli 等，2007）。对患有腹泻的荷斯坦奶牛施用多品种 DFM 可缩短腹泻持续时间（Renaud 等，2019）。对以芽孢杆菌为基础的 DFM 与口服电解质联合治疗断奶前犊牛腹泻进行了评估。与单独的电解质相比，相关添加剂确实减少了产气荚膜梭菌的粪便排出，减轻了腹泻的严重程度并降低了治疗成本（Wehnes 等，2009）。

最近的研究强调了分布在代乳粉和开食料中的 *S. boulardii* strain CNCM I-1079 对幼龄奶犊牛肠道发育的潜在积极作用（Fomenky 等，2017）。结肠形态的改变和中性黏蛋白产量的增加表明接受添加剂的犊牛肠道成熟的更早。在断奶时，饲喂益生菌的犊牛血清中急性期蛋白如触珠蛋白和 C-反应蛋白也有所增加。同时也刺激了从血浆中分离的多形核中性粒细胞的吞噬活性。这些数据表明 *S. boulardii* CNCM I-1079 在应激性断奶期增强犊牛先天免疫和炎症反应中的潜在作用。

在同一研究中，发现两种益生菌（*S. boulardii* CNCM I-1079 和 *L. acidophilus* BT-1386）在断奶前阶段对肠道细菌群落有显著影响，特别是回肠中（Fomenky 等，2018）。两种益生菌均显著降低了潜在致病菌 *Streptococcus* 和 *Tyzzerella* 的相对丰度，增加了有益纤维降解菌 *Fibrobacter* 的相对丰度。补充活酵母似乎比细菌 DFM 对肠道微生物有更大的影响。DFM 可能会影响胆汁酸分泌，而胆汁酸会影响脂肪的消化和吸收、蛋白质和能量代谢以及肠道微生物调节。胆汁酸调节途径的激活增强了肠道对细菌感染的抵抗和相关液体和电解质的分泌，减少了结肠炎症，并增加了生长激素 FGF19 的水平（Ipharraguerre 等，2018）。布拉迪酵母不同类型的作用机制已有报道，尽管大多数只在体外或实验室动物试验中被证实（Pothoulakis，2009；McFarland，2010；Stier 和 Bischoff，2016）。布拉迪酵母被分类为具有三种作用方式：管腔、营养和黏膜抗炎信号作用。在肠腔内，布拉迪酵母可能会通过酵母细胞壁甘露寡糖-细菌菌毛相互作用干扰病原的黏附。它们能够中和毒素，保护细胞生理，与肠道微生物互作并诱导短链脂肪酸谱的改变。

5 益生菌的益处和作用方式：成年反刍动物的饲料效率

提高饲料效率（FE）对可持续畜牧生产至关重要。最近宏基因组学研究表明瘤胃微生物组成/功能和 PE 之间存在关联（Mizrahi 和 Jami，2018）。几种植物细胞壁降解细菌菌群及其纤维降解功能与 FE 之间存在着特殊的联系（Delgado 等，2019）。使用 16S rDNA 测序方法，McGovern 等（2018）发现产琥珀酸丝状杆菌（*Fibrobacter succinogenes*）丰度和剩余采食量（RFI）呈负相关，表明这种纤维分解和半纤维分解细菌通过向宿主和其他微生物种群提供底物而对 FE 有所贡献。因此，提高瘤胃纤维消化率是益生菌的一个靶点。可能通过下列方式实现：

(1) 减少不可消化纤维部分。

(2) 提高纤维消化率，例如，通过维持能够提升纤维消化细菌数量的瘤胃环境。

在草料中不可消化纤维（iNDF）通常与木质素含量，特别是被"困"在木质素中的结构碳水化合物（纤维素和半纤维素）有关。由于其生化降解过程涉及氧化途径，木质素在动物胃肠道内不能被完全消化。但是，木质素结合碳水化合物的释放可能会增加粗饲料的饲用价值。McSweeney 等（1994）观察到高达 33.6% 的高粱木质素降解可能与瘤胃厌氧真菌 *Neocallimastix patriciarum* 的活性有关。作者认为，降解的木质素部分是通过木聚糖从基质中解体而不是通过木质素解聚而溶解的木质素碳水化合物复合物。

微生物 FA 能够通过不同的方式改善瘤胃内的纤维降解。一种间接途径是通过稳定 pH 值的作用（在下一节中讨论）。另一种途径是通过增强厌氧条件和氧气清除特性改变瘤胃环境（Chaucheyras-Durand 等，2008；Chaucheyras-Durand 和 Fonty，2002；Jouany，2006；Jouany 和 Morgavi，2007；Marden 等，2008）。这些情况促进了纤维降解微生物和它们对植物细胞壁多糖的作用。细菌和真菌对维生素、肽类、氨基酸、氨、有机酸或支链脂肪酸的营养需求已有描述，微生物 FA 可能为纤维分解微生物提供这些组分。

研究发现活酵母 FA 具有促进纤维降解瘤胃微生物生长和活力的潜力（Chaucheyras-

Durand 等，2008）。其机制包括增加真菌游动孢子萌发和纤维素降解，促进纤维分解细菌产琥珀酸丝状杆菌（*Fibrobacter succinogenes*），白色瘤胃球菌（*Ruminococcus albus*），黄色瘤胃球菌（*Ruminococcus flavefaciens*）和溶纤维丁酸弧菌（*Butyrivibrio fibrisolvens*）的生长和/或活性。一些研究使用基于 DNA 的技术（qPCR，DNA 测序）报告了产琥珀酸丝状杆菌（AlZahal 等，2017；Pinloche 等，2013；Uyeno 等，2017），瘤胃球菌（Mosoni 等，2007；Pinloche 等，2013；Silberberg 等，2013；Sousa 等，2018）或瘤胃真菌（Ding 等，2014）丰度的增加。Jiang 等（2017）使用 Illumina MiSeq 和 qPCR 比较了活的和热灭活酵母 FA 对瘤胃微生物的影响。活酵母的补充增加了瘤胃球菌和产琥珀酸丝状杆菌的相对丰度。活酵母菌株（*S. cerevisiae* CNCM I-1077）在促进纤维分解细菌（*F. succinogenes*，*R. flavefaciens*，*B. fibrisolvens*）和真菌定植于纤维底物的作用得到了证实。这种刺激程度取决于底物的性质和靶向微生物种类（Chaucheyras-Durand 等，2016）。在补充酵母期间，木质素含量最高且因此易消化碳水化合物较少的饲料降解得更好，表明添加酵母对木质素-多糖键的微生物分解具有影响，且对瘤胃真菌可能最为活跃。

已发现活酵母可以增加纤维底物上溶纤维丁酸弧菌的丰度。已知这种细菌具有阿魏酸和对香豆酸酯酶，它们水解半纤维素部分中酚酸和木聚糖链之间的酯键，从而使更多的多糖暴露在微生物酶的攻击之下（McSweeney 等，1998）。Guedes 等（2008）对同一活酵母菌株增加玉米青贮饲料样品的纤维（NDF-中性洗涤纤维）降解进行了报道。与易消化玉米青贮饲料相比，这种酵母 FA 更强地增加了低消化性玉米青贮饲料 NDF 的降解。这些结果表明活酵母可能通过提升参与木质素-多糖键水解的细菌和真菌的作用来帮助减少难消化的 NDF。

活酵母添加剂在瘤胃酸中毒的情况下通过稳定瘤胃 pH 值间接促进纤维降解。使用 18S 和 ITS 测序，Ishaq 等（2017）的研究表明，饮食引起的亚急性瘤胃酸中毒（SARA）改变了瘤胃真菌和原虫的多样性并且选择性地针对纤维降解菌种。补充活酵母产品的牛具有稳定的瘤胃 pH 值和更大的微生物多样性变化，避免了原虫的减少。活酵母（*S. cerevisiae* Y1242）也增加了属于产琥珀酸丝状杆菌的一些优势厌氧 OTU 的丰度和一些编码特定微生物纤维分解酶的基因丰度（AlZahal 等，2017）。一直有报道称活酵母的补充能够改善体内瘤胃纤维消化（Chaucheyras-Durand 等，2016；Dias 等，2018；Ding 等，2014；Ferraretto 等，2012；Guedes 等，2008；Sousa 等，2018）。

瘤胃酸中毒仍然普遍存在于奶牛和肉牛中，这些牛的饲料中有与低纤维相关的大量易发酵碳水化合物，可能会由于产酸量高和缓冲能力降低而对瘤胃功能产生负面影响（Villot 等，2018）。虽然酸中毒不仅仅是一种"瘤胃疾病"，还会影响整个消化道（Plaizier 等，2018），但是干预策略侧重于瘤胃微生物平衡和瘤胃 pH 值稳定（Humer 等，2018）。已经有许多关于酵母 FA 预防 SARA 的研究（Chaucheyras-Durand 等，2016；Ipharraguerre 等，2018；Jiang 等，2017；Jouany，2006；Jouany 和 Morgavi，2007；Mizrahi 和 Jami，2018）。瘤胃传感器（Villot 等，2018）已被用于检测活酵母 FA 对瘤胃 pH 值的有益作用。

酵母 FA 促进参与发酵酸（特别是乳酸，比 VFAs 更强的酸）释放和/或涉及乳酸去除的微生物菌群的转变，从而优化乳酸生产者和乳酸利用者之间的平衡。体外

（Chaucheyras-Durand 等，2008）和体内（Pinloche 等，2013）试验均已观察到不同活酵母对乳酸利用菌如 *Megasphaera elsdenii* 或 *Selenomonas ruminantium* 生长和代谢的促进作用。供应生长因子、肽类、氨基酸或维生素被认为是一种作用机制（Fonty 和 Chaucheyras-Durand，2006）。对参与乳酸生成的主要菌种之一牛链球菌（*Streptococcus bovis*）的生长抑制作用已在体外试验进行了测定（Fonty 和 Chaucheyras-Durand，2006）。酵母 FA 对瘤胃乳酸浓度的影响已在体内研究中被证实（Chaucheyras-Durand 等，2008；Kumprechtová 等，2019；Reis 等，2018）。

酵母 FA 还能缓解丁酸性酸中毒（Brossard 等，2004；Lettat 等，2010）。Brossard 等（2006）报道了一株酿酒酵母对羊丁酸性瘤胃酸中毒的 pH 值稳定作用。该菌株能促进纤毛原虫 Entodiniomorphid，已知这些原虫可以迅速吞噬淀粉颗粒从而有效地与淀粉分解菌竞争（Owens 等，1998）。活酵母对纤毛原虫的影响在其他研究中也有报道（Chaucheyras-Durand 和 Fonty，2002；Silberberg 等，2013）。纤毛虫比淀粉分解菌消化淀粉的速度慢，并且它们发酵的主要终产物是 VFAs 而不是乳酸，这些事实可能解释了它们可以通过延缓发酵而在瘤胃中起到稳定作用。尽管纤毛原虫与甲烷生成有关，一项最近的瘤胃宏转录组学研究表明它们对纤维分解的贡献似乎比以前认为的要大，因此增加它们可以提高纤维消化率（Comtet-Marre 等，2017）。

更好的纤维消化和稳定的瘤胃 pH 值可以通过改善 FE 有益于动物瘤胃健康及其功能。De Ondarza 等（2010）通过从 14 次试验中收集性能数据研究了活酵母（*Saccharomyces cerevisiae* CNCM I-1077）对奶牛的影响，明确表明 FE 得到了改善。当针对饲喂 NDF 高于 30%（高纤维日粮，低 SARA 风险）的牛时，FE 高于整体平均值。活酵母处理的动物每千克干物质采食量（DMI）多生产了 40 克牛奶。在 SARA 风险较高（低纤维日粮，>25% 淀粉）的情况下，FE 甚至更高，每千克 DMI 额外产奶 80 克。其他研究表明当给牛饲喂这种活酵母产品时，其进食行为发生了改变，饮食间隔更短（Bach 等，2007；DeVries 和 Chevaux，2014），这表明饲料消化得到了改善因为采食量没有受到影响。瘤胃 pH 值的改善使得纤维降解菌的活性增加从而解释了采食频率的提高。

最近一项研究使用内窥镜收集瘤胃活组织评估了活酵母（*S. cerevisiae* CNCM I-1077）在产犊期的作用（Bach 等，2018）。产乳酸的链球菌和乳酸杆菌属（Derakhshani 等，2016）以及变形菌门的糖解成员（Zhu 等，2017）在产后增加，提高了影响胃肠道上皮的 SARA 风险（Steele 等，2011）。结果表明产犊前补充活酵母 FA 增加了瘤胃中调控炎症和上皮屏障基因（例如紧密连接编码基因）的表达。

很少有研究关注益生菌在酸中毒期间减缓炎症或控制免疫系统的潜能。在饲喂高挑战性日粮（淀粉/果糖）的小母牛中，离子载体莫能菌素和一种活酵母 FA（*S. cerevisiae* I-1077）的结合使用显著降低了瘤胃液中的组胺浓度（Golder 等，2014）。在其他关注于微生物 FA 对瘤胃或血浆炎症分子浓度影响的试验中，Garcia Diaz 等（2018）表明给肉用公牛补充活酵母（*S. cerevisiae* NCYC 996）后，其血浆 LPS 以及急性期蛋白血清淀粉样蛋白 A（SAA）的浓度降低了，但对瘤胃或十二指肠的 LPS 浓度没有影响。Silberberg 等（2013）检测到使用活酵母 FA 后血浆 SAA 水平显著降低，同样对瘤胃 LPS 浓度没有影响。这可能是因为反刍动物 LPS 易位最有可能发生的部位是大肠而不是瘤胃，可能相比于网状瘤胃上皮单层肠上皮更容易被酸性损伤（Khiaosa‑Ard 和 Zebeli，2018）。

图 3 总结了补充活酵母 FA 对有亚急性瘤胃酸中毒（SARA）风险的反刍动物的预期益处。

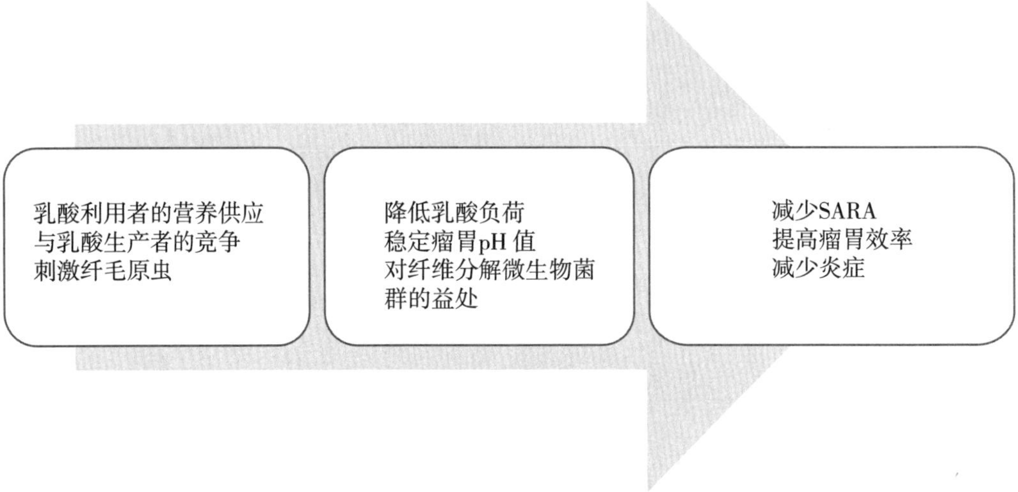

```
┌─────────────────┐   ┌─────────────────┐   ┌─────────────────┐
│ 乳酸利用者的营养供应 │   │ 降低乳酸负荷       │   │ 减少SARA         │
│ 与乳酸生产者的竞争  │   │ 稳定瘤胃pH值       │   │ 提高瘤胃效率      │
│ 刺激纤毛原虫      │   │ 对纤维分解微生物菌  │   │ 减少炎症         │
│                 │   │ 群的益处           │   │                │
└─────────────────┘   └─────────────────┘   └─────────────────┘
```

图 3　活酵母 FA 对反刍动物的益处

当牛遭受高温高湿条件（热应激）时，*Clostridium* coccoides‑*Eubacterium rectale* group 和 *Streptococcus* 在瘤胃中的相对比例增加，而纤维杆菌属的比例减少，导致微生物失衡（Uyeno 等，2010）。在这些情况下，微生物 FA 有助于稳定瘤胃生态系统并减轻热应激对牛生产性能的负面影响（Salvati 等，2015）。

埃氏巨型球菌（*Megasphaera elsdenii*）是瘤胃生态系统中一个重要的生态菌种，因为它可以去除乳酸从而防止酸中毒。最近的一项关注于低效或高效牛瘤胃微生物组的宏基因组学研究强调了在高效牛微生物组中富集最多的基因隶属于埃氏巨型球菌。参与将乳酸转化为丙酸的丙烯酸酯代谢途径的基因也在最高效的动物中得到了富集（Shabat 等，2016）。埃氏巨型球菌被认为是提高瘤胃内 *M. elsdenii* 浓度和加速瘤胃乳酸清除的一种 FA（Arik 等，2019；Muya 等，2015；Yohe 等，2018；Zebeli 等，2012）。然而，有益作用并不总是能够测量的（Yohe 等，2018；Zebeli 等，2012）。此外，准备和递送都具有挑战性。

其他来自丙酸杆菌（*Propionibacterium*）属的乳酸利用菌也被进行了用于减缓高谷物饲喂牛 SARA 严重程度的评估（Azad 等，2017；Philippeau 等，2017）。已有建议将它们和乳酸杆菌联合起来以促进乳酸生成，这样可以刺激利用乳酸的丙酸杆菌（Lettat 等，2012）。一项利用 16S rRNA 基因测序的研究报告了一种 *P. acidipropiici* P169 菌株对高谷物饲喂肉用公牛瘤胃富集微生物乳酸利用菌（*Veillonellaceae* 和 *Megasphaera*）以及纤维分解细菌 *Ruminococcaceae*、*Lachnospiraceae*、*Clostridiaceae*、*Christensenellaceae* 相对丰度的有益作用（Azad 等，2017）。微生物代谢产物也受到了影响（支链脂肪酸的摩尔比例更高，并且氨的浓度增加），表明纤维分解和蛋白水解活性状态有所改善。

6 益生菌的益处和作用方式：甲烷生成

使用微生物 FA 是减少反刍动物甲烷排放的一种可能性选择。潜在的作用机制包括（Jeyanathan 等，2014）：

（1）直接抑制甲烷生成；
（2）促进瘤胃中已经存在的替代途径如同型产乙酸；
（3）富马酸盐还原；
（4）通过丙烯酸酯途径的丙酸生成；
（5）硝酸盐/亚硝酸盐还原；
（6）二氧化碳固定；
（7）甲烷厌氧氧化（甲烷营养）。

这些途径中只有少数被探索过，可能是因为对这些复杂的微生物代谢及其与甲烷生成竞争的能力缺乏了解。已发现同型产乙酸途径可促进没有产甲烷菌的羔羊的功能性瘤胃活动，但产乙酸菌捕获 H_2 的效率远低于产甲烷菌（Fonty 等，2007）。候选微生物的生存也有问题因为大多数选择途径需要严格的厌氧条件。最后，可能会对动物发酵效率产生副作用或对动物产生毒性，例如，如果补充硝酸盐导致亚硝酸盐积累可能引起高铁血红蛋白血症。

已经有一些关于酿酒酵母作用于瘤胃甲烷生成以促进替代非产甲烷途径的研究。至少在体外已经表明对产乙酸菌的促进作用（Chaucheyras 等，1995；Nollet 等，1997）。使用无菌动物模型，Chaucheyras–Durand 等（2010b）表明纤维分解菌群（产氢菌与非产氢菌）组成可能会影响瘤胃生态系统中 H_2 的积累和随后的甲烷生成。促进不产生任何氢的纤维分解菌，如产琥珀酸丝状杆菌，可能有助于限制瘤胃中甲烷的排放。然而，迄今为止对酵母 FA 的研究未能证明其具有减排作用（Bayat 等，2015；Chung 等，2011）。活酵母 FA（请参阅前面的）存在时 FE 的增加应该会对甲烷排放产生间接作用，因为它会降低每产千克牛奶/肉的输出量（Jeyanathan 等，2014）。少数细菌被检测了抗产甲烷潜力。在绵羊上的结果显示了一株 *L. pentosus* 的某些功效（2 周后甲烷排放量减少 13%，持续 4 周）（Jeyanathan 等，2016）。三株丙酸杆菌混合物也在饲喂玉米精饲料的肉牛上进行了评估（Vyas 等，2014），但对肠道甲烷生成未有影响；作者认为高淀粉含量的日粮诱导了高水平的丙酸，这些情况可能降低了丙酸杆菌对甲烷减排的功效。

7 益生菌的益处和作用方式：病原体控制

畜牧生产中抗生素的使用与耐药菌群的发展和动物性食品中抗生素残留的持久性相关（Langford 等，2003；Ramatla 等，2017；Seymour 等，1988）。补充微生物 FA 被认为是一种降低牛的病原体负荷和人畜共患病病原体向人类传播风险的选择性方法。沙门氏菌病是 2017 年欧盟观察到的第二大人畜共患病（EFSA 和 ECDC，2018），并且沙门氏菌食源性暴发与牛肉产品的消费有关（EFSA 和 ECDC，2018）。沙门氏菌已被证明是由牛无症状携带的（Feye 等，2016），尽管它们也能引起动物的临床疾病。肠道沙门氏菌和大肠杆菌引发的腹泻和肠道疾病是养牛者经济损失的主要原因（USDA，2014a；Cho 和 Yoon，2014）。益生菌可以通过竞争性排斥、产生抗菌化合物和刺激宿主免疫防御来帮助消灭病原体，这将减少病原体在动物体内的定植和感染风险。

嗜淀粉乳杆菌 C94（*Lactobacillus amylovorus* C94）和唾液乳杆菌 C86（*L. salivarius* C86）菌株在体外试验中被证实具有对抗分离自牛的肠道沙门氏菌的良好潜力（Adetoye 等，2018）。研究表明 DFM（每天用 *L. acidophilus* LA51 和 *P. freudenreichii* PF24 混合处理）对肉用公牛具有预防沙门氏菌感染的积极作用（Tabe 等，2008）。对肉牛施用相同的 DFM 可显著降低屠宰时沙门氏菌的流行率，并且显著降低与牛肉糜污染相关的外周淋巴结（PLN）病原体浓度（Vipham 等，2015）。Soto 等（2015）测试了由干酪乳杆菌 DSPV318T（*Lactobacillus casei* DSPV318T）、唾液乳杆菌 DSPV315T（*L. salivarius* DSPV315T）和乳酸片球菌 DSPV006T（*Pediococcus acidilactici* DSPV006T）组成的接种物对幼龄犊牛的影响，观察到每天服用该接种物和乳糖会导致中性粒细胞/淋巴细胞比例下降，表明在感染急性期免疫反应增强（Soto 等，2016）。先前观察到该产品可降低感染沙门氏菌犊牛的腹泻严重程度（Soto 等，2015）。

产生志贺氏毒素的大肠杆菌（STEC）（如 *E. coli* O157:H7）因其对人类健康的影响而备受关注（Caprioli 等，2005；Chauret，2011；EFSA 和 ECDC，2018）。DFM 代表了减少肉牛 *E. coli* O157:H7 粪便脱落的一种有效策略（Brashears 和 Chaves，2017）。体外试验结果表明，嗜酸乳杆菌 BT-1386（*Lactobacillus acidophilus* BT-1386）对粪便中 *E. coli* O157:H7 具有剂量依赖性抑制作用（Chaucheyras-Durand 等，2006）。该菌株和一活酵母菌株 *Saccharomyces cerevisiae* CNCM I-1077 能显著降低绵羊瘤胃中致病性大肠杆菌的负荷（Chaucheyras-Durand 等，2010a）。*L. reuteri* LB1-7 联合甘油实现了对瘤胃内 *E. coli* O157:H7 的抑制。瘤胃液中暴露相同的混合物后减少了直肠内容物中病原体的生长（Bertin 等，2017）。这些体外结果强调了 DFM 对于降低消化系统病原体含量的潜力。

最近一项 meta 分析表明 LAB（特别是 *Lactobacillus acidophilus* NP51 和 *Propionibacterium freudenreichii* NP24 的 DFM 组合）降低了 *E. coli* O157:H7 粪便脱落的流行率（Wisener 等，2015）。在 2013 年，美国的牛肉业食品安全委员会（BIFSCO）已经认识到一些细菌菌株在减少肉牛 *E. coli* O157:H7 方面的功效并将其作为生产最佳实践（Production Best Practice）的一部分（Beef Industry Food Safety Council subcommittee on pre-harvest，2013）。对试验感染的小牛施用竞争性-排斥共生大肠杆菌的混合物显示 O157:H7 以及非 O157 血清型大肠杆菌的粪便脱落显著减少（Tkalcic 等，2003）。另一项研究发现，使用相同的益生菌混合物处理过的犊牛脱落的非 O157 大肠杆菌显著减少，但对大肠杆菌 O157 没有作用（Zhao 等，2003）。在试验感染的羔羊中，每日给予 *S. faecium* 或 *S. faecium*、*L. acidophilus*、*L. casei*、*L. fermentum* 和 *L. plantarum* 的混合物可显著降低粪便中 *E. coli* O157:H7 的含量（Lema 等，2001）。含 *Lactobacillus acidophilus* LC10、*Lactobacillus helveticus* LC3、*Lactobacillus bulgaricus* LC182、*Lactobacillus lactis*、*Streptococcus thermophilus* LC201 和 *Enterococcus faecium* LAT E-253 的益生菌混合物能显著降低绵羊致病性大肠杆菌的粪便脱落（Rigobelo 等，2015）。然而，限制 STEC 携带取决于许多难以控制的参数如日粮、环境因素、应激水平、间歇性和季节性脱落、畜群中出现几种血清型等（Bertin 等，2011，2013；Chaucheyras-Durand 等，2010a，2006；Dunière

等，2011；Fremaux 等，2006）。最有效策略的确定和实施应基于对牛 GIT 中 STEC 生态学和生理学的充分理解（Segura 等，2018）。

益生菌也可用于治疗牛的非消化性疾病。乳腺炎是乳制品行业最有害的疾病之一，由于其会导致产量下降和治疗成本增加（Hogeveen 和 Østerås，2005）。益生菌如 LAB 可以通过定植在乳房并建立一个有益的生物膜来防止病原体定植从而潜在地减少乳腺炎。*Lactobacillus brevis* 1595 和 1597 以及 *Lactobacillus plantarum* 1610 三株菌在与病原体竞争乳腺定植中显示出好的定植能力。它们还具有抗炎特性，使得大肠杆菌刺激下的牛乳腺上皮细胞（bMEC）分泌的 IL-8 较低（Bouchard 等，2015）。*Lactobacillus rhamnosus* ATCC7469 和 *L. plantarum* 2/37 已被证明可以形成生物膜来取代葡萄球菌（Wallis 等，2018，2019）。一种产细菌素的 *Lactococcus lactis* DPC3174 被证明在治疗奶牛乳腺炎方面与常规抗生素一样有效（Klostermann 等，2008）。两种 LAB 菌株（*Lactobacillus perolens* CRL 1724 和 nisin Z producer *Lactococcus lactis* sub. *Lactis* CRL1655）被证明可以调节奶牛在干奶期的宿主乳房免疫系统并刺激体内局部和全身防御系统（Pellegrino 等，2017）。在使用益生菌用于治疗牛乳腺炎的最新文献的综述中，Rainard 和 Foucras（2018）指出大多数 LAB 试验是在体外而不是在体内进行的。然而，他们证实乳房内给药是使用益生菌控制乳腺炎的最佳方法。

益生菌也可以用于提高繁殖性能。牛生殖系统疾病可导致产奶量降低以及繁殖性能受损（Bellows 等，2002）。急性子宫炎是产犊后由于细菌感染引起的子宫炎症，大肠杆菌是引发该疾病的主要病原体（Kassé 等，2016）。与对照动物相比，使用 LAB 益生菌的阴道治疗可降低子宫炎患病率多达 58%（Genís 等，2018）。活酵母日粮补充可以通过改变激素和卵巢卵泡动力来改善奶牛热应激期间的繁殖性能（Nasiri 等，2018）。

牛呼吸道疾病（BRD）会给牛生产商造成重大损失（Fulton 等，2002；Griffin 等，2010）。BRD 是多因素的但已确认的主要细菌病原体之一是溶血性曼氏杆菌（*Mannheimia haemolytica*）。体外试验已经表明隶属于乳杆菌、乳球菌和类芽孢杆菌属的益生菌菌株能黏附于牛呼吸细胞并能通过竞争和取代抑制溶血性曼氏杆菌（Amat 等，2017）。研究显示给小鼠灌胃 *Enterococcus faecalis* CECT7121 可以诱导体液免疫应答的增加以抵抗多杀巴斯德杆菌和溶血性曼氏杆菌。益生菌处理过的小鼠显示出更高的干扰素-γ 产量，这表明发生了更强的细胞免疫反应（Díaz 等，2018）。作者建议这种益生菌可以用作增强反刍动物免疫应答疫苗策略中可能的佐剂。在接受细菌和病毒攻击（溶血性曼氏杆菌和牛疱疹病毒-1）的肉用小母牛中，补充 *Saccharomyces boulardii* CNCM I-1079 改变了白细胞血像（在 *Mannheimia* 攻击后第 2 天中性粒细胞%显著增加，并且单核细胞%增加），表明固有免疫应答得到了增强（Kayser 等，2019）。

8 益生菌的益处和作用方式：对免疫系统的影响

如本章前面所述，益生菌干预以增强新生和断奶前反刍动物免疫功能已成为研究热点。研究证实生命早期肠道微生物对于生命后期免疫能力的塑造起着关键作用（Gensollen 等，2016）。在成年反刍动物中，益生菌可能在调节免疫功能中促炎和抗炎信号产生方面发挥作用（Raabis 等，2019）。然而，人们对益生菌参与宿主-微生物互作和调节免疫功能的方式还不是很清楚（Ma 等，2018）。与日粮补充益生菌相关的免疫反应调节不仅可以通过先天和适应性免疫系统进行，还可以通过调节肠道上皮通透性、黏液分泌和微生物生态系统内的竞争进行，在微生物生态系统中益生菌可以分泌抗微生物化合物（La Fata 等，2018）。这使得状况非常复杂（图4）。

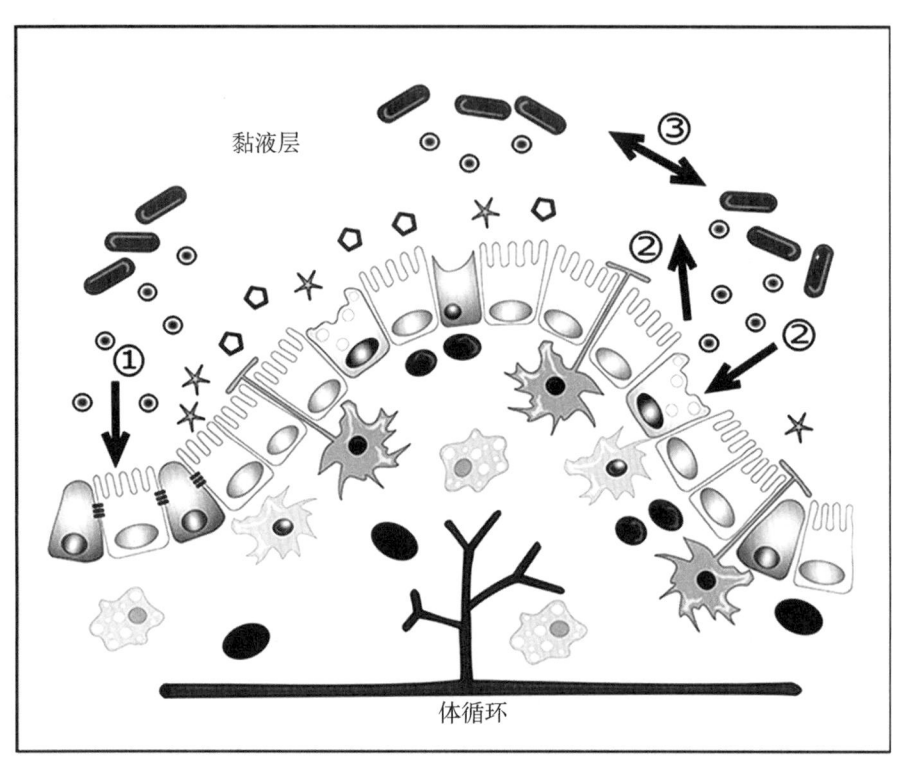

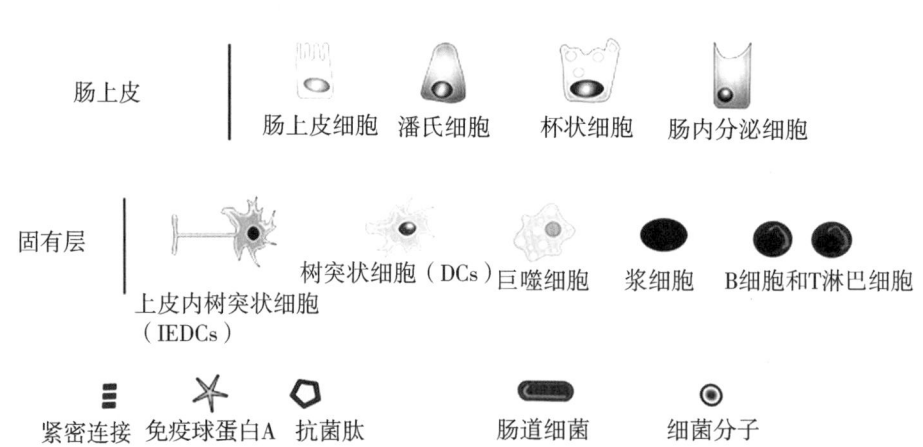

图4 肠道屏障和微生物相关主体（肠道细菌或益生菌及其代谢产物）—肠上皮互作。①代表紧密连接（TJ）蛋白的调节；②表示杯状细胞黏液分泌的调节；③表示肠道微生物不同成员之间可能的相互作用（例如抗菌性能）。改编自 La Fata 等（2018）。

Bach 等（2018）评估了一种活酵母 FA（*S. cerevisiae* I-1077）对奶牛产犊前后编码免疫反应相关蛋白基因表达的影响。奶牛对活酵母反应迅速，在补充 7d 后，参与免疫反应的基因表达发生了可测的变化，包括瘤胃中抗炎基因 IL-10 和结肠中 β-防御素基因的表达。β-葡聚糖或甘露寡糖（如酿酒酵母）也被认为是免疫刺激剂或病原体结合剂（Ganner 和 Schatzmayr，2012；Goodridge 等，2009；Li 等，2016，2018；Yuan 等，2015）。由于细胞壁 β-葡聚糖似乎在其他动物物种中诱导显著的免疫反应（Angulo 等，2018，2019），其他酵母种类如 *Debaryomyces hansenii* 也越来越受到人们关注。使用原子力显微镜等技术来提高酵母细胞壁成分对免疫系统调节和宿主反应作用的理解已经引起了极大的兴趣（Schiavone 等，2015，2017）。

有大量关于所选细菌对肠上皮细胞（IEC）反应影响的文献，主要来自人类细胞系或啮齿类动物模型（La Fata 等，2018）。Villena 等（2018）建议使用牛 IEC 来研究免疫生物制剂，如双歧杆菌或乳酸杆菌，对抗病毒免疫的影响。他们报道了免疫生物制剂通过激活干扰素调节因子-3（IRF-3），增加抗病毒因子生成和调节炎症介质来发挥有益作用。细菌益生菌（*Lactobacillus casei* BL23，*Lactococcus lactis* V7）对金黄色葡萄球菌刺激的牛乳腺细胞先天免疫反应的影响已被研究（Assis 等，2015；Souza 等，2018）。类酵母细菌细胞壁组成和结构很可能参与引发宿主反应（Lebeer 等，2018）。

通过微生物-肠-脑轴将产生神经化学物质的益生菌用于健康和疾病的治疗越来越受到关注（Lyte，2011）。这对反刍动物来说至关重要，因为越来越多的证据表明应激会增加肠道感染的易感性（Freestone 和 Lyte，2010）。最近的研究报告了在胃肠道作用的 *Enterococcus faecium* 产生了多巴胺（Villageliú 和 Lyte，2018）。其他被用作益生菌的微生物菌种也能产生一系列参与肠-脑轴交流的分子（Lyte，2011）（表 2）。

表 2 潜在益生菌属分离的神经化学物质

微生物属	神经化学物质
双歧杆菌，乳杆菌	γ-氨基丁酸（GABA）
乳杆菌	乙酰胆碱
芽孢杆菌，酵母菌	去甲肾上腺素
肠球菌，芽孢杆菌	多巴胺
肠球菌	血清素

资料来源：改编自 Lyte（2011），Villageliú 和 Lyte（2018）。

9 结论和未来趋势

正如本章讨论的研究所表明的那样，在牲畜中使用微生物 FA 可以提高盈利能力同时保持动物福利和农场可持续性。商用益生菌的发展将取决于监管约束。例如，在欧盟，聚焦于为一个目标物种批准单一的、定义明确的微生物菌株。如果监管机构能够接受微生物活性物质的组合甚至是已经证明有好处的未鉴定微生物群，那将会有所帮助。目前不允许使用创新技术，例如微生物移植（瘤胃或粪便），转基因益生菌也没有得到批准。

需要更多的研究来更好地了解微生物 FA 与肠道微生物和宿主动物相互作用的机制，以便更好地选择农场主易于使用的安全、强健、环保和高效的添加剂。多组学方法将有助于破译益生菌利用的代谢途径，这可以解释它们在宿主动物中的益处。一旦确定了这些途径的特征，就可以更容易地利用这些理解来更准确和适当地筛选和选择具有预期益处的最佳候选者。大多数反刍动物研究都集中于为了 FE 和甲烷减排改善瘤胃功能。然而，需要对益生菌如何调节宿主-微生物互作和宿主免疫进行更多的研究以优化益生菌干预策略来改善动物肠道健康和生产性能。

其他潜在的研究领域包括基于微生物 FA 的母体印记（关于子代 GIT 微生物群落的建立和免疫发育），尽管其作用机制在很大程度上仍然未知并且需要进一步研究。另一个令人兴奋的领域是微生物内分泌学，它是微生物学、内分泌学和神经生理学的交叉学科。此领域有助于更好地了解应激是如何影响反刍动物健康的。引用 Freestone 的名言："快乐的，应激较少的反刍动物可能是更有营养的动物和更安全的肉类来源"（Freestone 和 Lyte，2010）。

10　致　谢

我们非常感谢 Raphaële Gresse 女士（Lallemand/UMR MEDIS）在图 4 编制过程中所作出的技术贡献。

11　更多信息

进一步阅读以获得本章涵盖主题的更深入信息：

- Chaucheyras-Durand, F. and Durand, H. 2010. Probiotics in animal nutrition and health. Benef Microbes. 1（1），3-9. doi：10.3920/BM2008.1002（Review）.
- McCann, J. C., Elolimy, A. A. and Loor, J. J. 2017. Rumen microbiome, probiotics, and fermentation additives. *Vet. Clin. North Am. Food. Anim. Pract*. 33（3），539-53. doi：10.1016/j.cvfa.2017.06.009.
- McSweeney, C. and Mackie, R. 2012. Micro-organisms and ruminant digestion: state of knowledge, trends and future prospects. Commission on Genetic Resources for Food and Agriculture. Background study paper NO. 61.
- Probiotics in animal nutrition. FAO report 2016. Available at：http://www.fao.org/3/a-i5933e.pdf.

紧跟最新趋势的重要协会、专业组织或其他值得访问的网站：

- www.lallemandanimalnutrition.com.
- www.ruminantdigestivesystem.com.
- www6.ara.inrae.fr/medis.

重要期刊或会议（例如每年或几年定期举行的重要会议）：

- Beneficial Microbes Conference (https://www.bastiaanse-communication.com/BMC2018/) and associated journal.
- Congress on Gastro-intestinal Function (https://www.congressgastrofunction.org/).
- International Probiotic Conference (https://www.probiotic-conference.net/).
- Joint INRA-Rowett Symposium on gut microbiota (https://colloque.inrae.fr/inra-rowett-2016).

12 参考文献

Abaker, J. A., Xu, T. L., Jin, D., et al., 2017. Lipopolysaccharide derived from the digestive tract provokes oxidative stress in the liver of dairy cows fed a high-grain diet. J. Dairy Sci. 100 (1), 666-78.

Abecia, L., Martín-García, A. I., Martínez, G., et al., 2013. Nutritional intervention in early life to manipulate rumen microbial colonization and methane output by kid goats postweaning. J. Anim. Sci. 91 (10), 4832-40.

Abecia, L., Martínez-Fernandez, G., Waddams, K., et al., 2014. An antimethanogenic nutritional intervention in early life of ruminants modifies ruminal colonization by Archaea. Archaea 2014, 841463.

Abecia, L., Martínez-Fernandez, G., Waddams, K., et al., 2018. Analysis of the rumen microbiome and metabolome to study the effect of an antimethanogenic treatment applied in early life of kid goats. Front. Microbiol. 9, 2227.

Adetoye, A., Pinloche, E., Adeniyi, B. A., et al., 2018. Characterization and anti-salmonella activities of lactic acid bacteria isolated from cattle faeces. BMC Microbiol. 18 (1), 96.

Alugongo, G. M., Xiao, J., Wu, Z., et al., 2017. Review: utilization of yeast of *Saccharomyces cerevisiae* origin in artificially raised calves. J. Anim. Sci. Biotechnol. 8, 34.

AlZahal, O., Li, F., Guan, L. L., et al., 2017. Factors influencing ruminal bacterial community diversity and composition and microbial fibrolytic enzyme abundance in lactating dairy cows with a focus on the role of active dry yeast. J. Dairy Sci. 100 (6), 4377-93.

Amachawadi, R. G., Giok, F., Shi, X., et al., 2018. Antimicrobial resistance of *Enterococcus faecium* strains isolated from commercial probiotic products used in cattle and swine. J. Anim. Sci. 96 (3), 912-20.

Amat, S., Subramanian, S., Timsit, E., et al., 2017. Probiotic bacteria inhibit the bovine respiratory pathogen Mannheimia haemolytica serotype 1 *in vitro*. Lett. Appl. Microbiol. 64 (5), 343-9.

Angulo, M., Reyes-Becerril, M., Tovar-Ramírez, D., et al., 2018. Debaryomyces hansenii CBS 8339 β-glucan enhances immune responses and down-stream gene signaling pathways in goat peripheral blood leukocytes. Dev. Comp. Immunol. 88, 173-82.

Angulo, M., Reyes-Becerril, M., Cepeda-Palacios, R., et al., 2019. Probiotic effects of marine Debaryomyces hansenii CBS 8339 on innate immune and antioxidant parameters in newborn goats. Appl. Microbiol. Biotechnol. 103 (5), 2339-52.

Arik, H. D., Gulsen, N., Hayirli, A., et al., 2019. Efficacy of Megasphaera elsdenii inoculation in subacute ruminal acidosis in cattle. J. Anim. Physiol. Anim. Nutr. 103 (2), 416-26.

Ashenafi, D., Yidersal, E., Hussen, E., et al., 2018. The effect of long distance transportation stress on cattle: a review. Biomed. J. Sci. Tech. Res. 3.

Assis, B. S., Germon, P., Silva, A. M., et al., 2015. *Lactococcus lactis* V7 inhibits the cell invasion of bovine mammary epithelial cells by Escherichia coli and Staphylococcus aureus. Benef Microbes 6 (6), 879-86.

Azad, E., Narvaez, N., Derakhshani, H., et al., 2017. Effect of *Propionibacterium acidipropionici* P169 on the rumen and faecal microbiota of beef cattle fed a maize-based finishing diet. Benef. Microbes 8 (5), 785-99.

Bach, A., Iglesias, C. and Devant, M. 2007. Daily rumen pH pattern of loose-housed dairy cattle as affected by feeding pattern and live yeast supplementation. Anim. Feed Sci. Technol. 136 (1-2), 146-53.

Bach, A., Guasch, I., Elcoso, G., et al., 2018. Changes in gene expression in the rumen and colon epithelia during the dry period through lactation of dairy cows and effects of live yeast supplementation. J. Dairy Sci. 101 (3), 2631-40.

Baldwin, R. L., McLeod, K. R., Klotz, J. L., et al., 2004. Rumen development, intestinal growth and hepatic metabolism in the pre-and postweaning ruminant. J. Dairy Sci. 87, E55-65.

Bayat, A. R., Kairenius, P., Stefański, T., et al., 2015. Effect of Camelina oil or live yeasts (*Saccharomyces cerevisiae*) on ruminal methane production, rumen fermentation, and milk fatty acid composition in lactating cows fed grass silage diets. J. Dairy Sci. 98 (5), 3166-81.

Beef Industry Food Safety Council subcommittee on pre-harvest. 2013. Production Best Practices (PBP) to aid in the control of foodborne pathogens in groups of cattle. BIFSCO, USA.

Bellows, D. S., Ott, S. L. and Bellows, R. A. 2002. Review: Cost of reproductive diseases and conditions in cattle. Prof. Anim. Sci. 18 (1), 26-32.

Bertin, Y., Girardeau, J. P., Chaucheyras-Durand, F., et al., 2011. Enterohaemorrhagic *Escherichia coli* gains a competitive advantage by using ethanolamine as a nitrogen source in the bovine intestinal content. Environ. Microbiol. 13 (2), 365-77.

Bertin, Y., Chaucheyras-Durand, F., Robbe-Masselot, C., et al., 2013. Carbohydrate utilization by enterohaemorrhagic *Escherichia coli* O157:H7 in bovine intestinal content. Environ. Microbiol. 15 (2), 610-22.

Bertin, Y., Habouzit, C., Dunière, L., et al., 2017. *Lactobacillus reuteri* suppresses E. coli O157:H7 in bovine ruminal fluid: toward a pre-slaughter strategy to improve food safety? PLoS ONE 12 (11), e0187229.

Blecha, F., Boyles, S. L. and Riley, J. G. 1984. Shipping suppresses lymphocyte blastogenic responses in Angus and Brahman × Angus feeder calves. J. Anim. Sci. 59 (3), 576-83.

Bouchard, D. S., Seridan, B., Saraoui, T., et al., 2015. Lactic acid bacteria isolated from bovine mammary microbiota: potential allies against bovine mastitis. PLoS ONE 10 (12), e0144831.

Brashears, M. M. and Chaves, B. D. 2017. The diversity of beef safety: a global reason to strengthen our current systems. Meat Sci. 132, 59-71.

Brossard, L., Martin, C., Chaucheyras-Durand, F., et al., 2004. Protozoa involved in butyric rather than lactic fermentative pattern during latent acidosis in sheep. Reprod. Nutr. Dev. 44 (3), 195-206.

Brossard, L., Chaucheyras-Durand, F., Michalet-Doreau, B., et al., 2006. Dose effect of live yeasts on rumen microbial communities and fermentations during butyric latent acidosis in sheep: new type of interaction. Anim. Sci. 82 (6), 829-36.

Bush, S. J., McCulloch, M. E. B., Muriuki, C., et al., 2019. Comprehensive transcriptional profiling of the gastrointestinal tract of ruminants from birth to adulthood reveals strong developmental stage specific gene expression. G3 Bethesda MD 9 (2), 359–73.

Caprioli, A., Morabito, S., Brugère, H., et al., 2005. Enterohaemorrhagic *Escherichia coli*: emerging issues on virulence and modes of transmission. Vet. Res. 36 (3), 289–311.

Chaucheyras, F., Fonty, G., Bertin, G., et al., 1995. *In vitro* H_2 utilization by a ruminal acetogenic bacterium cultivated alone or in association with an archaea methanogen is stimulated by a probiotic strain of *Saccharomyces cerevisiae*. Appl. Environ. Microbiol. 61 (9), 3466–7.

Chaucheyras-Durand, F. and Fonty, G. 2001. Establishment of cellulolytic bacteria and development of fermentative activities in the rumen of gnotobiotically-reared lambs receiving the microbial additive *Saccharomyces cerevisiae* CNCM I-1077. Reprod. Nutr. Dev. 41 (1), 57–68.

Chaucheyras-Durand, F. and Fonty, G. 2002. Influence of a probiotic yeast (*Saccharomyces cerevisiae* CNCM I-1077) on microbial colonization and fermentations in the rumen of newborn lambs. Microb. Ecol. Health Dis. 14 (1), 30–6.

Chaucheyras-Durand, F., Madic, J., Doudin, F., et al., 2006. Biotic and abiotic factors influencing *in vitro* growth of *Escherichia coli* O157: H7 in ruminant digestive contents. Appl. Environ. Microbiol. 72 (6), 4136–42.

Chaucheyras-Durand, F., Walker, N. D. and Bach, A. 2008. Effects of active dry yeasts on the rumen microbial ecosystem: past, present and future. Anim. Feed Sci. Technol. 145 (1-4), 5–26.

Chaucheyras-Durand, F., Faqir, F., Ameilbonne, A., et al., 2010a. Fates of acid-resistant and non-acid-resistant Shiga toxin-producing Escherichia coli strains in ruminant digestive contents in the absence and presence of probiotics. Appl. Environ. Microbiol. 76 (3), 640–7.

Chaucheyras-Durand, F., Masséglia, S., Fonty, G., et al., 2010b. Influence of the composition of the cellulolytic flora on the development of hydrogenotrophic microorganisms, hydrogen utilization, and methane production in the rumens of gnotobiotically reared lambs. Appl. Environ. Microbiol. 76 (24), 7931–7.

Chaucheyras-Durand, F., Ameilbonne, A., Bichat, A., et al., 2016. Live yeasts enhance fibre degradation in the cow rumen through an increase in plant substrate colonization by fibrolytic bacteria and fungi. J. Appl. Microbiol. 120 (3), 560–70.

Chauret, C. 2011. Survival and control of Escherichia coli O157: H7 in foods, beverages, soil and water. Virulence 2 (6), 593–601.

Chen, Y., Penner, G. B., Li, M., et al., 2011. Changes in bacterial diversity associated with epithelial tissue in the beef cow rumen during the transition to a high-grain diet. Appl. Environ. Microbiol. 77 (16), 5770–81.

Chen, Y., Oba, M. and Guan, L. L. 2012. Variation of bacterial communities and expression of toll-like receptor genes in the rumen of steers differing in susceptibility to subacute ruminal acidosis. Vet. Microbiol. 159 (3-4), 451–9.

Cho, Y. I. and Yoon, K. J. 2014. An overview of calf diarrhea-infectious etiology, diagnosis, and intervention. J. Vet. Sci. 15 (1), 1–17.

Chung, Y. H., Walker, N. D., McGinn, S. M., et al., 2011. Differing effects of 2 active dried yeast (*Saccharomyces cerevisiae*) strains on ruminal acidosis and methane production in nonlactating dairy cows. J. Dairy Sci. 94 (5), 2431–9.

Comtet-Marre, S., Parisot, N., Lepercq, P., et al., 2017. Metatranscriptomics reveals the active bacterial and eukaryotic fibrolytic communities in the rumen of dairy cow fed a mixed diet. Front. Microbiol.

8, 67.

Comtet-Marre, S., Chaucheyras-Durand, F., Bouzid, O., et al., 2018. FibroChip, a functional DNA microarray to monitor cellulolytic and hemicellulolytic activities of rumen microbiota. Front. Microbiol. 9, 215.

Delgado, B., Bach, A., Guasch, I., et al., 2019. Whole rumen metagenome sequencing allows classifying and predicting feed efficiency and intake levels in cattle. Sci. Rep. 9 (1), 11.

de Ondarza, M. B., Sniffen, C. J., Dussert, L., et al., 2010. Case Study: multiple-study analysis of the effect of live yeast on milk yield, milk component content and yield, and feed efficiency. Prof. Anim. Sci. 26 (6), 661-6.

Derakhshani, H., Tun, H. M., Cardoso, F. C., et al., 2016. Linking peripartal dynamics of ruminal microbiota to dietary changes and production parameters. Front. Microbiol. 7, 2143.

DeVries, T. J. and Chevaux, E. 2014. Modification of the feeding behavior of dairy cows through live yeast supplementation. J. Dairy Sci. 97 (10), 6499-510.

Dias, A. L. G., Freitas, J. A., Micai, B., et al., 2018. Effect of supplemental yeast culture and dietary starch content on rumen fermentation and digestion in dairy cows. J. Dairy Sci. 101 (1), 201-21.

Díaz, A. M., Almozni, B., Molina, M. A., et al., 2018. Potentiation of the humoral immune response elicited by a commercial vaccine against bovine respiratory disease by *Enterococcus faecalis* CECT7121. Benef Microbes 9 (4), 553-62.

Ding, G., Chang, Y., Zhao, L., et al., 2014. Effect of *Saccharomyces cerevisiae* on alfalfa nutrient degradation characteristics and rumen microbial populations of steers fed diets with different concentrate-to-forage ratios. J. Anim. Sci. Biotechnol. 5 (1), 24.

Du, R., Jiao, S., Dai, Y., et al., 2018. Probiotic *Bacillus amyloliquefaciens* C-1 improves growth performance, stimulates GH/IGF-1, and regulates the gut microbiota of growth-retarded beef calves. Front. Microbiol. 9, 2006.

Dunière, L., Gleizal, A., Chaucheyras-Durand, F., et al., 2011. Fate of *Escherichia coli* O26 in corn silage experimentally contaminated at ensiling, at silo opening, or after aerobic exposure, and protective effect of various bacterial inoculants. Appl. Environ. Microbiol. 77 (24), 8696-704.

Durand-Chaucheyras, F., Fonty, G., Bertin, G., et al., 1998. Fate of *Levucell* SC I-1077 yeast additive during digestive transit in lambs. Reprod. Nutr. Dev. 38 (3), 275-80.

EFSA, ECDC. 2018. The European Union summary report on trends and sources of zoonoses, zoonotic agents and food-borne outbreaks in 2017 [WWW Document]. European Food Safety Authority. Available at: https://www.efsa.europa.eu/fr/efsajournal/pub/5500 (accessed on 03 June 2019).

Ferraretto, L. F., Shaver, R. D. and Bertics, S. J. 2012. Effect of dietary supplementation with live-cell yeast at two dosages on lactation performance, ruminal fermentation, and total-tract nutrient digestibility in dairy cows. J. Dairy Sci. 95 (7), 4017-28.

Feye, K. M., Anderson, K. L., Scott, M. F., et al., 2016. Abrogation of *Salmonella* and *E. coli* O157: H7 in feedlot cattle fed a proprietary Saccharomyces cerevisiae fermentation prototype. J. Vet. Sci. Technol. 7 (4).

Fomenky, B. E., Chiquette, J., Bissonnette, N., et al., 2017. Impact of *Saccharomyces cerevisiae boulardii* CNCMI-1079 and *Lactobacillus acidophilus* BT1386 on total lactobacilli population in the gastrointestinal tract and colon histomorphology of Holstein dairy calves. Anim. Feed Sci. Technol. 234, 151-61.

Fomenky, B. E., Do, D. N., Talbot, G., et al., 2018. Direct-fed microbial supplementation influences the bacteria community composition of the gastrointestinal tract of pre-and post-weaned calves. Sci. Rep. 8 (1), 14147.

Fonty, G. and Chaucheyras-Durand, F. 2006. Effects and modes of action of live yeasts in the rumen. Biol. (Bratisl.) 61 (6), 741-50.

Fonty, G., Joblin, K., Chavarot, M., et al., 2007. Establishment and development of ruminal hydrogenotrophs in methanogen-free lambs. Appl. Environ. Microbiol. 73 (20), 6391-403.

Freestone, P. and Lyte, M. 2010. Stress and microbial endocrinology: prospects for ruminant nutrition. Animal 4 (7), 1248-57.

Fremaux, B., Raynaud, S., Beutin, L., et al., 2006. Dissemination and persistence of Shiga toxin-producing *Escherichia coli* (STEC) strains on French dairy farms. Vet. Microbiol. 117 (2-4), 180-91.

Fulton, R. W., Cook, B. J., Step, D. L., et al., 2002. Evaluation of health status of calves and the impact on feedlot performance: assessment of a retained ownership program for postweaning calves. Can. J. Vet. Res. 66 (3), 173-80.

Gaggìa, F., Mattarelli, P. and Biavati, B. 2010. Probiotics and prebiotics in animal feeding for safe food production. Int. J. Food Microbiol. 141 (Suppl. 1), S15-28.

Galvão, K. N., Santos, J. E. P., Coscioni, A., et al., 2005. Effect of feeding live yeast products to calves with failure of passive transfer on performance and patterns of antibiotic resistance in fecal *Escherichia coli*. Reprod. Nutr. Dev. 45 (4), 427-40.

Ganner, A. and Schatzmayr, G. 2012. Capability of yeast derivatives to adhere enteropathogenic bacteria and to modulate cells of the innate immune system. Appl. Microbiol. Biotechnol. 95 (2), 289-97.

Garcia Diaz, T., Ferriani Branco, A., Jacovaci, F. A., et al., 2018. Inclusion of live yeast and mannan-oligosaccharides in high grain-based diets for sheep: ruminal parameters, inflammatory response and rumen morphology. PLoS ONE 13 (2), e0193313.

Genís, S., Cerri, R. L. A., Bach, À., et al., 2018. Pre-calving intravaginal administration of lactic acid bacteria reduces metritis prevalence and regulates blood neutrophil gene expression after calving in dairy cattle. Front. Vet. Sci. 5, 135.

Gensollen, T., Iyer, S. S., Kasper, D. L., et al., 2016. How colonization by microbiota in early life shapes the immune system. Science 352 (6285), 539-44.

Golder, H. M., Celi, P., Rabiee, A. R., et al., 2014. Effects of feed additives on rumen and blood profiles during a starch and fructose challenge. J. Dairy Sci. 97 (2), 985-1004.

Goodridge, H. S., Wolf, A. J. and Underhill, D. M. 2009. Beta-glucan recognition by the innate immune system. Immunol. Rev. 230 (1), 38-50.

Griffin, D., Chengappa, M. M., Kuszak, J., et al., 2010. Bacterial pathogens of the bovine respiratory disease complex. Vet. Clin. North Am. Food Anim. Pract. 26 (2), 381-94.

Griffin, D., Chengappa, M. M., Kuszak, J., et al., 2010. Bacterial pathogens of the bovine respiratory disease complex. Vet. Clin. North Am. Food Anim. Pract. 26 (2), 381-94.

Guedes, C. M., Gonçalves, D., Rodrigues, M. A. M., et al., 2008. Effects of a Saccharomyces cerevisiae yeast on ruminal fermentation and fibre degradation of maize silages in cows. Anim. Feed Sci. Technol. 145 (1-4), 27-40.

Guzman, C. E., Bereza-Malcolm, L. T., De Groef, B., et al., 2015. Presence of selected methanogens, Fibrolytic bacteria, and proteobacteria in the gastrointestinal tract of neonatal dairy calves from birth to 72 hours. PLoS ONE 10 (7), e0133048.

Guzman, C. E., Bereza-Malcolm, L. T., De Groef, B., et al., 2016. Uptake of milk with and without solid feed during the monogastric phase: effect on fibrolytic and methanogenic microorganisms in the gastrointestinal tract of calves. Anim. Sci. J. 87 (3), 378-88.

Hare, K. S., Leal, L. N., Romao, J. M., et al., 2019. Preweaning nutrient supply alters mammary gland transcriptome expression relating to morphology, lipid accumulation, DNA synthesis, and RNA expression in Holstein heifer calves. J. Dairy Sci. 102 (3), 2618-30.

He, Z. X., Ferlisi, B., Eckert, E., et al., 2017. Supplementing a yeast probiotic to pre-weaning Holstein calves: feed intake, growth and fecal biomarkers of gut health. Anim. Feed Sci. Technol. 226, 81-7.

Hogeveen, H. and Østerås, O. 2005. Mastitis management in an economic framework. In: Mastitis in Dairy Production. Wageningen Academic Publishers, Maastrich, Netherlands.

Humer, E., Petri, R. M., Aschenbach, J. R., et al., 2018. Invited review: practical feeding management recommendations to mitigate the risk of subacute ruminal acidosis in dairy cattle. J. Dairy Sci. 101 (2), 872-88.

Hutcheson, D. P. and Cole, N. A. 1986. Management of transit-stress syndrome in cattle: nutritional and environmental effects. J. Anim. Sci. 62 (2), 555-60.

Ingvartsen, K. L., Dewhurst, R. J. and Friggens, N. C. 2003. On the relationship between lactational performance and health: is it yield or metabolic imbalance that cause production diseases in dairy cattle? A position paper. Livest. Prod. Sci. 83 (2-3), 277-308.

Ipharraguerre, I. R., Pastor, J. J., Gavaldà-Navarro, A., et al., 2018. Antimicrobial promotion of pig growth is associated with tissue-specific remodeling of bile acid signature and signaling. Sci. Rep. 8 (1), 13671.

Ishaq, S. L., AlZahal, O., Walker, N., et al., 2017. An investigation into rumen fungal and protozoal diversity in three rumen fractions, during high-fiber or grain-induced sub-acute ruminal acidosis conditions, with or without active dry yeast supplementation. Front. Microbiol. 8, 1943.

Jami, E., Israel, A., Kotser, A., et al., 2013. Exploring the bovine rumen bacterial community from birth to adulthood. ISME J. 7 (6), 1069-79.

Jami, E., White, B. A. and Mizrahi, I. 2014. Potential role of the bovine rumen microbiome in modulating milk composition and feed efficiency. PLoS ONE 9 (1), e85423.

Jeyanathan, J., Martin, C. and Morgavi, D. P. 2014. The use of direct-fed microbials for mitigation of ruminant methane emissions: a review. Animal 8 (2), 250-61.

Jeyanathan, J., Martin, C. and Morgavi, D. P. 2016. Screening of bacterial direct-fed microbials for their antimethanogenic potential *in vitro* and assessment of their effect on ruminal fermentation and microbial profiles in sheep. J. Anim. Sci. 94 (2), 739-50.

Jiang, Y., Ogunade, I. M., Qi, S., et al., 2017. Effects of the dose and viability of *Saccharomyces cerevisiae*. 1. Diversity of ruminal microbes as analyzed by Illumina MiSeq sequencing and quantitative PCR. J. Dairy Sci. 100 (1), 325-42.

Jiao, J., Li, X., Beauchemin, K. A., et al., 2015. Rumen development process in goats as affected by supplemental feeding v. grazing: age-related anatomic development, functional achievement and microbial colonisation. Br. J. Nutr. 113 (6), 888-900.

Johnsen, J. F., Viljugrein, H., Bøe, K. E., et al., 2019. A cross-sectional study of suckling calves' passive immunity and associations with management routines to ensure colostrum intake on organic dairy farms. Acta Vet. Scand. 61 (1), 7.

Jouany, J. P. 2006. Optimizing rumen functions in the close-up transition period and early lactation to drive dry matter intake and energy balance in cows. Anim. Reprod. Sci. 96 (3-4), 250-64.

Jouany, J. P. and Morgavi, D. P. 2007. Use of 'natural' products as alternatives to antibiotic feed additives in ruminant production. Animal 1 (10), 1443-66.

Kassé, F. N., Fairbrother, J. M. and Dubuc, J. 2016. Relationship between *Escherichia coli* virulence factors and postpartum metritis in dairy cows. J. Dairy Sci. 99 (6), 4656-67.

Kayser, W. C., Carstens, G. E., Washburn, K. E., et al., 2019. Effects of combined viral-bacterial challenge with or without supplementation of *Saccharomyces cerevisiae boulardii* strain CNCM I-1079 on immune upregulation and DMI in beef heifers. J. Anim. Sci. 97 (3), 1171-84.

Khafipour, E., Krause, D. O. and Plaizier, J. C. 2009. A grain-based subacute ruminal acidosis challenge causes translocation of lipopolysaccharide and triggers inflammation. J. Dairy Sci. 92 (3), 1060-70.

Khan, M. A., Bach, A., Weary, D. M., et al., 2016. Invited review: transitioning from milk to solid feed in dairy heifers. J. Dairy Sci. 99 (2), 885-902.

Khiaosa-Ard, R. and Zebeli, Q. 2018. Diet-induced inflammation: from gut to metabolic organs and the consequences for the health and longevity of ruminants. Res. Vet. Sci. 120, 17-27.

Kleen, J. L., Hooijer, G. A., Rehage, J., et al., 2003. Subacute ruminal acidosis (SARA): a review. J. Vet. Med. A 50 (8), 406-14.

Klostermann, K., Crispie, F., Flynn, J., et al., 2008. Intramammary infusion of a live culture of *Lactococcus lactis* for treatment of bovine mastitis: comparison with antibiotic treatment in field trials. J. Dairy Res. 75 (3), 365-73.

Kumprechtová, D., Illek, J., Julien, C., et al., 2019. Effect of live yeast (*Saccharomyces cerevisiae*) supplementation on rumen fermentation and metabolic profile of dairy cows in early lactation. J. Anim. Physiol. Anim. Nutr. 103 (2), 447-55.

La Fata, G., Weber, P. and Mohajeri, M. H. 2018. Probiotics and the gut immune system: indirect regulation. Probiotics Antimicrob. Proteins 10 (1), 11-21.

Langford, F. M., Weary, D. M. and Fisher, L. 2003. Antibiotic resistance in gut bacteria from dairy calves: a dose response to the level of antibiotics fed in milk. J. Dairy Sci. 86 (12), 3963-6.

Lebeer, S., Bron, P. A., Marco, M. L., et al., 2018. Identification of probiotic effector molecules: present state and future perspectives. Curr. Opin. Biotechnol. 49, 217-23.

Lema, M., Williams, L. and Rao, D. R. 2001. Reduction of fecal shedding of enterohemorrhagic *Escherichia coli* O157:H7 in lambs by feeding microbial feed supplement. Small Rumin. Res. 39 (1), 31-9.

Lettat, A., Nozière, P., Silberberg, M., et al., 2010. Experimental feed induction of ruminal lactic, propionic, or butyric acidosis in sheep. J. Anim. Sci. 88 (9), 3041-46.

Lettat, A., Nozière, P., Silberberg, M., et al., 2012. Rumen microbial and fermentation characteristics are affected differently by bacterial probiotic supplementation during induced lactic and subacute acidosis in sheep. BMC Microbiol. 12, 142.

Li, Z., You, Q., Ossa, F., et al., 2016. Assessment of yeast *Saccharomyces cerevisiae* component binding to *Mycobacterium avium* subspecies paratuberculosis using bovine epithelial cells. BMC Vet. Res. 12, 42.

Li, Z., Kang, H., You, Q., et al., 2018. *In vitro* bioassessment of the immunomodulatory activity of *Saccharomyces cerevisiae* components using bovine macrophages and *Mycobacterium avium* ssp. paratuberculosis. J. Dairy Sci. 101 (7), 6271-86.

Lyte, M. 2011. Probiotics function mechanistically as delivery vehicles for neuroactive compounds: microbial endocrinology in the design and use of probiotics. Bioessays 33 (8), 574-81.

Ma, T., Suzuki, Y. and Guan, L. L. 2018. Dissect the mode of action of probiotics in affecting host-microbial interactions and immunity in food producing animals. Vet. Immunol. Immunopathol. 205, 35-48.

Malmuthuge, N., Li, M., Fries, P., et al., 2012. Regional and age dependent changes in gene expression of toll-like receptors and key antimicrobial defence molecules throughout the gastrointestinal tract of dairy calves. Vet. Immunol. Immunopathol. 146 (1), 18-26.

Malmuthuge, N., Chen, Y., Liang, G., et al., 2015. Heat-treated colostrum feeding promotes beneficial bacteria colonization in the small intestine of neonatal calves. J. Dairy Sci. 98 (11), 8044-53.

Marden, J. P., Julien, C., Monteils, V., et al., 2008. How does live yeast differ from sodium bicarbonate to stabilize ruminal pH in high-yielding dairy cows? J. Dairy Sci. 91 (9), 3528-35.

McFarland, L. V. 2010. Systematic review and meta-analysis of *Saccharomyces boulardii* in adult patients. World J. Gastroenterol. 16 (18), 2202-22.

McGovern, E., Kenny, D. A., McCabe, M. S., et al., 2018. 16S rRNA sequencing reveals relationship Between potent cellulolytic genera and feed efficiency in the rumen of bulls. Front. Microbiol. 9, 1842.

McGuirk, S. M. 2008. Disease management of dairy calves and heifers. Vet. Clin. North Am. Food Anim. Pract. 24, 139-53.

McGuirk, S. M. and Collins, M. 2004. Managing the production, storage, and delivery of colostrum. Vet. Clin. North Am. Food Anim. Pract. 20, 593-603.

McSweeney, C. S., Dulieu, A., Katayama, Y., et al., 1994. Solubilization of lignin by the ruminal anaerobic fungus *Neocallimastix patriciarum*. Appl. Environ. Microbiol. 60 (8), 2985-89.

McSweeney, C. S., Dulieu, A. and Bunch, R. 1998. *Butyrivibrio* spp. and other xylanolytic microorganisms from the rumen have Cinnamoyl esterase activity. Anaerobe 4 (1), 57-65.

Meale, S. J., Li, S., Azevedo, P., et al., 2016. Development of ruminal and fecal microbiomes are affected by weaning but not weaning strategy in dairy calves. Front. Microbiol. 7, 582.

Meale, S. J., Chaucheyras-Durand, F., Berends, H., et al., 2017. From pre-to postweaning: transformation of the young calf's gastrointestinal tract. J. Dairy Sci. 100 (7), 5984-95.

Minuti, A., Palladino, A., Khan, M. J., et al., 2015. Abundance of ruminal bacteria, epithelial gene expression, and systemic biomarkers of metabolism and inflammation are altered during the peripartal period in dairy cows. J. Dairy Sci. 98 (12), 8940-51.

Mizrahi, I. and Jami, E. 2018. Review: The compositional variation of the rumen microbiome and its effect on host performance and methane emission. Animal 12 (s2), s220-32.

Mokhber-Dezfouli, M. R., Tajik, P., Bolourchi, M., et al., 2007. Effects of probiotics supplementation in daily milk intake of newborn calves on body weight gain, body height, diarrhea occurrence and health condition. Pak. J. Biol. Sci. 10 (18), 3136-40.

Morelli, L. and Capurso, L. 2012. FAO/WHO guidelines on probiotics: 10 years later. J. Clin. Gastroenterol. 46 (Suppl.), S1-2.

Mosoni, P., Chaucheyras-Durand, F., Béra-Maillet, C., et al., 2007. Quantification by real-time PCR of cellulolytic bacteria in the rumen of sheep after supplementation of a forage diet with readily fermentable carbohydrates: effect of a yeast additive. J. Appl. Microbiol. 103 (6), 2676-85.

Mõtus, K., Viltrop, A. and Emanuelson, U. 2018. Reasons and risk factors for beef calf and youngstock on-farm mortality in extensive cow-calf herds. Animal 12 (9), 1958-66.

Muya, M. C., Nherera, F. V., Miller, K. A., et al., 2015. Effect of *Megasphaera elsdenii* NCIMB 41125 dosing on rumen development, volatile fatty acid production and blood β-hydroxybutyrate in neonatal dairy calves. J. Anim. Physiol. Anim. Nutr. 99 (5), 913-8.

Nasiri, A. H., Towhidi, A., Shakeri, M., et al., 2018. Effects of live yeast dietary supplementation on hormonal profile, ovarian follicular dynamics, and reproductive performance in dairy cows exposed to high ambient temperature. Theriogenology 122, 41-6.

Nollet, L., Demeyer, D. and Verstraete, W. 1997. Effect of 2-bromoethanesulfonic acid and Peptostreptococcus productus ATCC 35244 addition on stimulation of reductive acetogenesis in the ruminal ecosystem by selective inhibition of methanogenesis. Appl. Environ. Microbiol. 63 (1), 194-200.

Owens, F. N., Secrist, D. S., Hill, W. J., et al., 1998. Acidosis in cattle: a review. J. Anim. Sci. 76 (1), 275-86.

Pellegrino, M., Berardo, N., Giraudo, J., et al., 2017. Bovine mastitis prevention: humoral and cellular response of dairy cows inoculated with lactic acid bacteria at the dry-off period. Benef. Microbes 8 (4), 589-96.

Petri, R. M., Schwaiger, T., Penner, G. B., et al., 2013a. Changes in the rumen epimural bacterial diversity of beef cattle as affected by diet and induced ruminal acidosis. Appl. Env. Microbiol. 79, 3744-55.

Petri, R. M., Schwaiger, T., Penner, G. B., et al., 2013b. Characterization of the core rumen microbiome in cattle during transition from forage to concentrate as well as during and after an acidotic challenge. PLoS ONE 8 (12), e83424.

Petri, R. M., Kleefisch, M. T., Metzler-Zebeli, B. U., et al., 2018. Changes in the rumen epithelial microbiota of cattle and host gene expression in response to alterations in dietary carbohydrate composition. Appl. Environ. Microbiol. 84 (12).

Philippeau, C., Lettat, A., Martin, C., et al., 2017. Effects of bacterial direct-fed microbials on ruminal characteristics, methane emission, and milk fatty acid composition in cows fed high- or low-starch diets. J. Dairy Sci. 100 (4), 2637-50.

Pinloche, E., McEwan, N., Marden, J. P., et al., 2013. The effects of a probiotic yeast on the bacterial diversity and population structure in the rumen of cattle. PLoS ONE 8 (7), e67824.

Pitta, D. W., Indugu, N., Baker, L., et al., 2018. Symposium review: understanding diet-microbe interactions to enhance productivity of dairy cows. J. Dairy Sci. 101 (8), 7661-79.

Plaizier, J. C., Khafipour, E., Li, S., et al., 2012. Subacute ruminal acidosis (SARA), endotoxins and health consequences. Anim. Feed Sci. Technol. 172, 9-21.

Plaizier, J. C., Danesh Mesgaran, M., Derakhshani, H., et al., 2018. Review: enhancing gastrointestinal health in dairy cows. Animal 12 (s2), s399-418.

Pothoulakis, C. 2009. Review article: anti-inflammatory mechanisms of action of *Saccharomyces boulardii*. Aliment. Pharmacol. Ther. 30 (8), 826-33, Article: Anti-Inflammatory.

Priestley, D., Bittar, J. H., Ibarbia, L., et al., 2013. Effect of feeding maternal colostrum or plasma-derived or colostrum-derived colostrum replacer on passive transfer of immunity, health, and performance of preweaning heifer calves. J. Dairy Sci. 96 (5), 3247-56.

Raabis, S., Li, W. and Cersosimo, L. 2019. Effects and immune responses of probiotic treatment in ruminants. Vet. Immunol. Immunopathol. 208, 58-66.

Rainard, P. and Foucras, G. 2018. A critical appraisal of probiotics for mastitis control. Front. Vet. Sci.

5, 251.

Ramatla, T., Ngoma, L., Adetunji, M., et al., 2017. Evaluation of antibiotic residues in raw meat using different analytical methods. Antibiot. Basel Switz 6 (4).

Reis, L. F., Sousa, R. S., Oliveira, F. L. C., et al., 2018. Comparative assessment of probiotics and monensin in the prophylaxis of acute ruminal lactic acidosis in sheep. BMC Vet. Res. 14 (1), 9.

Renaud, D. L., Kelton, D. F., Weese, J. S., et al., 2019. Evaluation of a multispecies probiotic as a supportive treatment for diarrhea in dairy calves: a randomized clinical trial. J. Dairy Sci. 102 (5), 4498-505.

Rey, M., Enjalbert, F. and Monteils, V. 2012. Establishment of ruminal enzyme activities and fermentation capacity in dairy calves from birth through weaning. J. Dairy Sci. 95 (3), 1500-12.

Rey, M., Enjalbert, F., Combes, S., et al., 2014. Establishment of ruminal bacterial community in dairy calves from birth to weaning is sequential. J. Appl. Microbiol. 116 (2), 245-57.

Reynolds, J. D. and Morris, B. 1984. The effect of antigen on the development of Peyer's patches in sheep. Eur. J. Immunol. 14 (1), 1-6.

Riaz, Q. U. A. and Masud, T. 2013. Recent trends and applications of encapsulating materials for probiotic stability. Crit. Rev. Food Sci. Nutr. 53 (3), 231-44.

Rigobelo, E. E. C., Karapetkov, N., Maestá, S. A., et al., 2015. Use of probiotics to reduce faecal shedding of Shiga toxin-producing *Escherichia coli* in sheep. Benef. Microbes 6 (1), 53-60.

Roth, B. A., Keil, N. M., Gygax, L., et al., 2009. Influence of weaning method on health status and rumen development in dairy calves. J. Dairy Sci. 92 (2), 645-56.

Russell, J. B. and Wilson, D. B. 1996. Why are ruminal cellulolytic bacteria unable to digest cellulose at low pH? J. Dairy Sci. 79 (8), 1503-9.

Sadet, S., Martin, C., Meunier, B., et al., 2007. PCR-DGGE analysis reveals a distinct diversity in the bacterial population attached to the rumen epithelium. Animal 1 (7), 939-44.

Salvati, G. G. S., Morais Junior, N. N., Melo, A. C. S., et al., 2015. Response of lactating cows to live yeast supplementation during summer. J. Dairy Sci. 98 (6), 4062-73.

Schiavone, M., Sieczkowski, N., Castex, M., et al., 2015. Effects of the strain background and autolysis process on the composition and biophysical properties of the cell wall from two different industrial yeasts. FEMS Yeast Res. 15 (2).

Schiavone, M., Déjean, S., Sieczkowski, N., et al., 2017. Integration of biochemical, biophysical and transcriptomics data for investigating the structural and nanomechanical properties of the yeast cell wall. Front. Microbiol. 8, 1806.

Schofield, B. J., Lachner, N., Le, O. T., et al., 2018. Beneficial changes in rumen bacterial community profile in sheep and dairy calves as a result of feeding the probiotic *Bacillus amyloliquefaciens* H57. J. Appl. Microbiol. 124 (3), 855-66.

Segura, A., Bertoni, M., Auffret, P., et al., 2018. Transcriptomic analysis reveals specific metabolic pathways of enterohemorrhagic *Escherichia coli* O157:H7 in bovine digestive contents. BMC Genomics 19 (1), 766.

Seymour, E. H., Jones, G. M. and McGilliard, M. L. 1988. Persistence of residues in milk following antibiotic treatment of dairy cattle. J. Dairy Sci. 71 (8), 2292-6.

Shabat, S. K. B., Sasson, G., Doron-Faigenboim, A., et al., 2016. Specific microbiome-dependent mechanisms underlie the energy harvest efficiency of ruminants. ISME J. 10 (12), 2958-72.

Signorini, M. L., Soto, L. P., Zbrun, M. V., et al., 2012. Impact of probiotic administration on the health and fecal microbiota of young calves: A meta-analysis of randomized controlled trials of lactic acid bacteria. Res. Vet. Sci. 93 (1), 250-8.

Silanikove, N. 2000. Effects of heat stress on the welfare of extensively managed domestic ruminants. Livest. Prod. Sci. 67 (1-2), 1-18.

Silberberg, M., Chaucheyras-Durand, F., Commun, L., et al., 2013. Repeated acidosis challenges and live yeast supplementation shape rumen microbiota and fermentations and modulate inflammatory status in sheep. Animal 7 (12), 1910-20.

Soberon, F. and Van Amburgh, M. E. 2013. Lactation Biology Symposium: the effect of nutrient intake from milk or milk replacer of preweaned dairy calves on lactation milk yield as adults: a meta-analysis of current data. J. Anim. Sci. 91 (2), 706-12.

Soto, L. P., Frizzo, L. S., Signorini, M. L., et al., 2015. Faecal culturable microbiota, growth and clinical parameters of calves supplemented with lactic acid bacteria and lactose prior and during experimental infection with Salmonella Dublin DSPV 595T. Arch. Med. Vet. 47 (2), 237-44.

Soto, L. P., Astesana, D. M., Zbrun, M. V., et al., 2016. Probiotic effect on calves infected with Salmonella Dublin: haematological parameters and serum biochemical profile. Benef. Microbes 7 (1), 23-33.

Sousa, D. O., Oliveira, C. A., Velasquez, A. V., et al., 2018. Live yeast supplementation improves rumen fibre degradation in cattle grazing tropical pastures throughout the year. Anim. Feed Sci. Technol. 236, 149-58.

Souza, R. F. S., Rault, L., Seyffert, N., et al., 2018. *Lactobacillus casei* BL23 modulates the innate immune response in *Staphylococcus aureus*-stimulated bovine mammary epithelial cells. Benef. Microbes 9 (6), 985-95.

Steele, M. A., Kroom, J., Kahler, M., et al., 2011. Bovine rumen epithelium undergoes rapid structural adaptations during grain-induced subacute ruminal acidosis. Am. J. Physiol. Regul. Integr. Comp. Physiol. 300 (6), R1515-23.

Steele, M. A., Penner, G. B., Chaucheyras-Durand, F., et al., 2016. Development and physiology of the rumen and the lower gut: targets for improving gut health. J. Dairy Sci. 99 (6), 4955-66.

Stelwagen, K., Carpenter, E., Haigh, B., et al., 2009. Immune components of bovine colostrum and milk. J. Anim. Sci. 87 (13 Suppl.), 3-9.

Stier, H. and Bischoff, S. C. 2016. Influence of *Saccharomyces boulardii* CNCM I-745 on the gut-associated immune system. Clin. Expl. Gastroenterol. 9, 269-79.

Sullivan, M. L. and Bradford, B. J. 2011. Viable cell yield from active dry yeast products and effects of storage temperature and diluent on yeast cell viability. J. Dairy Sci. 94 (1), 526-31.

Sun, X., Yuan, X., Chen, L., et al., 2017. Histamine induces bovine rumen epithelial cell inflammatory response via NF-κB pathway. Cell. Physiol. Biochem. 42 (3), 1109-19.

Tabe, E. S., Oloya, J., Doetkott, D. K., et al., 2008. Comparative effect of direct-fed microbials on fecal shedding of *Escherichia coli* O157:H7 and Salmonella in naturally infected feedlot cattle. J. Food Prot. 71 (3), 539-44.

Terré, M., Maynou, G., Bach, A., et al., 2015. Effect of *Saccharomyces cerevisiae* CNCM I-1077 supplementation on performance and rumen microbiota of dairy calves. Prof. Anim. Sci. 31 (2), 153-8.

Tkalcic, S., Zhao, T., Harmon, B. G., et al., 2003. Fecal shedding of enterohemorrhagic *Escherichia* coli in weaned calves following treatment with probiotic *Escherichia coli*. J. Food Prot. 66 (7), 1184-9.

Trevisi, E. and Minuti, A. 2018. Assessment of the innate immune response in the periparturient cow. Res. Vet. Sci. 116, 47-54.

Trevisi, E., Riva, F., Filipe, J. F. S., et al., 2018. Innate immune responses to metabolic stress can be detected in rumen fluids. Res. Vet. Sci. 117, 65-73.

USDA. 2014a. Dairy 2014: health and management practices on U. S dairy operations.

USDA 2014b. Dairy 2014: dairy cattle management practices in the United States.

Uyeno, Y., Sekiguchi, Y., Tajima, K., et al., 2010. An rRNA-based analysis for evaluating the effect of heat stress on the rumen microbial composition of Holstein heifers. Anaerobe 16 (1), 27-33.

Uyeno, Y., Akiyama, K., Hasunuma, T., et al., 2017. Effects of supplementing an active dry yeast product on rumen microbial community composition and on subsequent rumen fermentation of lactating cows in the mid-to-late lactation period. Anim. Sci. J. 88 (1), 119-24.

Villageliú, D. and Lyte, M. 2018. Dopamine production in Enterococcus faecium: a microbial endocrinology-based mechanism for the selection of probiotics based on neurochemical-producing potential. PLoS ONE 13 (11), e0207038.

Villena, J., Aso, H., Rutten, V. P. M. G., et al., 2018. Immunobiotics for the bovine host: their interaction with intestinal epithelial cells and their effect on antiviral immunity. Front. Immunol. 9, 326.

Villot, C., Meunier, B., Bodin, J., et al., 2018. Relative reticulo-rumen pH indicators for subacute ruminal acidosis detection in dairy cows. Animal 12 (3), 481-90.

Vipham, J. L., Lonergan, G. H., Guillen, L. M., et al., 2015. Reduced burden of *Salmonella enterica* in bovine subiliac lymph nodes associated with administration of a direct-fed microbial. Zoonoses Public Health 62 (8), 599-608.

von Buenau, R., Jaekel, L., Schubotz, E., et al., 2005. *Escherichia coli* strain Nissle 1917: significant reduction of neonatal calf diarrhea. J. Dairy Sci. 88 (1), 317-23.

Vyas, D., McGeough, E. J., Mohammed, R., et al., 2014. Effects of *Propionibacterium* strains on ruminal fermentation, nutrient digestibility and methane emissions in beef cattle fed a corn grain finishing diet. Animal 8 (11), 1807-15.

Wallis, J. K., Krömker, V. and Paduch, J. H. 2018. Biofilm formation and adhesion to bovine udder epithelium of potentially probiotic lactic acid bacteria. AIMS Microbiol. 4 (2), 209-24.

Wallis, J. K., Krömker, V. and Paduch, J. H. 2019. Biofilm challenge: lactic acid bacteria isolated from bovine udders versus staphylococci. Foods 8 (2).

Wang, D. S., Zhang, R. Y., Zhu, W. Y., et al., 2013. Effects of subacute ruminal acidosis challenges on fermentation and biogenic amines in the rumen of dairy cows. Livest. Sci. 155 (2-3), 262-72.

Wehnes, C. A., Novak, K. N., Patskevich, V., et al., 2009. Benefits of supplementation of an electrolyte scour treatment with a Bacillus-based direct fed microbial for calves. Probiotics Antimicrob. Proteins 1 (1), 36-44.

Wisener, L. V., Sargeant, J. M., O'Connor, A. M., et al., 2015. The use of direct-fed microbials to reduce shedding of *Escherichia coli* O157 in beef cattle: a systematic review and meta-analysis. Zoonoses Public Health 62 (2), 75-89.

Yáñez-Ruiz, D. R., Abecia, L. and Newbold, C. J. 2015. Manipulating rumen microbiome and fermentation through interventions during early life: a review. Front. Microbiol. 6, 1133.

Yeoman, C. J., Ishaq, S. L., Bichi, E., et al., 2018. Biogeographical differences in the influence of maternal microbial sources on the early successional development of the bovine neonatal gastrointestinal tract. Sci. Rep. 8 (1), 3197.

Yohe, T. T., Enger, B. D., Wang, L., et al., 2018. Short communication: does early-life administration of a Megasphaera elsdenii probiotic affect long-term establishment of the organism in the rumen and alter rumen metabolism in the dairy calf? J. Dairy Sci. 101 (2), 1747-51.

Yuan, K., Mendonça, L. G. D., Hulbert, L. E., et al., 2015. Yeast product supplementation modulated humoral and mucosal immunity and uterine inflammatory signals in transition dairy cows. J. Dairy Sci. 98 (5), 3236-46.

Zebeli, Q., Terrill, S. J., Mazzolari, A., et al., 2012. Intraruminal administration of *Megasphaera elsdenii* modulated rumen fermentation profile in mid-lactation dairy cows. J. Dairy Res. 79 (1), 16-25.

Zebeli, Q., Ghareeb, K., Humer, E., et al., 2015. Nutrition, rumen health and inflammation in the transition period and their role on overall health and fertility in dairy cows. Res. Vet. Sci. 103, 126-36.

Zhao, T., Tkalcic, S., Doyle, M. P., et al., 2003. Pathogenicity of enterohemorrhagic *Escherichia coli* in neonatal calves and evaluation of fecal shedding by treatment with probiotic Escherichia coli. J. Food Prot. 66 (6), 924-30.

Zhu, Z., Noel, S. J., Difford, G. F., et al., 2017. Community structure of the metabolically active rumen bacterial and archaeal communities of dairy cows over the transition period. PLoS ONE 12 (11), e0187858.

Zhu, Z., Kristensen, L., Difford, G. F., et al., 2018. Changes in rumen bacterial and archaeal communities over the transition period in primiparous Holstein dairy cows. J. Dairy Sci. 101 (11), 9847-62.

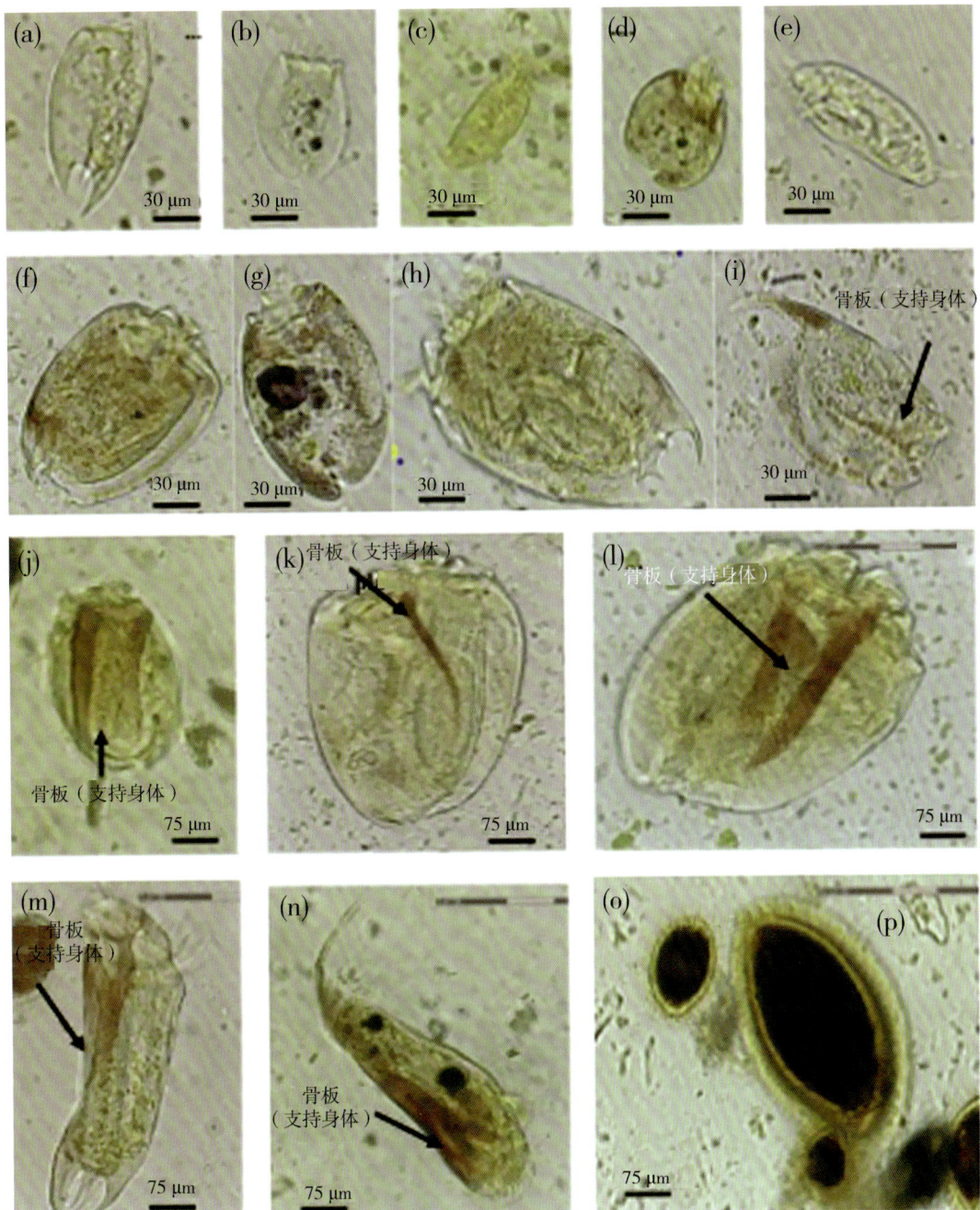

彩图1 鲁哥氏碘液染色瘤胃纤毛虫的光学显微镜照片（a）~（e）是内毛属（*Entodinium* spp.），（f）~（h）是双毛属（*Diplodium* spp.），（i）单甲属（*Eremoplastron* spp.），（j）硬甲属（*Ostracodinium* spp.），（k）真双毛属（*Endiplodinium* spp.），（l）后毛属（*Metadinium* spp.），（m）和（n）前毛属（*Epidimiu* spp.），（o）厚毛虫属（*Dasytricha* spp.），（p）等毛虫属（*Isotricha* spp.）。图像均来自 Williams（2018），图中原虫均属于 B 型种群。

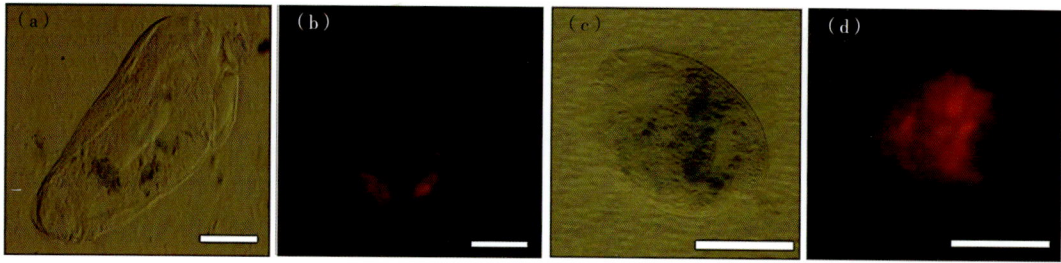

彩图 2　前毛虫和内毛虫的纤维成像（a）前毛虫光学显微镜图像。（比例尺 10 μm），内含胞内叶绿体，分离自饲喂草 2 h 后的肉牛瘤胃；（b）前毛虫荧光图像（比例尺 10 μm），内含胞内叶绿体，分离自饲喂草 2 h 后的肉牛瘤胃；（c）内毛虫光学显微镜图像（比例尺 10 μm），内含胞内叶绿体，分离自饲喂草 2 h 后的肉牛瘤胃（d）内毛虫光学显微镜图像（比例尺 10 μm），内含胞内叶绿体，分离自饲喂草 2 h 后的肉牛瘤胃。

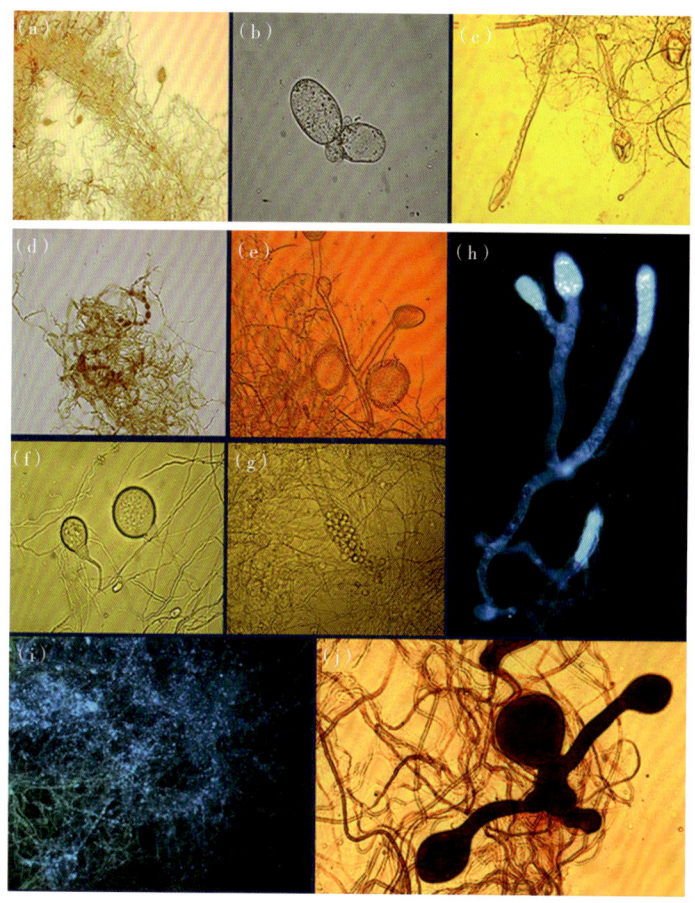

彩图 3　光镜下厌氧真菌的特征。（a）*Anaeromyces*：位于不分支孢子囊柄的孢子囊具尖形（短尖形）的顶端；（b）*Caecomyces*：球根状的假根；（c）*Piromyces*：长的、无分支的孢子囊柄；（d）*Anaeromyces*：具有大量缢缩的菌丝，呈香肠状或串珠状外观；（e）*Piromyces*：分叉的孢子囊柄；（f）*Piromyces*：在孢子囊下方的孢子囊柄形成一个卵杯状膨胀；（g）*Piromyces*：游动孢子从孢子囊释放；（h）*Piromyces*：分支的孢子囊柄，细胞核集中在顶端（DAPI 染色）；（i）*Orpinomyces*：普遍分支的菌丝上存在密集的多核根状菌丝体（DAPI 染色）；（j）*Piromyces*：孢子囊形状不规则（卢戈氏染色）。

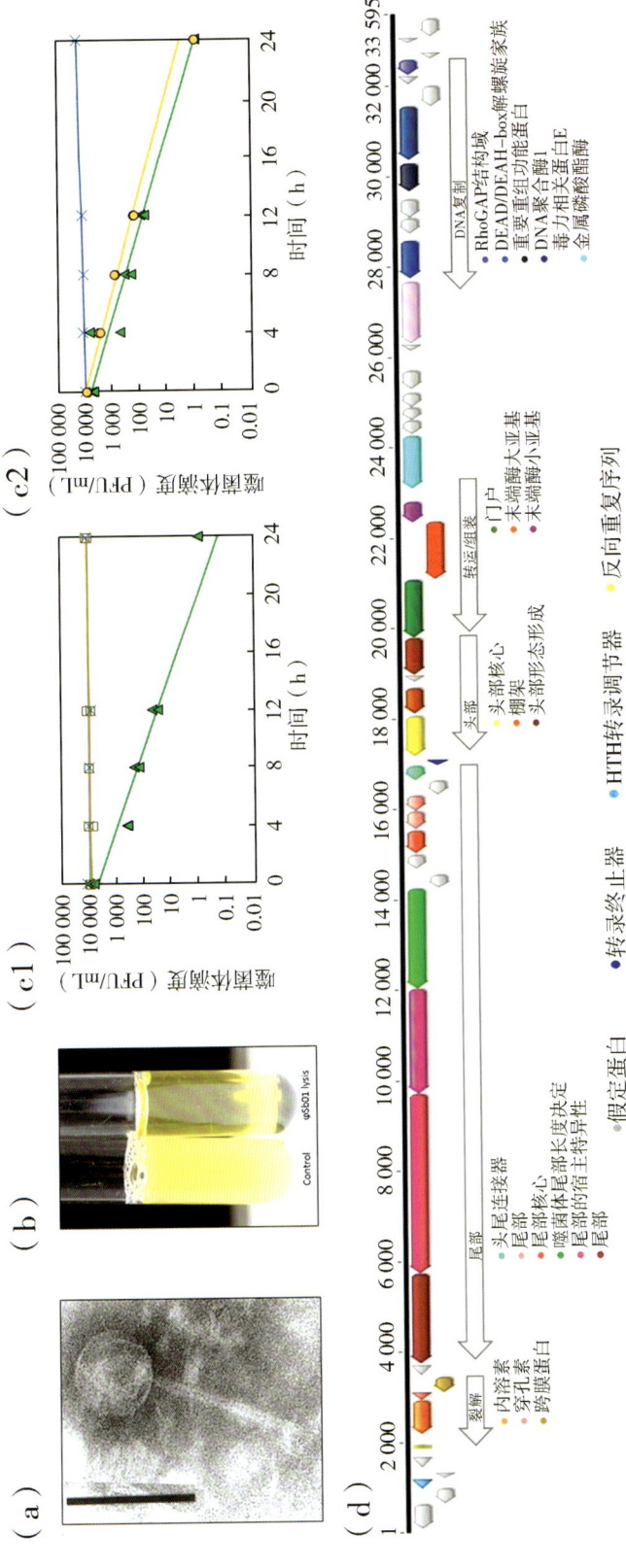

彩图 4 基于培养和遗传学技术解析瘤胃球菌马兽肠球菌噬菌体 φSb01 感染瘤胃细菌马粪肠球菌 2B 的特征。（a）透射电子显微镜观测 φSb01 的典型长尾噬菌体形态（比例尺 100nm）；（b）马粪肠球菌 2B 厌氧条件下纯培养生长 6h 后（对照）和感染 φSb01 后 3h 内（φSb01 裂解）；（c）φSb01 颗粒的存活率测定，用蚀斑试验测定剩余的活体噬菌体颗粒的滴度 [每毫升空斑形成单位（PFU）及其回归曲线]。（c1）分别在绵羊瘤胃液（▲），热处理（高压灭菌 107℃，103KPa，45min）瘤胃液对照培养基（×）和无瘤胃液对照培养基（×）的复制管中培养孵育；（c2）分别在绵羊瘤胃液（▲），澄清，0.22μm 过滤（低蛋白残留过滤器 HV, Millipore）的瘤胃液（●）和无瘤胃液的对照培养基（×）；（d）φSb01 的全基因组序列显示了模块化的基因排列和相关功能基因的聚类。改编自 Gilbert 等（2017）。

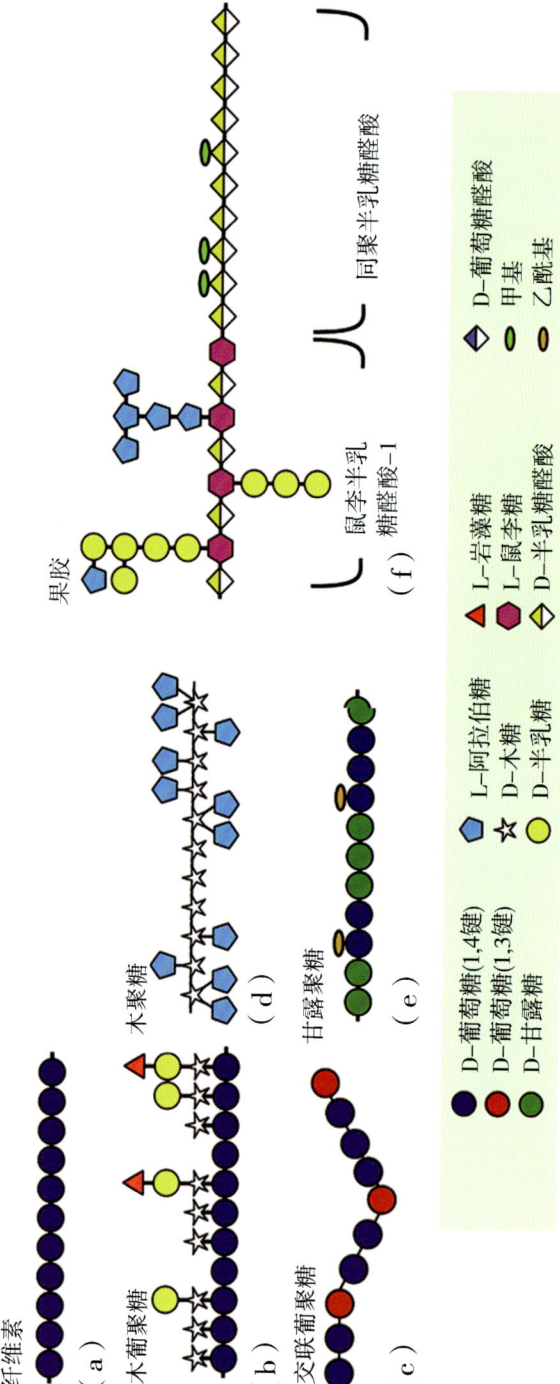

彩图 5 植物多糖结构和组成的类型。包括（a）纤维素、（b-e）半纤维素和（f）果胶。具体底物包括（a）纤维素、（b）木葡聚糖、（c）交联葡聚糖、（d）木聚糖、（e）甘露聚糖、（f）果胶（包括鼠李糖醛酸）。

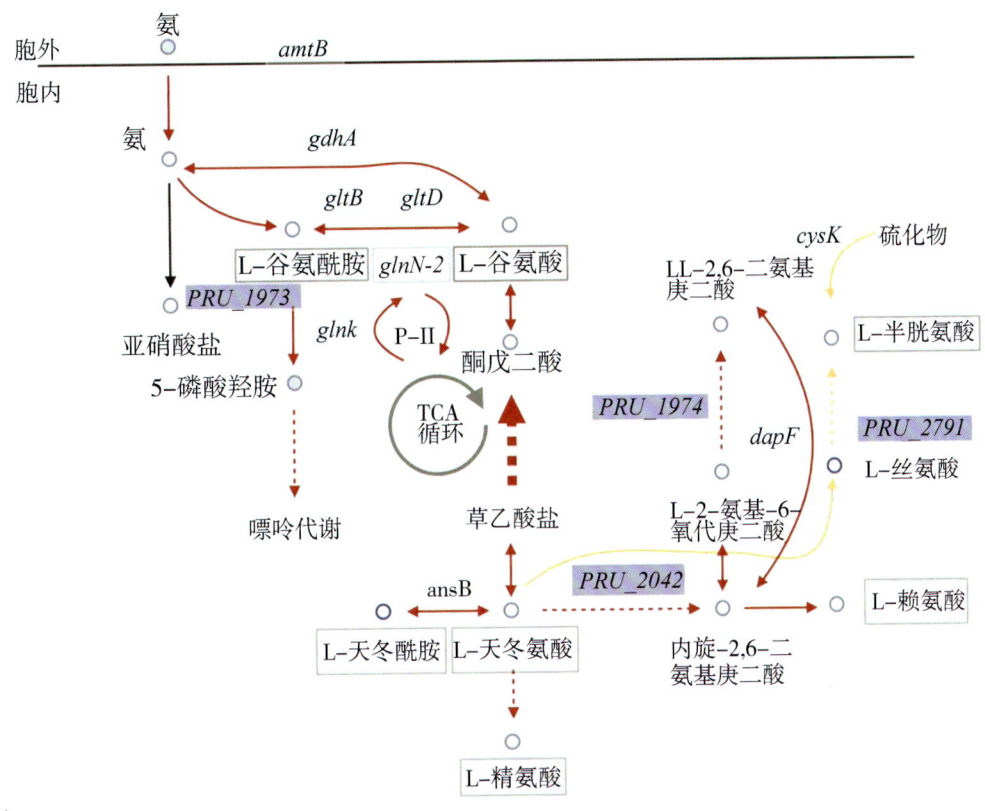

关键基因/位点	编码蛋白	途径
amtB	铵转运蛋白	
glnK	氮调控蛋白P-Ⅱ	
gdhA	谷氨酸脱氢酶（NADP-依赖型）	谷氨酸、谷氨酰胺代谢
glnN-2	谷氨酰胺合成酶Ⅲ-2型	谷氨酸、谷氨酰胺代谢
gltB	谷氨酸合成酶，大亚基	谷氨酸、谷氨酰胺代谢
gltD	谷氨酸合成酶，小亚基	谷氨酸、谷氨酰胺代谢
PRU_1974	O-乙酰高丝氨酸氨丙基转移酶	赖氨酸合成
cysK	半胱氨酸合成酶A	半胱氨酸合成
dapF	二氨基庚二酸异构酶	赖氨酸合成
PRU_1974	同源转氨酶	赖氨酸合成
PRU_2042	二氨基庚二酸脱氢酶	赖氨酸合成
PRU_1973	谷氨酰胺转氨酶	谷氨酸代谢
asnB	天冬酰胺合成酶	天冬酰胺合成

彩图6 普雷沃氏菌23（P.ruminicola 23）的氮代谢网（Kim等，2017）。当P.ruminicola 23在高铵浓度下生长时，GDH和CS/GOGAT的氨同化效率达到最大。化合物用灰色圆圈表示，含氮物用绿色表示，含硫物用蓝色表示。红色箭头代表氨诱导途径，黄色箭头代表氨基酸或肽诱导途径。

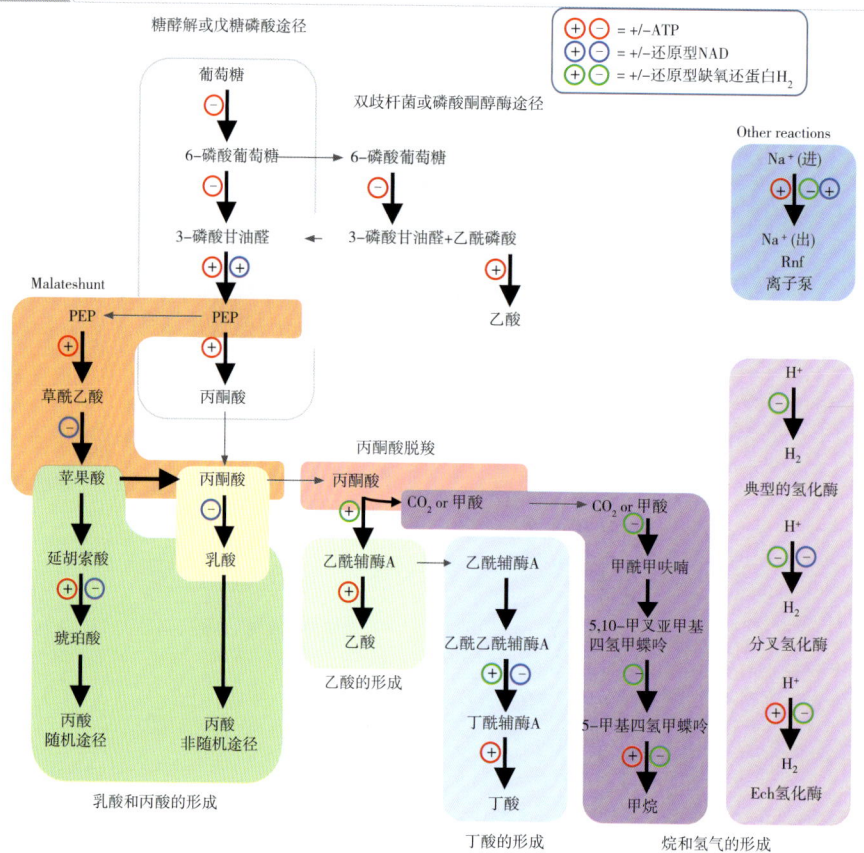

彩图7 瘤胃细菌和产甲烷菌的发酵途径概述。途径已被简化，每个箭头可能代表多种反应。详见 Hackmann 等（2017）。不同颜色代表参与的不同反应，红色：ATP 的生成或水解；蓝色：NAD 的还原或氧化；绿色：铁氧还蛋白或氢气（H_2）的形成。甲烷形成途径由 6 种瘤胃产甲烷菌（*Methanobrevibacter boviskoreani* JH-1、*Methanobrevibacter millerae* ZA-10、*Methanobrevibacter olleyae* 1H5-1P、*Methanobrevibacter ruminantium* M1、*Methanomicrobium mobile* 1、*Methanosarcina* sp.Ms 97）编码。一种产甲烷菌（*Methanosarcina* sp.Ms 97）不编码从甲酸盐形成甲烷的途径。缩写：-P=磷酸盐，-3P=3-磷酸盐，-6P=6-磷酸盐，CoA=辅酶 A，CoM=辅酶 M，H_4MPT=四氢甲蝶呤或四氢糖蝶呤，MFR=甲呋喃，PEP=磷酸烯醇式丙酮酸。

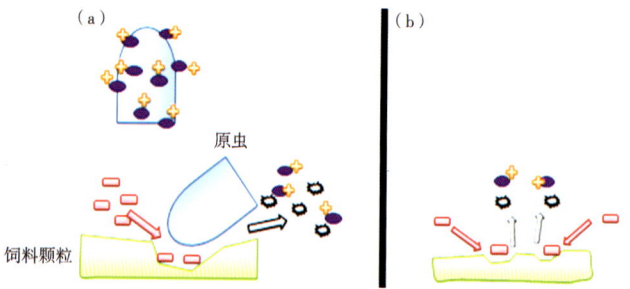

彩图8 具原虫和去原虫瘤胃之间差异的详细示意图。（a）具原虫瘤胃。原虫（蓝色）降解饲料颗粒（绿色）。在降解过程中，细菌（红色）进入饲料颗粒的可消化部分，细菌和原虫产生的发酵终产物（黑色）被产甲烷菌（紫色）利用以产生甲烷（橙）。产甲烷菌附着于原虫以利用其氢小体产生的过量氢。（b）去原虫瘤胃。在没有原生动物的情况下，摄入的饲料颗粒的降解导致发酵产物和甲烷产量减少。